# AFRICAN HANDBOOK OF BIRDS
## SERIES ONE

# AFRICAN HANDBOOK OF BIRDS
### SERIES ONE

# BIRDS OF EASTERN AND NORTH EASTERN AFRICA

## VOLUME II

C. W. MACKWORTH-PRAED
and
C. H. B. GRANT

**LONGMAN**
LONDON AND NEW YORK

**Longman Group Limited** London

*Associated companies, branches and representatives
throughout the world*

*Published in the United States of America
by Longman Inc., New York*

© Second edition C. W. Mackworth-Praed 1960

All rights reserved. No part of this publication may be reproduced, stored in a retrieval system, or transmitted in any form or by any means, electronic, mechanical, photocopying, recording, or otherwise, without the prior permission of the Copyright owner.

*First published 1955
Second edition 1960
Reprinted 1980*

British Library Cataloguing in Publication Data
Mackworth-Praed, Cyril Winthrop
  Birds of Eastern and North Eastern Africa. –
2nd ed. – (African handbook of birds; series 1).
Vol. 2
  1. Birds – Africa, East – Identification
  2. Birds – Africa, Northeast – Identification
  I. Title II. Grant, Claude Henry Baxter
  III. Series
  598.2'967    QL692.E18    79-40701

ISBN 0582 46083 2

Printed in Great Britain by
Butler & Tanner Ltd,
Frome and London

# CONTENTS

| | |
|---|---|
| Publisher's Note | vii |
| Biographical Note | vii |
| Authors' Note | xi |
| Maps | xii |

## GROUPS

| | |
|---|---|
| Larks | 1 |
| Wagtails and Pipits | 47 |
| Babblers | 82 |
| Bulbuls | 107 |
| Flycatchers | 153 |
| Thrushes | 227 |
| Warblers | 335 |
| Swallows | 521 |
| Cuckoo Shrikes | 555 |
| Drongos | 562 |
| Shrikes | 567 |
| Hypocolius | 645 |
| Tits | 646 |
| Orioles | 660 |
| Crows | 670 |
| Starlings | 680 |
| White-eyes | 724 |
| Sunbirds | 736 |
| Creepers | 821 |
| Weavers and Waxbills | 824 |
| Finches | 1055 |
| Buntings | 1086 |
| Indexes | 1115 |

## PUBLISHER'S NOTE

FOR more than two decades the *African Handbook of Birds* has been widely used and consulted by ornithologists the world over, and since the volumes went out of print some years ago there has been a steady demand for their reissue. It has not been possible to carry out the extensive revision required to bring them up to date, or even to incorporate the notes for the future editions left by Colonel Mackworth-Praed, but it has been decided to reprint the text and illustrations as they stand. There are two additions however. Firstly, on pp. xii, xiii can be found two maps showing the political boundaries of the African continent—one dating from the end of World War II, the other giving the current situation. It is hoped that the juxtaposition of these will assist the reader in translating the information on species distribution given in the text into modern terms. Secondly, we are printing below a biographical note on the two authors of the *Handbook*. This has been prepared with the kind assistance of Mrs. Pat Hall on behalf of the executors of the estate of the late Colonel C. W. Mackworth-Praed.

## BIOGRAPHICAL NOTE

The *African Handbook of Birds* was begun in 1932, Claude Grant being then 54 and Cyril Praed 41. The earliest entries in the joint account dated April 1933, are for postage stamps, with the addressees carefully listed, representing a considerable roll-call of the ornithologists of the time.

Many years of work followed, with enforced interruptions, until the publication in 1952 of 4990 copies of the first East Africa volume, and of 5000 copies of the second in 1955. Claude Grant died in 1958, having largely completed the taxonomic and nomenclatural groundwork for the whole series. Cyril Praed continued the project, publishing a second edition of 5000 each of the East Africa volumes in 1957 and 1960. The two volumes for Southern Africa, also 5000 each, followed in 1962 and 1963, with a second impression of 2000 of the first in 1969. The 3250 copies of the first volume of Western Africa were published in 1970, and the 2926 of the second in 1973. All publication costs were borne by Cyril Praed throughout.

Captain Claude Henry Baxter Grant was born on 24th December 1878, in London. He inherited from his father an interest in birds, and after studying field surveying, he worked as a taxidermist on the staff of the Natural History Museum in London, preparing himself for collecting in the field. These plans were interrupted by the Boer War, which gave him his first introduction to Southern Africa, where after hostilities ceased he made several collections for the Museum. From 1908 to 1910 he collected in South America, and later in New Guinea. He returned to the Museum in 1913 to work on a large East African collection until in 1914 war service took him to East Africa.

After the war, Grant transferred to the Colonial Administration of Tanganyika Territory, where he remained from 1919 to 1932, devoting a good deal of his spare time to surveying and map-making, his photographs and planetable sheets being now with the Royal Geographical Society. His other collections from all parts of the world, birds, mammals, insects, plant and mineral specimens are in the Natural History Museum. His personal experiences of course were legion; few men have been juggled with by a wounded bull elephant between its tusks and its trunk and then thrown over its shoulder without much harm. As a measure of his popularity in Tanganyika, the local African authority, some time after he left, asked that a street in Ujiji should be named after him, which was done.

During his years at the Museum's Bird Room, working on these books, he became well known to many visiting ornithologists, and he had an enormous corresponding circle overseas. A meticulously careful worker, he combined his first-hand knowledge of African birds with a particular study of the routes taken and collections made by the early travellers in Africa. His knowledge also of general ornithological literature was exhaustive and in nomenclature he was an uncompromising purist. He was a man of unfailing cheerfulness and always ready to help others, however busy he was himself.

Grant was a member of many scientific societies and was also an Honorary Associate of the Zoological Department of the British Museum. He contributed to a number of ornithological publications —*Ibis*, the Bulletin of the British Ornithologists' Club, *Ostrich*, *Tanganyika Notes and Records* and others. He was a member of the

B.O.U. Committee from 1933 to 1935 and Editor of *Ibis* from 1941 to 1947, Editor of the B.O.C. Bulletin from 1935 to 1940 and 1947 to 1952 and Vice-Chairman of the B.O.C. from 1940 to 1943. He was a member of the B.O.U. List Committee from 1935 to 1952 and Honorary Secretary from 1937 to 1952.

He married in 1915 Lena Harriett, daughter of Henry Priestley, who accompanied him on many of his expeditions in Tanganyika, and was survived by her and two daughters.

Lt. Colonel Cyril Winthrop Mackworth-Praed, O.B.E., was born on 21st September 1891 at Mickleham, Surrey, in the chalk country of the North Downs, from where his parents moved to Herefordshire in the Welsh Borders. A rather lonely childhood developed his interest in shooting, fishing, and natural history study, at that time chiefly birds, butterflies and moths, but later to extend to microlepidoptera, and to plants, of which he amassed a considerable herbarium. This, like all his collections, was carefully labelled and identified. After leaving Cambridge he went out to East Africa with a view to farming and was there when the First World War started. After some service in Africa he was invalided home with fever and then joined the Scots Guards, with which regiment he also served in the Second World War. In 1919 he married the daughter of Colonel Stephenson R. Clarke, the African naturalist and collector. Helping his father-in-law to identify birds from his collections kept Praed's interest in African birds alive and also introduced him to the study of taxonomy and to the coterie of ornithologists then working in the Bird Room at the Natural History Museum. His work on the London Stock Exchange, including later its Council, restricted the amount of time he could spend on birds and it was only in the 1930s when the opportunity arose to collaborate with Grant that it was possible to put in hand his project for the *African Handbook*.

To appreciate just what a mammoth undertaking this was, one must remember that when it was conceived such African bird literature as existed was widely scattered through diverse publications: it was before Bannerman, Roberts, Chapin, Archer and Godman, or Cave and Macdonald had published their regional works. While Grant did most of the taxonomic work, Praed provided the financial backing, laid down the policy and format and undertook much of the collation of field notes and data. It says much for his wisdom and foresight that it was possible to maintain the original format with only minor modifications through all six volumes and forty years of changing ornithological opinions and conventions.

While the name Mackworth-Praed is inevitably associated with African birds, he was active in many other spheres: in particular his interest in wildfowl preservation led him in the thirties to renovate the old duck decoy and re-start systematic ringing at Orielton, Pembrokeshire, on a property leased by his family. He was for a long time a prominent member of the British Section of the Wild Fowl Inquiry Committee and was responsible for the imaginative Duck Adoption Scheme whereby for 5/- anyone could adopt a ringed duck and be kept informed of its movements, the money raised being used to finance further ringing.

He was Chairman and a benefactor of the Bird Exploration Fund Committee for over twenty years, a past Chairman of the British Ornithologists' Club and a regular attender at the club meetings. He was a familiar figure among the British contingent at international conferences, making his first of several return visits to Africa at the 1st Pan-African Ornithological Congress in 1957. His services to ornithology were recognized by the award of the O.B.E., the Stamford Raffles Award of the Zoological Society and the Union Medal.

Outside bird circles, he won world honours with both rifle and shotgun, including a silver medal at the 1924 Olympics and first prize at the 1937 International Shooting Championship at Berlin. He was a member of the Athenaeum, and of many other London clubs and societies. His modesty and courteousness concealed acute observation of all that grew or moved around him, including his fellow men. His discovery, while commanding a training centre for Allied commandos in Invernesshire in 1942, of an unexpected Scottish race of the Chequered Skipper butterfly created an entomological furore, since this species was previously only known in Britain from the English Midlands, where it is now believed extinct.

Like many of his generation he became a keen conservationist in his later years, subscribing, or leaving bequests, to some forty organizations concerned with the study and preservation of threatened wildlife. He was largely responsible for financing a book on the insect fauna of Hampshire, and at his house in the New Forest maintained records for nearly thirty years of the moths coming to his light-trap before releasing them. Until incapacitated by increasing illness, he travelled widely in Europe from the Arctic to the Mediterranean, and was seldom at a loss to identify any species of plant, bird or insect. He was that nowadays rare person, an all-round naturalist. He died in July 1974.

## AUTHORS' NOTE

A PERUSAL of this Volume and also of Volume One will reveal that there are many items, particularly of habits, habitat and breeding data, which need to be filled in and it is to be hoped that further study in the field will bring to light new facts of the systematic and ecological status of certain groups. It is, however, essential that they should be published. We know of many cases where such biological data are known to local residents but to no one else.

Pending further knowledge we have in some groups leaned more towards retaining the larger numbers of probable species. In some cases these may prove to be but local representative races, but it is not easy to draw the line, and in any case divergence in the future is likely and will be accelerated by man's interference with natural conditions. The Swifts, the Bulbuls, the Paradise Flycatchers and the Seed-eaters are examples.

We should like to express our thanks for the comments and suggestions that Volume One has called forth as well as for much additional useful information. We should also like to apologize for the size and weight of this Volume, but the number of species dealt with renders it unavoidable.

C. W. M.-P.
C. H. B. G.

1953

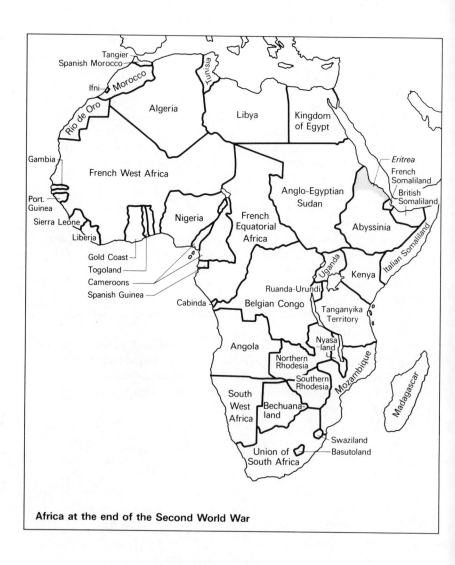

Africa at the end of the Second World War

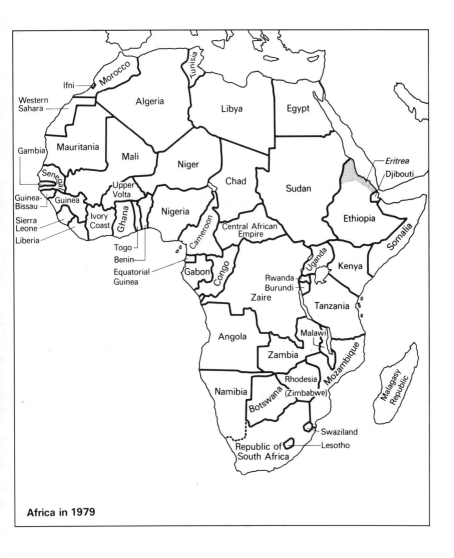

# VOLUME TWO

FAMILY—**ALAUDIDÆ**. **LARKS**. Genera: *Mirafra, Pinarocorys, Heteromirafra, Certhilauda, Alæmon, Ammomanes, Galerida, Heliocorys, Pseudalæmon, Eremopterix, Melanocorypha, Calandrella, Spizocorys* and *Aëthocorys*.

Thirty-six species occur in Eastern Africa, some of which are migrants from southern Europe, western Asia and north Africa. Larks are found nearly everywhere outside woodland, in the deserts, on grasslands and in sparsely timbered country. Their life is spent mainly on the ground, where they nest and find their food, which consists of grain, weed seeds, small molluscs and insects. The nest is usually a well made cup of grass or any other material available concealed among vegetation on flat ground, but certain groups, as for instance the Sand and Sun Larks have distinctive nesting habits.

Local movements occur in many species according to the seasons and supply of food, as for instance that of the Sparrow-Larks to Port Sudan from the surrounding areas, when the grain is being shipped. Being ground birds they are liable to earth staining, and several species have more than one colour phase.

KEY TO THE ADULT LARKS OF EASTERN AFRICA

1 Feathers on occiput long and forming a distinct crest:  CRESTED LARK
   *Galerida cristata*   675

2 Feathers on occiput of normal length and not forming a distinct crest:   3–72
3 Mantle plain:   5–13
4 Mantle variously marked:   14–72
5 Ends and bases of secondaries and inner primaries white:  HOOPOE LARK
   *Alæmon alaudipes*   670

6 Ends of secondaries dingy white:   7–8

| | | |
|---|---|---|
| 7 Tail dark sooty brown: | DUSKY BUSH-LARK *Pinarocorys nigricans* | **666** |
| 8 Basal half of tail chestnut: | RED-TAILED BUSH-LARK *Pinarocorys erythropygia* | **667** |
| 9 Ends and bases of secondaries uniform with rest of feathers: | | 10–72 |
| 10 Bill long and thin: | LESSER HOOPOE-LARK *Alæmon hamertoni* | **671** |
| 11 Bill stout and not long: | | 12–13 |
| 12 Above, sandy buff: | BLACK-TAILED SAND-LARK *Ammomanes cinctura* | **672** |
| 13 Above, greyish and isabelline: | SAND-LARK *Ammomanes deserti* | **673** |
| 14 Ear-coverts and cheeks white: | | 16–20 |
| 15 Ear-coverts and cheeks not white: | | 21–72 |
| 16 Mantle chestnut: | CHESTNUT-BACKED SPARROW-LARK, male *Eremopterix leucotis* | **679** |
| 17 Mantle pale stone colour: | WHITE-FRONTED SPARROW-LARK, male *Eremopterix nigriceps* | **680** |
| 18 Mantle ashy brown: | | 19–20 |
| 19 Top of head dark chocolate with a white centre: | CHESTNUT-HEADED SPARROW-LARK, male *Eremopterix signata* | **681** |
| 20 Top of head brown: | FISCHER'S SPARROW-LARK, male *Eremopterix leucopareia* | **682** |
| 21 Belly black or blackish: | | 23–26 |
| 22 Belly not black or blackish: | | 27–72 |
| 23 Above, blackish: | CHESTNUT-BACKED SPARROW-LARK, female *Eremopterix leucotis* | **679** |
| 24 Above, ash brown: | | 25–26 |

# LARKS 3

| | | | |
|---|---|---|---|
| 25 | Bill large: | CHESTNUT-HEADED SPARROW-LARK, female *Eremopterix signata* | **681** |
| 26 | Bill small: | FISCHER'S SPARROW-LARK, female *Eremopterix leucopareia* | **682** |
| 27 | Black patch at base of neck in front: | COLLARED LARK *Mirafra collaris* | **663** |
| 28 | Black patches at sides of neck at base: | | 30–37 |
| 29 | No black patch or black patches at base of neck: | | 38–72 |
| 30 | Size larger, wing over 111 mm.: | CALANDRA LARK *Melanocorypha bimaculata* | **683** |
| 31 | Size smaller, wing under 101 mm.: | | 32–37 |
| 32 | Above, dark, broad blackish centres to feathers: | RUFOUS SHORT-TOED LARK *Calandrella rufescens* | **685** |
| 33 | Above, pale, brown centres to feathers and broad sandy edges: | | 34–37 |
| 34 | Bill long: | SHORT-TAILED LARK *Pseudalæmon fremantlii* | **678** |
| 35 | Bill short: | | 36–37 |
| 36 | Wing over 86 mm.: | SHORT-TOED LARK *Calandrella brachydactila* | **684** |
| 37 | Wing under 86 mm.: | BLANFORD'S LARK *Calandrella blanfordi* | **687** |
| 38 | No spots or streaks on chest: | | 40–43 |
| 39 | Chest spotted or streaked: | | 44–72 |
| 40 | Lores and moustachial stripe black: | MASKED LARK *Aëthocorys personata* | **689** |

B

# LARKS

| | | |
|---|---|---|
| 41 | No black on lores or moustachial stripe: | 42–43 |
| 42 | Underwing-coverts white: DUNN'S LARK *Ammomanes dunni* | **674** |
| 43 | Underwing-coverts black: WHITE-FRONTED SPARROW-LARK, female *Eremopterix nigriceps* | **680** |

| | | |
|---|---|---|
| 44 | Above, various shades of russet or cinnamon brown, with or without blackish centres to feathers: | 46–55 |
| 45 | Above, various shades of buff or brown, with or without blackish centres to feathers: | 56–72 |
| 46 | Bill long: SOMALI LONG-BILLED LARK *Certhilauda somalica* | **669** |
| 47 | Bill normal: | 48–57 |
| 48 | Light pattern on underside of tail feathers on both inner and outer webs: | 50–51 |
| 49 | Light pattern on underside of outer tail confined to outer web: | 52–55 |
| 50 | Pattern on outer tail feathers white: KORDOFAN BUSH-LARK *Mirafra cordofanica* | **656** |
| 51 | Pattern on outer tail feathers brown: FLAPPET-LARK *Mirafra rufocinnamomea* | **660** |
| 52 | Some chestnut on sides of chest: | 54–55 |
| 53 | No chestnut on sides of chest: | 56–72 |
| 54 | Top of head chestnut not streaked: RED-CAPPED LARK *Calandrella cinerea* | **686** |

## LARKS

55 Top of head chestnut, or
brown, streaked:     FAWN-COLOURED LARK
                                   *Mirafra africanoides*    **661**

56 Below, white, with broadish russet brown streaks
on chest:     GILLETT'S LARK
                  *Mirafra gilletti*    **664**

57 Below, buffish:    58–59

58 Broad cinnamon brown
streaks on chest:     RED SOMALI LARK    **658**
                  *Mirafra sharpei*

59 Narrow blackish brown
streaks on chest:     RUSTY BUSH-LARK
                  *Mirafra rufa*    **662**

60 No distinct dark centres
to feathers of upperparts
and no dark streaks or
spots on chest:     PINK-BREASTED LARK
                  *Mirafra pœcilosterna*    **665**

61 Distinct dark centres to
feathers of upperparts and
dark streaks or spots on
chest:    62–72

62 Flight feathers edged with
dark brown or chestnut:    64–66

63 Flight feathers edged with
pale brown or buff:    67–72

64 Size small, wing under 86
mm.:     NORTHERN WHITE-
                  TAILED BUSH-LARK
                  *Mirafra albicauda*    **655**

65 Size larger, wing over 88
mm., tail 66 mm. or under:     RUFOUS-NAPED LARK
                  *Mirafra africana*    **659**

66 Size large, tail 71 mm. or
over:     RED-WINGED BUSH-LARK
                  *Mirafra hypermetra*    **657**

67 Above, mottled blackish
   brown, brown and pale
   buff:                          LONG-CLAWED LARK
                                  *Heteromirafra ruddi*       668
68 Above, broadly marked
   brown and blackish:            SHORT-CRESTED LARK
                                  *Galerida malabarica*       676
69 Bill rather heavy, feathers
   of mantle edged whitish:       OBBIA LARK
                                  *Spizocorys obbiensis*      688
70 Bill normal, feathers of
   mantle dark brown and
   buff:                                                            71-72
71 Wing-coverts edged brown: SUN LARK
                             *Heliocorys modesta*             677
72 Wing-coverts edged buff:  SINGING BUSH-LARK
                             *Mirafra cantillans*             654

**654 SINGING BUSH-LARK.  *MIRAFRA CANTILLANS*** Blyth.

*Mirafra cantillans marginata* Hawker. Pl. 54.

*Mirafra marginata* Hawker, Bull. B.O.C. 7, p. 55, 1898: Ujawagi, eastern Abyssinia.

**Distinguishing characters:** Above, mottled and streaked dark and light brown; outermost tail feathers mainly white; below, buff and white with narrow streaks on upper chest. The sexes are alike. Wing 71 to 84 mm. The young bird has light buff tips and edges to the feathers of the upperparts, wings and tail.

**Range in Eastern Africa:** Central and coastal Eritrea, Abyssinia, British and Italian Somalilands, the south-eastern Sudan and Uganda to northern and central Tanganyika Territory.

**Habits:** A bird of open country, either of grass plains or grass with scattered bushes. Sings as a rule while hovering, but occasionally from the top of a bush. When alarmed flies low and fast with a whirring sound, checking abruptly and diving into grass. In Tanganyika Territory is definitely migratory, arriving in the low country during the rains (March to August) probably to breed, and chooses low sandy or alkaline ground. It does not usually 'flappet' but has a fluttering sailing courtship flight.

**Nest and Eggs:** A typical Lark's nest in a grass-tuft, well covered over, of grass and grass stems with a finer lining. Eggs two to four, dull greenish white with heavy dull olive brown stippling and flecking, sometimes in a cap at the larger end; about 19 × 15 mm.

**Recorded breeding:** Central Abyssinia, October. Eastern Abyssinia, May and June. Kenya Colony, Marsabit, May to July; southern areas, April and May. Tanganyika Territory, March to August, also December.

**Call:** Rather a Bunting-like chirruping song, a ragged trill of four notes. Also pipes and twitters during courtship flight.

*Mirafra cantillans chadensis* Alex.

*Mirafra chadensis* Alexander, Bull. B.O.C. 21, p. 80, 1908: Kowa Baga, Lake Chad.

**Distinguishing characters:** General colour very much paler, more sandy buff on the upperparts, than the preceding race. Wing 72 to 83 mm.

**General distribution:** French Sudan and Nigeria to the Sudan and Eritrea.

**Range in Eastern Africa:** Western Sudan to western Eritrea.

**Habits:** As for the preceding race. Its hovering flight with the wings pointing downward is somewhat peculiar. Abundant in Kordofan and Darfur.

**Recorded breeding:** Nigeria, June to August. Air, French Sahara, August and September. Sudan, July to September.

**Distribution of other races of the species:** Arabia and India, the nominate race being described from Bengal.

**655 NORTHERN WHITE-TAILED BUSH-LARK.** *MIRAFRA ALBICAUDA* Reichenow. **Pl. 54.**

*Mirafra albicauda* Reichenow, J.f.O. p. 223, 1891: Gonda, Tabora District, Tanganyika Territory.

**Distinguishing characters:** Darker above than the last species, black or blackish with light edges to feathers; spots on chest usually larger and more distinct; feet and toes inclined to be larger. The sexes are alike. Wing 75 to 86 mm. The young bird is rather blacker above and has the feathers tipped and edged with pale buff.

**General distribution:** French Equatorial Africa to the White Nile and Tanganyika Territory.

**Range in Eastern Africa:** The southern Sudan to central Tanganyika Territory.

**Habits:** A bird of open country, plentiful in the Kenya highlands on the Athi and Kapiti Plains and also in Turkana. In the Sudan it seems to be confined to black cotton soil. It has a habit of rising and circling into the air while singing, and it also makes a purring or drumming noise with its wings. Said to be often found with the Harlequin Quail.

**Nest and Eggs:** Nest, a shallow scrape lined with grass, in the shelter of a tuft of grass or at the side of a rock. Eggs two, greyish or buffish white with blotches and freckling of dark brown; about 17 × 14 mm.

**Recorded breeding:** Sudan (breeding condition), May. Doinyo Erok, Kenya Colony, March. North-eastern Tanganyika Territory (breeding condition), May.

**Call:** A musical chattering song, uttered as it hovers or circles about a hundred feet up.

**656 KORDOFAN BUSH-LARK.** *MIRAFRA CORDOFANICA* Strickland. **Pl. 54.**

*Mirafra cordofanica* Strickland, P.Z.S. for 1850, p. 218, pl. 23, 1852: Kordofan, Sudan.

**Distinguishing characters:** Above, pale cinnamon brown with pale fawn edges to feathers; below, buff with cinnamon brown spots on upper chest; central tail feathers pale cinnamon brown, contrasting with the other black tail feathers; outermost tail feathers mainly white, penultimate one black with outer webs white. The sexes are alike. Wing 78 to 88 mm. The young bird has the feathers of the upperparts subterminally tipped and edged with black and tipped and edged with pale buff.

**General distribution:** French Sudan to Darfur and Kordofan.

**Range in Eastern Africa:** Darfur and Kordofan, Sudan.

**Habits:** A rare species inhabiting firm sandy soil and usually associated with feathery prairie-grass 'Heskanit' (*Aristida*).

**Nest and Eggs:** Undescribed.

**Recorded breeding:** Western Sudan, believed to breed from June to September.

**Call:** A sweet musical little song in the breeding season.

## LARKS

### 657 RED-WINGED BUSH-LARK. *MIRAFRA HYPERMETRA* (Reichenow).

*Mirafra hypermetra hypermetra* (Reichw.). Pl. 54.

*Spilocorydon hypermetrus* Reichenow, O.C. p. 155, 1879: Kibaradja, Tana River, Kenya Colony.

**Distinguishing characters:** A large Lark with a slight crest and dark tawny webs to the flight feathers; above, mottled blackish, brown and greyish brown, most of the feathers edged with tawny white; below, deepish buff, throat whiter, with black spots on upper chest. The sexes are alike. Wing 112 to 118 mm. Grey and rufous colour phases also occur. The young bird is much darker than the adult; has buff tips to the feathers of the upperparts; spots on upper chest dull black.

**Range in Eastern Africa:** Italian Somaliland and Kenya Colony to eastern Tanganyika Territory, as far south as the Ruvu River.

**Habits:** A very large Lark with a curious little soaring flight to a comparatively low altitude of ten to twenty feet where it hovers for a considerable period. Sings while hovering and also from a perch on a stump or on an ant-hill.

**Nest and Eggs:** A well built nest of grass and rootlets well covered over in a grass tuft. Eggs probably two to four, rather well marked with brown or olive brown on a dull white ground; about 23 × 18·5 mm.

**Recorded breeding:** Italian Somaliland, April to June. Northern Guaso Nyiro probably February and March.

**Call:** Song, short, loud and flute-like 'du-diau-did-dido' (Erlanger), also a loud clear call of two notes.

*Mirafra hypermetra gallarum* Hart.

*Mirafra hypermetra gallarum* Hartert, Bull. B.O.C. 19, p. 84, 1907: Bouta, Hawash Valley, eastern Abyssinia.

**Distinguishing characters:** Greyer above than the nominate race. Wing 113 to 125 mm.

**Range in Eastern Africa:** Eastern and southern Abyssinia.

**Habits:** Those of the nominate race, and is found in arid sparsely grassed country.

**Recorded breeding:** Eastern Abyssinia, probably April to June.

*Mirafra hypermetra kidepoensis* Macd.

*Mirafra hypermetra kidepoensis* Macdonald, Bull. B.O.C. 60, p. 59, 1940: Kidepo Plain, south of the Didinga Mts., south-eastern Sudan.

**Distinguishing characters:** Rather smaller than the preceding races and having the bases of the head feathers chestnut as in the Rufous-naped Lark, but differs from that bird in having a longer tail and rather heavier bill. Wing 102 to 111 mm.

**Range in Eastern Africa:** Southern Sudan to northern Uganda from the Kidepo River Valley to the plains near Mt. Maroto and the Nakwai Hills.

**Habits:** As for other races.

**Recorded breeding:** No records.

*Mirafra hypermetra kathangorensis* Cave.

*Mirafra hypermetra kathangorensis* Cave, Bull. B.O.C. 60, p. 96, 1940: Kathangor, Equatoria, Sudan.

**Distinguishing characters:** Above, blacker, darker than the other races, and with paler under wing-coverts and bases to the flight feathers. Wing 103 mm.

**Range in Eastern Africa:** South-eastern Sudan at Kathangor, north-east of Kapoeta.

**Habits:** No information.

**Recorded breeding:** No records.

**658 RED SOMALI LARK.** *MIRAFRA SHARPEI* Elliot.

*Mirafra sharpei* Elliot, Field Col. Mus. Pub. Orn. 1, p. 37, 1897: Silo Plain, British Somaliland. **Pl. 54.**

**Distinguishing characters:** Above, rich cinnamon brown streaked with buffish white; below, buffish white; chin and throat whiter, with cinnamon brown streaks on upper chest and a few cinnamon brown spots on lower neck. The sexes are alike. Wing 92 to 103 mm. Juvenile plumage unrecorded.

**Range in Eastern Africa:** British Somaliland.

**Habits:** No information.

**Nest and Eggs:** Undescribed.

**Recorded breeding:** No records.

**Call:** Unrecorded.

### 659 RUFOUS-NAPED LARK. *MIRAFRA AFRICANA* A. Smith.

*Mirafra africana nigrescens* Reichw.

*Mirafra nigrescens* Reichenow, O.M. p. 39, 1900: Elton Plateau, Ukinga, about 35 miles north of Lake Nyasa, south-western Tanganyika Territory.

**Distinguishing characters:** General colour above more black than tawny; below, tawny with black streaks on chest and flanks; flight feathers cinnamon brown with greyish black tips and ends; hind claw long and almost straight, 16 mm. The sexes are alike. Wing 92 mm. Somewhat similar to but smaller than the Red-winged Bush-Lark and having a shorter tail. The young bird has lighter and broader tawny tips and edges to the feathers of the upperparts, wings and tail.

**Range in Eastern Africa:** South-western Tanganyika Territory.

**Habits, Nest, Eggs, etc.:** See under other races.

**Recorded breeding:** No records.

*Mirafra africana tropicalis* Hart. **Pl. 54.**

*Mirafra africana tropicalis* Hartert, Nov. Zool. 7, p. 45, 1900: Bukoba, north-western Tanganyika Territory.

**Distinguishing characters:** Warmer tone above and with much less black than the preceding race; hind claw short and curved, 11 to 13 mm. Wing 86 to 109 mm.

**Range in Eastern Africa:** Southern Uganda to Tanganyika Territory from about Bukoba to the Mbulu, Kasulu and Iringa Districts.

**Habits:** As for other races, plentiful in cultivated or grass country. The song is said to be more varied than in other races and is often delivered in hovering flight.

**Recorded breeding:** Uganda, February to June, also November and December. Tanganyika Territory, September and December.

*Mirafra africana athi* Hart. **Ph. vii.**

*Mirafra africana athi* Hartert, Nov. Zool. 7, p. 46, 1900: Athi Plains, Kenya Colony.

**Distinguishing characters:** Colder and greyer in tone on upperparts than the two preceding races. Wing 87 to 102 mm.

**Range in Eastern Africa:** Kenya Colony to north-eastern Tanganyika Territory at Loliondo, Lake Manyara, the crater area and Ngare Nairobi.

**Habits:** A common and widespread stockily-built species of grass or open bush country. Its flight is rather short and jerky with quivering wings, but it also has a flapping gliding courtship flight with a little hover. It runs very fast when on the ground. It has a habit of calling from a perch in the breeding season and is often noticed sitting on a bush or rock in the early morning. It is difficult to flush.

**Nest and Eggs:** Nest of dried grass usually in a tussock. Eggs two or three, brownish white with dark brown or yellowish brown and grey markings mostly at larger end, or uniform greyish brown with grey and yellowish brown speckling; about 24 × 16 mm.

**Recorded breeding:** Kenya Colony, April to June, also November to January in wet seasons. North-eastern Tanganyika Territory (breeding condition), March to May.

**Call:** A mournful but clear whistled call of four or five notes 'chiwicki-chiwi.' Also a very sweet little song in the breeding season.


*Mirafra africana zuluensis* (Rob.).

*Africorys africana zuluensis* Roberts, Ann. Trans. Mus. 18, p. 215, 1936: Maputa, northern Zululand.

**Distinguishing characters:** Differs from the Ukinga and Bukoba races in being greyer above with narrower, less black, markings. Bill and hind claw shorter. Wing 89 to 99 mm. Hind claw curved, 9 to 11 mm.

**General distribution:** Tanganyika Territory and northern Portuguese East Africa to Zululand.

**Range in Eastern Africa:** South-western Tanganyika Territory, Portuguese East Africa and Nyasaland.

**Habits:** As for other races.

**Recorded breeding:** No records.

# LARKS

*Mirafra africana harterti* Neum.
**Mirafra africana harterti** Neumann, Bull. B.O.C. 23, p. 45, 1908: Kiboko River, south Ukamba, south-eastern Kenya Colony.

**Distinguishing characters:** Warmer and richer in colour than the other races, edges of feathers of upperparts cinnamon brown, or deep tawny brown. Wing 94 to 107 mm.

**Range in Eastern Africa:** South Ukamba to Taita, south-eastern Kenya Colony.

**Habits:** As for other races.

**Recorded breeding:** Southern Kenya Colony, March.

*Mirafra africana kurrœ* Lynes.
**Mirafra africana kurrœ** Lynes, Bull. B.O.C. 43, p. 95, 1923: Kurra, Jebel Marra, western Sudan.

**Distinguishing characters:** Differs from all the preceding races in having a vinous wash on the mantle. Wing 85 to 99 mm.

**General distribution:** Bauchi Plateau, Nigeria, to the Sudan.

**Range in Eastern Africa:** Western Sudan.

**Habits:** Local and in our area confined to one of the dry terraces to the north of the Jebel Marra massif. Often noticed sitting on bushes in the early morning, and occasionally sings sweetly from a perch.

**Recorded breeding:** Western Sudan, probably June to September.

**Distribution of other races of the species:** West to South Africa, the nominate race being described from south-eastern Bechuanaland.

### 660 FLAPPET-LARK. *MIRAFRA RUFOCINNAMOMEA* (Salvadori).

*Mirafra rufocinnamomea rufocinnamomea* (Salvad.).
**Megalophoneus rufocinnamomeus** Salvadori, Atti. Soc. Ital. Sci. Nat. 8, p. 378, 1865: Northern Abyssinia.

**Distinguishing characters:** Above, brick brown with blackish centres to feathers and with rusty brown edges to flight feathers; below, brown, more brick brown on chest with black spots on upper chest and lower neck; edges of inner webs of flight feathers tawny brown; outer tail feathers mainly pale tawny, central feathers blackish with tawny edges. A brighter cinnamon brown phase occurs rarely. The sexes are alike. Wing 80 to 87 mm. The young

bird has the upperparts duller and the blackish centres to the feathers have a barred appearance.

**Range in Eastern Africa:** Northern to central Abyssinia.

**Habits, Nest and Eggs:** As for the following race. Usually three 'flappets' are given but occasionally only single ones.

**Recorded breeding:** Abyssinia, August to October.

*Mirafra rufocinnamomea fischeri* (Reichw.). **Pl. 54.**
*Megalophoneus fischeri* Reichenow, J.f.O. p. 266, 1878: Rabai, near Mombasa, eastern Kenya Colony.

**Distinguishing characters:** Differs from the nominate race in having less black on the upperparts and more buffish-brown below. Both cinnamon brown and earth brown phases occur, the former usually in more open country; and a brighter cinnamon brown phase occurs rarely. Wing 68 to 88 mm.

**General distribution:** Gabon and Angola to Kenya Colony and south to the Transvaal and northern Natal.

**Range in Eastern Africa:** Coastal areas of Kenya Colony to southern areas of Tanganyika Territory, Portuguese East Africa and Nyasaland.

**Habits:** One of the most characteristic birds of tropical Eastern Africa, the muffled rattling flight attracting immediate attention. Its habit is to rise to a considerable height into the wind then to descend almost vertically to within twenty feet of the ground before swerving off and carrying on some time before alighting. The loud crackling flaps are usually three to five in number in each flight; they seem to be made by the males only, but one bird's flight will start off any other within hearing. It is a bird of bush or open woodland, which runs fast on the ground and does not take wing easily if alarmed. It is not certain whether the cracks are produced by the wings meeting below as well as above the body.

**Nest and Eggs:** A Pipit-like nest of dry grass well hidden in a tussock. Eggs two to four, buff or greyish white densely mottled with dark brown or pinkish greyish brown; about 21 × 14 mm.

**Recorded breeding:** Kenya Colony, April to July, also December. Tanganyika Territory, November to January. Nyasaland, November to January, also April.

**Call:** Uttered from a perch or from the ground, a soft whistled 'tuee-tui.'

# LARKS

*Mirafra rufocinnamomea torrida* Shelley.
*Mirafra torrida* Shelley, P.Z.S. p. 308, pl. 17, 1882: Ugogo, Dodoma district, central Tanganyika Territory.

**Distinguishing characters:** General colour above brighter vinous brown than the preceding race, and a warm brown buff below; usually brick brown centres to feathers of chest. Wing 70 to 87 mm.

**Range in Eastern Africa:** Southern Abyssinia, south-eastern Sudan, Kenya Colony, east of the Rift Valley, but not the coastal areas, and Tanganyika Territory as far south as Iringa.

**Habits:** As for other races.
**Recorded breeding:** No records.

*Mirafra rufocinnamomea tigrina* Oust.
*Mirafra tigrina* Oustalet, Le Naturaliste, 6, 2nd ser. p. 231, 1892: Poste de la Mission, Haut Kemo, Ubangi, French Equatorial Africa.

**Distinguishing characters:** In the earth brown phase much darker in general tone above than the Kenya race; in the cinnamon brown phase duller, rather less cinnamon, more vinous on the upperparts and more heavily spotted on the chest. Wing 71 to 82 mm.

**General distribution:** Northern Belgian Congo to the Sudan and Uganda.
**Range in Eastern Africa:** Southern Sudan to western and south-western Uganda.
**Habits:** As for other races, its crackling flaps are occasionally very loud indeed.
**Recorded breeding:** Uganda, May and June, also October.

*Mirafra rufocinnamomea sobatensis* Lynes.
*Mirafra sobatensis* Lynes, Bull. B.O.C. 33, p. 129, 1914: White Nile at lat. 10° N., eastern Sudan.

**Distinguishing characters:** Very dark, almost black above, with broad fawn edges to the wing feathers; heavily spotted on lower neck and chest. Wing 80 to 87 mm.

From the Northern White-tailed Bush Lark it may be at once distinguished by being darker above, more tawny below and having the edges of the tail feathers tawny, not white.

**Range in Eastern Africa:** Eastern Sudan.
**Habits:** No information.
**Recorded breeding:** No records.

*Mirafra rufocinnamomea kavirondensis* Van Som.

*Mirafra fischeri kavirondensis* Van Someren, Bull. B.O.C. 41, p. 125, 1921: Kisumu, western Kenya Colony.

**Distinguishing characters:** Markings on head and mantle more distinct, less diffused, than in the Mombasa race, cinnamon phase indistinguishable. Wing 71 to 82 mm.

**Range in Eastern Africa:** Uganda, Kenya Colony west to the Rift Valley and northern Tanganyika Territory as far south as Kigoma, Shinyanga and the Arusha area.

**Habits:** As for other races.

**Recorded breeding:** Uganda, May and June, also October.

*Mirafra rufocinnamomea furensis* Lynes.

*Mirafra fischeri furensis* Lynes, Bull. B.O.C. 43, p. 95, 1923: Kulme, Darfur, western Sudan.

**Distinguishing characters:** Pale cinnamon brown above with few dark markings; has no earth brown phase, but has rarely a deep cinnamon brown phase which is not distinguishable in colour from the same rare phase in the nominate and Kenya races. Wing 72 to 82 mm.

**Range in Eastern Africa:** Western Sudan.

**Habits:** As for other races, the crackling flaps are usually in threes.

**Recorded breeding:** Sudan, July and August, its eggs are usually two only.

*Mirafra rufocinnamomea omoensis* Neum.

*Mirafra fischeri omoensis* Neumann, J.f.O. p. 787, 1928: Lange Tombara dist. Djimma, south-western Abyssinia.

**Distinguishing characters:** Very similar to the nominate race, but usually lacks the vinous wash. Wing 78 to 86 mm.

**Range in Eastern Africa:** Omo River area. south-western Abyssinia.

**Habits:** As for other races.

**Recorded breeding:** No records.

**Distribution of other races of the species:** West Africa, and north-western Northern Rhodesia.

## 661 FAWN-COLOURED LARK. *MIRAFRA AFRICANOIDES* A. Smith.

*Mirafra africanoides intercedens* Reichw. **Pl. 54.**

*Mirafra intercedens* Reichenow, O.M. p. 96, 1895: Loeru, Kondoa Irangi District, central Tanganyika Territory.

**Distinguishing characters:** Above, deep tawny with broad blackish streaks; neck lighter in colour; below, buffish white; upper chest tawny with short black streaks; outer web of outermost tail feather and tip buffish white. The sexes are alike. Wing 78 to 95 mm. The young bird is paler in colour, has broader dark centres and more fawn coloured edges to the feathers of the upperparts.

**Range in Eastern Africa:** British Somaliland to eastern Uganda and central Tanganyika Territory.

**Habits:** A common species of the Masai steppe-country with the same 'flappeting' crackling flight as the Flappet Lark. There is a curious 'stalling' display flight, and the birds perch freely on trees.

**Nest and Eggs:** A typical Lark's nest of grass, well concealed. South African eggs of the nominate race are whitish, heavily streaked and mottled with light brown and slate blue; about 21·5 × 14·5 mm.

**Recorded breeding:** Tanganyika Territory (breeding condition), March, May and October.

**Call:** A short sweet song uttered from a perch.

*Mirafra africanoides alopex* Sharpe.

*Mirafra alopex* Sharpe, Cat. Bds. Brit. Mus. 13, p. 617, 1890: Haud (approx. long. 45° × lat. 8°), eastern Abyssinia. **Pl. 54.**

**Distinguishing characters:** Rather smaller and richer coloured than the preceding race, with narrower blackish streaks above; deeper buffish white below, and with small dark spots on upper chest. The sexes are alike. Wing 81 to 87 mm.

**Range in Eastern Africa:** South eastern British Somaliland and eastern Abyssinia.

**Habits:** A bush country species perching on the top bough of a tree and making a short soaring rather quivering flight while uttering a sweet fluty song reminiscent of a Woodlark in Europe.

**Recorded breeding:** No records.

**Distribution of other races of the species:** South Africa, the nominate race being described from south-eastern Bechuanaland.

## 662 RUSTY BUSH-LARK. *MIRAFRA RUFA* Lynes.

*Mirafra rufa rufa* Lynes. **Pl. 54.**

*Mirafra rufa* Lynes, Bull. B.O.C. 41, p. 15, 1920: Juga Juga, near El Fasher, western Sudan.

**Distinguishing characters:** Above, warm tawny, with black streaks mainly confined to the head and neck, less distinct on mantle; edges of wing-coverts deep buff; below, creamy buff with tawny streaks on upper chest. The sexes are alike. Wing 78 to 88 mm. The young bird has the feathers of the upperparts tipped with buff and subterminally tipped with black; upper chest more spotted than streaked with dark brown.

**Range in Eastern Africa:** Western Sudan.

**Habits:** A bird of rather Pipit-like characteristics; it has a relatively long tail and usually perches on bushes when flushed. It is a common species in the open bush of many parts of the western Sudan. Its courtship flight is a prolonged aerial cruise at some height while singing, terminated by a sudden vertical drop to a bush or to the ground.

**Nest and Eggs:** Undescribed.

**Recorded breeding:** Sudan, June to September.

**Call:** Unrecorded. Song not particularly described.

*Mirafra rufa lynesi* Grant and Praed.

*Mirafra rufa lynesi* C. Grant and Mackworth-Praed, Bull. B.O.C. 53, p. 246, 1933: Delami, Koalib, Kordofan Province, south central Sudan.

**Distinguishing characters:** Above, more cinnamon brown than the nominate race and lacking, or almost lacking, the dark centres to the feathers of the mantle. Wing 83 to 86 mm.

**Range in Eastern Africa:** Kordofan Province of the Sudan.

**Habits:** As for the nominate race.

**Recorded breeding:** No records.

## 663 COLLARED LARK. *MIRAFRA COLLARIS* Sharpe.

*Mirafra collaris* Sharpe, Bull. B.O.C. 5, p. 24, 1896: Aimola, southern Abyssinia. **Pl. 54.**

**Distinguishing characters:** Above, rich cinnamon brown, streaked with white; neck mottled blackish and white; flight feathers

# LARKS

black with whitish tips and white edging at base of primaries; upper tail-coverts grey with whitish edges; below, pale tawny; throat white; a large black patch at base of neck. The sexes are alike. Wing 78 to 84 mm. The young bird has black spots on the upperparts and wing-coverts and the patch at the base of the neck replaced by dull blackish spots.

**Range in Eastern Africa:** British and Italian Somaliland to south-eastern Abyssinia and north-eastern Kenya Colony.

**Habits:** Inhabits patches of brick red soil which it matches exactly, and is exceedingly hard to see on the ground. It prefers to run and squat and does not fly if it can avoid it. Usually occurs singly, and appears to be shy and uncommon, except on the Tana River, Kenya Colony, where Serle found it in some numbers. In the breeding season it has the normal short ascending flight with clapping wings.

**Nest and Eggs:** A typical Lark's nest in the shelter of a weed or grass tuft. Eggs three, white spotted with olive and reddish brown; about 20 × 15 mm.

**Recorded breeding:** Garra-Liwin country, southern Abyssinia, May.

**Call:** A curiously plaintive whistling song in an ascending scale made during the descent from a flight or after the bird has perched on a tree.

**664 GILLETT'S LARK.** *MIRAFRA GILLETTI* Sharpe.
*Mirafra gilletti* Sharpe, Bull. B.O.C. 4, p. 29, 1895: Sibbe, Ogaden, eastern Abyssinia. **Pl. 54.**

**Distinguishing characters:** Not unlike the Fawn-coloured Lark in appearance but differs in being more chestnut brown above, dark streaks less distinct, having the rump and upper tail-coverts greyish, and streaks on upper chest pale chestnut. The sexes are alike. Wing 79 to 90 mm. The young bird is much paler above, more blobbed than streaked with blackish, and the feathers broadly edged and tipped with buff; below, upper chest buff with blackish spots.

**Range in Eastern Africa:** Eastern Abyssinia to British and Italian Somalilands.

**Habits:** A bird of rather better watered hilly country, not of dry steppe, locally common.

**Nest and Eggs:** A well-made nest of fine stems and rootlets, on the ground in a grass tuft well covered over. Eggs three, dull white

with heavy olive-brown and sparse reddish-brown flecking, and grey undermarkings; about 20 × 15 mm.

**Recorded breeding:** Eastern Abyssinia, April and May.

**Call:** Song usually from a bough, not very loud, but also in soaring flight, a pleasant little 'zi-zi-zi-de-ti-e-o' (Erlanger).

### 665 PINK-BREASTED LARK. *MIRAFRA PŒCILOSTERNA* (Reichenow).

*Mirafra pœcilosterna pœcilosterna* (Reichw.). **Pl. 54.**
*Alauda pœcilosterna* Reichenow, O.M. p. 155, 1879: Kibaradje, Tana River, Kenya Colony.

**Distinguishing characters:** Above, brown with lighter edges to some of the feathers, especially inner secondaries and wing-coverts; wings and tail ashy brown, tail longish; below, whitish, tawny on flanks, chest and sides of head; lightish edges to feathers of upper chest giving a streaked appearance. The sexes are alike. Wing 84 to 94 mm. The young bird has the upperside mottled dusky and sandy; throat and breast rather more rusty coloured with blackish spots on lower throat and upper breast.

**Range in Eastern Africa:** Eastern Uganda to Kenya Colony south of about lat. 1° N., southern Italian Somaliland and north-eastern Tanganyika Territory.

**Habits:** A bird of thorn scrub country, locally common. It has few of the habits of a Flappet-Lark and may not be one. It perches freely on trees and bushes and usually sings from a perch on the topmost branch. Abundant in the Turkana country of Kenya Colony and has more the appearance of a Pipit.

**Nest and Eggs:** Undescribed.

**Recorded breeding:** Northern Guaso Nyiro, Kenya Colony, March.

**Call:** A thin grasshopper-like 'tweet' usually uttered three times, and the song is a flat unmusical trill (Moreau).

*Mirafra pœcilosterna australoabyssinica* Benson.
*Mirafra pœcilosterna australoabyssinicus* Benson, Bull. B.O.C. 63, p. 13, 1942: Mega, southern Abyssinia.

**Distinguishing characters:** Differs from the nominate race in being greyer above and rather paler below. Wing 85 to 97 mm.

**Range in Eastern Africa:** Southern Abyssinia to south-eastern

Sudan, northern Kenya Colony as far south as Lokitaung, northern
Turkana, and Lasamis on Marsabit-Archer's Post road.

**Habits:** As for the nominate race. Found in arid country at
3,000 feet.

**Recorded breeding:** South Abyssinia (breeding condition),
March.

### 666 DUSKY BUSH-LARK. *PINAROCORYS NIGRICANS* (Sund.).

*Alauda nigricans* Sundevall, Œfv. Vet. Akad, Förh. p. 99, 1850:
Aapies River, Pretoria district, Transvaal.

**Distinguishing characters:** Above, dark sooty brown including wings and tail; some light or tawny edges to flight feathers and wing-coverts; outer web of outer tail feathers pale tawny; lores, some patches round eye and on sides of face and throat white; below, buffish white with broad sooty brown spots and streaks from chest to upper belly; inner webs of primaries mainly pale chestnut. The sexes are alike. Wing 107 to 125 mm. The young bird is more fawn coloured above with tawny edges to the flight feathers and wing-coverts.

**General distribution:** Angola to southern Belgian Congo, Tanganyika Territory, Nyasaland, eastern Transvaal and Natal.

**Range in Eastern Africa:** Kigoma and Mpanda, western Tanganyika Territory and Shiré River, Nyasaland in non-breeding season.

**Habits:** A rather Thrush-like bird of open ground among trees, whose habits are little recorded.

**Nest and Eggs:** Nest neatly made and well bedded into the ground, of broad grasses unlined. Eggs two, glossy greenish blue with pronounced hair lines of dark brown or umber; about 24·5 × 15·5 mm.

**Recorded breeding:** Zululand, September.

**Call:** Unrecorded.

### 667 RED-TAILED BUSH-LARK. *PINAROCORYS ERYTHROPYGIA* (Strickland).

*Alauda erythropygia* Strickland, P.Z.S. for 1850, p. 219, pl. 24, 1852:
Kordofan, south central Sudan.

**Distinguishing characters:** Adult male, differs from the Dusky Bush-Lark in having the upper tail-coverts, basal half of tail and outer

tail feathers tawny chestnut. Birds in worn dress are generally much browner. The female has the inner webs of the primaries tawny chestnut, but this character is sometimes found in the male. Wing 99 to 116 mm. The young bird is similar to the adult female but browner above with light tips to the feathers and white tips to the wing-coverts.

**General distribution:** Ivory Coast to the Sudan and Uganda.

**Range in Eastern Africa:** Western and southern Sudan to north-western Uganda.

**Habits:** An interesting species with fine powers of flight and song, which appears to be confined to dry bush country. On the wing they appear very dark, with broad wings and a buoyant undulating flight; on the ground they look tall birds, run quickly and are very shy. Frequently perch in trees. In the breeding season they make a fine circling soaring flight, often to a great height, and the song is loud and very beautiful. Occasionally congregate in numbers after a grass fire.

**Nest and Eggs:** Undescribed.

**Recorded breeding:** Nigeria (breeding condition), March. Bahr-el-Ghazal, February and March. South-eastern Sudan (nestling), January. Uganda (nestling), February.

**Call:** Song as above.

**668** LONG-CLAWED LARK. *HETEROMIRAFRA RUDDI* (C. Grant).

*Heteromirafra ruddi archeri* S. Clarke. **Pl. 55.**

*Heteromirafra archeri* Stephenson Clarke, Bull. B.O.C. 40, p. 64, 1920: Jifa, western border of British Somaliland.

**Distinguishing characters:** A large-headed, short-tailed bird; above, mottled and streaked black and buffish white; a semblance of a lighter collar on the hind neck; hind claw long and almost straight; tail feathers narrower and pointed; below, pale buff with small brown spots on chest and lower neck. The sexes are alike. Wing 74 to 83 mm. The young bird has broader buffish white edges and tips to the feathers of the upperparts.

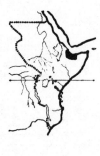

**Range in Eastern Africa:** Western British Somaliland.

**Habits:** A distinct looking Lark of rather unusual appearance owing to its large head and short tail, of which little is known.

**Nest and Eggs:** The nest of the South African race is well made of grass, usually in or under a tussock. Eggs three, pale pinkish drab, heavily stippled with light brown and with undermarkings of slate colour; about 21 × 15 mm.

**Recorded breeding:** No records.

**Call:** The South African race has a distinctive whistling bubbling song, uttered while hovering high in the air.

**Distribution of other races of the species:** South Africa, the nominate race being described from the Transvaal.

**669** SOMALI LONG-BILLED LARK. *CERTHILAUDA SOMALICA* Witherby. **Pl. 55.**

*Certhilauda somalica* Witherby, Bull. B.O.C. 14, p. 29, 1903: Dibbit, near Galkayu, northern Italian Somaliland.

**Distinguishing characters:** Similar in appearance to the Red Somali Lark, but is not quite so rich in colour, has a longer bill, longer hind claw, outer web and tip of the outermost tail feather white, and the under tail-coverts streaked with brown. The sexes are alike. Wing 91 to 105 mm. Juvenile plumage unrecorded.

**Range in Eastern Africa:** Central British Somaliland to northern Italian Somaliland.

**Habits:** No information.

**Nest and Eggs:** One clutch only known. Eggs four, white, speckled with shades of brown, somewhat zoned; about 22 × 16 mm.

**Recorded breeding:** British Somaliland, no date given. Italian Somaliland, September.

**Call:** Unrecorded.

**670** HOOPOE-LARK. *ALÆMON ALAUDIPES* (Desfontaine). *Alæmon alaudipes desertorum* (Stan.).
*Alauda desertorum* Stanley, in Salt's Trav. Abyss. App. p. 60, 1814: Amphila Island, Eritrea.

**Distinguishing characters:** A long-billed, long-tailed Lark, greyish brown above; below, white with large black spots on upper chest; tips of secondaries and inner primaries, and outer web of outermost tail feather white; bases of secondaries and inner primaries white, forming a large white patch in wing. The sexes are alike. Wing 105 to 125 mm. The young bird has black subterminal spots on the wing-coverts and scapulars with pale tips.

**General distribution:** The coastlands and islands of the Red Sea from Suez-Cairo desert to Eritrea, Abyssinia, British and Italian Somalilands and south-western Arabia.

**Range in Eastern Africa:** Coastlands of the Sudan, Eritrea, north-eastern Abyssinia, British Somaliland and northern Italian Somaliland.

**Habits:** Desert country birds usually seen in pairs and small family parties. They might easily be mistaken for Coursers on the ground, and they run swiftly and fly as little as possible. In flight the broad white wing-bars are very noticeable. They are usually very tame birds. Their strong bills are used for digging and they destroy large numbers of locust pupæ. Courtship flight of the male is a little upward soar with a melodious note becoming tremulous as the flight finishes, then a downward floating on outstretched wings.

**Nest and Eggs:** Nest, at times large and untidy, of any material, usually in the shade of a bush—at other times practically no nest is made and the eggs are laid on bare ground. Eggs two, white with rufous spots or blotches and lilac undermarkings; about $22 \times 16$ mm.

**Recorded breeding:** Red Sea coast, February to April.

**Call:** A long-drawn melodious call, uttered from the ground or from a bush, or in the breeding season during a display flight. The call has what is described as an 'instrumental' or clarion tone.

*Alæmon alaudipes meridionalis* (Brehm.).
*Certhilauda meridionalis* A. E. Brehm, J.f.O. p. 77, 1854: Dongola, Sudan.

**Distinguishing characters:** More sandy buff on upperparts than the preceding race. Wing 108 to 131 mm.

**Range in Eastern Africa:** The Nile Valley from Dongola to Khartoum, Sudan.

**Habits:** As for other races.

**Recorded breeding:** Omdurman (breeding condition), November.

**Distribution of other races of the species:** Cape Verde Islands and Rio de Oro to Egypt and Arabia, the nominate race being described from Tunisia.

# LARKS

**671 LESSER HOOPOE-LARK.** *ALÆMON HAMERTONI* Witherby.

*Alæmon hamertoni hamertoni* With. **Pl. 55.**

*Alæmon hamertoni* Witherby, Ibis, p. 513, 1905: Obbia, Italian Somaliland.

**Distinguishing characters:** Very similar above to the Hoopoe-Lark, but smaller with a shorter bill and tail; no distinct spots on upper chest and no white in wings. The sexes are alike. Wing 101 mm. Juvenile plumage unrecorded.

**Range in Eastern Africa:** Northern Italian Somaliland.

**Habits:** No information.

**Nest and Eggs:** Undescribed.

**Recorded breeding:** No records.

**Call:** Unrecorded.

*Alæmon hamertoni altera* With.

*Alæmon hamertoni altera* Witherby, Ibis, p. 514, 1905: El Afweena, Warsangeli, eastern British Somaliland.

**Distinguishing characters:** General colour more sandy buff than the Lesser Hoopoe-Lark. Wing 91 to 108 mm.

**Range in Eastern Africa:** Eastern British Somaliland.

**Habits:** No information.

**Recorded breeding:** No records.

*Alæmon hamertoni tertia* S. Clarke.

*Alæmon hamertoni tertia* Stephenson Clarke, Bull. B.O.C. 40, p. 219, 1919: Arori Plain, near Burao, central British Somaliland.

**Distinguishing characters:** General colour more tawny buff than either of the preceding races. Wing 93 to 108 mm.

**Range in Eastern Africa:** Central British Somaliland.

**Habits:** No information.

**Recorded breeding:** No records.

**672 BLACK-TAILED SAND-LARK.** *AMMOMANES CINCTURA* (Gould).

*Ammomanes cinctura kinneari* Bates. **Pl. 55.**

*Ammomanes cinctura kinneari* Bates, Bull. B.O.C. 55, p. 140, 1935: 50 miles south of Omdurman, Sudan.

**Distinguishing characters:** Above, plain biscuit colour with dusky tips to flight feathers and subterminal dusky spots on ends of

tail feathers; below, pale biscuit and white. The sexes are alike. Wing 84 to 93 mm. The young bird has underlying dull black spots and broken barring on the upperparts, wing-coverts, throat and chest.

**General distribution:** French Sudan to Red Sea coast of the Sudan.

**Range in Eastern Africa:** Northern areas of the Sudan.

**Habits:** Seen in small flocks in desert country. They run very fast and stop suddenly and frequently. Their flight is fairly swift and at times darting and jerky, and the call is reminiscent of a child's trumpet. Locally common.

**Nest and Eggs:** Nest in rock crevices or under stones making a little garden of pebbles outside the entrance. Eggs undescribed.

**Recorded breeding:** Hills of Red Sea Province, Sudan, March and April.

**Call:** As above.

**Distribution of other races of the species:** Cape Verde Islands, North Africa, Arabia to Iran, the nominate race being described from Cape Verde Islands.

**673** SAND-LARK. *AMMOMANES DESERTI* (Lichtenstein).
*Ammomanes deserti akeleyi* Ell. **Pl. 55.**
*Ammomanes akeleyi* Elliott, Field Colum. Mus. Publ. Orn. 1, p. 39, 1897: Deragodlet, Berbera plains, British Somaliland.

**Distinguishing characters:** General colour plain buff brown; flight feathers and tail dusky with light tips; rump, tail-coverts and base of tail more tawny; dull blackish spots on upper chest. The sexes are alike. Wing 85 to 95 mm. The young bird has narrow light tips to the feathers of the upperparts including wing-coverts and flight feathers; below, the spots on the upper chest are just indicated.

**Range in Eastern Africa:** Central and eastern British Somaliland and northern Italian Somaliland.

**Habits:** As for other races, an inhabitant of stony rocky ground.

**Nest and Eggs:** As for other races. The nests of all this group are extremely hard to find, as they are placed behind stones or in crevices and in consequence are little known.

**Recorded breeding:** British Somaliland, February.

# LARKS

*Ammomanes deserti samharensis* Shell.

*Ammomanes samharensis* Shelley, Bds. Afr. 3, p. 99, pl. 21, 1902: Amba, near Massowa, Eritrea.

**Distinguishing characters:** General colour above more ashy brown than the preceding race. Wing 92 to 103 mm.

**Range in Eastern Africa:** Eastern Sudan and northern Eritrea.

**Habits:** Inhabits rocks or stony ground among scattered bushes.

**Recorded breeding:** No records.

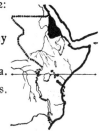

*Ammomanes deserti assabensis* Salvad.

*Ammomanes assabensis* Salvadori, Boll. Mus. Zool. Anat. Comp. Torino 17, No. 454, p. 2, 1902: Assab, Danakil coast, southern Eritrea.

**Distinguishing characters:** General colour above darker, more sooty brown than in the preceding races. Wing 88 to 94 mm.

**Range in Eastern Africa:** North-western British Somaliland and French Somaliland to the Danakil country of eastern Abyssinia and southern Eritrea.

**Habits:** Those of other races, it is locally common in the foothills along the coast.

**Recorded breeding:** No records.

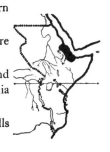

*Ammomanes deserti erythrochroa* Reichw.

*Ammomanes lusitana erythrochroa* Reichenow, J.f.O. p. 307, 1904: Ambukol, Dongola, northern Sudan.

**Distinguishing characters:** General colour above more tawny than in any of the preceding races. Wing 89 to 102 mm.

**Range in Eastern Africa:** Western to central Sudan.

**Habits:** In pairs or small parties among low rocky jebels or wadis and broken ground. Usually shy.

**Nest and Eggs:** The nest is usually in a hole or crevice with a little bank of pebbles outside. Eggs probably three or four, white or cream spotted and speckled with brown and with purplish grey undermarkings; about 22 × 16 mm.

**Recorded breeding:** Shendi, Sudan, February.

**Call:** A plaintive 'tweet' as they flit about among rocks.

**Distribution of other races of the species:** Sahara to Sinai, the nominate race being described from Upper Egypt.

## LARKS

### 674 DUNN'S LARK. *AMMOMANES DUNNI* (Shelley).
*Ammomanes dunni dunni* (Shell.). **Pl. 55.**
*Calendula dunni* Shelley, Bull. B.O.C. 14, p. 82, 1904: Ogageh Wells, Kordofan, Sudan.

**Distinguishing characters:** Very similar to the Black-tailed Sand-Lark, but has a stouter bill; darker biscuit-coloured streaks above; the tail feathers, with the exception of the four central ones, are wholly black with light edges. The sexes are alike. Wing 74 to 83 mm. The young bird has small white spots on the upperside.

**General distribution:** Damergu and Air to the Sudan.

**Range in Eastern Africa:** Western Sudan to the Kordofan Province.

**Habits:** Locally common on open grass plains in a very restricted habitat. Tame and difficult to flush or to see on the ground. Habits little noted. It can be distinguished in flight from the local Flappet Larks by the absence of white on the outer tail feathers.

**Nest and Eggs:** Undescribed.

**Recorded breeding:** Darfur (breeding condition), February and March, also young in February.

**Call:** Unrecorded.

**Distribution of other races of the species:** Arabia.

### 675 CRESTED LARK. *GALERIDA CRISTATA* (Linnæus).
*Galerida cristata isabellina* Bp.
*Galerida isabellina* Bonaparte, Consp. Av. 1, p. 245, 1850: Khartoum, Sudan.

**Distinguishing characters:** Above, pale isabelline with slightly darker centres to feathers; head distinctly crested; first primary shorter than primary coverts; below, buffish white and pale isabelline, with dull blackish brown streaks on upper chest; under wing-coverts tawny. The sexes are alike. Wing 99 to 110 mm. The young bird has the feathers of the crown, crest and mantle with subterminal dark spots and buff tips.

**Range in Eastern Africa:** Western and central Sudan from Jebel Meidob and El Fasher to Khartoum.

**Habits:** Birds of open country, preferring cultivation and the neighbourhood of water when available. Usually tame. They often have a very restricted distribution locally. Abundant along the Nile near Khartoum. They have a pleasing little song.

**Nest and Eggs:** Nest, a little circular hollow in bare ground as a rule, lined with a few pieces of fine dry grass or weed leaves. Eggs two or three, buff or yellowish white, densely speckled with various shades of pale brown and with pale purplish undercloudings; probably about 22 × 16 mm.

**Recorded breeding:** Western Sudan, October to December. Khartoum, Sudan, December to March.

**Call:** In the breeding season the song is often uttered from the ground, and is short and simple.

*Galerida cristata altirostris* Brehm.

*Galerida altirostris* C. L. Brehm, Vögelfang, p. 124, 1855: Akasheh, Dongola Province, Sudan.

**Distinguishing characters:** General colour above rather darker in tone than in the preceding race and having a longer and narrower bill. Wing 95 to 109 mm.

**General distribution:** South and south-eastern Asia Minor to Palestine, the Sudan and Eritrea.

**Range in Eastern Africa:** Eastern Sudan and Eritrea.

**Habits:** As for other races.

**Recorded breeding:** Red Sea coast, February.

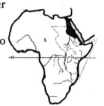

*Galerida cristata somaliensis* Reichw.

*Galerida cristata somaliensis* Reichenow, J.f.O. p. 49, 1907: Zeila, British Somaliland.

**Distinguishing characters:** General colour above rather darker and colder in tone than in either of the preceding races. Wing 93 to 106 mm.

**Range in Eastern Africa:** British Somaliland to north-western Kenya Colony.

**Habits:** Occurring in open desert country, but otherwise with the same habits as other races.

**Recorded breeding:** British Somaliland, January.

*Galerida cristata zalingei* Grant and Praed.
*Galerida cristata zalingei* Claude Grant and Mackworth-Praed, Bull.
B.O.C. 59, p. 141, 1939: Zalingei, western Sudan.

**Distinguishing characters:** General tone above rather richer isabelline than in neighbouring races and having a smaller bill. Wing 93 to 103 mm.

**Range in Eastern Africa:** Jebel Marra, Zalingei and Kallokitting, Darfur, western Sudan.

**Habits:** As for other races, but again a desert form.

**Recorded breeding:** Darfur, October to December.

**Distribution of other races of the species:** Europe to Korea and to Arabia, the nominate race being described from Austria.

**676** SHORT-CRESTED LARK. *GALERIDA MALABARICA* (Scopoli).

*Galerida malabarica prætermissa* (Blanf.). **Pl. 55.**
*Alauda prætermissa* Blanford, Ann. Mag. N.H. (4) 4, p. 330, 1869: Senafe, Acchele Guzai District, Eritrea.

**Distinguishing characters:** Similar in appearance to the Crested Lark; above, broadly mottled black and brown; below, buff with large black spots on upper chest; head slightly crested; first primary longer than primary coverts; under wing-coverts dusky. The sexes are alike. Wing 94 to 104 mm. The young bird is browner above, with buffish white tips and edges to the feathers.

**Range in Eastern Africa:** Eritrea to central Abyssinia.

**Habits:** Local and uncommon on the plateau of north-western Abyssinia, at 8,000 feet and over. Habits very much those of the preceding species.

**Nest and Eggs:** Nest of dried grass, roots and weeds on the ground. Eggs two or three, greyish white with brown markings and underlying purplish grey or light brown; about 22·5 × 16·5 mm.

**Recorded breeding:** Abyssinia, February and March, also a nestling in November.

**Call:** Unrecorded, the call of other races is a low whistle.

*Galerida malabarica elliotti* Hart.
*Galerida elliotti* Hartert, Nov. Zool. 4, p. 144, 1897: Dagahbur, British Somaliland.

**Distinguishing characters:** General colour much paler and more sandy than the preceding race. Wing 87 to 103 mm.

**Range in Eastern Africa:** British Somaliland to northern Italian Somaliland.

**Habits:** As for the Crested Lark, found among short grass and boulders.

**Recorded breeding:** Northern frontier of Kenya Colony (breeding condition), April.

*Galerida malabarica huriensis* Benson.

*Galerida theklæ huriensis* Benson, Bull. B.O.C. 68, p. 9, 1947: Huri Hills, northern Kenya Colony, 35 miles south of Mega in southern Abyssinia.

**Distinguishing characters:** Similar to the Eritrean race in the colour of the upperside; but below, whiter, less buff. Wing 90 to 101 mm.

**Range in Eastern Africa:** Northern Kenya Colony, east of Lake Rudolf.

**Habits:** As for other races.

**Recorded breeding:** No records.

**Distribution of other races of the species:** South-western Europe, north Africa and India, the nominate race being described from Malabar.

**677 SUN LARK.** *HELIOCORYS MODESTA* (Heuglin).

*Heliocorys modesta modesta* (Heugl.). **Pl. 55.**

*Galerida modesta* Heuglin, J.f.O. p. 74, 1864: Bongo, Bahr-el-Ghazal, southern Sudan.

**Distinguishing characters:** A small darkish Lark, warm brown above streaked with black; below, pale brown; throat whitish; black streaks on upper chest. The sexes are alike. Wing 75 to 84 mm. The young bird has white spots on the tips of the feathers of the upperparts.

**Range in Eastern Africa:** Southern Sudan to northern Uganda.

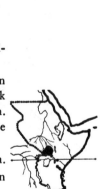

**Habits:** Birds of bare ground or short grass, not uncommon locally in pairs or small parties. They are often seen on roads or ant-hills and are not unlike Short-toed Larks in habits. They have a delightful silvery song uttered either on the ground or while hovering in the air. Usually quite tame.

**Recorded breeding:** No records.

**Nest, Eggs, etc.:** See under the western race.

*Heliocorys modesta bucolica* (Hartl.).

*Miraffra bucolica* Hartlaub, Zool. Jahrb. 2, p. 327, 1887: Tamayá, south-western Sudan.

**Distinguishing characters:** General colour above much darker than the preceding race and the black streaks on chest much broader and blacker. Wing 78 to 86 mm.

**General distribution:** Northern Belgian Congo and southern French Equatorial Africa to the Sudan.

**Range in Eastern Africa:** South-western Sudan at Wau, Aza, Tamayá, Kuderma and Kabayendi.

**Habits:** As for other races.

**Recorded breeding:** Wau (breeding condition), July.

*Heliocorys modesta giffardi* Hart.

*Heliocorys modesta giffardi* Hartert, Bull. B.O.C. 10, p. 5, 1899: Gambaga, Gold Coast, West Africa.

**Distinguishing characters:** Light edges to feathers of mantle much paler, more sandy than in either of the preceding races. Wing 77 to 86 mm.

**General distribution:** Northern Gold Coast to the Sudan.

**Range in Eastern Africa:** Darfur, western Sudan.

**Habits:** As for the nominate race, and somewhat migratory. They inhabit the western basins of Darfur from May to August, but breed elsewhere.

**Nest and Eggs:** Nest frequently among stones, occasionally in the open, a cup-shaped depression in the ground, scantily lined with grass or fibre. Eggs probably one or two, glossy creamy white, almost completely covered with reddish brown spots; about 20 × 15 mm., but only one egg measured.

**Recorded breeding:** Gold Coast, January and February. Nigeria, November to February.

**Call:** A small piping note and the song noted previously.

**Distribution of other races of the species:** West Africa.

### 678 SHORT-TAILED LARK. *PSEUDALÆMON FREMANTLII* (Phillips).

*Pseudalæmon fremantlii fremantlii* (Phill.). **Pl. 55.**

*Calendula fremantlii* Phillips, Bull. B.O.C. 6, p. 46, 1897: Gedais, British Somaliland.

**Distinguishing characters:** Above, streaked and mottled light and dark brown; bill long and heavy for size of bird; tail short; black streak from gape to under eye and along moustachial area; below, buffish white, throat whiter; upper chest with small brown spots; a small black patch on each side of upper chest. The sexes are alike. Wing 83 to 93 mm. The young bird has the feathers of the upperparts, secondaries and primaries tipped with white; only an indication of brown spots on chest.

**Range in Eastern Africa:** British and northern Italian Somaliland.

**Habits:** No information, a courtship flight as in the Hoopoe Lark is recorded.

**Recorded breeding:** British Somaliland (fledgling), January.

**Nest, Eggs, Call, etc.:** See under following race.

*Pseudalæmon fremantlii delamerei* Sharpe.

*Pseudalæmon delamerei* Sharpe, Bull. B.O.C. 10, p. 102, 1900: Athi River, Kenya Colony.

**Distinguishing characters:** General colour warmer in tone than in the nominate race and with darker and broader markings on upperparts. Wing 80 to 87 mm.

**Range in Eastern Africa:** South central Kenya Colony to the Arusha area, north-eastern Tanganyika Territory.

**Habits:** In Kenya Colony, a rare bird of the Athi and Kapiti plains. It utters a curious sharp unmistakable note on rising.

**Nest and Eggs:** A small grass nest in no way differing from those of many other Larks. Eggs two, white, well speckled and spotted at the larger end with irregular markings of brown and dark grey and with violet grey undermarkings; about 23 × 15 mm.

**Recorded breeding:** Kenya Colony, Athi Plains, May. Tanganyika Territory, Arusha (breeding condition), December.

**Call:** As above.

*Pseudalæmon fremantlii megaensis* Benson.

*Pseudalæmon fremantlii megaensis* Benson, Bull. B.O.C. 67, p. 25, 1946: 15 miles north of Mega, southern Abyssinia.

**Distinguishing characters:** Similar to the preceding race but larger. Wing 85 to 92 mm. The young bird is much darker than that of the nominate race.

**Range in Eastern Africa:** Southern Abyssinia.

**Habits:** Not uncommon in short grass country in flocks of ten to twenty when not breeding.

**Recorded breeding:** No records, young about two months old were seen in September.

### 679 CHESTNUT-BACKED SPARROW-LARK. *EREMOPTERIX LEUCOTIS* (Stanley).

*Eremopterix leucotis leucotis* (Stan.).

*Loxia leucotis* Stanley in Salt's Trav. Abyssinia, App. p. 60, 1814: Coast of Eritrea.

**Distinguishing characters:** Adult male, a small Finch-like Lark; head, neck all round, wing shoulders and underside, also under wing-coverts, black; cheeks, ear-coverts, and a half collar on occiput white; mantle, wing-coverts, other than those on shoulder, and innermost secondaries chestnut; rump greyish white. The female has the chestnut replaced by black except on the lesser wing-coverts, the ear-coverts and cheeks are black. A cinnamon brown phase occasionally occurs. Wing 73 to 82 mm. The young bird has the feathers of the upperparts tipped with white or buff.

**Range in Eastern Africa:** Kordofan, Nile Valley and Mongalla, Sudan to Abyssinia and Eritrea.

**Habits:** Nice cheerful little birds of sandy or bare stony ground, usually seen in pairs or small parties, tame and at times abundant. Occasionally they gather in large flocks and raid crops like Weaver birds, but as a rule they prefer uninhabited country. They fly round in a circle at some height when alarmed. On the ground all species of this genus appear Finch-like, but they run and do not hop.

**Nest and Eggs:** Nest a shallow cup or pad of dry grass with a few bits of cotton down or hair in a depression in the ground. Egg usually only one, greyish white thickly speckled with olive brown and grey with a tendency to zoning; about 19 × 14 mm.

Plate 54

Singing Bush-Lark (p. 6)  
Mirafra cantillans marginata

Fawn-coloured Lark (p. 17)  
Mirafra africanoides intercedens

Pink-breasted Lark (p. 20)  
Mirafra poecilosterna poccilosterna

Short-toed Lark (p. 39)  
Calandrella brachydactila brachydactila

Collared Lark (p. 18)  
Mirafra collaris

Fawn-coloured Lark  
Mirafra a. alopex (p. 17)

Rusty Bush-Lark (p. 18)  
Mirafra rufa rufa

Gillett's Lark  
Mirafra gilletti (p.19)

Kordofan Bush-Lark (p. 8)  
Mirafra cordofanica

Flappet-Lark (p. 14)  
Mirafra rufocinnamomea fischeri

Northern White-tailed Bush-Lark  
Mirafra albicauda (p. 7)

Red-winged Bush-Lark (p. 9)  
Mirafra hypermetra

Rufous-naped Lark (p. 11)  
Mirafra africana tropicalis

Red Somali Lark (p. 10)  
Mirafra sharpei

R. Green and C. E. Talbot Kelly

**Long-clawed Lark** (p. 22)
*Heteromirafra ruddi*
**Black-tailed Sand-Lark** (p. 25)
*Ammomanes cinctura kinneari*
Sun Lark (p. 31)
*Heliocorys modesta modesta*
Red-capped Lark (p. 42)
*Calandrella cinerea saturatior*
Obbia Lark (p. 44)
*Spizocorys obbiensis*

Somali Long-billed Lark
*Certhilauda somalica* (p. 23)
Sand-Lark (p. 26)
*Ammomanes deserti akeleyi*
Short-tailed Lark (p. 33)
*Pseudalaemon fremantlii fremantlii*
Rufous Short-toed Lark (p. 41)
*Calandrella rufescens athensis*
Masked Lark (p. 45)
*Aethocorys personata personata*

Lesser Hoopoe-Lark (p. 25)
*Alaemon hamertoni hamertoni*
Short-crested Lark (p. 30)
*Galerida malabarica praetermissa*
Dunn's Lark (p. 28)
*Ammomanes dunni dunni*
Blanford's Lark (p. 44)
*Calandrella blanfordi blanfordi*
Fischer's Sparrow-Lark (p. 38)
*Eremopterix leucopareia*

# LARKS

**Recorded breeding:** Nile Valley, May, also December and January. Eastern Abyssinia (nestling), December.

**Call:** A sharp rattling little call or song in the breeding season.

*Eremopterix leucotis melanocephala* (Licht.).

*Alauda melanocephala* Lichtenstein, Verz. Doubl. p. 28, 1823: Nubia.

**Distinguishing characters:** Differs from the nominate race in having the wing shoulder white and chestnut, instead of black. Wing 72 to 78 mm.

**General distribution:** Senegal to the middle Nile Valley.

**Range in Eastern Africa:** The southern, western and central Sudan to the Nile Valley from Berber to Khartoum.

**Habits:** Of the preceding race with which it intergrades.

**Recorded breeding:** West Africa, November to February. Sudan, mostly December to March.

*Note:* Birds from the Nile Valley north of about Fashoda are intermediate in wing character between this and the nominate race.

*Eremopterix leucotis madaraszi* (Reichw.).

*Pyrrhulauda leucotis madaraszi* Reichenow, O.M. p. 78, 1902: Tabora, Tanganyika Territory.

**Distinguishing characters:** Differs from the nominate race in having the rump more dusky, not dull white. Wing 75 to 84 mm.

**General distribution:** Kenya Colony to northern Portuguese East Africa and Nyasaland.

**Range in Eastern Africa:** Kenya Colony and Manda Island to Tanganyika Territory, Portuguese East Africa and Nyasaland.

**Habits:** Of the other races. Commonly found on recently burnt ground. Possibly migratory to some extent. Vincent notes that a pair will frequently fly round at a considerable height in single file and return to the same place.

**Recorded breeding:** Northern Tanganyika Territory, May and June. Nyasaland, April and May, also September. Portuguese East Africa, March and April.

**Distribution of other races of the species:** South Africa.

### 680 WHITE-FRONTED SPARROW-LARK. *EREMOPTERIX NIGRICEPS* (Gould).

*Eremopterix nigriceps albifrons* (Sund.).
*Coraphites albifrons* Sundevall, Œfv. K. Vet. Förh. 7, p. 127, 1850: Nubia.

**Distinguishing characters:** Adult male, differs from the Chestnut-backed Sparrow-Lark in having a white forehead and the whole upperside pale stone colour. The female is pale tawny above, buffish white below, with black under wing-coverts and outer tail feathers. In life the long cheek feathers of the adult males stand out like whiskers. Wing 72 to 82 mm. The young bird has blackish subterminal tips to the feathers of the crown; mantle with terminal buff tips.

**General distribution:** French Sudan to the Nile Valley.

**Range in Eastern Africa:** Western Sudan to the Nile Valley.

**Habits:** A common and locally abundant species inhabiting open country or sparse bush. Habits much those of the last species with which it often associates, but it is a distinctly shyer bird as a rule. By day they keep rather to the shade of bushes when possible. Curious little courtship flight, the male soaring up with a little twittering note and parachuting down in a butterfly-like fashion with wings over back. Makes considerable local movements in the non-breeding season.

**Nest and Eggs:** The nest consists of a tiny cup or pad of dry grass stems etc. in a hollow in the ground usually by a tuft of grass. Eggs two, greyish white sparsely speckled with olive brown and grey.

**Recorded breeding:** North of Timbuktu, French Sudan, apparently September and October. Darfur, Sudan, December to March. Omdurman, Sudan, December and January.

**Call:** A little chirping rather Pipit-like song during the breeding season flight.

*Eremopterix nigriceps melanauchen* (Cab.).
*Coraphites melanauchen* Cabanis, Mus. Hein. 1, p. 124, 1851: Dahlak Island, Red Sea.

**Distinguishing characters:** Differs from the preceding race in having the black of the crown of the head extending on to the upper back. Wing 72 to 82 mm.

**General distribution:** Egypt to Italian Somaliland and Socotra Island.

**Range in Eastern Africa:** Eastern Sudan and Dahlak Island to Abyssinia, British and Italian Somalilands and Socotra Island.

**Habits:** Of the last race, locally abundant and somewhat migratory, occasionally in very large flocks. The song is a tinkling whistle made during a spiral flight, and the call is a liquid chirrup.

**Recorded breeding:** Red Sea coast, February. British Somaliland, April. Socotra Island, February.

**Distribution of other races of the species:** Cape Verde Islands, North Africa, Iraq, Arabia and India, the nominate race being described from St. Jago, Cape Verde Islands.

### 681 CHESTNUT-HEADED SPARROW-LARK. *EREMOPTERIX SIGNATA* (Oustalet).

*Eremopterix signata signata* (Oust.).

*Pyrrhulauda signata* Oustalet, Bibl. école haut. Etud. 31, art. 10, p. 9, 1886: northern Italian Somaliland.

**Distinguishing characters:** Adult male, head, base of bill, chin to chest chestnut and black; chest to under tail-coverts and under wing-coverts black; spot on crown of head, ear-coverts, cheeks and flanks white; rest of upperparts brown. The female is mottled grey and blackish above; below, whitish, with black confined to under wing-coverts, lower belly and under tail-coverts; sides of face and chest washed with tawny, the latter with small dull blackish spots. Wing 74 to 81 mm. The young bird is very similar to the adult female, but is white from breast to under tail-coverts.

**Range in Eastern Africa:** Northern Kenya Colony to British Somaliland and northern Italian Somaliland.

**Habits:** Locally not uncommon in small parties and flocks up to fifty in stony desert scrub country. Quite common in southern Turkana, Kenya Colony.

**Nest and Eggs:** Undescribed.

**Recorded breeding:** South Abyssinia (breeding condition), March.

**Call:** A sharp 'chip-op' (Tomlinson).

*Eremopterix signata harrisoni* (O. Grant).
*Pyrrhulauda harrisoni* O. Grant, Bull. B.O.C. 11, p. 30, 1900: South end of Lake Rudolf.

**Distinguishing characters:** Differs from nominate race in being darker above, more earth brown, less tawny. The female has the chin to lower neck vinous brown. Wing 76 to 81 mm.

**Range in Eastern Africa:** South-eastern Sudan to north-western Kenya Colony and Lake Rudolf.

**Habits:** As for the nominate race.

**Recorded breeding:** No records.

### 682 FISCHER'S SPARROW-LARK. *EREMOPTERIX LEUCOPAREIA* (Fischer and Reichenow). Pl. 55, Ph. vii.

*Coraphites leucopareia* Fischer and Reichenow, J.f.O. p. 55, 1884: Klein Arusha, north-eastern Tanganyika Territory.

**Distinguishing characters:** Adult male, very similar to the Chestnut-headed Sparrow-Lark, but has the crown of the head brown, and the space between eye and base of bill, and chin to chest black. The female differs from that of that species in having the upperside browner and the whole underside duskier with dull blackish streaks. Wing 71 to 80 mm. The young bird has whitish tips to the feathers of the upperparts.

**General distribution:** Kenya Colony to Nyasaland.

**Range in Eastern Africa:** Kenya Colony and Tanganyika Territory to Likoma Island, eastern Lake Nyasa.

**Habits:** A bird of short grass plains or bare patches in grass country. Very often seen on roads in little flocks of five to eight. Common in parts of Kavirondo and in dry short grass country in Tanganyika Territory.

**Nest and Eggs:** Nest a shallow scrape in the ground lined with short pieces of dry grass. Eggs two or three, creamy white or greyish white spotted or blotched with brown or sepia and with mauve grey undermarkings; about 19 × 12 mm.

**Recorded breeding:** Kenya Colony, March to June. Tanganyika Territory, May and June. Nyasaland (breeding condition), April and May.

**Call:** A low 'tweezee' and a soft twitter.

# LARKS

**683 CALANDRA LARK.** *MELANOCORYPHA BIMACULATA* (Ménétriés).

*Melanocorypha bimaculata bimaculata* (Mén.).

*Alauda bimaculata* Ménétriés, Cat. Rais. Zool. Caucase, p. 37, 1832: Talych Mts., south of Baku, Black Sea.

**Distinguishing characters:** A large Lark with large black patches on the sides of the upper chest; mottled brown above, mainly white below. The sexes are alike. Wing 113 to 128 mm.

**General distribution:** Kirghiz Steppes to Transcaspia, northern Iran and Afghanistan; in non-breeding season to the Sudan, Arabia and India.

**Range in Eastern Africa:** The Sudan as far south as lat. 14° N. in non-breeding season.

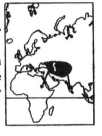

**Habits:** A palæarctic winter migrant to north-eastern Africa usually keeping in flocks. They are shy birds, but occur in some numbers in most years, usually in open country. In the field they appear stockily built, without any white in the tail or wing.

*Melanocorypha bimaculata rufescens* Brehm.

*Melanocorypha rufescens* C. L. Brehm, Vögelfang, p. 120, 1855: Blue Nile, Sudan.

**Distinguishing characters:** General tone above rather more buff, warmer in tone, than in the nominate race. Wing 111 to 127 mm.

**General distribution:** Asia Minor and Syria; in non-breeding season to Egypt, Sinai, the Sudan and Eritrea; once recorded from Swakopmund, South-West Africa.

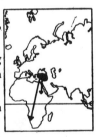

**Range in Eastern Africa:** The Sudan, as far south as Khartoum and Eritrea in non-breeding season.

**Habits:** As for the preceding race, and indistinguishable in the field.

**684 SHORT-TOED LARK.** *CALANDRELLA BRACHYDACTILA* (Leisler).

*Calandrella brachydactila brachydactila* (Leisl.). **Pl. 54.**

*Alauda brachydactila* Leisler, Ann. Wett. Ges. 3, p. 357, 1814: Montpellier, Herault, south of France.

**Distinguishing characters:** Above, buffish brown, streaked and mottled with blackish brown; below, white, buffish on chest with no

streaks or spots; a small blackish patch on each side of upper chest; first primary very small and rudimentary or missing, second longest. The sexes are alike. Wing 86 to 96 mm. The young bird has subterminal black ends and terminal whitish tips to the feathers of the upperparts.

**General distribution:** Southern Europe, Crete, Asia Minor and North Africa; in non-breeding season to the Sudan, British Somaliland, south-western Arabia and India.

**Range in Eastern Africa:** The Sudan, Eritrea and British Somaliland to about lat. 12° N. in non-breeding season.

**Habits:** A common palæarctic winter migrant to the Sudan, sometimes in vast numbers, with individual flocks composed of from fifty birds to many thousands.

*Calandrella brachydactila hermonensis* Trist.

*Calandrella hermonensis* Tristram, P.Z.S. p. 434, 1864: Hermon, Palestine.

**Distinguishing characters:** General tone above warmer, more buff and with more indication of chestnut on the crown than in the nominate race. Wing 87 to 101 mm.

**General distribution:** North Africa to Palestine; in non-breeding season to the southern Sahara, the Sudan, Eritrea and south-western Arabia.

**Range in Eastern Africa:** The Sudan and Eritrea to about lat. 12° N. in non-breeding season.

**Habits:** Not uncommon palæarctic winter visitor to the Sudan. In the southern French Sahara Bates met with it commonly and noted that it was usually in pairs.

*Calandrella brachydactila longipennis* (Evers.).

*Alauda longipennis* Eversmann, Bull. Soc. Imp. Nat. Moscow 21, p. 219, 1848: Lake Ala Kul, Dzungaria, Semiryechensk Province, Turkestan, eastern Russia.

**Distinguishing characters:** General tone above greyer and colder than the nominate race. Wing 85 to 103 mm.

**General distribution:** Transcaspia, Turkestan and Afghanistan; in non-breeding season to Lake Chad, the Sudan, Egypt, Sinai, south-western Arabia and north-western India.

# LARKS

**Range in Eastern Africa:** Western Sudan in non-breeding season.

**Habits:** Fairly common western Asiatic visitor to the Sudan and north-eastern Africa.

## 685 RUFOUS SHORT-TOED LARK. *CALANDRELLA RUFESCENS* Vieillot.

*Calandrella rufescens somalica* (Sharpe).
*Alaudula somalica* Sharpe, P.Z.S. p. 472, 1895: Haud, British Somaliland.

**Distinguishing characters:** Above, mottled dark and light brown; below, buffish white; narrow blackish brown spots on upper chest. The sexes are alike. Wing 81 to 93 mm. Not unlike the Singing Bush-Lark in appearance but paler in bill and general colour, and has a long second primary, equal in length to the other primaries. Juvenile plumage unrecorded.

**Range in Eastern Africa:** Borama, eastern Abyssinia and British Somaliland.

**Habits:** Often found in long grass in open glades in woodland.

**Recorded breeding:** No records.

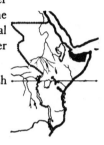

*Calandrella rufescens athensis* (Sharpe). **Pl. 55.**
*Spizocorys athensis* Sharpe, Bull. B.O.C. 10, p. 101, 1900: Athi River, Kenya Colony.

**Distinguishing characters:** Differs from the Somaliland race in being less sandy coloured and with blacker markings on the upperparts. Wing 81 to 91 mm. Similar in general appearance to the Northern White-tailed Bush-Lark, but may be distinguished by the larger black spots on the upper chest, by the first primary being very small and rudimentary and the second primary being of normal length, not very short as in that species, and there is no brown or chestnut edging to the flight feathers.

**Range in Eastern Africa:** Southern Kenya Colony to north-eastern Tanganyika Territory.

**Habits:** Local but not uncommon on the Athi Plains, Kenya Colony.

**Nest and Eggs:** Nest a rather deep cup, lined with grass, at the base of a grass tuft. Eggs two or three, rather Pipit-like of whitish ground colour, closely speckled with sepia and darker brown, particularly at the larger end; about 18 × 14·5 mm.

**Recorded breeding:** Athi Plains, probably March and April. Arusha, Tanganyika Territory (breeding condition), December.

**Call:** A low-pitched 'piri-pip.' A breeding season song is also uttered in flight.

*Calandrella rufescens megaensis* Benson.

*Calandrella somalica megaensis* Benson, Bull. B.O.C. 67, p. 26, 1946: Ten miles north of Mega, southern Abyssinia.

**Distinguishing characters:** Differs from the preceding race in being warmer, more buffish in tone, and from the Somali race in having blacker markings on the upperparts. Wing 82 to 91 mm.

**Range in Eastern Africa:** Southern Abyssinia.

**Habits:** Found in open grass country with scattered low thorn trees at 3,000 to 4,500 feet, also on open arid plains.

**Recorded breeding:** No records.

**Distribution of other races of the species:** Europe, Asia and North Africa, the nominate race being described from Tenerife, Canary Islands.

**686 RED-CAPPED LARK.** *CALANDRELLA CINEREA* (Gmelin).

*Calandrella cinerea saturatior* Reichw. **Pl. 55, Ph. vii.**

*Calandrella cinerea saturatior* Reichenow, Vög. Afr. 3, p. 376, 1904: Kondeland, south-western Tanganyika Territory.

**Distinguishing characters:** Above, mottled dark brown and blackish; head chestnut; below, buffish white; chest darker; sides of chest and flanks chestnut. The sexes are alike. Wing 85 to 98 mm. The young bird is almost black above with white spots at tips of feathers.

**General distribution:** Uganda, Kenya Colony and the Belgian Congo to the Rhodesias and Nyasaland.

**Range in Eastern Africa:** Uganda and Kenya Colony to Nyasaland.

**Habits:** A locally common species of open plains and dry sandy country, congregating into flocks in the non-breeding season. They are tame little birds, running fast on the ground and twittering and fluttering their wings as they feed. Strong undulating flight not usually protracted, but they are locally migratory to a considerable extent. The male displays in the breeding season, and sings from high in the air.

**Nest and Eggs:** A small nest of dry grass at the side of a grass tuft, often with a little bank of earth round it. Eggs two, whitish, heavily mottled and clouded with brown or rufous brown and with dusky purple undermarkings; about 21 × 15 mm.

**Recorded breeding:** Uganda, June. Kenya Colony, May and June, also November and December. Tanganyika Territory, March to May, also (breeding condition) January. Nyasaland, June to October. Rhodesia, June to September.

**Call:** A continuous twittering is made by a feeding flock, and they call as they rise. The male has a musical song uttered high in the air while hovering or during a flight of alternate soarings and dives.

*Calandrella cinerea erlangeri* (Neum.).
*Tephrocorys cinerea erlangeri* Neumann, J.f.O. p. 239, 1906: Sheikh
    Mohamed, headwaters of Juba River, south-central Abyssinia.

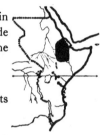

**Distinguishing characters:** Differs from the preceding race in having darker centres to the feathers of the mantle; the underside more diffused with chestnut; and black patches on each side of the lower neck. Wing 80 to 96 mm.

**Range in Eastern Africa:** Northern and central Abyssinia.

**Habits:** A local race on grassland of high plateaux, habits presumably similar to those of the preceding race.

**Recorded breeding:** No records.

*Calandrella cinerea asmaraensis* Smith.
*Calandrella cinerea asmaraensis* K. D. Smith, Bull. B.O.C. 71, p. 55,
    1951: Asmara, Eritrea.

**Distinguishing characters:** Differs from the preceding races in having the forehead and crown paler chestnut; dark markings of mantle and scapulars smaller and the brown edges to the feathers broader, showing much less black; chest and flanks paler, more deep buff than chestnut. Wing 87 to 90 mm.

**Range in Eastern Africa:** The plateau of central and southern Eritrea.

**Habits:** A local race frequenting open agricultural land and stony moorlands. In non-breeding season in parties up to forty.

**Nest and Eggs:** Nest on bare ploughed fields and grassy meadows, nest scanty, of dry grass bents only, and sometimes in a hollow on bare ground, eggs two.

**Recorded breeding:** May and June, also October.

**Distribution of other races of the species:** South Africa, the habitat of the nominate race.

### 687 BLANFORD'S LARK. *CALANDRELLA BLANFORDI* (Shelley).

*Calandrella blanfordi blanfordi* (Shell.). **Pl. 55.**

*Tephrocorys blanfordi* Shelley, Bds. Afr. 3, p. 128, pl. 21, fig. 2, 1902: Senafé, Eritrea.

**Distinguishing characters:** Not unlike the last species but smaller in size and paler in colour; red cap much less distinct; no chestnut on sides of chest or down flanks; small black patches on sides of upper chest. The sexes are alike. Wing 75 to 85 mm. Juvenile plumage unrecorded.

**Range in Eastern Africa:** Southern Eritrea, northern Abyssinia and British Somaliland.

**Habits:** No information.

**Nest and Eggs:** Undescribed.

**Recorded breeding:** No records.

**Call:** Unrecorded.

**Distribution of other races of the species:** Arabia.

### 688 OBBIA LARK. *SPIZOCORYS OBBIENSIS* Witherby.

*Spizocorys obbiensis* Witherby, Ibis, p. 514, 1905: Obbia, Italian Somaliland. **Pl. 55.**

**Distinguishing characters:** A small Lark with a rather heavy bill and short tail; above, mottled blackish brown and whitish buff; below, white, streaked with blackish brown on chest and flanks; first primary very short and narrow. Presumably the sexes are alike. Wing 67 mm. Juvenile plumage unrecorded.

**Range in Eastern Africa:** Northern Italian Somaliland.

**Habits:** No information.

**Nest and Eggs:** Undescribed.

**Recorded breeding:** No records.

**Call:** Unrecorded.

# LARKS

**689 MASKED LARK.** *AËTHOCORYS PERSONATA* (Sharpe).

*Aëthocorys personata personata* (Sharpe). **Pl. 55.**

Spizocorys personata Sharpe, P.Z.S. p. 471, 1895: Sassabana, near Milmil, eastern Abyssinia.

**Distinguishing characters:** Lores, under eyes and moustachial area black; above, pale brown, sparingly striped with blackish brown; below, pale cinnamon brown; chest more brown; throat white; bill and feet pale yellow or pale brown. Presumably the sexes are alike. Wing 86 mm. Juvenile plumage unrecorded.

**Range in Eastern Africa:** Eastern Abyssinia.

**Habits:** As for other races.

**Nest and Eggs:** Undescribed.

**Recorded breeding:** No records.

**Call:** Unrecorded.

*Aëthocorys personata intensa* Roths.

Aëthocorys personata intensa Rothschild, Bull. B.O.C. 51, p. 100, 1931: Chanler's Falls, Northern Guaso Nyiro, Kenya Colony.

**Distinguishing characters:** Differs from the nominate race in being much darker and in having the upperparts more chocolate brown. Wing 85 to 92 mm.

**Range in Eastern Africa:** Central Kenya Colony.

**Habits:** As for other races.

**Recorded breeding:** No records.

*Aëthocorys personata yavelloensis* Benson.

Aëthocorys personata yavelloensis Benson, Bull. B.O.C. 68, p. 9, 1947: 10 miles west of Yavello, southern Abyssinia.

**Distinguishing characters:** Above, darker than the nominate race, greyer with blacker centres to feathers; chest rather greyer; much colder and greyer in tone than the Kenya Colony race. The sexes are alike. Wing 85 mm.

**Range in Eastern Africa:** Southern Abyssinia.

**Habits:** An apparently scarce or local species of which little is known. In southern Abyssinia they were found on bare or scantily grassed ground covered with small lava boulders and with a few scattered thorn-bushes.

**Recorded breeding:** South Abyssinia (breeding condition), July.

Names in Sclater's *Syst. Av. Æthiop.* 2, 1930, which have been changed, or have become synonyms in this work:

*Mirafra cheniana schillingsi* Reichenow, treated as synonymous with *Mirafra cantillans marginata* Hawker.
*Mirafra africana ruwenzoria* Kinnear, treated as synonymous with *Mirafra africana tropicalis* Hartert.
*Mirafra africana dohertyi* Hartert, treated as synonymous with *Mirafra africana athi* Hartert.
*Mirafra fischeri zombae* O. Grant, treated as synonymous with *Mirafra rufocinnamomea fischeri* (Reichenow).
*Mirafra fischeri degeni* O. Grant, treated as synonymous with *Mirafra rufocinnamomea rufocinnamomea* (Salvadori).
*Mirafra africanoides longonotensis* Van Someren, treated as synonymous with *Mirafra africanoides intercedens* Reichenow.
*Mirafra pœcilosterna massaica* (Fischer and Reichenow), treated as synonymous with *Mirafra pœcilosterna* (Reichenow).
*Galerida theklæ* Brehm, now a race of *Galerida malabarica* (Scopoli).
*Tephrocorys cinerea anderssoni* (Tristram), treated as synonymous with *Calandrella cinerea cinerea* (Gmelin).
*Tephrocorys cinerea ruficeps* (Rüppell), now *Calandrella cinerea erlangeri* (Neumann).

Names introduced since 1930, and which have become synonyms in this work:

*Mirafra candida* Friedmann, 1930, treated as synonymous with *Mirafra cantillans marginata* Hawker.
*Mirafra pulpa* Friedmann, 1930, treated as synonymous with *Mirafra cantillans marginata* Hawker.
*Tephrocorys cinerea fuertesi* Friedmann, 1932, treated as synonymous with *Calandrella cinerea erlangeri* (Neumann).
*Eremopterix signata cavei* Grant and Praed, 1941, treated as synonymous with *Eremopterix signata harrisoni* (O. Grant).

## Addenda and Corrigenda

**654** *Mirafra c. marginata*. *Recorded breeding:* add Eritrea February to June.
*M.c. chadensis*. *Recorded breeding:* add Eritrea.

**655** *Mirafra albicauda*. *Recorded breeding:* add south-western Tanganyika Territory, April and May.

**658** *Mirafra sharpei*. *Distinguishing characters:* add Wing 90 to 107 mm. *Habits:* Not uncommon south of Hargeisa, on plains, calling from the tops of low bushes or anthills. Display flight a short run of a few feet into the air then a swerving glide to the ground. Produces a hollow-sounding noise while 'flappeting' like two pieces of wood being knocked together (G. Clarke).

**661** *Mirafra africanoides macdonaldi* White.
*Mirafra africanoides macdonaldi* White, Bull. B.O.C. 73, p. 88, 1953: Yavello, southern Abyssinia. *Distinguishing characters:* Darker and warmer in colour than the eastern Abyssinian race, with more distinct streaks; but streaking above narrower than in the Tanganyika race. Wing 80 to 92 mm. *Range in Eastern Africa:* Yavello and Mega area, southern Abyssinia.

**662** *Mirafra rufa*. Add *Distribution of other races of the species:* West Africa.

**670** *Alæmon a. desertorum*. *Recorded breeding:* add Probably commences to breed in November.

*Continued on p.* 1101

# WAGTAILS

FAMILY—**MOTACILLIDÆ. WAGTAILS.** Genera: *Motacilla* and *Budytes*.

Twelve species occur in Eastern Africa, some of which are migrants from Europe and Asia.

The Wagtails are divisible into two groups, the White and Grey Wagtails which frequent water, and are often found round human habitations, and the shorter-tailed Yellow Wagtails (*Budytes*) which frequent and breed in open cultivated land or marshes. In Africa the habits of the two groups converge to some extent, as on migration the Yellow Wagtails are largely found near water also; but in the breeding season the latter have far more the habits of Pipits. Their calls also are quite distinct from the 'chissick' of the other Wagtails, and they usually nest on flat ground. They are liable to earth staining and some of the females and young birds, especially those of the Blue-headed Wagtail, fade to a brownish or brownish-grey on the upperside, having in this state little or no green colour, and the underside becomes very white, with little or no yellow.

## KEY TO THE ADULT WAGTAILS OF EASTERN AFRICA

1 Mantle black: AFRICAN PIED WAGTAIL *Motacilla aguimp* **691**

2 Mantle olivaceous grey: WELLS' WAGTAIL *Motacilla capensis wellsi* **693**

3 Mantle green or with a greenish wash: YELLOW WAGTAIL GROUP *Budytes* species **695** to **701**

4 Mantle grey: 5–8

5 Below, yellow: GREY WAGTAIL *Motacilla cinerea* **694**

6 Below, white: 7–8

7 Four outermost tail feathers white, side of head white or whitish: WHITE WAGTAIL *Motacilla alba* **690**

8 Eight outermost tail feathers white, side of head grey: MOUNTAIN WAGTAIL *Motacilla clara* **692**

## WAGTAILS

**690** WHITE WAGTAIL. *MOTACILLA ALBA* Linnæus.

*Motacilla alba alba* Linn.
*Motacilla alba* Linnæus, Syst. Nat. 10th ed. p. 185, 1758: Sweden.

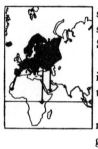

**Distinguishing characters:** Adult male, crown of head, nape and broad band across chest black; mantle, rump and wing-coverts grey; forehead, face, chin, throat and belly white. The female usually has some grey feathers in the black of the crown of head and nape. In breeding dress the chin to chest band is black. Wing 81 to 94 mm. The young bird has the crown of head and nape grey and much less white on the forehead.

**General distribution:** Iceland, Europe to the Ural Mts. and south to the Mediterranean; in non-breeding season to Africa as far south as Gambia, Uganda and Kenya Colony; southern Arabia and Socotra Island.

**Range in Eastern Africa:** As far south as southern Kenya Colony in non-breeding season, probably a straggler to Tanganyika Territory and Nyasaland.

**Habits:** Regular palæarctic winter visitor to the Sudan, found round wells and other suitable places. They arrive rather late, generally mid-October, sometimes in very large numbers. They do not go far south as a rule and are only uncommon visitors to Kenya Colony and Uganda.

*Motacilla alba forwoodi* O. Grant and Forbes.
*Motacilla forwoodi* O. Grant and Forbes, Bull. Liverpool Mus. 2, p. 3, 1899: Abd-el-Kuri Island, near Socotra Island.

**Distinguishing characters:** Differs from the nominate race in having the forehead grey and the black of the chin to chest including the chest band slightly brownish or bronzy, not so deep a black. Wing 82 to 84 mm. The young bird also has the forehead grey and the narrower chest band slightly washed with brownish.

**Range in Eastern Africa:** Island of Abd-el-Kuri, near Socotra Island.

**Habits:** Indistinguishable from those of the nominate race.

**Nest and Eggs:** Undescribed.

**Recorded breeding:** No records.

**Call and Food:** No information.

**Distribution of other races of the species:** Siberia, Iran and China.

## WAGTAILS

**691 AFRICAN PIED WAGTAIL.** *MOTACILLA AGUIMP* Dumont.

*Motacilla aguimp vidua* Sund.

*Motacilla vidua* Sundevall, Œfv. Vet. Ak. Förh. 7, p. 128, 1850: Assouan, Upper Egypt.

**Distinguishing characters**: Whole upperside, base of bill to ear-coverts and broad band across chest black; rest of plumage white. The sexes are alike. Wing 85 to 101 mm. The young bird has the black of the adult replaced by ashy grey. It may be at once distinguished from the young of the White Wagtail by the white on the outer webs at the base of the primaries. Occasionally subject to albinism.

**General distribution**: Sierra Leone and Upper Egypt southwards to Angola and the eastern Cape Province.

**Range in Eastern Africa**: Throughout, except British and French Somalilands.

**Habits**: The common Wagtail of Africa, found in nearly every village. They are tame and fearless of man and are very generally regarded by natives as bringing good luck, in fact it is considered rather ominous if a pair of Wagtails do not colonize any new group of huts. They occur along rivers or in any other well-watered localities, and like the European Wagtails roost in numbers together in reed beds.

**Nest and Eggs**: Nest a cup of leaves and dry grass lined with hair and fine rootlets, in huts, buildings or river banks, etc. Eggs two or three, sometimes more, pale brownish, greyish or greenish ground colour thickly freckled with darker brown; about 22 × 15 mm. They are parasitized by several species of Cuckoo.

**Recorded breeding**: Uganda, April to July, October to December, February and March, etc. Kenya Colony in most months of the year. Tanganyika Territory, January to March, August to October, November to December, etc., possibly three broods a year. Nyasaland, in most months, commonest September and October.

**Food**: Almost entirely insects.

**Call**: A typical Wagtail 'chizzic.' They also have a very pleasant song and are far superior performers to the European Wagtails.

The song may be uttered in flight, while running, or from a perch, and is almost as fine as that of a Canary.

**Distribution of other races of the species:** South Africa, the nominate race being described from the lower Orange River.

### 692 MOUNTAIN WAGTAIL. *MOTACILLA CLARA* Sharpe.

*Motacilla clara clara* Sharpe.

*Motacilla clara* Sharpe, Ibis, p. 341, 1908: Simien, northern Abyssinia.

**Distinguishing characters:** Above, pale clear grey including lores and ear-coverts; flight feathers black with broad pure white edges to inner secondaries; tail rather long, and all except four central feathers pure white; below, pure white with a narrow black band across chest. The sexes are alike. Wing 84 to 90 mm. The young bird is not as pure white below and has a rather obsolete chest band.

**Range in Eastern Africa:** Abyssinia.

**Habits, Nest and Eggs:** See under the following race.

**Recorded breeding:** No records.

*Motacilla clara torrentium* Ticehurst.

*Motacilla clara torrentium* C. B. Ticehurst, Bull. B.O.C. 60, p. 80, 1940: Ngoye Forest, Zululand, South Africa.

**Distinguishing characters:** Differs from the nominate race in being smaller. Wing 74 to 84 mm.

**General distribution:** Liberia to Uganda and Kenya Colony to Angola, eastern Cape Province and Natal.

**Range in Eastern Africa:** Uganda and Kenya Colony to the Zambesi River.

**Habits:** A distinctive species with the habits of the Grey Wagtail in Europe, preferring fast rocky streams and waterfalls. They are generally to be seen on rocks in running water and are much shyer than the black and white Wagtails and even more graceful in appearance.

**Nest and Eggs:** A deep cup nest of grass lined with rootlets in a cliff or on a ledge of rock by a river, occasionally in a crevice on a tree. Eggs two or three, pale brown or greyish, speckled to almost a uniform hue of light yellowish brown; about $20 \cdot 5 \times 15 \cdot 5$ mm.

Plate 56

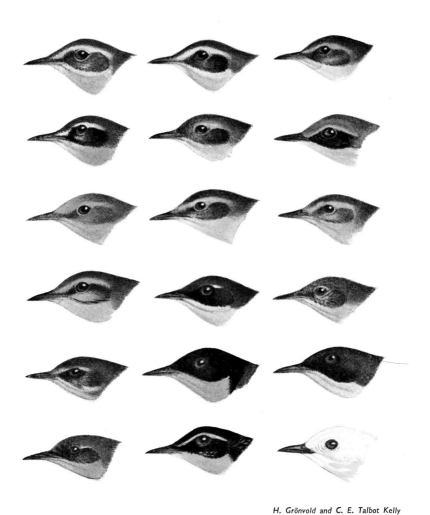

H. Grönvold and C. E. Talbot Kelly

Budytes flavus flavus, male (p. 53)
Budytes flavus dombrowskii. male (p. 53)
Budytes luteus luteus, male (p. 54)
Budytes luteus flavissima, male (p. 55)
Budytes thunbergi thunbergi, young (p. 55)
Budytes feldegg, young (p. 56)

Budytes flavus flavus, female (p. 53)
Budytes flavus dombrowskii, female (p. 53)
Budytes luteus luteus, female (p. 54)
Budytes thunbergi thunbergi, male (p. 55)
Budytes feldegg, male (p. 56)
Budytes superciliaris, male (p. 57)

Budytes flavus flavus, young (p. 53)
Budytes perconfusus. male (p 54)
Budytes luteus luteus, young (p. 54)
Budytes thunbergi thunbergi, female (p. 55)
Budytes feldegg, female (p. 56)
Budytes leucocephalus male (p. 57)

Plate 57

Sandy Plain-backed Pipit (p. 66)
*Anthus vaalensis saphiroi*
Sokoke Pipit (p. 71)
*Anthus sokokensis*
Richard's Pipit (p. 67)
*Anthus novaeseelandiae cinnamomeus*
Short-tailed Pipit (p. 72)
*Anthus brachyurus leggei*

Tree Pipit (p. 70)
*Anthus trivialis trivialis*
Striped Pipit (p. 74)
*Anthus lineiventris*
Tawny Pipit (p. 61)
*Anthus campestris campestris*
Plain-backed Pipit (p. 64)
*Anthus leucophrys omoensis*

Long-billed Pipit (p. 62)
*Anthus similis hararensis*
Little Tawny Pipit (p. 72)
*Anthus caffer blayneyi*
Malindi Pipit (p. 71)
*Anthus melindae*
Jackson's Pipit (p. 69)
*Anthus latistriatus*

**Recorded breeding:** Mt. Elgon, June. Kenya Colony, May and June. Tanganyika Territory, September and October. Nyasaland, July to September, also April.

**Food:** Insects, both land and water, small molluscs, etc.

**Call:** A sharp 'trit-trit' and a pleasing low warbling song.

**693 CAPE WAGTAIL. *MOTACILLA CAPENSIS* Linnæus.**
*Motacilla capensis wellsi* O. Grant. Wells' Wagtail.
*Motacilla wellsi* O. Grant, Bull. B.O.C. 29, p. 30, 1911: Kigesi, south-western Uganda.

**Distinguishing characters:** Above, dark olivaceous grey; head and rump darker grey; edges of flight feathers and wing-coverts dull yellowish olive; below, white washed with pale yellow, chin and throat whiter; a dull black band across chest. The sexes are alike. Wing 78 to 85 mm. The young bird is browner above, and buffish below. May be distinguished at any age from the young of the African Pied Wagtail by the absence of white on the upperside of the wings.

**General distribution:** Eastern Belgian Congo to Uganda, Kenya Colony and Tanganyika Territory.

**Range in Eastern Africa:** Southern Uganda to Kenya Colony and western Tanganyika Territory as far south as Kasulu.

**Habits:** The nominate race of this bird is South African and it is there a very tame, almost domesticated, species of farms and buildings. The present race is much scarcer and is found round swamps or the shallow bays of lakes, where it is to be seen running about on the lily pads. It is shy and restless also, and quite unlike its southern relative. Rare in Kenya Colony, rather more common in Uganda.

**Nest and Eggs:** Nest on the ground under a tuft of grass and probably also in banks or low cliffs, of dry grass lined with finer grass. Eggs three, grey or stone, heavily mottled and freckled with brown; about 21 × 15 mm.

**Recorded breeding:** Nandi, Kenya Colony, April.

**Food:** As for other species.

**Call:** A shrill distinctive chirrup.

**Distribution of other races of the species:** South Africa, the habitat of the nominate race.

### 694 GREY WAGTAIL. *MOTACILLA CINEREA* Tunstall.

*Motacilla cinerea cinerea* Tunst.

*Motacilla cinerea* Tunstall, Orn. Brit. p. 2, 1771: Wycliffe, Yorkshire, England.

**Distinguishing characters:** Above, grey; upper tail-coverts yellow, tail long; below, chin and throat white; chest to under tail-coverts yellow. In breeding dress the chin and throat are black in the male. The sexes are otherwise alike. Wing 82 to 111 mm. The young bird has the chest buff not yellow.

**General distribution:** Western Europe; in non-breeding season to Africa as far south as Gambia and Tanganyika Territory.

**Range in Eastern Africa:** To Uganda and northern Tanganyika Territory in non-breeding season.

**Habits:** Common palæarctic winter visitors to Uganda, Kenya Colony and Tanganyika Territory, mostly frequenting highland streams, and found in much the same localities as the Mountain Wagtail. A rare passage migrant in the Sudan. Heuglin considers that they may breed in some of the mountain valleys of Abyssinia.

**Distribution of other races of the species:** Madeira and the Azores, northern India and Siberia.


## YELLOW WAGTAILS. *BUDYTES* SPECIES.

**Habits:** Seven species of Yellow Wagtails visit Eastern Africa in the non-breeding season, mostly by the Nile Valley route. They congregate into large scattered flocks and several races or even species may occur in the same party. Young birds are hardly distinguishable and too little is known as yet to define their migrations or their routes. The number of Yellow Wagtails which congregate in the Nile Valley on the migration northward in the spring must be seen to be believed, but the banks of the river may be yellow with them for mile after mile and to a considerable depth. The most abundant is probably the eastern race of the Yellow Wagtail, or at times the Blue-headed. Their wintering range does not extend much south of Nyasaland.

The call of most species is a shrill 'peep-pip' and the alarm note a repeated 'peep.' In the breeding season they extend over almost the whole palæarctic region. Food very largely insects.

# WAGTAILS

## 695 BLUE-HEADED YELLOW WAGTAIL. *BUDYTES FLAVUS* (Linnæus).

*Budytes flavus flavus* (Linn.). **Pl. 56.**

*Motacilla flava* Linnæus, Syst. Nat. 10th ed. p. 185, 1758: South Sweden.

**Distinguishing characters:** Adult male, above, forehead to neck and sides of face grey; some white usually mixed with grey of cheeks; a distinct white line from bill over eye to well behind the eye; mantle to rump yellowish green; underparts rich lemon yellow, occasionally chrome yellow; usually some white on chin and between yellow throat and grey cheeks. The female is duller, having the forehead to neck more olivaceous grey and the underparts whiter, the yellow being more confined to the belly and under tail-coverts; the mantle is darker than the head. Wing 73 to 86 mm. The young bird is generally more olivaceous above; the eye stripe is buff and there are some dusky spots on upper chest; the head is slightly greyer than the mantle. Occasionally subject to albinism.

**General distribution:** Western and middle Europe to the Caspian Sea; in non-breeding season to Africa throughout and southern Arabia.

**Range in Eastern Africa:** Throughout in non-breeding season.

*Budytes flavus beema* Sykes.

*Budytes beema* Sykes, P.Z.S. p. 90, 1832: Deccan, India.

**Distinguishing characters:** Adult male, differs from the nominate race in having the grey of the head pale french grey. The female and young bird are indistinguishable from those of the nominate race. Wing 74 to 86 mm.

**General distribution:** Western Siberia to Tomsk and Turkestan; in non-breeding season to Kenya Colony, Nyasaland, Arabia and India.

**Range in Eastern Africa:** The Sudan and one record from Nairobi, Kenya Colony, in non-breeding season.

*Budytes flavus dombrowskii* Tschusi. **Pl. 56.**

*Budytes flavus dombrowskii* Tschusi, Orn. Jahrb. 14, p. 161, 1903: Pantelimon, Rumania.

**Distinguishing characters:** Adult male, similar to the nominate race, but differs from it in having the ear-coverts much darker.

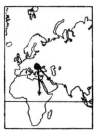

The female is very similar to that of the nominate race but has much darker ear-coverts. Wing 73 to 85 mm. The young bird is similar to that of the nominate race, but has darker ear-coverts.

**General distribution:** Rumania; in non-breeding season to Palestine, Iraq, north-eastern Africa as far south as the Sudan and Abyssinia.

**Range in Eastern Africa:** The Sudan and Abyssinia.

**Distribution of other races of the species:** South-western Europe, Siberia, Russia, Turkestan and Alaska.

### 696 YELLOW-BROWED YELLOW WAGTAIL. *BUDYTES PERCONFUSUS* Grant and Praed. Pl. 56.

*Budytes perconfusus* C. Grant and Mackworth-Praed, Bull. B.O.C. 69, p. 130, 1950: Khartoum, Sudan.

**Distinguishing characters:** Differs from the Blue-headed Yellow Wagtail in having the eye stripe from the base of the bill to well behind the eye bright yellow; chin and throat yellow. It differs from the Yellow Wagtail in having the forehead and nape grey and the lores and ear-coverts slightly olivaceous dusky grey. The female has the top of the head uniform with the mantle; eye stripe yellowish; below, chin yellow; chest buffish with some dusky spots; belly and under tail-coverts yellow. Wing 78 to 85 mm. The young bird has the eye stripe washed with yellow.

**General distribution:** From an unknown breeding area to the British Cameroons, the Sudan, Abyssinia, the Transvaal and western Arabia in the non-breeding season. Passage migrant through Denmark, Holland and Germany.

**Range in Eastern Africa:** The Sudan and Abyssinia in non-breeding season.

*Note:* The breeding area of this species is as yet unknown.

### 697 YELLOW WAGTAIL. *BUDYTES LUTEUS* (Gmelin).
*Budytes luteus luteus* (Gmel.). **Pl. 56.**
*Parus luteus* S. G. Gmelin, Reise Russ. 3, p. 101, pl. 20, fig. 1, 1774: Astrakhan, southern Russia.

**Distinguishing characters:** Adult male, differs from the Blue-headed Yellow Wagtail in having the head and cheeks yellowish-green;

and the eye stripe which extends to behind the eye, chin and throat bright yellow; the forehead is more yellow than the crown. The female has the forehead and crown olivaceous brown uniform with the mantle; eye stripe, throat and neck buffish yellow. Wing 75 to 87 mm. The young bird has the head and mantle uniform olivaceous brown, the head not being greyer or differing in colour from the mantle; eye stripe buff.

**General distribution:** The Volga River to the headwaters of the Yenesei River; in non-breeding season to south-western Europe, Malta, Asia Minor, Iran, India, Ceylon, Arabia and western and eastern Africa as far south as Sierra Leone, Gold Coast, Northern Rhodesia and the Transvaal, also Socotra Island.

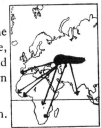

**Range in Eastern Africa:** Throughout in non-breeding season.

*Budytes luteus flavissimus* (Blyth). **Pl. 56.**
*Motacilla flavissima* Blyth, Loudons' Mag. 7, p. 342, 1834: England.

**Distinguishing characters:** Adult male, similar to the nominate race but has the forehead yellowish green uniform with the crown, with usually no pure yellow. The female and young bird are indistinguishable from those of the nominate race. Wing 72 to 87 mm.

**General distribution:** Southern Norway, British Isles, Heligoland, Holland, Belgium, northern France and the Channel Islands; in non-breeding season to Africa as far south as the Belgian Congo and Southern Rhodesia.

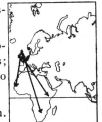

**Range in Eastern Africa:** Throughout in non-breeding season.

**Distribution of other races of the species:** Formosa Island.

**698** DARK-HEADED YELLOW WAGTAIL. *BUDYTES THUNBERGI* (Billberg).
*Budytes thunbergi thunbergi* (Bill.). **Pl. 56.**
*Motacilla thunbergi* Billberg, Syn. Faun. Scand. 1, pt. 2, Aves, p. 50, 1828: Lapland.

**Distinguishing characters:** Adult male, differs from the Blue-headed Yellow Wagtail in having the forehead to nape darker, sides of face blacker and in having no eye stripe, but sometimes a white fleck behind the eye. The female is much greyer above and whiter below than the male, and differs from the female of the Blue-headed

and Yellow Wagtails in usually having the white or buffish white eye stripe less broad; top of head and ear-coverts darker and more uniform with mantle. Wing 73 to 87 mm. The young bird has the top of the head yellowish green uniform with the mantle.

**General distribution:** Northern Scandinavia to Estonia and northern Russia as far east as the lower Yenesei River; in non-breeding season to Africa as far south as Damaraland and the Transvaal, India, Burma and the Malay Peninsula.

**Range in Eastern Africa:** Throughout in non-breeding season.

*Budytes thunbergi cinereocapillus* (Savi.).
*Motacilla cinereocapilla* Savi, N. Giorn. Dei. Lett. 22, p. 190, 1831: Tuscany, Italy.

**Distinguishing characters:** Adult male, differs from the nominate race in having the chin to neck in front white, or rarely some yellow mixed with the white of the neck and occasionally there is a short white mark behind the eye. The female is duller, less green above and less yellow below, but usually has a white or whitish chin and front of neck. Wing 76 to 83 mm. The young bird is similar to the adult female.

**General distribution:** Eastern southern France to Switzerland, Italy, Dalmatia and Algeria; in non-breeding season to northern, western and eastern Africa as far south as Senegal and Uganda; also Arabia.

**Range in Eastern Africa:** One record from Entebbe, Uganda, in non-breeding season.

**Distribution of other races of the species:** Eastern Russia, Turkestan and Egypt.

**699** BLACK-HEADED YELLOW WAGTAIL. *BUDYTES FELDEGG* (Michahelles). **Pl. 56.**
*Motacilla feldegg* Michahelles, Isis, p. 812, 1830: Split, Dalmatia, Yugo-Slavia.

**Distinguishing characters:** Adult male, differs from the preceding species in having the head from forehead to neck above and sides of face deeper black; no eye stripe; often a variable white line between the black face and yellow throat; the yellow and green of the mantle is intermixed with the black of the head on the hinder neck, and the black extends on to the upper mantle. The female also has the forehead to crown and sides of face black or ash brown, mantle ash brown with a faint wash of green; below, pale yellow, almost white on chin and throat to breast. Wing 76 to 88 mm. The young bird is very similar to that of the Dark-headed Yellow Wagtail, but usually has some indications of black on the forehead.

**General distribution:** The Balkans, Greece, Turkey, northwest Iran, Syria, to the Black and Caspian Seas and Turkestan; in non-breeding season to Eastern Africa as far south as Uganda and Kenya Colony, also southern Arabia, Socotra Island and India.

**Range in Eastern Africa:** The Sudan, Eritrea, Abyssinia, British Somaliland, Uganda, Kenya Colony and Socotra Island in non-breeding season.

**700 WHITE-EYE-BROWED YELLOW WAGTAIL.** *BUDYTES SUPERCILIARIS* Brehm. **Pl. 56.**

*Budytes superciliaris* A. E. Brehm, J.f.O. p. 74, 1854: Khartoum, Sudan.

**Distinguishing characters:** Adult male, differs from the Black-headed Yellow Wagtail in having a white stripe over the eye extending well behind the eye, the eye stripe is sometimes yellow or particoloured white and yellow. The female differs from that of the Black-headed Yellow Wagtail in having a stripe over the eye. Wing 80 to 84 mm.

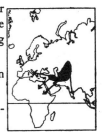

**General distribution:** Iran to eastern Russian Turkestan; in non-breeding season to the Sudan and Arabia.

**Range in Eastern Africa:** The Sudan and Eritrea in non-breeding season.

**701 WHITE-HEADED YELLOW WAGTAIL.** *BUDYTES LEUCOCEPHALUS* Przevalski. **Pl. 56.**

*Budytes leucocephala* Przevalski Zap. Ump. Akad. Nauk. St. Petersb. 55, p. 85, 1887: Dzungaria, northern Turkestan.

**Distinguishing characters:** Similar to other Yellow Wagtails but has the whole head to nape, chin and usually part of throat white or white washed with grey; sometimes a white eye stripe is distinguishable. The sexes are alike. Wing 77 to 85 mm. The

immature bird has the head and ear-coverts olivaceous grey; a white stripe from base of bill to over and behind eye; mantle washed with greyish; below, chin and throat whitish, washed with yellow; broken spotted collar at base of neck; chest to under tail-coverts paler yellow than adult.

**General distribution:** Eastern Russia to Turkestan and western Mongolia; in non-breeding season to Africa as far south as north-eastern Northern Rhodesia, northern Nyasaland, Arabia and India.

**Range in Eastern Africa:** The Sudan in non-breeding season and should occur as far south as south-western Tanganyika Territory.

Names in Sclater's *Syst. Av. Æthiop.* 2, 1930, which have been changed or have become synonyms in this work:

*Budytes flavus rayi* Bonaparte, now *Budytes luteus flavissimus* (Blyth).
*Budytes flavus campestris* (Pallas), now *Budytes luteus luteus* (Gmelin).
*Budytes flavus melanogriseus* (Homeyer), treated as synonymous with *Budytes feldegg* (Michahelles).

**PIPITS.** Family—**MOTACILLIDÆ.** Genera: *Anthus, Tmetothylacus* and *Macronyx.*

Twenty species occur in Eastern Africa, some of which are migrants from Europe, Asia and North Africa. Most of them are birds of the grasslands or savannah country, though one inhabits open spaces in woods. The majority spend most of their time on the ground and are, therefore, liable to a certain amount of earth staining. Their food consists almost exclusively of insects, particularly *Diptera*, small *Coleoptera* and *Orthoptera*, with a few seeds and small molluscs. Like all birds they take termites when available. Their nests are usually well-made cups of grass lined with finer material, and well concealed on the ground in herbage or on banks.

### Key to the Adult Pipits of Eastern Africa

| | | | |
|---|---|---|---|
| 1 | A black band across chest: | | 3–11 |
| 2 | No black band across chest: | | 12–36 |
| 3 | Underside rich salmon: | ROSY-BREASTED LONG-CLAW, male *Macronyx ameliæ* | 721 |
| 4 | Underside buff, throat saffron: | ABYSSINIAN LONG-CLAW *Macronyx flavicollis* | 719 |
| 5 | Underside yellow: | | 6–11 |

# PIPITS

| | | | |
|---|---|---|---|
| 6 | Outer tail feathers yellow: | GOLDEN PIPIT, male *Tmetothylacus tenellus* | **715** |
| 7 | Outer tail feathers blackish and white: | | **8–11** |
| 8 | Flanks and under tail-coverts white, with black streaks: | PANGANI LONG-CLAW *Macronyx aurantiigula* | **720** |
| 9 | Flanks and under tail-coverts yellow or buffish yellow: | | **10–11** |
| 10 | Underside bright yellow: | YELLOW-THROATED LONG-CLAW *Macronyx croceus* | **716** |
| 11 | Underside buffish yellow: | FÜLLEBORN'S LONG-CLAW *Macronyx fülleborni* | **717** |
| 12 | Belly pale salmon: | ROSY-BREASTED LONG-CLAW, female *Macronyx ameliæ* | **721** |
| 13 | Whole underside bright yellow: | SHARPE'S LONG-CLAW *Macronyx sharpei* | **718** |
| 14 | Underside of wings and centre of belly yellow, no streaks on chest: | GOLDEN PIPIT, female *Tmetothylacus tenellus* | **715** |
| 15 | Underside not bright yellow or salmon, chest streaked: | | **16–36** |
| 16 | Size smaller, wing under 70 mm.: | | **18–21** |
| 17 | Size larger, wing over 77 mm.: | | **22–36** |
| 18 | Broad black streaks on chest: | SOKOKE PIPIT *Anthus sokokensis* | **710** |

19 Narrow blackish streaks on chest: 20–21

20 Above, almost black with darkish centres to feathers: SHORT-TAILED PIPIT *Anthus brachyurus* 711

21 Above, brown with dark centres to feathers: LITTLE TAWNY PIPIT *Anthus caffer* 712

22 Green edges to flight and tail feathers: STRIPED PIPIT *Anthus lineiventris* 714

23 No green on edges to flight and tail feathers: 24–36

24 General colour above olivaceous brown: TREE PIPIT *Anthus trivialis* 708

25 General colour above warm brown: 26–27

26 Top of head and mantle streaked and mottled: RICHARD'S PIPIT *Anthus novæseelandiæ* 706

27 Top of head and mantle practically plain: SANDY PLAIN-BACKED PIPIT *Anthus vaalensis* 705

28 General colour above pale fawn brown: TAWNY PIPIT *Anthus campestris* 702

29 General colour above mainly black: 30–31

30 Edges of mantle feathers pale brown: RED-THROATED PIPIT *Anthus cervinus* 713

31 Edges of mantle feathers dark brown: JACKSON'S PIPIT *Anthus latistriatus* 707

32 General colour above dusky or smoky brown or earth brown: 33–36

# PIPITS

33 Upperside practically plain,
no defined edges to feathers
of head and mantle:     PLAIN-BACKED PIPIT
*Anthus leucophrys*    **704**

34 Upperside not practically
plain, feathers of head and
mantle with defined edges:    35–36

35 Streaks on underside confined to chest:     LONG-BILLED PIPIT
*Anthus similis*    **703**

36 Streaks on underside extend from chest to throat and flanks:     MALINDI PIPIT
*Anthus melindæ*    **709**

**702** TAWNY PIPIT. *ANTHUS CAMPESTRIS* (Linnæus).
*Anthus campestris campestris* (Linn.). **Pl. 57.**
*Alauda campestris* Linnæus, Syst. Nat. 10th ed. p. 166, 1758: Sweden.

**Distinguishing characters:** General colour sandy buff with darker centres to feathers of upperparts, especially noticeable on the head; edges of wing-coverts and inner secondaries more tawny; below, some small narrow blackish streaks on the chest and a narrow blackish malar stripe; second, third and fourth primaries emarginated on outer web; hind claw short and slightly curved. The sexes are alike. Wing 81 to 100 mm. The young bird has light sandy edges to the feathers of the upperparts and small brown spots on the chest.

**General distribution:** Central Sweden to the Mediterranean, Asia Minor, Turkestan, North Africa and Iran; in non-breeding season to West Africa, Lake Chad, Kenya Colony, southern Arabia and India.

**Range in Eastern Africa:** Eritrea, northern Kenya Colony and Socotra Island in non-breeding season.

**Habits:** A palæarctic winter visitor to Africa, not uncommon in the Sudan where it remains throughout the non-breeding season, and also in Somaliland. An uncommon migrant to Kenya Colony and does not appear to go further south. Usually seen singly or in pairs.

**Distribution of other races of the species:** Southern Siberia, eastern Mongolia and Manchuria.

### 703 LONG-BILLED PIPIT. *ANTHUS SIMILIS* Jerdon.

*Anthus similis nivescens* Reichw.

*Anthus nivescens* Reichenow, O.M. p. 179, 1905: Kismayu, southern Italian Somaliland.

**Distinguishing characters:** General tone above dusky with paler edges to the feathers; below, pale buff with dull dusky streaks on chest; second, third, fourth and fifth primaries emarginated on the outer web; hind claw short and slightly curved. The sexes are alike. Wing 85 to 100 mm. The young bird has more distinct dark buff edges to the feathers of the upperparts, and streaks on chest sharper and smaller, inclining to spots.

**Range in Eastern Africa:** Red Sea Province of the Sudan to British and Italian Somalilands.

**Habits, Nest, Eggs, etc.:** See under the following races.

**Recorded breeding:** Red Sea hills, April.

*Anthus similis hararensis* Neum. **Pl. 57.**

*Anthus nicholsoni hararensis* Neumann, J.f.O. p. 233, 1906: Abu Bekr, near Harar, eastern Abyssinia.

**Distinguishing characters:** General colour above darker, more dusky brown than in the preceding race, with narrow lighter edges to the feathers; below, pale brown or buffish brown with dusky streaks on chest. Wing 88 to 103 mm.

**General distribution:** Eritrea, Abyssinia and the Sudan to north-eastern Belgian Congo, Uganda, Kenya Colony and Tanganyika Territory.

**Range in Eastern Africa:** Eritrea, Abyssinia, south-eastern Sudan, Uganda, Kenya Colony and north-eastern Tanganyika Territory.

**Habits:** As for other races, a very dark looking bird in the field. A common resident in north-western Abyssinia and occurs over most of Kenya Colony. Prefers rocky grassy situations but also perches in trees. It is easily overlooked among the numbers of Richard's Pipits.

**Recorded breeding:** Uganda, May and June. Northern Tanganyika Territory (breeding condition), March.

## PIPITS

*Anthus similis nyassæ* Neum.

*Anthus nicholsoni nyassæ* Neumann, J.f.O. p. 233, 1906: Between Sangesi and Songea, south-western Tanganyika Territory.

**Distinguishing characters:** Very similar to the last race, but centres of feathers of upperparts darker and edges of feathers broader and more tawny especially top of head, mantle, wing-coverts and secondaries. Wing 85 to 104 mm.

**General distribution:** Tanganyika Territory to eastern Belgian Congo, Northern Rhodesia and Nyasaland.

**Range in Eastern Africa:** Tanganyika Territory as far north as Njombe to Nyasaland.

**Habits:** Probably only a breeding visitor to Nyasaland, when they inhabit woodland or light bush, and perch freely on trees.

**Nest and Eggs:** A deep cup nest of grass, well concealed among herbage, lined with fine grass and often with a little barrier of dead leaves and small sticks round the rim. Eggs two or three, creamy white, mottled all over with sandy brown; about 21 × 15 mm.

**Recorded breeding:** Tanganyika Territory, September to November. Nyasaland, September to November. Northern Rhodesia, September and October.

**Call:** A loud chirrup and the song is described as an unmusical 'kliddh-kliddh' or 'chwee-chwa-chwee-chwe' while soaring (Benson).

*Anthus similis sokotræ* Hart.

*Anthus sordidus sokotræ* Hartert, Nov. Zool. 24, p. 457, 1917: Alilo Pass, Socotra Island.

**Distinguishing characters:** Very similar to the neighbouring mainland race, but more distinctly marked, dark centres and light edges to feathers of upperparts more defined. Wing 82 to 91 mm.

**Range in Eastern Africa:** Socotra Island.

**Habits:** Abundant on the island, habits as for other races. The male has a very sweet song in the breeding season which is uttered either from a perch or while rising into the air.

**Recorded breeding:** November to February.

*Anthus similis jebelmarræ* Lynes.

*Anthus sordidus jebelmarræ* Lynes, Bull. B.O.C. 41, p. 16, 1920: Jebel Marra, western Sudan.

**Distinguishing characters**: Very similar to the Nyasa race but with a greyish tinge on hind neck, and streaks on the chest less defined. Wing 86 to 99 mm.

**Range in Eastern Africa**: Western Sudan.

**Habits**: Abundant on middle and higher levels of Jebel Marra and Jebel Meidob, and wanders to other hills outside the main range. It has here the habits of a Rock Pipit and is found in rocky country as long as grass is present.

**Recorded breeding**: Darfur, June to September.

**Distribution of other races of the species**: Sahara, South Africa, Arabia and India, the nominate race being described from the Nilgiris, India.

**704 PLAIN-BACKED PIPIT.** *ANTHUS LEUCOPHRYS* Vieillot.

*Anthus leucophrys omoensis* Neum. **Pl. 57.**

*Anthus leucophrys omoensis* Neumann, J.f.O. p. 234, 1906: Ergino Valley, between Gofa and Doka, south-western Abyssinia.

**Distinguishing characters**: Above, dark earth brown; faint darker streaks on top of head; buff superciliary stripe extending to well behind eye; flight feathers, wing-coverts and tail edged with buff or pale brown, outer web of outer feathers and inner web near end lighter; second, third, fourth and fifth primaries emarginated on outer webs; below, dark buff or pale brown with dusky streaks on chest; chin and throat paler; a more or less distinct dusky streak through and below eyes and on each side of throat. The sexes are alike. Wing 96 to 110 mm.; hind claw long and curved, 8 to 17 mm. The young bird has paler buff edges to the feathers of the nape, rump, wings and tail.

**Range in Eastern Africa**: Abyssinia.

**Habits**: As for other races.

**Recorded breeding**: No records.

*Anthus leucophrys zenkeri* Neum.

*Anthus leucophrys zenkeri* Neumann, J.f.O. p. 235, 1906: Jaunde, Cameroons.

**Distinguishing characters:** Differs from the Abyssinian race in being generally lighter in colour. Wing 85 to 104 mm.; hind claw 8 to 12 mm.

**General distribution:** Senegal, Gambia, Portuguese Guinea and Gold Coast to Nigeria, Cameroons, French Equatorial Africa, the Sudan, Uganda and Kenya Colony.

**Range in Eastern Africa:** Southern Sudan, Uganda and western Kenya Colony.

**Habits:** Owing to the confusion of this and the following species it is remarkably difficult to give any field notes which are specifically reliable. It is to be hoped that further observations of the birds in life may throw some light on the problem of these extremely closely related and, in the field, almost indistinguishable species, and that some differences of habits, voice, etc., may become apparent.

**Nest and Eggs:** Believed to be indistinguishable from those of the Sandy Plain-backed Pipit, a grass nest lined with finer material well concealed. Eggs three, freckled and speckled with various shades of brown and slate; about $22 \times 16$ mm.

**Recorded breeding:** See above.

**Call:** Not particularly described.


*Anthus leucophrys bohndorffi* Neum.

*Anthus leucophrys bohndorffi* Neumann, J.f.O. p. 236, 1906: Kasongo, eastern Belgian Congo.

**Distinguishing characters:** Differs from the Abyssinian race in being paler below and from the Cameroons race in having a longer hind claw. Wing 89 to 100 mm.; hind claw 12 to 17 mm.

**General distribution:** Angola to Belgian Congo, Northern Rhodesia, Nyasaland and Tanganyika Territory.

**Range in Eastern Africa:** South-western Tanganyika Territory at the Songwe River.

**Habits:** As for other races.

**Recorded breeding:** No records.

**Distribution of other races of the species:** Sierra Leone to the Ivory Coast, Angola, the Belgian Congo and South Africa, the nominate race being described from the Cape of Good Hope.

### 705 SANDY PLAIN-BACKED PIPIT. *ANTHUS VAALENSIS* Shelley.

*Anthus vaalensis saphiroi* Neum. **Pl. 57.**

*Anthus leucophrys saphiroi* Neumann, J.f.O. p. 235, 1906: Belassiri, near Harar, eastern Abyssinia.

**Distinguishing characters:** Differs from the very similar Plain-backed Pipit in being warmer brown above, including wings and tail, not dark nor dusky earth brown; below, streaks on chest often less sharp and clear; second, third, fourth and fifth primaries emarginated on outer webs, last rather indistinctly. The sexes are alike. Wing 88 to 99 mm.; hind claw long and curved, 9 to 12 mm. The young bird has pale edges to the feathers of the nape, wing-coverts and innermost secondaries.

**Range in Eastern Africa:** Abyssinia and British Somaliland.

**Habits, Nest and Eggs:** See under the following race.

**Recorded breeding:** No records.

*Anthus vaalensis goodsoni* Meinertz.

*Anthus leucophrys goodsoni* Meinertzhagen, Bull. B.O.C. 41, p. 23, 1920: Nakuru, Kenya Colony.

**Distinguishing characters:** Differs from the Abyssinian race in being much paler below, pale buff or buffish white. Wing 93 to 101 mm.; hind claw 9 to 10 mm.

**Range in Eastern Africa:** Kenya Colony east of the Rift Valley.

**Habits:** A common species of open country where the grass is short, congregating into small flocks in the non-breeding season. The breeding season flight is a soar to some thirty feet and then a swoop to the ground.

**Nest and Eggs:** A typical Pipit's nest of grass lined with finer grass and rootlets and well concealed under a grass tuft. Eggs usually three, bluish white finely but densely speckled with brown and with underlying grey blotches; about 23 × 15 mm.

**Recorded breeding:** Kenya Colony, April to June, also probably September and October.

**Call:** A little twittering breeding song and various other calls not particularly described.

**Distribution of other races of the species:** Angola to Nyasaland and South Africa, the nominate race being described from Natal.

### 706 RICHARD'S PIPIT.* *ANTHUS NOVÆSEELANDIÆ* Gmelin.

*Anthus novæseelandiæ cinnamomeus* Rüpp. **Pl. 57.**

*Anthus cinnamomeus* Rüppell, N. Wirbelt. Vög. p. 103, 1840: Simen Province, northern Abyssinia.

**Distinguishing characters:** General colour above, tawny brown, centre of feathers blackish brown; below, tawny, throat and centre of belly paler; black spots on chest; second, third and fourth primaries emarginated on outer web; hind claw rather long and curved. The sexes are alike. Wing 81 to 97 mm. The young bird has pale buffish white edges to the feathers of upperparts and wing-coverts.

**Range in Eastern Africa:** Eritrea, Abyssinia and the Sudan.

**Habits, Nest and Eggs:** As for the following race; it is locally common in parts of Abyssinia.

**Recorded breeding:** North-western Abyssinia, April.

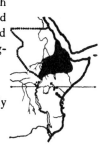

*Anthus novæseelandiæ lacuum* Meinertz. **Ph. vii.**

*Anthus richardi lacuum* Meinertzhagen, Bull. B.O.C. 41, p. 22, 1920: Lake Naivasha, Kenya Colony.

**Distinguishing characters:** General colour above duller, more olivaceous brown than in the preceding race; below, paler. Wing 79 to 93 mm.

**General distribution:** Uganda and Kenya Colony to eastern Belgian Congo, Tanganyika Territory and northern Portuguese East Africa.

**Range in Eastern Africa:** Uganda and Kenya Colony to Tanganyika Territory and Portuguese East Africa, also Pemba, Zanzibar and Mafia Islands.

* *Note:* This bird has always been known as Richard's Pipit, and we do not like to change the vernacular name, although the scientific name for the species has to be Gmelin's name for the New Zealand race.

**Habits:** A common and abundant species over most of its range and found at any elevation. In the non-breeding season it occurs in scattered flocks up to a dozen strong and prefers situations in woodland and semi-open country. It perches freely on trees and usually sings from a perch, though it also sings while soaring.

**Nest and Eggs:** Nest well concealed in a tuft or in the side of a bank. Eggs usually three, whitish ground colour heavily streaked and mottled with pale brown, and with purple or grey undermarkings; about 21 × 16 mm.

**Recorded breeding:** Uganda and Kenya Colony, all the year round, chiefly April to June, and October to December. Tanganyika Territory, April, also September to November. Mafia Island (breeding condition), June. Portuguese East Africa, March and April.

**Call:** A sharp 'prip' or a double 'pip-pit' also quite a pleasant song in the breeding season.

*Anthus novæseelandiæ annæ* Meinertz.

*Anthus richardi annæ* Meinertzhagen, Ibis, p. 656, 1921: Lake Megago, British Somaliland.

**Distinguishing characters:** General colour paler than in either of the preceding races with a greyish wash on upperparts. Wing 81 to 90 mm.

**General distribution:** British Somaliland and south-western Arabia.

**Range in Eastern Africa:** British Somaliland.

**Habits:** Presumably as for other races.

**Recorded breeding:** No records.

*Anthus novæseelandiæ lynesi* Bann. and Bates.

*Anthus rufulus lynesi* Bannerman and Bates, Ibis, p. 802, 1926: near Bamenda, Cameroons.

**Distinguishing characters:** General colour darker than the other races; centres of feathers of upperparts black; below, more tawny including throat and centre of belly; black streaks on chest broader and larger. Wing 90 to 100 mm.

**General distribution:** Cameroons; in non-breeding season to western Sudan.

## PIPITS

**Range in Eastern Africa:** Western Sudan in non-breeding season.

**Habits:** Common as a non-breeding visitor to Jebel Marra and Darfur in May and June, spending December to March further north and returning to breed to the south-west apparently in Cameroons.

**Recorded breeding:** Cameroons probably February.

*Anthus novæseelandiæ lichenya* Vinc.
*Anthus richardi lichenya* Vincent, Bull. B.O.C. 53, p. 131, 1933: Mlanje Mt. southern Nyasaland.

**Distinguishing characters:** Similar to the Kenya race but with a warmer tone above. Wing 81 to 92 mm.

**General distribution:** North-eastern Northern Rhodesia as far west as the Kafue River, Nyasaland and north-western Northern Portuguese East Africa.

**Range in Eastern Africa:** Portuguese East Africa at Unangu to Nyasaland.

**Habits:** As for other races, occurring in pairs, small parties or loose flocks in any fairly open situation. Vincent mentions that like other Pipits this species will immediately come to investigate smoke, and also that in Nyasaland and Portuguese East Africa all races of Richard's Pipit alight on the ground and do not perch in trees. There is a distinctive song-flight, undulating, and as the bird drops it utters a resonant 'ching-ching-ching-ching-ching' (Benson).

**Recorded breeding:** Nyasaland, August and December, also possibly April to July.

**Distribution of other races of the species:** Siberia to Turkestan, China, West and South Africa, India, the Malay States, Dutch East Indies and New Zealand, the nominate race being described from New Zealand.

**707 JACKSON'S PIPIT.** *ANTHUS LATISTRIATUS* Jackson.
*Anthus latistriatus* Jackson, Ibis, p. 628, 1899: Kavirondo, western Kenya Colony. **Pl. 57.**

**Distinguishing characters:** General colour above, dark blackish brown, the feathers narrowly edged with dark brown; below, buffish

brown; chin and throat lighter; black streaks on chest; some blackish brown streaks on flanks; second, third and fourth primaries emarginated on outer web; hind claw rather long and curved. The sexes are alike. Wing 87 to 94 mm. The young bird is more strongly mottled above and on chest.

**General distribution:** Eastern Belgian Congo (between Lake Edward and the north end of Lake Tanganyika) to Kenya Colony.

**Range in Eastern Africa:** Southern Uganda and western Kenya Colony.

**Habits:** No information, easily overlooked among other Pipits.

**Nest and Eggs:** Undescribed.

**Recorded breeding:** No records.

**Call:** Unrecorded.

**708 TREE PIPIT.** *ANTHUS TRIVIALIS* (Linnæus).

*Anthus trivialis trivialis* (Linn.). **Pl. 57.**

*Alauda trivialis* Linnæus, Syst. Nat. 10th ed. p. 166, 1758: Sweden.

**Distinguishing characters:** Above, olivaceous brown, with darker centres to feathers of head and mantle; edges of wing-coverts buff; below, throat to chest and flanks buff; chest with broad blackish streaks and a few streaks down flanks; belly white; second, third and fourth primaries emarginated on outer web; hind claw short and curved. The sexes are alike. Wing 77 to 95 mm.

**General distribution:** Europe and Asia; in non-breeding season to Africa as far south as the Transvaal and north-western India.

**Range in Eastern Africa:** Throughout in non-breeding season.

**Habits:** A very common passage migrant or palæarctic winter visitor to most of Eastern Africa, occasionally in flocks of some size, but more usually singly. It prefers open bush or cultivated country. It is most abundant in Kenya Colony and stragglers reach as far south as Southern Rhodesia and the Transvaal. Usually very tame.

**Distribution of other races of the species:** Eastern Siberia to China, and north-western India.

## PIPITS

**709 MALINDI PIPIT.** *ANTHUS MELINDÆ* Shelley. **Pl. 57.**
*Anthus melindæ* Shelley, Bds. Afr. 2, p. 305, 1900: Malindi, eastern Kenya Colony.

**Distinguishing characters:** Above, very similar to the Cameroons race of the Plain-backed Pipit, but size smaller; below, dull white with broad brownish-black streaks on chest and flanks; often some streaks on upper belly; second, third, fourth and fifth primaries emarginated on outer web; hind claw short and curved. The sexes are alike. Wing 83 to 85 mm. The young bird is rather pale above with broad sandy tips to feathers; streaks on chest larger.

**Range in Eastern Africa:** Coastal districts of Kenya Colony.

**Habits:** A rare and local species of the Kenya Colony coastal belt. Some were noticed on the Mombasa waterfront in August by H. F. I. Elliott who says they have a very dipping flight and a loud screeching call.

**Nest and Eggs:** Undescribed.
**Recorded breeding:** No records.
**Call:** As above.

**710 SOKOKE PIPIT.** *ANTHUS SOKOKENSIS* Van Someren.
*Anthus sokokensis* Van Someren, Bull. B.O.C. 41, p. 124, 1921: Sokoke Forest, near Mombasa, eastern Kenya Colony. **Pl. 57.**

**Distinguishing characters:** A small Pipit; above, broadly streaked with black, buff and tawny; wing-coverts tipped with buffish white; below, white, tinged with pale yellow; large and broad black streaks on chest; narrow black streaks on flanks; second, third and fourth primaries emarginated on outer web; hind claw short and very slightly curved. The sexes are alike. Wing 68 to 69 mm. The young bird is rather more sandy above; streaks on chest more distinct.

**Range in Eastern Africa:** Sokoke Forest, eastern Kenya Colony to the Pugu Hills, near Dar-es-Salaam, eastern Tanganyika Territory.

**Habits:** A local species of unusual habitat. Found in scrubby forest or undergrowth of forest or in clearings. Little has been reported about its habits.

**Nest and Eggs:** Undescribed.
**Recorded breeding:** No records.
**Call:** Unrecorded.

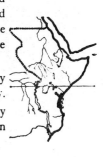

**711 SHORT-TAILED PIPIT.** *ANTHUS BRACHYURUS* Sundevall.

*Anthus brachyurus leggei* O. Grant. **Pl. 57.**

*Anthus leggei* O. Grant, Bull. B.O.C. 19, p. 26, 1906: Mokia, south-east Ruwenzori Mts., Uganda.

**Distinguishing characters:** A small, short-tailed, dark Pipit, mainly black above with dark olivaceous brown edges to feathers; below buff; belly white; chest closely streaked with black; narrower blackish streaks on flanks; second, third and fourth primaries emarginated on outer web; hind claw very short and slightly curved. The sexes are alike. Wing 60 to 67 mm. The young bird has browner, less olivaceous brown, edges to the feathers of the upperparts.

**General distribution:** Uganda, Tanganyika Territory, the Belgian Congo, north-eastern Northern Rhodesia and Angola.

**Range in Eastern Africa:** Western Uganda to north-western and south central Tanganyika Territory.

**Habits:** Apparently a highland species of local distribution and nowhere common. Usually seen singly or in pairs and is said to be found in long grass and to be difficult to flush. In the Belgian Congo, however, Lynes found it in pasture and short grass country. It has a short aerial flight and song in the breeding season.

**Nest and Eggs:** A typical Pipit's nest of grass, well concealed. Eggs three, yellowish white with heavy speckling and large confluent blotches of yellowish brown and with purplish grey undercloudings; about 15·5 × 13 mm.

**Recorded breeding:** South-east Belgian Congo, January and February. Tanganyika Territory highlands, November to February.

**Call:** Unrecorded, except as above.

**Distribution of other races of the species:** Natal and Zululand, the nominate race being described from Durban.

**712 LITTLE TAWNY PIPIT.** *ANTHUS CAFFER* Sundevall.

*Anthus caffer blayneyi* Van Som. **Pl. 57.**

*Anthus blayneyi* Van Someren, Bull. B.O.C. 40, p. 53, 1919: Olgerei, south-western Kenya Colony.

**Distinguishing characters:** A small Pipit, size as in the Short-tailed Pipit, but markings above broader with wider light edges; tail longer and light part of outer tail feathers pure white, not buffish white; below, whiter with broader blackish streaks; hind claw very short and very slightly curved. The sexes are alike. Wing 65 to 71 mm. The young bird is more sandy coloured above than the adult.

**Range in Eastern Africa:** Southern Kenya Colony from south Ukamba to north-eastern Tanganyika Territory from Ikoma to Monduli.

**Habits:** Another rare and local species from the dryer parts of southern Kenya Colony and northern Tanganyika Territory. Inhabits bush country and perches on branches on occasions. It prefers open sandy patches with grass tufts and sparse vegetation.

**Nest and Eggs:** Nest of fine grass in a hollow at the base of a grass tuft. Eggs two, occasionally three, greyish white, irregularly blotched and speckled with grey, blue grey and dark grey, and with sandy brown surface markings; about $17 \cdot 5 \times 13 \cdot 5$ mm.

**Recorded breeding:** March to June.

**Call:** A bleating call of two notes 'see-ip' and a short twittering song usually uttered from the top of a bush.

*Anthus caffer australoabyssinicus* Benson.
*Anthus caffer australoabyssinicus* Benson, Bull. B.O.C. 63, p. 12, 1942: Yavello, southern Abyssinia.

**Distinguishing characters:** Differs from the preceding race in being paler tawny in general colour. Wing 66 to 71 mm.

**Range in Eastern Africa:** Southern Abyssinia.

**Habits:** Not uncommon in arid park-like country at Yavello, at 4,500 to 5,000 feet.

**Recorded breeding:** No records.

**Distribution of other races of the species:** South Africa, the nominate race being described from the Transvaal.

**713** RED-THROATED PIPIT. *ANTHUS CERVINUS* (Pallas).
*Motacilla cervina* Pallas, Zoog. Rosso-Asiat. 1, p. 511, 1811: Kovyma River, Kamchatka, Siberia.

**Distinguishing characters:** Above, feathers edged with pale brown with well marked and broad black centres; below, creamy white, with broad black streaks on sides of neck, chest and flanks;

second, third and fourth primaries emarginated on outer web. In breeding dress the side of the head and chin to throat is vinous brown; hind claw rather long and curved. The sexes are alike. Wing 77 to 91 mm.

**General distribution:** Europe, Asia and north-western America; in non-breeding season to Africa as far south as Nigeria and Tanganyika Territory, East Indian Islands and California.

**Range in Eastern Africa:** The Sudan and Eritrea to Tanganyika Territory in non-breeding season.

**Habits:** An abundant palæarctic winter visitor to most of Eastern Africa, being particularly numerous on migration, when it consorts with Yellow Wagtails near water, but if disturbed it rises high and flies right away. It is common in Kenya Colony throughout the non-breeding season at most elevations up to high moorland, but mainly a migrant through Uganda. A number remain to winter in the Sudan in wetter places. Its call is a distinctive high-pitched metallic squeak and the flight is jerky and Finch-like. A few remain in Africa throughout the year, but these are non-breeding birds.

### 714 STRIPED PIPIT. *ANTHUS LINEIVENTRIS* Sundevall.

*Anthus lineiventris* Sundevall, Œfv. Vet. Akad. Förh. 7, p. 100, 1850: Dwar's Berg, Marico district, w. Transvaal. **Pl. 57.**

**Distinguishing characters:** A large Pipit, above olivaceous brown with a yellowish tinge and darker centres to the feathers; edges of upper and under wing-coverts, flight and tail feathers green; outer tail feathers partly white; below, buffish, closely streaked with blackish; second, third, fourth and fifth primaries emarginated on outer web; hind claw rather short and well curved. The sexes are alike. Wing 80 to 90 mm. The young bird is slightly browner above than the adult, less olivaceous. In flight the white on the outer tail feathers is conspicuous.

**General distribution:** Angola to Kenya Colony, Tanganyika Territory, Nyasaland and eastern South Africa.

**Range in Eastern Africa:** Taita Hills, Kenya Colony and north-eastern Tanganyika Territory to the Zambesi River.

**Habits:** Confined to rocky wooded hills and rocky hillsides, preferably near water and widely distributed. Generally seen singly, and perches in trees when flushed.

## PIPITS

**Nest and Eggs:** Nest a cup of grass near a tussock on a rocky hillside. Eggs three, whitish ground colour heavily freckled with red brown and lilac; about 22 × 16 mm.

**Recorded breeding:** Nyasaland, at 5,000 feet, October. Portuguese East Africa, probably September and October. Southern Rhodesia, December.

**Call:** The song is a startlingly loud Thrush-like whistling warble, but is rarely uttered.

715 GOLDEN PIPIT. *TMETOTHYLACUS TENELLUS* (Cabanis). Pl. 58.

*Macronyx tenellus* Cabanis, J.f.O. p. 205, 1878: Taita, south-eastern Kenya Colony.

**Distinguishing characters:** Adult male, above, head and mantle blackish brown with yellow edges to the feathers; flight feathers bright yellow with black ends to primaries; tail bright yellow except central feathers; below, wholly bright yellow with a black band across chest; hind claw long and curved. The female is dusky brown above with browner edges to the feathers; below, buff brown; edges of outer webs of flight feathers, greater part of inner webs, under wing-coverts, centre of belly and outermost tail feathers yellow. Wing 77 to 88 mm. The young bird has narrow whitish edges to the feathers of the upperparts and sometimes a band of small brown spots on the chest, otherwise similar to the female.

**General distribution:** The Sudan, Abyssinia and British Somaliland to Tanganyika Territory; once recorded from Irene, near Pretoria, Transvaal, but record doubtful.

**Range in Eastern Africa:** Southern Sudan, eastern Abyssinia and British Somaliland to Kenya Colony and eastern Tanganyika Territory as far south as the Central Railway Line.

**Habits:** A brightly coloured bird with the habits of both a Pipit and a Wagtail. It inhabits dry open scrub country in family parties or small flocks, but it perches freely on trees and bushes and wags its tail like a Wagtail when perched. Very local and usually a shy wild species, but pugnacious in the breeding season. Its breeding season flight is a plane towards the ground from a tree top, with wings in a V over its back.

**Nest and Eggs:** Nest well hidden in the grass, but not actually on the ground and made of thick stems and grasses with a lining of rootlets. Eggs two to four, rosy white or greenish white, heavily spotted and mottled with darkish clay colour; about 20 × 14·5 mm.

**Recorded breeding:** Eastern Abyssinia, May. Tanganyika Territory, May to July.

**Food:** As for other Pipits.

**Call:** The male has a high, thin, sibilant whistle, uttered with fluttering wings and spread tail.

### 716 YELLOW-THROATED LONG-CLAW. *MACRONYX CROCEUS* (Vieillot). Pl. 58, Ph. vii.

*Alauda crocea* Vieillot, N. Dict. d'Hist. Nat. 1, p. 365, 1816: Senegal.

**Distinguishing characters:** Above, mottled black and tawny; below, rich yellow; superciliary stripe yellow; a broad black collar on chest extending to gape; ends of flank feathers buff brown; a few black streaks on sides of chest and flanks; hind claw very long and curved. The sexes are alike. Wing 85 to 107 mm. The young bird is mainly buff below with some yellow on belly and flanks, and the black collar on chest is replaced by blackish spots.

**General distribution:** Senegal and the Sudan to Angola and Natal.

**Range in Eastern Africa:** Southern Sudan, Uganda and Kenya Colony to the Zambesi River.

**Habits:** Common and conspicuous birds of grass or cultivated open country, open glades in woods or even in swamps. Benson remarks that in Kenya Colony and Tanganyika Territory this bird frequents short grass in dry country or cultivated land, while in Nyasaland it prefers short grassed swampy ground. Usually tame and in pairs. The flight consists of a few Lark-like flaps and then a dive into the grass, but in the breeding season there is a slower flapping flight with tail expanded while singing.

**Nest and Eggs:** Nest of grass and root fibres, rather loosely made and well concealed under a tuft of grass, often in tallish grass. Eggs normally three, whitish or faintly greenish mottled with purplish grey and speckled or spotted with brown, chiefly in a zone at the large end; about 25 × 18 mm.

**Recorded breeding:** Southern Nigeria, June to August. Uganda, March to June, also September to November. Kenya Colony, most of the year, especially April to July. Tanganyika Territory, November to June. Nyasaland, December to March. Northern Rhodesia, December.

**Food:** As for the Pipits, largely insects, with seeds, small molluscs and a little vegetable matter.

**Call:** A monotonous cry from the top of a bush 'tuewhee,' also a cheerful song uttered on the wing or from a perch.

### 717 FÜLLEBORN'S LONG-CLAW. *MACRONYX FÜLLEBORNI* Reichenow.

*Macronyx fülleborni* Reichenow, O.M. p. 39, 1900: Unjika Highlands, south-western Tanganyika Territory.

**Distinguishing characters:** Similar to the Yellow-throated Long-claw, but yellow below suffused with buff; flanks and under-tail-coverts practically wholly buff brown. The sexes are alike. Wing 96 to 109 mm. The young bird has the chin and throat buff brown; an indication only of blackish below; a broken black collar on chest; chest to belly suffused with buff brown.

**General distribution:** Angola and Tanganyika Territory, central and southern Belgian Congo and Northern Rhodesia.

**Range in Eastern Africa:** South central and south-western Tanganyika Territory.

**Habits:** Much those of the preceding species inhabiting highland downs or grassy glades near water.

**Nest and Eggs:** Nest as for other Long-claws, a cup of grass lined with rootlets on top of a tuft of grass, open to view or concealed. Eggs three, whitish with brown and pale purple mottlings tending to a zone at the larger end; about 25 × 18 mm.

**Recorded breeding:** Tanganyika Territory, November and December, also May. South-eastern Belgian Congo, November.

**Food:** As for other species.

**Call:** The breeding call is a monotonous whistled 'jee-o-wee' or 'chee-er' uttered from the top of a bush. There is also a Sparrow-like chirping 'weee.'

**718 SHARPE'S LONG-CLAW.** *MACRONYX SHARPEI* Jackson. Pl. 58.

*Macronyx sharpei* Jackson, Bull. B.O.C. 14, p. 74, 1904: Mau Plateau, Kenya Colony.

**Distinguishing characters:** Adult male, smaller than the Yellow-throated Long-claw, but similar above; below, paler yellow and black collar on chest replaced by black streaks which extend down flanks; hind claw very long and curved. The female is rather less bright yellow and has a buff superciliary stripe. Wing 83 to 87 mm. The young bird has the underparts sandy buff, some yellow on belly, and a few black spots on breast.

**Range in Eastern Africa:** Highlands at 7,000 to 8,000 feet on both sides of the Rift Valley, Kenya Colony.

**Habits:** Widespread and locally common at high elevations on suitable downland on either side of the Rift Valley. Prefers open grass downs without trees and settles on the ground. It has much more Pipit-like habits than the Yellow-throated Long-claw, but like that species is found singly or in pairs.

**Nest and Eggs:** As for other species of the genus, a grass nest lined with rootlets. Eggs two or three, pale greenish white mottled with pale yellowish brown and purplish grey; about 23 × 17 mm.

**Recorded breeding:** Kenya Colony highlands, May to July.

**Food:** As for other species.

**Call:** Said to be quite distinct from that of the Yellow-throated Long-claw and one series suggests the call of a Cloud-Scraper Cisticola.

**719 ABYSSINIAN LONG-CLAW.** *MACRONYX FLAVICOLLIS* Rüppell. Pl. 58.

*Macronyx flavicollis* Rüppell, N. Wirbelt. Vög. p. 102, pl. 38, fig. 2, 1840: Simen, northern Abyssinia.

**Distinguishing characters:** Above, similar to the Yellow-throated Long-claw; below, throat saffron yellow and breast to belly buff, with a touch of yellow in centre of belly; hind claw very long and curved. The sexes are alike. Wing 83 to 95 mm. The young bird has dark tawny brown edges to the feathers of the upperparts; below, buff including throat, with a collar on chest of black spots; edge of wing yellow. Albinistic examples occur.

**Range in Eastern Africa:** Northern, central and south-western Abyssinia.

**Habits:** A bird of high mountain plateaux, abundant but local and usually in pairs, tame and friendly while breeding. It frequents open grass country and is usually noticed sitting on a stone and uttering a call in the manner of a Stonechat.

**Nest and Eggs:** Nest in young growing corn or grass, like that of a Lark, of grass lined with fibre or horsehair but with the cup raised a little from the ground. Eggs two or three, glossy, pale greenish white, finely speckled and flecked with dull brown; about 24 × 17 mm.

**Recorded breeding:** Central Abyssinia, June to August.

**Food:** As for other species.

**Call:** A clear trilling little song from a perch or on the wing, and a piping call note.

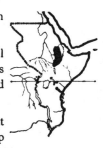

**720 PANGANI LONG-CLAW.** *MACRONYX AURANTII-GULA* Reichenow. **Pl. 58.**

*Macronyx aurantiigula* Reichenow, J.f.O. p. 222, 1891: Pangani River, eastern Tanganyika Territory.

**Distinguishing characters:** Adult male, similar above to the Yellow-throated Long-claw, but edges of feathers paler; superciliary stripe yellow; below, throat usually more orange; chest collar narrow; sides of chest, flanks and under tail-coverts white slightly tinged with buff; black streaks on chest below narrow black collar and also on flanks; hind claw very long and curved. The female has the posterior half of the superciliary stripe white. Wing 87 to 96 mm. The young bird is buff below with a wash of yellow on the breast.

**Range in Eastern Africa:** Central and eastern Kenya Colony to north-eastern Tanganyika Territory.

**Habits:** A shy and wary species, found in open parts of bush among aloe patches or in open thorn country.

**Nest and Eggs:** Nest similar to that of the Yellow-throated Long-claw, placed under a grass tuft. Eggs usually two, chalky white streaked and mottled with yellowish brown, grey and rufous; about 24 × 16·5 mm.

**Recorded breeding:** Kenya Colony, February to July, double brooded.

**Food:** As for other species, particularly small moths and their larvæ, small grasshoppers and termites.

**Call:** Very like that of the Yellow-throated Long-claw, but fuller and the song richer.

### 721 ROSY-BREASTED LONG-CLAW. *MACRONYX AMELIÆ* de Tarragon.

*Macronyx ameliæ wintoni* Sharpe. **Pl. 58, Ph. vii.**
*Macronyx wintoni* Sharpe, Ibis, p. 444, 1891: Kavirondo, western Kenya Colony.

**Distinguishing characters:** Adult male, above, black with tawny edges to the feathers; below, rich salmon; a broad black collar across chest and curving upwards to gape; hind claw very long and curved. The female is buffish brown below, with the salmon colour much paler and confined to the belly and a few spots on the throat; chest band replaced by black streaks. Wing 81 to 94 mm. The young bird has narrow, very pale buff edges to the feathers of the mantle, and only a trace of pale salmon on the belly.

**General distribution:** Western Kenya Colony to Tanganyika Territory, the Rhodesias, Bechuanaland, western Nyasaland and Portuguese East Africa north of the Zambesi River.

**Range in Eastern Africa:** Western Kenya Colony to south central Tanganyika Territory and Portuguese East Africa.

**Habits:** A shy and local species of open grass country preferring the neighbourhood of water and swampy places. It has a conspicuous habit of soaring, sometimes to a great height, while uttering a plaintive little song, and hovers before settling with legs stretched downwards. On the ground it runs fast and freely. It does not perch on trees.

**Nest and Eggs:** Nest of grass, lined with rootlets in or under a tuft of grass in marshy ground. Eggs normally three or four, very pale green mottled with brown and with pale lilac undermarkings; about 23 × 16 mm.

**Recorded breeding:** Kenya Colony, May and June, also December. Tanganyika Territory, December, January and June. Nyasaland, December and January. Portuguese East Africa, March. Southern Rhodesia, January.

**Food:** As for other species.

**Call:** The song is a squeaking little whistle of several notes with a wheezing last syllable and there is also a sharp plaintive call of 'chuit-chuit.'

**Distribution of other races of the species:** Natal and Zululand, the nominate race being described from Durban.

Names in Sclater's *Syst. Av. Æthiop.* 2, 1930, which have been changed or have become synonyms in this work:

*Anthus campestris griseus* Nicoll, treated as synonymous with *Anthus campestris campestris* (Linnæus).
*Anthus sordidus* Rüppell, now *Anthus similis* Jerdon.
*Anthus sordidus longirostris* Neumann, now *Anthus similis neumannianus* Collins and Hartert.
*Anthus similis neumannianus* Collins and Hartert, treated as synonymous with *Anthus similis hararensis* Neumann.
*Anthus gouldi turneri* Meinertzhagen, treated as synonymous with *Anthus leucophrys zenkeri* Neumann.
*Anthus richardi* Vieillot now a race of *Anthus novæseelandiæ* Gmelin.

Names introduced since 1930 and which have become synonyms in this work:

*Anthus nicholsoni chyuluensis* Van Someren, 1939, treated as synonymous with *Anthus similis hararensis* Neumann.
*Macronyx aurantiigula subocularis* Friedmann, 1930, treated as synonymous with *Macronyx aurantiigula* Reichenow.

## Addenda and Corrigenda

**694** Now *Motacilla caspica* S. G. Gmelin, Reise durch Russl. 3, p. 104, 1774: Enzeli, Iran.

**699** *Budytes feldegg. General distribution:* add to African localities, Nigeria and north-eastern Belgian Congo.

**704** *Anthus l. zenkeri. Range in Eastern Africa:* add to Abyssinia, Lake Stephanie area.

**705** *Anthus v. goodsoni. Range in Eastern Africa:* add Also occurs at Eldoret and Nakuru.

**706** *Anthus n. lacuum. General distribution:* add to Beira area.

*Continued on p. 152*

## Family—TURDOIDIDÆ. BABBLERS and CHATTERERS.
Genera: *Turdoides* and *Argya*.

Nine species of Babblers and three species of Chatterers occur in Eastern Africa. The Babblers are noisy birds, keeping up a continual babbling chatter and are thus easily observed. They are usually found in parties on the outskirts of woods and thickets, scratching and searching for food in low bushes or among debris on the ground.

The Chatterers are long-tailed, plain-coloured birds usually found in parties, and spend much of their time on the ground, where the tail is carried in an elevated position. They are noisy and usually noticeable birds, and their food consists mainly of insects with a little fruit.

KEY TO THE ADULT BABBLERS AND CHATTERERS OF EASTERN AFRICA

1 Tail long and graduated: 3-6
2 Tail medium length and rounded: 7-19
3 Chin to chest having a scaly appearance: SCALY CHATTERER *Argya aylmeri* 733
4 Chin to chest not having a scaly appearance: 5-6
5 General colour buff: FULVOUS CHATTERER *Argya fulva* 731
6 General colour brown and russet brown: RUFOUS CHATTERER *Argya rubiginosa* 732

7 Rump white or whitish: WHITE-RUMPED BABBLER *Turdoides leucopygia* 728
8 Rump not white or whitish: 9-19
9 Chin to belly white: PIED BABBLER *Turdoides hypoleuca* 729
10 Head white: WHITE-HEADED BABBLER *Turdoides leucocephala* 722
11 Chin and neck in front white: KENYA BLACK-LORED BABBLER *Turdoides melanops vepres* 726

## BABBLERS and CHATTERERS

12 Head and chin to belly not white: 13–19
13 General colour dark olivaceous brown: DUSKY BABBLER
*Turdoides tenebrosa* 727
14 General colour blackish brown and tawny: HINDE'S PIED BABBLER
*Turdoides hindei* 730
15 General colour mouse brown, or mouse grey; chin to chest speckled with white: or chin to chest with lighter edges to feathers giving a scaly appearance: 16–19
16 Under wing-coverts pale ashy brown, flanks pale mouse brown, chin whitish: BROWN BABBLER
*Turdoides plebeja* 723
17 Under wing-coverts tawny, chin sooty: ARROW-MARKED BABBLER
*Turdoides jardinei* 725
18 Lores to ear-coverts black, under wing-coverts pale brown: SCALY BABBLER
*Turdoides squamulata* 724
19 Lores black, under wing-coverts buff and mouse grey: BLACK-LORED BABBLER
*Turdoides melanops* 726

**722 WHITE-HEADED BABBLER.** *TURDOIDES LEUCOCEPHALA* Cretzschmar.

*Turdoides leucocephala leucocephala* Cretz.

*Turdoides leucocephala* Cretzschmar, in Rüppell's Atlas zu Reise, Vög. p. 6, pl. 4, 1826: Wellea Medina, Sennar, eastern Sudan.

**Distinguishing characters:** General colour mouse grey; top of head, cheeks and chin white; belly paler; very pale tips to feathers of throat and chest, giving a scaly appearance. The sexes are alike.

G

Albinistic examples occur. Wing 101 to 112 mm. The young bird has the top of the head mouse grey very slightly paler than the mantle; cheeks and chin white as in adult; flanks and belly browner.

**Range in Eastern Africa:** Nile Valley south of Khartoum to the Eritrean coast.

**Habits:** Common in the eastern Sudan in noisy parties a dozen or so strong. Inquisitive scolding birds following from bush to bush behind an intruder with a noise which A. L. Butler likens to that of a policeman's rattle.

**Nest and Eggs:** Undescribed.
**Recorded breeding:** No records.
**Call:** As above.

*Turdoides leucocephala abyssinica* (Neum.).

*Crateropus leucocephalus abyssinicus* Neumann, J.f.O. p. 550, 1904: Anseba Valley, Eritrea.

**Distinguishing characters:** General colour browner than the nominate race. Wing 105 to 110 mm.

**Range in Eastern Africa:** Anseba River Valley, Eritrea.
**Habits:** As for the nominate race.
**Recorded breeding:** No records.

**723 BROWN BABBLER.** *TURDOIDES PLEBEJA* (Cretzschmar).

*Turdoides plebeja plebeja* (Cretz.).

*Ixos plebejus* Cretzschmar, in Rüppell's Atlas zu Reise, Vög. p. 35, pl. 23, 1826: Kordofan, Sudan.

**Distinguishing characters:** General colour mouse brown with lighter edges to feathers; centres of feathers on head blacker giving a scaly appearance; below, chin to breast paler mouse brown, with subterminal blackish and terminal white tips, giving a speckled appearance; breast to belly creamy white; wings and tail bronzy brown; under wing-coverts pale or ashy brown; chin whitish. The sexes are alike. Wing 100 to 115 mm. The young bird is rather browner above than the adult, especially the wings and tail.

**General distribution:** Lake Chad area to the central Sudan.

**Range in Eastern Africa:** Darfur and Kordofan areas of the Sudan.

## BABBLERS and CHATTERERS

**Habits:** A typical Babbler in habits occurring mainly in low country in small noisy chattering parties. They fly from bush to bush with a flapping skimming flight, and set up a loud babbling chattering on sighting a human being. Feed usually on the ground creeping about among bushes, but do not hop. When drinking they are said to sidle down on an overhanging branch that dips into the water.

**Nest and Eggs:** The nest is a large shallow cup of hard rootlets lined with finer materials in a thick, usually thorny, bush. Eggs two to four, glossy and usually deep turquoise blue; about 27 × 19 mm., but in West Africa bright salmon pink, grey-blue and mauvish stone-coloured eggs are also reported, and they are somewhat smaller.

**Recorded breeding:** Nigeria, May to September. Sudan, August and September.

**Call:** Various grating babbling cries, including a chattering not unlike a Wood-Hoopoe.

*Turdoides plebeja cinerea* (Heugl.).
*Crateropus cinereus* Heuglin, J.f.O. p. 300, 1862: Upper White Nile, between lat. 5° and 10° N., Sudan.

**Distinguishing characters:** General colour much darker than in the nominate race. Wing 94 to 114 mm.

**General distribution:** Eastern Nigeria to Abyssinia, the Sudan and Kenya Colony.

**Range in Eastern Africa:** Western Abyssinia and the southern Sudan to Uganda and western Kenya Colony.

**Habits:** Of the Brown Babbler, somewhat subject to local migrations. The bright yellow iris is noticeable at some distance.

**Recorded breeding:** Southern Sudan, November.

**Distribution of other races of the species:** West Africa.

**724** SCALY BABBLER. *TURDOIDES SQUAMULATA* (Shelley).
*Turdoides squamulata squamulata* (Shell.).
*Crateropus squamulatus* Shelley, Ibis, p. 45, 1884: Mombasa, eastern Kenya Colony.

**Distinguishing characters:** Top and sides of head black, feathers of former with white edges and tips giving a scaly appearance; rest of upperparts mouse grey; wings and tail darker; below,

more ashy, with white tips to the feathers, but also scaly; under wing-coverts pale brown. The sexes are alike. Wing 101 to 102 mm. The young bird is paler with a whiter throat.

**Range in Eastern Africa:** Coastal area of Kenya Colony from Lamu to Vanga.

**Habits:** A bird of the very dense bush of the coastal belt with the same habits as others of the genus. Rarely collected and little is recorded of its habits.

**Nest and Eggs:** Nest undescribed; eggs pale blue and glossy; about 26 × 19 mm.

**Recorded breeding:** No records.

**Call:** Unrecorded.

*Turdoides squamulata jubaensis* Van Som.

*Turdoides squamulata jubaensis* Van Someren, Journ. E. A. & U. Nat. Hist. Soc. No. 37, for Jan. 1930, p. 196, July 1931: Upper Juba River, southern Italian Somaliland.

**Distinguishing characters:** Differs from the nominate race in being lighter olive grey above, and in not having the ear-coverts dark greyish. Wing 98 to 104 mm.

**Range in Eastern Africa:** Upper Juba River from Dolo to Serenli, southern Italian Somaliland.

**Habits:** As for the nominate race.

**Recorded breeding:** No records.

**725 ARROW-MARKED BABBLER.** *TURDOIDES JARDINEI* (Smith).

*Turdoides jardinei kirki* (Sharpe).

*Crateropus kirki* Sharpe, in Layard's Bds. S. Afr. 2nd ed. p. 213, 1876: Mazaro, on Zambesi River, about 75 miles west of mouth of Shiré River.

**Distinguishing characters:** Above, bronzy brown, with a few light flecks and blackish centres to feathers of head; below, throat ashy; chest to belly buffish white; white tips to feathers of chin to upper belly, giving a speckled appearance; under wing-coverts pale tawny; chin sooty. The sexes are alike. Wing 88 to 107 mm. The young bird is brown, paler on underparts and with slight indications of speckling on the chin to chest.

## BABBLERS and CHATTERERS

**General distribution:** Kenya Colony and eastern areas of Tanganyika Territory to Nyasaland, north-eastern Northern Rhodesia, eastern Southern Rhodesia and Portuguese East Africa.

**Range in Eastern Africa:** Eastern areas of Kenya Colony from Lamu to Tanganyika Territory, as far west as Monduli; Mpapwa and Iringa, Nyasaland and Portuguese East Africa.

**Habits:** Noisy small flocks among creepers, rank grass and thorny bush, especially in rocky country or along stream beds. Typical Babbler in habits with low straight flight from bush to bush.

**Nest and Eggs:** Nest of roots, grass stems, etc. lined with rootlets, in a thick bush, or in driftwood piled up in trees along a stream. Eggs usually three, deep turquoise blue; about 23 × 19 mm. It is the commonest host of Levaillant's Cuckoo, whose eggs are a little less glossy and slightly larger.

**Recorded breeding:** Tanganyika Territory, October to March, etc. Nyasaland, at any season, usually October to April. Portuguese East Africa, most months of year.

**Call:** The usual chattering babbling calls, as Moreau says 'their usual intercourse sounds like the bandying of filthy language in a harsh voice.' Also a call of 'kaa-kaa' not unlike the distant call of a Rook.

*Turdoides jardinei emini* (Neum.).
*Crateropus plebeius emini* Neumann, J.f.O. p. 549, 1904: Wala River, Uniamwesi, Tabora Province, central Tanganyika Territory.

**Distinguishing characters:** Differs from the preceding race in being generally more ashy brown than bronzy brown. Wing 97 to 107 mm.

**Range in Eastern Africa:** Uganda and western and southern Kenya Colony to western and central Tanganyika Territory at Kigoma, Kasulu, Tabora and Mbulu districts, also Ukerewe Island, south-eastern Lake Victoria.

**Habits:** As for the preceding race.

**Recorded breeding:** Northern Tanganyika Territory (breeding condition), January.

**Distribution of other races of the species:** Angola and the Belgian Congo to South Africa, the nominate race being described from the north-western Transvaal.

### 726 BLACK-LORED BABBLER. *TURDOIDES MELANOPS* (Hartlaub).

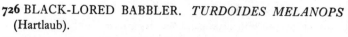

*Turdoides melanops sharpei* (Reichw.).

*Crateropus sharpei* Reichenow, J.f.O. p. 432, 1891: Kakoma, Tabora Province, Tanganyika Territory.

**Distinguishing characters:** General colour mouse grey; under wing-coverts buff and mouse grey; a black spot in front of eye; indistinct lighter tips to feathers of chin to chest; head to back somewhat scaly in certain lights; usually fine black streaks on chin to chest; wings and tail dusky bronze. The sexes are alike. Wing 102 to 116 mm. The young bird is plain mouse grey above with no lighter tips to feathers of head and neck; below, throat to chest not so scaly in appearance. Sir F. Jackson remarks that this bird can be identified in the field by its white eye.

**Range in Eastern Africa:** Uganda, western Kenya Colony, and western and central Tanganyika Territory.

**Habits:** Noisy inquisitive birds in typical Babbler parties, locally common. They are occasionally seen in company with other Babblers, and usually keep to thick cover. Their calls have earned them the local name of 'Cat-Bird.'

**Nest and Eggs:** Nest of grass and roots rather roughly made and not unlike a Thrush's, placed in a dense bush. Eggs two or three, darkish blue; about 26 × 20 mm.

**Recorded breeding:** Uganda, April, May, July, September. Mt. Elgon, April.

**Call:** Various harsh grating cries, also one like the yelling of a cat, and another which is a kind of bleating squeak.

*Turdoides melanops vepres* Meinertz. Kenya Black-faced Babbler.

*Turdoides melanops vepres* Meinertzhagen, Bull. B.O.C. 57, p. 69, 1937: Nanyuki, north of Mt. Kenya, central Kenya Colony.

**Distinguishing characters:** Generally darker than the preceding race; chin, or chin and whole neck in front, white, variable in amount. Wing 110 to 115 mm.

**Range in Eastern Africa:** Nanyuki, central Kenya Colony.

**Habits:** No information.

# BABBLERS and CHATTERERS

**Recorded breeding:** No records.

**Distribution of other races of the species:** Angola and Damaraland, the nominate race being described from Damaraland.

**727 DUSKY BABBLER.** *TURDOIDES TENEBROSA* (Hartlaub).

*Turdoides tenebrosa tenebrosa* (Hartl.).
*Crateropus tenebrosus* Hartlaub, J.f.O. p. 425, 1883: Kudurma, southeastern Bahr-el-Ghazal, southern Sudan.

**Distinguishing characters:** General colour dark olivaceous brown; forehead greyish; black spot in front of eye; feathers of chin to chest with blackish centres and olive brown edges, giving a scaly appearance; flanks and under tail-coverts browner; wings and tail dark bronzy brown; under wing-coverts olive brown and black. The sexes are alike. Wing 106 to 122 mm. The young bird is rather browner above, wings and tail less dark; below, much browner on belly and chin to breast grey with faint streaking, not having the scaly appearance of the adult.

**Range in Eastern Africa:** Western Abyssinia to southern Sudan.

**Habits:** A very secretive species of dense cover near water.

**Nest and Eggs:** The nest is similar to that of the preceding species, though rather more lightly built. It is made of dried leaves and grasses with a fairly deep cup lined with rootlets, and is usually placed in thick cover or creepers a few feet from the ground. Eggs two or three, pale blue and extremely glossy; about $25 \times 17$ mm.

**Recorded breeding:** Equatoria, April and July. Uganda, March and June.

**Call:** A hoarse 'chow' or a more nasal 'what-cow' (Chapin).

**Distribution of other races of the species:** The Belgian Congo.

**728 WHITE-RUMPED BABBLER.** *TURDOIDES LEUCOPYGIA* (Rüppell).

*Turdoides leucopygia leucopygia* (Rüpp.).
*Ixos leucopygius* Rüppell, N. Wirbelt, Vög. p. 82, pl. 30, fig. 1, 1840: Coast of Eritrea.

**Distinguishing characters:** General colour bronzy brown and white; whole head, rump and belly white; top of head pale grey. The sexes are alike. Wing 115 to 117 mm. The young bird is duller

than the adult and has the top of the head brownish rather lighter than the mantle, the sides of the face and chin being whitish.

**Range in Eastern Africa:** Eritrea and north-eastern Abyssinia from Massowa to Adigrat.

**Habits:** Typical Babblers, found in noisy chattering parties, with the habits of other races.

**Nest and Eggs:** See under other races.

**Recorded breeding:** No records.

*Turdoides leucopygia limbata* (Rüpp.).
*Crateropus limbatus* Rüppell, Syst. Uebers, p. 48, 1845: Ali Amba, Shoa, central Abyssinia.

**Distinguishing characters:** Differs from the nominate race in having the top of the head blackish with silvery grey edges to the feathers giving a scaly appearance. Wing 114 to 122 mm.

**Range in Eastern Africa:** Central Eritrea in the Anseba Valley to central Abyssinia and Adau, Danakil, eastern Abyssinia.

**Habits:** As for the nominate race.

**Recorded breeding:** No records.

*Turdoides leucopygia hartlaubii* (Boc.).
*Crateropus hartlaubii* Bocage, J. Lisboa, 2, No. 5, p. 48, 1868: Huilla, southern Angola.

**Distinguishing characters:** Very little white on rump; no white on head, and no black on throat; lores blackish. Wing 107 to 120 mm.

**General distribution:** Angola and eastern Belgian Congo to Tanganyika Territory, northern Bechuanaland and the Transvaal.

**Range in Eastern Africa:** Ufipa Plateau, south-western Tanganyika Territory.

**Habits:** Locally plentiful in thick bush or reed beds, with the general habits of the family, of which it is one of the noisiest members. When disturbed, a flock closes up, curses the intruder in chorus, and then the individuals silently slip away.

**Nest and Eggs:** Nest a large untidy heap of grass twigs and leaves in the fork of a tree with a centre cup lined with fibres. Eggs usually three, deep greenish blue; about 27 × 20 mm.

**Recorded breeding**: Northern Rhodesia, April. Northern Transvaal, September to November.

**Call**: A varied selection of squawks, squeals and cackles.

*Turdoides leucopygia smithii* (Sharpe).
*Crateropus smithii* Sharpe, Bull. B.O.C. 4, p. 41, 1895: Sheik Hussein, Arussi, central Abyssinia.

**Distinguishing characters**: Feathers of throat dusky and broadly edged with white; feathers of chest to upper belly broadly edged with white. Wing 109 to 121 mm.

**Range in Eastern Africa**: Central and west-central Abyssinia east of the Didessa River to central British Somaliland.

**Habits**: As for other races.

**Recorded breeding**: British Somaliland (nestling), August.

*Turdoides leucopygia omoensis* (Neum.).
*Crateropus smithi omoensis* Neumann, Bull. B.O.C. 14, p. 15, 1903: Senti River, affluent of Omo River, south-western Abyssinia.

**Distinguishing characters**: Head, mantle and chest blackish brown with broad grey or white edges to feathers which look like scales; in front of eye, cheeks, chin and throat black; ear-coverts white. Wing 99 to 118 mm.

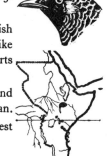

**Range in Eastern Africa**: Lake Abaya to the Omo River and Boran area, Abyssinia, and the Boma Plateau, south-eastern Sudan.

**Habits**: Common among thick juniper scrub or secondary forest in noisy chattering parties near Yavello in southern Abyssinia.

**Recorded breeding**: South Abyssinia, March and November.

*Turdoides leucopygia clarkei* Macdonald.
*Turdoides leucopygia clarkei* Macdonald, Bull. B.O.C. 60, p. 10, 1939: Bello, about 20 miles north-east of Goré, western Abyssinia.

**Distinguishing characters**: Similar to the preceding race, but differs in having white in front of and under eye; black of chin and throat sooty black or blackish brown. Wing 98 to 109 mm.

**Range in Eastern Africa**: Upper Baro River, Goré and Buré, western Abyssinia, as far east as Bello and Dabana

**Habits**: As for other races.

**Recorded breeding**: No records.

# BABBLERS and CHATTERERS

### 729 NORTHERN PIED BABBLER. *TURDOIDES HYPOLEUCA* (Cabanis).

*Turdoides hypoleuca hypoleuca* (Cab.).

*Crateropus hypoleucus* Cabanis, J.f.O. pp. 205, 226, 1878: Kitui, Ukamba, Kenya Colony.

**Distinguishing characters:** Above, including wings and tail, sides of head and flanks dark brown with a bronzy wash; below, chin to belly white. The sexes are alike. Wing 102 to 118 mm. The young bird is slightly more olivaceous than the adult and is streaked on the chest and breast.

**Range in Eastern Africa:** Central and south-eastern Kenya Colony to the Mt. Kilimanjaro and Mt. Meru areas of Tanganyika Territory.

**Habits:** Typical Babblers proceeding from bush to bush in family parties with harsh persistent cries. One would at least suppose from the noise that they had located a leopard or something menacing in each bush. Common in woodland scrub and along the outskirts of forests.

**Nest and Eggs:** Nest roughly made with twigs and leaves as a platform, a coarse grass cup lined with finer grass and fibre. Nest usually hidden in a thorn bush though often in an isolated one. Eggs three, occasionally four, of a rich uniform dark blue without much gloss; about $26 \times 19$ mm.

**Recorded breeding:** Kenya Colony, February to May, also August, October and November. Tanganyika Territory, November and February.

**Call:** A series of harsh cries, mainly an insistent 'quorr-quorr,' not perhaps so discordant as those of some other species of Babblers.

*Turdoides hypoleuca rufuensis* (Neum.).

*Crateropus hypoleuca rufuensis* Neumann, O.M. p. 148, 1906: Usegua, Pangani District, north-eastern Tanganyika Territory.

**Distinguishing characters:** Differs from the nominate race in having the upperparts distinctly greyer and head more scaly. Wing 107 to 113 mm.

**Range in Eastern Africa:** Eastern Tanganyika Territory from Lake Manyara to the Usambara, Pangani and Kilosa districts.

**Habits:** As for the nominate race.

**Recorded breeding:** No records.

## BABBLERS and CHATTERERS

**730 HINDE'S PIED BABBLER.** *TURDOIDES HINDEI* (Sharpe).

*Crateropus hindei* Sharpe, Bull. B.O.C. 11, p. 29, 1900: Athi River, southern Kenya Colony.

**Distinguishing characters:** A bird of curiously mixed plumage, mainly blackish or sooty brown, with more or less broad white tips and ends to most of the feathers, including in some specimens the wings and tail; flanks, rump and under tail-coverts usually tawny and in some specimens tawny markings occur anywhere on the head and body. The sexes are alike. Wing 100 to 101 mm. The young bird is similar to the adult, but more fluffy in appearance and has browner, less bronzy, wings and tail.

**Range in Eastern Africa:** Kikuyu and Ukamba Provinces, central Kenya Colony.

**Habits:** A rare and local species, with typical Babbler habits.

**Nest and Eggs:** Nest undescribed. Eggs light blue and glossy; about 26 × 18 mm.

**Recorded breeding:** No records.

**Call:** Unrecorded.

**731 FULVOUS CHATTERER.** *ARGYA FULVA* (Desfontaines).
*Argya fulva acaciæ* (Licht.).

*Sphenura acaciæ* Lichtenstein, Verz. Doubl. p. 40, 1823: Nubia.

**Distinguishing characters:** General colour greyish buff; blackish streaks on head and nape. The sexes are alike. Wing 86 to 101 mm. The young bird is more tawny above than the adult.

**General distribution:** Upper Egypt, the Sudan and Eritrea.

**Range in Eastern Africa:** The western and central Sudan, as far south as Darfur and Kordofan to Eritrea.

**Habits:** Gregarious birds always found in small parties, but much more quiet and furtive than Babblers, though they have the habit of chattering angrily at an intruder. This species is confined to desert scrub and spend much time on the ground, proceeding along it at great speed in long hops with uplifted tail. Bates describes their gait exactly when he says they 'scud along the ground like weasels.'

**Nest and Eggs:** A loose cup nest of twigs and dry grass placed in a dense thorn bush. Eggs three to five, glossy greenish or turquoise blue; about 24 × 17·5 mm.

**Recorded breeding:** Sudan, Darfur, May. Kordofan, April. White Nile, March. Red Sea coast, March and November.

**Food:** Insects and small seeds, or grain with berries at certain seasons.

**Call:** A series of chattering alarm notes, and a thin mewing crescendo sort of song of 'wee-wee-wee.'

**Distribution of other races of the species:** Tunisia and the French Sudan, the nominate race being described from Tunisia.

### 732 RUFOUS CHATTERER. *ARGYA RUBIGINOSA* (Rüppell).

*Argya rubiginosa rubiginosa* (Rüpp.).
*Crateropus rubiginosus* Rüppell, Syst. Uebers, Vög., p. 47, pl. 19, 1845: Shoa, central Abyssinia.

**Distinguishing characters:** General colour above brown; below, russet brown; narrow silvery white streaks on forehead. The sexes are alike. Wing 78 to 89 mm. The young bird is more russet brown above and has few of the silvery white streaks of the adult.

**Range in Eastern Africa:** Southern Sudan, central and southern Abyssinia and eastern Uganda to central Kenya Colony.

**Habits:** Much those of the last species, keeping in small bands or family parties and playing follow-my-leader from bush to bush. They have the same habit of hopping rapidly along with their tails cocked up. They are shy retiring birds of thick bush along rivers or dense cover, and very noisy with queer growling cries.

**Nest and Eggs:** Nest of thin twigs, grass stems etc. often with living stems woven in, with an internal grass cup lined with roots and fine grass. The nest is usually placed one to four feet from the ground in thick bushes. Eggs two to four, oval and glossy blue; about 22 × 17 mm.

**Recorded breeding:** Southern Sudan, August. Central and southern Abyssinia, March to July, probably double brooded. Karamoja, Uganda, February and April to October.

**Food:** Largely insects.

**Call:** A variety of guttural and chattering calls, with a curious little soft cry of 'kweer,' and a long plaintive quavering whistle.

*Argya rubiginosa heuglini* Sharpe.
*Argya heuglini* Sharpe, Cat. Bds. Brit. Mus. 7, p. 391, 1883: Zanzibar Island.

**Distinguishing characters:** Above, general colour tone darker than in the nominate race, with narrow blackish rather indistinct streaks from head to mantle. Wing 75 to 87 mm.

**Range in Eastern Africa:** Coastal areas from the Juba River, Italian Somaliland to Dar-es-Salaam, Tanganyika Territory, as far west as Taveta, Sinya, Kilimanjaro area, Moshi, Ngerengere and northern Paré Mts., also Zanzibar Island.

**Habits:** As for the nominate race.

**Recorded breeding:** Kenya Colony, Malindi, December; Mombasa, June; southern areas, March and April. Northern Tanganyika Territory, July.

*Argya rubiginosa sharpii* O. Grant.
*Argya sharpii* O. Grant, Ibis, p. 662, 1901: Webi Shebeli, southern Abyssinia.

**Distinguishing characters:** Differs from the nominate race in being larger. Wing 91 to 96 mm.

**Range in Eastern Africa:** Upper Webi Shebeli in the Dolo and Unsi areas, southern Abyssinia.

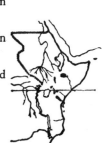

**Habits:** As for other races.

**Recorded breeding:** No records.

*Argya rubiginosa emini* Reichw.
*Argya rubiginosa emini* Reichenow, O.M. p. 30, 1907: Scamuje, Tabora Province, Tanganyika Territory.

**Distinguishing characters:** Above, browner and warmer in tone than the nominate race and having the silvery white streaks on forehead extending on to crown. Wing 86 mm.

**Range in Eastern Africa:** The Tabora and Mkalama areas, central Tanganyika Territory.

**Habits:** Probably differing little from those of the nominate race.

**Recorded breeding:** No records.

**733 SCALY CHATTERER.** *ARGYA AYLMERI* Shelley.
*Argya aylmeri aylmeri* Shell.
*Argya aylmeri* Shelley, Ibis, p. 404, pl. 2, 1885: Somaliland.

**Distinguishing characters:** Above, ashy brown; head brown; below, chin to chest dark grey with buff edges to feathers, giving a scaly appearance; breast to belly buff. The sexes are alike. Wing 70 to 78 mm. The young bird is rather browner above than the adult, less ashy; scaly appearance of chin to chest not so clear; bill bronzy brown lighter at base, not wholly pale horn yellow.

**Range in Eastern Africa:** Eastern Abyssinia and British and Italian Somalilands.

**Habits:** Typical of the genus. Very active among thick scrub, grass and aloes, keeping to bushes as much as possible. Continually makes a little squeaking cry like a mouse.

**Nest and Eggs:** A strongly made nest of twigs and stems lined with fine grass and placed in bushes among aloes and Euphorbias. Eggs two or three, pale bluish green and glossy; about $21 \cdot 5 \times 16$ mm.

**Recorded breeding:** Abyssinia, April.

**Food:** Insects.

**Call:** Many piping and grating calls, but chiefly a noise described by Couchman and Elliott as a 'squeaking wood-screw.'

*Argya aylmeri keniana* Jack.
*Argya keniana* Jackson, Bull. B.O.C. 27, p. 7, 1910: Emberre, east of Embu, Kenya Province, central Kenya Colony.

**Distinguishing characters:** General colour browner than in the nominate race; above russet brown; head more chestnut brown; below, pale brown. Wing 68 to 78 mm.

**Range in Eastern Africa:** Central Kenya Colony.

**Habits:** As for the nominate race.

**Recorded breeding:** No records.

*Argya aylmeri loveridgei* Hart.
*Argya aylmeri loveridgei* Hartert, Bull. B.O.C. 43, p. 118, 1923: Campi-ya-Bibi, near Samburu Station, south-eastern Kenya Colony.

**Distinguishing characters:** General colour darker and warmer brown than in the preceding race. Wing 76 to 80 mm.

**Range in Eastern Africa:** South-eastern Kenya Colony and north-eastern Tanganyika Territory from Tsavo to Gelai Mt. and Moshi.

**Habits:** Common in small parties among sansevieria and thorn scrub in the country south-west of Tsavo near the Kenya Colony-Tanganyika Territory boundary.

**Recorded breeding:** Northern Tanganyika Territory (breeding condition), December.

*Argya aylmeri mentalis* Reichw.

*Argya mentalis* Reichenow, J.f.O. p. 75, 1887: Soboro, Kondoa Irangi district, central Tanganyika Territory.

**Distinguishing characters:** General colour darker ash brown than the nominate race, especially below. Wing 73 to 78 mm.

**Range in Eastern Africa:** Northern and central Tanganyika Territory from Lolbene and Lossogonoi Mts. to Kondoa Irangi and Dodoma districts.

**Habits:** As for other races.

**Recorded breeding:** Northern Tanganyika Territory (breeding condition), December.

*Argya aylmeri boranensis* Benson.

*Argya aylmeri boranensis* Benson, Bull. B.O.C. 68, p. 10, 1947: 10 miles south of Yavello, southern Abyssinia.

**Distinguishing characters:** Differs from the preceding race in having blacker centres to the feathers of the chin and throat; breast to belly more russet brown; and from the nominate race in being darker both above and below. Wing 72 to 78 mm.

**Range in Eastern Africa:** Southern Abyssinia from the Boran area to the headwaters of the Webi Shebeli.

**Habits:** As for other races.

**Recorded breeding:** Southern Abyssinia, April.

Names in Sclater's *Syst. Av. Æthiop.* 2, 1930, which have been changed or have become synonyms in this work:

*Turdoides melanops clamosa* (Van Someren), treated as synonymous with *Turdoides melanops sharpei* (Reichenow).

Names introduced since 1930 and which have become synonyms in this work:

*Turdoides hypoleuca kilosa* Vincent, 1935, treated as synonymous with *Turdoides hypoleuca rufuensis* Neumann.

# 98 THRUSH-BABBLERS, ILLADOPSES, HILL-BABBLERS

FAMILY—**PELLORNEUMIDÆ. THRUSH-BABBLERS, ILLADOPSES and HILL-BABBLERS**. Genera: *Ptyrticus*, *Malacocincla* and *Pseudoalcippe*.

One Thrush-Babbler, five species of Illadopsis and one Hill-Babbler occur in Eastern Africa.

The Thrush-Babbler is a very secretive bird, inhabiting dense bush. The general colour and spotting on the chest gives it the appearance of a Thrush, for which it may quite easily be mistaken.

The Illadopses are birds of thick undergrowth, and spend most of their time either on the ground searching for food, or in the undergrowth near the ground.

The Hill-Babblers are birds of dense cover and are rather secretive.

### KEY TO THE ADULT THRUSH-BABBLER, ILLADOPSES AND HILL-BABBLERS OF EASTERN AFRICA

1 Triangular spots on chest: THRUSH-BABBLER *Ptyrticus turdinus* **734**

2 Distinct black streaks on chin to chest: STIERLING'S HILL-BABBLER *Pseudoalcippe abyssinicus stierlingi* **740**

3 No spots or streaks on chin, throat or chest: 4–13

4 Top of head black: RUWENZORI HILL-BABBLER *Pseudoalcippe abyssinicus atriceps* **740**

5 Top of head grey: 6–7

6 Belly and under tail-coverts grey: ABYSSINIAN HILL-BABBLER *Pseudoalcippe abyssinicus abyssinicus* **740**

7 Belly and under tail-coverts brown: MOUNTAIN ILLADOPSIS *Malacocincla pyrrhopterus* **738**

8 Top of head more olivaceous grey: 9–13

# THRUSH-BABBLERS, ILLADOPSES, HILL-BABBLERS 99

| | | | |
|---|---|---|---|
| 9 Below brownish: | BROWN ILLADOPSIS *Malacocincla fulvescens* | **735** | |
| 10 Below white: | | | 11–12 |
| 11 Breast feathers plain: | PALE-BREASTED ILLADOPSIS *Malacocincla rufipennis* | **736** | |
| 12 Breast feathers tipped with olivaceous brown: | SCALY-BREASTED ILLADOPSIS *Malacocincla albipectus* | **737** | |
| 13 Below grey: | GREY-CHESTED ILLADOPSIS *Malacocincla poliothorax* | **739** | |

**734 THRUSH-BABBLER.** *PTYRTICUS TURDINUS* Hartlaub.
*Ptyrticus turdinus turdinus* Hartl.
*Ptyrticus turdinus* Hartlaub, J.f.O. p. 425, 1883: Tamaja (Tomaya), south-western Sudan.

**Distinguishing characters:** Above, deep russet brown; head and upper tail-coverts richer brown; below, white with blackish brown triangular spots on chest. The sexes are alike. Wing 94 to 110 mm. The young bird is rather more deep tawny above than the adult, especially the tail feathers and the outer webs of the flight feathers.

**General distribution:** North-eastern Belgian Congo to the Sudan.

**Range in Eastern Africa:** South-western Sudan.

**Habits:** A very secretive species living in dense bush and grass thickets along the edge of forest streams and with skulking habits, little known.

**Nest and Eggs:** Undescribed.

**Recorded breeding:** Belgian Congo, June to August or later.

**Food:** Largely insects.

**Call:** A fluid and beautiful bell-like note 'i-din—i-din,' with the second note higher than the first (Cave).

**Distribution of other races of the species:** West Africa and southern Belgian Congo.

H

### 735 BROWN ILLADOPSIS. *MALACOCINCLA FULVESCENS* (Cassin).

*Malacocincla fulvescens ugandæ* (Van Som.). **Pl. 59.**

*Turdinus ugandæ* Van Someren, Bull. B.O.C. 35, p. 125, 1915: Sezibwa River Forest, southern Uganda.

**Distinguishing characters:** Above, dark russet brown, head darker; below, practically wholly brownish; throat dirty white; cheeks grey. The sexes are alike. Wing 71 to 82 mm. The young bird is rather paler brown above, including the head.

**General distribution:** The Sudan, Kenya Colony, Tanganyika Territory and northern Belgian Congo.

**Range in Eastern Africa:** Southern Sudan, Uganda, western Kenya Colony and western Tanganyika Territory.

**Habits:** A shy and retiring species, inhabiting dense forest undergrowth, of which little is recorded within our limits. They are birds of the forest floor, hunting among debris for insects and small molluscs.

**Nest and Eggs** (West African races): Nests are loose shallow cups made of dead leaves and lined with a few fibres, the whole thing almost decaying from damp. Eggs two, white or cream or pinkish white either densely mottled with maroon or else boldly blotched with the same colour; undermarkings purplish grey; very variable in colour, about 22 × 16 mm.

**Recorded breeding:** No records, probably at any time of year.

**Food:** Mainly insects taken on the ground, largely ants.

**Call:** A long drawn soft whistle first in a higher then in a lower key.

**Distribution of other races of the species:** West Africa, the nominate race being described from Gabon.

### 736 PALE-BREASTED ILLADOPSIS. *MALACOCINCLA RUFIPENNIS* (Sharpe).

*Malacocincla rufipennis rufipennis* (Sharpe). **Pl. 59.**

*Trichostoma rufipennis* Sharpe, Ann. Mag. N.H. (4), 10, p. 451, 1872: Gabon.

**Distinguishing characters:** Above, similar to the Brown Illadopsis but head and sides of face more greyish and olivaceous; below, paler, throat and belly whiter. The sexes are alike. Wing 65 to 80 mm. The young bird has the top of the head brownish; tawny tips to the lesser wing-coverts.

# THRUSH-BABBLERS, ILLADOPSES, HILL-BABBLERS

**General distribution:** Cameroons to Gabon, Spanish Muni, north-eastern Belgian Congo, the Sudan and Kenya Colony.

**Range in Eastern Africa:** Southern Sudan to Uganda and western Kenya Colony.

**Habits:** A little-known bird inhabiting dense forest undergrowth.

**Nest and Eggs:** See under the following race.

**Recorded breeding:** Southern Sudan (breeding condition), March.

*Malacocincla rufipennis distans* (Friedm.).
*Turdinus rufipennis distans* Friedmann, Proc. N. Engl. Zool. Cl. 10, p. 48, 1928: Amani, Usambara Mts., north-eastern Tanganyika Territory.

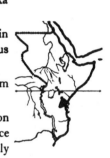

**Distinguishing characters:** Differs from the nominate race in having the chest to belly pale grey and flanks more olivaceous brown. Wing 68 to 78 mm.

**Range in Eastern Africa:** Eastern Tanganyika Territory from the Usambara Mts. to the Uluguru Mts., also Zanzibar Island.

**Habits:** A secretive species living entirely in undergrowth or on the ground. It has, as described by Moreau, a strong resemblance to a *Bradypterus*. Usually found in small parties moving steadily through thick undergrowth close to the ground.

**Nest and Eggs:** A deep cup nest of dead leaves lined with horsehair-like mycelium, well hidden among leaves on stumps, etc. Eggs two, greyish white, blotched or speckled with reddish brown; about $23 \cdot 5 \times 16$ mm.

**Recorded breeding:** Eastern Tanganyika Territory, November and December.

**Food:** Insects of all kinds.

**Call:** A slow meditative whistle 'hoooit-hooooee' sometimes answered by another bird, also much churring conversation between members of a flock. Alarm note a grating 'ka-a-a' (Moreau).

*Malacocincla rufipennis puguensis* Grant and Praed.
*Illadopsis rufipennis puguensis* C. Grant and Mackworth-Praed, Bull. B.O.C. 60, p. 61, 1940: Pugu Hills, 20 miles west of Dar-es-Salaam, eastern Tanganyika Territory.

**Distinguishing characters:** Differs from the preceding race in being paler above, more olivaceous buff, and lacking the chestnut wash on the wings and tail. Wing 64 to 76 mm.

**Range in Eastern Africa:** Pugu Hills, eastern Tanganyika Territory.

**Habits:** As for other races, but the alarm call is distinctive and not so grating.

**Recorded breeding:** No records.

**Distribution of other races of the species:** Fernando Po.

### 737 SCALY-BREASTED ILLADOPSIS. *MALACOCINCLA ALBIPECTUS* (Reichenow).

*Turdinus albipectus* Reichenow, J.f.O. p. 307, 1887: Near Stanley Falls, Belgian Congo.

**Distinguishing characters:** Similar in size and colour to the Pale-breasted Illadopsis, but breast feathers tipped with olivaceous brown giving a scaly appearance. The sexes are alike. Wing 72 to 80 mm. The young bird differs from the young of that species in having the breast and flank feathers tipped with brown.

**General distribution:** The Belgian Congo from Lukolela to the Sudan, Uganda and Kenya Colony.

**Range in Eastern Africa:** The Imatong Mts., southern Sudan to Uganda and western Kenya Colony as far east as Kakamega.

**Habits:** Very similar to those of the preceding species.

**Nest and Eggs:** Undescribed.

**Recorded breeding:** No records, probably in many months.

**Food:** Ants, beetles and other insects.

**Call:** A chirp, followed by several short whistles.

### 738 MOUNTAIN ILLADOPSIS. *MALACOCINCLA PYRRHOPTERUS* (Reichenow and Neumann).

*Malacocincla pyrrhoptera pyrrhoptera* (Reichw. and Neum.). **Pl. 59.**
*Callene pyrrhoptera* Reichenow and Neumann, O.M. p. 75, 1895: Mau, Kenya Colony.

**Distinguishing characters:** Top of head olivaceous grey with indistinct darker tips, giving a faint scaly appearance; below, grey; flanks and under tail-coverts brown. The sexes are alike. Wing

68 to 70 mm. The young bird has the whole upperside, including the head, more russet brown than the adult.

**Range in Eastern Africa:** Western Kenya Colony.

**Habits:** Another species of dense or impenetrable forest undergrowth, usually moving through it in a constant direction with a steady twittering call. Little is known of its habits or breeding.

**Nest and Eggs:** Undescribed.

**Recorded breeding:** No records.

**Food:** Insects, berries and small snails.

**Call:** Song a melodious warble 'chee-cherilu-cherilu-chi' (Van Someren).

*Malacocincla pyrrhoptera kivuensis* (Neum.).
*Turdinus pyrrhopterus kivuensis* Neumann, Bull. B.O.C. 21, p. 55, 1908: Mt. Sabinio, Uganda-Belgian Congo Boundary.

**Distinguishing characters:** Differs from the nominate race in having the top of the head grey, not olivaceous. Wing 66 to 76 mm.

**General distribution:** Eastern Belgian Congo, Uganda and Tanganyika Territory.

**Range in Eastern Africa:** Ruanda and Urundi, eastern Belgian Congo, western Uganda and western Tanganyika Territory.

**Habits:** As for the nominate race.

**Recorded breeding:** No records.

**Distribution of other races of the species:** Nyasaland.

### 739 GREY-CHESTED ILLADOPSIS. *MALACOCINCLA POLIOTHORAX* (Reichenow).

*Alethe poliothorax* Reichenow, O.M. p. 6, 1900: Bangwa, north-west Cameroons.

**Distinguishing characters:** Above, warm chestnut brown; below, grey; centre of breast to belly whitish. The sexes are alike. Wing 80 to 87 mm. Juvenile plumage unrecorded.

**General distribution:** Fernando Po, Cameroon Mt., Uganda and the mountains north-west of Lake Tanganyika.

**Range in Eastern Africa:** Ruwenzori Mts. and Mt. Elgon.

**Habits:** A mountain forest species occurring in thick undergrowth and creeping quietly about, of which little is known. It appears to have a wide distribution in its particular localities.

**Nest and Eggs:** Undescribed.

**Recorded breeding:** Cameroon Mt., April and May.

**Food:** No information.

**Call:** A warbling call 'chee-cheerlee-cheerlee-chii-chee-cher-chii-ri-uu' the last note trilled (Van Someren). Woosnam also records a harsh call.

### 740 ABYSSINIAN HILL-BABBLER. *PSEUDOALCIPPE ABYSSINICUS* (Rüppell).

*Pseudoalcippe abyssinicus abyssinicus* (Rüpp.). **Pl. 59.**
*Drymophila abyssinica* Rüppell, N. Wirbelt, Vögel, p. 108, pl. 40, fig. 2, 1840: Simen, northern Abyssinia.

**Distinguishing characters:** Head grey; upperparts olive brown; below, paler grey; belly lighter. The sexes are alike. Wing 64 to 75 mm. The young bird has the top of the head olive brown, uniform with the mantle.

**Range in Eastern Africa:** Abyssinia and south-eastern Sudan, eastern Uganda, Kenya Colony and north-eastern to western Tanganyika Territory.

**Habits:** Birds of dense forest undergrowth and forest edge, found in most if not all of the mountain rain-forests of East Africa. Locally abundant but with definite seasonal movements, and are, in some localities at least, only found in damp forest. According to Granvik this bird is the finest songster in Africa. Its usual range is from 6,500 to 8,500 feet.

**Nest and Eggs:** A small cup nest of lichen, rootlets and fine stems lined with horsehair or fibres, in the fork of a shrub a few feet from the ground. Eggs two, but up to five recorded, pale pinkish or buff with darker clouding and with fine reddish brown spots; about 21 × 15 mm.

**Recorded breeding:** Southern Sudan (breeding condition),

July and November. Abyssinia, April. Mt. Elgon (breeding condition), March. Tanganyika Territory, January and February, also young in July.

**Food:** Insects, small molluscs and berries.

**Call:** A fine clear Nightingale-like song, often uttered after dusk, and much guttural chatter by day as well as a two-syllabled call.

*Pseudoalcippe abyssinicus stierlingi* (Reichw.). Stierling's Hill-Babbler.

*Turdinus stierlingi* Reichenow, O.M., p. 82, 1898: Iringa, south-central Tanganyika Territory. **Pl. 59.**

**Distinguishing characters:** Differs from the nominate race in having forehead and chin blacker and black streaks on throat. The sexes are alike. Wing 64 to 70 mm. The young bird has the top of the head brown, the sides of face paler than the adult, only an indication of the streaks on the throat, and some brown on the chest.

**Range in Eastern Africa:** East-central to south-western Tanganyika Territory from Kilosa and Uluguru Mts. to the Njombe, Rungwe and Songea areas and western Portuguese East Africa.

**Habits:** Usually in small parties in highland forest or bamboo zone, but are often found at lower levels. They do not normally feed high up in trees, and are more often heard than seen. The pale colour of the bill is very noticeable in forest. They have a little Robin-like chant of 'meeting to-morrow' (Lynes).

**Recorded breeding:** Tanganyika Territory, December to February.

*Pseudoalcippe abyssinicus atriceps* (Sharpe). Ruwenzori Hill-Babbler.

*Turdinus atriceps* Sharpe, Bull. B.O.C. 13, p. 10, 1902: Mubuku Valley, Ruwenzori Mts., western Uganda.

**Distinguishing characters:** Head black; rest of upperparts warm brown; below, throat to under tail-coverts dark grey; ends of flank feathers brown. The sexes are alike. Wing 65 to 76 mm. The young bird has the head brownish black, and the underside mainly olivaceous brown.

**General distribution:** Cameroons to Uganda.

**Range in Eastern Africa:** Western Uganda, at elevations above 5,000 feet, to Ruanda and Urundi.

**Habits:** A mountain forest species seen usually in small parties of eight to ten among ferns and tangled forest growth busily engaged in searching for insects, and moving in one direction like a flock of Helmet-Shrikes. They utter a continuous twittering rather Swallow-like note, and do not confine themselves to undergrowth, but follow creepers high up among trees. Most frequently seen near water. They have a beautiful clear song usually uttered high up in some dense mass of creepers (Woosnam), also the twittering note mentioned above.

**Recorded breeding:** No records.

*Pseudoalcippe abyssinicus stictigula* (Shell.).
*Alcippe stictigula* Shelley, Bull. B.O.C. 13, p. 61, 1903: Mwenembe, north-western Nyasaland.

**Distinguishing characters:** Differs from the nominate race in being warmer brown above. Wing 63 to 70 mm.

**General distribution:** North-western to southern Nyasaland.

**Range in Eastern Africa:** Nyasaland.

**Habits:** Of the nominate race, confined to evergreen forest, and living on or near the forest floor.

**Recorded breeding:** Nyasaland, October and November.

**Distribution of other races of the species:** West Africa to Angola.

Names in Sclater's *Syst. Av. Æthiop.* 2, 1930, which have been changed or have become synonyms in this work:
*Illadopsis rufipennis barakæ* Jackson, treated as synonymous with *Malacocincla albipectus* (Reichenow).

Names introduced since 1930, and which have become synonyms in this work:
*Pseudoalcippe abyssinicus chyulu* Van Someren, 1939, treated as synonymous with *Pseudoalcippe abyssinicus abyssinicus* (Rüppell).

FAMILY—**PYCNONOTIDÆ. BULBULS.** Genera: *Pycnonotus, Tricophorus, Bleda, Thescelocichla, Pyrrhurus, Bœopogon, Ixonotus, Phyllastrephus, Suaheliornis, Arizelocichla, Chlorocichla, Stelgidillas, Andropadus, Eurillas, Stelgidocichla* and *Neolestes.*

Thirty-seven species occur in Eastern Africa. The majority are inhabitants of forest and woodland, but some occur in more open country and around habitations. In some the females are usually appreciably smaller and have shorter bills than the males, this being especially so in the genus *Phyllastrephus*. All are mainly frugivorous and many are good songsters. The Greenbuls are not easy to identify either in the field or in the hand until one has had some experience of them.

KEY TO THE ADULT BULBULS OF EASTERN AFRICA

| | | | |
|---|---|---|---|
| 1 | White spots on wings and rump: | SPOTTED GREENBUL *Ixonotus guttatus* | 752 |
| 2 | No white spots on wings and rump: | | 3–60 |
| 3 | Black band across chest: | BLACK-COLLARED BULBUL *Neolestes torquatus* | 777 |
| 4 | No black band across chest: | | 5–60 |
| 5 | Above, golden green: | JOYFUL GREENBUL *Chlorocichla lætissima* | 770 |
| 6 | Above, pale olive green: | | 12–13 |
| 7 | Above, olive brown or russet brown: | | 14–15 |
| 8 | Above, ashy grey green: | | 16–17 |
| 9 | Above, mouse brown: | | 18–23 |
| 10 | Above, dull or dark olive green or brownish olive green: | | 24–45 |
| 11 | Above, bright olive green: | | 46–60 |
| 12 | Size larger, wing over 96 mm.: | YELLOW-BELLIED GREENBUL *Chlorocichla flaviventris* | 769 |

| | | | |
|---|---|---|---|
| 13 | Size smaller, wing under 95 mm.: | ZANZIBAR SOMBRE GREENBUL *Andropadus importunus* | 773 |

| | | | |
|---|---|---|---|
| 14 | Above, olive brown, chest white: | BROWNBUL *Phyllastrephus terrestris* | 753 |
| 15 | Above, russet brown, chest ashy brown: | NORTHERN BROWNBUL *Phyllastrephus strepitans* | 754 |

| | | | |
|---|---|---|---|
| 16 | Forehead and cheeks grey: | LEAF-LOVE *Pyrrhurus scandens* | 750 |
| 17 | Forehead and cheeks ashy grey green: | GREY-OLIVE GREENBUL *Phyllastrephus cerviniventris* | 759 |

| | | | |
|---|---|---|---|
| 18 | Under tail-coverts white: | WHITE-VENTED BULBUL *Pycnonotus barbatus* | 744 |
| 19 | Under tail-coverts yellow: | | 20–21 |
| 20 | White feathers at side of neck: | WHITE-EARED BULBUL *Pycnonotus dodsoni* | 743 |
| 21 | No white feathers at side of neck: | | 22–23 |
| 22 | Top of head dark brown: | DARK-CAPPED BULBUL *Pycnonotus tricolor* | 742 |
| 23 | Top of head black: | BLACK-CAPPED BULBUL *Pycnonotus xanthopygos* | 741 |

| | | | |
|---|---|---|---|
| 24 | Dark olive streaks on chest: | DAPPLED MOUNTAIN-GREENBUL *Phyllastrephus oreostruthus* | 762 |
| 25 | No dark olive streaks on chest: | | 26–45 |
| 26 | Tail chestnut: | BRISTLE-BILL *Bleda syndactyla* | 746 |

# BULBULS

27 Tail not chestnut: 28–45
28 Throat yellow: YELLOW-THROATED
LEAF-LOVE
*Pyrrhurus flavicollis* **749**
29 Underside mainly grey: SLENDER-BILLED
GREENBUL
*Stelgidillas gracilirostris* **771**
30 Underside mainly dark olive green: 31–37
31 Yellow streak down each side of throat: YELLOW-WHISKERED
GREENBUL
*Stelgidocichla latirostris* **776**
32 Bill short and broad at base: LITTLE GREENBUL
*Eurillas virens* **775**
33 Bill not short and broad: 34–45
34 Throat pale olive grey: TORO OLIVE GREENBUL
*Phyllastrephus hypochloris* **760**
35 Throat dark grey: 36–37
36 Inner edges of flight feathers warm yellowish buff, first primary short: LITTLE GREY GREENBUL
*Andropadus gracilis* **772**
37 Inner edges of flight feathers cold greyish buff, first primary long: CAMEROON SOMBRE
GREENBUL
*Andropadus curvirostris* **774**
38 Underside pale greyish or creamy white with lemon yellow streaks: 39–42
39 Head grey: YELLOW-STREAKED
GREENBUL
*Phyllastrephus flavostriatus* **755**
40 Feathers of head having a scaly appearance: 41–42

| | | | |
|---|---|---|---|
| 41 | White ends to tail feathers: | WHITE-TAILED GREENBUL *Thescelocichla leucopleura* | 748 |
| 42 | No white ends to tail feathers: | WHITE-THROATED GREENBUL *Phyllastrephus albigularis* | 761 |
| 43 | Head olive brown: | | 44–45 |
| 44 | Mantle olive brown: | SHARPE'S GREENBUL *Phyllastrephus alfredi* | 756 |
| 45 | Mantle greenish olive brown: | FISCHER'S GREENBUL *Phyllastrephus fischeri* | 758 |
| 46 | Size smaller, wing under 71 mm.: | | 47–48 |
| 47 | Inner edges of flight feathers buff: | KRETSCHMER'S GREENBUL *Suaheliornis kretschmeri* | 764 |
| 48 | Inner edges of flight feathers lemon yellow: | SMALLER YELLOW-STREAKED GREENBUL *Phyllastrephus debilis* | 757 |
| 49 | Size larger, wing over 72 mm.: | | 50–60 |
| 50 | Throat pure white: | RED-TAILED GREENBUL *Tricophorus calurus* | 745 |
| 51 | Throat not white: | | 52–60 |
| 52 | Below, bright sulphur yellow: | GREEN-TAILED BRISTLE-BILL *Bleda eximia* | 747 |
| 53 | Below, lemon yellow: | XAVIER'S GREENBUL *Phyllastrephus xavieri* | 763 |
| 54 | Below, olive green or olive yellow: | STRIPE-CHEEKED GREENBUL *Arizelocichla milanjensis* | 767 |

| | | | |
|---|---|---|---|
| 55 | Below, olivaceous yellow, throat grey: | OLIVE-BREASTED MOUNTAIN GREENBUL *Arizelocichla tephrolæma* | 765 |
| 56 | Below, olivaceous and grey: | | 57–60 |
| 57 | Outer tail feathers white: | HONEY-GUIDE GREENBUL *Bœopogon indicator* | 751 |
| 58 | Outer tail feathers green: | | 59–60 |
| 59 | Bill deep and heavy: | MOUNTAIN-GREENBUL *Arizelocichla nigriceps* | 766 |
| 60 | Bill not deep and heavy: | SHELLEY'S GREENBUL *Arizelocichla masukuensis* | 768 |

**741 BLACK-CAPPED BULBUL.** *PYCNONOTUS XANTHOPYGOS* (Hemprich and Ehrenberg).

*Pycnonotus xanthopygos layardi* Gurn.

*Pycnonotus layardi* Gurney, Ibis, p. 390, 1879: Rustenberg, western Transvaal.

**Distinguishing characters:** General colour above also chest dark mouse brown; sides of face and throat darker; top of head black, sharply contrasting with mantle; below, breast and belly white; flanks mouse brown; under tail-coverts lemon yellow. Albinistic examples occur. The sexes are alike. Wing 87 to 107 mm. The young bird has the head dull black.

**General distribution:** Northern Rhodesia, Nyasaland also the Zambesi Valley to the Transvaal, eastern Cape Province and Natal.

**Range in Eastern Africa:** Nyasaland and the Zambesi Valley.

**Habits:** Probably the commonest and best known bird of Nyasaland. Fearless lively birds of every garden, with habits as for other races.

**Nest and Eggs:** A cup of dried grass lined with fine grass and fibre, usually low and well hidden. Eggs three, salmon pink to purplish pink, freckled and blotched with reddish brown and dark purplish brown with grey undermarkings, often zoned; about 23 × 16·5 mm.

**Recorded breeding:** Nyasaland, September to January.

**Food:** Largely fruit and berries.

**Call:** Mellow and cheery, one is described as 'Well, George, how's the wife?' another as 'Back to Calcutta,' and the alarm call is a scolding 'chit-chit.'

*Pycnonotus xanthopygos micrus* Oberh.

*Pycnonotus layardi micrus* Oberholser, Proc. U.S. Nat. Mus., 28, p. 891, 1905: Taveta, south-eastern Kenya Colony.

**Distinguishing characters:** Similar to the preceding race but upperside rather darker; sometimes a few whitish feathers on side of neck. Wing 84 to 98 mm.

**Range in Eastern Africa:** Coastal areas of Kenya Colony from Malindi to Tanganyika Territory as far west as the Arusha District and Tabora, and Portuguese East Africa, but not reaching the Zambesi River; also Zanzibar and Mafia Islands.

**Habits:** Common, tame and conspicuous bird, both on mainland and on Zanzibar Island. Habits as for other races; almost ubiquitous.

**Recorded breeding:** Coastal Kenya Colony, April to July and again October to February. Tanganyika Territory, September to April. Zanzibar Island, October to January. Portuguese East Africa, October to February.

*Pycnonotus xanthopygos spurius* Reichw.

*Pycnonotus spurius* Reichenow, Vög., Afr. 3, p. 841, 1905: Ennia, Gallaland, eastern Abyssinia.

**Distinguishing characters:** Differs from the preceding race in having the neck in front blacker and the chest appreciably darker; usually a few whitish feathers on side of neck. Wing 86 to 96 mm.

**Range in Eastern Africa:** Southern and eastern Abyssinia.

**Habits:** As for other races.

**Recorded breeding:** Southern Abyssinia, April and May.

**Distribution of other races of the species:** Arabia to Palestine, the nominate race being described from Arabia.

### 742 DARK-CAPPED BULBUL. *PYCNONOTUS TRICOLOR* (Hartlaub).

*Pycnonotus tricolor tricolor* (Hartl.).

*Ixos tricolor* Hartlaub, Ibis, p. 341, 1862: Congo.

**Distinguishing characters:** General colour above, dark mouse brown; head and throat darker, almost blackish brown; chest rather paler; breast and belly white; flanks mouse brown, under tail-coverts lemon yellow. The sexes are alike. Wing 83 to 104 mm. The young bird is rather browner than the adult and has the head uniform with the mantle.

**General distribution:** Northern Cameroons to the Sudan, Tanganyika Territory, Angola, northern Damaraland, Belgian Congo and Northern Rhodesia.

**Range in Eastern Africa:** Southern Sudan, Uganda, Ruanda, Urundi and western Tanganyika Territory.

**Habits:** Common, widespread and noticeable bird of gardens and woodland. It is tame and one of the most persistent songsters. Cheerful confiding little bird, but among the first to mob and pursue any snake or predatory animal.

**Nest and Eggs:** As for the following race.

**Recorded breeding:** Sudan, August to October. Uganda (double-brooded), February to June, and again in September to January. South-eastern Belgian Congo, October.

*Pycnonotus tricolor fayi* Mearns. **Ph. viii.**

*Pycnonotus layardi fayi* Mearns, Smiths. Misc. Coll. Wash. 56, No. 20, p. 7, 1911: Fays Farm, Njabini, near Mt. Kinankop, Aberdare Mts., western Kenya Colony.

**Distinguishing characters:** Similar to the nominate race, but top of head, sides of face and throat usually darker, though not the black of the Black-capped Bulbul. Wing 84 to 102 mm.

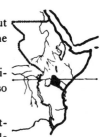

**Range in Eastern Africa:** Kenya Colony and Tanganyika Territory from Mt. Elgon and Mt. Kenya to Athi and Loliondo, also Rusinga Island, Lake Victoria.

**Habits:** Common widespread bird of woodland and forest outskirts. Its pleasant warbling song is characteristic. Cheerful little birds usually quite confiding, but will mob an animal or predatory bird in a determined manner with crest erected and lowered repeatedly, and with shrill angry chattering cries.

**Nest and Eggs:** Nest in a tree fork or in creepers, bushes or large herbs, made of small twigs, grass stalks or fibres, lined hair and fibres. Eggs two, white to pink, spotted and blotched with dark brown and dark violet with ash-grey undermarkings; about 22 × 16 mm. (larger measurements are also noted).

**Recorded breeding:** Kenya Colony, February to July but probably double brooded in most localities, and may breed in any month of year. Tanganyika Territory, November to February.

**Food:** Berries and insects, takes some cultivated fruit, but probably more than makes up for it by the destruction of insects.

**Call:** A pleasing cheerful song and a variety of mellow call notes. One call is like that of the preceding species, 'Back to Calcutta-back to Calcutta.'

**Distribution of other races of the species:** Northern Rhodesia.

### 743 WHITE-EARED BULBUL. *PYCNONOTUS DODSONI* Sharpe.

*Pycnonotus dodsoni* Sharpe, P.Z.S p. 488, 1895: Sillul, Ogaden, eastern Abyssinia.

**Distinguishing characters:** Differs from the Dark-capped and Black-capped Bulbuls in having a more pronounced white patch at the side of the neck; the feathers of the upperparts are also edged with lighter colour giving a rather scaly appearance; breast mottled with dark ashy. The sexes are alike. Wing 77 to 93 mm. The young bird has the head browner, not so black.

**Range in Eastern Africa:** Central Abyssinia to British and Italian Somaliland, the country east of Mt. Kenya as far south as Tsavo, north Paré Mts. and Mkomasi, north-eastern Tanganyika Territory.

**Habits:** Not easy to tell from the races of the Dark-capped Bulbul in the field, consequently most field notes may refer to either species. This species is unlikely to differ in habits in any way but is more likely to be found in arid conditions.

**Nest and Eggs:** Cup nest usually in a low bush, made of grass and fibre and lined with fine grass, rather like a Warbler's nest. Eggs two or three, very variable, white or pinkish ground colour with much grey blotching and clouding, and with chestnut or chocolate spots and flecks; about 20 × 15 mm., but many eggs are larger.

**Recorded breeding:** Eastern Abyssinia, April to June. South Abyssinia, October and January. Kenya Colony, Turkana, May. southern areas, March to May. Mombasa, December.

**Food:** Berries and insects.

**Call:** A clear call of 'pii-chi-chi-re,' but with many other notes and variations.

Plate 58

Green-breasted Pitta
*Pitta reichenowi* (Vol. I., p. 797)

African Broadbill (Vol. I., p. 793)
*Smithornis capensis capensis*

African Pitta (Vol. I., p. 796)
*Pitta angolensis*

Rosy-breasted Long-claw (p. 80)
*Macronyx ameliae wintoni*
male     female

Sharpe's Long-Claw
*Macronyx sharpei* (p. 78)

Golden Pipit (p. 75)
*Tmetothylacus tenellus*
male     female

Pamgani Long-claw (p. 79)
*Macronyx aurantiigula*

Abyssinian Long-claw
*Macronyx flavicollis* (p. 78)

Yellow-throated Long-claw
*Macronyx croceus* (p. 76)
adult     young

Plate 59

Mountain Illadopsis (p. 102)  Brown Illadopsis (p. 100)  Pale-breasted Illadopsis (p. 100)
Stierling's Hill-Babbler (p. 105)  Abyssinian Hill-Babbler (p. 104)  Green-tailed Bristle-bill (p. 117)
Yellow Longbill (p. 446)
Bristle-bill (p. 117)  Yellow-bellied Greenbul (p. 141)  Sharpe's Greenbul (p. 126)
Joyful Greenbul (p. 142)
Northern Brownbul (p. 123)  Brownbul (p. 122)
Fischer's Greenbul (p. 123)  Yellow-streaked Greenbul (p. 124)

## BULBULS

**744 WHITE-VENTED BULBUL.** *PYCNONOTUS BARBATUS* (Desfontaines).

*Pycnonotus barbatus arsinoe* (Licht.).

*Turdus Arsinoe* Lichtenstein, Verz. Doubl. p. 39, 1823: Faiyum, Egypt.

**Distinguishing characters:** Similar to the Dark-capped Bulbul, but head blacker contrasting with mantle and under tail-coverts white, not yellow. Albinism is not uncommon. The sexes are alike. Wing 85 to 99 mm. The young bird has the head brown, not black.

**General distribution:** Egypt to the Sudan.

**Range in Eastern Africa:** The Sudan, except south-eastern area.

**Habits:** Tame cheery little bird, well distributed in suitable localities, which are many, as it seems to thrive in any form of bush or woodland, whether near cultivation or in the desert. It is equally common in the gardens of Khartoum or the arid belt of Kordofan.

**Nest and Eggs:** A neat but light cup nest of grass with occasionally a few dead leaves, and often suspended between two forking twigs. Eggs two or three, white or pinkish heavily speckled and spotted with reddish brown and with pale lilac undermarkings; about 23 × 16 mm.

**Recorded breeding:** Sudan, Kordofan, April, May and July. Khartoum, March and April. Nile Valley, June. Red Sea coast, March and April.

**Food:** Equally fruit and insects.

**Call:** Cheerful but rather monotonous song, also usual alarm notes.

*Pycnonotus barbatus schoanus* Neum.

*Pycnonotus barbatus schoanus* Neumann, O.M. p. 77, 1905: Kilbe, Kollo Province, central Abyssinia.

**Distinguishing characters:** Differs from the nominate race in being much darker above. Wing 84 to 98 mm.

**Range in Eastern Africa:** Eritrea, Abyssinia and south-eastern Sudan.

**Habits:** As for the preceding race.

**Recorded breeding:** Eastern Abyssinia and Eritrea, May, also July and August. Coastal Eritrea, January to March.

*Pycnonotus barbatus somaliensis* Reichw.

*Pycnonotus arsinoe somaliensis* Reichenow, Vög. Afr. 3, p. 840, 1905: Zeila, British Somaliland.

**Distinguishing characters:** Differs from the preceding race in having a distinct white patch at side of neck; breast feathers edged with whitish, giving a mottled appearance. Wing 80 to 95 mm.

**Range in Eastern Africa:** British Somaliland.

**Habits:** As for other races.

**Recorded breeding:** British Somaliland, February, also probably later.

**Distribution of other races of the species:** North and West Africa, the nominate race being described from Algiers.

745 RED-TAILED GREENBUL. *TRICOPHORUS CALURUS* Cassin.

*Tricophorus calurus ndussumensis* (Reichw.).

*Criniger verreauxi ndussumensis* Reichenow, Vög. Afr. 3, p. 383, 1904: Ndussuma country, west of Lake Albert, north-eastern Belgian Congo.

**Distinguishing characters:** Above, olive green; head brown; sides of face greyer; tail feathers russet, edged with olive green; below, chin and throat white; chest and under tail-coverts olivaceous yellow; flanks dark olive green. The sexes are alike. Wing 82 to 92 mm. The young bird has the top of the head brownish olive; mantle browner; below duller yellow.

**General distribution:** North-eastern Belgian Congo to Uganda.

**Range in Eastern Africa:** Uganda.

**Habits:** Common in most of the forests of Uganda, but little is noted of it. It is a bird of dense forest undergrowth which also ranges to the tree tops, and is often a member of mixed bird hunting parties. It has a habit of flipping its wings and puffing out its throat.

**Nest and Eggs:** In West Africa the nest is a platform of twigs heaped with damp moss and an inner cup of fine black fibre. Eggs two, pinkish, but so densely speckled with chocolate brown as to be almost uniform in colour; about 23 × 16 mm.

**Recorded breeding:** Not yet recorded in our area.

**Food:** Berries and insects.

**Call:** A little song of 'chit-chu-cherry' and a loud distinctive 'cheep,' as well as a soft 'zut' and a harsher 'churr.'

**Distribution of other races of the species:** West Africa, the nominate race being described from Gabon.

**746 BRISTLE-BILL.** *BLEDA SYNDACTYLA* Swainson.
*Bleda syndactyla woosnami* O. Grant. **Pl. 59.**
*Bleda woosnami* O. Grant, Bull. B.O.C. 19, p. 87, 1907: Mpanga Forest, western Uganda.

**Distinguishing characters:** Above, dark olive green; a few white feathers over eye; tail chestnut; below, sulphur yellow; flanks olive green. The sexes are alike. Wing 97 to 113 mm. The young bird has the mantle and breast washed with rusty; belly buff; flanks rusty; under tail-coverts buffish yellow.

**General distribution:** The Sudan, north-eastern Belgian Congo, Uganda and Kenya Colony.

**Range in Eastern Africa:** Southern Sudan, Uganda and western Kenya Colony at Kakamega.

**Habits:** Not common and always in tall dense undergrowth, not in trees, a very shy bird apparently, alert and difficult to observe.

**Nest and Eggs:** Nest a slight shallow cup of decaying leaves and a few sticks with fibre and rootlets, a few feet from the ground in a leafy bush. Eggs two, buff almost obscured by dense blotches and suffusions of vandyke brown and pale brown; about 26·5 × 18 mm. (West African race).

**Recorded breeding:** Southern Sudan (breeding condition), October and March. Eastern Uganda, April and May.

**Food:** No information.

**Call:** Monotonous 'tur-tur-tur-tur' and a little song of descending notes 'se-e-e-e-er-r-r.' Alarm note 'chirit-chirit.'

**Distribution of other races of the species:** West Africa, the habitat of the nominate race.

**747 GREEN-TAILED BRISTLE-BILL.** *BLEDA EXIMIA* Hartlaub.
*Bleda eximia ugandæ* Van Som. **Pl. 59.**
*Bleda eximia ugandæ* Van Someren, Bull. B.O.C. 35, p. 116, 1915: Mabira Forest, southern Uganda.

**Distinguishing characters:** Very similar to the Bristle-bill, but smaller, and has the tail olive green with broad yellow tips to the outer feathers. The sexes are alike. Wing 93 to 104 mm. The young bird is duller, has some tawny feathers in the wing-coverts and flesh-coloured, not slate, toes.

**General distribution:** The Sudan, Uganda and the northern Belgian Congo.

**Range in Eastern Africa:** South-western Sudan to western and southern Uganda.

**Habits:** A local forest species of which little is recorded from within our area. In Cameroons an allied race is said to keep to low undergrowth and to be restless and agile. It feeds largely on ants, or appears to be dependent on them for its food.

**Nest and Eggs:** Undescribed.

**Recorded breeding:** No records.

**Food:** One of several species which sit on branches overlooking a column of ants and occasionally dive down on it from time to time. It is uncertain whether they actually eat ants or whether they take from them the larvæ or other food they are carrying.

**Call:** A mournful whistle varied by a flute-like note (Sjostedt).

**Distribution of other races of the species:** West Africa, the nominate race being described from the Gold Coast.

### 748 WHITE-TAILED GREENBUL. *THESCELOCICHLA LEUCOPLEURA* (Cassin).

*Phyllostrophus leucopleurus* Cassin, Proc. Acad. Sci. Philad. 8, p. 328, 1855: Moonda River, Gabon.

**Distinguishing characters:** Above, dark olive; feathers of forehead and crown with darker edges giving a scaly appearance; sides of face streaked olive and whitish; all except central tail feathers with broad white ends; below, throat and chest mottled olive and whitish; rest of underparts pale yellow. The sexes are alike. Wing 102 to 115 mm. The young bird is browner olive above; throat to chest pale yellow; tail feathers more pointed.

**General distribution:** Senegal to Gabon and Uganda.

**Range in Eastern Africa:** The Bwamba Valley of western Uganda.

**Habits:** Shy restless birds in small parties, found as a rule along swampy streams in forest, especially among low Raphia palms, and

# BULBULS

remarkable for their noisy and incessant chattering. Bates remarks that their ringing guttural notes run together so as to resemble excited talking. They are locally migratory.

**Nest and Eggs:** Undescribed.

**Recorded breeding:** Cameroons, February.

**Food:** Probably largely fruit.

**Call:** A characteristic chattering ending with a call like 'kwe-poor-kwaper' (Marchant). They are also excellent songsters in a conversational sort of way.

### 749 YELLOW-THROATED LEAF-LOVE. *PYRRHURUS FLAVICOLLIS* Swainson.

*Pyrrhurus flavicollis flavigula* (Cab.).

*Trichophorus flavigula* Cabanis, O.C. p. 174, 1880: Angola.

**Distinguishing characters:** Above, dull olive, including wings and tail; feathers of head edged with grey, giving a scaly appearance; below, paler dull yellowish olive; fine grey streaks on chest; throat pale yellow. The sexes are alike. Wing 101 to 118 mm. The young bird is duller, more brownish olive; throat dull white; toes pale bluish brown not lead black.

**General distribution:** North-eastern Belgian Congo, as far north as Welle River to Uganda, Kenya Colony, Tanganyika Territory, Angola and Northern Rhodesia.

**Range in Eastern Africa:** Western and southern Uganda, western Kenya Colony and western Tanganyika Territory; also Kome, Sesse and Ukerewe Islands, Lake Victoria.

**Habits:** A bird of woodland, gallery forest and the wooded banks of streams as well as of true forest, restless, shy and almost boisterously noisy. Usually seen singly or in small parties, and obtaining a good deal of its food on the ground. It has a typical habit of flicking open its wings while uttering a clucking note.

**Nest and Eggs:** The nest is a cup of small twigs, rootlets and skeleton leaves, often attached to a fork by spiders' webs and placed in a bush or thicket a few feet from the ground. Eggs two, pale pink or dirty white, heavily blotched with reddish brown and with pale purple undermarkings; about 25 × 17 mm. and rather oval.

**Recorded breeding:** Uganda, April to August, also December, breeding season irregular.

**Food:** Fruit, berries and insects.

**Call:** The normal call is a shrill harsh scolding chatter 'chiro-chiro.' There is also a Thrush-like warbling song, and a cat-like mew.

*Pyrrhurus flavicollis soror* (Neum.).
*Xenocichla flavicollis soror* Neumann, O.M. p. 9, 1914: Ogowe River, Gabon.

**Distinguishing characters:** Differs from the preceding race in having the throat pale creamy white. Wing 92 to 114 mm.

**General distribution:** Gabon to the Ubangi area of French Equatorial Africa, northern Belgian Congo and the Sudan.

**Range in Eastern Africa:** South-western Sudan.

**Habits:** As for the preceding race.

**Recorded breeding:** Nigeria and Cameroons in most months of the year.

**Distribution of other races of the species:** West Africa, the nominate race being described from Gambia.

750 LEAF-LOVE. *PYRRHURUS SCANDENS* Swainson.
*Pyrrhurus scandens orientalis* (Hartl.). **Pl. 60.**
*Xenocichla orientalis* Hartlaub, J.f.O. p. 425, 1883: Tomajá, south-western Sudan.

**Distinguishing characters:** Above, ashy grey green; forehead and sides of face grey; wings dusky edged with fawn brown; tail pale tawny; below, creamy white with pale yellow streaks; chin and throat white, chest greyer. The sexes are alike. Wing 96 to 110 mm. The young bird is rather duller and paler than the adult.

**General distribution:** Ubangi-Welle area of French Equatorial Africa to northern Belgian Congo, the Sudan, Uganda and Tanganyika Territory.

**Range in Eastern Africa:** South-western Sudan to western Tanganyika Territory as far south as Kabogo Head.

**Habits:** A bird of local distribution frequenting thickets along stream banks. It climbs about in the foliage rather than hops.

## BULBULS

Little is known of its habits, but it is noisy when alarmed, and usually travels in small parties, keeping up an incessant chatter.

**Nest and Eggs:** The nest of the western race is a cup of fine grass and leaves, slung between the stems of a creeper. Eggs two, buff, finely speckled with brown and mottled with a zone of purplish brown spots at the larger end. No measurements available.

**Recorded breeding:** Welle district (breeding condition) October onwards.

**Food:** Mainly insects.

**Call:** Described as distinctive with the accent on the last syllable.

**Distribution of other races of the species:** West Africa, the habitat of the nominate race.

**751 HONEY-GUIDE GREENBUL. *BÆOPOGON INDICATOR* (Verreaux).**
*Bæopogon indicator chlorosaturata* (Van Som.).
*Chlorocichla indicator chlorosaturata* Van Someren, Bull. B.O.C. 35, p. 127, 1915: Kyetume Forest, southern Uganda.

**Distinguishing characters:** Above, dark olive green; sides of face and below, greyish brown, streaked with olive green; belly and under tail-coverts buff; outer tail feathers white with sulphur yellow edges and black brown tips. This tail pattern is very similar to that of the Honey-Guides. The sexes are alike. The male has a cream coloured eye noticeable at some distance. Wing 99 to 111 mm. The young bird is above duller; top of head washed with brownish; belly greyish white; flanks buffish; outer tail feathers wholly white.

**General distribution:** Central Belgian Congo to the Sudan and Uganda.

**Range in Eastern Africa:** Southern Sudan and western Uganda.

**Habits:** A bird of high forest tree tops, usually keeping beyond ordinary gunshot from the ground. Not uncommon, but owing to its habitat little known. Several observers have noted its resemblance to the Honey-Guides, and Col. Cave notes that it has also the same sort of musty smell.

**Nest and Eggs:** Undescribed.

**Recorded breeding:** No records.

**Food:** Mainly berries and small fruits.

**Call:** A monotonous whistle 'keeto-keeto-keeto-keeto' and the West African race has a song, rather a pleasing song according to

Bates, of long-drawn slurred notes. There are also many other calls, some cat-like.

**Distribution of other races of the species:** West Africa, the nominate race being described from Gabon.

### 752 SPOTTED GREENBUL. *IXONOTUS GUTTATUS* Verreaux.

*Ixonotus guttatus* J. and E. Verreaux, Rev. Mag. Zool. p. 306, 1851: Gabon.

**Distinguishing characters:** Above, ashy with a green wash; broad white spots at tips of wing-coverts, innermost secondaries, rump and upper tail-coverts; outer tail feathers white; below, white with a creamy yellow wash. The sexes are alike. Wing 84 to 95 mm. The young bird has less green wash above.

**General distribution:** Gold Coast to Gabon and Uganda.

**Range in Eastern Africa:** Western Uganda.

**Habits:** A species of both low bushes and of high forest tree tops, of wide distribution but locally confined to small areas. Usually in restless twittering flocks with curious wing-flirting movements and much more gregarious than most Greenbuls.

**Nest and Eggs:** Nest, a slight rough shallow cup of leaf stalks and bark lined with black rootlets. Eggs yellowish buff, densely speckled with dark brown spots obscuring the ground colour. No measurements available.

**Recorded breeding:** Cameroons, young in October.

**Food:** Insects, berries and fruit.

**Call:** A twittering chirp and a clicking note, said to be like the chinking of coins.

### 753 BROWNBUL. *PHYLLASTREPHUS TERRESTRIS* Swainson. Pl. 59.

*Phyllastrephus terrestris suahelicus* Reichw.

*Phyllastrephus capensis suahelicus* Reichenow, Vög. Afr. 3, p. 405, 1904: Msua, Bagamoyo District, eastern Tanganyika Territory.

**Distinguishing characters:** Above, dull olivaceous brown; tail pale russet; below, white; flanks and under tail-coverts pale olivaceous brown. The sexes are alike. Wing 73 to 97 mm. The young bird is rather more earth brown in colour above, and the wing-coverts are narrowly edged with russet.

# BULBULS

**General distribution:** Kenya Colony to northern Bechuanaland and Southern Rhodesia.

**Range in Eastern Africa:** Coastal area of Kenya Colony from Malindi to eastern and southern Tanganyika Territory and the Zambesi River.

**Habits:** A bird of scrub and thick bush in low-lying country. Usually to be seen scratching and rustling about in dead leaves under bushes in little parties of three or four individuals.

**Nest and Eggs:** Nest, a flimsy frail untidy cup of small twigs and stems lined with rootlets and a few leaves, generally among low bushes on the outskirts of a thicket. Eggs two, white with streaks and blotches of sepia or olive brown and with grey undermarkings; about 24 × 17 mm.

**Recorded breeding:** Nyasaland, October to March. Portuguese East Africa, March. Rhodesia, December.

**Food:** Believed to be largely ants.

**Call:** A harsh grating chirping chatter. Also a short pleasant warbled song 'whicherwer-whicherwer.'

*Phyllastrephus terrestris bensoni* Van Som.

*Phyllastrephus terrestris bensoni* Van Someren, Bull. B.O.C. 46, p. 11, 1945: Lower Meru Forest, Kenya Colony.

**Distinguishing characters:** Differs from the preceding race in being darker above and on sides of face. Wing 95 to 98 mm.

**Range in Eastern Africa:** Meru and Chuka Forests, Mt. Kenya, Kenya Colony.

**Habits:** As for the preceding race.

**Recorded breeding:** No records.

**Distribution of other races of the species:** South Africa, the nominate race being described from the Knysna, Cape Province.

754 NORTHERN BROWNBUL. *PHYLLASTREPHUS STREPITANS* (Reichenow). Pl. 59.

*Criniger strepitans* Reichenow, O.C. p. 139, 1879: Malindi, eastern Kenya Colony.

**Distinguishing characters:** Similar to the Brownbul, but differs in being above warm russet, not olivaceous brown; tail darker in

tone, and bill rather smaller and darker coloured, more blackish brown. The sexes are alike. Wing 68 to 85 mm. The young bird has the top of the head and mantle warmer in tone.

**Range in Eastern Africa:** Western and southern Abyssinia and southern Sudan, Kenya Colony and eastern Tanganyika Territory, as far south as Dar-es-Salaam.

**Habits:** Found on or near ground in dense scrub or lowland forest, either near water or in quite dry country. Shy but inquisitive birds, with a habit of twinkling the wings and tail.

**Nest and Eggs:** Undescribed.

**Recorded breeding:** Tanganyika Territory (breeding condition), March.

**Food:** Insects.

**Call:** A cheerful chattering cry 'che-che-che-cha-cha-che (Fuggles Couchman), also a call like that of a Reed Warbler.

### 755 YELLOW-STREAKED GREENBUL. *PHYLLASTREPHUS FLAVOSTRIATUS* (Sharpe).

*Phyllastrephus flavostriatus tenuirostris* (Fisch. and Reichw.). **Pl. 59.**
*Xenocichla tenuirostris* Fischer and Reichenow, J.f.O. p. 262, 1884: Lindi, south-eastern Tanganyika Territory.

**Distinguishing characters:** Head grey; rest of upperparts olive green; below, chin and throat greyish white; chest to belly very pale grey with clear yellow streaks; flanks olive green; under wing-coverts yellow. The sexes are alike. Wing 80 to 103 mm. The young bird has a dusky olive green chest and the centre of the belly is wholly yellow.

**Range in Eastern Africa:** Taita area of south-eastern Kenya Colony and coastal areas of Tanganyika Territory as far west as the south Paré Mts., Kilosa, the Pugu Hills and the coastal areas of Portuguese East Africa opposite Mozambique Island.

**Habits:** Birds of scrub or forest, usually seen running along limbs of trees like Creepers in search of insects and hanging on head downwards like Tits. They have a curious habit of flipping one wing open from time to time. Tame birds and often members of mixed bird parties.

**Nest and Eggs:** Nest a plain cup of fibres, usually unlined, in a low shrub, often between branches. Eggs two, grey or pinkish mauve with a zone of Bunting-like scrawls in deep purple and dark brown or slate at larger end; about 23 × 16 mm.

**Recorded breeding:** Amani, Tanganyika Territory, October to January.

**Food:** Mainly insects.

**Call:** Long loud and not unpleasing song, with a nasal fluting tone at the beginning 'blo-bli-blew-tu-tu-tu-tu.' Also a clear vibrant call of 'bwa-bi-hi-hi-hi,' almost whinnying, and a metallic chattering alarm note (Moreau).

*Phyllastrephus flavostriatus olivaceo-griseus* Reichw.
*Phyllastrephus olivaceo-griseus* Reichenow, O.M. p. 47, 1908: Rugege Forest, Ruanda, near Lake Kivu, eastern Belgian Congo.

**Distinguishing characters:** Differs from the preceding race in having a shorter bill; brighter green wings and tail; inner webs of flight feathers bright yellow. Wing 87 to 105 mm.

**General distribution:** Uganda to eastern Belgian Congo between Lakes Kivu and Tanganyika, west of the Ruzizi River.

**Range in Eastern Africa:** South-western Uganda at Kigezi to the Rugege Forest, Ruanda.

**Habits:** Another rare forest bird of which nothing is known. It is possibly more local than rare.

**Recorded breeding:** No records.

*Phyllastrephus flavostriatus vincenti* Grant and Praed.
*Phyllastrephus flavostriatus vincenti* C. Grant and Mackworth-Praed, Bull. B.O.C. 60, p. 62, 1940: Namuli Mts., northern Portuguese East Africa.

**Distinguishing characters:** Differs from the preceding race in having the mantle washed with grey; below greyer, and the yellow streaks on the underside much less bright. Wing 78 to 100 mm.

**Range in Eastern Africa:** Namuli and Chiperoni Mts., Portuguese East Africa to Nyasaland.

**Habits:** Birds of mountain forest, but with the same characteristics as the other races, and notes and nesting habits identical.

**Recorded breeding:** Nyasaland, October to January.

*Phyllastrephus flavostriatus kungwensis* Mor.

*Phyllastrephus flavostriatus kungwensis* Moreau, Bull. B.O.C. 62, p. 29, 1941: Forest above Ujamba, Kungwe-Mahare Mts., western Tanganyika Territory.

**Distinguishing characters:** Differs from the preceding race in being whiter, less grey on chest; under wing-coverts and inner edging to flight feathers brighter and clearer lemon yellow without any buffish wash; bill at tip straighter, less curved downwards. Wing 88 to 105 mm.

**Range in Eastern Africa:** Kungwe-Mahare Mts. area, western Tanganyika Territory.

**Habits:** A mountain forest bird, locally common.

**Recorded breeding:** Kungwe-Mahare Mts., Tanganyika Territory, August.

**Distribution of other races of the species:** Belgian Congo and South Africa, the nominate race being described from the Transvaal.

### 756 SHARPE'S GREENBUL. *PHYLLASTREPHUS ALFREDI* (Shelley). Pl. 59.

*Bleda alfredi* Shelley, Bull. B.O.C. 13, p. 61, 1903: Mewenembe, Nyika Plateau, northern Nyasaland.

**Distinguishing characters:** Below, similar to the Yellow-streaked Greenbul, but above, including wings and tail, brown with an olive wash. The sexes are alike. Wing 76 to 100 mm. Juvenile plumage unrecorded.

**General distribution:** Tanganyika Territory to north-eastern Northern Rhodesia and western Nyasaland.

**Range in Eastern Africa:** Ufipa Plateau, south-western Tanganyika Territory.

**Habits:** A relatively common species of evergreen forest where the trees are tall, but does not seem to occur where the timber is less well grown.

**Nest and Eggs:** A slight cup nest of skeleton leaves, grass stems and moss, with a lining of fine grass placed in or suspended from the fork of a bough. Eggs two, grey mauve with Bunting-like scrawls of dark grey and blackish in a ring at the larger end; about 23 × 16 mm.

**Recorded breeding:** Nyasaland, September to December.

**Food:** No information.

**Call:** A single shrill rather bleating call 'fwa.'

## 757 SMALLER YELLOW-STREAKED GREENBUL. *PHYLLASTREPHUS DEBILIS* (Sclater).

*Phyllastrephus debilis debilis* (Scl.). Pl. 60.

*Xenocichla debilis* W. L. Sclater, Ibis, p. 284, 1899: Near Inhambane, Portuguese East Africa.

**Distinguishing characters:** Top of head grey with a slight olive wash; sides of face pale grey; rest of upperparts bright olive green; wing feathers edged yellowish green; tail brownish olive edged yellowish green; below, chin and throat white; rest of underparts very pale greyish white, broadly streaked with bright lemon yellow; under wing-coverts and edges of inner webs of secondaries bright lemon yellow. The sexes are alike. Wing 62 to 70 mm. The young bird has the top of the head greener; sides of face washed with green.

**General distribution:** Tanganyika Territory to southern Portuguese East Africa as far south as Inhambane.

**Range in Eastern Africa:** Southern Tanganyika Territory at Liwale and Rondo Plateau to the Zambesi River.

**Habits, Nest, etc.:** As for other races.

**Recorded breeding:** No records.

*Phyllastrephus debilis albigula* (Grote).

*Macrosphenus albigula* Grote, O.M. p. 62, 1919: Mlalo, near Lushoto, north-eastern Tanganyika Territory.

**Distinguishing characters:** Differs from the nominate race in having the top of the head olive green and grey, and the underside much greyer with less yellow streaking. Wing 62 to 68 mm.

**Range in Eastern Africa:** Usambara Mts., north-eastern Tanganyika Territory, between Lushoto and Amani to upper Sigi River below Amani above 1,000 feet.

**Habits:** In appearance the smallest and slenderest of the local Greenbuls, and quite plentiful in the mountain forests of the Usambara Mts. Habits much the same as a *Phylloscopus*. Intermediates occur between this race and the next, this race inhabiting the higher ground.

**Nest and Eggs:** The nest is a little cup of fibres in the fork of a tree. Eggs undescribed.

**Recorded breeding:** No records.

**Food:** Insects mainly.

**Call:** A loud warbling song not often heard, also a throaty gurgling alarm note.

*Phyllastrephus debilis rabai* Hart. and V. Som.
*Phyllastrephus rabai* Hartert and Van Someren, Bull. B.O.C. 41, p. 64, 1921: Rabai Hills, near Mombasa, eastern Kenya Colony.

**Distinguishing characters:** Differs from the nominate race in being rather greyer on chest, breast and flanks and from the last race in being brighter green above, with grey forehead and crown; paler, less grey below. Wing 60 to 71 mm. The young bird has the grey head washed with green.

**Range in Eastern Africa:** Coastal areas of south-eastern Kenya Colony and north-eastern Tanganyika Territory from Rabai and Shimba Hills to area east of Amani as far west as middle Sigi River at 500 feet, and south to the Nguru and Uluguru Mts. and Pugu Hills.

**Habits:** Those of the last race except that it inhabits low coastal scrub and low-lying forest patches. Its call is said to be very ventriloquial and sibilant on a rising scale. Elliott says the song has an extraordinary explosive quality.

**Recorded breeding:** Pugu Hills (breeding condition), November.

## 758 FISCHER'S GREENBUL. *PHYLLASTREPHUS FISCHERI* (Reichenow).

*Phyllastrephus fischeri fischeri* (Reichw.). **Pl. 59.**
*Criniger fischeri* Reichenow, O.C. p. 139, 1879: Muniumi, near mouth of the Juba River, southern Italian Somaliland.

**Distinguishing characters:** Above, olive brown, with a slight greenish tinge; tail dull brown; below, creamy white with rather indistinct yellow streaks from chest to belly; throat practically white; under tail-coverts pale brown. The sexes are alike. Wing, male 78 to 98, female 75 to 86 mm. The young bird has the top of the head greener; tail more rusty in tone.

**Range in Eastern Africa:** Coastal areas of southern Italian Somaliland to northern Portuguese East Africa from the Juba River to Netia, Mozambique district and as far west as the eastern Usambara Mts., Morogoro, eastern Uluguru Mts. and Mahenge district.

**Habits, Nest and Eggs:** See under other races.

**Recorded breeding:** No records.

*Phyllastrephus fischeri placidus* (Shell.). **Ph. viii.**

*Xenocichla placida* Shelley, P.Z.S. p. 363, 1889: Kilimanjaro, north-eastern Tanganyika Territory.

**Distinguishing characters:** Differs from the nominate race in being darker above, more olive green, and below rather greyer. Wing, male 75 to 94, female 72 to 90 mm.

**General distribution:** Kenya Colony, Tanganyika Territory, Nyasaland and northern Portuguese East Africa.

**Range in Eastern Africa:** Kenya Colony east of the Rift Valley and Tanganyika Territory east of a line through Mbulu to Iringa and Njombe to southern Nyasaland and Portuguese East Africa, but not the coastal areas below about 1,200 feet.

**Habits:** A common species of highland forest and dense riparian scrub, keeping mostly to undergrowth and creepers, not ascending to the tree tops. It frequently flirts its tail and flicks its wings. Usually seen singly or in pairs hunting for insects in dense cover in evergreen forest. In Nyasaland there is some evidence of a migration to lower levels in the cold season.

**Nest and Eggs:** Nest a cup of stems and leaves lined with rootlets and a little moss, placed low in forest undergrowth. Eggs two, glossy pale greenish grey or pinkish white with bold brown and sepia blotches tending to form a zone or cap; about 23 × 15·5 mm.

**Recorded breeding:** Kenya Colony highlands, March to July, also November and December. Tanganyika Territory, December to February. Nyasaland, October to January.

**Food:** Mainly insects, some fruit.

**Call:** Sharp bleating alarm note 'prip-prip-pre-pre-pre-pre.' The song is a slight fluty whistle.

*Phyllastrephus fischeri cabanisi* (Sharpe).

*Criniger cabanisi* Sharpe, Cat. Bds. B.M. 6, p. 83, 1881: Eastern Angola.

**Distinguishing characters:** Similar to the preceding race, but below pale yellow including under wing-coverts. Wing, male 83 to 97, female 72 to 87 mm.

**General distribution:** The Sudan, Uganda and Kenya Colony to Tanganyika Territory, eastern Angola, southern Belgian Congo and Northern Rhodesia.

**Range in Eastern Africa:** Southern Sudan, Uganda and Kenya Colony west of the Rift Valley to western Tanganyika Territory as far south as the Ufipa Plateau.

**Habits:** Those of other races but seems a scarcer bird over most of its range. Apparently occurs rarely if at all under 4,000 feet.

**Recorded breeding:** Southern Sudan (breeding condition), July.

### 759 GREY-OLIVE GREENBUL. *PHYLLASTREPHUS CERVINIVENTRIS* Shelley. Pl. 60.

*Phyllostrophus cerviniventris* Shelley, Ibis, p. 10, pl. 2, 1894: Zomba, southern Nyasaland.

**Distinguishing characters:** Above, ashy grey green, head rather greyer; tail dull brown; below, pale greyish brown; centre of belly paler; under tail-coverts pale tawny. The sexes are alike. Wing 75 to 89 mm. The contrast of the olive back and brown tail is noticeable in the field, and at rest its whitish feet are conspicuous. The young bird is browner, less olive, above; and has deep tawny edges to the secondary wing-coverts.

**General distribution:** Central Kenya Colony to southern Belgian Congo, Northern Rhodesia, Nyasaland and Portuguese East Africa.

**Range in Eastern Africa:** Mt. Kenya to the Zambesi River.

**Habits:** In Eastern Africa a bird of hot semi-deciduous forest with thick undergrowth from which it spreads along gallery strips of forest up the slopes of hills to higher altitudes. In Nyasaland it also occurs in highland forest. It is usually seen singly or in small parties searching undergrowth, or occasionally fluttering out after an insect. It also has the wing-flicking, tail-flirting habit of so many of the Greenbuls.

**Nest and Eggs:** Nest a deep cup of leaves and moss in the fork of a shrub with a well-made inner lining of rootlets and fibres. Eggs

Plate 60

Grey-olive Greenbul (p. 130)
Leaf-Love (p. 120)
White-throated Greenbul (p. 131)
Xavier's Greenbul (p. 133)
Toro Olive-Greenbul (p. 131)
Smaller-Yellow-streaked Greenbul (p. 127)
Kretschmer's Greenbul (p. 134)
Mountain Greenbul (p. 136)
Olive-breasted Mountain Greenbul (p. 135)
Stripe-cheeked Greenbul (p. 139)
Shelley's Greenbul (p. 140)
Slender-billed Greenbul (p. 143)
Zanzibar Sombre Greenbul (p. 145)
Cameroon Sombre Greenbul (p. 147)
Little Grey Greenbul (p. 145)

Plate 61

| Little Yellow Flycatcher (p. 192) | Yellow-bellied Flycatcher (p. 198) | | Chestnut-cap Flycatcher (p. 193) |
| --- | --- | --- | --- |
| | Male | Female | |
| Ashy Flycatcher (p. 171) | Dusky Flycatcher (p. 165) | Swamp Flycatcher (p. 168) | Spotted Flycatcher (p. 162) |
| Grey Tit-Flycatcher (p. 173) | Red Sea Warbler (p. 352) | Abyssinian Catbird (p. 176) | Pale Flycatcher (p. 177) |
| White-eyed Slaty Flycatcher (p. 180) | Grey Flycatcher (p. 179) | Yellow Flycatcher (p. 191) | Silver Bird (p. 187) |

two, greyish cream well covered with speckles and blotches of brown and bluish grey; about 23·5 × 16 mm.

**Recorded breeding:** Nyasaland, May, October and November. Portuguese East Africa, May.

**Food:** Mainly insects.

**Call:** Distinctive, a somewhat rasping chatter, 'jhiddie-jweck-jweck' described by Vincent as the sort of sound a leather hinge makes under strain.

### 760 TORO OLIVE GREENBUL. *PHYLLASTREPHUS HYPOCHLORIS* (Jackson). Pl. 60.

*Stelgidillas hypochloris* Jackson, Bull. B.O.C. 19, p. 20, 1906: Kibirau, Toro, western Uganda.

**Distinguishing characters:** Above and below, dull olive green; but paler below and with some narrow paler yellowish streaks on chest and breast; tail more bronzy and green. Generally much duller, especially below, than the Angolan race of Fischer's Greenbul, and tail more bronzy brown than dull brown and with narrow olive green edges to the outer webs. The sexes are alike. Wing 74 to 83 mm. The young bird is duller than the adult.

**General distribution:** Eastern Belgian Congo to the Sudan and Uganda.

**Range in Eastern Africa:** Southern Sudan to western and southern Uganda.

**Habits:** A rare and little-known species of Equatorial forest, usually seen among tangled undergrowth along streams or at the forest edge in pairs or small parties.

**Nest and Eggs:** Undescribed.

**Recorded breeding:** Southern Sudan (near breeding), March.

**Food:** No information.

**Call:** Unrecorded.

### 761 WHITE-THROATED GREENBUL. *PHYLLASTREPHUS ALBIGULARIS* (Sharpe). Pl. 60.

*Xenocichla albigularis* Sharpe, Cat. Bds. Brit. Mus. 6, p. 103, pl. 7, 1881: Fantee, Gold Coast Colony.

**Distinguishing characters:** Upperside and wings olive green; forehead to crown darker with dark centres to feathers, giving a scaly appearance; tail russet with green edges to outer webs; below, throat white; chest greyish with pale yellow streaks; breast to belly whitish with pale yellow streaks; lower belly and under tail-coverts pale yellow. The sexes are alike. Wing 68 to 86 mm. The young bird has the top of the head much greener; less grey on side of head.

**General distribution:** Gold Coast, Cameroons and Upper Congo, the Sudan and Uganda.

**Range in Eastern Africa:** Southern Sudan and Uganda.

**Habits:** A forest species inhabiting most of the forests of Uganda, but little known or noticed. It is reputed to be very shy and difficult to see.

**Nest and Eggs:** Undescribed.

**Recorded breeding:** Uganda, April.

**Food:** Insects, largely ants.

**Call:** A soft 'trit-trit' but with many variations of emphasis.

### 762 DAPPLED MOUNTAIN-GREENBUL. *PHYLLASTRE-PHUS OROSTRUTHUS* Vincent.

*Phyllastrephus orostruthus orostruthus* Vinc.

*Phyllastrephus orostruthus* Vincent, Bull. B.O.C. 53, p. 133, 1933: Namuli Mts., Portuguese East Africa.

**Distinguishing characters:** Adult male, above, dark brownish olive green; flight feathers browner; tail very dark chestnut; upper tail-coverts dark cinnamon brown; a faint superciliary stripe; below, yellowish white; chest and upper breast mottled dark olive green; flanks dark olive green. Female probably similar to the male. Wing 84 mm. Juvenile plumage unrecorded.

**Range in Eastern Africa:** Namuli Mts., Portuguese East Africa.

**Habits:** A bird of dense high mountain forest, only known from Namuli Mt. at about 4,800 feet. The call is said to be a very distinctive three-syllabled whistling call with a high middle note like a *Cossypha*.

**Nest and Eggs:** Undescribed.

**Recorded breeding:** No records.
**Food:** No information.
**Call:** See above.

*Phyllastrephus orostruthus amani* Scl. and Mor.
*Phyllastrephus orostruthus amani* Sclater and Moreau, Bull. B.O.C. 56, p. 16, 1935: Amani Forest, north-eastern Tanganyika Territory.

**Distinguishing characters:** Similar to the nominate race, but generally paler; upperside greener, and the mottling on chest and upper breast sharper and clearer. Wing 88 mm.

**Range in Eastern Africa:** Amani Forest, Usambara Mts., north-eastern Tanganyika Territory.

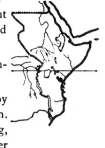

**Habits:** A very rare forest species, but recognizable at once by its song from the other Greenbuls of its area. Habits unknown. The male has a loud clear song of five notes, the first three ascending, then the fourth as low as the first, then a pause, and the fifth higher than any.

**Recorded breeding:** No records.

**763 XAVIER'S GREENBUL.** *PHYLLASTREPHUS XAVIERI* (Oustalet).

*Phyllastrephus xavieri sethsmithi* Hart. and Neum. **Pl. 60.**
*Phyllastrephus icterinus sethsmithi* Hartert and Neumann, O.M. p. 81, 1910: Budongo Forest, Unyoro, western Uganda.

**Distinguishing characters:** Above, bright olive green; wings and tail more olive brown; below, pale yellow; flanks olivaceous; a faint yellowish superciliary stripe over eye. The sexes are alike. Wing 75 to 90 mm. Can be distinguished from the Angolan race of Fischer's Greenbul by the olive green, not olive brown, upper tail-coverts and edging to the tail feathers. The young bird has the wing-coverts rather more brownish olive.

**General distribution:** Eastern Belgian Congo and Uganda.

**Range in Eastern Africa:** Western Uganda.

**Habits:** Common locally in forest both in undergrowth and high trees, often in company with other Bulbuls.

**Nest and Eggs:** Undescribed.

**Recorded breeding:** No records.

**Food:** Insects and small berries.

**Call:** Unrecorded.

**Distribution of other races of the species:** Cameroons to eastern Belgian Congo, the nominate race being described from the Ubangi River, Belgian Congo.

### 764 KRETSCHMER'S GREENBUL. *SUAHELIORNIS KRETSCHMERI* (Reichenow and Neumann).

*Suaheliornis kretschmeri kretschmeri* (Reichw. and Neum.). **Pl. 60.**

*Phyllastrephus kretschmeri* Reichenow and Neumann, O.M. p. 75, 1895: Kiboscho, Mt. Kilimanjaro, north-eastern Tanganyika Territory.

**Distinguishing characters:** Very similar to the Smaller Yellow-treaked Greenbul in size and general colour, but generally darker; tail more olive, less green; bill much longer, throat not so white; whiskers shorter and not so conspicuous. The sexes are alike. Wing 62 to 67 mm. The young bird has the top of the head greener.

**Range in Eastern Africa:** Eastern Tanganyika Territory from Mt. Kilimanjaro and the Usambara Mts. to the Uluguru Mts. and the Pugu Hills.

**Habits:** Common locally, and looks like a small slender *Phyllastrephus*. Persistent songster but usually from the depths of a bush and the song is unmusical and emphatic. Mostly found in fringing forest not in high trees.

**Nest and Eggs:** Undescribed.

**Recorded breeding:** Usambara Mts. Tanganyika Territory, at any time of year.

**Food:** Mainly insects.

**Call:** A monotonous song 'eet-i-riid,' or 'eet-itti-rid.' Two alarm notes, a low 'charr' and a clear call 'ker-ip.'

*Suaheliornis kretschmeri griseiceps* (Grote).

*Macrosphenus griseiceps* Grote, O.M. p. 162, 1911: Mikindani, south-eastern Tanganyika Territory.

**Distinguishing characters:** Differs from the nominate race in being paler and with a greyer top to head. Wing 63 to 67 mm. The young bird has some green in the grey head and some yellow on the throat.

**Range in Eastern Africa:** South-eastern Tanganyika Territory to Portuguese East Africa, from Mikindani to Netia, Mozambique district.

**Habits:** As for the nominate race, found in small parties, often scratching for insects among leaves under thick bushes, and commonly members of a mixed bird party. Quite common in coastal scrub and extending inland in suitable localities. The call is a cheery chirruping whistled song 'cheer-up-cheer-up,' but more usually a throaty gurgling warble.

**Recorded breeding:** No records.

## 765 OLIVE-BREASTED MOUNTAIN-GREENBUL. *ARIZELOCICHLA TEPHROLÆMA* (Gray).

*Arizelocichla tephrolæma kikuyuensis* (Sharpe). Pl. 60.

*Xenocichla kikuyuensis* Sharpe, Ibis, p. 118, 1891: Kikuyu, central Kenya Colony.

**Distinguishing characters:** Head and neck all round and upper chest grey; forehead to nape darker; some light streaks on ear-coverts; rest of plumage bright olive green, more yellowish in centre of belly. The sexes are alike. Wing 84 to 98 mm. The young bird is generally duller in colour, and has a green wash on the grey of the head and neck.

**Range in Eastern Africa:** Uganda, Ruanda, Belgian Congo and Kenya Colony as far east as the Aberdares and Mt. Kenya.

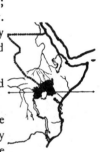

**Habits:** Common and often abundant species, certainly the commonest bird of the forest zone of Ruwenzori Mts. and probably also of the higher forests fringing the Rift Valley. Found in dense undergrowth, in bamboo, in tree tops, in fact everywhere, and ranges up to 10,000 feet or more, feeding mainly on fruit.

**Nest and Eggs:** Nest of grass, fine roots and moss, placed in a low shrub. Eggs one or two, pinkish-white, heavily blotched and clouded with dark brown and leaden grey; about 25 × 18 mm.

**Recorded breeding:** Ruwenzori Mts., March.

**Call:** Rather silent, but is said to have a varied and delightful Thrush-like song at times.

**Distribution of other races of the species:** West Africa, the nominate race being described from Cameroon Mt.

**766 MOUNTAIN-GREENBUL.** *ARIZELOCICHLA NIGRICEPS* (Shelley).

*Arizelocichla nigriceps nigriceps* (Shell.). **Pl. 60.**

*Xenocichla nigriceps* Shelley, P.Z.S. p. 362, 1889: Mt. Kilimanjaro, north-eastern Tanganyika Territory.

**Distinguishing characters:** Adult male, forehead to crown dull black; rest of upperparts bright olive green; sides of face, stripe over eye and chin to breast grey; centre of belly grey with bright olive green streaks; flanks and under tail-coverts bright olive green. The female has a duller black head than the male. Wing 82 to 95 mm. The young bird is similar to the adult female, but has the chin to breast more dusky grey.

**Range in Eastern Africa:** Hills south of Narossera Plateau, southern Kenya Colony to the Mt. Kilimanjaro area, Arusha district and south-west to Mt. Gerui, Tanganyika Territory.

**Habits:** A common mountain forest species, rather noisy and obtrusive for a Greenbul, otherwise it does not differ in habits from others of the group.

**Nest and Eggs:** Nest a cup of moss with a fibre lining built on twigs of forest undergrowth occasionally at some height from the ground. Eggs one or two, pinkish or purplish white, heavily streaked and blotched with dark purplish brown which has a tendency to form a cap at the larger end. Very handsome eggs; about 25 × 18 mm.

**Recorded breeding:** Mt. Kilimanjaro, January to March, also September. Oldeani, January.

**Food:** Seeds, berries and insects.

**Call:** Moreau gives a wide range of song for this bird—first a 'talking' song 'kwew-ki-kwew-ki-kwew,' secondly a deliberate song 'kwo-kwer-kwer-kwee-kwo' and thirdly a rich passionate warbling song. Alarm call a growling 'churr.'


*Arizelocichla nigriceps fusciceps* (Shell.).

*Xenocichla fusciceps* Shelley, Ibis, p. 13, 1893: Mlanji Plateau, southern Nyasaland.

**Distinguishing characters:** Very similar to the preceding race but has the top of the head grey. Wing 86 to 98 mm.

**General distribution:** Tanganyika Territory to north-eastern Northern Rhodesia, Nyasaland and Portuguese East Africa.

**Range in Eastern Africa:** South-western Tanganyika Territory to Namuli Mts., Portuguese East Africa and Nyasaland.

**Habits:** A local species among patches of mountain forest, keeping to the higher levels; noisy and inquisitive but occasionally flying round an intruder's head with fluttering flight and uttering loud harsh cries; the most noticeable call is a throaty babble.

**Recorded breeding:** Nyasaland, August to November.

*Arizelocichla nigriceps usambarœ* (Grote).
*Phyllastrephus tephrolœmus usambarœ* Grote, O.M. p. 62, 1919: Mlalo, near Lushoto, Usambara Mts., north-eastern Tanganyika Territory.

**Distinguishing characters:** Differs from the preceding race in having a black streak between the grey crown and the grey stripe over the eye. Wing 84 to 91 mm.

**Range in Eastern Africa:** South Paré Mts. to west Usambara Mts., north-eastern Tanganyika Territory.

**Habits:** As for other races, locally common. Calls given as 'hee-hee-hee' also a Thrush-like chatter and a nasal call.

**Recorded breeding:** North-eastern Tanganyika Territory, December.

*Arizelocichla nigriceps neumanni* Hart.
*Arizelocichla neumanni* Hartert, Bull. B.O.C. 42, p. 50, 1922: Uluguru Mts., Morogoro District, eastern Tanganyika Territory.

**Distinguishing characters:** Differs from the preceding race in having the top of the head dull black and no grey stripe over eye. Wing 91 to 100 mm.

**Range in Eastern Africa:** Uluguru Mts.

**Habits:** As for other races. The song according to Fuggles-Couchman is a 'chipchuck-chuck-choo-kweoo,' the third syllable being the lowest.

**Recorded breeding:** December.

*Arizelocichla nigriceps kungwensis* Mor.

*Arizelocichla tephrolæma kungwensis* Moreau, Bull. B.O.C. 61, p. 60, 1941: Kungwe-Mahare Mts., western Tanganyika Territory.

**Distinguishing characters:** Differs from the other races in having the sides of the face, lores and throat pale grey. Wing 90 to 94 mm.

**Range in Eastern Africa:** Kungwe-Mahare Mt. area, western Tanganyika Territory.

**Habits:** As for other races.

**Recorded breeding:** No records.

*Arizelocichla nigriceps chlorigula* (Reichenow).

*Xenocichla chlorigula* Reichenow, O.M. p. 8, 1899: Kalinga, Iringa District, south-central Tanganyika Territory.

**Distinguishing characters:** Differs from other races in having a greyer head, white eyelids both above and below the eye and a bright olive green throat joining up with the green on the mantle. Wing 85 to 97 mm. The young bird is duller and has much less bright olive green on the throat.

**Range in Eastern Africa:** Eastern and southern Tanganyika Territory from Nguru Mts., to eastern Dodoma, Kilosa, Iringa, Njombe and Songea districts.

**Habits:** Very common species of highland forest undergrowth. The call is a curious distinctive song of two deep notes followed by two falsetto ones described as 'tit-tirio' constantly repeated. Alarm call a chatter.

**Recorded breeding:** Tanganyika Territory, September to December.

**767 STRIPE-CHEEKED GREENBUL.** *ARIZELOCICHLA MILANJENSIS* (Shelley).

*Arizelocichla milanjensis milanjensis* (Shell.).

*Xenocichla milanjensis* Shelley, Ibis, p. 9, pl. I, fig. I, 1894: Mlanji, southern Nyasaland.

**Distinguishing characters:** Similar to some races of the Mountain-Greenbul, but is slightly longer billed and rather darker in colour, with the chin green and grey and neck and upper chest olive green. The sexes are alike. Wing 86 to 98 mm. The young bird has a greenish wash on the grey head.

**General distribution:** Portuguese East Africa, Nyasaland and eastern Southern Rhodesia.

**Range in Eastern Africa:** Namuli and Chiperone Mts., Portuguese East Africa to Mt. Mlanje, Nyasaland.

**Habits:** Mostly found in intermediate and lower highland forest or even in dense scrub forest. Rather silent birds, but the call is unmistakable and while calling the birds sidle along a bough with little hops. Fairly common and widely distributed but apparently rare on Mt. Mlanje itself.

**Nest and Eggs:** Nest a neat rounded cup of twigs, grass stems and roots, occasionally at some little height. Eggs two, whitish densely freckled and marbled with streaks and spots of brown, chocolate and grey; about 24·5 × 17 mm.

**Recorded breeding:** Nyasaland, September and October. Southern Rhodesia, November and December.

**Food:** Seeds, fruit, berries and worms.

**Call:** A throaty 'ukkeri-ukkeri-ukkeri,' occasionally followed by a low trill, also some attempt at a scrambling song.

*Arizelocichla milanjensis striifacies* (Reichw. and Neum.). **Pl. 60.**
*Xenocichla striifacies* Reichenow and Neumann, O.M. p. 74, 1895: Marangu, Kilimanjaro, north-eastern Tanganyika Territory.

**Distinguishing characters:** Differs from the nominate race in having the forehead to crown olive green, uniform with the mantle. Wing 89 to 106 mm. The young bird is duller and has only faint streaks on the ear-coverts.

**General distribution:** Kenya Colony, Tanganyika Territory, Portuguese East Africa and Nyasaland as far south and east as Mt. Chiradzulu.

**Range in Eastern Africa:** South-eastern Kenya Colony at Taveta and Chyulu Hills to Tanganyika Territory, Portuguese East Africa at Unangu and Nyasaland north and west of Mts. Chiradzulu and Cholo.

**Habits:** As for the other races, common in canopy and mid-strata of forest and in places there is a marked seasonal migration to lower levels.

**Recorded breeding:** Kenya Colony, Taveta, May; Chyulu Hills, December to February. Tanganyika Territory, Mt. Kilimanjaro, January and February. Usambara Mts. probably October and November. Nyasaland, October.

### 768 SHELLEY'S GREENBUL. *ARIZELOCICHLA MASUKUENSIS* (Shelley).

*Arizelocichla masukuensis masukuensis* (Shell.).
*Andropadus masukuensis* Shelley, Ibis, p. 534, 1897: Masuku Hills, just south of Songwe River, North Nyasa District, Nyasaland.

**Distinguishing characters:** Above and below, olive green, somewhat darker green above. The sexes are alike. Wing 82 to 85 mm. The young bird is like the adult but duller in colour.

**General distribution:** Tanganyika Territory to northern Nyasaland.

**Range in Eastern Africa:** Rungwe Mt., south-western Tanganyika Territory.

**Habits:** As for other races.

**Nest and Eggs:** As for other races.

**Recorded breeding:** Matipa, Northern Nyasaland, May and November.

*Arizelocichla masukuensis roehli* (Reichw.). **Pl. 60.**
*Andropadus roehli* Reichenow, O.M. p. 181, 1905: Mlalo, near Lushoto, Usambara Mts., north-eastern Tanganyika Territory.

**Distinguishing characters:** Differs from the nominate race in being paler below and in having the throat greyer. Wing 77 to 88 mm.

**Range in Eastern Africa:** Eastern to southern Tanganyika Territory from the Paré and Usambara Mts., to the Uluguru Mts., Mahenge, Njombe and Songea areas.

**Habits:** Not uncommon in forest and riverside timber. It is generally seen creeping about in dense vegetation but will also cling to tree trunks while searching for food like a Woodpecker. Probably not as skulking as most Greenbuls and occurs in more open forest. Rather shy silent birds.

**Nest and Eggs:** Nest a cup of fibre and rootlets lined with fine black horsehair-like material with a little moss round edge of nest. A light strong nest usually in the fork of a sapling. Eggs two, pinkish buff or brownish cream, densely scrawled and blotched with sepia, chocolate and grey; about 22·5 × 16 mm.

**Recorded breeding:** Tanganyika Territory, September to December.

**Food:** Insects and fruit.

**Call:** A soft nasal 'kwew-kwa-kwew.'

*Arizelocichla masukuensis kakamegæ* (Sharpe).
*Xenocichla kakamegæ* Sharpe, Bull. B.O.C. 11, p. 29, 1900: Kakamega Forest, Nandi, western Kenya Colony.

**Distinguishing characters:** Differs from the preceding races in having a grey wash on the top of the head; and the sides of the face and throat grey. Wing 82 to 84 mm. The greyish head and grey throat closely resembles the local form of the Mountain Greenbul, but that bird is larger, brighter olive green, and has a heavier bill.

**General distribution:** Kenya Colony to Tanganyika Territory and south-eastern Belgian Congo.

**Range in Eastern Africa:** Western Kenya Colony to western Tanganyika Territory.

**Habits:** As for other races.

**Recorded breeding:** No records.

**769 YELLOW-BELLIED GREENBUL.** *CHLOROCICHLA FLAVIVENTRIS* (Smith).
*Chlorocichla flaviventris occidentalis* Sharpe. **Pl. 59.**
*Chlorocichla occidentalis* Sharpe, Cat. Bds. B.M. 6, p. 113, pl. **7,** 1881: Ovaquenyama, Damaraland.

**Distinguishing characters:** Above, pale olive green; head slightly darker; below, including under wing-coverts, sulphur yellow; olivaceous wash on chest. The sexes are alike. Wing 95 to 108 mm. The young bird has the head uniform with the mantle.

**General distribution:** Tanganyika Territory to southern Angola, Damaraland, the Rhodesias, Nyasaland and Portuguese East Africa south of the Zambesi River and west of the mouth of the Shiré River.

**Range in Eastern Africa:** Tanganyika Territory west of Arusha and Mpapwa to Nyasaland.

**Habits:** As for other races and mostly found at low levels.

**Nest and Eggs:** The nest is a flimsy but neat cup of tendrils, grass and a few fine twigs and stems, lined with grass blades, tendrils and rootlets, often in dense cover. Eggs two, cream or pale olive heavily marbled and smudged with dark olive brown and with a few darker spots and streaks; about 25 × 17 mm.

**Recorded breeding:** South-eastern Belgian Congo, November. Nyasaland, November to February.

**Food:** Seeds, berries and some insects.

**Call:** A persistent 'pao-pao-pao' or 'barac-barac-barac' nasal, tinny and querulous, also a little 'twe-twe-twe.' The directional call is a mournful double note 'kerr-quar' answered and repeated by the other bird of the pair (Van Someren).

*Chlorocichla flaviventris centralis* Reichw.
*Chlorocichla centralis* Reichenow, J.f.O. p. 74, 1887: Loeru, Pangani District, Tanganyika Territory.

**Distinguishing characters:** Generally much darker than the preceding race, more olive brown above, head much darker; under wing-coverts usually chrome yellow. Wing 97 to 111 mm.

**Range in Eastern Africa:** Kenya Colony east of Mt. Kenya and south of Lamu to Tanganyika Territory east of Arusha and Mpapwa and northern Portuguese East Africa.

**Habits:** Bird of dense bush always difficult to see or move, on the sea coast it inhabits dense windswept scrub. Shy retiring species with particularly skulking habits generally in pairs or small parties.

**Recorded breeding:** Kenya Colony highlands, April to June, also December. Mombasa, July and August. Portuguese East Africa, November and December.

**Distribution of other races of the species:** Natal and Zululand, the nominate race being described from Durban.

**770 JOYFUL GREENBUL.** *CHLOROCICHLA LÆTISSIMA* (Sharpe). Pl. 59.
*Andropadus lætissimus* Sharpe, Bull. B.O.C. 10, p. 27, 1899: Nandi, western Kenya Colony.

**Distinguishing characters:** Above, golden green; feathers of head with darker centres, giving a scaly appearance; below, golden yellow. The sexes are alike. Wing 101 to 107 mm. The young bird has the mantle washed with brownish and a greenish wash on the underside.

**General distribution:** North-eastern Belgian Congo to the Sudan and Kenya Colony.

**Range in Eastern Africa:** South-eastern Sudan, Uganda and western Kenya Colony.

**Habits:** Very local but common where they occur, often in small flocks with other species. They have a beautiful clear song which R. B. Woosnam considered was the finest bird song heard in Africa.

**Nest and Eggs:** Undescribed.

**Recorded breeding:** No records.

**Food:** Berries.

**Call:** Song as above, also loud chattering calls.

### 771 SLENDER-BILLED GREENBUL. *STELGIDILLAS GRACILIROSTRIS* (Strickland).

*Stelgidillas gracilirostris percivali* (Neum.). **Pl. 60.**

*Criniger gracilirostris percivali* Neumann, O.M. p. 185, 1903: Fort Smith, south-central Kenya Colony.

**Distinguishing characters:** Similar to the Toro Olive Greenbul but bill shorter and smaller than that species; above, dull olive green; below, grey with a paler throat; under wing-coverts and edges to inner webs of flight feathers chrome yellow. The sexes are alike. Wing 82 to 90 mm. The young bird is more olive brown above.

**Range in Eastern Africa:** The southern Sudan at Talanga Forest, Lotti Forest, Bendere 190 miles south of Wau, Aloma Plateau and Imatong Mts. to central Kenya Colony east of the Rift Valley.

**Habits:** Of the Uganda race, locally plentiful in thick forest feeding on fruit.

**Nest and Eggs, Call, etc.:** See under other races.

**Recorded breeding:** No records.

*Stelgidillas gracilirostris chagwensis* (Van Som.).

*Chlorocichla gracilirostris chagwensis* Van Someren, Bull. B.O.C. 35, p. 127, 1915: Nazigo Hill, Chagwe, southern Uganda.

**Distinguishing characters:** Differs from the preceding race in being darker below, more olivaceous grey with a few pale yellow streaks in centre of breast to belly. Wing 81 to 89 mm.

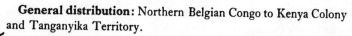

**General distribution:** Northern Belgian Congo to Kenya Colony and Tanganyika Territory.

**Range in Eastern Africa:** Southern Uganda and Kenya Colony west of the Rift Valley to the Biharamulo district, north-western Tanganyika Territory.

**Habits:** Common in dense forest in either undergrowth or tree tops according to season. They frequent the highest tree tops when the figs are ripe and are found in most of the forests of Uganda. They appear in any given locality when certain fruit is ready for them and then vanish again. Usually in small parties of three or four and are excellent songsters.

**Nest and Eggs:** Undescribed.

**Recorded breeding:** No records.

**Food:** Fruit and berries also small bean-like white seeds and grasshoppers.

**Call:** An excitable chattering alarm note, and a warbling song often heard late at night. Van Someren describes the call as a plaintive 'pleet-till-uu.'

*Stelgidillas gracilirostris congensis* (Reichw.).
*Andropadus gracilirostris congensis* Reichenow, O.M. p. 181, 1916: Leopoldville, south-western Belgian Congo.

**Distinguishing characters:** Differs from the preceding race in having the under wing-coverts, axillaries and inner edges of flight feathers darker; in having the under tail-coverts grey, not buff; and the inner edges of the tail feathers more olivaceous, less buff. Wing 90 to 91 mm.

**General distribution:** Northern Angola to southern Belgian Congo and Tanganyika Territory.

**Range in Eastern Africa:** The Kungwe-Mahare area of western Tanganyika Territory.

**Habits:** No information.

**Recorded breeding:** No records, probably in many months.

**Distribution of other races of the species:** West Africa, the nominate race being described from Fernando Po.

# BULBULS

## 772 LITTLE GREY GREENBUL. *ANDROPADUS GRACILIS* Cabanis.

*Andropadus gracilis gracilis* Cab. **Pl. 60.**

*Andropadus gracilis* Cabanis, O.C. p. 147, 1880: Angola.

**Distinguishing characters:** Above, dark olive-green; head darker and duller; lower rump washed with golden; tail brown washed with olive, narrow light tips to outer feathers; below, throat grey; upper chest olivaceous grey; rest of underparts bright olive or olive brown, with some yellow in centre of belly. The sexes are alike. Wing, 67 to 76 mm. The young bird is more olive brown than olive green above.

**General distribution:** Cameroons to Uganda and north Angola.

**Range in Eastern Africa:** Western and southern Uganda.

**Habits:** A bird of wide range, very common in many parts of the Equatorial forests, but of whose habits little has been recorded, though it probably differs in no respect from its congeners.

**Nest and Eggs:** Undescribed.

**Recorded breeding:** No records, might breed in any month.

**Food:** Said to be largely caterpillars, with some fruit.

**Call:** A chirping note of 'twit' and a low sweet song.

*Andropadus gracilis kavirondensis* (Van Som.).

*Charitillas gracilis kavirondensis* Van Someren, Bull. B.O.C. 40, p. 95, 1920: Kakamega Forest, Kavirondo, south-western Kenya Colony.

**Distinguishing characters:** Differs from the nominate race in being darker above and paler below. Wing 70 to 80 mm.

**Range in Eastern Africa:** Western Kenya Colony.

**Habits:** As for the nominate race.

**Recorded breeding:** No records.

**Distribution of other races of the species:** Sierra Leone to Nigeria.

## 773 ZANZIBAR SOMBRE GREENBUL. *ANDROPADUS IMPORTUNUS* (Vieillot).

*Andropadus importunus insularis* Hartl. **Pl. 60.**

*Andropadus insularis* Hartlaub, Orn. Beitr. Faun. Mad. p. 44, 1861: Zanzibar Island.

**Distinguishing characters:** Very similar, especially above, to the Yellow-bellied Greenbul but smaller and less yellow below.

The sexes are alike. Wing 77 to 92 mm. The pale yellow eye is conspicuous in the adult. The young bird is more olivaceous brown on the upperparts; eye brown.

**Range in Eastern Africa:** Coastal areas of southern Italian Somaliland to Kenya Colony and Tanganyika Territory as far south as Dar-es-Salaam and as far west as Mt. Kilimanjaro and Kilosa, also Manda and Zanzibar Islands.

**Habits:** Locally common in any thick vegetation, feeding in scrub or often among dead leaves on the ground. Sings freely even at midday from a prominent perch, but when not singing is a solitary skulking bird which is generally difficult to observe. It may be found either in bush or among cultivation, but only at lower levels.

**Nest and Eggs:** Nest of twigs and creeper stems lined with hairlike fibre and fine grass seedheads, a rather flimsy shallow cup often semi-transparent in the fork of a branch. Eggs two, dull white or cream spotted and streaked with reddish brown or purple and with ashy undermarkings often in the form of a zone; about $20 \cdot 5 \times 15 \cdot 5$ mm.

**Recorded breeding:** Kenya Colony, Malindi, December and January; Mombasa, April to July, also October. Zanzibar Island, October to January.

**Food:** Berries and insects.

**Call:** A persistent cheerful songster with a warbling twitter ending in a metallic 'chink.'

*Note:* Some specimens from the Dar-es-Salaam areas and Mafia Island are intermediate between this and the following race.

*Andropadus importunus hypoxanthus* Sharpe.
*Andropadus hypoxanthus* Sharpe, Bds. S. Afr. p. 205, 1876: Tete, Rivi River, southern Nyasaland.

**Distinguishing characters:** Differs from the preceding race in being rather paler above and yellower below, and in having the under wing-coverts lemon yellow instead of ochre yellow. It is very similar to the Yellow-bellied Greenbul, but is smaller and has an olivaceous wash across the chest and breast. Wing 82 to 95 mm.

**General distribution:** Tanganyika Territory to Nyasaland and northern Portuguese East Africa.

**Range in Eastern Africa:** South-eastern Tanganyika Territory to northern Portuguese East Africa and Nyasaland, also Mafia Island.

# BULBULS

**Habits:** As for the nominate race, the song ends with a sound almost like the intake of a deep breath and starts with a shrill double whistle. The alarm call is a double whistle followed by three quicker ones 'tee-wheet-triri-ri-rit' (Vincent).

**Recorded breeding:** Nyasaland, November to January. Portuguese East Africa, November to February.

*Andropadus importunus fricki* Mearns. **Ph. viii.**

*Andropadus fricki* Mearns, Smiths, Misc. Coll. 61, No. 25, p. 4, 1914: Ndoto Mt., central Kenya Colony.

**Distinguishing characters:** Differs from the Zanzibar race in having a yellow ring round the eye. Wing 87 to 91 mm.

**Range in Eastern Africa:** Central Kenya Colony from the Ndoto Mts. to about Nairobi.

**Habits:** Those of the preceding races but found at somewhat higher elevations. Van Someren describes the call as 'pii-a-roo' as the birds move about, with a longer call from the top of a bush of 'cheerit-cho-chiro-cho-chiro.'

**Recorded breeding:** Southern Kenya Colony, March and April.

**Distribution of other races of the species:** South Africa, the nominate race being described from Knysna.

### 774 CAMEROON SOMBRE GREENBUL. *ANDROPADUS CURVIROSTRIS* Cassin.

*Andropadus curvirostris curvirostris* Cass. **Pl. 60.**

*Andropadus curvirostris* Cassin, Proc. Philad, Acad. p. 46, 1859: Camma River, near Gabon, West Africa.

**Distinguishing characters:** General colour similar to the Little Grey Greenbul, but usually larger and bill longer and heavier. The sexes are alike. Wing 73 to 87 mm. The young bird is similar to the adult, but the wings are slightly browner.

**General distribution:** Fernando Po to Liberia, north Angola, the Sudan and Kenya Colony.

**Range in Eastern Africa:** Southern Sudan, Uganda and western Kenya Colony.

L

**Habits:** A bird of the tops of medium-sized trees and of secondary growth in true forest. A shy silent species of which little is recorded.

**Nest and Eggs:** Unrecorded, but believed to nest in the lower strata of forest.

**Recorded breeding:** Southern Sudan (breeding condition), October. Western Uganda (breeding condition), July.

**Food:** Berries.

**Call:** Said to have a sweet song at times, and a call of four or five syllables, with the last note trilled.

**Distribution of other races of the species:** Sierra Leone.

**775 LITTLE GREENBUL.** *EURILLAS VIRENS* (Cassin).
*Eurillas virens virens* (Cass.).
*Andropadus virens* Cassin, Proc. Ac. Philad. p. 3, 1857: Cape Lopez, Gabon.

**Distinguishing characters:** Size and general colour very similar to the Cameroon Sombre Greenbul, but bill shorter and broader at base; chin and throat olive green; rarely the mantle and rump are barred. The sexes are alike. Wing 67 to 84 mm. The young bird has the wing-coverts and lower back brown, and the chest and breast dark olive brown contrasting with the belly. In this latter respect it is easily distinguished from the young of the Cameroon Sombre Greenbul.

**General distribution:** Gabon to north Angola, the Sudan and Uganda.

**Range in Eastern Africa:** South-western Sudan to western and southern Uganda.

**Habits:** A common and often abundant species among dense bushes and heavy undergrowth. In forest it ascends to mid-strata but not often to high tree-tops. A bird of wide ecological range, from mountain forest to thick low-lying scrub. It may be found in any little isolated patch of thick cover. A fine songster and sings all the year round. Probably the tamest of the Greenbuls.

**Nest and Eggs:** Nest a cup of twigs, leaf shreds, pieces of bark and grass lined with fine grass or fibre in a bush or tree a few feet from the ground. Eggs two, pinkish or greyish white and rather glossy, spotted or speckled all over with brownish red and with slate grey undermarkings; about 22 × 15·5 mm.

**Recorded breeding:** Uganda, February to May, also October.

**Food:** Largely tree seeds and berries.

**Call:** A series of guttural rather explosive notes followed by a warble of clear notes ending with a whistle, also a rattling trill.

*Eurillas virens zombensis* (Shell.).

*Andropadus zombensis* Shelley, Ibis, p. 10, 1894: Zomba, southern Nyasaland.

**Distinguishing characters:** Differs from the nominate race in being generally rather paler and brighter olive green above. Wing 77 to 89 mm.

**General distribution:** Kenya Colony and Tanganyika Territory, south-eastern Belgian Congo, eastern Angola, Northern Rhodesia, Portuguese East Africa north of the Zambesi River and Nyasaland.

**Range in Eastern Africa:** South-eastern Kenya Colony to Tanganyika Territory, Nyasaland and Portuguese East Africa, also Mafia Island.

**Habits:** As for the nominate race and very widely distributed. The nests are described as very rough and shallow in some instances.

**Recorded breeding:** Tanganyika Territory, probably September to February. Nyasaland, October and November. Portuguese East Africa, November.

*Eurillas virens zanzibaricus* Pak.

*Eurillas virens zanzibaricus* Pakenham, Bull. B.O.C. 55, p. 111, 1935: Jozani Forest, Zanzibar Island.

**Distinguishing characters:** Differs from the nominate race in being purer olive green above, but not so bright as the preceding race. Wing 78 to 85 mm.

**Range in Eastern Africa:** Zanzibar Island.

**Habits:** As for the other races.

**Recorded breeding:** In breeding condition from November to February.

**Distribution of other races of the species:** Gambia to Nigeria.

**776 YELLOW-WHISKERED GREENBUL.** *STELGIDO-CICHLA LATIROSTRIS* (Strickland).

*Stelgidocichla latirostris eugenia* (Reichw.). **Ph. viii.**

*Andropadus eugenius* Reichenow, J.f.O. p. 53, 1892: Bukoba, north-western Tanganyika Territory.

**Distinguishing characters:** Above, dark olive green; tail dull brown washed with olivaceous; below, pale olive green; bright yellow stripe down each side of throat. The sexes are alike. Wing 76 to 95 mm. The young bird lacks the yellow stripe down each side of throat, but is otherwise similar to the adult.

**General distribution:** Eastern Belgian Congo to the Sudan, Kenya Colony and Tanganyika Territory.

**Range in Eastern Africa:** Southern Sudan to Uganda, Kenya Colony as far east as Matthew's Range and Mt. Kenya, and northern and western Tanganyika Territory as far east as Loliondo and as far south as Kungwe-Mahare Mts. and the Ufipa Plateau.

**Habits:** A common wide-ranging species found in both lowland and mountain forest or in any dense patch of vegetation clad with vines and creepers. Rather noisy birds with a loud monotonous call.

**Nest and Eggs:** Nest a cup of twigs, leaves and rootlets lined with fine fibre or vegetable horsehair, placed in a low shrub or tangled vegetation. Eggs, two, smooth and glossy dull white or dirty pink with purplish brown or liver coloured spots, and with greyish markings. Size very variable, about 24 × 16 mm.

**Recorded breeding:** South-eastern Sudan (near breeding), May. Uganda, April to August. Kenya Colony, Mt. Elgon (breeding condition), June; Highlands, March to June, also December.

**Food:** Fruit, especially small figs, and insects.

**Call:** A rather monotonous chuckling twittering song-call of about eight notes uttered all day long, usually from the thick top of a tree.

**Distribution of other races of the species:** West Africa, the nominate race being described from Fernando Po.

**777 BLACK-COLLARED BULBUL.** *NEOLESTES TORQUATUS* Cabanis.

*Neolestes torquatus* Cabanis, J.f.O. p. 237, pl. 1, fig. 1, 1875: Chinchoxo, Portuguese Congo.

**Distinguishing characters:** Above, forehead to hind neck grey; a glossy black streak from lores through eye to side of crown and neck; rest of upperparts green; below, chin to throat and ear-coverts buffish white; a broad glossy blue-black band across chest joining up to streak from lores to neck; breast to belly whitish; flanks greenish, under wing-coverts and axillaries golden yellow; bill and feet black. The sexes are alike. Wing 69 to 76 mm. The young bird has the forehead to hind neck dull olive green, uniform with rest of upperparts; mantle and wing-coverts tipped with buff; below, chest band dull black; under wing-coverts and axillaries dingy white washed with yellow.

**General distribution:** Lower Congo and Angola to the eastern Belgian Congo.

**Range in Eastern Africa:** Not yet recorded, but should occur in the Kivu area of Ruanda.

**Habits:** Usually seen in small parties feeding on fruits in grassy woodland country. This bird was originally believed to be a Shrike, but its habits and affinities are with the Bulbuls.

**Nest and Eggs:** The nest is a frail cup of grass and plant stems in a bush. Eggs two, pinkish-white with indistinct spots of darker pink and with a faint rufous zone at the larger end; about 20·5 × 14 mm.

**Recorded breeding:** Belgian Congo, December, also a nestling in September.

**Food:** As above.

**Call:** A twittering cry, not unlike that of a *Pycnonotus*.

Names in Sclater's *Syst. Av. Æthiop.* 2, 1930, which have been changed or have become synonyms in this work:

*Pycnonotus tricolor minor* Heuglin, treated as synonymous with *Pycnonotus tricolor tricolor* (Hartlaub).
*Atimastillas flavicollis shelleyi* (Neumann), treated as synonymous with *Pyrrhurus flavicollis flavigula* (Cabanis).
*Atimastillas flavicollis pallidigula* (Sharpe), treated as synonymous with *Pyrrhurus flavicollis flavigula* (Cabanis).
*Phyllastrephus cabanisi sucosus* Reichenow, treated as synonymous with *Phyllastrephus fischeri cabanisi* (Sharpe).
*Phyllastrephus sucosus sylvicultor* (Neave), treated as synonymous with *Phyllastrephus fischeri cabanisi* (Sharpe).
*Arizelocichla nigriceps percivali* Hartert, treated as synonymous with *Arizelocichla nigriceps usambaræ* Grote.
*Andropadus insularis kilimandjaricus* (Sjostedt), treated as synonymous with *Andropadus importunus insularis* Hartlaub.
*Andropadus insularis somaliensis* Reichenow, treated as synonymous with *Andropadus importunus insularis* Hartlaub.

# BULBULS

Names introduced since 1930, and which have become synonyms in this work:

*Pycnonotus tricolor naumanni* Meise, 1934, treated as synonymous with *Pycnonotus xanthopygos micrus* Oberholser.

*Pycnonotus tricolor chyulu* Van Someren, 1939, treated as synonymous with *Pycnonotus dodsoni* Sharpe.

*Phyllastrephus fischeri marsabit* Van Someren, 1930, treated as synonymous with *Phyllastrephus fischeri fischeri* (Reichenow).

*Phyllastrephus alfredi itoculo* Vincent, 1933, treated as synonymous with *Phyllastrephus fischeri fischeri* (Reichenow).

*Phyllastrephus flavostriatus litoralis* Vincent, 1933, treated as synonymous with *Phyllastrephus flavostriatus tenuirostris* Fischer and Reichenow.

*Phyllastrephus albigula shimbanus* Van Someren, 1943, treated as synonymous with *Phyllastrephus debilis rabai* Hartert and Van Someren.

*Phyllastrephus fischeri chyuluensis* Van Someren, 1939, treated as synonymous with *Phyllastrephus fischeri placidus* (Shelley).

*Arizelocichla milanjensis chyulu* Van Someren, 1939, treated as synonymous with *Arizelocichla milanjensis striifacies* (Reichenow & Neumann).

*Chlorocichla flaviventris chyuluensis* Van Someren, 1939, treated as synonymous with *Chlorocichla flaviventris centralis* Reichenow.

*Charitillas minor* Van Someren, 1922, treated as synonymous with *Andropadus gracilis gracilis* Cabanis.

*Eurillas virens shimba* Van Someren, 1931, treated as synonymous with *Eurillas virens zombensis* Shelley.

*Stelgidocichla latirostris australis* Moreau, 1941, treated as synonymous with *Stelgidocichla latirostris eugenia* Reichenow.

## Addenda and Corrigenda

**735** etc., pp. 98 to 103 and 107, for *Malacocincla* read *Trichastoma*.

**736** *Trichostoma* should be *Trichastoma*.

**743** *Pycnonotus dodsoni*. *Recorded breeding:* add British Somaliland, May.

**744** *Pycnonotus b. schoanus*. *Recorded breeding:* add Eritrea, February and March.

*Pycnonotus b. somaliensis*. *Recorded breeding:* add British Somaliland, May to August.

**745** For *Tricophorus* read *Criniger*. And No. 745 should be *Criniger calurus emini* Chapin, Auk, p. 444, 1948: Lukolela, Congo River, Belgian Congo. Wing 81 to 95 mm. *General distribution:* delete North-eastern.

**753** (p. 123). *Phyllastrephus t. bensoni* Bull. B.O.C. 46 should be *16*.

**758** *Phyllastrephus fischeri*. Size of wing should read Wing male 78 to 89: female 75 to 80 mm. *Range in Eastern Africa:* delete all localities south of the Pangani River.

**761** Should be *Phyllastrephus albigularis albigularis*. Add: *Distribution of other races of the species:* Angola.

**770** Should be *Chlorocichla lætissima lætissima*.

**771** *Charitillas g. kavirondensis* Van Som. is a race of *Andropadus ansorgei*, not of *A. gracilis*.

**773** *Andropadus importunus*. Add to young birds: A pale yellow ring round eye. *Recorded breeding:* add Dar-es-Salaam, January.

*A.i. fricki*. *Distinguishing characters:* delete, and substitute: Darker above than the Zanzibar race and yellower below.

*Continued on p.* 1101

## FLYCATCHERS

FAMILY—**MUSCICAPIDÆ. FLYCATCHERS.** Genera: *Muscicapa, Alseonax, Parisoma, Lioptilornis, Parophasma, Bradornis, Dioptrornis, Melænornis, Empidornis, Myopornis, Pedilorhynchus, Artomyias, Chloropeta, Chloropetella, Stizorhina, Megabyas, Bias, Hyliota, Batis, Platysteira, Dyaphorophyia, Erythrocercus, Erannornis, Trochocercus* and *Tchitrea*.

Sixty-two species have so far been found in Eastern Africa, of which three are migrants from Europe and Asia. They are found in all types of country wherever there are trees or scrub and at all altitudes. Their food consists mainly of insects of all kinds, and the majority of species are expert at catching them on the wing. Most of the larger species sit on some low bough or other vantage point from which they can see and chase passing insects, but many of the smaller species have also the leaf-hunting habits of Warblers.

### KEY TO THE ADULT FLYCATCHERS OF EASTERN AFRICA

| | | |
|---|---|---|
| 1 Chest with arrow-shaped black markings: | BOEHM'S FLYCATCHER *Myopornis böhmi* | 802 |
| 2 No arrow-shaped markings on chest: | | 3–128 |
| 3 Above, bright cærulean blue: | BLUE FLYCATCHER *Erannornis longicauda* | 827 |
| 4 Above, not bright blue: | | 5–128 |
| 5 General colour sooty grey: | ABYSSINIAN CATBIRD *Parophasma galinieri* | 791 |
| 6 General colour blackish brown: | SOOTY FLYCATCHER *Artomyias fuliginosa* | 804 |
| 7 General colour above, white: | PARADISE FLYCATCHER (white-backed phase of males) *Tchitrea viridis* | 832 |
| 8 General colour wholly black: | | 17–18 |
| 9 General colour above, dull black: | | 19–20 |

10 General colour above, warm brown: 21–23
11 Above, blue-black and grey: 24–27
12 General colour above, except head, bright chestnut: 28–35
13 General colour above, green, yellow-green, or olivaceous brown: 36–49
14 General colour above, glossy black, violet, or green: 50–63
15 General colour above, ashy brown, or mouse-brown: 64–84
16 General colour of mantle grey or blue-grey: 85–128

| | | |
|---|---|---|
| 17 Glossy blue or violet black: | SOUTH AFRICAN BLACK FLYCATCHER *Melænornis pammelaina* | 799 |
| 18 Dull black or slaty black: | BLACK FLYCATCHER *Melænornis edolioides* | 798 |

| | | |
|---|---|---|
| 19 White collar on hind neck: | WHITE-COLLARED FLYCATCHER, male *Muscicapa albicollis* | 780 |
| 20 No white collar on hind neck: | PIED FLYCATCHER, male *Muscicapa hypoleuca* | 779 |

| | | |
|---|---|---|
| 21 White streaks on underside: | SHRIKE-FLYCATCHER, female *Megabyas flammulatus* | 811 |
| 22 Below, wholly chestnut brown: | RUFOUS FLYCATCHER *Stizorhina fraseri* | 810 |
| 23 Below, pale whitish and tawny: | BLACK-AND-WHITE FLYCATCHER, female *Bias musicus* | 812 |

## FLYCATCHERS

| | | | |
|---|---|---|---|
| 24 | Throat and chest blue black: | BLACK-THROATED WATTLE-EYE, female *Platysteira peltata* | **823** |
| 25 | Throat white, black band across chest: | | 26–27 |
| 26 | White patch on wing: | WATTLE-EYE, male *Platysteira cyanea* | **822** |
| 27 | No white patch on wing: | BLACK-THROATED WATTLE-EYE, male *Platysteira peltata* | **823** |
| 28 | Tail very short: | CHESTNUT WATTLE-EYE, female *Dyaphorophyia castanea* | **824** |
| 29 | Tail normal length, or central tail-feathers greatly elongated: | | 30–35 |
| 30 | Chest to belly saffron or light chestnut: | BLACK-HEADED PARADISE FLYCATCHER *Tchitrea nigriceps* | **835** |
| 31 | Chest to belly grey: | | 32–35 |
| 32 | Top of head slightly glossy violet grey: | ZAMBESI PARADISE FLYCATCHER *Tchitrea plumbeiceps* | **834** |
| 33 | Top of head glossy blue-black: | | 34–35 |
| 34 | White on wing, under tail-coverts slaty or chestnut: | PARADISE FLYCATCHER *Tchitrea viridis races* | **832** |
| 35 | No white on wing, under tail-coverts whitish or buffish white: | RED-WINGED PARADISE FLYCATCHER *Tchitrea suahelica* | **833** |
| 36 | Throat to chest saffron, belly yellow: | YELLOW-BELLIED WATTLE-EYE, female *Diaphorophyia concreta* | **826** |

| | | |
|---|---|---|
| 37 | Throat white, belly pale yellow: | LIVINGSTONE'S FLYCATCHER *Erythrocercus livingstonei* **809** |
| 38 | Throat and belly yellow: | 39–43 |
| 39 | Eye wattled: | YELLOW-BELLIED WATTLE-EYE, male *Dyaphorophyia concreta* **826** |
| 40 | Eye not wattled: | 41–44 |
| 41 | Top of head brownish: | YELLOW FLYCATCHER *Chloropeta natalensis* **805** |
| 42 | Top of head green: | 43–44 |
| 43 | Size large, wing over 55 mm.: | MOUNTAIN YELLOW FLYCATCHER *Chloropeta similis* **806** |
| 44 | Size smaller, wing under 51 mm.: | LITTLE YELLOW FLYCATCHER *Chloropetella holochlora* **807** |
| 45 | Below, white and chestnut: | 46–48 |
| 46 | Tail chestnut: | CHESTNUT-CAP FLYCATCHER *Erythrocercus mccallii* **808** |
| 47 | Tail blackish: | 48–49 |
| 48 | Throat to breast suffused pale chestnut: | ZULULAND FLYCATCHER, female *Batis fratrum* **821** |
| 49 | Throat, chest and flanks chestnut: | PUFF-BACK FLYCATCHER female *Batis capensis* **815** |

---

| | | |
|---|---|---|
| 50 | Head crested: | 51–52 |
| 51 | Tail very short: | BLACK AND WHITE FLYCATCHER, male *Bias musicus* **812** |
| 52 | Tail normal length or central tail-feathers very long: | 53–54 |

# FLYCATCHERS

| | | |
|---|---|---|
| 53 White patch on wings: | PARADISE FLYCATCHER (western race) black phase of males *Tchitrea viridis speciosa* | **832** |
| 54 No white on wings: | BLUE-HEADED CRESTED-FLYCATCHER, male *Trochocercus nitens* | **830** |
| 55 Head not crested: | | 56–63 |
| 56 Rump white: | | 57–58 |
| 57 Eye wattled: | CHESTNUT WATTLE-EYE, male *Dyaphorophyia castanea* | **824** |
| 58 Eye not wattled: | SHRIKE-FLYCATCHER, male *Megabyas flammulatus* | **811** |
| 59 Rump not white: | | 60–63 |
| 60 Eye wattled, above, bottle-green: | JAMESON'S WATTLE-EYE *Dyaphorophyia jamesoni* | **825** |
| 61 Eye not wattled: | | 62–63 |
| 62 Above, blue black or violet blue-black: | YELLOW-BELLIED FLYCATCHER, male *Hyliota flavigaster* | **813** |
| 63 Above, velvety black: | SOUTHERN YELLOW-BELLIED FLYCATCHER *Hyliota australis* | **814** |
| 64 A white patch on wing: | WHITE-COLLARED FLYCATCHER, female *Muscicapa albicollis* | **780** |
| 65 No white patch on wing: | | 66–89 |
| 66 Flanks tawny: | BANDED TIT-FLYCATCHER *Parisoma böhmi* | **787** |
| 67 Flanks brownish-ashy: | RED-COLLARED FLYCATCHER *Lioptilornis rufocinctus* | **790** |
| 68 Top of head streaked: | | 69–73 |
| 69 Chest streaked: | SPOTTED FLYCATCHER *Muscicapa striata* | **778** |
| 70 Size small, wing under 67 mm.: | DUSKY FLYCATCHER *Alseonax adustus* | **781** |

| | | |
|---|---|---|
| 71 | Chest not streaked: | 72–73 |
| 72 | Size larger, wing over 81 mm.: | GREY FLYCATCHER *Bradornis microrhynchus* **793** |
| 73 | Size smaller, wing under 81 mm.: | LITTLE GREY FLYCATCHER *Bradornis pumilus* **794** |
| 74 | Top of head not streaked: | 75–84 |
| 75 | White at base of primaries: | PIED FLYCATCHER, female *Muscicapa hypoleuca* **779** |
| 76 | No white at base of primaries: | 77–84 |
| 77 | White on outermost tail feathers: | BROWN TIT-FLYCATCHER *Parisoma lugens* **789** |
| 78 | No white on outermost tail feathers: | 79–84 |
| 79 | Size smaller, wing under 72 mm.: | SWAMP FLYCATCHER *Alseonax aquaticus* **782** |
| 80 | Size larger, wing over 73 mm.: | 81–84 |
| 81 | General colour darker, more sooty: | ABYSSINIAN SLATY FLYCATCHER *Dioptrornis chocolatinus* **797** |
| 82 | General colour paler, more mouse-brown: | 83–84 |
| 83 | Bill deeper and thicker: | PALE FLYCATCHER *Bradornis pallidus* **792** |
| 84 | Bill less deep and thinner: | WAJHEIR GREY FLYCATCHER *Bradornis bafirawari* **795** |

| | | |
|---|---|---|
| 85 | Head crested: | 86–93 |
| 86 | Head and throat glossy blue-black: | 87–91 |
| 87 | Breast and belly white: | CRESTED FLYCATCHER, male *Trochocercus cyanomelas* **828** |

# FLYCATCHERS

| | | | |
|---|---|---|---|
| 88 | Breast to belly grey: | BLUE-HEADED CRESTED FLYCATCHER, male *Trochocercus nitens* | 830 |
| 89 | Top of head only glossy blue-black: | | 90–91 |
| 90 | Throat grey: | BLUE-HEADED CRESTED FLYCATCHER, female *Trochocercus nitens* | 830 |
| 91 | Throat speckled white and slate: | CRESTED FLYCATCHER female *Trochocercus cyanomelas* | 828 |
| 92 | Head and throat dull black: | WHITE-TAILED CRESTED FLYCATCHER *Trochocercus albonotatus* | 829 |
| 93 | Throat slate-blue: | DUSKY CRESTED FLYCATCHER *Trochocercus nigromitratus* | 831 |
| 94 | Head not crested: | | 95–128 |
| 95 | Below, white, irregularly barred slate: | FOREST FLYCATCHER *Fraseria ocreata* | 800 |
| 96 | Throat rich deep chestnut; breast and belly white: | WATTLE-EYE, female *Platysteira cyanea* | 822 |
| 97 | Below, tawny or buff: | | 98–99 |
| 98 | Below, tawny: | SILVER BIRD *Empidornis sempipartitus* | 801 |
| 99 | Below, buff: | YELLOW-BELLIED FLYCATCHER, female *Hyliota flavigaster* | 813 |
| 100 | Below, white, chestnut or tawny; band across chest: | | 101–106 |
| 101 | Chestnut patch on throat: | CHIN-SPOT PUFF-BACK FLYCATCHER, female *Batis molitor* | 817 |
| 102 | Throat white: | | 103–106 |

| | | |
|---|---|---|
| 103 | Top of head black: | BLACK-HEADED PUFF-BACK FLYCATCHER, female *Batis minor* **820** |
| 104 | Top of head grey: | 105–106 |
| 105 | Under wing-coverts white: | ZULULAND FLYCATCHER, male *Batis fratrum* **821** |
| 106 | Under wing-coverts black: | GREY-HEADED PUFF-BACK* FLYCATCHER, female *Batis orientalis* **818** |
| 107 | Below, white, black band across chest: | 108–115 |
| 108 | Under wing-coverts white: | 109–110 |
| 109 | White spots on forehead: | RUWENZORI PUFF-BACK FLYCATCHER *Batis diops* **816** |
| 110 | No white spots on forehead: | PUFF-BACK FLYCATCHER, male *Batis capensis* **815** |
| 111 | Under wing-coverts black: | 112–115 |
| 112 | Top of head black: | BLACK-HEADED PUFF-BACK FLYCATCHER, male *Batis minor* **820** |
| 113 | Top of head grey: | 114–115 |
| 114 | Little or no white at nape: | CHIN-SPOT PUFF-BACK FLYCATCHER, male *Batis molitor* **817** |
| 115 | Large patch of white at nape: | GREY-HEADED PUFF-BACK* FLYCATCHER, male *Batis orientalis* **818** |
| 116 | Below, mainly grey: | 117–128 |

\* *Note:* **821** Pigmy Puff-back Flycatcher, *Batis perkeo*, is similar but smaller, wing 47–52 mm. against 53–63 mm.

| | | |
|---|---|---|
| 117 Outermost tail feathers white: | GREY TIT-FLYCATCHER *Parisoma plumbeum* | **788** |
| 118 Outermost tail-feathers grey: | | 119–128 |
| 119 Feet yellow: | YELLOW-FOOTED FLYCATCHER *Alseonax seth-smithi* | **783** |
| 120 Feet not yellow: | | 121–128 |
| 121 Wings black: | | 122–123 |
| 122 Above, dark grey: | DUSKY BLUE FLYCATCHER *Pedilorhynchus comitatus* | **803** |
| 123 Above, pale grey: | CASSIN'S GREY FLYCATCHER *Alseonax cassini* | **784** |
| 124 Wings grey: | | 125–128 |
| 125 Tail black: | GREY-THROATED FLYCATCHER *Alseonax griseigularis* | **786** |
| 126 Tail grey: | | 127–128 |
| 127 Whitish edges to secondaries: | ASHY FLYCATCHER *Alseonax cinereus* | **785** |
| 128 No whitish edges to secondaries: | WHITE-EYED SLATY FLYCATCHER *Dioptrornis fischeri* | **796** |

**778** SPOTTED FLYCATCHER. *MUSCICAPA STRIATA* (Pallas).

*Muscicapa striata striata* (Pall.).

*Motacilla striata* Pallas, in Vroeg's Cat. Adum. p. 3, 1764: Holland.

**Distinguishing characters:** Above, ashy brown; blackish streaks on forehead and crown; below, white; sides of throat and chest streaked dark ashy; flanks buffish with some ashy streaks. The sexes are alike. Wing 80 to 92 mm.

**General distribution:** Europe and northern Africa; in non-breeding season to greater part of Africa as far south as Cape Province and Arabia.

**Range in Eastern Africa:** Throughout in the non-breeding season.

**Habits:** The Spotted Flycatcher is an extensive palæarctic winter visitor to Africa, and is found throughout Eastern Africa from October to April. A number appear to remain throughout Uganda and parts of Kenya Colony though moving about a good deal locally, but the main body pass on to Nyasaland and the Rhodesias. In Tanganyika Territory it is both a bird of passage and a visitor. Its habits are characteristic of the family as a whole, and it spends its time on a low bare bough or perch from which it makes dashes at passing insects, very rarely missing them. Its note is a thin sibilant 'sep' which hardens in alarm into a sharp 'tec-tec.' It has also a habit of constantly flicking its wings while perched, and it does so more repeatedly than most of the native African Flycatchers, with some of whom it can easily be confused in the field.

*Muscicapa striata neumanni* Poche.
*Muscicapa grisola neumanni* Poche, O.M. p. 26, 1904: Loita Mts., Kenya Colony.

**Distinguishing characters:** Differs from the nominate race in being paler above, more ashy grey, less brown; streaks on chest usually narrower and paler. Wing 81 to 92 mm.

**General distribution:** Western Asia to Lake Baikal and Iran; in non-breeding season to Eastern Africa and Arabia.

**Range in Eastern Africa:** Abyssinia to Uganda and Tanganyika Territory; also Zanzibar and Pemba Islands in non-breeding season.

**Habits:** The eastern race of the Spotted Flycatcher, indistinguishable in the field but in Africa occurring more particularly on the east coast and on Zanzibar and Pemba Islands. It has not so far been noticed south of Tanganyika Territory.

*Muscicapa striata gambagæ* (Alex.). **Pl. 61.**
*Alseonax gambagæ* Alexander, Bull. B.O.C. 12, p. 11, 1901: Gambaga, Gold Coast Colony.

**Distinguishing characters:** Considerably smaller than the nominate race; bill shorter; upperside mouse-brown; streaks on head and chest smaller and much less distinct. The sexes are alike. Wing 71 to 76 mm. The young bird has whitish-buff spots above and blackish streaking below.

# FLYCATCHERS

**General distribution:** Gold Coast Colony to British Somaliland and Arabia.

**Range in Eastern Africa:** The Sudan to British Somaliland.

**Habits:** Much like those of the nominate race. Frequents light woodland in the more fertile parts of the Sudanese and Somali semi-arid belts.

**Nest and Eggs:** Nest usually in a hollow of a dead stump, and composed of fine grass lined with seed down and bound round with cobwebs. Eggs two, greenish white speckled with faint grey and yellowish brown; about 16·5 × 13 mm.

**Recorded breeding:** Nigeria, March to April. Darfur, Sudan, June.

**Call:** Unrecorded.

## 779 PIED FLYCATCHER. *MUSCICAPA HYPOLEUCA* (Pallas).

*Muscicapa hypoleuca semitorquata* Hofm.
*Muscicapa semitorquata* Hofmeyer, Zeitsch, ges. Orn. 2, p. 185, pl.10, 1885: Caucasus.

**Distinguishing characters:** Adult male, upperside black; forehead, base and outer webs of primaries, inner secondaries and outer tail-feathers white; below, entirely white. The female has the black of the male replaced by ashy brown. Wing 77 to 84 mm. The young bird is similar to the female, but the white on the wings and tail is buffish.

**General distribution:** Greece to Asia Minor, the Caucasus and Iran; in non-breeding season to Iraq, Egypt, the Sudan, British Somaliland, Uganda and Tanganyika Territory.

**Range in Eastern Africa:** Eritrea and the Sudan, British Somaliland, western Uganda and southern Tanganyika Territory in non-breeding season.

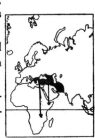

**Habits:** A palæarctic passage migrant in some numbers through the Sudan and has recently been recorded from Tanganyika Territory. Its habits are somewhat like those of the Spotted Flycatcher, but it often uses a perch higher up in a tree and not always a dead one, nor does it return so constantly to the same place. The call is a sharp 'whit-whit.'

**Distribution of other races of the species :** Western and northern Europe, the nominate race being described from Holland.

M

# FLYCATCHERS

### 780 WHITE-COLLARED FLYCATCHER. *MUSCICAPA ALBICOLLIS* Temminck.

*Muscicapa albicollis* Temminck, Man. d'Orn, p. 100, 1815: Waltershausen, Thuringia, Germany.

**Distinguishing characters:** Adult male, differs from the Pied Flycatcher in having a broad white collar on hind neck. The female has white, or whitish, bases to the feathers of the hind neck, which are not found in the female of the Pied Flycatcher. Wing 79 to 82 mm. The young bird has no white at the base of the feathers of the hind neck, but has the upperparts paler, less brown, than the young bird of the Pied Flycatcher.

**General distribution:** Central and southern Europe to Iran; in non-breeding season to the western Sahara, Egypt, the Sudan, and south to Northern and Southern Rhodesia and Nyasaland.

**Range in Eastern Africa:** The Sudan in the non-breeding season, and should occur in Uganda and Tanganyika Territory.

**Habits:** A comparatively rare palæarctic winter visitor to Africa, passing through the Sudan in September and occurring as far south as Nyasaland and Northern Rhodesia. In habits indistinguishable from the Pied Flycatcher.

### 781 DUSKY FLYCATCHER. *ALSEONAX ADUSTUS* (Boie).
*Alseonax adustus minimus* (Heugl.).

*Muscicapa minima* Heuglin, J.f.O. p. 301, 1862: Gondar, northern Abyssinia.

**Distinguishing characters:** Above, brownish mouse grey; below, throat and belly whitish; chest and flanks ash brown. The sexes are alike. Wing 58 to 67 mm. The immature bird is umber brown; throat and belly washed with brown. The young bird is spotted buff above and streaked with dark ash brown below.

**Range in Eastern Africa:** Eritrea and Abyssinia.

**Habits:** A common resident on the Abyssinian plateau. Habits as for other races.

**Nest and Eggs:** See under other races.

**Recorded breeding:** Abyssinia, June and July.

# FLYCATCHERS

*Alseonax adustus murinus* Fisch. & Reichw.
*Alseonax murinus* Fischer & Reichenow, J.f.O. p. 54, 1884: Mt. Meru, northern Tanganyika Territory.

**Distinguishing characters:** Above, greyer than the preceding race; chest and flanks ash grey. The immature bird has the lower flanks and belly pale brown. Wing 57 to 67 mm.

**Range in Eastern Africa:** Taita Hills, south-eastern Kenya Colony to Mt. Meru, Essimingor, Lolkissale, Nou Mbulu and Mt. Kilimanjaro in north-eastern Tanganyika Territory.

**Habits:** As for other races.

**Recorded breeding:** Tanganyika Territory, December and January, also (breeding condition) October.

*Alseonax adustus pumilus* Reichw. **Pl. 61, Ph. viii.**
*Alseonax pumila* Reichenow, J.f.O. pp. 32, 218, 1892: Bukoba, north-western Tanganyika Territory.

**Distinguishing characters:** Above, similar to the Abyssinian race, but more extensively ashy brown below, with less whitish on throat and belly. The immature bird is rather deeper brown than that of the Abyssinian race. Wing 59 to 66 mm.

**General distribution:** North-eastern Belgian Congo and the southern Sudan to Uganda, Kenya Colony and Tanganyika Territory.

**Range in Eastern Africa:** Southern Sudan, Uganda, Kenya Colony south of lat. 0.30 N. (except south-eastern areas) and northern Tanganyika Territory as far east as Loliondo.

**Habits:** A common resident of well-wooded country or open forest, often becoming very tame in gardens. It may be found on any wooded hill top or in woodland at all elevations. Its habits are much those of the Spotted Flycatcher as described under that species, and it also has the characteristic flicking of the wings to some degree. Its choice of perch is more varied and it may be seen at any height in a tree. It occasionally picks insects off the ground.

**Nest and Eggs:** The nest is a small open cup of grass stems, fibres and rootlets, often lined with plant down or feathers. It may be placed on a house or among creepers on a rock ledge or on a lichen covered bough or stump, and is usually well concealed and often bound with cobwebs. It may be at any height. Eggs usually two or three, pale blue, green or dull white with brown or red brown spotting and speckling and with pale ashy undermarkings; about 17 × 12·5 mm.

**Recorded breeding:** Uganda, April to July, and September to January. Kenya Colony, January, February and April to June.

**Call:** A short sibilant 'seet-seet.' Alarm note a ticking chatter.

*Alseonax adustus subadustus* Shell.
*Alseonax subadustus* Shelley, Ibis, p. 452, 1897; Nyika Plateau, northern Nyasaland.

**Distinguishing characters:** Above, paler ash-grey than other races; below, white with pale ash-grey chest and flanks. Wing 61 to 67 mm.

**General distribution:** Southern Belgian Congo and Northern Rhodesia to Nyasaland, eastern Southern Rhodesia and Portuguese East Africa.

**Range in Eastern Africa:** Nyasaland and Portuguese East Africa.

**Habits:** As for other races, common and widely distributed in suitable woodland or at forest edges, preferring shade and the proximity of water and generally seen in pairs.

**Recorded breeding:** Nyasaland, September to December.

*Alseonax adustus fülleborni* (Reichw.).
*Muscicapa fülleborni* Reichenow, O.M. p. 122, 1900: Rupira, Ukinga, Tukuyu district, south-western Tanganyika Territory.

**Distinguishing characters:** Differs from the last race in being darker ash grey. Wing 61 to 65 mm.

**Range in Eastern Africa:** Mpapwa to the Uluguru Mts., Njombe, Ifakara, Ukinga, upper Nyamansi River and Kungwe-Mahare Mts., Tanganyika Territory.

**Habits:** As for other races, common among trees along streams and in wooded country.

**Recorded breeding:** Tanganyika Territory, November and December.

*Alseonax adustus roehli* Grote.
*Alseonax murinus roehli* Grote, O.M. p. 62, 1919: Mlalo, Lushoto, north-eastern Tanganyika Territory.

**Distinguishing characters:** Warmer in general colour than the Ukinga race. Wing 58 to 63 mm.

**Range in Eastern Africa:** Paré and Usambara Mts. to the Nguru Hills, Ukuguru Hills and Kilosa, eastern Tanganyika Territory.

**Habits:** As for other races.

**Recorded breeding:** No records.

*Alseonax adustus marsabit* V. Som.
*Alseonax minimus marsabit* Van Someren, J.E.A & U.N.Hist.Soc. 37, p. 1, 1931: Mt. Marsabit, northern Kenya Colony.

**Distinguishing characters:** General colour very brown, and very similar to the immature bird of the Bukoba race, but slightly greyer above. The immature bird is deeper brown than the adult and is, therefore, deeper brown than the immature bird of the neighbouring race. Wing 61 to 63 mm.

**Range in Eastern Africa:** Northern Kenya Colony, from Turkana and Marsabit to Mt. Elgon.

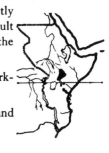

**Habits:** As for other races but usually seen in dense forest and often at the tops of trees.

**Recorded breeding:** Mt. Elgon, May.

*Alseonax adustus chyulu* V. Som.
*Alseonax minimus chyulu* Van Someren, J.E.A. & U.N.Hist.Soc. 14, p. 71, 1939: Chyulu Hills, south-eastern Kenya Colony.

**Distinguishing characters:** Similar to the race from north-eastern Tanganyika Territory in the greyness of the upperside, but underside suffused with pale brown. The immature bird is rather browner above and deepish brown below. Wing 60 to 67 mm.

**Range in Eastern Africa:** Chyulu Hills, south-eastern Kenya Colony to Endulen, Oldeani, Nguruka and Ketumbeine, north-eastern Tanganyika Territory.

**Habits:** As for other races, common at forest edges and tame.

**Recorded breeding:** Chyulu Hills, Kenya Colony 'early in year, over by May.' Oldeani, Tanganyika Territory, January.

**Distribution of other races of the species:** West and South Africa, the nominate race being described from Knysna, Cape Province.

## 782 SWAMP FLYCATCHER. *ALSEONAX AQUATICUS* (Heuglin).

*Alseonax aquaticus aquaticus* (Heugl.).

*Muscicapa aquatica* Heuglin, J.f.O. p. 256, 1864: Wau, Bahr-el-Ghazal, south-western Sudan.

**Distinguishing characters:** Above, ash-colour; below, white; chest and flanks washed with ash-colour. Larger than the Dusky Flycatcher; bill and tail appreciably longer. The sexes are alike. Wing 66 to 70 mm. The young bird is spotted with buff above and has a few dark streaks on the chest.

**General distribution:** Gambia to the Sudan.

**Range in Eastern Africa:** South-western Sudan.

**Habits:** Local and uncommon or overlooked, among papyrus or high reeds.

**Nest and Eggs, Call, etc.:** See under the following race.

**Recorded breeding:** Adamawa, Cameroons, April.

*Alseonax aquaticus infulatus* (Hartl.). **Pl. 61, Ph. viii.**

*Muscicapa infulata* Hartlaub, P.Z.S. p. 626, 1880: Wadelai, north-western Uganda.

**Distinguishing characters:** Differs from the nominate race in being much darker, more sooty brown above, and with a more distinct ash-coloured chest. Wing 61 to 72 mm.

**General distribution:** The Sudan, Uganda, Kenya Colony, south-eastern Belgian Congo and north-eastern Northern Rhodesia.

**Range in Eastern Africa:** Southern Sudan (except south-western areas) to Uganda, western Kenya Colony, and western Tanganyika Territory.

**Habits:** Not at all an uncommon species in its very restricted local habitats, which are elephant grass, papyrus, and reed beds in lakes and along the banks of large rivers. Its habits are those of a Flycatcher, but it has also been noted running about on lily pads and picking insects off them.

**Nest and Eggs:** Nests generally in the old nests of Weaver-birds, that of any species available may be used, whether the nest has a spout or not. Other nests are recorded from the crevices of rough bark on trees. Old nests are relined with fine grass or feathers to form a cup for the eggs. Eggs usually two, pale blue thickly spotted with reddish brown especially at larger end; about 17 × 12 mm.

**Recorded breeding:** Uganda and western Kenya Colony, March to June. Southern Belgian Congo, January.

**Call:** A sharp alarm cry 'pzitt.' Also a rather distinctive song 'tweet-a-weet-tit-weet-a-weet' (Elliott).

**783 YELLOW-FOOTED FLYCATCHER.  *ALSEONAX SETH-SMITHI* (Van Someren).**
*Pedilorhynchus epulatus seth-smithi* Van Someren, Nov. Zool. 29, p. 96, 1922: Budongo Forest, western Uganda.

**Distinguishing characters:** General colour dark slate; flight feathers and tail black; chin, belly and under tail-coverts white; feet yellow. The sexes are alike. Wing 52 to 59 mm. The young bird is spotted with deep tawny above; below, dusky with some tawny spotting. The immature bird retains the tawny edges to the inner secondaries and wing-coverts.

**General distribution:** Cameroons and Gabon to Uganda.

**Range in Eastern Africa:** Western Uganda.

**Habits:** A small dark forest Flycatcher perching on dead boughs wherever there is sufficient of a clearing to get a view of passing insects. Typical Flycatcher habits.

**Nest and Eggs:** Nest a loose pile of moss with a cup-shaped cavity lined with lichen. Eggs two, dull greenish white indistinctly clouded and mottled with rufous and grey; about 17 × 13·5 mm. (O. Grant).

**Recorded breeding:** Cameroons, June and July. Belgian Congo, April and July.

**Call:** Unrecorded.

**784 CASSIN'S GREY FLYCATCHER. *ALSEONAX CASSINI* (Heine).**
*Muscicapa cassini* Heine, J.f.O. p. 428, 1859: Camma River, Gabon.

**Distinguishing characters:** Above, dark grey; indistinct darker streaks on head; flight feathers and tail black; below, grey; belly white. The sexes are alike. Wing 67 to 74 mm. The young bird is spotted with buff and white on chest and upperside. The immature bird retains the buff tips to the wing-coverts.

**General distribution:** Sierra Leone and northern Angola, to Uganda, northern areas of north-eastern Northern Rhodesia and southern Belgian Congo.

**Range in Eastern Africa:** South-western Uganda.

**Habits:** A strictly forest species with a strong predilection for water, and is nearly always seen on a dead bough over a forest stream. Usually tame and with normal Flycatcher habits.

**Nest and Eggs:** Nest on a stump or snag projecting from or hanging over a stream. It is a shallow cup of rootlets and moss lined with bark fibre or fine grass tops. Eggs two, whitish olive speckled with fine reddish brown and with liver coloured spots mostly at large end; measurements according to Reichenow 18·5 × 13·75 mm.

**Recorded breeding:** Nigeria, January to April. Cameroons, November to March. Ankole, Uganda, November. South-eastern Belgian Congo, September.

**Call:** A short vigorous song, and a normal call of 'pink.' Alarm call a shriller 'tseet, tseet, tseet.'

### 785 ASHY FLYCATCHER. *ALSEONAX CINEREUS* (Cassin).

*Alseonax cinereus cinereus* (Cass.).

*Eopsaltria cinerea* Cassin, Proc. Ac. N. Sci. Philad. p. 253, 1856: Moonda River, Gabon.

**Distinguishing characters:** Very similar to Cassin's Grey Flycatcher, but paler grey above; flight feathers and tail dark grey, not black; white eye-streak from base of bill to over eye; white streak on lower eyelid. The sexes are alike. Wing 67 to 78 mm. The young bird is spotted above with tawny and black; below, marbled with blackish, pale buff and white. The immature bird retains the tawny tips to the wing-coverts.

**General distribution:** Cameroons, Gabon and northern Angola to the Sudan and Uganda.

**Range in Eastern Africa:** Southern Sudan to western and southern Uganda.

**Habits:** Said to be not uncommon locally in open forest, with typical Flycatcher habits. They are shy and easily overlooked, as they remain motionless for long periods.

**Nest and Eggs:** See under the following race, but the nests are said to be commonly in old nests of Weavers.

**Recorded breeding:** Uganda, June.

# FLYCATCHERS

*Alseonax cinereus cinereolus* (Finsch & Hartl.). **Pl. 61.**
*Muscicapa cinereola* Finsch & Hartlaub, Vög. Ostafr. in von der Decken's Reisen, 4, p. 302, 1870: Usaramo, Dar-es-Salaam District, eastern Tanganyika Territory.

**Distinguishing characters:** Differs from the nominate race in being appreciably paler grey. Wing 71 to 86 mm.

**General distribution:** Kenya Colony to the Mossamedes area of southern Angola, Damaraland, Bechuanaland, eastern Southern Rhodesia and Portuguese East Africa as far south as Inhambane.

**Range in Eastern Africa:** South-central Kenya Colony to eastern half of Tanganyika Territory and the Zambesi River.

**Habits:** Apparently a bird of rather diverse situations, but it occurs freely in the mountain forests of Tanganyika Territory at the edges of clearings, and down to sea-level in suitable woodland. Elsewhere it is found in thorn scrub or riverside belts of trees. Typical Flycatchers' habits and is often a member of a mixed bird party. Somewhat sluggish for a Flycatcher, but a beautiful blue-grey bird in life. After a flight it frequently gives a flick with its wings on returning to a perch.

**Nest and Eggs:** Normally a cup of fine fibres and roots in a crevice of bark or a shallow hole of a tree or stump, also a very small pad of a nest in the cupped fork of a tree. Larger cup nests of twigs, moss, etc. with a grass lining in the forks of trees are noted from Nyasaland. Eggs two or three, cream or whitish, thickly spotted or speckled with brown or chocolate brown and grey; about 18 × 14 mm.

**Recorded breeding:** Tanganyika Territory, September. Kafue, Northern Rhodesia, November. Nyasaland, October to December. Portuguese East Africa, September and October.

**Call:** A thin short whistle, or a weak stuttering sibilant cry 'twit-it-it-tweet.'

**Distribution of other races of the species:** Gold Coast and Natal.

**786** GREY-THROATED FLYCATCHER. *ALSEONAX GRISEIGULARIS* Jackson.
*Alseonax griseigularis griseigularis* Jack.
*Alseonax griseigularis* Jackson, Bull. B.O.C. 19, p. 19, 1906: Kibirau, Toro, western Uganda.

**Distinguishing characters:** Similar to the Ashy Flycatcher, but

smaller and tail blackish; wings dark grey. The sexes are alike. Wing 63 to 64 mm. Juvenile plumage unrecorded.

**General distribution:** Eastern Belgian Congo to Uganda.

**Range in Eastern Africa:** Western and southern Uganda.

**Habits:** Confined to wooded water courses and forest streams, little known but probably of typical Flycatcher habits.

**Nest and Eggs:** Apparently undescribed.

**Recorded breeding:** Uganda (breeding condition), March and July.

**Call:** Plaintive, and said to be similar to that of the Grey Tit-Flycatcher.

**Distribution of other races of the species:** West Africa.

**787 BANDED TIT-FLYCATCHER.** *PARISOMA BÖHMI* Reichenow.

*Parisoma böhmi böhmi* Reichw.

*Parisoma böhmi* Reichenow, J.f.O. p. 209, pl. 2, fig. 2, 1882: Seke, Ugogo, Dodoma district, central Tanganyika Territory.

**Distinguishing characters:** Above, ash-grey; broad white edges to inner secondaries and wing-coverts; tail blackish, outer tail-feathers edged with white; below, throat to belly white, former with blackish spots, coalescing on lower neck and joining a broken collar; flanks and under tail-coverts rich tawny. The sexes are alike. Wing 58 to 68 mm. The young bird has buff edges to the wing-coverts; no black spots on throat and no collar.

**Range in Eastern Africa:** Eastern and southern Kenya Colony to central Tanganyika Territory as far south as the Iringa district.

**Habits:** Cheerful little birds with somewhat the habits of Tits, though with Flycatcher affinities. They are locally common in acacia country usually searching leaves, and often utter a pretty little trilling song while feeding.

**Nest and Eggs:** Nest in the fork of a small bough, a neat little cup of grass and rootlets lined with grass heads, some feet from the ground in a thorn thicket. Eggs two or three, whitish, spotted with yellowish brown, greyish brown and pale grey; about $17 \times 12 \cdot 5$ mm.

**Recorded breeding:** Southern Kenya Colony, March and April. Northern Tanganyika Territory (near breeding condition), January.

**Call:** A song as above, also a rather squeaky call 'chicky-wurrah-chick-wurr.'

*Parisoma böhmi somalicum* Friedm.

*Parisoma böhmi somalicum* Friedmann, Proc. N. Eng. Zool. Cl. 10, p. 51, 1928: Sogsoda, British Somaliland.

**Distinguishing characters:** Differs from the nominate race in being more ashy above and in usually having a wash of tawny on the chest below the broken black collar. Wing 58 to 63 mm.

**Range in Eastern Africa:** Abyssinia and British Somaliland.

**Habits:** Locally common in thorn-acacia country.

**Recorded breeding:** No records.

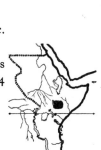

*Parisoma böhmi marsabit* V. Som.

*Parisoma böhmi marsabit* Van Someren, J.E.A. & U.N.Hist. Soc. No. 37, p. 194, 1930: Marsabit, northern Kenya Colony.

**Distinguishing characters:** Differs from the preceding races in having pale tawny flanks and under tail-coverts. Wing 58 to 64 mm.

**Range in Eastern Africa:** Northern Kenya Colony.

**Habits:** As for other races.

**Recorded breeding:** No records.

**788** GREY TIT-FLYCATCHER. *PARISOMA PLUMBEUM* (Hartlaub).

*Parisoma plumbeum plumbeum* (Hartl.). **Pl. 61.**

*Stenostira plumbea* Hartlaub, J.f.O. p. 41, 1858: Casamanse, Senegal.

**Distinguishing characters:** Very similar in general appearance to the Ashy Flycatcher, but tail longer, more blackish, and outermost tail feathers white; under tail-coverts buff. The sexes are alike. Wing 61 to 75 mm. The young bird has buff edges to the wing-coverts and inner secondaries; sometimes buff spots on scapulars and mantle; below, dirty brown with some lighter streaks on chest.

**General distribution:** Senegal to Abyssinia and Uganda.

**Range in Eastern Africa:** Central Sudan to Abyssinia and Uganda.

**Habits:** As for the eastern race.

**Nest and Eggs:** See under the following race.

**Recorded breeding:** Lake Chad, March to August. Uganda, April to June.

*Parisoma plumbeum orientale* Reichw. & Neum.

*Parisoma orientale* Reichenow & Neumann, O.M. p. 74, 1895: Kibwesi, south-central Kenya Colony.

**Distinguishing characters:** Differs from the nominate race in having the under tail-coverts white or only faintly buff. Wing 61 to 75 mm.

**General distribution:** Kenya Colony to Angola and Natal.

**Range in Eastern Africa:** Central Kenya Colony to the Zambesi River.

**Habits:** Birds of open bush or light woodland, or occasionally of open forest clearings, and locally not uncommon, with most of the habits of Flycatchers. The white outer tail feathers are said to be noticeable in the field as the birds have a distinctive habit of raising the back and flirting the tail like a fan.

**Nest and Eggs:** Nest unlike those of the preceding species, normally in some kind of hole or depression in a tree or stump, often in old holes of Woodpeckers or Barbets, which are scantily lined with grass, rootlets, feathers or lichens. Eggs two or three, dull white or whitish green spotted and blotched with various shades of brown; about 17 × 13 mm. (South Africa).

**Recorded breeding:** Nyasaland, May and June, also October and November. Rhodesia, October. South Africa, December.

**Call:** A little rising and falling song call, rather musical 'tee-ry-tee,' described as plaintive and ventriloquial.

### 789 BROWN TIT-FLYCATCHER. *PARISOMA LUGENS* (Rüppell).

*Parisoma lugens lugens* (Rüpp.).

*Sylvia (Curruca) lugens* Rüppell, N. Wirbelt. p. 113, pl. 42, fig. 2, 1840: Simen, northern Abyssinia.

**Distinguishing characters:** Above, smoky brown; below, white tinged with smoky brown; chin and sides of throat mottled black and whitish; lower flanks and under tail-coverts buffish; outermost tail feathers edged and tipped with white. The sexes are alike. Wing 60 to 70 mm. The young bird has the head darker than the mantle.

**Range in Eastern Africa:** Abyssinia.

**Habits, Nest, etc.:** See under the following race.

**Recorded breeding:** No records.

# FLYCATCHERS

*Parisoma lugens jacksoni* Sharpe.
*Parisoma jacksoni* Sharpe, Bull. B.O.C. 10, p. 28, 1899: Mt. Elgon, Kenya Colony–Uganda boundary.

**Distinguishing characters:** Differs from the nominate race in being rather less smoky in tone above and rather more dusky on chest and flanks. Wing 61 to 70 mm.

**General distribution:** The Sudan to Kenya Colony, Tanganyika Territory and south-eastern Belgian Congo.

**Range in Eastern Africa:** Imatong Mts. south-eastern Sudan to south-western Kenya Colony, northern Tanganyika Territory west of Mt. Kilimanjaro.

**Habits:** Birds of open forest and woodland, also of bush clad slopes, feeding among tree tops and singing like Warblers. Locally common.

**Nest and Eggs:** Nest a light cup of moss and rootlets lined with grass and bark-fibre on a stump or in the thin fork of a tree. Eggs two, whitish olive or cream, speckled or spotted with fine reddish brown and liver and grey spots; about 18 × 14 mm.

**Recorded breeding:** Uganda, November. Kenya Colony, June also April.

**Call:** A musical little song. No other calls recorded.


*Parisoma lugens clara* Meise.
*Parisoma lugens clara* Meise, O.M. p. 16, 1934: Mahuka, north-west of Lipumba, Matengo highlands, south-western Tanganyika Territory.

**Distinguishing characters:** Rather darker above than the preceding race; below, chest more dusky and chin darker. Wing 63 to 65 mm.

**General distribution:** Tanganyika Territory to Nyasaland.

**Range in Eastern Africa:** South-western Tanganyika Territory.

**Habits:** As for other races, but seems to be confined to *Acacia woodii* woodland at 4,500–5,000 feet.

**Recorded breeding:** Dedza, Nyasaland, October and November.

**790 RED-COLLARED FLYCATCHER.** *LIOPTILORNIS RUFOCINCTUS* (Rothschild).

*Lioptilus rufocinctus* Rothschild, Bull. B.O.C. 23, p. 6, 1908: Rugege Forest, south-east of Lake Kivu, western Ruanda.

**Distinguishing characters:** Forehead to occiput and round eyes black; sides of face, nape, chin to chest, rump and upper and under tail-coverts cinnamon rufous; mantle and flanks brownish ashy; wings and tail brownish black; chest to belly greyish brown washed with pale rufous; bill yellowish white. The sexes are alike. Wing 99 to 106 mm. The young bird has a stronger cinnamon rufous wash from chest to belly.

**Range in Eastern Africa:** Western Ruanda.

**Habits:** Nothing is known of this species except that it frequents high trees in forest and occasionally enters the bamboo-zone.

**Nest and Eggs:** Undescribed.

**Recorded breeding:** No records.

**Call:** A sharp 'chuck.'

**791 ABYSSINIAN CATBIRD.** *PAROPHASMA GALINIERI* (Guérin). **Pl. 61.**

*Parisoma galinieri* Guerin, Rev. Zool. p. 162, 1843: Abyssinia.

**Distinguishing characters:** General colour sooty grey; black spot in front of eye; forehead dirty white; lower belly and under tail-coverts chestnut. The sexes are alike. Wing 83 to 91 mm. The young bird is paler sooty grey; lower belly and under tail-coverts paler, more tawny than chestnut.

**Range in Eastern Africa:** Central and southern Abyssinia.

**Habits:** A species of unknown affinities, which is now treated as a Flycatcher. It is usually found in pairs or small parties in thick cover at high elevations, 8,000–10,000 feet, and each pair occupies a small territory, but little is known of their habits which do not appear to be particularly Flycatcher-like.

**Nest and Eggs:** Undescribed.

**Recorded breeding:** Abyssinia, probably March to June.

**Call:** The male has a loud clear ringing call and is answered by the female with a churring note. There is also a very fine Nightingale-like song, which may be heard up to late dusk.

# FLYCATCHERS

**792 PALE FLYCATCHER.** *BRADORNIS PALLIDUS* (Müller).
*Bradornis pallidus pallidus* (Müll.).
*Muscicapa pallida* von Müller, Naumannia, pt. 4, p. 28, 1851: Melpess, Kordofan, Sudan.

**Distinguishing characters:** Above, mouse brown; below, pale clear buff; throat and centre of belly white. The sexes are alike. Wing 75 to 89 mm. The immature bird is rather darker. The young bird has white streaks on the head; buffish spots on the mantle and wing-coverts; blackish streaks on chest and breast.

**Range in Eastern Africa:** North-western Abyssinia to the Sudan (except south-eastern area) and north-western Uganda.

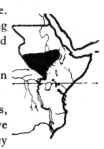

**Habits:** Common and usually tame bird of cultivation, gardens, woodland or scrub. In habits slower moving and more unobtrusive than most Flycatchers, and it takes a good deal of its insect prey from the ground. It is not unlike a Spotted Flycatcher in appearance, but is a somewhat larger, stouter bird with a rounder head. Perches on low boughs and spends much time motionless.

**Nest and Eggs:** Nest a very neat little cup of wiry rootlets well woven and almost transparent in the fork of a tree or bush. Eggs usually three, pale bluish green, spotted and blotched with reddish brown and with violet undermarkings; about 20 × 15 mm.

**Recorded breeding:** Sudan, Darfur, May, White Nile, April and May.

**Call:** Rarely uttered, a little chirp or churr and occasionally a warbling twitter of six or seven notes. Alarm note a rasping 'tsek.'

*Bradornis pallidus subalaris* Sharpe. **Pl. 61.**
*Bradyornis pallidus subalaris* Sharpe, P.Z.S. p. 713, pl. 58, fig. 1, 1873: Mombasa, eastern Kenya Colony.

**Distinguishing characters:** Size as in the nominate race, but darker both above and below. Wing 77 to 100 mm.

**General distribution:** Eastern Kenya Colony, Tanganyika Territory, south-eastern Belgian Congo, north-eastern Northern Rhodesia, and south to the eastern Transvaal, Natal and Zululand.

**Range in Eastern Africa:** Coastal areas of Kenya Colony from Lamu and Manda Island and eastern Tanganyika Territory inland as far as Amani and Kilosa to Portuguese East Africa and Nyasaland.

**Habits:** As for the nominate race. The nest is often loosely built and rather flat.

**Recorded breeding:** Kenya Colony, Mombasa—where it is abundant—August and September. Tanganyika Territory, September to December. South-eastern Belgian Congo, September to November. Nyasaland, October and November. Rhodesia, September and October.

*Bradornis pallidus griseus* Reichw.
*Bradyornis grisea* Reichenow, J.f.O. p. 221, 1882: Mgunda Mkali, south-eastern Tabora district, central Tanganyika Territory.

**Distinguishing characters:** Size as large as in the Mombasa race from which it differs in being darker above. Wing 83 to 104 mm.

**General distribution:** North-eastern Belgian Congo to the Sudan, Uganda, Kenya Colony and Tanganyika Territory.

**Range in Eastern Africa:** South-eastern Sudan to Uganda (except north-western), Kenya Colony (except coastal area), northern and western Tanganyika Territory as far east as the Paré Mts. and Kilosa, and as far south as Mpanda.

**Habits:** As for other races.

**Recorded breeding:** Kenya Colony, November to February, also March to May. Tanganyika Territory, September to December.

*Bradornis pallidus neumanni* Hilg.
*Bradornis griseus neumanni* Hilgert, Kat. Coll. Erl. p. 250, 1908: Are-Dare, Webbe Mana, southern Abyssinia.

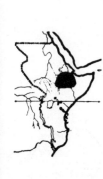

**Distinguishing characters:** Size as in the nominate race, but darker and warmer in tone than either that race or the Mombasa race. Wing 79 to 92 mm.

**Range in Eastern Africa:** Central, southern and south-western Abyssinia.

**Habits:** As for other races.

**Recorded breeding:** No records.

*Bradornis pallidus bowdleri* Coll. & Hart.
*Bradyornis pallidus bowdleri* Collins & Hartert, Nov. Zool. 34, p. 52, 1927: Eritrea.

**Distinguishing characters:** Larger and darker than the nominate race, but not nearly so dark nor so warm in tone as the preceding race. Wing 90 to 95 mm.

**Range in Eastern Africa:** Eritrea to north-eastern Abyssinia north of lat. 10° N.

**Habits:** As for other races.

**Recorded breeding:** No records.

**Distribution of other races of the species:** West Africa and Angola.

### 793 GREY FLYCATCHER. *BRADORNIS MICRORHYNCHUS* Reichenow. Pl. 61.

*Bradyornis microrhynchus* Reichenow, J.f.O. p. 62, 1887: Irangi, Kondoa-Irangi district, central Tanganyika Territory.

**Distinguishing characters:** Above, ash grey with dusky streaks on head; below, paler ash grey; throat and centre of belly whitish. Albinistic examples occur. The sexes are alike. Wing 77 to 89 mm. The young bird has whitish streaks on the head; buff spots on the mantle and wing-coverts; blackish streaks on chest and breast. This species may be distinguished from the Pale Flycatcher by the general grey appearance, more rounded wing and smaller bill.

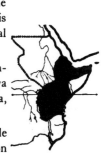

**Range in Eastern Africa:** Abyssinia, British Somaliland, northern Italian Somaliland as far south as lat. 9° N., Uganda, Kenya Colony and Tanganyika Territory as far south as Kilosa, Iringa, the Ufipa Plateau and the north end of Lake Nyasa.

**Habits:** Insufficiently distinguished in the past from the Pale Flycatcher, but unlikely to differ much in habits. It is not uncommon in the Ukamba country of Kenya Colony.

**Nest and Eggs:** The nest is rather like that of the Pale Flycatcher and is always lined with feathers. Eggs two, uniform green, usually with white excremental smears; about 19 × 15 mm.

**Recorded breeding:** Kenya Colony, April, also August to October. Tanganyika Territory (nestling) February.

**Call:** A low 'tweety' (Benson) is the only call recorded.

### 794 LITTLE GREY FLYCATCHER. *BRADORNIS PUMILUS* Sharpe.

*Bradyornis pumilus* Sharpe, P.Z.S. p. 480, 1895: Hargeisa, British Somaliland.

**Distinguishing characters:** Similar to the Grey Flycatcher but lighter below, whiter on throat and belly. Wing 74 to 81 mm.

**Range in Eastern Africa:** The southern Sudan east of the Nile to Abyssinia, British and Italian Somalilands, eastern Uganda and Kenya Colony.

**Habits:** Not easily distinguished in the field from the Grey Flycatcher, and records are therefore uncertain.

**Nest and Eggs:** In the same situation as the preceding species, but nests are said to be much more deeply cupped, and to have a few feathers mixed with the lining. Eggs three, pale clay colour with darker shading; about $18 \cdot 5 \times 13 \cdot 5$ mm.

**Recorded breeding:** South-western Abyssinia and Italian Somaliland, April. Turkana, Kenya Colony, May.

**Call:** Unrecorded.

### 795 WAJHEIR GREY FLYCATCHER. *BRADORNIS BAFIRAWARI* Bannerman.

*Bradornis bafirawari* Bannerman, Bull. B.O.C. 45, p. 41, 1924: Wajheir, eastern Kenya Colony.

**Distinguishing characters:** Similar to the Grey Flycatcher in length and shape of wing, but bill as long as in the larger races of the Pale Flycatcher, and differs from both in having the under wing-coverts and edges of inner webs of flight feathers white. The sexes are alike. Wing, male 77 to 79, female 73 mm. The young bird has the head streaked with buff; mantle and wing-coverts more mottled; breast mottled with irregular black markings.

**Range in Eastern Africa:** Eastern Kenya Colony.

**Habits:** As for the Pale Flycatcher, of which it is a near relative, usually seen in pairs in open thorn scrub.

**Nest and Eggs:** Undescribed.

**Recorded breeding:** No records.

**Call:** Unrecorded.

### 796 WHITE-EYED SLATY FLYCATCHER. *DIOPTRORNIS FISCHERI* Reichenow.

*Dioptrornis fischeri fischeri* Reichw. **Pl. 61. Ph. ix.**

*Dioptrornis fischeri* Reichenow, J.f.O. p. 53, 1884: Mt. Meru, Arusha district, north-eastern Tanganyika Territory.

**Distinguishing characters:** Above, slate-grey; a white ring round eye; below, throat and flanks pale grey; chest and under tail-coverts white with a buffish wash. The sexes are alike. Wing 85 to 96 mm. The young bird has whitish spots on the upperparts, more especially on the head and rump; below, feathers of chest and belly tipped with blackish.

**Range in Eastern Africa:** South-eastern Sudan to south-eastern and central Kenya Colony, as far west as Mt. Elgon, and north-eastern Tanganyika Territory in the area of Mts. Monduli, Meru, Kilimanjaro and the Paré.

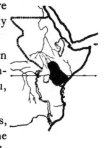

**Habits:** A common species, inhabiting woodland, forest edges, cultivated land, and thorn scrub. Generally very active in the evenings, and is much in evidence during a swarm of ants. Feeds mostly by hawking insects, including butterflies, but also searches foliage for them and will on occasion eat berries. They have even been known to eat nestlings of other birds. Often abundant and usually tame, feeding on till late dusk.

**Nest and Eggs:** A neat cup nest of moss with some dry leaves and fibre and a lining of fibre or hair, placed in the fork of a tree often at some height and in thick foliage or among lichen. Eggs two or three, oval, pale dull green heavily spotted and blotched with brown with a tendency to form a zone at the larger end; about 21 × 15·5 mm.

**Recorded breeding:** Kenya Colony, March to May and June, also October, December and January. Northern Tanganyika Territory, March and April, also September and January.

**Call:** A loud 'zit' and a chattering trill on a descending scale. Alarm or anxiety call, a hoarse throaty squeal, 'tchuiskere-cheeweet' (Van Someren).


*Dioptrornis fischeri nyikensis* (Shell.).
*Muscicapa nyikensis* Shelley, Bull. B.O.C. 8, p. 35, 1899: Nyika Plateau, northern Nyasaland.

**Distinguishing characters:** Differs from the nominate race in being duller slate grey above and with less white round the eye. Wing 84 to 95 mm.

**General distribution:** Tanganyika Territory to northern Nyasaland.

**Range in Eastern Africa:** Tanganyika Territory from Mbulu to Uluguru Mts., Njombe, and north end of Lake Nyasa.

**Habits:** As for the nominate race but mostly found in highland forest, and have a habit, according to Admiral Lynes, of cocking up their tails. The two eggs are described as turquoise blue spotted and blotched with reddish brown and with purplish grey undermarkings, no measurements available.

**Recorded breeding:** Tanganyika Territory, September to December. Northern Nyasaland, October.

*Dioptrornis fischeri toruensis* (Hart.).
*Muscicapa toruensis* Hartert, Nov. Zool. 7, p. 37, 1900: Fort Gerry, Uganda.

**Distinguishing characters:** Similar to the preceding race but smaller. Wing 79 to 83 mm.

**General distribution:** Eastern Belgian Congo to Uganda.

**Range in Eastern Africa:** Ruwenzori Mts. to south-western Uganda.

**Habits:** As for the nominate race, local and little recorded, not uncommon in the higher forests of the western side of the Ruwenzori range.

**Recorded breeding:** No records.


*Dioptrornis fischeri ufipæ* Mor.
*Dioptrornis fischeri ufipæ* Moreau, Bull. B.O.C. 62, p. 41, 1942: Mbisi Forest, Sumbawanga, Ufipa Plateau, south-western Tanganyika Territory.

**Distinguishing characters:** Similar to the nominate race in having a complete white ring round the eye, but is paler slate above, very close to the two preceding races in this respect, but has the chin, throat and belly much whiter than in any other race. Wing 90 to 93 mm.

**Range in Eastern Africa:** Ufipa Plateau, south-western Tanganyika Territory.

**Habits:** As for other races.

**Recorded breeding:** No records.

**Distribution of other races of the species:** North-eastern Belgian Congo.

## 797 ABYSSINIAN SLATY FLYCATCHER. *DIOPTRORNIS CHOCOLATINUS* (Rüppell).

*Dioptrornis chocolatinus chocolatinus* (Rüpp.).
*Muscicapa chocolatina* Rüppell, N. Wirbelt. Vög. p. 107, 1840: Semien, northern Abyssinia.

**Distinguishing characters:** Above, sooty brown; below, throat to chest and flanks grey, with indication of streaks on throat; breast to under tail-coverts whitish or buffish white. The sexes are alike. Occasionally subject to albinism. Wing 80 to 90 mm. The young bird has broad buff streaks on head; large buff spots on mantle, wing-coverts and rump; below, buffish white with blackish tips to feathers.

**Range in Eastern Africa:** Southern Eritrea and Abyssinia as far south as Alghe and west to Charada Forest, Kaffa and Maji.

**Habits:** A common resident of the northern Abyssinian plateau, but little has been recorded of its habits. It is usually associated with the edge of forest or heavy woodland.

**Nest and Eggs:** A neatly made cup nest of moss and rootlets, lined with the latter, placed on a horizontal bough. Eggs probably three, pale greenish or turquoise blue with regular umber or clay coloured flecks and spots and underlying blue-grey blotches; about 22 × 15·5 mm.

**Recorded breeding:** Abyssinia, March to June.

**Call:** Unrecorded.


*Dioptrornis chocolatinus reichenowi* (Neum.).
*Muscicapa reichenowi* Neumann, O.M. p. 10, 1902: Budda, Gimirra, south-western Abyssinia.

**Distinguishing characters:** Differs from the nominate race in being darker. Wing 82 to 85 mm.

**Range in Eastern Africa:** Western Abyssinia from Wallega to Gimirra.

**Habits:** As for the nominate race.

**Recorded breeding:** No records.

## FLYCATCHERS

### 798 BLACK FLYCATCHER. *MELÆNORNIS EDOLIOIDES* (Swainson).

*Melænornis edolioides lugubris* (v. Müll.). **Pl. 96.**
*Muscicapa lugubris* J. W. von Müller, Naumannia, pt. 4, p. 28, 1851: Kolla, north of Gondar, northern Abyssinia.

**Distinguishing characters:** General colour wholly blackish slate. The sexes are alike, but the female is sometimes washed with paler slate colour below. Wing 89 to 105 mm. The young bird has numerous tawny spots.

**General distribution:** Bamingui and Shari Rivers, French Equatorial Africa to Abyssinia, Uganda and Tanganyika Territory.

**Range in Eastern Africa:** Northern and western Abyssinia (as far east as the Didessa and Omo River valleys) to the Sudan, Uganda, western Kenya Colony and northern Tanganyika Territory at Mwanza.

**Habits:** A locally common species in cultivated areas and open woodland. Somewhat crepuscular in habits and rather lethargic in the daytime; it has, however, the habits of a typical Flycatcher. May easily be passed over as a Drongo at first sight. Apparently has some seasonal migrations or movements.

**Nest and Eggs:** The nest is very often in banana plantations among the fruit of the banana palm. Also, and apparently freely, in the old nests of other species of birds which they reline with rootlets. Eggs two or three, white, greenish-white, or rufous-white heavily blotched with rufous and with a few pale purple undermarkings; about $21 \cdot 5 \times 16$ mm.

**Recorded breeding:** Uganda, March to May.

**Call:** Some attempt at a song in breeding season, a few scattered clear notes.

*Melænornis edolioides schistacea* Sharpe.
*Melænornis schistacea* Sharpe, P.Z.S. p. 481, 1895: Darro Mts., Arussi, south-central Abyssinia.

**Distinguishing characters:** Differs from the preceding race in being generally greyer in tone. Wing 94 to 105 mm.

**Range in Eastern Africa:** Abyssinia east of the Omo and Didessa River valleys to Eritrea and northern Kenya Colony at Moyale.

**Habits:** As for the preceding race.

# FLYCATCHERS

**Recorded breeding:** No records.

**Distribution of other races of the species:** West Africa, the nominate race being described from Senegal.

### 799 SOUTH AFRICAN BLACK FLYCATCHER. *MELÆNORNIS PAMMELAINA* (Stanley).

*Melænornis pammelaina pammelaina* (Stan.). **Pl. 96.**
*Sylvia pammelaina* Stanley, in Salt's Abyssinia, App. p. 59, 1814: Mozambique, Portuguese East Africa.

**Distinguishing characters:** Similar in size to the Black Flycatcher, but general colour glossy blue black. The sexes are alike. Wing 97 to 117 mm. The immature bird is much duller below, and has tawny tips to the inner secondaries, wing-coverts and tail. The young bird is also dull with tawny spots on the body.

**General distribution:** Damaraland to Northern Rhodesia, Nyasaland, Portuguese East Africa, the Transvaal and Natal.

**Range in Eastern Africa:** Nyasaland and Portuguese East Africa.

**Habits:** A common species over most of its range, and possibly the commonest Flycatcher of Nyasaland. It is exceedingly easy to confuse with a Drongo, and it has a curious liking for associating with them. It is, however, a much less bold bird and is more lethargic and has a somewhat slower flight, as well as hiding in foliage much more than a Drongo. Both species will, however, come to a grass fire to feed on insects and they are said to eat berries also. In the non-breeding season it collects into small parties. Usually shy and wary, and does not perch on dead boughs like a Flycatcher.

**Nest and Eggs:** Nest usually in the old nests of other birds or in old stumps, or in a hollow at the base of a bough, in each case lined with rootlets. Eggs two or three, whitish, or pale green, heavily mottled with greenish brown or reddish brown, with lilac undermarkings; about 21·5 × 15·5 mm.

**Recorded breeding:** South-eastern Belgian Congo, September to November. Nyasaland, September to December. Rhodesia, September and October.

**Call:** A faint piping, with a few clear notes in the early morning.

*Melænornis pammelaina tropicalis* (Cab.).
*Melanopepla tropicalis* Cabanis, J.f.O p. 241, 1884: Ikanga, Ukamba, southern Kenya Colony.

**Distinguishing characters:** Differs from the nominate race in being washed with violet. Wing 100 to 110 mm.

**Range in Eastern Africa:** Kenya Colony from the Ndoto Mts., to Tanganyika Territory.

**Habits:** As for the nominate race. Moreau noted that when this species and a Drongo were together on the ground the Drongo always held its tail up to a greater degree. The nest is often found among bananas. There is said to be a sweet feeble sort of song, produced apparently with great effort with the head thrown back.

**Recorded breeding:** Tanganyika Territory, September to December.

**800 FOREST FLYCATCHER.   *FRASERIA OCREATA*** (Strickland).
*Fraseria ocreata ocreata* (Strick.).
*Tephrodornis ocreatus* Strickland, P.Z.S. p. 102, 1844: Fernando Po.

**Distinguishing characters:** Above, blackish grey; below, white with dark grey concentric barring; centre of belly white; flanks blackish grey. The sexes are alike. Wing 89 to 102 mm. The immature bird has some tawny spots on the wing-coverts. The young bird has the upperparts spotted with tawny.

**General distribution:** Fernando Po to southern Nigeria, Cameroons, Gabon, southern French Equatorial Africa, northern Belgian Congo and Uganda.

**Range in Eastern Africa:** Western Uganda.

**Habits:** A somewhat Shrike-like rather aberrant Flycatcher, inhabiting the tops of trees in open cultivated land, or along the outskirts of forest. It usually hawks from a perch like a Flycatcher, but is distinctly noisy and is often found in small parties.

**Nest and Eggs:** In West Africa the nest is a collection of skeleton leaves in a shallow hole in a tree or stump. Eggs two, light olive green longitudinally streaked and blotched with umber brown and dark grey; about 21 × 17 mm.

**Recorded breeding:** Southern Cameroons, December.

**Call:** A curious harsh buzzing sound. Also a song of surprising suddenness and sweetness, interspersed with notes mimicking other species.

**Distribution of other races of the species:** Sierra Leone to Gold Coast.

### 801 SILVER-BIRD. *EMPIDORNIS SEMIPARTITUS* (Rüppell).

*Empidornis semipartitus semipartitus* (Rüpp.).
*Muscicapa semipartita* Rüppell, N. Wirbelt. Vög. p. 107, pl. 40, fig. 1, 1840: Gondar, northern Abyssinia.

**Distinguishing characters:** Above, french-grey; below, rich tawny. The sexes are alike. Wing 84 to 89 mm. The young bird is spotted with pale buff and black on the french-grey upperparts; below, buff mottled with black.

**Range in Eastern Africa:** Northern Abyssinia.

**Habits:** As for other races, common in dry thorn country.

**Nest and Eggs:** See under other races. Appears invariably to breed in old nests of Weaver birds, usually *Dinemellia*.

**Recorded breeding:** No records.

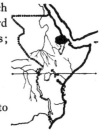

*Empidornis semipartitus kavirondensis* (Neum.). **Pl. 61.**
*Bradyornis kavirondensis* Neumann, J.f.O. p. 257, 1900: Kwa Kissero, Kavirondo, western Kenya Colony.

**Distinguishing characters:** Differs from the nominate race in being paler grey above and having longer wings and tail. Wing, male 97 to 101; female, 91 to 96 mm.

**Range in Eastern Africa:** South-eastern Sudan in the Dongotona and Didinga Mts. area, Uganda, the western half of Kenya Colony, and northern Tanganyika Territory from Mwanza to Kahama and Shinyanga.

**Habits:** As for other races, fairly common in open acacia bush.

**Recorded breeding:** Uganda, June.

*Empidornis semipartitus orleansi* Roths.
*Empidornis semipartitus orleansi* Rothschild, Bull. B.O.C. 43, p. 45, 1922: Rejaf, southern Sudan.

**Distinguishing characters:** Similar in colour to the preceding race, but with shorter wings and tail. Wing, male 82 to 95, female 86 to 93 mm.

**Range in Eastern Africa:** Western, central and southern Sudan (except Dongotona to Didinga) from Darfur to the Sobat and Baro River valleys in western Abyssinia.

**Habits:** A distinctly aberrant Flycatcher with the habits of a *Bradornis*, both hawking from a perch and picking insects from the ground. Locally common in open acacia country where it spends some time running about on the ground. Shy and wary as a rule, sitting on the tops of bushes in full sunlight.

**Nest and Eggs:** A substantial nest of dry grass lined with grass heads and with a roof or dome of thorny twigs. The bird also freely uses the nests of other species, particularly those of the larger Weavers. Eggs two, sea-green or pale green stippled and speckled with reddish brown, slightly glossy and rather long and oval; about 21·5 × 15 mm.

**Recorded breeding:** Darfur, May to August.

**Call:** A sweet rather Thrush-like song.

### 802 BOEHM'S FLYCATCHER. *MYOPORNIS BÖHMI* (Reichenow).

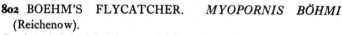

*Bradyornis böhmi* Reichenow, J.f.O. p. 253, 1884: Kakoma, Tabora district, west central Tanganyika Territory.

**Distinguishing characters:** Above, brown streaked with blackish; slight indications of a collar on hind neck; broad buff edges to wing-coverts and inner secondaries; below, white with arrow shaped black spots from chin to breast and down flanks. The sexes are alike. Wing 76 to 82 mm. The young bird is spotted above with buff and blackish; below, white with crescentic black bars, but showing the distinctive arrow-shaped spots.

**General distribution:** Tanganyika Territory to Angola, southern Belgian Congo, Northern Rhodesia and western Nyasaland.

**Range in Eastern Africa:** Western Tanganyika Territory to the Tabora district.

**Habits:** An apparently rare species of which little has been recorded within our area. It is found in open bush or woodland and has the stance and habits of a Flycatcher. Not uncommon and tame in Northern Rhodesia where it has the habits of a rather sluggish *Muscicapa* or *Alseonax* and takes at least part of its food from branches or on the ground. Much of its food consists of black tree-ants (C. M. N. White).

**Nest and Eggs:** Apparently usually in the old nests of Weavers, *Anaplectes* or *Plocepasser*, which are relined with vegetable down. Eggs four, dull pale green with faint but close freckling of pale pinkish brown; about 19 × 14 mm.

**Recorded breeding:** Nyasaland, October and November, also young in March. South-eastern Belgian Congo, September.

**Call:** Very silent birds. The only call recorded is a faint 'siiiii.'

**803** DUSKY BLUE FLYCATCHER. *PEDILORHYNCHUS COMITATUS* (Cassin).
*Pedilorhynchus comitatus stuhlmanni* Reichw.
*Pedilorhynchus stuhlmanni* Reichenow, J.f.O. pp. 34, 132, pl. 1, fig. 1, 1892: Manjonjo, southern Uganda.

**Distinguishing characters:** Above, dark bluish slate; wings and tail black; below, throat white; sides of face and chest dark blue slate; flanks grey; belly paler grey. Differs from Cassin's Grey Flycatcher in the broader bill and very much darker coloration below, and from the Yellow-footed Flycatcher in being larger and having a black bill and feet. The sexes are alike. Wing 61 to 67 mm. The young bird is similar to the adult, but the plumage is more fluffy in texture.

**General distribution:** Eastern Belgian Congo to the Sudan and Uganda.

**Range in Eastern Africa:** South-western Sudan to western and southern Uganda.

**Habits:** Locally not uncommon in true forest wherever it is sufficiently open for it to see and catch insects. It seems to take most of its food on the wing.

**Nest and Eggs:** Nests very often in old nests of Weaver birds which it relines with dry grass. Eggs usually two, four have been recorded, uniform light olive brown or else greenish grey, densely mottled with yellowish brown or reddish brown; about 20 × 13·5 mm., though they seem extraordinarily large for a bird of this size.

**Recorded breeding:** Uganda, June and July.

**Call:** Silent bird, but has a little song in the breeding season, which it utters with spread out tail.

**Distribution of other races of the species:** West Africa, the nominate race being described from Gabon.

## FLYCATCHERS

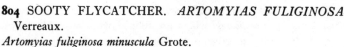

**804 SOOTY FLYCATCHER.** *ARTOMYIAS FULIGINOSA* Verreaux.
*Artomyias fuliginosa minuscula* Grote.
*Artomyias fuliginosa minuscula* Grote, Anz. Orn. Ges. Bayern, No. 7, p. 58, 1922: Beni, Semliki Valley, Belgian Congo.

**Distinguishing characters:** General colour dark sooty brown; below, paler with dark streaks. The sexes are alike. Wing 78 to 85 mm. The young bird has faint light tips to the feathers of the mantle, rump and wing-coverts and smaller, more distinct, streaks below.

**General distribution:** North-eastern Belgian Congo to the Sudan and Uganda.

**Range in Eastern Africa:** Southern Sudan to western and southern Uganda.

**Habits:** Silent birds of open spaces in tall forest, usually seen on a dead bough at a considerable height. True Flycatchers in habits, returning to the same perch after a flight. Often seen in small parties and Woosnam noted that their flight was very Martin-like.

**Nest and Eggs:** Nest, a shallow cup of moss, grass and rootlets at the end of a bough. Eggs two, undescribed.

**Recorded breeding:** Uganda (breeding condition), July and August.

**Call:** Unrecorded.

**Distribution of other races of the species:** West Africa and Angola, the nominate race being described from Gabon.

**805 YELLOW FLYCATCHER.** *CHLOROPETA NATALENSIS* Smith.
*Chloropeta natalensis natalensis* Smith.
*Chloropeta natalensis* A. Smith, Ill. Zool. S. Afr., Aves, pl. 112, fig. 2, 1847: Durban, Natal, South Africa.

**Distinguishing characters:** Above, yellow and olive brown; head olive brown; below, bright lemon yellow. The sexes are alike. Wing 57 to 64 mm. The young bird is washed with buff above and below.

**General distribution:** Tanganyika Territory to Nyasaland, north-eastern Northern Rhodesia, eastern Southern Rhodesia, eastern Transvaal, Natal and Zululand.

**Range in Eastern Africa:** Southern Tanganyika Territory from Ufipa and Njombe to Nyasaland.

**Habits, Nest and Eggs:** See under the following race, usually found in rank grass mixed with bracken or other dense short cover.

**Recorded breeding:** Nyasaland, October to December. Portuguese East Africa, November to January. Southern Rhodesia, October to February.

*Chloropeta natalensis massaica* Fisch. & Reichw. **Pl. 61. Ph. ix.**
*Chloropeta massaica* Fischer & Reichenow, J.f.O. p. 54, 1884: Tschaga, Kilimanjaro, north-eastern Tanganyika Territory.

**Distinguishing characters:** Differs from the nominate race in being darker above. Wing 57 to 66 mm.

**General distribution:** North-eastern Belgian Congo to Abyssinia and Tanganyika Territory.

**Range in Eastern Africa:** Southern Sudan to western and central Abyssinia south of Lake Tana, Uganda, Kenya Colony and northern Tanganyika Territory as far south as the Kasulu district and the Uluguru Mts.

**Habits:** An aberrant Flycatcher with the general habits of a Sedge-Warbler or Eremomela, inhabiting dense nettle-beds, bracken and bush herbage; usually found in swampy wooded localities, occasionally in the open. Locally common and rather noisy for a Flycatcher, often seen sidling up stems and uttering the harsh alarm note of a Sedge-Warbler. Usually found at lower elevations than the Mountain Yellow Flycatcher. Moreau notes that the pose of these birds and the raised feathers of the crown are reminiscent of a Bulbul.

**Nest and Eggs:** A bulky nest of Warbler type, thick and well made of broad grass with a deep cup lined with finer grass and seed heads, usually in the fork of a large plant or small shrub. Eggs two to four, oval, creamy white or pinkish with or without sparse liver coloured spots and streaks, often in a cap at the larger end, and a few faint grey or purplish undermarkings; about 17 × 13 mm.

**Recorded breeding:** Southern Sudan, June. Kenya Colony highlands, April and June. Tanganyika Territory, December to February.

**Call:** Song brief and clear, almost a warble 'twee-twee-twee' followed by a trill. Sings most in the evenings. Alarm note a harsh 'churr.'

**Distribution of other races of the species:** Cameroons to Angola and the Belgian Congo.

**806 MOUNTAIN YELLOW FLYCATCHER.** *CHLOROPETA SIMILIS* Richmond.

*Chloropeta similis* Richmond, Auk. p. 163, 1897: Mt. Kilimanjaro at 10,000 feet, north-eastern Tanganyika Territory.

**Distinguishing characters:** Very similar to the Yellow Flycatcher, but differs in being green above including head, and in having twelve instead of ten tail feathers. The sexes are alike. Wing 55 to 62 mm. The young bird is faintly washed with buff.

**General distribution:** Eastern Belgian Congo as far south as Baraka to the Sudan, Kenya Colony, Tanganyika Territory and northern Nyasaland.

**Range in Eastern Africa:** Southern Sudan to Uganda, Ruanda, Belgian Congo, western Kenya Colony and Tanganyika Territory from the Uluguru Mts. to the south-western areas.

**Habits:** Very much as for the last species and their relationship has been the subject of some discussion. They are now treated as separate species, and in the field they seem to keep to little communities all of one species or the other. Their ranges are similar and though usually this species occurs at higher elevations and more usually in open forest, this is not always the case.

**Nest and Eggs:** Nest a deep thick-walled cup of moss and leaves lined with hair and a few feathers in the fork of a shrub. Eggs two, darkish cream, sparsely spotted with brown and ashy grey; about 18 × 13·5 mm.

**Recorded breeding:** Kinangop, Kenya Colony, August and November. Northern Tanganyika Territory (breeding condition), January.

**Call:** A short musical song, said to be far better than, and easily distinguishable from, that of the Yellow Flycatcher.

**807 LITTLE YELLOW FLYCATCHER.** *CHLOROPETELLA HOLOCHLORA* (Erlanger). **Pl. 61.**

*Erythrocercus holochlorus* Erlanger, O.M. p. 181, 1901: Salole, lower Juba River swamps, southern Italian Somaliland.

**Distinguishing characters:** Above, golden green; below, bright canary yellow. The sexes are alike. Wing 44 to 51 mm. The young bird is slightly greener above and of a more fluffy appearance.

# FLYCATCHERS

**Range in Eastern Africa:** Eastern Kenya Colony and southern Italian Somaliland to eastern Tanganyika Territory as far west as Amani, Nguru and the Uluguru Mts.

**Habits:** Lively dainty little golden-coloured birds hunting among leaves like *Phylloscopi*, and with no resemblance in habits to Flycatchers at all unless it is to the genus *Erythrocercus*. Locally not uncommon, as in the central hill area of Dar-es-Salaam district.

**Nest and Eggs:** Undescribed.

**Recorded breeding:** No records.

**Call:** A plaintive 'zee-zee' or 'tit-tit-chee' and also a jumbled Canary-like song.

## 808 CHESTNUT-CAP FLYCATCHER. *ERYTHROCERCUS MCCALLII* (Cassin).

*Erythrocercus mccallii congicus* O. Grant. **Pl. 61.**
*Erythrocercus congicus* O. Grant, Bull. B.O.C. 19, p. 41, 1907: Irumu, Ituri Forest, eastern Belgian Congo.

**Distinguishing characters:** Forehead and crown chestnut with white streaks; tail chestnut; rest of upperparts olivaceous brown; below, chin to chest tawny; rest of underparts buffish white. The sexes are alike. Wing 50 mm. The young bird has the forehead, crown and throat pale brownish.

**General distribution:** Eastern Belgian Congo and Uganda.

**Range in Eastern Africa:** Western Uganda.

**Habits:** Found in small flocks among the dense foliage of medium sized trees, and are in habits and appearance more like the Long-tailed Tits of Europe than anything else. They have none of the habits of Flycatchers but are active and continuously on the move with a frequent spreading out of their tails.

**Nest and Eggs:** Nest (of West African race) a deep cup or pocket suspended from twigs and stuck on with cobwebs, composed chiefly of dry leaves with a little plant down as a lining. Eggs undescribed.

**Recorded breeding:** No records.

**Call:** A little high-pitched call like that of a White-eye.

**Distribution of other races of the species:** West Africa, the nominate race being described from Gabon.

**809** LIVINGSTONE'S FLYCATCHER. *ERYTHROCERCUS LIVINGSTONEI* Gray.

*Erythrocercus livingstonei livingstonei* Gray. **Pl. 63.**
*Erythrocercus livingstonei* G. R. Gray, in Finsch & Hartlaub, Vög. Ostafr. p. 303, 1870: Zambesi River.

**Distinguishing characters:** Top of head and sides of face pale grey; tail pale chestnut with black penultimate blobs on the six central tail feathers; rest of upperparts yellowish green; below, throat white; rest of underparts lemon yellow. The sexes are alike. Wing 46 to 52 mm. The young bird has the hind crown and nape yellowish green; and the blobs on the tail feathers very small or absent.

**General distribution:** North-eastern Northern Rhodesia to Nyasaland and Portuguese East Africa as far south as Inhambane.

**Range in Eastern Africa:** Portuguese East Africa at Netia to southern Nyasaland at Chikwakwa and Chiromo.

**Habits, Nest and Eggs:** As for the following race.

**Recorded breeding:** Southern Nyasaland, January to March.

*Erythrocercus livingstonei thomsoni* Shell.
*Erythrocercus thomsoni* Shelley, P.Z.S. p. 303, pl. 16, fig. 2, 1882: Rovuma River, Tanganyika Territory–Portuguese East Africa boundary.

**Distinguishing characters:** Differs from the nominate race in having the top of the head yellowish green uniform with the mantle, and the tail has a black sub-terminal bar on the six central feathers. Wing 46 to 52 mm.

**General distribution:** Tanganyika Territory and northern Portuguese East Africa to Nyasaland from Fort Johnston to Kotakota.

**Range in Eastern Africa:** Southern Tanganyika Territory to Portuguese East Africa as far south as the Lurio River and Nyasaland.

**Habits:** Restless tireless little birds hopping and darting about among creepers and foliage in any sort of thick scrub or woodland, and generally in higher thick-foliaged trees. They feed by searching, but occasionally pursue an insect on the wing, all the while fanning

## FLYCATCHERS

their tails. Their habits are much those of the Long-tailed Tits of Europe and they are usually in small parties.

**Nest and Eggs:** The nest is placed some feet up in dense scrub, and is described as dome-shaped with a side entrance, and made of leaves bound together with spiders' webs. It is lined inside up to the entrance level with soft grass fibres. Eggs of this species are given as white, finely but densely spotted with chestnut and rufous with lilac undershading; about 13 × 10·5 mm.

**Recorded breeding:** Nyasaland, December to March.

**Call:** Little sharp calls of 'zert' or 'chip-chip' with the occasional outburst of warbling notes, also a Sunbird-like 'tweet' on the wing. They also make a snapping noise with their bills.

### 810 RUFOUS FLYCATCHER. *STIZORHINA FRASERI* (Strickland).

*Stizorhina fraseri vulpina* Reichw.
*Stizorhina vulpina* Reichenow, J.f.O. p. 125, 1902: Bundeko, Semliki, eastern Belgian Congo.

**Distinguishing characters:** General colour chestnut brown; rump and tail much brighter; head and throat rather duskier. The sexes are alike. Wing 92 to 108 mm. The young bird is duller than the adult; rather mottled on throat and upper breast.

**General distribution:** Eastern Belgian Congo to the Sudan, and Uganda.

**Range in Eastern Africa:** South-western Sudan to western and southern Uganda.

**Habits:** A fairly common species of forest tree-tops, where it feeds by searching among foliage like a Cuckoo-Shrike. It occasionally makes a flight after an insect but in no way hawks from a perch, and its food consists largely of caterpillars. Usually single or in pairs. It may very easily be mistaken for the Red-tailed Ant-Thrush (*Neocossyphus rufus*) but it is a smaller bird and has a shorter tail.

**Nest and Eggs:** Undescribed. Suspected of breeding in holes of trees.

**Recorded breeding:** Uganda (breeding condition), July and August but season irregular.

**Call:** A low pleasing whistle 'sweet-swee-sweet' with an occasional break or hoarseness in it (C. G. Young).

**Distribution of other races of the species:** West Africa, the nominate race being described from Fernando Po.

### 811 SHRIKE-FLYCATCHER. *MEGABYAS FLAMMULATUS* Verreaux.

*Megabyas flammulatus æquatorialis* Jacks.
*Megabias æquatorialis* Jackson, Bull. B.O.C. 15, p. 11, 1904: Entebbe, southern Uganda.

**Distinguishing characters:** Adult male, above, glossy blue black; rump and underside white. The female has the head and mantle earth brown; edges of wing-coverts, flight feathers and rump deep tawny rufous; below, broadly streaked with white and earth brown; lower belly and under tail-coverts pale tawny rufous. Wing 85 to 94 mm. The young bird of both sexes is similar to the adult female, but has some white flecks on the wing-coverts and upper tail-coverts.

**General distribution:** Northern Angola to north-eastern Belgian Congo, the Sudan and Uganda.

**Range in Eastern Africa:** South-western Sudan at Bendere to western and southern Uganda.

**Habits:** A bird of forest clearings and forest edges or of high forest with dense undergrowth. It is a rare bird within our limits. The nominate race from West Africa is described as a striking bird whose musical notes attract attention. The male has an up and down Pipit-like flight while singing. Both sexes have a habit of flying slowly with rapid wing beats as if in very low gear.

**Nest and Eggs:** In West Africa, a small shallow cup nest of rotten wood and fibres bound with cobwebs, high in a tree. Eggs three, bluish or greenish grey with small spots and blotches of umber brown and lilac grey tending to form a zone round the middle of the egg; about 21 × 16 mm.

**Recorded breeding:** Uganda, April to June.

**Call:** Presumably as for the West African race, a musical 'chuick,' a whistling call and song as above. The female has a churring note.

**Distribution of other races of the species:** West Africa, the nominate race being described from Gabon.

## FLYCATCHERS

**812 BLACK-AND-WHITE FLYCATCHER.** *BIAS MUSICUS* (Vieillot).

*Bias musicus femininus* Jacks.
*Bias feminina* Jackson, Bull. B.O.C. 16, p. 87, 1906: Toro, western Uganda.

**Distinguishing characters:** Adult male, whole head to chest, flanks and upperparts black with a bottle green iridescent wash; centre of breast to under tail-coverts, basal half of primaries and sub-terminal tips of rump feathers white. The female is cinnamon brown above; head dusky brown with a bronzy wash; below, whitish and tawny. Wing 86 to 92 mm. The young bird is similar to the adult female.

**Range in Eastern Africa:** South-western Sudan and Uganda.

**Habits, Nest and Eggs:** See under the following race.

**Recorded breeding:** Uganda, March to May, also July and September.

*Bias musicus changamwensis* V. Som.
*Bias musicus changamwensis* Van Someren, Bull. B.O.C., 40, p. 24, 1919: Changamwe, near Mombasa, eastern Kenya Colony.

**Distinguishing characters:** Differs from the preceding race in being smaller. Wing 81 to 86 mm.

**General distribution:** Eastern Kenya Colony to southern Nyasaland and southern Portuguese East Africa.

**Range in Eastern Africa:** Eastern Kenya Colony as far west as Mt. Kenya to eastern Tanganyika Territory as far west as Amani and Kilosa, also Nyasaland and Portuguese East Africa.

**Habits:** Usually found in pairs or small parties in fairly high trees, the males often making slow circling flights round the tree. Curiously grotesque little birds in silhouette. Pugnacious, and will drive away any other bird approaching the nesting site. Normal flight is slow and flapping, but in courtship a 'dithering' effect is produced. Distinctly noisy and noticeable, but not common.

**Nest and Eggs:** The nest is a small open cup of fibres and rootlets bound with cobwebs in a fork of a tree. Eggs two or three, greyish white speckled and blotched with brown and grey; about $19 \times 15$ mm.

**Recorded breeding:** Western Uganda, July and August. Pangani River, Tanganyika Territory, January.

**Call:** A harsh 'churr' uttered on the wing. The song is a sharp startling series of notes of some variety with a whistled 'wit-tu-wit-tu-tui-tu-tu-' also uttered on the wing, at first ascending in scale and then descending.

**Distribution of other races of the species:** West Africa to northern Angola, the nominate race being described from the Portuguese Congo.

### 813 YELLOW-BELLIED FLYCATCHER. *HYLIOTA FLAVIGASTER* Swainson.

*Hyliota flavigaster flavigaster* Swains.
*Hyliota flavigaster* Swainson, Bds. W. Afr. 2, p. 47, 1837: Senegal.

**Distinguishing characters:** Adult male, above, iridescent blue black, with a violet wash; lesser wing-coverts and sometimes base of outer web of innermost secondaries white; below, pale tawny. The female has the upperparts grey. Wing 65 to 78 mm. The young bird is similar to the adult female, but has ashy tips to the feathers of the upperparts.

**General distribution:** Senegal to Abyssinia, Uganda and Kenya Colony.

**Range in Eastern Africa:** Abyssinia to Uganda and Kenya Colony.

**Habits:** As for the southern race, not uncommon, but little is recorded of its habits.

**Nest and Eggs:** See under the following race.

**Recorded breeding:** Belgian Congo, May and June.

*Hyliota flavigaster barbozæ* Hartl. **Pl. 61.**
*Hyliota barbozæ* Hartlaub, J.f.O. p. 329, 1883: Caconda, Angola.

**Distinguishing characters:** Differs from the nominate race in having the white on outer web of the innermost secondaries extending further down the feathers, sometimes to the tip, and breast usually bright tawny. Wing 70 to 75 mm.

**General distribution:** Angola to Tanganyika Territory, Nyasaland and Portuguese East Africa.

**Range in Eastern Africa:** Tanganyika Territory to Portuguese East Africa and Nyasaland.

**Habits:** Inconspicuous birds of tree tops in spite of their coloration, and in habits they have been variously described as similar to Eremomelas, small Shrikes, Sunbirds, *Phylloscopi* or Tits. They are often members of a mixed bird party, and obtain their food by searching leaves, flowers, and branches in open woodland, appearing to be constantly travelling as they feed and never staying long in a tree. According to Zimmer, easily mistaken for an *Anthreptes* in the field.

**Nest and Eggs:** A small open cup nest with a rounded edge made of moss stalks and lichen, at any height in a tree. Eggs two, dull white with a dense band or zone of brown and lilac spots; about 17 × 13 mm.

**Recorded breeding:** South-eastern Belgian Congo, September. Nyasaland and Portuguese East Africa, October to December.

**Call:** A slight two-syllabled whistle freely uttered and rather Tit-like.

**814 SOUTHERN YELLOW-BELLIED FLYCATCHER.** *HYLIOTA AUSTRALIS* Shelley.
*Hyliota australis australis* Shell.
*Hyliota australis* Shelley, Ibis, p. 258, 1882: Umvuli River, Southern Rhodesia.

**Distinguishing characters:** Adult male differs from the Yellow-bellied Flycatcher in being dull velvety black above, not blue black. The female has the upperparts much duller, more brownish grey. Wing 66 to 74 mm. The young bird is similar to the adult female, but is whiter, very pale tawny below.

**General distribution:** Eastern and southern Belgian Congo to Northern Rhodesia, Uganda, Nyasaland, northern Portuguese East Africa and Southern Rhodesia.

**Range in Eastern Africa:** Western Uganda to western Kenya Colony.

**Habits:** A quiet easily-overlooked bird of the upper branches of trees, with Tit-like habits, very similar to those of the Yellow-bellied Flycatcher, moving in pairs or small parties in open woodland.

**Nest and Eggs:** A small neat cup nest covered with lichens and lined with fine grass and feathers. Eggs two or three, whitish grey spotted with yellowish or reddish brown and with purple under-markings; about 17·5 × 12 mm.

**Recorded breeding:** Nyasaland, October to December. Southern Rhodesia, September to December.

**Call:** A series of little squeaking whistles with an occasional loud trilling warble.

*Hyliota australis usambaræ* Scl.
*Hyliota australis usambaræ* Sclater, Bull. B.O.C. 52, p. 104, 1932: Amani, north-eastern Tanganyika Territory.

**Distinguishing characters:** Adult male, smaller than the nominate race; below, darker, more washed with chestnut. Wing 62 to 65 mm.

**Range in Eastern Africa:** Eastern Tanganyika Territory from Amani to the Ruvu River.

**Habits:** As for the southern race.

**Recorded breeding:** No records.


**815 PUFF-BACK FLYCATCHER. *BATIS CAPENSIS* (Linnæus).**
*Batis capensis mixta* (Shell.). **Pl. 62.**
*Pachyprora mixta* Shelley, P.Z.S. p. 359, pl. 40, 1889: Mt. Kilimanjaro, north-eastern Tanganyika Territory.

**Distinguishing characters:** Adult male, above, slate grey, with some black markings on scapulars; a broad black stripe from forehead under eye towards hind neck, including ear-coverts; median wing-coverts and outer edge of innermost secondaries white: below, white with a very broad slightly iridescent black band across chest; under wing-coverts black near edge of wing, remainder white. The female differs from the male in having the mantle and wings olivaceous brown, the latter with chestnut coverts and edges to flight feathers; below, throat and chest chestnut brown, latter with white tips giving a hoary appearance; belly white; flanks chestnut. Wing 58 to 65 mm., tail 31 to 38 mm. The young bird is streaked with buff above. The immature bird is similar to the adult female but has the head olivaceous brown.

**General distribution:** Kenya Colony and Tanganyika Territory to the Masuku Mts. of northern Nyasaland.

**Range in Eastern Africa:** South-eastern Kenya Colony from the Tana River to north-eastern, central and south-western Tanganyika Territory.

**Habits:** A common species of highland forest, mostly keeping to the tree top canopy, but calling attention to itself by a persistent hollow whistling note uttered with great regularity.

**Nest and Eggs:** See under the following race.

**Recorded breeding:** Tanganyika Territory, October to December.

*Batis capensis dimorpha* (Shell.).
*Pachyprora dimorpha* Shelley, Ibis, p. 18, 1893: Milanje Plateau, southern Nyasaland.

**Distinguishing characters:** Adult male, differs from the preceding race in having a longer tail; the female differs in having a longer tail and no hoary appearance on the chest. Wing 59 to 65 mm., tail 37 to 46 mm. The immature male is similar to the adult female, and the immature female has the head olivaceous brown uniform with the mantle.

**General distribution:** North-eastern Northern Rhodesia to western and southern Nyasaland and southern Portuguese East Africa.

**Range in Eastern Africa:** Portuguese East Africa from the Namuli Mts. to the Zambesi River and Nyasaland.

**Habits:** Birds of mountain forest, and common in their restricted habitat. Their habits are those of true Flycatchers, but they also do much searching among leaves and are often members of a mixed bird party. Usually tame and fearless of man.

**Nest and Eggs:** Nest a deep cup of moss lined with fibre and plastered outside with lichens and well concealed on a horizontal bough. Eggs two or three, creamy white or greenish white, with a ring of reddish brown and grey spots, zoned round the largest part of the egg; about 19 × 14 mm., large for the bird.

**Recorded breeding:** Nyasaland, September to January. Portuguese East Africa, October to December.

**Call:** A short rasping squeak, also a curious clicking buzz, but its chief call is a monotonous whistled 'keep-keep' which it maintains for long periods.

*Batis capensis reichenowi* Grote.
*Batis reichenowi* Grote, O.M. p. 162, 1911: Mikindani, south-eastern Tanganyika Territory.

**Distinguishing characters:** Similar to the preceding race, but the male has a rather narrower black chest-band, and the female has the mantle greyer, less olivaceous brown; below, chin and throat

white; breast and belly chestnut brown, washed with grey. Wing 56 to 60 mm., tail 30 to 37 mm.

**Range in Eastern Africa:** South-eastern Tanganyika Territory.

**Habits:** As for other races.

**Recorded breeding:** No records.

**Distribution of other races of the species:** South Africa, the nominate race being described from the Cape of Good Hope.

### 816 RUWENZORI PUFF-BACK FLYCATCHER. *BATIS DIOPS* Jackson.

*Batis diops* Jackson, Bull. B.O.C. 15, p. 38, 1905: Ruwenzori, western Uganda.

**Distinguishing characters:** Above, slate-grey; some black markings on scapulars; white spots on each side of forehead; head and face glossy blue black; tips of median wing-coverts and edge of innermost secondaries white; below, white with a glossy blue black band across chest; under wing-coverts black near edge of wing, remainder white. The sexes are alike. Wing 61 to 65 mm. Juvenile plumage unrecorded.

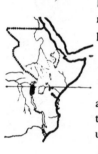

**General distribution:** Kivu district, Belgian Congo to Uganda.

**Range in Eastern Africa:** Mt. Ruwenzori, western Uganda.

**Habits:** Quite common in the forest zone of the Ruwenzori Mts. and on the Virunga Volcanoes, Belgian Congo, where it occurs up to the bamboo zone. Is equally at home among high trees or dense undergrowth. Usually extremely tame.

**Nest and Eggs:** Undescribed.

**Recorded breeding:** Western Uganda (breeding condition), July and August also November.

**Call:** A rather deep 'tcweer-tcweer' repeatedly uttered by both birds of a pair, also a scolding alarm note (Van Someren). There is an audible clicking of wings in flight, especially in the breeding season.

### 817 CHIN-SPOT PUFF-BACK FLYCATCHER. *BATIS MOLITOR* (Hahn & Küster).

*Batis molitor molitor* (Hahn & Küst.). **Ph. ix.**

*Muscicapa molitor* Hahn & Küster, Vög. aus Asien etc. Lief. 20, pl. 2, 1850: Kaffirland, eastern Cape Province, South Africa.

**Distinguishing characters:** Adult male, very similar to the two preceding species, but may be distinguished from them by having

# FLYCATCHERS

the under wing-coverts wholly black and the whole outer web of the outermost tail-feathers white. The female has the mantle grey like the head; throat rich chestnut, surrounded by white; chest band rich chestnut. Wing 55 to 66 mm., tail 40 to 47 mm. The young bird is similar to the adult female, but has buffish speckling on the head, back and chest band. The immature male exactly resembles the adult female; and the immature female has a brown mantle and tawny coloured median wing-coverts.

**General distribution:** The Sudan to Kenya Colony, Angola, Damaraland, Portuguese East Africa and eastern Cape Province.

**Range in Eastern Africa:** South-eastern Sudan to Uganda, Kenya Colony and Tanganyika Territory, but not to the coastal areas.

**Habits:** Locally common among acacias in open woodland or forest edge, keeping mainly to the lower boughs of trees and searching for its food like a Tit. It also hawks from a perch like a true Flycatcher as well as hovering in the air while searching among leaves. The wings make a distinct rustling noise in flight, and in the breeding season there is a wing clapping display flight.

**Nest and Eggs:** A small cup nest of moss fibre and lichens lashed with cobwebs, usually in a small fork and well concealed. Eggs two, rarely three, greyish and greenish white with a girdle of brownish, blackish or sepia spots and with grey undermarkings; about 17 × 13 mm. but larger measurements are recorded.

**Recorded breeding:** Ruwenzori Mts., January. Kenya Colony, December to May or later. Tanganyika Territory, October to December. South-eastern Belgian Congo, September and October. Rhodesia, September to December. Tete, Portuguese East Africa, March and April.

**Call:** A little squeaking 'chi-chirr'; alarm note a harsh 'purk-purk.' Song a flute-like whistle of three notes on a descending scale surprisingly loud for the size of the bird, made by the male from a prominent perch.

*Batis molitor soror* Reichw.
*Batis puella soror* Reichenow, Vög. Afr. 2, p. 485, 1903: Quelimane, Portuguese East Africa.

**Distinguishing characters:** Differs from the nominate race in being paler grey above, and the female has the throat and chest band tawny. Wing 50 to 58 mm., tail 35 to 42 mm.

## FLYCATCHERS

**General distribution:** Kenya Colony to Tanganyika Territory, southern Nyasaland and Portuguese East Africa as far west as the Gorongoza Mts. and as far south as Inhambane.

**Range in Eastern Africa:** Coastal areas of Kenya Colony from Lamu to eastern Tanganyika Territory as far west as Amani and Morogoro, also Portuguese East Africa and Nyasaland, Zanzibar and Mafia Islands.

**Habits:** In all respects as for the nominate race, common in scrub and open woodland, usually in pairs and tame. Moreau believes the females do the courting and Winterbottom mentions this also, but they also do much of the incubation. The nest is very small and neat, camouflaged with lichens. Moreau compares the call to the noise made by a sharply pushed bicycle pump.

**Recorded breeding:** Tanganyika Territory, August to December. Zanzibar Island, January. Mafia Island (breeding condition) July and October. Nyasaland, September to November.

### 818 GREY-HEADED PUFF-BACK FLYCATCHER. *BATIS ORIENTALIS* (Heuglin).

*Batis orientalis orientalis* (Heugl.).
*Platystira orientalis* Heuglin, Orn. Nordost. Afr. 2, p. 449, 1871: Modat Valley, Bogosland, Eritrea.

**Distinguishing characters:** Adult male, similar to the last species, but the male has a more distinct patch of white on the nape, and the female has the throat white, not tawny; chest band chestnut; the white superciliary stripe and patch on hind neck are washed with tawny. Wing 53 to 63 mm. The young bird is speckled with white on the upperparts and has a blackish chest band; the immature male is similar to the adult female; and the immature female has tawny, not white, median wing-coverts, and there is a brownish tinge on the mantle.

**Range in Eastern Africa:** Eritrea, Abyssinia, eastern Sudan from Roseires to the Didinga Mts., the Somalilands and northern Kenya Colony at Moyale.

**Habits:** As these birds are by no means easy to tell in the field from others of the group, and as there has been much confusion in the naming of them, it is not possible to give any distinctive field notes. They inhabit thorn scrub and open woodland and are generally in pairs; are tame and show great affection for one another.

**Nest and Eggs:** See above.

**Recorded breeding:** Eastern Abyssinia, probably February and March.

**Call:** Described by Benson as 'weet-weet-weet-seerr.'

*Batis orientalis chadensis* Alex.
*Batis chadensis* Alexander, Bull. B.O.C. 21, p. 105, 1908: Arrège, west of Lake Chad, north-eastern Nigeria.

**Distinguishing characters:** Adult male, similar to the nominate race though occasionally the top of the head is slightly blackish; the female has the grey mantle washed with brown and an indistinct tawny collar on hind neck. Wing 54 to 59 mm.

**General distribution:** Southern Niger River to Lake Chad and the Sudan.

**Range in Eastern Africa:** Western Sudan as far east as Jebel Melbis and Chakchak.

**Habits:** See remarks under the nominate race.

**Recorded breeding:** Darfur, April and May.

*Batis orientalis lynesi* Grant & Praed.
*Batis orientalis lynesi* C. Grant & Mackworth-Praed, Bull. B.O.C. 60, p. 92, 1940: Sinkat, Red Sea Province, eastern Sudan.

**Distinguishing characters:** Differs from the preceding race in the female having a bright tawny, not chestnut, chest band. Wing 55 to 58 mm.

**Range in Eastern Africa:** Red Sea province of the Sudan.

**Habits:** See remarks under the nominate race.

**Recorded breeding:** No records.

819 PIGMY PUFF-BACK FLYCATCHER. *BATIS PERKEO* Neumann.
*Batis perkeo* Neumann, J.f.O. p. 352, 1907: Darassam, near junction of Webi Mana and Gonele Dorya Rivers, south-eastern Abyssinia.

**Distinguishing characters:** Similar to the Red Sea race of the last species in that the female has a tawny not chestnut chest band, but is smaller in size; the male is smaller than the male of the nominate race of the last species, but is otherwise similar. Wing 47 to 55 mm.

**Range in Eastern Africa:** Eildab, 100 miles south-east of Berbera, British Somaliland to southern Abyssinia, Turkana, the northern Guaso Nyiro, Kenya Colony and southern Italian Somaliland.

**Habits:** See remarks under the last species, this bird appears to be confined to thorn scrub and acacia country. The male has a soaring breeding season flight.

**Nest and Eggs:** Undescribed.

**Recorded breeding:** Mesille, northern Kenya Colony, probably November.

**Call:** A harsh 'churr' and a series of four weak 'peeps,' wings noisy in flight.

## 820 BLACK-HEADED PUFF-BACK FLYCATCHER. *BATIS MINOR* Erlanger.

*Batis minor minor* Erl.
*Batis orientalis minor* Erlanger, O.M. p. 181, 1901: Salole, middle Juba River, southern Italian Somaliland.

**Distinguishing characters:** Adult male, similar to the Grey-headed Puff-back Flycatcher, but the top of the head is black with a slight gloss; and the female has a dark chestnut chest band. Wing 50 to 55 mm. The young bird is blackish above with buffish speckling.

**Range in Eastern Africa:** Juba River Valley, southern Italian Somaliland.

**Habits, Nest and Eggs:** See under other races.

**Recorded breeding:** Probably this species—Italian Somaliland, April.

*Batis minor erlangeri* Neum.
*Batis minor erlangeri* Neumann, J.f.O. p. 353, 1907: Gara Mulata, near Harar, eastern Abyssinia.

**Distinguishing characters:** Similar to the nominate race but larger. The male may be distinguished from the male of the Chinspot and Grey-headed Puff-back Flycatchers by the black or blackish, not grey or dark grey, top of head. Wing 54 to 64 mm.

**General distribution:** Cameroons and northern Angola to Eritrea and British Somaliland.

**Range in Eastern Africa:** Southern Sudan to Eritrea, Abyssinia, Uganda and British Somaliland.

**Habits:** As for the coastal race, generally in pairs and in any kind of woodland, but more particularly associated with evergreen scrub.

**Recorded breeding:** Uganda, June to November.

*Batis minor suahelica* Neum. **Pl. 62.**
*Batis minor suahelicus* Neumann, J.f.O. p. 353, 1907: Kahé, near Moshi, north-eastern Tanganyika Territory.

**Distinguishing characters:** Differs from the nominate race in having the head a greyer black. Wing 52 to 59 mm.

**Range in Eastern Africa:** Coastal areas of Kenya Colony and Tanganyika Territory from Lamu to Dar-es-Salaam and inland as far as North Paré Mts. and Kilosa, also Manda Island.

**Habits:** A common species of the coastal belt in any situation. Restless little birds searching foliage like Tits or catching insects like Flycatchers, moving from tree to tree with undulating flight.

**Nest and Eggs:** Nest, a small neat cup of fibre and lichen in the fork of a tree, bound together with cobwebs. Eggs pale bluish or greenish grey with dark grey, brown or umber spots and blotches tending to form a zone at the larger end; about $15 \cdot 5 \times 12 \cdot 5$ mm.

**Recorded breeding:** Manda Island, June. Mombasa, October. Tanganyika Territory, June.

**Call:** A few long piping notes clear and ringing with a cry of 'ploop ploop.' Alarm note a sort of buzz.

**821  ZULULAND FLYCATCHER.  *BATIS FRATRUM*** (Shelley). **Pl. 62.**
*Pachyprora fratrum* Shelley, Ibis, p. 522, 1900: Zululand.

**Distinguishing characters:** Adult male, above, slate grey; mantle slightly olivaceous; broad stripe from lores, through eye and ear-coverts to down side of neck; wing shoulders and outer scapulars black; median wing-coverts white; flight feathers narrowly edged with white; tail black tipped with white, outer web of outermost tail feathers mainly white; below, white; broad band of tawny across chest spreading down flanks; under wing-coverts black near edge of wing, remainder white. The female has the grey head suffused with brown; mantle and wing shoulders olivaceous brown; median wing-coverts and edges of flight feathers tawny; much less black on face;

throat as well as chest and flanks tawny. The female is much paler than the female of the Nyasaland race of the Puff-back Flycatcher; has much less black on side of face; throat, chest and flanks tawny with a rather frosted appearance, not chestnut brown, and there is no whitish division between the throat and chest. Wing 58 to 62 mm. The immature male has a tawny throat and median wing-coverts. The young bird differs from the adult female in having a tawny stripe over the eye; lores to ear-coverts brownish.

**General distribution:** Northern Portuguese East Africa to Nyasaland and Zululand.

**Range in Eastern Africa:** Portuguese East Africa and Nyasaland from the Lurio River to the Zambesi River.

**Habits:** A species which has recently been discovered within our limits. It is locally common in the coastal jungle and scrub of Portuguese East Africa, and is said to have the same habits as others of the genus but to be somewhat more shy and retiring. It is frequently a member of a mixed bird party.

**Nest and Eggs:** Nest, a cup of fine grass stems ornamented with leaves and bound with cobwebs in the fork of a tree or bush. Eggs two, white and glossy, spotted and blotched with brown, chocolate and grey, which tends to form a zone at the larger end; about $17 \cdot 5 \times 12 \cdot 5$ mm.

**Recorded breeding:** South Africa, November.

**Call:** Alarm note is a jerky rasping 'prer-rer-rert.'

### 822 WATTLE-EYE. *PLATYSTEIRA CYANEA* (Müller).
*Platysteira cyanea nyanzæ* Neum. **Pl. 62.**
*Platysteira cyanea nyanzæ* Neumann, J.f.O. p. 210, 1905: Bukoba, north-western Tanganyika Territory.

**Distinguishing characters:** Adult male, above, glossy blue black; sides of face glossy blue black; median wing-coverts and outer edge of inner secondaries white; narrow white edge and tip to outermost tail feather; wattle over eye vermilion; below, white; chest band glossy blue black. The female has the head and mantle darker grey glossed with blue black; below, chin white; throat to chest rich chestnut. Wing 63 to 70 mm. The young bird is speckled with buff above; median wing-coverts and outer edge of inner secondaries buff or buffish white; below, white, slightly buffish on chest. The immature male is brownish grey above; eye-wattle small; median wing-coverts and outer edge of inner secondaries buff; below, sides of neck and broken chest band tawny; whole outer web of outermost tail feathers white. The immature female has the chin white and the throat to chest tawny.

**General distribution:** Northern Belgian Congo to southern French Equatorial Africa, the Sudan, Uganda and Kenya Colony.

**Range in Eastern Africa:** Southern Sudan, Uganda and north-western Tanganyika Territory to western Kenya Colony, also Kome and Ukerewe Islands, Lake Victoria.

**Habits:** Would hardly be classed as a Flycatcher in the field, its habits being more those of a Tit or a *Phylloscopus*. Common in forest among secondary growth or in forest strips, feeding by searching among branches. Restless little birds, with some seasonal migrational movements, usually in noisy little parties. Most species of this genus and the next flick their wings audibly in flight.

**Nest and Eggs:** A small cup nest of fine grass, fibre and lichen bound with spiders' webs in the fork of a tree or bush. Eggs two, cream or pale olive with brownish spots and purple undermarkings chiefly at larger end; about $18 \cdot 5 \times 14$ mm. (West African measurements).

**Recorded breeding:** South-eastern Sudan, July. Uganda, April to July, also November and December.

**Call:** A sort of song of three or four descending clear whistles with some interval between them. Also a cheery whistling cry and a buzzing note from the female. Much snapping of mandibles in breeding season and a low mating song of rather churring nature by the male.

*Platysteira cyanea æthiopica* Neum.
*Platysteira cyanea æthiopica* Neumann, J.f.O. p. 210, 1905: Banka, Malo, south-western Abyssinia.

**Distinguishing characters:** Very similar to the preceding race but smaller and male rather greyer on mantle. Wing 58 to 62 mm.

**Range in Eastern Africa:** Western areas of Abyssinia to Lake Tana and Lake Zwai.

**Habits:** As for the preceding race.

**Recorded breeding:** No records.

**Distribution of other races of the species:** West Africa, the nominate race being described from Senegal.

## 823 BLACK-THROATED WATTLE-EYE. *PLATYSTEIRA PELTATA* Sundevall.

*Platysteira peltata peltata* Sund. **Pl. 62.**

Platystira peltata Sundevall, Œfv. Vet Akad. Förh. Stockholm, 7, p. 105, 1850: Umlalazi River, Zululand.

**Distinguishing characters:** Adult male, very similar to the Wattle-eye, but has a much narrower chest band and no white on the upperside of the wing; mantle rather greyer than head and both washed with green; wattle over eye scarlet. The female has the whole head and throat to chest glossy blue-black; chin white. Wing 61 to 68 mm. The young and immature birds differ from those of the Wattle-eye in having very narrow tawny edges to the median wing-coverts and inner secondaries, not a broad tawny patch, and in having a narrow white edge and tip to the outermost tail feather, not outer web wholly white.

**General distribution:** Kenya Colony to Tanganyika Territory, Nyasaland, Portuguese East Africa, eastern Southern Rhodesia, Natal and Zululand.

**Range in Eastern Africa:** Eastern Kenya Colony as far west as Thika and south of the Tana River, central and eastern Tanganyika Territory as far west as Mwanza to the Zambesi River; also Ukerewe Island, south-east Lake Victoria, and Mafia Island.

**Habits:** Birds of forest outskirts and strips, and of riverside evergreen scrub, almost always near water. Silent little birds, but tame and fearless, searching leaves and branches for food. The red eye-wattle is quite noticeable.

**Nest and Eggs:** Nest a very small cup of grass and fibres lined with vegetable down, usually in a small fork to which it is attached by cobwebs. In Rhodesia the only nests recorded were unlined. The male does at least some of the incubation. Eggs two, pale greenish white heavily mottled with various shades of brown and with slaty blue undermarkings; about 17 × 13 mm.

**Recorded breeding:** Kenya Colony, March to June and December to February. Tanganyika Territory, December and January, also (breeding condition) August. Nyasaland, September to February. Rhodesia, October and November.

**Call:** A loud 'aing' or 'zee' from the male, and a harsh low rattling alarm note. A little tinkling song 'er-er-fee-er-er-fee-fee'.

Plate 62

Chestnut Wattle-eye (p. 211)
Male    Female
Yellow-bellied Wattle-eye (p. 213)
Male    Female
Black-headed Puff-back Flycatcher (p. 207)
Male    Female
Wattle-eye (p. 203)
Male    Female

Jameson's Wattle-eye (p. 212)
Young    Adult
Puff-back Flycatcher (p. 200)
Female    Male
Zululand Flycatcher (p. 207)
Female    Male
Black-throated Wattle-eye (p. 210)
Female    Male

Plate 63

**Black-headed Paradise Flycatcher**
*Tchitrea nigriceps emini* (p. 224)

**Grey-headed Paradise Flycatcher**
*Tchitrea plumbeiceps violacea* (p. 223)

**Paradise Flycatcher** (p. 219;
*Tchitrea viridis ferreti*
Male        Female

**Livingstone's Flycatcher** (p. 194)
*Erythrocercus livingstonei livingstonei*

**Dusky Crested Flycatcher** (p. 219)
*Trochocercus nigromitratus toroensis*

**Blue-headed Crested Flycatcher**
*Trochocercus nitens nitens* (p. 218)
Male        Female

**Blue Flycatcher** (p. 214)
*Erannornis longicauda teresita*

**White-tailed Crested Flycatcher** (p. 217)
*Trochocercus albonotatus albonotatus*

**Crested Flycatcher** (p. 216)
*Trochocercus cyanomelas vivax*
Male        Female

# FLYCATCHERS

*Platysteira peltata mentalis* Boc.
*Platystira mentalis* Bocage, Jorn. Lisbon, p. 256, 1878: Caconda, Angola.

**Distinguishing characters:** Differs from the nominate race in having the whole upperside glossy deep blue black, not washed with green. Wing 64 to 70 mm.

**General distribution:** Uganda, Kenya Colony and Tanganyika Territory to northern Nyasaland, southern Belgian Congo, Northern Rhodesia and Angola.

**Range in Eastern Africa:** Uganda and western Kenya Colony, Tanganyika Territory west of Emin Pasha Gulf south to the Songwe River on the Nyasaland boundary.

**Habits:** As for the nominate race. Hale-Carpenter says the call sounds like 'how are you? I'm pretty well, thanks.'

**Recorded breeding:** Northern Rhodesia, September to December.

## 824 CHESTNUT WATTLE-EYE. *DYAPHOROPHYIA CASTANEA* (Fraser).

*Dyaphorophyia castanea castanea* (Fraser). **Pl. 62.**
*Platysteira castanea* Fraser, P.Z.S. for 1842, p. 141, 1843: Clarence, Fernando Po.

**Distinguishing characters:** Adult male, head, sides of face, mantle, wings, tail and broad band across chest glossy blue black; remainder of plumage white; tail very short; feathers of flanks and rump very fluffy and silky white; eye wattle purplish black or mauve. The female has the head and sides of face blue grey; rest of plumage bright chestnut except chest to belly which is white; tail black. Wing 55 to 63 mm. The young bird has the head mixed grey and pale chestnut; rest of upperparts pale chestnut; tail blackish; below, white with a broad band of pale grey and chestnut across chest, more or less barred with blackish.

**General distribution:** Fernando Po and Southern Nigeria to the Sudan, Kenya Colony and northern Angola.

**Range in Eastern Africa:** Southern Sudan to Uganda and western Kenya Colony.

**Habits:** Best described by Sir Frederick Jackson as curious restless little birds like animated powder-puffs with heads in constant movement on the look out for food. They are mainly a forest species, silent and somewhat shy, with a curious habit of diving vertically into the undergrowth for no apparent reason. They make a sharp clicking noise with their wings in flight. Often join in mixed bird parties.

**Nest and Eggs:** Nest a cup on the branch of a shrub in dense shade. Eggs two, bluish white with brownish spots and grey undermarkings, mainly at larger end; about $18 \times 13 \cdot 5$ mm.

**Recorded breeding:** Uganda, April to July. South-eastern Belgian Congo, October.

**Call:** Rather low, harsh and complaining, with a clear shrill alarm whistle, and the wing clicking previously referred to. There is also a whippy swishing sound in the breeding season, but it is not clear how it is produced.

**Distribution of other races of the species:** Sierra Leone to Togoland.

**825** JAMESON'S WATTLE-EYE. *DYAPHOROPHYIA JAMESONI* Sharpe. **Pl. 62.**

*Diaphorophyia jamesoni* Sharpe, in Jameson's Story of the Rear
    Column, p. 414, 1890: Yambuya, Aruwimi River, north-eastern Belgian Congo.

**Distinguishing characters:** Adult male, upperside and below from chin to chest glossy bottle green; broad dark chestnut patch from below eye down side of neck; breast and belly white; eye-wattle turquoise blue. The female is dark grey above washed with glossy bottle green. Wing 52 to 58 mm. The young bird is similar to the adult female but has the chin to chest chestnut.

**General distribution:** Northern and north-eastern Belgian Congo to Kenya Colony.

**Range in Eastern Africa:** Lotti Forest area of southern Sudan to Uganda and Nandi, western Kenya Colony.

**Habits:** Common in true equatorial forest in secondary growth, and generally found in the darkest places among tall undergrowth. Does not seem to ascend into trees at all.

**Nest and Eggs:** Undescribed.

**Recorded breeding:** Southern Sudan, January. Uganda (breeding condition), July.

**Call:** Little recorded; in the breeding season, a whippy swishing sound similar to that made by the Chestnut Wattle-eye.

# FLYCATCHERS

## 826 YELLOW-BELLIED WATTLE-EYE. *DYAPHOROPHYIA CONCRETA* (Hartlaub).

*Dyaphorophyia concreta graueri* Hart.
*Diaphorophyia graueri* Hartert, Bull. B.O.C. 23, p. 7, 1908: Forest, 90 km. west of Lake Albert Edward, eastern Belgian Congo.

**Distinguishing characters:** Adult male, above, olive green; head darker; below, yellow; eye-wattle green. The female is above dull olive green; below, chin to chest rich chestnut; breast to belly ochraceous buff; flanks pale chestnut. Wing 61 to 63 mm. The young bird is grey above with a slight bottle green wash; buffish white tips to secondaries; pale yellow below.

**General distribution:** Eastern Belgian Congo to Uganda.

**Range in Eastern Africa:** Bwamba area, western Uganda.

**Habits:** A rare and little known forest species.

**Nest and Eggs:** Undescribed.

**Recorded breeding:** No records.

**Call:** Several whistles followed by guttural clucks.

*Dyaphorophyia concreta silvæ* Hart. & V. Som.
*Diaphorophyia graueri silvæ* Hartert & Van Someren, Bull. B.O.C. 43, p. 79, 1923: Silwa, Kaimosi, western Kenya Colony.

**Distinguishing characters:** Adult male differs from the Belgian Congo race in being light olive green above; nape washed with grey. The female has the chin to chest deep saffron. Wing 63 mm.

**Range in Eastern Africa:** Kakemega Forest, western Kenya Colony.

**Habits:** No information.

**Recorded breeding:** No records.

*Dyaphorophyia concreta kungwensis* Moreau. **Pl. 62.**
*Dyaphorophyia ansorgei kungwensis* Moreau, Bull. B.O.C. 61, p. 25, 1941: Forest above Ujamba, Kungwe-Mahare Mts., Kigoma district, western Tanganyika Territory.

**Distinguishing characters:** Adult male, differs from the Belgian Congo race in being dark slate above with a greenish-blue gloss. The female is paler slate above with a very slight greenish-blue gloss, and with chin to chest deep saffron. Wing 58 to 61 mm. The young bird is similar to the adult female above and to the adult male below; edges and tips of innermost secondaries buffish white.

**Range in Eastern Africa:** Kungwe-Mahare Mts., western Tanganyika Territory.

**Habits:** No information.

**Recorded breeding:** Kungwe-Mahare Mts., Tanganyika Territory (breeding condition), August.

**Distribution of other races of the species:** West Africa to Angola, the nominate race being described from the Gold Coast.

**827 BLUE FLYCATCHER.** *ERANNORNIS LONGICAUDA* (Swainson).
*Erannornis longicauda teresita* (Ant.). **Pl. 63.**
*Elminia teresita* Antinori, Cat. Descr. Ucc., p. 50, 1864: Djur, Bahr-el-Ghazal, southern Sudan.

**Distinguishing characters:** Above, cærulean blue; below, chin to chest whitish and cærulean blue; breast to belly white; tail long and graduated. The sexes are alike. Wing 63 to 71 mm. The young bird is much paler than the adult and has buff edges to the feathers of the upperparts.

**General distribution:** Cameroons to the Sudan and Kenya Colony.

**Range in Eastern Africa:** Southern Sudan to Uganda, western Kenya Colony and north-western Tanganyika Territory.

**Habits:** A lovely little bird of forest clearings and woodland, tame and often common locally. Usually seen playing round low tree tops or among bushes continually flirting its tail and turning its body. It feeds mainly by searching, but also makes flights at insects and is an adept at this method if necessary. Quite delightful little birds, and are freely found in gardens.

**Nest and Eggs:** The nest is an exquisite rather shallow cup of cobwebs and grey lichen lined with hair-like fibres and an odd feather or two, often moulded on to a branch at a fork. Eggs two, oval, greyish white with a wide zone of greenish or brownish grey and lilac mottlings; about 16 × 12 mm.

**Recorded breeding:** Uganda, March to August, also October and November. North-western Tanganyika Territory, October.

**Call:** Various chirping cries, also a fine twittering though disjointed song.

# FLYCATCHERS

*Erannornis longicauda kivuensis* (Grote).
*Elminia albicauda kivuensis* Grote, J.f.O. p. 485, 1922: Kwidschwi Island, Lake Kivu, Belgian Congo.

**Distinguishing characters:** Differs from the preceding race in having the mantle duller, more grey blue; head streaked with pale blue and two outermost tail feathers almost wholly white. Wing 64 to 71 mm.

**General distribution:** Southern and eastern Belgian Congo as far north as Lake Kivu to Tanganyika Territory, north-eastern Northern Rhodesia and Nyasaland.

**Range in Eastern Africa:** Western to north-eastern Tanganyika Territory.

**Habits:** Very much those of the preceding race, not uncommon in bush country or cultivated land at higher elevations. Generally among higher trees in pairs or small parties, and always tame, particularly noticeable near water. The tail is fanned perpendicularly while the wings are spread downward.

**Recorded breeding:** Nyasaland, October to December.

**Distribution of other races of the species:** Gambia to Nigeria, the nominate race being described from Sierra Leone.

## 828 CRESTED FLYCATCHER. *TROCHOCERCUS CYANOMELAS* (Vieillot).

*Trochocercus cyanomelas bivittatus* Reichw.
*Trochocercus bivittatus* Reichenow, O.C. p. 108, 1879: Muniumi, Tana River, eastern Kenya Colony.

**Distinguishing characters:** Head crested; adult male, head and crest and chin to chest glossy blue black; rest of upperparts bluish slate; narrow white bands and a white patch on wing; below, breast to belly white. The female is duller and greyer, lacks the white patch on the wing, and has the chin to chest mottled blackish and white. Wing 63 to 70 mm. The young bird is similar to the adult female but has buffish tips and edges to the wing-coverts and flight feathers.

**General distribution:** Italian Somaliland to Kenya Colony, Nyasaland and eastern Southern Rhodesia.

**Range in Eastern Africa:** Southern Italian Somaliland from the Juba River and eastern Kenya Colony to Nyasaland and the Zambesi River, also Zanzibar Island.

**Habits:** A rather retiring restless species inhabiting dense undergrowth or other thick cover, usually at lower levels. Shy excitable birds found in pairs and small parties, most noticeable when they dash out of cover at a passing insect in true Flycatcher fashion. On Mt. Kilimanjaro it occurs up to 5,000 feet in fringing forest.

**Nest and Eggs:** A small cup nest of moss and lichens bound with cobwebs, placed in the fork of a shrub. Eggs two, pale cream or pinkish white blotched and speckled with purplish brown or greenish olive brown often in the form of a zone; about 16 × 13 mm.

**Recorded breeding:** Southern Italian Somaliland, May and June. Rhodesia, June to January. Zanzibar Island, probably June to November but might breed in any month. Nyasaland, December and February.

**Call:** A rasping 'zi-zerdt-zerdt.' The song is a most curious high pitched 'kwew-ew-ew-ew' followed by four loud clicks (Moreau).

*Trochocercus cyanomelas vivax* Neave. **Pl. 63.**
*Trochocercus vivax* Neave, Ann. Mag. N.H. (8), 4, p. 129, 1909: Bunkeya, south-eastern Belgian Congo.

**Distinguishing characters:** Adult male, differs from the preceding race in lacking the white patch on the wing. The female is not distinguishable from that of the last race. Wing 68 to 71 mm.

**General distribution:** Uganda, Tanganyika Territory and the southern Belgian Congo.

**Range in Eastern Africa:** South-western Uganda at Mubendi to western Tanganyika Territory as far south as Karema.

**Habits:** As for the preceding race.

**Recorded breeding:** No records.

*Trochocercus cyanomelas kikuyuensis* V. Som.
*Trochocercus bivittatus kikuyuensis* Van Someren, J.E.A. & U. N.Hist. Soc. No. 37, p. 194, 1931: Kyambu Forest, south-central Kenya Colony.

**Distinguishing characters:** Similar to the Tana River race in having white in the wing, but is larger. Wing 68 to 75 mm.

**Range in Eastern Africa:** Central Kenya Colony from Mt. Kenya to Ngong.

**Habits:** Do not differ from those of other races, but this race occurs at slightly higher elevations.

**Recorded breeding:** No records.

**Distribution of other races of the species:** South Africa, the nominate race being described from Knysna, Cape Province.

### 829 WHITE-TAILED CRESTED FLYCATCHER. *TROCHOCERCUS ALBONOTATUS* Sharpe.

*Trochocercus albonotatus albonotatus* Sharpe. **Pl. 63.**
*Trochocercus albonotatus* Sharpe, Ibis, p. 121, 1891: Mt. Elgon.

**Distinguishing characters:** Head, crest and throat black; rest of upperparts dark grey; below, centre of breast to belly white; flanks dark grey; apical third of all except central tail feathers white. The sexes are alike. Wing 59 to 70 mm. The young bird has the sides of the head and throat dark grey.

**General distribution:** Uganda, Kenya Colony, Tanganyika Territory, north-eastern Northern Rhodesia, Nyasaland and Portuguese East Africa.

**Range in Eastern Africa:** Uganda, western Kenya Colony, Tanganyika Territory, Nyasaland and south-western Portuguese East Africa.

**Habits:** Tame, common and attractive little birds of upland or highland forests, flitting about with restless jerky flight and much fanwise spreading of the tail. They are very local and remain in a very small radius, so that they are easily overlooked except for their incessant agitated cries. They do, however, wander down from the mountain forests after the breeding season. Their movements are very rapid as they search foliage or dash out after passing insects. In courtship the male hops round the female with tail raised vertically and wings drooped.

**Nest and Eggs:** The nest is a delightful little structure of closely woven green moss bound with cobwebs and lined with lichen built round a fork of a low tree or shrub. Eggs two, white or pale buff with brown or olive green spotting and bluish grey undermarkings often in the form of a zone but both ground colour and markings are variable; about 17 × 12 mm.

**Recorded breeding:** Tanganyika Territory, September to

February, mostly October and November. Nyasaland, October and November, also February and young in March.

**Call:** A high thin twitter. The song consists of a series of thin sharp notes uttered with stuttering eagerness, rather like a twittering whistle.

**Distribution of other races of the species:** West Africa to Southern Rhodesia.

**830 BLUE-HEADED CRESTED FLYCATCHER.** *TROCHO-CERCUS NITENS* Cassin.
*Trochocercus nitens nitens* Cass. **Pl. 63.**
*Trochocercus nitens* Cassin, Proc. Acad. Sci. Philad. p. 50, 1859: Camma River, Gabon.

**Distinguishing characters:** Adult male, above, chin to chest glossy blue black; rest of underparts grey. The female is dark grey above; top of head and crest glossy blue black; below, paler grey. Wing 61 to 68 mm. The young bird is similar to the adult female but has less glossy blue black on the head.

**General distribution:** Southern Nigeria to Uganda and northern Angola.

**Range in Eastern Africa:** Western and southern Uganda.

**Habits:** A shy but attractive bird of dense forest or bush, with the same habit as the White-tailed Crested Flycatcher of continually spreading the tail. Little recorded of its habits except that in the breeding season it has been seen to dart round and round a bush with spread tail making a rattling sound.

**Nest and Eggs:** Nest, a cup of bark fibre and cobwebs firmly attached to the fork of a small tree and lined with fine tendrils and fibres. Eggs two, cream colour with brownish or chestnut spots and mauve undermarkings; about 20 × 13 mm.

**Recorded breeding:** Lagos, Nigeria (breeding condition), February. Uganda, July and August.

**Call:** A shrill sharp call of five or six notes 'tir-i-ti-ti-turt.' Also a harsher alarm cry, and a short bubbling song.

**Distribution of other races of the species:** Sierra Leone to Gold Coast.

## 831 DUSKY CRESTED FLYCATCHER. *TROCHOCERCUS NIGROMITRATUS* (Reichenow).

*Trochocercus nigromitratus toroensis* Jacks. **Pl. 63.**

*Trochocercus toroensis* Jackson, Bull. B.O.C. 19, p. 20, 1906: Kibirau, Toro, western Uganda.

**Distinguishing characters:** Adult male, general colour wholly pale slaty blue; top of head and crest dull black; wings and tail blackish. The female has a duller black head and crest. Wing 54 to 63 mm. The young bird is generally duller in plumage than the adult.

**General distribution:** North-eastern Belgian Congo to Uganda.

**Range in Eastern Africa:** Western and southern Uganda.

**Habits:** Another dense forest Flycatcher of little known habits, but presumably differing little from others of the group.

**Nest and Eggs:** A small cup of moss and cobwebs. Eggs two, dull white with a grey zonal ring and a few faint brown spots; about 17 × 12 mm.

**Recorded breeding:** Uganda, September.

**Call:** Unrecorded.

**Distribution of other races of the species:** Cameroons, the habitat of the nominate race.

## 832 PARADISE FLYCATCHER. *TCHITREA VIRIDIS* (Müller).

*Tchitrea viridis ferreti* Guér. **Pl. 63. Ph. ix.**

*Tchitrea ferreti* Guérin, Rev. Zool. p. 162, 1843: Northern Abyssinia.

**Distinguishing characters:** Head and neck all round to chest glossy violet blue-black; rest of upperparts varying from chestnut to white, rarely glossy blue black; flight feathers black; wing-coverts and edges of secondaries mainly white; two central tail feathers considerably elongated and varying chestnut and white; below, breast to belly dark grey, grey and white or white; under tail-coverts usually pale chestnut to white. The female has less white on the wing; two central tail feathers not considerably elongated; mantle, inner secondaries and tail always dark chestnut; and under tail and wing-coverts grey or washed with grey. Wing 73 to 91 mm. The young bird is similar to the adult female, but has a rather less glossy head, shorter crest and paler under wing-coverts.

**General distribution:** Eritrea and British Somaliland to eastern and south-western Belgian Congo, north of lat. 6° S. and Tanganyika Territory.

**Range in Eastern Africa:** Eritrea and British Somaliland to the Sudan except south-western area, Uganda except area north end of Lake Victoria, Kenya Colony and Tanganyika Territory.

**Habits:** Owing to the similarity of this species to the Red-winged Paradise Flycatcher, the females and young being almost impossible to distinguish in the field, it is not easy to attribute any one characteristic habit to either species. It seems probable however, that this species is the one which occurs most freely in open country, and in strips of bush along rivers away from forests. They are graceful and delightful birds at all times, and in the breeding season particularly tame. They catch much of their food on the wing, and have all the habits of true Flycatchers, they are also fearless and will attack any large bird of prey that appears. The food includes quite large insects.

**Nest and Eggs:** A small neat cup nest usually in the fork of a low tree, made of twigs, moss, lichens or grass fibre according to locality, lined with hair or fine grass and often decorated outside with lichen. The male assists in the incubation. Eggs two or three, white or creamy white with a well marked zone of reddish or reddish brown and purplish grey spots and blotches; about 19 × 14 mm.

**Recorded breeding:** Abyssinia, June. Uganda, March to July, also October to January. Kenya Colony, November to June, multi-brooded, and probably other months as well. Tanganyika Territory, April also September to December.

**Call:** A guttural rasping cry also a hissing note. The song is cheery and rippling 'whee-wheeo whit-whit.'

*Tchitrea viridis speciosa* (Cass.).
*Muscipeta speciosa* Cassin, Proc. Acad. Sci. Philad. for 1859, p. 48, 1860: Camma River, Gabon.

**Distinguishing characters:** Adult male, differs from the Abyssinian race in having the head and neck more greenish than violet; mantle usually darker chestnut; under tail-coverts usually dark grey, sometimes chestnut or white. Wing 71 to 91 mm.

**General distribution:** Ivory Coast to Nigeria, Gabon, French Equatorial Africa and the Sudan.

**Range in Eastern Africa:** Western Sudan.

**Habits:** Presumably as for the last race. A breeding visitor to the Darfur area between April and October.

**Recorded breeding:** Cameroons, all months of the year. Darfur, June and July.

*Tchitrea viridis restricta* Salom.
*Tchitrea viridis restricta* Salomonsen, Bull. B.O.C. 54, p. 48, 1933: Nkose Island, southernmost of Sesse Islands, Lake Victoria, Uganda.

**Distinguishing characters:** Adult male, differs from the Abyssinian race in having chestnut under tail-coverts and in being generally darker, elongated central tail feathers varying from chestnut to white and top of head sometimes white; breast to belly often a mixture of chestnut and violet blue-black or chestnut and grey. Wing 80 to 89 mm.

**Range in Eastern Africa:** North end of Lake Victoria and Nkose Island.

**Habits:** As for other Paradise Flycatchers.

**Recorded breeding:** No records.

**Distribution of other races of the species:** West Africa and Arabia, the nominate race being described from Senegal.

*Note:* This species and the next are closely related and some authors still treat them as conspecific. We now know, however, that there is an overlap in distribution in Uganda and Tanganyika Territory. Close study of the birds in the field would be of great value.

**833** RED-WINGED PARADISE FLYCATCHER. *TCHITREA SUAHELICA* (Reichenow).
*Tchitrea suahelica suahelica* (Reichw.).
*Terpsiphone perspicillata suahelica* Reichenow, Werther Mittl. Hochl. D. Ostafr. p. 275, 1898: Mpondi River, Usandawe, Tanganyika Territory.

**Distinguishing characters:** Adult male, head, chin and throat glossy violet blue-black; rest of upperparts including tail, tawny chestnut; flight feathers dusky with tawny edges; two central tail feathers very long; below, chest to belly grey; lower belly, under tail-coverts and under wing-coverts white. Rarely there is a little white on the upper wing-coverts and edges of flight feathers. The female has the top of the head violet blue-black; central tail feathers

not considerably elongated. Wing 75 to 83 mm. The young bird is similar to the adult female but is more deep tawny than chestnut above and has the top of the head dark grey with a very faint gloss.

**Range in Eastern Africa:** Eastern Tanganyika Territory from south-eastern Mt. Kilimanjaro and the Usambara Mts. to Kilosa, Dar-es-Salaam, Kisiju, forty miles south of Dar-es-Salaam, Usandawe, Iringa and Njombe areas, also Zanzibar Island.

**Habits:** As for the preceding species.

**Nest and Eggs:** Nest a cup of varying depth, made of moss bound on with cobwebs, generally placed in the upward branching fork of a small bough, and usually over water. Eggs two, occasionally three, probably indistinguishable from those of the Grey-headed Paradise Flycatcher.

**Recorded breeding:** Njombe, Tanganyika Territory, November onwards. Amani, November to February. Zanzibar Island all the year round.

**Call:** A clear impetuous rippling song with a falling cadence (Moreau), also a normal call of 'zi-zk-zk,' and an alarm call of 'zwä-i-zwer.'

*Tchitrea suahelica ruwenzoriæ* Grant & Praed.
*Tchitrea perspicillata ruwenzoriæ* C. Grant & Mackworth-Praed, Bull. B.O.C. 60, p. 93, 1940: Mokia, south-eastern Ruwenzori Mts., Uganda.

**Distinguishing characters:** Differs from the preceding race in the male having the head greenish blue-black, chin and throat grey, slightly glossy. The female has the top of the head less glossy. Rarely there is some white on the upper wing-coverts and edges of flight feathers. Wing 75 to 82 mm. From the Paradise Flycatcher it may be distinguished by the paler chestnut above, and white lower belly and under tail-coverts.

**Range in Eastern Africa:** Western and south-western Uganda from the Ruwenzori Mts. to near Entebbe and northern and western Tanganyika Territory as far south as the headwaters of the Nyamansi River and Mpanda and as far east as Kahe and Moshi; also Kome Island, Lake Victoria.

**Habits:** As for the other races.

**Recorded breeding:** Uganda (breeding condition), July and August.

*Tchitrea suahelica granti* Roberts.
*Tchitrea granti* Roberts, Bull. B.O.C. 68, p. 129, 1948: Duivenhoek River, Swellendam, Cape Province, South Africa.

**Distinguishing characters:** Differs from the nominate race in the male having the head, neck, chin and throat glossy blue-black washed with greenish. Wing 73 to 85 mm.

**General distribution:** Nyasaland, the lower Zambesi River, southern Portuguese East Africa, eastern Transvaal, Cape Province and Natal.

**Range in Eastern Africa:** Southern Nyasaland.

**Habits:** Much those of the Paradise Flycatcher, but it seems probable that this is the species which is more often found in true forest and gallery forest.

**Recorded breeding:** Nyasaland, November to February. South Africa, October to December.

## 834 GREY-HEADED PARADISE FLYCATCHER. *TCHITREA PLUMBEICEPS* Reichenow.

*Tchitrea plumbeiceps violacea* Grant & Praed. **Pl. 63.**
*Tchitrea plumbeiceps violacea* C. Grant & Mackworth-Praed, Bull. B.O.C. 60, p. 93, 1940: Fort Hill, North Nyasa district, Nyasaland.

**Distinguishing characters:** Differs from the various races of the last species in having the top of the head violet grey with very little gloss and the chin and throat grey with a very faint gloss. There is less variation in colour; the two central tail feathers in adult males are exceptionally long. Wing 73 to 88 mm. The young bird is similar to the adult female but is paler tawny above; head paler violet grey with less gloss.

**General distribution:** River Ja, Cameroons to the Belgian Congo, Kenya Colony, Northern and Southern Rhodesia, Bechuanaland and Portuguese East Africa.

**Range in Eastern Africa:** Coastal area of Kenya Colony from the Tana River to Tanganyika Territory, Portuguese East Africa and Nyasaland, also Pemba and Mafia Islands.

**Habits:** Usually common and conspicuous species of gardens and woodland, a cheerful friendly bird. There seems to be considerable evidence of local migration. Its habits are indistinguishable from those of the last two species and it obtains most of its food by catching insects on the wing. Particularly fond of shady rivers and woodland streams. The males are very pugnacious and both sexes incubate.

**Nest and Eggs:** Nest of fine grass lined with rootlets and lashed with cobwebs and lichens to the fork of a bush or tree. Sometimes the nest is made wholly of web-like substance and there is no lining, sometimes also there is a sort of tail to the nest underneath. Eggs two or three, creamy white finely speckled or spotted with red or reddish brown in the form of a zone at the larger end; about 20 × 14 mm.

**Recorded breeding:** Cameroons, nestling, November. Kenya Colony, July. Tanganyika Territory, March to May, also December. South-eastern Belgian Congo, October to December. Nyasaland, September to February. Portuguese East Africa, November and December. Northern Rhodesia, breeding migrant, occurring from September to March.

**Call:** A clear ringing song call of three sections, first two or three notes, then four, then about eight. Harsh clear call 'zeet-zeet.'

**Distribution of other races of the species:** Angola and Damaraland, the nominate race being described from northern Angola.

### 835 BLACK-HEADED PARADISE FLYCATCHER. *TCHITREA NIGRICEPS* (Hartlaub).

*Tchitrea nigriceps emini* (Reichenow). **Pl. 63.**
*Muscipeta emini* Reichenow, O.M. p. 31, 1893: Bukoba, north-western Tanganyika Territory.

**Distinguishing characters:** Adult male, head and neck all round glossy deep blue-black, with a slight greenish wash; rest of plumage above and below, dark or light saffron or light chestnut; primaries black; two central tail feathers elongated, but not so long as in the other Paradise Flycatchers. The female lacks the elongated central tail feathers and has a duller, often greyish, throat. Wing 73 to 83 mm. The young bird has the throat dark slate grey.

**Range in Eastern Africa:** Western and southern Uganda from the Mpanga River, southern Ruwenzori Mts. to Lake George and areas north of Lake Victoria and south of lat. 2° N. to the Bukoba district, north-western Tanganyika Territory.

**Habits:** Common and often abundant species of forest, gardens or woodland. Habits of the genus, taking insects on the wing, and fearlessly attacking any large bird of prey that comes near them. Frequently members of a mixed bird party, watching for insects to be disturbed.

**Nest and Eggs:** A small cup nest of fibres, etc. bound with cobwebs and decorated with lichens often in a fairly exposed situation. Eggs two or three, pinkish white with a ring or zone of reddish brown blotches and spots; about 20 × 15 mm.

**Recorded breeding:** Uganda, March to June and October to January.

**Call:** Described as a loud insistent ringing cry.

*Tchitrea nigriceps somereni* (Chapin).
*Terpsiphone rufiventer somereni* Chapin, Evol. II, 2, p. 114, 1948: Budongo Forest, western Uganda.

**Distinguishing characters:** Differs from the Bukoba race in being darker and richer in general colour and having a violet, not greenish wash on the glossy head. Wing 75 to 83 mm.

**Range in Eastern Africa:** Western Uganda from the Budongo Forest, between Lake Albert and Masindi to the areas north of the Ruwenzori Mts. including the Semliki Valley.

**Habits:** As for the other race.

**Recorded breeding:** No records.

**Distribution of other races of the species:** West Africa, the nominate race being described from Guinea.

Names in Sclater's *Syst. Av. Æthiop.* 2, 1930, which have been changed, or have become synonyms in this work:

*Muscicapa gambagæ somaliensis* Bannerman, treated as synonymous with
   *Muscicapa striata gambagæ* Alexander.
*Alseonax minimus djamdjamensis* Neumann, treated as synonymous with
   *Alseonax adustus minimus* (Heuglin).
*Alseonax aquaticus ruandæ* Gyldenstolpe, treated as synonymous with
   *Alseonax aquaticus infulatus* (Hartlaub).
*Alseonax flavipes* Bates, now *Alseonax seth-smithi* Van Someren.
*Alseonax cinereus kikuyuensis* Van Someren, treated as synonymous with
   *Alseonax cinereus cinereolus* Finsch & Hartlaub.
*Bradornis taruensis* Van Someren, treated as synonymous with *Bradornis*
   *pumilus* Sharpe.
*Bradornis pallidus granti* Bannerman, treated as synonymous with *Bradornis*
   *pallidus neumanni* Hilgert.
*Bradornis pallidus sharpei* Rothschild, now *Bradornis pallidus bowdleri*
   Collins & Hartert.
*Bradornis pallidus suahelicus* Van Someren, treated as synonymous with
   *Bradornis pallidus griseus* Reichenow.
*Bradornis griseus erlangeri* Reichenow, treated as synonymous with *Bradornis*
   *pumilus* Sharpe.

*Melænornis edolioides ugandæ* Van Someren, treated as synonymous with *Melænornis edolioides lugubris* (von Müller).
*Chloropetella holochlorus suahelica* Roberts, treated as synonymous with *Chloropetella holochlora* (Erlanger).
*Stizorhina fraseri intermedia* S. Clarke, treated as synonymous with *Stizorhina fraseri vulpina* Reichenow.
*Hyliota slatini* Sassi, treated as synonymous with *Hyliota australis australis* Shelley.
*Batis molitor puella* Reichenow, treated as synonymous with *Batis molitor molitor* (Hahn & Küster).
*Batis minor nyanzæ* Neumann, treated as synonymous with *Batis minor erlangeri* Neumann.
*Batis orientalis bella* (Elliot) treated as synonymous with *Batis orientalis orientalis* (Heuglin).
*Batis mystica* Neumann, treated as synonymous with *Batis molitor molitor* (Hahn & Küster).
*Platysteira peltata jacksoni* Sharpe, treated as synonymous with *Platysteira peltata mentalis* Bocage.
*Trochocercus nigromitratus kibaliensis* Alexander, treated as synonymous with *Trochocercus nigromitratus toroensis* Jackson.
*Tchitrea perspicillata* Swainson, found to be pre-occupied, now *Tchitrea suahelica granti* Roberts.
*Tchitrea poliothorax* Reichenow, treated as synonymous with *Tchitrea viridis ferreti* Guérin.

Names introduced since 1930 and which have become synonyms in this work:

*Alseonax minimus interpositus* Van Someren, 1932, treated as synonymous with *Alseonax adustus pumilus* Reichenow.
*Bradornis pallidus chyuluensis* Van Someren, 1939, treated as synonymous with *Bradornis pallidus griseus* Reichenow.
*Bradornis pallidus leucosoma* Grote, 1937, treated as synonymous with *Bradornis pallidus subalaris* Sharpe.
*Bradornis griseus ukamba* Van Someren, 1932, treated as synonymous with *Bradornis microrhynchus* Reichenow.
*Dioptrornis fischeri amani* Sclater, 1931, treated as synonymous with *Alseonax cinereus cinereola* Finsch & Hartlaub.
*Hyliota australis inornata* Vincent, 1933, treated as synonymous with *Hyliota australis australis* Shelley.
*Erythrocercus livingstonei monapo* Vincent, 1933, treated as synonymous with *Erythrocercus livingstonei livingstonei* Gray.
*Tchitrea perspicillata ungujaensis* Grant & Praed, 1947, treated as synonymous with *Tchitrea suahelica suahelica* Reichenow.

## Addenda and Corrigenda

See *Addenda*, pp. 1101, 1102.

Plate 64

Kurrichane Thrush (p. 240)
*Turdus libonyanus tropicalis*

Taita Olive Thrush
*Turdus helleri* (p. 246)

Mottled Rock-Thrush (p. 258)
  Male          Female

White-winged Cliff-Chat
  Male       Female
    (p. 281)

African Thrush (p. 242)
*Turdus pelios centralis*

Bare-eyed Thrush (p. 247)
*Turdus tephronotus*

Cliff-Chat (p. 280)
  Male    Female

Olive Thrush (p. 243)
*Turdus olivaceus abyssinicus*

Abyssinian Ground Thrush (p. 249)
*Geokichla piaggiae piaggiae*

Sooty Chat (p. 284)
  Female    Male

White-headed Black Chat (p. 283)
  Male       Female

Plate 65

White-under-winged Wheatear (p. 266)

Hooded Wheatear (p. 267)
Female     Male

Abyssinian Black Wheatear (p. 268)
Female          Male
Male

Capped Wheatear (p. 270)    Red-breasted Wheatear (p. 269)

Wheatear (p. 259)
Female     Male

Desert Wheatear (p. 262)
Female     Male

Red-rumped Wheatear (p. 265)

Somali Wheatear (p. 261)

Red-tailed Wheatear (p. 266)

# THRUSHES, WHEATEARS, CHATS, AKALATS, etc.

## Family—TURDIDÆ.

Of this large Family which contains the Thrushes, Wheatears, Chats, Akalats, Alethes, Morning Warblers, Robins, Redstarts, Bluethroats and Nightingales, ninety species are known to occur in Eastern Africa, and several of these are migrants from Europe and Asia in the non-breeding season. They are found in all types of country, some in woodland and scrub, others such as the Wheatears and some of the Chats in open country or desert. Their food is mainly insects, though in most cases mollusca form part of their diet, and the woodland species consume a good deal of fruit. This family comprises species that have very different habits, appearance and habitat, although all are relatively closely related. Introductions have been given to those groups that appear to form the most compact divisions.

Key to the Adult Thrushes, Wheatears, Chats, Akalats, Alethes, Morning Warblers, Robins, Redstarts, Bluethroats and Nightingales of Eastern Africa

1 White band across lower neck:
   RING-OUZEL
   *Turdus torquatus*     837
2 Black band across chest:
   CAPPED WHEATEAR
   *Œnanthe pileata*     868
3 Wholly slate blue:
   BLUE ROCK-THRUSH, male
   *Monticola solitaria*     851
4 Chin to chest tawny, breast to belly grey:
   ROBIN-CHAT
   *Cossypha caffra*     893
5 Chin to chest white, sometimes tipped with black, rest of underparts black:
   WHITE-HEADED BLACK CHAT, female
   *Thamnolæa arnotti*     879
6 Chin to chest bright blue:
   BLUETHROAT, male
   *Cyanosylvia svecica*     919
7 Lower neck to belly golden yellow:
   WHITE-STARRED BUSH-ROBIN
   *Pogonocichla stellata*     915

| | | |
|---|---|---|
| 8 | Throat and neck chestnut, breast to belly white: | RED-THROATED ALETHE *Alethe poliophrys* 904 |
| 9 | Chin to neck buff or whitish buff surrounded by a broken black line: | 27–28 |
| 10 | Chin and neck grey: | 29–30 |
| 11 | Chin to belly sooty black or sooty brown: | 31–32 |
| 12 | Chest streaked: | 33–36 |
| 13 | Below, mottled and barred: | 37–40 |
| 14 | Chest and flanks pale tawny: | 41–46 |
| 15 | Chin to chest brownish grey, or brown washed with chestnut or grey: | 47–51 |
| 16 | Underparts with large dark spots: | 52–57 |
| 17 | Below, various shades of russet: | 58–63 |
| 18 | Chin and throat white contrasting with sides of face: | 64–69 |
| 19 | Throat streaked: | 70–74 |
| 20 | Below, dusky brown or dusky grey: | 75–85 |
| 21 | Below, mainly grey: | 86–93 |
| 22 | Chin and throat black: | 94–106 |
| 23 | Chin to chest black or blackish: | 107–118 |
| 24 | Chin to belly black: | 119–131 |
| 25 | Below, isabelline, buff or buff-grey: | 132–149 |
| 26 | Below, wholly or partly deep tawny, rich orange brown, orange buff, rich chestnut brown, rich cinnamon rufous or olivaceous brown: | 150-179 |

| | | | |
|---|---|---|---|
| 27 | Mantle brown: | MORNING WARBLER *Cichladusa arquata* | 908 |
| 28 | Mantle sooty: | BLUE-THROAT, female *Cyanosylvia svecica* | 919 |

| | | | |
|---|---|---|---|
| 29 | Above, grey with white tips: | ROCK-THRUSH, male *Monticola saxatilis* | 850 |
| 30 | Above, mottled grey and black: | MOTTLED ROCK-THRUSH, male *Monticola angolensis* | 852 |

| | | | |
|---|---|---|---|
| 31 | Chin to belly sooty black: | ANTEATER-CHAT *Myrmecocichla æthiops* | 881 |
| 32 | Chin to belly sooty brown: | SOOTY CHAT, female *Myrmecocichla nigra* | 880 |

| | | | |
|---|---|---|---|
| 33 | Chest streaked with grey: | | 34–35 |
| 34 | Mantle blackish brown: | BROWN-BACKED SCRUB-ROBIN *Erythropygia hartlaubi* | 914 |
| 35 | Mantle warm brown: | WHITE-WINGED SCRUB-ROBIN *Erythropygia leucoptera* | 911 |
| 36 | Chest streaked with black: | RED-BACKED SCRUB-ROBIN *Erythropygia zambesiana* | 910 |

| | | | |
|---|---|---|---|
| 37 | Base of primaries white: | WHITE-WINGED CLIFF-CHAT, female *Thamnolæa semirufa* | 877 |
| 38 | Base of primaries not white: | | 39–40 |
| 39 | Tail chestnut: | ROCK-THRUSH, female *Monticola saxatilis* | 850 |
| 40 | Tail black: | BLUE ROCK-THRUSH, female *Monticola solitaria* | 851 |

| | | | |
|---|---|---|---|
| 41 | Head and mantle mottled: | MOTTLED ROCK-THRUSH, female *Monticola angolensis* | 852 |
| 42 | Head and mantle plain: | | 43–46 |
| 43 | Upper tail-coverts chestnut: | | 44–45 |
| 44 | Duller in colour, less distinctly marked: | BEARDED SCRUB-ROBIN *Erythropygia barbata* | 912 |
| 45 | Brighter in colour, more distinctly marked: | EASTERN BEARDED SCRUB-ROBIN *Erythropygia quadrivirgata* | 913 |
| 46 | Upper tail-coverts grey: | WHITE-THROATED ROBIN, female *Irania gutturalis* | 920 |

| | | | |
|---|---|---|---|
| 47 | Chin to chest brownish grey: | LITTLE ROCK-THRUSH, female *Monticola rufocinerea* | 853 |
| 48 | Chin to chest brown washed with chestnut: | WHITE-CROWNED CLIFF-CHAT, female *Thamnolæa coronata* | 878 |
| 49 | Chin to chest grey: | | 50–51 |
| 50 | Breast to belly tawny: | LITTLE ROCK-THRUSH, male *Monticola rufocinerea* | 853 |
| 51 | Breast to belly chestnut: | TROPICAL CLIFF-CHAT, female *Thamnolæa cinnamomeiventris* | 876 |

| | | | |
|---|---|---|---|
| 52 | White tips to wing-coverts: | SPOTTED GROUND-THRUSH *Psophocichla guttata* | 846 |
| 53 | No white tips to wing-coverts: | | 54–57 |

# THRUSHES, WHEATEARS, CHATS, AKALATS, etc.

| | | |
|---|---|---|
| 54 Tail chestnut: | SPOTTED MORNING WARBLER | |
| | *Cichladusa guttata* | 909 |
| 55 Tail earth colour: | | 56–57 |
| 56 Inner webs of flight feathers, excepting ends, dark buff: | GROUND-SCRAPER THRUSH | |
| | *Psophocichla litsitsirupa* | 847 |
| 57 Inner webs of flight feathers not dark buff: | SONG-THRUSH | |
| | *Turdus ericetorum* | 836 |
| 58 White on wing: | | 59–60 |
| 59 Upper tail-coverts white: | STONECHAT, female | |
| | *Saxicola torquata* | 882 |
| 60 Upper tail-coverts not white: | WHINCHAT, male | |
| | *Saxicola rubetra* | 883 |
| 61 No white on wing: | | 62–63 |
| 62 Wing 95 mm. and over: | RED-BREASTED WHEATEAR | |
| | *Œnanthe bottæ* | 866 |
| 63 Wing 94 mm. and under: | HEUGLIN'S RED-BREASTED WHEATEAR | |
| | *Œnanthe heuglini* | 867 |
| 64 Bases of head feathers tawny: | FIRE-CREST ALETHE | |
| | *Alethe castanea* | 901 |
| 65 Bases of head feathers not tawny: | | 66–69 |
| 66 Ends of tail feathers white: | CHOLO MOUNTAIN ALETHE | |
| | *Alethe choloensis* | 902 |
| 67 Ends of tail feathers not white: | | 68–69 |
| 68 White stripe over eye: | BROWN-CHESTED ALETHE | |
| | *Alethe poliocephala* | 903 |
| 69 No white stripe over eye: | WHITE-CHESTED ALETHE | |
| | *Alethe fülleborni* | 905 |
| 70 Chest olivaceous: | OLIVE THRUSH | |
| | *Turdus olivaceus* | 841 |

| | | | |
|---|---|---|---|
| 71 | Chest buff: | KURRICHANE THRUSH *Turdus libonyanus* | **839** |
| 72 | Chest greyish: | | 73–74 |
| 73 | Under wing-coverts pale tawny: | AFRICAN THRUSH *Turdus pelios* | **840** |
| 74 | Under wing-coverts dark tawny: | BARE-EYED THRUSH *Turdus tephronotus* | **843** |

| | | | |
|---|---|---|---|
| 75 | White on tail: | | 76–77 |
| 76 | Upper tail-coverts white: | ABYSSINIAN BLACK WHEATEAR, female *Œnanthe lugubris* | **865** |
| 77 | Upper tail-coverts not white: | HILL CHAT *Pinarochroa sordida* | **873** |
| 78 | No white in tail: | | 79–85 |
| 79 | Tail chestnut: | BLACK REDSTART, female *Phœnicurus ochruros* | **918** |
| 80 | Tail olivaceous brown: | | 81–82 |
| 81 | Superciliary streak russet brown: | USAMBARA ALETHE *Alethe montana* | **906** |
| 82 | Superciliary streak white: | IRINGA ALETHE *Alethe lowei* | **907** |
| 83 | Tail russet brown: | | 84–85 |
| 84 | Chest mottled: | SPROSSER *Luscinia luscinia* | **922** |
| 85 | Chest not mottled: | NIGHTINGALE *Luscinia megarhynchos* | **921** |

| | | | |
|---|---|---|---|
| 86 | Chin to neck black: | SOMALI BLACKBIRD *Turdus ludoviciæ* | **838** |
| 87 | Chin to neck white: | OLIVE-FLANKED FOREST-CHAT *Dessonornis anomala* | **895** |
| 88 | Chin and neck grey or greyish, uniform with rest of underparts: | | 89–93 |

# THRUSHES, WHEATEARS, CHATS, AKALATS, etc.

| | | | |
|---|---|---|---|
| 89 | Tail chestnut: | RED-TAILED CHAT *Cercomela familiaris* | 871 |
| 90 | Tail ashy grey: | BROWN-TAILED ROCK-CHAT *Cercomela scotocerca* | 870 |
| 91 | Tail black: | | 92–93 |
| 92 | Under tail-coverts white: | BLACK-TAILED ROCK-CHAT *Cercomela melanura* | 869 |
| 93 | Under tail-coverts blackish brown: | SOMBRE ROCK-CHAT *Cercomela dubia* | 872 |

| | | | |
|---|---|---|---|
| 94 | Rump and tail (except central tail feathers), chestnut: | REDSTART, male *Phœnicurus phœnicurus* | 917 |
| 95 | Tail tawny and black: | RED-RUMPED WHEATEAR *Œnanthe xanthoprymna* | 861 |
| 96 | Tail black and white: | | 97–106 |
| 97 | White in wing-coverts: | | 98–99 |
| 98 | Head and mantle sandy: | DESERT WHEATEAR, male *Œnanthe deserti* | 857 |
| 99 | Head and mantle black or black edged with sandy or dark brown: | STONECHAT *Saxicola torquata* | 882 |
| 100 | No white in wing-coverts: | | 101–106 |
| 101 | Mantle buffish white: | BLACK-EARED WHEATEAR, male (black-throated form) *Œnanthe hispanica* | 858 |
| 102 | Mantle black: | | 102–106 |
| 103 | Inner webs of primaries white: | WHITE-UNDER-WINGED WHEATEAR *Œnanthe lugens* | 863 |
| 104 | Inner webs of primaries grey: | | 105–106 |
| 105 | Rump and upper tail-coverts buff: | ABYSSINIAN BLACK WHEATEAR, male *Œnanthe lugubris* | 865 |

| | | |
|---|---|---|
| 106 Rump and upper tail-coverts white: | PIED WHEATEAR, male (black-throated form) *Œnanthe leucomela* | 859 |

| | | |
|---|---|---|
| 107 Tail chestnut: | BLACK REDSTART, male *Phœnicurus ochruros* | 918 |
| 108 Tail tawny and black or white and black: | | 109–110 |
| 109 Mantle grey: | SOMALI WHEATEAR *Œnanthe phillipsi* | 856 |
| 110 Mantle black: | HOODED WHEATEAR, male *Œnanthe monacha* | 864 |
| 111 Tail black: | | 112–118 |
| 112 Wing shoulder white: | | 113–115 |
| 113 Upper tail-coverts chestnut: | | 114–115 |
| 114 Top of head black: | TROPICAL CLIFF-CHAT, male *Thamnolæa cinnamomeiventris subrufipennis* | 876 |
| 115 Top of head white: | WHITE-CROWNED CLIFF-CHAT, male *Thamnolæa coronata* | 878 |
| 116 No white on wing shoulder: | | 117–118 |
| 117 Base of primaries white: | WHITE-WINGED CLIFF-CHAT, male *Thamnolæa semirufa* | 877 |
| 118 Base of primaries black: | ABYSSINIAN CLIFF-CHAT, female *Thamnolæa cinnamomeiventris albiscapulata* | 876 |

| | | |
|---|---|---|
| 119 Wing shoulder white: | | 120–122 |
| 120 Forehead white: | WHITE-SHOULDERED BLACK CHAT, male *Pentholæa albifrons clericalis* | 874 |
| 121 Top of head white: | WHITE-HEADED BLACK CHAT, male *Thamnolæa arnotti* | 879 |

# THRUSHES, WHEATEARS, CHATS, AKALATS, etc.

| | | | |
|---|---|---|---|
| 122 | Top of head and forehead black: | SOOTY CHAT, male *Myrmecocichla nigra* | **880** |
| 123 | Wing shoulder not white: | | 124–131 |
| 124 | White in tail: | | 125–128 |
| 125 | White confined to tips of outer tail feathers: | BLACK BUSH-ROBIN *Cercotrichas podobe* | **916** |
| 126 | Greater part of tail white: | | 127–128 |
| 127 | Top of head white: | WHITE-RUMPED WHEATEAR, male *Œnanthe leucopyga* | **860** |
| 128 | Top of head black: | WHITE-RUMPED WHEATEAR, female *Œnanthe leucopyga* | **860** |
| 129 | No white in tail: | | 130–131 |
| 130 | Wing 85 mm. or over: | RÜPPELL'S CHAT *Pentholæa melæna* | **875** |
| 131 | Wing 85 mm. or under: | WHITE-FRONTED BLACK CHAT *Pentholæa albifrons albifrons* | **874** |
| 132 | White on wings: | WHINCHAT, female *Saxicola rubetra* | **883** |
| 133 | No white on wings: | | 134–149 |
| 134 | Tail chestnut: | REDSTART, female *Phœnicurus phœnicurus* | **917** |
| 135 | Tail tawny and black: | RED-TAILED WHEATEAR *Œnanthe chrysopygia* | **862** |
| 136 | Rump and upper tail-coverts buff: | | 137–138 |
| 137 | Apical half of tail black: | DESERT WHEATEAR, female *Œnanthe deserti* | **857** |
| 138 | Tail mainly buff: | HOODED WHEATEAR, female *Œnanthe monacha* | **864** |
| 139 | Upper tail-coverts and base of tail white: | | 140–149 |

# THRUSHES, WHEATEARS, CHATS, AKALATS, etc.

140 Ear-coverts black: 141–143
141 Head and mantle buffish white with dusky edges: BLACK-EARED WHEATEAR, male (plain-throated form) *Œnanthe hispanica* **858**
142 Head buffish white with dusky edges, mantle black with brownish edges: PIED WHEATEAR, male (white-throated form) *Œnanthe leucomela* **859**
143 Head and mantle grey with brown edges: WHEATEAR, male *Œnanthe œnanthe* **854**
144 Ear-coverts brown or isabelline: 145–149
145 Under wing-converts white with dusky bases: ISABELLINE WHEATEAR *Œnanthe isabellina* **855**
146 Under wing-coverts dusky: PIED WHEATEAR, female *Œnanthe leucomela* **859**
147 Under wing-coverts dusky with whitish edges: 148–149
148 Head and mantle stone brown: BLACK-EARED WHEATEAR, female *Œnanthe hispanica* **858**
149 Head and mantle warm brown: WHEATEAR, female *Œnanthe œnanthe* **854**

---

150 Chin and throat white: WHITE-THROATED ROBIN, male *Irania gutturalis* **920**
151 Chin and throat black: TAITA OLIVE THRUSH *Turdus helleri* **842**
152 Chin and throat dusky: WHITE-TAILED ANT-THRUSH *Neocossyphus poensis* **849**

# THRUSHES, WHEATEARS, CHATS, AKALATS, etc.

153 Chin and throat deep
 fawn colour:     FOREST-ROBIN
 *Stiphrornis erythrothorax* **901**

154 Chin and throat brown
 spotted with black:     SPOT-THROAT
 *Modulatrix stictigula*     **894**

155 Chin and throat orange,
 rich orange brown or
 orange yellow:     156–179

156 White tips to wing-coverts:     157–158

157 Forehead olivaceous
 brown:     ORANGE GROUND-THRUSH
 *Geokichla gurneyi*     **844**

158 Forehead orange brown:     ABYSSINIAN
 GROUND-THRUSH
 *Geokichla piaggiæ*     **845**

159 No white tips to wing-
 coverts:     160–179

160 Forehead to occiput
 white:     SNOWY-HEADED
 ROBIN-CHAT
 *Cossypha niveicapilla*     **892**

161 Forehead to occiput black
 and white giving a scaly
 appearance:     WHITE-CROWNED
 ROBIN-CHAT
 *Cossypha albicapilla*     **891**

162 White stripe over eye;
 sometimes concealed by
 coloured tips to feathers:     163–173

163 Wing shoulder bright
 blue:     BLUE-SHOULDERED
 ROBIN-CHAT
 *Cossypha cyanocampter*     **889**

164 Wing shoulder not bright
 blue:     165–173

165 Tail olivaceous:     SHARPE'S AKALAT
 *Sheppardia sharpei*     **897**

166 Tail russet:     AKALAT
 *Sheppardia cyornithopsis*     **899**

238  THRUSHES, WHEATEARS, CHATS, AKALATS, etc.

| | | | |
|---|---|---|---|
| 167 | Tail dusky washed and edged with chestnut: | MOREAU'S ROBIN-CHAT *Cossypha insulana kungwensis* | 888 |
| 168 | Tail tawny: | | 169–170 |
| 169 | Central tail feathers black: | RÜPPELL'S ROBIN-CHAT *Cossypha semirufa* | 885 |
| 170 | Central tail feathers olivaceous: | WHITE-BROWED ROBIN-CHAT *Cossypha heuglini* | 884 |
| 171 | Tail chestnut: | | 172–173 |
| 172 | Top of head blackish grey: | GREY-WINGED ROBIN-CHAT *Cossypha polioptera* | 887 |
| 173 | Top of head chestnut brown: | ARCHER'S ROBIN-CHAT *Cossypha archeri* | 886 |
| 174 | No white stripe over eye: | | 175–179 |
| 175 | Below, chestnut brown: | RED-TAILED ANT-THRUSH *Neocossyphus rufus* | 848 |
| 176 | Below, cinnamon rufous: | RED-CAPPED ROBIN-CHAT *Cossypha natalensis* | 890 |
| 177 | Below, orange buff or orange brown: | | 178–179 |
| 178 | Rump and tail olive green: | EAST COAST AKALAT *Sheppardia gunningi* | 896 |
| 179 | Rump and tail russet brown: | EQUATORIAL AKALAT *Sheppardia æquatorialis* | 898 |

**THRUSHES and OUZELS.** FAMILY—**TURDIDÆ.** Genera: *Turdus, Geokichla* and *Psophocichla*.

Twelve species occur in Eastern Africa, of which two are migrants from Europe. They inhabit forest, woodland, and gardens, and are not usually shy. They spend much of their time on the ground searching for their insect food among the leaves and debris under trees and bushes, but they also eat fruit and berries.

THRUSHES, WHEATEARS, CHATS, AKALATS, etc. 239

## 836 SONG-THRUSH. *TURDUS ERICETORUM* (Turton).
*Turdus ericetorum philomelos* Brehm.
*Turdus philomelos* Brehm, Handb. Nat. Vög. Deutschl. p. 382, 1831: Middle Germany.

**Distinguishing characters:** Above, earthy brown; below, chin to breast buff; belly white; blackish brown spots down sides of throat and on breast, chest and flanks; under wing-coverts yellowish buff. The sexes are alike. Wing 110 to 123 mm.

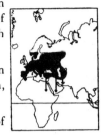

**General distribution:** Europe, except British Isles, to Asia; in non-breeding season to Canary Islands, North Africa, the Sudan, Arabia and Iran.

**Range in Eastern Africa:** Khartoum, the Red Sea Province of the Sudan and Eritrea in the non-breeding season.

**Habits:** A comparatively rare palæarctic winter migrant to the Sudan and Eritrea, unlikely to occur further south.

**Distribution of other races of the species:** Great Britain, the nominate race being described from Kent.

## 837 RING-OUZEL. *TURDUS TORQUATUS* Linnæus.
*Turdus torquatus alpestris* (Brehm).
*Merula alpestris* Brehm. Handb. Nat. Vög. Deutschl. p. 377, 1831: Alpine Tirol.

**Distinguishing characters:** Adult male, above, dull blackish, feathers edged with dusky white; whitish edges to wing-coverts and flight feathers; below, a broad white collar across lower neck; rest of underparts dull blackish, feathers edged and centred with white. The female is duller than the male and the collar across the lower neck is dusky white. Wing 136 to 148 mm.

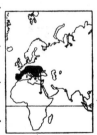

**General distribution:** Central to southern Europe; in non-breeding season to Cyprus, Asia Minor and the Sudan.

**Range in Eastern Africa:** The Sudan as far as Khartoum in non-breeding season.

**Habits:** A rare and probably exceptional migrant to the Sudan.

**Distribution of other races of the species:** Northern Europe, Iran and Caucasus, the nominate race being described from Sweden.

## 240 THRUSHES, WHEATEARS, CHATS, AKALATS, etc.

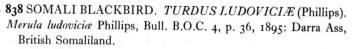

**838 SOMALI BLACKBIRD.** *TURDUS LUDOVICIÆ* (Phillips).
*Merula ludoviciæ* Phillips, Bull. B.O.C. 4, p. 36, 1895: Darra Ass, British Somaliland.

**Distinguishing characters:** General colour brownish grey; upperparts darker; lores, ear-coverts, chin and throat black with white and grey streaks on chin and throat; bill dull yellow. The sexes are alike. Wing 111 to 122 mm. The young bird is spotted with buffish white and blackish and the bill is horn-brown.

**Range in Eastern Africa:** The plateau of British Somaliland.

**Habits:** No information.

**Nest and Eggs:** Undescribed.

**Recorded breeding:** No records.

**Food:** No information.

**Call:** Unrecorded.

**839 KURRICHANE THRUSH.** *TURDUS LIBONYANUS* (Smith).
*Turdus libonyanus tropicalis* Pet. **Pl. 64.**
*Turdus tropicalis* Peters, J.f.O. p. 50, 1881: Inhambane, Portuguese East Africa.

**Distinguishing characters:** Above, brownish grey; below, throat and belly white; black streaks on sides of throat and neck; chest greyish buff; flanks and under wing-coverts rich yellowish buff. The sexes are alike. Wing 101 to 125 mm. The young bird has blackish spots on the chest and down the flanks; and buff tips to the upper wing-coverts.

**General distribution:** Tanganyika Territory to south-eastern Belgian Congo, Portuguese East Africa, the Rhodesias, southern Nyasaland, Swaziland and Natal.

**Range in Eastern Africa:** Tanganyika Territory to the Zambesi River.

**Habits:** Common in light bush, woodlands and rocky hills: in Nyasaland the most common Thrush. Tame round gardens but usually shy away from cultivation. Feeds largely on the ground in typical Thrush fashion and darts under bushes if disturbed.

**Nest and Eggs:** Nest in the fork of a tree or on a stump, large, made of twigs, roots and grasses, with leaves or lichens outside on a mud base. Eggs usually three, dull bluish ground colour with

# THRUSHES, WHEATEARS, CHATS, AKALATS, etc.

more or less dense brown speckling and pale lilac underblotches; about 26 × 18·5 mm.

**Recorded breeding:** Tanganyika Territory, December to February. South-eastern Belgian Congo, September to December. Nyasaland, August to January. Portuguese East Africa, September and October. Rhodesia, September to December.

**Food:** Insects and fruit.

**Call:** A single Thrush-like squeak and a low song in the breeding season, also a musical whistling 'tchi-cheee' at all times of year.

**Distribution of other races of the species:** Transvaal, the nominate race being described from Kurrichane.

## 840 AFRICAN THRUSH. *TURDUS PELIOS* Bonaparte.

*Turdus pelios pelios* Bp.
*Turdus pelios* Bonaparte, Consp. Gen. Av. 1, p. 273, 1850: Fazogli, eastern Sudan.

**Distinguishing characters:** Differs from the Kurrichane Thrush in being rather browner above; chest ashy; chin to throat and sides of neck streaked with ashy brown. The sexes are alike. Wing 105 to 120 mm. The young bird is spotted with ash-grey on the chest, breast and upperparts of the flanks, and has tawny tips to the upper wing-coverts.

**General distribution:** Ubangi River Valley to the Sudan, Eritrea and Abyssinia.

**Range in Eastern Africa:** Western and south-western Sudan to Eritrea and central Abyssinia as far south as Neghelli.

**Habits:** As for other races, but nowhere numerous, and may be only a migrant over part of its range. It is usually a quiet little bird seen searching for food in thickets and old stream beds.

**Nest and Eggs:** See under other races.

**Recorded breeding:** Darfur, June to August. Abyssinia, April to June.

*Turdus pelios schuetti* (Cab.).
*Pelocichla schuetti* Cabanis, J.f.O. p. 319, 1882: Malanje, northern Angola.

**Distinguishing characters:** Differs from the nominate race in being generally darker, with a warmer tone of colour on flanks and under wing-coverts and a wash of buff on the grey chest. Wing 107 to 120 mm.

**General distribution:** Northern Angola to Tanganyika Territory.

**Range in Eastern Africa:** North-western Tanganyika Territory.

**Habits:** As for the nominate race.

**Recorded breeding:** Northern Angola (nestling), March.

*Turdus pelios centralis* Reichenow.
*Turdus pelios centralis* Reichenow, Vög. Afr. 3, p. 690, 1905: Wadelai, north-western Uganda. **Pl. 64.**

**Distinguishing characters:** Differs from the nominate race in being darker above, but similar below. Wing 105 to 122 mm.

**General distribution:** Welle River, north-eastern Belgian Congo, to the Sudan, Uganda, Abyssinia, Kenya Colony and Tanganyika Territory.

**Range in Eastern Africa:** The south-eastern and eastern Sudan, western and southern Abyssinia as far east as the Didessa River, Alghe, Yavello and Mega to north-western Kenya Colony, Uganda and western Tanganyika Territory.

**Habits:** The common Thrush of Uganda gardens, behaving exactly like a Song-Thrush in England, feeding on the ground and making a series of quick hops followed by a short listening period, but is usually somewhat shy and darts under a bush if alarmed. It is found in true forest as well as in cultivated areas.

**Nest and Eggs:** Nest of coarse grass mixed with mud and lined with fibres and rootlets, placed in trees, bushes or banana palms. Eggs two or three, pale blue with more or less rufous speckling, sometimes with so much as to appear rufous all over; about 26 × 20 mm.

**Recorded breeding:** Uganda, April to August, and October to December, but in other months also.

**Food:** Insects, with a good deal of fruit in season.

**Call:** A clear loud song in breeding season, also a clear whistle uttered as the bird leaves its perch. A very common call is a little twitter of alarm very freely uttered.

**Distribution of other races of the species:** West Africa.

## THRUSHES, WHEATEARS, CHATS, AKALATS, etc.

### 841 OLIVE THRUSH. *TURDUS OLIVACEUS* (Linnæus).

*Turdus olivaceus abyssinicus* Gmel. **Pl. 64. Ph. ix.**

*Turdus abyssinicus* Gmelin, Syst. Nat. 1, pt. 2, p. 824, 1789: Abyssinia.

**Distinguishing characters:** Above, dark brownish grey; below, sides of face and chin to chest buffish grey; chin and throat streaked with blackish; breast to belly, flanks and under wing-coverts deep tawny; under tail-coverts dark ashy and white. The sexes are alike. Wing 104 to 124 mm. The young bird has black spots from the chin to breast and down flanks.

**Range in Eastern Africa:** Eritrea and Abyssinia to Kenya Colony and north-eastern Tanganyika Territory at Loliondo and Olonoti.

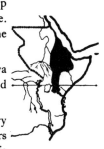

**Habits:** A common resident of the Abyssinian high plateau, very destructive to fruit in gardens, and found at all elevations. It occurs freely on open hillsides along streams, in fields or in forest, and is usually shy with a fast flight.

**Nest and Eggs:** Nest composed mainly of grass and rather soft, with a substantial foundation of twigs, rootlets, earth and coarse grass, well hidden close to the trunks of big trees in cover, on stumps or in thick bushes. Eggs two, bluish green speckled with fine brown spots or blotched with rufous and with occasional purplish under-markings; about 27 × 21 mm.

**Recorded breeding:** Abyssinia, April to July. Kenya Colony highlands, April to July, also September to December.

**Food:** Insects and fruit, worms, snails, etc.

**Call:** A thin whistle, also a low sweet song, and a sharp clacking alarm note.

*Turdus olivaceus deckeni* Cab.
*Turdus deckeni* Cabanis, J.f.O. p. 412, 1868: Mt. Kilimanjaro, north-eastern Tanganyika Territory.

**Distinguishing characters:** Generally darker than the preceding race, and tawny of underside duller and less clear. Wing 109 to 124 mm.

**Range in Eastern Africa:** Longido to Ketumbeine, Monduli and Mt. Kilimanjaro.

**Habits:** As for the other races, found at all elevations from 4,500 to 10,000 feet; both in cultivated land and forest.

**Recorded breeding:** Mt. Kilimanjaro, Tanganyika Territory, January to March.

*Turdus olivaceus milanjensis* Shell.
*Turdus milanjensis* Shelley, Ibis, p. 12, 1893: Milanji Plateau, southern Nyasaland.

**Distinguishing characters:** Very similar to the last race but chin and throat buffish white with distinct black stripes. Wing 115 to 125 mm.

**General distribution:** Nyasaland from Dedza to southern areas and northern Portuguese East Africa.

**Range in Eastern Africa:** Portuguese East Africa to southern Nyasaland.

**Habits:** As for other races, common but shy and usually solitary in forests and woodland at higher elevations.

**Recorded breeding:** Nyasaland, September to December, apparently double-brooded.


*Turdus olivaceus baraka* (Sharpe).
*Merula baraka* Sharpe, Bull. B.O.C. 14, p. 19, 1903: Ruwenzori, Uganda.

**Distinguishing characters:** Similar to the Abyssinian race, but upperside and chest darker. Wing 107 to 120 mm.

**General distribution:** The Kivu district of the eastern Belgian Congo and Uganda.

**Range in Eastern Africa:** Western and southern Uganda from Mt. Ruwenzori to Entebbe.

**Habits:** As for other races, occurring up to the snow line of the Ruwenzori mountains and among tree heaths, as well as down to 6,000 feet in hot tropical valleys.

**Recorded breeding:** Ruwenzori (breeding condition), December.

## THRUSHES, WHEATEARS, CHATS, AKALATS, etc.

### Turdus olivaceus nyikæ Reichw.

Turdus nyikæ Reichenow, O.M. p. 95, 1904: Nyika Plateau, northern Nyasaland.

**Distinguishing characters:** Similar to the southern Nyasaland race but differs in being more olivaceous above and the throat is darker. Wing 108 to 123 mm.

**General distribution:** Tanganyika Territory to northern Nyasaland.

**Range in Eastern Africa:** Nguru and Uluguru Mts. east-central Tanganyika Territory to south-western Tanganyika Territory.

**Habits:** As for other races, with a fine varied song at dawn and dusk.

**Recorded breeding:** Tanganyika Territory, October and January. Nyasaland, October to December.

### Turdus olivaceus roehli Reichw.

Turdus roehli Reichenow, O.M. p. 182, 1905: Mlalo, near Lushoto, Usambara Mts., north-eastern Tanganyika Territory.

**Distinguishing characters:** Very similar to the Abyssinian race, but darker above and centre of breast to belly white. Wing 105 to 112 mm.

**Range in Eastern Africa:** North Paré to Usambara Mts., north-eastern Tanganyika Territory.

**Habits:** Abundant in forest, hunting about on the ground and feeding on berries. Has the same habit as the Song-Thrushes of Europe of cracking snails against a stone. Has a squealing alarm note and a soft varied little whistling song, rather melancholy in character.

**Recorded breeding:** Eastern Tanganyika Territory (breeding condition), October to December.

### Turdus olivaceus oldeani Scl. & Mor.

Turdus olivaceus oldeani Sclater & Moreau, Bull. B.O.C. 56, p. 13, 1935: Oldeani Forest, Mbulu district, northern Tanganyika Territory.

**Distinguishing characters:** General colour darker and duller than either the Kilimanjaro or northern Nyasaland races, more dusky with only a faint wash of brownish on the flanks; under wing-coverts tawny. Wing 106 to 116 mm.

**Range in Eastern Africa:** Embagai, Elanairobi, Ngorongoro, Oldeani, Mt. Meru, Mbulu, Lolkissale, Mt. Gerui (Hanang) and Ufiome, north-eastern Tanganyika Territory.

**Habits:** As for other races, occurring mainly in forest. The song is not as melancholy as that of the Usambara race.

**Recorded breeding:** Northern Tanganyika Territory (breeding condition), January.

*Turdus olivaceus graueri* Neum.
*Turdus graueri* Neumann, Bull. B.O.C. 21, p. 56, 1908: Nsasa, Ruanda, eastern Belgian Congo.

**Distinguishing characters:** Differs from the Uganda, Oldeani and northern Nyasaland races in being much brighter tawny below, and from the Abyssinian race in having the chin and throat tawny white showing distinct streaks; chest only washed with greyish. Wing 116 to 125 mm.

**Range in Eastern Africa:** Ruanda and Urundi, eastern Belgian Congo and western Tanganyika Territory from the Bukoba and Kasulu areas to the upper Nyamanse River, Ushamba area of Ubende.

**Habits:** As for other races.

**Recorded breeding:** No records.

**Distribution of other races of the species:** Northern and Southern Rhodesia and South Africa, the nominate race being described from the Cape of Good Hope.

**842 TAITA OLIVE THRUSH.** *TURDUS HELLERI* (Mearns). **Pl. 64.**

*Planesticus helleri* Mearns, Smiths, Misc. Coll. 61, No. 10, p. 1, 1913: Mt. Mbololo, east of Mt. Kilimanjaro, south-eastern Kenya Colony.

**Distinguishing characters:** Similar to the Olive Thrush in having centre of breast and belly white, but differs in having the chest dark grey and the head all round black. Wing 117 to 122 mm.

**Range in Eastern Africa:** The Taita Hills area of south-eastern Kenya Colony.

**Habits:** Presumably as for the Olive Thrush, but no information available.

**Nest and Eggs:** Undescribed.

**Recorded breeding:** No records.

**Call:** Unrecorded.

### 843 BARE-EYED THRUSH.  *TURDUS TEPHRONOTUS* Cabanis. Pl. 64.

*Turdus tephronotus* Cabanis, J.f.O. pp. 205, 218, pl. 3, fig. 2, 1878: Ndi, Taita district, south-eastern Kenya Colony.

**Distinguishing characters:** Above, ashy grey; eyes encircled by patch of bare skin bright chrome yellow in colour; below, throat white streaked with black; chest grey; rest of underparts and under wing-coverts tawny; centre of lower belly white. The sexes are alike. Wing 100 to 115 mm. The young bird has dusky spots on the chest and breast.

**Range in Eastern Africa:** Central and southern Abyssinia and southern Italian Somaliland to central and eastern Kenya Colony, as far west as the Endoto Mts. and the Taita Hills, and eastern Tanganyika Territory as far west as Mt. Kilimanjaro and the Dodoma district.

**Habits:** A fairly common resident along the Eastern African coast, especially plentiful among cultivations at Lamu and Witu and again common in the southern Masai area. Habits typically Thrush-like, feeding a good deal on the ground. Sings from the top of tall trees.

**Nest and Eggs:** Nest of twigs, grasses, roots and dead leaves, lined with fine grass and rootlets, placed in a tree or bush. Eggs two or three, pale bluish green spotted at the larger end with dark rufous and with grey or violet undermarkings; about 26 × 19 mm.

**Recorded breeding:** Southern Abyssinia, March. Italian Somaliland, May. Kenya Colony coast, April to June.

**Food:** Fruit, caterpillars, beetles and other insects.

**Call:** Has a typical short Thrush-like song, 'throo-throo-thrit' rather throaty in character (Benson) also a rattling call and a sibilant alarm note. Moreau notes also a soft repeated whinnying call 'tu' made by birds on the ground.

248 *THRUSHES, WHEATEARS, CHATS, AKALATS, etc.*

**844 ORANGE GROUND-THRUSH.** *GEOKICHLA GURNEYI*
(Hartlaub).

*Geokichla gurneyi gurneyi* (Hartl.).

*Turdus gurneyi* Hartlaub, Ibis, p. 350, pl. 9, 1864: near Pietermaritzburg, Natal, South Africa.

**Distinguishing characters:** Adult male, above, olive brown; head slightly darker; large white tips to wing-coverts; below, chin to chest and flanks orange brown; centre of belly and under tail-coverts white; under wing-coverts olive brown. The female is usually paler orange brown below. Wing 103 to 115 mm. The young bird is streaked with dark buff above, and mottled with black below.

**General distribution:** Nyasaland, Portuguese East Africa, eastern Transvaal, eastern Cape Province and Natal.

**Range in Eastern Africa:** Nyasaland to Portuguese East Africa.

**Habits:** Common in hill and mountain forest in restricted localities, chiefly among big timber. Shy, wild, and difficult to observe, but it appears to feed largely on the ground.

**Nest and Eggs:** Rather conspicuous large cup nests of green moss with black fibre lining in the forks of small trees or shrubs. Eggs usually three, turquoise blue either plain or spotted with reddish brown; about 29 × 20 mm.

**Recorded breeding:** Nyasaland, September to January. Southern Rhodesia, January. Zululand, January.

**Food:** Insects, worms, fruit, etc., picked up from the ground.

**Call:** Described as a thin whistling note, or a hissing trill. It also has a rather wild breeding season song, uttered often at dusk.

*Geokichla gurneyi raineyi* Mearns.

*Geocichla gurneyi raineyi* Mearns, Smiths Misc. Coll. 41, No. 10, p. 4, 1913: Mt. Mbololo, east of Mt. Kilimanjaro, south-eastern Kenya Colony.

**Distinguishing characters:** Differs from the nominate race in having the top of the head grey, and throat and chest rather paler. Wing 105 to 116 mm.

**General distribution:** Kenya Colony, Tanganyika Territory and Nyasaland as far south as Mzumara.

**Range in Eastern Africa:** Taita area of south-eastern Kenya Colony to Mt. Kilimanjaro, Paré Mts., Usambara Mts., Oldeani,

Mt. Meru, Nguru Hills, Uluguru Mts., Njombe and Tukuyu, Tanganyika Territory.

**Habits:** Much as for the nominate race, a terrestrial Thrush of the damper parts of the forests.

**Recorded breeding:** Northern Tanganyika Territory, November onwards.

*Geokichla gurneyi chuka* V. Som.
*Geocichla gurneyi chuka* Van Someren, J.E.A. & U. Nat. Hist. Soc. No. 37, p. 195, 1930: Chuka, Mt. Kenya, central Kenya Colony.

**Distinguishing characters:** Larger, rather darker above, and bill longer and less stout than the last race. Wing 116 to 125 mm.

**Range in Eastern Africa:** Mt. Kenya, Kenya Colony.

**Habits:** Presumably as for the nominate race, but field notes have not in the past been reliable owing to the occurrence of both this and the following species in the same locality.

**Recorded breeding:** No records.

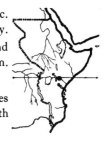

*Geokichla gurneyi chyulu* V. Som.
*Geokichla gurneyi chyulu* Van Someren, Journ. E.A.U. Nat. Hist. Soc. 14, p. 77, 1939: Chyulu Range, south-eastern Kenya Colony.

**Distinguishing characters:** Differs from the neighbouring Usambara race in having the top of head rather less grey and the orange brown below less rich in colour. Wing 104 to 112 mm.

**Range in Eastern Africa:** Chyulu Mts.

**Habits:** As for other races.

**Recorded breeding:** No records.

**845 ABYSSINIAN GROUND-THRUSH.** *GEOKICHLA PIAGGIÆ* (Bouvier).
*Geokichla piaggiæ piaggiæ* (Bouv.). **Pl. 64.**
*Turdus piaggiæ* Bouvier, Bull. Soc. Zool. France, 2, p. 456, 1877: Lake Tana, northern Abyssinia.

**Distinguishing characters:** Similar to the Orange Ground-Thrush, but forehead orange brown or russet instead of olive brown; crown washed with the same colour; and upperside generally more russet brown. The sexes are alike. Wing 96 to 110 mm. The young bird differs from the young bird of the Orange Ground-Thrush in being tawny above, especially on the head, instead of olive brown.

**General distribution:** Abyssinia, Kenya Colony, Uganda and eastern Belgian Congo.

**Range in Eastern Africa:** Abyssinia, northern and western Kenya Colony and Uganda.

**Habits:** Very much those of the last species, with which it has been confused hitherto. Field notes may refer to either species, so it is impossible to give any distinguishing field characterizations. Found mostly in the upper parts of the forest zone.

**Nest and Eggs:** See above. Eggs from Mt. Ruwenzori are said to be two in number, pale greenish blue spotted and blotched with chestnut and purplish grey; about 27 × 20 mm.

**Recorded breeding:** Boma Plateau, south-eastern Sudan (nestling), August. Southern Abyssinia (2 months old young), November. Ruwenzori Mts., March to June.

**Food:** Fruit, molluscs and insect larvæ.

**Call:** Van Someren describes the call from western Uganda as 'hii-o-cie-tchuu—hii-o-cie-tchuu—hii-o-whii-whi-whi' in fact not unlike the word *Geokichla*.

*Geokichla piaggiæ kilimensis* Neum.
*Geocichla gurneyi kilimensis* Neumann, J.f.O. pp. 188 & 310, 1900: Kifinika, Kilimanjaro, north-eastern Tanganyika Territory.

**Distinguishing characters:** Differs from the nominate race in having the colour of the throat and chest darker and deeper. Wing 97 to 107 mm.

**Range in Eastern Africa:** Mt. Kenya and the Aberdares, Kenya Colony, to Mt. Kilimanjaro, north-eastern Tanganyika Territory.

**Habits:** See remarks under the nominate race.

**Recorded breeding:** No records.

*Geokichla piaggiæ rowei* Grant and Praed.
*Geokichla piaggiæ rowei* C. Grant & Mackworth-Praed, Bull. B.O.C. 57, p. 101, 1937: Loliondo Forest, Arusha district, north-eastern Tanganyika Territory.

**Distinguishing characters:** Differs from the preceding races in having the throat and chest paler and the upperside generally more olivaceous. Wing 99 to 106 mm.

# THRUSHES, WHEATEARS, CHATS, AKALATS, etc.

**Range in Eastern Africa:** Loliondo and Magaidu Forests, Arusha district, northern Tanganyika Territory.

**Habits:** See remarks under the nominate race.

**Recorded breeding:** No records.

*Geokichla piaggiæ hadii* Macd.
*Geokichla piaggiæ hadii* Macdonald, Bull. B.O.C. 60, p. 98, 1940: Emogadung, Dongotona Mts., south-eastern Sudan.

**Distinguishing characters:** Differs from the nominate race in being generally darker, more olivaceous above, and the dark parts of wing-coverts and flight feathers blacker. Wing 96 to 106 mm.

**Range in Eastern Africa:** Dongotona and Imatong Mts. to Boma Plateau, south-eastern Sudan.

**Habits:** As for other races.

**Recorded breeding:** No records.

*Geokichla piaggiæ williamsi* Macd.
*Geocichla piaggiæ williamsi* Macdonald, Bull. B.O.C. 69, p. 16, 1948: Mt. Muhavura, Kigezi, south-western Uganda-Ruanda boundary.

**Distinguishing characters:** Differs from the other races in having the whole top of the head to neck warm russet brown; lower rump and upper tail-coverts russet brown; mantle and scapulars suffused with russet. Wing 95 to 105 mm.

**Range in Eastern Africa:** Muhavura Mt., south-western Uganda.

**Habits:** As for other races.

**Recorded breeding:** No records.

**Distribution of other races of the species:** Belgian Congo.

**846** SPOTTED GROUND - THRUSH.  *PSOPHOCICHLA GUTTATA* (Vigors).

*Psophocichla guttata guttata* (Vigors).
*Turdus guttatus* Vigors, P.Z.S. p. 92, 1831: Durban, Natal, South Africa.

**Distinguishing characters:** Above, olivaceous brown; black bars on ear-coverts; wing-coverts tipped with white; below, white, buffish on chest; large blackish spots on chest, breast, and flanks;

under wing-coverts white and blackish; base of inner webs of flight feathers white. The sexes are alike. Wing 117 to 125 mm. Juvenile plumage unrecorded.

**General distribution:** Nyasaland to eastern Cape Province and Natal.

**Range in Eastern Africa:** Soche and Cholo Mts., Nyasaland.

**Habits:** A scarce and local species of damp bush country or of low evergreen forest.

**Nest and Eggs:** The nest is a bulky cup of leaves, roots and moss, lined with dark fibre and placed in a bush or small tree. Eggs two or three, greenish blue, thickly covered with dark reddish brown or greenish brown markings; about 28 × 19 mm.

**Recorded breeding:** Nyasaland, November. Pondoland, eastern Cape Province, January.

**Food:** Insects taken mainly on the ground, especially ants.

**Call:** Is reputed to have a wonderful song.

*Psophocichla guttata fischeri* (Hellm.).
*Turdus guttatus fischeri* Hellmayr, O.M. p. 54, 1901: Pangani River, eastern Tanganyika Territory.

**Distinguishing characters:** Smaller than the nominate race, and spots on the underside smaller and sparser. Wing 108 mm. Juvenile plumage unrecorded.

**Range in Eastern Africa:** Coastal area of Kenya Colony and Tanganyika Territory from Kipini to the Pangani River.

**Habits:** A woodland species feeding on the ground, rare in collections and little known.

**Recorded breeding:** No records.

*Note:* Probably now extinct.

**847 GROUND - SCRAPER THRUSH.** *PSOPHOCICHLA LITSIPSIRUPA* (Smith).
*Psophocichla litsitsirupa simensis* (Rüpp.).
*Merula (Turdus) simensis* Rüppell, N. Wirbelt. Vög. p. 81, pl. 29, fig. 1, 1840: Angethat, northern Abyssinia.

**Distinguishing characters:** Above, ashy brown; below, buff with large black drop-shaped markings; under wing-coverts and greater part of inner webs of flight feathers deep buff. The sexes

# THRUSHES, WHEATEARS, CHATS, AKALATS, etc. 253

are alike. Wing 122 to 142 mm. The young bird has buff tips to the wing-coverts and some indistinct dusky spots on the mantle. Albinistic examples occur.

**Range in Eastern Africa:** Abyssinia as far east as Harar and as far south as the Arussi area.

**Habits:** Common on the high plateau of Abyssinia, particularly near cultivated ground, where it is tame and is said to have all the characteristics of an English Song-Thrush. It feeds mainly on the ground but spends most of its time in trees, and sings from a perch.

**Nest and Eggs:** A base of coarse sticks with a cup nest of grass and fibres without mud; placed in a tree or bush. Eggs three, occasionally four, sea-green covered with purplish spots which tend to form a zone and with a few overlying reddish brown flecks; about 28 × 20 mm.

**Recorded breeding:** Abyssinia, March to May, later in the north.

**Food:** Insects, worms, crustacea, etc.

**Call:** A deep low 'pit-it-it-it-it' or 'tü-tü-tü-tü' also a loud clear and rather monotonously repeated song 'dia-dö-dia-dö' (Erlanger). Alarm note 'chee-chee.'

*Psophocichla litsipsirupa stierlingi* (Reichw.).
*Geocichla litsitsirupa stierlingi* Reichenow, O.M. p. 5, 1900: Iringa, south-central Tanganyika Territory.

**Distinguishing characters:** Differs from the preceding race in being more ashy grey above, and white below with the same large spots. Wing 121 to 136 mm.

**General distribution:** Northern Angola to Tanganyika Territory, Northern Rhodesia and Nyasaland.

**Range in Eastern Africa:** Western and south-central Tanganyika Territory from the Ufipa Plateau to Iringa and Mbeya.

**Habits:** Common in the middle uplands of Tanganyika Territory but uncommon in most parts of Nyasaland. A much shyer bird by all reports than the Abyssinian race and found mostly in open woodland. Zimmer states they have very much the habits of Starlings, living in small flocks and making a murmuring call. The flight is fast and undulating. There is evidence of some migrational movement.

**Recorded breeding:** Tanganyika Territory, October to December. Nyasaland, September to November.

**Distribution of other races of the species:** South Africa, the nominate race being described from between the Orange River and lat. 20° 23′ S.

## ANT-THRUSHES. Family—TURDIDÆ. Genus: *Neocossyphus*.

Two species occur in Eastern Africa, and are found in scrub and undergrowth in woodlands and forest. Little is known of their general habits and breeding.

### 848 RED-TAILED ANT-THRUSH.  *NEOCOSSYPHUS RUFUS* (Fischer and Reichenow).

*Neocossyphus rufus rufus* (Fisch. & Reichw.). **Pl. 66.**
*Pseudocossyphus rufus* Fischer and Reichenow, J.f.O. p. 58, 1884: Pangani River, eastern Tanganyika Territory.

**Distinguishing characters:** Adult male, above, earthy-brown; rump and tail chestnut brown; below, chestnut brown. The sexes are alike. Wing 119 to 125 mm. Juvenile plumage unrecorded.

**Range in Eastern Africa:** The lower Tana River Kenya Colony to the Usambara Mts., Pangani and the Uluguru Mts., Tanganyika Territory, also Zanzibar Island.

*Note:* The superficial resemblance of this bird to the Rufous Flycatcher (*Stizorhina fraseri*) is remarkable, but as they inhabit two different areas within our geographical limits they are not likely to be confused.

**Habits:** A shy bird of forest or coastal scrub, rare and little known. It is believed to feed largely on the ground, and ants form a considerable part of its food.

**Nest and Eggs:** Nest in a hole in a tree. Eggs pale greenish white, heavily freckled with reddish and with red-brown blotches; measurements of an oviduct egg 26·5 × 19·5 mm.

**Recorded breeding:** Zanzibar Island (breeding condition), September.

**Food:** Largely ants.

**Call:** A clicking note followed by a thin high whistle. Alarm note a high-pitched grating noise.

*Neocossyphus rufus gabunensis* Neumann.
*Neocossyphus rufus gabunensis* Neumann, Bull. B.O.C. 21, p. 77, 1908: Ohumbe, Lake Onange, Ogowi River, Gabon.

**Distinguishing characters:** Differs from the nominate race in being smaller and darker; head and neck all round more dusky; mantle darker earthy brown; rump and upper tail-coverts deeper chestnut brown. Wing 109 to 117 mm.

**General distribution:** Cameroons and Gabon to Uganda.

**Range in Eastern Africa:** The Bwamba area of western Uganda.

**Habits:** A scarce species usually seen hunting among fallen leaves on the ground in fairly dense undergrowth, with low chuckling calls.

**Recorded breeding:** Cameroons (nestling), November. Western Uganda (near breeding condition), October. Belgian Congo, April.

### 849 WHITE-TAILED ANT-THRUSH. *NEOCOSSYPHUS POENSIS* (Strickland).

*Neocossyphus poensis præpectoralis* Jacks.
*Neocossyphus præpectoralis* Jackson, Bull. B.O.C. 16, p. 90, 1906: Kibera, Toro, western Uganda.

**Distinguishing characters:** Above, sooty brown; below, tawny chestnut; throat sooty brown; three outermost tail feathers white on inner webs and tips. The sexes are alike. Wing 105 to 110 mm. Juvenile plumage undescribed.

**Range in Eastern Africa:** Western Uganda.

**Habits:** Nothing is known about the race occurring in our area. The West African race is a bird of low lying dense forest usually found in parties and feeding on the ground.

**Nest and Eggs:** Undescribed.

**Recorded breeding:** Belgian Congo (breeding condition), April onwards.

**Food:** No information.

**Call:** A high-pitched whistle. Alarm call a sharp 'sip-sip.'

**Distribution of other races of the species:** West Africa to the Belgian Congo, the nominate race being described from Fernando Po.

## ROCK-THRUSHES. Family—TURDIDÆ. Genus: *Monticola*.

Four species occur in Eastern Africa, two of which are visitors in the non-breeding season from Europe and Asia. They are found in both open and woodland country and are mainly ground birds, though the Mottled Rock-Thrush is more of a tree species.

The Rock-Thrush is remarkable in that the whole population spend the non-breeding season in Africa and southern Arabia, even the far eastern birds not migrating south to India or the Malay States as might be expected, but all moving westwards to Eastern Africa.

### 850 ROCK-THRUSH. *MONTICOLA SAXATILIS* (Linnæus).
*Turdus saxatilis* Linnæus, Syst. Nat. 12th ed. 1, p. 294, 1766: Switzerland.

**Distinguishing characters:** Adult male, head and neck all round and mantle blue grey, feathers tipped with buff and subterminally with black; lower back white tipped with blue grey; wings black, edged with whitish; tail and upper tail-coverts rich tawny; below, from chest to under tail-coverts and including under wing-coverts rich tawny, tipped with white. The female differs from the male in being ashy brown above with buffish feather tips; below, buff or pale tawny with black concentric barring; throat whiter. Wing 113 to 131 mm. The young bird is similar to the adult female but is browner above.

**General distribution:** South and Middle Europe to Asia Minor, Central Asia, Mongolia, China, Morocco and Algeria; in non-breeding season to southern Arabia, northern, western and eastern Africa, as far south as Portuguese Guinea and Tanganyika Territory.

**Range in Eastern Africa:** Northern areas to Mpanda and Iringa, Tanganyika Territory, in non-breeding season, also Zanzibar Island.

**Habits:** A common palæarctic winter migrant to Africa, extending commonly as a winter visitor to about the middle of Tanganyika Territory. To be found wherever there are suitable rocky outcrops or hills and usually singly and tame. They pass through the Sudan in September, where some stay for the winter, the main body reaching Kenya Colony and Uganda in the end of October. The young birds head the migration in the autumn and the old males seem to go first in the spring. The call is a rather soft 'chak-chak.'

# THRUSHES, WHEATEARS, CHATS, AKALATS, etc. 257

## 851 BLUE ROCK-THRUSH. *MONTICOLA SOLITARIA* (Linnæus).

*Monticola solitaria solitaria* (Linn.).
*Turdus solitarius* Linnæus, Syst. Nat. 10th ed. p. 170, 1758: Italy.

**Distinguishing characters:** Adult male, general colour blue-slate with a slight brighter blue gloss, feathers edged with ashy. The female is browner above; below, chin to chest mottled buff and grey; breast to under tail-coverts buff barred blackish and grey. Wing 119 to 131 mm.

**General distribution:** Central and southern Europe, Asia Minor and North Africa; in non-breeding season from Senegal to British Somaliland and south-western Arabia.

**Range in Eastern Africa:** Western Sudan to Eritrea, Abyssinia and British Somaliland in non-breeding season.

**Habits:** A common palæarctic winter visitor to the Sudan in suitable localities, such as rocky hills, ruins, etc. but appears to be a very rare vagrant further south. It has occasionally been recorded in Abyssinia.

*Monticola solitaria longirostris* (Blyth).
*Petrocincla longirostris* Blyth, J.A.S.B. 16, p. 150, 1847: bet. Sind and Ferozepore, India.

**Distinguishing characters:** The male differs from the nominate race in being paler blue-slate. The female is more ashy brown above with little or no slate blue coloration; below, paler. Wing 114 to 128 mm.

**General distribution:** Iran to Kurdistan and Transcaspia; in non-breeding season to the Sudan, British Somaliland, south-western Arabia, Baluchistan and Sind.

**Range in Eastern Africa:** Western Sudan to Eritrea, Abyssinia and British Somaliland in non-breeding season.

**Habits:** A palæarctic winter migrant probably indistinguishable from the nominate race in the field; it is impossible to say how freely it occurs, but it appears to be not uncommon.

**Distribution of other races of the species:** Tibet to China.

258 THRUSHES, WHEATEARS, CHATS, AKALATS, etc.

**852 MOTTLED ROCK-THRUSH or ANGOLA THRUSH. *MONTICOLA ANGOLENSIS* (Sousa). Pl. 64.**

*Monticola angolensis* Sousa, J. Lisboa, 12, pp. 225, 233, 1888: Caconda, Angola.

**Distinguishing characters:** Adult male, above, mottled pale blue grey and black; upper tail-coverts and tail bright tawny; central tail feathers brown; below, sides of face and throat pale blue grey; rest of underparts tawny, darker and brighter on chest; whiter on lower belly. The female is mottled buff and black above; below, throat buff; sides of throat and sides of face mottled buff and black; chest to belly tawny, darker at base of neck. Wing 91 to 104 mm. The young bird is spotted above with buff, grey and black; white mottled with black below; under wing-coverts and tail as in adult.

**General distribution:** Angola to the southern Belgian Congo, Northern Rhodesia, Tanganyika Territory, eastern Southern Rhodesia, Nyasaland, and northern Portuguese East Africa.

**Range in Eastern Africa:** Western and southern Tanganyika Territory from Kungwe-Mahare Mts. and Mpapwa to Nyasaland and Portuguese East Africa.

**Habits:** Widely distributed in Nyasaland and Portuguese East Africa in suitable localities, which are small hills in woodland country or open bush. Usually shy and solitary and settles in trees and not on rocks, with which it seems to have little or no connection.

**Nest and Eggs:** In shallow holes in trees, usually a slight layer of twigs without much lining, but occasionally a nest of coarse grass lined with fine grass and without twigs. Eggs three or four, turquoise blue, sometimes with a few small brown spots; about 23 × 18 mm.

**Recorded breeding:** Belgian Congo, October. Nyasaland, September to December. Southern Rhodesia, October.

**Food:** No information.

**Call:** A small fluty two-noted whistle of which the second note is higher, also an alarm note of a chattering nature.

**853 LITTLE ROCK-THRUSH. *MONTICOLA RUFO-CINEREA* (Rüppell).**

*Monticola rufocinerea rufocinerea* (Rüpp.).

*Saxicola rufocinerea* Rüppell, N. Wirbelt. Vög. p. 76, pl. 27, 1837: Simen Province, northern Abyssinia.

**Distinguishing characters:** Adult male, above, brownish grey; rump, upper tail-coverts and tail deep tawny; central tail feathers and tips of others dusky brown; below, chin to chest pale slate; rest of underparts rich tawny; belly whitish. The female is generally duller and browner and has some white on the chin and throat. Wing 77 to 90 mm. The young bird is spotted with buff and black above and below, but has the under wing-coverts and upper tail-coverts as in the adult.

**Range in Eastern Africa:** Eritrea, Abyssinia, British Somaliland, south-eastern Sudan, eastern Uganda, western half of Kenya Colony and north-eastern Tanganyika Territory at Longido.

**Habits:** A forest country bird in Abyssinia, where it is found on the edges of big forested ravines, not on the high plateau. It occurs fairly freely in many suitable localities, but is distinctly crepuscular in its habits and is easily overlooked. Apparently somewhat migratory over at least the southern part of its range. In the field a constant shivering of the tail is noticeable.

**Nest and Eggs:** Undescribed.

**Recorded breeding:** Southern Sudan (near breeding), November.

**Food:** No information.

**Call:** Unrecorded.

**Distribution of other races of the species:** Arabia.

**WHEATEARS.** FAMILY—TURDIDÆ. Genus: *Œnanthe*.

Fifteen species occur in Eastern Africa, some of which are migrants from Europe and Asia. They are birds of open country, and the white or light markings on the rump and tail are conspicuous characters in flight. They are not always tame, but can be easily observed. Their food consists of insects, small molluscs and occasionally small seeds.

**854** WHEATEAR. *ŒNANTHE ŒNANTHE* (Linnæus).
*Œnanthe œnanthe œnanthe* (Linn.). **Pl. 65.**
*Motacilla Œnanthe* Linnæus, Syst. Nat. 10th ed. p. 186, 1758: Sweden.

**Distinguishing characters:** Adult male, above, grey and brown; forehead and stripe over eye, lower rump and upper tail-coverts white; wings black; whitish edges and tips to wing-coverts and secondaries; whitish tips to primaries; lores to ear-coverts black; below, buffish white, deeper and darker on throat to breast; under wing-coverts black, edged and tipped with white; tail, basal three-quarters white; apical quarter black, tipped with white; central feathers black, base white. The female is generally warm brown throughout and lacks the grey and black of the male. Wing, male 94 to 103, female 87 to 98 mm. The immature bird is very similar to the adult female but rather warmer in colour. Albinistic examples are not uncommon.

**General distribution:** Northern and central Europe to northern Asia, Alaska and the Pribyloff Islands; in non-breeding season to Arabia and Africa as far as Senegal, northern Nigeria, northern Nyasaland, Tanganyika Territory and the Zambesi River, also Vulcan Island and North Borneo.

**Range in Eastern Africa:** Northern areas to Tanganyika Territory and Zanzibar Island in non-breeding season.

**Habits:** A common palæarctic winter visitor to most of Eastern Africa, particularly to Kenya Colony and northern Tanganyika Territory, where it is found in all types of open country. In the Sudan it is a passage migrant as well as a palæarctic winter visitor and there appears to be a regular migration through Darfur from east to west. Although it is a bird of open country it does not like hot sun, and Sir F. Jackson notes that in Kenya Colony it is always in a shady spot in the middle of the day. Its habits are as in Europe and it flits from one vantage post to another, though often it may be from a rock or mound a few inches high to the next. A silent bird in Africa and usually seen singly.

*Œnanthe œnanthe libanotica* (Hemp. & Ehr.).
*Saxicola libanotica* Hemprich & Ehrenberg, Symb. Phys. Aves. fol. b.b.(5) 1828: Lebanon.

**Distinguishing characters:** Differs from the nominate race in being generally paler. Wing, male 90 to 105, female 89 to 101 mm.

**General distribution:** Southern Europe and southern Asia, to Afghanistan and Mongolia; in non-breeding season to Arabia and Africa as far as the Gold Coast, French Sudan, Lake Chad and Tanganyika Territory.

**Range in Eastern Africa:** As far south as Tanganyika Territory in non-breeding season.

**Distribution of other races of the species:** Canada, Greenland, Iceland and Algeria.

# THRUSHES, WHEATEARS, CHATS, AKALATS, etc.

## 855 ISABELLINE WHEATEAR. *ŒNANTHE ISABELLINA* (Temminck and Langier).

*Saxicola isabellina* Temminck & Laugier, Pl. Col. livr. 79, pl. 472, fig. 1, 1829: Nubia.

**Distinguishing characters:** Very similar to the adult female of the Wheatear but appreciably paler in general colour and having the under wing-coverts white with dusky bases. The sexes are alike. Wing 93 to 103 mm.

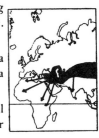

**General distribution:** Southern Russia to Mongolia, Tibet, Asia Minor and Iran; in non-breeding season to Africa, Arabia, Socotra Island and India.

**Range in Eastern Africa:** Northern areas as far south as central Uganda, northern Tanganyika Territory, Pemba and Zanzibar Islands in non-breeding season.

**Habits:** A common palæarctic winter visitor with very much the same habits as the Wheatear, but preferring more desert country on the whole. Found fairly commonly at all elevations on open plains in Kenya Colony, Uganda and Tanganyika Territory, but only extends in numbers to the north of the last-named country. Common in sandy deserts in the Sudan and also on the high Abyssinian plateau.

## 856 SOMALI WHEATEAR. *ŒNANTHE PHILLIPSI* (Shelley). Pl. 65.

*Saxicola phillipsi* Shelley, Ibis, p. 404, pl. 12, 1885: Mountains near Berbera, British Somaliland.

**Distinguishing characters:** Adult male, above, grey; wing-coverts lighter; forehead, stripe over eye and upper tail-coverts white; flight feathers black; sides of face and neck, and chin to breast black; under wing-coverts and flanks black; rest of underparts white; tail white, edges and tips black, central feathers black. The female has the sides of face, neck and chin to breast grey, and tail feathers broadly tipped with black. Wing 77 to 86 mm. The young bird is speckled with buff and blackish.

**Range in Eastern Africa:** British and Italian Somalilands to the Ogaden area of eastern Abyssinia.

**Habits:** Exactly those of the Wheatear, of which it appears to be the local representative.

**Nest and Eggs:** Undescribed.
**Recorded breeding:** No records.
**Call:** Unrecorded.

### 857 DESERT WHEATEAR. *ŒNANTHE DESERTI* (Temminck).

*Œnanthe deserti deserti* (Temminck). **Pl. 65.**
Saxicola deserti Temminck, Pl. Col. livr. 60, pl. 359, fig. 2, 1825: Nubia.

**Distinguishing characters:** Adult male, above, isabelline; forehead, stripe over eye, lower rump and upper tail-coverts buffish white; wings black; white patch on shoulder; tail black, basal third white; below, sides of face, chin to neck black with white tips; chest to under tail-coverts whitish isabelline; under wing-coverts and axillaries mainly black; inner webs of flight feathers white. The female is paler, lacks the black of the male, and has the under wing-coverts and axillaries white with dusky bases. Wing 84 to 98 mm.

**General distribution:** Russian Turkestan to Iran, Palestine and Egypt; in non-breeding season to Lake Chad, the Sudan, Eritrea, British Somaliland, Arabia, Iraq and north-western India.

**Range in Eastern Africa:** The Sudan to Eritrea and British Somaliland in non-breeding season.

**Habits:** Common palæarctic winter visitor to sandy desert and scrub country of the Sudan and across to British Somaliland, but it does not appear to be known south of the equator. Breeds in Egypt, nesting in a hole in the sand. The birds have quite a sweet little song.

*Œnanthe deserti oreophila* (Ober.).
Saxicola oreophila Oberholser, Proc. U.S. Nat. Mus. 22, p. 221, 1900: Tibet.

**Distinguishing characters:** Differs from the preceding race in that the male has the top of the head greyer. Wing 91 to 103 mm.

**General distribution:** Tibet to Baltistan and Chinese Turkestan; in non-breeding season to Socotra Island, southern Arabia and Baluchistan.

**Range in Eastern Africa:** Islands of Socotra and Abd-el-kuri in non-breeding season. Not yet recorded from the mainland.

# THRUSHES, WHEATEARS, CHATS, AKALATS, etc.

**Habits:** An eastern palæarctic race which just comes into our area in the non-breeding season and is indistinguishable in the field from the nominate race.

**Distribution of other races of the species:** The Sahara, Egypt and Palestine.

## 858 BLACK-EARED WHEATEAR. ŒNANTHE HISPANICA (Linnæus).

*Œnanthe hispanica melanoleuca* (Güld.).
*Muscicapa melanoleuca* Güldenstadt, Nov. Com. Petrop. 19, p. 468, 1775: Georgia, Caucasus.

**Distinguishing characters:** The black-throated phase of the adult male is similar below to the Desert Wheatear, but above, the head to rump is buffish white and the wings wholly black. The light-throated form of the male is similar but has the chin to throat pale isabelline uniform with chest to under tail-coverts. The female is darker above, browner than the Desert Wheatear, and has distinctly dusky under wing-coverts and axillaries; there is also a dark and a light-throated phase. Wing 81 to 96 mm. The immature male is very similar to the adult female.

**General distribution:** Southern Italy to the Balkan peninsula, Crete, the Caucasus and Iran, in non-breeding season to Algeria and the eastern side of Africa as far south as the Sudan and Eritrea, also south-western Arabia.

**Range in Eastern Africa:** The Sudan to Eritrea and northern Abyssinia in non-breeding season.

**Habits:** A common palæarctic winter visitor of rather late appearance to the Sudan, Abyssinia and the Red Sea coast, not usually arriving before October. It appears to be most active in the evening and has been noted hawking flies over water after dusk like a Flycatcher.

**Distribution of other races of the species:** Southern Europe to Iran, the nominate race being described from Gibraltar.

## 859 PIED WHEATEAR. ŒNANTHE LEUCOMELA (Pallas).

*Œnanthe leucomela leucomela* (Pall.).
*Motacilla leucomela* Pallas, Nov. Com. Petrop. 14, p. 584, pl. 22, 1771: Lower Volga, southern Russia.

**Distinguishing characters:** Adult male, similar to the Black-eared Wheatear, but has the mantle black tipped with buff. The

## 264 THRUSHES, WHEATEARS, CHATS, AKALATS, etc.

female is darker, less brown, more dark ash colour than the female of the Black-eared Wheatear. Rarely both sexes have no black on the chin and throat, which are uniform with the rest of the underside. Wing, male 89 to 101, female 86 to 98 mm.

**General distribution:** Southern Russia to Turkestan, south-eastern Siberia and northern China; in non-breeding season to the Sudan, Eritrea, British Somaliland, Tanganyika Territory and southern Arabia.

**Range in Eastern Africa:** The Sudan and Eritrea, British Somaliland, Kenya Colony and northern Tanganyika Territory in non-breeding season, one record from Zanzibar Island.

**Habits:** A common, at times very common, palæarctic migrant to the high plateau of Abyssinia, and south as far as northern Tanganyika Territory. Generally single and usually conspicuous as it is fond of perching on any eminence. Has a sweet little call of two notes in spring.

*Œnanthe leucomela cypriaca* (Hom.).
*Saxicola cypriaca* v. Homeyer, Zeitsch. ges. Orn. 1, p. 397, 1884: Cyprus.

**Distinguishing characters:** Differs from the nominate race in being smaller. Wing, male 82 to 88, female 79 to 83 mm.

**General distribution:** Cyprus to Syria; in non-breeding season to the Sudan, Kenya Colony and Arabia.

**Range in Eastern Africa:** The Sudan to central Kenya Colony in non-breeding season.

**Habits:** Of the nominate race and indistinguishable in the field.

860 WHITE-RUMPED WHEATEAR. *ŒNANTHE LEUCOPYGA* (Brehm.).
*Œnanthe leucopyga leucopyga* (Brehm.).
*Vitiflora leucopyga* Brehm. Vogelf. p. 225, 1855: Nubia.

**Distinguishing characters:** Adult male, general colour glossy blue black; crown of head and upper and under tail-coverts white; tail feathers, except central pair, white with black subterminal spots. The sexes are alike. Wing 93 to 113 mm. The immature bird has the crown of the head black. The young bird is brownish black, with lighter tips to the wing-coverts and flight feathers.

**General distribution:** Egypt to Abyssinia and French Somaliland.

**Range in Eastern Africa:** The eastern Sudan as far south as lat. 15° N. to eastern Abyssinia north of the Adis Ababa–Jibuti railway line, and French Somaliland.

**Habits:** A locally common resident of the eastern Sudan, Eritrea and eastern Abyssinia with the habits of a typical Wheatear, frequenting bare rocky or stony ground.

**Nest and Eggs:** In houses, walls, rocky ravines or under stones, a nest of dried grass stems lined with any form of soft material. Eggs three or four, pale greenish-blue with spots or blotches of pale reddish brown, mostly at larger end; about $20 \cdot 5 \times 16$ mm.

**Recorded breeding:** Sudan, April.

**Call:** A delightful subdued little song.

*Œnanthe leucopyga ægra* Hart.
*Œnanthe leucopyga ægra* Hartert, Nov. Zool. 20, p. 55, 1913: Gara Klima, southern Algeria.

**Distinguishing characters:** Differs from the nominate race in being rather smaller and in being slightly glossy black, not glossy blue black, and having less black on forehead in adult males. Wing 97 to 107 mm.

**General distribution:** Algeria to Tunis, Air and Darfur.

**Range in Eastern Africa:** Jebels Marra and Meidob, western Sudan.

**Habits:** Of the nominate race, breeding in the western Sudan and the central Sahara from December to February. Probably indistinguishable in the field.

**Recorded breeding:** Air, central Sahara, 'as late as June.' Darfur (nestlings), April and December.

**Distribution of other races of the species:** Sinai.

**861 RED-RUMPED WHEATEAR.** *ŒNANTHE XANTHOPRYMNA* (Hemprich & Ehrenberg). **Pl. 65.**
*Saxicola xanthoprymna* Hemprich & Ehrenberg, Symb. Phys. fol. d.d. 1828: Nubia.

**Distinguishing characters:** Adult male, above, ashy brown; rump russet; white stripe over eye; sides of face and neck black; below, chin to throat, under wing-coverts and axillaries black; chest to belly buffish white; lower belly and under wing-coverts washed with russet; tail tawny and black. The female is duller, with duller black chin and throat. Wing 86 to 95 mm. The young bird has the dull chin and throat of the adult female and dusky not black under wing-coverts and axillaries.

**General distribution:** Iraq and western Iran; in non-breeding season to Egypt, the Sudan, Eritrea and western Arabia.

**Range in Eastern Africa:** The coastal area of the Sudan and Eritrea in non-breeding season.

**Habits:** A rare and little known species, with typical Wheatear habits. It is not known whether this species ever breeds within our limits, and its breeding habits appear to be undescribed.

**862 RED-TAILED WHEATEAR.** *ŒNANTHE CHRYSO-PYGIA* (De Filippi). **Pl. 65.**

*Dromolæa chrysopygia* De Filippi, Arch. Zool. Genova, 2, p. 381, 1863: Dernavend, Iran.

**Distinguishing characters:** Adult male, very similar above and below, including rump and tail, to the Red-rumped Wheatear; but sides of face and neck, chin and throat buffish white; lores dusky; under wing-coverts and axillaries white. The female is slightly browner above than the male. Wing 85 to 98 mm.

**General distribution:** Armenia to northern Iran and southern Russia; in non-breeding season to the Sudan, Arabia and India.

**Range in Eastern Africa:** Coastal areas of Eritrea and the Sudan in non-breeding season.

**Habits:** Local, and often found at the bases of rocky hills among acacia scrub, but with much the same range as the Red-rumped Wheatear, and has been freely confused with it in the past.

**863 WHITE-UNDERWINGED WHEATEAR.** *ŒNANTHE LUGENS* (Lichtenstein).

*Œnanthe lugens persica* (Seeb.). **Pl. 65.**

*Saxicola persica* Seebohm, Cat. Bds. B.M. 5, p. 372, 1881: Shiraz, Iran.

**Distinguishing characters:** Very similar to the adult male Pied Wheatear, but bill rather heavier; inner webs of flight feathers much lighter and black of sides of face and throat extending more on to the sides of the chest; chest to belly white. The sexes are alike. Wing 91 to 105 mm.

**General distribution:** Syria and southern Iran; in non-breeding season to Egypt and the Sudan.

**Range in Eastern Africa:** Northern Sudan in non-breeding season.

**Habits:** A palæarctic winter migrant to the extreme north of our area frequenting rocky outcrops and low rocky hills. Not common and usually shy, with a habit of diving under rocks if alarmed.

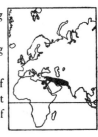

**Distribution of other races of the species:** Egypt to Palestine, the nominate race being described from north Nubia, Upper Egypt.

## 864 HOODED WHEATEAR. *ŒNANTHE MONACHA* (Temminck). Pl. 65.

*Saxicola monacha* Temminck, Pl. Col. livr. 60, p. 359, fig. 1, 1825: Nubia.

**Distinguishing characters:** Adult male, very similar to the White-underwinged Wheatear, but black of chin and throat extends well down on to the chest; the white on the rump extends on to the lower back; the tail is almost wholly white, the central pair of feathers having the apical half sooty brown, and there are some similar coloured markings at the ends of some of the other feathers. The female is buff to ashy brown above; lower back to upper tail-coverts and underside isabelline; throat lighter; tail isabelline with more extensive sooty brown than the male, especially on the central feathers. Wing 100 to 108 mm. Juvenile plumage unrecorded.

**General distribution:** Egypt and the Sudan, Palestine, Iraq, Arabia and north-western India.

**Range in Eastern Africa:** The Sudan from the Nile Valley to the Red Sea coast north of about lat. 18° N.

**Habits:** A bird of desolate wadis, or desert bush country, preferring cliffs and ravines, and rarely seen near cultivation. It appears to be nowhere common.

**Nest and Eggs:** Undescribed.

**Recorded breeding:** Egypt (young), June.

**Call:** A subdued but very sweet song in the spring.

## 865 ABYSSINIAN BLACK WHEATEAR. *ŒNANTHE LUGUBRIS* (Rüppell).

*Œnanthe lugubris lugubris* (Rüpp.). **Pl. 65.**

*Saxicola lugubris* Rüppell, N. Wirbelt. Vög. p. 77, pl. 28, fig. 1, 1837: Simen, northern Abyssinia.

**Distinguishing characters:** Adult male, general colour black, sometimes chest to belly white or mixed black and white; top of head sooty brown with black streaks; upper and under tail-coverts buff; basal two-thirds of all except central tail feathers white; lower belly white. Birds with white from chest to belly sometimes have the basal two-thirds of all except central tail feathers tawny not white. The female is sooty black above; below, dusky white; more sooty brown on chin to chest; flanks darker, streaked with sooty black; white in the tail tinged with tawny. Wing 78 to 90 mm. Juvenile plumage unrecorded.

**Range in Eastern Africa:** Eritrea to central Abyssinia and British Somaliland.

**Habits:** Local resident of the ravines and gullies of the high plateau of Abyssinia, rare and little known.

**Recorded breeding:** Eritrea, July and August.

*Œnanthe lugubris schalowi* (Fisch. & Reichw.). *Schalow's Wheatear.* **Pl. 65. Ph. xi.**

*Saxicola schalowi* Fischer & Reichenow, J.f.O. p. 57, 1884: Lake Naivasha, western Kenya Colony.

**Distinguishing characters:** Adult male, differs from the nominate race in having the chest to belly constantly white and the basal two-thirds of all except the central tail feathers always tawny. The female is sooty brown above; below, chin to chest paler with darker brown streaks; belly dusky white; flanks streaked sooty brown. Wing 84 to 93 mm.

**Range in Eastern Africa:** Southern Kenya Colony from Molo to Naivasha, and eastern Tanganyika Territory from Ngorongoro Crater to Mt. Meru.

**Habits:** The resident Wheatear of the Rift Valley, inhabiting steep boulder-strewn slopes. Always singly or in pairs.

**Nest and Eggs:** In a deep crevice in rocks, stone walls or banks, a cup nest of dry grass stems and roots lined with rootlets and hair.

Eggs one to three, very pale blue or blue green to almost pure white, smooth and glossy, occasionally spotted or blotched pale brown, reddish brown and purplish brown; about 20 × 15·5 mm.

**Recorded breeding:** Rift Valley, April to July. Northern Tanganyika Territory, March.

**Call:** Unrecorded.

*Œnanthe lugubris vauriei* Meinertzhagen.
*Œnanthe lugens vauriei* Meinertzhagen, Bull. B.O.C. 69, p. 107, 1949: Erigavo, British Somaliland.

**Distinguishing characters:** Adult male, differs from other races in having the crown whiter; basal half of tail and rump pale orange. The female is paler and browner. Wing 80 to 84 mm.

**Range in Eastern Africa:** Eastern British Somaliland.

**Habits:** As for other races, local and inhabiting rocky ravines.

**Recorded breeding:** No records.

**866 RED-BREASTED WHEATEAR.** *ŒNANTHE BOTTÆ*
(Bonaparte).
*Œnanthe bottæ frenata* (Heugl.). **Pl. 65.**
*Saxicola frenata* Heuglin, J.f.O. p. 158, 1869: Mensa, Abyssinia.

**Distinguishing characters:** Above, brown washed with ashy; upper tail-coverts white; lores to ear-coverts blackish; below, russet; throat and belly whiter; under wing-coverts white or white tinged with russet; basal half and tips of all except central tail feathers white. The sexes are alike. Wing 95 to 106 mm. Juvenile plumage unrecorded.

**Range in Eastern Africa:** Eritrea to central Abyssinia.

**Habits:** A common resident of the Abyssinian high plateau occurring in downland meadows of short grass at high elevations. Habits very much those of a typical Wheatear.

**Nest and Eggs:** Undescribed.

**Recorded breeding:** Abyssinia probably May and June.

**Call:** A twittering song heard all night in April, especially on moonlight nights.

**Distribution of other races of the species:** Arabia, the nominate race being described from the Yemen.

**867** HEUGLIN'S RED-BREASTED WHEATEAR. *ŒNANTHE HEUGLINI* (Finsch. & Hartlaub).

*Saxicola heuglini* Finsch & Hartlaub, Vög. Ost-Afr. in V. der Decken Reisen, 4, p. 259, 1870: Sudan.

**Distinguishing characters:** Very similar to the Red-breasted Wheatear but smaller and darker, especially above. The sexes are alike. Wing 82 to 94 mm. The young bird is mottled with blackish below, and spotted above with tawny.

**General distribution:** French Sudan and northern Gold Coast, to the Sudan, Eritrea and Kenya Colony.

**Range in Eastern Africa:** The Sudan from Darfur to Khartoum, Wad Medani and Gedaref, western Eritrea and Kavirondo in western Kenya Colony where probably a straggler.

**Habits:** A bird of cotton soil and low hillsides, especially among recently burned grass, tame and conspicuous. Common in many parts of the Sudan, being locally abundant in the Jebel Marra, and an occasional visitor to Kenya Colony. Has been noted by many observers for its fondness for burnt ground, where it resorts to breed before the grass grows too high again. It also visits grass fires from a considerable distance. In the breeding season the cock has a pretty little mating flight.

**Nest and Eggs:** Undescribed.

**Recorded breeding:** Western Sudan, February to May.

**Call:** A harsh 'chak' also a sweet little song in the breeding season 'twee-twee-twee-twee.'

**868** CAPPED WHEATEAR. *ŒNANTHE PILEATA* (Gmelin) *Œnanthe pileata livingstonii* (Trist.). **Pl. 65.**

*Campicola livingstonii* Tristram, P.Z.S. p. 888, 1867: Murchison Flats, southern Nyasaland.

**Distinguishing characters:** Above, dark russet or blackish brown; forehead and stripe over eye white; upper forehead and crown black or blackish merging into colour of back on nape; upper tail-coverts white; basal half of all except central tail feathers white; lores to ear-coverts and down side of neck black; below, white including under wing-coverts, broad black band across chest joining the black on sides of neck; flanks buff. The sexes are alike. Wing 86 to 100 mm. The young bird is brown above, with a few buff spots; buffish brown below; chin to breast mottled with dusky.

**General distribution:** Angola to Kenya Colony, the Rhodesias, south-eastern Belgian Congo, Nyasaland, Bechuanaland and Damaraland.

**Range in Eastern Africa:** Central Kenya Colony to the Zambesi River, also Zanzibar Island.

**Habits:** A tame and conspicuous bird of open sandy country or grass land. Widely distributed throughout its range but not usually common, except possibly in the Kenya Colony highlands. In the breeding season resorts to various devices much as Plovers or Waders do to lure intruders away from its nesting site. Unlike the true Wheatears quite often perches on a bush, and has a far wider range of calls. Fond of newly burned ground.

**Nest and Eggs:** Nest about a foot underground in a rat-hole or ant heap, made of dry grass lined with hair or rootlets. Eggs three, white, with slight gloss and a faint bluish tinge sometimes with pale red spots; about $24 \cdot 5 \times 16 \cdot 5$ mm.

**Recorded breeding:** Kenya Colony, April to June, also December and January. South-eastern Belgian Congo, August. Nyasaland, May to November, also January. Rhodesia, July to September.

**Call:** Many and varied calls and the birds are excellent mimics. They have a little soaring flight from which they descend with fluttering wings and a little jingling song.

**Distribution of other races of the species:** South Africa, the nominate race being described from the Cape of Good Hope.

## CHATS. FAMILY—TURDIDÆ. Genera: *Cercomela, Pinarochroa, Pentholæa, Thamnolæa, Myrmecocichla* and *Saxicola*.

Fifteen species occur in Eastern Africa. They are found in open country, mainly in rocky areas, although one or two are woodland birds. They are usually lively and conspicuous, and though they are shy can always be seen or heard. Their food consists mainly of insects picked up from bare ground.

Two races of the Stonechat are migrants from Europe and Asia and there are a number of resident races. Stonechats are usually found in open rough country and are perhaps most plentiful on open highlands with sparse vegetation and rocky outcrops. The migrants may be found in more diverse types of country, although all are seen perched on the tops of bushes or other tall herbage, and the male particularly is very conspicuous.

**869 BLACK-TAILED ROCK-CHAT.** *CERCOMELA MELANURA* (Temminck).

*Cercomela melanura lypura* (Hemp. & Ehr.). **Pl. 66.**

*Sylvia lypura* Hemprich & Ehrenberg, Symb. Phys. fol. e.e., 1828: Eastern Eritrea.

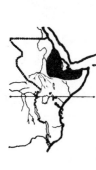

**Distinguishing characters:** Above, ash grey; tail and upper tail-coverts bronzy black, rarely barred ashy; below, greyish white; under wing-coverts and under tail-coverts white. The sexes are alike. Wing 70 to 81 mm. The young bird is generally slightly browner; throat greyish; some buffish tips to wing-coverts.

**Range in Eastern Africa:** Eastern Sudan, Eritrea, eastern Abyssinia and British Somaliland.

**Habits:** Birds of dry and rocky country, tame but restless little creatures, ceaselessly flitting about with much opening and closing of wings and tails. Lynes describes this action well as 'twinkling.'

**Nest and Eggs:** Nest in a hole under a rock, usually with a little platform or surround of pebbles outside, made of grass lined with hair or other soft material. Eggs three, pale blue with rusty brown spots, and mauve grey undermarkings; about 18 × 13 mm.

**Recorded breeding:** Sudan, Nile Valley, August. Red Sea Hills, April and May. Abyssinia, April.

**Call:** A subdued chirp and in the breeding season a sweet little monotonous song.

*Cercomela melanura airensis* Hart.

*Cercomela melanura airensis* Hartert, Nov. Zool. 28, p. 114, 1921: Mt. Baguezan, French Sudan.

**Distinguishing characters:** Differs from the preceding race in being generally warmer in tone. Wing 75 to 84 mm.

**General distribution:** Asben in French Sudan to the Sudan.

**Range in Eastern Africa:** Western Sudan from Darfur to Kordofan.

**Habits:** As for other races.

**Recorded breeding:** Darfur, May.

*Cercomela melanura aussæ* Thes. & Mey.
*Cercomela melanura aussæ* Thesiger & Meynell, Bull. B.O.C. 55, p. 79, 1934: Aussa, Danakil, eastern Abyssinia.

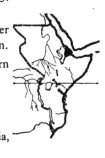

**Distinguishing characters:** General colour darker and rather greyer than either of the two preceding races. Wing 71 to 82 mm.

**Range in Eastern Africa:** The Danakil country of eastern Abyssinia to French Somaliland.

**Habits:** As for other races.

**Recorded breeding:** Aussa, Abyssinia (nestling), April.

**Distribution of other races of the species:** Sinai and Arabia, the nominate race being described from Arabia.

## 870 BROWN-TAILED ROCK-CHAT. *CERCOMELA SCOTOCERCA* (Heuglin).

*Cercomela scotocerca scotocerca* (Heugl.). **Pl. 66.**
*Saxicola scotocerca* Heuglin, Orn. Nordost Afr. 1, p. 363, 1869: near Keren, Bogosland, Eritrea.

**Distinguishing characters:** Similar to the Black-tailed Rock-Chat, but smaller and tail and upper tail-coverts brown to ash grey instead of black. The sexes are alike. Wing 67 to 73 mm. Juvenile plumage unrecorded.

**Range in Eastern Africa:** The Red Sea province of the Sudan and Eritrea.

**Habits:** Almost identical in habits with the Black-tailed Rock-Chat and occasionally found with it, but on the whole inhabits more fertile country and is freely found in bush.

**Nest and Eggs:** Like that of the last species, a grass nest lined with hair, in a hole or deep crevice in rocks with a little slope of pebbles outside. Eggs undescribed.

**Recorded breeding:** Sudan, Red Sea Hills, April and May.

**Call:** Of Turkana race, a sweet strong 'chuke-chuke' and also a short sweet trill.

*Cercomela scotocerca spectatrix* Clarke.
*Cercomela spectatrix* Stephenson Clarke, Bull. B.O.C. 40, p. 49, 1919: Las Khorai, eastern British Somaliland.

**Distinguishing characters:** Larger and greyer than the nominate race. Wing 75 to 84 mm.

**Range in Eastern Africa:** Eastern Abyssinia and northern British Somaliland.

**Habits:** Presumably as for the nominate race.

**Recorded breeding:** No records.

*Cercomela scotocerca turkana* V. Som.
*Cercomela turkana* Van Someren, Bull. B.O.C. 40, p. 91, 1920: Turkana, north-western Kenya Colony.

**Distinguishing characters:** General colour darker brown than the nominate race. Wing 71 to 80 mm.

**Range in Eastern Africa:** South-western Abyssinia and northern Kenya Colony.

**Habits:** As for the nominate race, having a wide range in desert country, and feeding on the ground among rocks. Often flies into a bush and creeps down like a mouse.

**Recorded breeding:** No records.

*Cercomela scotocerca furensis* Lynes.
*Cercomela scotocerca furensis* Lynes, Ibis, p. 391, 1926: Jebel Marra, Darfur, western Sudan.

**Distinguishing characters:** Similar to the preceding race, but browner, warmer, in tone, not so dark. Wing 71 to 80 mm.

**Range in Eastern Africa:** Darfur area, western Sudan.

**Habits:** As for the nominate race.

**Recorded breeding:** No records.

## 871 RED-TAILED CHAT. *CERCOMELA FAMILIARIS* (Stephens).

*Cercomela familiaris falkensteini* (Cab.). **Pl. 66.**
*Saxicola falkensteini* Cabanis, J.f.O. p. 235, 1875: Chinchocho, Portuguese Congo.

**Distinguishing characters:** Very similar to the Brown-tailed Rock-Chat, but tail chestnut with central tail feathers and tips of the others blackish brown. The sexes are alike. Wing 73 to 86 mm. The young bird is mottled with blackish below, and spotted above with dusky and dull buff.

## THRUSHES, WHEATEARS, CHATS, AKALATS, etc.

**General distribution:** Gold Coast, northern Nigeria and Portuguese Congo, to Abyssinia, Kenya Colony, Tanganyika Territory, Northern Rhodesia and northern Portuguese East Africa.

**Range in Eastern Africa:** Kordofan area of the Sudan to northern Abyssinia, and south to Portuguese East Africa and Nyasaland.

**Habits:** A very widespread species in many types of country, and at various elevations from rocky highlands to cultivated lowlands. Usually common and tame, and especially to be looked for in rocky hills among woodland, where it lives among trees but feeds on the ground. Habits much those of a Redstart, pursuing insects on the wing or on the ground with much flirting of tail and wings.

**Nest and Eggs:** Nest large and bulky, often very large, of mud, bark, wood or anything available with a thick cup of wool, hair, feathers or other soft material. The nest is either in a shallow hole in a bank, wall or tree, or in a rock crevice or other situation, very like a Robin's nest in Europe. Eggs three or four, bright greenish-blue with rufour or chestnut markings at the larger end, often forming a cap; about 20 × 15 mm.

**Recorded breeding:** Northern Nigeria, April. South-western Kenya Colony, October. Tanganyika Territory, October to December. North-eastern Northern Rhodesia, October. Nyasaland, August to November.

**Call:** A deep whistled 'sweep-sweep-sweep,' alarm note a squeaky scolding 'whee-chuck-chuck.'

*Cercomela familiaris omoensis* (Neum.).
*Saxicola galtoni omoensis* Neumann, O.M. p. 163, 1904: Baka, south-western Abyssinia.

**Distinguishing characters:** Differs from the preceding race in being darker and greyer above, less brown. Wing 75 to 85 mm.

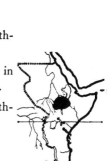

**Range in Eastern Africa:** South-eastern Sudan and south-western Abyssinia.

**Habits:** As for the last race.

**Recorded breeding:** No records.

**Distribution of other races of the species:** West to South Africa, the nominate race being described from South Africa.

T

276  THRUSHES, WHEATEARS, CHATS, AKALATS, etc.

**872 SOMBRE ROCK-CHAT.** *CERCOMELA DUBIA* (Blundell & Lovat). Pl. 66.
*Myrmecocichla dubia* Blundell & Lovat, Bull. B.O.C. 10, p. 22, 1899: Fontaly, central Abyssinia.

**Distinguishing characters:** Very similar to the Black-tailed Rock-Chat but bill rather stouter; upper tail-coverts and tail not so dark, the former with ashy brown edges; under tail-coverts brownish black, edged with either brown or grey. The sexes are alike. Wing 80 to 83 mm. Juvenile plumage unrecorded.

**Range in Eastern Africa:** Central Abyssinia to British Somaliland.

**Habits:** Described as a large upstanding bird with the same 'twinkling' habits as the Black-tailed Rock-Chat. Rare and little observed.

**Nest and Eggs:** Undescribed.

**Recorded breeding:** No records.

**Call:** Unrecorded.

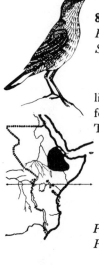

**873 HILL CHAT.** *PINAROCHROA SORDIDA* (Rüppell).
*Pinarochroa sordida sordida* (Rüpp.).
*Saxicola sordida* Rüppell, N. Wirbelt. Vög. p. 75, pl. 26, fig. 2, 1837: Simen, northern Abyssinia.

**Distinguishing characters:** Above and below, dull brown; rather lighter from breast to flanks, and belly paler still; all except central tail feathers white with dull brown markings on outer webs near and at tip. The sexes are alike. Wing 65 to 78 mm. Juvenile plumage unrecorded.

**Range in Eastern Africa:** Abyssinia.

**Habits:** As for other races.

**Nest and Eggs:** See under other races.

**Recorded breeding:** Eastern Abyssinia, probably March.

*Pinarochroa sordida hypospodia* Shell.
*Pinarochroa hypospodia* Shelley, P.Z.S. p. 226, pl. 13, 1885: Mt. Kilimanjaro, north-eastern Tanganyika Territory.

**Distinguishing characters:** Above, darker, more blackish brown than the nominate race; bill dark with a pale yellow patch on lower mandible. Wing 73 to 76 mm. The young bird is indistinctly barred above and has the chest spotted with dusky.

**Range in Eastern Africa:** Mt. Kilimanjaro, north-eastern Tanganyika Territory.

**Habits:** Quiet little birds, usually extremely tame, of very high altitudes in rocky moorland. Their usual range is from 12,000 to 16,000 feet. They flit about among rocks and boulders with a pleasant chirping call. They have the same habit of flirting their tails and wings as the other Rock-Chats, and usually occur in small parties. Nest in clefts of rocks. Eggs described as white with faint tawny spots and streaks. No measurements available.

**Recorded breeding:** Mt. Kilimanjaro at 12,500 feet, June.

*Pinarochroa sordida ernesti* Sharpe.
*Pinarochroa ernesti* Sharpe, Bull. B.O.C. 10, p. 36, 1900: Mt. Kenya, Kenya Colony.

**Distinguishing characters:** Differs from the nominate race in being darker and warmer brown above, but not the blackish brown of the preceding race. Wing 72 to 78 mm. The young bird has darker tips to the feathers both above and below, giving a slightly mottled appearance, especially on the underside.

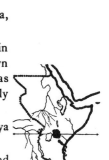

**Range in Eastern Africa:** Mountain areas of central Kenya Colony from Mt. Elgon to Mt. Kenya and the Aberdares.

**Habits:** Those of the last race, but is found at lower elevations and occurs in bamboo glades and open spaces down to 7,500 feet. Its general behaviour is that of a Stonechat and it perches on flower stems or twigs. The tail opens fanwise when jerked. They are often very tame.

**Nest and Eggs:** The only nest so far described was a grass cup lined with moss and lichens and placed in a cleft at the top of a bole of Giant Groundsel at about 13,000 feet. Eggs three, blue, spotted and streaked with black and violet; probably about 22 × 16 mm.

**Recorded breeding:** Mount Kenya, August, January and February.

**Food:** Largely small beetles.

**Call:** A metallic little call of 'werp-werp' and a more sibilant alarm call.

*Pinarochroa sordida olimotiensis* Ell.
*Pinarochroa sordida olimotiensis* Elliott, Bull. B.O.C. 66, p. 19, 1945: Olimoti Mt., north of Ngorongoro, north-eastern Tanganyika Territory.

**Distinguishing characters:** Differs from the neighbouring Kilimanjaro race in being less sooty, rather browner above; bill wholly black. Wing 73 to 75 mm. The young bird is also less sooty above.

278   THRUSHES, WHEATEARS, CHATS, AKALATS, etc.

**Range in Eastern Africa:** Highlands of north-eastern Tanganyika Territory between long. 35° and 36° E., north of Lakes Eyasi and Manyara.

**Habits:** As for other races, found at about 9,000 feet on the summit crater rim of Mt. Olmoti in tussocky grass land and Artemisia-covered sub-alpine scrub. Very tame. Nest in side of tussock or similar situation a little above ground.

**Recorded breeding:** December and January.

**874   WHITE-FRONTED BLACK CHAT. *PENTHOLÆA ALBIFRONS* (Rüppell).**

*Pentholæa albifrons albifrons* (Rüpp.).
*Saxicola albifrons* Rüppell, N. Wirbelt. Vög. p. 78, 1837: Mt. Takeragiro, Temben Province, northern Abyssinia.

**Distinguishing characters:** Adult male, general colour slightly glossy black; forehead white; inner webs of flight feathers silvery whitish below. The female has ashy tips to the feathers of the forehead, chin and neck. Wing 73 to 81 mm. Juvenile plumage unrecorded.

**Range in Eastern Africa:** Eritrea and Abyssinia as far south as Yavello.

**Habits:** Presumably as for other races, but nothing so far recorded.

**Nest and Eggs:** See under the next race.

**Recorded breeding:** No records.

*Pentholæa albifrons clericalis* Hartl.   *White-shouldered Black Chat.*
*Pentholæa clericalis* Hartlaub, O.C. p. 91, 1882: Langomeri, north-western Uganda.

**Distinguishing characters:** Adult male, differs from the nominate race in having a smaller bill; more white on forehead; a white patch on the wing shoulder; no silvery white inner webs to the flight feathers. The female lacks the white on the forehead and wing shoulder. Wing 70 to 80 mm. The immature male has no white on forehead, but some white on wing shoulder, and tawny tips and edges to the wing-coverts and innermost secondaries; the immature female has the tawny tips and edges to the wing-coverts and innermost secondaries. The young bird is spotted above and below with tawny.

## THRUSHES, WHEATEARS, CHATS, AKALATS, etc.

**General distribution:** North-eastern Belgian Congo to the Sudan and Uganda.

**Range in Eastern Africa:** Southern Sudan and northern half of Uganda.

**Habits:** Little known, though not uncommon and rather conspicuous birds. In appearance they look like Black Tits, except for the shoulders and forehead of the male, and among trees they have something of a Tit's habits. Usually in light bush or thin strips of woodland.

**Nest and Eggs:** The nest, of dry grass, is placed in crevices under large boulders. Eggs three, pale green with light rufous spots.

**Recorded breeding:** Belgian Congo, January and February.

**Call:** The West African race has a call of 'tweet' or a clear little burst of song, uttered while flicking the tail, also other chattering notes.

*Pentholæa albifrons pachyrhyncha* Neum.
*Pentholæa albifrons pachyrhyncha* Neumann, O.M. p. 8, 1906: Uba, Omo district, south-western Abyssinia.

**Distinguishing characters:** Adult male, differs from the nominate race in being more glossy blue black; and in having the white extending from the forehead to the crown. The female has an ashy grey forehead and throat. Wing 75 to 85 mm. The immature male has the white restricted to near the bill and some ashy tips to the belly feathers and upper and under tail-coverts.

**Range in Eastern Africa:** Gofa and Uba areas, south-western Abyssinia.

**Habits:** As for the other races.

**Recorded breeding:** No records.

**Distribution of other races of the species:** West Africa.

**875 RÜPPELL'S CHAT.** *PENTHOLÆA MELÆNA* (Rüppell).
*Saxicola melæna* Rüppell, N. Wirbelt. Vög. p. 77, pl. 28, 1837: Alegua Mt., Agami Province, northern Abyssinia.

**Distinguishing characters:** General colour wholly black, except for inner webs of primaries and innermost secondaries which are white, sharply contrasting with the black of the rest of the wing; larger than the White-fronted Black Chat. The sexes are alike. Wing 85 to 94 mm. Juvenile plumage unrecorded.

## 280 THRUSHES, WHEATEARS, CHATS, AKALATS, etc.

**Range in Eastern Africa:** Southern Eritrea and northern and central Abyssinia.

**Habits:** A rare and local species of the high Abyssinian plateau. Usually met with in pairs among waterfalls and wet rocks at the tops of precipitous ravines. They have a habit of flirting the tail high over the back.

**Nest and Eggs:** Undescribed.

**Recorded breeding:** No records.

**Call:** Unrecorded.

### 876 CLIFF-CHAT. *THAMNOLÆA CINNAMOMEIVENTRIS* (Lafresnaye).

*Thamnolæa cinnamomeiventris subrufipennis* Reichw. Tropical Cliff-Chat. **Pl. 64.**

*Thamnolæa subrufipennis* Reichenow, J.f.O. p. 78, 1887: near Ussure, Kondoa Irangi district, north-central Tanganyika Territory.

**Distinguishing characters:** Adult male, head all round to chest, mantle, wings above and below, and tail glossy blue-black; white patch on wing shoulder; rump and breast to under tail-coverts chestnut; sometimes white on upper breast. The female has the black in the male replaced with grey; the wings and tail washed with grey; and there is no white patch on the wing shoulder. Wing 103 to 122 mm. In the female a rare albinistic phase occurs having the grey replaced by silvery grey with grey shaft streaks. The young male is rather duller and is more fluffy in appearance than the adult, and has less white on the wing shoulder. The young female is duller than the adult and has a wash of chestnut on the grey chest.

**General distribution:** North-eastern Belgian Congo to the Sudan, Abyssinia, north-eastern Northern Rhodesia and Nyasaland.

**Range in Eastern Africa:** Southern Sudan, south-western Abyssinia, Uganda, Kenya Colony and the Zambesi River.

**Habits:** A widespread but local species confined to rocky hill tops and boulder strewn hill-sides, but occasionally found on buildings. Usually in pairs. They have a habit of raising and lowering the tail slowly but almost continuously. Delightful and attractive birds, good songsters, and excellent mimics. They have been noted taking long jumps downhill, feet foremost and with wings closed until they almost reach the ground.

**Nest and Eggs:** A flat cup nest of rootlets, twigs, and leaves, lined with finer material and a few feathers, placed in caves, clefts of rock, holes in walls, or among boulders, occasionally using the old nests of Swallows. Eggs three, cream or pale bluish-green spotted and blotched with pale chestnut brown, with a few pale mauve undermarkings; the spots often form a zone, rather glossy; about 27 × 19 mm.

**Recorded breeding:** Kenya Colony, Mt. Elgon, June; Kilima Theki, March and April. Tanganyika Territory, November to January, also May. Nyasaland, September to December. Portuguese East Africa, November.

**Call:** A clear double whistle and various other calls including a wild sweet warbling song. Alarm note a harsh 'krät-krät-krät' followed by a clear whistle. Also mimic freely at least half a dozen species of birds.

*Thamnolæa cinnamomeiventris albiscapulata* (Rüpp.). Abyssinian Cliff-Chat.

*Saxicola albiscapulata* Rüppell, N. Wirbelt. Vög. p. 74, pl. 26, fig. 1, 1837: Northern Abyssinia.

**Distinguishing characters:** Adult male, differs from the preceding race in being brighter blue black and in having the apical half of the upper and under tail-coverts blue black. The female is similar to the male, but lacks the white shoulder patch. Wing 105 to 122 mm.

**Range in Eastern Africa:** Eastern Sudan and Eritrea to eastern and central Abyssinia.

**Habits:** A common resident on the Abyssinian plateau, where it is tame and often breeds in or near houses in holes in masonry or rocks.

**Recorded breeding:** Abyssinia, February.

**Distribution of other races of the species:** West to South Africa, the nominate race being described from Cape Province.

**877 WHITE-WINGED CLIFF-CHAT.** *THAMNOLÆA SEMIRUFA* (Rüppell). **Pl. 64.**

*Saxicola semirufa* Rüppell, N. Wirbelt. Vög. p. 74, pl. 25, 1837: Lake Tana, north-western Abyssinia.

**Distinguishing characters:** Adult male, above, wholly black including head to chest; basal half of primaries white; breast to

# 282 THRUSHES, WHEATEARS, CHATS, AKALATS, etc.

under tail-coverts light chestnut. The female is black or brownish black above; rump washed with chestnut and with brownish black barring; basal half of primaries white; below, brownish grey mottled with blackish; a pale chestnut streak down centre of throat; belly and under tail-coverts chestnut. Wing 106 to 122 mm. The young bird is blackish, spotted above and below with dark buff and has the white wing patch of the adult.

**Range in Eastern Africa:** Eritrea to north-western, central and south-western Abyssinia as far south as Yavello.

**Habits:** A local but not uncommon species of the higher levels of the Abyssinian plateau in open country. The white basal half of the primaries is noticeable and distinguishes it from the last species.

**Nest and Eggs:** A compact nest of grass stems and moss with a lining of hair and feathers, placed in the crevice of a rock or other suitable situation. Eggs three, white or greenish white and glossy, covered with fine pale rust coloured speckling; about $25 \times 19$ mm.

**Recorded breeding:** Northern and central Abyssinia, June to August.

**Call:** A modulated flute-like song.

## 878 WHITE-CROWNED CLIFF-CHAT. *THAMNOLÆA CORONATA* (Reichenow).

*Thamnolæa coronata coronata* Reichw.

*Thamnolæa coronata* Reichenow, O.M. p. 157, 1902: Tapong, Togoland, West Africa.

**Distinguishing characters:** Adult male, similar to the Cliff-Chat, but whole top of head white or mixed black and white, and no white on upper chest. The female differs from the male in having a brown mantle; greyish brown head and sides of face; chin to chest brown washed with chestnut; normally no white on wing shoulder; wing-coverts edged with grey. Rarely the female has some white on the wing shoulder and top of head. Wing 101 to 116 mm. Juvenile plumage unrecorded.

**General distribution:** Borders of Togoland to northern Nigeria, northern Cameroons and the western Sudan.

**Range in Eastern Africa:** The Darfur area of the Sudan.

**Habits:** A bird of rocky gorges and cliffs, with much of the habits of the Rock-Thrushes, and usually seen on a rock point or a bough

# THRUSHES, WHEATEARS, CHATS, AKALATS, etc.

projecting from a cliff. Mostly found in inaccessible places and very local, usually in pairs.

**Nest and Eggs:** Undescribed.

**Recorded breeding:** Darfur, May to July.

**Call:** A clear whistled 'wheet-wheet' and a scraping alarm note of 'chak-chak.'

*Thamnolæa coronata kordofanensis* Wetts.
*Thamnolæa coronata kordofanensis* Wettstein, Anz. Akad. Wien. 53, p. 135, 1916: Gebel Rihal, Kordofan, central Sudan.

**Distinguishing characters:** Adult male, differs from the nominate race in having some white on the upper chest. The female has a browner chin, throat and chest than the nominate race, with practically no chestnut wash; a deeper coloured underside and under tail-coverts. Wing 107 to 115 mm.

**Range in Eastern Africa:** The Kordofan and Nuba areas of the Sudan.

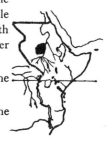

**Habits:** As for the last race; it is locally common among the granite hills and kopjes of the Nuba Province.

**Recorded breeding:** No records.

### 879 WHITE-HEADED BLACK CHAT. *THAMNOLÆA ARNOTTI* (Tristram).

*Thamnolæa arnotti arnotti* (Trist.). **Pl. 64.**
*Saxicola arnotti* Tristram, Ibis, p. 206, pl. 6, 1869: Victoria Falls, north-western Southern Rhodesia.

**Distinguishing characters:** Adult male, general colour black with a slight gloss; top of head and large patch on wing white. The female has the chin to ear-coverts and lower neck white, feathers sometimes tipped with black. Wing 87 to 107 mm. The young bird of both sexes has the head and neck all round black; the male sometimes has white bases to the feathers of the top of the head; the female sometimes has some white feathers on the chin and neck.

**General distribution:** Southern Belgian Congo and Tanganyika Territory to northern Bechuanaland, northern Portuguese East Africa, north-western Southern Rhodesia and north-eastern Transvaal.

**Range in Eastern Africa:** Tanganyika Territory from the Tabora district to Nyasaland.

**Habits:** A local and rather sedentary species of open woodland, with much of the habits of a Thrush, usually found in pairs or small parties hopping about among trees and apparently searching for food more on tree trunks than on the ground. They have a habit of raising and lowering the tails on settling, and their wing beats are audible in flight. Usually shy.

**Nest and Eggs:** Nest of soft grass, feathers, etc. in a shallow hole of a tree or not uncommonly on the beams of grass-roofed rest houses. Sometimes the hole is unlined. Eggs three or four, pale blue or greenish, with rufous or brown freckling and mauve undermarking, often forming a cap or zone at the larger end; about 22 × 16 mm.

**Recorded breeding:** Tanganyika Territory, July to September. Portuguese East Africa, October and November. South-eastern Belgian Congo, October. Northern Rhodesia, September and October. Nyasaland, September to November.

**Food:** Believed to be largely ants.

**Call:** A shrill sibilant alarm whistle of 'sweep-sweep.' Song a high squeaking whistle interspersed with rasping notes.

**Distribution of other races of the species:** Angola.

**880 SOOTY CHAT.** *MYRMECOCICHLA NIGRA* (Vieillot). **Pl. 64. Ph. ix.**

*Œnanthe nigra* Vieillot, N. Dict. d'Hist. Nat. 21, p. 431, 1818: Malimbe, Portuguese Congo.

**Distinguishing characters:** Adult male, very similar to the White-headed Black Chat, but with no white on the head, and shorter tail, 56 to 63 mm., against 64 to 75 mm.; the bill is slightly heavier; the white shoulder patch does not extend on to the greater primary coverts, and the flight feathers are broader. The female is wholly umber or sooty brown. Wing 88 to 103 mm. The young male is dull sooty black without gloss. The young female is similar to the adult female, but is darker, more sooty, from chin to chest.

**General distribution:** The Cameroons to Angola, Northern Rhodesia, the Sudan, Uganda, Kenya Colony and Tanganyika Territory.

**Range in Eastern Africa:** The southern Sudan and Uganda to south-western Kenya Colony and western Tanganyika Territory.

# THRUSHES, WHEATEARS, CHATS, AKALATS, etc. 285

**Habits:** A noticeable bird often locally common among termite hills in open country or light bush. Essentially a bird of short grass country and usually tame. Sings from a perch or a termite hill, swaying its body up and down as it does so; the tail also is cocked up and spread as the bird hops about.

**Nest and Eggs:** Nest of grass, lined with soft grass, flowers, and down, in hollows of old termite hills, sand pits, or stumps, but also bores a tunnel three feet or more long if no suitable hollow is available. The nesting tunnel is used for roosting. Eggs four or five, bluish-white, or white with occasional fine black spotting; about 24 × 17 mm.

**Recorded breeding:** Uganda, January to March and August to October, other months are mentioned.

**Call:** Whistling calls and a small Thrush-like song in breeding season.

## 881 ANTEATER-CHAT. *MYRMECOCICHLA ÆTHIOPS* Cabanis.

*Myrmecocichla æthiops cryptoleuca* Sharpe.
*Myrmecocichla cryptoleuca* Sharpe, Ibis, p. 445, 1891: Kikuyu, western Kenya Colony.

**Distinguishing characters:** General colour sooty brownish black; forehead, lores, chin and throat speckled with brown; basal three-quarters of inner webs of primaries white. The sexes are alike. Wing 107 to 122 mm. The young bird has brown tips to the wing-coverts. The adult is very similar to the female of the White-winged Cliff-Chat from which it can be distinguished by the longer bill, and by there being less white on the outer webs of the primaries and more on the inner webs.

**Range in Eastern Africa:** Mt. Elgon, the Mathews Range and Suk area of Kenya Colony to Eldoret, Nakuru and Kikuyu, and south to the Crater highlands of northern Tanganyika Territory.

**Habits:** A common species of the highlands of Kenya Colony, usually associated with termite nests into which it frequently enters. Occasionally perches on trees and bushes, a tame and conspicuous bird with a rather weak flight, and often seen in family parties of four or five.

**Nest and Eggs:** Nest of dry grass lined with rootlets in holes in banks, ant-bear holes, or large termite nests. Eggs usually three to five, white; about 25·5 × 19 mm. Sir C. Belcher found on at least two occasions that all the eggs in a nest of this Chat were dented, as if by a bird's beak, and that an egg of some parasitic species, possibly an *Indicator*, had been inserted.

**Recorded breeding:** Kenya Colony, March to August, and December and January.

**Food:** Insects, largely termites and beetles.

**Call:** Piping and whistling cries, with a whistling song at times. The piping notes are often accompanied by drooping of the wings and tail.

*Myrmecocichla æthiops sudanensis* Lynes.
*Myrmecocichla æthiops sudanensis* Lynes, Bull. B.O.C. 41, p. 18, 1920: El Fasher, Darfur, western Sudan.

**Distinguishing characters:** Differs from the preceding race in being browner and smaller. Wing 97 to 110 mm.

**Range in Eastern Africa:** Darfur and Kordofan areas of the Sudan.

**Habits:** As for the last race, but with a more desert habitat. Parties in the non-breeding season appear to engage in furious battles like a football scrimmage (Lynes). Locally common.

**Recorded breeding:** Darfur and Kordofan, June to August.

**Distribution of other races of the species:** West Africa, the nominate race being described from Senegal.

**882 STONECHAT.** *SAXICOLA TORQUATA* (Linnæus).
*Saxicola torquata variegata\** (Gmel.).
*Parus varietagus* S. G. Gmelin, Reise. durch. Russl. 3, p. 105, 1774: Shemakha, Azerbaijan, west of Baku, Caspian Sea, south Russia.

**Distinguishing characters:** Adult male, above, whole head and throat black with more or less rusty tips to feathers; patch on wing, sides of neck, upper tail-coverts and more than half of all except central tail feathers white; below, chest and breast chestnut; belly to under tail-coverts white washed with tawny; under wing-coverts and axillaries black tipped with white. The female has the black of the male replaced by brown or blackish brown, and has the white

---

\* *Note:* If original orthography is adhered to the spelling of this name should be *varietaga*.

# THRUSHES, WHEATEARS, CHATS, AKALATS, etc.

in the tail more confined to the base and in some replaced by buff. Wing 65 to 73 mm.

**General distribution:** Azerbaijan to Caucasus, lower Ural River valley (Uralsk to Caspian Sea) and Lake Aral; in non-breeding season to Egypt, the Sudan, British Somaliland, western Arabia and Iraq.

**Range in Eastern Africa:** Eastern and central Sudan to about lat. 8° N., Eritrea, Abyssinia and British Somaliland in non-breeding season.

**Habits:** A common palæarctic winter visitor to all types of open country, but not ranging far south. All Stonechats perch on the tops of small bushes or tall herbage, and fly down to pick up insects on the ground. The call is a 'chip' or 'clak.'

*Saxicola torquata albofasciata* Rüpp.
*Saxicola albofasciata* Rüppell, Syst. Uebers, p. 39, 1845: Simen Province, northern Abyssinia.

**Distinguishing characters:** Adult male, differs from the preceding race in being black and white; tail black with indication of white at extreme base; black extending well down to breast and sides of upper flanks; breast sometimes mixed chestnut and black; under wing-coverts and axillaries black. The female differs from the female of the preceding race in having no white in the base of the tail; under wing-coverts and axillaries dusky with tawny or buff edges. Wing 65 to 75 mm. The young bird is blackish brown above spotted with buff; below, buff, mottled with blackish from chin to breast.

**Range in Eastern Africa:** Abyssinia to south-eastern Sudan.

**Habits:** A common resident in Abyssinia from 7,000 to 10,000 feet with the habits of other races. The male has a cheerful little song.

**Nest and Eggs:** See under other races.

**Recorded breeding:** Sudan, young in April. Abyssinia, probably March to June.

*Saxicola torquata caffra* Keyserl. & Blas.
*Saxicola Rubicola* var *Caffra* Keyserling and Blasius, Die Wirbelt. Europas, p. 59, 1840; Uitenhage, eastern Cape Province.

**Distinguishing characters:** Adult male, differs from other races in having the chin and throat dark chestnut; centre of belly and under tail-coverts white washed with chestnut; under wing-coverts and axillaries black with white edges; tail as in the preceding race.

288  *THRUSHES, WHEATEARS, CHATS, AKALATS, etc.*

The female has the under wing-coverts and axillaries white or buffish with dusky bases. Wing 64 to 76 mm.

**General distribution:** Angola to Tanganyika Territory, southern Belgian Congo, northern Portuguese East Africa, Nyasaland, Cape Province except western areas, and Natal.

**Range in Eastern Africa:** Western Tanganyika Territory from Kungwe-Mahare Mts. to Nyasaland and Portuguese East Africa.

**Habits:** Confined to higher levels among bracken and short grass, with no difference in habits to other races.

**Recorded breeding:** South-eastern Belgian Congo, August. Nyasaland, August to November. Portuguese East Africa, probably August to October. Southern Rhodesia, August and September.

*Saxicola torquata axillaris* (Shell.). **Ph. x.**
*Pratincola axillaris* Shelley, P.Z.S. p. 556, 1884: Mt. Kilimanjaro, north-eastern Tanganyika Territory.

**Distinguishing characters:** Adult male, differs from the preceding race in having the chestnut below confined to a patch on the central chest adjoining the black throat, though there is considerable individual variation in the extent of the chestnut and, rarely, in some specimens, it is replaced by either black or white; rest of underparts white with sometimes a chestnut wash down flanks; under wing-coverts and axillaries black tipped with white, which, together with the restriction of the black to chin and neck, distinguishes this race from the Abyssinian race. The female is generally darker and more warmly coloured than the female of the preceding race. Very rarely the female has the greater part of the throat and neck in front black with a dull chestnut chest. Wing 61 to 75 mm.

**General distribution:** Eastern Belgian Congo to Uganda, Kenya Colony and Tanganyika Territory.

**Range in Eastern Africa:** Uganda to Kenya Colony and Tanganyika Territory, except the eastern area from Uluguru Mts. to Mpapwa and the western area from the Kungwe-Mahare Mts. to the Nyasaland boundary.

**Habits:** Common and widespread in open country, especially in wet or marshy places at all elevations up to 10,000 feet. Found also in glades in woodland or in high alpine moorland.

**Nest and Eggs:** Nest of grass and rootlets lined with hair, well hidden and often under overhanging cover on or near the ground.

Eggs two to four, greenish or dull greyish blue flecked or speckled with rufous; about 19 × 14·5 mm. A common host of the Red-chested Cuckoo.

**Recorded breeding**: Uganda, May, also September and October. Kenya Colony highlands, March to June, and probably in many months of the year. Tanganyika Territory, December and January.

**Food**: Mainly small insects, with a few small molluscs, worms, etc.

**Call**: The male has a cheerful little song in the breeding season which is uttered from a perch, or more rarely, when descending from a short flight. The normal call is that of the European Stonechat, a sharp but low 'chep-chep.'

*Saxicola torquata jebelmarræ* Lynes.
*Saxicola torquata jebelmarræ* Lynes, Bull. B.O.C. 41, p. 27, 1920: Jebel Marra, Darfur, western Sudan.

**Distinguishing characters**: Adult male, very similar to the preceding race but with chestnut extending on to breast and down flanks. The female is generally paler, more buffish in general colour. Wing 65 to 72 mm.

**Range in Eastern Africa**: Western Sudan.

**Habits**: Confined to middle and upper levels. Locally common with the typical habits of the species.

**Recorded breeding**: May.

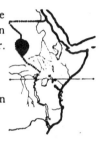

*Saxicola torquata promiscua* Hart.
*Saxicola torquata promiscua* Hartert, Bull. B.O.C. 42, p. 51, 1922: Uluguru Mts., eastern Tanganyika Territory.

**Distinguishing characters**: Adult male, similar to that of the Kilimanjaro race, but the female differs in being generally paler, especially the throat and breast to belly and under tail-coverts; chest colour rather contrasting with rest of underparts. Wing 64 to 68 mm.

**Range in Eastern Africa**: Mpapwa to Kilosa and the Uluguru Mts.

**Habits**: Of the species, common in open ground of the higher levels.

**Recorded breeding**: Tanganyika Territory, November to January.

*Saxicola torquata hemprichii* Ehr.
*Saxicola Hemprichii* Ehrenberg, Symb. Phys. fol. a.a. (8), 1832: Egypt.

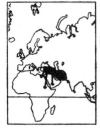

**Distinguishing characters:** Adult male, similar to the other migrant race from Asia, but with only about a third of all except the central tail feathers white. The female has very little or no white at the base of the tail and is very much paler, more sandy in general colour, than any resident race. Wing 67 to 78 mm.

**General distribution:** Eastern Asia Minor to Syria, Iraq and Iran; in non-breeding season to Egypt, the Sudan, Abyssinia and British Somaliland, also Arabia.

**Range in Eastern Africa:** The Sudan, Eritrea, Abyssinia and British Somaliland in non-breeding season.

**Habits:** A palæarctic winter visitor not easily distinguished in the field from the other migrant race first mentioned.

**Distribution of other races of the species:** Northern Europe and Asia, West and South Africa and Madagascar; the nominate race being described from the Cape of Good Hope.

**883 WHINCHAT.** *SAXICOLA RUBETRA* (Linnæus).
*Saxicola rubetra rubetra* (Linn.).
*Motacilla Rubetra* Linnæus, Syst. Nat. 10th ed. p. 186, 1758: Sweden.

**Distinguishing characters:** Adult male, above, buff or buffish brown, broadly streaked with black; white patch on wing; bases of primary coverts white; a white or buffish white streak from base of bill to over eye; ear-coverts blackish; below, chin and sides of neck white; throat to breast tawny; belly and under tail-coverts paler tawny to white. The female has brown ear-coverts, is paler below, and lacks the distinct white on the chin and sides of neck; in some the breast is streaked and spotted with blackish. Wing 70 to 84 mm.

**General distribution:** The greater part of Europe; in non-breeding season to Africa as far south as Damaraland, Nyasaland and Tanganyika Territory, also western Arabia.

**Range in Eastern Africa:** Northern areas south to Tanganyika Territory in non-breeding season.

**Habits:** A common and abundant palæarctic winter visitor to open marshy ground or damp clearings. A bird of somewhat lower levels than the Stonechat, though it may occur anywhere on migration, and with a slightly more western distribution. Has many of the habits of the Stonechat, but is less often seen on bushes and prefers to perch on grass or herbage.

**Distribution of other races of the species:** Siberia, Dalmatia and Iran.

**ROBIN-CHATS, AKALATS and ALETHES.** FAMILY— **TURDIDÆ.** Genera: *Cossypha, Modulatrix, Dessonornis, Sheppardia, Stiphrornis* and *Alethe*.

Twenty-three species occur in Eastern Africa. They are all inhabitants of woods and forests, and the majority are more often heard than seen.

The brighter colours of the Robin-Chats often catch the eye, but the Akalats, and especially the Alethes, are much more inconspicuous and difficult to observe. The food consists largely of insects picked up from the ground, and several species have a close association with ants. Very little is known of the nesting habits of many species of this group, particularly of the Akalats and Alethes.

**884** WHITE-BROWED ROBIN-CHAT. *COSSYPHA HEUGLINI* Hartlaub.
*Cossypha heuglini heuglini* Hartl. **Pl. 66. Ph. x.**
*Cossypha heuglini* Hartlaub, J.f.O. p. 36, 1866: Wau, Bahr-el-Ghazal, south-western Sudan.

**Distinguishing characters:** Adult male, top of head to nape and sides of face black; a broad white stripe over eye dividing black of head from black of face; mantle slaty blue to olivaceous brown; wings palish slate; lower rump, upper tail-coverts, all except central tail feathers and underside tawny orange; central tail feathers olivaceous brown. The female is usually browner on the mantle. Wing, male 92 to 111, female 85 to 97 mm.; tail, male 84 to 98, female, 77 to 86 mm. The young bird is spotted with tawny on the head and mottled tawny and black on rest of body; large tawny spots on upper wing-coverts.

**General distribution:** Ubangi-Shari area of French Equatorial Africa to Abyssinia, Uganda, Kenya Colony, eastern Angola, Belgian Congo, Tanganyika Territory, Nyasaland, Portuguese East Africa, Bechuanaland, Southern Rhodesia and north-eastern Zululand.

**Range in Eastern Africa:** Southern half of the Sudan to southern and western Abyssinia and south to the Zambesi River, but not the coastal areas of Kenya Colony, Tanganyika Territory and Portuguese East Africa.

**Habits:** A common and characteristic bird over most of its range, occurring at all levels in woodland or gardens. A fine songster and a remarkable mimic with a wide range of notes, which it generally utters while concealed in thick cover. It feeds largely on the ground and is rarely seen at any height in trees, except perhaps while singing. Normally shy and skulking but exceedingly pugnacious, attacking any other species of bird on its own territory, which appears to be very limited as a rule.

**Nest and Eggs:** Nest well concealed on a stump or bank, a mass of dead leaves on a twig foundation with a cup of roots and fibres. Eggs two or three, blue, but usually so heavily flecked and stippled as to be pale brown or chocolate, with darker smudges; about 22 × 17·5 mm.

**Recorded breeding:** South-western Sudan, probably June. Uganda and Kenya Colony, April to July and October to December. Tanganyika Territory, December to February, also May. Northern Rhodesia, October to December. Nyasaland, September to December, also April. Portuguese East Africa, October to January.

**Food:** Beetles and other insects, seeds, etc.

**Call:** A loud 'pip-ip-uree' with a fine if disconnected song of bubbling high pressure flute-like notes, increasing rapidly in volume. Wood describes it as 'Think of it' 'Think of it' 'Think of it' 'Think of it' working up into terrific volume. A fine mimic of many birds.

*Cossypha heuglini intermedia* (Cab.).
*Bessornis intermedia* Cabanis, J.f.O. p. 412, 1868: Coastal districts of Kenya Colony.

**Distinguishing characters:** Similar to the nominate race but smaller. Wing, male 81 to 97, female 86 to 87 mm.; tail, male 76 to 85, female 71 to 78 mm.

**Range in Eastern Africa:** Coastal areas of Italian Somaliland to Kenya Colony, Tanganyika Territory and Portuguese East Africa from the Juba River to the Zambesi River, and as far west as Amani in Tanganyika Territory, and Nhauela in northern Portuguese East Africa.

**Habits:** Indistinguishable from the nominate race.

**Recorded breeding:** No records.

**Distribution of other races of the species:** Portuguese Congo to Angola.

## 885 RÜPPELL'S ROBIN-CHAT. *COSSYPHA SEMIRUFA* (Rüppell).

*Cossypha semirufa semirufa* (Rüpp.).
*Petrocincla semirufa* Rüppell, N. Wirbelt. Vög. p. 81, 1840: Eritrea.

**Distinguishing characters:** Similar in colour pattern to the White-browed Robin-Chat, but smaller; mantle and wings olive brown, not slaty; two central tail feathers blackish brown. Wing, male 74 to 88, female 72 to 83 mm.; tail, male 63 to 81, female, 54 to 75 mm. The young bird is mottled and spotted with rufous above; no stripe over eye; throat, breast and flanks tawny mottled with dusky.

**Range in Eastern Africa:** Eritrea to central, western and south-western Abyssinia, the Boma Plateau of the south-eastern Sudan, and Moyale and Marsabit in northern Kenya Colony.

**Habits:** A common resident of wooded localities on the higher Abyssinian plateau, often seen in gardens, but always shy and skulking. A fine songster singing on after dark at times. General habits very much those of the last species, but the calls are distinctive, either a shrill 'chi-chee' or a tinkling note (Benson).

**Nest and Eggs:** See under the following race.

**Recorded breeding:** Abyssinia, May to August.

*Note:* Formerly called Black-tailed Robin-Chat, but this is somewhat of a misnomer as only the two central tail feathers are at all black.

*Cossypha semirufa intercedens* (Cab.).
*Bessornis intercedens* Cabanis, J.f.O. pp. 205, 219, 1878: Kitui, Ukamba, Kenya Colony.

**Distinguishing characters:** Differs from the nominate race in being larger and darker; mantle and wing-coverts slate washed with olivaceous, and from the White-browed Robin-Chat in being usually darker above and with central tail feathers much darker and blacker. Wing, male 86 to 95, female 82 to 89 mm.; tail, male 76 to 84, female 67 to 73 mm.

**Range in Eastern Africa:** Central and south-eastern Kenya Colony to northern Tanganyika Territory from Longido to the Arusha area, south Paré Mts. and Essimingor.

**Habits:** Much those of other Robin-Chats, a shy retiring species of forest or dense scrub.

**Nest and Eggs:** The nest is a solid bulky cup of twigs, leaves, roots and mud lined with rootlets and grass stems, well concealed in the fork of a shrub or tree. Eggs three, olive brown and glossy; about $23 \cdot 5 \times 16 \cdot 5$ mm.

**Recorded breeding:** Kenya Colony, March to June, also December and January in wet seasons. Northern Tanganyika Territory (breeding condition), October, December and February.

**Food:** Mainly insects, often seen feeding among a column of driver ants.

**Call:** Three whistled notes 'hoo-hoo-hiu' but much interspersed with mimicry of other birds' or squirrels' calls. Also a low warbling song.

*Cossypha semirufa donaldsoni* Sharpe.
*Cossypha donaldsoni* Sharpe, Bull. B.O.C. 4, p. 28, 1895: Sheikh Husein, Arussi, south-central Abyssinia.

**Distinguishing characters:** Adult male, similar to the preceding race but with a longer tail. The female is usually slightly browner on the mantle than the male. Wing, male 86 to 88, female 78 to 81 mm.; tail, male 84 to 86, female 75 mm.

**Range in Eastern Africa:** Eastern and south-eastern Abyssinia from Arussi to Harar.

**Habits:** In no way differing from those of other races.

**Recorded breeding:** Eastern Abyssinia, May.

**886** ARCHER'S ROBIN-CHAT. *COSSYPHA ARCHERI* Sharpe. Pl. 66.

*Cossypha archeri* Sharpe, Bull. B.O.C. 13, p. 9, 1902: Ruwenzori Mts.

**Distinguishing characters:** Adult male, above, chestnut brown; head browner; tail chestnut; white stripe from base of bill to over and behind eye; sides of face blackish; below, orange brown; centre of belly buffish white. The sexes are alike. Wing 70 to 80 mm. Juvenile plumage like adult, unspotted above.

# THRUSHES, WHEATEARS, CHATS, AKALATS, etc. 295

**General distribution:** Eastern Belgian Congo from the Ruwenzori Mts. to the west side of Lake Tanganyika, and Uganda.

**Range in Eastern Africa:** Western Uganda from the Ruwenzori Mts. to the Kigezi area.

**Habits:** A bird of forest undergrowth and scrub, and of shy retiring disposition. Very local indeed and easily overlooked. Little is known of its life history. On the Ruwenzori Mts. it is fairly plentiful in the bamboo zone and extends up to 13,000 feet. In life it is almost Warbler-like in its movements.

**Nest and Eggs:** Undescribed.

**Recorded breeding:** No records.

**Food:** Beetles and small insects.

**Call:** A sharp double croak, a melancholy piping, not unlike a creaking cart wheel (Woosnam); 'tchui-chuich-chuoh' (Van Someren).

### 887 GREY-WINGED ROBIN-CHAT. *COSSYPHA POLIOPTERA* Reichenow.

*Cossypha polioptera polioptera* Reichw.
*Cossypha polioptera* Reichenow, J.f.O. p. 59, 1892: Bukoba, north-western Tanganyika Territory.

**Distinguishing characters:** Above, top of head dark grey; a black streak through and behind the eye; a spotted black and white stripe over eye; rest of upperparts and wings olivaceous brown; wing-coverts slaty; upper tail-coverts and tail chestnut; below, orange brown; centre of belly white. The sexes are alike. Wing 72 to 81 mm. The young bird has the top of the head mottled with dusky; mantle slightly mottled; tawny tips to wing-coverts and secondaries; breast slightly mottled.

**General distribution:** Northern Angola to Uganda, Kenya Colony and Tanganyika Territory.

**Range in Eastern Africa:** Southern Sudan to Uganda, north-western and western Kenya Colony, and north-western Tanganyika Territory as far as the south end of Lake Victoria.

**Habits:** A rare and shy inhabitant of dense forest undergrowth, of which little is known or recorded. Occasionally seen along streams or in riparian gallery forest.

**Nest and Eggs:** Nest undescribed. Eggs olive green (Van Someren).

**Recorded breeding:** Uganda, believed to nest in June, also in October.

**Food:** Chiefly insects, for which it searches among fallen leaves on the ground.

**Call:** Nothing recorded, a closely allied race is said to have a brief pleasing song.

**Distribution of other races of the species:** West Africa.

**888** ALEXANDER'S ROBIN-CHAT. *COSSYPHA INSULANA* Grote.
*Cossypha insulana kungwensis* Mor. Moreau's Robin-Chat.
*Cossypha polioptera kungwensis* Moreau, Bull. B.O.C. 61, p. 60, 1941: Kungwe Mt., western Tanganyika Territory.

**Distinguishing characters:** Differs from the Grey-winged Robin-Chat in having the top of the head more olivaceous; a more or less concealed white eyebrow tipped with grey instead of black; stripe behind eye grey instead of black; tail shorter and darker, washed and edged with chestnut. The sexes are alike. Wing, 69 to 81 mm. Juvenile plumage unrecorded.

**General distribution:** Cameroons to Tanganyika Territory.

**Range in Eastern Africa:** Kungwe-Mahare Mts. to the Upper Nyamansi River area, western Tanganyika Territory.

**Habits:** No information.

**Nest and Eggs:** Undescribed.

**Recorded breeding:** No records.

**Food:** No information.

**Call:** Unrecorded.

**Distribution of other races of the species:** Fernando Po and northern Belgian Congo, the nominate race being described from Fernando Po.

**889** BLUE-SHOULDERED ROBIN-CHAT. *COSSYPHA CYANOCAMPTER* (Bonaparte).
*Cossypha cyanocampter bartteloti* Shell. **Pl. 66.**
*Cossypha bartteloti* Shelley, Ibis, p. 159, pl. 5, fig. 2, 1890: Yambuya, northern Belgian Congo.

**Distinguishing characters:** Adult male, above, forehead to nape and sides of face black; a white stripe from base of bill to over eye and down side of neck; rest of upperparts slaty to olivaceous brown;

wing shoulder blue; wings and central tail feathers almost black; rest of tail feathers tawny with black edges and tips; whole underside orange buff. The female is usually browner on the mantle. Wing 80 to 87 mm. The young bird is spotted above with tawny; breast with irregular black markings.

**General distribution:** Northern Belgian Congo to the Sudan and Kenya Colony.

**Range in Eastern Africa:** Southern Sudan, Uganda and the Kakamega, Kaimosi and Nandi areas of western Kenya Colony.

**Habits:** Another shy and retiring forest species, rather partial to stream sides and swampy places. Its call may often be heard but the bird is very rarely seen.

**Nest and Eggs:** Nest of the nominate West African race is a loose nest of dry leaves with a few fibres for lining. Eggs two, glossy greenish blue, more or less covered at the larger end with rufous and lilac grey mottlings; about $23 \times 15 \cdot 5$ mm.

**Recorded breeding:** Ituri (breeding condition), March to August.

**Food:** Insects taken on the ground.

**Call:** Many and varied and a good mimic. The song is said to be a mixture of curious long drawn whining notes with snatches of strange varied rather ventriloquial song. Van Someren describes the song as low and warbling and of little variation, but in other areas it starts with slow rising whistles and becomes more varied afterwards.

**Distribution of other races of the species:** West Africa, the nominate race being described from the Gold Coast.

**890** RED-CAPPED ROBIN-CHAT. *COSSYPHA NATALENSIS* Smith. **Pl. 66.**

*Cossypha natalensis* A. Smith, Ill. Zool. S. Afr. Aves, pl. 60, 1840: Durban, Natal, South Africa.

**Distinguishing characters:** Adult male, head cinnamon brown; mantle mixed slate and cinnamon brown; wings pale slate blue; rump and all except central tail feathers cinnamon rufous; central tail feathers and edge of outer tail feathers black; whole underside and sides of face cinnamon rufous. The female is usually rather smaller than the male. Wing 84 to 101 mm. The young bird is spotted with black and cinnamon brown on the head; mantle, wing-coverts and underside mottled with black.

## 298 THRUSHES, WHEATEARS, CHATS, AKALATS, etc·

**General distribution:** The Sudan, Abyssinia, Uganda and Italian Somaliland to Portuguese Congo, Angola, the eastern areas of Southern Rhodesia, eastern Transvaal and Natal.

**Range in Eastern Africa:** Southern Sudan, south-western Abyssinia, Uganda, and southern Italian Somaliland to the Zambesi River, also Ukerewe, Zanzibar and Mafia Islands.

**Habits:** A bird of extensive range though local, and not usually common. Mostly in forest undergrowth or coastal jungle, but widely distributed and found as high as 6,500 feet in parts of its range. An inquisitive but shy bird, and a most amazing mimic with considerable ventriloquial powers. Sings with slightly raised quivering wings.

**Nest and Eggs:** Nest on the ground or in the side of a stump or rock, a cup of leaves, twigs and moss, lined with root fibres. Eggs two or three, blue, overlaid with dense olive green, and heavily mottled with brown; about $23 \cdot 5 \times 17$ mm.

**Recorded breeding:** Uganda, April and May. Tanganyika Territory, November. Nyasaland, November.

**Food:** Grasshoppers, beetles, ants and some berries and fruit.

**Call:** Normally a long piping call, but its powers of mimicry are great and Pakenham gives lists of some twenty-five species mimicked in Zanzibar Island alone. It can also give a remarkably human whistle. The song is a sweet rather incoherent whistling and the alarm note is described as a sharp 'preemp-tik.'

### 891 WHITE-CROWNED ROBIN-CHAT. *COSSYPHA ALBICAPILLA* (Vieillot).

*Cossypha albicapilla omoensis* Sharpe.

*Cossypha omoensis* Sharpe, Bull. B.O.C. 11, p. 28, 1900: Omo River, south-western Abyssinia.

**Distinguishing characters:** Above, forehead to nape with white scale-like markings; sides of face, malar stripe, chin, mantle, wings, and central tail feathers black; lower rump, upper tail-coverts, remainder of tail feathers and underside (except chin) pale chestnut; part of outer web of outer tail-feather black. The sexes are alike. Wing 119 to 130 mm. Juvenile plumage unrecorded.

**General distribution:** Ba-mingui River, Ubangi-Shari area, French Equatorial Africa, to Abyssinia.

## THRUSHES, WHEATEARS, CHATS, AKALATS, etc. 299

**Range in Eastern Africa:** Southern Sudan and south-western Abyssinia.

**Habits:** Little recorded, allied races are found in dense dark bush in swampy places, usually in pairs.

**Nest and Eggs** (of the Nigerian race): Scanty nest of leaves, tendrils and rootlets in the top of a dead stump, in July. Eggs two, pale grey-green largely obscured by reddish spots and blotches and with a few violet undermarkings; about 27 × 17·5 mm.

**Recorded breeding:** No records.

**Food:** Mainly insects taken on the ground.

**Call:** A cheery song in breeding season, at other times a harsh monosyllabic call.

**Distribution of other races of the species:** West Africa, the nominate race being described from Senegal.

**892** SNOWY-HEADED ROBIN-CHAT. *COSSYPHA NIVEI-CAPILLA* (Lafresnaye). **Pl. 66.**
*Turdus niveicapillus* Lafresnaye, Eas. Nouv. man. group, Pass. p. 16, 1838: Senegal.

**Distinguishing characters:** Adult male, not unlike Rüppell's Robin-Chat, but mantle, wings, and central tail feathers darker; a cinnamon brown collar on hinder neck; the centre of the head from forehead to occiput white. Differs from the White-crowned Robin-Chat in being smaller and having the chin chestnut. The female is smaller than the male. Wing, male 94 to 108, female 87 to 102 mm. Often subject to degrees of melanism. The young bird is spotted with tawny and black on the head; mantle, wing-coverts and underside mottled tawny and black.

**General distribution:** Gambia to Gabon, Portuguese Congo, Abyssinia, Uganda, Kenya Colony and Tanganyika Territory.

**Range in Eastern Africa:** Southern Sudan and south-western Abyssinia to Uganda, western Kenya Colony and northern Tanganyika Territory from Bukoba and Kasulu districts to Shinyanga; also Ukerewe Island, Lake Victoria.

**Habits:** A distinguished looking bird of rather retiring disposition, but one of the finest songsters in Africa. Inhabits forest or thick bush of stream beds and water-courses and also, in certain districts, gardens. Sings freely even after dark and is an excellent mimic.

**Nest and Eggs:** Nest of leaves and moss lined with rootlets on a palm stump or some similar situation. Eggs two or three, glossy, dark greenish olive brown; about 23 × 16 mm.

**Recorded breeding:** Nigeria, May to July. Uganda, May, June and November.

**Food:** Largely insects, and is often seen attacking a column of soldier ants, possibly for the food they are carrying.

**Call:** A very fine songster and a noted mimic with a large repertory of calls, even extending to the notes of native musical instruments.

## 893 ROBIN-CHAT. *COSSYPHA CAFFRA* (Linnæus).

*Cossypha caffra iolæma* Reichw. **Pl. 66. Ph. x.**

*Cossypha caffra iolæma* Reichenow, O.M. p. 5, 1900: Kilimanjaro, north-eastern Tanganyika Territory.

**Distinguishing characters:** Adult male, head grey; sides of face blackish; broad white stripe over eye; mantle olivaceous brown; lower rump, upper tail-coverts and all except central tail feathers tawny chestnut; central tail feathers olivaceous brown; below, throat to chest deep tawny; black streak on each side of chin; breast and flanks grey; belly white; lower belly and under tail-coverts deep buff. The sexes are alike. Wing 72 to 92 mm. The young bird has blackish mottling both above and below; throat to chest washed with buff.

**General distribution:** The Sudan to Nyasaland and northern Portuguese East Africa.

**Range in Eastern Africa:** Southern Sudan, Uganda and Kenya Colony to the Zambesi River.

**Habits:** Not uncommon in woodland, forest edge, or on bush-clad slopes in the highlands of Kenya Colony or the hills of Nyasaland, and keeping to the higher levels in Tanganyika Territory and Portuguese East Africa. Has a notably sweet warbling song, but is a shy skulking bird as a rule, and when on the ground hops about with much bobbing and flirting of tail.

**Nest and Eggs:** Nest of moss, rootlets, etc. in a tree stump or a bank concealed by creepers or moss. Eggs two or three, bluish or greenish, densely speckled with dull brown which often forms a cap at the larger end, and with lilac undermarkings; about 23 × 17 mm.

**Recorded breeding:** Kenya Colony, April to July and October to December. Tanganyika Territory, October to December. Nyasaland, October to January.

**Food:** Mainly insects and berries.

**Call:** A low soft call of six or seven syllables and a short warbling song of great sweetness.

**Distribution of other races of the species:** South Africa, the nominate race being described from the Cape of Good Hope.

## 894 SPOT-THROAT. *MODULATRIX STICTIGULA* (Reichenow).

*Modulatrix stictigula stictigula* (Reichw.).
*Turdinus stictigula* Reichenow, O.M. p. 10, 1906: Mbaramo, Usambara Mts., north-eastern Tanganyika Territory.

**Distinguishing characters:** Above, dark olivaceous brown; tail dark chestnut; below, throat pale brown with black spots; rest of underparts rich chestnut brown, centre of belly white. The sexes are alike. Wing 78 to 84 mm. The young bird is considerably duller, especially below and the spots on the throat are not well defined.

**Range in Eastern Africa:** Eastern Tanganyika Territory from the Usambara to the Nguru Mountains.

**Habits:** Shy and skulking bird of Thrush-like habits which passes all its time within a few feet of the ground, and frequently drops thereon and searches for insects among fallen debris. Confined to the undergrowth of heavy forest.

**Nest and Eggs:** The nest is a neat cup of twigs lined with a thick pad of skeleton leaves in the fork of a sapling. Eggs two, pinkish white blotched and scrawled with crimson lake or chocolate, and occasionally with some bright green spots; about $23 \cdot 5 \times 18$ mm.

**Recorded breeding:** Tanganyika Territory, November.

**Food:** Insects and some berries.

**Call:** Most distinctive, a plaintive long-drawn whistle with almost a rattle in the middle. It has also a beautiful and varied Thrush-like song, plaintive and exquisitely clear (Moreau).

*Modulatrix stictigula pressa* (Bangs and Lov.)
*Illadopsis stictigula pressa* Bangs and Loveridge, Proc. N. Engl. Zool. Cl. 12, p. 94, 1931: Nkuka Forest, Rungwe Mt., south-western Tanganyika Territory.

**Distinguishing characters:** Differs from the nominate race in being generally darker and duller, and in having larger and blacker spots on the throat. Wing 75 to 82 mm.

**General distribution:** Tanganyika Territory and northern Nyasaland.

**Range in Eastern Africa:** Eastern to south-western Tanganyika Territory, from the Uluguru Mts., to Rungwe Mt., Njombe and Songea.

**Habits:** As for the nominate race. The only distinction is in the call which is said to be somewhat different, a bi-syllabic whistling.

**Recorded breeding:** No records.

### 895 OLIVE-FLANKED ROBIN-CHAT. *DESSONORNIS ANOMALA* (Shelley).

*Dessonornis anomala anomala* (Shell.). **Pl. 66.**
*Callene anomala* Shelley, Ibis, p. 14, 1893: Milanji Plateau, southern Nyasaland.

**Distinguishing characters:** Above, brown; forehead and stripe over eye grey; upper tail-coverts and tail russet; sides of face greyish; below, throat white; chest to belly grey; lower flanks and under tail-coverts pale brown. The sexes are alike. Wing 73 to 80 mm. Juvenile plumage undescribed.

**Range in Eastern Africa:** Milanji Plateau, southern Nyasaland.

**Habits:** Found in dense woodland patches, usually in pairs and locally common in its restricted habitat, feeding among bushes on the ground.

**Nest and Eggs:** Nest a deep cup of dead leaves and moss lined with vegetable fibre. Eggs two, pyriform, putty-colour with a darker shading at the larger end; about 26 × 16 mm.

**Recorded breeding:** October to January.

**Food:** No information.

**Call:** A low deep croak. The male has a little chirruping whistle in the breeding season, which Vincent describes for the Namuli Mts. race as 'Back-my-sweetie.'

*Dessonornis anomala macclounii* (Shell.). **Pl. 66.**
*Callene macclounii* Shelley, Bull. B.O.C. 13, p. 61, 1903: Mwenembe, Nyika Plateau, northern Nyasaland.

**Distinguishing characters:** Forehead and stripe over eye white; head, mantle and wings olivaceous grey; scapulars and lower back olivaceous brown; upper tail-coverts russet; tail brown; sides of face black; chin to throat white; chest to upper belly grey; flanks and lower belly and under tail-coverts brown. The sexes are alike. Wing 71 to 80 mm. The young bird has the upperparts and sides of face spotted with russet; below, pale russet mottled with black.

**General distribution:** Tanganyika Territory to northern Nyasaland.

**Range in Eastern Africa:** Tukuyu district, south-western Tanganyika Territory.

**Habits:** A shy and skulking inhabitant of dense evergreen forest, feeding on or near the ground. It has something of the appearance and habits of a Rock-Chat, continually flicking its wings and lowering its tail. The nest is an open cup of moss lined with fine rootlets on a steep bank or in an old stump.

**Recorded breeding:** Tanganyika Territory, December. Nyasaland, October to December.


*Dessonornis anomala grotei* Reichw.
*Bessonornis grotei* Reichenow, Verh. Orn. Ges. Bay. p. 584, 1932: Uluguru, Morogoro district, eastern Tanganyika Territory.

**Distinguishing characters:** Differs from the nominate race in having the head and mantle slate grey. Wing 69 to 80 mm. The immature bird is more olive brown above and sometimes has tawny tips to the wing-coverts and innermost secondaries.

**Range in Eastern Africa:** Morogoro to Njombe and Songea districts, Tanganyika Territory.

**Habits:** A scarce species only found in dense forest and would be overlooked except for its short melodious ringing song.

**Recorded breeding:** South-western Tanganyika Territory, probably September and October.

*Dessonornis anomala gurué* (Vinc.).
*Alethe anomala gurué* Vincent, Bull. B.O.C. 53, p. 138, 1933: Namuli Mt., Quelimane Province, Portuguese East Africa.

**Distinguishing characters:** Differs from the nominate race in being darker, especially on the upperside. Wing 74 to 81 mm.

**Range in Eastern Africa:** Namuli and Chiperone Mts., Portuguese East Africa.

**Habits:** As for the nominate race, but generally found in dense patches of weeds and herbage at the forest edge. Most difficult to see and generally skulks about close to the ground.

**Recorded breeding:** No records.

*Dessonornis anomala mbuluensis* Grant and Praed.
*Bessonornis macclounii mbuluensis* C. Grant & Mackworth-Praed, Bull. B.O.C. 57, p. 80, 1937: Nou Forest, Mbulu district, northern Tanganyika Territory.

**Distinguishing characters:** Differs from the two preceding races in having the head and mantle darker slate and in being much darker below. Wing 82 to 87 mm.

**Range in Eastern Africa:** Mbulu district, Tanganyika Territory.

**Habits:** As for the nominate race but the call is said to be softer.

**Recorded breeding:** Northern Tanganyika Territory (just after breeding), January.

## 896 EAST COAST AKALAT. *SHEPPARDIA GUNNINGI* (Haagner).

*Sheppardia gunningi sokokensis* (V. Som.). **Pl. 67.**
*Callene sokokensis* Van Someren, Bull. B.O.C. 41, p. 125, 1921: Sokoke Forest, Malindi, eastern Kenya Colony.

**Distinguishing characters:** Above, olive green; stripe over eye and wing-coverts pale slate; white spot in front of eye concealed; below, bright orange buff; centre of breast to belly white. The sexes are alike. Wing 65 to 74 mm. The young bird has the top of the head and mantle spotted tawny and black; below, tawny yellow with dark edges to feathers of chest; belly white; flanks pale yellow.

**Range in Eastern Africa:** Coastal areas of Kenya Colony and Tanganyika Territory from Malindi to the Pugu Hills.

**Habits:** No information, except that it is found in evergreen forest.

**Nest and Eggs:** Undescribed.

**Recorded breeding:** No records.

**Food:** Insects.

**Call:** Unrecorded.

**Distribution of other races of the species:** Western Nyasaland and southern Portuguese East Africa, the nominate race being described from Beira, Portuguese East Africa.

**897 SHARPE'S AKALAT.** *SHEPPARDIA SHARPEI* (Shelley). *Sheppardia sharpei sharpei* (Shell.). Pl. 67.
*Callene sharpei* Shelley, Bull. B.O.C. 13, p. 60, 1903: Masisi Hill, Nyika Plateau, northern Nyasaland.

**Distinguishing characters:** Above, similar to the East Coast Akalat but lacks the slate on the wings, and the white spot in front of eye is usually prominent; below, olivaceous brown brighter on throat, not bright orange buff. The sexes are alike. Wing 62 to 70 mm. The young bird is streaked with dark buff above; and mottled with black below.

**General distribution:** Tanganyika Territory to northern Nyasaland.

**Range in Eastern Africa:** The Uluguru Mts. to south-western areas of Tanganyika Territory.

**Habits:** An easily overlooked silent bird of forest or bamboo, little known, but occasionally seen turning over dead leaves on the ground. They are locally not uncommon and in certain areas appear to be the chief host of the Long-tailed Cuckoo (*Cercococcyx*).

**Nest and Eggs:** The nest is a deep cup of leaves, mostly skeleton leaves, about three feet from the ground in any suitable bushy growth. Eggs two, pale pinkish buff with reddish brown markings at the larger end; about 20 × 15 mm.

**Recorded breeding:** Tanganyika Territory, October and November. Nyasaland, November.

**Food:** Insects taken on the ground.

**Call:** Song brief and feeble—a sort of 'tee-tee-tuiddy'—the alarm note is a girding 'thathathatha' with an occasional slight metallic clink (Moreau).

*Sheppardia sharpei usambaræ* Macd.
*Sheppardia sharpei usambaræ* Macdonald, Ibis, p. 669, 1940: Amani Forest, Usambara Mts., north-eastern Tanganyika Territory.

**Distinguishing characters:** Differs from the nominate race in being paler below; the white of the belly extends further on to the chest. Wing 65 to 74 mm. The young bird is also paler than that of the nominate race.

**Range in Eastern Africa:** Usambara Mts. to the Nguru Hills, Tanganyika Territory.

**Habits:** Shy skulking forest bird spending most of its time turning over dead leaves, but it is probable that most of the notes under the last race refer equally to this one. In courtship Moreau notes that one bird sits on a twig while the other hovers above uttering Robin-like squeaks.

**Recorded breeding:** No records.

### 898 EQUATORIAL AKALAT. *SHEPPARDIA ÆQUATORIALIS* (Jackson).

*Sheppardia æquatorialis æquatorialis* (Jacks.).
*Callene æquatorialis* Jackson, Bull. B.O.C. 16, p. 46, 1906: Kericho, Lumbwa, western Kenya Colony.

**Distinguishing characters:** Rather similar to the East Coast Akalat, but rump and tail russet brown; below, darker, more orange brown. The sexes are alike. Wing 64 to 66 mm. The young bird is much darker than that of the East Coast Akalat, spotting blacker and deeper tawny both above and on chest.

**General distribution:** North of Lake Kivu, eastern Belgian Congo to western Kenya Colony.

**Range in Eastern Africa:** Uganda and western Kenya Colony.

**Habits:** A species of dense forest undergrowth of which little is known or recorded. It emerges into clearings in the mornings or late evenings, and has a low fast flight from cover to cover. It always keeps close to the ground.

**Nest and Eggs:** Undescribed.

**Recorded breeding:** No records.

**Food:** No information.

**Call:** A curious toad-like croak.

Plate 66

Olive-flanked Robin-Chat (p. 302)  Olive-flanked Robin-Chat (p. 303)  White-chested Alethe (p. 312)
Brown-chested Alethe (p. 310)  (Nyika Race)  Archer's Robin-Chat (p. 294)
  Usambara Alethe (p. 313)
White-browed Robin-Chat (p. 291)    Blue-shouldered Robin-Chat (p. 296)
Robin-Chat (p. 300)  Red-capped Robin-Chat (p. 297)  Snowy-headed Robin-Chat (p. 299)
Red-tailed Chat (p. 274)  Red-tailed Ant-Thrush (p. 254)  Sombre Rock-Chat (p. 276)
Brown-tailed Rock-Chat (p. 273)    Black-tailed Rock-Chat (p. 272)

Plate 67

Black Redstart (p. 327)
Female    Male

Redstart    (p. 326)
Male    Female

Blue-throat (p. 328)
Male    Female    Male

Nightingale (p. 330)   East-coast Akalat (p. 304)   Sharpe's Akalat (p. 305)
White-starred Bush-Robin (p. 323)  Eastern Bearded Scrub-Robin (p. 320)  Red-backed Scrub-Robin (p. 316)
Brown-backed Scrub-Robin (p. 321)  White-winged Scrub-Robin (p. 318)  Rufous Warbler (p. 358)

# THRUSHES, WHEATEARS, CHATS, AKALATS, etc. 307

*Sheppardia æquatorialis acholiensis* Macd.
*Sheppardia æquatorialis acholiensis* Macdonald, Ibis, p. 670, 1940: Kitibol, Acholi Hills, Imatong Mts., southern Sudan.

**Distinguishing characters:** Differs from the nominate race in having the head and mantle more olive brown. Wing 72 mm.

**Range in Eastern Africa:** Imatong Mts., southern Sudan.

**Habits:** Presumably as for the nominate race.

**Recorded breeding:** No records.

**Distribution of other races of the species:** Sierra Leone and Liberia.

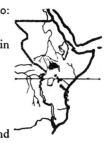

## 899 AKALAT. *SHEPPARDIA CYORNITHOPSIS* (Sharpe).
*Sheppardia cyornithopsis lopezi* (Alex.).
*Callene lopezi* Alexander, Bull. B.O.C. 19, p. 46, 1907: Libokwa, lower Welle district, north-eastern Belgian Congo.

**Distinguishing characters:** Similar in colour to the Equatorial Akalat and has a russet brown rump and tail, but differs in having the breast to belly white; flanks slightly olivaceous. The sexes are alike. Wing 66 to 68 mm.

**General distribution:** North-eastern Belgian Congo and Uganda.

**Range in Eastern Africa:** Western Uganda.

**Habits:** Another retiring Robin-like bird of high forest, keeping close to the ground. Shy but very inquisitive.

**Nest and Eggs:** Undescribed.

**Recorded breeding:** No records.

**Food:** Insects, including termites.

**Call:** A repeated series of short whistles.

**Distribution of other races of the species:** Cameroons, the habitat of the nominate race.

## 900 FOREST-ROBIN. *STIPHRORNIS ERYTHROTHORAX* Hartlaub.
*Stiphrornis erythrothorax mabiræ* Jacks.
*Stiphrornis mabiræ* Jackson, Bull. B.O.C. 25, p. 85, 1910: Mabira Forest, southern Uganda.

**Distinguishing characters:** Above, olivaceous; top of head greyer; a large white spot in front of eye; around eye and cheek black; below, chin to chest tawny; breast to under tail-coverts creamy white;

sides of chest and flanks grey; bill black. The sexes are alike. Wing 64 to 67 mm. The young bird is paler above with large tawny spots mainly on wings; spot in front of eye small; sides of face grey; below, chin and throat creamy white; lower neck and chest mixed tawny and grey; bill pale horn.

**General distribution:** North-eastern and eastern Belgian Congo to Uganda.

**Range in Eastern Africa:** Western and southern Uganda.

**Habits:** Little is recorded of this species except that it moves in a jerky Wren-like manner and is local in swampy forest, with the usual retiring habits of birds which live in dense shade, but it may be seen bathing freely at dusk.

**Nest and Eggs:** The only nest found so far is that of one of the races from the Cameroons, and was a shallow cup of moss at the base of a forest tree. Eggs undescribed.

**Recorded breeding:** Ituri (breeding condition), January to July.

**Food:** Insects taken on the ground, particularly ants.

**Call:** Said to have a clear delightful little song (Woosnam). There is also a croaking churr, a click, a low penetrating whistle, and another more trilled and sweeter whistle.

**Distribution of other races of the species:** West Africa, the nominate race being described from the Gold Coast.

**901 FIRE-CREST ALETHE.** *ALETHE CASTANEA* (Cassin).
*Alethe castanea woosnami* O. Grant.
*Alethe woosnami* O. Grant, Bull. B.O.C. 19, p. 24, 1906: Irumu, north-eastern Belgian Congo.

**Distinguishing characters:** Above, warm dark brown; tawny streak along crown; tail dusky; streak over eye and sides of face grey; below, white; sides of throat, chest, flanks and under wing-coverts grey. The sexes are alike. Wing 86 to 97 mm. The young bird has the head, mantle and wing-coverts blackish brown, broadly spotted with tawny; below, whitish washed with tawny; chest mottled with black; throat tawny or white.

**General distribution:** North-eastern Belgian Congo, the Sudan and Uganda.

**Range in Eastern Africa:** South-western Sudan to western and southern Uganda.

**Habits:** Another forest species found only in dense shade. Very definitely associated with ants; erects its crest if excited while spreading and flirting tail and is rather pugnacious. It also flutters its wings frequently while on a perch.

**Nest and Eggs:** In West African races a nest of moss and roots lined with rootlets on a stump or heap of refuse on the forest floor. Eggs two or three, white or pinkish, heavily blotched with deep maroon and lilac; about 24 × 17 mm.

**Recorded breeding:** No records.

**Food:** Ants or the insects they are carrying, also small frogs.

**Call:** A slowly uttered whistling song well known in deep forest, also a weak churring note and a soft double whistle.

**Distribution of other races of the species:** West Africa, the nominate race being described from the Moonda River, Gabon.

**902** CHOLO MOUNTAIN ALETHE. *ALETHE CHOLOENSIS* Sclater.
*Alethe choloensis choloensis* Scl.
*Alethe choloensis* W. Sclater, Bull. B.O.C. 47, p. 86, 1927: Cholo Mt., southern Nyasaland.

**Distinguishing characters:** Above, olivaceous brown; below, chin and throat white; rest of underparts greyish buff; all except central tail feathers with white ends. The sexes are alike. Wing 95 to 104 mm. The young bird is spotted with tawny above; some blackish mottling below; chest tawny and blackish.

**Range in Eastern Africa:** Mangoche, Chikala, Chiradzulu, Milanji, Soche and Cholo Mts., southern Nyasaland.

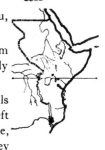

**Habits:** A local and probably not uncommon species found from about 4,000 to 4,500 feet in patches of primaeval forest and generally seen on or near the forest floor.

**Nest and Eggs:** A cup of green moss lined with fine dried tendrils and placed, in the case of the only nest so far recorded, in the cleft of a large forest tree some twelve feet from the ground. Eggs three, glossy green, mottled with shades of chestnut over underlying grey and lilac; about 26 × 19 mm.

**Recorded breeding:** Southern Nyasaland, November to January.

**Food:** Ants and beetles.

**Call:** A melodious whistle of two notes, the second lower and longer, also a low 'cheer' which is likely to develop into a duet between two birds; a croaking alarm note.

*Alethe choloensis namuli* Vinc.
*Alethe choloensis namuli* Vincent, Bull. B.O.C. 53, p. 138, 1933: Namuli Mt., Quelimane Province, Portuguese East Africa.

**Distinguishing characters:** Differs from the nominate race in having the chest and under tail-coverts whiter, not so distinctly greyish buff. Wing 97 to 102 mm.

**Range in Eastern Africa:** Namuli and Chiperone Mts., Portuguese East Africa.

**Habits:** A shy and uncommon species of dense herbage and bush along forest edge at about 5,800 feet, often heard but rarely seen.

**Recorded breeding:** No records.

**903** BROWN-CHESTED ALETHE. *ALETHE POLIOCEPHALA* (Bonaparte).
*Alethe poliocephala carruthersi* O. Grant. **Pl. 66.**
*Alethe carruthersi* O. Grant, Bull. B.O.C. 19, p. 25, 1906: Toro district, western Uganda.

**Distinguishing characters:** Above, warm dark brown; head, and sides of face darker, more sooty; white stripe from base of bill to above eye; below, throat white; chest buff; rest of underparts and under wing-coverts white, washed with buff. The sexes are alike. Wing 83 to 91 mm. The young bird is very similar to that of the Fire-crest Alethe, but grey not tawny below.

**General distribution:** North-western Belgian Congo to the Sudan and Kenya Colony.

**Range in Eastern Africa:** Southern Sudan, Uganda and western Kenya Colony.

**Habits:** A shy forest species of very much the same habits as the Cholo Mountain Alethe. Usually associated with ant columns on which it depends for food, and lives on or near the ground. It has the same habit of rapidly flirting wings and tail.

**Nest and Eggs:** Undescribed.

**Recorded breeding:** Southern Sudan (near breeding), March.

**Food:** Insects including beetles and termites, also frogs.

**Call:** A repeated short whistle.

*Alethe poliocephala akeleyæ* Dearb.
*Alethe akeleyæ* Dearborn, Field Mus. Publ. Orn. 1, p. 170, 1909: Mt. Kenya, central Kenya Colony.

**Distinguishing characters:** Differs from the preceding race in being larger. Wing 95 to 100 mm.

**Range in Eastern Africa:** Central Kenya Colony from Mt. Kenya to Kikuyu and Nairobi.

**Habits:** Much as for the last race, a thick undergrowth haunting species with the habits of a Bush-Robin, sitting on the lower boughs of a bush and dropping down on to an insect on the ground. Rather crepuscular and feeds largely on ants or on what the ants are carrying.

**Recorded breeding:** Kenya Colony, probably December to February.

*Alethe poliocephala kungwensis* Mor.
*Alethe poliocephala kungwensis* Moreau, Bull. B.O.C. 61, p. 46, 1941: Kungwe Mt., western Tanganyika Territory.

**Distinguishing characters:** Differs from the preceding races in having the upperside darker and richer chestnut brown; forehead to nape dusky olive. Wing 94 to 101 mm.

**Range in Eastern Africa:** Kungwe-Mahare Mts. to headwaters of the Nyamansi River, western Tanganyika Territory.

**Habits:** Presumably as for other races.

**Recorded breeding:** No records.

*Alethe poliocephala ufipæ* Mor.
*Alethe poliocephala ufipæ* Moreau, Bull. B.O.C. 62, p. 54, 1942: Mbisi Forest, Sambawanga, Ufipa Plateau, south-western Tanganyika Territory.

**Distinguishing characters:** Similar to the Uganda and Kenya Colony races but forehead to nape and sides of face olivaceous brown. Wing 93 to 97 mm.

**Range in Eastern Africa:** Ufipa Plateau, south-western Tanganyika Territory.

**Habits:** Presumably as for other races.

**Recorded breeding:** No records.

**Distribution of other races of the species:** West Africa, the nominate race being described from Fernando Po.

312  THRUSHES, WHEATEARS, CHATS, AKALATS, etc.

### 904 RED-THROATED ALETHE. *ALETHE POLIOPHRYS* (Sharpe).

*Alethe poliophrys* Sharpe, Bull. B.O.C. 13, p. 10, 1902: Ruwenzori Mts., Uganda-Belgian Congo boundary.

**Distinguishing characters:** Top of head and occiput black; broad streak over eye to nape grey; sides of face blackish grey; rest of upperparts warm russet brown; below, chin grey; throat and neck clear russet; chest grey washed with russet; rest of underparts and under wing-coverts white, washed with buff. The sexes are alike. Wing 93 to 99 mm. The young bird is clearer russet above, including head, with tawny and black markings; below, chin to breast streaked black, tawny and whitish; centre of lower belly white.

**General distribution:** Eastern Belgian Congo and Uganda.

**Range in Eastern Africa:** Ruwenzori Mts., western Uganda.

**Habits:** Another forest species intimately connected with ants. It inhabits the edge of the bamboo zone on the Ruwenzori Mts. and is locally not uncommon at between 7,000 and 9,000 feet. Habits little recorded.

**Nest and Eggs:** Undescribed.

**Recorded breeding:** No records.

**Food:** Ants, and any food they may be carrying.

**Call:** Alarm call a quick 'too-too-tut' (Van Someren), and a sort of song of 'what-whad-whad-whadid' the last double call higher.

### 905 WHITE-CHESTED ALETHE. *ALETHE FÜLLEBORNI* Reichenow.

*Alethe fülleborni fülleborni* Reichw. **Pl. 66.**
*Alethe fülleborni* Reichenow, O.M. p. 99, 1900: between Poroto Mts. and Tandala, south-western Tanganyika Territory.

**Distinguishing characters:** Upperside and sides of neck and chest warm olivaceous brown; lores and ear-coverts greyer; rump and tail more russet; below, white, including under wing-coverts; flanks grey and olivaceous. The sexes are alike. Wing 102 to 108 mm. The young bird is spotted above with black and tawny, and has crescentic black markings below, mainly on chest.

**General distribution:** Tanganyika Territory to northern Nyasaland.

**Range in Eastern Africa:** South-central and south-western Tanganyika Territory from Njombe to the Tukuyu district.

**Habits:** Shy skulking little known bird of evergreen forest and tangled undergrowth, where it hops about on the ground like a Thrush.

**Nest and Eggs:** Undescribed.

**Recorded breeding:** No records.

**Food:** Beetles, ants and berries.

**Call:** Said to be a low piping whistle.

*Alethe fülleborni usambaræ* Reichw.
*Alethe fülleborni usambaræ* Reichenow, O.M. p. 182, 1905: Mlalo, near Lushoto, north-eastern Tanganyika Territory.

**Distinguishing characters:** Differs from the nominate race in having the mantle more olivaceous, less brown. Wing 106 to 120 mm.

**Range in Eastern Africa:** Eastern Tanganyika Territory from the Usambara Mts. to the Uluguru Mts. and Mahenge.

**Habits:** As for the nominate race, it is also believed to have a lively whistling song 'fweer-her-hee-her-hee-her' (Moreau).

**Recorded breeding:** No records.

**906 USAMBARA ALETHE.** *ALETHE MONTANA* Reichenow.
**Pl. 66.**

*Alethe montana* Reichenow, O.M. p. 30, 1907: Usambara, north-eastern Tanganyika Territory.

**Distinguishing characters:** Adult male, above, olivaceous; streak between base of bill to above eye russet brown, partly concealed; sometimes lores also russet brown; below, olivaceous grey; centre of throat and centre of breast to belly whitish. The sexes are alike. Wing 75 to 80 mm. Juvenile plumage unrecorded.

**Range in Eastern Africa:** Usambara Mts., north-eastern Tanganyika Territory.

**Habits:** Inconspicuous little birds inhabiting dense shade and hunting for food among fallen leaves, tame with Thrush-like habits and generally associated with ants.

**Nest and Eggs:** Undescribed.

**Recorded breeding:** No records.

**Food:** Insects, largely ants.

**Call:** A soft chitter, and a brief warble rarely uttered.

**907 IRINGA ALETHE.** *ALETHE LOWEI* Grant and Praed.
*Alethe lowei* C. Grant & Mackworth-Praed, Bull. B.O.C. 61, p. 61,
1941: Njombe, Iringa Province, southern Tanganyika Territory.

**Distinguishing characters:** Size and general characters as in the Usambara Alethe; above, dark olivaceous brown; a concealed white stripe over eye tipped with pale brown; a small white spot below the eye; below, dull brown; centre of belly white. The sexes are alike. Wing 69 to 76 mm. The plumage of the young bird is unrecorded. This Alethe has a close superficial resemblance to Sharpe's Akalat, but the shorter and fewer rictal bristles, broader and longer first primary and stouter and darker coloured feet are distinguishing characters.

**Range in Eastern Africa:** The Njombe area of Iringa Province, southern Tanganyika Territory.

**Habits:** Only known from inside forest jungle.

**Nest and Eggs:** Undescribed.

**Recorded breeding:** Njombe (near breeding), December.

**Food:** Insects.

**Call:** Unrecorded.

**MORNING WARBLERS.** Family — **TURDIDÆ.** Genus: *Cichladusa*.

Two species occur in Eastern Africa. The Morning Warbler in particular is an early riser and is frequently seen and heard on the window-ledge or verandah, where its lively calls attract attention. They are entirely insectivorous as far as is known.

**908 MORNING WARBLER.** *CICHLADUSA ARQUATA* Peters.

*Cichladusa arquata* Peters, Monatb. Akad. Berlin (for 1863), p. 134, 1864: Sena, Zambesi River, Portuguese East Africa.

**Distinguishing characters:** Above, brown; edges of flight feathers, upper tail-coverts and tail chestnut; streak over eye, sides of face to nape and down sides of chest grey; below, chin to chest buff, surrounded with a broken black border; centre of breast to under tail-coverts buff; under wing-coverts tawny. The sexes are alike. Wing 83 to 98 mm. The young bird is mottled below with blackish, though there is an indication of the buff patch from chin to chest.

# THRUSHES, WHEATEARS, CHATS, AKALATS, etc.

**General distribution:** Eastern Belgian Congo to Uganda, Kenya Colony, Northern Rhodesia, Nyasaland and the Zambesi River.

**Range in Eastern Africa:** South-western Uganda and southern Kenya Colony to the Zambesi River.

**Habits:** A tame and common species of somewhat restricted distribution occurring freely in towns and villages as well as in scrub or palm plantations. Might possibly be mistaken for a Bulbul in appearance, but rather Thrush-like on the ground. A low-country bird commonest along the coast or the shores of Lake Nyasa, but widely spread in Tanganyika Territory. It breeds in numbers in the town of Dar-es-Salaam.

**Nest and Eggs:** In buildings or in borassus palms, etc., rather Swallow-like nests of coarse grass and mud lined with fine grass. Eggs two, white or bluish white, with small brownish spots and blotches and a few lavender undermarkings; about 24 × 17 mm.

**Recorded breeding:** Nyasaland, November to March.

**Call:** A piping 'sweet-sweet' followed by a rasping call of three syllables, or a rattling note followed by whistles. Alarm note a shrill 'pree.' This species has a fine clear song in the early morning from which it gets its English name.

## 909 SPOTTED MORNING WARBLER. *CICHLADUSA GUTTATA* (Heuglin).

*Cichladusa guttata guttata* (Heugl.). **Ph. x.**
*Crateropus guttatus* Heuglin, J.f.O. p. 300, 1862: Upper White Nile, Sudan.

**Distinguishing characters:** Smaller than the Morning Warbler, rather paler above and lacking the grey sides of the face, nape and sides of chest; chin and throat surrounded by large drop-shaped spots which extend down the flanks; under wing-coverts buff. The sexes are alike. Wing 79 to 92 mm. The young bird has smaller spots below.

**Range in Eastern Africa:** South-western Abyssinia and the southern Sudan to Uganda, central and south-eastern Kenya Colony and Tanganyika Territory to the Central Railway Line, but not east of Mombo and Dodoma.

**Habits:** Usually among palm-trees or scrub and generally shy in such situations. A sweet songster singing all day and much of the night, and has much of the habits and song of a Robin-Chat, being also something of a mimic. It has been called one of the finest songsters in Africa. Occasionally nests in houses and becomes tame.

**Nest and Eggs:** Nest almost entirely of mud with a few fibres as lining, and it is plastered on to the centre of a bough or other support in the form of a deep cup. Eggs usually two, uniform blue; about 22 × 15·5 mm.

**Recorded breeding:** Equatoria, April. Kenya Colony, Rift Valley, August; southern areas, March to May. Tanganyika Territory, January to April.

**Call:** A loud clear varied song and a rasping alarm note.

*Cichladusa guttata rufipennis* Sharpe.
*Cichladusa rufipennis* Sharpe, Bull. B.O.C., 12, p. 35, 1901: Lamu, eastern Kenya Colony.

**Distinguishing characters:** Differs from the nominate race in being generally darker, with a deeper tone of colour below, and usually more distinct streaks on head. Wing 75 to 83 mm.

**Range in Eastern Africa:** Southern Abyssinia, northern and eastern Kenya Colony as far west as the Orr Valley, and southern Italian Somaliland to eastern Tanganyika Territory east of Mombo and Dodoma.

**Habits:** As for the last race.

**Recorded breeding:** Taita, Kenya Colony, April. Tanganyika Territory (nestling), May.

## SCRUB-ROBINS and BUSH-ROBINS. FAMILY—TURDIDÆ.

Genera: *Erythropygia*, *Pogonocichla* and *Cercotrichas*.

Seven species occur in Eastern Africa. They are conspicuous and lively birds and are usually found in thick and dense vegetation and in mountain forest and bamboo. Their food consists mainly, if not entirely, of insects.

**910 RED-BACKED SCRUB-ROBIN.** *ERYTHROPYGIA ZAMBESIANA* Sharpe.

*Erythropygia zambesiana zambesiana* Sharpe. **Pl. 67.**
*Erythropygia zambesiana* Sharpe, P.Z.S. p. 588, pl. 45, fig. 2, 1882: Tete, Rivi River, a tributary of the Shiré, southern Nyasaland.

**Distinguishing characters:** Above, warm brown; darker on head; a white stripe from base of bill to over eye; wing-coverts tipped with white; upper tail-coverts and tail russet brown to chestnut with subterminal black ends and all except central tail feathers with white ends; below, chin and throat white; black moustachial stripe down side of throat; chest streaked with black; flanks pale brown; centre of breast to belly white. The sexes are alike. Wing 60 to 71 mm. The young bird is mottled black and tawny above, and mottled with blackish on chest and flanks and more sparsely on breast and upper belly; tips of wing-coverts and edges of inner secondaries tawny.

**General distribution:** Portuguese Congo to the Sudan, Uganda, Kenya Colony, Tanganyika Territory, Northern Rhodesia, Portuguese East Africa and eastern Southern Rhodesia.

**Range in Eastern Africa:** South-western Sudan at Rejaf, Yei and Kajo Kaji, Uganda, western Kenya Colony as far east as the Amala River and Mt. Leganisho, western and eastern Tanganyika Territory as far west as Amani and Arusha and south to the Zambesi River, but not central areas of Kenya Colony and Tanganyika Territory.

**Habits:** A cheerful little bird of bush, woodlands, or gardens with many of the habits of a Robin but with some of the characteristics of a Pipit, though rather shy and restless as a rule. Sings from a bush top or rock but dives into cover at slightest alarm. In display runs along a bough with drooping head and wings and erected tail. Apparently a local migrant in some localities.

**Nest and Eggs:** Nest in a grass tuft or thick herbage near the ground, a deep cup of fine grass lined with rootlets. Eggs two or three, white with rusty red spots and blotches and purplish undermarkings; about $18 \cdot 5 \times 14$ mm.

**Recorded breeding:** Tanganyika Territory, October to December. Rhodesia and Nyasaland, September to January, probably double-brooded. Portuguese East Africa, December to February.

**Call:** A cheerful if rather monotonous song of descending notes, rather ventriloquial. The tail is fanned and raised over the back as the call finishes. Alarm note a chattering rattle, also a loud clear call of 'hee-er-wi-er-wi' (Moreau).

*Erythropygia zambesiana brunneiceps* Reichw. **Ph. x.**
*Erythropygia brunneiceps* Reichenow, J.f.O. p. 63, 1891: Nguruman, north end Lake Natron, southern Kenya Colony.

**Distinguishing characters:** Differs from the nominate race in having the head and mantle darker and browner; white edges to inner secondaries. Wing 70 to 77 mm.

**Range in Eastern Africa:** Southern Kenya Colony west of Nairobi to Mt. Suswa and Lake Natron and north-eastern Tanganyika Territory west of Mt. Kilimanjaro to Loliondo.

**Habits:** As for the nominate race, but inhabiting low dry country wherever there is suitable scrubby bush.

**Recorded breeding:** Nguruman near Lake Natron, April. Tanganyika Territory (breeding condition), September and November.

*Erythropygia zambesiana sclateri* Grote.
*Erythropygia leucoptera sclateri* Grote, Bateleur, 2, p. 14, 1930: Iringa, south central Tanganyika Territory.

**Distinguishing characters:** Differs from the nominate race in having white edges to the inner secondaries and mantle rather more tawny in colour. Wing 65 to 74 mm.

**Range in Eastern Africa:** Tanganyika Territory from Shinyanga, Mbulu and Lolkisale to Kilosa and Iringa.

**Habits:** As for other races.

**Recorded breeding:** No records.

911 WHITE-WINGED SCRUB-ROBIN. *ERYTHROPYGIA LEUCOPTERA* (Rüppell).
*Erythropygia leucoptera leucoptera* (Rüpp.). **Pl. 67.**
*Salicaria leucoptera* Rüppell, Syst. Ueb. p. 38, pl. 15, 1845: Shoa, central Abyssinia.

**Distinguishing characters:** Very similar to the Red-backed Scrub-Robin, but differs in having the head greyish brown; white edges to the inner secondaries as well as white tips to the wing-coverts; comparatively more white at the end of the outermost tail feathers; grey not black streaks on chest, and an indistinct dusky moustachial stripe. The sexes are alike. Wing 62 to 74 mm. The young bird differs from that of the Red-backed Scrub-Robin in having white tips to the wing-coverts and white edges to the inner secondaries instead of tawny.

**Range in Eastern Africa:** Central Abyssinia to British Somaliland, the south-eastern Sudan, northern and south-eastern Kenya Colony and north-eastern Tanganyika Territory.

**Habits:** Rather more of a desert bush species than the last, being common in the arid thorn scrub of northern Kenya Colony. Usually a shy retiring bird, but in the breeding season may become very tame. As it flies from bush to bush it raises its tail to show the black and white tips to the feathers on settling. It is often noticed foraging among the dung of animals.

**Nest and Eggs:** A deep cup nest of grass lined with fine grass or hair among herbage in the base of a bush. Eggs two or three, pale greenish or bluish cream closely speckled with brown, and with pinkish mauve undermarkings, often in the form of a zone; about 19 × 15 mm. Common host of the Red-chested Cuckoo.

**Recorded breeding:** Abyssinia, March to May. Kenya Colony, January to May.

**Call:** A plaintive chattering alarm call, and a rich full-throated warbling song of eight or ten notes.

*Erythropygia leucoptera pallida* Benson.
*Erythropygia leucoptera pallida* Benson, Bull. B.O.C. 63, p. 14, 1942: Serenli, Juba River, southern Italian Somaliland.

**Distinguishing characters:** Differs from the nominate race in being less rufous on the flanks and the dusky streaking on chest less distinct. Wing 64 to 70 mm.

**Range in Eastern Africa:** Juba River, southern Italian Somaliland.

**Habits:** As for other races.

**Recorded breeding:** No records.

**912 BEARDED SCRUB-ROBIN.** *ERYTHROPYGIA BARBATA* (Finsch & Hartlaub).
*Cossypha barbata* Finsch & Hartlaub, Vög. Ost. Afr. p. 864, 1870: Caconda, Benguella, Angola.

**Distinguishing characters:** Above, earthy brown; blackish stripe on each side of crown from forehead to over eye with a white stripe below; lores blackish; ear-coverts brown; base of primaries white; bastard wing tipped with white; lower rump and upper tail-coverts chestnut; apical end of three outermost tail feathers white, outermost almost wholly white; below, chin to neck and malar stripe white; a blackish moustachial stripe from base of lower mandible down side of neck; rest of underparts tawny buff; centre of breast to belly white. The sexes are alike. Wing 74 to 87 mm. The young bird is more tawny above with darker tawny mottling.

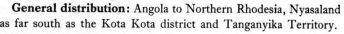

**General distribution:** Angola to Northern Rhodesia, Nyasaland as far south as the Kota Kota district and Tanganyika Territory.

**Range in Eastern Africa:** South-western Tanganyika Territory.

**Habits:** As for other species.

**Recorded breeding:** Rhodesia, September to October. Western Nyasaland, October to December.

### 913 EASTERN BEARDED SCRUB-ROBIN. *ERYTHROPYGIA QUADRIVIRGATA* (Reichenow).

*Erythropygia quadrivirgata quadrivirgata* (Reichw.). **Pl. 67.**

*Thamnobia quadrivirgata* Reichenow, O.C. p. 114, 1879: Kipini, lower Tana River, eastern Kenya Colony.

**Distinguishing characters:** Differs from the Bearded Scrub-Robin in being browner above; stripes on head blacker, and white more extensive on breast and belly. The sexes are alike. Wing 74 to 89 mm. In immature dress the stripe over the eye is buff; black stripe above it less sharp; sides of face buffish; malar stripe less distinct; greater wing-coverts tipped with buff. The young bird is mottled tawny and black above, with a black stripe over eye; below, chest and part of flanks mottled black and tawny.

**General distribution:** Kenya Colony and Tanganyika Territory to Nyasaland, Portuguese East Africa as far south as Inhambane and the Rhodesias.

**Range in Eastern Africa:** Eastern areas of Kenya Colony to Tanganyika Territory, Nyasaland and the Zambesi River.

**Habits:** A bird of thick scrub and bushy watercourses, or, in open country, of thickets and ant hills. Usually shy, and in pairs; has a delightful song and is something of a mimic. Feeds on the ground where it hops about cocking up its tail like a Robin-Chat.

**Nest and Eggs:** Nest a small cup in the fork of a shrub in dense undergrowth, or occasionally in a shallow hole in a tree, made of grass lined with rootlets. Eggs two or three, aquamarine heavily spotted and blotched with light russet brown or chocolate and with mauve undermarkings; about $19 \times 15$ mm.

**Recorded breeding:** Kenya Colony (nestling), January. Nyasaland, October to January. Portuguese East Africa, November.

**Call:** A very beautiful Lark-like song usually in the morning and evening only, also a loud whistle.

*Erythropygia quadrivirgata erlangeri* Reichw.

*Erythropygia quadrivirgata erlangeri* Reichenow, Vög. Afr. 3, p. 770, 1905: lower Juba River, southern Italian Somaliland.

**Distinguishing characters:** Differs from the nominate race in being paler above. Wing 70 to 79 mm.

**Range in Eastern Africa:** Southern Italian Somaliland.

**Habits:** A bird of dense shrubby undergrowth, feeding largely on the ground, which would be unnoticed except for its song.

**Recorded breeding:** No records.

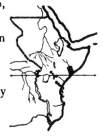

*Erythropygia quadrivirgata greenwayi* Moreau.

*Erythropygia barbata greenwayi* Moreau, Bull. B.O.C. 58, p. 64, 1938: Mafia Island, Tanganyika Territory.

**Distinguishing characters:** Differs from the nominate race in being more ashy above and with paler chest and flanks, which in some cases are merely pale buff. Wing 84 to 86 mm.

**Range in Eastern Africa:** Zanzibar and Mafia Islands.

**Habits:** As for the mainland races.

**Recorded breeding:** Zanzibar and Mafia Islands (breeding condition), November and December.

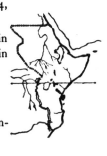

**914 BROWN-BACKED SCRUB-ROBIN.** *ERYTHROPYGIA HARTLAUBI* Reichenow.

*Erythropygia hartlaubi hartlaubi* Reichw. **Pl. 67.**

*Erythropygia hartlaubi* Reichenow, J.f.O. p. 63, 1891: Mutjora (Mtsora) on tributary of Semliki River, south-west of Ruwenzori Mts., Uganda-Belgian Congo boundary.

**Distinguishing characters:** Above, blackish brown; a white stripe from bill to over eye; white spots at ends of wing-coverts; upper tail-coverts and tail chestnut, apical third of latter black and all except central feathers tipped with white; below, white; diffused grey streaks on chest; flanks and under tail-coverts buff. The sexes are alike. Wing 61 to 70 mm. The young bird has the top of the head mottled with tawny; stripe over eye washed with tawny; chest mottled with dusky.

**General distribution:** Cameroons to Uganda, Tanganyika Territory and northern Angola.

**Range in Eastern Africa:** Western Uganda to north-western Tanganyika Territory.

**Habits:** As for the other species of the genus. Often found near villages and especially fond of elephant grass, feeding on the ground and singing its clear loud song from bush tops or grass stems. Occurs up to 6,000 feet on the Ruwenzori Mts.

**Nest and Eggs:** Nest a few inches from the ground in low herbage. Eggs two or three, rarely four, pinkish white to greyish yellow heavily speckled and freckled with varying shades of brown, often with a zone of spots toward the larger end; about 21 × 14 mm.

**Recorded breeding:** Uganda, February and July.

**Call:** A loud cheerful little song, rather challenging in tone but ventriloquial. Also mimics other birds. Annoyance call 'whii,' note of anxiety 'culit' (Van Someren).

*Erythropygia hartlaubi kenia* Van Som. **Ph. x.**
*Erythropygia hartlaubi kenia* Van Someren, J.E.A. & U.N.Hist. Soc. 37, p. 4, 1931: Mt. Kenya, central Kenya Colony.

**Distinguishing characters:** Differs from the nominate race in being darker above, and the streaks on the chest are more distinct. Wing 63 to 71 mm.

**Range in Eastern Africa:** Mt. Kenya to Mau and Kikuyu, central Kenya Colony.

**Habits:** Locally common and usually found in fairly dense cover in valleys between forest ridges and near water.

**Recorded breeding:** Kenya highlands, April to June, occasionally December.

**915 WHITE-STARRED BUSH-ROBIN.** *POGONOCICHLA STELLATA* (Vieillot).
*Pogonocichla stellata guttifer* (Reichw. & Neum.).
*Tarsiger guttifer* Reichenow & Neumann, O.M. p. 76, 1875: Kifinika, Mt. Kilimanjaro, north-eastern Tanganyika Territory.

**Distinguishing characters:** Head and neck all round dark blue slate; a spot above front of eye and at base of neck in front white; mantle dark golden olive green; upper tail-coverts and underside from neck to under tail-coverts golden yellow; wings slate; tail golden yellow with central tail feathers, tips of others and outer edge of outermost black. The sexes are alike. Wing 75 to 92 mm. This race has no immature dress. The young bird is black above, broadly spotted with pale golden yellow; below, mottled pale golden yellow and black.

Plate 68

Black-capped Warbler (p. 353)
Male    Female
Desert Warbler (p. 355)
Garden Warbler (p. 350)
Bonelli's Warbler (p. 385)
Willow Warbler (p. 382)

Subalpine Warbler (p. 354)
Female    Male
Upcher's Warbler (p. 359)
African Reed Warbler (p. 367)
Yellow-throated Woodland-Warbler (p. 386)

Lesser Whitethroat (p. 348)

Savi's Warbler (p. 361)
Reed Warbler (p. 365)
Brown Woodland-Warbler (p. 388)

Plate 69

Red-faced Woodland-Warbler
  *Seicercus laetus* (p. 392)
Red-capped Forest-Warbler
  *Artisornis metopias* (p. 419)
Black-breasted Apalis (p. 407)
  *Apalis flavida flavocincta*
Grey Apalis (p. 400)
  *Apalis cinerea cinerea*
**C**innamon Bracken-Warbler (p. 373)
  *Sathrocercus cinnamomeus cinnamomeus*

Long-billed Apalis (p. 418)
  *Apalis moreaui moreaui*
Bar-throated Apalis (p. 396)
  *Apalis murina murina*
Yellow Bar-throated Apalis (p. 398)
  *Apalis flavigularis griseiceps*
Buff-throated Apalis (p. 400)
  *Apalis rufogularis denti*
Little Rush-Warbler (p. 369)
  *Bradypterus baboecala abyssinicus*

Black-headed Apalis (p. 402)
  *Apalis melanocephala melanocephala*
Black-capped Apalis (p. 413)
  *Apalis nigriceps collaris*
Black-backed Apalis (p. 405)
  *Apalis nigrescens*
Bamboo Warbler (p. 372)
  *Bradypterus alfredi albicrissalis*
Evergreen-forest Warbler (p. 375)
  *Sathrocercus mariae mariae*

# THRUSHES, WHEATEARS, CHATS, AKALATS, etc. 323

**Range in Eastern Africa:** Southern Sudan to east of Lake Rudolf, central and south-western Kenya Colony, and the highlands of north-eastern Tanganyika Territory as far east as Mt. Kilimanjaro.

**Habits:** Lovely little birds with the habits of a Robin or a Redstart, but rather more retiring and often keeping to thick cover. Take insects on the wing or by foraging among dead leaves. Often quite tame and have a variety of sweet subdued calls. Tail frequently flirted and spread out. Mainly found in woodland and forest at higher elevations.

**Nest and Eggs:** Nest a cup-shaped hollow in a mass of leaves, moss and grass, lined with feathers, usually domed or semi-domed, and often in a hollow or in the cleft of a stump or tree, or on a bank. Eggs two or three, greenish white speckled and spotted with reddish brown, which usually forms a zone at one end; about 21 × 15 mm.

**Recorded breeding:** Kenya Colony, probably November and December. Tanganyika Territory, December to February.

**Food:** Insects taken both on the ground and on the wing. This species is also very often attendant on an ant column.

**Call:** A double piping squeak 'hoo-ee—hoo-ee' and a number of soft little song calls. Alarm note a grating squeak.

*Pogonocichla stellata orientalis* (Fisch. & Reichw.). **Pl. 67.**
*Tarsiger orientalis* Fischer & Reichenow, J.f.O. p. 57, 1884: Pangani, north-eastern Tanganyika Territory.

**Distinguishing characters:** Differs from the preceding race in lacking the golden wash on the mantle, and in having the edges of the secondaries olive. Wing 72 to 87 mm. The immature dress is plain green above with a few yellow spots and pale green below mottled with darker green.

**General distribution:** Kenya Colony to north-eastern Northern Rhodesia, Nyasaland and northern Portuguese East Africa.

**Range in Eastern Africa:** Taita district of south-eastern Kenya Colony to central, eastern and southern Tanganyika Territory from Arusha and the north Paré Mts. to Portuguese East Africa and Nyasaland.

## 324 THRUSHES, WHEATEARS, CHATS, AKALATS, etc.

**Habits:** As for the last race, common but, in spite of their colouring, inconspicuous birds of forest at higher elevations. Usually solitary when not breeding.

**Recorded breeding:** Nyasaland, September to February. Portuguese East Africa, October to December.

*Pogonocichla stellata ruwenzorii* (O. Grant).
*Tarsiger ruwenzorii* O. Grant, Bull. B.O.C. 19, p. 33, 1906: Mubuku Valley, south-east Ruwenzori, western Uganda.

**Distinguishing characters:** Differs from the Kilimanjaro race in having the mantle, head and neck paler. Wing 73 to 83 mm. This race has a similar immature dress to that of the last race.

**General distribution:** Eastern Belgian Congo to Uganda and Lake Kivu.

**Range in Eastern Africa:** Ruwenzori Mts. area, western Uganda.

**Habits:** As for other races, found freely in bamboo and heath zones at some height on mountains.

**Recorded breeding:** Ruwenzori Mt., 8,000 to 13,500 feet, November to January.

*Pogonocichla stellata elgonensis* (O. Grant).
*Tarsiger elgonensis* O. Grant, Bull. B.O.C. 27, p. 57, 1911: Mt. Elgon, Uganda-Kenya Colony boundary.

**Distinguishing characters:** Differs from the preceding races in having the yellow in the tail confined to the base of the feathers. Wing 82 to 84 mm. The young bird is similar to that of the Ruwenzori race and has the same amount of yellow in the tail, *i.e.* not confined to the base as in the adult. This race also has a similar immature dress.

**Range in Eastern Africa:** Mt. Elgon, Uganda-Kenya Colony boundary.

**Habits:** Numerous in forest and lower bamboo zone, 6,000 to 12,000 feet, with the habits of other races.

**Recorded breeding:** Mt. Elgon, January.

*Pogonocichla stellata macarthuri* Van Som.
*Pogonocichla stellata macarthuri* Van Someren, J.E.A. & U.N. Hist. Soc. 14, p. 83, 1939: Chyulu Range, south-eastern Kenya Colony.

**Distinguishing characters:** The adult and nestling are similar to the Pangani race; but the immature differs in having the sides of

# THRUSHES, WHEATEARS, CHATS, AKALATS, etc.

the face and throat olive green, the former with a few yellow streaks; rest of underparts olive green streaked with yellow; pale yellow patch at base of foreneck. Wing 76 to 86 mm.

**Range in Eastern Africa:** Chyulu Hills, south-eastern Kenya Colony.

**Habits:** As for other races. Tame and abundant. Song a repeated 'tu-we tu-we ti-ti.' The alarm note is a ratchet-like 'pirut-pirut.' (Van Someren).

**Recorded breeding:** May and June.

**Distribution of other races of the species:** South Africa, the nominate race being described from Plettenberg Bay, Cape Province.

**916 BLACK BUSH-ROBIN.** *CERCOTRICHAS PODOBE* (Müller).

*Cercotrichas podobe podobe* (Müll.).
*Turdus podobe* P. L. S. Müller, Syst. Nat. Suppl. p. 145, 1776: Senegal.

**Distinguishing characters:** Adult male, general colour sooty black; tail black with broad white tips to all except central tail feathers; inner webs of flight feathers, except ends, brown. The sexes are alike. Wing 84 to 101 mm. The young bird is generally duller and more sooty brown than the adult.

**General distribution:** Senegal to British Somaliland and western Arabia.

**Range in Eastern Africa:** Western Sudan to the Red Sea and northern British Somaliland.

**Habits:** A very pleasing active little bird of arid bush, often the most abundant desert species wherever there is bush or thorn scrub. Scrub-Robin-like habits with a varied sweet song in most months of the year, but in appearance a miniature Blackbird, hopping about with raised tail and feeding largely on the ground. Active pugnacious little birds often flirting their tails when settled on a tree and usually seen in pairs.

**Nest and Eggs:** Nest a cup with a bulky foundation of grass or any material handy in a bank, tree stump, or any similar situation. Eggs two or three, pale greenish dotted with greenish brown; no measurements available.

**Recorded breeding:** Sudan, April and May.

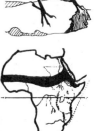

**Food:** Worms and insects.

**Call:** A pleasant varied little song, if rather scratchy.

**Distribution of other races of the species:** Arabia.

## REDSTARTS and BLUETHROAT.  FAMILY — TURDIDÆ.
Genera: *Phœnicurus* and *Cyanosylvia*.

Two species of Redstarts and the Bluethroat occur in Eastern Africa. All are insectivorous migrants from Europe and Asia during the northern winter.

**917 REDSTART. *PHŒNICURUS PHŒNICURUS* (Linnæus).**
*Phœnicurus phœnicurus phœnicurus* (Linn.). **Pl. 67.**
*Motacilla Phœnicurus* Linnæus, Syst. Nat. 10th ed. p. 187, 1758: Sweden.

**Distinguishing characters:** Adult male, forehead white; rest of top of head, mantle and wing-coverts grey or brownish; rump and tail chestnut; central pair of tail feathers blackish; below, sides of face and neck and chin to breast black, more or less tipped whitish; rest of underparts including under wing-coverts chestnut; centre of belly white. The female is brown above and paler brown below; centre of belly buffish white; no black on head; upper tail-coverts and tail as in male. Wing 75 to 87 mm. The female can be confused with the Red-tailed Chat, but lacks the black markings at the tips of the tail feathers and is brown not grey below. The female of the Redstart and White-winged Redstart can be distinguished from the female of the Kashmir Redstart as follows:—

Redstart and White-winged Redstart: Second primary equal to or longer than sixth primary.

Kashmir Redstart: Second primary equal to or shorter than seventh primary.

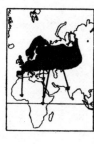

**General distribution:** Europe and Asia to Lake Baikal; in non-breeding season to Africa as far south as the Gold Coast and Tanganyika Territory, also western Arabia and north-western India.

**Range in Eastern Africa:** The Sudan, Eritrea, Abyssinia, Italian Somaliland, northern Uganda, Kenya Colony and north-western Tanganyika Territory in non-breeding season.

**Habits:** A common passage migrant and palæarctic winter visitor to most parts of the Sudan, also fairly common in Uganda, but only

## THRUSHES, WHEATEARS, CHATS, AKALATS, etc.

casual in Kenya Colony and very few records are known from Tanganyika Territory. Shy, inconspicuous bird in Africa but the flash of the red tail as it disappears into a bush is usually noticeable.

*Phœnicurus phœnicurus samamisicus* (Habl.). White-Winged Redstart.
*Motacilla samamisicus* Hablizl, N. Nord. Beit, 4, p. 60, 1783: Samamisch Alps, Gilan Province, North Iran.

**Distinguishing characters:** Adult male, differs from the nominate race in having white edges to the flight feathers; the female is generally darker both above and below, with a tendency to greyish on the mantle. Wing 73 to 84 mm.

**General distribution:** Asia Minor to the Crimea, Caucasus, Iran and Turkestan; in non-breeding season to north-eastern Africa, western Arabia and north-western India.

**Range in Eastern Africa:** Eastern Sudan, Eritrea and Abyssinia in non-breeding season.

**Habits:** As for the nominate race and not easy to distinguish in the field unless close.

**Distribution of other races of the species:** Algeria.

### 918 BLACK REDSTART. *PHŒNICURUS OCHRUROS* (Gmelin).

*Phœnicurus ochruros phœnicuroides* (Moore). Kashmir Redstart. Pl. 67.
*Ruticilla phœnicuroides* Moore, P.Z.S. for 1854, p. 25, 1855: Shikarpur, Sind, northern India.

**Distinguishing characters:** Adult male, head and mantle grey, or sometimes blackish; bases of grey feathers on forehead white; forehead and sides of face black; lower rump, tail-coverts and all tail feathers except central ones chestnut; below, chin to breast black; rest of underparts including axillaries and under wing-coverts chestnut. The female is brown. Wing 79 to 88 mm. The female may be confused with the Red-tailed Chat, but differs in a similar way to the previous species.

**General distribution:** Iran to central Asia, Kashmir and north-western India; in non-breeding season to Egypt, Eastern Africa, Arabia, Iraq and the plains of India.

**Range in Eastern Africa:** Eastern half of Sudan to Abyssinia and British Somaliland in non-breeding season.

**Habits:** Local palæarctic winter visitor, at times not uncommon, preferring woodlands at lower levels. Some examples have been obtained late in the year and it has been suspected of breeding in Abyssinia, but there is no evidence as yet of it having done so.

**Distribution of other races of the species:** Europe to Syria, Palestine, and Iran, the nominate race being described from Iran.

### 919 BLUETHROAT. *CYANOSYLVIA SVECICA* (Linnæus).

*Cyanosylvia svecica svecica* (Linn.). *Red-spotted Bluethroat.* **Pl. 67.**
*Motacilla svecica* Linnæus, Syst. Nat. 10th ed. p. 187, 1758: Sweden.

**Distinguishing characters:** Adult male, above, sooty brown; black streaks on head; buff or buffish white stripe over eye; basal half of all tail feathers except central ones chestnut; below, malar stripe and chin to neck blue; a chestnut spot on lower neck; a black and a chestnut band across lower chest below the blue, remainder of underparts buff. The female has the malar stripe, chin and a stripe down centre of neck white; sides of neck black joining a black band across lower neck. The immature male has the chin to throat buff with some black spotting. Wing, male 68 to 80, female 67 to 75 mm.

**General distribution:** Sweden, Lapland and northern Russia as far east as the Yenesei River; in non-breeding season to Palestine, Iran, Egypt, the Sudan, Abyssinia, western Arabia and India.

**Range in Eastern Africa:** The Sudan and Abyssinia in non-breeding season.

**Habits:** A rather uncommon passage migrant and palæarctic winter visitor to the northern parts of our area. The main body winters in Egypt and only stragglers reach the Sudan except on occasions and then a number may be met with together. Usually found in riparian bush and rank herbage.

*Cyanosylvia svecica cyanecula* (Meis.). *White-spotted Bluethroat.*
**Pl. 67.**
*Sylvia cyanecula* Meisner, Syst. Verz. Vög. Schweiz. p. 30, 1804: Ardennes, France.

**Distinguishing characters:** Adult male, differs from the nominate race in having the spot on the lower neck above the blue band white instead of chestnut; sometimes this white spot is lacking or concealed in which case there are white or black bases to the blue

# THRUSHES, WHEATEARS, CHATS, AKALATS, etc.

feathers. The female is indistinguishable from that of the Red-spotted Bluethroat. Wing, male 70 to 78, female 65 to 73 mm.

**General distribution:** Middle Europe; in non-breeding season to Egypt and the Sudan.

**Range in Eastern Africa:** The Sudan in non-breeding season.

**Habits:** As for the nominate race.

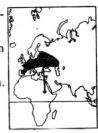

*Cyanosylvia svecica magna* Sar. & Loud. *Iran Bluethroat.*
*Cyanosylvia wolfi magna* Sarudny & Loudon, Orn. Jahrb. 15, p. 225, 1904: Bidesar, Arabistan, Iran.

**Distinguishing characters:** Similar to the White-spotted Bluethroat but larger, and has no black band between the blue and chestnut bands across the chest. Wing, male 78 to 84, female 75 to 80 mm.

**General distribution:** Iran; in non-breeding season to Iraq, Egypt, the Sudan, Abyssinia and western Arabia.

**Range in Eastern Africa:** The eastern Sudan to Eritrea and northern Abyssinia in non-breeding season.

**Habits:** As for the nominate race, but much rarer in Africa.

**Distribution of other races of the species:** India and eastern Asia to China.

## WHITE-THROATED ROBIN. FAMILY—TURDIDÆ. Genus: *Irania*.

Occurs in Eastern Africa as a winter migrant from Asia. Its skulking habits make it difficult to observe, although it is probably fairly plentiful.

**920 WHITE-THROATED ROBIN.** *IRANIA GUTTURALIS* (Guérin).

*Cossypha gutturalis* Guérin, Rev. Zool. p. 162, 1843: Northern Abyssinia.

**Distinguishing characters:** Adult male, above, grey; tail black: white stripe over eye; sides of face and neck black; below, chin and neck, lower belly and under tail-coverts white; rest of underparts including under wing-coverts tawny; more rarely the whole underside is creamy white. The female is more ashy grey above; sides of face and neck ashy grey; underside less tawny. Wing 88 to 102 mm.

## 330 THRUSHES, WHEATEARS, CHATS, AKALATS, etc.

**General distribution:** Asia Minor, Syria, Palestine, Iraq, Iran and western Turkestan; in non-breeding season to Eastern Africa and Arabia.

**Range in Eastern Africa:** Eritrea, Abyssinia, British Somaliland and Kenya Colony to central Tanganyika Territory in non-breeding season.

**Habits:** A passage migrant and palæarctic winter visitor. A shy species with much the habits of a Nightingale, keeping to tangled undergrowth in ravines, or to dense scrub.

### NIGHTINGALE and SPROSSER.    FAMILY — TURDIDÆ.
Genus: *Luscinia*.

These two species both occur in Eastern Africa as migrants from Europe and Asia. Skulking in habits, they are easily overlooked and the Sprosser can only be distinguished from the Nightingale by the marbled markings on the chest when seen at close quarters. Mainly insectivorous.

### 921 NIGHTINGALE.    *LUSCINIA MEGARHYNCHOS* Brehm.
*Luscinia megarhynchos megarhynchos* Brehm. **Pl. 67.**
*Luscinia megarhynchos* Brehm, Handb. Nat. Vög. Deutsch. p. 356, 1831: Germany.

**Distinguishing characters:** Above, dull russet brown; upper tail-coverts and tail russet; a slight whitish stripe over eye; below, ashy brown; chin to throat and centre of breast to belly dull creamy white; under tail-coverts pale buff. The sexes are alike. Wing 76 to 91 mm.

**General distribution:** Southern half of Europe to Asia Minor and North Africa; in non-breeding season to Sierra Leone, Gold Coast, Nigeria, Uganda and Kenya Colony.

**Range in Eastern Africa:** The Sudan, Uganda and Kenya Colony in non-breeding season.

**Habits:** A palæarctic winter visitor to as far south as Kenya Colony, but nowhere common; it also occurs in Uganda where it has been noted as singing freely for some days, unlike most winter migrants. Keeps to dense bush along rivers or in ravines, and is very liable to be overlooked unless its unmistakable deep notes can be heard. Alarm note a harsh churr.

*Luscinia megarhynchos hafizi* Sew.
*Luscinia hafizi* Sewertzov, Vert. gouz. Rasp. Turk. jirot, p. 120, 1872: Turkestan.

**Distinguishing characters:** Differs from the other races in being larger and perhaps rather paler above, more ashy, and with little or no russet on wings. Wing, male 91 to 95, female 87 to 93 mm.

**General distribution:** Transcaspia and Turkestan to western China; in non-breeding season to British Somaliland, Kenya Colony, Tanganyika Territory, southern Arabia and India.

**Range in Eastern Africa:** British Somaliland to coastal area of Kenya Colony and north-eastern Tanganyika Territory in non-breeding season.

**Habits:** As for the nominate race but with a more easterly breeding range. In Africa often found in coastal scrub, though it also occurs inland. Noted as in full song in northern Tanganyika Territory at the end of March or even earlier, and is usually in pairs early in the year.

*Luscinia megarhynchos africana* (Fisch. & Reichw.).
*Lusciola africana* Fischer & Reichenow, J.f.O. p. 182, 1884: Lower Arusha, near Mt. Kilimanjaro, north-eastern Tanganyika Territory.

**Distinguishing characters:** Differs from the nominate race in being duller russet, more earth-brown above; no russet edging to flight feathers; wing-coverts considerably less bright, but without the ashy coloration of the preceding race. Wing, male 85 to 90, female 83 to 86 mm.

**General distribution:** Transcaucasia to Iran and Iraq; in non-breeding season to British Somaliland, Kenya Colony, Tanganyika Territory and southern Arabia.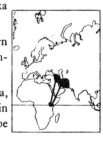

**Range in Eastern Africa:** British Somaliland to south-eastern Kenya Colony and north-eastern Tanganyika Territory in non-breeding season.

**Habits:** As for other races, but a decidedly rarer bird in Africa, only known from the east coast. Also recorded as singing early in the year. Its racial name is a misnomer as it is now known to be only a palæarctic winter visitor to Africa from Asia.

## 922 SPROSSER. *LUSCINIA LUSCINIA* (Linnæus).

*Motacilla Luscinia* Linnæus, Syst. Nat. 10th ed. p. 184, 1758: Sweden.

**Distinguishing characters:** Differs from the Nightingale in having the lower throat and chest mottled. Wing 84 to 96 mm.

**General distribution:** Northern Europe to western Siberia; in non-breeding season to Arabia and Africa as far south as Northern Rhodesia, Nyasaland and the Transvaal.

**Range in Eastern Africa:** The Sudan, British Somaliland and Kenya Colony to Nyasaland.

**Habits:** Indistinguishable from those of the Nightingale but the mottling of throat and chest is fairly noticeable. Probably commoner in most localities than the Nightingale and with a wider range. Sings fairly freely, especially in the evenings. Its usual call is a croaking noise.

Names in Sclater's *Syst. Av. Æthiop.* 2, 1930, which have been changed, or have become synonyms in this work:

*Turdus libonyanus niassæ* Rensch, treated as synonymous with *Turdus libonyanus tropicalis* Peters.
*Turdus libonyanus cinerascens* Reichenow, treated as synonymous with *Turdus libonyanus tropicalis* Peters.
*Turdus libonyanus costæ* Rensch, treated as synonymous with *Turdus libonyanus tropicalis* Peters.
*Turdus olivaceus uluguru* Hartert, treated as synonymous with *Turdus olivaceus nyikæ* Reichenow.
*Turdus olivaceus elgonensis* (Sharpe) treated as synonymous with *Turdus olivaceus abyssinicus* Gmelin.
*Geokichla gurneyi otomitra* Reichenow, treated as synonymous with *Geokichla gurneyi gurneyi* Hartlaub.
*Geokichla gurneyi usambaræ* Neumann, treated as synonymous with *Geokichla gurneyi raineyi* Mearns.
*Geokichla gurneyi keniensis* Mearns treated as synonymous with *Geoklicha piaggiæ kilimensis* Neumann.
*Œnanthe deserti atrogularis* (Blyth) treated as synonymous with *Œnanthe deserti deserti* (Temminck).
*Cercomela scotocerca enigma* Neumann & Zedlitz, treated as synonymous with *Cercomela dubia* Blundell & Lovat.
*Cercomela familiaris sennaarensis* (Seebohm) treated as synonymous with *Cercomela familiaris falkensteini* (Cabanis).
*Cercomela familiaris gambagæ* Hartert, treated as synonymous with *Cercomela familiaris falkensteini* (Cabanis).
*Pinarochroa sordida erlangeri* Reichenow, treated as synonymous with *Pinarochroa sordida sordida* (Rüppell).
*Pinarochroa sordida schoana* Neumann, treated as synonymous with *Pinarochroa sordida sordida* (Rüppell).
*Thamnolæa argentata* Reichenow, treated as synonymous with *Thamnolæa cinnamomeiventris subrufipennis* Reichenow.
*Thamnolæa arnotti leucolæmia* (Finsch & Reichenow) treated as synonymous with *Thamnolæa arnotti arnotti* (Tristram).
*Thamnolæa arnotti collaris* (Reichenow) treated as synonymous with *Thamnolæa arnotti arnotti* (Tristram).

# THRUSHES, WHEATEARS, CHATS, AKALATS, etc. 333

*Saxicola torquata robusta* (Tristram) now *Saxicola torquata caffra* Keyserling & Blasius.
*Saxicola torquata orientalis* W. L. Sclater, treated as synonymous with *Saxicola torquata caffra* Keyserling & Blasius.
*Cossypha niveicapilla melanoptera* (Cabanis) treated as synonymous with *Cossypha niveicapilla* (Lafresnaye).
*Bessonornis albigularis* (Reichenow) now *Dessonornis anomala grotei* Reichenow.
*Erythropygia leucophrys ruficauda* Sharpe, treated as synonymous with *Erythropygia zambesiana zambesiana* Sharpe.
*Erythropygia leucophrys vansomereni* Sclater, treated as synonymous with *Erythropygia zambesiana zambesiana* Sharpe.
*Erythropygia leucophrys soror* Reichenow, treated as synonymous with *Erythropygia zambesiana zambesiana* Sharpe.
*Erythropygia vulpina* Reichenow, treated as synonymous with *Erythropygia leucoptera leucoptera* (Rüppell).
*Erythropygia barbata rovumæ* Grote, treated as synonymous with *Erythropygia quadrivirgata quadrivirgata* (Reichenow).
*Pogonocichla stellata johnstoni* Shelley, treated as synonymous with *Pogonocichla stellata orientalis* (Fischer & Reichenow).
*Pogonocichla intensa* Sharpe, treated as synonymous with *Pogonocichla stellata guttifer* (Reichenow & Neumann).
*Luscinia megarhyncha golzii* Cabanis, now *Luscinia megarhynchos hafizi* Sewertzov.

Names introduced since 1930 and which have become synonyms in this work:

*Turdus olivaceus chyuluensis* Van Someren, 1939, treated as synonymous with *Turdus olivaceus abyssinicus* Gmelin.
*Turdus pelios ubendeensis* Moreau, 1944, treated as synonymous with *Turdus olivaceus graueri* Neumann.
*Turdus fischeri natalicus* Grote, 1938, treated as synonymous with *Psophocichla guttata guttata* (Vigors).
*Turdus tephronotus australoabyssinicus* Benson, 1942, treated as synonymous with *Turdus tephronotus* Cabanis.
*Psophocichla guttata belcheri* (Benson), 1950, treated as synonymous with *Psophocichla guttata guttata* (Vigors).
*Petrophila rufocinerea tenis* Friedmann, 1930, treated as synonymous with *Monticola rufocinerea rufocinerea* (Rüppell).
*Saxicola torquata stonei* Bowen, 1931, treated as synonymous with *Saxicola torquata caffra* Keyserling & Blasius.
*Cossypha heuglini euronota* Friedmann, 1930, treated as synonymous with *Cossypha heuglini intermedia* (Cabanis).
*Cossypha insulana granti* Serle, 1949, treated as synonymous with *Cossypha insulana kungwensis* Moreau.
*Bessonornis albigularis porotoensis* Bangs & Loveridge 1931, treated as synonymous with *Dessonornis anomala macclounii* Shelley.
*Alethe macclouniei njombe* Benson, 1936, treated as synonymous with *Dessonornis anomala grotei* Reichenow.
*Sheppardia cyornithopsis bangsi* Friedmann, 1930, treated as synonymous with *Sheppardia sharpei sharpei* Shelley.

# THRUSHES, WHEATEARS, CHATS, AKALATS, etc.

## Addenda and Corrigenda

**838** *Turdus ludoviciæ.* Add *Habits:* Only found in Juniper forest between 3500 and 8000 ft., keeping in the shade as much as possible. Add: *Nest and Eggs:* Nest in fork of tree or branch, well hidden, generally 15 to 20 ft. up; made mainly of Juniper bark fibres with slight lining of grass and fine moss. Eggs normally three of clear blue ground colour dusted with reddish brown; about 27 × 20·5 mm. Add: *Food:* Berries of Juniper, Olea and other trees (Tribe).

**839** For *Turdus l. tropicalis* read *T.l. cinerascens* Reichenow, O.M. p. 82, 1898: Tabora, Tanganyika Territory. Add *General distribution:* For the Rhodesias read Northern Rhodesia and eastern Southern Rhodesia, and add Portuguese East Africa as far south as Beira. Delete Swaziland and Natal.

**840** *Pelocichla,* correctly *Peliocichla, schuetti* Cabanis, is a synonym of *Turdus libonyanus verreauxi* Cabanis 1882, and not a race of *Turdus pelios.* The correct name of the race under *T.p. schuetti* is *T.p. bocagei. Peliocichla bocagei* Cabanis, J.f.O., p. 320, 1882: Angola.

**846** *Psophocichla guttata guttata* is confined to South Africa and extra limital. The race occurring in Nyasaland is *Psophocichla guttata belcheri* Benson, Ostrich, p. 58, 1950: Socha, Nyasaland.

**847** *Psophocichla l. litsipsirupa.* Add to *Range in Eastern Africa:* Eritrea. Add to *Recorded breeding:* Eritrea, June and July.

**853** *Monticola r. rufocinerea.* Add to *Recorded breeding:* Eritrea, nestling September.
*Monticola rufocinerea.* Add *Habits:* Always found in Juniper forest areas on high plateau. Add *Nest and Eggs:* Nest constructed of fibres from Juniper bark and light grasses. May be in hollow tree or side of mud and wattle hut beneath eaves; one nest has been found actually inside the hut on a beam. Nests found and breeding observed in December 1954, May and June 1956. Eggs three, glossy bluish-green; about 22 × 17 mm. Add *Call:* A very sweet little song can be heard in the evenings, usually from the top of a tree or shrub in full sight. More like a whistle than anything else, starting high, down, high, down, so ∵ ∴ (Tribe).

**856** *Œnanthe phillipsi.* Add: *Nest and Eggs:* Nest a small cup-shaped structure made of grass, sometimes with a hair lining, at ground level, usually at base of Aloe or other small ground shrub. Eggs three, glossy bluish-green sparsely spotted and blotched with brown; about 22 × 16 mm. Add *Recorded Breeding:* British Somaliland, April, May and June. Add *Call:* A very silent bird, and I have never recorded any call nor remember one (Tribe).

**859** *Œnanthe leucomela* now *Œnanthe pleschanka* (Lepechin).
*Œnanthe pleschanka pleschanka* (Lep).
*Motacilla pleschanka* Lepechin, Nov. Comm. Acad. Sci. Imp. Petrop. 14, p. 503, 1770: Saratov, Volga River.

**864** *Œnanthe monacha.* Add: *Distinguishing characters:* The young males are similar to the adult female. Immature males have brownish, *not* black, wings.

**865** *Œnanthe l. lugubris. Recorded breeding:* Should be May to August.
*Œnanthe l. vaurei.* Add: *Recorded breeding:* British Somaliland, April and May.

**867** *Œnanthe heuglini.* Reference and type locality should be: *Saxicola Heuglini* Finsch & Hartlaub, Orn. N. Ost Afr. 1, p. 346, 1869: Gondar, Abyssinia.

*Continued on p. 1102*

## WARBLERS. Family—SYLVIIDÆ.

At least one hundred and thirty-two species are known from Eastern Africa, of which thirty belong to the Cisticola group. Warblers occur in all types of country and at all altitudes. Many are migrants from Europe and Asia. They show considerable diversity of colour and form, habits and habitat. All are insectivorous and may eat fruit as well at certain times of the year. For convenience we have divided them into a number of groups under separate headings, and into two keys, the second key containing the Cisticolas only.

We regret that it appears to us impossible to give any definition useful to the man in the field of what is a Cisticola and what is not. Luckily their habits are mostly distinctive.

### Key to the Adult Warblers of Eastern Africa

| | | | |
|---|---|---|---|
| 1 | Head and mantle spotted and streaked: | | 3–6 |
| 2 | Head and mantle not spotted and streaked: | | 7–189 |
| 3 | Tail feathers with subterminal blackish bars: | STRIPED-BACK PRINIA *Prinia gracilis* | 1047 |
| 4 | Tail feathers plain: | | 5–6 |
| 5 | Top of head spotted: | GRASSHOPPER WARBLER *Locustella nævia* | 941 |
| 6 | Top of head streaked: | SEDGE WARBLER *Acrocephalus schœnobænus* | 947 |
| 7 | Above, black, or sooty black: | | 13–17 |
| 8 | Above, ashy: | | 18–27 |
| 9 | Above, olivaceous, or olivaceous grey or brown: | | 28–52 |
| 10 | Above, various shades of brown, but not olivaceous brown: | | 53–88 |
| 11 | Above, various shades of green: | | 89–121 |
| 12 | Above, various shades of grey, but not olivaceous grey: | | 122–189 |

13 Above, glossy black, chest and belly yellow: WHITE-WINGED APALIS, male *Apalis chariessa* **984**

14 Above, dull black, chest and belly white; tail feathers tipped white: BLACK-HEADED APALIS *Apalis melanocephala* **975**

15 Above, sooty black: 16–17

16 Breast to belly white; outer tail feathers wholly white: BLACK-BACKED APALIS *Apalis nigrescens* **976**

17 Breast to belly barred black and white; tail feathers with subterminal black bars and white tips: BANDED PRINIA *Prinia bairdii* **1049**

18 Wings green: GREY-BACKED CAMAROPTERA non-breeding dress *Camaroptera brevicaudata* **1011**

19 Wings of same colour as the mantle: 20–27

20 No white in tail: OLIVACEOUS WARBLER *Hippolais pallida* **938**

21 Outer tail feathers edged and tipped, or only tipped, with white: 22–27

22 Upper tail-coverts chestnut: DESERT WARBLER *Sylvia nana* **932**

23 Upper tail-coverts of same colour as mantle: 24–27

24 Size larger, wing 70 mm. or over: UPCHER'S WARBLER *Hippolais languida* **937**

| | | |
|---|---|---|
| 25 Size smaller, wing 61 mm. or under: | | 26–27 |
| 26 Tail long and narrow: | PALE PRINIA<br>*Prinia somalica* | 1046 |
| 27 Tail shorter and broader: | BLACK-CAPPED WARBLER, female<br>*Sylvia melanocephala* | 930 |
| 28 Below, barred black: | BARRED WREN-WARBLER<br>*Calamonastes fasciolatus* | 967 |
| 29 Below, not barred: | | 30–52 |
| 30 Above, olivaceous grey: | BASRA REED WARBLER<br>*Acrocephalus griseldis* | 943 |
| 31 Above, olivaceous: | | 32–39 |
| 32 Below, pale yellow or cream: | | 33–36 |
| 33 Chin to chest pale yellow: | WILLOW WARBLER<br>*Phylloscopus trochilus* | 959 |
| 34 Chin to under tail-coverts pale yellow or cream: | | 35–36 |
| 35 Size larger, wing over 69 mm.: | ICTERINE WARBLER<br>*Hippolais icterina* | 935 |
| 36 Size smaller, wing under 49 mm.: | BUFF-BELLIED WARBLER<br>*Phyllolais pulchella* | 995 |
| 37 Below, white: | BONELLI'S WARBLER<br>*Phylloscopus bonelli* | 962 |
| 38 Below, chin to chest buffish, faintly streaked with yellow: | CHIFF-CHAFF<br>*Phylloscopus collybita* | 960 |
| 39 Below, wholy pale buff: | MARSH WARBLER<br>*Acrocephalus palustris* | 945 |
| 40 Above, olivaceous brown: | | 41–52 |
| 41 Neck in front streaked: | RIVER WARBLER<br>*Locustella fluviatilis* | 940 |
| 42 Neck in front not streaked: | | 43–52 |

43 Size larger, wing 90 mm.
or over: GREAT REED WARBLER
*Acrocephalus*
*arundinaceus* **942**

44 Size smaller, wing 85 mm.
or under: 45-52

45 Bill short: GARDEN WARBLER
*Sylvia borin* **925**

46 Bill long: 47-52

47 Feet and toes black and
large: LESSER SWAMP-WARBLER
*Calamocichla*
*leptorhyncha* **957**

48 Feet and toes not black
and not large: 49-52

49 Chest and flanks olivaceous brown: SAVI'S WARBLER
*Locustella luscinioides* **939**

50 Chest and flanks buff: 51-52

51 Size larger, wing 63 to
69 mm.; 2nd primary
equal to or longer than
4th: REED WARBLER
*Acrocephalus scirpaceus* **944**

52 Size smaller, wing 52 to
64 mm.; 2nd primary
appreciably shorter than
4th: AFRICAN REED WARBLER
*Acrocephalus bæticatus* **946**

---

53 A moustachial stripe: MOUSTACHE WARBLER
*Melocichla mentalis* **1052**

54 No moustachial stripe: 55-88

55 Wing shoulder and bar
across wing white: WHITE-WINGED WARBLER
*Bradypterus carpalis* **950**

56 Wings green: 57-58

57 Broad white stripe over
eye, tail very short: WHITE-BROWED CROMBEC
*Sylvietta leucophrys* **1002**

# WARBLERS

| | | |
|---|---|---|
| 58 | Narrow brown stripe over eye, tail not short: | BROWN WOODLAND-WARBLER *Seicercus umbrovirens* **965** |
| 59 | Wing-coverts black or blackish, distinctly edged and tipped with white: | 60–61 |
| 60 | Top of head streaked with black: | CRICKET WARBLER *Spiloptila clamans* **1052** |
| 61 | Top of head not streaked: | RED-FACED APALIS *Apalis rufifrons* **987** |
| 62 | Tail feathers with subterminal black or blackish bar: | 63–66 |
| 63 | Size larger, wing 75 mm. or over: | RUFOUS WARBLER *Agrobates galactotes* **934** |
| 64 | Size smaller, wing 63 mm. or under: | 65–66 |
| 65 | Wings chestnut: | RED-WINGED WARBLER (except breeding dress) *Heliolais erythroptera* **1050** |
| 66 | Wings not chestnut: | TAWNY-FLANKED PRINIA *Prinia subflava* **1045** |
| 67 | Tail not subterminally barred: | 67–88 |
| 68 | Feet and toes large and high claw distinctly long: | 69–70 |
| 69 | Size larger, wing 83 mm. or over: | FOX'S SWAMP-WARBLER *Calamocichla foxi* **956** |
| 70 | Size smaller, wing 82 mm. or under: | GREATER SWAMP-WARBLER *Calamocichla gracilirostris* **955** |
| 71 | Feet and toes normal: | 72–88 |
| 72 | Under tail-coverts thick and nearly as long as tail: | FAN-TAILED WARBLER *Schœnicola brevirostris* **969** |

| | | |
|---|---|---|
| 73 | Under tail-coverts normal length: | 74–88 |
| 74 | Forehead, sides of face to chest black: | BLACK-FACED RUFOUS WARBLER, male *Bathmocercus rufus* **1053** |
| 75 | No black on head to chest: | 76–88 |
| 76 | Some white on outer tail feathers: | WHITETHROAT *Sylvia communis* **924** |
| 77 | No white in tail: | 78–88 |
| 78 | Tail feathers broad: | 79–82 |
| 79 | Chest mottled grey and white: | BAMBOO WARBLER *Bradypterus alfredi* **951** |
| 80 | Chest streaked with blackish: | 81–82 |
| 81 | Size larger, wing 65 mm. or over: | GRAUER'S WARBLER *Bradypterus graueri* **949** |
| 82 | Size smaller, wing 60 mm. or under: | LITTLE RUSH-WARBLER *Bradypterus babæcala* **948** |
| 83 | Tail feathers narrow: | 84–88 |
| 84 | Size smaller, wing 51 mm. or under: | RED-CAPPED FOREST-WARBLER *Artisornis metopias* **991** |
| 85 | Size larger, wing 55 mm. or over: | 86–88 |
| 86 | Upperside, sides of face and chest chestnut: | LOPEZ'S WARBLER *Sathrocercus lopezi* **954** |
| 87 | Upperside and chest russet brown: | CINNAMON BRACKEN-WARBLER *Sathrocercus cinnamomeus* **952** |

| | | |
|---|---|---|
| 88 | Upperside dark brown, below olive brown: | EVERGREEN-FOREST WARBLER *Sathrocercus mariæ* 953 |
| 89 | White stripe over eye: | UGANDA WOODLAND-WARBLER *Seicercus budongoensis* 965 |
| 90 | Stripe over eye not white, or no stripe over eye: | 91–122 |
| 91 | Forehead, sides of face and chin to foreneck pale chestnut: | MRS. MOREAU'S WARBLER *Scepomycter winifredæ* 992 |
| 92 | Throat and chest buffish brown: | 93–94 |
| 93 | Above, grass green: | RED-FACED WOODLAND-WARBLER *Seicercus lætus* 966 |
| 94 | Above, dull olive green: | OLIVE-GREEN CAMAROPTERA *Camaroptera chloronota* 1010 |
| 95 | Dark chestnut patch in centre of throat: | GREY-CAPPED WARBLER *Eminia lepida* 993 |
| 96 | No chestnut on head or throat: | 97–122 |
| 97 | Black band across upper chest: | YELLOW BAR-THROATED APALIS *Apalis flavigularis* 971 |
| 98 | No black band across chest: | 99–122 |
| 99 | Belly mainly grey: | 100–102 |
| 100 | Below, wholly olive grey: | GREY LONGBILL *Macrosphenus concolor* 1015 |
| 101 | Throat and chest dusky brown: | GREEN CROMBEC *Sylvietta virens* 1001 |

| | | | |
|---|---|---|---|
| 102 | Throat white, black or grey patch on lower neck: | BLACK-CAPPED APALIS *Apalis nigriceps* | 983 |
| 103 | Belly various shades of yellow: | | 104–110 |
| 104 | Throat white, grey patch at base of foreneck: | WHITE-WINGED APALIS, female *Apalis chariessa* | 984 |
| 105 | Chin to chest white: | | 106–107 |
| 106 | Top of head grey: | GREEN-BACKED EREMOMELA *Eremomela canescens* | 1005 |
| 107 | Top of head dull olive green: | YELLOW LONGBILL *Macrosphenus flavicans* | 1014 |
| 108 | Line down centre of throat and patch at base of foreneck black or grey: | BLACK-THROATED APALIS *Apalis jacksoni* | 982 |
| 109 | Throat to belly dull yellow, flanks brownish: | YELLOW SWAMP-WARBLER *Calamonastides gracilirostris* | 958 |
| 110 | Belly mainly white: | | 111–121 |
| 111 | Throat and chest black: | MASKED APALIS *Apalis binotata* | 981 |
| 112 | Throat white: | | 113–118 |
| 113 | Upper chest yellow or greenish with or without a black patch: | | 114–115 |
| 114 | Outer tail feathers mainly yellow: | BLACK-BREASTED APALIS *Apalis flavida* | 979 |
| 115 | Outer tail feathers only tipped with yellow: | GREEN-TAILED APALIS *Apalis caniceps* | 980 |
| 116 | Upper chest white: | | 117–118 |

| | | |
|---|---|---|
| 117 | Above, bright green, sides of face yellow: | YELLOW-BROWED CAMAROPTERA *Camaroptera superciliaris* **1012** |
| 118 | Above, olive green, sides of face grey: | GREEN-BACKED CAMAROPTERA *Camaroptera brachyura* **1009** |
| 119 | Throat yellow: | 120–121 |
| 120 | Above, pale yellow green: | WOOD WARBLER *Phylloscopus sibilatrix* **961** |
| 121 | Above, head saffron brown, mantle green: | YELLOW-THROATED WOODLAND-WARBLER *Seicercus ruficapillus* **963** |

| | | |
|---|---|---|
| 122 | Throat to belly lemon yellow: | GREEN-CAP EREMOMELA *Eremomela scotops* **1006** |
| 123 | Top of head clearly contrasting with mantle: | 124–140 |
| 124 | Top of head chestnut: | BROWN-CROWNED EREMOMELA *Eremomela badiceps* **1007** |
| 125 | Top of head pale brown: | BLACKCAP, female *Sylvia atricapilla* **926** |
| 126 | Top of head ashy or sooty brown: | 127–130 |
| 127 | Black band across chest: | BAR-THROATED APALIS *Apalis murina* **970** |
| 128 | No black band across chest: | 129–130 |
| 129 | Outer tail feathers white: | GREY APALIS *Apalis cinerea* **973** |
| 130 | Outer tail feathers tipped with white: | BROWN-HEADED APALIS *Apalis alticola* **974** |
| 131 | Top of head black or blackish: | 132–140 |
| 132 | Chin to chest black: | 133–134 |

133 Outer tail feathers white: RÜPPELL'S WARBLER, male
*Sylvia rüppelli* **929**

134 Outer tail feathers ashy grey: BLACK-FACED RUFOUS WARBLER, female
*Bathmocercus rufus* **1053**

135 Chin to chest greyish pinkish white or white: 136–139

136 Chin to chest greyish: BLACKCAP, male
*Sylvia atricapilla* **926**

137 Chin to chest white: 138–140

138 Size larger, wing 72 mm. or over: ORPHEAN WARBLER, male
*Sylvia hortensis* **927**

139 Size medium, wing 62 to 70 mm.: RED SEA WARBLER
*Sylvia leucomelæna* **928**

140 Size smaller, wing 58 mm. or under: BLACK-CAPPED WARBLER, male
*Sylvia melanocephala* **930**

141 Top of head not clearly contrasting with mantle: 142–189

142 Outer webs of secondaries white: KARAMOJA WARBLER
*Apalis karamojæ* **988**

143 Outer webs of all flight feathers chestnut: RED-WINGED WARBLER, breeding dress
*Heliolais erythroptera* **1050**

144 Outer webs of primaries tawny: RED-WINGED GREY WARBLER
*Drymocichla incana* **994**

145 No white, chestnut or tawny in wings: 146–189

146 Black band across chest: 147–148

147 Outer tail feathers mainly white: BLACK-COLLARED APALIS
*Apalis pulchra* **977**

## WARBLERS

| | | |
|---|---|---|
| 148 | No white in outer tail feathers: | COLLARED APALIS *Apalis ruwenzorii* **978** |
| 149 | No black band across chest: | 150–189 |
| 150 | Tail very short: | 151–159 |
| 151 | Some yellow on belly: | 152–153 |
| 152 | Yellow confined to vent: | YELLOW-VENTED EREMOMELA *Eremomela flavicrissalis* **1005** |
| 153 | Yellow extending to belly or breast: | YELLOW-BELLIED EREMOMELA *Eremomela icteropygialis* **1003** |
| 154 | No yellow on underside: | 155–162 |
| 155 | Below, white or pale isabelline: | 156–157 |
| 156 | Ear-coverts buffish: | SOMALI LONG-BILLED CROMBEC *Sylvietta isabellina* **999** |
| 157 | Ear-coverts chestnut: | RED-CAPPED CROMBEC *Sylvietta ruficapilla* **1000** |
| 158 | Below, deep buff or russet: | 159–162 |
| 159 | Blackish stripe between base of bill and eye: | CROMBEC *Sylvietta brachyura* **996** |
| 160 | No blackish stripe between base of bill and eye: | 161–162 |
| 161 | Throat greyish contrasting with breast colour: | RED-FACED CROMBEC *Sylvietta whytii* **997** |
| 162 | Throat uniform with breast colour: | LONG-BILLED CROMBEC *Sylvietta rufescens* **998** |
| 163 | Tail not very short: | 164–189 |
| 164 | Below more or less barred: | 165–166 |

| | | | |
|---|---|---|---|
| 165 | Size larger, wing 80 mm. or over: | BARRED WARBLER *Sylvia nisoria* | **933** |
| 166 | Size smaller, wing 66 mm. or under: | GREY WREN-WARBLER *Calamonastes simplex* | **968** |
| 167 | No barring below: | | 168–189 |
| 168 | Tail feathers tipped and edged with white: | OLIVE-TREE WARBLER *Hippolais olivetorum* | **936** |
| 169 | No white in tail feathers and edges of wing feathers green: | GREY-BACKED CAMAROPTERA breeding dress *Camaroptera brevicaudata* | **1011** |
| 170 | Tail feathers tipped with white and edges of wing feathers not green: | | 171–178 |
| 171 | Subterminal black bar to outer tail feathers: | SOCOTRA WARBLER *Incana incana* | **1013** |
| 172 | No subterminal black bar to outer tail feathers: | | 173–189 |
| 173 | Chin and throat white, neck and breast grey: | WHITE-CHINNED PRINIA *Prinia leucopogon* | **1048** |
| 174 | Chin and throat tawny, neck and breast grey: | CHESTNUT-THROATED APALIS *Apalis porphyrolæma* | **985** |
| 175 | Chin whitish, throat and chest chestnut: | BAMENDA APALIS *Apalis bamendæ* | **986** |
| 176 | No white in tail: | | 177–178 |
| 177 | Below, greyish white: | LONG-BILLED APALIS *Apalis moreaui* | **989** |

| | | | |
|---|---|---|---|
| 178 | Below, creamy buff | BURNT-NECK EREMOMELA *Eremomela usticollis* | **1008** |
| 179 | Outer tail feathers mainly white: | | 180–189 |
| 180 | Chin to breast chestnut: | SUBALPINE WARBLER, male *Sylvia cantillans* | **931** |
| 181 | Chin to breast deep buff: | BUFF-THROATED APALIS *Apalis rufogularis* | **972** |
| 182 | Whole underside silvery grey: | KUNGWE APALIS *Apalis argentea* | **990** |
| 183 | Whole underside white or buffish white: | | 184–185 |
| 184 | Size larger, wing 72 mm. or over: | ORPHEAN WARBLER, female *Sylvia hortensis* | **927** |
| 185 | Size smaller, wing 71 mm. or under: | | 186–189 |
| 186 | Grey of face darker contrasting with white throat: | LESSER WHITETHROAT *Sylvia curruca* | **923** |
| 187 | Grey of face paler, not contrasting so strongly with white throat: | | 188–189 |
| 188 | Bill shorter, underparts distinctly buffish white: | SUBALPINE WARBLER, female *Sylvia cantillans* | **931** |
| 189 | Bill longer, underparts not distinctly buffish white: | RÜPPELL'S WARBLER, *Sylvia rüppelli* female | **929** |

*Note:* The Cisticola group is not included in this Key.

## WARBLERS

### WHITETHROATS, BLACKCAP, GARDEN, ORPHEAN, RÜPPELLS, SUBALPINE, DESERT and BARRED WARBLERS. FAMILY—SYLVIIDÆ. Genus: *Sylvia*.

A group of palæarctic Warblers of which nine species migrate to our area in the non-breeding season and one is resident. They are birds of bushes, hedges, and dense vegetation; mostly greyish or brownish in colour above and whitish below. Food mainly insects, but also a great deal of fruit when in season. Songs varied, but mostly of some power and quality.

**923** LESSER WHITETHROAT. *SYLVIA CURRUCA* (Linnæus).
*Sylvia curruca curruca* (Linn.). **Pl. 68.**
*Motacilla Curruca* Linnæus, Syst. Nat. 10th ed. p. 184, 1758: Sweden.

**Distinguishing characters:** Above, ash grey; head paler grey; below, white; iris brownish white; bill and feet lead colour. The sexes are alike. Wing 61 to 70 mm.

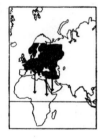

**General distribution:** Europe to the Caucasus and northern Iran; in non-breeding season to Africa as far south as Lake Chad, the Sudan and Abyssinia, also western Arabia.

**Range in Eastern Africa:** The Sudan, Eritrea and Abyssinia in non-breeding season.

**Habits:** A common, often abundant, palæarctic winter visitor to the Sudan, but so far has not been recorded from further south. It is a rather skulking, easily overlooked little bird with an alarm note of 'tac-tac.' Feeds in Europe largely on insects, but Morrison-Scott's observations on migrating Warblers in Arabia make it clear that this species, in common with others, feeds largely on seeds while on migration.

*Sylvia curruca blythi* Ticeh. & Whist.
*Sylvia curruca blythi* Ticehurst & Whistler, Ibis, p. 556, 1933: Siberia.

**Distinguishing characters:** Similar in colour and size to the nominate race, but differs in having the second primary shorter than the sixth, the opposite being the case in the nominate race. Wing 60 to 71 mm.

**General distribution:** Russia in Asia and Manchuria; in non-breeding season to Eastern Africa, Arabia, Iran, India and Ceylon.

# WARBLERS

**Range in Eastern Africa:** So far only noted from Abyssinia in non-breeding season.

**Habits:** Indistinguishable from the nominate race in the field.

**Distribution of other races of the species:** Transcaspia and Turkestan.

## 924 WHITETHROAT. *SYLVIA COMMUNIS* Latham.

*Sylvia communis communis* Lath.
*Sylvia communis* Latham, Gen. Syn. Suppl. 1, p. 287, 1787: England.

**Distinguishing characters:** Adult male, above, brown; wings brighter brown; head slightly grey; below, chin and throat white; rest of underparts buffish white, lighter in centre of belly; often a wash of pink on the lower neck and upper chest. The female has the head brown and the lower neck and upper chest buff. Wing 65 to 78 mm.

**General distribution:** Europe to Siberia and North Africa from Morocco to Tunis; in non-breeding season to western and eastern Africa as far south as Damaraland, Southern Rhodesia and western Nyasaland; also southern Arabia.

**Range in Eastern Africa:** Throughout in non-breeding season at least as far south as Tanganyika Territory.

**Habits:** Locally common palæarctic winter visitor to most parts of Eastern Africa, usually seen singly, but in flocks or parties when on migration. Has been noted in fair numbers as far south as Southern Rhodesia, and is said to sing freely in Kenya Colony before the spring migration. Keeps to cover as much as possible.

*Sylvia communis icterops* Méné.
*Sylvia icterops* Ménétriés, Cat. rais. Caucase. p. 34, 1832: Talysch, Caspian Sea.

**Distinguishing characters:** Differs from the nominate race in being rather greyer above, with paler edges to the flight feathers. Wing 68 to 79 mm.

**General distribution:** Asia Minor, Palestine, the Caucasus and Iran; in non-breeding season to western and eastern Africa as far south as western Nyasaland and Southern Rhodesia, also Arabia.

**Range in Eastern Africa:** Eritrea, Abyssinia and British Somaliland to Tanganyika Territory in non-breeding season.

**Habits:** Indistinguishable in the field from the nominate race.

*Sylvia communis rubicola* Stres.
*Sylvia communis rubicola* Stresemann, J.f.O. p. 378, 1928: Kuldja, western Chinese Turkestan.

**Distinguishing characters:** Differs from the preceding races in being more buff or ashy buff above, not brown as in the nominate race, nor grey as in the last race. Wing 67 to 78 mm.

**General distribution:** Russian and Chinese Turkestan to Mongolia and the Altai; in non-breeding season to northern and eastern Africa, north-eastern Angola and western Nyasaland, Arabia and India.

**Range in Eastern Africa:** The Sudan and Eritrea to Uganda, central Kenya Colony and Tanganyika Territory in non-breeding season.

**Habits:** Indistinguishable in the field from the nominate race.

**925** GARDEN WARBLER. *SYLVIA BORIN* (Boddaert). **Pl. 68.**
*Motacilla borin* Boddaert, Tabl. Pl. Enl. p. 35, 1783: France.

**Distinguishing characters:** Above, ashy olivaceous; sides of neck slightly grey; below, creamy olivaceous; under tail-coverts paler; centre of belly white. The sexes are alike. Wing 72 to 85 mm.

**General distribution:** Europe to western Asia and Iran; in non-breeding season to Africa throughout, and western Arabia.

**Range in Eastern Africa:** Throughout in non-breeding season.

**Habits:** A common palæarctic winter visitor to all parts of Eastern Africa, but mainly a passage migrant in the Sudan and the north. Collect into flocks for migration, but are otherwise found singly or in twos and threes in bush, woodland, or cultivation, being especially fond of Lantana bushes, of which they eat the fruit. Sing freely on or before the spring migration, and have a cheerful bubbling little sub-song at all times.

**926** BLACKCAP. *SYLVIA ATRICAPILLA* (Linnæus).
*Sylvia atricapilla atricapilla* (Linn.).
*Motacilla atricapilla* Linnæus, Syst. Nat. 10th ed. p. 187, 1758: Sweden.

**Distinguishing characters:** Adult male, above, ashy grey; whole top of head black; below, sides of face, neck and chest greyish; flanks brownish; centre of belly white. The female is rather browner than the male and has the whole top of the head russet brown. Wing 68 to 81 mm.

**General distribution:** Europe to western Siberia, North Africa and Iran; in non-breeding season to Africa occasionally as far south as Nyasaland.

**Range in Eastern Africa:** Northern areas to southern Tanganyika Territory in non-breeding season.

**Habits:** One of the commonest palæarctic winter visitors to all parts of Eastern Africa. Rather more of a shady woodland, or even forest, bird than most migrants, and penetrates into high forest on mountains. Moves about in scattered flocks during the non-breeding season and sings freely before the spring migration, from December onwards. Probably feeds to a considerable extent on fruit as well as insects.

*Sylvia atricapilla dammholzi* Stres.
*Sylvia atricapilla dammholzi* Stresemann, J.f.O. p. 377, 1928: Pish Kuh, Gilan, northern Iran.

**Distinguishing characters:** Adult male, paler than the nominate race; greyer above, especially on the rump. The female is also paler above, and the top of the head is duller brown. Wing 71 to 82 mm.

**General distribution:** Southern Asia Minor, Iraq and Iran; in non-breeding season to the Sudan, Abyssinia, Tanganyika Territory and Arabia.

**Range in Eastern Africa:** Abyssinia and the Sudan to Tanganyika Territory in non-breeding season.

**Habits:** As for the nominate race, and would be doubtfully distinguishable in the field.

**Distribution of other races of the species:** Madeira, Canary Islands.

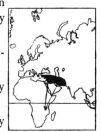

**927 ORPHEAN WARBLER.** *SYLVIA HORTENSIS* (Gmelin).
*Sylvia hortensis crassirostris* Cretz.
*Sylvia crassirostris* Cretzschmar, Atlas zu der Reise, Vög. p. 49, pl. 33, fig. a, 1830: Nubia.

**Distinguishing characters:** Adult male, above, grey; head and sides of face blackish; tail black; outer web of outermost tail feathers and tips of all except central pair white; below, white or creamy white. The female has the head greyer and the mantle browner than the male. Wing 72 to 86 mm.

# WARBLERS

**General distribution:** South-eastern Europe and western Asia from Dalmatia and the Ionian Islands to Afghanistan and Cyrenaica; in non-breeding season to Kenya Colony, Arabia and India.

**Range in Eastern Africa:** The Sudan, Eritrea, British Somaliland and Kenya Colony in non-breeding season.

**Habits:** A palæarctic winter visitor to the northern parts of our area, being especially numerous on migration on the Red Sea coast. In the field the white on the outer tail feathers is generally sufficient to distinguish it from the Blackcap.

**Distribution of other races of the species:** South-western Europe and West Africa, the nominate race being described from France.

**928 RED SEA WARBLER.** *SYLVIA LEUCOMELÆNA* (Hemprich & Ehrenberg).
*Sylvia leucomelæna blanfordi* Seeb. **Pl. 61.**
*Sylvia blanfordi* Seebohm, P.Z.S. p. 979, 1878: Rairo, Eritrea.

**Distinguishing characters:** Adult male, head and sides of face brownish black; mantle and wings ash brown; tail black, outermost tail feather edged and tipped with white; below, creamy white. The female has the head and sides of face duller and browner. Wing 64 to 70 mm. Juvenile plumage undescribed.

**Range in Eastern Africa:** Red Sea coastal area and Eritrea.

**Habits:** No information.

**Nest and Eggs:** Undescribed.

**Recorded breeding:** No records.

**Call:** Unrecorded.

*Sylvia leucomelæna somaliensis* (Scl. & Praed).
*Parisoma blanfordi somaliensis* W. Sclater & Mackworth-Praed, Ibis, p. 707, 1918: Mundara, British Somaliland.

**Distinguishing characters:** Differs from the preceding race in having a greater extent of white on the outer web of the outermost tail feather and more white on the tips of the two outermost. Wing 62 to 68 mm.

**Range in Eastern Africa:** British Somaliland to northern Italian Somaliland as far south as lat. 9° N.

**Habits:** No information.

**Recorded breeding:** No records.

**Distribution of other races of the species:** Arabia, the habitat of the nominate race.

### 929 RÜPPELL'S WARBLER. *SYLVIA RÜPPELLI* Temminck.

*Sylvia rüppelli* Temminck, Pl. Col. livr. 41, pl. 245, fig. 1, 1823: Crete.

**Distinguishing characters:** Adult male, forehead and crown black; rest of upperparts grey; tail blackish, outermost feathers mainly white; light edges to wing-coverts and innermost secondaries; moustachial streak white; chin to chest black; rest of underparts pinky white. The female is rather browner above; much less black on head or none; below, chin to chest blackish and buffish white, or wholly buffish white. The female is very similar to the Lesser Whitethroat, but has a longer and sharper bill, a much shorter and narrower first primary; iris reddish brown; legs red brown. Wing 66 to 71 mm. The immature bird of both sexes is similar to the adult female, but has no black on head or chin to breast, and is paler above.

**General distribution:** Greece, Crete, Cyprus, Asia Minor and Palestine; in non-breeding season to the Sudan.

**Range in Eastern Africa:** The Sudan in non-breeding season.

**Habits:** A locally common palæarctic winter visitor, not ranging further south than the Sudan. Lynes notes that its winter distribution is closely correlated with that of the Tundub bush (*Capparis*) which grows in cotton soil, and that the birds' heads are usually discoloured by the pollen from the flowers. They start singing in February and leave for the north in March.

### 930 BLACK-CAPPED WARBLER. *SYLVIA MELANO-CEPHALA* (Gmelin).

*Sylvia melanocephala momus* (Hemp. & Ehr.). **Pl. 68.**
*Curruca momus* Hemprich & Ehrenberg, Symb. Phys. fol. bb., 1828: Egypt.

**Distinguishing characters:** Adult male, top of head and sides of face sooty black; rest of upperparts brownish grey; white tip to outer tail feathers, outer edge of outermost white; below, chin and throat white; rest of underparts pinky white; eye ring red. The female is dull brown above; top of head greyer; below, buffish white; throat white; eye ring buff. Wing 50 to 58 mm.

**General distribution:** Syria and Palestine; in non-breeding season to Egypt, Sinai, the Sudan and southern Arabia.

**Range in Eastern Africa:** The eastern Sudan in non-breeding season.

**Habits:** Skulking, restless, and extremely inquisitive little birds, usually found among bushes or in gardens and cultivated land. The constant spreading of the graduated tail is a noticeable characteristic in the field. They have a rattling churring alarm note.

*Sylvia melanocephala mystacea.* Méné.
*Sylvia mystacea* Ménétriés, Cat. rais. Caucase. p. 34, 1832: Salyany, Transcaucasia.

**Distinguishing characters:** Adult male, differs from the preceding race in having the black on the head not so well defined and less intense in colour. The female is paler than the female of the preceding race. Wing 55 to 63 mm. The immature male is similar to the adult female, but has a greyer tone above and a pinkish wash below.

**General distribution:** Southern Caucasus and Transcaucasia to northern Iran, Turkestan and Afghanistan; in non-breeding season to the Sudan, British Somaliland and southern Arabia.

**Range in Eastern Africa:** Eastern Sudan, Eritrea and British Somaliland in non-breeding season.

**Habits:** As for other races.

**Distribution of other races of the species:** Spain, Asia Minor and North Africa, the nominate race being described from Sardinia.

931 SUBALPINE WARBLER. *SYLVIA CANTILLANS* (Pallas).
*Sylvia cantillans albistriata* (Brehm). **Pl. 68.**
*Curruca albistriata* C. L. Brehm, Vögelfang, p. 229, 1855: Egypt.

**Distinguishing characters:** Adult male, above, ashy grey including top of head; outermost tail feathers mainly white; moustachial streak white; underparts vinous brown, darker from chin to breast, centre of belly whitish; fleshy ring round eye orange. The female is pale brown above, and creamy white below; flanks buffish. This bird has a similar short and narrow first primary to Rüppell's Warbler, but the female can be distinguished from that species by the shorter bill, yellow eye and pale brown feet. Wing 59 to 63 mm.

# WARBLERS

**General distribution:** Balkan peninsula to Asia Minor, Sicily, Cyprus and Palestine; in non-breeding season to Egypt, the French Sudan, Lake Chad, Eastern Africa and south-western Arabia.

**Range in Eastern Africa:** Eastern Sudan and British Somaliland to the coastal areas of Tanganyika Territory in non-breeding season.

**Habits:** A rare and local palæarctic winter visitor, but very skulking and secretive and easily overlooked. The call is a harsh 'tac-tac.'

**Distribution of other races of the species:** Western Mediterranean countries, the nominate race being described from Italy.

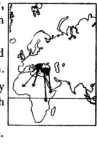

**932 DESERT WARBLER.** *SYLVIA NANA* (Hemprich & Ehrenberg).
*Sylvia nana nana* (Hemp. & Ehr.). **Pl. 68.**
*Curruca nana* Hemprich & Ehrenberg, Symb. Phys. fol. cc, 1828: El Tor, Sinai peninsula.

**Distinguishing characters:** Adult male, above, sandy ash-grey; tail-coverts tawny; tail tawny and blackish, outermost feathers mainly white; below, pale creamy white. The sexes are alike. Wings, male 54 to 62, female 54 to 58 mm.

**General distribution:** Turkey to Iran, Transcaspia and Turkestan; in non-breeding season to Palestine, Egypt, Eastern Africa, Arabia and northern India.

**Range in Eastern Africa:** Coastal areas of the Sudan to British Somaliland in non-breeding season.

**Habits:** A local palæarctic winter visitor to a restricted area and found mostly in coastal scrub.

**Distribution of other races of the species:** Algeria and Tripoli.

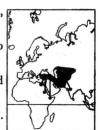

**933 BARRED WARBLER.** *SYLVIA NISORIA* (Bechstein).
*Motacilla nisoria* Bechstein, Gem. Nat. Deutschl. 4, p. 580, 1795: Germany.

**Distinguishing characters:** Adult male, above, ashy grey; whitish edges to wing-coverts, inner secondaries, rump and tail-coverts; some white at tip of outer tail feathers; below, white or creamy white, barred with ashy grey. The female is plain below or with only a few bars. Wing 80 to 94 mm. The young bird is paler above than the adult female and has less distinct edges to the wing-coverts and inner secondaries. The female and young bird might be

2A

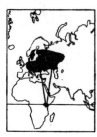

confused with those of the Orphean Warbler, but can be distinguished by the light edges to the wing-coverts and inner secondaries, and by the first primary being small and shorter than the primary coverts, instead of large and longer than the primary coverts. They are also larger.

**General distribution:** Northern and central Europe to Turkestan; in non-breeding season to Eastern Africa and Arabia.

**Range in Eastern Africa:** Eastern Sudan, Eritrea and Abyssinia to south-eastern Kenya Colony in non-breeding season.

**Habits:** A palæarctic winter migrant to Kenya Colony arriving mainly by the Red Sea route, and rare or absent from the Nile Valley.

**RUFOUS WARBLER.** FAMILY—**SYLVIIDÆ.** Genus: *Agrobates*.

The only species of the genus, with five races, two resident in Eastern Africa, the other three migrants from Europe and Asia in the non-breeding season. They frequent bush country and woodlands and are largely ground birds. The tail pattern is distinctive and when singing from the top of a bush the tail is frequently spread.

**934 RUFOUS WARBLER.** *AGROBATES GALACTOTES* (Temminck).

*Agrobates galactotes galactotes* (Temm.).
*Sylvia galactotes* Temminck, Man. d'Orn. 2nd ed. p. 182, 1820: Algeciras, southern Spain.

**Distinguishing characters:** Above, tawny; light stripe from bill to over eye; tail pale chestnut with white tips and subterminal black bar; below, sandy. The sexes are alike. Wing 83 to 92 mm.

**General distribution:** Southern Spain and Portugal, North Africa and Palestine; in non-breeding season to the Sahara and Eastern Africa.

**Range in Eastern Africa:** The Sudan and British Somaliland in non-breeding season.

**Habits:** Warblers of rather Thrush-like habits, with something of the behaviour of Robins, common in bush country at lower levels and feeding mainly on the ground. They have a shrill rather Thrush-like song uttered from a bough generally in an exposed position, and a habit of frequently spreading their tails. They do not occur in the Sudan much south of the latitude of Khartoum.

*Agrobates galactotes familiaris* (Méné.).   *Caucasus Rufous Warbler.*
*Sylvia familiaris* Ménétriés, Cat. rais. Caucase, p. 32, 1832: Saliane, Transcaucasia.

**Distinguishing characters:** Differs from the nominate race in being ash brown above. Wing 81 to 92 mm.

**General distribution:** Transcaucasia, Iraq, Iran and Turkestan; in non-breeding season to Abyssinia, Kenya Colony, Arabia and north-western India.

**Range in Eastern Africa:** Eastern Abyssinia to northern and eastern Kenya Colony, and north-eastern Tanganyika Territory.

**Habits:** As for the other migrant races, apparently not as common in Africa as the Syrian race.

*Agrobates galactotes syriacus* (Hemp. & Ehr.).
*Curruca galactotes* var *syriaca* Hemprich & Ehrenberg, Symb. Phys. fol. bb. 1833: Beyrout, Syria.

**Distinguishing characters:** Above, darker and browner than the nominate race, and rather greyer below. Wing 78 to 90 mm.

**General distribution:** Balkan peninsula, Asia Minor and Syria; in non-breeding season to Eastern Africa as far west as the eastern Belgian Congo, and as far south as Tanganyika Territory.

**Range in Eastern Africa:** Eritrea, eastern Abyssinia, French Somaliland, Kenya Colony and north-eastern Tanganyika Territory in non-breeding season.

**Habits:** As for the nominate race, and occurs not uncommonly as far south as the northern areas of Tanganyika Territory, mostly in thorn bush at low levels from late October to April.

*Agrobates galactotes minor* (Cab.).
*Ædon minor* Cabanis, Mus. Hein. 1, p. 39, 1850: Eritrea.

**Distinguishing characters:** Similar to the nominate race, but rather smaller. Wing 75 to 86 mm. The young bird is slightly mottled on the chest; paler, more fawn colour above; black spots on tail practically absent.

**General distribution:** Senegal and northern Nigeria to the Sudan and British Somaliland.

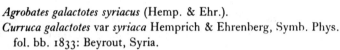

**Range in Eastern Africa:** The central areas of the Sudan, Eritrea and British Somaliland as far east as Bulhar and Hargeisa.

**Habits:** A resident African race preferring dry dusty bush country, but may be found anywhere in bush. Rather a conspicuous bird owing to its restlessness and its loud song, but is shy and vanishes into thick cover if alarmed.

**Nest and Eggs:** A rather loosely constructed cup nest of grass, leaves, and plant stems lined with finer material, placed in a thick bush when available. Eggs three, pale grey or pale bluish white with spots and blotches of umber brown and lavender grey; about 21 × 15·5 mm.

**Recorded breeding:** Timbuktu, French Sudan, probably June and July. Sokoto, Nigeria, April. Air, French Sahara, July and August.

**Food:** Largely insects taken mainly on the ground.

**Call:** The song is Thrush-like but rather shrill and is uttered from a perch, the bird flirting and spreading its tail while singing.

*Agrobates galactotes hamertoni* (O. Grant). **Pl. 67.**
*Erythropygia hamertoni* O. Grant, Bull. B.O.C. 19, p. 24, 1906: Bera, 120 miles north-west of Obbia, Italian Somaliland.

**Distinguishing characters:** Darker above and smaller than the Eritrean race. Wing 70 to 71 mm.

**Range in Eastern Africa:** British and Italian Somaliland as far west as Waghar and Bera.

**Habits:** No information, but unlikely to differ from those of other races.

**Recorded breeding:** No records.

## ICTERINE, OLIVE-TREE, UPCHER'S and OLIVACEOUS WARBLERS. Family—SYLVIIDÆ. Genus: *Hippolais*.

These four species are migrants from Europe and Asia during the non-breeding season, though a race of one has recently been found breeding in the Sudan. None appear to be common, but they may easily be overlooked. Compared with the Phylloscopine Warblers the bill is rather long and broad, but not so broad as in the Reed and Swamp-Warblers. Their colours are sober olives and yellows. Purely insectivorous as far as is known.

## WARBLERS

**935 ICTERINE WARBLER. *HIPPOLAIS ICTERINA*** (Vieillot).

*Sylvia icterina* Vieillot, N. Dict. d'Hist. Nat. 11, p. 194, 1817: France.

**Distinguishing characters:** Above, pale olive green, with a yellowish wash; a faint yellowish stripe over eye; below, pale yellow. The sexes are alike. Wing 69 to 83 mm.

**General distribution:** The greater part of Europe, except western areas, to the Urals and Caspian Sea; in non-breeding season to Africa as far south as Kenya Colony, Nyasaland, Damaraland and Bechuanaland, also western Arabia.

**Range in Eastern Africa:** Eritrea, the Sudan, Uganda and Kenya Colony in non-breeding season.

**Habits:** A typical Warbler in habits, easily overlooked among other species and not common anywhere. The song is rather striking with almost an explosive quality, but otherwise it is a retiring little bird.

**936 OLIVE-TREE WARBLER. *HIPPOLAIS OLIVETORUM*** (Strickland).

*Salicaria olivetorum* Strickland, in Gould's Bds. Europe, 2, pl. 107, 1837: Zante, Ionian Islands, Greece.

**Distinguishing characters:** Above, ash-grey; tips of outer tail-feathers and outer edge of outermost white; below, white washed with cream colour; first primary very small and sharply pointed. The sexes are alike. Wing 82 to 91 mm.

**General distribution:** Balkan peninsula, Asia Minor, Syria and northern Palestine; in non-breeding season to Egypt, Eastern Africa, Southern Rhodesia and the Transvaal.

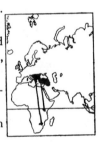

**Range in Eastern Africa:** Eritrea and eastern Sudan to south-eastern Kenya Colony and Nyasaland in non-breeding season.

**Habits:** Much as for the Icterine Warbler; most records have been from the dry bush country of north-eastern Kenya Colony.

**937 UPCHER'S WARBLER. *HIPPOLAIS LANGUIDA*** (Hemprich & Ehrenberg). **Pl. 68.**

*Curruca languida* Hemprich & Ehrenberg, Symb. Phys. fol. cc, 4, 1833: Syria.

**Distinguishing characters:** Above, ash brown; a faint light stripe from bill to above eye; below, creamy white; tips of tail feathers and outer edge of outermost whitish; first primary broader

and longer than in the Olive-Tree Warbler. The sexes are alike. Wing 70 to 80 mm.

**General distribution:** Palestine to Iran, Transcaspia and Afghanistan; in non-breeding season to Eastern Africa and Arabia.

**Range in Eastern Africa:** Red Sea Province of the Sudan, Eritrea, and British Somaliland to Kenya Colony and north-eastern Tanganyika Territory in non-breeding season.

**Habits:** An occasional palæarctic winter visitor to as far south as Kenya Colony and northern Tanganyika Territory, arriving in British Somaliland as early as July and August. Not readily distinguished in the field from other Warblers of this group.

### 938 OLIVACEOUS WARBLER. *HIPPOLAIS PALLIDA* (Hemprich & Ehrenberg).

*Hippolais pallida pallida* (Hemp. & Ehr.).
*Curruca pallida* Hemprich & Ehrenberg, Symb. Phys. fol. bb. 1833: Egypt.

**Distinguishing characters:** Very similar to Upcher's Warbler, but smaller, and with a buffish wash above. The sexes are alike. Wing 57 to 69 mm. The young bird is rather paler than the adult and has buff edges to the flight feathers and buff tips to the tail feathers.

**General distribution:** Northern Nigeria to Egypt and the Suez Canal; in non-breeding season to Eastern Africa.

**Range in Eastern Africa:** The Sudan to Abyssinia in non-breeding season.

**Habits:** A common, at times abundant, visitor to the Sudan staying occasionally as late as June and singing freely from January onwards. They have a churring irregular rather Reed Warbler-like song.

*Hippolais pallida elæica* (Linder.).
*Salicaria elæica* Lindermayer, Isis, p. 343, 1843: Greece.

**Distinguishing characters:** Differs from the nominate race in being more olivaceous, less buffish above. Wing 60 to 71 mm. The young bird is paler and very similar to the adult of the nominate race.

**General distribution:** Balkan peninsula, Asia Minor, Crete, Cyprus, Sinai, Palestine, the Sudan, Iran and Turkestan; in non-breeding season to Eastern Africa and Arabia.

# WARBLERS

**Range in Eastern Africa:** Eritrea, the Sudan, Abyssinia, British Somaliland, Kenya Colony and north-eastern Tanganyika Territory in non-breeding season; also breeds in the Sudan.

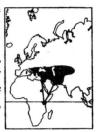

**Habits:** A common palæarctic winter migrant to Eastern Africa, ranging as far south as Tanganyika Territory, and it is also now known to breed in the Sudan near Khartoum. Tame and quite fearless of man, frequently found in gardens or cultivated land, gathering into considerable flocks at times of migration.

**Nest and Eggs:** Nest a neat cup of thin stems and fibres bound together and lined with a plentiful supply of plant down or other soft material, placed in any sort of low undergrowth or bush. Eggs two or three, pale mauvish white with sepia spots and speckles mainly at the larger end; about 19 × 13 mm.

**Recorded breeding:** Khartoum, May and June.

**Food:** Entirely insects.

**Call:** A short, loud, cheerful song in the breeding season and a low rambling sub-song heard at other times of year.

**Distribution of other races of the species:** Senegal and the Sahara.

## SAVI'S, RIVER and GRASSHOPPER WARBLERS. FAMILY—SYLVIIDÆ. Genus: *Locustella*.

These three species are migrants to Eastern Africa during the non-breeding season. They frequent swampy localities and in such situations can be easily overlooked. Their food consists of insects.

**939** SAVI'S WARBLER. *LOCUSTELLA LUSCINIOIDES* (Savi).

*Locustella luscinioides luscinioides* (Savi). **Pl. 68.**

*Sylvia luscinioides* Savi, N. Giorn. Litter. 7, p. 341, 1824: Pisa, Italy.

**Distinguishing characters:** Above, brown, slightly olivaceous; tail feathers graduated; below, pale brown; whiter from chin to throat and centre of belly; first primary small and sharply pointed. The sexes are alike. Wing 64 to 72 mm.

**General distribution:** Western Europe, Russia, North Africa, Palestine and Syria; in non-breeding season to the Sudan and Eritrea.

**Range in Eastern Africa:** The Sudan and Eritrea in non-breeding season.

**Habits:** Unobtrusive, secretive birds confined to swampy places. They were noticed in western Darfur in moderate numbers by Admiral Lynes, but were very local. The alarm note is a harsh chatter.

*Locustella luscinioides fusca* (Severtz.).
*Cettia fusca* Severtzov, Turk. Jevotn. (in Izv. Obshch. Moskov. 8, 2), p. 131, 1873: River Oxus, Turkestan, southern Russia.

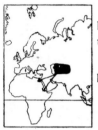

**Distinguishing characters:** Differs from the nominate race in being generally paler, more olive above and lighter below. Wing 68 to 71 mm.

**General distribution:** Transcaspia and Turkestan; in non-breeding season to Egypt, Trans-Jordan and Eritrea.

**Range in Eastern Africa:** Eritrea in non-breeding season.

**Habits:** As for the nominate race.

**940** RIVER WARBLER.   *LOCUSTELLA FLUVIATILIS* (Wolf).
*Sylvia fluviatilis* Wolf, in Meyer & Wolf, Taschenb. 1, p. 229, 1810: River Danube, near Vienna, Austria.

**Distinguishing characters:** Above, olivaceous brown; a faint stripe over eye; below, paler olivaceous brown; throat and chest streaked; centre of breast to belly whitish; tail graduated; under tail-coverts long and tipped with whitish; first primary small and sharply pointed. The sexes are alike. Wing 67 to 80 mm.

**General distribution:** Southern Sweden to central Germany, Hungary, Austria, Rumania, the Ural Mts. and Transcaspia; in non-breeding season to east and south Africa as far south as the Transvaal and Arabia.

**Range in Eastern Africa:** Kenya Colony in non-breeding season, and should occur elsewhere in Eastern Africa.

**Habits:** A skulking secretive species, mostly confined to reed beds, difficult to observe or identify. It has been noticed as a rare palæarctic winter visitor to Kenya Colony.

## WARBLERS

**941 GRASSHOPPER WARBLER.** *LOCUSTELLA NÆVIA* (Boddaert).

*Locustella nævia straminea* Seeb.
*Locustella straminea* Seebohm, Cat. Bds. Brit. Mus. 5, p. 117, 1881: Orenburg, Ural Mts., Russia.

**Distinguishing characters:** Adult male, above, olivaceous buff or brown, spotted and streaked with blackish; tail graduated; below, white or creamy white; flanks buffish; under tail-coverts with blackish streaks; often some spots on lower throat and neck. The sexes are alike. Wing 55 to 62 mm. The young bird is washed with yellow below.

**General distribution:** Europe and Asia from the southern Ural Mts. to the Caucasus and the Altai Mts.; in non-breeding season to India, Arabia and Africa.

**Range in Eastern Africa:** Only recorded so far from Danakil, eastern Abyssinia, in non-breeding season.

**Habits:** A palæarctic winter migrant which has rarely been recorded from Africa. A very shy species slinking among dense vegetation where it may readily escape observation. A bird of slim build, at close quarters the graduated tail and streaking on the back may attract attention. The reeling call is unlikely to be heard in Africa.

**Distribution of other races of the species:** Europe and Asia, the nominate race being described from Italy.

**REED WARBLERS.** FAMILY—**SYLVIIDÆ.** Genus: *Acrocephalus*.

A mainly palæarctic genus of which five species visit Africa and one is resident; very largely confined in the breeding season to marshy ground and river banks. General colour brownish or streaky above and pale buff below, usually with a light coloured stripe above the eye. Bill rather long and broad. Restless birds with grating calls. Food, insects.

**942 GREAT REED WARBLER.** *ACROCEPHALUS ARUNDINACEUS* (Linnæus).

*Acrocephalus arundinaceus arundinaceus* (Linn.).
*Turdus arundinaceus* Linnæus, Syst. Nat. 10th ed. p. 170, 1758: Danzig, Baltic Sea.

**Distinguishing characters:** Above, olive brown with a rufous wash; rump rather paler; a buffish stripe between bill and top of eye; below, pale buff; throat and centre of belly whitish; some greyish streaks on lower neck; tail graduated; first primary small, narrow and sharply pointed. The sexes are alike. Wing 90 to 102 mm.

**General distribution:** Southern and central Europe to the Ural Mts., Asia Minor and Palestine; in non-breeding season to Africa as far south as the Portuguese Congo, Northern Rhodesia, western Nyasaland and Tanganyika Territory, also south-western Arabia.

**Range in Eastern Africa:** Eritrea and the Sudan to Kenya Colony and Tanganyika Territory in non-breeding season.

**Habits:** A relatively large and easily identifiable species, quite common as a passage migrant or palæarctic winter visitor. On migration it is not by any means confined to reeds or even to wet country. The song is low and vibrant with rather ventriloquial lower notes and is often heard in Africa after December.

*Acrocephalus arundinaceus zarudnyi* Hart.
*Acrocephalus arundinaceus zarudnyi* Hartert, Bull. B.O.C. 21, p. 26, 1907: Djarkent, Turkestan.

**Distinguishing characters:** Differs from the nominate race in being less rufous above and paler below, especially on chest. Wing 90 to 103 mm.

**General distribution:** Transcaspia to Turkestan; in non-breeding season to Iraq, Iran, Arabia and Eastern Africa as far south as Bechuanaland, the Transvaal and Natal.

**Range in Eastern Africa:** Eastern Abyssinia and Eritrea, French and British Somalilands and eastern Tanganyika Territory, also Zanzibar Island in non-breeding season.

**Habits:** Those of the nominate race, though with a somewhat more eastern distribution. Not uncommon in Zanzibar Island and Tanganyika Territory, and has been obtained in Nyasaland at quite high elevations.

**Distribution of other races of the species:** Japan and Korea.

943 BASRA REED WARBLER. *ACROCEPHALUS GRISELDIS* (Hartlaub).

*Calamoherpe griseldis* Hartlaub, Abh. Nat. Ver. Bremen, 12, p. 7, 1891: Nguru, Morogoro district, eastern Tanganyika Territory.

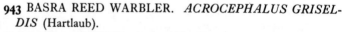

**Distinguishing characters:** Long-billed; above, olivaceous grey, light stripe over eye; below, cream or creamy white; first primary small, narrow and sharply pointed. The sexes are alike. Wing 77 to 84 mm. The greyer and duller coloured upperside serves to distinguish this Reed Warbler from the Great Reed, the Reed and the Marsh Warblers; it is also appreciably smaller than the Great Reed and appreciably larger than the Reed and Marsh Warblers.

**General distribution:** Iraq; in non-breeding season to eastern and south-eastern Africa.

**Range in Eastern Africa:** Kenya Colony and eastern Tanganyika Territory in non-breeding season with at least one record from Nyasaland.

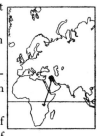

**Habits:** A rare and little known migrant to the east coast of Africa. It has mostly been noticed in rank herbage at the edge of forest, or in secondary growth forest itself.

**944** REED WARBLER. *ACROCEPHALUS SCIRPACEUS* (Hermann).
*Acrocephalus scirpaceus scirpaceus* (Herm.). **Pl. 68.**
*Turdus scirpaceus* Hermann, Obs. Zool. p. 202, 1804: Alsace.

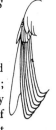

**Distinguishing characters:** Above, warmish brown; rump and upper tail-coverts paler; a faint light stripe from bill to over eye; below, buff; throat and centre of belly usually whiter; first primary small, narrow and sharply pointed. The notch on the inner web of the second primary comes within the tip of the eighth primary, but sometimes this is not clearly defined and colour is the only distinguishing feature between this bird and the Marsh Warbler. The sexes are alike. Wing 60 to 70 mm.

**General distribution:** Western Europe to the Ural River; in non-breeding season to Africa as far south as the Cameroons and the Zambesi River, also Arabia.

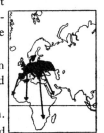

**Range in Eastern Africa:** Throughout in non-breeding season.

**Habits:** A common palæarctic migrant and in Africa not confined to reeds as it is in its breeding quarters. It may be met with in dry long grass in any type of country. It has a short song of rather squeaking tone in the early months of the year.

*Acrocephalus scirpaceus fuscus* (Hemp. & Ehr.).
*Curruca fusca* Hemprich & Ehrenberg, Symb. Phys. Aves. fol. cc., 1833: northern Arabia.

**Distinguishing characters:** Differs from the nominate race in being more olive brown above. Wing 62 to 72 mm.

**General distribution:** Southern Russia to Turkestan, Iran and Baluchistan; in non-breeding season to Egypt, Eastern Africa and Arabia.

**Range in Eastern Africa:** The Sudan, Abyssinia, Eritrea, Kenya Colony and north-eastern Tanganyika Territory as far south as Monduli in non-breeding season.

**Habits:** As for the nominate race.

### 945 MARSH WARBLER. *ACROCEPHALUS PALUSTRIS* (Bechstein).

*Sylvia palustris* Bechstein, Orn. Taschenb. p. 186, 1803: Germany.

**Distinguishing characters:** Similar to the eastern race of the Reed Warbler in colour and size, and can only be distinguished by the notch on the inner web of the second primary being outside the tip of the eighth primary. The sexes are alike. Wing 62 to 72 mm. This species can hardly be distinguished from the Reed Warbler in the field, but the colour of the upperparts is somewhat colder and more olive brown. It would be indistinguishable from the eastern race of that species. Its breeding habits and song are very distinct.

**General distribution:** Europe to Iran; in non-breeding season to eastern half of Africa as far south as eastern Cape Province and Natal, also Arabia.

**Range in Eastern Africa:** Throughout in non-breeding season.

**Habits:** Not easily identifiable in the field from the Reed Warbler, and apparently a fairly common visitor with the same range as that species and the same habitat. A rich fine song totally different to that of the Reed Warbler, but it is not very often heard in Africa. Moreau notes a curious January movement through Tanganyika Territory, presumably northward.

### 946 AFRICAN REED WARBLER. *ACROCEPHALUS BÆTICATUS* (Vieillot).

*Acrocephalus bæticatus cinnamomeus* Reichw.
*Acrocephalus cinnamomeus* Reichenow, O.M. p. 161, 1908: North end of Lake Edward, south-western Uganda.

**Distinguishing characters:** Size small; above, pale brown; below, buff; chin, throat and centre of belly whiter. First primary short, narrow and sharply pointed. The sexes are alike. Wing 52 to 56 mm. The young bird is more buff in general colour than the adult.

**General distribution:** Lake Chad to the Sudan, north-eastern and eastern Belgian Congo and Uganda.

**Range in Eastern Africa:** Western and south-western Sudan to north-western Uganda.

**Habits:** A common but local species with a pleasant clear song, usually in reeds or long grass. Migratory in places, and in western Darfur only a visitor to breed from August to October.

**Nest and Eggs:** A typical Reed Warbler's cup nest of grass and fibres slung between stems of thick grass or reeds, or occasionally in small shrubs. Eggs two or three, greyish green, with dark grey mottlings; about 16·5 × 12·5 mm.

**Recorded breeding:** Darfur, August and September. Southern Sudan, July and August.

**Call:** A pleasant scratchy light-pitched song of typical Reed Warbler nature 'with slight acceleration and decrescendo.' It also has a sub-song.

*Acrocephalus bæticatus suahelicus* Grote. **Pl. 68.**
*Acrocephalus bæticatus suahelicus* Grote, O.M. p. 145, 1926: Zanzibar Island.

**Distinguishing characters:** Differs from the preceding race in being larger and a darker and warmer tone of brown above. Wing 52 to 64 mm. The small overlap in wing measurement with the Reed Warbler may cause confusion, but both this race and the last have the second primary appreciably shorter than the fourth, whereas the Reed Warbler has the second primary equal to or longer than the fourth.

**General distribution:** Kenya Colony to south-western Nyasaland.

**Range in Eastern Africa:** Kenya Colony to Tanganyika Territory, also Pemba, Zanzibar and Mafia group islands.

**Habits:** A bird of very mixed habitat, but distinctly local, found in gardens or mangrove swamps, or on sand-banks crowned with vegetation, or in reed beds. As a general rule it does not appear to frequent the larger fresh water swamps much; it is usually very tame.

**Recorded breeding:** Coastal East Africa in any month in the year. Mafia Island (breeding condition), July to October.

**Distribution of other races of the species:** Southern Africa, the nominate race being described from Knysna, Cape Province.

### 947 SEDGE WARBLER. *ACROCEPHALUS SCHŒNOBÆNUS* (Linnæus).

*Motacilla Schœnobænus* Linnæus, Syst. Nat. 10th ed. p. 184, 1758: South Sweden.

**Distinguishing characters:** Adult male, above, head and mantle brown, streaked with black; rump plain warmer brown; white stripe from bill to over and well behind eye; below, brownish or buffish; whiter on throat and belly. The sexes are alike. Wing 61 to 73 mm.

**General distribution:** Europe to western Siberia but not Spain or Greece; in non-breeding season to Africa as far south as Nigeria, Damaraland and Natal, occasional Madeira.

**Range in Eastern Africa:** Throughout in non-breeding season.

**Habits:** A common, often abundant, palæarctic winter visitor usually in reed beds and swampy places, with a jerky cheerful song and a grating call. Generally calls or sings if disturbed, and the song is heard all the year round. In the north mainly a passage migrant only.

*Note:* The Clamorous Reed-Warbler. *Acrocephalus stentoreus* (Hemprich & Ehrenberg). *Curruca stentorea* Hemprich & Ehrenberg, Symb. Phys. Aves., fol. b.b. 1833: Damietta, Egypt, which is similar to the Great Reed-Warbler but has a longer and slenderer bill and shorter wing (74 to 86 mm.), and inhabits Egypt and Palestine, was recorded from the Blue Nile and Eritrean coast over 100 years ago. Being normally a resident and sedentary species its occurrence in Eastern Africa requires confirmation.

## RUSH, BAMBOO, BRACKEN, FOREST and SWAMP WARBLERS. FAMILY—SYLVIIDÆ. Genera: *Bradypterus*, *Sathrocercus*, *Calamocichla* and *Calamonastides*.

Eleven species occur in Eastern Africa. They mainly inhabit swamps and swampy situations, though some species are found in forest or in the rough herbage along the edge of the forests. The *Calamocichla* and *Calamonastides* groups can be distinguished from the *Bradypterus* and *Sathrocercus* groups by the long and large feet and toes, and all four can be distinguished from the Reed Warbler (*Acrocephalus*) group by the long and broad first primary; the Reed Warblers having a short, narrow, and sharply pointed first primary. The food consists of insects, small mollusca, etc.

## WARBLERS

**948 LITTLE RUSH WARBLER.** *BRADYPTERUS BABŒ-CALA* (Vieillot).

*Bradypterus babœcala abyssinicus* (Blund. & Lov.). **Pl. 69.**
*Lusciniola abyssinica* Blundell & Lovat, Bull. B.O.C. 10, p. 19, 1899: Lake Chercher, eastern Abyssinia.

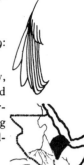

**Distinguishing characters:** Above, olivaceous brown; below, whitish; blackish streaks on lower throat and foreneck; flanks and under tail-coverts olivaceous brown; tail feathers broad with underlying barring. The sexes are alike. Wing 55 to 60 mm. The young bird is warmer and richer brown on the chest, flanks and under tail-coverts and with smaller streaks on the foreneck.

**Range in Eastern Africa:** Abyssinia.

**Habits:** As for other races but breeding in colonies in reeds in deep water.

**Nest and Eggs:** See under other races.

**Recorded breeding:** Abyssinia, August and September.

*Bradypterus babœcala centralis* Neum.
*Bradypterus brachypterus centralis* Neumann, Bull. B.O.C. 21, p. 55, 1908: between Mkingo and Muhera, Ruanda, eastern Belgian Congo.

**Distinguishing characters:** Similar to the preceding race, but general colour warmer in tone. Wing 55 to 56 mm.

**General distribution:** Southern Cameroons to the Belgian Congo.

**Range in Eastern Africa:** Ruanda, eastern Belgian Congo.

**Habits:** As for other races, inhabiting papyrus or dense reeds.

**Recorded breeding:** Cameroons, June to August.

*Bradypterus babœcala elgonensis* Mad.
*Bradypterus elgonensis* Madarász, O.M. p. 175, 1912: Buchungu, Mt. Elgon.

**Distinguishing characters:** Darker and warmer in tone of colour than any of the preceding races. Wing 51 to 58 mm.

**General distribution:** Lake Chad to Mt. Elgon and Kenya Colony.

**Range in Eastern Africa:** Uganda to central Kenya Colony.

**Habits:** Shy but inquisitive birds of reed beds or long swampy grass, or of dense herbage near water. During their short flights over or among the vegetation their wings make a curious snapping noise. The rather long tail is noticeable in the field.

**Nest and Eggs:** The nest is a deep cup of coarse grass lined with fibre and well hidden low down in the reeds or herbage. Eggs two, pinkish white with reddish brown spots and ashy violet undermarkings; about 19 × 14 mm.

**Recorded breeding:** Kenya Colony highlands, April to June.

**Call:** A jerky and rather unmelodious song, and a somewhat ventriloquial call 'thri' or at times a 'churr.' The song is compared by Belcher to that of *Nectarinia famosa*.

*Bradypterus babœcala moreaui* Scl.
*Bradypterus brachypterus moreaui* W. L. Sclater, Bull. B.O.C. 52, p. 57, 1931: Amani, Usambara Mts., north-eastern Tanganyika Territory.

**Distinguishing characters:** Similar above to the Ruanda race, but below, flanks and under tail-coverts rather brighter in tone, less olivaceous; streaks on foreneck variable, *i.e.* smaller, fewer or almost absent. Wing 52 to 60 mm.

**General distribution:** Kenya Colony, Tanganyika Territory, Nyasaland and the Tete Province of northern Portuguese East Africa.

**Range in Eastern Africa:** South-eastern Kenya Colony from the southern Tana River to south-western Tanganyika Territory.

**Habits:** As for the other races, the call is a series of notes sounding like 'thri' on an accelerating descending scale.

**Recorded breeding:** Southern Nyasaland, March to June.

*Bradypterus babœcala sudanensis* Grant & Praed.
*Bradypterus babœcala sudanensis* C. Grant & Mackworth-Praed, Bull. B.O.C. 61, p. 25, 1941: Lat. 9° 30' N.; long. 30° 40' E.; White Nile, southern Sudan.

**Distinguishing characters:** Similar in colour to the Abyssinian race but smaller. Wing 50 to 54 mm.

**Range in Eastern Africa:** Upper White Nile from Lake No to Tonga, southern Sudan.

**Habits:** As for other races.

**Recorded breeding:** No records.

**Distribution of other races of the species:** Angola, Belgian Congo and South Africa, the nominate race being described from the Cape Province.

### 949 GRAUER'S WARBLER. *BRADYPTERUS GRAUERI* Neumann.

*Bradypterus graueri* Neumann, Bull. B.O.C. 21, p. 56, 1908: Western Kivu Volcanoes, Ruanda, eastern Belgian Congo.

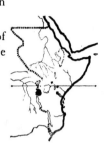

**Distinguishing characters:** Very similar to the Ruanda race of the little Rush Warbler, but larger. Wing 59 to 65 mm. Juvenile plumage unrecorded.

**Range in Eastern Africa:** Eastern Belgian Congo.

**Habits:** No information.

**Nest and Eggs:** Undescribed.

**Recorded breeding:** No records.

**Call:** Unrecorded.

### 950 WHITE-WINGED WARBLER. *BRADYPTERUS CARPALIS* Chapin.

*Bradypterus carpalis* Chapin, Bull. Amer. Mus. Nat. Hist. 35, fig. 4, p. 27, 1916: Faradje, north-eastern Belgian Congo.

**Distinguishing characters:** Adult male, above, dark black brown; wing-shoulder and bar across wing white; below, white; large arrow-shaped black marks on throat; flanks brownish black. The female is less boldly marked on the throat. Wing 67 to 71 mm. Juvenile plumage unrecorded.

**General distribution:** North-eastern Belgian Congo to Uganda.

**Range in Eastern Africa:** Western and southern Uganda.

**Habits:** Found among reeds along the Sezibwa River in Uganda, and in papyrus swamp in north-eastern Belgian Congo.

**Nest and Eggs:** Undescribed.

**Recorded breeding:** Belgian Congo (breeding condition), February to September.

**Call:** A succession of chirping notes of peculiar resonant quality, which start slowly, then quickening and dying gradually away, often to be followed by four or five loud explosive wing-beats (Chapin).

# WARBLERS

### 951 BAMBOO WARBLER. *BRADYPTERUS ALFREDI* Hartlaub.

*Bradypterus alfredi albicrissalis* Neum. **Pl. 69.**

*Bradypterus alfredi albicrissalis* Neumann, O.M. p. 10, 1914: Mubuku Valley, south-eastern Ruwenzori Mts., western Uganda.

**Distinguishing characters:** Above, brown; below, white; neck, chest, flanks and under tail-coverts grey; centre of neck and chest mottled grey and white. The sexes are alike. Wing 59 to 60 mm. Juvenile plumage unrecorded.

**Range in Eastern Africa:** Western Uganda from south-eastern Ruwenzori Mts., to Bugoma.

**Habits:** Found among grass and dense vegetation below the forest line on the Ruwenzori Mts. at about 5,000 feet. R. B. Woosnam noted that it has great development of leg muscles, but little is recorded of its habits.

**Nest and Eggs:** Undescribed.

**Recorded breeding:** No records.

**Call:** Unrecorded.


*Bradypterus alfredi kungwensis* Mor.

*Bradypterus alfredi kungwensis* Moreau, Bull. B.O.C. 62, p. 42, 1942: Ujamba, Kungwe-Mahare Mts., western Tanganyika Territory.

**Distinguishing characters:** Differs from the preceding race in being darker olivaceous brown above and with darker grey flanks. Wing 57 to 60 mm.

**General distribution:** Tanganyika Territory to north-western Northern Rhodesia.

**Range in Eastern Africa:** Kungwe-Mahare Mt. area, western Tanganyika Territory.

**Habits:** Inhabits mixed forest and bamboo at 6,000 to 7,000 feet.

**Recorded breeding:** No records.

**Distribution of other races of the species:** North-eastern Belgian Congo, the habitat of the nominate race.

## 952 CINNAMON BRACKEN-WARBLER. *SATHROCERCUS CINNAMOMEUS* (Rüppell).

*Sathrocercus cinnamomeus cinnamomeus* (Rüpp.). **Pl. 69.**
*Sylvia* ? (*Salicaria*) *cinnamomea* Rüppell, N. Wirbelt. Vög. p. 3, pl. 42, 1840: Entschetqab, Simen Province, northern Abyssinia.

**Distinguishing characters:** Above, warm russet brown; stripe over eye brown; below, brighter russet brown; chin, throat and belly white or whitish; tail feathers narrow and decomposed in appearance. The sexes are alike. Wing 58 to 65 mm. The young bird is blackish brown above; below, olive brown and yellow.

**General distribution:** Abyssinia and Kenya Colony to the Kivu area of the eastern Belgian Congo.

**Range in Eastern Africa:** Northern, central and southern Abyssinia to Uganda and western Kenya Colony.

**Habits:** A highland species of dense herbage, bushes, reeds or bracken, common and widely distributed. Occurs up to 12,000 or 13,000 feet on mountains, sometimes inside forest. Shy but inquisitive bird, breeding on or near the ground. Has a short song of the same explosive nature as that of Cetti's Warbler in Europe, 'hi-chwi-chwi-chwi-chwi.' Generally to be found round forest outskirts, or in small densely bushed ravines. This species also has a very large development of the leg muscles.

**Nest and Eggs:** Nest a deep cup of wool, down, and feathers, lined with hair in brambles or other herbage near the ground, often on a rather large untidy base of leaves. Eggs two, white or pinkish, speckled and blotched with purplish mauve or ash grey. No measurements available.

**Recorded breeding:** Kenya Colony, February, also probably May to July.

**Call:** Besides the song mentioned above there is a chirping, rasping alarm note 'twee-twee' and a melodious babbling.

*Sathrocercus cinnamomeus nyassæ* (Shell.).
*Bradypterus nyassæ* Shelley, Ibis, p. 16, 1893: Milanji Plateau, southern Nyasaland.

**Distinguishing characters:** Differs from the nominate race in being more blackish brown above; duller brown below, with a buffish wash mainly on chest; occasional dark streaks on foreneck. Wing 58 to 66 mm.

**General distribution:** Tanganyika Territory to Nyasaland and north-eastern Northern Rhodesia.

**Range in Eastern Africa:** East-central Tanganyika Territory from the Nguru Mts. and Njombe to Nyasaland.

**Habits:** As for the nominate race, but also occurs inside forests among secondary growth, and is common in damp forest patches. The song is described as 'tee-tee-tee-chit-chit-chit-chit' loud, sudden and trilling.

**Recorded breeding:** Tanganyika Territory (about to breed), January. Nyasaland, August to October.

*Sathrocercus cinnamomeus rufoflavidus* (Reichw. & Neum.).

*Bradypterus rufoflavidus* Reichenow & Neumann, O.M. p. 75, 1895: Kifinika Hut, Mt. Kilimanjaro, north-eastern Tanganyika Territory.

**Distinguishing characters:** Differs from the nominate race in being less bright above and paler below; stripe over eye lighter. Wing 57 to 65 mm.

**Range in Eastern Africa:** South-eastern Kenya Colony and northern Tanganyika Territory.

**Habits:** As for the nominate race, but generally met with outside forest in nettles, weeds or bracken. Also occurs in the heath zone of Mt. Kilimanjaro.

**Recorded breeding:** Northern Tanganyika Territory, August, also (breeding condition), December and January.

*Sathrocercus cinnamomeus cavei* (Macd.).

*Bradypterus cinnamomea cavei* Macdonald, Bull. B.O.C. 60, p. 9, 1939: Kifia, Imatong Mts., southern Sudan.

**Distinguishing characters:** Differs from the nominate race in being darker above, usually with a chocolate wash. Wing 60 to 65 mm.

**Range in Eastern Africa:** Imatong Mts., southern Sudan.

**Habits:** As for other races.

**Recorded breeding:** Imatong Mts. (near breeding), May.

*Sathrocercus cinnamomeus macdonaldi* Grant & Praed.
*Sathrocercus cinnamomea macdonaldi* C. Grant & Mackworth-Praed, Bull. B.O.C. 61, p. 26, 1941: Gummaro stream, three miles west of Goré, western Abyssinia.

**Distinguishing characters:** Differs from the nominate race in being warm rich chocolate brown above and chestnut below; chin, throat and belly white or white washed with chestnut. Wing 60 to 65 mm.

**Range in Eastern Africa:** Western Abyssinia.

**Habits:** As for other races.

**Recorded breeding:** South-western Abyssinia (breeding condition), August.

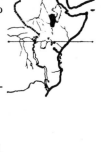

*Sathrocercus cinnamomeus ufipæ* Grant & Praed.
*Sathrocercus cinnamomea ufipæ* C. Grant & Mackworth-Praed, Bull. B.O.C. 62, p. 30, 1941: Sumbawanga, Ufipa Plateau, south-western Tanganyika Territory.

**Distinguishing characters:** Very similar to the Nyasaland race but bronzy brown above, including wings and tail, and with indistinct darker edges to the feathers of the head and mantle; below, throat suffused with olivaceous brown, not whitish. Wing 63 to 67 mm.

**Range in Eastern Africa:** Ufipa Plateau, south-western Tanganyika Territory.

**Habits:** No information.

**Recorded breeding:** No records.

**Distribution of other races of the species:** Cameroons.

953 EVERGREEN-FOREST WARBLER. *SATHROCERCUS MARIÆ* (Madarász).
*Sathrocercus mariæ mariæ* (Mad.). **Pl. 69.**
*Bradypterus mariæ* Madarász, Ann. Mus. Nat. Hung. 3, pp. 401, 402, 1905: Kibosho, Mt. Kilimanjaro, north-eastern Tanganyika Territory.

**Distinguishing characters:** General appearance dull and dingy; above, dark brown; below, dull olive brown; throat dusky white; centre of belly paler olive brown; black streaks on foreneck; tail as in the Cinnamon Bracken Warbler but shorter. The sexes are alike. Wing 59 to 65 mm. The young bird has a yellowish wash on the underside; it is darker and has less yellow than the young bird of the Cinnamon Bracken Warbler.

**Range in Eastern Africa:** Central and southern Kenya Colony from Mt. Kenya, Molo, Aberdares, Mau and Escarpment, to the Chyulu Hills and the Lumi River; also north-eastern Tanganyika Territory in the area of Oldeani, Mt. Meru and Mt. Kilimanjaro.

**Habits:** An inhabitant of dense undergrowth inside rain forest, not uncommon locally but difficult to observe, and the tail is almost always threadbare; very lively little birds with a habit of singing a duet to each other. Generally found only in the damper parts of the forest where they creep about picking up insects. Occur up to 8,000 feet. This species might easily be mistaken for the Brown Illadopsis.

**Nest and Eggs:** The nest is a collection of leaves or other matter with a hair lining placed in dense low cover, not readily distinguishable from that of the Cinnamon Bracken-Warbler. Eggs two or three, plain white, or pinkish white with liver coloured spots and lilac undermarkings; about $19 \cdot 5 \times 15$ mm.

**Recorded breeding:** North-central Tanganyika Territory, November to January.

**Call:** Described as a chirping or rattling 'tiku-tiku-tiku-tik' and a high-pitched 'cheu-cheat-cheat,' but Moreau notes that both this species and the Bracken-Warbler have a 'duetting' habit, one bird producing the song and the other chiming in with high-pitched notes.

*Sathrocercus mariæ usambaræ* (Reichw.).
*Bradypterus usambaræ* Reichenow, J.f.O. p. 391, 1917: Usambara, north-eastern Tanganyika Territory.

**Distinguishing characters:** Differs from the nominate race in being warmer in tone, less grey below. Wing 57 to 65 mm. It can be distinguished from the Nyasaland race of the Cinnamon Bracken-Warbler by its dark, duller, more olive brown appearance below, less buffish. It has a superficial resemblance to the Usambara race of the Pale-breasted Illadopsis, which however, is larger, heavier billed and grey below. A grey semi-albinistic form occurs in the Uluguru Mts. with white alulæ and white claws.

# WARBLERS

**General distribution:** Kenya Colony to northern Nyasaland at Masuku and Nyankhowa and Portuguese East Africa.

**Range in Eastern Africa:** Taita Hills, south-eastern Kenya Colony to the Paré Mts., Samé, Mlalo near Lushoto, Usambara Mts., Nguru Hills, Mpapwa, Uluguru Mts., Njombe and Rungwe in Tanganyika Territory, and Portuguese East Africa at Unangu.

**Habits:** As for the nominate race. It utters a sharp 'chew' or 'chi-chew' with a rattling vociferous little chirping song. The alarm note is a constant twittering.

**Recorded breeding:** Amani, Tanganyika Territory, July to October.

*Sathrocercus mariæ granti* (Benson).
*Bradypterus usambaræ granti* Benson, Bull. B.O.C. 59, p. 110, 1939: Lichenya Plateau, Mlanje Mt., southern Nyasaland.

**Distinguishing characters:** Differs from the preceding race in having the underside warmer in tone. It can be distinguished from the Cinnamon Bracken-Warbler by the much darker, olive brown, not brown, flanks and under tail-coverts, and by the shorter tail. Wing 59 to 65 mm.

**General distribution:** Western and southern Nyasaland as far north as the Vipya Plateau to Portuguese East Africa.

**Range in Eastern Africa:** Southern Nyasaland to south-western Portuguese East Africa.

**Habits:** Much as for the other races, occurring only inside evergreen forest.

**Recorded breeding:** Nyasaland, October to December.

**Distribution of other races of the species:** Cameroons and Angola.

**954 LOPEZ'S WARBLER.** *SATHROCERCUS LOPEZI* (Alexander).
*Sathrocercus lopezi barakæ* (Sharpe).
*Bradypterus barakæ* Sharpe, Ibis, p. 546, 1906: Eastern Ruwenzori Mts., western Uganda.

**Distinguishing characters:** Above, dark chestnut brown, less dark on head; below, rich chestnut brown; centre of lower belly white or whitish; tail feathers narrow and greatly decomposed in

appearance. The sexes are alike. Wing 55 to 61 mm. Juvenile plumage unrecorded.

**Range in Eastern Africa:** Ruwenzori Mts., western Uganda.

**Habits:** Another inhabitant of damp forest undergrowth, frequenting the densest cover and always appearing wet and bedraggled. Rarely seen to fly. Occurs between 6,500 and 8,000 feet.

**Nest and Eggs:** Undescribed.

**Recorded breeding:** No records.

**Call:** A short strident song and a faint double chirp.

**Distribution of other races of the species:** Fernando Po, the habitat of the nominate race.

**955 GREATER SWAMP-WARBLER.** *CALAMOCICHLA GRACILIROSTRIS* (Hartlaub).
*Calamocichla gracilirostris parva* (Fisch. & Reichw.). **Pl. 71.**
*Phyllostrephus parvus* Fischer & Reichenow, J.f.O. p. 262, 1884: Murentat, near Lake Naivasha, central Kenya Colony.

**Distinguishing characters:** Adult male, above, blackish earth brown; below, dull buffish white; flanks and under tail-coverts olivaceous brown; feet large; claws well developed. The female is slightly smaller. Wing, male 71 to 77, female 66 to 70 mm. The young bird is warmer in tone and more russet above; breast and flanks washed with russet.

**Range in Eastern Africa:** Kenya Colony from Laikipia, Naivasha, Nairobi, Taveta and Simba.

**Habits:** Very much those of a Reed Warbler, locally common along the reedy shores of lakes or rivers. Shy inconspicuous birds except in the breeding season, when their song is noticeable.

**Nest and Eggs:** Nest a deep cup of grass and reed blades lined with fibre and with feathers usually round the rim, slung between reed stems some feet above water. Eggs two or three, pale bluish white heavily spotted with several shades of brown; about 20 × 15 mm.

**Recorded breeding:** Naivasha, May to August.

**Call:** Quite a loud song, almost Thrush-like in tone. Other calls rather harsh and scolding.

*Calamocichla gracilirostris nilotica* Neum. **Ph. xi.**
*Calamocichla ansorgei nilotica* Neumann, Nov. Zool. 15, p. 246, 1908: Wadelai, north-western Uganda.

**Distinguishing characters:** Differs from the preceding race in being duller earth brown above, less warm in tone. Wing, male 76 to 82, female 72 to 78 mm. The young bird is warmer brown above, and is thus similar to the adult of the preceding race, but has a fluffier appearance.

**General distribution:** The Sudan, Uganda, western Kenya Colony, eastern Belgian Congo and north-eastern Northern Rhodesia at Lake Bangweulu.

**Range in Eastern Africa:** Lake No and Mongalla areas of the southern Sudan to Uganda and western Kenya Colony at Kisumu.

**Habits:** Seldom seen but not uncommon in papyrus swamps or in elephant grass in wet places. It would escape notice altogether except for its harsh call. The nest is similar to that of the preceding race but is in many cases placed high up in among the flowers of papyrus.

**Recorded breeding:** Kisumu, Kenya Colony, July.

**Distribution of other races of the species:** West and South Africa, the nominate race being described from Natal.

## 956 FOX'S SWAMP-WARBLER. *CALAMOCICHLA FOXI* (Sclater).

*Calamornis foxi* W. L. Sclater, Bull. B.O.C. 47, p. 118, 1927: Lake Maraye, Kivu district, eastern Belgian Congo.

**Distinguishing characters:** Similar to the western race of the last species but larger. Wing, male 85, female 83 mm. The young bird is russet brown above; below, mainly paler russet brown; chin, throat, breast and belly white.

**Range in Eastern Africa:** South-western Uganda at Lake Mutanda to northern Ruanda and eastern Belgian Congo at Lake Maraye.

**Habits:** Presumably those of the other members of the genus. Only known from a very limited area.

**Nest and Eggs:** Undescribed.
**Recorded breeding:** No records.
**Call:** Unrecorded.

**957** LESSER SWAMP-WARBLER. *CALAMOCICHLA LEPTORHYNCHA* (Reichenow).

*Calamocichla leptorhyncha leptorhyncha* (Reichw.). **Pl. 71.**

*Turdirostris leptorhyncha* Reichenow, O.C. p. 155, 1879: Tschra, mouth of the Tana River, eastern Kenya Colony.

**Distinguishing characters:** Above, warmish russet brown to darker brown; rump usually brighter; below, white or creamy white; flanks and under tail-coverts pale brown; feet large and claws well developed. The sexes are alike. Wing, male 58 to 72, female 58 to 69 mm. The young bird is warmer in colour and more russet above than the adult. In general appearance this Swamp-Warbler is very similar to the Reed Warbler, but can be at once distinguished by its heavy black, or blackish, feet and claws and large first primary.

**General distribution:** Kenya Colony to Tanganyika Territory, south-eastern Belgian Congo, Northern Rhodesia, Nyasaland, Portuguese East Africa, Bechuanaland and South Africa.

**Range in Eastern Africa:** Coastal areas of Kenya Colony from the mouth of the Tana River to southern Tanganyika Territory and the Zambesi River, also Pemba and Zanzibar Islands.

**Habits:** Locally common in marshes and reed beds, with the appearance and habits of a Reed Warbler. Shy bird as a rule but quite pugnacious in the breeding season. Common in Pemba and Zanzibar Islands, also in Nyasaland. May also be found in thickets or long grass.

**Nest and Eggs:** Nest of coarse grass and strips of reeds unlined but solidly made, placed in thick vegetation or shrubs, or slung on to two or three reed stems above water. Eggs usually two, occasionally three or even four, whitish or bluish green heavily spotted or freckled with brown, grey or purple; about $19 \cdot 5 \times 15$ mm. on the east coast, $18 \times 13 \cdot 5$ mm. in Nyasaland.

**Recorded breeding:** Tanganyika Territory (breeding condition), January. Pemba and Zanzibar Islands, mostly May to July. Nyasaland, October to July.

**Call:** A short song of powerful notes rather harsh in tone. The bird 'shivers' its tail when singing. The usual call is an occasional harsh 'churr.'

*Calamocichla leptorhyncha jacksoni* Neum.
*Calamocichla jacksoni* Neumann, O.M. p. 185, 1901: Entebbe, southern Uganda.

**Distinguishing characters:** Above darker and duller, more olivaceous, than the nominate race; below, duller dusky brown. Wing, male 61 to 69, female 56 to 69 mm.

**General distribution:** Eastern Belgian Congo to the Sudan and Uganda.

**Range in Eastern Africa:** Upper White Nile to Uganda, including Sesse Islands.

**Habits:** Locally common among reeds along lake shores or among shrubs and creepers. Habits of the genus.

**Recorded breeding:** Entebbe, Uganda, March.

*Calamocichla leptorhyncha macrorhyncha* (Jack.).
*Bradypterus macrorhynchus* Jackson, Bull. B.O.C. 27, p. 8, 1910: Il-polossat (=Ol Bolossat), Laikipia, central Kenya Colony.

**Distinguishing characters:** Above, very similar to the nominate race, but rather more dusky below. Wing, male 65 to 73, female 63 to 71 mm.

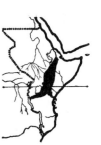

**Range in Eastern Africa:** Lake Zwai and southern Abyssinia to central, western and southern Kenya Colony and Tanganyika Territory at Kaserazi, the Kome Islands of southern Lake Victoria, the Monduli and Mbulu areas and Kabogo Head.

**Habits:** Unrecorded, but presumably as for the other races. Has a clear liquid song.

**Recorded breeding:** Northern Tanganyika Territory (breeding condition), June.

*Calamocichla leptorhyncha tsanæ* (Bann.).
*Calamœcetor leptorhyncha tsanæ* Bannerman, Bull. B.O.C. 57, p. 71, 1936: Achera Marian, Lake Tana, northern Abyssinia.

**Distinguishing characters:** Darker and duskier above than other races. Wing, male 67 to 75, female 69 to 75 mm.

**Range in Eastern Africa:** Lake Tana area, northern Abyssinia.
**Habits:** Unrecorded, but presumably those of other races.
**Recorded breeding:** No records.
**Distribution of other races of the species:** Lake Chad area.

## WARBLERS

### 958 YELLOW SWAMP-WARBLER. *CALAMONASTIDES GRACILIROSTRIS* (O. Grant). Pl. 71.

*Chloropeta gracilirostris* O. Grant, Bull. B.O.C. 19, p. 33, 1906: Mokia, south-eastern Mt. Ruwenzori (3,400 feet), western Uganda.

**Distinguishing characters:** Above, olivaceous green; rump slightly russet; below, yellow, flanks brownish; feet, toes and claws dark and large. The sexes are alike. Wing 60 to 64 mm. The young bird is rather browner on the flanks.

**General distribution:** Eastern Belgian Congo and western Uganda from the Ruwenzori Mts. area to Lake Edward and Lake Mweru.

**Range in Eastern Africa:** South-eastern Ruwenzori Mts., Uganda.

**Habits:** This bird has for many years been regarded as one of the *Chloropetas*, a genus of aberrant Flycatchers, but its similarity is only in colour, and in structure and affinities it undoubtedly belongs to the present group. Its habits are those of a Reed or Swamp-Warbler and it is found in papyrus swamps and reed beds.

**Nest and Eggs:** Undescribed.

**Recorded breeding:** No records.

**Call:** A series of loud half-whistled notes 'chwee-chwee-chwee' (Chapin).

### WILLOW, WOOD and BONELLI'S WARBLERS and CHIFF-CHAFF. FAMILY—SYLVIIDÆ. Genus: *Phylloscopus*.

A Warbler family of four species with very similar habits, all of which are migrants from Europe and Asia during the non-breeding season. They merit their name 'Leaf-searchers' and are always in trees generally searching for food among the thinner twigs. Their food is entirely insects and their general coloration greenish above and pale yellowish below.

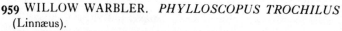

### 959 WILLOW WARBLER. *PHYLLOSCOPUS TROCHILUS* (Linnæus).

*Phylloscopus trochilus trochilus* (Linn.). **Pl. 68.**

*Motacilla Trochilus* Linnæus, Syst. Nat. 10th ed. p. 188, 1758: England.

**Distinguishing characters:** Above, pale olivaceous green; pale yellow streak from bill to over and behind eye; below, sulphur yellow; belly whiter. Third and fourth primaries usually equal and longest, sometimes third slightly longer than fourth; fifth primary appreciably shorter than third and fourth, giving wing a more pointed appearance; sixth primary never emarginated on outer web; first primary longer than coverts. The sexes are alike. Wing 60 to 75 mm. The immature bird has the whole of the underparts suffused with yellow as in the Icterine Warbler, but it is much smaller, has a smaller and narrower bill and has a long first primary compared to the short narrow first primary of the Icterine Warbler.

**General distribution:** Western, central and southern Europe as far as south Poland and Hungary; in non-breeding season to the Canary Islands, Africa and south-western Arabia.

**Range in Eastern Africa:** Throughout in non-breeding season.

**Habits:** An abundant passage migrant in the north, and a common visitor to the remainder of Eastern Africa during the palæarctic winter, though a considerable number pass on to South Africa. Sings fairly freely in Africa in the early months of the year, a soft sweet unmistakable song rising at first and then falling away. May be found in any type of country including forest, and up to considerable elevations.

*Phylloscopus trochilus acredula* (Linn.).
*Motacilla acredula* Linnæus, Syst. Nat. 10th ed. p. 189, 1758: Upsala, Sweden.

**Distinguishing characters:** Generally duller than the nominate race, especially below; chest and flanks buffish with little or no yellow; throat and belly whiter. Wing 61 to 74 mm.

**General distribution:** Eastern Greenland and northern Europe to the Yenisei River; in non-breeding season to eastern half of Africa as far south as Natal, also Iran and Arabia.

**Range in Eastern Africa:** Abyssinia and the Sudan to the Zambesi River in non-breeding season.

**Habits:** Hardly distinguishable in the field from the nominate race, but appears to winter mainly in tropical East Africa, and does not pass on much further south.

**Distribution of other races of the species:** Eastern Siberia, Taimyr to Anadyr.

**960** CHIFF - CHAFF.   *PHYLLOSCOPUS   COLLYBITA* (Vieillot).

*Phylloscopus collybita collybita* (Vieill.).

Sylvia collybita Vieillot, N. Dict. d'Hist. Nat. 11, p. 235, 1817: France.

**Distinguishing characters:** Above, similar to the Willow Warbler, but perhaps generally slightly browner; below, chin and throat pale buffish brown, not yellowish or yellowish white; flanks buffish brown; centre of belly white; some faint yellowish streaks on underparts; feet dark brown, almost blackish, especially toes, not pale brown or horn-colour as in the Willow Warbler, and bill more blackish. Third, fourth and fifth primaries usually equal, but fifth sometimes slightly, but not very appreciably, shorter than third and fourth, giving wing a more rounded appearance; sixth primary usually emarginated on outer web, but not constantly so. The sexes are alike. Wing 51 to 68 mm.

**General distribution:** Western and middle Europe to western Germany; in non-breeding season to Africa as far south as Gambia, French Sudan and the southern Sudan; also Iraq and Arabia.

**Range in Eastern Africa:** The Sudan and Abyssinia, in non-breeding season.

**Habits:** An abundant palæarctic winter visitor to the Sudan and Abyssinia. In the western Sudan it is rarer and only a scarce visitor to as far south as the Jebel Marra. Unmistakable onomatopœic call note, but otherwise most difficult to tell in the field from the Willow Warbler.

*Phylloscopus collybita abietinus* (Nils.).

Sylvia abietina Nilsson, K. Sv. Vet.-Akad. Handl. p. 113, 1819: Sweden.

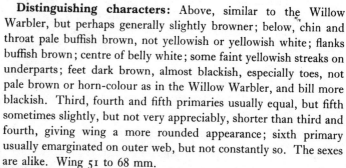

**Distinguishing characters:** Differs from the nominate race in being usually rather paler and greyer. Wing 53 to 66 mm.

**General distribution:** Scandinavia to eastern Germany and Russia as far east as the Yenisei River; in non-breeding season to eastern half of Africa and Arabia.

**Range in Eastern Africa:** The Sudan to Abyssinia and British Somaliland in non-breeding season.

**Habits:** As for the nominate race with a rather more eastern distribution in Africa. Indistinguishable from the nominate race in the field.

**Distribution of other races of the species:** Canary Islands, Portugal and Spain to Algiers, Armenia to Caucasus, eastern Siberia and northern India.

### 961 WOOD WARBLER. *PHYLLOSCOPUS SIBILATRIX* (Bechstein).

*Motacilla sibilatrix* Bechstein, Naturforch. 27, p. 47, 1793: Thuringia, Germany.

**Distinguishing characters:** Above, yellow green; stripe from bill to over and behind eye, throat, neck, breast and upper flanks bright lemon yellow; breast to under tail-coverts white. The sexes are alike. Wing 69 to 80 mm. Appreciably brighter colour and whiter below than the Willow Warbler. Third primary always longer than second and fourth which are about equal in length; first primary shorter than coverts, not longer as in Willow Warbler.

**General distribution:** Greater part of Europe; in non-breeding season to west and eastern Africa as far south as Cameroons and Uganda, also Arabia.

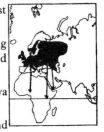

**Range in Eastern Africa:** The Sudan to Uganda and Kenya Colony in non-breeding season.

**Habits:** A palæarctic winter migrant not usually common and appears to be only a passage migrant in the Sudan. Its bright yellow breast and white underparts are easily distinguished, and in the early months of the year its sibilant song is noticeable.

### 962 BONELLI'S WARBLER. *PHYLLOSCOPUS BONELLI* (Vieillot).

*Phylloscopus bonelli orientalis* (Brehm.). **Pl. 68.**

*Phyllopneuste orientalis* C. L. Brehm, Vögelf. p. 232, 1855: Wadi Halfa, northern Sudan.

**Distinguishing characters:** Above, olivaceous, with usually a yellowish wash; a slight light stripe over eye; rump and upper tail-coverts brighter yellow; edges of flight feathers, wing-coverts and tail feathers greenish yellow; below, whitish, under wing-coverts and axillaries yellow. The sexes are alike. Wing 62 to 70 mm. This Warbler can be distinguished from the Willow Warbler, Chiff-Chaff and Wood-Warbler by the paler head and mantle, with which the

yellower rump and upper tail-coverts show a contrast, and by the uniform whitish underside.

**General distribution:** South-eastern Europe, Asia Minor and Palestine; in non-breeding season to Egypt and the Sudan.

**Range in Eastern Africa:** The Sudan as far south as about lat. 9° N. in non-breeding season.

**Habits:** A scarce palæarctic winter migrant to the Sudan, but an easily overlooked species, keeping to thicker trees. Its song, if heard, has something of the sibilant quality of the Wood Warbler, but is easily recognizable.

**Distribution of other races of the species:** Germany, Portugal, Spain, Italy and North Africa, the nominate race being described from Italy.

**WOODLAND-WARBLERS.**   Family—SYLVIIDÆ.   Genus: *Seicercus*.

Four species occur in Eastern Africa. Small non-migratory Warbler-like birds, whose habitat is woodland and forest and the undergrowth in such situations. In addition to the leaf-searching habits of the true Warblers they have the Flycatcher habit of flying out from a perch to catch passing insects.

### 963 YELLOW-THROATED WOODLAND-WARBLER. *SEICERCUS RUFICAPILLUS* (Sundevall).

*Seicercus ruficapillus minullus* (Reicnw.). **Pl. 68.**

*Chloropeta minulla* Reichenow, O.M. p. 181, 1905: Mlalo, near Lushoto, north-eastern Tanganyika Territory.

**Distinguishing characters:** Above, top of head and hind neck saffron brown; rest of upperparts olive green; stripe over eye, sides of face and chin to upper chest, underside of wings and under tail-coverts greenish lemon yellow; chest pale grey; flanks greenish yellow and grey; belly creamy white. The sexes are alike. Wing 48 to 57 mm. The young bird has the chest washed with yellow green.

**Range in Eastern Africa:** Taita Hills district of south-eastern Kenya Colony and Mt. Mbololo to north-eastern and eastern Tanganyika Territory, as far south as Kilosa and the Uluguru Mts.

**Habits:** These birds are mostly found in woodland or evergreen forest, usually in pairs, but congregating into little flocks in the non-breeding season which pass from tree to tree like Long-tailed Tits in Europe. They also flutter out to catch insects in flight and in general behave very much like the Phylloscopine Warblers, of which they might almost be considered to be the African representatives.

**Nest and Eggs:** Nest domed, of skeleton leaves, moss and hair-like mycelium placed low down in herbage and generally near tracks in forest. Eggs usually two or three, white—pink when fresh—with reddish spots; about 17 × 13 mm.

**Recorded breeding:** Tanganyika Territory, Amani, November. Dar-es-Salaam, September.

**Food:** Entirely insects.

**Call:** A rather sibilant 'see-eer' harsh and ringing. An occasional little bell-like trill is believed to come from this species also.

*Seicercus ruficapillus johnstoni* Scl.
*Seicercus ruficapilla johnstoni* W. Sclater, Bull. B.O.C. 48, p. 13, 1927: Kombi, Masuku Range, north Nyasaland.

**Distinguishing characters:** Differs from the preceding race in having the top of the head and hinder neck darker, tinged with saffron, less green, and in having an admixture of grey on the mantle. Wing 51 to 59 mm.

**General distribution:** Tanganyika Territory to Nyasaland and Portuguese East Africa.

**Range in Eastern Africa:** South-western Tanganyika Territory to Nyasaland and Chiperone Mt., south-western Portuguese East Africa.

**Habits:** As for the preceding race, birds of evergreen forest. The call, however, is said to be notably less harsh.

**Recorded breeding:** Nyasaland, Masuku and Cholo Mts. (breeding condition), August. Mangoche Mt. (breeding condition), October.

*Seicercus ruficapillus quelimanensis* Vinc.
*Seicercus ruficapilla quelimanensis* Vincent, Bull. B.O.C. 53, p. 136, 1933: Namuli Mt., Quelimane Province northern Portuguese East Africa.

**Distinguishing characters:** Differs from the two preceding races in having the mantle grey with only traces of green. Wing 52 to 58 mm.

**Range in Eastern Africa:** Namuli Mts., northern Portuguese East Africa.

**Habits:** Local in forest at from 4,500 to 6,000 feet, keeping to the higher branches of trees. An incessant squeaking note, and an occasional trill are noticeable, otherwise the habits are those of the last race.

**Recorded breeding:** No records.

*Seicercus ruficapillus ochrogularis* Mor.
*Seicercus ruficapilla ochrogularis* Moreau, Bull. B.O.C. 61, p. 24, 1941: Kungwe Mt., Kigoma district, western Tanganyika Territory.

**Distinguishing characters:** Differs from the other three races in having the stripe over the eye, sides of face and chin to upper chest yellow washed with saffron, not greenish lemon yellow. Wing 54 to 59 mm.

**Range in Eastern Africa:** Kungwe-Mahare Mts., western Tanganyika Territory.

**Habits:** Unrecorded, but unlikely to differ from those of other races.

**Recorded breeding:** Kungwe-Mahare Mts. (breeding condition), August.

**Distribution of other races of the species:** South Africa, the nominate race being described from Durban, Natal.

### 964 BROWN WOODLAND-WARBLER. *SEICERCUS UMBROVIRENS* (Rüppell).

*Seicercus umbrovirens umbrovirens* (Rüpp.). **Pl. 68.**
*Sylvia (Ficedula) umbrovirens* Rüppell, N. Wirbelt. Vög. p. 112, 1840: Simen, northern Abyssinia.

**Distinguishing characters:** Above, warm olivaceous brown; edges of flight feathers, wing-coverts and tail feathers yellow green; stripe over eye, sides of face and underside pale brown; centre of belly whitish. The sexes are alike. Wing 56 to 61 mm. The young bird is washed with yellow below.

**Range in Eastern Africa:** Eritrea, northern and central Abyssinia and British Somaliland.

**Habits:** As for other races and locally plentiful in Abyssinia above 7,500 feet, inhabiting undergrowth among groves of high

## WARBLERS

trees. The nest is usually among low bushes or hanging moss-covered branches rather than on the ground.

**Nest and Eggs:** As for the following race.

**Recorded breeding:** Abyssinia, March and April, also (breeding condition), November.

*Seicercus umbrovirens mackenzianus* (Sharpe).
*Cryptolopha mackenziana* Sharpe, Ibis, p. 153, 1892: Kikuyu, central Kenya Colony.

**Distinguishing characters:** Differs from the nominate race in being darker, warmer brown above; below, whitish with the sides of the head and flanks warmer brown. Wing 52 to 64 mm.

**Range in Eastern Africa:** Southern Sudan to Uganda and Kenya Colony.

**Habits:** Fairly common mainly at higher levels in forest or bamboo. Habits of a Willow Warbler, equally as tame and very easily mistaken for it.

**Nest and Eggs:** Nest a domed structure of dry grass and leaves lined with feathers and plant down placed in the side of a bank or against a fallen log or root. Eggs two or three, white or pinkish-white with small rusty red spots occasionally forming a cap at the larger end; about $17 \cdot 5 \times 13$ mm.

**Recorded breeding:** Kenya Colony, March to August and November to February.

**Food:** Insects, often caught on the wing.

**Call:** A series of clear notes rather trilled and very varied. A constantly repeated 'pee-piri' is the most usual. Alarm note a sharp 'swee-vik.' The song is not unlike that of a Willow Warbler.

*Seicercus umbrovirens dorcadichrous* (Reichw. & Neum.).
*Camaroptera dorcadichroa* Reichenow & Neumann, O.M. p. 76, 1895: Kifinika, Mt. Kilimanjaro, north-eastern Tanganyika Territory.

**Distinguishing characters:** Differs from the last race in being more of an olivaceous deep warm brown above; throat and chest washed with brown. Wing 52 to 60 mm.

**Range in Eastern Africa:** North-eastern Tanganyika Territory from Oldeani, Mbulu and Mt. Gerui (Hanang) to Mt. Kilimanjaro and north Paré Mts.

**Habits:** As for other races, but a more persistent songster with an enormously varied range of song, comprising those of most of the migrant European Warblers and of other species as well.

**Recorded breeding:** Northern Tanganyika Territory (breeding condition), October and December.

*Seicercus umbrovirens omoensis* (Neum.).
*Cryptolopha umbrovirens omoensis* Neumann, J.f.O. p. 208, 1905: Banka, Malo, south-western Abyssinia.

**Distinguishing characters:** Differs from the nominate race in being greenish brown above; underside pale brown, centre of belly white. Wing 52 to 61 mm.

**Range in Eastern Africa:** Western and southern Abyssinia south of the Great Abai Valley to the Omo River Valley and the Boran area.

**Habits:** As for other races.

**Recorded breeding:** No records.

*Seicercus umbrovirens alpinus* (O. Grant).
*Cryptolopha alpina* O. Grant, Bull. B.O.C. 16, p. 117, 1906: Mubuku Valley, Ruwenzori Mts., western Uganda.

**Distinguishing characters:** Above, similar to the Kilimanjaro race, but below, chin to breast dull russet brown; rest of underparts washed with pale brown. Wing 61 to 67 mm.

**Range in Eastern Africa:** Ruwenzori Mts., western Uganda.

**Habits:** As for other races, but mostly found among Tree-Heaths at from 10,000 feet to 14,000 feet. Has a short cheerful song. Nest usually in moss dependent from Tree-Heaths and most difficult to see. Eggs up to three, white spotted all over with light reddish and lavender grey spots; about $18 \times 12 \cdot 5$ mm.

**Recorded breeding:** Ruwenzori Mts., December and January.

*Seicercus umbrovirens wilhelmi* (Gyld.).
*Cryptolopha wilhelmi* Gyldenstolpe, Bull. B.O.C. 43, p. 37, 1922: Mt. Muhavura, northern Ruanda, eastern Belgian Congo-Uganda boundary.

**Distinguishing characters:** Very similar to the last race, but differs in being duller above, more olivaceous brown, not so warm in colour; below, chin to breast duller brown. Wing 63 to 64 mm.

**Range in Eastern Africa:** Virunga Volcanoes, northern Ruanda, Belgian Congo.

**Habits:** Those of other races, found among the high forest and heath zones of its habitat.

**Recorded breeding:** No records.

*Seicercus umbrovirens fuggles-couchmani* Mor.
*Seicercus umbrovirens fuggles-couchmani* Moreau, Bull. B.O.C. 61, p. 24, 1941: Chenzema, 7,000 feet, Uluguru Mts., Morogoro district, eastern Tanganyika Territory.

**Distinguishing characters:** Similar to the Kilimanjaro race, but above colder olivaceous brown. Wing 54 to 60 mm.

**Range in Eastern Africa:** Uluguru Mts., Morogoro district, eastern Tanganyika Territory.

**Habits:** As for other races.

**Recorded breeding:** No records.

**Distribution of other races of the species:** South-western Arabia.

**965 UGANDA WOODLAND - WARBLER.** *SEICERCUS BUDONGOENSIS* (Seth-Smith).
*Cryptolopha budongoensis* D. Seth-Smith, Bull. B.O.C. 21, p. 12, 1907: Budongo Forest, western Uganda.

**Distinguishing characters:** Above, olive green; broad whitish stripe over eye; a blackish stripe from bill through eye; wing-coverts and edges of flight feathers olive yellow; below, whitish, tinged olive green, darker on flanks; under wing-coverts yellowish white. The sexes are alike. Wing 54 mm. The young bird is duller than the adult; breast washed with olive.

**General distribution:** Ituri Forest, eastern Belgian Congo to Kenya Colony.

**Range in Eastern Africa:** Uganda and western Kenya Colony.

**Habits:** Little known species occurring among dense undergrowth of heavy forest.

**Nest and Eggs:** Undescribed.

**Recorded breeding:** No records.

**Food:** No information.

**Call:** A short high-pitched song freely uttered from dense cover.

## WARBLERS

**966 RED-FACED WOODLAND-WARBLER.** *SEICERCUS LÆTUS* (Sharpe). **Pl. 69.**

*Cryptolopha læta* Sharpe, Bull. B.O.C. 13, p. 9, 1902: Ruwenzori Mts., western Uganda.

**Distinguishing characters**: Above, green; forehead, stripe over eye and sides of face buffish brown; below, chin to chest buff brown, paler on chest; belly white; under wing-coverts lemon yellow. The sexes are alike. Wing 54 to 60 mm. The young bird is duller than the adult; less buffish brown on sides of face; breast washed with greyish.

**Range in Eastern Africa**: Western Uganda and Ruanda from the Ruwenzori Mts. to the Virunga Volcanoes.

**Habits**: Not uncommon in small parties in higher forest or bamboo zones, passing along in little scattered flocks from tree to tree.

**Nest and Eggs**: Undescribed.

**Recorded breeding**: Uganda, probably April to June.

**Food**: Of the genus.

**Call**: A faint chirping 'sip-sip,' and a low warbling song.

**WREN-WARBLERS.** FAMILY—**SYLVIIDÆ.** Genus: *Calamonastes*.

Two species occur in Eastern Africa and are distinguished by the barred underparts and the habit of fanning the tail. They occur in both woodland and thorn-scrub. The food is believed to be entirely insects, taken largely on or near the ground.

**967 BARRED WREN-WARBLER.** *CALAMONASTES FASCIOLATUS* (Smith).

*Calamonastes fasciolatus stierlingi* Reichw.

*Calamonastes stierlingi* Reichenow, O.M. p. 39, 1901: Songea, south-western Tanganyika Territory.

**Distinguishing characters**: Above, olivaceous brown; wing-coverts tipped with white; below, white, barred with black. The sexes are alike. Wing 54 to 68 mm. The young bird is rather paler above; below, faintly washed with yellow and bars dull blackish.

**General distribution**: Tanganyika Territory to Rhodesia, Nyasaland and northern Portuguese East Africa.

# WARBLERS

**Range in Eastern Africa:** East-central to south-western Tanganyika Territory and Portuguese East Africa.

**Habits:** Among woodlands, particularly in grassy patches in woods, in pairs or small parties, local but not uncommon. The wings make a rattling noise as the bird rises.

**Nest and Eggs:** The nest is a remarkable 'Tailor-bird' like construction of dry grass enclosed within the living leaves of a shrub or plant which latter are pierced and sewn on to the nest with strong spiders' webs or other material. It is oval or round and lightly made with the entrance hole at the side. Eggs two or three, white, closely speckled with purplish red and bluish grey, or turquoise blue with sepia spots and pale purple undermarkings; about 18 × 13 mm.

**Recorded breeding:** Nyasaland, October to January. Rhodesia, June and October to December.

**Call:** A penetrating trill, almost a bleat, 'peep-peep.' Song a whistling flute-like trill of short duration.

**Distribution of other races of the species:** Angola and South Africa, the nominate race being described from Bechuanaland.

## 968 GREY WREN-WARBLER. *CALAMONASTES SIMPLEX* (Cabanis).

*Calamonastes simplex simplex* (Cab.).
*Thamnobia simplex* Cabanis, J.f.O. pp. 205, 221, 1878: Ndi, Taita district, south-eastern Kenya Colony.

**Distinguishing characters:** Above, ash brown with a grey wash; below, darkish grey with more or less indistinct light barring. The sexes are alike. Wing 52 to 66 mm. The young bird is paler below with fainter barring; bill horn, not black as in adult.

**Range in Eastern Africa:** Central Abyssinia and British Somaliland to south-eastern Sudan, Uganda, Kenya Colony (except the south-western area), southern Italian Somaliland and northern Tanganyika Territory from the south-east corner of Lake Natron to Pangani.

**Habits:** Bird of dry low level country mainly keeping to thick thorn scrub and usually difficult to see. The tail is constantly fanned up and down, and in the breeding season the male makes a little aerial hop over the nest with spread tail while building is in progress.

**Nest and Eggs:** A spherical nest of fine silk-like fibre with a side entrance and a fibre cup inside. Nest is placed low in herbage with live leaves stitched on to it. Eggs three, greyish white thickly spotted with minute dark dots forming a zone at larger end; about 19 × 12·5 mm.

**Recorded breeding:** Taru Desert, Kenya Colony, December and January. Tanganyika Territory, November and April.

**Call:** A remarkable far-reaching metallic chinking exactly like that produced by knocking two stones together, also a soft 'oo-tu-tu-tu-tu.'

*Calamonastes simplex undosus* (Reichw.).
*Drymoica undosa* Reichenow, J.f.O. p. 211, 1882: Kakoma, Tabora district, west-central Tanganyika Territory.

**Distinguishing characters:** Differs from the nominate race in having chest to belly and flanks whiter, with more distinct barring. Wing 55 to 68 mm.

**General distribution:** Kenya Colony, Tanganyika Territory and north-eastern Northern Rhodesia.

**Range in Eastern Africa:** South-western Kenya Colony and Tanganyika Territory from southern end of Lake Victoria to Tabora, Uluguru, Iringa and the Ufipa Plateau.

**Habits:** As for the nominate race.

**Recorded breeding:** Tanganyika Territory, October to December, but probably double-brooded.

**FAN-TAILED WARBLER.**   FAMILY—SYLVIIDÆ.   Genus: *Schœnicola*.

This species, the only one of the genus, is an inhabitant of high grass, reeds and bush along streams. The heavy tail with its long under tail-coverts is a clear distinctive character which is noticeable in life. It is believed to be entirely insectivorous.

**969** FAN-TAILED WARBLER. *SCHŒNICOLA BREVIROSTRIS* (Sundevall).
*Schœnicola brevirostris brevirostris* (Sund.).
*Bradypterus brevirostris* Sundevall, Œfv. K. Sv. Vet-Akad. Forhandl. 7, p. 103, 1850: Umlazi River, Natal, South Africa.

**Distinguishing characters:** Above, olivaceous to russet brown; tail long and broad, dark blackish brown, tipped whitish; below, white or buffish white; sides of body and flanks browner; under tail-coverts long and fluffy, brown to blackish brown, tipped with whitish. The sexes are alike. Wing 57 to 64 mm. The young bird has the underparts washed with yellow.

**General distribution:** Southern Nyasaland and the Tete Province of northern Portuguese East Africa to Southern Rhodesia, southern Portuguese East Africa, eastern Cape Province and Natal.

**Range in Eastern Africa:** Southern Nyasaland.

**Habits:** As for the northern race, the male in display makes short circular cruises some twenty feet from the ground with loudly 'burring' and quickly beating wings, but as Vincent remarks, makes slow progress.

**Nest and Eggs:** See under the following race.

**Recorded breeding:** Nyasaland, November to February. Portuguese East Africa, March. Natal, March.

*Schœnicola brevirostris alexinæ* (Heugl.).
*Sphenœcus alexinæ* Heuglin, J.f.O. p. 166, 1863: Gazelle River southern Sudan.

**Distinguishing characters:** Differs from the nominate race in being generally rather darker in tone of colour. Wing 55 to 67 mm.

**General distribution:** Sierra Leone to Abyssinia, Tanganyika Territory, northern and eastern Angola, Northern Rhodesia and northern Nyasaland.

**Range in Eastern Africa:** Southern Sudan and Abyssinia to Uganda, Kenya Colony and Tanganyika Territory.

**Habits:** Birds of high grass lands usually confined to damp hollows or reeds and bush fringing mountain streams; locally common and occasionally found in long grass and scrub well away from water. Conspicuous in flight, which is short and jerky above the grass tops with an abrupt dive into the grass, and difficult to flush again as they creep away through the grass at once. The broad tail is rather noticeable in flight, very noticeable in the display flight, and the wings make a distinct rustle at times.

**Nest and Eggs:** Nest, close to ground in grass tuft, a shallow frail cup of broad leaved grass lined with finer grass. Eggs two, white or creamy white mottled with pale reddish brown and with sparse lilac undermarkings; about 18 × 13 mm.

**Recorded breeding:** Uganda, September, also February to June. Tanganyika Territory, November and February. Northern Nyasaland, January to March.

**Call:** Uttered from a grass top, a loud whistle often followed by a rattle. A piping and clicking 'pink' is also heard in the breeding season.

**APALIS WARBLERS.** FAMILY—**SYLVIIDÆ.** Genera: *Apalis, Artisornis* and *Scepomycter*.

Twenty-two species occur in Eastern Africa. They have rather long bills and the *Apalis* group have as a rule rather long tails. Many of the *Apalis* have also a black band across the chest. They are usually inhabitants of undergrowth in woodlands, but some frequent the tops of forest trees. Their food consists of insects. They are as remarkable as any group of birds in the variation of the colour of their eggs, and it must be understood that the various examples mentioned in this work are not by any means the only forms likely to be met with.

**970 BAR-THROATED APALIS.** *APALIS MURINA* Reichenow.
*Apalis murina murina* Reichw. **Pl. 69.**
*Apalis murina* Reichenow, O.M. p. 28, 1904: Mararupia, Tukuyu district, south-western Tanganyika Territory.

**Distinguishing characters:** Forehead to nape and sides of head ash brown to dusky brown; mantle and upperside of wings slate grey, sometimes washed with olive; rump more or less olive green; below, white with a broad black band across lower neck; lower belly yellow. The sexes are alike. Wing 50 to 55 mm. The young bird has the mantle more washed with green, and the black band across lower neck is much narrower than in the adult.

**General distribution:** Tanganyika Territory to northern Nyasaland and the Mafinga Mts. of north-eastern Northern Rhodesia.

**Range in Eastern Africa:** North-eastern Tanganyika Territory to the Iringa and Tukuyu districts.

**Habits:** Fairly common forest or forest jungle species in the highlands of Tanganyika Territory. Restless inquisitive little birds found both in forest canopy and near the ground in undergrowth.

**Nest and Eggs:** Nest a compact oval or round ball of moss and lichen often placed in a mass of leaves and cobwebs, or occasionally woven on to a pendent leaf. It is lined with vegetable down or soft material. Eggs two or three, light turquoise blue with light freckles and cloudings of dull red; about 17 × 13 mm. White eggs with red speckling are also reported, but it is not clear whether they are of this or of the following race.

**Recorded breeding:** Tanganyika Territory, December to February.

**Call:** A bleating note which is probably an alarm call, and various trilling calls.

*Apalis murina youngi* Kinn.
*Apalis thoracica youngi* Kinnear, Bull. B.O.C. 57, p. 8, 1936: Vipya, northern Nyasaland.

**Distinguishing characters:** Above, darker than the nominate race; back and wings slate colour; below, whiter with little or no yellow on lower belly. Wing 49 to 56 mm.

**General distribution:** Tanganyika Territory to the Nyika and Vipya Plateaux and Mwanjati Hill, Nyasaland.

**Range in Eastern Africa:** Ufipa Plateau, south-western Tanganyika Territory.

**Habits:** As for the nominate race. Eggs said to be very pale blue with the usual markings.

**Recorded breeding:** Nyika Plateau, October.

*Apalis murina whitei* Grant & Praed.
*Apalis murina whitei* C. Grant & Mackworth-Praed, Bull. B.O.C. 57, p. 114, 1937: Dedza Mt., Dedza district, Nyasaland.

**Distinguishing characters:** Differs from the nominate race in being paler above; top of head paler ashy; below, throat buffish; lemon yellow more confined to lower belly. Wing 49 to 55 mm.

**General distribution:** Tanganyika Territory to north-eastern Northern Rhodesia and south-western Nyasaland.

**Range in Eastern Africa:** Luwiri-Kitessi, south-west of Songea.

**Habits:** As for other races. The call of this race and the preceding one is a low trill followed by a more distinct call of 'pirri-pirri.'

**Recorded breeding:** Nyasaland, October and November.

*Apalis murina fuscigularis* Moreau.
*Apalis murina fuscigularis* Moreau, Bull. B.O.C. 58, p. 48, 1937: Taita Hills, south-eastern Kenya Colony.

**Distinguishing characters**: Differs from the nominate race in having the throat black, forming with the band one black patch across the lower neck; chin ashy brown. Wing 52 to 55 mm.

**Range in Eastern Africa**: Taita Hills, south-eastern Kenya Colony.

**Habits**: As for the nominate race.

**Recorded breeding**: No records.

**Distribution of other races of the species**: Southern Rhodesia.

### 971 YELLOW BAR-THROATED APALIS. *APALIS FLAVIGULARIS* Shelley.

*Apalis flavigularis flavigularis* Shell.
*Apalis flavigularis* Shelley, Ibis, p. 16, 1893: Milanji, southern Nyasaland.

**Distinguishing characters**: Above, head black with an admixture of green; mantle, wings and rump green; tail grey, outer feathers edged and tipped with white; below, bright lemon yellow; sides of face and band across lower neck black; flanks green. The sexes are alike. Wing 49 to 56 mm. The young bird has the head green.

**Range in Eastern Africa**: Nyasaland.

**Habits**: Locally common in hill forest or forest outskirts, rather stocky little birds for the genus, noisy and inquisitive.

**Nest and Eggs**: See under the following race.

**Recorded breeding**: Nyasaland, October to January.

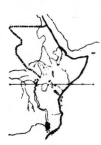

*Apalis flavigularis griseiceps* Reichw. & Neum. **Pl. 69.**
*Apalis griseiceps* Reichenow & Neumann, O.M. p. 75, 1895: Kifinika Hut, Mt. Kilimanjaro, north-eastern Tanganyika Territory.

**Distinguishing characters**: Differs from the nominate race in having the head and sides of face ash brown; chin to black band and immediately below black band white. Wing 50 to 56 mm. The young bird has the green of the mantle extending towards the crown of the head, and a much less distinct band across the lower neck.

**Range in Eastern Africa**: Chyulu Hills, south-eastern Kenya Colony to north-eastern and south-central Tanganyika Territory.

**Habits:** Attractive noisy little birds common in forest or bush in the highest parts of wooded mountains.

**Nest and Eggs:** A rather untidy nest of grass, lichen, etc. plastered over with moss, with the entrance hole at the side of the top, lined with cobwebs and vegetable down, and usually placed in a low shrub. Eggs two or three, white with brownish red speckling at larger end. Blue eggs also occur. Variable in size and shape; about $17 \cdot 5 \times 13$ mm.

**Recorded breeding:** Tanganyika Territory, September to January.

**Call:** A trilling 'pirri-pirri' rather more trilled than in the Bar-Throated Apalis, also a harsh bleating 'tsewi.'

*Apalis flavigularis uluguru* Neum.
*Apalis griseiceps uluguru* Neumann, O.M. p. 10, 1914: East Uluguru Mts., Morogoro district, Tanganyika Territory.

**Distinguishing characters:** Differs from the nominate race in having the chin and throat above the black band white, and from the last race in having the yellow of the underparts extending right up to the black band. Wing 51 to 52 mm.

**Range in Eastern Africa:** Uluguru Mts., Morogoro district, Tanganyika Territory.

**Habits:** As for the nominate race.

**Recorded breeding:** No records.

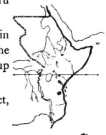

*Apalis flavigularis lynesi* Vinc.
*Apalis lynesi* Vincent, Bull. B.O.C. 53, p. 142, 1933: Namuli Mts., northern Portuguese East Africa.

**Distinguishing characters:** Differs from the nominate race in having the head and sides of face grey, and chin to chest band black. Wing 51 to 55 mm.

**Range in Eastern Africa:** Namuli Mts., northern Portuguese East Africa.

**Habits:** Locally common, generally in pairs in primæval forest, both in low undergrowth and in forest canopy. Noticeable and restless little birds, with a variety of chirruping calls.

**Recorded breeding:** Probably November and December.

**Distribution of other races of the species:** South Africa.

# WARBLERS

**972 BUFF-THROATED APALIS.** *APALIS RUFOGULARIS* (Fraser).
*Apalis rufogularis denti* O. Grant. **Pl. 69.**
*Apalis denti* O. Grant, Bull. B.O.C. 19, p. 86, 1907: Mpanga Forest, western Uganda.

**Distinguishing characters:** Above, dull olive; below, chin to chest deep buff; rest of underparts, including four outer tail feathers white. The sexes are alike. Wing 45 to 52 mm. The young bird is slightly greener above, with indications of the buff on chin to chest, and centre of breast to belly yellowish.

**General distribution:** Ituri Forest of eastern Belgian Congo and Uganda.

**Range in Eastern Africa:** South-western Sudan and Uganda.

**Habits:** Little recorded except that it is locally not uncommon among forest tree tops.

**Nest and Eggs:** Undescribed.

**Recorded breeding:** No records, breeding season probably prolonged.

**Call:** Unrecorded.

**Distribution of other races of the species:** West Africa and Angola, the nominate race being described from Fernando Po.

**973 GREY APALIS.** *APALIS CINEREA* (Sharpe).
*Apalis cinerea cinerea* (Sharpe). **Pl. 69.**
*Euprinoides cinerea* Sharpe, Ibis, p. 120, 1891: Mt. Elgon.

**Distinguishing characters:** Top of head and sides of face ashy to dusky brown; mantle and rump grey; three outer tail feathers mainly white; below, creamy white. The sexes are alike. Wing 51 to 60 mm. The young bird is washed with green above and with yellow below.

**General distribution:** Cameroon Mt. to the Sudan, Kenya Colony and north-eastern Tanganyika Territory.

**Range in Eastern Africa:** Southern Sudan and Uganda to central Kenya Colony and north-eastern Tanganyika Territory at Loliondo.

**Habits:** Common in dense forest and forest outskirts, both in tree tops and in undergrowth, darting about searching for insects and hovering with outspread tail, in which the white feathers are very noticeable.

**Nest and Eggs:** Nests, usually built in thick but fine-foliaged trees on the outskirts of forest, are rather bulky domed structures of lichen and cobwebs, lined with feathers. Eggs normally three, pale blue finely speckled with reddish brown; about $15 \times 11 \cdot 5$ mm.

**Recorded breeding:** Mpumu, Uganda (breeding condition), June. Kenya Colony highlands, March to June, also November and December.

**Call:** A variety of notes, normally a high pitched 'pirrrr-pik-pik-pik,' also a repeated 'chip-it' or 'puik' which becomes almost a song.

**Distribution of other races of the species:** Fernando Po and Sao Thomé Island.

974 BROWN-HEADED APALIS. *APALIS ALTICOLA* (Shelley).
*Apalis alticola brunneiceps* (Reichw.).
*Burnesia brunneiceps* Reichenow, O.M. p. 122, 1900: Rupira, Rungwe, south-western Tanganyika Territory.

**Distinguishing characters:** Very similar to the Grey Apalis, but head browner and outer tail feathers grey with white ends, not wholly white. Wing 51 to 61 mm. The young bird is slightly washed with olivaceous above, including top of head and sides of face; a faint wash of yellow below; base of lower mandible horn colour, not black.

**General distribution:** Tanganyika Territory to south-eastern Belgian Congo, eastern Northern Rhodesia and northern Nyasaland.

**Range in Eastern Africa:** North-eastern to south-western Tanganyika Territory, Oldeani, Mbulu, Mt. Gerui (Hanang), Njombe, Ufipa Plateau and Rungwe.

**Habits:** In south-western Tanganyika Territory reported to be a bird of the woodland edges of streams and of woodlands on downs, but in the central part of the country it is definitely a bird of forest canopy.

**Nest and Eggs:** Undescribed, though a nest was observed fifty feet up in a lichen covered streamer.

**Recorded breeding:** Northern and central Tanganyika Territory, probably October to February.

**Call:** Loud and varied, the most usual being a two syllabled 'chip-it-chip-it.'

**Distribution of other races of the species:** North-eastern Northern Rhodesia, the habitat of the nominate race.

## 975 BLACK-HEADED APALIS. *APALIS MELANOCEPHALA* (Fischer & Reichenow).

*Apalis melanocephala melanocephala* (Fisch. & Reichw.). **Pl. 69.**

*Burnesia melanocephala* Fischer & Reichenow, J.f.O. p. 56, 1884: Pangani, eastern Tanganyika Territory.

**Distinguishing characters:** Adult male, head including sides of face black; mantle, rump, wings and central tail feathers rather more dusky; below, creamy white; chest often rather buffish; tail feathers grey with whitish ends. The female is similar but slightly olivaceous grey above. Wing 44 to 51 mm. The young bird is washed with olive above and with yellow below; bill brown not black as in adult.

**Range in Eastern Africa:** East Usambara Mts. and Amani to Tanga and Pangani, Tanganyika Territory.

**Habits:** Common in all types of forest at any height from the ground, usually in sunshine not in dense shade, searching leaves for insects like a Willow-Warbler. Often associates with other insect hunting birds.

**Nest and Eggs:** See under other races.

**Recorded breeding:** No records.


*Apalis melanocephala nigrodorsalis* Granvik.

*Apalis melanocephala nigrodorsalis* Granvik, J.f.O. Sond. p. 244, 1923: Kyambu, near Nairobi, south-central Kenya Colony.

**Distinguishing characters:** Adult male, differs from the nominate race in being more velvety blackish brown above. The female is more olivaceous above than the female of the nominate race. Wing 47 to 55 mm.

**Range in Eastern Africa:** Central Kenya Colony from south of Mt. Kenya to Kiambu and Nairobi.

**Habits:** As for the nominate race, but not usually common.

**Recorded breeding:** No records.

*Apalis melanocephala moschi* V. Som.
*Apalis melanocephala moschi* Van Someren, J. E. A. & U. N. Hist. Soc. 9, p. 195, 1931: Moshi, north-eastern Tanganyika Territory.

**Distinguishing characters:** Adult male, differs from the nominate race in having a longer tail; in being dusky grey above, including head; the sides of the face being black. The female is paler, especially on the head, than the female of the nominate race. Wing 45 to 53 mm.

**Range in Eastern Africa:** Taita Hills, south-eastern Kenya Colony to Mts. Kilimanjaro and Meru, west Usambara Mts., Mafi Mt., Handeni and the Nguru and Uluguru Mts., Tanganyika Territory.

**Habits:** As for the nominate race, but not as common nor as noisy, and has mostly been noted from secondary growth forest and clearings.

**Recorded breeding:** No records.

*Apalis melanocephala tenebricosa* Vinc.
*Apalis melanocephala tenebricosa* Vincent, Bull. B.O.C. 53, p. 141, 1933: Namuli Mt., Quelimane Province, Portuguese East Africa.

**Distinguishing characters:** Differs from the nominate race in being washed with sooty brown below. Wing 47 to 52 mm.

**Range in Eastern Africa:** Unangu to Namuli and Chiperone Mts., Portuguese East Africa.

**Habits:** Common in its restricted habitat in the larger patches of forest among the highest branches of tall trees, restless and constantly calling with a chirruping cry and a trilling song of bell-like three-syllabled whistles quickly repeated, 'chiririt-chiririt-chiririt,' and usually answered by another bird.

**Recorded breeding:** July and August.

*Apalis melanocephala fuliginosa* Vinc.
*Apalis melanocephala fuliginosa* Vincent, Bull. B.O.C. 53, p. 141, 1933: Cholo Mt., southern Nyasaland.

**Distinguishing characters:** Adult male, differs from the preceding race in being less intense black above, rather duller, and sometimes greyish on mantle. The female is indistinguishable from the female of the last race. Wing 47 to 52 mm.

**Range in Eastern Africa:** Milanji and Cholo Mts., southern Nyasaland.

**Habits:** As for other races.

**Nest and Eggs:** A rounded and thick-walled structure of lichen lined with plant down with the entrance rather high on one side; it is usually well hidden among leaves at some height from the ground and bound on to twigs with spiders' webs. Eggs two or three, pale greenish, boldly spotted and speckled with brown and chestnut and with purplish slate undermarkings; about 15·5 × 12 mm.

**Recorded breeding:** Southern Nyasaland, October and November.

**Call:** A repeated 'chirrit' also a low trill and an almost inaudible 'peep.'

*Apalis melanocephala ellinoræ* V. Som.
*Apalis melanocephala ellinoræ* Van Someren, Bull. B.O.C. 64, p. 50, 1944: Meru, north of Mt. Kenya, central Kenya Colony.

**Distinguishing characters:** Adult male, nearest to the Kyambu race but differs in being clearer black above, less brownish black. The female is also darker above than the female of that race. Wing 53 to 55 mm.

**Range in Eastern Africa:** Meru and Mt. Kenya, central Kenya Colony.

**Habits:** As for other races.

**Recorded breeding:** No records.

*Apalis melanocephala lightoni* Roberts.
*Apalis chirindensis lightoni* Roberts, Ostrich 9, p. 119, 1938: Beira, Portuguese East Africa.

**Distinguishing characters:** Very similar to the Moshi race but darker above, although grey in tone when compared to the Cholo race. Wing 47 to 52 mm.

**General distribution:** Nyasaland to Beira, Portuguese East Africa.

**Range in Eastern Africa:** Namizimu, Mangoche, Chikala, Ndirande, Mpingwe and Soche Mts., and the Zomba Plateau, southern Nyasaland.

**Habits:** As for other races.

**Recorded breeding:** No records, probably October to December.

*Apalis melanocephala muhuluensis* Grant & Praed.

*Apalis melanocephala muhuluensis* C. Grant & Mackworth-Praed, Bull. B.O.C. 67, p. 43, 1947: Mahenge Boma, southern Tanganyika Territory.

**Distinguishing characters:** Adult male, above, including sides of face, wings and central tail feathers, deep black. Female plumage unrecorded. Wing 51 to 55 mm. The young bird is sooty black above; lower rump, upper tail-coverts and lower flanks washed with yellow.

**Range in Eastern Africa:** Mahenge to the Luwiri-Kitessa Forest, Songea district, southern Tanganyika Territory.

**Habits:** As for other races.

**Recorded breeding:** No records.

**Distribution of other races of the species:** Southern Rhodesia.

**976 BLACK-BACKED APALIS.** *APALIS NIGRESCENS* (Jackson). **Pl. 69.**

*Euprinoides nigrescens* Jackson, Bull. B.O.C. 16, p. 90, 1906: Ankole, south-western Uganda.

**Distinguishing characters:** Very similar to the Black-headed Apalis but has the four outer tail feathers wholly white. The sexes are alike. Wing 47 to 52 mm. The young bird has the throat and breast slightly washed with yellowish.

**General distribution:** Eastern Belgian Congo to western Kenya Colony.

**Range in Eastern Africa:** Uganda to western Kenya Colony from Mt. Elgon to Kavirondo.

**Habits:** Locally common among forest tree tops, singly or in pairs in most of the larger forests of Uganda. Occurs also in secondary growth, especially the tops of medium sized trees.

**Nest and Eggs:** Nest as for other species but rather elongate with a side entrance near top, made chiefly of beard moss. Eggs undescribed.

**Recorded breeding:** Western Uganda, July and August.

**Call:** The only sound recorded is a short cheep.

### 977 BLACK-COLLARED APALIS. *APALIS PULCHRA* Sharpe.

*Apalis pulchra pulchra* Sharpe. **Ph. xi.**
*Apalis pulchra* Sharpe, Ibis, p. 119, 1891: Mt. Elgon.

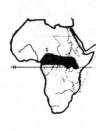

**Distinguishing characters:** Upperside and sides of head dark grey; tail with white ends, outermost feathers almost wholly white; below, throat creamy white; a broad black band across upper chest; flanks and belly rich chestnut; centre of breast and under wing-coverts white. The sexes are alike. Wing 50 to 59 mm. The young bird has the chest band dusky grey.

**General distribution:** Cameroons highlands to the Sudan and Kenya Colony.

**Range in Eastern Africa:** Southern Sudan to Uganda and central Kenya Colony.

**Habits:** Cheerful noisy little birds usually seen sidling up and down plant stems or searching foliage. They carry their tails very much cocked up and move them rapidly from side to side, occasionally fanning them out also. A species inhabiting highland forest nearly always in undergrowth and not in high trees.

**Nest and Eggs:** Nest when made by the bird itself an elongated purse-like structure, roofed over, and well concealed among upright stems or twigs. It is made of moss and lichen and plentifully lined with feathers. Very often, however, the birds adopt the nests of other species, particularly of *Eminia*, *Camaroptera*, various Weavers, or other species of Apalis. Eggs two, rarely three, pale greenish white with liver or reddish brown spots and violet undermarkings, rather elongated and large for the bird; about 18 × 12 mm. (Van Someren).

**Recorded breeding:** Cameroons, January and February. Kenya Colony highlands, March to July, irregularly also in December and January.

**Call:** A shrill 'tseeye-tseeye' or 'tuit-tuit' with an occasional pleasant little song.

**Distribution of other races of the species:** South-eastern Belgian Congo.

**978 COLLARED APALIS.** *APALIS RUWENZORII* Jackson.
*Apalis ruwenzorii* Jackson, Bull. B.O.C. 15, p. 11, 1904: Ruwenzori Mts., western Uganda.

**Distinguishing characters:** Similar to the Black-collared Apalis, but differs in having no white in the tail; paler grey above; chest band greyish black; throat and flanks tawny. The sexes are alike. Wing 47 to 53 mm. The young bird is duller than the adult; lower mandible horn colour, not black, as in the adult.

**General distribution:** Eastern Belgian Congo to the Ruwenzori Mts.

**Range in Eastern Africa:** Ruwenzori Mts., western Uganda.

**Habits:** Plentiful on the Ruwenzori Mts. in the forest zone and lower margin of the bamboo zone, always skulking in dense undergrowth or in tangled masses of creeper, never in the tree-tops.

**Nest and Eggs:** Undescribed.

**Recorded breeding:** No records.

**Call:** Low and scolding, Chapin also records a double 'tooting' sound.

**979 BLACK-BREASTED APALIS.** *APALIS FLAVIDA* (Strickland).
*Apalis flavida flavocincta* (Sharpe). **Pl. 69. Ph. xii.**
*Euprinoides flavocinctus* Sharpe, J.f.O. p. 346, 1882: Athi River, south-central Kenya Colony.

**Distinguishing characters:** Adult male, above, green; forehead towards crown and sides of face grey; below, white or creamy white; broad yellow band across chest; black spot in centre of chest; tail green, with broad yellow ends to all except central feathers. The female usually lacks the black spot on the chest. Wing 47 to 57, tail 42 to 60 mm. The young bird is paler; ring round eye and lores whitish; top of head green uniform with mantle; bill horn, not black as in adult.

**Range in Eastern Africa:** South-western Uganda to central, southern and eastern Kenya Colony and Manda Island.

**Habits:** Birds of very Phylloscopine habits, diligently searching leaves and small branches for insects. They are locally plentiful in forest, gallery forest or woodland, in parties of three to five, either among tree-tops or undergrowth.

**Nest and Eggs:** Nest either fairly high in a tree or a few feet from the ground in a bush or herb, a small semi-domed cup with one side higher than the other, made of lichen and moss with a lining of vegetable down. Eggs three, bluish green or pinkish white, spotted with liver and orange brown and with ashy violet undermarkings. It seems likely that the ground colour may vary considerably; about 15 × 11·5 mm.

**Recorded breeding:** Central Kenya Colony, April to June, also December. Southern areas, March and April.

**Call:** A churring alarm note 'chieeer' and a number of freely uttered chittering cries, rather like the clicking of pebbles.

*Apalis flavida golzi* (Fisch. & Reichw.).
*Euprinoides golzi* Fischer & Reichenow, J.f.O. p. 182, 1884: Great Arusha, north-eastern Tanganyika Territory.

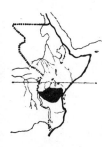

**Distinguishing characters:** Differs from the preceding race in having the grey of the forehead extending to the occiput; black spot in centre of chest in male usually larger. Wing 46 to 55, tail 45 to 55 mm.

**Range in Eastern Africa:** South-eastern Kenya Colony to north-eastern, central and western Tanganyika Territory, Ukerewe Island, Ketumbeine, Mbulu, Monduli, Mt. Kilimanjaro, south Paré Mts., Mkomasi, Kigoma, Mkalama and Iringa.

**Habits:** As for the last race, but more a bird of bush and thorn scrub. The nest is a minute cup of fine fibres with one side higher than the other, placed in a thick bush. The call is a ringing repeated 'tsirri-tsirri' and there is also a churring alarm note.

**Recorded breeding:** South-eastern Kenya Colony, January to March. Tanganyika Territory, November to March.

*Apalis flavida viridiceps* Hawk.
*Apalis viridiceps* Hawker, Bull. B.O.C. 7, p. 55, 1895: Sheik Woofli, western British Somaliland.

**Distinguishing characters:** Differs from the preceding races in having the chest band yellow and green; no black spot in centre of chest in male; above, more olivaceous green; ends of tail feathers almost white or faintly tinged with yellow; grey confined to forehead. Wing 47 to 53, tail 52 to 62 mm.

# WARBLERS

**Range in Eastern Africa:** British Somaliland and Italian Somaliland as far as Afgoi, lower Webi Shebeli.

**Habits:** As for the last race but the nests are usually in low bushes.

**Recorded breeding:** Italian Somaliland, May.

*Apalis flavida malensis* Neum.
*Apalis malensis* Neumann, O.M. p. 78, 1905: Shambala River, Bazala, south-western Abyssinia.

**Distinguishing characters:** Similar to the Somaliland race but brighter green above, and ends of tail feathers pale yellow not almost white. Wing 46 to 53, tail 46 to 62 mm.

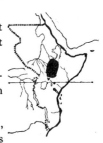

**Range in Eastern Africa:** South-western Abyssinia, south-eastern Sudan, Karamoja area of eastern Uganda and northern Kenya Colony.

**Habits:** A common bush country species of sandy stony thorn-bush, particularly common in Turkana, Kenya Colony, with the habits of other races. The nest is small and purse-shaped, usually placed in the middle of a large thorn bush. Eggs variable, one type is warm olive with a zone of fine russet spots. No measurements available.

**Recorded breeding:** Karamoja, May.

**Distribution of other races of the species:** South Africa, the nominate race being described from Damaraland.

980 GREEN-TAILED APALIS. *APALIS CANICEPS* (Cassin).
*Apalis caniceps caniceps* (Cass.).
*Camaroptera caniceps* Cassin, Proc. Ac. Nat. Sci. Philad. 2, p. 38, 1859: Camma River, Gabon.

**Distinguishing characters:** Differs from the Black-breasted Apalis in having the head grey to the occiput; no black on centre of chest in male, or rarely a very faint indication of black; and in having a shorter tail with narrow pale yellow or whitish tips, not broad yellow ends. The sexes are alike. Wing 48 to 55, tail 31 to 44 mm. The young bird is paler than the adult and has the head and sides of face green uniform with the mantle; the chin and throat pale yellow uniform with the chest band.

**General distribution:** Cameroons to the Sudan, Uganda and Kenya Colony.

**Range in Eastern Africa:** The southern Sudan to Uganda and western Kenya Colony at Kisumu.

**Habits:** Little recorded, but habits are probably those of the Black-breasted Apalis.

**Nest and Eggs:** See under the following race.

**Recorded breeding:** No records.

*Apalis caniceps neglecta* (Alex.).
*Chlorodyta neglecta* Alexander, Bull. B.O.C. 10, p. 17, 1900: near Zumbo, northern Portuguese East Africa.

**Distinguishing characters:** Differs from the nominate race in having the hinder part of the crown and occiput green; the male has a black spot in centre of chest; tail tips yellow. Wing 45 to 55, tail 39 to 54 mm.

**General distribution:** Northern Angola to southern Belgian Congo, the Rhodesias, Nyasaland, Portuguese East Africa, the Transvaal and Zululand.

**Range in Eastern Africa:** Portuguese East Africa from Netia to Nyasaland.

**Habits:** Local and not usually common in bush or thick river scrub.

**Nest and Eggs:** The nest which is domed and pear-shaped is made of moss, lichen and spiders' webs and bound on to the supporting twigs of a bough of a tree. The bird has also been reported using an old domed nest, probably that of a Pytilia. Eggs two or three, greyish green or whitish, densely speckled with reddish brown, especially at larger end; about $15 \cdot 5 \times 11$ mm.

**Recorded breeding:** Nyasaland, March and May, also September. South-eastern Belgian Congo, October.

**Call:** A two syllabled 'chirrer' and a harsh scolding alarm note, also a Sparrow-like chittering.

*Apalis caniceps tenerrima* Grote.
*Apalis flavida tenerrima* Grote, O.M. p. 119, 1935: Mikindani, south-eastern Tanganyika Territory.

**Distinguishing characters:** Differs from the last race in having the forehead to occiput grey. Wing 45 to 50, tail 38 to 45 mm.

**Range in Eastern Africa:** Eastern areas of Kenya Colony and Tanganyika Territory from Mombasa to Mikindani and inland to Korogwe, Dodoma, the Uluguru Mts., Nguru Mts., and Mahenge, also Zanzibar Island.

**Habits:** Little recorded, but not uncommon in bush country on Zanzibar Island.

**Recorded breeding:** Zanzibar Island, probably November to January.

**Distribution of other races of the species:** South Africa.

## 981 MASKED APALIS. *APALIS BINOTATA* Reichenow.

*Apalis binotata binotata* Reichw.
*Apalis binotata* Reichenow, O.M. p. 113, 1895: Jaunde, Cameroons.

**Distinguishing characters:** Adult male, head grey with a slight wash of green; lores and ear-coverts blackish; rest of upperparts green; below, chin and throat to chest black; small white patch at base of sides of neck; sides of chest yellow green; breast to belly buffish white; flanks olivaceous; tail green with small yellowish tips. The female has the white on sides of neck extending to base of bill, and chin usually also white. Wing 44 to 52 mm. The young bird has the head green uniform with the mantle, and no black in centre of throat and chest.

**General distribution:** Cameroons to Uganda.

**Range in Eastern Africa:** Mpanga Forest, Fort Portal and Mt. Elgon, Uganda.

**Habits:** Locally abundant in forest, and found both in tree tops and undergrowth, rather noisy.

**Nest and Eggs:** Nests are said to be large pocket-shaped structures of grey *Usnea* lichen bound with cobwebs. Eggs long narrow ovals with a slight gloss, dull greenish blue washed with rufous and with darker red spots and smears. No measurements given.

**Recorded breeding:** Cameroons, March and April, also June to August.

**Call:** A variety of hoarse cheeping notes.

*Apalis binotata personata* Sharpe.
*Apalis personata* Sharpe, Bull. B.O.C. 13, p. 9, 1902: Ruwenzori Mts., western Uganda.

**Distinguishing characters:** Differs from the nominate race in having the top of the head, sides of face, and chin to chest black, with a slight wash of green on the crown; a larger white patch on sides of neck, which does not extend up sides of throat in the female. Wing 52 to 58 mm.

**General distribution:** The Kivu area of the eastern Belgian Congo, and Uganda.

**Range in Eastern Africa:** Western Uganda from the southern Ruwenzori Mts. to the Kigezi and Kivu areas and Ruanda, eastern Belgian Congo.

**Habits:** As for the nominate race, being more usually found in tree tops. Distinctly sociable and often seen in small parties.

**Recorded breeding:** Ruwenzori Mts. (breeding condition), November.

### 982 BLACK-THROATED APALIS. *APALIS JACKSONI* Sharpe.

*Apalis jacksoni jacksoni* Sharpe. Pl. 70.

*Apalis jacksoni* Sharpe, Ibis, p. 119, 1891: Mt. Elgon, Kenya Colony-Uganda boundary.

**Distinguishing characters:** Adult male, forehead to crown, wings and tail grey; white edges to inner secondaries and white tips to all tail feathers, outermost with outer web white; mantle to rump bright green; below, sides of face and centre of chin and throat black, broadening to a patch on lower neck; broad white streak on side of neck; chest to belly yellow; flanks olivaceous. The female has the sides of the face and centre of chin to lower neck grey. Wing 48 to 55 mm. The young bird is duller than the adult female; top of head and lower throat greenish yellow in female, greyer in male; streak on side of neck washed with yellow; bill paler horn colour than in adult.

**General distribution:** The Sudan and Uganda, to Kenya Colony and northern Angola.

**Range in Eastern Africa:** South-eastern Sudan to Uganda and central Kenya Colony.

**Habits:** Frequents open sunny glades in the interior of dark forests, being usually seen among tall trees, but occasionally coming down to the undergrowth. Generally in family parties searching for insects.

**Nest and Eggs:** Nest a frail transparent but deep cup of *Usnea* lichen bound with spiders' webs. Eggs two, light bluish green evenly covered with small reddish spots. No measurements available.

**Recorded breeding:** Entebbe, Uganda, April and June.

**Call:** Prolonged calls of 'churng-churng-churng' almost in a duet (Chapin).

**Distribution of other races of the species:** Cameroons.

## 983 BLACK-CAPPED APALIS. *APALIS NIGRICEPS* (Shelley).

*Apalis nigriceps collaris* V. Som. **Pl. 69.**
*Apalis nigriceps collaris* Van Someren, Bull. B.O.C. 35, p. 107, 1915: Bugoma Forest, western Uganda.

**Distinguishing characters:** Adult male, head, sides of face to patch on lower neck in front black; band between black head and mantle golden yellow; mantle green; tail black, outer tail feathers white; chin and throat white; breast to under tail-coverts whitish. The female has the head, sides of face and patch on lower neck grey. Wing 45 to 46 mm. The young bird is washed with yellowish below; the young male having some grey in crown.

**General distribution:** Eastern Belgian Congo and Uganda.

**Range in Eastern Africa:** Western and southern Uganda.

**Habits:** Little known and comparatively rare in collections. Reported as a bird of forest tree-tops in parties of four to six by Van Someren. The West African race is usually associated with small trees and ferns in ravines.

**Nest and Eggs:** Undescribed.

**Recorded breeding:** Belgian Congo (breeding condition), April also October.

**Call:** A little cry of 'whit-whit' uttered with cocked up tail and puffed out feathers.

**Distribution of other races of the species:** West Africa, the nominate race being described from the Gold Coast.

## 984 WHITE-WINGED APALIS. *APALIS CHARIESSA* Reichenow. **Pl. 70.**

*Apalis chariessa* Reichenow, O.C. p. 114, 1879: Mitole, Tana River, eastern Kenya Colony.

**Distinguishing characters:** Adult male, above, including sides of face, wings, tail and a large patch in centre of lower neck, glossy blue black; edges of inner secondaries and ends of all except central

tail feathers white; below, chin, throat, sides of neck and under wing-coverts white; chest to belly bright yellow, wash of saffron on chest. The female has the head, sides of face, wings, tail and patch in centre of lower neck grey; white edges to inner secondaries and ends of tail feathers; mantle to rump green. Wing 45 to 54 mm. The young bird is similar to the adult female, but the centre of the lower neck is yellow.

**Range in Eastern Africa:** Tana River, eastern Kenya Colony to the Uluguru Mts., eastern Tanganyika Territory, the Blantyre and Cholo districts, southern Nyasaland and Chiperone Mt., south-western Portuguese East Africa.

**Habits:** Very local and in the north of its range a bird of high trees fringing rivers in open country; in Nyasaland it inhabits strips of forest leading down from mountain plateaux.

**Nest and Eggs:** Nest usually made in a hanging piece of 'Old-man's-beard' lichen—*Usnea barbata*, which is simply hollowed out to form a bag. Frequently placed at some height from the ground and very difficult to locate. Eggs three, pale green, spotted with yellowish brown over dull slate; about 17 × 11 mm.

**Recorded breeding:** Nyasaland, October to January.

**Call:** A repeated 'tweety' rather slow and not unmelodious, also a quicker 'teety-teetup' or 'tweety-chy' (Benson).

### 985 CHESTNUT-THROATED APALIS. *APALIS PORPHY-ROLÆMA* Reichenow & Neumann.

*Apalis porphyrolæma porphyrolæma* Reichw. & Neum. Pl. 70.
*Apalis porphyrolæma* Reichenow & Neumann, O.M. p. 75, 1895: Eldoma, south-western Kenya Colony.

**Distinguishing characters:** Above, ash grey; sides of face darker; below, chin, throat and malar stripe chestnut, sharply defined; rest of underparts whitish washed with grey; flanks grey; light tips to tail feathers. The sexes are alike. Wing 49 to 55 mm. The young bird is slightly paler above and washed with olivaceous; chin, throat and malar stripe yellowish; rest of underparts washed with yellowish.

**Range in Eastern Africa:** Ruanda and Ruwenzori Mts., western Uganda, to central Kenya Colony and north-eastern Tanganyika Territory at Loliondo.

**Habits:** Lively active little birds living among the topmost branches of trees, and extending from 6,000 feet upwards to the bamboo zone. They are said to come down to lower elevations in November and December on Mt. Elgon. Described as flitting from twig to twig swaying the body with drooping half-open wings, tail erect and spread out flicking from side to side. It is not clear whether this is only a breeding season activity.

**Nest and Eggs:** Undescribed.

**Recorded breeding:** No records.

**Call:** A chirping note, and also a faint whistle.

**Distribution of other races of the species:** Cameroons to Belgian Congo.

**986 BAMENDA APALIS.** *APALIS BAMENDÆ* Bannerman.
*Apalis bamendæ chapini* Fried. Pl. 70.
*Apalis chapini* Friedmann, Proc. N. Engl. Zool. Cl. 10 p. 47, 1928: Nyingwa, Uluguru Mts., eastern Tanganyika Territory.

**Distinguishing characters:** Above, grey; forehead and sides of face chestnut; crown washed with chestnut; below, chin white; throat to chest and thighs pale chestnut; breast to belly buffish white; chestnut of chest merging into breast colour. The sexes are alike. Wing 47 to 52 mm. Juvenile plumage unrecorded.

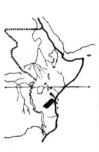

**Range in Eastern Africa:** The Nguru and Uluguru Mts. to Dabaga and the Uzungwe Mts., eastern to south-central Tanganyika Territory.

**Habits:** Not uncommon among evergreen foliage of trees at forest edge or in forest undergrowth. They are not easily observed and little is recorded of their life history.

**Nest and Eggs:** A long purse-shaped nest with a tail of lichen hanging from a high bough with an elongated entrance at the side. Eggs undescribed.

**Recorded breeding:** Eastern Tanganyika Territory, December to February.

**Call:** 'Chirrit-chirrit' not very loud but far carrying, also a sharp squeak repeated two or three times.

*Apalis bamendæ strausæ* Boul.

*Apalis bamendæ strausæ* Boulton, Ann. Carn. Mus. 21, p. 53, 1931: Mt. Rungwe, south-western Tanganyika Territory.

**Distinguishing characters:** Differs from the preceding race in having the chin chestnut, uniform with the throat. Wing 48 to 53 mm.

**General distribution:** Tanganyika Territory to northern Nyasaland as far south as Mzimba and the Ncheu district.

**Range in Eastern Africa:** South-western Tanganyika Territory from Njombe to the Nyasaland boundary.

**Habits:** As for the last race.

**Recorded breeding:** Njombe, Tanganyika Territory, December and January. Nyasaland, October.

**Distribution of other races of the species:** Cameroons, the habitat of the nominate race.

### 987 RED-FACED APALIS. *APALIS RUFIFRONS* (Rüppell).

*Apalis rufifrons rufifrons* (Rüpp.).

*Prinia rufifrons* Rüppell, N. Wirbelt. Vög. p. 110, pl. 41, fig. 2, 1840: Eritrea.

**Distinguishing characters:** Above, ash brown; forehead chestnut; wing-coverts and inner secondaries narrowly edged with whitish; tail blackish with white ends to all except central feathers; below, creamy white. The sexes are alike. Wing 42 to 49 mm. The young bird has only a very faint indication of chestnut on the forehead.

**Range in Eastern Africa:** The Sudan from Kordofan to the Red Sea and French Somaliland.

**Habits:** In the Sudan a bird of low sterile hills covered with poor grass and sparse scrub, in small parties in the non-breeding season, usually very active in search of food. Its most noticeable characteristic is the way it erects its tail and waves it from side to side.

**Nest and Eggs:** See under the following race.

**Recorded breeding:** Eritrea (nestling), June. Sudan, probably June to August.

*Apalis rufifrons smithii* (Sharpe). **Pl. 70.**
Dryodromas smithii Sharpe, Bull. B.O.C. 4, p. 29, 1895: Webi-Shebeli River, south-eastern Abyssinia.

**Distinguishing characters:** Differs from the nominate race in having the chestnut of the forehead extending to the occiput and the wing-coverts and inner secondaries broadly edged and tipped with white. Wing 41 to 50 mm.

**Range in Eastern Africa:** Eastern Abyssinia and British and Italian Somalilands, Kenya Colony and north-eastern Tanganyika Territory as far as Gelai, Ketumbeine and Mkomasi.

**Habits:** As for the last race, tame and fearless, the erect dark tail being very noticeable. A most systematic searcher of bushes for insects.

**Nest and Eggs:** A neat long pocket-like nest, open on a slant at the top and often decorated with cocoons or plant wool, hanging from twigs in a thin bush, usually amongst grass. Eggs four or five, white or greenish white, with reddish spots and flecks and pale undermarkings; about $14 \cdot 5 \times 11$ mm.

**Recorded breeding:** British Somaliland, January. Eastern Abyssinia, April and May.

**Call:** A clear chirping song 'tick-tick-tick-tick-tick-tick' and an alarm call of 'seep-seep.'

*Apalis rufifrons rufidorsalis* (Sharpe).
Dryodromas rufidorsalis Sharpe, Bull. B.O.C. 6, p. 48, 1897: Tsavo River, south-eastern Kenya Colony.

**Distinguishing characters:** Similar to the last race in having white edges and tips to the wing-coverts and inner secondaries, but has the chestnut of the head extending to the mantle and rump. Wing 49 mm.

**Range in Eastern Africa:** The Tsavo area of south-eastern Kenya Colony.

**Habits:** Those of the other races and in their movements very reminiscent of a *Cercomela*. The call is a loud ringing 'sippe-sippe' and a little bubbling trill 'spi-spi-hee-hee-hee' (Moreau).

**Recorded breeding:** No records.

### 988 KARAMOJA APALIS. *APALIS KARAMOJÆ* (Van Someren).

*Euprinoides karamojæ* Van Someren, Bull. B.O.C. 41, p. 120, 1921: Mt. Kamalinga, Karamoja, north-eastern Uganda.

**Distinguishing characters:** Above, ash grey, darker on rump; tail black, outer tail feathers wholly white; outer webs of innermost secondaries white; below, creamy white; bill long. The sexes are alike. Wing 50 to 51 mm. Juvenile plumage unrecorded.

**Range in Eastern Africa:** Eastern Uganda.

**Habits:** No information, apparently a rare and local species.

**Nest and Eggs:** Undescribed.

**Recorded breeding:** No records.

**Call:** Unrecorded.

### 989 LONG-BILLED APALIS. *APALIS MOREAUI* Sclater.

*Apalis moreaui moreaui* Scl. **Pl. 69.**

*Apalis moreaui* W. L. Sclater, Bull. B.O.C. 51, p. 109, 1931: Amani, Usambara Mts., north-eastern Tanganyika Territory.

**Distinguishing characters:** Bill long; above, olivaceous grey; forehead and lores faintly washed with tawny; tail grey; below, pale grey. The sexes are alike. Wing 42 to 47 mm. Juvenile plumage undescribed.

**Range in Eastern Africa:** Amani forests, Usambara Mts., north-eastern Tanganyika Territory.

**Habits:** A very inconspicuous bird except for its call note, keeping to the undergrowth of dense forest and diligently searching foliage for insects.

**Nest and Eggs:** Undescribed.

**Recorded breeding:** No records.

**Call:** Remarkable, a sound like a series of deliberate blows with a mallet on an iron peg (Moreau).

*Apalis moreaui sousæ* Bens.

*Apalis moreaui sousæ* Benson, Bull. B.O.C. 66, p. 19, 1945: Njesi Plateau, 10 miles north of Unangu, northern Portuguese East Africa.

**Distinguishing characters:** Differs from the nominate race in being slightly darker olivaceous above; forehead, lores and ear-coverts more chestnut brown; below, greyer, especially on chest and flanks. Wing 45 to 50 mm.

**Range in Eastern Africa:** Njesi Plateau, northern Portuguese East Africa.

**Habits:** An evergreen forest species, found in the forest canopy. It has a song call very like that of the Black-headed Apalis.

**Recorded breeding:** No records.

**990 KUNGWE APALIS.** *APALIS ARGENTEA* Moreau.
*Apalis argentea* Moreau, Bull. B.O.C. 61, p. 47, 1941: Kungwe Mt., western Tanganyika Territory.

**Distinguishing characters:** Upperside and sides of head grey; wings blackish with an olivaceous wash on edges of flight feathers; tail darker grey; below, white, including under wing-coverts; three outer tail feathers wholly white, fourth white with sooty edging to outer web. The sexes are alike. Wing 48 to 52 mm. The young bird differs from the adult in having the mantle and rump olivaceous.

**Range in Eastern Africa:** Kungwe-Mahare and Ubende areas, western Tanganyika Territory as far south as the Mpanga, Katuma and upper Nyamanzi Rivers.

**Habits:** Unrecorded, is said to inhabit the tops of forest trees or of tall trees on forest outskirts or in gallery forest.

**Nest and Eggs:** Undescribed.

**Recorded breeding:** No records.

**Call:** Said to be a 'pe-pe-pe-' rather like that of the Black-headed Apalis, but softer.

**991 RED-CAPPED FOREST-WARBLER.** *ARTISORNIS METOPIAS* (Reichenow). **Pl. 69.**
*Prinia metopias* Reichenow, O.M. p. 30, 1907: Usambara, northeastern Tanganyika Territory.

**Distinguishing characters:** Bill long and tail rather short; head and sides of face, neck and throat chestnut, darker on crown and occiput; mantle and rump olivaceous brown; wings and tail greyer; below, chin to belly white; flanks olivaceous brown. The sexes are alike. Wing 45 to 51 mm. The young bird is duller and with a wash of yellow on the throat, sides of neck and belly. In the colour and markings of the head this Warbler has a resemblance to the Bamenda Apalis, but it can be distinguished by its longer bill, olivaceous

brown instead of grey mantle, the shorter narrow feathered tail without light tips, and the black not yellowish brown feet.

**Range in Eastern Africa:** Tanganyika Territory from the Usambara Mts., to the Nguru Hills, Uluguru Mts., the Luwiri-Kitessi Forest, Songea district and Unangu, Portuguese East Africa.

**Habits:** Little known bird of dense undergrowth of forests, usually hunting for insects close to the forest floor. The call is freely heard but is ventriloquial, and as the birds move about without shaking the leaves, they are very difficult to see. Mr. Moreau, who, as far as we know, is the only person who has ever seen a nest, remarks on the correctness of Dr. Friedmann's deduction that this species is related to the Tailor-birds of India.

**Nest and Eggs:** Nest, an exquisitely delicate little cup of lichen scraps lined with horsehair-like mycelium, sewn on to leaves which had holes pierced in them with some moss-like silky substance of a yellowish-green colour. Eggs undescribed.

**Recorded breeding:** Usambara Mts., January.

**Food:** Insects.

**Call:** Various little soft calls. The song is a flat trill, like the latter part of a Wood Warbler's song, and preceded by two loud preliminary notes.

### 992 MRS. MOREAU'S WARBLER. *SCEPOMYCTER WINIFREDÆ* (Moreau). Pl. 70.

*Artisornis winifredæ* Moreau, Bull. B.O.C. 58, p. 139, 1938: Uluguru Mts., Morogoro district, eastern Tanganyika Territory.

**Distinguishing characters:** Adult male, head all round to nape and chest chestnut; mantle and rump dull olivaceous; wings and tail dusky edged with dull olivaceous; below, dull olivaceous paler than mantle and rump; centre of breast and upper belly rather buffish. The female differs from the male in having the chestnut paler. Wing 55 to 58 mm. The young bird is more olivaceous green above and paler olivaceous green below; forehead and crown mixed pale chestnut and olivaceous, sides of face and chin to upper chest pale tawny chestnut.

**Range in Eastern Africa:** Uluguru Mts., eastern Tanganyika Territory.

# WARBLERS

**Habits:** A little known species, usually seen singly or in pairs in dense evergreen undergrowth, moving about on the ground or very close to it.

**Nest and Eggs:** Undescribed.

**Recorded breeding:** No records.

**Food:** No information.

**Call:** The call is a long drawn out 'weeeee,' but a soft uninflected whistle, continually repeated, sometimes for nearly an hour on end, is also said to emanate from this bird.

## GREY-CAPPED WARBLER, RED-WINGED GREY WARBLER and BUFF-BELLIED WARBLER.   FAMILY—SYLVIIDÆ.

Genera: *Eminia*, *Drymocichla* and *Phyllolais*.

These are somewhat aberrant Warblers, which respectively prefer swampy woodland, savanna, and acacia country; the last may be easily overlooked as it is partial to the tops of acacia trees. Food, insects.

**993 GREY-CAPPED WARBLER.** *EMINIA LEPIDA* Hartlaub Ph. xi.

*Eminia lepida* Hartlaub, P.Z.S. for 1880, p. 625, pl. 60, fig. 1, 1881 Majungo, north end of Lake Albert, north-western Uganda.

**Distinguishing characters:** Above, forehead and crown grey surrounded by a black band from lores through eye to occiput; rest of upperparts bright green including wings and tail; wing shoulders chestnut; below, pale grey; a chestnut patch in centre of throat; under wing-coverts chestnut and buff; flanks green. The sexes are alike. Wing 63 to 75 mm. The young bird is slightly duller and has paler and rather less chestnut in centre of throat.

**Range in Eastern Africa:** Uganda to central Kenya Colony and north-eastern Tanganyika Territory; also Kome Island, south-western Lake Victoria.

**Habits:** Locally common over a fairly wide area in scrub or forest undergrowth, or among dense vegetation along stream beds. Rather shy little birds with a short but powerful song, now common in gardens at Entebbe and elsewhere, usually in pairs.

**Nest and Eggs:** Nest like a loosely made Sunbird's, domed, usually with a projecting porch over the entrance at the side of the top. It is made of dry grass-like fibre, leaves and feathers, usually with a fibre lining, and often has a draggled appearance like a mass of rubbish; it is usually suspended from or placed in creepers or dense vegetation. Eggs two or three, either white with an occasional purplish-brown spot or blotch, or glossy pale blue, long and ovate; about 19·5 × 13·5 mm.

**Recorded breeding:** Uganda, April to July. Kenya Colony, April to August, double brooded, and in some areas also in November and December.

**Call:** A repeated 'twi-twi' in a crescendo trill with some clear loud whistles, also a little 'tk-tk' uttered with head thrown back and tail cocked up.

### 994 RED-WINGED GREY WARBLER. *DRYMOCICHLA INCANA* Hartlaub.

*Drymocichla incana* Hartlaub, P.Z.S. for 1880, p. 626, pl. 60, fig. 2, 1881: Majungo, north end of Lake Albert, north-western Uganda.

**Distinguishing characters:** General colour pale grey; wings and tail darker; basal three-quarters of primaries tawny; below, paler grey, almost whitish; bill black. The sexes are alike. Wing 50 to 60 mm. The young bird is rather paler above and has the upper mandible dark brown, lower pale horn.

**General distribution:** Cameroons to the Sudan and Uganda.

**Range in Eastern Africa:** South-western Sudan and north-western Uganda.

**Habits:** Little recorded, apparently a bird of low woodland or of trees in well watered grass country.

**Nest and Eggs:** Undescribed.

**Recorded breeding:** Lake Albert (breeding condition), April to October.

**Call:** Said to have a short monotonous double-noted song.

### 995 BUFF-BELLIED WARBLER. *PHYLLOLAIS PULCHELLA* (Cretzschmar). **Pl. 70. Ph. xi.**

*Malurus pulchellus* Cretzschmar, Atlas zu der Reise, Vög. p. 53, pl. 35, 1830: Kordofan, central Sudan.

**Distinguishing characters:** Above, pale powdery olive green; below, pale creamy yellow; tail blackish, with white ends and white outer webs to outer feathers. The sexes are alike. Wing 42 to 49 mm. The young bird is rather darker above and not so delicately coloured below.

**General distribution:** Lake Chad to Eritrea, eastern Belgian Congo, Uganda, Kenya Colony and Tanganyika Territory.

**Range in Eastern Africa:** Western Sudan to Eritrea, Uganda, central Kenya Colony and northern Tanganyika Territory.

**Habits:** Charming little birds inhabiting the tops of small acacias in open woodland or bush, and locally common. In the non-breeding season they move in small close flying flocks of five or six and seem to feed exclusively in the tree tops.

**Nest and Eggs:** A closely woven purse-shaped nest mainly of vegetable down, hanging from trees and bushes at all heights up to thirty feet. Eggs two or three, pale greenish blue with reddish brown, grey brown, or chocolate spots at the larger end; about 14·5 × 10 mm.

**Recorded breeding:** Southern Kenya Colony, March and April. Darfur, probably June to September. Karamoja, Uganda, April to July.

**Call:** Song a sort of dry trill of 'zit-zit-zit-char-char-chip' (Van Someren), and a churring alarm call.

## CROMBECS or STUMP-TAILS. FAMILY—SYLVIIDÆ. Genus: *Sylvietta*.

Called Crombecs by the early Dutch settlers in South Africa because of their curved bills. Their very short tails are also a distinctive character. They are birds of the rather dryer but wooded country, and are usually seen searching trunks and branches for their insect food, the bill being especially adapted to probe the crannies and hollows in the bark. Seven species occur in Eastern Africa.

## 996 CROMBEC. *SYLVIETTA BRACHYURA* Lafresnaye.
*Sylvietta brachyura brachyura* Laf.
*Sylvietta brachyura* Lafresnaye, Rev. Zool. p. 258, 1839: Senegambia.

**Distinguishing characters:** Tail very short; above, pale grey with a slight brownish wash; stripe over eye pale brown; brownish grey streak from base of bill through eye to ear-coverts; below, tawny, paler on chin and centre of belly. The sexes are alike. Wing 51 to 58 mm. The young bird has tawny tips to the wing-coverts.

**General distribution:** Senegal to Eritrea and the Sudan.

**Range in Eastern Africa:** Northern Eritrea to the Sudan from the Red Sea coast to Waliko, Upper Nile, and the Sobat River.

**Habits:** A. L. Butler called this bird the 'Nuthatch Warbler' which is a very good description of it both in appearance and in its habit of running up a bough like a mouse. They are birds of dry bush country searching for insects along boughs in a very systematic manner. Usually solitary or in small parties and often members of a mixed bird party.

**Nest and Eggs:** A small deep cup or purse nest with the back higher than the front, made of grass or other material, bound with silk and usually ornamented with cocoons and lined with grass heads. It is suspended by its higher side to a twig, but is not roofed in and is frequently right out in the open. Eggs two, white with scattered spots of olive brown, rufous, and grey; about $17 \cdot 5 \times 12$ mm.

**Recorded breeding:** French Sudan (breeding condition), June and July. Northern Nigeria, June. Sudan, April to June.

**Call:** A varied shivering little song something like that of a Willow Warbler but shorter. Alarm note a sharp clicking 'chit-chit.'

*Sylvietta brachyura leucopsis* Reichw. **Pl. 70.**
*Sylvietta leucopsis* Reichenow, O.C. p. 114, 1879: Kibaradja, Tana River, eastern Kenya Colony.

**Distinguishing characters:** Differs from the nominate race in having the stripe over the eye, chin and throat white. Wing 49 to 60 mm.

**Range in Eastern Africa:** Southern Eritrea as far north as Bogos, to Abyssinia, south-eastern Sudan, the Somalilands, central and eastern Kenya Colony and north-eastern Tanganyika Territory.

**Habits:** As for the nominate race.

**Recorded breeding:** Southern Abyssinia, April. Kenya Colony, February to April. Tanganyika Territory, November.

*Sylvietta brachyura dilutior* Reichw.
*Sylvietta carnapi dilutior* Reichenow, O.M. p. 154, 1916: Ruwenzori Mts., Uganda.

**Distinguishing characters:** Similar to the nominate race, but greyer above and rather warmer tawny below. Wing 51 to 60 mm.

**Range in Eastern Africa:** South-eastern Sudan as far west as

Gondokoro, Kajo Kaji and Mongalla, to Uganda and western Kenya Colony in the eastern Trans Nzoia area.

**Habits:** As for the nominate race, found exclusively in acacias.

**Recorded breeding:** Sudan, May and June, also (nestling) October. Uganda, February and March.

**Distribution of other races of the species:** West Africa.

## 997 RED-FACED CROMBEC. *SYLVIETTA WHYTII* Shelley.

*Sylvietta whytii whytii* Shell. Pl. 70.
*Sylviella whytii* Shelley, Ibis, p. 13, 1894: Zomba, southern Nyasaland.

**Distinguishing characters:** Above, grey; sides of face and underside pale buffish brown; throat greyish caused by base of feathers showing through the tawny brown colour; no streak through eye. The sexes are alike. Wing 54 to 61 mm. The young bird is tawny grey above and with slight tawny tips to wing-coverts.

**General distribution:** Portuguese East Africa to southern Nyasaland and eastern Southern Rhodesia.

**Range in Eastern Africa:** Portuguese East Africa and Nyasaland.

**Habits:** Common in the highlands of Nyasaland in open bush or woodland searching branches for insects, but also hunting among fine twigs and leaves like a *Phylloscopus*. Frequent members of a mixed bird party.

**Nest and Eggs:** A deep purse-like nest fastened by its higher side to a bough or twig, usually in a shady place and often attached to a pendent bough and made of fine grass ornamented with rotten wood or other material, bound on with cobwebs and with a lining of fine grass and rootlets. It looks very like some old insect nest. Eggs two, white, finely freckled with reddish and pale brown, often in the form of a zone with some grey undermarkings; about 19 × 12 mm.

**Recorded breeding:** Rhodesia and Nyasaland, September to December. Portuguese East Africa, October and November.

**Call:** The alarm note is a sharp 'tip' uttered so continuously as to become a penetrating rattling trill. There is also a small twittering call 'si-si-si-see.'

*Sylvietta whytii jacksoni* Sharpe. **Ph. xii.**
*Sylviella jacksoni* Sharpe, Bull. B.O.C. 7, p. 7, 1897: Kamassia, western Kenya Colony.

**Distinguishing characters:** Similar to the nominate race but differs in having the sides of the face and the underside warm tawny brown. Wing 56 to 66 mm.

**General distribution:** Uganda, Kenya Colony and Tanganyika Territory to Nyasaland, north of lat. 12° S.

**Range in Eastern Africa:** Southern and eastern Uganda to Kenya Colony and Tanganyika Territory, but not coastal areas nor the north-western area.

**Habits:** Of the genus, generally found in fairly dense bush or river fringing forest strips.

**Recorded breeding:** Uganda, June and July. Kenya Colony, March to June, also December. Tanganyika Territory, October to January, also April.

*Sylvietta whytii minima* O. Grant.
*Sylviella minima* O. Grant, Ibis, p. 156, 1900: Manda Island, Kenya Colony.

**Distinguishing characters:** Similar to the nominate race, but general colour paler. Wing 52 to 60 mm.

**Range in Eastern Africa:** Coastal areas of Kenya Colony and Tanganyika Territory as far west as Maji-ya-Chumvi, Ngomeni and Morogoro, also Manda Island.

**Habits:** As for other races.

**Recorded breeding:** No records.

*Sylvietta whytii abayensis* Mearns.
*Sylvietta whytii abayensis* Mearns, Smiths, Misc. Coll. 61, No. 20, p. 4, 1913: Gato River, Lake Abaya, southern Abyssinia.

**Distinguishing characters:** Very similar to the Kamassia race but upperside more olive brown. Wing 56 to 63 mm.

**Range in Eastern Africa:** South-eastern Sudan, southern Abyssinia and north-western Kenya Colony.

**Habits:** As for other races.

**Recorded breeding:** No records.

# WARBLERS

**998 LONG-BILLED CROMBEC.** *SYLVIETTA RUFESCENS* (Vieillot).

*Sylvietta rufescens pallida* Alex.

*Sylviella pallida* Alexander, Bull. B.O.C. 8, p. 48, 1899: between Tete and Chicowa, Zambesi River, Portuguese East Africa.

**Distinguishing characters:** Similar to the Red-faced Crombec, but differs in having a longer, heavier and slightly more down-curved bill; in having the ear-coverts greyish, thus showing an eye-stripe over eye, especially behind the eye; and in having the throat uniform with the rest of the underparts, showing no grey. The sexes are alike. Wing 54 to 62 mm. The young bird is whiter below.

**General distribution:** Northern Rhodesia and Nyasaland to Damaraland and Bechuanaland.

**Range in Eastern Africa:** Nyasaland.

**Habits:** Almost indistinguishable in the field from the Red-faced Crombec and of similar habits, but as a rule prefers low bushes in hilly country.

**Nest and Eggs:** Nest like that of the Red-faced Crombec, but distinctly larger and usually quite low down. Eggs two, white, often heavily marked with brown or reddish spots and with lilac grey undermarkings; about $19 \times 12 \cdot 5$ mm.

**Recorded breeding:** Northern Rhodesia, October. Nyasaland, September to April. Southern Rhodesia, October to December.

**Call:** A shrill alarm call of three notes, with the emphasis on the first syllable and rather distinctive.

**Distribution of other races of the species:** Belgian Congo to South Africa, the nominate race being described from western Cape Province.

**999 SOMALI LONG-BILLED CROMBEC.** *SYLVIETTA ISABELLINA* Elliott.

*Sylviella isabellina* Elliott, Field-Columbian Mus. Publ. Orn. ser. 1, p. 44, 1897: Le Gud, Ogaden, eastern Abyssinia.

**Distinguishing characters:** Above, pale grey; slight whitish stripe over eye; bill much longer than the Red-faced Crombec; below, pale isabelline, paler on throat. The sexes are alike. Wing 53

to 63 mm. The young bird is duller than the adult; throat slightly mottled.

**Range in Eastern Africa:** Eastern and southern Abyssinia to British Somaliland, Italian Somaliland and south-eastern Kenya Colony.

**Habits:** A bird of dry acacia steppe and bush country of which little is recorded.

**Nest and Eggs:** Nest as for other species, hanging from a twig of an acacia bush, bottle-shaped with a slanting opening at the top, decorated outside with spiders' webs and plant down and lined with fibre and fine grass. Eggs two, white with considerable olive brown freckling and with ashy grey undermarkings; about 17·5 × 12 mm.

**Recorded breeding:** Southern Abyssinia, March.

**Call:** Unrecorded.

**1000** RED-CAPPED CROMBEC. *SYLVIETTA RUFICAPIL-LA* Bocage.

*Sylvietta ruficapilla chubbi* O. Grant.

*Sylviella chubbi* O. Grant, Bull. B.O.C. 27, p. 10, 1910: Broken Hill, Northern Rhodesia.

**Distinguishing characters:** Above, pale grey; forehead and crown of head slightly brownish; lores and behind eye, behind ear-coverts and underside white; ear-coverts and patch on upper chest chestnut. The sexes are alike. Wing 63 to 70 mm. The young bird has the forehead and top of head more distinctly brownish.

**General distribution:** South-eastern Belgian Congo to Northern Rhodesia, Nyasaland and Portuguese East Africa.

**Range in Eastern Africa:** Not yet recorded, but should occur in south-western areas.

**Habits:** Apparently those of other species of the genus.

**Nest and Eggs:** Nest similar to that of the last species. Eggs two, white with a zone of sepia spots; about 18 × 11·5 mm.

**Recorded breeding:** Northern Rhodesia, August to November. Nyasaland, October.

**Call:** Loud and ringing: 'richi, chichi, chichir' repeated about half a dozen times.

# WARBLERS

**1001 GREEN CROMBEC.** *SYLVIETTA VIRENS* Cassin.
*Sylvietta virens baraka* Sharpe. **Pl. 70.**
*Sylviella baraka* Sharpe, Bull. B.O.C. 7, p. 6, 1897: Entebbe, southern Uganda.

**Distinguishing characters:** Above, olive green; head browner; a faint brown stripe over eye; sides of face and chin to chest dusky brown; rest of underparts dusky grey; centre of belly whitish; under wing-coverts pale yellow. The sexes are alike. Wing 47 to 53 mm. The young bird is rather paler than the adult and has a whitish throat.

**General distribution:** Eastern Belgian Congo to the Sudan and Uganda.

**Range in Eastern Africa:** Southern Sudan and Uganda.

**Habits:** Not uncommon locally among dense undergrowth at the edge of forest or in gardens. Habits practically unrecorded but probably those of other members of the genus.

**Nest and Eggs:** Nest pear-shaped and suspended from a twig, strongly and compactly made of grasses and fibres woven with cobwebs and ornamented with cocoons, bark, etc. Eggs two, white or very pale blue faintly spotted with rufous; about $16 \cdot 5 \times 11$ mm.

**Recorded breeding:** Semliki River, Belgian Congo, July. Uganda, February to June.

**Call:** The usual call is of three notes followed by a trill and is rather high-pitched.

**Distribution of other races of the species:** West Africa to Angola, the nominate race being described from Gabon.

**1002 WHITE-BROWED CROMBEC.** *SYLVIETTA LEUCOPHRYS* Sharpe.
*Sylvietta leucophrys leucophrys* Sharpe. **Ph. xi.**
*Sylviella leucophrys* Sharpe, Ibis, p. 120, 1891: Mt. Elgon, Uganda-Kenya Colony boundary.

**Distinguishing characters:** Above, top of head and ear-coverts chocolate brown; mantle grey and olivaceous; edges of wing feathers bright green; a broad white stripe over eye; sides of face and chin to throat white; chest to belly grey; under wing-coverts yellow. The sexes are alike. Wing 55 to 63 mm. The young bird has more brown on the sides of the face; chin, throat and underside brownish grey; centre of belly whitish.

**Range in Eastern Africa:** Uganda (except south-western areas) to central Kenya Colony.

**Habits:** A bird of dense undergrowth and rank herbage found either in forest or outside it.

**Nest and Eggs:** Nests of hanging purse shape slung from forks of shrubs or between grasses or plant stems about three feet from the ground, built of soft grass, cobwebs and moss, and lined with tendrils or a few feathers. Eggs two, white and glossy with fine rusty red spots rather tending to form a cap; also according to Van Someren, a beautiful pink sparsely blotched and spotted with maroon; about 15 × 12 mm.

**Recorded breeding:** Mt. Elgon, March and April. Kenya highlands, March to June also December and January.

**Call:** A trilling 'chit-chit-chirrrrrr' (Van Someren).

*Sylvietta leucophrys chloronota* Hart.
*Sylvietta leucophrys chloronota* Hartert, Nov. Zool. 27, p. 460, 1920: North-west of Baraka, eastern Belgian Congo.

**Distinguishing characters:** Similar to the nominate race but yellowish-green above; top of head and ear-coverts brighter chestnut. Wing 55 to 59 mm.

**General distribution:** Eastern Belgian Congo to south-western Uganda and western Tanganyika Territory.

**Range in Eastern Africa:** South-western Uganda to Kungwe-Mahare Mts., western Tanganyika Territory.

**Habits:** As for the nominate race. It occurs on the Ruwenzori Mts. at least up to 8,500 feet.

**Recorded breeding:** No records.

**EREMOMELAS.** FAMILY—**SYLVIIDÆ.** Genus: *Eremomela*.

Six species occur in Eastern Africa. Warbler-like birds, active in habits, frequenting wooded country, though not apparently either forest or thick woods.

Their general colour is similar to the Willow and Wood Warblers and for them they might be mistaken, though their black bills and brighter coloration should distinguish them. They feed entirely on insects as far as is known.

The Brown-crowned Eremomela has a similar chest band to that of many species of the *Apalis* group.

## 1003 YELLOW-BELLIED EREMOMELA. *EREMOMELA ICTEROPYGIALIS* (Lafresnaye).

*Eremomela icteropygialis griseoflava* Heugl. Pl. 70.
*Eremomela griseoflava* Heuglin, J.f.O. p. 40, 1862: Bogosland, Eritrea.

**Distinguishing characters:** Above, ash grey; tail short; a dusky streak through eye, and a faint white stripe over eye; below, chin to upper belly white or buffish white; sides of chest ash brown; lower belly yellow; under tail-coverts yellow and white. The sexes are alike. Wing 50 to 56 mm. The young bird is duller than the adult; mantle often washed with greenish; less yellow on belly.

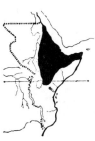

**Range in Eastern Africa:** Eritrea and the eastern and southern Sudan, Abyssinia, British Somaliland, Italian Somaliland as far south as lat. 9° N. and western, northern and south-western Kenya Colony.

**Habits:** Birds of dry open bush, found well out in true desert country as well as near water, generally not more than two or three together. Habits very Phylloscopine, searching twigs and flowers for insects and their larvæ, restless and untiring. Common among *Calotropis* (Sodom Apple) bushes.

**Nest and Eggs:** A small and deep, but delicate, cup of grass and cobwebs, often semi-transparent, slung by its rim to two twigs or the fork of a bough. Eggs two or three, white and unglossed, with dark red, occasionally blackish, spots and small grey undermarkings; about 15 × 11 mm.

**Recorded breeding:** Southern Abyssinia, March to May and October and November. Kenya Colony, February to April.

**Call:** A chittering jingling little song of four syllables—also a plaintive call 'see-see,' and a scolding 'chee-chiri-chea-chiri-chit.'

*Eremomela icteropygialis polioxantha* Sharpe.
*Eremomela polioxantha* Sharpe, Cat. Bds. B.M. 7, p. 160, 1883: Swaziland, South Africa.

**Distinguishing characters:** Differs from the preceding race in being darker above; chin to chest greyish white; yellow of the belly extending nearly to the breast. Wing 53 to 63 mm.

**General distribution:** Tanganyika Territory to Nyasaland, Northern Rhodesia, southern Belgian Congo, eastern Southern Rhodesia and eastern Transvaal.

**Range in Eastern Africa:** South-central to south-western Tanganyika Territory, Portuguese East Africa and Nyasaland.

**Habits:** As for the preceding race but not found in desert.

**Recorded breeding:** Tanganyika Territory, November to January. Belgian Congo, September. Nyasaland, October.

*Eremomela icteropygialis abdominalis* Reichw.
*Eremomela flaviventris abdominalis* Reichenow, Vög. Afr. 3, p. 635, 1905: Igonda, Tabora district, central Tanganyika Territory.

**Distinguishing characters:** Differs from the preceding race in having a rather smaller bill and no whitish stripe over eye. Wing 50 to 57 mm.

**Range in Eastern Africa:** Tanganyika Territory from Longido to the Tabora district.

**Habits:** As for the preceding race, but found in woodland and thick bush as well as in open country. Inconspicuous little birds chiefly active in the morning and evening feeding among branches like Tits.

**Recorded breeding:** No records.

*Eremomela icteropygialis alexanderi* Scl. & Praed.
*Eremomela flaviventris alexanderi* W. Sclater & Mackworth-Praed, Ibis, p. 673, 1918: Bara, Kordofan, Sudan.

**Distinguishing characters:** Similar to the Eritrean race but upperparts and sides of the chest paler ash grey. Wing 49 to 57 mm.

**General distribution:** French Sudan and Lake Chad area to Kordofan, Sudan.

**Range in Eastern Africa:** Western and southern Sudan from Darfur to Kordofan and the While Nile.

**Habits:** As for other races.

**Recorded breeding:** Darfur, June and July.

*Eremomela icteropygialis belli* Grant & Praed.
*Eremomela griseoflava belli* C. Grant & Mackworth-Praed, Bull. B.O.C. 67, p. 44, 1947: Liwale, south-eastern Tanganyika Territory.

**Distinguishing characters:** Differs from the other races in having the chin to throat white, with the yellow of the belly extending in a wash over the lower neck in front. Wing 68 mm.

**Range in Eastern Africa:** Liwale area of south-eastern Tanganyika Territory.

**Habits:** As for other races.

**Recorded breeding:** No records.

**Distribution of other races of the species:** South Africa.

**1004 YELLOW-VENTED EREMOMELA.** *EREMOMELA FLAVICRISSALIS* Sharpe.

*Eremomela flavicrissalis* Sharpe, P.Z.S. p. 48, 1895: Webi Shebeli, southern Abyssinia.

**Distinguishing characters:** Similar to the Yellow-bellied Eremomela, but smaller and the yellow confined to the lower belly. The sexes are alike. Wing 48 to 51 mm. The young bird has the lower belly white or only slightly tinged with yellow.

**Range in Eastern Africa:** British Somaliland to southern Italian Somaliland and north-eastern and eastern Kenya Colony as far south as the northern Guaso Nyiro and the Tana River.

**Habits:** Those of the last species, found in thin open bush of dry or desert country. Little is recorded of its characteristics, except that it is tame, and sits very close when breeding.

**Nest and Eggs:** The nest is said to be much shallower than that of other Eremomelas, a flattish cup hanging from a forked twig of low acacia bushes. It is made of the usual materials and ornamented with cocoons and spiders' webs. Eggs two, white with a few dark brown or sepia spots and fine speckling of the same colour; about 14 × 11 mm.

**Recorded breeding:** Southern Abyssinia, May. Eastern Kenya Colony (breeding condition), November.

**Call:** A low piping call.

**1005 GREEN-BACKED EREMOMELA.** *EREMOMELA CANESCENS* Antinori.

*Eremomela canescens canescens* Ant. **Pl. 70.**

*Eremomela canescens* Antinori, Cat. Coll. Ucc. p. 38, 1864: Djur River, Bahr-el-Ghazal, south-western Sudan.

**Distinguishing characters:** Head pale ashy grey; rest of upperparts bright greenish yellow; a black streak from base of bill through eye to ear-coverts; below, chin to upper chest white; chest to under

tail-coverts and under wing-coverts canary yellow. The sexes are alike. Wing 51 to 59 mm. The young bird has the mantle slightly olivaceous.

**Range in Eastern Africa:** Southern Sudan, Uganda and western Kenya Colony.

**Habits:** Common in small parties in scrub and acacia country and extending into woodland or thick bush on the slopes of hills. Habits of the genus, searching for insects restlessly among bushes or low shrubs and generally shy.

**Nest and Eggs:** Nest a cup of leaves, twigs and stems bound together with cobwebs and suspended from a fork or from between two twigs. Eggs one or two, bright bluish-green with a zone of brown spots at the larger end, about 14 × 10·5 mm.

**Recorded breeding:** Western Kenya Colony, May and June.

**Call:** A rather unmusical chittering song. A buzzing chirp is also recorded for allied races.

*Eremomela canescens elegans* Heugl.
*Eremomela elegans* Heuglin, J.f.O. p. 259, 1864: Sennar-Abyssinian boundary.

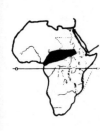

**Distinguishing characters:** Differs from the nominate race in being paler above and having a slight white stripe over eye. Wing 50 to 57 mm.

**General distribution:** Cameroons to the Sudan.

**Range in Eastern Africa:** The Sudan from Darfur to Sennar.

**Habits:** A woodland species confined to the better wooded valleys and usually in trees not bushes, otherwise its habits are those of the genus.

**Recorded breeding:** Darfur, May to July.

*Eremomela canescens abyssinica* Bann.

*Eremomela elegans abyssinica* Bannerman, Bull. B.O.C. 29, p. 38, 1911: Omo River, south-western Abyssinia.

**Distinguishing characters:** Differs from the preceding races in being darker above, more olivaceous, less yellow; no stripe over eye. Wing 54 to 59 mm.

**Range in Eastern Africa:** Central Eritrea and Abyssinia to the Sobat Valley of the eastern Sudan as far as Ayod, south-eastern Upper Nile Province.

**Habits:** Little recorded and probably not differing from those of other races. They are described as inhabiting the tops of tall trees in valleys and thick scrub along water courses.

**Recorded breeding:** Southern Abyssinia, February, probably other times of year also.

**Distribution of other races of the species:** West Africa.

### 1006 GREEN-CAP EREMOMELA. *EREMOMELA SCOTOPS* Sundevall.

*Eremomela scotops scotops* Sund. **Pl. 71.**

*Eremomela scotops* Sundevall, Œfv. K. Sv.Vet.-Akad. Förh. 7, p. 103, 1850: Mohapoani, Witfontein Mts., western Transvaal.

**Distinguishing characters:** Above, pale ash grey with a slight greenish wash; forehead to crown and sides of face pale green; below, chin white; rest of underparts lemon yellow; under wing-coverts white. The sexes are alike. Wing 53 to 61 mm. The young bird is paler.

**General distribution:** Tanganyika Territory to northern Portuguese East Africa, southern Nyasaland, Southern Rhodesia, Bechuanaland and the Transvaal.

**Range in Eastern Africa:** Songea and Liwale districts of Tanganyika Territory to Portuguese East Africa and southern Nyasaland.

**Habits:** Common inhabitants of open woodland, usually in small groups and often with a mixed bird party. Restless and noisy little birds always chasing and fighting with each other as they hunt for insects among the leafy boughs of the higher trees.

**Nest and Eggs:** A delicate neatly made cup nest suspended or attached by its rim, made of pieces of leaf and flowers bound together and lined with white silky threads of vegetable matter. Eggs two or three, pale blue lightly spotted with rufous and with lilac undermarkings; about 16 × 11·5 mm.

**Recorded breeding:** Nyasaland, September to December.

**Call:** A throaty twittering call 'nyum-nyum-nyum' also a rasping alarm note and some attempt at a liquid song of five or six notes (Swynnerton). Benson records a repeated monotonous 'chip.'

*Eremomela scotops pulchra* (Boc.).
*Tricholais pulchra* Bocage, Jorn. Lisboa, 6, p. 257, 1878: Caconda, Angola.

**Distinguishing characters:** Differs from the nominate race in having the breast to belly white or faintly washed with yellow. Wing 52 to 64 mm.

**General distribution:** Southern Angola and southern Belgian Congo to Northern Rhodesia and Nyasaland.

**Range in Eastern Africa:** South Nyasa district west of Fort Johnston and Blantyre district, Nyasaland.

**Habits:** As for the nominate race.

**Recorded breeding:** Nyasaland, October.

*Eremomela scotops citriniceps* (Reichw.).
*Tricholais citriniceps* Reichenow, J.f.O. p. 210, 1882: Kakoma, Tabora district, west-central Tanganyika Territory.

**Distinguishing characters:** Differs from the nominate race in having the breast to belly whiter; green of head brighter and extending to nape. Wing 58 to 63 mm.

**General distribution:** Eastern Belgian Congo to Uganda, Kenya Colony and Tanganyika Territory.

**Range in Eastern Africa:** Uganda to western Kenya Colony and western and central Tanganyika Territory as far south as the Ufipa Plateau and Iringa district.

**Habits:** As for the other races, common in upland country. They are said to have a short loud trilling call and a monotonous double syllabled breeding call.

**Recorded breeding:** Tanganyika Territory, October to December.

*Eremomela scotops occipitalis* (Fisch. & Reichw.).
*Tricholais occipitalis* Fischer & Reichenow, J.f.O. p. 181, 1884: Pangani, north-eastern Tanganyika Territory.

**Distinguishing characters:** Similar to the nominate race in having the whole underside pale yellow, but this colour is brighter on chin to chest; top of head also similar to that race. Wing 50 to 59 mm.

**Range in Eastern Africa:** Central Kenya Colony from Nairobi to eastern Tanganyika Territory as far south as Kilosa.

**Habits:** As for the other races.

**Recorded breeding:** No records.

**Distribution of other races of the species:** Lower Congo Valley, Belgian Congo and northern Angola.

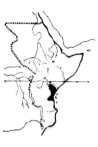

### 1007 BROWN-CROWNED EREMOMELA. *EREMOMELA BADICEPS* Fraser.

*Eremomela badiceps badiceps* Fras.
*Sylvia badiceps* Fraser, P.Z.S. p. 144, 1842: Clarence, Fernando Po.

**Distinguishing characters:** Adult male, forehead and crown chestnut; mantle and rump grey; a black stripe from lores under eye to lower ear-coverts; below, chin and throat creamy white; a black band across lower neck; chest to belly grey, sometimes creamy white down centre. The sexes are alike, but the female does not usually have the chestnut extending so far towards the nape. Wing 52 to 58 mm. The young bird is olive green above, with a wash of chestnut on the head; below, pale yellow with an ill-defined neck band.

**General distribution:** Fernando Po to Uganda and northern Angola.

**Range in Eastern Africa:** Bwamba valley of western Uganda.

**Habits:** Pretty little active birds inhabiting tree-tops of secondary forest growth, feeding on insects and chasing each other with twittering cries.

**Nest and Eggs:** Undescribed.

**Recorded breeding:** Belgian Congo (breeding condition), February, April and November. Western Uganda (breeding condition), July to September.

**Call:** As above.

*Eremomela badiceps turneri* V. Som.
*Eremomela badiceps turneri* Van Someren, Bull. B.O.C. 40, p. 92, 1920: Yala River, Nandi, western Kenya Colony.

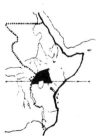

**Distinguishing characters:** Differs from the nominate race in having the chestnut confined to the forehead, but extending back as a superciliary stripe. Wing 48 mm.

**Range in Eastern Africa:** Uganda to western Kenya Colony.

**Habits:** Unrecorded: apparently local and rare.

**Recorded breeding:** Uganda (breeding condition), July and August.

*Eremomela badiceps latukæ* Hall.

*Eremomela badiceps latukæ* Hall, Bull. B.O.C. 69, p. 76, 1949: Katire, Imatong Mts., southern Sudan.

**Distinguishing characters:** Differs from the nominate race in having the chestnut of the head less bright, not extending to the nape. Wing 52 to 57 mm.

**Range in Eastern Africa:** Southern Sudan.

**Habits:** As for other races.

**Recorded breeding:** No records.

**1008** BURNT-NECK EREMOMELA. *EREMOMELA USTICOLLIS* Sundevall.

*Eremomela usticollis rensi* Ben. **Pl. 70.**

*Eremomela (Magalilais) usticollis rensi* Benson, Ostrich, 13, p. 241, 1943: Fort Johnston, southern Nyasaland.

**Distinguishing characters:** Above, ash grey, including wings and tail; below, pale creamy buff; throat whitish; usually a rusty short streak below ear-coverts and a rusty bar on chest, but often only indicated. The sexes are alike. Wing 50 to 54 mm. The young bird is similar to the adult.

**General distribution:** Portuguese East Africa and Nyasaland.

**Range in Eastern Africa:** Nyasaland.

**Habits:** A local species within our area found in pairs or small parties among acacias.

**Nest and Eggs:** Nest a shallow cup in a tree or bush at some height from the ground made of spiders' webs and other insect made material. Egg one, white, sparingly spotted with pale brown; about 17 × 11·5 mm.

**Recorded breeding:** Nyasaland, March, April and November.

**Call:** A thin musical 'di-di-di-di' (Benson) or 'dyup-dyup-dyup-dyup.'

**Distribution of other races of the species:** South Africa, the nominate race being described from the Transvaal.

## WARBLERS

### CAMAROPTERAS or GLASS-EYES and SOCOTRA WARBLER.

Family—SYLVIIDÆ. Genera: *Camaroptera and Incana*.

Four species occur in Eastern Africa. They inhabit most types of wooded country and are frequently seen in gardens and cultivated areas, feeding on insects. Some of the races have a distinctive breeding and non-breeding dress; some have a perennial dress and but one complete moult a year. The Socotra Warbler is an insular species formerly placed among the Cisticolas.

**1009** GREEN-BACKED CAMAROPTERA. *CAMAROPTERA BRACHYURA* (Vieillot). 
*Camaroptera brachyura pileata* Reichw. **Pl. 71.**
*Camaroptera pileata* Reichenow, J.f.O. p. 66, 1891: Zanzibar Island.

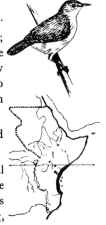

**Distinguishing characters:** Forehead to crown dark grey; mantle, rump, wings and tail olive green; sides of face and underside pale grey; centre of chest to belly whitish; under wing-coverts yellow and white; legs yellowish brown. The sexes are alike. Wing 52 to 59 mm. The young bird has the forehead and crown green uniform with the mantle and is washed with yellow below.

**Range in Eastern Africa:** Coastal area of Kenya Colony and Tanganyika Territory, also Zanzibar and Mafia Islands.

**Habits:** Common in small patches of scrub forest or coastal thickets, and on Zanzibar Island in any type of country where sufficient cover exists. A curious presumably nuptial flight is described by Pakenham, the bird swerving, swooping, and diving, all the time uttering a quick little snapping or clicking noise. There is a curious four-syllabled call uttered from a thicket 'tchweep' rather bleated, also a slight reedy note.

**Nest and Eggs:** See under other races.

**Recorded breeding:** Coastal Tanganyika Territory (breeding condition), October and January.

*Camaroptera brachyura bororensis* Gunn. & Rob.
*Camaroptera brachyura bororensis* Gunning & Roberts, Ann. Trans. Mus. 3, p. 117, 1911: Ngamwe, Boror, Quelimane Province, northern Portuguese East Africa.

**Distinguishing characters:** Differs from the preceding race in being paler, whiter, below. Wing 52 to 60 mm.

**General distribution:** Southern Belgian Congo between Lualaba and Kaluli Rivers to southern Nyasaland, Portuguese East Africa as far south as Inhambane and eastern Southern Rhodesia.

**Range in Eastern Africa:** Portuguese East Africa from the Lurio River to Nyasaland.

**Habits:** Widely distributed in low country and on the central plateau of Portuguese East Africa, wherever there is open country with thickets and dense undergrowth in patches. In Nyasaland it also inhabits dense patches of undergrowth, but is more local in distribution. A secretive and skulking species, rarely observable, but with a remarkable call.

**Nest and Eggs:** A deep or semi-domed cup of fine grass or of a white silky textured substance with leaves sewn on to its sides, usually well concealed. Eggs two to four, white, or white with small reddish spots; about 16·5 × 12·5 (South African measurements).

**Recorded breeding:** Nyasaland, January to April, also October.

**Call:** A bleated kid-like 'maa-maa' very ventriloquial is presumably the alarm note, but it is freely uttered when the bird has no reason to be alarmed. There is also a loud clucking noise in the breeding season as of two stones struck together, made apparently by the male.

*Camaroptera brachyura fuggles-couchmani* Mor.

*Camaroptera brachyura fuggles-couchmani* Moreau, Bull. B.O.C. 60, p. 15, 1939: Kibungo Forest, Uluguru Mts., eastern Tanganyika Territory.

**Distinguishing characters:** Differs from the coastal race in having a dusky throat and the lower flanks olive. Wing 47 to 55 mm.

**General distribution:** Tanganyika Territory to northern Nyasaland as far south as Kota-kota.

**Range in Eastern Africa:** The Uluguru Mts. to Mahenge, south-eastern Tanganyika Territory.

**Habits:** Presumably as for other races, but little recorded.

**Recorded breeding:** No records.

**Distribution of other races of the species:** South Africa, the nominate race being described from Knysna, Cape Province.

# WARBLERS

**1010 OLIVE-GREEN CAMAROPTERA.** *CAMAROPTERA CHLORONOTA* Reichenow.

*Camaroptera chloronota toroensis* (Jacks.). **Pl. 71.**
*Sylviella toroensis* Jackson, Bull. B.O.C. 15, p. 38, 1905: Kibera River, Toro, western Uganda.

**Distinguishing characters:** Above, dull olive green; below, chin, throat and belly white or dusky white; chest and flanks dusky and buff. The sexes are alike. Wing 50 to 57 mm. The young bird is olivaceous and yellow below, and has little or no green edges to the tail feathers.

**General distribution:** Eastern Belgian Congo to Kenya Colony.

**Range in Eastern Africa:** Southern half of Uganda and western Kenya Colony.

**Habits:** Not uncommon locally among thick undergrowth in most of the Uganda forests. It is described as very noisy but practically nothing is recorded of its life history.

**Nest and Eggs:** Nest (of nominate race) formed by uniting the leaves of a spray as a framework, with two or three other leaves as a roof. Nest of soft white down with a few spiders' webs. Eggs normally two, pale greenish blue ground colour, well marked with spots and small blotches of pale brown and lilac; about $17 \times 12 \cdot 5$ mm.

**Recorded breeding:** No records.

**Call:** A short whistled note repeated for minutes at a time (Chapin), also a plaintive 'wheet-wheet,' clearly distinguishable to an expert ear from the call of other Camaropteras.

**Distribution of other races of the species:** West Africa, the nominate race being described from Togoland.

**1011 GREY-BACKED CAMAROPTERA.** *CAMAROPTERA BREVICAUDATA* (Cretzschmar).

*Camaroptera brevicaudata brevicaudata* (Cretz.). **Pl. 71. Ph. xii.**
*Sylvia brevicaudata* Cretzschmar, Atlas zu der Reise, Vög. p. 53, pl. 35, fig. b, 1831: Kordofan, Sudan.

**Distinguishing characters:** Head, mantle and rump grey; scapulars, wing-coverts and edges of flight feathers green; tail ashy with an indistinct subterminal bar and indistinct barring; below, pale grey; centre of belly whitish; legs buffish or yellowish brown; under wing-coverts yellow and white; bill black. In non-breeding

dress (October to April or May) the grey of the upperparts is replaced by ashy brown and the colour of the underparts is much paler, in some cases white with buffish chest and flanks; bill, upper mandible black, lower pale horn. The sexes are alike. Wing 50 to 61 mm. The young bird is very similar to the adult in non-breeding dress, but is paler and has some yellow on the chest and belly.

**General distribution:** Senegal to Portuguese Guinea, the Sudan and Abyssinia.

**Range in Eastern Africa:** Darfur, Kordofan and the White Nile areas of the Sudan to north-western Abyssinia, north of Lake Tana.

**Habits:** Very common inquisitive little birds, inhabiting any type of country but generally seen creeping among the denser patches of bush and behaving very much like a European Hedge-Sparrow. It may also be found in woodland, forest, or in bush patches far out in the desert. The tail is often carried at a sharp angle over the back.

**Nest and Eggs:** A well made purse-shaped nest of fine grass, fibre and cocoons placed between sewn together leaves of some large foliaged plant or shrub, and usually placed in a dense thicket near the ground. It may also have a leaf or leaves sewn on as a roof. Eggs two or three, generally two, white and glossy, spotted with rusty brown and lilac; about 16 × 12 mm.

**Recorded breeding:** Nigeria, July to September. Darfur, July and August.

**Call:** A querulous little note of 'squee' or 'bzeeb' continuously uttered from which it gets its name of 'Bush-goat.' Other calls are a low broken whistle, a loud 'chink' like pebbles struck together and a sharp 'psitt' (Elliott). There is also a short pleasant little song, rarely heard.

*Camaroptera brevicaudata tincta* (Cass.).
*Syncopta tincta* Cassin, Proc. Ac. Nat. Sci. Philad. p. 325, 1855: Moonda River, Gabon.

**Distinguishing characters:** Differs from the nominate race in being very much darker grey both above and below, and in having no non-breeding dress. Wing 48 to 61 mm. The young bird is also much darker, but has some yellow on the chest and belly.

**General distribution:** Sierra Leone and Gabon to Uganda, Kenya Colony, Tanganyika Territory and southern Belgian Congo.

**Range in Eastern Africa:** Uganda (except north-western area) to western Kenya Colony and western Tanganyika Territory at Kasulu, Kigoma and the Kungwe-Mahare Mts.

**Habits:** As for the nominate race, locally common.

**Recorded breeding:** Cameroon Mt., October to January. Uganda, April to June, also December and January. Southern Belgian Congo, September and October.

*Camaroptera brevicaudata griseigula* Sharpe.
*Camaroptera griseigula* Sharpe, Ibis, p. 158, 1892: Voi, south-eastern Kenya Colony.

**Distinguishing characters:** Differs from the nominate race in being darker above and greyer below, but not so dark above as the Gabon race. The non-breeding dress, June to November, is only slightly browner above than the breeding dress, and the base of the lower mandible is seldom brown, but usually remains black. Wing 49 to 61 mm. The young bird does not always have an indication of yellow on the chest and belly.

**Range in Eastern Africa:** Kenya Colony (except north-eastern area) and northern Tanganyika Territory from Mt. Kilimanjaro to Mbulu, Lossogonoi and Kibaya; also Ukerewe Island, south-eastern Lake Victoria.

**Habits:** As for other races, common but rather skulking in habits; a forest species, but it may be met with in strips of bush or gardens, and feeds largely on the ground.

**Recorded breeding:** Kenya Colony highlands, April to June, also December. Southern areas, March and April. Northern Tanganyika Territory, January to March.

*Camaroptera brevicaudata abessinica* Zedl.
*Camaroptera griseoviridis abessinica* Zedlitz, J.f.O. p. 338, 1911: Harar, eastern Abyssinia.

**Distinguishing characters:** Differs from the nominate race in being darker grey above, but not so dark as the Gabon race. The non-breeding dress, August, September or October to April, is only slightly browner above. Wing 50 to 61 mm. The young bird is also darker above than that of the Gabon race.

**Range in Eastern Africa:** Red Sea Province of the Sudan, Eritrea, Abyssinia (except north-western area north of Lake Tana), the Somalilands, southern Sudan, north-western Uganda and north-eastern Kenya Colony.

**Habits:** As for other races, but found at higher levels as well, and also inhabits high trees as well as bushes.

**Recorded breeding:** Southern Sudan, November. Northern Uganda, March to June. Baringo, Kenya Colony, September.

*Camaroptera brevicaudata sharpei* Zedl.
*Camaroptera griseoviridis sharpei* Zedlitz, J.f.O. p. 342, 1911: Damaraland.

**Distinguishing characters:** Above, very similar to the last two races but breast to belly with a wash of buff. In non-breeding dress, June to October, the upperparts are pale brown and the underparts buff. Wing 49 to 59 mm. The young bird is similar to the adult in non-breeding dress, but has a slight olive wash on the mantle and a faint wash of yellow on the throat.

**General distribution:** Tanganyika Territory to Nyasaland, the Rhodesias, South-West Africa, Bechuanaland, southern Portuguese East Africa and northern Transvaal.

**Range in Eastern Africa:** Mt. Gerui (Hanang), the headwaters of the Nyamansi River and Iringa, Tanganyika Territory, to Nyasaland near Fort Johnston.

**Habits:** As for other races. Its call is as for the northern races, but its persistent squeak is often followed by a shrill note, and when excited the squeaks are punctuated with clicks. The ground colour of the eggs is said to be normally pale blue.

**Recorded breeding:** Tanganyika Territory, January to March. Northern Rhodesia, February. Nyasaland, March.

*Camaroptera brevicaudata erlangeri* Reichw.
*Camaroptera erlangeri* Reichenow, Vög. Afr. 3, p. 617, 1905: Solole, lower Juba River, southern Italian Somaliland.

**Distinguishing characters:** Above, rather darker grey than in the nominate race, but below, white or creamy white with a silky appearance, and in some cases with a grey tinge on throat, sides of chest and flanks. In non-breeding dress, June to November, the upperparts are only very slightly browner and the base of lower

# WARBLERS

mandible pale horn. Wing 50 to 59 mm. The young bird is browner above than the adult; green of scapulars and wings duller and washed with brown; lower mandible wholly pale horn; no yellow on chest and belly.

**Range in Eastern Africa:** Coastal areas of Italian Somaliland, Kenya Colony and Tanganyika Territory from the Juba River to as far south and west as Mpapwa.

**Habits:** As for other races, common but skulking with a persistent call, described as a rapid squeaky 'cleb-cleb-cleb-cleb' and a curious little note like that of a tin trumpet.

**Recorded breeding:** Tanganyika Territory, January.

**Distribution of other races of the species:** Angola.

### 1012 YELLOW-BROWED CAMAROPTERA. *CAMAROPTERA SUPERCILIARIS* Fraser.

*Camaroptera superciliaris ugandæ* S. Clarke. **Pl. 71.**
*Camaroptera superciliaris ugandæ* Stephenson Clarke, Bull. B.O.C. 33, p. 136, 1914: Uganda.

**Distinguishing characters:** Above, bright green, including tail and edges of flight feathers; stripe from base of bill to over eye and sides of face bright yellow; lores and stripe through eye green; below, white, under wing-coverts yellowish white; flanks greenish; under tail-coverts bright yellow; bare patch on each side of throat. The sexes are alike. Wing 50 to 56 mm. The young bird is duller than the adult; no yellow on sides of face; breast often washed with yellow.

**General distribution:** Eastern Belgian Congo and Uganda.

**Range in Eastern Africa:** Southern and western Uganda.

**Habits:** Common but local in equatorial forest, a small skulking bird creeping about in dense undergrowth or low trees. The male has a display of throat swelling and wing flickering as he flies round the female, with a call of 'pweep-pweep-pweep' (Van Someren).

**Nest and Eggs:** Nest (of Upper Guinea race), similar to that of other species of the genus, sewn into the leaves of a bush. Eggs three, very pale blue spotted and speckled with dark brown.

**Recorded breeding:** Western Uganda (breeding condition), July and August. Probably breeds at any time of year.

**Call:** A chirping note and a short melodious song, but see above. Other races have a distinctive double whining call.

**Distribution of other races of the species:** West Africa and Angola, the nominate race being described from Fernando Po.

**1013 SOCOTRA WARBLER.** *INCANA INCANA* (Sclater & Hartlaub).

*Cisticola incana* P. L. Sclater & Hartlaub, P.Z.S. p. 166, pl. 15, fig. 1, 1881: Socotra Island.

**Distinguishing characters:** Above, brownish grey; head browner; whitish edges to primaries; tail feathers with broad subterminal blackish band and whitish ends; below, dusky white; centre of belly pure white. The sexes are alike. Wing 48 to 51 mm. The young bird is browner above.

**Range in Eastern Africa:** Socotra Island.

**Habits:** Common in bush country or on grassy plains up to at least 4,500 feet, restless active little birds described as of a noisy, fussy disposition. The tail is carried in a semi-erect position when hopping about.

**Nest and Eggs:** A neat dome-shaped nest of fine grass ornamented with lichen and placed in a thick bush a few feet from the ground. Eggs undescribed.

**Recorded breeding:** Socotra Island, December to February or later.

**Call:** A scolding 'chip-chip-chip-chip-chip-it-chip-it-chip-it' very much like the noise made by striking a flint and steel together. The male also has a pretty little Stonechat-like song.

## LONGBILLS. FAMILY—SYLVIIDÆ. Genus: *Macrosphenus*.

Two species occur in Eastern Africa. They inhabit forests and dense undergrowth, where they search for their insect food in a Warbler-like manner among the lower branches of the bushes.

**1014 YELLOW LONGBILL.** *MACROSPHENUS FLAVICANS* Cassin.

*Macrosphenus flavicans hypochondriacus* (Reichw.). **Pl. 59.**

*Rectirostrum hypochondriacum* Reichenow, O.M. p. 32, 1893: Kinjawanga, near Beni, eastern Belgian Congo.

**Distinguishing characters:** Small, short-tailed and long-billed; above, olive-green; head darker and duller; wing-coverts grey; below, chin to chest dusky white; rest of underparts silky lemon yellow, more golden on flanks. The sexes are alike. Wing 53 to 64 mm. The young bird has the head olive green uniform with the mantle; white tips to the bastard wing.

**General distribution:** Eastern Belgian Congo and Uganda.

**Range in Eastern Africa:** Western and southern Uganda.

**Habits:** A remarkable bird which in life assumes a curious appearance with the body feathers puffed out while the neck is thin and appears disproportionately long. A bird of dense forest undergrowth among tangled masses of creepers, but sings from a perch high up in the trees, though remaining well concealed.

**Nest and Eggs:** Undescribed.

**Recorded breeding:** No records.

**Food:** Insects.

**Call:** A repeated 'tziss' high pitched and far-carrying but not loud. The song is plaintive and melodious of four notes in a descending scale.

**Distribution of other races of the species:** West Africa to Angola, the nominate race being described from Gabon.

**1015 GREY LONGBILL.** *MACROSPHENUS CONCOLOR* (Hartlaub).

*Camaroptera concolor* Hartlaub, Syst. Orn. Westafr. p. 62, 1857: Guinea.

**Distinguishing characters:** In size and colour of the upperside similar to the Yellow Longbill, but below, wholly dull olive grey. The sexes are alike. Wing 54 to 61 mm. The young bird is rather browner above and greyer below.

**General distribution:** Fernando Po to Uganda.

**Range in Eastern Africa:** Western Uganda.

**Habits:** Another little known inhabitant of dense forest undergrowth, which probably has the same puffed out appearance in life as the Yellow Longbill. Generally in thick secondary growth but also found among tree tops.

**Nest and Eggs:** Undescribed.

**Recorded breeding:** West Africa (breeding condition), October and December. Ituri (breeding condition) August to November.

**Call:** Jingling rapid high and low notes (Bates). A plaintive double whining cry (Allen). A sweet silvery rippling song not easy to locate (Marchant).

## CISTICOLAS or GRASS WARBLERS. Family—SYLVIIDÆ.
Genus: *CISTICOLA*.

Thirty species occur in Eastern Africa. Whatever their general habitat, whether downland, marshland, or woodland, grass of some sort is essential to their existence. They are insectivorous, and of generally inconspicuous habits, except in the case of breeding males. The tail is always of twelve feathers, and unless very short, well graduated. In most species there is a subterminal black spot on the tail feathers, while the coloration is almost entirely made up of brown tints and shades in a variety of colour patterns with greys, buffs and tawny.

Many of the species are so much alike that identification is difficult, until one becomes familiar with the moults from the breeding to the non-breeding season and *vice versa*, and from immature to adult, as well as with the more clearly defined differences of proportion found in the feet and wing feathers, and other points such as the sexual disparity of size.

As will be seen from the key, the shape of the first primary in a Cisticola has considerable diagnostic value, and these have been differentiated into five types, *i.e.*, broad-blade, blade, narrow-blade, scimitar and acute.

The females are always smaller than the males, but in greatly varying degree according to the species, and they invariably show less black pigmentation, especially in the bill and palate; the latter is of importance for in the breeding season sexes can be thereby determined with certainty. The juvenile birds are browner and rustier than the adult, and commonly—but not invariably—yellow below.

The grass nests are divisible into three types :—Ball-type, which is found in all the species except the Singing, Red-faced and Zitting Cisticolas; Tailor-bird type, found in the Singing and Red-faced Cisticolas, and soda-bottle type found only in the Zitting Cisticola. Most of these nests are lined with plant down, and it is remarkable that all the birds have the habit of continuing to add to the lining throughout the period of incubation.

Usually the breeding season coincides with the rains but this is not an invariable rule. Changes of plumage in the breeding and non-breeding season occur in the more northern and southern areas, but in the central areas a perennial dress with only one moult a year is the general rule.

Through the researches of the late Admiral Lynes more is known about the Grass Warblers than about any comparable group of African birds.

### KEY TO THE ADULT CISTICOLAS OF EASTERN AFRICA

1 Top of head various shades of brown, grey brown, buff or tawny, plain or streaked not contrasting with the rest of upperparts: 3–27

2 Top of head various shades of brown or chestnut brown, plain or streaked contrasting with the rest of upperparts: 28–72

3 Top of head and mantle plain: 5–9

4 Top of head and mantle more or less streaked: 10–27

5 Above, tawny rufous: FOXY CISTICOLA *Cisticola troglodytes* 1038

6 Above, isabelline fawn: RED-FACED CISTICOLA *Cisticola erythrops* non-breeding dress 1032

7 Above, dusky, or dusky brown: 8–9

8 Bill, feet and toes large and strong: WHISTLING CISTICOLA *Cisticola lateralis* 1025

9 Bill, feet and toes small and weak: SIFFLING CISTICOLA *Cisticola brachyptera* breeding dress 1037

| | | |
|---|---|---|
| 10 Above, pale sandy: | DESERT CISTICOLA *Cisticola aridula* | 1018 |
| 11 Above, greyish: | | 12–15 |
| 12 Size smaller, wing 50 mm. and under: | SOCOTRA CISTICOLA *Cisticola hæsitata* | 1017 |
| 13 Size larger, wing 54 mm. and over: | | 14–15 |
| 14 Bill, feet and toes thick and strong: | CROAKING CISTICOLA *Cisticola natalensis argentea* | 1036 |
| 15 Bill, feet and toes thinner and weak: | ASHY CISTICOLA *Cisticola cinereola* | 1042 |
| 16 Above, dusky or dusky brown: | | 17–20 |
| 17 Bill, feet and toes thick and strong: | CROAKING CISTICOLA *Cisticola natalensis* breeding dress | 1036 |
| 18 Bill, feet and toes thinner and weaker: | | 19–20 |
| 19 Below, buffish: | SIFFLING CISTICOLA *Cisticola brachyptera* non-breeding dress | 1037 |
| 20 Below, grey: | HUNTER'S CISTICOLA *Cisticola hunteri* | 1030 |
| 21 Above, tawny: | | 22–27 |
| 22 White tips to tail feathers: | | 23–24 |
| 23 Tail short, first primary blade-shaped: | ZITTING CISTICOLA *Cisticola juncidis* | 1016 |
| 24 Tail very short, first primary short and acute: | WING-SNAPPING CISTICOLA *Cisticola ayresii* | 1019 |
| 25 Buff tips to tail feathers: | | 26–27 |

Plate 70

Green Crombec (p. 429)  Chestnut-throated Apalis (p. 414)  Yellow-bellied Eremomela (p. 431)
Bamenda Apalis (p. 415)  White-winged Apalis (p. 413)
  Male  Female
Black-throated Apalis (p. 412)   Red-faced Apalis (p. 417)
  Male  Female
Red-faced Crombec (p. 425)  Green-backed Eremomela (p. 433)  Crombec (p. 424)
Mrs. Moreau's Warbler (p. 420)  Buff-bellied Warbler (p. 422)  Burnt-neck Eremomela (p. 438)

Plate 71

Tawny-flanked Prinia (p. 505)
Breeding dress   Non-breeding dress
White-chinned Prinia      Green-backed
   (p. 510)            Camaroptera (p. 439)
Red-wing Warbler (p. 512)
Breeding dress   Non-breeding dress
Banded Prinia (p. 511)
   Lesser Swamp-Warbler (p. 380)

Striped-back Prinia
     (p. 509)
  Olive-green
Camaroptera (p. 441)

Yellow-browed Camaroptera (p. 445)
Yellow Swamp-Warbler (p. 382)

Green-cap Eremomela
      (p. 435)
Grey-backed Camaroptera (p. 441)
    Breeding dress
    Non-breeding dress

Greater Swamp-Warbler (p. 378)

| | | |
|---|---|---|
| 26 Tail long, bill, feet and toes thick and strong: | CROAKING CISTICOLA *Cisticola natalensis* non-breeding dress | **1036** |
| 27 Tail short, bill, feet and toes thinner and weaker: | PECTORAL-PATCH CISTICOLA *Cisticola brunnescens* non-breeding dress | **1020** |

| | | |
|---|---|---|
| 28 Top of head more or less streaked and mantle streaked: | | 29–42 |
| 29 Edges of flight feathers bright tawny: | | 30–31 |
| 30 Mantle streaked grey and black: | STOUT CISTICOLA *Cisticola robusta* breeding dress | **1035** |
| 31 Mantle streaked tawny and black: | WINDING CISTICOLA *Cisticola galactotes* non-breeding dress | **1033** |
| 32 Edges of flight feathers buffish or brownish, not bright tawny: | | 33–44 |
| 33 Rump tawny: | | 34–35 |
| 34 Bill and toes longer and stronger: | PECTORAL-PATCH CISTICOLA *Cisticola brunnescens* breeding dress | **1020** |
| 35 Bill and toes shorter and weaker: | BLACK-BACKED CISTICOLA *Cisticola eximia* non-breeding dress | **1021** |
| 36 Rump greyish or brownish: | | 37–44 |

| | | |
|---|---|---|
| 37 Top of head darker, tawny or slightly chestnut: | | 38–43 |
| 38 Dark streaks on mantle narrow: | SINGING CISTICOLA *Cisticola cantans* non-breeding dress | 1031 |
| 39 Black streaks on mantle broad: | | 40–43 |
| 40 Top of head streaked: | | 41–42 |
| 41 Bill, feet and toes larger: | STOUT CISTICOLA *Cisticola robusta* non-breeding dress | 1035 |
| 42 Bill, feet and toes smaller: | WAILING CISTICOLA *Cisticola lais* | 1022 |
| 43 Top of head very indistinctly streaked, almost plain: | CHURRING CISTICOLA *Cisticola njombe* | 1023 |
| 44 Top of head paler, more brownish or buffish: | RATTLING CISTICOLA *Cisticola chiniana* | 1024 |

| | | |
|---|---|---|
| 45 Top of head plain: | | 46–72 |
| 46 Streaks on mantle sharp and clear: | | 47–51 |
| 47 Mantle streaked pale sandy and black: | RED-PATE CISTICOLA *Cisticola ruficeps* non-breeding dress | 1043 |
| 48 Mantle streaked grey and black, rump grey: | WINDING CISTICOLA *Cisticola galactotes* breeding dress | 1033 |
| 49 Mantle streaked tawny and black: | | 50–51 |
| 50 Rump blackish: | TINKLING CISTICOLA *Cisticola tinniens* | 1044 |
| 51 Rump tawny: | BLACK-BACKED CISTICOLA *Cisticola eximia* breeding dress | 1021 |

| | | |
|---|---|---|
| 52 Streaks on mantle not sharp and clear: | | 53-54 |
| 53 Tail blackish: | CARRUTHER'S CISTICOLA *Cisticola carruthersi* | 1034 |
| 54 Tail olivaceous brown: | RATTLING CISTICOLA (some races) *C. chiniana heterophrys*, *C. c. procera*, and *C. c. emendata* | 1024 |
| 55 Mantle plain not streaked: | | 56-72 |
| 56 Lores black: | BLACK-LORED CISTICOLA *Cisticola nigriloris* | 1028 |
| 57 Lores not black: | | |
| 58 Mantle and tail darkish grey: | TABORA CISTICOLA *Cisticola angusticauda* | 1040 |
| 59 Mantle and tail greyish isabelline: | RED-FACED CISTICOLA *Cisticola erythrops* breeding dress | 1032 |
| 60 Mantle dull olivaceous grey: | | 61-72 |
| 61 Edges of flight feathers tawny: | SINGING CISTICOLA *Cisticola cantans* breeding dress | 1031 |
| 62 Edges of flight feathers not tawny: | | 63-72 |
| 63 Size smaller, wing 49 mm. and under: | | 64-66 |
| 64 Tail shorter, top of head more russet: | TINY CISTICOLA *Cisticola nana* | 1041 |
| 65 Tail longer, top of head more chestnut: | PIPING CISTICOLA *Cisticola fulvicapilla* | 1039 |
| 66 Size larger, wing 49 mm. and over: | | 67-72 |

| | | |
|---|---|---|
| 67 End of tail feathers clear white: | RED-PATE CISTICOLA *Cisticola ruficeps* breeding dress | **1043** |
| 68 End of tail feathers not clear white: | | 69–72 |
| 69 Top of head dull brownish: | TRILLING CISTICOLA *Cisticola woosnami* | **1026** |
| 70 Top of head brightish russet: | | 71–72 |
| 71 Feet and toes large and strong: | CHUBB'S CISTICOLA *Cisticola chubbi* | **1029** |
| 72 Feet and toes small and weak: | ROCK-LOVING CISTICOLA *Cisticola emini* | **1027** |

**1016 ZITTING CISTICOLA.  *CISTICOLA JUNCIDIS*** (Rafinesque).

*Cisticola juncidis terrestris* (Smith). **Pl. 72.**

*Drymoica terrestris* A. Smith, Ill. Zool. S. Afr. Aves. pl. 74, fig. 2, 1842: near Kurrichane, Bechuanaland.

**Distinguishing characters:** A small species, first primary blade-shaped; adult male, above, brown streaked with blackish; rump tawny, below, whitish; tail with subterminal black spots above and below and white tips. In non-breeding dress the top of the head and mantle have broader buff edges to the feathers with rather less buff suffusion below. The female in breeding dress is more like the male in non-breeding dress and does not differ to such a marked degree from breeding dress to non-breeding dress. Wing, male 49 to 53, female 46 to 50 mm. The young bird is a rusty edition of the adult in non-breeding dress; below, pale yellow.

**General distribution:** Gabon, Angola, Tanganyika Territory, southern Belgian Congo, the Rhodesias, Portuguese East Africa and South Africa.

**Range in Eastern Africa:** Southern Tanganyika Territory, Nyasaland and Portuguese East Africa.

**Habits:** Found in open grass country or waste lands, or on the edge of cultivation, and spends most of its time on the ground. In the non-breeding season it skulks in the grass and makes only short flights when flushed. The cock bird, however, may be readily identified in the breeding season, for the courting antics consist of a rapid dipping flight at a considerable height in the air above the area which his mate inhabits. This species does not snap its wings.

**Nest and Eggs:** Nest soda-bottle shape, a type found only in this species, with entrance facing upwards, in grass, six inches to two feet from ground so that the grass overtops the nest by a few inches; composed of soft living grass stems bound with a fine network of cobwebs, lined copiously with plant down. Eggs four or five, occasionally three or six, white or light blue, freckled and finely spotted with reds or browns, sometimes though rarely plain; about 15 × 11 mm.

**Recorded breeding:** Nyasaland about November to March. Southern Rhodesia, December to April.

**Call:** The call uttered when courting is characteristic, consisting of one syllable, a sharp 'zit,' one utterance at each upward swoop of the quick and undulating progress.

*Cisticola juncidis uropygialis* (Fraser).
*Drymoica uropygialis* Fraser, Proc. Zool. Soc. p. 17, 1843: Accra, Gold Coast.

**Distinguishing characters:** Paler and more sandy coloured than the Bechuanaland race. In non-breeding dress more sandy than in breeding dress, especially on the rump; whiter below. Wing 44 to 53 mm.

**General distribution:** Senegal, Nigeria, the French Congo, the Sudan, Eritrea, Abyssinia and south-west Arabia.

**Range in Eastern Africa:** The Sudan and southern Eritrea and Abyssinia.

**Habits:** As for the South African race. Eggs are described as usually pale blue; about 13·5 × 10·5 mm.

**Recorded breeding:** Sudan, about June to September in the northern, and May to October in the southern part of the range.

*Cisticola juncidis perennia* Lynes.
*Cisticola juncidis perennia* Lynes, Ibis, Suppl. p. 105, 1930: Mokia, near Lake George, western Uganda.

**Distinguishing characters:** Top of head and mantle rather more heavily streaked than in the Gold Coast race. There is no non-breeding dress. Wing 44 to 53 mm. In north Pemba Island a dark phase exists which breeds indiscriminately with birds of its own colour or with the normal form. It is warm rusty brown above with almost black streaking; throat dark grey washed with pale rusty; rest of underparts rust coloured over dark grey.

**General distribution:** North-eastern Belgian Congo from the Ituri and Welle districts to Kenya Colony and Tanganyika Territory.

**Range in Eastern Africa:** Uganda, Kenya Colony and northern Tanganyika Territory, also Pemba, Zanzibar and Mafia Islands.

**Habits:** As for the species.

**Recorded breeding:** Uganda, Lake Albert, August. Tanganyika Territory, June, also Rufigi River delta, January. Pemba Island, March and August to January. Zanzibar Island, March to May and September to December. Nesting activity commences soon after the rains begin, and varies from place to place accordingly.

**Distribution of other races of the species:** Southern Europe, east to China and Australia, the nominate race being described from Sicily.

**1017** SOCOTRA CISTICOLA.   *CISTICOLA HÆSITATA* (Sclater & Hartlaub).
*Drymœca hæsitata* Sclater & Hartlaub, P.Z.S. p. 106, 1881: Socotra Island.

**Distinguishing characters:** First primary blade-shaped. Differs from the Zitting Cisticola in being greyer; rump greyish brown. The non-breeding dress is more buffish. The sexes are alike. Wing, male 49 to 50, female 46 mm. Juvenile plumage unrecorded.

**Range in Eastern Africa:** Socotra Island.

**Habits:** Little known, but it is found on open grass-covered country, interspersed with bushes. Like the Zitting Cisticola the male cruises in the air in the breeding season.

**Nest and Eggs:** Undescribed.

**Recorded breeding:** No records, but nesting is said to occur in January and February.

**Call:** The song is said to consist of a little series of several notes uttered either in flight or when perched on the summit of a bush.

### 1018 DESERT CISTICOLA. *CISTICOLA ARIDULA* Witherby.

*Cisticola aridula aridula* With.
*Cisticola aridula* Witherby, Bull. B.O.C. 11, p. 13, 1900: Gerazi, White Nile, Sudan.

**Distinguishing characters:** First primary narrow blade-shaped. Adult male, pale fawn colour; differs from the Zitting Cisticola in having the subterminal black spots only on the underside of the tail, which is plain black above with white tips. In non-breeding dress the dusky streaking of the top of the head and mantle is narrower. The female is more like the male in non-breeding dress with mantle paler and less heavily streaked; bill paler. Wing, male 48 to 52, female 46 to 48 mm. The young bird is deeper coloured than the adult in non-breeding dress.

**General distribution:** Northern Nigeria to the Sudan.

**Range in Eastern Africa:** The Sudan west of the Nile.

**Habits:** Found on open dry and short grass covered country spending most of its time on the ground. Inconspicuous in movements even in the case of breeding males. The male is noticeable only when alarmed and will then flit low and erratically over his mate with rapid steep swoops, snapping his wings and uttering sharp cries. The wing snapping sound is similar to that made with the finger and thumb, and in this species only seems to be made on alarm.

**Nest and Eggs:** Nest of ball type, elliptical, about six by three inches externally with side-top entrance; built of dry grass at from twelve inches above to nearly on ground, according to whether built in long grass or very small tufts. Builds in last year's dry grasses. Eggs of the species are white or pale blue, plain or finely speckled with dots of purplish red often an ill-defined ring of spots round the big end; about 14·5 × 11·5 mm.

**Recorded breeding:** Normally July, August and September, but season variable according to whether rains are good or bad, early or late.

**Call:** The song of the breeding male consists of small high-pitched tinkling notes of a single syllable several times repeated, at half-second intervals, from a low or grass top perch.

*Cisticola aridula lavendulæ* O. Grant.

*Cisticola lavendulæ* O. Grant & Reid, Ibis, p. 650, 1901: Aroharlaise, British Somaliland.

**Distinguishing characters:** Similar to the nominate race but more grey-sandy in colour with conspicuous white bases to the feathers of the hind neck. Wing, male 46 to 55, female 45 to 47 mm.

**Range in Eastern Africa:** Coastal Eritrea, southern Abyssinia and British Somaliland.

**Habits:** Said to be confined to *Panicum turgidum* areas.

**Recorded breeding:** Most probably from November to December, but no actual records.

*Cisticola aridula tanganyika* Lynes.

*Cisticola aridula tanganyika* Lynes, Ibis, Supp. p. 126, 1930: Morogoro, eastern Tanganyika Territory.

**Distinguishing characters:** Similar to the nominate race but smaller and darker. There is no non-breeding dress. Wing, male 47 to 51, female 45 to 47 mm.

**Range in Eastern Africa:** Kenya Colony and Tanganyika Territory as far south as Morogoro and Dar-es-Salaam.

**Habits:** As for the nominate race, a very common species throughout Masailand.

**Recorded breeding:** The principal nesting season is April to July; there is a secondary breeding season in November, December and January.

**Distribution of other races of the species:** Angola and South Africa.

## 1019 WING-SNAPPING CISTICOLA. *CISTICOLA AYRESII* Hartlaub.

*Cisticola ayresii ayresii* Hartl. **Pl. 72.**
*Cisticola ayresii* Hartlaub, Ibis, p. 325, 1863: Natal, South Africa.

**Distinguishing characters:** First primary short and acute. Adult male very similar to the Zitting Cisticola but tail much shorter and lacking the subterminal black spots. In non-breeding season the plumage is paler, more buff and black above. The female is more like the male in non-breeding dress. Wing, male 47 to 51, female 43 to 47 mm. The young bird is paler above than the adult; yellowish below.

**General distribution:** Gabon and Angola to Tanganyika Territory, Southern Rhodesia, the Transvaal, eastern Cape Province and Natal.

**Range in Eastern Africa:** Southern Tanganyika Territory.

**Habits:** Found on open grass plains or downland and, like the Desert Cisticola, inconspicuous in habits and difficult to flush from the ground during the non-breeding season. This species, together with the Pectoral-patch Cisticola and the Black-backed Cisticola, belong to a group in which breeding males carry out the most remarkable courting antics, from which Lynes named them 'Cloud-scrapers.' Rising from the grass the bird mounts quickly on rapidly beating wings to a great height, certainly to several hundred feet and often out of ordinary sight, there to commence a high cruising flight and song usually accompanied by loud wing snapping over its breeding area. After several minutes there comes an almost vertical plunge earthwards, usually too quick to follow, with an excitedly uttered single note then a sudden check close to the grass tops, a rapid and vertical 'jink' upwards for a number of feet, a short traversing flight and a final flop into the grass. On the descent it makes a twang like a bow string. In this species the call does not commence until the upward climb is completed. The song when up aloft is a shrill whistle of five or six notes accompanied by volleys of wing snaps and many swerves to and fro. The descent is accompanied by a chattering call and the volleys of wing snappings, with erratic darts in several directions when close to the grass.

**Nest and Eggs:** Nest of the characteristic ball-type described under the Desert Cisticola, placed on, or almost on, the ground in a tuft of short grass with a few soft green blades bent over and woven in a bar over the roof of the dry grass, plant-down lined nest. Eggs three or four, variable in coloration but with fine speckling as the dominant type of marking; about 16·5 × 11·5 mm.

**Recorded breeding:** About November to March.

**Call:** See Habits.

*Cisticola ayresii mauensis* V. Som.
*Cisticola terrestris mauensis* Van. Someren, Nov. Zool. 29, p. 207, 1922: Mau, south-central Kenya Colony.

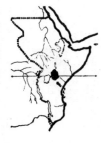

**Distinguishing characters:** Size similar to the nominate race, but rather more deeply coloured. There is no non-breeding dress. Wing, male 48 to 51, female 44 to 49 mm.

**Range in Eastern Africa:** Kenya Colony from Eldoret and Mt. Kenya to Ngong.

**Habits:** As for the nominate race; the male in cruising flight in the breeding season has a noticeably metallic clinking call interspersed with wing-snaps.

**Recorded breeding:** Principally April to July, also November and December.

*Cisticola ayresii entebbe* Lynes.
*Cisticola ayresii entebbe* Lynes, Ibis, Supp. p. 154, 1930: Entebbe, Uganda.

**Distinguishing characters:** Smaller than the nominate race and lighter coloured, but very similar to the non-breeding dress of that race. There is no non-breeding dress. Wing, male 46 to 48, female 43 to 45 mm.

**General distribution:** Kivu district, north-eastern Belgian Congo, to Kenya Colony and Tanganyika Territory.

**Range in Eastern Africa:** Uganda, north-western Tanganyika Territory, the Ruanda Province of the Belgian Congo and western Kenya Colony.

**Habits:** As for the nominate race.

**Recorded breeding:** More or less all the year round, but mainly April to November, not usually in the dry season.

*Cisticola ayresii imatong* Cave.
*Cisticola ayresii imatong* Cave, Bull. B.O.C. 59, p. 8, 1938: Imatong Mts., Equatorial Province, southern Sudan.

**Distinguishing characters:** Similar to the Kenya Colony race, but edges of feathers of mantle and rump of a darker shade; below, deeper coloured; chin and throat white. In non-breeding dress darker above; deeper toned below. Wing, male 51 to 53, female 48 to 49 mm.

**Range in Eastern Africa:** Imatong Mts., southern Sudan.

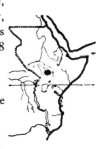

**Habits:** Found in mountain grasslands in the upper zone of the Imatong Mts.

**Recorded breeding:** April to June.

## 1020 PECTORAL-PATCH CISTICOLA. *CISTICOLA BRUN-NESCENS* Heuglin.

*Cisticola brunnescens brunnescens* Heugl.
*Cisticola brunnescens* Heuglin, J.f.O. p. 289, 1862: Godofelasi, Seraweh Province, Eritrea.

**Distinguishing characters:** Adult male, similar to the Wing-snapping Cisticola, but larger, with the first primary narrow blade-shaped; feet and toes big and strong and top of head practically plain buff; black patches at side of neck. In non-breeding dress brighter buff above, broadly streaked with black; below, deeper tone of buff. The female does not differ so markedly in the two dresses and is more like the male in non-breeding dress; upperside brighter; sub-loral spot and pectoral patches absent or inconspicuous. Wing, male 49 to 57, female 47 to 52 mm. The young bird is similar to the adult in non-breeding dress, chin to chest washed with yellow.

**Range in Eastern Africa:** Eritrea, Abyssinia and British Somaliland.

**Habits:** Found in open grass plains and with habits which are in general similar to those of the Wing-snapping Cisticola. In this, however, there are a few dissimilarities which the field ornithologist will soon learn to perceive since the two birds are frequently performing in the same area. The song which is usually a small and oft-repeated high-pitched note commences whilst the ascent is in progress, is continued throughout a more steady high cruise, with less erratic jerkings to and fro and with less erratic wing snappings, these last being at regular intervals and as though an accompaniment to the song. The rapid descent is made with the chattering call but without wing snaps. This species is parasitized by the Pintailed Whydah.

**Nest and Eggs:** As for the Wing-snapping Cisticola. Eggs three or four, variable, but usually pale blue with small scattered purplish dots and spots, least plentiful towards the small end; about 17·5 × 12·5 mm.

**Recorded breeding:** Abyssinia, June to October, probably about a month earlier in extreme south.

**Call:** See Habits.

*Cisticola brunnescens hindii* Sharpe.
*Cisticola hindii* Sharpe, Bull. B.O.C. 6, p. 7, 1896: Machakos, southern Kenya Colony.

**Distinguishing characters:** Similar to the breeding dress of the Nakuru race and the same size, but much paler. There is no non-breeding dress. Wing, male 48 to 55, female 46 to 48 mm.

**Range in Eastern Africa:** The middle highlands of Kenya Colony, 4,500 to 6,000 feet elevation, east of Kikuyu to Mts. Kilimanjaro, Meru and Gerui (Hanang), Tanganyika Territory.

**Habits:** As for the nominate race. Eggs appear to average smaller, about 16 × 11 mm.

**Recorded breeding:** Principally April to July, also November and December.

*Cisticola brunnescens cinnamomea* Reichw.
*Cisticola cinnamomea* Reichenow, O.M. p. 28, 1904: Ngomingi, Uhehe, Iringa district, Tanganyika Territory.

**Distinguishing characters:** More warmly coloured than other races with less black pigmentation and slightly smaller. There is no non-breeding dress. Wing, male 52 to 56, female 46 to 51 mm.

**General distribution:** The eastern Belgian Congo to Tanganyika Territory and the Rhodesias.

**Range in Eastern Africa:** Tanganyika Territory from the Iringa district to south-western areas.

**Habits:** As for the nominate race.

**Recorded breeding:** From about lat. 8° S., November to April, but further north more or less throughout the year.

# WARBLERS

*Cisticola brunnescens nakuruensis* V. Som.
*Cisticola terrestris nakuruensis* Van Someren, Nov. Zool. 29, p. 207, 1922: Nakuru Plains, Kenya Colony highlands.

**Distinguishing characters:** Similar to the nominate race in breeding dress, but slightly smaller and colour pattern less boldly contrasting. There is no non-breeding dress. Wing, male 53 to 55, female 47 to 51 mm.

**Range in Eastern Africa:** Kenya Colony from Kisumu, Eldoret, and Laikipia to Narossura Plain and northern Tanganyika Territory at the Crater Highlands.

**Habits:** As for the nominate race.

**Recorded breeding:** As for the Machakos race.

*Cisticola brunnescens wambera* Lynes.
*Cisticola brunnescens wambera* Lynes, Ibis, Suppl. p. 162, 1930: Wanbera, 8,000 feet, western Abyssinia.

**Distinguishing characters:** Similar to the nominate race but colour much deeper in tone and generally more tawny. In non-breeding dress brighter and lighter above; below, deeper buff. Wing, male 56 to 58, female 51 mm.

**Range in Eastern Africa:** The Wanbera Plateau, western Abyssinia.

**Habits:** As for other races.

**Recorded breeding:** No records.

**Distribution of other races of the species:** West and South Africa.

## 1021 BLACK-BACKED CISTICOLA. *CISTICOLA EXIMIA* (Heuglin).

*Cisticola eximia eximia* (Heugl.).
*Drymœca eximia* Heuglin, Ibis, p. 106, 1869: Upper Bahr-el-Ghazal, south-western Sudan.

**Distinguishing characters:** Adult male, very similar to the Wing-snapping Cisticola, but differs in having the first primary narrow blade-shape, not very short and acute, and differs from the Pectoral-patch Cisticola in having smaller and weaker feet. In non-breeding dress the tail has subterminal black spots below and the top of the head is streaked. The female in breeding dress is similar

to the male in non-breeding dress, but has some indistinct streaks on the top of the head. The non-breeding dress is very similar to the breeding dress. Wing, male 48 to 54, female 44 to 48 mm. The young bird is more buffish above than the adult in non-breeding dress; below, yellow.

**General distribution:** North-eastern Belgian Congo to the Sudan, Eritrea and Abyssinia.

**Range in Eastern Africa:** Eritrea, Abyssinia, southern Sudan and Uganda.

**Habits:** Found in open grass plains and with habits similar to those of the Pectoral-patch Cisticola. The only difference between the two seems to lie in the actual twittering song which has been called more 'twinkling' or varied. The difference is largely comparative, almost impossible to describe, but clearly distinguishable where both species happen to be performing over the same ground.

**Nest and Eggs:** Nest as described for the Wing-snapping Cisticola. Eggs three, plain white or white well marked with rusty and purplish brown dots, small and scattered, least plentiful on the smaller end; about 16 × 11·5 mm.

**Recorded breeding:** Commences in June and is probably continued until about October.

**Call:** See Habits.

**Distribution of other races of the species:** West Africa.

**1022 WAILING CISTICOLA.** *CISTICOLA LAIS* (Finsch & Hartlaub).
*Cisticola lais semifasciata* Reichw. **Pl. 72.**
*Cisticola semifasciata* Reichenow, Vög. Afr. 3, p. 544, 1905: Mlanje Plateau, southern Nyasaland.

**Distinguishing characters:** First primary blade-shaped. Adult, top of head deep chestnut brown streaked with dusky; mantle greyish brown narrowly streaked with blackish; tail feathers pointed with subterminal black spots on the inner webs. The sexes are alike and there is very little seasonal difference, adults in non-breeding dress having the colour tints rather lighter. Wing, male 55 to 60, female 48 to 53 mm. The young bird is similar above to the adult; below, yellow.

**General distribution:** Tanganyika Territory to Nyasaland and northern Portuguese East Africa.

**Range in Eastern Africa:** Southern Tanganyika Territory, Portuguese East Africa and southern Nyasaland.

**Habits:** Found on dry and open long-grass-covered mountain sides at from 5,000 to 7,000 feet, frequently those which are also boulder strewn, the slimly built but rather long tailed birds perching readily on the grass stems or on the rocks. Like most other members of the genus they poke about unobtrusively in family parties in the long grass during the non-breeding season. In the breeding season the males do on occasion hover excitedly low over their mates, but the species cannot be said to possess any distinctive courting antics.

**Nest and Eggs:** Ball-type nest, as described under the Desert Cisticola, built in grass among low shrubs, constructed of dry grass with a living grass bower. Eggs three or four, light to medium blue fairly well marked with freckles of claret red, the spots rather thicker towards the larger end; about 17 × 13 mm.

**Recorded breeding:** About October to March, commences to breed earlier than most other species.

**Call:** The call or song is of a few notes mostly in the nature of a shrill 'sweeping' whistle 'piet-piet-piet' which seems to echo and carry far round the hillsides in wailing fashion; the alarm, a high-pitched scolding bleat, is also unmistakable.

*Cisticola lais distincta* Lynes.
*Cisticola distincta* Lynes, Ibis, Supp. p. 240, 1930: Kedong Valley, south-central Kenya Colony.

**Distinguishing characters:** Differs from the Nyasaland race in being larger. There is no non-breeding dress. Wing, male 59 to 65, female 56 to 58 mm.

**Range in Eastern Africa:** Eastern Uganda to northern and central Kenya Colony.

**Habits:** This bird is found in bush on dry rocky hillsides or in grassy glades on the outskirts of mountain forest and does not call from one stance, but keeps hopping about in the undergrowth. The nest is a somewhat large ball-type domed structure placed on the ground with a circular entrance, and is made of fine grass well lined with creamy white seed-down.

**Recorded breeding:** Longonot Escarpment, April.

**Distribution of other races of the species:** South Africa, the nominate race being described from Natal.

**1023** CHURRING CISTICOLA. *CISTICOLA NJOMBE* Lynes.

*Cisticola njombe njombe* Lynes. **Pl. 72.**

*Cisticola aberrans njombe* Lynes, Bull. B.O.C. 53, p. 170, 1933: Njombe, southern Tanganyika Territory.

**Distinguishing characters:** First primary broad blade-shaped. Adult, top of head dull chestnut; mantle earth brown, indistinctly streaked with dusky; feet large; below, dull buffish white; tail graduated with rather indistinct subterminal spots on inner webs of tail feathers. The sexes are alike. In non-breeding dress the top of head is slightly brighter and the mantle warmer brown. Wing, male 51 to 57, female 47 to 51 mm. In general appearance this species is very similar to the Wailing Cisticola, but differs mainly in having an almost plain top to the head. The young bird is similar to the adult in non-breeding dress, streaking on mantle less distinct, below washed with yellow.

**Range in Eastern Africa:** Iringa and Njombe areas of southern Tanganyika Territory.

**Habits:** Found on wooded hillsides and most commonly those where boulders are prominent and where long grass growing round rocks affords ideal cover. Noisy but wary birds, the males on a boulder top or prominent perch utter a loud 'swink-swink' but at the least sign of danger dive down into the grass where for a while they move about with Tit-like churring calls. When hard pressed they take to the upper foliage of woodland trees. The courting male has a number of utterances, most commonly a quick rasp followed by a shrill 'swip' or a musical rattling warble, at the same time he will bob his body quickly up and down, flip his wings and jerk his tail up to the vertical in excited demonstrations towards the hen bird as she moves about beneath his perch. The tail is invariably held at a slight angle above the back.

**Nest and Eggs:** Nest ball-type, generally low down in a tangle of grass and weed growth. Eggs normally three, pale blue finely spotted with shades of red and brown; about 17·5 × 12·5 mm.

**Recorded breeding:** About November to April, eggs mostly recorded from December to February.

**Call:** See Habits.

Plate 72

H. Grönvold and C. E. Talbot Kelly

| Zitting Cisticola (p. 454) | Wing-snapping Cisticola (p. 459) | Rattling Cisticola (p. 467) | Wailing Cisticola (p. 464) |

Whistling Cisticola (p. 471)  Trilling Cisticola (p. 472)  Rock-loving Cisticola (p. 474)  Black-lored Cisticola (p. 476)

Singing Cisticola (p. 479)  Red-faced Cisticola (p. 482)  Winding Cisticola (p. 484)
Breeding dress  Non-breeding dress

Stout Cisticola (p. 488)  Croaking Cisticola (p. 491)  Siffling Cisticola (p. 495)
  Non-breeding dress  Breeding dress

Foxy Cisticola (p. 498)  Piping Cisticola (p. 499)  Red-pate Cisticola (p. 503)
  Non-breeding dress  Breeding dress

Ashy Cisticola (p. 502)  Tinkling Cisticola (p. 504)  Churring Cisticola (p. 466)  Tabora Cisticola (p. 500)

Plate 73

Blue Rough-wing Swallow (p. 551)  
Black Rough-wing Swallow (p. 549)  
Brown Rough-wing Swallow (p. 552)  
Eastern Rough-wing Swallow (p. 550)  
Western Rough-wing Swallow (p. 550)  
European Crag Martin (p. 545)  
African Sand Martin (p. 542)  
Pale Crag Martin (p. 545)  
Egyptian Sand Martin (p. 541)  
African Rock Martin (p. 547)  
Red-rumped Swallow (p. 534)  
Mascarene Martin (p. 540)  
Swallow (p. 524)  
Wire-tailed Swallow (p. 531)  
Blue Swallow (p. 530)  
Mosque Swallow (p. 535)  
Angola Swallow (p. 526)  
Rufous-chested Swallow (p. 536)  
Ethiopian Swallow (p. 528)  
Grey-rumped Swallow (p. 539)

*Cisticola njombe mariæ* Benson.
*Cisticola lais mariæ* Benson, Bull. B.O.C. 66, p. 16, 1945: Nyika Plateau, northern Nyasaland.

**Distinguishing characters:** Differs from the nominate race in being darker, especially above. Wing, male 55 to 59, female 51 to 54 mm.

**General distribution:** South-western Tanganyika Territory to the Nyika Plateau, northern Nyasaland.

**Range in Eastern Africa:** Mbeya, Rungwe and Tukuyu areas of south-western Tanganyika Territory.

**Habits:** As for the nominate race, mainly found from 7,000 to 8,000 feet in open grassland with scattered bushes.

**Recorded breeding:** Nyika Plateau, November.

**1024 RATTLING CISTICOLA.** *CISTICOLA CHINIANA* (A. Smith).
*Cisticola chiniana simplex* (Heugl.). **Pl. 72.**
*Drymœca simplex* Heuglin, Ibis, p. 105, 1869: Upper Nile, Sudan.

**Distinguishing characters:** First primary blade-shaped. Adult, very similar to the Wailing Cisticola and with the same sized feet and toes, but top of head less chestnut; subterminal black spots on both webs of the tail feathers. In non-breeding dress the plumage is generally darker; streaks on head and mantle broader. The female is smaller than the male. Wing, male 63 to 69, female 51 to 57 mm. The young bird has more suffused streaks above than the adult; below, chin to chest yellow.

**Range in Eastern Africa:** Southern Sudan and northern Uganda.

**Habits:** Usually found in thorn veldt but is widely distributed in open woodland type of country. Unobtrusive in the non-breeding season although an occasional male may be heard to utter a somewhat weak edition of the courting call. The breeding males advertise themselves almost more than any other species and their identity cannot be mistaken. Perched on the very top of a bush or tree they utter a loud rattling call of 'chaa-chaa-churr-chee-chee,' the rattling flourish in the middle very considerably prolonged at the commencement of the season but becoming somewhat enfeebled later. On the least sign of intrusion the male descends in a trilling glide towards the hen, either to drive her furiously away from the nest, or to perch near by, quite beside himself with rage, bowing up and down, flicking and fanning his tail and vigorously uttering a 'chaa-chaa' swearing of alarm. The female has a similar but softer warning. When the female has been driven well away from the nest both birds usually become silent.

**Nest and Eggs:** The nest of this species is of the ball type, about five inches high and three-and-a-half inches wide with top-side entrance, placed low down, from nearly on the ground to about a foot above it, in grass or low herbage; it is made of the usual dry grass bound with cobweb, and lined with plant down. The eggs are inclined to vary, blue or bluish white, plain, or with varied degrees of spotting; about 17 × 13 mm.

**Recorded breeding:** Lake Albert, March to May. Upper Nile Valley, Sudan, June to September.

**Call:** See Habits.

*Cisticola chiniana fischeri* Reichw.
*Cisticola fischeri* Reichenow, J.f.O. p. 162, 1891: Tura, Tabora district, Tanganyika Territory.

**Distinguishing characters:** Differs from the Nile race in having the top of the head and wing edges darker. The non-breeding dress is paler and more sandy buff, especially on the head; top of head more heavily striped. Wing, male 61 to 67, female 51 to 55 mm. The young bird is more striped above and more buffish in general colour; face, throat and breast yellowish.

**Range in Eastern Africa:** Urundi-Ruanda districts of Belgian Congo to northern and south-central Tanganyika Territory.

**Habits:** As for the Nile race, much parasitized by the Pin-tailed Whydah.

**Recorded breeding:** Probably November to May.

*Cisticola chiniana humilis* Mad.
*Cisticola humilis* Madarász, O.M. p. 168, 1904: Settima Mts., south-eastern Kenya Colony.

**Distinguishing characters:** Larger and darker and slightly browner than the Nile race. There is no non-breeding dress. Wing, male 68 to 74, female 56 to 60 mm.

**Range in Eastern Africa:** The Kenya Colony highlands, 6,000 to 7,000 feet.

**Habits:** As for the Nile race.

**Recorded breeding:** April to July, also November and December.

*Cisticola chiniana heterophrys* Ober.
*Cisticola heterophrys* Oberholser, Ann. Carneg. Mus. 3, p. 496, 1906: Mombasa, eastern Kenya Colony.

**Distinguishing characters:** Differs from other races in having the top of the head plain; streaks on mantle chestnut brown; edges of flight feathers more chestnut. There is no non-breeding dress. Wing, male 60 to 66, female 50 to 54 mm. The young bird is paler brown above than the adult or the young bird of other races.

This race can be distinguished from the Trilling Cisticola and the Rock-loving Cisticola by the chestnut edging to the flight feathers.

**Range in Eastern Africa:** Coastal areas of Kenya Colony to eastern Tanganyika Territory as far west as headwaters Sanya River, Lushoto, Korogwe, Handeni and Kilosa.

**Habits:** As for the Nile race.

**Recorded breeding:** April, also probably January.

*Cisticola chiniana bodessa* Mearns.
*Cisticola subruficapilla bodessa* Mearns, Smiths, Misc. Coll. 61, No. 11, p. 2, 1913: Bodessa, southern Abyssinia.

**Distinguishing characters:** Similar to the Mombasa race but larger; top of head darker and usually streaked; streaks on mantle more suffused. Irregular seasonal changes of dress; in the north of its range there is a breeding and non-breeding dress, the latter darker, browner and the streaks bolder; in the south a perennial dress, but which varies in individuals over as wide a range as those in the north. Wing, male 65 to 71, female 55 to 59 mm.

**Range in Eastern Africa:** Eritrea to the eastern Sudan at the Boma Plateau, southern Abyssinia and the northern Frontier Province of Kenya Colony.

**Habits:** As for the Nile race.

**Recorded breeding:** May to September, also about January and February; in the Hawash Valley, Abyssinia, it appears to breed in January and February, but no doubt breeding depends upon the rains.

*Cisticola chiniana victoria* Lynes.
*Cisticola chiniana victoria* Lynes, Ibis, Supp. p. 264, 1930: Sio River, northern Kavirondo, western Kenya Colony.

**Distinguishing characters:** Larger and more boldly marked than the Tabora race and less warm in colour, and slightly smaller than the Kenya Colony highlands race. There is no non-breeding dress. Wing, male 63 to 70, female 55 to 57 mm.

**Range in Eastern Africa:** Western Kenya Colony and northern Tanganyika Territory within the Lake Victoria basin, also Kaserazi and Ukerewe Islands of southern Lake Victoria, but not the Mwanza area.

**Habits:** As for the Nile race.

**Recorded breeding:** April to July, also November and December.

*Cisticola chiniana ukamba* Lynes.
*Cisticola chiniana ukamba* Lynes, Ibis, Supp. p. 267, 1930: Masongaleni, southern Ukamba Province, south-eastern Kenya Colony.

**Distinguishing characters:** Very like the non-breeding dress of the nominate race. There is no non-breeding dress. Wing, male 63 to 69, female 51 to 58 mm.

**Range in Eastern Africa:** Eastern Masai, Taita, southern Ukamba and Tana River areas of Kenya Colony from Loita to Simba, the middle Tana, Taveta and Mwatate, and to Gelai, Longido, Monduli, Mt. Meru and Kibaya in Tanganyika Territory.

**Habits:** As for the Nile race.

**Recorded breeding:** Principal season April to July, also November and December.

*Cisticola chiniana procera* Peters.
*Cisticola procera* Peters, J.f.O. p. 132, 1868: Tete, Zambesi River, Portuguese East Africa.

**Distinguishing characters:** Differs from the Tabora race in being much darker and in having the head top quite plain; mantle less distinctly streaked; the breeding dress is very like that of the

Mombasa race, though slightly darker and edges of flight feathers more chestnut. In non-breeding dress the plumage is generally browner; mantle clearly streaked; below, buff. Wing, male 63 to 67, female 51 to 53 mm.

**General distribution:** Nyasaland and Portuguese East Africa.

**Range in Eastern Africa:** Tete area of Portuguese East Africa to southern Nyasaland.

**Habits:** As for the Nile race.

**Recorded breeding:** About November to May.

*Cisticola chiniana emendata* Vincent.
*Cisticola chiniana emendata* Vincent, Bull. B.O.C. 64, p. 63, 1944: Mirrote, Mozambique Province, northern Portuguese East Africa.

**Distinguishing characters:** Differs from the preceding race in being darker above; streaks on mantle less distinct. In non-breeding dress general colour browner than the preceding race and below buff. Wing, male 61 to 68, female 51 to 55 mm.

**General distribution:** Tanganyika Territory to north-eastern Northern Rhodesia, western Nyasaland and northern Portuguese East Africa.

**Range in Eastern Africa:** Southern Tanganyika Territory from south-eastern Morogoro district to the Mozambique area of Portuguese East Africa.

**Habits:** As for other races.

**Recorded breeding:** No records.

**Distribution of other races of the species:** Angola and Northern Rhodesia to South Africa, the nominate race being described from Bechuanaland.

**1025 WHISTLING CISTICOLA.** *CISTICOLA LATERALIS* (Fraser).

*Cisticola lateralis antinorii* (Heugl.). **Pl. 72.**
*Drymœca antinorii* Heuglin, Ibis, p. 102, 1869: Djur, Bahr-el-Ghazal, south-western Sudan.

**Distinguishing characters:** Adult, differs from the Rattling Cisticola by the more robust bill; top of the head and back uniform sooty brown; below, creamy white; flanks dusky. The female is often paler and browner. There is normally no non-breeding dress

though it is possible that there may be one in the extreme north of its range. Wing, male 62 to 68, female 50 to 56 mm. The young bird is dull russet above; sides of face and chin to breast pale yellow.

**General distribution:** North-eastern Belgian Congo, the Sudan and Uganda.

**Range in Eastern Africa:** Southern Sudan and Uganda.

**Habits:** Found in surroundings which are difficult to describe but soon recognized when one is familiar with the species; it inhabits savannah woodland, but not in open dry portions like the Rattling Cisticola, rather in hot moist hollows where rank soft vegetation grows, such as on the edge of forest-like patches. The birds are wary and unobtrusive, creeping about in the thick cover with occasional shrill alarm notes. Breeding males when undisturbed take up a very high and prominent tree top perch, whence they utter a loud, rich whistling song unlike any other Cisticola, an opening note followed by even full syllables.

**Nest and Eggs:** Nest ball-type, of dry grass, and about a foot or less from the ground in clumps of green grass. Eggs, only one clutch of two known, plain blue; about $15 \cdot 5 \times 13$ mm.

**Recorded breeding:** The season seems to be principally from April to October but like other Cisticolas in these areas birds may be found breeding at any time.

**Call:** See Habits.

**Distribution of other races of the species:** West Africa and Angola, the nominate race being described from Liberia.

### 1026 TRILLING CISTICOLA. *CISTICOLA WOOSNAMI* O. Grant.

*Cisticola woosnami woosnami* O. Grant. **Pl. 72.**
*Cisticola woosnami* O. Grant, Bull. B.O.C. 21, p.72, 1908: Mokia, near Lake George, western Uganda.

**Distinguishing characters:** First primary blade-shaped. Adult male similar to the Whistling Cisticola but with the bill considerably weaker and more curved. Top of the head dull chestnut brown. There is no non-breeding dress. The female is smaller and lighter coloured than the male. Wing, male 64 to 68, female 52 to 60 mm. The young bird is dull russet above; yellow below.

**General distribution:** Eastern Belgian Congo and Uganda to Tanganyika Territory and Nyasaland north of Mzimba.

**Range in Eastern Africa:** Central and southern Uganda to central and southern Tanganyika Territory as far east as Mpapwa, but not to the Ufipa Plateau.

**Habits:** Found in typical woodland or orchard-bush country of Brachystegia-type timber where, if disturbed, they keep to the grass and bushes with harsh swearing alarm notes. Like the Whistling Cisticola breeding males take up a tree top perch and remain there for hours at a time. The song is one which renders identification an easy matter, for it is a soft reeling trill, high pitched and slightly crescendo, and lasting sometimes fully five seconds; it may be likened to the call of some Cicadas. When thus shivering out its pleasant reel it moves its head from side to side rendering the note most ventriloquistic and difficult to locate.

**Nest and Eggs:** Ball-type nest of dry grass lined with plant down, bowered over and low down in green grass. Eggs two or three, turquoise blue well sprinkled with dots of browns and purples; about $17 \cdot 5 \times 12 \cdot 5$ mm.

**Recorded breeding:** As for the Whistling Cisticola, but January to April is the season in the south of its range.

**Call:** See Habits.

*Cisticola woosnami schusteri* Reichw.
*Cisticola schusteri* Reichenow, J.f.O. p. 557, 1913: Uluguru Mts., eastern Tanganyika Territory.

**Distinguishing characters:** Differs from the nominate race in being darker above, more blackish. There is no non-breeding dress. Wing, male 64 to 67, female 54 to 55 mm.

**Range in Eastern Africa:** Northern and eastern Tanganyika Territory from Moshi to the Uluguru Mts.

**Habits:** As for the nominate race.

**Recorded breeding:** About January to April.

*Cisticola woosnami lufira* Lynes.
*Cisticola woosnami lufira* Lynes, Ibis, Suppl. p. 300, 1930: Lufira River, Haut Luapula district, south-eastern Belgian Congo.

**Distinguishing characters:** Differs from the nominate race in being warmer and browner in tone above. The non-breeding dress is more russet brown above. Wing 64 to 70 mm.

General distribution: South-eastern Belgian Congo, Northern Rhodesia and Tanganyika Territory.

Range in Eastern Africa: Malagarasi River delta to upper Nyamanse River and Ufipa Plateau, western Tanganyika Territory.

Habits: As for other races.

Recorded breeding: Southern Belgian Congo, December and January.

**1027** ROCK-LOVING CISTICOLA.  *CISTICOLA EMINI* Reichenow.

*Cisticola emini emini* Reichw. **Pl. 72.**

*Cisticola emini* Reichenow, J.f.O. p. 56, 1892: Busissi, south end Lake Victoria, northern Tanganyika Territory.

**Distinguishing characters:** First primary blade-shaped. Adult, differs from the Whistling and Trilling Cisticolas by its smaller size and very conspicuous chestnut brown top of the head. Probably no non-breeding dress. Wing, male 61 to 63, female 56 mm. The young bird is warmer brown above; the top of the head being uniform with the mantle.

**Range in Eastern Africa:** Central Kenya Colony at Limoru to Tanganyika Territory from Mwanza district to the Mkalama and Kilosa districts.

**Habits:** Little has been recorded concerning this species. One reason for the lack of information seems to be due to the fact that it haunts flat bare rocks in situations that most people would consider unlikely for any form of bird life.

**Nest, Eggs and Call:** See under other races.

**Recorded breeding:** No records.

*Cisticola emini petrophila* Alex.

*Cisticola petrophila* Alexander, Bull. B.O.C. 19, p. 104, 1907: Pettu, northern Nigeria.

**Distinguishing characters:** Differs from the nominate race in having the upperparts dark sepia with the top of the head brighter chestnut brown; wings uniform with mantle, with no distinct edging; face suffused with buff; below, chin pure white; rest of underparts strongly washed with buff. In non-breeding dress the chestnut brown of the top of the head extends on to the upper mantle; below, pale fawn. The sexes are alike. Wing, male 58 to 64, female 52 to 56 mm.

**General distribution:** Northern Nigeria to north-eastern Belgian Congo and the Sudan.

**Range in Eastern Africa:** Western and southern Sudan from Darfur to the Dongotona Mts.

**Habits:** Found among bare rocks with a little low herbage, unobtrusive and rather silent, although often hopping about on the open rock with some tail movement.

**Recorded breeding:** About July to October.

*Cisticola emini teitensis* Van Som.
*Cisticola teitensis* Van Someren, Nov. Zool. 29, p. 217, 1922: Sagala, Taita district, south-eastern Kenya Colony.

**Distinguishing characters:** Similar to the non-breeding dress of the Nigerian race, but warmer colour above and less suffused with buff below; the chin, throat and lower belly whitish. Wing, male 59 to 60 mm.

**Range in Eastern Africa:** South-eastern Kenya Colony at Taita to Mkomasi, north-eastern Tanganyika Territory.

**Habits:** As for other races.

**Recorded breeding:** No records.

*Cisticola emini lurio* Vin.
*Cisticola (emini) lurio* Vincent, Bull. B.O.C. 53, p. 173, 1933: Mirrote, Mozambique Province, Portuguese East Africa.

**Distinguishing characters:** Differs from other races in being darker above and streaks on mantle rather more distinct; whiter below. In non-breeding dress the chestnut brown of the head extends on to the upper mantle; pale tawny below. Wing, male 58 to 62, female 51 to 53 mm. It differs from the Mozambique race of the Rattling Cisticola in being much darker above.

**Range in Eastern Africa:** Northern Portuguese East Africa and Nyasaland.

**Habits:** Found on rock slopes where aloes abound but grass is of the scantiest, and are wild and difficult to approach. The piping whistle as well as many of the movements are reminiscent of the Wailing Cisticola.

**Nest and Eggs:** Nest ball-shaped with a side entrance of coarse grass outside with finer grass beneath, and lined with vegetable down. The only two nests known were concealed in patches of coarse grass eighteen inches high growing on a rocky slab. Eggs three, whitish or pale bluish green finely freckled with reddish brown and pale lilac in somewhat of a zone at the larger end; about 18·5 × 13·5 mm.

**Recorded breeding:** January, the season is probably November to March.

**Call:** On alarm a shrill complaining bi-syllabic whistle of 'pee-u' is uttered by both sexes, the male somewhat harsher. The breeding male also emits a little high-pitched trill, bell-like but of short duration. A small alarm note heard in the non-breeding season is nasal and rasping.

**1028 BLACK-LORED CISTICOLA.** *CISTICOLA NIGRILORIS* Shelley. **Pl. 72.**

*Cisticola nigriloris* Shelley, Ibis, p. 536, pl. 2, fig. 2, 1897: Kombi, Masuku Range, northern Nyasaland.

**Distinguishing characters:** First primary blade-shaped. Adult, lores black; above, dusky earth brown; top of the head and hind neck chestnut brown; below, buffish cream; breast and sides dusky; legs russet. There is no non-breeding dress. The sexes are alike. Wing, male 68 to 72, female 64 to 68 mm. The young bird has the mantle and rump chestnut brown; forehead washed with dull black.

**General distribution:** Southern Tanganyika Territory to northern Nyasaland.

**Range in Eastern Africa:** The highlands of southern Tanganyika Territory.

**Habits:** Found in patches of thick bushes, grass and luxuriant soft herbage, usually on the fringes of hill forest, where the birds are inclined to be more often heard than seen. This species and Chubb's and Hunter's Cisticolas are extremely similar in habits. A pair will range themselves alongside one another on some prominent perch with tails fanned, and then commence a chorus of piping babbles in duet and with a variety of notes, the female with one set and the male with another, all the while turning and bowing.

Sometimes other birds will appear and join in the chorus. Often the duet takes place with the sexes situated at some little distance apart, the hen uttering her flute-like enquiring notes, and the male replying with his crescendo and diminuendo bell-like trillings; but they usually range up together before the chorus is completed.

**Nest and Eggs:** Ball-type nest of dry grass lined with plant down, low down among tall grass. Eggs two or three, whitish to blue, marked in varying degrees with reds and secondary purplish greys. No measurements available.

**Recorded breeding:** November to February.

**Call:** See Habits.

1029 CHUBB'S CISTICOLA. *CISTICOLA CHUBBI* Sharpe. *Cisticola chubbi* Sharpe, Ibis, p. 157, 1892: Kimangitchi, Mt. Elgon.

**Distinguishing characters:** First primary blade-shaped. Adult, very similar to the Black-lored Cisticola, but smaller and whiter below; top of head pale chestnut red; lores slightly dusky. There is no non-breeding dress. The sexes are alike. Wing, male 63 to 67, female 58 to 62 mm. The young bird is duller than the adult; top of head pale russet.

**General distribution:** North-eastern Belgian Congo to Kenya Colony and Tanganyika Territory.

**Range in Eastern Africa:** Highlands of Uganda, Urundi-Ruanda, western Kenya Colony and north-western Tanganyika Territory.

**Habits:** As for the Black-lored Cisticola, secretive birds but noisy inside cover.

**Nest and Eggs:** Nest of the usual ball-type. Eggs one or two, larger than most Cisticolas, pale blue freckled with rust red inclining to a ring on the larger end; about 19 × 13·5 mm.

**Recorded breeding:** Nearly all the year round but mostly about September to November.

**Call:** A series of piping babbling notes often in a duet or chorus, very much as for the Black-lored Cisticola, but even more noisy.

**1030** HUNTER'S CISTICOLA.   *CISTICOLA HUNTERI*
Shelley.

*Cisticola hunteri hunteri* Shell.
*Cisticola hunteri* Shelley, P.Z.S. p. 364, 1889: Mt. Kilimanjaro.

**Distinguishing characters:** First primary blade-shaped. Adult, above, top of head dark brown, with suffused darker streaks; rest of upperparts dusky brown with broad suffused darker streaks; below, dusky, throat slightly paler; bill thin. There is no non-breeding dress. The sexes are alike. Wing, male 60 to 65, female 57 to 61 mm. The young bird is browner above, less dusky than the adult; below, more dusky buff.

**Range in Eastern Africa:** Mt. Kilimanjaro.

**Habits:** As for the Black-lored Cisticola and Chubb's Cisticola but even more secretive and only seen out of cover during the mating season.

**Recorded breeding:** No records but probably about October to December.

**Nest, Eggs and Call:** See under other races.

*Cisticola hunteri prinioides* Neum. **Ph. xii.**
*Cisticola prinioides* Neumann, J.f.O. p. 304, 1900: Mau, Kenya Colony.

**Distinguishing characters:** Paler and browner above than the nominate race; usually less distinctly streaked on head and mantle; below, less dusky. Wing 57 to 64 mm.

**Range in Eastern Africa:** Kenya Colony highlands from Mt. Kenya to Meru and Mau.

**Habits:** As for the nominate race, found in bush near streams and swamps.

**Nest and Eggs:** Nest large, of elongated ball-type with a large opening near the top, made of wide grass and lined with down or soft material. It is usually placed in a grass tussock or shrub and is not usually well-concealed. Eggs two, rarely three, white, pale blue or green ground colour with small spots or freckles of brown or blackish brown; about 17·5 × 12·5 mm.

**Recorded breeding:** Kenya Colony highlands, June to December, possibly in other months also.

# WARBLERS

**Call:** Much as for the last species. The duet consists of a soft flute-like 'twee-twee-twee' from one bird answered by a succession of silvery trills from the other. There are also a number of lesser call notes.

*Cisticola hunteri masaba* Lynes.
*Cisticola hunteri masaba* Lynes, Ibis, Supp. p. 343, 1930: Mt. Elgon.

**Distinguishing characters:** Very similar to the nominate race but darker below. Wing, male 63 to 65, female 61 mm.

**Range in Eastern Africa:** Mt. Elgon from about 9,000 feet to the summit.

**Habits:** As for other races.

**Recorded breeding:** Probably as for the nominate race, variable but mostly October to December.

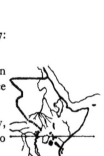

*Cisticola hunteri hypernephala* Elliott.
*Cisticola hunteri hypernephala* Elliott, Bull. B.O.C. 68, p. 10, 1947: Olosirwa Mt., north-eastern Tanganyika Territory.

**Distinguishing characters:** Above, darker, more mottled than the nominate race and paler below, and differs from the Mau race in the grey tone and heavy mottling above. Wing 58 to 63 mm.

**Range in Eastern Africa:** North-eastern Tanganyika Territory, Loliondo, Ketumbeine, Elanairobi, Olosirwa, Mbagai, Ngorongoro and Oldeani.

**Habits:** As for other races, the song is a duet consisting of a repeated 'whirdle whirdle' answered by 'eee.'

**Recorded breeding:** Normally November to February.

**1031** SINGING CISTICOLA. *CISTICOLA CANTANS* (Heuglin).
*Cisticola cantans cantans* (Heugl.). **Pl. 72.**
*Drymœca cantans* Heuglin, Ibis, p. 96, 1869: Gondar, northern Abyssinia.

**Distinguishing characters:** First primary blade-shaped. Adult, top of head and edges of flight feathers chestnut brown; sides of face and stripe over eye whitish; rest of upperparts earth brown; subterminal black spots on both webs of tail feathers; below, creamy

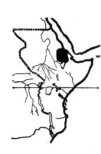

buff. In non-breeding dress the mantle is streaked with black. The sexes are alike. Wing, male 57 to 63, female 52 to 56 mm. The young bird is duller above, paler brown with indistinct streaking on the mantle.

**Range in Eastern Africa:** The highlands of southern Eritrea and Abyssinia at 6,000 to 8,000 feet.

**Habits:** A well-known bird where suitable environment is to be found, namely thick bushes and rank undergrowth, but always where it is of a soft and luxuriant nature, and is particularly fond of the profuse growth of grass and bracken to be met with in the gullies on the lower slopes of mountains. Found round forest but does not enter the big timber. In general movements it is like the Red-faced Cisticola, and utters various stuttering notes before creeping away low down in the tangle. There are several small and sibilant alarm calls, but now and again a male will hop on to a prominent perch just above the undergrowth and with vibrating tail well cocked up will utter a metallic chippering song in bursts of three or four notes which is strangely like the chippering of *Cinnyris* Sunbirds. On occasion the female has a similar but less shrill and voluble call. This species is much parasitized by the Pin-tailed Whydah.

**Nest and Eggs:** Nest of Tailor-bird type, of dry grass lined with plant down sewn into the leaves of a broad and soft-leaved herb or sapling, generally two to three feet from the ground. Eggs two or three, pale blue, plain or covered with small or large reddish spots; about 17 × 12·5 mm.

**Recorded breeding:** June to October.

**Call:** See Habits.

*Cisticola cantans concolor* (Heugl.).
*Drymœca concolor* Heuglin, Ibis, p. 97, 1869: Upper Nile, Sudan.

**Distinguishing characters:** Differs from the nominate race in having the top of the head and hind neck less chestnut red and in being generally paler. The non-breeding dress is lighter and brighter. Wing, male 55 to 59, female 51 to 58 mm.

**General distribution:** Nigeria to the Sudan.

**Range in Eastern Africa:** The Sudan from Darfur to the Bahr-el-Ghazal.

**Habits:** As for the nominate race.

**Recorded breeding:** June to September and October.

*Cisticola cantans pictipennis* Mad. **Ph. xii.**
*Cisticola pictipennis* Madarász, Ann. Mus. Hung. 2, p. 205, 1904:
   Moshi, north-eastern Tanganyika.

**Distinguishing characters:** Differs from the nominate race in having the top of the head rather darker chestnut brown. There is no non-breeding dress. Wing, male 58 to 62, female 55 to 59 mm.

**Range in Eastern Africa:** Kenya Colony and northern Tanganyika Territory from Mt. Elgon to Taita and Arusha, but not coastal areas.

**Habits:** As for the nominate race.

**Recorded breeding:** Principally April to August, also November and December.

*Cisticola cantans belli* O. Grant.
*Cisticola belli* O. Grant, Bull. B.O.C. 21, p. 71, 1909: Mokia, Lake
   George, Uganda.

**Distinguishing characters:** Differs from the nominate race in being darker; mantle and rump dark greyish brown. There is no non-breeding dress. Wing, male 54 to 60, female 48 to 52 mm.

**General distribution:** North-eastern Belgian Congo, the Sudan, Abyssinia, Uganda and Tanganyika Territory.

**Range in Eastern Africa:** Southern Sudan, southern Abyssinia, Uganda (but not Mt. Elgon), and north-western and western Tanganyika Territory as far south as the Kigoma area; also Kaserazi Island, Lake Victoria.

**Habits:** As for the nominate race.

**Recorded breeding:** Principally about April to October.

*Cisticola cantans munzneri* Reichw.
*Cisticola munzneri* Reichenow, J.f.O. p. 163, 1916: Sanya, Mahenge
   district, southern Tanganyika Territory.

**Distinguishing characters:** Differs from the nominate race in being smaller and whiter below. In non-breeding dress the top of the head is brighter chestnut brown; mantle warmer brown. Wing, male 54 to 58, female 48 to 55 mm.

**General distribution:** Tanganyika Territory to Portuguese East Africa and Southern Rhodesia.

**Range in Eastern Africa:** Southern Tanganyika Territory, Portuguese East Africa and Nyasaland.

**Habits:** As for the nominate race.

**Recorded breeding:** November to April.

**Distribution of other races of the species:** West Africa.

### 1032 RED-FACED CISTICOLA. *CISTICOLA ERYTHROPS* Hartlaub.

*Cisticola erythrops nilotica* Mad.
*Cisticola nilotica* Madarász, Ann. Mus. Hung. 12, p. 591, 1904: Blue Nile, Sudan.

**Distinguishing characters:** First primary blade-shaped. Adult, forehead, sides of face and eye stripe pale russet; rest of upperparts greyish isabelline; edges of flight feathers slightly russet; below, creamy buff, throat paler; subterminal black spots on both webs of tail feathers. In non-breeding dress more isabelline fawn colour above; top of head clearer pale tawny. The sexes are alike. Wing, male 56, female 51 mm. The young bird is similar to the adult in non-breeding dress but paler above.

**Range in Eastern Africa:** Middle Blue Nile in the Fung Province of the Sudan.

**Habits:** Although with few exceptions found in damper situations, this species favours similar vegetation to the Singing Cisticola to which species it is not unlike in general movements and call. A very elusive bird at all times, and even breeding males will drop down into cover at the least sign of intrusion.

**Nest, Eggs and Call:** See under other races.

**Recorded breeding:** June to October.

*Cisticola erythrops sylvia* Reichw. **Pl. 72. Ph. xii.**
*Cisticola sylvia* Reichenow, O.M. p. 28, 1904: Ulegga, Lake Albert, eastern Belgian Congo.

**Distinguishing characters:** Differs from the Nile race in having the top of the head and sides of face brighter russet. There is no non-breeding dress. Wing, male 59 to 65, female 53 to 57 mm. The young bird is duller than the adult and has a pale russet mantle.

**General distribution:** North-eastern and central Belgian Congo, to the Sudan, Uganda, Kenya Colony and Tanganyika Territory.

# WARBLERS

**Range in Eastern Africa:** South-western Sudan, Uganda, Kenya Colony and northern and central Tanganyika Territory.

**Habits:** As for the Nile race, inhabiting dense herbage near streams or swamps or overgrown forest clearings.

**Nest and Eggs:** Nest of Tailor-bird-type as for the Singing Cisticola. Eggs three, occasionally two or four, greenish blue to white, strongly marked with rusty reds and sometimes secondary greys; about 17·2 × 13 mm.

**Recorded breeding:** Principally April to August, also November and December.

**Call:** Uttered from a prominent weed or bush top, very loud and striking, a string of rich metallic whistles emitted in jerking fashion. It varies to some extent, but the commonest strains may be written as 'chip-wip, chip-wip, chip-wip' or as 'chewip-che, we-we-we.' A slightly smaller but harsher burst of conversational notes is also common.

*Cisticola erythrops pyrrhomitra* Reichw.
*Cisticola pyrrhomitra* Reichenow, J.f.O. p. 162, 1916: Gallaland, Abyssinia.

**Distinguishing characters:** Similar in size to the Belgian Congo race but rather darker above; forehead and sides of face inclining to tawny. The non-breeding dress is paler; chin, throat and middle of belly whitish. Wing, male 59 to 65, female 54 to 58 mm.

**Range in Eastern Africa:** Abyssinia to the Imatong Mts. and Boma Plateau areas of the southern Sudan.

**Habits:** As for the Nile race.

**Recorded breeding:** Probably June to October.

*Cisticola erythrops nyasa* Lynes.
*Cisticola erythrops nyasa* Lynes, Ibis, Supp. p. 374, 1930: Chiromo, southern Nyasaland.

**Distinguishing characters:** Similar to the Belgian Congo race, but smaller and rather paler coloured and slightly browner above; rather less russet on forehead and sides of face. The non-breeding dress is browner; forehead and top of head more uniform with mantle. Wing, male 55 to 59, female 49 to 56 mm. The young bird often has the chin to belly pale yellow.

**General distribution:** South-eastern Belgian Congo and Tanganyika Territory to the Rhodesias, Portuguese East Africa, Nyasaland, the Transvaal and Zululand.

**Range in Eastern Africa:** Southern Tanganyika Territory to Portuguese East Africa and Nyasaland.

**Habits:** As for the Nile race.

**Recorded breeding:** November to April.

**Distribution of other races of the species:** West Africa and Angola, the nominate race being described from between Cape Palmas and Calabar.

**1033** WINDING CISTICOLA. *CISTICOLA GALACTOTES* (Temminck).
*Cisticola galactotes galactotes* (Temm.). **Pl. 72.**
*Malurus galactotes* Temminck, in Temm. & Laug. Pl. Col. livr. No. 3, pl. 65, fig. 1, 1823: Natal, South Africa.

**Distinguishing characters:** First primary blade-shaped. Adult, top of head warm russet brown; buffish stripe over eye; mantle greyish, broadly streaked with black; rump greyish; edges of flight feathers tawny; black subterminal spots on both webs of tail feathers, less distinct on the upperside; below, creamy buff; feet and toes large. In non-breeding dress tawny above; top of head indistinctly streaked with black; mantle broadly streaked with black. The sexes are alike. Wing, male 59 to 65, female 51 to 55 mm. The young bird is somewhat similar to the adult in non-breeding dress, but has the top of the head broadly streaked with black and some pale yellow on chin to chest.

**General distribution:** South-eastern Belgian Congo, Northern Rhodesia, Nyasaland, Portuguese East Africa, northern Bechuanaland and Natal.

**Range in Eastern Africa:** Portuguese East Africa and Nyasaland.

**Habits:** With a few rare exceptions, a typical marsh bird, frequenting luxuriant grass and sedge of fringes of swamps and lakes, or of sodden ground. In the non-breeding season silent and unobtrusive, but when breeding the males perch on grass stems and utter a characteristic single rasping repeated note which is very like the winding-up of a clock. They not infrequently fly back and forth over the nesting area with the same sharp note, a quick succession of 'trit-trits' to which the female, if away from the nest, replies with two longer drawn out 'treets'; whilst occasionally a loud piping call of 'pirrit-pirrit' is heard.

**Nest and Eggs:** Nest ball-type, as described under the Desert Cisticola, generally two to four feet up in the grasses, sedges or water weeds, sometimes among herbage on dry grasslands; it is often decorated exteriorly with plant and cocoon fluff and warmly lined. Eggs three, glossy, from white with red or liver brown blotches to deep terra-cotta or brick red; about $17 \times 12 \cdot 5$ mm.

**Recorded breeding:** Generally about November to April, in Nyasaland in any month from October to June.

**Call:** See Habits.

*Cisticola galactotes lugubris* Rüpp.
*Sylvia (Cisticola) lugubris* Rüppell, N. Wirbelt. Aves. p. 111, 1840: Gondar, northern Abyssinia.

**Distinguishing characters:** The male differs from the nominate race in having the top of the head striped in breeding dress. In non-breeding dress the top of the head is distinctly streaked with black; mantle tawny, streaked with black. The female has paler coloured lores; bill less black. Wing, male 60 to 66, female 53 to 57 mm.

**Range in Eastern Africa:** Eritrea and Abyssinia.

**Habits:** As for the nominate race.

**Recorded breeding:** June to October, a month or so earlier in the south.

*Cisticola galactotes hæmatocephala* Cab.
*Cisticola hæmatocephala* Cabanis, J.f.O. p. 412, 1868: Mombasa, eastern Kenya Colony.

**Distinguishing characters:** Differs from the nominate race in having sparse dull blackish markings on the mantle; sides of face whiter. There is no non-breeding dress. Wing, male 55 to 59, female 49 to 53 mm.

**Range in Eastern Africa:** Coastal areas of southern Italian Somaliland, Kenya Colony and Tanganyika Territory.

**Habits:** As for the nominate race.

**Recorded breeding:** Mombasa, Kenya Colony, May.

*Cisticola galactotes marginata* (Heugl.).

*Drymœca marginata* Heuglin, Ibis, p. 94, pl. 1, 1869: Upper Nile, lat. 6°–7° N., southern Sudan.

**Distinguishing characters:** Differs from the nominate race in being smaller; top of head greyish brown. The non-breeding dress is similar to that of the Abyssinian race. Wing, male 53 to 60, female 49 to 55 mm.

**Range in Eastern Africa:** The Sudan south of lat. 12° N. to northern Uganda.

**Habits:** As for the nominate race.

**Recorded breeding:** June to October.

*Cisticola galactotes nyansæ* Neum. **Ph. xiii.**

*Cisticola lugubris nyansæ* Neumann, O.M. p. 78, 1905: Sesse Island, Lake Victoria, Uganda.

**Distinguishing characters:** Differs from the nominate race in having broader black markings on the mantle. Usually no non-breeding dress, but occasionally one is assumed and is similar to that of the Abyssinian race. Wing, male 54 to 65, female 50 to 59 mm.

**General distribution:** North-eastern Belgian Congo to Kenya Colony and Tanganyika Territory.

**Range in Eastern Africa:** Uganda, Kenya Colony from Kisumu, Nakuru and middle Tana River to Doinyo Erok and Mwatate, and Tanganyika Territory at north-western areas and Mwanza.

**Habits:** As for the nominate race.

**Recorded breeding:** Principally between March and August with rarely a secondary season in November and December.

*Cisticola galactotes suahelica* Neum.

*Cisticola lugubris suahelica* Neumann, O.M. p. 78, 1905: Begu, Usegua district, Tanganyika Territory.

**Distinguishing characters:** Differs from the nominate race in being paler. There is no non-breeding dress, except south of lat. 7° S. where it is similar to the Abyssinian race in having the mantle fawn colour streaked with black. Wing, male 57 to 61, female 51 to 55 mm.

# WARBLERS

**General distribution:** East central Belgian Congo to Tanganyika Territory.

**Range in Eastern Africa:** North-eastern to south-western Tanganyika Territory from Lake Natron, Moshi to Tabora, Morogoro, Iringa and Lake Rukwa.

**Habits:** As for the nominate race.

**Recorded breeding:** Probably more or less all the year round but mostly April to June.

*Cisticola galactotes zalingei* Lynes.
*Cisticola galactotes zalingei* Lynes, Ibis, Suppl. p. 390, 1930: Zalingei, Darfur Province, Sudan.

**Distinguishing characters:** Differs from the southern Sudan race in being rather paler above. In non-breeding dress paler with much more distinct streaks on top of head. Wing, male 58 to 64, female 52 to 56 mm.

**General distribution:** Northern Nigeria to the Sudan.

**Range in Eastern Africa:** The Darfur Province of the Sudan.

**Habits:** As for the nominate race. Eggs are three to five, white with orange and brown blotches or rich pinkish cream, heavily marked with purple madder blotches and secondary greys; about 16·5 × 12 mm.

**Recorded breeding:** July to October.

**Distribution of other races of the species:** West Africa.

**1034** CARRUTHER'S CISTICOLA. *CISTICOLA CARRUTHERSI* O. Grant.

*Cisticola carruthersi* O. Grant, Bull. B.O.C. 23, p. 94, 1909: Mokia, Lake George, Uganda.

**Distinguishing characters:** First primary blade-shaped. Adult, similar to the Winding Cisticola in breeding dress, but top of head chestnut; mantle darker; edges of flight feathers brown not tawny; sides of face washed with russet; below, flanks dusky; subterminal black in tail extending well up feathers; bill finer and straighter. There is no non-breeding dress. The sexes are alike. Wing, male 56 to 62, female 52 to 56 mm. Juvenile plumage unrecorded.

**General distribution:** North-eastern Belgian Congo to Kenya Colony.

**Range in Eastern Africa:** Uganda and western Kenya Colony at Kisumu.

**Habits:** Found in papyrus swamps, but its behaviour is undescribed.

**Nest and Eggs:** Nest a loosely made oval of grass blades and felty material with a side of top entrance hole. Eggs three or four, pinkish-red like those of the Winding Cisticola, glossy and thickly spotted with maroon and with mauve underspotting; about 16 × 12 mm.

**Recorded breeding:** Western Uganda, April. Kisumu, Kenya Colony, July.

**Call:** Unrecorded.

**1035** STOUT CISTICOLA. *CISTICOLA ROBUSTA* (Rüppell).
*Cisticola robusta robusta* (Rüpp.). **Pl. 72.**
*Drymoica robusta* Rüppell, Syst. Uebers, p. 35, 1845: Shoa, central Abyssinia.

**Distinguishing characters:** First primary blade-shaped. Adult male, top of head dull brown streaked with suffused blackish; hind neck slightly tawny; lightish stripe over eye; mantle greyish brown, broadly streaked with dusky black; rump plainer; subterminal black spots on both webs of tail feathers; edges of flight feathers tawny; sides of face dusky; below, creamy buff; sides of chest dusky; bill heavy. In non-breeding dress the head and hind neck are dark tawny streaked with black; mantle broadly streaked with black; edges of flight feathers brighter tawny. The female is much smaller than the male. Wing, male 75 to 81, female 62 to 66 mm. The young bird is pale tawny streaked with blackish above; bright yellow below.

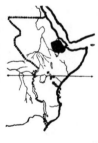

**Range in Eastern Africa:** The Abyssinian plateau to the Harar Range.

**Habits:** Found in open grass lands or clearings, with or without scattered trees or shrubs, where, like the Croaking Cisticola, it is a conspicuous bird on account of its size. It generally prefers moist spots and hollows, probably only because the type of grass is more suitable for its nesting. The male bird occupies a prominent perch on grass stems or shrub tops, and utters a distinctive call which can only be described as a piping and rippled or tremulo whistle; sometimes it is not rippling and is not unlike the piped whistle of a Wader. Wary and difficult to approach and quickly warns the hen of the approach of danger.

**Nest and Eggs:** Nest of ball-type, on or near the ground, somewhat flimsy and with a bower of living green grass-blades bowered over its dome. Eggs two or three, light turquoise blue, occasionally white, plain, or freckled strongly towards the big end with pale red; about 19 × 14 mm.

**Recorded breeding:** About April to October or November.

**Call:** See Habits.

*Cisticola robusta nuchalis* Reichw.
*Cisticola nuchalis* Reichenow, O.M. p. 61, 1903: Kagera River, Bukoba district, north-western Tanganyika Territory.

**Distinguishing characters:** Differs from the nominate race in being smaller; usually brighter in colour; black streaks on mantle less broad; head and neck brighter. There is no non-breeding dress. Wing, male 62 to 68, female 55 to 59 mm. The young bird is paler yellow below.

**General distribution:** North-eastern Belgian Congo and southern Sudan, Kenya Colony and Tanganyika Territory.

**Range in Eastern Africa:** Uganda, the southern Sudan as far north as Shambé, western Kenya Colony and north-western Tanganyika Territory.

**Habits:** As for the nominate race.

**Recorded breeding:** March to July and November to December.

*Cisticola robusta ambigua* Sharpe. **Ph. xiii.**
*Cisticola ambigua* Sharpe, Bull. B.O.C. 11, p. 28, 1900: Ravine, Kenya Colony highlands.

**Distinguishing characters:** Similar to the Bukoba race but larger. Wing, male 68 to 72, female 58 to 62 mm.

**Range in Eastern Africa:** Central and southern Kenya Colony from Eldoret, Laikipia and the upper Tana River to Mts. Meru and Kilimanjaro, north-eastern Tanganyika Territory.

**Habits:** As for the nominate race, eggs are somewhat smaller, about 17·5 × 12·5 mm.

**Recorded breeding:** Principally April to July; also November and December.

*Cisticola robusta schraderi* Neum.
*Cisticola robusta schraderi* Neumann, J.f.O. p. 265, 1906: Senafé, Eritrea.

**Distinguishing characters:** Very similar in size to the nominate race, but differs in being lighter and brighter above, and the hind neck is somewhat lighter coloured. Wing, male 75, female 62 mm.

**Range in Eastern Africa:** The highlands of southern Eritrea.

**Habits:** As for the nominate race.

**Recorded breeding:** No records.

*Cisticola robusta omo* Neum. & Lynes.
*Cisticola robusta omo* Neumann & Lynes, Bull. B.O.C. 48, p. 136, 1928: Kaukati, south-western Abyssinia.

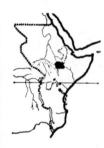

**Distinguishing characters:** Similar in size to the nominate race but general coloration richer and darker; top of head and mantle very black. Wing, male 75 to 78, female 69 mm.

**Range in Eastern Africa:** South-western Abyssinia, on the watershed of the upper and middle Omo River east to Alghe.

**Habits:** As for the nominate race.

**Recorded breeding:** April to July.

*Cisticola robusta aberdare* Lynes.
*Cisticola robusta aberdare* Lynes, Ibis, Supp. p. 426, 1930: Aberdare Range, Kenya Colony.

**Distinguishing characters:** A very dark coloured race and heavily streaked on the mantle; tail plain black on the upperside. Wing, male 72, female 59 to 63 mm.

**Range in Eastern Africa:** Molo to the Aberdare Mts., Kenya Colony.

**Habits:** As for the nominate race.

**Recorded breeding:** No records.

*Cisticola robusta awemba* Lynes.

*Cisticola robusta awemba* Lynes, Bull. B.O.C. 53, p. 169, 1933: Luwingu, Awemba Province, north-eastern Northern Rhodesia.

**Distinguishing characters:** Similar in size to the Aberdare race. The breeding dress is similar to the Bukoba race but warmer in tone and more strongly washed with buff below. In non-breeding dress the top of the head is streaked with black. Wing, male 71 to 76, female 57 to 62 mm.

**General distribution:** South-western Tanganyika Territory to north-eastern Northern Rhodesia.

**Range in Eastern Africa:** South-western Tanganyika Territory at Ufipa and Mbeya.

**Habits:** As for other races.

**Recorded breeding:** No records. Breeding dress in March.

**Distribution of other races of the species:** West Africa and Angola.

**1036** CROAKING CISTICOLA. *CISTICOLA NATALENSIS* (Smith).

*Cisticola natalensis natalensis* (Smith). **Pl. 72.**

*Drymoica natalensis* A. Smith, Ill. Zool. S. Afr. pl. 80, 1843: Durban, South Africa.

**Distinguishing characters:** First primary blade-shaped. Adult, differs from the Stout Cisticola in having a shorter and thicker bill; hind-neck not tawny but uniform with mantle; streaks on mantle and head less distinct; below, creamy white. In non-breeding dress the tail is longer; mantle tawny buff streaked with black; below, buff. The female is smaller than the male. Wing, male 70 to 78, female 58 to 64 mm. The young bird is similar above to the adult in non-breeding dress; below, yellow. It differs from the young of the Stout Cisticola by its shorter and thicker bill.

**General distribution:** Tanganyika Territory, Nyasaland, southern Portuguese East Africa, Southern Rhodesia, the Transvaal to Cape Province and Natal.

**Range in Eastern Africa:** Southern Tanganyika Territory to Portuguese East Africa and Nyasaland.

**Habits:** Found in open grass or waste lands or clearings in savannah woodland, usually where there is some growth of low shrub or sapling trees. In general appearance, movements and size, and in the male's wary guarding of the female, this species is remarkably like the Stout Cisticola. It is rather more conspicuous as the males occupy a more prominent perch, such as on the tip-top of a tree out in the open from which they continually advertise their presence. They also perform haphazard and short courting flights, reminiscent of some Whydahs, with flapping, butterfly-like wings.

**Nest and Eggs:** As for the Stout Cisticola. Eggs two, three or four, white or pale blue, plain or speckled with light reds or browns; about $19 \cdot 5 \times 14$ mm.

**Recorded breeding:** About November to April.

**Call:** An oft-repeated and unmusical vibrating single syllable; the alarm note, which is the most distinctive and is loudly uttered at the slightest sign of alarm, is a frog-like croak of two syllables.

*Cisticola natalensis valida* (Heugl.).
*Drymœca valida* Heuglin, J.f.O. p. 258, 1864: Wau, Bahr-el-Ghazal, south-western Sudan.

**Distinguishing characters:** This race has a perennial mode of dress, also a variety of different breeding dresses varying from one like the nominate race in breeding dress to a half-and-half mixture of its breeding and non-breeding dresses. Wing, male 68 to 74, female 54 to 60 mm.

**General distribution:** Portuguese Congo, north and central Belgian Congo, the Sudan, Kenya Colony and Tanganyika Territory.

**Range in Eastern Africa:** The southern Sudan, Uganda, western Kenya Colony and north and central Tanganyika Territory as far south as Kilosa.

**Habits:** As for the nominate race.

**Recorded breeding:** More or less all the year round, but mainly March to July; northern Uganda, May to November.

*Cisticola natalensis inexpectata* Neum.
*Cisticola natalensis inexpectata* Neumann, J.f.O. p. 268, 1906: Lake Abassi, southern Abyssinia.

**Distinguishing characters:** Differs from the nominate race in having the top of the head warmer coloured; streaks on mantle wider and darker. In non-breeding dress the streaking on the head and mantle is broader than in the nominate race. Wing, male 68 to 78, female 58 to 68 mm.

*Note:* Moults are rather irregular in the north of its range.

**Range in Eastern Africa:** Southern Eritrea to south-western and southern Abyssinia as far south as Adola and Alghe.

**Habits:** As for the nominate race.

**Recorded breeding:** About June to October.

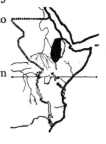

*Cisticola natalensis argentea* Reichw.
*Cisticola argentea* Reichenow, O.M. p. 25, 1905: Umfudu, lower Juba River, Italian Somaliland.

**Distinguishing characters:** Differs from all other races in being pale greyish or pale sandy above with broad black streaking on head and mantle. In non-breeding dress the feathers of the head and mantle have paler edges and the bill is horn colour, not blackish. Wing, male 70 to 76, female 60 to 62 mm.

**Range in Eastern Africa:** The Yavello area of the southern Abyssinia, northern Kenya Colony at Mt. Marsabit and southern Italian Somaliland.

**Habits:** As for the nominate race.

**Recorded breeding:** No records but probably about February to September.

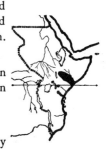

*Cisticola natalensis kapitensis* Mearns.
*Cisticola strangei kapitensis* Mearns, Smiths Misc. Coll. 61, No. 25, p. 4, 1911: Potha, Kapiti Plains, Kenya Colony.

**Distinguishing characters:** Differs from the nominate race in having the top of the head and hind neck brown streaked with dusky black; mantle streaking broad. The female has the mantle rather more buff and black. There is no non-breeding dress. Wing, male 67 to 71, female 55 to 59 mm.

**Range in Eastern Africa:** The central plateau of Kenya Colony from about 3,000 to 5,000 feet.

**Habits:** As for the nominate race.

**Recorded breeding:** Probably as for other Cisticolas in the same region, namely April to July; also November and December more irregularly.

*Cisticola natalensis tonga* Lynes.
*Cisticola natalensis tonga* Lynes, Ibis, Supp. p. 448, 1930: Kodok, White Nile, Sudan.

**Distinguishing characters:** Differs from the nominate race in non-breeding dress in being smaller, paler, and mantle cream coloured streaked with black. Breeding dress not known. Wing, male 68 to 72, female 59 to 61 mm.

**Range in Eastern Africa:** White and Blue Nile areas of the Sudan between about lat. 9° 30′ and 12° N.

**Habits:** As for the nominate race, but there is some evidence of seasonal movement and they do not appear to breed along the rivers where they are found in the non-breeding season.

**Recorded breeding:** No records.

*Cisticola natalensis katanga* Lynes.
*Cisticola natalensis katanga* Lynes, Ibis, Suppl. p. 443, 1930: Kambove, Haut Luapula district, south-eastern Belgian Congo.

**Distinguishing characters:** In breeding dress differs from the nominate race in having the top of the head rather warmer brown; in non-breeding dress general colour above darker and warmer brown than in the nominate race. Wing 59 to 72 mm.

**General distribution:** Northern Rhodesia, southern Belgian Congo, Nyasaland, and Tanganyika Territory.

**Range in Eastern Africa:** South-western Tanganyika Territory.

**Habits:** As for other races.

*Cisticola natalensis littoralis* V. Som.
*Cisticola natalensis littoralis* Van Someren, Bull. B.O.C. 64, p. 23, 1943: Rabai, near Mombasa, eastern Kenya Colony.

**Distinguishing characters:** Similar to the Wau race, but differs in having no non-breeding dress. Wing 58 to 71 mm.

**Range in Eastern Africa:** Coastal area of Kenya Colony and Tanganyika Territory from the Tana River to Dar-es-Salaam.

**Habits:** As for other races.

**Recorded breeding:** No records.

**Distribution of other races of the species:** West Africa and Angola to Northern Rhodesia.

## 1037 SIFFLING CISTICOLA. *CISTICOLA BRACHYPTERA* Sharpe.

*Cisticola brachyptera brachyptera* Sharpe. **Pl. 72.**
*Cisticola brachyptera* Sharpe, Ibis, p. 476, 1890: pl. 24, fig. 1: Volta River, Gold Coast.

**Distinguishing characters:** A small species; first primary scimitar-shaped. Adult male, above, sepia brown with very faint streaks; rump lighter; flight feathers faintly edged with brown; tail darker sepia brown with dusky subterminal spots; below, white; flanks buff. In non-breeding dress general colour warmer and more snuff brown; mantle with dusky centres to feathers, tail rather longer. The female has more streaking on the mantle in breeding dress than the male. In non-breeding dress the head and mantle are streaked with black. Wing, male 48 to 50, female 43 to 45 mm. The young bird has a rather longer first primary and is yellow below.

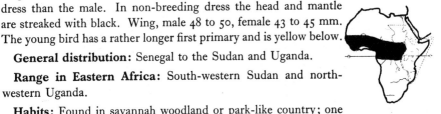

**General distribution:** Senegal to the Sudan and Uganda.

**Range in Eastern Africa:** South-western Sudan and north-western Uganda.

**Habits:** Found in savannah woodland or park-like country; one of the most ubiquitous Cisticolas, and common enough everywhere. Males in the non-breeding season and females at all times are very quiet and unobtrusive in their movements among the grass, although when disturbed they fly up into a tree to watch the intruder. Breeding males are quite unmistakable in their manner of taking up a commanding tree-top perch at varying heights, but are always seen in silhouette on a tip-top bare twig where they sit quietly and at intervals utter their wispy little song. This consists of a short chirruping or siffling twitter, best described as 'mississippi-ing.' Occasionally a male will carry out a high aerial cruise followed by a bullet-like dive to earth, but the habit seems to be a rare one and the urge for it is not entirely understood.

**Nest and Eggs:** Normal ball-type, small and well lined with plant down, without a bower, placed at about nine inches from the ground in grass or low herbage. Eggs three, exceptionally two or four, white to greenish blue, nearly always well marked with rusty red, occasionally plain; about 16 × 11·5 mm. The eggs of the species seem to be similarly variable in all races.

**Recorded breeding:** Southern Sudan, July.

**Call:** See Habits.

*Cisticola brachyptera hypoxantha* Hartl.
*Cisticola hypoxantha* Hartlaub, P.Z.S. p. 624, 1880: Magungu, Lake Albert, Uganda.

**Distinguishing characters:** Similar to the nominate race but mantle darker with broader dusky streaks in non-breeding dress. Wing, male 49 to 53, female 44 to 51 mm.

**General distribution:** North-eastern Belgian Congo, the Sudan and Uganda.

**Range in Eastern Africa:** South-eastern Sudan and northern Uganda.

**Habits:** As for the nominate race.

**Recorded breeding:** Season variable, generally about May to October according to rains from year to year; perhaps August to April in the unusual climate of the Lake Albert trough.

*Cisticola brachyptera katonæ* Mad.
*Cisticola katonæ* Madarász, Ann. Mus. Hung. p. 204, 1904: near Moshi, north-eastern Tanganyika Territory.

**Distinguishing characters:** Differs from the nominate race in being larger and having the mantle distinctly marked with dusky. There is no non-breeding dress. Wing, male 51 to 57, female 45 to 51 mm.

**Range in Eastern Africa:** Central and southern Kenya Colony from Nandi, Laikipia, Taita and Chyulu to northern Tanganyika Territory in the Arusha, Kilimanjaro and Moshi areas.

**Habits:** As for the nominate race.

**Recorded breeding:** Principally April to July and November to December.

*Cisticola brachyptera isabellina* Reichw.
*Cisticola isabellina* Reichenow, O.M. p. 60, 1907: Songea, south-western Tanganyika Territory.

**Distinguishing characters:** Similar to the nominate race but paler, duller and yellower. In non-breeding dress the yellowish colour is more pronounced and the top of head and mantle less distinctly streaked. Wing, male 48 to 52, female 42 to 46 mm.

**General distribution:** Tanganyika Territory, Portuguese East Africa, Nyasaland, and Southern Rhodesia to about lat. 20° S.

**Range in Eastern Africa:** Central and southern Tanganyika Territory to Portuguese East Africa and Nyasaland.

**Habits:** As for the nominate race.

**Recorded breeding:** From about December to May according to rains.

*Cisticola brachyptera zedlitzi* Reichw.
*Cisticola zedlitzi* Reichenow, O.M. p. 42, 1909: Mareb River, Eritrea.

**Distinguishing characters:** Differs from the nominate race in being well marked above with black. In non-breeding dress the top of head and mantle are broadly streaked with black; tail longer. Wing, male 51 to 56, female 45 to 50 mm.

**Range in Eastern Africa:** Eritrea and Abyssinia, except the dry south-eastern area.

**Habits:** As for the nominate race.

**Recorded breeding:** Abyssinia, April and July but probably from April to October or November.

*Cisticola brachyptera reichenowi* Mearns.
*Cisticola hypoxantha reichenowi* Mearns, Smiths Misc. Coll. 61, No. 25, p. 6, 1911: Changamwe, near Mombasa, eastern Kenya Colony.

**Distinguishing characters:** Differs from the nominate race in being duller, with the top of head of warmer colour. There is no non-breeding dress. Wing, male 47 to 51, female 42 to 44 mm.

**Range in Eastern Africa:** The coastal areas of southern Italian Somaliland and Kenya Colony, but not Mombasa Island.

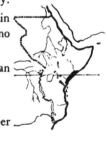

**Habits:** As for other races.

**Recorded breeding:** Principally April to July, also November and December.

*Cisticola brachyptera ankole* Lynes.
*Cisticola brachyptera ankole* Lynes, Ibis, Supp. p. 489, 1930: Ankole district, south-western Uganda.

**Distinguishing characters:** Differs from the nominate race in being browner above. Wing, male 48 to 52, female 44 to 50 mm.

**Range in Eastern Africa:** South-western Uganda and north-western Tanganyika Territory.

**Habits:** As for other races.

**Recorded breeding:** More or less all the year round, but principally about April to October.

*Cisticola brachyptera kericho* Lynes.
*Cisticola brachyptera kericho* Lynes, Ibis, Supp. p. 491, 1930: Kericho, south-western Kenya Colony.

**Distinguishing characters:** Differs from the nominate race in having a longer first primary; above, warm brown. Wing, male 54 to 55, female 49 to 50 mm.

**Range in Eastern Africa:** Central and southern Kenya Colony from Kericho to Nairobi.

**Habits:** As for other races.

**Recorded breeding:** No records, but no doubt breeds during the rains, *i.e.*, April to August and October to December.

**Distribution of other races of the species:** Angola to Northern Rhodesia.

**1038 FOXY CISTICOLA.   *CISTICOLA TROGLODYTES* (Antinori).**
*Cisticola troglodytes troglodytes* (Ant.).   **Pl. 72.**
*Drymoica troglodytes* Antinori, Cat. Ucc. p. 38, 1864: Djur, Bahr-el-Ghazal, south-western Sudan.

**Distinguishing characters:** A small species; first primary scimitar-shaped; adult, above, bright russet; below, warm buff; no subterminal spots in tail. The female is smaller. There is no non-breeding dress. Wing, male 46 to 48, female 42 to 44 mm. The young bird is duller than the adult; below, yellow.

**General distribution:** Eastern Cameroons and Nigeria to north-eastern Belgian Congo, the Sudan, Uganda and Kenya Colony.

**Range in Eastern Africa:** Western and southern Sudan, northern Uganda and north-western Kenya Colony.

**Habits:** Little is known of this bird in life except that it seems to divide its time equally between the grass and the trees. It seems most similar to the Siffling Cisticola although far more unobtrusive. The males only occasionally perch on a tree top to utter a few non-descript and wispy little call notes. There are two moults in the year.

# WARBLERS

**Nest and Eggs:** Nest a roundish oval with the entrance at the top of the side, of broad leaved grass lined with plant wool. Eggs apparently undescribed.

**Recorded breeding:** Uganda, about June to October.

**Call:** See Habits.

*Cisticola troglodytes ferruginea* Heugl.
*Cisticola ferruginea* Heuglin, J.f.O. p. 259, 1864: Sawakwo district, Rahad River watershed, western Abyssinia.

**Distinguishing characters:** Differs from the nominate race in being larger; above, browner and less bright russet; below, more whitish. Wing, male 49 to 51, female 45 to 50 mm.

**Range in Eastern Africa:** The middle Blue Nile, eastern Sudan to western and south-western Abyssinia.

**Habits:** As for the nominate race.

**Recorded breeding:** About May or June to October.

**1039 PIPING CISTICOLA. *CISTICOLA FULVICAPILLA*** (Vieillot).
*Cisticola fulvicapilla muelleri* Alex. **Pl. 72.**
*Cisticola muelleri* Alexander, Bull. B.O.C. 8, p. 49, 1899: Mesanangue, lower Zambesi River, Portuguese East Africa.

**Distinguishing characters:** First primary scimitar-shaped; adult, top of head warm rust red brown; mantle plain dusky brown; below, white washed with buff; flanks washed with smoky grey; rather indistinct subterminal dusky spots in tail feathers. In non-breeding dress the colour of the upperside is warmer with less contrast between the top of head and mantle. The sexes are alike. Wing, male 44 to 48, female 42 to 44 mm. The young bird is more rusty in colour than the adult in non-breeding dress. Some in the south of its range show faint indications of yellow below.

**General distribution:** North-eastern Northern Rhodesia, to Tanganyika Territory, Portuguese East Africa and Nyasaland.

**Range in Eastern Africa:** Central and southern Tanganyika Territory to Portuguese East Africa and Nyasaland.

**Habits:** Found in light woodland country where its general habits are exactly like those of the Siffling Cisticola, although its calls are quite different. Males take up a similar commanding perch where

their longer tails form a distinguishing character, but the song consists of a plaintive and monotonous single-noted piping of 'keep-keep-keep' repeated from ten to thirty times at the rate of about two a second. It is apt to be ventriloquistic in utterance, and although the song is smaller and more piping, it is not unlike the monotonous repetition of a Tinker-bird. The alarm call most often heard is a small harsh and rapid rattle of twittering clicks, uttered both when settled or in flight, and is just like the wing-rattle of a large tree-locust in flight.

**Nest and Eggs:** Ball-type nest lined with plant down, placed low in the grass with no living grass bower over it. Eggs three or four, plain or deeply freckled with rusty red, the ground colour varying from greenish white, bluish or pinkish white to turquoise; about 15 × 11 mm.

**Recorded breeding:** About October to March.

**Call:** See Habits.

**Distribution of other races of the species:** Angola and South Africa, the nominate race being described from Cape Province.

### 1040 TABORA CISTICOLA. *CISTICOLA ANGUSTICAUDA*\* Reichenow. Pl. 72.

*Cisticola augusticauda* Reichenow, J.f.O. p. 69, 1891: Ugunda, Tabora district, Tanganyika Territory.

**Distinguishing characters:** First primary scimitar-shaped. Adult very similar to the Piping Cisticola but differs in having the top of the head paler russet; primaries narrower; tail feathers longer and narrower. In non-breeding dress the tail is shorter. The sexes are alike. Wing, male 45 to 50, female 42 to 47 mm. The young bird has the top of the head dull russet and the mantle earth brown.

**General distribution:** Kenya Colony, Tanganyika Territory, south-eastern Belgian Congo and north-eastern Northern Rhodesia.

**Range in Eastern Africa:** South-western Kenya Colony to central and southern Tanganyika Territory.

**Habits:** Habitat and habits as for the Piping Cisticola.

\**Note:* If original orthography is adhered to, the spelling of this name should be *augusticauda*.

# WARBLERS

**Nest and Eggs:** Nest similar to that of the Piping Cisticola; eggs three, white with small spots and speckles of Indian Red; about 14·5 × 11 mm.

**Recorded breeding:** Iringa, Tanganyika Territory, Northern Rhodesia and south-eastern Belgian Congo, December to March.

**Call:** Very much like that of the Piping Cisticola.

### 1041 TINY CISTICOLA. *CISTICOLA NANA* Fischer & Reichenow.

*Cisticola nana* Fischer & Reichenow, J.f.O. p. 260, 1884: Ngaruka, Arusha district, north-eastern Tanganyika Territory.

**Distinguishing characters:** First primary blade-shaped. Adult similar to the Piping Cisticola in size and coloration; top of the head russet brown; mantle slightly darker with more or less distinct darker streaks; tail shorter, averaging 30 against 37 mm.; below, creamy white; flanks slightly dusky. The sexes are alike. There is no non-breeding dress. Wing, male 45 to 49, female 41 to 45 mm. The young bird has the mantle washed with russet.

**Range in Eastern Africa:** Southern Abyssinia to eastern Kenya Colony and northern and central Tanganyika Territory.

**Habits:** Found in thorn bush or typical park-like acacia country where its habits are distinctive. Appearing as a tiny white-breasted dark-backed bird, the male perches on the ends of higher branches or bush tops uttering its brief clear song; it darts off on whirring wings to another perch to repeat the notes. It carries out occasional courting antics on the way, short bursts of quickly whirring wings and sudden swerves; movements which, except that they are carried out quite close to the ground level, are apt to remind the observer of the Wing-snapping, Pectoral-patch and Black-backed Cisticolas. The song is of sharp whistled notes preceded by half a dozen small clicks, 'tic-tic-twee-twee-chick-chick' repeated at intervals. The call is a repeated 'churr-it-it.'

**Nest and Eggs:** Nest of ball type, woven of dry grass and placed in green grass at about six inches from the ground; well lined with plant down. Eggs four, pale greenish blue freckled with pale red; about 14·5 × 11·5 mm.

**Recorded breeding:** Principally April to July, also November and December.

**Call:** See Habits.

### 1042 ASHY CISTICOLA. *CISTICOLA CINEREOLA* Salvadori.

*Cisticola cinereola cinereola* Salvad.

*Cisticola cinereola* Salvadori, Ann. Mus. Genova, 26, p. 254, 1888: Farré, southern Abyssinia.

**Distinguishing characters:** First primary blade-shaped. Above, ashy grey broadly streaked with dusky; below, creamy white; subterminal black spots on both webs of tail feathers. The sexes are alike. In non-breeding dress the streaking on the head and mantle is broader and more blackish. Wing, male 60 to 66, female 51 to 54 mm. The young bird is washed with buff above; a very faint wash of yellow on sides of face and chin to chest.

**Range in Eastern Africa:** Eastern and southern Abyssinia from Danakil, Lake Stephanie and Mega to the Somalilands north of lat. 5° N.

**Habits:** Found in similar country to the Tiny Cisticola, among thorn bushes or in acacia woodland where in outline and general movements it is very like a Whitethroat, keeping mostly to bushes with the male coming to the uppermost twigs to utter a sweet and plaintive warbling song.

**Nest and Eggs:** Nest (of the southern race) ball-shaped with side of top entrance, made of fine grass and fibre lined with plant down. Living grass is plaited down on to the dome of the nest. Eggs two, very pale blue either unspotted or with a few dark brown spots at the larger end; about 17·5 × 12 mm.

**Recorded breeding:** No records.

**Call:** A sweet warbling plaintive song 'twi-twi-twi-twee-oo' delivered from a perch, and a three-syllabled and agitated alarm call 'chi-weet-oo.'

*Cisticola cinereola schillingsi* Reichw. **Pl. 72.**

*Cisticola schillingsi* Reichenow, in Schill. mit. Blitz. und Büchse, p. 556, 1905: Doinyo Erok, Masailand, northern Tanganyika Territory.

**Distinguishing characters:** Differs from the nominate race in being slightly browner above, especially on top of head. There is no non-breeding dress. Wing, male 60 to 66, female 51 to 55 mm.

**Range in Eastern Africa:** Italian Somaliland and northern, eastern and southern Kenya Colony.

**Habits:** As for the nominate race.

**Recorded breeding:** March to July are certainly breeding months and also probably November and December.

### 1043 RED-PATE CISTICOLA. *CISTICOLA RUFICEPS* (Cretzschmar).

*Cisticola ruficeps ruficeps* (Cretz.). **Pl. 72.**
*Malurus ruficeps* Cretzschmar, in Rüpp. Atlas Vög. p. 54, pl. 36a, 1826: Kordofan, Sudan.

**Distinguishing characters:** First primary blade-shaped. Very similar to the Tiny Cisticola but larger; feet larger; inner toe shorter than outer toe; top of head and hind neck bright russet; mantle drab brown; below, creamy white washed with cream on chest and flanks; subterminal black spots on both webs of feathers. The sexes are alike. In non-breeding dress the tail is rather longer; top of head bright tawny; mantle light sandy buff streaked with dark brown to blackish; upper tail-coverts tawny. Wing, male 54 to 58, female 49 to 53 mm. The young bird is similar to the adult in non-breeding dress but has the top of the head uniform with the mantle; some yellow below.

**General distribution:** Shari River, French Equatorial Africa to the Sudan.

**Range in Eastern Africa:** The Sudan from Darfur to Kordofan.

**Habits:** Found principally in bush country of scattered thorn trees and other thorny vegetation, but occasionally in softer herbage where the range of the species encroaches on the savannah. Like the Ashy Cisticola it has more the habits of a Whitethroat and keeps to the bushes rather than to the grass. It has no conspicuous call nor aerial courting antics, and a courting male merely utters a small musical song only audible at close range. The common call is a small rattling trill rather more harsh in utterance when the bird is alarmed.

**Nest and Eggs:** Ball-type nest, placed low in grass and foot high herbage. Eggs three and four, white to medium blue marked in all degrees with freckles and blotches of light and dark reds; about 16 × 12 mm.

**Recorded breeding:** Usually July to October.

**Call:** See Habits.

*Cisticola ruficeps scotoptera* (Sund.).
*Drymoica scotoptera* Sundevall, Œfv. K. Sv. Vet.-Akad. Förh. 7, No. 5, p. 129, 1850: Blue Nile, Sudan.

**Distinguishing characters**: Differs from the nominate race in being less bright russet, more ashy brown, on top of head and mantle. In non-breeding dress the top of the head is more chestnut brown; buff streaking on mantle paler. Wing, male 54 to 61, female 49 to 54 mm.

**Range in Eastern Africa**: The White and Blue Nile areas of the Sudan to southern Eritrea and northern Abyssinia.

**Habits**: As for other races.

**Recorded breeding**: As for the nominate race.

*Cisticola ruficeps mongalla* Lynes.
*Cisticola ruficeps mongalla* Lynes, Ibis, Supp. p. 451, 1930: Malek, Mongalla Province, southern Sudan.

**Distinguishing characters**: Differs from the nominate race in being darker. In breeding dress the mantle is smoke grey streaked with dusky. Wing, male 53 to 57, female 48 to 51 mm.

**Range in Eastern Africa**: The Upper Nile Province of the Sudan and northern Uganda.

**Habits**: As for the nominate race.

**Recorded breeding**: During and varying with the rains, mainly about July to October.

**Distribution of other races of the species**: West Africa.

**1044 TINKLING CISTICOLA.** *CISTICOLA TINNIENS* (Lichtenstein).
*Cisticola tinniens oreophila* Van Som. **Pl. 72.**
*Cisticola tinniens oreophila* Van Someren, Nov. Zool. 29, p. 214, 1922: Mt. Kenya, central Kenya Colony.

**Distinguishing characters**: First primary blade-shaped. Adult, top of head chestnut red; mantle with broad black and narrow buff streaks; subterminal black spots on both webs of tail feathers; below, creamy white. The sexes are alike. There is no non-breeding dress. Wing, male 57 to 59, female 52 to 54 mm. The young bird is more tawny above with less deep black streaking; dusky streaks on top of head.

**Range in Eastern Africa:** Central Kenya Colony from Eldoret, Nandi and Mt. Kenya to Mau, Naivasha, Aberdares and Kikuyu.

**Habits:** Always found at or near swampy ground, or banks of streams, where the birds flit about low over the grass or sedge with jerky and apparently rather feeble flight and flirtings of their tail. Males also indulge in low dancing pirouettes a few feet above their mates.

**Nest and Eggs:** Nest ball-type, some six or seven inches from the ground in long grass, made of grass stems bound with cobweb, often covered externally with fine dark coloured swamp rootlets. Eggs usually four, very variable from white to blue and spotted; the dominant type being greenish white, freckled mostly at the larger end with rusty red; about 16 × 12 mm.

**Recorded breeding:** April to July, also about November and December.

**Call:** Distinctive and repeated at intervals, a far-carrying chirruped whistle quite high-pitched, and preceded by some small opening notes which are inaudible unless at close quarters.

**Distribution of other races of the species:** Angola and South Africa, the nominate race being described from eastern Cape Province.

## PRINIAS or LONG-TAILS. FAMILY—SYLVIIDÆ. Genus: *Prinia*.

Five species occur in Eastern Africa. They are very closely related to the Cisticolas and are found in similar situations. The Tawny-flanked Prinia in its short-tailed breeding dress can easily be mistaken for a Cisticola. Insectivorous as far as is known.

**1045** TAWNY-FLANKED PRINIA. *PRINIA SUBFLAVA* (Gmelin).
*Prinia subflava subflava* (Gmel.). **Pl. 71.**
*Motacilla subflava* Gmelin, Syst. Nat. 1, pt. 2, p. 982, 1789: Senegal.

**Distinguishing characters:** Above, grey brown; lores blackish; edges of primaries pale brown; tail feathers with subterminal band and light tips; stripe from bill to over eye and underparts creamy white; flanks, legs and under tail-coverts pale tawny; bill black. In non-breeding dress, November to May, paler, more hair-brown above, and tail longer; bill horn colour. The sexes are alike. Wing, 44 to 55 mm. The young bird is more or less yellow below.

General distribution: Senegal, Portuguese Guinea, central areas of the Gold Coast, Nigeria, Cameroons and French Equatorial Africa to the Sudan, Abyssinia, Uganda and north-eastern Belgian Congo.

Range in Eastern Africa: The Sudan, south of lat. 9° 30′ N., Eritrea, Abyssinia (except the upper Baro River east of Gambela and the Wallega area) and northern Uganda.

Habits, Nest, Eggs, etc.: As for other races.

Recorded breeding: Southern Sudan, July to October.

*Prinia subflava affinis* (Smith).
*Drymoica affinis* A. Smith, Ill. Zool. S. Afr. Aves. pl. 77, 1843: Rustenberg, western Transvaal.

Distinguishing characters: Above, more russet less grey brown; rump and tail brown; edges of flight feathers russet brown; bill black. In non-breeding dress, usually May to October, tawny brown above and tail longer; bill paler horn colour. Wing 44 to 55 mm.

General distribution: South-eastern Belgian Congo, Northern Rhodesia, except perhaps south-western area, Nyasaland, south-western Tanganyika Territory, Portuguese East Africa, Southern Rhodesia and the Transvaal.

Range in Eastern Africa: South-western Tanganyika Territory, Portuguese East Africa and Nyasaland.

Habits: As for the other races, and is the most abundant Warbler of Nyasaland at all levels.

Recorded breeding: Nyasaland, November to April, but most months recorded. South-eastern Belgian Congo, January to April.

*Prinia subflava tenella* (Cab.). **Ph. xiii.**
*Drymoica tenella* Cabanis, in Von der Decken's Reise, 3, p. 23, 1869: Mombasa, south-eastern Kenya Colony.

Distinguishing characters: Similar to the nominate race but rather browner above, less grey, and has no seasonal change of dress; bill at all seasons black or blackish. Wing 44 to 58 mm.

General distribution: Eastern Belgian Congo to Ruwenzori Mts., Uganda (except northern areas), Kenya Colony and Tanganyika Territory.

**Range in Eastern Africa:** Central Uganda to Kenya Colony and Tanganyika Territory as far south as the Tongwe area of Kigoma district, Iringa and Dar-es-Salaam.

**Habits:** A common widespread species of varied habitat, usually among dense low herbage, creepers and secondary forest growth. A noticeable little bird owing to its long tail, which often seems to get rather in the way and usually tame; and in the non-breeding season found in parties of five or six individuals moving about with a cheerful piping call. The eggs are exceedingly variable, green, blue, pink or red, usually with reddish spots or scrawls and highly glossed.

**Recorded breeding:** Uganda in any month. Kenya Colony highlands, March to July and again later in the year. Witu, May. Tanganyika Territory, January to April, June and August.

*Prinia subflava desertæ* Macd.
*Prinia superciliosa desertæ* Macdonald, Bull. B.O.C. 62, p. 27, 1941:
    Kulme, Darfur, western Sudan.

**Distinguishing characters:** Differs from the nominate race in being paler above; upper back and head pale greyish buff; bill black; in non-breeding dress, October to May, pale tawny buff above and tail longer; bill horn colour. In non-breeding dress strikingly different to the nominate race in the same dress. Wing 43 to 54 mm.

**General distribution:** French Sudan from about Mopti on the Niger to the Sudan, Abyssinia and Eritrea.

**Range in Eastern Africa:** The Sudan between lat. 9° 30' and 16° N. as far east as the Sennar area and Gallabat and western Abyssinia in the Wallega area and the Baro River Valley east of Gambela, and south-western Eritrea.

**Habits:** Much those of other races, but owing to its habitat less of a forest species and found in woodland or open bush. Usually tame and inquisitive.

**Nest and Eggs:** A beautifully woven deep purse or bottle-shaped nest of grass stems and seed heads, often with quite a defined pattern made by the use of different materials, and sewn on to, often into, a plant so that the top leaf of the plant forms a roof over the nest. Eggs usually three but often two or four, of very great variety. The following among others are recorded: white stippled with salmon pink; white with sharp red brown spots; plain turquoise blue; blue blotched with rufous; blue with faint pink spots; salmon pink with purplish spots or heavily stippled all over to a uniform reddish brown; about 15 × 11·5 mm.

**Recorded breeding:** Nigeria, September. Darfur, August and September. Abyssinia, August and September.

**Call:** Has a sweet song and a considerable variety of call notes, including a cheerful piping and a sharp persistent alarm note of 'sbee' uttered while jerking the tail.

**Distribution of other races of the species:** West and South Africa.

**1046 PALE PRINIA.** *PRINIA SOMALICA* (Elliot).
*Prinia somalica somalica* (Ell.).
*Burnesia somalica* Elliot, Publ. Field Col. Mus. Orn. 1, p. 45, 1897: Las Durban, near Berbera, British Somaliland.

**Distinguishing characters:** Differs from the Tawny-flanked Prinia in being ashy grey above and pale creamy white below. The sexes are alike. Wing 48 to 51 mm. The young bird is more sandy above; breast and flanks buffish.

**Range in Eastern Africa:** Coastal areas of British Somaliland.

**Habits:** Quiet easily overlooked little birds of grassy steppe country, usually seen low among bushes and often on the ground.

**Nest and Eggs:** See under the following race.

**Recorded breeding:** No records.

*Prinia somalica erlangeri* Reichw.
*Prinia somalica erlangeri* Reichenow, O.M. p. 24, 1905: Gurra, southern Abyssinia.

**Distinguishing characters:** Differs from the nominate race in being darker above. Wing 44 to 51 mm.

**Range in Eastern Africa:** Southern Abyssinia, south-eastern Sudan, southern Italian Somaliland and northern, eastern and south-eastern Kenya Colony.

**Habits:** Little recorded, but appears to have been found mostly in pairs or small parties in the tops of tall acacias in grassy country.

**Nest and Eggs:** Nest bottle-shaped and domed, with an opening at side near top, made of grass and lined with wool and plant down, and fixed or tied on to grass stems. Eggs three or four, pale green or pale rufous with darker spots and blotches; about 15·5 × 11 mm.

**Recorded breeding:** Southern Abyssinia and Italian Somaliland, April and May. Northern Kenya Colony (breeding condition), March.

**Call:** A short churring call uttered from a high point and a little chipping song of five or six notes.

**1047** STRIPED-BACK PRINIA. *PRINIA GRACILIS* (Lichtenstein).

*Prinia gracilis gracilis* (Licht.). **Pl. 71.**
*Sylvia gracilis* Lichtenstein, Verz. Doubl. p. 34, 1823: Nubia.

**Distinguishing characters:** Differs from the last species in being streaked from forehead to rump with blackish brown. The sexes are alike. Wing 43 to 46 mm. Juvenile plumage unrecorded.

**General distribution:** Nile Valley, from about Aswan in Upper Egypt to Khartoum.

**Range in Eastern Africa:** Nile Valley as far south as Khartoum.

**Habits:** Abundant in scrub and cultivated ground near Khartoum with the habits of other members of the genus.

**Nest and Eggs:** Nest a deep egg-shaped cup of grass lined with soft material, placed in small shrubs near the ground. Eggs apparently undescribed.

**Recorded breeding:** Sudan, February and March.

**Call:** Unrecorded.

*Prinia gracilis carlo* Zedl.
*Prinia gracilis carlo* Zedlitz, J.f.O. p. 610, 1911: Dadab, eastern Abyssinia.

**Distinguishing characters:** Differs from the nominate race in being darker above. Wing 43 to 48 mm.

**Range in Eastern Africa:** Red Sea Province of the Sudan to northern British Somaliland.

**Habits:** Locally common on cultivated ground and among bushes.

**Recorded breeding:** Port Sudan, July.

**Distribution of other races of the species:** Arabia.

### 1048 WHITE-CHINNED PRINIA. *PRINIA LEUCOPOGON* (Cabanis).

*Prinia leucopogon reichenowi* (Hartl.). **Pl. 71.**

*Burnesia reichenowi* Hartlaub, J.f.O. p. 151, 1890: Njangalo, north-eastern Belgian Congo.

**Distinguishing characters:** Above, grey; forehead feathers tipped with paler grey giving a scaly appearance; lores black; below, chin and throat creamy white; lower neck to breast grey; belly and under tail-coverts pale buff, whiter in centre of upper belly; tail feathers tipped with white. The sexes are alike. Wing 53 to 61 mm. The young bird is very similar to the adult, but has the throat whiter and lower neck to breast much paler grey.

**General distribution:** North-eastern Belgian Congo to the southern Sudan, Kenya Colony and Tanganyika Territory.

**Range in Eastern Africa:** Southern Sudan, Uganda, western Kenya Colony in the Kaimosi and Nandi areas and western Tanganyika Territory as far south as the Kungwe-Mahare Mts.

**Habits:** Locally common in pairs or family parties in open bush, light forest or cultivated ground. Usually noticeable as a constantly twittering little party of birds searching for insects. On Mt. Elgon common among dense riparian vegetation. In courtship display the male throws his tail right over his back.

**Nest and Eggs:** Nest a grass bottle-shaped structure sewn between leaves drawn together. Eggs two or three, pale bluish white with rusty brown spots and lilac undermarkings; about $17 \cdot 5 \times 12$ mm.

**Recorded breeding:** Uganda, May to August, also January, April and other months.

**Call:** A rather harsh two-syllabled chattering cry constantly uttered, and also a very beautiful little song in the evening and early morning.

**Distribution of other races of the species:** Cameroons to northern Angola, the nominate race being described from Loango.

**1049 BANDED PRINIA.** *PRINIA BAIRDII* (Cassin).
*Prinia bairdii melanops* (Reichw. & Neum.). **Pl. 71.**
Burnesia melanops Reichenow & Neumann, O.M. p. 75, 1895: Mau, Kenya Colony.

**Distinguishing characters:** Above, sooty black with head rather darker than the body; white ends to wing-coverts, secondaries and tail; below, chin to neck dark sooty black; chest, breast and flanks barred black and white; this barring extends in some cases on to neck and throat; centre of breast and belly white. The sexes are alike. Wing 52 to 58 mm. The young bird is brownish above; grey brown below; belly whitish; flanks slightly barred.

**General distribution:** Eastern Belgian Congo to western Kenya Colony.

**Range in Eastern Africa:** Uganda to Ruanda-Urundi and western Kenya Colony from Mt. Elgon to Mau.

**Habits:** A local species of tangled forest undergrowth and old clearings around native villages. It is probably a good deal more common than is supposed, but prefers rough ground and dense vegetation. The male, and possibly the female also, cock the tail at an acute angle, and are restless little birds moving about with much wing-flicking.

**Nest and Eggs** (West African race): A deep pocket nest partly roofed over of rather large size, made of grass lined with finer grass and seed heads. Eggs three, greenish or bluish white blotched and clouded with light rufous or grey, and with fine freckling tending to form a zone at the larger end; about 16 × 12·5 mm.

**Recorded breeding:** No records.

**Call:** A shrill rapid 'plee-plee-plee' or 'chip-chip' while feeding.

**Distribution of other races of the species:** West Africa, the nominate race being described from Gabon.

# RED-WING WARBLER, MOUSTACHE-WARBLER, CRICKET-WARBLER and BLACK-FACED WARBLER. FAMILY—SYLVIIDÆ. Genera: *Heliolais*, *Melocichla*, *Spiloptila* and *Bathmocercus*.

All four are examples of monotypic genera and all are completely insectivorous as far as is known. The Red-wing Warbler has a seasonal change of dress throughout its distribution.

## 1050 RED-WING WARBLER. *HELIOLAIS ERYTHROPTERA* (Jardine).

*Heliolais erythroptera jodoptera* (Heugl.).
*Drymœca jodoptera* Heuglin, J.f.O. p. 258, 1864: Bongo, Bahr-el-Ghazal, south-western Sudan.

**Distinguishing characters:** Upperside and sides of head vinous grey; wings chestnut; below, creamy white; flanks, belly, under tail-coverts and legs pale tawny; tail feathers with subterminal dusky ends and white tips; bill horn. In non-breeding dress, October to April, the upperparts are vinous chestnut; bill horn. The sexes are alike. Wing 56 to 60 mm. The young bird is a paler and more fluffy version of the adult in non-breeding dress.

**General distribution:** Cameroons to the Sudan.

**Range in Eastern Africa:** South-western Sudan.

**Habits:** Presumably as for the other races, but little recorded within our area. Has a varied pleasing little song, and a high rather rattling call.

**Recorded breeding:** Belgian Congo, probably June to November.

*Heliolais erythroptera rhodoptera* (Shell.). **Pl. 71.**
*Cisticola rhodoptera* Shelley, Ibis, p. 333, 1880: Usambara Mts., north-eastern Tanganyika Territory.

**Distinguishing characters:** Head greyer and mantle more olivaceous brown than in the preceding race; bill horn, sometimes black. In non-breeding dress, June to October, the upperparts are vinous brown, less chestnut than in the preceding race; bill horn. Wing 47 to 57 mm.

**General distribution:** Kenya Colony to Nyasaland and Portuguese East Africa as far south as the Gorongoza Mts.

**Range in Eastern Africa:** Western Kenya Colony to the Zambesi River.

**Habits:** Birds of open grassy woodland, particularly of Brachystegia woodland where long grass grows among trees. Rather conspicuous birds as a rule, usually flying up to a tree top when flushed, common in Nyasaland and the central plateau of Portuguese East Africa, less common further south.

**Nest and Eggs:** The nest is a flattish ball of fine grass with a side entrance, sewn between the leaves of a shrub. Eggs usually two, green, mottled and blobbed with pinkish; about 17·5 × 12·5 mm.

**Recorded breeding:** Tanganyika Territory, October. Nyasaland, December to April. Portuguese East Africa, December to February.

**Call:** A thin twittering 'pee-pee-pee' which can be heard for some distance. Another call noted is one of 'chair-chair-chair' followed by several shrill cheeping whistles. Believed to have a song in the breeding season.

*Heliolais erythroptera major* (Blund. & Lovat).
*Orthotomus major* Blundell & Lovat, Bull. B.O.C. 10, p. 20, 1899: Getemma, western Abyssinia.

**Distinguishing characters:** Larger and longer-billed than the preceding races; above, vinous brown; bill horn. In non-breeding dress, March to April, paler vinous brown above; bill horn. Wing 59 to 63 mm.

**Range in Eastern Africa:** Abyssinia.

**Habits:** Found in sheltered valleys up to 6,000 feet. Habits little recorded, but the male is said to have a fine song which it utters from a tree-top.

**Recorded breeding:** No records.

**Distribution of other races of the species:** West Africa, the nominate race being described from the Gold Coast.

**1051 MOUSTACHE-WARBLER.** *MELOCICHLA MENTALIS* (Fraser).
*Melocichla mentalis grandis* (Boc.).
*Drymoica (Cisticola) grandis* Bocage, Jorn. Lisboa, 8, p. 56, 1880: Caconda, Angola.

**Distinguishing characters:** Above, forehead and ear-coverts chestnut; top of head and mantle warm dark brown; white stripe over eye; rump russet brown; wings and tail blackish brown, former with russet brown edges to flight feathers; below, cheeks and throat white; malar stripe black; breast to under tail-coverts russet brown; centre of belly whitish; ends of tail feathers sooty buff; bill, upper

mandible blackish brown, lower whitish horn. The female is usually smaller than the male. Wing 75 to 83 mm. The young bird has no chestnut on the forehead and the innermost secondaries are tipped with buff.

**General distribution:** Angola to Nyasaland and northern Portuguese East Africa.

**Range in Eastern Africa:** Nyasaland and Portuguese East Africa.

**Habits:** As for other races. The male has a fine rich song which can be heard at almost any time of year.

**Nest and Eggs:** As for other races.

**Recorded breeding:** South-eastern Belgian Congo, January, also April. Nyasaland, November and December.

*Melocichla mentalis amauroura* (Pelz.).
Argya amauroura Pelzeln, Verh. Zool. Bot. Ges. Wien. 32, p. 503, 1883: Fadibek, northern Uganda.

**Distinguishing characters:** Darker above than the preceding race, less warm brown. Wing 70 to 82 mm.

**General distribution:** The Sudan and north-eastern Belgian Congo to Kenya Colony and Tanganyika Territory.

**Range in Eastern Africa:** Southern Sudan to Uganda, except the Ruwenzori area, Ruanda, Urundi, Kenya Colony and western Tanganyika Territory as far south as the Ufipa Plateau.

**Habits:** A common but shy grass-loving species found always in dense low cover or rank herbage. Very little in evidence during the day, but in the early morning or late evening it sidles up an elephant grass stem, or some other tall herb, and utters a cheerful throaty little song. When suspicious it flirts its tail like a Wheatear and dives into cover if alarmed. Very fond of thick grass and scrub fringing streams, and is usually seen in pairs moving from cover to cover with a low skimming flight.

**Nest and Eggs:** A large flat cup nest of rough grass lined with fine grass and rootlets usually low down and well concealed in a tuft of high grass. Eggs two, rarely three, pale cream to brownish pink, spotted and zoned with brownish cloudings and lilac undermarkings; about 23 × 16 mm.

**Recorded breeding:** Uganda, April to August, also November to January. Southern Kenya Colony, March and April. Tanganyika Territory, December to February.

**Call:** Alarm note a sharp rasping 'ti-ti-ti-ti-ti' or a very rapidly uttered querulous 'cheep-cheep-cheep-cheep.' Song short but cheerful and loud 'tip-tip-twiddle-iddle-ee.'

*Melocichla mentalis orientalis* (Sharpe).
*Cisticola orientalis* Sharpe, Cat. Bds. B.M. 7, p. 245, 1883: Pangani River, Usambara, eastern Tanganyika Territory.

**Distinguishing characters:** Paler than the preceding races; more russet brown above. Wing 73 to 77 mm.

**Range in Eastern Africa:** Tanganyika Territory from the Usambara Mts. to the Pangani River, Pugu Hills, Kilosa and Njombe.

**Habits:** As for other races.

**Recorded breeding:** Amani, August.

*Melocichla mentalis atricauda* Reichw.
*Melocichla atricauda* Reichenow, O.M. p. 61, 1893: Nkondjo, Ndussuma, Semliki Valley, eastern Belgian Congo.

**Distinguishing characters:** Darker above than the preceding races, rather more blackish brown. Wing 74 to 81 mm. The young bird is also more blackish brown above.

**General distribution:** Semliki Valley, eastern Belgian Congo and western Uganda.

**Range in Eastern Africa:** Ruwenzori area, western Uganda.

**Habits:** As for other races.

**Recorded breeding:** No records.

*Melocichla mentalis granviki* Grant & Praed.
*Melocichla mentalis granviki* C. Grant & Mackworth-Praed, Bull. B.O.C. 62, p. 31, 1941: Wardji, Limmu, Jimma, south-western Abyssinia.

**Distinguishing characters:** The darkest race of all, above more uniform blackish brown, and rump darker brown. Wing 76 to 82 mm.

**Range in Eastern Africa:** Western Abyssinia from Wallega to the Omo River Valley.

**Habits:** As for other races.

**Recorded breeding:** No records.

**Distribution of other races of the species:** West Africa, the nominate race being described from Accra, Gold Coast.

### 1052 CRICKET - WARBLER.    *SPILOPTILA CLAMANS* (Temminck).

*Malurus clamans* Temminck, Pl. Col. livr. 78, pl. 466, fig. 2, 1828: Nubia.

**Distinguishing characters:** Adult male, top of head, wing-coverts and innermost secondaries mottled black and white; occiput and hind neck grey; mantle vinous tawny; rump pale yellow; wings and tail grey, latter with white ends and subterminal black bars; below, pale creamy buff. The female has the occiput and hind neck vinous tawny uniform with the mantle. Wing 44 to 49 mm. The young bird has the top of the head streaked with dark brown.

**General distribution:** Timbuktoo, French Sahara, to the Sudan and Eritrea.

**Range in Eastern Africa:** Darfur and Kordofan to western Eritrea.

**Habits:** A desert edge species, locally common among low acacia scrub, but avoiding grass or dense bush. It is found also in lesser numbers in true desert country wherever there is suitable low scrub. Usually noticed as a party of five or six active restless little birds jerking their tails up and down and from side to side, and uttering a loud monotonous call. When alarmed the bird dives into a bush and prefers to escape by running instead of flying.

**Nest and Eggs:** A deep egg-shaped nest domed, or partly domed, of grass or dry herbage lined with down, feathers or other soft material. The nest is usually in a dwarf shrub a few inches from the ground. Eggs three, white with fine red speckling. No measurements available.

**Recorded breeding:** Eritrea, July and August. Sudan, March. Southern Sahara, November, also June and July.

**Call:** A monotonous 'du-du-du-du-du-du' rather tinkling and insect-like. Has a pretty little song uttered from the top of a tree in the breeding season.

## 1053 BLACK-FACED RUFOUS WARBLER. *BATHMOCERCUS RUFUS* (Reichenow).

*Bathmocercus rufus jacksoni* Sharpe.

*Bathmocercus jacksoni* Sharpe, Bull. B.O.C. 13, p. 10, 1902: Kibera, Toro, western Uganda.

**Distinguishing characters:** Adult male, forehead and sides of face black; rest of upperparts rich russet brown; below, chin to upper belly black; sides of chest russet brown; lower belly and flanks grey. The female has the russet brown replaced by olivaceous grey; sides of chest pale buff. Wing 55 to 60 mm. The young bird is wholly olivaceous grey, except tail and edges of flight feathers which are dull russet brown.

**General distribution:** North-eastern Belgian Congo to the Sudan and Kenya Colony.

**Range in Eastern Africa:** Southern Sudan to Uganda and western Kenya Colony as far east as Mau.

**Habits:** Locally plentiful, generally in marshy places, amongst low forest undergrowth, but occurs up to 9,000 feet on Mt. Elgon. Little recorded as to habits, but cocks up its tail at right angles as it moves about. It is said to use nests, either old or 'cocks' nests for roosting purposes. It is believed to feed largely on the ground among the debris of the forest floor.

**Nest and Eggs:** Nest large and bulky, of grass lined with fine grass tops. Eggs undescribed.

**Recorded breeding:** Ituri (breeding condition) March to September.

**Food:** Insects.

**Call:** A sharp little alarm note, and a persistent call 'iipit-iipit-iipit' (Van Someren).

**Distribution of other races of the species:** West Africa, the nominate race being described from Cameroons.

---

Names in Sclater's *Syst. Av. Æthiop.* 2, 1930, which have been changed, or have become synonyms in this work:

*Acrocephalus scirpaceus crassirostris* (Brehm) now *Acrocephalus scirpaceus griseus* (Hemprich and Ehrenberg).

*Phylloscopus trochilus eversmani* (Bonaparte) now *Phylloscopus trochilus acredula* (Linnæus).

*Seicercus umbrovirens erythreæ* (Salvadori) treated as synonymous with *Seicercus umbrovirens umbrovirens* (Rüppell).

*Bradypterus brachypterus brachypterus* (Vieillot) now *Bradypterus babæcala babæcala* (Vieillot).

*Bradypterus brachypterus fraterculus* Mearns, treated as synonymous with *Sathrocercus mariæ mariæ* (Madarász).
*Bradypterus altumi* Van Someren, treated as synonymous with *Sathrocercus mariæ mariæ* (Madarász).
*Bradypterus roehli* Grote, treated as synonymous with *Sathrocercus mariæ usambaræ* (Reichenow).
*Bradypterus alfredi sjöstedti* Neumann, treated as synonymous with *Sathrocercus mariæ mariæ* (Madarász).
*Bradypterus yokanæ* Van Someren, treated as synonymous with *Bradypterus carpalis* Chapin.
*Calamornis palustris* (Reichenow) treated as synonymous with *Calamocichla leptorhyncha leptorhyncha* (Reichenow).
*Calamornis gracilirostris zuluensis* Neumann, treated as synonymous with *Calamocichla leptorhyncha leptorhyncha* (Reichenow).
*Apalis flavida æquatorialis* Neumann, treated as synonymous with *Apalis flavida flavocincta* (Sharpe).
*Apalis porphyrolæma affinis* O. Grant, treated as synonymous with *Apalis porphyrolæma porphyrolæma* Reichenow and Neumann.
*Apalis porphyrolæma vulcanorum* Gyldenstolpe, treated as synonymous with *Apalis porphyrolæma porphyrolæma* Reichenow and Neumann.
*Apalis ruficeps* Reichenow, treated as synonymous with *Artisornis metopias* (Reichenow).
*Artisornis metopias altus* (Friedmann) treated as synonymous with *Artisornis metopias* (Reichenow).
*Sylvietta brachyura micrura* (Rüppell) treated as synonymous with *Sylvietta brachyura brachyura* Lafresnaye.
*Sylvietta brachyura hilgerti* Zedlitz, treated as synonymous with *Sylvietta brachyura leucopsis* Reichenow.
*Eremomela griseoflava* Heuglin, now *Eremomela icteropygialis* (Lafresnaye).
*Eremomela griseoflava karamojensis* Stoneham, treated as synonymous with *Eremomela icteropygialis griseoflava* Heuglin.
*Eremomela griseoflava crawfurdi* S. Clarke, treated as synonymous with *Eremomela icteropygialis griseoflava* Heuglin.
*Eremomela griseoflava archeri* W. Sclater, treated as synonymous with *Eremomela icteropygialis griseoflava* Heuglin.
*Eremomela pusilla tessmani* Grote, treated as synonymous with *Eremomela canescens elegans* Heuglin.
*Eremomela pusilla elgonensis* Van Someren, treated as synonymous with *Eremomela canescens canescens* Antinori.
*Camaroptera brevicaudata noomei* Gunning & Roberts, treated as synonymous with *Camaroptera brevicaudata sharpei* Zedlitz.
*Heliolais erythroptera kavirondensis* Van Someren, treated as synonymous with *Heliolais erythroptera jodoptera* (Heuglin).
*Heliolais erythroptera kirbyi* Haagner, treated as synonymous with *Heliolais erythroptera rhodoptera* (Shelley).
*Prinia mistacea mistacea* Rüppell, now *Prinia subflava subflava* Gmelin.
*Prinia mistacea superciliosa* Swainson, treated as synonymous with *Prinia subflava subflava* Gmelin.
*Prinia mistacea immutabilis* Van Someren, treated as synonymous with *Prinia subflava tenella* (Cabanis).

Names introduced since 1930, and which have become synonyms in this work:

*Seicercus ruficapilla mbololo* Van Someren 1939, treated as synonymous with *Seicercus ruficapilla minulla* (Reichenow).
*Seicercus umbrovirens chyulu* Van Someren, 1939, treated as synonymous with *Seicercus umbrovirens mackenziana* (Sharpe).
*Sathrocercus cinnamomeus chyuluensis* (Van Someren) 1939, treated as synonymous with *Sathrocercus cinnamomeus rufoflavidus* (Reichenow and Neumann).

# WARBLERS 519

*Schœnicola brevirostris æquatorialis* Granvik, 1934, treated as synonymous with *Schœnicola brevirostris alexinæ* (Heuglin).

*Schœnicola brevirostris chyulu* Van Someren, 1939, treated as synonymous with *Schœnicola brevirostris alexinæ* (Heuglin).

*Apalis thoracica injectiva* Bangs and Loveridge 1931, treated as synonymous with *Apalis flavigularis griseiceps* Reichenow and Neumann.

*Apalis griseiceps chyulu* Van Someren 1939, treated as synonymous with *Apalis flavigularis griseiceps* Reichenow and Neumann.

*Apalis melanocephala songeaensis* Grant & Praed, 1947, treated as synonymous with *Apalis melanocephala muhuluensis* Grant & Praed.

*Apalis chariessa macphersoni* Vincent, 1934, treated as synonymous with *Apalis chariessa* Reichenow.

*Apalis bamendæ bensoni* Vincent, 1935, treated as synonymous with *Apalis bamendæ strausæ* Boulton.

*Eremomela scotops kikuyuensis* Van Someren, 1931, treated as synonymous with *Eremomela scotops occipitalis* (Fischer and Reichenow).

*Camaroptera brevicaudata aschani* Granvik, 1934, treated as synonymous with *Camaroptera brevicaudata brevicaudata* Cretzschmar.

*Camaroptera brevicaudata albiventris* Granvik, 1934, treated as synonymous with *Camaroptera brevicaudata erlangeri* Reichenow.

*Cisticola chiniana mocuba* Vincent, 1933, treated as synonymous with *Cisticola chiniana procera* Peters.

*Cisticola natalensis matengorum* Meise, 1934, treated as synonymous with *Cisticola natalensis natalensis* (Smith).

*Heliolais castanopsis* Vincent, 1933, treated as synonymous with *Heliolais erythroptera rhodoptera* (Shelley).

*Melocichla mentalis chyulu* Van Someren, 1939, treated as synonymous with *Melocichla mentalis amauroura* (Pelzeln).

## Addenda and Corrigenda

**934** *Agrobates g. galactotes.* Range in Eastern Africa: delete 'British Somaliland'.

*A.g. minor.* Recorded breeding: add British Somaliland, January.

**943** *Acrocephalus griseldis.* Has recently been found in British Somaliland.

**955, 956 & 957** *Swamp Warblers.* Genus *Calamocichla*.

Since the original publication of this Volume much additional information has come to hand which has caused us to alter our views considerably.

We now recognise only two species in Africa, the southern *Calamocichla gracilirostris* (*Calamoherpe gracilirostris* Hartlaub, Ibis, p. 348, 1864: Natal) with races in Eastern Africa, *C.g. parva*, *C.g. jacksoni*, *C.g. macrorhyncha* (doubtfully distinct from *C.g. parva*), *C.g. leptorhyncha* and *C.g. tsanæ*. This species has a white or buffish streak above the eye and a clear resonant song.

The other species is *Calamocichla rufescens* (*Bradypterus rufescens* Sharpe & Bouvier, Bull. Soc. Zool. France 1, p. 347, 1876: Landana, Congo), with races in Eastern Africa, *C.r. nilotica* and *C.r. foxi*.

This species has no eye stripe and its song is hoarser and more throaty. It is usually found in taller herbage, although both may inhabit the same swamp.

**955** *Calamocichla gracilirostris nilotica* should be *Calamocichla rufescens nilotica*.

**956** *Calamocichla foxi* should be *Calamocichla rufescens foxi*. Range in Eastern Africa: delete 'eastern Belgian Congo at Lake Mutanda'.

Continued on p. 1103

# SWALLOWS and MARTINS

FAMILY—**HIRUNDINIDÆ.** **SWALLOWS and MARTINS.**
Genera: *Hirundo, Phedina, Riparia, Ptyonoprogne, Delichon,* and *Psalidoprocne.*

Twenty-seven species occur in Eastern Africa, some being migrants from Europe and Asia. They inhabit all types of country and all altitudes. They feed on insects taken on the wing, and the different resident species have divergent and very interesting breeding habits.

The Rough-wing Swallows have the outer edge of the first primary rough, or saw-edged, caused by the barbules being turned backwards. The reason for this is not apparent.

KEY TO THE ADULT SWALLOWS AND MARTINS OF EASTERN AFRICA

1 Top of head white or mixed white and sooty black, throat white:    WHITE-HEADED ROUGH-WING SWALLOW *Psalidoprocne albiceps* **1080**

2 Top of head and throat not white:    3–46

3 Rump brownish grey contrasting with mantle:    GREY-RUMPED SWALLOW *Hirundo griseopyga* **1066**

4 Rump white contrasting with mantle:    HOUSE MARTIN *Delichon urbica* **1074**

5 Rump tawny:    6–11

6 Top of head tawny:    SMALLER STRIPED SWALLOW *Hirundo abyssinica* **1065**

7 Top of head blue:    8–11

8 Sides of face blue black:    RUFOUS-CHESTED SWALLOW *Hirundo semirufa* **1064**

9 Sides of face tawny:    10–11

10 Size larger, wing over 137 mm.:    MOSQUE SWALLOW *Hirundo senegalensis* **1063**

| | | |
|---|---|---|
| 11 | Size smaller, wing under 133 mm. | RED-RUMPED SWALLOW *Hirundo daurica* **1062** |
| 12 | Rump uniform in colour with mantle: | 13–44 |
| 13 | Above, bronzy brown: | BROWN ROUGH-WING SWALLOW *Psalidoprocne antinorii* **1079** |
| 14 | Above, oily green: | 15–18 |
| 15 | Under wing-coverts white: | EASTERN ROUGH-WING SWALLOW *Psalidoprocne orientalis* **1077** |
| 16 | Under wing-coverts not white: | 17–18 |
| 17 | Under wing-coverts pale ash-grey: | WESTERN ROUGH-WING SWALLOW *Psalidoprocne chalybea* **1076** |
| 18 | Under wing-coverts ashy brown: | BLACK ROUGH-WING SWALLOW *Psalidoprocne holomelæna* **1075** |
| 19 | Above, glossy blue or steel blue: | 20–33 |
| 20 | Forehead and top of head blue: | 21–22 |
| 21 | White in tail feathers: | WHITE-TAILED SWALLOW *Hirundo megaensis* **1059** |
| 22 | No white in tail feathers: | PEARL-BREASTED SWALLOW *Hirundo dimidiata* **1058** |
| 23 | Forehead and top of head chestnut: | WIRE-TAILED SWALLOW *Hirundo smithii* **1061** |
| 24 | Forehead only chestnut: | 25–33 |
| 25 | Chin and throat white or buffish white: | 26–27 |

| | | |
|---|---|---|
| 26 | Chin and throat white, blue band across chest, size larger, wing over 124 mm.: | WHITE-THROATED SWALLOW *Hirundo albigularis* **1056** |
| 27 | Chin and throat buffish white; blue on sides of chest only, size smaller, wing under 119 mm. | ETHIOPIAN SWALLOW *Hirundo æthiopica* **1057** |
| 28 | Chin and throat chestnut: | 29-33 |
| 29 | Blue-black band across chest, breast to under tail-coverts white or white washed with chestnut: | SWALLOW *Hirundo rustica* **1054** |
| 30 | Broken blue-black band across chest, breast to under tail-coverts ashy brown: | ANGOLA SWALLOW *Hirundo angolensis* **1055** |
| 31 | Above, and below, blue; no chestnut on forehead or head: | 32-33 |
| 32 | Under wing-coverts white: | BLUE ROUGH-WING SWALLOW *Psalidoprocne pristoptera* **1078** |
| 33 | Under wing-coverts blue: | BLUE SWALLOW *Hirundo atrocærulea* **1060** |
| 34 | Above, varying shades of brown or grey brown: | 35-46 |
| 35 | Below, streaked with black: | MASCARENE MARTIN *Phedina borbonica* **1067** |
| 36 | Not streaked below: | 37-46 |
| 37 | Band across chest: | 38-39 |

38 Size larger, wing over
122 mm.:  BANDED MARTIN
 *Riparia cincta*  **1070**

39 Size smaller, wing under
112 mm.:  SAND MARTIN
 *Riparia riparia*  **1068**

40 No band across chest:  41–46

41 No white spots on inner
webs of tail feathers:  AFRICAN SAND MARTIN
 *Riparia paludicola*  **1069**

42 White spots on inner webs
of tail feathers:  43–46

43 Spots on throat:  EUROPEAN
 CRAG-MARTIN
 *Ptyonoprogne rupestris*  **1071**

44 No spots on throat:  45–46

45 Chin and neck in front
pale russet, rest of under-
parts sooty brown:  AFRICAN ROCK-MARTIN
 *Ptyonoprogne fuligula*  **1073**

46 Below, uniform pale
brownish:  PALE CRAG-MARTIN
 *Ptyonoprogne obsoleta*  **1072**

**1054** SWALLOW. *HIRUNDO RUSTICA* Linnæus.
*Hirundo rustica rustica* Linn. European Swallow. **Pl. 73.**
*Hirundo rustica* Linnæus, Syst. Nat. 10th ed. p. 191, 1758: Sweden.

**Distinguishing characters:** Adult male, forehead dark chestnut; rest of upperparts glossy blue black; tail feathers with large white spots on inner webs, outermost considerably elongated; below, chin and throat dark chestnut; broad glossy blue-black band across chest; breast and under tail-coverts white washed with chestnut. The female has shorter outer tail feathers. Wing 115 to 133 mm. The young bird is duller above; forehead, chin and throat pale chestnut; chest band dull blackish; outer tail-feathers only slightly elongated and much broader than adult.

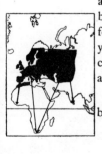

**General distribution:** Europe, Asia and North Africa; in non-breeding season to Africa, India and rarely to Ceylon.

**Range in Eastern Africa:** Throughout in non-breeding season.

**Habits:** A very abundant migrant and a common palæarctic winter visitor to most parts of our area. It would appear from ringing experiments that British breeding Swallows go almost entirely to the eastern half of South Africa, particularly Natal; while central European birds remain further north, being rather more numerous on the western side, in the southern Belgian Congo. Numbers, however, appear to spend their stay in Africa as far north as Abyssinia, where they have been seen in December, and in the highlands of Kenya Colony and Tanganyika Territory where they move about locally. The majority of birds appear to begin moving north through Kenya Colony in late January and February, but others do not move until May, and a certain proportion of non-breeding birds remain all the year round. At times on migration they occur in vast numbers; I have myself heard the desert thorn trees crack under the weight of birds settling to roost (C.W.M.-P.). The song is a soft warble and the alarm note a sharp 'chip.'

*Hirundo rustica transitiva* (Hart.).
*Chelidon rustica transitiva* Hartert, Vög. pal. Fauna, 1, p. 802, 1910: Esdraelon, Palestine.

**Distinguishing characters:** Similar in size to the European Swallow, but breast to under tail-coverts and under wing-coverts darker, more deeply washed with chestnut. Wing 119 to 127 mm.

**General distribution:** Palestine and the Jordan Valley; in non-breeding season to Egypt and as far south as Tanganyika Territory.

**Range in Eastern Africa:** The Sudan, Abyssinia, Uganda and Tanganyika Territory in non-breeding season.

**Habits:** At present only known as a rare non-breeding visitor but probably occurs in some numbers among the vast quantities of the European Swallow. Would only be noticeable if settled overhead or seen at very close range.

*Hirundo rustica rothschildi* Neum.
*Hirundo rothschildi* Neumann, O.M. p. 143, 1904: Schubba, Kaffa, south-western Abyssinia.

**Distinguishing characters:** Similar to the European race, but above more violet blue, and underside from breast to under tail-coverts white; outer tail feathers in the male not so long as in the European Swallow. The very bright sheen on the back is said to be noticeable in the field. Wing 119 to 123 mm.

526 SWALLOWS and MARTINS

**Range in Eastern Africa:** Northern to south-western Abyssinia.

**Habits:** A not uncommon resident of the high plateau of Abyssinia. Its habits are those of the European race and it might be easily overlooked.

**Nest and Eggs:** A cup-shaped mud nest with grass built into it, lined with grass and feathers and resting on a support of some kind; generally placed near streams or rocks, steep banks, or the arches of bridges. Eggs three, white with reddish brown spots. No measurements available.

**Recorded breeding:** Abyssinia, May.

**Call:** Song much like that of the European Swallow, a twittering warble, but appreciably different to a trained ear.

**Distribution of other races of the species:** Siberia to China, Japan and North Africa.

**1055** ANGOLA SWALLOW.  *HIRUNDO ANGOLENSIS* Bocage.
*Hirundo angolensis angolensis* Boc. **Pl. 73.**
*Hirundo angolensis* Bocage, Jorn. Lisboa, 2, p. 47, 1868: Huilla, southern Angola.

**Distinguishing characters:** Adult male, above, similar to the Swallow, but outer tail feathers only slightly elongated; below, chestnut from chin to chest with a much broken glossy blue black narrow band; rest of underparts ashy brown. The female has shorter outer tail feathers. Wing 114 to 128 mm. The young bird is duller above than the adult and the chestnut of chin to chest appreciably duller.

**General distribution:** Tanganyika Territory to northern Nyasaland, southern Belgian Congo, Angola and northern Damaraland.

**Range in Eastern Africa:** Western Tanganyika Territory from Bukoba to Kasulu, Kigoma and Njombe.

**Habits:** Believed to be a migrant from Angola, visiting Tanganyika Territory to breed, and is found there from November to May. Rather a more sluggish species than the European Swallow, or at least is more often seen resting.

**Nest and Eggs:** A half cup-shaped nest of mud and grass lined with feathers, usually plastered on to a cliff, or close to the roof of a cave or verandah, usually not supported below. Eggs three, white spotted with rusty brown; about 18·5 × 13·5 mm.

**Recorded breeding:** Tanganyika Territory, November to February. Northern Nyasaland, October and November.

**Call:** Song as for the following race, a warble often uttered on the wing.

*Hirundo angolensis arcticincta* Sharpe. *Uganda Swallow.*
*Hirundo arcticincta* Sharpe, Ibis, p. 119, 1891: Mt. Elgon.

**Distinguishing characters:** Similar to the nominate race but breast to under tail-coverts paler, whiter in centre of belly. Wing 113 to 122 mm.

**Range in Eastern Africa:** Uganda from Hoima, Mt. Elgon, Mpumu and Entebbe to western Kenya Colony as far east as Escarpment.

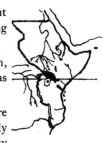

**Habits:** Very much those of the European Swallow but more sluggish in flight and more frequently seen settled. Quite easily distinguishable in the field, and locally common. It has a very pretty little song often uttered on the wing, and with it executes a curious little quivering sailing flight with wing-tips pointed downwards.

**Recorded breeding:** Uganda, February to July and October to December. Mt. Elgon in caves, June.

**1056** WHITE-THROATED SWALLOW. *HIRUNDO ALBI-GULARIS* Strickl.
*Hirundo albigularis albigularis* Strickl.
*Hirundo albigularis* Strickland, Contr. Orn. p. 17, pl. 1849: Cape Peninsula, South Africa.

**Distinguishing characters:** Adult male, above, similar to the Swallow, tail with white patches on inner webs; outer tail-feathers elongated; below, throat white; blue black band across chest, often broken in centre; rest of underparts greyish white. The female has rather shorter outer tail feathers. Wing 125 to 139 mm. The young bird has less and paler chestnut on forehead; rather duller above; pale edges to flight feathers; chest band dull black.

**General distribution:** Southern Nyasaland to South Africa; in non-breeding season to Angola and Northern Rhodesia and Nyasaland.

**Range in Eastern Africa:** The Shiré River, Nyasaland, in non-breeding season.

**Habits:** A common southern species which only just occurs within our limits. Those birds which breed in the Union between September and January migrate northwards in the non-breeding season.

**Nest and Eggs:** Nest of mud lined with feathers; open cup shape, plastered on to a wall or cliff or under the eave of a verandah or similar situation. Eggs three or four, pinkish white with spots and speckles of reddish brown and slate; about 21·5 × 14·5 mm.

**Recorded breeding:** Nyasaland, October and November. Southern Rhodesia, November to January. Union of South Africa, October to December.

**Call:** Unrecorded.

**Distribution of other races of the species:** Angola and north-western Northern Rhodesia.

### 1057 ETHIOPIAN SWALLOW.   *HIRUNDO ÆTHIOPICA* Blanford. Pl. 73.

*Hirundo æthiopica* Blanford, Ann. Mag. Nat. Hist. (4), 4, p. 329, 1869: Barakit, Tigré, northern Abyssinia.

**Distinguishing characters:** Above, similar to the Swallow, but rather more steel blue; chestnut of forehead extending on to fore-crown; tail as in the Swallow but outer feathers only slightly elongated; below, dusky white; chin to chest washed with buff; blue of upperparts extending on to sides of chest. The sexes are alike. Wing 100 to 118 mm. The young bird is ashy above with some steel blue gloss; forehead buff.

**General distribution:** Timbuktu to Abyssinia and south to Tanganyika Territory.

**Range in Eastern Africa:** Sudan, south of Shendi, and Abyssinia to Pangani, north-eastern Tanganyika Territory.

**Habits:** The common House Swallow of its area, generally found round human habitations. It has its own local migrations which are not completely clear, but it certainly visits Darfur and Kordofan in the Sudan to breed, and is very rarely seen there at other times of the year. Apparently a resident breeding species on the coast of Kenya Colony. Always tame and domesticated and starts building as soon as the rains soften the mud.

**Nest and Eggs:** A half-cup nest of mud in caves or verandahs of houses, lined with feathers, hair, etc., and not supported below. Eggs three or four, occasionally more, pinkish-white spotted with rufous or chestnut; about 19 × 12·5 mm.

**Recorded breeding:** Nigeria, April to June. Western Sudan, May to July. Southern Sudan, April to June. Abyssinia, July. Italian Somaliland, April and May. Uganda, March. Kenya Colony (breeding condition), July.

**Call:** A remarkably sweet Canary-like song in the breeding season, and the usual alarm cry of 'chip' or 'cheut.'

**1058 PEARL-BREASTED SWALLOW.** *HIRUNDO DIMIDIATA* Sundevall.

*Hirundo dimidiata marwitzi* Reichw.

*Hirundo dimidiata var marwitzi* Reichenow, Vög. Afr. 2, p. 404, 1903: Malangali, Usafua, south-western Tanganyika Territory.

**Distinguishing characters:** Adult male, above, blue black with a violet wash including sides of face and sides of chest, wings and tail; no white in tail; below, white; chest greyish. The female has shorter outer tail feathers. Wing 94 to 107 mm. The young bird has a greenish wash above.

**General distribution:** Angola to Northern Rhodesia, northern Bechuanaland, Nyasaland and Tanganyika Territory.

**Range in Eastern Africa:** South-western Tanganyika Territory.

**Habits:** A relatively uncommon but widely distributed species which has lately been found within our limits.

**Nest and Eggs:** Nest bowl-shaped, made of mud plastered on to a wall and generally supported by a ledge or projection of some sort. They have been recorded as nesting deep in wells as well as in the more ordinary situations of barns and outhouses. Eggs three, pure white; about 17 × 12 mm.

**Recorded breeding:** Southern Belgian Congo, September and October. Northern Rhodesia, September. Portuguese East Africa (feeding young), October. Nyasaland, August to October.

**Call:** No information.

**Distribution of other races of the species:** South Africa, the nominate race being described from the Transvaal.

**1059 WHITE-TAILED SWALLOW.** *HIRUNDO MEGAENSIS* Benson.

*Hirundo megaensis* C. W. Benson, Bull. B.O.C. 63, p. 10, 1942: Mega, southern Abyssinia.

**Distinguishing characters:** Adult male, above, sides of face, neck and wings glossy steel blue; tail white with dusky tips, outer feathers edged with steel blue, outermost slightly elongated; below, white. The female has shorter outermost tail feathers. The large amount of white in the tail is very noticeable in flight. Wing 100 to 105 mm. The young bird is duller above than the adult with some brownish tips to the feathers; tail dull steel blue with white on outer webs and with white spots on inner webs, outer feathers with black edges and outermost slightly elongated and more rounded at tip than in the adult female.

**Range in Eastern Africa:** Southern Abyssinia.

**Habits:** Found locally in open short grass country with scattered low bushes in southern Abyssinia, and apparently limited in altitude to between 4,000 and 4,500 feet.

**Nest and Eggs:** Undescribed. Suspected of nesting in tall anthills.

**Recorded breeding:** Probably January and February.

**Call:** No call heard.

**1060 BLUE SWALLOW.** *HIRUNDO ATROCÆRULEA* Sundevall.

*Hirundo atrocærulea atrocærulea* Sund. **Pl. 73.**
*Hirundo atrocærulea* Sundevall, Œfv. K. Sv. Vet.-Akad. Förh. 7, p. 107, 1850: Umvoti, Natal, South Africa.

**Distinguishing characters:** Adult male, wholly glossy steel blue; some white feathers on flanks and sides of rump; shafts of primaries white; outer tail feathers narrow and considerably elongated. The female has the outer tail feathers much shorter. Wing 101 to 120 mm. The young bird is sooty black with some blue gloss above; throat brownish.

**General distribution:** Southern Rhodesia to eastern Transvaal and Natal; in non-breeding season to Uganda.

**Range in Eastern Africa:** Uganda, in non-breeding season.

**Habits:** Locally common birds with swift erratic flight full of swooping dives, mainly found in grassy valleys, or by the edges of woods or swamps. This race is a migrant over a considerable distance, breeding in South Africa where it stays from November to April and spending the rest of the year in Uganda, passing through western Nyasaland en route. It perches rather freely for a Swallow.

**Nest and Eggs:** An open cup or half-cup, lined with moss and rootlets or feathers, on an overhanging bank or wall, or in the roof of an antbear's burrow, and not supported below. In the Transvaal it has been seen to enter holes in the ground. Eggs two or three, white spotted with rufous or yellowish brown; 18·5 × 13·5 mm.

**Recorded breeding:** Southern Rhodesia, November. Natal, December to March.

**Call:** A musical twittering, and the male is said to have a sweet warble, also a plaintive whistle of 'cheek-cheek.'

*Hirundo atrocærulea lynesi* Grant and Praed.
*Hirundo atrocærulea lynesi* C. Grant & Mackworth-Praed, Bull. B.O.C. 62, p. 45, 1942: Njombe, southern Tanganyika Territory.

**Distinguishing characters:** Similar to the nominate race but with a violet wash on the upperparts. Wing 106 to 119 mm.

**General distribution:** Tanganyika Territory to Nyasaland.

**Range in Eastern Africa:** Njombe area of southern Tanganyika Territory and the Mlanji Plateau of Nyasaland.

**Habits:** As for the preceding race and appears to occur in Nyasaland during the breeding season only, but its non-breeding quarters are as yet unknown. Indistinguishable on the wing, it is extremely difficult to differentiate from the other race when on migration.

**Recorded breeding:** Central and southern Tanganyika Territory, November to February. Nyasaland, October to December.

**1061 WIRE-TAILED SWALLOW.  *HIRUNDO SMITHII*** Leach.
*Hirundo smithii smithii* Leach. **Pl. 73.**
*Hirundo smithii* Leach, App. Tuckey's Voy. Congo, p. 407, 1818: Chisalla Island, about 75 miles up the Congo River, Belgian Congo.

**Distinguishing characters:** Adult male, above, top of head chestnut; rest of upperparts glossy violet blue; a slight bronzy wash on wing-coverts and upper tail-coverts; lores and ear-coverts black,

latter with a violet blue wash; tail with white patches on inner webs; outer tail feathers elongated and very narrow; below, white. The female has much shorter outer tail feathers. Wing 102 to 125 mm. The young bird is much duller; top of head ashy brown; outer tail feathers very slightly elongated; below, washed with buff.

**General distribution:** Gold Coast to Abyssinia, and south to Angola and Natal.

**Range in Eastern Africa:** The Sudan to Eritrea, Abyssinia and the Zambesi River, also Pemba and Zanzibar Islands.

**Habits:** Probably the commonest resident Swallow of most of our area and a tame delightful little bird. Rather uncommon in the Sudan and confined there to the neighbourhood of water, but very widely distributed in eastern and south-eastern Africa, and found in nearly all towns and villages as well as in open country.

**Nest and Eggs:** A small open cup or half-cup of mud lined with a few feathers, generally rather flat and often placed as close as possible to the roof and not supported below. Nest in verandahs, houses, under arches or on rocks, occasionally adopting old nests of other Swallows. Eggs usually three, white, rather heavily marked with reddish or dark brown spots, and with lilac undermarkings; about 17·5 × 13 mm.

**Recorded breeding:** Nigeria, February. Abyssinia, April and May. Sudan, February to April, also November. Kenya Colony, March to June, also October to December. Tanganyika Territory, February, April, also October to December and other months irregularly. Zanzibar Island, any month. Nyasaland, January to September. Northern Rhodesia, February and March, also June to September.

**Call:** A low twitter not often heard.

**Distribution of other races of the species:** Southern Russia, Iran and India.

**1062 RED-RUMPED SWALLOW.** *HIRUNDO DAURICA* Linnæus.
*Hirundo daurica rufula* Temm.
*Hirundo rufula* Temminck, Man. d'Orn. 2nd ed. 3, p. 298, 1835: Egypt.

**Distinguishing characters:** Top of head, mantle, wings, tail, upper and under tail-coverts glossy blue-black; lores, stripe over eye, sides of face to nape chestnut; usually a very narrow line of chestnut on forehead; rump dark tawny shading to buff on lower rump; below, pale tawny white usually narrowly streaked with pale brown; no white in tail. The sexes are alike. Wing 120 to 127 mm. The young bird has the crown of the head duller in tone; tawny ends to the wing-coverts and inner secondaries.

**General distribution:** Morocco, Greece, Asia Minor, Cyprus, Syria, Palestine, Iraq, Iran and Russian Turkestan; in non-breeding season to Libya, Egypt, the Sudan, Abyssinia, Eritrea, British Somaliland and Arabia.

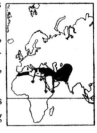

**Range in Eastern Africa:** Eastern Sudan, Eritrea, Abyssinia as far south as Adis-Ababa and British Somaliland in non-breeding season.

**Habits:** A non-breeding migrant to the northern parts of our area, but a single pair has been found breeding in British Somaliland.

**Nest and Eggs:** As for other races.

**Recorded breeding:** British Somaliland, February.

*Hirundo daurica scullii* Seeb.
*Hirundo scullii* Seebohm, Ibis, p. 167, 1883: Gilgit, Kashmir, India.

**Distinguishing characters:** Similar to the Egyptian race but smaller. Wing 111 to 121 mm.

**General distribution:** Afghanistan and India; in non-breeding season to Palestine, Iran, south-western Arabia, British Somaliland and Abyssinia.

**Range in Eastern Africa:** British Somaliland and Abyssinia in non-breeding season.

**Habits:** As for other races.

*Hirundo daurica melanocrissa* (Rüpp.).
*Cecropis melanocrissus* Rüppell, Syst. Uebers, p. 17, pl. 5, 1845: Temben, northern Abyssinia.

**Distinguishing characters:** Differs from the preceding race in rarely having any chestnut on forehead; stripe over eye much narrower; rump the same chestnut colour as the sides of the face and nape, and not shading into paler coloration on the lower rump; below, streaking confined to throat and chest or absent. Wing 117 to 133 mm. The young bird has a paler rump and tawny ends to the innermost secondaries.

**Range in Eastern Africa:** Eritrea and Abyssinia.

**Habits:** As for other races.

**Recorded breeding:** Abyssinia, June and July, also apparently December.

*Hirundo daurica domicella* Hartl. & Finsch.
*Hirundo domicella* Hartlaub & Finsch, Vög. Ost. Afr. p. 143, 1870: Casamanse, Senegal.

**Distinguishing characters:** Differs from the last race in being creamy white below with no streaking, or with only a few streaks on the chest. Wing 108 to 121 mm. The young bird is washed with tawny below and has tawny tips to the innermost secondaries.

**General distribution:** Senegal, Gambia, northern Gold Coast, Nigeria and Cameroons to the Sudan and Abyssinia.

**Range in Eastern Africa:** Darfur to the Baro River area of Abyssinia and the south-western Sudan.

**Habits:** Little known from our area. It appears to migrate southward through western Darfur in October and November and returns in April and May. A very common species in northern West Africa. Eggs are white, occasionally with a few orange spots; about $20 \cdot 5 \times 14$ mm.

**Recorded breeding:** Nigeria, December and January.

*Hirundo daurica emini* Reichw. **Pl. 73.**
*Hirundo emini* Reichenow, J.f.O. p. 30, 1892: Bussissi, near Mwanza, south end Lake Victoria, Tanganyika Territory.

**Distinguishing characters:** Differs from the two preceding races in being larger and in having the whole underside chestnut. Wing 121 to 132 mm.

**General distribution:** Eastern Belgian Congo to the Sudan and south to Nyasaland.

**Range in Eastern Africa:** Uganda, south-eastern Sudan and Kenya Colony to Nyasaland.

**Habits:** A common species in no way differing from other Swallows in habits but with considerable local migrations not yet clearly defined.

**Nest and Eggs:** A retort-shaped nest with a half-tubular entrance spout, lined with feathers, in caves, houses, etc., and plastered to an overhanging projection. Often nesting in a little colony. Eggs two or three, white; about 22·5 × 15·5 mm.

**Recorded breeding:** Uganda, March to December. Kenya Colony, March, April and June to December. Tanganyika Territory, November to March. Nyasaland, August, also October to January and about to breed at the end of February.

**Call:** A sweet metallic twitter.

**Distribution of other races of the species:** Siberia to India, China and Japan, the nominate race being described from Siberia.

### 1063 MOSQUE SWALLOW. *HIRUNDO SENEGALENSIS* Linnæus.

*Hirundo senegalensis senegalensis* Linn. **Pl. 73.**
*Hirundo senegalensis* Linnæus, Syst. Nat. 12th ed. 1, p. 345, 1766: Senegal.

**Distinguishing characters:** Similar to the Red-rumped Swallow but larger; under wing-coverts paler and whiter; breast, belly and under tail-coverts chestnut or tawny, with sometimes a black spot at the tip of the feathers of the under tail-coverts. The sexes are alike. Wing 137 to 158 mm. The young bird is duller and has some tawny edges and tips to wing-coverts, innermost secondaries and upper tail-coverts.

**General distribution:** Senegal and the Gold Coast to the Sudan, Uganda and Kenya Colony.

**Range in Eastern Africa:** The Sudan, Abyssinia, Uganda and Kenya Colony (except southern areas).

**Habits:** A large Swallow with more lethargic flight than the smaller species, and much more often seen settled on a tree. It has a curious flight, consisting of a quick flutter followed by a sail, and looks very like a small Hawk on the wing. Much more a species of large timber and woodland, and only rarely builds in houses. Has considerable local movements and is usually found in small parties. In the western Sudan it is mainly a visitor to breed.

**Nest and Eggs:** Usually in a hole in a tree, but occasionally in a house, on a cliff, or below a bridge, and at times adapts nests of other species to its use. In a tree it fills up a hole to a suitable size with mud, but if it makes a nest it is a thick walled retort-shaped structure with a wide mouth, takes a long time to make, and is plastered to an overhanging projection. In a house it merely makes a rim of mud lined with feathers on a board. Eggs usually three or four, white; about 22 × 15 mm., rather variable in size.

**Recorded breeding:** Northern Cameroons, July and August. Darfur, June to September. Southern Sudan, December to February. Abyssinia, July and November. Uganda, February to July, also October and November.

**Call:** A single tin-trumpet-like call 'harp' and a guttural croak, a trilling mew is also reported.

*Hirundo senegalensis monteiri* Hartl.
*Hirundo monteiri* Hartlaub, Ibis, p., 340, pl. 11, 1862: Angola.

**Distinguishing characters:** Differs from the nominate race in having white in the inner webs of all except the central tail feathers. Wing 137 to 152 mm.

**General distribution:** Angola to the southern Belgian Congo, Kenya Colony, Tanganyika Territory, Northern Rhodesia, Portuguese East Africa as far south as Inhambane and Damaraland.

**Range in Eastern Africa:** Central Kenya Colony to the Zambesi River, also Mafia Island.

**Habits:** As for other races.

**Recorded breeding:** South-eastern Kenya Colony, April and May, also December. Tanganyika Territory and Nyasaland, very varied, February, May, October and December. Rhodesia, March and April, also November and December.

**Distribution of other races of the species:** The Accra area of the Gold Coast.

**1064** RUFOUS-CHESTED SWALLOW. *HIRUNDO SEMIRUFA* Sundevall.
*Hirundo semirufa gordoni* Jard. **Pl. 73.**
*Hirundo gordoni* Jardine, Contr. Orn. for 1851, p. 141, 1852: Gold Coast.

**Distinguishing characters:** Differs from the Red-rumped and Mosque Swallows in having the lores, under the eyes, and the ear-coverts glossy blue black, while the chestnut of the underside does not extend on to the nape; a considerable amount of white in the tail. The sexes are alike. Wing 103 to 126 mm. The young bird has fawn coloured tips to the innermost secondaries and a very pale breast.

**General distribution:** Senegal to Uganda, the Sudan, Kenya Colony and Angola.

**Range in Eastern Africa:** Western and southern Sudan to Uganda and western Kenya Colony.

**Habits:** Locally common and in places abundant, a bird of slow powerful flight with much gliding, usually in pairs and not mixing with other Swallows. Easily mistaken in the field for the Mosque Swallow, and is mostly found in level open country. In the north it is a migrant to breed from May to September.

**Nest and Eggs:** Nest a mud retort with a long entrance funnel when a full nest is made, lined with grass and with an inner lining of feathers, and it is plastered to an overhanging projection. Much more usually, however, the nest is in a drainpipe, ant-hill, hollow tree, or under a roof, and the construction is much reduced. Eggs two to four, white with some gloss; about 22 × 16 mm.

**Recorded breeding:** Sierra Leone, June. Nigeria, April and May. Darfur, July and August. Uganda, May and June.

**Call:** A soft gurgling song and a plaintive call of 'sēeur-sēeur.' Alarm call 'weet-weet.'

**Distribution of other races of the species:** South Africa, the nominate race being described from the Transvaal.

**1065 STRIPED SWALLOW.**   *HIRUNDO ABYSSINICA* Guérin.

*Hirundo abyssinica abyssinica* Guér.
*Hirundo abyssinica* Guérin, Rev. Zool. p. 322, 1843: Northern Abyssinia.

**Distinguishing characters:** Whole top and sides of head and rump deep tawny; mantle, wings and tail blue black, latter with white spots on inner webs of outer tail feathers and outermost elongated; below, chin to belly white profusely streaked with black; under wing-coverts pale tawny; under tail-coverts white streaked with tawny. The female usually has the outermost tail feathers rather shorter. Wing 105 to 112 mm. The young bird is duller than the adult, has some black on the head; wing-coverts and innermost secondaries tipped with tawny; chest and flanks washed with tawny.

**Range in Eastern Africa:** Eritrea and Abyssinia to the southern Sudan at Kajo Kaji, Opari and the Boma Plateau.

**Habits:** As for other races.

**Recorded breeding:** Abyssinia, May to July. Southern Abyssinian border, building in March.

*Hirundo abyssinica unitatis* Scl. & Praed.
*Hirundo puella unitatis* W. Sclater & Mackworth-Praed, Ibis, p. 718, 1918: Pinetown, Natal, South Africa.

**Distinguishing characters:** More heavily striped below than the nominate race. Wing 101 to 117 mm.

**General distribution:** The Sudan to Uganda, Kenya Colony, the Belgian Congo, the Rhodesias, eastern Cape Province and Natal.

**Range in Eastern Africa:** Southern Sudan, Uganda, and Kenya Colony to the Zambesi River, also Pemba, Zanzibar and Mafia Islands.

**Habits:** A common and widely distributed species found in very diverse localities, usually tame and with a 'flutter and sail' flight. It takes an incredible time building its nest—seven weeks in some instances—and is then easily and freely evicted by Swifts or other Swallows. Pitman suggests that both its nest building and its eggs show that it has only recently taken to building its present type of nest, and that its ancestors made cup nests and laid spotted eggs. Some local migration, nest usually in colonies. There is one interesting record of large flocks of this species eating caterpillars off standing crops.

**Nest and Eggs:** In caves or under roofs of metal or concrete, also on rocks. The nest is of a retort shape with a long entrance spout, lined with feathers and plastered to an overhanging projection, but the spout is often broken off short. Eggs three, occasionally two or four, pure white or more rarely white with occasional pale reddish spots or speckling; about 19·5 × 13·5 mm., rather larger in the south.

**Recorded breeding:** Southern Sudan, young in December. Kenya Colony and Uganda, February to August, also October to December. Tanganyika Territory, January to April, June to August, October to December; mostly October to March. Pemba Island,

March to July. South-east Belgian Congo, September and October. Nyasaland, October to January. Portuguese East Africa, May to August, also October and November. Southern Rhodesia, August to November or later.

**Call:** Two or three tinny mewing chirps followed by nasal squeaks, also a pleasant little song of almost organ-like pipings.

*Hirundo abyssinica bannermani* Grant & Praed.
*Hirundo abyssinica bannermani* C. Grant & Mackworth-Praed, Bull. B.O.C. 62, p. 54, 1942: Aribo Valley, Darfur, western Sudan.

**Distinguishing characters:** Differs from the preceding races in having much finer and narrower striping below. Wing 99 to 112 mm.

**Range in Eastern Africa:** Western Sudan.

**Habits:** As for other races.

**Recorded breeding:** Darfur, July and August.

**Distribution of other races of the species:** West Africa.

**1066 GREY-RUMPED SWALLOW.** *HIRUNDO GRISEO-PYGA* Sundevall.
*Hirundo griseopyga griseopyga* Sund. **Pl. 73.**
*Hirundo griseopyga* Sundevall, Œfv. K. Sv. Vet.-Akad. Förh. 7, p. 107, 1850: Durban, Natal, South Africa.

**Distinguishing characters:** Head ashy brown with a slight gloss; lores black, also ear-coverts; mantle, wings and tail glossy blue black; rump and upper tail-coverts ashy brown; tail well forked and outer feathers elongated; below, white; throat and upper chest washed with pale buff; under wing-coverts ashy. The sexes are alike. Wing 88 to 105 mm. The young bird is duller above; rump and upper tail-coverts browner; outer tail feathers shorter; buffish white tips to innermost secondaries.

**General distribution:** The Sudan and Abyssinia to Bechuanaland and Natal.

**Range in Eastern Africa:** Central Sudan to Abyssinia and south to the Zambesi River.

**Habits:** An aberrant species of Swallow looking in flight like a grey-rumped House Martin and preferring open fields, clearings in woods or swampy places. Usually seen in small parties flying rather

aimlessly about and occasionally settling on trees or grass tufts. Its movements are not well known, but it is distinctly local and appears to be migratory in a number of districts.

**Nest and Eggs:** Nest usually in a small hole in open level country, or occasionally in banks, either made by the bird or by some mammal. The nest is a dry grass pad in a hollowed chamber in the soil, about a foot or eighteen inches down. They are difficult to find, as the bird passing over in flight closes its wings and dives straight in. Eggs two to four, white, delicate and thin shelled; about 17 × 12 mm., two hens occasionally using the same nest.

**Recorded breeding:** Abyssinian plateau, March and April. Tanganyika Territory, August. Portuguese East Africa, September. Nyasaland, July to October. Rhodesia, August and September. South Africa, July and August.

**Call:** A weak nasal twitter, rather reminiscent of a Sand Martin.

**Distribution of other races of the species:** West Africa.

**1067** MASCARENE MARTIN.   *PHEDINA BORBONICA* (Gmelin).
*Phedina borbonica madagascariensis* Hartl. **Pl. 73.**
*Phedina madagascariensis* Hartlaub, J.f.O. p. 83, 1860: Madagascar.

**Distinguishing characters:** Above, ashy or sooty brown, finely streaked with black; wings and tail sooty black; below, white; chin to belly streaked with sooty black; sides of neck and flanks sooty. The sexes are alike. Wing 109 to 123 mm. The young bird has broad white tips to the innermost secondaries.

**General distribution:** Madagascar and possibly Pemba Island; a migrant to the mainland.

**Range in Eastern Africa:** Pemba Island to Nyasaland.

**Habits:** A bird of slow flight and easily recognizable. Little is recorded of its habits and it appears to be only a migrant from Madagascar between March to October or November. There are, however, records from Pemba Island from November to March as well as in August and September. Great numbers recorded at Lake Chilwa, Nyasaland, in June 1944.

**Nest and Eggs:** Nest (in Madagascar) a shallow cup of twigs, stems and other materials on a rocky ledge of a cliff or in a cave. Eggs believed to be white with pale brown spotting. No measurements given.

## SWALLOWS and MARTINS

**Recorded breeding:** Madagascar, October and November.

**Call:** Unrecorded.

**Distribution of other races of the species:** Réunion and Mauritius, the nominate race being described from Réunion.

**1068** SAND MARTIN. *RIPARIA RIPARIA* (Linnæus).
*Riparia riparia riparia* (Linn.)
Hirundo riparia Linnæus, Syst. Nat. 10th ed. p. 192, 1758: Sweden.

**Distinguishing characters:** Above, including wings and tail, mouse-brown; below, white; band across chest, under wing-coverts and flanks mouse-brown. The sexes are alike. Wing 99 to 112 mm. The young bird has light edges to wing- and tail-coverts, innermost secondaries and rump.

**General distribution:** Europe to Siberia, North America and North Africa; in non-breeding season to Sierra Leone and from the Sudan to the Transvaal, south-western Arabia, India and Brazil, also one record from Madagascar.

**Range in Eastern Africa:** Throughout in non-breeding season.

**Habits:** A passage migrant and a locally common non-breeding visitor, mostly in the northern part of the area, where it frequents rivers and lakes and assembles to any rise of fly. An occasional migrant only to Nyasaland and Portuguese East Africa. The normal call is a little thin 'svee.'

*Riparia riparia shelleyi* (Sharpe). *Egyptian Sand Martin.* **Pl. 73.**
*Cotile shelleyi* Sharpe, Cat. Bds. B.M. 10, p. 100, 1885: Egypt.

**Distinguishing characters:** Similar to the nominate race but smaller. Wing 88 to 95 mm. The young bird is browner above and has dark buffish tips to the wing-coverts, innermost secondaries and rump.

**General distribution:** Egypt to the northern Sudan; in non-breeding season to the Gulf of Suez and the rest of the Sudan.

**Range in Eastern Africa:** Northern Sudan at Argo; in non-breeding season to as far south as lat. 12° N.

**Habits:** As for the nominate race, settling in flocks on telegraph wires and roosting in reed beds. Occasionally common in small flocks at Khartoum, feeding over water or cultivated land.

**Nest and Eggs:** In a tunnel in a sand or earth bank or even in flat ground. Eggs white, about 17 × 12 mm.

**Recorded breeding:** Egypt, March and April.

**Call:** As for the nominate race.

**Distribution of other races of the species:** Siberia and Turkestan.

**1069** AFRICAN SAND MARTIN. *RIPARIA PALUDICOLA* (Vieillot).

*Riparia paludicola paludicola* (Vieill.). **Pl. 73.**

*Hirundo paludicola* Vieillot, N. Dict. d'Hist. Nat. 14, p. 511, 1817: South Africa.

**Distinguishing characters:** Smaller than the Sand Martin, and similar above, but below, chin to upper belly pale mouse-brown, lower belly and under tail-coverts white; sometimes wholly mouse-brown below. The sexes are alike. Wing 99 to 112 mm. The young bird has the wing-coverts, secondaries, rump and upper tail-coverts edged and tipped with dark buff.

**General distribution:** Angola to Nyasaland, Tanganyika Territory and South Africa.

**Range in Eastern Africa:** One record from Njombe, southern Tanganyika Territory, in December.

**Habits:** A rare straggler to the south of our area, habits as for other races. It appears to be migratory in Nyasaland.

**Recorded breeding:** Nyasaland, July and August. South Africa, breeding at any time of year but mostly from May to September.

*Riparia paludicola minor* (Cab.).

*Cotyle minor* Cabanis, Mus. Hein. 1, p. 49, 1850: Dongola, northern Sudan.

**Distinguishing characters:** Darker above than the nominate race. Wing 88 to 100 mm.

**General distribution:** French Sudan to the Anglo-Egyptian Sudan and Eritrea.

**Range in Eastern Africa:** The Sudan and south-western Eritrea.

**Habits:** Numerous and generally in company with Sand Martins and not easily distinguished on the wing; usually seen feeding over or near water. Has considerable migrations in the non-breeding season.

**Nest and Eggs:** Nest normally in a hole two feet or more long in a vertical bank, with a well-felted pad of grass and feathers at the end, but occasionally in a mere depression in a cliff face. Usually nests in small colonies. Eggs one to five, white; about 16·5 × 12·5 mm.

**Recorded breeding:** Nigeria, December to February. Western Sudan, November and December. Eastern Sudan, November to March.

**Call:** A Sand Martin like 'svee-svee' with a twittering warble in the breeding season.

*Riparia paludicola ducis* Reichw. **Ph. xiii.**
*Riparia ducis* Reichenow, O.M. p. 81, 1908: Western Ruanda, eastern Belgian Congo.

**Distinguishing characters:** Rather larger and much darker dusky or greyish brown than the last race, with only the under tail-coverts white. Wing 92 to 110 mm.

**Range in Eastern Africa:** Uganda and Kenya Colony to Ruanda and Tanganyika Territory.

**Habits:** Gregarious species not by any means confined to water, and nesting colonies may be some way from any river or lake. Habits much as for the Sand Martin, and is migratory in the non-breeding season.

**Recorded breeding:** Kenya Colony highlands, March to June, also August and September. Mwanza, Tanganyika Territory, June.

*Riparia paludicola schoensis* Reichw.
*Riparia minor schoensis* Reichenow, J.f.O. p. 88, 1920: Adis-Ababa, central Abyssinia.

**Distinguishing characters:** Similar to the preceding race but larger. Wing 109 to 117 mm.

**Range in Eastern Africa:** Abyssinia.

**Habits:** As for other races.

**Recorded breeding:** Abyssinia, July to September.

**Distribution of other races of the species:** Madagascar.

**1070 BANDED MARTIN.** *RIPARIA CINCTA* (Boddaert).
*Riparia cincta erlangeri* Reichw.

*Riparia cincta erlangeri* Reichenow, J.f.O. p. 673, 1905: Irma, Adis-Ababa-Harar Road, eastern Abyssinia.

**Distinguishing characters:** Above, dark mouse-brown; a white streak from nostril to over eye; below, white with a dark mouse-brown chest band which usually extends down centre of breast. The sexes are alike. Wing 131 to 143 mm. The larger size, white streak in front of eye and whiter under wing-coverts distinguishes this species from the Sand Martin. The young bird has the wing-coverts, innermost secondaries, rump and part of the mantle tipped with buff.

**Range in Eastern Africa:** Eritrea and Abyssinia to the southern Sudan and north-eastern Uganda at Wadelai.

**Habits:** As for other races.

**Nest and Eggs:** See under following race.

**Recorded breeding:** Abyssinia, June to August.

*Riparia cincta suahelica* V. Som. **Ph. xiii.**
*Riparia cincta suahelica* Van Someren, Nov. Zool. 29, p. 90, 1922: Escarpment, southern Kenya Colony.

**Distinguishing characters:** Darker above, more blackish brown than the Abyssinian race, and rather smaller. Wing 122 to 136 mm.

**General distribution:** Uganda, Kenya Colony and Tanganyika Territory to Northern Rhodesia and Nyasaland.

**Range in Eastern Africa:** Uganda, except the north-western area; to Kenya Colony, Tanganyika Territory and Nyasaland at Lake Chilwa.

**Habits:** A comparatively sluggish bird for a Martin and distinctly larger; it has been noted picking insects off grass tops, and settles freely on grass stems. Usually in pairs and not gregarious. Common locally in highlands of Kenya Colony and of wide distribution, but not common elsewhere. Frequently found in grassy country well away from water.

**Nest and Eggs:** Nest a hole of some length in a bank or cliff, the chamber lined with grass stems and a few feathers. Nests are not necessarily in a colony. Eggs normally three, up to five recorded, white and glossy; about 23 × 16 mm.

**Recorded breeding**: Uganda and western Kenya Colony, March to June. Southern Kenya Colony, March and April. Tanganyika Territory, January onwards.

**Call**: A varied warbling twitter, occasionally ending in a little trill.

**Distribution of other races of the species**: South Africa, the nominate race being described from the Cape of Good Hope.

**1071** EUROPEAN CRAG MARTIN.  *PTYONOPROGNE RUPESTRIS* (Scopoli). **Pl. 73.**
*Hirundo rupestris* Scopoli, Annus I, Hist. Nat. p. 167, 1769: Tyrol, southern Europe.

**Distinguishing characters**: General colour mouse-brown, paler below; belly more russet; chin and throat spotted; white spots on inner webs of all except central and outermost tail feathers, though latter sometimes with small spots. The sexes are alike. Wing 124 to 143 mm. The young bird has russet or whitish edges and tips to the feathers of the upperparts, secondaries and wing-coverts and is more russet below.

**General distribution**: Southern Europe to China and northern Africa; in non-breeding season to the Sudan, Abyssinia, Arabia and India.

**Range in Eastern Africa**: Sudan as far south as Koalib about 250 miles south of Khartoum and Abyssinia in non-breeding season.

**Habits**: A scarce non-breeding visitor to the north of our area. A bird of relatively slow flight likely to be found near cliffs or ravines. The white tail-spots are conspicuous in flight.

**1072** PALE CRAG MARTIN. *PTYONOPROGNE OBSOLETA* (Cabanis).
*Ptyonoprogne obsoleta obsoleta* (Cab.). **Pl. 73.**
*Cotyle obsoleta* Cabanis, Mus. Hein. 1, p. 50, 1850: Egypt.

**Distinguishing characters**: Smaller and greyer than the European Crag Martin, with less russet on belly and no spots on chin and throat. The sexes are alike. Wing 112 to 122 mm. The young bird has ashy edges and tips to the feathers of the upperparts.

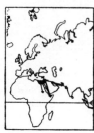

**General distribution:** Libya, Egypt to Sinai and Palestine; in non-breeding season to the Sudan, British Somaliland, Socotra Island, Arabia and Sind.

**Range in Eastern Africa:** Nile Valley to eastern Sudan, British Somaliland and Socotra Island in non-breeding season.

**Habits:** Usually seen on or near high cliffs in pairs or small parties. They also inhabit ruins or caves and make a mud nest attached to the roof. There is a record of this species breeding on a cliff on the Red Sea coast in May, but it may have been a misidentification. As far as is known this race does not nest within our area, but it is closely allied to the African Rock Martin, which has a wide distribution.

*Ptyonoprogne obsoleta arabica* (Reichw.).
*Riparia arabica* Reichenow, Vög. Afr. 3, p. 828, 1905: Lahej, Aden, south-west Arabia.

**Distinguishing characters:** Darker than the nominate race and more suffused with brown below. Wing 110 to 126 mm.

**General distribution:** South-eastern Egypt, eastern Sudan and Arabia; in non-breeding season to British and Italian Somaliland, Socotra Island and Sind.

**Range in Eastern Africa:** Red Sea Province of the Sudan, also British and northern Italian Somaliland and Socotra Island in the non-breeding season.

**Habits:** As for the last race.

**Recorded breeding:** Erkowit, Red Sea Province, Sudan, August.

*Ptyonoprogne obsoleta reichenowi* (Zedl.).
*Riparia rupestris reichenowi* Zedlitz, O.M. p. 177, 1908: North Arabia.

**Distinguishing characters:** Very much paler than the nominate race. Wing 112 to 122 mm.

**General distribution:** Northern and eastern Arabia to south-western Iran; Afghanistan and northern India; in non-breeding season to Egypt at Suez, British Somaliland and southern Arabia.

**Range in Eastern Africa:** British Somaliland in non-breeding season.

**Habits:** As for other races.

**Distribution of other races of the species:** French Sahara and India.

**1073** AFRICAN ROCK MARTIN. *PTYONOPROGNE FULI-GULA* (Lichtenstein).
*Ptyonoprogne fuligula rufigula* (Fisch. & Reichw.). **Pl. 73.**
*Cotyle rufigula* Fischer & Reichenow, J.f.O. p. 53, 1884: Lake Naivasha, southern Kenya Colony.

**Distinguishing characters:** Size and tail markings as in the Pale Crag Martin, but dark sooty brown above and below, except chin and neck below which are pale russet. The sexes are alike. Wing 105 to 120 mm. The young bird has russet edging to the wing, mantle, rump, and to the tail-coverts.

**General distribution:** Northern Nigeria to Abyssinia, and south to Nyasaland, northern Portuguese East Africa, Northern Rhodesia and eastern Southern Rhodesia.

**Range in Eastern Africa:** Southern Sudan and south-western Abyssinia, Uganda, Kenya Colony, Tanganyika Territory, northern Portuguese East Africa and Nyasaland.

**Habits:** Commonly found wherever precipitous cliffs occur over all Eastern Africa. They are silent birds of relatively weak gliding flight and are not usually seen very far from their habitat, though they will go many miles to visit a grass fire. Occasionally they are found in towns and nest on houses. Incubation is by both sexes.

**Nest and Eggs:** Nest either a cup placed on a ledge or a half-cup plastered on to a wall or cliff, in an overhung angle if possible. Nest of mud mixed with grass and lined with fine grass or fibre and feathers. Eggs two or three, white freely spotted or speckled with light or dark rusty brown and with a few violet undermarkings mostly at larger end; about 20 × 13·5 mm.

**Recorded breeding:** Kenya Colony, February to July and occasionally in other months. Tanganyika Territory, September to May, most commonly in October, February and March. Nyasaland, October to April.

**Call:** A low twitter or little high-pitched 'twee' also occasionally a clacking noise like a Chat.

*Ptyonoprogne fuligula pusilla* (Zedl.).
*Riparia rupestris pusilla* Zedlitz, O.M. p. 177, 1908: Asmara, Eritrea.

**Distinguishing characters:** Paler than the preceding race both above and below, especially from chest to belly. Wing 110 to 128 mm.

**Range in Eastern Africa:** Darfur to Eritrea and Abyssinia except south-western areas.

**Habits:** Silent inconspicuous birds common among the rocky tops of hills, but only wanderers elsewhere, and in the field are only found where suitable cliffs occur: they have, however, adapted themselves to a town life in many places. Some evidence of local migration.

**Recorded breeding:** Darfur, December to March. Abyssinia, January and February, May, July, August and October. Eritrea, February and March.

**Distribution of other races of the species:** West and South Africa, the nominate race being described from eastern Cape Province.

**1074 HOUSE MARTIN.** *DELICHON URBICA* (Linnæus).
*Delichon urbica urbica* (Linn.).
*Hirundo urbica* Linnæus, Syst. Nat. 10th ed. p. 192, 1758: Sweden.

**Distinguishing characters:** Above, glossy blue black; rump and upper tail-coverts white, latter with glossy blue black ends; wings and tail dull black, latter forked; below, white; feet and toes feathered white. The female has a less deeply forked tail than the male. Wing 99 to 117 mm. The young bird is duller above with some glossy blue black on mantle and head; tips of innermost secondaries white; dark centres to feathers of rump and upper tail-coverts; and chin to chest washed with dusky. The young bird has a superficial resemblance to the Grey-rumped Swallow, but the usually lighter coloured rump and much less forked tail, together with the larger size should serve to distinguish the House Martin.

**General distribution:** Europe and Asia to the Petchora, Altai and Kashmir; in non-breeding season to the Canary Islands, Africa as far south as Angola, Transvaal and Natal, occasional to India.

**Range in Eastern Africa:** Throughout in non-breeding season.

**Habits:** A common passage migrant through Kenya Colony and Tanganyika Territory, also to a certain extent through Nyasaland.

On the other hand it appears to be rarely seen on passage in the Sudan or Abyssinia. Its quarters during the northern winter months are not well-defined and it is still somewhat of a mystery where the large numbers of migrants seen passing south go to in the non-breeding season.

*Note:* House Martins have recently been found nesting in small numbers in widely separated districts of South Africa. It is not known to what race they belong.

**Distribution of other races of the species:** Eastern Siberia and northern Africa.

**1075 BLACK ROUGH-WING SWALLOW.** *PSALIDOPROCNE HOLOMELÆNA* (Sundevall).
*Psalidoprocne holomelæna holomelæna* (Sund.). **Pl. 73. Ph. xiii.**
*Hirundo holomelas* Sundevall, Œfv. K. Sv. Vet.-Akad. Förh. 7, p. 108, 1850: Durban, Natal, South Africa.

**Distinguishing characters:** General colour black with a bottle green gloss; under wing-coverts ashy brown; tail forked. The female has the tail less deeply forked. Wing 98 to 113 mm. The young bird is rather less glossy above and browner below.

**General distribution:** Uganda and Kenya Colony to Natal.

**Range in Eastern Africa:** Uganda (except Ruwenzori and south-western areas) Kenya Colony and Tanganyika Territory to the Zambesi River.

**Habits:** A common but local species found at all elevations but preferring high ground. Frequently seen in forest glades in some numbers flying low over the ground but rising to tree-top level in the evenings. Normally found from about the 4,000 feet level upwards, they have been recorded in numbers up to 13,500 feet. They seem to sleep or shelter in thick trees during the heat of the day, and appear suddenly in numbers with weak silent flight.

**Nest and Eggs:** In a short hole in a bank, natural or excavated. Nest is substantially made of grass with lichen or moss. Occasionally nest on a ledge in a cave. Eggs two, white; about 19 × 13 mm., thin shelled, as are the eggs of all Rough-winged Swallows.

**Recorded breeding:** Mt. Elgon, February. Kenya Colony in many months, apparently irregular but usually in rains. Amani, Tanganyika Territory, all months except April and May.

**Call:** A soft chirp seldom uttered.

*Psalidoprocne holomelæna ruwenzori* Chap.
*Psalidoprocne holomelæna ruwenzori* Chapin, Ann. Mus. Nov. p. 13, 1932: Kalongi, western Ruwenzori Mts., eastern Belgian Congo.

**Distinguishing characters:** Similar to the nominate race, but tail much less deeply forked. Wing 103 to 119 mm.

**General distribution:** Eastern Belgian Congo to Uganda.

**Range in Eastern Africa:** Ruwenzori Mts. and south-western Uganda.

**Habits:** As for the preceding race.

**Recorded breeding:** West Ruwenzori, February.

### 1076 WESTERN ROUGH-WING SWALLOW. *PSALIDOPROCNE CHALYBEA* Reichenow. Pl. 73.

*Psalidoprocne chalybea* Reichenow, J.f.O. p. 442, 1892: Victoria, Cameroons.

**Distinguishing characters:** General colour dull oily green with a slight sheen; tail well forked; under wing-coverts and axillaries pale ash grey. The sexes are alike. Wing 88 to 101 mm. The young bird is sooty brown, with a slight green wash.

**General distribution:** Cameroons to eastern Belgian Congo.

**Range in Eastern Africa:** Not yet recorded, but should occur in Ruanda on the eastern side of Lake Kivu.

**Habits:** An inhabitant of forest clearings, perching occasionally on bare twigs of high trees. The forked tail is noticeable.

**Nest and Eggs:** Undescribed.

**Recorded breeding:** Ituri (breeding condition), May to September.

**Call:** An occasional soft mellow little call.

### 1077 EASTERN ROUGH-WING SWALLOW. *PSALIDOPROCNE ORIENTALIS* Reichenow.

*Psalidoprocne orientalis orientalis* Reichw. Pl. 73.
*Psalidoprocne petiti orientalis* Reichenow, J.f.O. p. 277, 1889: Lewa, 12 miles north-west of Pangani, north-eastern Tanganyika Territory.

**Distinguishing characters:** General colour oily green; under wing-coverts and axillaries white; tail well forked, outer feathers narrow. The sexes are alike. Wing 98 to 114 mm. The young bird is duller and the under wing-coverts and axillaries are tinged with dusky.

**General distribution:** Tanganyika Territory to Nyasaland, north-eastern Northern Rhodesia, eastern Southern Rhodesia and the Beira area of Portuguese East Africa.

**Range in Eastern Africa:** North-eastern and central Tanganyika Territory to Nyasaland.

**Habits:** Locally common in highlands round forest edges or on grass plateaux at high elevations. Usually in small parties, perching freely on dead tree tops, but hawking back and fore over the same ground with silent flight, rather wavering and usually low over the ground. Some evidence of migration.

**Nest and Eggs:** Nest a pad of lichen at the end of a short tunnel in a bank. Eggs two, white; about $18 \cdot 5 \times 13$ mm.

**Recorded breeding:** Tanganyika Territory, December. Nyasaland, December to June. Rhodesia, November.

**Call:** A low twittering, also a little whistling mewing cry.

*Psalidoprocne orientalis oleaginea* Neum.
*Psalidoprocne orientalis oleaginea* Neumann, O.M. p. 144, 1904: Schubba, Kaffa, south-western Abyssinia.

**Distinguishing characters:** General colour richer and darker, especially below, and tail much less deeply forked than the nominate race. Wing 97 to 109 mm.

**Range in Eastern Africa:** Western and south-western Abyssinia to the Yei area, southern Sudan.

**Habits:** A bird of forest edge or clearings of which little is so far recorded.

**Recorded breeding:** No records.

**Distribution of other races of the species:** Angola.

**1078** BLUE ROUGH-WING SWALLOW. *PSALIDOPROCNE PRISTOPTERA* (Rüppell).
*Psalidoprocne pristoptera pristoptera* (Rüpp.). **Pl. 73.**
*Hirundo pristoptera* Rüppell, N. Wirbelt. Vög. p. 105, pl. 39, fig. 2, 1836: Simen Province, northern Abyssinia.

**Distinguishing characters:** Differs from the last species in being blue black with a slight gloss above and below; wings and

tail washed with green, the latter forked and the feathers much broader; under wing-coverts and axillaries white; the primary coverts have dusky ends. The female has the tail less deeply forked. Wing 93 to 108 mm. The young bird is sooty brown and blue, with dusky white under wing-coverts.

**Range in Eastern Africa:** Northern, western and south-western Abyssinia and Eritrea.

**Habits:** A common but local resident on the high plateau of Eritrea and Abyssinia, usually seen in pairs hawking over streams.

**Nest and Eggs:** Nest in a tunnel about two feet long excavated in a bank or cliff near water, the chamber of which is padded with fine grass. Eggs three, white and glossy. No measurements available.

**Recorded breeding:** Eritrea and Abyssinia, May and June.

**Call:** Unrecorded.

*Psalidoprocne pristoptera blanfordi* Blund. & Lov.
*Psalidoprocne pristoptera blanfordi* Blundell & Lovat, Bull. B.O.C. 10, p. 20, 1899: Bilo, 120 miles west of Adis Ababa, central Abyssinia.

**Distinguishing characters:** Differs from the nominate race in having a greenish wash on the head, mantle and wing-coverts over the blue; below, darker blue with a very slight wash of green. Wing 106 to 110 mm.

**Range in Eastern Africa:** Upper Blue Nile and Bilo to Adis Ababa, Abyssinia.

**Habits:** As for the nominate race.

**Recorded breeding:** No records.

**1079** BROWN ROUGH-WING SWALLOW. *PSALIDO-PROCNE ANTINORII* Salvadori. **Pl. 73.**
*Psalidoprocne antinorii* Salvadori, Ann. Mus. Civ. Genova (2) 1, p. 123, 1884: Denz, Shoa, central Abyssinia.

**Distinguishing characters:** Similar to the preceding species but differs in the general colour being slightly glossy dark bronzy brown; also the tail is well forked and the outer feathers not quite so broad. The female has the tail less forked. Wing 95 to 110 mm. The young bird is duller and below brownish; under wing-coverts and axillaries mixed dusky and white.

**Range in Eastern Africa:** Central and southern Abyssinia.

**Habits:** Locally common in the Abyssinian highlands, from 4,000 feet upwards, usually round the fringes of evergreen forest.

**Nest and Eggs:** Undescribed.

**Recorded breeding:** Abyssinia (breeding condition), July.

**Call:** Unrecorded.

**1080 WHITE-HEADED ROUGH-WING SWALLOW. *PSALIDOPROCNE ALBICEPS* Sclater.**
*Psalidoprocne albiceps* P. L. Sclater, P.Z.S. p. 108, pl. 14, 1864: Uzinza, northern Tabora Province, central Tanganyika Territory.

**Distinguishing characters:** Adult male, general colour sooty black with a slight oily green gloss; head and throat white; sooty black stripe from base of bill through eye to ear-coverts; under wing-coverts and axillaries ashy; tail forked. The female has the throat white, but the top of the head mixed white and sooty black. Wing 95 to 110 mm. The young bird is wholly sooty black, except that the young male has an ashy brown throat.

**General distribution:** The Sudan to Tanganyika Territory, Northern Rhodesia and Nyasaland.

**Range in Eastern Africa:** Southern Sudan to Tanganyika Territory.

**Habits:** Locally common and at times very numerous, particularly so in Uganda. In flight they have rather the appearance of Swifts, but the white heads of the males are noticeable. Frequent open country and the lower slopes of hills as well as forest glades. Stoneham remarks that as the birds chase one another he has frequently seen large butterflies of the genus *Charaxes* join in the chase, without any attention being paid to them by the birds. Perch freely on bushes or grass stems and on tree tops in the evening. Some evidence of migration in the south.

**Nest and Eggs:** Nest in holes in banks, generally a rodent hole adapted by the bird, chamber lined with grass and a few feathers. Eggs two to four, white; about $17·5 \times 12·5$ mm.

**Recorded breeding:** Southern Sudan, August and September. Uganda, April to June, also August. Kavirondo, Kenya Colony, June to October. Tanganyika Territory, November to March. Nyasaland, February.

**Call:** Unrecorded.

Names in Sclater's *Syst. Av. Æthiop.* 2, 1930, which have been changed, or have become synonyms in this work.

*Hirundo senegalensis hybrida* Van Someren, treated as synonymous with *Hirundo senegalensis monteiri* Hartlaub.

*Psalidoprocne holomelæna massaica* Neumann, treated as synonymous with *Psalidoprocne holomelæna holomelæna* Sundevall.

Names introduced since 1930, and which have become synonyms in this work:

*Hirundo senegalensis aschani* Granvik, 1934, treated as synonymous with *Hirundo senegalensis senegalensis* Linnæus.

*Ptyonoprogne rufigula fusciventris* Vincent, 1933, treated as synonymous with *Ptyonoprogne fuligula rufigula* (Fischer and Reichenow).

## Addenda and Corrigenda

**1057** *Hirundo æthiopica.* *Recorded breeding:* add Eritrea, May to August. British Somaliland, May and June.

**1065** *H.a. abyssinica.* *Recorded breeding:* add Eritrea, April to August.

*Continued on p.* 644

FAMILY—**CAMPEPHAGIDÆ**. **CUCKOO SHRIKES**. Genera: *Campephaga* and *Coracina*.

Six species occur in Eastern Africa. In the genus *Campephaga* the sexes differ remarkably and might easily be taken for different species in the field. The nostrils are partly or wholly covered by feathers. The feathers of the rump have strong shafts, but the apical quarter is weak; the effect is that of a spine when handled.

They are birds of wooded country and forests, and are usually seen singly or in pairs.

### KEY TO THE ADULT CUCKOO-SHRIKES OF EASTERN AFRICA

1 General colour wholly grey: GREY CUCKOO-SHRIKE *Coracina cæsia* 1086

2 General colour above, grey, also throat; rest of underparts white: WHITE-BREASTED CUCKOO-SHRIKE *Coracina pectoralis* 1085

3 General colour glossy blue black and purple or blue black: 4–8

4 Sides of face and underparts purple: PURPLE-THROATED CUCKOO-SHRIKE, male *Campephaga quiscalina* 1084

5 Sides of face and underparts blue black, inner webs of flight feathers grey or yellow: 6–8

6 Wing shoulder blue black, inner webs of flight feathers dark grey: PETIT'S CUCKOO-SHRIKE, male *Campephaga petiti* 1082

7 Wing shoulder blue black or lemon yellow, inner webs of flight feathers yellow:    BLACK CUCKOO-SHRIKE, male *Campephaga sulphurata* **1081**

8 Wing shoulder red or orange yellow, inner webs of flight feathers yellow:    RED-SHOULDERED CUCKOO-SHRIKE, male *Campephaga phœnicea* **1083**

9 General colour olive brown and yellow:    10–15

10 Below, mainly yellow:    11–12

11 Indistinct barring below:    PURPLE-THROATED CUCKOO-SHRIKE, female *Campephaga quiscalina* **1084**

12 Black specks on breast:    PETIT'S CUCKOO-SHRIKE, female *Campephaga petiti* **1082**

13 Below, mainly white, with black crescentic markings and bars:    14–15

14 Above, brown or yellowish brown:    BLACK CUCKOO-SHRIKE, female *Campephaga sulphurata* **1081**

15 Above, olive brown:    RED-SHOULDERED CUCKOO-SHRIKE, female *Campephaga phœnicea* **1083**

**1081** BLACK CUCKOO-SHRIKE. *CAMPEPHAGA SULPHURATA* (Lichtenstein). **Pls. 74, 96. Ph. xiv.**
*Cuculus sulphuratus* Lichtenstein, Cat. Rer. Nat. Rar. p. 15, 1793: Eastern Cape Province, South Africa.

**Distinguishing characters:** Adult male, general colour glossy blue black; wings and tail black, edged with glossy blue black; wing shoulder either glossy blue black or bright yellow; under wing-coverts blue black or blue black and yellow; inner webs of flight feathers washed with yellow. The female is brown or yellowish brown above, barred with black on rump and upper tail-coverts and more indistinctly barred on lower mantle; wings brown with bright yellow edges to all the feathers; tail olive brown and blackish, outer feathers half to quarter tipped with bright yellow; underside, white sometimes with a yellowish wash with black bars and crescentic marks; throat occasionally buffish. Wing 95 to 108 mm. The young bird differs from the female in having the mantle and sometimes the head barred, more markings below, and the tail feathers pointed and not rounded.

**General distribution:** Uganda, the Sudan, Abyssinia, and Italian Somaliland to Angola and South Africa.

**Range in Eastern Africa:** South-eastern Sudan, southern Abyssinia, Uganda, Kenya Colony, and Italian Somaliland to the Zambesi River, also Zanzibar and Mafia Islands.

**Habits:** Silent inconspicuous birds hunting through the tops of bushes and trees in wooded country. Restless and always on the move, usually in pairs, and shy; occasionally seen feeding on the ground. Locally common in forest or forest strips, but with a wide ecological range and occurring also in lowland scrub in the coastal belt. The males often consort with Drongos and may then be easily overlooked. They also take part in mixed bird hunting parties.

**Nest and Eggs:** Nest usually at some height, either a flat pad of moss or lichen on a bough, or a deeper cup in the fork of a tree very small for the bird and extremely hard to find. Eggs two, pale greyish green or yellowish green, plentifully blotched and spotted with brown or sepia and with lilac undermarkings; about 24 × 16·5 mm.

**Recorded breeding:** Kenya Colony, March and April. Tanganyika Territory, northern areas (breeding condition), May and January. Amani, October. Zanzibar Island, probably October. Southern Belgian Congo, December to February. Rhodesia, October and November. Nyasaland, October to January. Portuguese East Africa, September to January.

**Food:** Largely caterpillars, fruit and tree seeds.

**Call:** A low Flycatcher-like 'chup' or 'mutt' also a low tinkling trill 'wheeo-wheeo-eee-ee-e,' but as a rule very silent birds.

### 1082 PETIT'S CUCKOO-SHRIKE. *CAMPEPHAGA PETITI* Oustalet.

*Campophaga petiti* Oustalet, Ann. Sci. Nat. (6) 17, Art 8, p. 1, 1884: Landana, Portuguese Congo.

**Distinguishing characters:** Adult male, similar to the Black Cuckoo-Shrike but lacks the yellow wash on the inner webs of the flight feathers which are grey; wing shoulder glossy blue black. The female is very similar to that of the Black Cuckoo-Shrike but is wholly bright yellow below with a few black flecks on the breast. Wing 92 to 104 mm. The young bird is bright yellow below with black arrow-head markings.

**General distribution:** Gabon to Kenya Colony and northern Angola.

**Range in Eastern Africa:** Uganda and western Kenya Colony.

**Habits:** A rare and little known species with probably much the same habits as the Black Cuckoo-Shrike, mostly seen in the tops of forest trees.

**Nest and Eggs:** Undescribed.

**Recorded breeding:** No records.

**Food:** Caterpillars and other insects.

**Call:** Unrecorded.

### 1083 RED-SHOULDERED CUCKOO-SHRIKE. *CAMPEPHAGA PHŒNICEA* (Latham).

*Ampelis phœnicea* Latham, Ind. Orn. 1, p. 367, 1790: Gambia, West Africa.

**Distinguishing characters:** Adult male, differs from the Black Cuckoo-Shrike in having the wing shoulder red or orange yellow. The female is more olive brown above than the female of the Black Cuckoo-Shrike and this is especially noticeable on the rump and upper tail-coverts. Wing 91 to 107 mm. The young bird differs from the young of the Black Cuckoo-Shrike in the same way. Albinistic examples are known.

**General distribution:** Gambia and Sierra Leone to the Sudan, Eritrea, Abyssinia, Uganda and Kenya Colony.

**Range in Eastern Africa:** Southern half of the Sudan, Eritrea, Abyssinia, Uganda and western Kenya Colony.

**Habits:** Not usually common but with a wide distribution. Local in the woods of the west basin of Darfur, Sudan. In Abyssinia a rare and local resident among tall trees, in Uganda mostly found in forest or woodland. Habits of the genus, but usually tame, and if insects are disturbed from foliage pursues and catches them like a Flycatcher, or follows them to the ground if necessary.

**Nest and Eggs:** A cup or saddle-shaped nest of moss and lichen in the fork of a tree, bound round with spiders' webs. Eggs two, yellowish green with purplish spots and blotches; about 21 × 16 mm.

**Recorded breeding:** Darfur, July. Equatoria, June. Uganda, March and April.

**Food:** Caterpillars and other insects.

**Call:** A soft whistled call of two notes by the male, often uttered on the wing, also various harsher calls.

**1084** PURPLE-THROATED CUCKOO-SHRIKE. *CAMPEPHAGA QUISCALINA* Finsch.
*Campephaga quiscalina martini* Jacks. **Pl. 74.**
*Campophaga martini* Jackson, Bull. B.O.C. 31, p. 18, 1912: Ravine, Nandi, western Kenya Colony.

**Distinguishing characters:** Adult male, upperside and edges of wings and tail glossy steel blue with a purple wash from top of head to upper mantle; below, glossy purple; inner webs of flight feathers black. The female has the head and hind neck grey; ear-coverts grey and white; rest of upperparts yellowish green including wing-coverts; below, throat buffish white; rest of underparts bright yellow with faint barring on chest and flanks. Wing 94 to 101 mm. The young bird differs from the adult female in having a black sub-terminal bar and yellow tips to the wing-coverts, innermost secondaries and rump, and from the young of the Black and Red-Shouldered Cuckoo-Shrikes by lacking the broad yellow edges to the wing-coverts and flight feathers, and from the young bird of Petit's Cuckoo-Shrike in having barred, not arrow-shaped, markings below.

**General distribution:** North-eastern Belgian Congo to the Sudan, Uganda, Kenya Colony and Tanganyika Territory.

**Range in Eastern Africa:** Southern Sudan to Uganda, western Kenya Colony and northern Tanganyika Territory.

**Habits:** Locally common and at times abundant in dense forest particularly on Mt. Elgon. Habits as for the genus.

**Nest and Eggs:** Undescribed.

**Recorded breeding:** Mt. Elgon (breeding condition), June and November.

**Food:** Insects, particularly caterpillars.

**Call:** A modulated whistle.

*Campephaga quiscalina münzneri* Reichw.
*Campephaga quiscalina münzneri* Reichenow, O.M. p. 91, 1915: Mahenge, south-eastern Tanganyika Territory.

**Distinguishing characters:** From the preceding race the male differs in having the purple wash confined to the upper mantle, the top of the head being steel blue; the females are indistinguishable. Wing 92 to 98 mm.

**Range in Eastern Africa:** Eastern and southern Tanganyika Territory from the Morogoro to Mahenge districts.

**Habits:** Of the genus, a typical forest species at higher elevations.

**Recorded breeding:** No records.

**Distribution of other races of the species:** West Africa to Angola, the nominate race being described from Fantee, Gold Coast.

**1085** WHITE-BREASTED CUCKOO-SHRIKE. *CORACINA PECTORALIS* (Jardine and Selby).
*Graucalus pectoralis* Jardine & Selby, Ill. Orn. 2, pl. 57, 1828: Sierra Leone.

**Distinguishing characters:** Adult male, above, pale grey lighter on forehead and over eye; flight feathers blackish; outer tail feathers black; below, sides of face and chin to neck dark slate; rest of underparts white, including underside of wings. The female has the chin to neck pale grey or white and grey. Wing 133 to 152 mm. The young bird is similar to the adult female, but has some of the feathers of the upperparts with subterminal black ends and white tips, broad white edges to the flight feathers and primary coverts; pointed outer tail feathers and black spot at end of under tail-coverts.

# CUCKOO SHRIKES

**General distribution:** Senegal to Abyssinia and south to Northern Rhodesia, the Transvaal and Damaraland.

**Range in Eastern Africa:** Southern Sudan and Abyssinia to the Zambesi River.

**Habits:** Inhabits large trees in open woodland, or lower trees in open bush country, generally in pairs. A conspicuous bird with slow flapping or gliding flight, rather Hawk-like. Occasionally hunts on the wing like a Drongo, or feeds on the ground, also sometimes a member of mixed bird parties. Moves unobtrusively from branch to branch with long hops, and is said to peer at leaves and twigs in a curiously short-sighted manner.

**Nest and Eggs:** The nest is a shallow cup of weed stems and lichen plastered on to a bough high up in a tree, usually difficult or impossible to see from the ground. Eggs one or two, green, freely speckled with pale brown and with ashy grey undermarkings; about 27 × 20 mm.

**Recorded breeding:** Southern Sudan (near breeding), February and March. Rhodesia, October to November. Nyasaland, September to November. Portuguese East Africa, December.

**Food:** Insects, largely caterpillars.

**Call:** Male a soft whistled 'duid-duid,' female a long trilled 'che-e-e-e-e.' There is also a song and a weak squeaking cry. Calls uttered perched or on the wing.

## 1086 GREY CUCKOO-SHRIKE. *CORACINA CÆSIA* (Lichtenstein).

*Coracina cæsia pura* (Sharpe).
*Graucalus purus* Sharpe, Ibis, p. 121, 1891: Mt. Elgon.

**Distinguishing characters:** Adult male, wholly bluish grey; wings and tail darker; lores and sometimes whole throat, black. The female is rather paler and lacks the black on the lores and throat. Wing 115 to 127 mm. The young bird has the head and body barred black and white; wings and tail dark bluish grey with whitish edges to the wing-coverts, tips of flight feathers and tail; upper tail-coverts bluish grey with subterminal black bars and white ends.

**Range in Eastern Africa:** Southern Sudan, central and southern Abyssinia, eastern Uganda, Kenya Colony, Tanganyika Territory and Nyasaland.

**Habits:** Locally common in forest outskirts or high woodland, usually seen among the top boughs of high trees where it remains motionless for long periods or sidles quietly along a branch. Weak flight, but is capable of catching insects on the wing like a Drongo.

**Nest and Eggs:** Nest a cupped pad of moss or lichen in a fork or on a bough at some height from the ground. Eggs one or two, clear green or dull blue with elongated olive spots, about 26 × 19·5 mm. (South African race).

**Recorded breeding:** Southern Sudan, August and (near breeding) November. Kenya Colony, December.

**Food:** Insects.

**Call:** A long drawn feeble squeak like a mammal, also a whistle.

**Distribution of other races of the species:** South Africa, the nominate race being described from eastern Cape Province.

Names in Sclater's *Syst. Av. Æthiop.* 2, 1930, which have been changed or have become synonyms in this work:

*Campephaga flava* Vieillot, now *Campephaga sulphurata* (Lichtenstein).
*Campephaga hartlaubi* Salvadori, a colour phase, treated as synonymous with *Campephaga sulphurata* (Lichtenstein).
*Campephaga xanthornoides* (Lesson) treated as synonymous with *Campephaga phœnicea* (Latham).

**FAMILY—DICRURIDÆ. DRONGOS.** Genus: *Dicrurus*.

Three species occur in Eastern Africa. They are black birds with the nostrils wholly covered by feathers and bristles. The 'fish-tailed' species are birds of both lightly timbered country and forest, and are conspicuous, being specially noticeable for their aggressiveness to any birds entering their territory. The 'square-tailed' species are forest birds and are often seen in mixed bird parties.

### Key to the Adult Drongos of Eastern Africa

1 Tail practically square, inner webs of flight feathers black:  SQUARE-TAILED DRONGO
*Dicrurus ludwigii*  **1089**

2 Tail 'fish-tail' shape, outer feathers slightly splayed outwards at tip, inner webs of flight feathers blackish or ashy:  3–4

3 Mantle glossy velvety blue black tinged with violet, inner webs of flight feathers blackish: VELVET-MANTLED DRONGO *Dicrurus modestus* 1087

4 Mantle glossy blue black, no velvety appearance, inner webs of flight feathers ashy: DRONGO *Dicrurus adsimilis* 1088

## 1087 VELVET-MANTLED DRONGO. *DICRURUS MODESTUS* Hartlaub.

*Dicrurus modestus coracinus* Verr. **Pl. 96.**
*Dicrurus coracina* Verreaux, Rev. Mag. Zool. p. 311, 1851: Gabon.

**Distinguishing characters:** General colour slightly glossy velvety blue black with a violet wash on head and mantle; wings and tail more intense in colour, former with inner webs of flight feathers blackish, and latter deeply 'fish-tailed' with outer feathers slightly splayed outwards at tip. The female has a less deeply forked 'fish-tail.' Wing 118 to 137 mm. The young bird is duller than the adult and has much less gloss on upperparts.

**General distribution:** Fernando Po to southern Nigeria, northern Angola, the Sudan, Uganda and Kenya Colony.

**Range in Eastern Africa:** South-western Sudan, Uganda and western Kenya Colony.

**Habits:** Essentially forest birds usually found in clearings on dead trees or other good vantage points.

**Nest and Eggs:** Nest believed to be slung in a fork like that of the Drongo, but no records of a close examination exist. Eggs undescribed.

**Recorded breeding:** No records.

**Food:** Said to take caterpillars, but mostly beetles and an occasional butterfly.

**Call:** Harsh and scolding, with a chattering song at times.

**Distribution of other races of the species:** Principe Island, Gulf of Guinea, the habitat of the nominate race.

**1088 DRONGO.** *DICRURUS ADSIMILIS* (Bechstein).
*Dicrurus adsimilis adsimilis* (Bechst.). **Pl. 96**
*Corvus adsimilis* Bechstein, Latham, Allgem. Ueber. Vög. 2, p. 362, 1794: Duywen-hock River, southern Cape Province.

**Distinguishing characters:** Differs from the Velvet-mantled Drongo in being glossy blue black without any velvety appearance, and in having the inner webs of the flight feathers ashy, not blackish. The female has a less deeply forked 'fish-tail.' Wing 111 to 140 mm. The immature dress of both sexes is similar to the adult, but with ashy white edges to the feathers of the breast to belly and ashy white along the bend of the wing. The young bird is dull black with ashy edges to the feathers; wings and tail less glossy than the adult. It can be distinguished from the young of the other Drongos by the ash-coloured inner webs to the flight feathers.

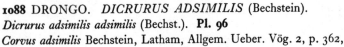

**General distribution:** Senegal, Sierra Leone and northern Nigeria to Abyssinia, and south to Angola and South Africa.

**Range in Eastern Africa:** The Southern Sudan, Abyssinia, Eritrea, and British Somaliland to the Zambesi River, also Pemba and Zanzibar Islands.

**Habits:** A common species of open bush or woodland, usually noticed sitting on a bough in a clearing and hawking insects from a perch. Has a rapid twisting tumbling flight and a habit of chasing fearlessly all large birds, particularly large Birds of Prey and Hornbills, at which it dives and frequently actually strikes about the back and head. A very typical African species widely distributed and almost universal in open country. Frequently a member of a mixed bird party, and often consorts with Helmet-Shrikes.

**Nest and Eggs:** A flat saucer-shaped nest built across a horizontal fork or wedged into it, made of stems lined with rootlets, and of flimsy almost transparent appearance but of relatively strong construction and beautifully woven. Eggs usually three, either white or cream spotted with chestnut or purplish brown, or cream with pinkish brown blotches, or a pale pinkish orange suffused with darker orange and with pale stone undermarkings; about 25 × 17·5 mm., but very variable.

**Recorded breeding:** Nigeria, March and April. Abyssinia, April to June. Sudan, April. Equatoria, November. Uganda, June and July. Kenya Colony highlands, June also August to October. Witu, May. Tanganyika Territory, October to December. Zanzibar

Island, October and November. Nyasaland, September to November. Portuguese East Africa, August to October. Northern and Southern Rhodesia, September and October. South-eastern Belgian Congo, September and October.

**Food:** Insects, largely taken on the wing. It is one of the few African birds which habitually takes butterflies.

**Call:** Harsh call of 'wurchee-wurchee,' also a sort of tin trumpet note. Has a chattering song and some clear liquid notes and whistles, but the bird is an excellent mimic of a number of species. It calls freely on moonlight nights.

*Dicrurus adsimilis jubaensis* V. Som.
*Dicrurus adsimilis jubaensis* Van Someren, J.E.A. & U.N.Hist. Soc. No. 37, p. 196 (for 1930) 1931: Upper Juba River, southern Italian Somaliland.

**Distinguishing characters:** Differs from the nominate race in having the tail less deeply 'fish-tailed' and in being more glossy greenish blue black above. Wing 110 to 120 mm.

**Range in Eastern Africa:** Juba River valley, southern Italian Somaliland.

**Habits:** Little recorded, but presumably as for the nominate race.

**Recorded breeding:** No records.

**Distribution of other races of the species:** Sierra Leone to Togoland.

### 1089 SQUARE-TAILED DRONGO. *DICRURUS LUDWIGII* (Smith).

*Dicrurus ludwigii ludwigii* (Smith).
*Edolius ludwigii* A. Smith, S. Afr. Q. Journ. 2nd ser. p. 144, 1834: Durban, Natal, South Africa.

**Distinguishing characters:** Adult male, smaller and more glossy blue black than the Drongo, with an almost square tail and blackish inner webs to flight feathers. The female is duller and slaty below with only a faint gloss. Wing 97 to 107 mm. The young bird is similar to the adult female, but is more or less washed with bottle green above.

**General distribution:** South-eastern Belgian Congo, Northern Rhodesia, southern Nyasaland, eastern Southern Rhodesia, Portuguese East Africa, eastern Transvaal, Natal and Zululand.

**Range in Eastern Africa:** Portuguese East Africa and Nyasaland.

**Habits:** The forest and woodland counterpart of the Drongo, abundant locally but rarely seen out of densely wooded country. Common in bird parties and is usually the leader. Equally fearless in attack on large birds and will actually ride on a bird's back while attacking it. Same habit of hawking from a perch as the Drongo, but in general is a much more secretive bird. It has a notably orange red iris, and a habit of twitching its tail sideways.

**Nest and Eggs:** A small strongly made nest of tendrils, rootlets, lichen and cobwebs, built across a horizontal fork of a tree usually in dense shade. Eggs three, white, either plain or well spotted and blotched with pale brown and lilac; about 21 × 16 mm.

**Recorded breeding:** Northern Rhodesia, October. Nyasaland, September to November. South Africa, November.

**Food:** Insects taken mainly on the wing.

**Call:** A strident alarm note, almost a shriek of 'chidder-chick,' also a chirruping 'chuit' and many other calls.

*Dicrurus ludwigii münzneri* Reichw.
*Dicrurus münzneri* Reichenow, O.M. p. 91, 1915: Sanyi, Mahenge district, south-eastern Tanganyika Territory.

**Distinguishing characters:** Adult male, differs from the nominate race in being brighter in colour and more glossy; the female is more glossy below, especially on chest. Wing 97 to 110 mm.

**Range in Eastern Africa:** Southern Italian Somaliland to eastern Kenya Colony and eastern Tanganyika Territory from the Juba River to the Mahenge and Lindi districts.

**Habits:** As for the nominate race and locally not uncommon.

**Recorded breeding:** No records.

*Dicrurus ludwigii sharpei* Oust. **Pl. 96.**
*Dicrurus sharpei* Oustalet, N. Arch. Mus. Paris, p. 97, 1879: Gabon.

**Distinguishing characters:** Generally duller than the preceding race, more velvety blue black, with a violet wash. Wing 100 to 112 mm. Not unlike the Velvet-mantled Drongo but it is smaller and has a practically square, not 'fish-tailed' tail.

# HELMET-SHRIKES and RED-BILLED SHRIKES

**General distribution:** Senegal and Gabon, to northern Angola, the Sudan, north-eastern Belgian Congo, Uganda and Kenya Colony.

**Range in Eastern Africa:** Southern Sudan to Uganda and western Kenya Colony.

**Habits:** As for the nominate race.

**Recorded breeding:** No records.

Names in Sclater's *Syst. Av. Æthiop.* 2, 1930, which have been changed or have become synonyms in this work:

*Dicrurus adsimilis divaricatus* (Lichtenstein) treated as synonymous with *Dicrurus adsimilis adsimilis* (Bechstein).
*Dicrurus modestus ugandensis* Van Someren, treated as synonymous with *Dicrurus modestus coracinus* Verreaux.
*Dicrurus elgonensis* Van Someren, treated as synonymous with *Dicrurus ludwigii sharpei* Oustalet.

## FAMILY—PRIONOPIDÆ. HELMET-SHRIKES and RED-BILLED SHRIKES. Genera: *Prionops* and *Sigmodus*.

Seven species occur in Eastern Africa. A very distinctive group with marked characteristics, of which intense sociability at all seasons is the most pronounced. They are always in small flocks of six to twenty in wooded or forested country, do everything in concert, breed close together and it has been noted more than once that several birds share in the building of each nest and even in the feeding of the young. A further characteristic is a curious snapping noise made by the bill, which, with a sort of communal chattering at times, and the conspicuous colour pattern, renders them unmistakable in the field. They search branches and leaves for food like Tits, and are probably entirely insectivorous.

### KEY TO THE ADULT HELMET-SHRIKES AND RED-BILLED SHRIKES OF EASTERN AFRICA

| | | |
|---|---|---|
| 1 Bill red or crimson: | | 3–6 |
| 2 Bill black: | | 7–12 |
| 3 A chestnut-coloured 'cushion' on forehead: | CHESTNUT-FRONTED SHRIKE *Sigmodus scopifrons* | 1096 |
| 4 No chestnut-coloured 'cushion' on forehead: | | 5–6 |

# HELMET-SHRIKES and RED-BILLED SHRIKES

| | | |
|---|---|---|
| 5 Top of head blue grey: | RED-BILLED SHRIKE *Sigmodus caniceps* | 1094 |
| 6 Top of head blue black: | RETZ'S RED-BILLED SHRIKE *Sigmodus retzii* | 1095 |
| 7 Top of head yellow: | YELLOW-CRESTED HELMET-SHRIKE *Prionops alberti* | 1093 |
| 8 Top of head grey or grey and white: | | 9–12 |
| 9 Crest curly: | | 10–11 |
| 10 Yellow wattle round eye: | CURLY-CRESTED HELMET-SHRIKE *Prionops cristata* | 1091 |
| 11 No wattle round eye: | GREY-CRESTED HELMET-SHRIKE *Prionops poliolopha* | 1092 |
| 12 Crest not curly: | STRAIGHT-CRESTED HELMET-SHRIKE *Prionops plumata* | 1090 |

**1090 STRAIGHT-CRESTED HELMET-SHRIKE.** *PRIONOPS PLUMATA* (Shaw).

*Prionops plumata poliocephala* (Stanley).
*Lanius poliocephalus* Stanley, in Salt's Trav. Abyssinia, App. p. 50, 1814: Mozambique, Portuguese East Africa.

**Distinguishing characters:** Head and short straight crest grey; blackish bar behind ear-coverts; broad white bar on hind neck; mantle black with an oily green gloss; wings black with white tips to middle coverts and white edges and tips to inner secondaries forming a longitudinal bar; white spots on inner webs of primaries; tail black, outermost feather and ends of others white; below, white; throat greyish; under wing-coverts and axillaries black; eye and eye-wattle yellow; feet and toes orange or orange red. The sexes are alike. Wing 100 to 116 mm. The young bird has the head brownish or ashy and whitish tips and edges to the primary coverts.

**General distribution:** Uganda and Kenya Colony to Angola, Damaraland and Zululand.

**Range in Eastern Africa:** Southern Uganda and Kenya Colony to the Zambesi River, probably a non-breeding visitor only to Uganda and Kenya Colony.

**Habits:** Tame and delightful little birds, intensely sociable at all seasons. Usually in small flocks of eight to twelve or so, always close together, silent when travelling, but as soon as they stop or are feeding break into a low skirling chatter. Usually found in open woodland or bush at lower elevations travelling through the trees and bushes and often with other birds in a mixed foraging party. Paget-Wilkes has described them as 'falling from a tree like a shower of snow-flakes and making off through the bush together.' They are gregarious even when nesting, and several different birds may share in the brooding and feeding of the young at any one nest.

**Nest and Eggs:** Nest a neat compact cup of grass and fibre, lined with bark shreds and plastered with cobwebs, beautifully made and camouflaged, usually on a lateral bough or in a horizontal fork some feet from the ground. Eggs three or four, occasionally more, pinkish white, bluish green, or pale bluish stone spotted with rufous and purplish brown, usually in a zone at the larger end, with blue or pale purple undermarkings; about 21 × 17 mm. Two clutches in one nest, eight eggs, have been recorded.

**Recorded breeding:** Northern Rhodesia, March, also September and October. Tanganyika Territory, March, also September to December. Nyasaland, September to December. South-eastern Belgian Congo, September and October. Southern Rhodesia, October to December.

**Call:** A skirling chatter made by whole flock together with occasional flute-like notes. Characteristic snapping of bill often heard, and is also used as an alarm signal. Normal alarm call, a low far-carrying 'chow-chow.'

*Prionops plumata vinaceigularis* Richmond.
*Prionops vinaceigularis* Richmond, Auk. 14, p. 162, 1897: Plains east of Kilimanjaro, south-eastern Kenya Colony.

**Distinguishing characters:** Similar to the preceding race in having a straight crest and the mantle black with an oily green gloss, but differs in having the crest of forehead and crown white, and only a bare trace of a white longitudinal bar on wing or none at all. Wing 107 to 125 mm.

# HELMET-SHRIKES and RED-BILLED SHRIKES

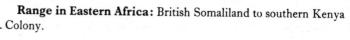

**Range in Eastern Africa:** British Somaliland to southern Kenya Colony.

**Habits:** As for other species. Quite common locally in bush country.

**Recorded breeding:** Southern Abyssinia and northern Kenya Colony, February.

**Distribution of other races of the species:** West Africa, the nominate race being described from Senegal.

### 1091 CURLY-CRESTED HELMET-SHRIKE. *PRIONOPS CRISTATA* Rüppell.

*Prionops cristata cristata* Rüpp.

*Prionops (Lanius) cristatus* Rüppell, N. Wirbelt. Vög. p. 30, pl. 12, fig. 2, 1836: Massaua, Eritrea.

**Distinguishing characters:** Similar to the Straight-crested Helmet-Shrike, but crest of forehead and crown longer, distinctly curly and white or faintly greyish; mantle blacker, with a more bottle green gloss; longitudinal bar on wings narrow or absent. The sexes are alike. Wing 114 to 130 mm. The young bird has a shorter curly crest and the hind crown and nape is sooty brown, not grey.

**Range in Eastern Africa:** Eritrea, Abyssinia and south-eastern Sudan.

**Habits:** As for the Straight-crested Helmet-Shrike, found in small parties throughout the year in open bush or woodland. They are also said to follow much the same route day after day. They are quite pugnacious and combine to attack an enemy and at least one instance is recorded of their attacking a man when he attempted to pick up a wounded member of a flock. In certain areas erratic in appearance and with local migratory movements. Rather shyer and noisier than the last species.

**Nest and Eggs:** Nest exactly as for the Straight-crested Helmet-Shrike. Eggs four, pale bluish green with a zone of confluent brown and lilac spots; about 22 × 16·5 mm.

**Recorded breeding:** Southern Abyssinia, March.

**Call:** Nothing noted except a noisy chattering.

## HELMET-SHRIKES and RED-BILLED SHRIKES

*Prionops cristata concinnata* Sund.

*Prionops concinnatus* Sundevall, Œfv. K. Sv. Vet.-Akad. Förh. 7, p. 130, 1850: Roseirés, eastern Sudan.

**Distinguishing characters:** Differs from the preceding race in having a broad white longitudinal bar on wings. Wing 110 to 124 mm.

**General distribution:** Cameroons to the Sudan and Uganda.

**Range in Eastern Africa:** Southern areas of the Sudan and northern half of Uganda.

**Habits:** As for the last race. At least four birds have been seen to help in building one nest.

**Recorded breeding:** Nigeria, March and April. Darfur, probably September to November. Southern Sudan (near breeding), March.

**1092 GREY-CRESTED HELMET-SHRIKE.** *PRIONOPS POLIOLOPHA* Fischer & Reichenow.

*Prionops poliolophus* Fischer & Reichenow, J.f.O. p. 180, 1884: Naivasha, Kenya Colony.

**Distinguishing characters:** Differs from the Curly-crested Helmet-Shrike in being larger and in having a still longer grey crest, especially on hind crown; sides of face also grey; eye bright yellow; feet and toes red or orange red; no eye wattles. The sexes are alike. Wing 132 to 135 mm. Juvenile plumage undescribed.

**Range in Eastern Africa:** Central and south-western Kenya Colony to the Tabora district, Tanganyika Territory.

**Habits:** As for the other members of the genus, travelling from tree to tree in flocks and feeding with much chattering and bill-snapping. Somewhat erratic in appearance in Kenya Colony.

**Nest and Eggs:** A neat but shallow cup nest of grass fibre and spiders' webs in a fork of a tree. Eggs greenish blue with spots of reddish and greyish brown; about 21 × 17 mm.

**Recorded breeding:** Naivasha, Kenya Colony, May.

**Call:** As above, nothing distinctive noted.

# HELMET-SHRIKES and RED-BILLED SHRIKES

**1093 YELLOW-CRESTED HELMET-SHRIKE.** *PRIONOPS ALBERTI* Schouteden.

*Prionops alberti* Schouteden, Rev. Zool. et Bot. Afr. 24, p. 211, 1933: Mt. Mikeno, eastern Belgian Congo.

**Distinguishing characters:** General colour wholly black, with head to nape, including the crest, lemon yellow. The sexes are alike. Wing 126 to 136 mm. The young bird has a paler crest and nape.

**General distribution:** Eastern Belgian Congo, north of Lake Kivu.

**Range in Eastern Africa:** Kivu area, north-western Ruanda.

**Habits:** No information.

**Nest and Eggs:** Nest of lichens and rootlets in the fork of a tree.

**Recorded breeding:** Eastern Belgian Congo, February and March.

**Call:** Unrecorded.

**1094 RED-BILLED SHRIKE.** *SIGMODUS CANICEPS* Bonaparte.

*Sigmodus caniceps mentalis* Sharpe. **Pl. 74.**

*Sigmodus mentalis* Sharpe, J. Linn. Soc. Zool. 17, p. 425, 1884: Sassa, near Semmio, south-eastern French Equatorial Africa.

**Distinguishing characters:** Head and sides of face blue grey; mantle, wings and tail black with a slight greenish gloss; white spots on inner webs of primaries; below, throat and neck black, and also a narrow collar on hind neck; chest white; breast to under tail-coverts chestnut; under wing-coverts black; eye yellow; bill crimson; feet and toes red. The sexes are alike. Wing 106 to 117 mm. The young bird has the forehead and crown white; a black stripe from lores through eye to nape; throat white; neck in front either black or buff; bill black.

**General distribution:** South-eastern French Equatorial Africa to north-eastern Belgian Congo and western Uganda.

**Range in Eastern Africa:** Western Uganda.

**Habits:** Very much those of the Helmet Shrikes, passing from tree to tree in small flocks and chattering while feeding, but they inhabit larger timber as a rule and keep more to the tree-tops. They also make a curious snapping swishing noise. Locally not uncommon, especially in Bugoma Forest, Uganda.

**Nest and Eggs:** Undescribed.

**Recorded breeding:** No records.

**Call:** As above.

**Distribution of other races of the species:** West Africa, the nominate race being described from the Gold Coast.

**1095 RETZ'S RED-BILLED SHRIKE.** *SIGMODUS RETZII* (Wahlberg).

*Sigmodus retzii tricolor* (Gray).

*Prionops tricolor* Gray, P.Z.S. p. 45, 1864: Tete, Rivi River, southern Nyasaland.

**Distinguishing characters:** Whole head, neck and underside slightly glossy blue black; mantle, rump and wing-coverts earth grey; white spots on inner webs of primaries; lower belly, under wing-coverts and ends of all except central tail feathers white; eyes orange; eye wattles red; bill red, tip yellow; feet and toes orange red. The sexes are alike. Wing 123 to 141 mm. The young bird has the head, neck and underside earth grey, uniform with the mantle, rump and wing-coverts.

**General distribution:** Tanganyika Territory to north-eastern Northern Rhodesia, Nyasaland and northern Portuguese East Africa.

**Range in Eastern Africa:** Western and central Tanganyika Territory to the Zambesi River, but not eastern areas except near Kilosa.

**Habits:** Usually noticeable in small flocks of six to ten flying from tree top to tree top and mostly keeping to the larger trees. They are often accompanied by other birds, frequently by a Drongo, and are very often in the same flock as Helmet-Shrikes. Rather noisy little birds continually chattering as they search for insects.

**Nest and Eggs:** See under the following race.

**Recorded breeding:** Tanganyika Territory, probably October to December. Nyasaland, September to December. Portuguese East Africa, October and November.

*Sigmodus retzii graculinus* (Cab.). **Pl. 74.**
*Prionops graculinus* Cabanis, J.f.O. p. 412, pl. 3, 1868: Mombasa, eastern Kenya Colony.

**Distinguishing characters:** Differs from the preceding race in having the mantle, rump and wing-coverts much paler; spots on inner webs of primaries vary in size and are almost absent in some specimens. Wing 114 to 131 mm.

**Range in Eastern Africa:** Southern Italian Somaliland to Kenya Colony and Tanganyika Territory as far south as Kisiju, 40 miles south of Dar-es-Salaam, and as far west as Mwanza, the south Paré Mts., and Kilosa.

**Habits:** Almost exactly those of Helmet Shrikes, with which they often associate, and usually extremely tame. Common in small parties in coastal bush and scrub, searching foliage and hawking insects. The vivid red and yellow of bill and eye show up noticeably as the birds allow of close approach. Highly gregarious, and even while nesting three or four birds may be round the nest together, though there is as yet no evidence that they share the duties as in *Prionops*.

**Nest and Eggs:** The nest is a broad shallow cup of grass stems, lichens, and tendrils, well made and woven on to or into the upper fork of a tree or bush, and usually difficult to see. Eggs three or four, pale bluish green or greenish cream with brownish, reddish or blackish lines and spots usually in a zone, and with grey underspots. The yolk is blood red; about 22 × 18 mm.

**Recorded breeding:** No records.

**Call:** As for others of the group, a noisy chattering interspersed with whistles, but contains, according to Moreau, a peculiar humming quality 'seeoo-gooria-gooria.' Alarm note a sharp 'tche-wrik.'

**Distribution of other races of the species:** Angola and South Africa, the nominate race being described from Damaraland.

### 1096 CHESTNUT-FRONTED SHRIKE. *SIGMODUS SCOPIFRONS* Peters.

*Sigmodus scopifrons scopifrons* Pet. **Pl. 74.**
*Sigmodus scopifrons* Peters, J.f.O. p. 422, 1854: Mozambique, Portuguese East Africa.

**Distinguishing characters:** General colour smoke grey; a chestnut coloured 'cushion' on forehead with a palish grey band behind it; rest of head, throat, wings, and tail blacker; a white spot on inner webs of primaries; white ends to tail feathers decreasing to a spot at tip of central pair; lower belly and under tail-coverts white; eye-wattle blue; bill red; feet and toes coral red. The sexes are alike. Wing 99 to 104 mm. The young bird has the forehead blackish uniform with rest of the head; feathers of the bastard wing with white tips.

**General distribution:** Tanganyika Territory to Portuguese East Africa as far south as Beira.

**Range in Eastern Africa:** Eastern Tanganyika Territory from the Usambara Mts. to the Morogoro and Lindi districts and the Zambesi River.

**Habits:** As for other species of the group, small parties passing from tree top to tree top and intensely sociable at all seasons. The flight is rather feeble.

**Nest and Eggs:** A shallow well made cup saddled on to a fork of a tree. They build in company, and of three birds which were seen to feed one nest of young, at least one was a young male. Eggs undescribed.

**Recorded breeding:** Amani, Tanganyika Territory, February. Portuguese East Africa, November.

**Call:** A nasal humming like that made by a comb on taut tissue paper (Moreau).

*Sigmodus scopifrons keniensis* V. Som.
*Sigmodus scopifrons keniensis* Van Someren, Bull. B.O.C. 43, p. 80, 1923: Meru, north-east of Mt. Kenya, central Kenya Colony.

**Distinguishing characters:** Differs from the nominate race in having the band behind the 'cushion' obscure, and the white on inner webs of secondaries reduced to a small spot. Wing 105 to 107 mm.

**Range in Eastern Africa:** Meru area, north-west of Mt. Kenya.

**Habits:** As for the nominate race.

**Recorded breeding:** No records.

*Sigmodus scopifrons kirki* Scl.
*Sigmodus scopifrons kirki* W. L. Sclater, Bull. B.O.C. 44, p. 92, 1924: Lamu, eastern Kenya Colony.

**Distinguishing characters:** Differs from the two preceding races in having the band behind the 'cushion' much paler grey, almost white; white on inner webs of primaries less than the

nominate race, but more extensive than the preceding race. Wing 97 to 106 mm.

**Range in Eastern Africa:** Coastal areas of Kenya Colony and Tanganyika Territory from Lamu to Pangani and the Dar-es-Salaam district.

**Habits:** As for other races.

**Recorded breeding:** No records.

Names in Sclater's *Syst. Av. Æthiop.* 2, 1930, which have been changed, or have become synonyms in this work:

*Prionops cristata melanoptera* Sharpe, treated as synonymous with *Prionops plumata vinaceigularis* Richmond.
*Prionops cristata omoensis* Neumann, treated as synonymous with *Prionops cristata cristata* Rüppell.
*Sigmodus retzii intermedius* Neumann, treated as synonymous with *Sigmodus retzii graculinus* (Cabanis).
*Sigmodus retzii neumanni* Zedlitz, treated as synonymous with *Sigmodus retzii graculinus* (Cabanis).

FAMILY—**LANIIDÆ. SHRIKES.** Genera: *Eurocephalus, Nilaus, Lanius, Corvinella, Urolestes, Laniarius, Dryoscopus, Tchagra, Bocagia, Chlorophoneus, Telophorus, Malaconotus, Rhodophoneus* and *Nicator*.

Fifty-two species of Shrikes are known to occur in Eastern Africa, some of which are migrants from Europe and Asia. They inhabit all types of timbered country from forest to the most sparsely timbered semi-desert areas, and many are conspicuous objects of the countryside. Their food consists of insects of all kinds, young birds, eggs, small mammals and reptiles. The Slate-coloured, Black and Sooty Boubous bear a close superficial resemblance to the Square-tailed Drongos, but these latter have small feet and longer, more pointed, wings. The Brubru group (*Nilaus*) are somewhat distinctive, and probably have affinities with the Flycatchers. The White-crowned Shrike (*Eurocephalus*) is also aberrant.

### KEY TO THE ADULT SHRIKES OF EASTERN AFRICA

| | | |
|---|---|---|
| 1 Top of head yellow: | GONOLEK *Laniarius barbarus* | 1120 |
| 2 Top of head white, mantle grey: | MANY-COLOURED BUSH SHRIKE, male *Chlorophoneus multicolor* | 1137 |

3 Top of head white, upper tail-coverts white: WHITE-CROWNED SHRIKE *Eurocephalus anguitimens* 1097

4 Top of head ash brown, head and mantle streaked: YELLOW-BILLED SHRIKE *Corvinella corvina* 1117

5 Top of head and upper mantle black: BLACK-CAP BUSH-SHRIKE *Bocagia minuta* 1136

6 Top of head blackish brown, mantle earth brown: PUFF-BACK, female *Dryoscopus gambensis* 1131

7 Top of head slate or grey; mantle ash grey, below grey or fawn colour: PINK-FOOTED PUFF-BACK *Dryoscopus angolensis* 1132

8 Top of head and mantle ash brown: PRINGLE'S PUFF-BACK, female *Dryoscopus pringlii* 1130

9 Top of head ashy brown, black streak down centre of head: THREE-STREAKED BUSH-SHRIKE *Tchagra jamesi* 1135

10 Top of head ashy brown washed with rosy; rump rose colour: ROSY-PATCHED SHRIKE *Rhodophoneus cruentus* 1147

11 Top of head and mantle isabelline grey, tail chestnut: RED-TAILED SHRIKE *Lanius cristatus* 1114

12 Top of head grey, mantle wholly grey, underside white with or without pinkish or peach wash: 24–33

| | | |
|---|---|---|
| 13 Top of head black, mantle black, rump white or grey: | | 34–43 |
| 14 Top of head black, mantle grey or blackish grey: | | 44–51 |
| 15 Top of head to upper mantle grey: | | 52–59 |
| 16 Top of head black, mantle and rump black: | | 60–63 |
| 17 Top of head and mantle green: | | 64–69 |
| 18 Top of head brown or blackish brown: | | 70–73 |
| 19 Above and below, black, blackish slate or slate blue: | | 74–79 |
| 20 Top of head grey or brownish grey, mantle chestnut: | | 80–83 |
| 21 Top of head chestnut, mantle black, grey or brown: | | 84–86 |
| 22 Top of head black, mantle black with a central streak of buff or white: | | 87–88 |
| 23 Top of head black, mantle brown or tawny: | | 89–90 |
| 24 Scapulars black: | GREY-BACKED FISCAL *Lanius excubitorius* | 1102 |
| 25 Scapulars grey: | | 26–27 |
| 26 A large amount of white at base of primaries: | LESSER GREY SHRIKE *Lanius minor* | 1103 |
| 27 A very small amount of white at base of primaries: | BOGDANOW'S SHRIKE *Lanius bogdanowi* | 1116 |
| 28 Scapulars white: | | 29–34 |
| 29 Bill black, lores black: | | 30–33 |
| 30 No black on forehead: | MACKINNON'S SHRIKE *Lanius mackinnoni* | 1110 |

## SHRIKES

| | | |
|---|---|---|
| 31 Forehead black: | | 32–33 |
| 32 Rump and upper tail-coverts white, whitish, or grey: | GREY SHRIKE* <br> *Lanius elegans* races | 1100 |
| 33 Rump and upper tail-coverts nearly uniform in colour with mantle: | JEBEL MARRA GREY SHRIKE <br> *Lanius jebelmarræ* | 1101 |

| | | |
|---|---|---|
| 34 Scapulars black: | ZANZIBAR PUFF-BACK <br> *Dryoscopus affinis* | 1129 |
| 35 Scapulars grey: | | 36–37 |
| 36 Size smaller, wing under 70 mm.: | PRINGLE'S PUFF-BACK, male <br> *Dryoscopus pringlii* | 1130 |
| 37 Size larger, wing over 80 mm.: | PUFF-BACK, male <br> *Drvoscopus gambensis* | 1131 |
| 38 Scapulars black with broad white edges: | BLACK-BACKED PUFF-BACK <br> *Dryoscopus cubla* | 1128 |
| 39 Scapulars wholly white: | | 40–43 |
| 40 Forehead and stripe over eye white: | UHEHE FISCAL <br> *Lanius marwitzi* | 1105 |
| 41 No white on forehead or over eye: | | 42–43 |
| 42 Below, white: | FISCAL <br> *Lanius collaris* | 1104 |
| 43 Below, black: | MAGPIE-SHRIKE <br> *Urolestes melanoleucus* | 1118 |

\* *Note: Lanius elegans pallidirostris* has the bill horn colour or dusky:

| | | |
|---|---|---|
| 44 Scapulars grey or blackish: | | 45–46 |
| 45 Forehead and stripe over eye white: | JACKSON'S BUSH-SHRIKE *Chlorophoneus bocagei* | 1141 |
| 46 No white on forehead or over eye: | LONG-TAILED FISCAL *Lanius cabanisi* | 1108 |
| 47 Scapulars white: | | 48–51 |
| 48 Forehead white: | NUBIAN SHRIKE, female *Lanius nubicus* | 1109 |
| 49 Forehead black: | | 50–51 |
| 50 Outer tail feathers wholly white: | SOMALI FISCAL *Lanius somalicus* | 1106 |
| 51 Outer tail feathers with white outer webs and ends: | TAITA FISCAL *Lanius dorsalis* | 1107 |

| | | |
|---|---|---|
| 52 Forehead and stripe over eye white: | MANY-COLOURED BUSH-SHRIKE, female *Chlorophoneus multicolor* | 1137 |
| 53 Forehead and stripe over eye yellow: | SULPHUR-BREASTED BUSH-SHRIKE *Chlorophoneus sulfureopectus* | 1138 |
| 54 Forehead black, no white or yellow stripe over eye: | BLACK-FRONTED BUSH-SHRIKE *Chlorophoneus nigrifrons* | 1140 |
| 55 Forehead grey: | | 56–59 |
| 56 Ear-coverts and sides of neck black: | RUFOUS-BREASTED BUSH-SHRIKE *Chlorophoneus rubiginosus* | 1139 |
| 57 Ear-coverts and sides of neck grey: | | 58–59 |

# SHRIKES

| | | |
|---|---|---|
| 58 Wing-coverts green with small yellow tips: | GREY-HEADED BUSH-SHRIKE *Malaconotus blanchoti* | **1144** |
| 59 Wing-coverts black with large yellow tips: | LAGDEN'S BUSH-SHRIKE *Malaconotus lagdeni* | **1146** |

| | | |
|---|---|---|
| 60 Forehead and stripe over eye white: | NUBIAN SHRIKE, male *Lanius nubicus* | **1109** |
| 61 No white on forehead or over eye: | | 62–63 |
| 62 Below, crimson: | BLACK-HEADED GONOLEK *Laniarius erythrogaster* | **1119** |
| 63 Below, white: | BOUBOU *Laniarius æthiopicus* | **1125** |

| | | |
|---|---|---|
| 64 Forehead crimson: | DOHERTY'S BUSH-SHRIKE *Telophorus dohertyi* | **1143** |
| 65 Forehead green or grey: | | 66–69 |
| 66 Black band across chest, no yellow spots on tips of wing-coverts: | FOUR-COLOURED BUSH-SHRIKE *Telophorus quadricolor* | **1142** |
| 67 No black band across chest, yellow spots on tips of wing-coverts: | | 68–69 |
| 68 Throat olivaceous grey or buff: | NICATOR *Nicator chloris* | **1148** |
| 69 Throat yellow green: | YELLOW-THROATED NICATOR *Nicator vireo* | **1149** |

| | | |
|---|---|---|
| 70 Mantle streaked black and white or black and buff: | | 71–72 |
| 71 Stripe over eye extending to nape: | NORTHERN BRUBRU, female *Nilaus afer* | **1098** |

72 Stripe over eye not extending to nape or absent: BLACK-BROWED BRUBRU, female *Nilaus nigritemporalis* **1109**

73 Mantle plain brown, brown of top of head bordered with black: BROWN-HEADED BUSH-SHRIKE *Tchagra australis* **1134**

74 General colour slate blue, some white in rump feathers: SLATE-COLOURED BOUBOU *Laniarius funebris* **1121**

75 No white in rump feathers: 76–79

76 General colour blackish slate: FÜLLEBORN'S BLACK BOUBOU *Laniarius fülleborni* **1124**

77 General colour black: 78–79

78 General colour dull black, with a slight greenish gloss; bill shorter and smaller: MOUNTAIN SOOTY BOUBOU *Laniarius poensis* **1123**

79 General colour slightly glossy deep black; bill longer and heavier: SOOTY BOUBOU *Laniarius leucorhynchus* **1122**

80 Scapulars white: SOUSA'S SHRIKE *Lanius souzæ* **1111**

81 Scapulars chestnut: 82–83

82 Rump and upper tail-coverts grey or brownish grey: RED-BACKED SHRIKE *Lanius collurio* **1112**

# SHRIKES

| | | |
|---|---|---|
| 83 Rump and upper tail-coverts chestnut: | EMIN'S SHRIKE *Lanius gubernator* | 1113 |
| 84 Scapulars white: | WOODCHAT SHRIKE *Lanius senator* | 1115 |
| 85 Scapulars white with black ends: | RED-NAPED BUSH-SHRIKE *Laniarius ruficeps* | 1126 |
| 86 Scapulars black: | LUHDER'S BUSH-SHRIKE *Laniarius lühderi* | 1127 |
| 87 White stripe over eye extending to nape: | NORTHERN BRUBRU, male *Nilaus afer* | 1098 |
| 88 White stripe over eye not extending to nape or absent: | BLACK-BROWED BRUBRU, male *Nilaus nigritemporalis* | 1099 |
| 89 Below, white, creamy white or buffish white: | BLACK-HEADED BUSH-SHRIKE *Tchagra senegala* | 1133 |
| 90 Below, tawny: | BLACK-CAP BUSH-SHRIKE *Malaconotus alius* | 1145 |

**1097 WHITE-CROWNED SHRIKE.** *EUROCEPHALUS ANGUITIMENS* Smith.

*Eurocephalus anguitimens rüppelli* Bp.

*Eurocephalus rüppelli* Bonaparte, Rev. Mag. Zool. p. 440, 1853: White Nile, southern Sudan.

**Distinguishing characters:** Forehead to nape, chin to chest, lower rump and upper tail-coverts white; lores, through eye to ear-coverts black, extending to nape below the white; sides of neck and mantle ash colour; wings and tail blackish; flanks ashy brown with blackish centres to feathers; centre of belly and under tail-coverts white. The sexes are alike. Wing 120 to 137 mm. The young bird has the top of the head and nape ash coloured with buffish tips and buffish edges and tips to the mantle, wing-coverts and flight feathers.

**Range in Eastern Africa:** Southern Sudan and Abyssinia to Tanganyika Territory.

**Habits:** Noticeable birds in noisy parties of six to twelve following each other from tree top to tree top in bush or open woodland. Rather shy as a rule, but their white rumps are conspicuous and their flight between trees is gliding and butterfly-like. They usually take off all together with a curious little squeak. A good deal of their food is said to be picked up from the ground, but it probably depends on the season. Sociable towards other species.

**Nest and Eggs:** Nest very small for the bird and quite amazingly neatly and compactly built of various materials woven over with spiders' webs. Moreau describes them as so solid and well made of fine materials that they look like little flat yellow cheeses stuck on to boughs. They are saddled on to a bough or a fork usually towards the end of the branch but are also recorded as stuck to the side of trees. Eggs probably only two or three, but two hens often share a nest. Eggs are white or faint lilac handsomely blotched with amber, ochre or violet; about 27 × 21 mm.

**Recorded breeding:** Southern Sudan (breeding condition), August. Southern and eastern Abyssinia, April and May, also October. Kenya Colony, Turkana, September, southern areas December to March. Tanganyika Territory, June to December.

**Call:** A sharp 'kak-kak' and other high-pitched squawking notes, also the squeaks above mentioned.

**Distribution of other races of the species:** South Africa, the nominate race being described from Bechuanaland.

**1098** NORTHERN BRUBRU. *NILAUS AFER* (Latham).
*Nilaus afer afer* (Lath.).
*Lanius afer* Latham, Suppl. Ind. Orn. p. 19, 1801: Senegal.

**Distinguishing characters:** Adult male, above, black; forehead and stripe over eye to nape white; broad broken streak down centre of back, edges of wing-coverts and inner secondaries tawny or tawny white; rump mottled white and tawny; outer edges of all except central tail feathers and tips white or tawny white; underside white, but sides of chest and flanks very dark chestnut. The female has the black replaced by brown or blackish brown, and some blackish streaks on throat and neck. Wing 80 to 84 mm. The young bird has the head, mantle and rump mottled and barred white, tawny or

# SHRIKES

blackish brown; tips of flight feathers and outer edges and tips to tail feathers buff or tawny; below, white irregularly barred brown.

**General distribution:** Senegal to the Sudan, Eritrea, Abyssinia, eastern Belgian Congo and Uganda.

**Range in Eastern Africa:** The Sudan and Eritrea, Abyssinia as far south as Lake Zwai and Arussi, and Uganda north of the Lango swamps.

**Habits:** Common and widely distributed in open bush, woodland, or forest edge, searching leaves for food singly or in pairs. Silent unobtrusive birds of rather Tit-like habits, generally made noticeable by the call of the male which is instantly answered by the female.

**Nest and Eggs:** Nest a neat cup of soft vegetable matter plastered with cobwebs twenty to thirty feet up in the fork of a thorn tree. Eggs two, greyish or whitish ground colour with blackish or sepia spots and with dark greyish undermarkings; about $20 \times 17$ mm.

**Recorded breeding:** French Sudan, June, July and November. White Nile, April and May, also November. Northern Uganda, March to July.

**Call:** A clear ventriloquial whistle of male answered by female with a lower note. In breeding season said to have a low monotonous song like the noise made by an iron file (Cheesman). A churring alarm call.

*Nilaus afer minor* Sharpe.
*Nilaus minor* Sharpe, P.Z.S. p. 479, 1895: Okoto, near Upper Webi Shebeli, eastern Abyssinia.

**Distinguishing characters:** Differs from the nominate race in having pale chestnut sides to chest and flanks. Wing 71 to 81 mm.

**Range in Eastern Africa:** South-eastern Sudan, southern and south-eastern Abyssinia, British and Italian Somalilands, and from north-western Kenya Colony to north-eastern Tanganyika Territory east of Mt. Kilimanjaro.

**Habits:** As for the nominate race; eggs are rather smaller, about $18 \times 15$ mm.

**Recorded breeding:** Southern Abyssinia, April. Kenya Colony, Turkana, May; Kibwesi, March. Somaliland, April.

*Nilaus afer massaicus* Neum.
*Nilaus afer massaicus* Neumann, J.f.O. p. 363, 1907: Doinyo Erok, Kenya Colony-Tanganyika Territory boundary.

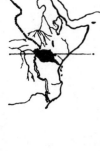

**Distinguishing characters:** Differs from the nominate race in having the sides of chest and flanks paler chestnut, although darker than in the last race; streak down back and edges to wing-coverts and inner secondaries varying from white to tawny. Wing 78 to 87 mm. The young bird is darker above than that of the nominate race.

**Range in Eastern Africa:** Ruwenzori Mts. and Uganda as far north as the Parosa River north of the Lango Swamp to south-western Kenya Colony and northern Tanganyika Territory west of Mt. Kilimanjaro.

*Note :* The nominate race occurs at Kibusi which is north of the Lango Swamp and south-west of the Parosa River.

**Habits:** As for other races, seen singly or in pairs in acacias along dry water courses.

**Recorded breeding:** Southern Kenya Colony, March and April.

**Distribution of other races of the species:** South Africa.

**1099 BLACK-BROWED BRUBRU.** *NILAUS NIGRITEM-PORALIS* Reichenow.
*Nilaus nigritemporalis* Reichenow, J.f.O. pp. 36, 218, 1892: Ngoma (Ngome), about 15 miles south of Mwanza, northern Tanganyika Territory.

**Distinguishing characters:** Adult male, differs from the Northern Brubru in having much paler chestnut on sides of chest and flanks, in having the streak down the back and the edges of wing-coverts and inner secondaries white, and in having no white stripe over the eye, or only the barest indication which never extends to the nape. The female often has an eye-stripe, but this does not extend to the nape either, and has black streaks and specks on throat and breast. Wing 75 to 89 mm. The young bird is more streaked than barred below, unlike the young bird of the Masai race of the Northern Brubru, and the eye-stripe does not extend to the nape.

**General distribution:** Tanganyika Territory, Portuguese East Africa, Northern and Southern Rhodesia, south-eastern Belgian Congo, Nyasaland and Zululand.

## SHRIKES

**Range in Eastern Africa:** Tanganyika Territory, northern Portuguese East Africa and Nyasaland.

**Habits:** Widely distributed and not uncommon in woodland country, singly or in pairs. Frequently a member of a mixed bird party keeping to the denser foliage of tree tops and searching leaves for food; generally silent and inconspicuous in the non-breeding season.

**Nest and Eggs:** Nest a very neat and closely woven cup in a bare fork, very inconspicuous. Eggs two to four, greenish white thickly mottled and blotched with brown and pale grey, sometimes spotted and speckled with black or dark brown; about 21 × 16 mm.

**Recorded breeding:** Tanganyika Territory uplands, October to December. Portuguese East Africa and Nyasaland, September to December. Rhodesia, December.

**Call:** A shrill harsh 'greek' and a mellow whistled courting or breeding note of three syllables.

**1100 GREY SHRIKE.** *LANIUS ELEGANS* Swainson.
*Lanius elegans elegans* Swains.
*Lanius elegans* Swainson, Faun. Bor. Amer. 2, p. 122, 1831: Algeria.

**Distinguishing characters:** Above, pale grey; narrow band on forehead and lores to ear-coverts black; wings black, basal half of primaries and secondaries white; ends of scapulars white; rump and upper tail-coverts white to greyish white; tail, central feathers black, others mainly white; below, white, including underside of wings; ends of flight feathers dusky; bill black. The sexes are alike. Wing 93 to 111 mm. The young bird is isabelline; ear-coverts, wings and tail dull black; no black on forehead; lores dusky; bill horn colour.

**General distribution:** Northern Nigeria to French Sahara, southern Algeria, Egypt, the Sudan, Eritrea and Palestine.

**Range in Eastern Africa:** The Sudan and Eritrea.

**Habits:** A fairly common species of desert bush country often far from water. Habits of the genus, perching on the tops of low trees and bushes from which it swoops down on to its prey. Flight fast but undulating with a pronounced sweep upwards to a perch and generally a drop to nearly ground level on leaving it. All Shrikes seem to have keen eyesight and can detect the movement of a small insect on the ground at twenty yards. Most also have the habit of impaling their larger prey on thorns in a sort of larder. The true Shrikes flick their tails up and down freely when perched.

**Nest and Eggs:** Rather roughly made nest of twigs, dry grass and other material, lined with rootlets, hair, etc. in a bush or tree. Eggs two to three, creamy white spotted with pale brown and with pale mauve undermarkings; about 24·5 × 18 mm.

**Recorded breeding:** French Sahara, March. French Sudan, September to November. Sudan, Darfur, February to May, Kordofan, February and March, Red Sea Province, March to May and possibly earlier. Eritrea, December to March.

**Call:** A harsh 'chack' and various other calls including a melodious whistle.

*Lanius elegans lahtora* Sykes.
*Lanius lahtora* Sykes, P.Z.S. p. 86, 1832: Deccan, India.

**Distinguishing characters:** Above, tone of grey darker than in the preceding race; below, white, often with a buff wash on chest, breast and flanks; bill black. Wing 105 to 111 mm.

**General distribution:** India; in non-breeding season to Persian Gulf, Arabia, Iraq, Iran, Palestine and British Somaliland.

**Range in Eastern Africa:** British Somaliland in non-breeding season.

**Habits:** As for other races.

*Lanius elegans pallidirostris* Cass.
*Lanius pallidirostris* Cassin, Proc. Acad. Nat. Sci. Philad. 5, p. 244, 1852: Eritrea.

**Distinguishing characters:** Differs from the other races in having no black on forehead; mantle more or less washed with isabelline; bill dusky to horn colour, not black; below, white, with usually a pinkish or buff wash on chest. Wing 102 to 117 mm.

**General distribution:** Lake Aral and Trans-Caspian areas in Russian Turkestan; in non-breeding season to Iraq, Palestine, Iran, Afghanistan, Punjab, Arabia, Egypt, the Sudan, Eritrea and Abyssinia.

**Range in Eastern Africa:** The Sudan, Eritrea and Abyssinia in non-breeding season, October to March.

**Habits:** As for other races.

# SHRIKES

*Lanius elegans aucheri* Bon.
*Lanius aucheri* Bonaparte, Rev. Mag. Zool. p. 294, 1853: Iran.

**Distinguishing characters:** Differs from the Algerian race in having the rump and upper tail-coverts greyer, more uniform with the mantle; a greyish wash on the chest. Wing 98 to 115 mm. The young of the two races are indistinguishable.

**General distribution:** Eastern Palestine to Iraq, Iran, Sinai and Arabia (except hills of southern areas); in non-breeding season to south-eastern Egypt, the Sudan, Eritrea, Abyssinia, British Somaliland and India.

**Range in Eastern Africa:** The Sudan, Eritrea, eastern Abyssinia and British Somaliland in non-breeding season.

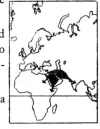

**Habits:** A migrant to north-eastern Africa from October to March. Not distinguishable in habits from other races, and differing little in appearance in the field. Its nearest known nesting habitat is Arabia, where it breeds in April and May.

*Lanius elegans uncinatus* Scl. & Hartl.
*Lanius uncinatus* P. L. Sclater & Hartlaub, P.Z.S. p. 168, 1881: Socotra Island.

**Distinguishing characters:** Differs from the preceding race in having a longer bill and rather less white in the scapulars. Wing 98 to 102 mm. The young bird is whiter below than the young bird of the last race, and has a buffish grey chest and flanks.

**Range in Eastern Africa:** Socotra Island.

**Habits:** Described as a very tame bird, usually solitary, with the habits of the genus, and is unlikely to differ in behaviour from other Grey Shrikes. The call is a sharp 'clink-clink' and there is also what is described as a 'harsh mournful song.'

**Recorded breeding:** No records.

*Lanius elegans buryi* Lor. & Hellm.
*Lanius buryi* Lorenz & Hellmayr, O.M. p. 39, 1901: Yeshbum, southern Arabia.

**Distinguishing characters:** Much darker above, more dusky grey, than in the preceding races, and usually has chest and flanks grey and under primary coverts dark grey. Wing 99 to 111 mm.

**General distribution:** Abyssinia and hills of southern Arabia.

**Range in Eastern Africa:** Eastern Abyssinia.
**Habits:** As for other races.
**Recorded breeding:** No records.

*Lanius elegans dubarensis* Grant & Praed.
*Lanius excubitor dubarensis* C. Grant & Mackworth-Praed, Bull. B.O.C. 71, p. 55, 1951: Dubar, about seven miles south of Berbera, British Somaliland.

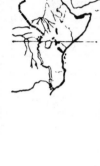

**Distinguishing characters:** Above, similar to the Arabian race; below, whiter and lacking the grey on the chest and flanks, bill black. Wing 105 mm.
**Range in Eastern Africa:** British Somaliland.
**Habits:** No particular information.
**Recorded breeding:** No records.
**Distribution of other races of the species:** Tunisia.

*Note:* Many authors have treated all these Grey Shrikes as races of the Great Grey Shrike, *Lanius excubitor* Linnaeus. We have kept them apart under the name *Lanius elegans* Swainson, as in our opinion the similarity of the sexes, the absence of barring below in the female, the narrower tail feathers, and the fact that the young are never barred below, warrant specific distinction.

**1101 JEBEL MARRA GREY SHRIKE.** *LANIUS JEBEL-MARRÆ* Lynes.
*Lanius excubitor jebelmarræ* Lynes, Bull. B.O.C. 43, p. 94, 1923: Jebel Marra, Darfur, western Sudan.

**Distinguishing characters:** Differs from the resident African races of the Grey Shrike in being smaller and darker above; rump and upper tail-coverts nearly uniform with mantle. The sexes are alike. Wing 96 to 100 mm.
**Range in Eastern Africa:** Jebel Marra, Darfur, western Sudan.
**Habits:** Confined to the higher levels of the Jebel Marra, and not known to occur below 7,000 feet. As far as is known its habits do not differ from those of other Grey Shrikes.
**Nest and Eggs:** Undescribed.
**Recorded breeding:** March and April.
**Call:** Unrecorded.

## SHRIKES

**1102 GREY-BACKED FISCAL.** *LANIUS EXCUBITORIUS*
Prévost & Des Murs.

*Lanius excubitorius excubitorius* Prév. and Des Murs.
*Lanius excubitorius* Prévost & Des Murs, in Lefeb. Voy. Abyss. pp. 99, 170, pl. 8, 1847: Upper Nile, southern Sudan.

**Distinguishing characters:** Adult male, very similar to a light-rumped Grey Shrike, but differs in having the black from the forehead, lores and ear-coverts extending down the side of the neck; in having the wing shoulder, scapulars, under wing-coverts and axillaries black; and in having a longer tail with the white confined to the basal two-thirds and the tips of the feathers. The female has some dark chestnut on the flanks. Wing 106 to 121 mm. The young bird has the grey of the upperside replaced by pale brown with narrow dusky bars; some brownish and dusky bars on the wing-coverts and scapulars; below, buffish white with faint dusky barring on chest.

**General distribution:** The Sudan to eastern Belgian Congo and Uganda; in non-breeding season to Kenya Colony.

**Range in Eastern Africa:** Western and southern Sudan to Uganda; in non-breeding season to western Kenya Colony.

**Habits:** Very different to those of the true Grey Shrikes. These birds are most gregarious, often dancing about all over a tree in a party, like Babblers, or flying screaming from one tree to another. The flight is rather lazy and the bird's tail seems too long for it. They inhabit bushy country as a rule and occasionally assemble into large parties in chase of flying ants. They have, however, the true Shrike habit of attacking Crows and Birds of Prey, and they are remarkably sharp-sighted, detecting insect prey at quite considerable distances. Somewhat migratory, appearing in western Kenya Colony in December and January.

**Nest and Eggs:** A compact cup of twigs and interlaced rootlets lined with fibres and grass, generally in a low thorn bush or thick thorn tree. Eggs two to four, pale yellowish grey or creamy pink with a few brown spots and grey undermarkings; about 25 × 19 mm.

**Recorded breeding:** Sudan, April and May, also (nestling) November. Uganda, April to July, also November to February and (nestling) September.

**Call:** Very varied; a flock often utter low musical notes in harmony, but there are also harsher cries.

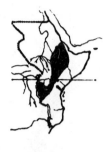

Lanius excubitorius böhmi Reichw.

Lanius böhmi Reichenow, J.f.O. p. 258, 1902: Boga Katani, Ujiji district, western Tanganyika Territory.

**Distinguishing characters:** Differs from the nominate race in being rather larger and rather more dusky grey above. Wing 111 to 130 mm.

**Range in Eastern Africa:** Abyssinia to western Kenya Colony and western Tanganyika Territory.

**Habits:** As for the nominate race.

**Recorded breeding:** Kenya Colony, May to July.

**1103 LESSER GREY SHRIKE. *LANIUS MINOR* Gmelin.**

Lanius minor Gmelin, Syst. Nat. 1, pt. 1, p. 308, 1788: Italy.

**Distinguishing characters:** Adult male, similar to the Grey Shrike, but differs in having a broad black band on forehead, chest and flanks pale pink and a shorter and narrower first primary. The female is duller grey above and has the chest and flanks more peach-coloured. Wing 110 to 126 mm. The immature bird is ashy grey above, with no black band on forehead. The young bird lacks the black forehead and has the upperparts, including the innermost secondaries, more or less barred with dusky. The shorter and narrower first primary distinguishes this Shrike from the races of the Grey Shrike at all ages.

**General distribution:** Middle and southern Europe to the Altai and Turkestan; in non-breeding season to eastern and southern Africa and Arabia.

**Range in Eastern Africa:** Throughout in non-breeding season.

**Habits:** A common migrant to Eastern Africa, often occurring in small flocks. Certain individuals appear to pass the breeding season in Africa also, but do not breed. This species appears to migrate south mostly by the Egypt-Sinai route, but is comparatively rare there in the early months of the year and probably travels north further east. Its habits and flight are much those of the Grey Shrike, but it is a somewhat smaller neater bird with a good deal of white noticeable on the wings in flight. Its range extends further south in Africa than that of any of the large Grey Shrikes. The call is a harsh note of 'tchek.'

## SHRIKES

**1104 FISCAL.** *LANIUS COLLARIS* Linnæus.
*Lanius collaris humeralis* Stan. **Ph. xiv.**
*Lanius humeralis* Stanley, in Salt's Voy. Abyss. App. p. 51, 1814: Chelicut, northern Abyssinia.

**Distinguishing characters:** Adult male, upperside and sides of face, slightly glossy black; rump and upper tail-coverts grey; scapulars, basal third of primaries and underside white; axillaries black and white; tail black, all except central pair of feathers with white ends, outermost mainly white; bill black. The female has more or less chestnut flanks, and is occasionally more brownish black above. Wing 86 to 106 mm. The young bird is narrowly barred black and tawny above; some white in scapulars; below, whitish, more or less barred with dusky black; ends of tail feathers buff with subterminal black crescent-shaped markings; bill blackish horn.

**General distribution:** Eritrea and the Sudan to the Transvaal.

**Range in Eastern Africa:** Eritrea, Abyssinia, the south-eastern Sudan and eastern Uganda to the Zambesi River.

**Habits:** A generally common and well known bird to be seen on telegraph wires or other posts of vantage in all types of country, including towns. It has acquired various local names of which 'Bullhead' and 'Jacky Hangman' are examples, the latter from its habits of impaling its prey in a larder—often in this case on a sisal plant, though this habit is by no means universal. In spite of their destruction of young birds they are undoubtedly beneficial in the large numbers of insects, mostly *Orthoptera*, and of mice which they kill. They also feed greedily on flying ants when available, eating only the abdomen. There is some evidence of seasonal movements.

**Nest and Eggs:** A neat and compact cup nest of grass, roots, cotton, or other materials lined with fibre, fine grass or hair, placed in a bush or the fork of a tree. Eggs usually three, though two or four are not uncommon, creamy or pinkish white with sepia, brown, or chestnut spots and pale lilac undermarkings, often mainly in a zone round the eggs; about 23 × 18 mm.

**Recorded breeding:** Abyssinia, April to July. Kenya Colony, all the year, especially in the rainy seasons. Tanganyika Territory, March to May and August to December. Nyasaland, October to January. Rhodesia, September to February. Portuguese East Africa, November to February.

**Call:** A typical sharp Shrike call answered by the female with a soft piping. Alarm note a clear whistle. Very rarely there is a short sweet plaintive song, particularly in the southern parts of the bird's range.

*Lanius collaris smithii* (Fraser).
*Collurio smithii* Fraser, P.Z.S. p. 16, 1843: Cape Coast Castle, Gold Coast, West Africa.

**Distinguishing characters:** Differs from the preceding race in having the outer tail feathers black with only the ends white, not mainly white. Wing 81 to 96 mm.

**General distribution:** Sierra Leone and Liberia to south-western Sudan, western Uganda and Tanganyika Territory.

**Range in Eastern Africa:** South-western Sudan to western Uganda and the Kigoma area, western Tanganyika Territory.

**Habits:** As for the preceding race.

**Recorded breeding:** Sierra Leone, February to May. Uganda irregularly. Western Tanganyika Territory, September and October.

*Note:* Uganda and western Tanganyika Territory specimens show intermediate characters between this and the preceding race.

**Distribution of other races of the species:** Angola and South Africa; the nominate race being described from the Cape of Good Hope.

**1105 UHEHE FISCAL.** *LANIUS MARWITZI* Reichw.
*Lanius marwitzi* Reichenow, O.M. p. 90, 1901: Ngomingi, Iringa district, south-central Tanganyika Territory.

**Distinguishing characters:** Differs from the Fiscal in having a white superciliary stripe and some white on forehead. Wing 91 to 95 mm. The young bird is darker than that of the Fiscal and much more heavily barred below.

**Range in Eastern Africa:** Mpapwa to Njombe and the country north of Lake Nyasa, Tanganyika Territory.

**Habits:** Exactly those of the Fiscal and might easily be mistaken for that species. It is said to perch for choice on the highest trees in bush country and to be silent and rather shy. Local birds with a very restricted range.

## SHRIKES

**Nest and Eggs:** Undescribed.

**Recorded breeding:** Tanganyika Territory, Uzungwe Mts., December and January. Njombe, probably October to December.

**Call:** As for the Fiscal, and also a faint dry warbling song, desultory but sweet.

**1106 SOMALI FISCAL.** *LANIUS SOMALICUS* Hartlaub.
*Lanius somalicus* Hartlaub, Ibis, p. 342, 1859: Bender Gam (Gaan), north-eastern British Somaliland.

**Distinguishing characters:** Adult male, head and sides of face slightly glossy black; mantle grey; scapulars and upper tail-coverts white; wings black, basal half to two-thirds of primaries and ends of secondaries white; outermost pair of tail feathers white, remainder with white ends; below, white; axillaries black or ashy; bill black. The female has the axillaries duller black. Wing 93 to 106 mm. The young bird is isabelline brown or isabelline grey above, with buff edges to wing-coverts and secondaries; some faint barring on scapulars; rump and upper tail-coverts buff with faint bars; below, buffish white; greyish brown on chest; axillaries ashy with faint barring; tail markings as in adult, but black replaced by blackish brown; bill horn colour. The immature bird is similar to the adult but has a duller black head and axillaries ashy to blackish; bill not so deep a black as adult, and often horn colour at base.

**Range in Eastern Africa:** Eastern Abyssinia, British Somaliland, northern Italian Somaliland and northern Kenya Colony.

**Habits:** A local and apparently scarce species of dry bush country over most of its range, but by far the commonest Shrike of north-eastern Italian Somaliland. A rather sluggish bird sitting about on bushes and occasionally diving to the ground to pick up an insect; often tame.

**Nest and Eggs:** A typical Shrike's nest of grass fibre, etc. lined with rootlets and finer material, placed in a tree or bush, sometimes at some height. Eggs normally four, dull yellowish white with pale brown and pale grey spots forming a zone round the egg; about 24 × 18 mm.

**Recorded breeding:** Eastern Abyssinia and British Somaliland, June. Southern Abyssinia, March. Italian Somaliland, January, February and May.

**Call:** A song of short phases, rather complicated and variable, of which a common form is 'bur-er-er' followed by a quick 'lit-it-it' (North). A low churring alarm note.

**1107 TAITA FISCAL.** *LANIUS DORSALIS* Cabanis.
*Lanius (Fiscus) dorsalis* Cabanis, J.f.O. pp. 205, 225, 1878: Ndi, Taita district, south-eastern Kenya Colony.

**Distinguishing characters:** Adult male, differs from the Somali Fiscal in having no white ends to the secondaries, and in the female having some dark chestnut on the flanks; the bill is also rather more robust. Wing 94 to 106 mm. Albinistic examples occur. The young bird has the top of the head and nape sandy greyish, finely barred with black; wing-coverts and inner secondaries edged with buff, latter with subterminal crescent-shaped black and buff markings; below, some sparse black bars on chest and flanks; bill dusky horn, not black as in adult.

**Range in Eastern Africa:** Eastern Uganda, southern Abyssinia, southern Italian Somaliland, Kenya Colony and north-eastern Tanganyika Territory.

**Habits:** Locally common in desert bush country at lower levels in northern and eastern Kenya Colony and in the country round Meru, with the habits of the Fiscal, and appearing to be subject to local migrations.

**Nest and Eggs:** A typical Shrike's nest of twigs and grass in the centre of a thorn bush. Eggs three or four, creamy or greyish white spotted and blotched with brown and grey mostly at the larger ends; about 23 × 14 mm.

**Recorded breeding:** Wajheir, northern Kenya Colony, March. Southern Italian Somaliland, March.

**Call:** A low chuckling, a flute-like whistle, and a harsh grating alarm call.

**1108 LONG-TAILED FISCAL.** *LANIUS CABANISI* Hartert.
*Lanius cabanisi* Hartert, Nov. Zool. 13, p. 404, 1906: Mombasa, eastern Kenya Colony.

**Distinguishing characters:** Differs from the Taita Fiscal in being larger and longer tailed; the mantle and scapulars greyish black; the white in tail confined to the tips and to the bases of the feathers; bill black. The sexes are alike. Wing 104 to 118 mm.

## SHRIKES

The young bird is greyish brown above, narrowly barred with dusky; rump and upper tail-coverts buffish white, barred blackish; below, speckled with dusky on chest and flanks; bill blackish horn.

**Range in Eastern Africa:** Southern Italian Somaliland to central Kenya Colony and eastern Tanganyika Territory as far south as Dar-es-Salaam.

**Habits:** Locally common in coastal and lower level districts of Kenya Colony and generally seen in small parties among low scrub. It could be described as sociable, if continuous chattering and bickering in a party come under that heading. A conspicuous noisy bird not easily overlooked.

**Nest and Eggs:** Nest large, of rootlets and coarse grass with a finer lining of the same materials, in a thick tree or bush. Eggs three or four, creamy white or pale olive spotted and blotched with brown, and with lavender undermarkings, mostly at larger end; about $25 \times 19$ mm.

**Recorded breeding:** Kenya Colony mainly March to June, also August and September and November and December. Tanganyika Territory, January.

**Call:** A large variety of chattering scolding cries, of which the commonest is a harsh 'chit-er-row' (Van Someren); also a mellow whistle.

**1109 NUBIAN SHRIKE.** *LANIUS NUBICUS* Lichtenstein.
*Lanius nubicus* Lichtenstein, Verz. Doubl. p. 47, 1823: Nubia.

**Distinguishing characters:** Adult male, above, crown to rump, wings and tail, slightly glossy bluish black; forehead, stripe over eye, scapulars, basal half of primaries, edge of inner secondaries, two outermost tail feathers and ends of others except the central pair white; below, palish russet brown, with chin, throat, centre of belly and under tail-coverts white. The female is duller above, much less black and white, and the mantle is often more ashy. Wing 83 to 96 mm. The young bird is ashy, barred with black above; below, sides of chest and flanks white, barred black.

**General distribution:** Asia Minor and Palestine to Iran and Egypt; in non-breeding season to Lake Chad area, Eastern Africa as far south as Lake Rudolf, and Arabia.

**Range in Eastern Africa:** The Sudan, Abyssinia and British Somaliland in non-breeding season.

**Habits:** A migrant from Eastern Europe and Asia, but there is a strong possibility that it may be found nesting in the northern Sudan. A. L. Butler believed that it did so, and Rothschild and Wollaston saw young birds hardly able to fly being fed by their parents near Shendi. It is a species of gardens and shaded cultivated land, when possible settling on the lower boughs of trees rather in the manner of a Woodchat Shrike.

**Nest and Eggs:** Nest (in Asia) a cup of moss and lichens lined with fine rootlets. Eggs four or five, olive green with a zone of spots round the larger end; about 22 × 16 mm.

**Recorded breeding:** Not yet recorded in Africa.

**Call:** Alarm note a harsh low 'shrek.'

### 1110 MACKINNON'S SHRIKE. *LANIUS MACKINNONI* Sharpe.

*Lanius mackinnoni* Sharpe, Ibis, p. 444, pl. 13, 1891: Bugemaia, Kavirondo, western Kenya Colony.

**Distinguishing characters:** Adult male, above, sooty grey; stripe over eye, scapulars and tips of tail feathers white; wings wholly black with no white; black streak from lores to ear-coverts; below, white. The female has a patch of chestnut on the flanks. Wing 82 to 92 mm. The young bird is grey above with narrow blackish or tawny bars; below, white with narrow wavy blackish markings mainly on chest and flanks.

**General distribution:** Cameroons to northern Angola, northern Belgian Congo, Uganda, Kenya Colony and Tanganyika Territory.

**Range in Eastern Africa:** Uganda, western Kenya Colony and north-western and central Tanganyika Territory.

**Habits:** Not unlike the Grey-backed Fiscal in appearance but much less noisy and excitable. It is usually seen sitting motionless on some vantage point from which it drops on to its prey. Mainly found in open woodland or on forest outskirts solitary or in pairs; not common and distinctly local.

**Nest and Eggs:** Nest of fibre grass and weed stalks lined with rootlets and fine fibres, usually low down in a thick thorn-bush. Eggs two, buff or cream speckled and spotted with yellowish brown and various shades of grey; about 23·5 × 17·5 mm.

## SHRIKES

**Recorded breeding:** Cameroons, March and April, also (nestlings) September. Belgian Congo (nestling) April. Uganda, April and May, also July and August.

**Call:** The male has quite a sweet song, varied and with some mimicry of other species. A rather musical 'chickarea' (Elliott). A low 'churr' while feeding young.

**1111 SOUZA'S SHRIKE.** *LANIUS SOUZÆ* Bocage.
*Lanius souzæ burigi* Chap.
*Lanius souzæ burigi* Chapin, Auk. 67, p. 241, 1950: between Usuvi, north-western Tanganyika Territory and the Kasaki district, eastern Ruanda.

**Distinguishing characters:** Adult male, head, upper mantle and rump ash or brownish grey; mantle and wings russet brown, latter with blackish bars; scapulars white; lightish stripe over eye; dull blackish stripe from lores to ear-coverts; tail feathers narrow, dusky, often barred, outer pair of feathers mainly white; below, buffish white. The female has tawny flanks. Wing 78 to 87 mm. In immature dress the head and mantle are warm brown. The young bird is dark russet and chestnut above including scapulars, with narrow black bars; below, buffish white narrowly barred with blackish; tail dark russet with light tips. Vincent notes that this species is immediately distinguishable from the Red-backed Shrike in the hand by the narrow tail feathers.

**General distribution:** Northern Rhodesia and south-eastern Belgian Congo to Tanganyika Territory, Nyasaland and Portuguese East Africa at Furancungo.

**Range in Eastern Africa:** Western and south-western Tanganyika Territory.

**Habits:** A woodland species, in appearance like the European Red-backed Shrike, but with distinctive barring on the wings and tail. Apparently migratory and a visitor to Nyasaland from June to January to breed, and while there is a bird of middle levels. Little is recorded of its habits, except that it appears to prefer tall perches like the Fiscal Shrikes.

**Nest and Eggs:** Nest a fairly deep well-built cup of twigs, grass and lichens bound with insect webs and placed in the fork of a shrub or small tree. Eggs two or three, cream or greenish white with spots and freckles of brown and purple and pale ashy dusting, rather in a zone round the egg; about 21 × 16·5 mm.

**Recorded breeding:** Nyasaland, October to December. South-eastern Belgian Congo, September to November.

**Call:** The only recorded calls are a low scraping note, and a chattering made by the female in alarm for the safety of the nest.

**Distribution of other races of the species:** Angola and probably north-western Northern Rhodesia, the nominate race being described from Angola.

1112 RED-BACKED SHRIKE. *LANIUS COLLURIO* Linnæus.
*Lanius Collurio* Linnæus, Syst. Nat. 10th ed. p. 94, 1758: Sweden.

**Distinguishing characters:** Adult male, narrow band on forehead and lores to ear-coverts black; head, neck, rump and upper tail-coverts grey; mantle, scapulars, edges of wing-coverts, and inner secondaries chestnut; base of primaries white; all except central tail feathers white with broad black ends, central feathers black; below, pinkish; throat white; bill black. The female has the head and ear-coverts ash brown; mantle, scapulars and wing-coverts brown; no white at base of primaries; below, pale buffish white with crescent-shaped blackish markings on sides of neck, chest and flanks; outer tail feathers dusky bordered and tipped with white; bill horn colour. Occasionally there is some grey on the head and upper tail-coverts and some of the tail feathers have black ends. Wing 84 to 99 mm. Similar in appearance to Souza's Shrike, but is larger and has broader tail feathers. The immature bird is brighter above than the adult female. The young bird is browner above than the adult female, with more or less sparse blackish bars; below, buffish white, with blackish crescent-shaped markings.

**General distribution:** Europe and Asia; in non-breeding season to Africa throughout and Arabia, but rare in tropical West Africa.

**Range in Eastern Africa:** Throughout in non-breeding season.

**Habits:** A common and at times abundant passage migrant and visitor. It may be found in any type of country and at all altitudes, but prefers bush country and grassy glades when available. Perches on top of a low bush or on bare lower boughs of trees and drops on to its prey on the ground, occasionally hovering in the air if it cannot see it at once. A passage migrant only to the Sudan, Uganda and Kenya Colony, spending the non-breeding season in parts of Tanganyika Territory and further south. The usual call is a harsh 'chak-chak.'

# SHRIKES

**1113 EMIN'S SHRIKE.** *LANIUS GUBERNATOR* Hartert. Pl. 74.

*Lanius gubernator* Hartert, O.C. p. 91, pl. 1, fig. 2, 1882: Langomeri, north-western Uganda.

**Distinguishing characters:** Adult male, differs from the Red-backed Shrike in having the rump and upper tail-coverts chestnut; tail less black; below, chest to under tail-coverts tawny; centre of belly white. The female has less or no black on the forehead, and the mantle dull brown or ashy brown; below, similar to male. Wing 72 to 87 mm. The young bird is brownish barred with blackish above; below, pale tawny barred with blackish.

**General distribution:** Cameroons to north-eastern Belgian Congo and the Sudan.

**Range in Eastern Africa:** Southern Sudan to northern Uganda.

**Habits:** An uncommon species within our limits, usually seen in small parties flying from tree to tree. It has been seen to take insects on the wing.

**Nest and Eggs:** Undescribed.

**Recorded breeding:** No records.

**Call:** A low hissing note, and the male has a loud clear call also.

**1114 RED-TAILED SHRIKE.** *LANIUS CRISTATUS* Linnæus. *Lanius cristatus isabellinus* Hemp. & Ehr. *Pale Red-tailed Shrike.*
*Lanius isabellinus* Hemprich & Ehrenberg, Symb. Phys. fol. e, note 2, 1828: Kumfuda, Arabia.

**Distinguishing characters:** Adult male, differs from the Red-backed Shrike in being wholly isabelline grey above; upper tail-coverts and tail pale chestnut without any markings. The female is paler above; ear-coverts ashy brown; tail as in male, but paler. Wing 91 to 101 mm. The immature bird is similar to the adult female but warmer coloured above, and often has lightish tips to the tail feathers with a subterminal narrow black bar. The young bird has some narrow blackish barring on the upperside. The adult female, immature and young can be distinguished from the Red-backed Shrike by the more tawny chestnut colouring of the underside of the tail feathers as against the darker, more ashy coloration

on the underside of the tail feathers in the Red-backed Shrike. In this species also the second primary is shorter than the fifth, whereas in the Red-backed Shrike the reverse is the case.

**General distribution:** South-eastern Russian Turkestan to Manchuria; in non-breeding season to the Lake Chad area, Eastern Africa, Arabia and India.

**Range in Eastern Africa:** Central, eastern and southern Sudan, Eritrea, Abyssinia, Kenya Colony and eastern Tanganyika Territory as far south as Dar-es-Salaam in non-breeding season.

**Habits:** Visitor to as far south as Tanganyika Territory, a common and at times abundant passage migrant further north. It appears to have a slightly more eastern range on migration than the Rufous Red-tailed Shrike from which it is not easily distinguished in the field. It is described as partial to small patches of cultivated ground which it can watch from some point of vantage. It has been heard singing in its non-breeding quarters, a low soft warble, interspersed with harsher notes.

*Lanius cristatus phœnicuroides* (Schal.). *Rufous Red-tailed Shrike.*
*Otomela phœnicuroides* Schalow, J.f.O. p. 148, 1875: Chimkent, Russian Turkestan.

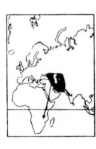

**Distinguishing characters:** Differs from the Pale Red-tailed Shrike in being darker above, and in having the top of the head pale chestnut. Wing 85 to 99 mm. The immature and young bird are also darker above, and show some contrast between the top of the head and the mantle.

**General distribution:** Southern Russia to Iran; in non-breeding season to Eastern Africa as far south as Nyasaland, Arabia and occasionally to India.

**Range in Eastern Africa:** Southern Sudan, Abyssinia, Eritrea, Uganda, British Somaliland, Kenya Colony as far west as Loita and north-eastern Tanganyika Territory as far south as Monduli, Moshi and Mkomasi in non-breeding season.

**Habits:** As for the last species.

**Distribution of other races of the species:** Asia, Dutch East Indies and Borneo; the nominate race being described from Bengal.

# SHRIKES

**1115 WOODCHAT SHRIKE.** *LANIUS SENATOR* Linnæus.
*Lanius senator niloticus* (Bp.).
*Enneoctonus niloticus* Bonaparte, Rev. Mag. Zool. p. 439, 1853: White Nile, Sudan.

**Distinguishing characters:** Adult male, forehead, lores to ear-coverts and down side of neck and mantle black; crown to upper mantle tawny chestnut; scapulars, rump and upper tail-coverts white; wing black, base of primaries white; tail black, basal part and tips of all the feathers white; below, white. The female has a grey mantle. Wing 93 to 105 mm. The immature bird has the mantle ash-coloured and the head pale rufous with dusky barring. The young bird is white or creamy white above, barred with blackish.

**General distribution:** Transcaucasia and Iran; in non-breeding season to the Sudan, British Somaliland, Uganda and Kenya Colony.

**Range in Eastern Africa:** The Sudan, Abyssinia, British Somaliland, Uganda and Kenya Colony as far south as Mt. Elgon and Lake Rudolf in non-breeding season.

**Habits:** A passage migrant and visitor to the northern half of Africa, passing through the Sudan and northern Abyssinia and spending the non-breeding season probably in the Abyssinian lowlands and Kenya Colony. Habits of other Shrikes, but much more prone to settle in a tree and not on it.

**Distribution of other races of the species:** Europe and North Africa, the nominate race being described from the River Rhine.

**1116 BOGDANOW'S SHRIKE.** *LANIUS BOGDANOWI* (Bianchi).
*Otomela bogdanowi* Bianchi, Bull. Ac. Imp. Sci. St. Petersburg, 30, p. 514, 1886: Astrabad, northern Iran.

**Distinguishing characters:** Above, grey; forehead whiter; lower mantle to rump washed with isabelline; base of bill to lores and ear-coverts black; stripe over eye white; wings dull blackish; base of primaries white; tail black, outer feathers mainly white; below, including under wing-coverts white; flanks buff; first primary rather narrow; bill black. The sexes are alike. Wing 90 to 92 mm. The immature dress is darker and more isabelline above; upper tail-coverts pale tawny; tail pale fawn colour instead of white; bill horn.

**General distribution:** Caspian Sea to north-western China.

**Range in Eastern Africa:** Once recorded from Umbugwe, south of Arusha, north-eastern Tanganyika Territory in November.

**Habits:** A species of which there is at present only one African record.

### 1117 YELLOW-BILLED SHRIKE. *CORVINELLA CORVINA* (Shaw).

*Corvinella corvina affinis* Hartl.

*Corvinella affinis* Hartlaub, Syst. Orn. Westafr. p. 104, 1857: White Nile, south of lat. 7° N., southern Sudan.

**Distinguishing characters:** A long-tailed bird; adult male, above, ash brown, broadly streaked with black; wings blackish brown, basal two-thirds of primaries tawny; tail long and blackish brown with narrow subterminal blackish ends and narrow buffish tips; lores and ear-coverts blackish; below, buffish white with black streaks; patch on flanks pale chestnut; bill yellow. The female has no chestnut patch on flanks. Wing 113 to 130 mm. The young bird is barred and mottled above with blackish, not streaked; below, white with blackish barring on breast and flanks.

**Range in Eastern Africa:** Southern Sudan east of the River Nile to eastern Uganda as far west as Kabuzi and western Kenya Colony.

**Habits:** A noisy and usually tame species of sociable habits, mostly seen in parties of ten to fifteen, flying from tree to tree. They have the appearance of Shrikes, constantly jerking their tails up and down, but have most of the habits of Babblers. Birds of the woodland and bush frequently entering thick foliaged trees. Subject to at least local migrations. The food consists of fruit, insects and small reptiles and they are particularly fond of lizards.

**Nest and Eggs:** A fairly large cup of twigs, rootlets and fibre lined with fine grass, in small trees or bushes. Eggs four or five, cream with irregular spots and speckles of yellowish brown and with pale grey undermarkings often in a zone round the egg; about 25 × 18 mm.

**Recorded breeding:** Uganda, December to March.

**Call:** A constant repeated 'scis-scis' from which it gets the local name of Scissor-Bird. Also a shrill call described as 'may we-may we wait-may we' and various rippling and Parrot-like calls when in a flock.

*Corvinella corvina togoensis* Neum.

*Corvinella corvina togoensis* Neumann, J.f.O. p. 263, 1900: Kete Krachi, Togoland.

**Distinguishing characters:** Differs from the preceding race in being more rufescent above. Wing 116 to 130 mm.

**General distribution:** Sierra Leone to Nigeria, Cameroons and the Sudan.

**Range in Eastern Africa:** Darfur and Kordofan areas, Sudan.

**Habits:** As for other races. It appears to be migratory, visiting the western Sudan to breed.

**Recorded breeding:** Darfur, May to July.

*Corvinella corvina caliginosa* Fried. & Bowen.

*Corvinella corvina caliginosa* Friedmann & Bowen, Proc. Biol. Soc. Washington, 46, p. 122, 1933: Rangu, southern Bahr-el-Ghazal, south-western Sudan.

**Distinguishing characters:** Above, greyer, less brown, than the other races. Wing 121 to 127 mm.

**General distribution:** North-eastern Belgian Congo to the Sudan and Uganda.

**Range in Eastern Africa:** The Sudan, west of the River Nile to western Uganda as far east as Kabuzi.

**Habits:** As for other races.

**Recorded breeding:** No records.

**Distribution of other races of the species:** West Africa, the nominate race being described from Senegal.

**1118** MAGPIE-SHRIKE. *UROLESTES MELANOLEUCUS* (Jardine).

*Urolestes melanoleucus æquatorialis* Reichw.

*Urolestes æquatorialis* Reichenow, J.f.O. p. 65, 1887: Gasa Mts., 12 to 14 miles south-west of Kondoa, Kondoa-Irangi district, north-central Tanganyika Territory.

**Distinguishing characters:** General colour slightly glossy black and white; tail long; scapulars, basal half of primaries and tips of secondaries white; flanks usually black and white; rump greyish white; tail feathers usually with white tips. The sexes are alike. Wing 126 to 136 mm. The young bird is brownish black, tips of secondaries buffish white; ashy tips to feathers of underparts and upper tail-coverts.

**General distribution:** Kenya Colony to northern Nyasaland and northern Portuguese East Africa.

**Range in Eastern Africa:** South-western Kenya Colony to Tanganyika Territory and the northern area of northern Portuguese East Africa.

**Habits:** Conspicuous birds with true Shrike habits, sitting on the tops of bushes and dropping on to insects on the ground. They are, however, often found in parties and may be accompanied by other species of birds. The flight is strong and dipping and they fly at a higher level than most Shrikes. They are reputed to have the Shrike habit of spitting its prey on thorns. Rather shy birds as a rule.

**Nest and Eggs:** Nest of thorny twigs, rather large and bulky, lined with grass and rootlets, and usually placed in a particularly thorny situation. Eggs three or four, cream buff or pinkish with umber, rufous and grey spots and blotches, mainly at larger end; about 27 × 19 mm.

**Recorded breeding:** Tanganyika Territory, January to April.

**Call:** A loud and not unmusical whistle, described (for the southern race) as 'needle-boom-needle-boom-come here-come here.' Has also harsh scolding alarm notes which it utters on the approach of any large bird.

**Distribution of other races of the species:** Angola and South Africa, the nominate race being described from the Orange River.

**1119 BLACK-HEADED GONOLEK.** *LANIARIUS ERYTHROGASTER* (Cretzschmar). **Pl. 75.**

*Lanius erythrogaster* Cretzschmar, in Rüpp. Atlas, Vög. p. 43, pl. 29: 1829: Kordofan, Sudan.

**Distinguishing characters:** Above, including sides of head, wings and tail, slightly glossy black; below, crimson; under tail-coverts buff. The sexes are alike. Wing 91 to 114 mm. The immature bird is similar to the adult, but is duller black above and has buff tips to the feathers of the whole underside. The young bird has the feathers of the upperside tipped with buff; below, yellowish buff narrowly barred with black.

**General distribution:** Lake Chad to Eritrea, and south to Tanganyika Territory and the eastern Belgian Congo.

**Range in Eastern Africa:** Western Sudan and Uganda to Eritrea, Abyssinia, Kenya Colony and northern Tanganyika Territory.

**Habits:** An inhabitant of dense bush or other form of thick cover, never seen for more than a few seconds in the open, but very noticeable from its call wherever it occurs. Like other species of this group it is inquisitive and can be called up to establish its identity. It is common, widely distributed, and even locally plentiful, almost invariably in pairs and usually very shy, though in places it has become a garden bird and has lost a little of its shyness.

**Nest and Eggs:** Nest in a thick bush, a cup of rootlets interlaced with grass and lined with fine fibres and rootlets. Eggs two, pale bluish green spotted and blotched with umber and various shades of grey; about 25 × 18 mm.

**Recorded breeding:** Shari River, French Equatorial Africa, July. Darfur, March and April. Uganda, March to July, and September to November.

**Call:** A fine clear double whistle 'whatt-ho' or 'yochayoo.' In certain localities this is answered by a grating call from the female as a duet, but it is not as usual as it is in other species of the genus. Sir F. Jackson compares the grating note of the female to the noise made by tearing calico.

**1120** GONOLEK. *LANIARIUS BARBARUS* (Linnæus).
*Laniarius barbarus mufumbiri* O. Grant. **Pl. 75.**
*Laniarius mufumbiri* O. Grant, Bull. B.O.C. 29, p. 30, 1911: Vichumbi, south end Lake Edward, eastern Belgian Congo.

**Distinguishing characters:** Differs from the Black-headed Gonolek in having the forehead to nape golden yellow; white tips to the wing-coverts, and a thinner bill. Wing 87 to 92 mm. The young bird has the top of the head yellow.

**General distribution:** Eastern Belgian Congo and Uganda.

**Range in Eastern Africa:** Uganda.

**Habits:** A shy retiring bird not often seen, inhabiting swamps and patches of papyrus. It is, however, very inquisitive and will come to inspect any unusual noise. Mainly insectivorous.

**Nest and Eggs:** Nest (of the nominate ace) a transparent cup of small rootlets in a tree or bush. Eggs two, pale bluish green sparingly spotted and blotched with dark brown and lilac; about 23·5 × 17 mm.

**Recorded breeding:** No records.

**Call:** A loud mellow call of 'yo-yo' or 'yoo-yo' (F. J. Jackson).

**Distribution of other races of the species:** West Africa, the nominate race being described from Senegal.

### 1121 SLATE-COLOURED BOUBOU. *LANIARIUS FUNEBRIS* (Hartlaub). Pl. 96.

*Dryoscopus funebris* Hartlaub, P.Z.S. p. 105, 1863: Meninga, headwaters Gombe River, Tabora district, Tanganyika Territory.

**Distinguishing characters:** Adult male, head and throat blackish slate to bluish black; mantle and underside slightly glossy deep slate blue; wings and tail black with a slight greenish gloss; inner webs of flight feathers ashy; concealed white spots on rump feathers. The female is paler slate blue from breast to belly. Wing 81 to 98 mm. The young bird is dull blackish above, more or less barred with tawny; below, olivaceous, narrowly barred with blackish. Albinistic examples occur.

**Range in Eastern Africa:** Southern Sudan and Abyssinia, the Somalilands, Uganda, Kenya Colony and Tanganyika Territory as far south as Iringa and Lake Rukwa.

**Habits:** A bird of very similar habits to the Black-headed Gonolek, inhabiting thick dense cover, but in this case usually in dryer country. Again it is the distinctive call and the answer of the female that notifies the bird's presence. Usually in pairs and not uncommon in dry thorn bush country.

**Nest and Eggs:** Nest of grass and rootlets rather small and flat, placed low in a bush. Eggs two or three, pale bluish green spotted with reddish brown and underlying lavender spots and speckles often in the form of a zone; about 23 × 16 mm.

**Recorded breeding:** Southern Sudan (breeding condition), October. Southern Abyssinia, April. Kenya Colony, northern areas, May and June; Mombasa (nestlings), October; southern areas, March and April. Tanganyika Territory, December to March, also October and November.

**Call:** A duet, variously described by many authors. It consists of three or four clear notes by the male answered instantly by the female with a snarling noise, but this procedure may be reversed.

# SHRIKES

Elliott gives the following notes: Call, 'plonk-plonk-plonk,' answer 'wheck,' or 'cuhunk-cuhunk,' answer 'whee-er' or 'squeesquaw-squosquee,' answer 'plonk.' There are also other calls including a recognizable song, and a vibrating churring alarm note.

**1122 SOOTY BOUBOU.** *LANIARIUS LEUCORHYNCHUS* (Hartlaub). **Pl. 96.**

*Telephonus leucorhynchus* Hartlaub, Rev. Zool. p. 108, 1848: Elmina, Gold Coast Colony.

**Distinguishing characters:** Differs from the Slate-coloured Boubou in being slightly glossy deep black; inner webs of flight feathers blackish; rump feathers long and very fluffy with no concealed white spots. The sexes are alike. Wing 90 to 102 mm. The young bird is duller and has a white bill, though the adult has a black one.

**General distribution:** Sierra Leone and Gabon to the Sudan and Uganda.

**Range in Eastern Africa:** South-western Sudan to western and southern Uganda.

**Habits:** An inhabitant of thick forest undergrowth and tangled vegetation of old forest clearings. Shy, furtive and little noted, and would be overlooked but for its distinctive calls.

**Nest and Eggs:** Nest a flat platform of thorn and twigs lined with rootlets and almost as scanty as a Dove's. Eggs two, greenish grey closely spotted with various shades of brown and with stone grey undermarkings; about 25 × 17·5 mm.

**Recorded breeding:** Uganda, probably June and July.

**Call:** A clear vibrating whistle of six or seven notes by male with female interposing a low vibrant long drawn 'kwee' between the third and fourth notes—though at times they utter alternate notes. Also many gurgling and babbling calls while feeding.

**1123 MOUNTAIN SOOTY BOUBOU.** *LANIARIUS POENSIS* (Alexander).

*Laniarius poensis holomelas* (Jacks.). **Pl. 96.**

*Dryoscopus holomelas* Jackson, Bull. B.O.C. 16, p. 90, 1906: Ruwenzori Mts., western Uganda.

**Distinguishing characters:** Adult male, general colour wholly dull black, slightly slaty and with a very slight greenish gloss; no

white in rump feathers; tail shortish as in the sooty Boubou. The female is duller than the male. Wing 78 to 87 mm. The young bird is duller than the adult female.

**General distribution:** Eastern Belgian Congo and Uganda.

**Range in Eastern Africa:** Western Uganda from Lake Albert to Lake Kivu.

**Habits:** A bird of the high forest zone from 6,000 to 10,000 feet coming lower occasionally. It occurs fairly freely on Mt. Ruwenzori and also on the Birunga Volcanoes. An inhabitant of tangled masses of creepers, shy and difficult to see, but with a marvellous variety of notes.

**Nest and Eggs:** Undescribed.

**Recorded breeding:** No records.

**Call:** Usually a curious clinking sound followed by piping notes, but it has such a variety of calls that it seems remarkable that one bird can make them.

**Distribution of other races of the species:** West Africa, the nominate race being described from Fernando Po.

**1124 FÜLLEBORN'S BLACK BOUBOU.** *LANIARIUS FÜLLEBORNI* (Reichenow). **Pl. 96.**
*Dryoscopus fülleborni* Reichenow, O.M. p. 39, 1900: Usafua, Tukuyu district, south-western Tanganyika Territory.

**Distinguishing characters:** Adult male, general colour dark slaty black; no white in rump feathers. The female has the chest to belly washed with olivaceous. Wing 80 to 94 mm. The young bird is dark olivaceous grey above; below, olivaceous green or olivaceous grey.

**General distribution:** Central Tanganyika Territory to north-eastern Northern Rhodesia and northern Nyasaland.

**Range in Eastern Africa:** Tanganyika Territory from the Usambara Mts. to the Nguru and Uluguru Mts. and the south-western areas.

**Habits:** A forest species with remarkable song calls, but otherwise its habits have been little noted. It occurs fairly freely in the highland forest of Tanganyika Territory and in the evergreen forest of the Nyasaland highlands.

**Nest and Eggs:** A shallow cup nest of grass, tendrils and rootlets in the fork of a shrub or among creepers. Eggs two, pale greenish blue sparingly spotted or freckled with brown and olive brown and with slate or leaden grey undermarkings; about 25 × 19 mm.

**Recorded breeding:** Tanganyika Territory, December to February. Northern Nyasaland, December.

**Call:** Many and varied. The normal call is a deep three-syllabled flute-like call of the male followed by two notes from the female of which the second is shriller. Moreau also notes a deep liquid call of either two or three notes from the male, followed by a bubbling 'u-u-u-u' from the female. The alarm note is a hard click.

**1125 TROPICAL BOUBOU.** *LANIARIUS ÆTHIOPICUS* (Gmelin).
*Laniarius æthiopicus æthiopicus* (Gmel.).
*Turdus æthiopicus* Gmelin, Syst. Nat. 1, pt. 2, p. 824, 1789: Abyssinia.

**Distinguishing characters:** Upperside and sides of face slightly glossy blue black; middle and some secondary wing-coverts white; concealed white spots on rump; below, white with a pinkish or peach coloured wash on chest, breast and flanks. The sexes are alike. Wing 91 to 109 mm. The immature bird is duller above; chest, breast and flanks pale brown; some tawny tips to wing-coverts and bill not so black as in adult. The young bird is similar to the immature, but has tawny tips and bars to the feathers of the upperparts, and some dusky bars on chest.

**Range in Eastern Africa:** Kassala Province to the Boma Plateau, eastern Sudan to Eritrea, Abyssinia, British Somaliland and northern Kenya Colony.

**Habits:** As for other races, but possibly less skulking in its habits.

**Nest and Eggs:** See under the following race, though the nest is more frequently in open situations.

**Recorded breeding:** Northern Kenya Colony, April and May (breeding condition).

*Note:* Intergrades with the Kilimanjaro race in the Meru area, north-east Mt. Kenya.

*Laniarius æthiopicus major* (Hartl.).

*Telephonus major* Hartlaub, Rev. Mag. Zool. p. 108, 1848: Elmina, Gold Coast Colony.

**Distinguishing characters:** Differs from the preceding race in that the white on the wing extends down the edges of some of the secondaries and forms a longitudinal wing bar. Wing 90 to 114 mm. The young bird has a few tawny tips to the feathers of the upperparts and some dusky barring below.

**General distribution:** Sierra Leone, Nigeria, and the Cameroons to the Sudan, Uganda, Kenya Colony, Tanganyika Territory, northern Angola, Belgian Congo, northern Northern Rhodesia and northern Nyasaland.

**Range in Eastern Africa:** Central and southern Sudan to Uganda, western Kenya Colony as far east as the Aberdares and western Tanganyika Territory as far east as Loliondo, Esimingor, Njombe and Mahenge.

**Habits:** A skulking species of thick cover, almost always in pairs and with wonderful bell-like notes. Common and very widespread in any form of cover from highland forest to scrub vegetation in dry river beds. It is commonly known as 'Bell-bird' and although secretive is not usually shy if cautiously approached. Regrettably destructive of young birds of other species.

**Nest and Eggs:** Saucer-shaped nest of tendrils, dry stems and rootlets lined with fibre and fine rootlets, placed in a bush or in the fork of a shrub or tree, not necessarily well concealed. Eggs two, either blue green or greenish buff, rather glossy, with spots or speckles of brown, lilac and grey; about $23 \times 17$ mm., with some local variation.

**Recorded breeding:** Nigeria, June. Uganda, March to July. Kenya Colony highlands, March to June and October to December. Tanganyika Territory, October to April. Portuguese East Africa, December to April. Southern Belgian Congo, September to December. Rhodesia, September to November.

**Call:** Varied and remarkable duet of calls. Normally male gives three bell-like notes and female answers with a groaning 'huee' but so instantaneously that the call seems to be made by the one bird. There are very many combinations, however, of which Stoneham has listed at least seven, and these variations may be peculiar to one locality and locally constant without any change in the bird itself. The alarm calls are a harsh 'churr' and an explosive 'tchak' (Vincent).

**Note:** Intergrades with the Kilimanjaro race in the area between Nakuru, the Aberdares, Nairobi and Kiambu.

*Laniarius æthiopicus sublacteus* (Cass.).
*Dryoscopus sublacteus* Cassin, Proc. Acad. Sci. Philad. p. 246, 1851: Mombasa, eastern Kenya Colony.

**Distinguishing characters:** Differs from the Abyssinian race in having the wings black, with no white on coverts. Wing 85 to 98 mm. The young bird has no barring below. There is a wholly black colour phase on the coast of Kenya Colony and Manda Island.

**Range in Eastern Africa:** Eastern Kenya Colony and eastern Tanganyika Territory from Kipini to Dar-es-Salaam and as far west as Makindu, Lake Jipe, north Paré Mts., and Mpapwa; also Manda and Zanzibar Islands.

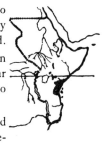

**Habits:** As for other races, the call being particularly fine and liquid. Moreau notes that the call of the male on the coast is triple-noted, inland double-noted, but there is no apparent change in the appearance of the bird and the division does not correspond to that of the observable races.

**Recorded breeding:** Amani, Tanganyika Territory, October. Zanzibar Island, May.

*Laniarius æthiopicus mossambicus* (Reichw.).
*Dryoscopus major mossambicus* Reichenow, J.f.O. p. 141, 1880: Mozambique, Portuguese East Africa.

**Distinguishing characters:** Differs from the Gold Coast race in having the underparts distinctly washed with pinkish or peach colour. Wing 84 to 102 mm.

**General distribution:** Northern Rhodesia, Bechuanaland, Southern Rhodesia, greater part of Nyasaland and Portuguese East Africa.

**Range in Eastern Africa:** Portuguese East Africa and Nyasaland.

**Habits:** As for other races.

**Recorded breeding:** Nyasaland, November to January and April.

**Note:** Intergrades with the Gold Coast race in north-western Northern Rhodesia to the Rovuma River area.

*Laniarius æthiopicus ambiguus* Mad. **Ph. xiv.**
*Laniarius ambiguus* Madarász, Ann. Mus. Nat. Hung. 2, p. 205, 1904: Kibosho, Kilimanjaro, north-eastern Tanganyika Territory.

**Distinguishing characters:** Differs from the Abyssinian race in having the white on the wings confined to the middle coverts, and often a slight green gloss on the upperparts. Wing 89 to 103 mm.

**Range in Eastern Africa:** Northern and central Kenya Colony from Mt. Kulal east of Lake Rudolf, the Aberdares, and Fort Hall to north-eastern Tanganyika Territory from Longido, Ketumbeine and Monduli to Mt. Kilimanjaro.

**Habits:** As for other races, but not necessarily a skulking bird. Very fine liquid notes, possibly the finest of all.

**Recorded breeding:** Kenya Colony, mainly November to January and April to June.

*Note:* Intergrades with the Gold Coast race in the area between Nakuru, the Aberdares, Nairobi and Kiambu; and with the Abyssinian race in the Meru area, north of Mt. Kenya.

*Laniarius æthiopicus erlangeri* Reichw. **Pl. 96.**
*Laniarius erlangeri* Reichenow, Vög. Afr. 3, p. 834, 1905: Umfudu, Juba River, southern Italian Somaliland.

**Distinguishing characters:** Similar to the last race but smaller. Wing 80 to 90 mm. There is a wholly black colour phase in the coastal area.

**Range in Eastern Africa:** Lower Juba River, southern Italian Somaliland.

**Habits, Nest and Eggs:** As for other race.

**Recorded breeding:** No records.

**Distribution of other races of the species:** Southern Rhodesia and the Transvaal.

**1126 RED-NAPED BUSH-SHRIKE.** *LANIARIUS RUFICEPS* (Shelley).

*Laniarius ruficeps ruficeps* (Shell.). **Pl. 74.**
*Dryoscopus ruficeps* Shelley, Ibis, p. 402, pl. 10, 1885: Burao, central British Somaliland.

**Distinguishing characters:** Adult male, white stripe from base of bill over eye to sides of neck; lores to ear-coverts black; forehead black; crown to nape orange red; mantle grey or grey and black; wings and tail black, former with a longitudinal white band, latter with outer edge of outer tail and tips of all except central feathers white; below, creamy or pinky white; throat and belly white. The female has the mantle more olivaceous grey. Wing 72 to 81 mm. The immature bird has the crown mottled blackish and olive brown; lores to ear-coverts dull blackish; bill horn colour not black as in adult. The young bird has the forehead, crown, mantle and tail olivaceous grey.

**Range in Eastern Africa:** Central to southern British Somaliland.

**Habits, Nest and Eggs:** See under other races.

**Recorded breeding:** No records.

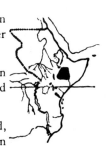

*Laniarius ruficeps rufinuchalis* (Sharpe).
*Dryoscopus rufinuchalis* Sharpe, P.Z.S. p. 479, 1895: Durro, near Abdula, southern Abyssinia.

**Distinguishing characters:** Differs from the nominate race in having the crown black with the chestnut confined to the hinder crown and occiput. Wing 76 to 78 mm.

**Range in Eastern Africa:** Southern Abyssinia and southern Italian Somaliland from Yavello and Dabuli to El Wak and Bardera.

**Habits:** Birds of thick undergrowth, rarely seen but often heard, which feed a good deal on the ground. Food largely insects taken on the ground or in the lower parts of bushes.

**Nest and Eggs:** A rather flat Dove-like nest of sticks and thick grass stems in small bushes. Eggs two or three, greenish white streaked and flecked with pale brown and with ash-coloured undermarkings; about 22 × 16 mm.

**Recorded breeding:** Southern Abyssinia, May.

**Call:** The male is said to give a growling call, answered by the female with a more liquid note. One call is described as Crow-like, another as a creak, and there are various chuckling calls as well as a sharp cracking sound almost like that of cracking a whip.

*Laniarius ruficeps kismayensis* (Erl.).
*Dryoscopus ruficeps kismayensis* Erlanger, O.M. p. 182, 1901: Kismayu, southern Italian Somaliland.

**Distinguishing characters:** Differs from the Abyssinian race in having a paler grey mantle. Wing 76 to 84 mm.

**Range in Eastern Africa:** Coastal areas of southern Italian Somaliland and Kenya Colony from Mogadishu to Taru.

**Habits:** A local species of low-lying country among thick bushes and undergrowth, usually in pairs, but calling to one another frequently as they hunt some distance apart, and moving furtively through thick patches of cover.

**Recorded breeding:** Jubaland, Italian Somaliland, May.

## 1127 LÜHDER'S BUSH SHRIKE. *LANIARIUS LÜHDERI* Reichenow.

*Laniarius lühderi* Reichenow, J.f.O. p. 101, 1874: Cameroons Delta, West Africa.

**Distinguishing characters:** Forehead and crown to nape chestnut; lores to ear-coverts and rest of upperparts black; white bar down the wing; below, chin to upper belly tawny; lower belly, lower flanks, under wing-coverts and under tail-coverts white. The sexes are alike. Wing 81 to 97 mm. The immature bird is olive brown above with a whitish or yellowish bar down the wing; tail tawny chestnut; below, tawny buff with some yellow from breast to belly; under tail-coverts yellowish white. The young bird is olivaceous brown above more or less barred with blackish; below, olivaceous yellow barred with dusky.

**General distribution:** Cameroons and Gabon to the Sudan, eastern Belgian Congo, Uganda, Kenya Colony and Tanganyika Territory.

**Range in Eastern Africa:** Southern Sudan to Uganda, western Kenya Colony and western Tanganyika Territory.

**Habits:** A secretive species rarely leaving the shelter of dense cover, often heard but seldom seen. Generally found in old overgrown forest clearings or in thick undergrowth along streams, occasionally in tree-tops. Inquisitive birds.

**Nest and Eggs:** Nest a cup of grass, weed stems and rootlets interlaced with creeper fibre, placed in a fork of a tree or bush.

Eggs two, whitish cream or very pale blue with irregular blotches of brown and grey at the larger end; about 25 × 18 mm.

**Recorded breeding:** Cameroons, July. Uganda (breeding condition), July. Mt. Elgon, June.

**Call:** Said to be a soft tremulous 'keow' from the male answered by the female with 'cha-cha.' C. G. Young describes the call of the male as a sharp 'tlk-tlk' and a sudden hollow 'co' followed by a rapid 'hohoho' to which the female answers with clicks. Alarm call a sharp 'tzowarr' like a released spring.

## 1128 BLACK-BACKED PUFF-BACK. *DRYOSCOPUS CUBLA* (Shaw).

*Dryoscopus cubla hamatus* Hartl. **Ph. xiv.**
*Dryoscopus hamatus* Hartlaub, P.Z.S. p. 106, 1863: Kaseh, Unyamwesi, Tanganyika Territory.

**Distinguishing characters:** Adult male, head, sides of face, mantle, wings and tail glossy blue black; edges of wing-coverts and inner secondaries white; edge of outer tail feathers and tips of all white; rump and lower back greyish white; below, white with a greyish tint; feet grey. The female differs from the male in having a white streak from the nostrils to half-way over the eye; lower back and rump grey; bill black in both sexes. Wing 73 to 89 mm. The immature bird has a distinct wash of buffish below and on the white of the wings. The young bird is similar to the adult female, but is dull black above with ashy tips and a buffish wash on the wings; lower mandible horn colour.

**General distribution:** Eastern Belgian Congo to Kenya Colony, Tanganyika Territory, northern Angola and the Limpopo River.

**Range in Eastern Africa:** Central Kenya Colony to Lamu and Manda Island, Ruanda, Urundi and Tanganyika Territory to the Zambesi River; also Ukerewe Island, Lake Victoria.

**Habits:** Differing widely in habits from the skulking habits of the Bush-Shrikes, the birds of this genus have many of the habits of Warblers and are distinctly sociable, being freely found as members of mixed bird parties. They are birds of light bush and open woodland and are usually seen hunting among leaves for caterpillars, quite often catching insects in flight. Their name arises from their curious habit of puffing out the feathers of the back till they almost encircle the bird, and in display they fly about making a buzzing crackling noise with their wings, and they also drop vertically while giving a little series of whistles or clacking noises. Delightful cheery little birds, usually tame, but regrettably destructive to the eggs and young of other small birds in the nesting season.

**Nest and Eggs:** A cup of bark fibre and rootlets well hidden and bound with cobwebs round the fork of a tree, usually at some height from the ground. Van Someren gives a most interesting account of how the birds tear off bark from a dead creeper stem. It is a considerable acrobatic performance and full use is made of leverage while the bird keeps up a continued whistling (Ibis, p. 394, 1916). Eggs two or three, white or cream with a light zone of brown sepia and lilac spotting at the larger end; about 20 × 15·5 mm.

**Recorded breeding:** Kenya Colony, March to June and December to February. Tanganyika Territory, January to March. Southern Belgian Congo, September and October. Nyasaland, September to January, also at other times of year. Portuguese East Africa, October to January. Southern Rhodesia, October and November.

**Call:** Many, but usually two clicks followed by a whistle. Alarm note is 'chak' followed by a swearing 'skurr' at times. Breeding season call as above.

**Distribution of other races of the species:** South Africa, the nominate race being described from Knysna.

**1129** ZANZIBAR PUFF-BACK. *DRYOSCOPUS AFFINIS* (Gray).
*Dryoscopus affinis affinis* (Gray).
*Hapalophus affinis* G. R. Gray, Charlesw. Mag. p. 489, 1837: Zanzibar Island.

**Distinguishing characters:** Adult male, differs from the Black-backed Puff-back in having the lower back, rump and underside pure white or creamy white, not greyish white, and a longer heavier bill; feet blue grey. The female has a white streak from the nostrils to half-way over the eyes; a black spot in front of eye; and the rump is greyer. Wing 74 to 85 mm. The immature bird is buffish below, and has the under mandible horn colour. The young bird is dusky black above; a white stripe over the lores; rump greyish white; below, white.

# SHRIKES

**Range in Eastern Africa:** Coastal areas of Kenya Colony and Tanganyika Territory from Lamu and Manda Island to Dar-es-Salaam, and as far inland as the Pugu Hills, also Zanzibar and Mafia Islands.

**Habits:** A common species of the East African coastal belt, in places abundant. It has a habit of snapping its wings together during a short flight, and also makes a planing flight with its rump feathers puffed out. Bold conspicuous birds of cultivated country and woodland.

**Nest and Eggs:** Apparently undescribed.

**Recorded breeding:** Zanzibar Island, probably mainly September to December.

**Call:** A variety of whistling chirps, and also a loud shrill note as it executes its downward planing flight.

*Dryoscopus affinis senegalensis* (Hartl.).
*Sigelus senegalensis* Hartlaub, Syst. Orn. West Afr. p. 112, 1857: Gabon.

**Distinguishing characters:** Differs from the nominate race in having black, not white, bases to the scapular feathers. Wing 73 to 86 mm. Albinistic examples are known.

**General distribution:** Southern Nigeria, Cameroons and Gabon to Uganda.

**Range in Eastern Africa:** Bwamba area, western Uganda.

**Habits:** A bird of forest edge, or to be found along forest streams in low-lying country. Little is recorded of its habits.

**Recorded breeding:** Cameroons (nestling), August. Western Uganda (breeding condition), July.

**1130 PRINGLE'S PUFF-BACK. *DRYOSCOPUS PRINGLII*** Jackson.
*Dryoscopus pringlii* Jackson, Bull. B.O.C. 3, p. 3, 1893: between Tsavo and Kufumika, southern Kenya Colony.

**Distinguishing characters:** Adult male, similar to the Black-backed Puff-back but smaller; scapulars grey; lower mantle and rump greyer; lower rump white; below, greyer, centre of belly white; feet blue. The female is ashy brown above; white edges to

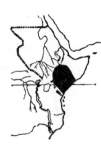

wing-coverts and flight feathers; below, white, chest creamy. Wing 68 to 70 mm. The immature bird is similar to the adult female. The young bird is duller than the adult female and has a slight buff wash on chest.

**Range in Eastern Africa:** Southern Abyssinia to southern Italian Somaliland and Kenya Colony.

**Habits:** An inhabitant of dry thorn scrub of the Eastern African lowlands. A small desert species with the habits of the genus, but of which little is recorded.

**Nest and Eggs:** A very neat cup moulded into the fork of a branch and well hidden although in the open. Eggs undescribed.

**Recorded breeding:** Tanganyika Territory, November.

**Call:** A low pitched churring sound and a short monotonous harsh song of rather low pitch.

**1131 PUFF-BACK.** *DRYOSCOPUS GAMBENSIS* (Lichtenstein).
*Dryoscopus gambensis malzacii* (Heugl.).
*Malaconotus malzacii* Heuglin, Orn. Nordost. Afr. p. 457, 1871: Upper White Nile, southern Sudan.

**Distinguishing characters:** Adult male, very similar to the Black-backed Puff-back but differs in having a rather heavier lower mandible, duller black mantle; greyer scapulars, and only a trace of light colour to the edge of the outer tail feathers and tips of the others; feet slate grey. Female differs in being earth brown above, darker on head, with buffish edges to wing-coverts and scapulars; below, buff or tawny buff, darker from chin to breast. Wing 85 to 97 mm. The immature male differs from the adult in having the light edges to the wings and the underside buffish; the immature female is similar to the adult, but is sometimes more tawny below. The young bird is similar to the adult female but has some ashy edges to the feathers of the head and mantle, and buff edges to wing-coverts and flight feathers.

**General distribution:** French Equatorial Africa to the Sudan, north-eastern Belgian Congo, Uganda, and Kenya Colony.

**Range in Eastern Africa:** Central Sudan to Uganda and Kenya Colony.

## SHRIKES

**Habits:** Typical small Puff-backs inhabiting thick-foliaged trees which they search diligently for insects. Common, and at times noisy in woodland and forest.

**Nest and Eggs:** In a bush or tree, a nest of rootlets and bark fibre, covered with lichen and cobwebs. Nest occasionally at some height from the ground. Eggs usually two, greyish white spotted and streaked with various shades of brown and grey; about 16 × 10 mm.

**Recorded breeding:** Darfur, June to August. Uganda, March to July, also December and January.

**Call:** A chattering alarm note of 'tuk-tak' also a repeated Sparrow-like chirp. The male also has a low quite melodious warbling song in the breeding season.

*Dryoscopus gambensis erythreæ* Neum.
*Dryoscopus malzacii erythreæ* Neumann, J.f.O. p. 412, 1899: Salamona, Eritrea.

**Distinguishing characters:** Differs from the preceding race in the female having the head, mantle and wings more brownish black, and a blacker tail. Males indistinguishable. Wing 79 to 94 mm.

**Range in Eastern Africa:** Eritrea to eastern Sudan from Sennar to the Boma Plateau and Abyssinia.

**Habits:** As for the other races but prefers isolated thick-leaved trees in low country, or bush at forest edges. Mostly found at 5,000 feet or over, in small parties calling to each other with a sharp single note.

**Recorded breeding:** Southern Abyssinia (near breeding condition), December.

*Dryoscopus gambensis erwini* Sassi.
*Dryoscopus gambensis erwini* Sassi, O.M. p. 109, 1923: Forest west of Lake Tanganyika, eastern Belgian Congo.

**Distinguishing characters:** Male similar to those of the two preceding races; female differs in having the top of the head greyer than the female of the last race. Wing 80 to 88 mm.

**General distribution:** Eastern Belgian Congo and northern Tanganyika Territory.

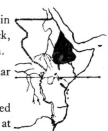

**Range in Eastern Africa:** The Bukoba and Mwanza districts of northern Tanganyika Territory.

**Habits:** As for other races.

**Recorded breeding:** No records.

**Distribution of other races of the species:** West Africa, the nominate race being described from Senegambia.

### 1132 PINK-FOOTED PUFF-BACK. *DRYOSCOPUS ANGOLENSIS* Hartlaub.

*Dryoscopus angolensis nandensis* Sharpe. **Pl. 74.**
*Dryoscopus nandensis* Sharpe, Bull. B.O.C. 11, p. 28, 1900: Nandi, western Kenya Colony.

**Distinguishing characters:** Adult male, head to upper back and including sides of face dark slate colour; mantle, wings and tail ash-grey; lower back and rump silky pale grey; below, very pale grey; feet fleshy pink. The female has the head, upper back and sides of face grey; mantle, wings and tail slightly olivaceous brown; lower back and rump silky greyish brown; below, tawny; centre of belly and under tail-coverts white. Wing 82 to 87 mm. The young bird is like the adult female, but duller.

**General distribution:** Eastern Belgian Congo to the Sudan, Uganda and Kenya Colony.

**Range in Eastern Africa:** South-eastern Sudan to Uganda and western Kenya Colony.

**Habits:** Rare and local forest species at heights of 4,000 to 8,000 feet.

**Nest and Eggs:** Undescribed.

**Recorded breeding:** No records.

**Call:** Both sexes have a harsh churring call.

*Dryoscopus angolensis kungwensis* Mor.
*Dryoscopus angolensis kungwensis* Moreau, Bull. B.O.C. 61, p. 45, 1941: Nganja, Kungwe area, western Tanganyika Territory.

**Distinguishing characters:** Differs from the preceding race in the greyer crown and hind neck of the male. Wing 87 mm.

**Range in Eastern Africa:** Kungwe-Mahare Mts. area, western Tanganyika Territory.

## SHRIKES

**Habits:** Nothing particularly recorded.

**Recorded breeding:** No records.

**Distribution of other races of the species:** West Africa and Angola, the nominate race being described from Angola.

### 1133 BLACK-HEADED BUSH-SHRIKE. *TCHAGRA SENEGALA* (Linnæus).

*Tchagra senegala senegala* (Linn.). **Ph. xiv.**
*Lanius Senegalus* Linnæus, Syst. Nat. 12th ed. 1, p. 137, 1766: Senegal.

**Distinguishing characters:** Centre of head from forehead to nape and stripe through eye from base of bill to above ear-coverts black; face and stripe over eye buffish; mantle brown; rump greyish brown; wings chestnut with black or blackish ends to primaries and greater part of inner secondaries; tail black with white ends; central tail feathers browner grey, more or less distinctly barred; below, white or creamy white with more or less grey on chest and flanks. The sexes are alike. Wing 79 to 92 mm. In the young bird the black of the head is replaced by dull blackish brown and the bill is horn colour, not black.

**General distribution:** Senegal to the Sudan, Kenya Colony, Italian Somaliland, Angola and Cape Province.

**Range in Eastern Africa:** Southern Sudan in the area of Torit, Didinga and Atoporopos, Uganda, Kenya Colony and southern Italian Somaliland to the Zambesi River.

**Habits:** A common and widespread species of desert scrub, bush, woodland and gardens. Rather shy and with skulking habits except in the breeding season. The courtship flight is a sharp mount into the air with crackling flapping wings and then a long slant or spiral down, often finishing with a glide up into a bush while uttering a clear piping call. This flight is not necessarily confined to the breeding season and is often high, the bird occasionally rising to two hundred feet. This species and the next are very alike except in the colour of the top of the head and the slightly larger size of the present one. They are often found together and might easily be mistaken for male and female. The food consists mainly of insects, especially beetles, but it also attacks mice and even rats as well as young birds.

**Nest and Eggs:** Nest a neatly made rather flat cup of weed stems, tendrils, and grass stalks, lined with rootlets, placed in a low tree or bush a few feet from the ground. Eggs two or three, variable, but usually white, well marked with spots, streaks and Bunting-like scrawls of various shades of brown and grey; about 25 × 17 mm. In coastal East Africa the Bunting-like scrawls appear to be absent.

**Recorded breeding:** Southern Italian Somaliland, June. Uganda, May and June, also September to November. Western Kenya Colony, October. Kenya Colony coast, July; southern areas, March and April. Tanganyika Territory, November to April. South-eastern Belgian Congo, September and October. Rhodesia, September to February. Nyasaland, September to January in highlands, later in the lowlands. Portuguese East Africa, December.

**Call:** A powerful but plaintive series of piping notes, rather rattling at first but becoming clear and bell-like. The emphasis and the length of the call differs considerably with locality, and it may be uttered from a perch or during flight. A churring alarm note.

*Tchagra senegala habessinica* (Hemp. & Ehr.).
*Lanius senegalus* var. *habessinica* Hemprich & Ehrenberg, Symb. Phys. 1, fol. e, 1823: Eritrea.

**Distinguishing characters:** Differs from the nominate race in having the mantle darker, more earth brown. Wing 78 to 91 mm.

**Range in Eastern Africa:** Eritrea and Abyssinia, British Somaliland and the southern Sudan as far west as the Imatong Mts.

**Habits:** Local and less common than the nominate race. It is recorded as breeding in coffee plantations at Harar, but otherwise seems to be a bird of light bush.

**Recorded breeding:** Eritrea, July. Harar, Abyssinia, May and June.

*Tchagra senegala remigialis* (Hartl. & Finsch.).
*Telephonus remigialis* Hartlaub & Finsch, Vög. Ost. Afr. in Van der Decken's Reise. 4, p. 34, 1870: Blue Nile, eastern Sudan.

**Distinguishing characters:** Differs from the nominate race in having the mantle tawny brown; below, whiter with sides of face, chest and flanks buff. Wing 83 to 92 mm.

**Range in Eastern Africa:** The Sudan from Darfur to the Blue Nile and Malakal.

## SHRIKES

**Habits:** Common locally in bush or woodland, a shy retiring bird with a quick flight from bush to bush. It has the same courtship flight and bell-like piping notes as the nominate race.

**Recorded breeding:** Sudan, May to August, also November.

**Distribution of other races of the species:** North and West Africa and Arabia.

### 1134 BROWN-HEADED BUSH-SHRIKE. *TCHAGRA AUSTRALIS* (Smith).

*Tchagra australis minor* (Reichw.).
*Telephonus minor* Reichenow, J.f.O. p. 64, 1887: Kagehi, Mwanza district, northern Tanganyika Territory.

**Distinguishing characters:** Differs from the Black-headed Bush-Shrike in having the centre of the head from forehead to nape brown, with a black stripe between this and the buffish stripe over the eye. The sexes are alike. Wing 63 to 80 mm. The young bird is duller than the adult and the stripe along the side of the crown and through the eye is dull black. There is no blackish in the brown of the head as in the young bird of the Black-headed Bush-Shrike.

**General distribution:** Eastern Kenya Colony to Tanganyika Territory, Nyasaland, and Portuguese East Africa as far south as Inhambane.

**Range in Eastern Africa:** Central and south-eastern Kenya Colony and central and eastern Tanganyika Territory to the Zambesi River.

**Habits:** A common species frequenting low bushes among damp grass, or tangled undergrowth at forest edge in low country. Rather local and usually in pairs with somewhat skulking habits, but in the breeding season the male rises into the air with whirring wings and then planes down like a Pipit, uttering a descending scale of musical notes. It also claps its wings loudly in short flights from bush to bush.

**Nest and Eggs:** A very neat cup of roots and fibres, rather like a Warbler's nest, usually in a low bush in long grass. Eggs two or three, white and rather round with irregular streaks and spots of purplish brown and lilac grey; about 22 × 16 mm.

**Recorded breeding:** Kenya Colony, Kikuyu, April to July, also January; southern areas, March and April. Tanganyika Territory,

January to March, also June. Nyasaland, October to March. Northern Portuguese East Africa (nestling), January. Rhodesia, September and October.

**Call:** A piping note, also a little whistling song 'chawee-chawy-chewo-chewoo' (Moreau), uttered as the bird descends to a perch. Alarm note a guttural 'charr.'

*Tchagra australis emini* (Reichw.). **Ph. xiv.**
*Telephonus australis emini* Reichenow, O.M. p. 60, 1893: Bukoba, north-western Tanganyika Territory.

**Distinguishing characters:** Differs from the preceding race in being darker brown above. Wing 73 to 83 mm.

**General distribution:** The Sudan, eastern Belgian Congo, Uganda, Western Kenya Colony and Tanganyika Territory.

**Range in Eastern Africa:** Southern Sudan and Uganda to western Kenya Colony and north-western Tanganyika Territory.

**Habits:** As for the last race, common and locally abundant in dense undergrowth or low bushes in long grass, but occurs more often in or around forest.

**Recorded breeding:** Uganda, December to February; (nestling) May. Western Kenya Colony, August to November.

**Distribution of other races of the species:** Western and southern Africa, the nominate race being described from Bechuanaland.

**1135** THREE-STREAKED BUSH-SHRIKE. *TCHAGRA JAMESI* (Shelley).
*Tchagra jamesi jamesi* (Shell.).
*Telephonus jamesi* Shelley, Ibis, p. 403, pl. 10, fig. 2, 1885: Goolis Mts., British Somaliland.

**Distinguishing characters:** Above, ashy brown; black stripes down crown from forehead to nape and through eye; wings chestnut, ends of feathers and innermost secondaries ash brown; rump and central tail feathers greyer, latter barred; rest of tail black with white ends to feathers; below, pale ashy grey, lighter on throat and belly. The sexes are alike. Wing 67 to 75 mm. The young bird is rather paler than the adult, sides of face washed with buff; stripe from forehead not extending to nape and duller black.

Plate 74

Black Cuckoo-Shrike (p. 556)
Male    Female
Purple-throated Cuckoo-Shrike (p. 559)
Female    Male
Red-naped Bush-Shrike (p. 614)    Emin's Shrike (p. 601)
Pink-footed Puff-Back (p. 622)
Female    Male
Chestnut-fronted Shrike (p. 574)
Grey-green Bush-Shrike (p. 633)
Red-billed Shrike (p. 572)    Retz's Red-billed Shrike (p. 574)
Grey-headed Bush-Shrike (p. 636)    Four-coloured Bush-Shrike (p. 634)    Rosy-patched Shrike (p. 639)
Female    Male

Gonolek (p. 607)
*Laniarius barbarus mfumbiri*

Black-headed Gonolek (p. 606)
*Laniarius erythrogaster*

Rufous-breasted Bush-Shrike (p. 631)
*Chlorophoneus rubiginosus bertrandi*

Black-fronted Bush-Shrike (3 phases)
*Chlorophoneus nigrifrons* (p. 632)

Black-fronted Bush-Shrike (p. 632) (2 phases)

Sulphur-breasted Bush-Shrike (p. 630)
*Chlorophoneus sulfureopectus similis*

Many-coloured Bush-Shrike (p. 629) (3 phases)
*Chlorophoneus multicolor batesi*

# SHRIKES

**Range in Eastern Africa:** British Somaliland to southern Abyssinia, south-eastern Sudan, Mt. Elgon, southern Italian Somaliland and Kenya Colony as far south as Taita.

**Habits:** As for other races, little recorded, but the bird is not uncommon in thorn scrub in the northern part of Kenya Colony in many localities.

**Nest and Eggs:** See under other races.

**Recorded breeding:** Southern Abyssinia, March to July.

*Tchagra jamesi mandanus* (Neum.).
*Telephonus jamesi mandanus* Neumann, O.M. p. 183, 1903: Manda Island, Kenya Colony.

**Distinguishing characters:** Differs from the nominate race in having the mantle more russet-brown, less ashy. Wing 72 mm.

**Range in Eastern Africa:** Eastern Kenya Colony at Witu, Lamu, Manda and Patta Islands.

**Habits:** Presumably as for other races but little is recorded. It inhabits dense low thorn scrub and Sansevieria.

**Recorded breeding:** Lamu, Kenya Colony, May.

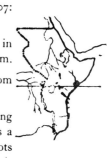

*Tchagra jamesi kismayensis* (Neum.).
*Telephonus jamesi kismayensis* Neumann, J.f.O. p. 369, 1907: Kismayu, southern Italian Somaliland.

**Distinguishing characters:** Differs from the nominate race in being lighter below with no ash-grey colour. Wing 63 to 68 mm.

**Range in Eastern Africa:** Southern Italian Somaliland from Umfudu to Kismayu.

**Habits:** A locally plentiful but shy secretive bird found among thick undergrowth with the habits of the genus. This bird has a remarkably coloured iris, said to be brown with a ring of white spots round the pupil which contract to minute specks if the bird is alarmed.

**Nest and Eggs:** Small flat cup nest of twigs and dry grass lined with rootlets placed in a thorn bush and usually fairly conspicuous. Eggs two or three, white and glossy with dark brown and violet spots and blotches; about 23 × 15 mm.

**Recorded breeding:** Jubaland, March and April.

**Call:** Very like that of the Brown-headed Bush-Shrike—a trill on a descending scale.

### 1136 BLACKCAP BUSH-SHRIKE. *BOCAGIA MINUTA* (Hartlaub).

*Bocagia minuta minuta* (Hartl.).

*Telephonus minutus* Hartlaub, P.Z.S. p. 292, 1858: Ashanti, Gold Coast, West Africa.

**Distinguishing characters:** Adult male, whole top of head slightly glossy black; rest of upperparts tawny with black markings on mantle and scapulars, forming more or less a black patch; wings chestnut; tail black with buffish ends; below, deep buff, throat white; bill black. The female has a white streak from the base of bill to over the eye. Wing 71 to 81 mm. The young bird has the centre of the crown tawny or mottled tawny and black, and the bill horn-colour, not black.

**General distribution:** West Africa from Sierra Leone and Portuguese Congo to Abyssinia, the Sudan, Uganda, Kenya Colony and Tanganyika Territory.

**Range in Eastern Africa:** Abyssinia, the southern Sudan, Uganda, Kenya Colony and the Kasulu, Kigoma and Tabora districts of Tanganyika Territory.

**Habits:** Those of other Bush-Shrikes, preferring in this case bushes in long swampy grass, or damp edges of forest. Widespread and not uncommon, but not easy to see. Frequently found in elephant grass climbing up a stem and chattering if alarmed. It has a short courtship flight as noted under the following race.

**Nest and Eggs:** Nest a neat well made cup of tendrils, stems and rootlets in the fork of a bush or woven on to upright stems, or in a stump, usually two or three feet from the ground. Eggs two occasionally three, white and glossy with streaks of reddish or purplish brown, and undermarkings of lavender, mostly in a zone at the larger end; about 23 × 17 mm.

**Recorded breeding:** Nigeria, July. Equatoria, October. Uganda, March, June and September to November. Kenya Colony, May and June. Tanganyika Territory, December to April.

**Call:** A sharp 'chup' probably of the male answered by a harsh bleating cry from the female. There is also a whistling song.

*Bocagia minuta anchietæ* (Boc.).
*Telephonus anchietæ* Bocage, Jorn. Lisboa, 2, p. 344, 1870: Pongo Andongo, Loanda district, northern Angola.

**Distinguishing characters:** Differs from the nominate race in having no black markings on the mantle. Wing 69 to 80 mm. The young bird has black centres to some of the feathers of the mantle.

**General distribution:** Angola to Tanganyika Territory, Nyasaland, Northern Rhodesia and eastern Southern Rhodesia.

**Range in Eastern Africa:** North-eastern and eastern Tanganyika Territory to Nyasaland.

**Habits:** As for the last race. It is not as shy a bird as most Bush Shrikes and more often sits on the top of a bush. It has a shrill whistling song of about four notes which is often uttered on the wing during the courtship flight, which is a sudden check in flight and a mount upwards on rapidly fluttering wings. On higher ground where swamps and long grass are not available it lives in bracken and low bush.

**Recorded breeding:** Tanganyika Territory, November to February. South-eastern Belgian Congo, February and March. Portuguese East Africa, December. Nyasaland, November. Rhodesia, October and November. Southern Belgian Congo, February and March.

**1137 MANY-COLOURED BUSH-SHRIKE. *CHLOROPHONEUS MULTICOLOR* (Gray).**
*Chlorophoneus multicolor batesi* Sharpe. **Pl. 75.**
*Chlorophoneus batesi* Sharpe, Ibis, p. 330, 1908: River Ja, Cameroon.

**Distinguishing characters:** Adult male, forehead, lores to around eye and ear-coverts black; forward part of crown white; stripe over eye white; centre of crown to mantle grey; lower mantle, rump and wings green, latter with yellow spots at ends of flight feathers; tail green or black with broad yellow or orange ends to feathers; below, lemon yellow, orange yellow or scarlet, but lower belly and under tail-coverts always more or less yellow; rarely chin to breast black uniform with forehead, lores and ear-coverts. The female differs from the male in having the forehead and lores white; ear-coverts grey; tail never wholly black. Wing 90 to 102 mm. The immature bird of both sexes is similar to the adult female but has yellow tips to the middle wing-coverts. The young bird has the

upper mantle green tipped with yellow, wing-coverts tipped with yellow, and underside yellow barred with dusky.

**General distribution:** Cameroons to Uganda and northern Angola.

**Range in Eastern Africa:** Western Uganda.

**Habits:** A species of forest or secondary growth, little seen and probably rare, though as it is an expert in keeping itself concealed it is easily overlooked.

**Nest and Eggs:** Undescribed.

**Recorded breeding:** West Africa (breeding condition), January.

**Call:** A harsh 'chak-chak' and a melodious whistling 'whoop' are the only calls described.

**Distribution of other races of the species:** West Africa, the habitat of the nominate race.

**1138** SULPHUR-BREASTED BUSH-SHRIKE. *CHLOROPHONEUS SULFUREOPECTUS* (Lesson).
*Chlorophoneus sulfureopectus similis* (Smith). **Pl. 75.**
*Melaconotus similis* A. Smith, Rep. Exp. Cent. Afr. p. 44, 1836: north of Kurrichane, Rustenburg district, western Transvaal.

**Distinguishing characters:** Adult male, forehead yellow or greenish yellow; crown to mantle grey; yellow stripe over eye; lores to ear-coverts black; rump, wings and tail green, latter with yellow ends to feathers; below, bright canary yellow; chest orange. The female has the lores and ear-coverts less black. Wing 81 to 97 mm. A colour phase has the grey extending over the rump to the central tail feathers; below, white, chest creamy buff; upper belly and flanks washed with creamy buff. The immature bird has the forehead and ear-coverts grey; lores and stripe over eye whitish; throat whitish or whitish and yellow; wing-coverts tipped with yellow; narrow yellow ends to tail feathers; breast yellow or washed with orange. The young bird is barred above and below with blackish.

**General distribution:** Abyssinia, Uganda and Kenya Colony to Angola, eastern Cape Province and Natal.

**Range in Eastern Africa:** Sudan and Abyssinia to the Zambesi River, also Mafia Island.

**Habits:** A rather shy and skulking species with a very wide distribution and not uncommon. Usually in pairs in fairly thick cover,

## SHRIKES

especially of riverside trees and bush, or among densely foliaged clumps of trees. Mainly a bird of low levels. Easily overlooked except for its call, it is surprisingly inconspicuous for its size and colour. Its food consists of insects of all kinds including hairy caterpillars.

**Nest and Eggs:** Nest a small frail and rather flat platform of twigs, tendrils and grass stems, not unlike a Dove's, placed at any height up to thirty feet in a fairly thick tree. Eggs two or three, green or greenish white spotted and blotched or streaked with dark brown and lilac and with pale grey undercloudings, often in a zone; about 23 × 16 mm.

**Recorded breeding:** Uganda, February and March. Kenya Colony, March and April. Tanganyika Territory, November to January. South-eastern Belgian Congo, September to November. Nyasaland, October to February. Portuguese East Africa, June, November to March. Rhodesia, September to November.

**Call:** A low piping whistle of about ten notes uttered with the head thrown back. In Tanganyika Territory and Portuguese East Africa the call is particularly loud and clear, and has been likened to a man whistling up a dog.

**Distribution of other races of the species:** West Africa, the nominate race being described from Senegal.

**1139  RUFOUS-BREASTED BUSH-SHRIKE.  *CHLOROPHONEUS RUBIGINOSUS* (Sundevall).**

*Chlorophoneus rubiginosus bertrandi* (Shell.). **Pl. 75.**

*Laniarius bertrandi* Shelley, Ibis, p. 15, pl. 2, fig. 2, 1894: Milanji Mt., southern Nyasaland.

**Distinguishing characters:** Forehead to hind neck grey; lores, faint stripe over eye and malar stripe white; spot above lores and through eye to ear-coverts and down sides of neck black; rest of upperparts including outer webs of flight feathers and tail green; below, chin to upper belly saffron; centre of belly white; flanks and under tail-coverts green; under wing-coverts, axillaries and inner edges of flight feathers bright yellow. The sexes are alike. Wing 81 to 87 mm. The young bird is grey around the eye, ear-coverts and down side of neck; underside from chin to upper belly greyish or buffish, narrowly barred with dusky.

**General distribution:** Nyasaland to eastern Southern Rhodesia.

**Range in Eastern Africa:** Nyasaland.

**Habits:** A bird of mountain forest, very shy and retiring and seldom seen. It has an extremely pleasant varied warbling song with some very Nightingale-like notes in it which reveal its presence at all times of year.

**Nest and Eggs:** Nest a small rather frail saucer of twigs lined with weed stems and tendrils in thick undergrowth a few feet from the ground. Eggs two, greenish or greyish white with speckles or streaks and blotches of reddish brown and lavender; about 24 × 18 mm.

**Recorded breeding:** Nyasaland, October and November. Portuguese East Africa, November and December.

**Call:** A series of clear whistling notes. Song as above.

**Distribution of other races of the species:** Western Nyasaland and South Africa, the nominate race being described from Durban, Natal.

1140 BLACK-FRONTED BUSH-SHRIKE. *CHLOROPHONEUS NIGRIFRONS* (Reichenow). **Pl. 75.**

*Laniarius nigrifrons* Reichenow, O.M. p. 95, 1896: Marangu, Kilimanjaro, north-eastern Tanganyika Territory.

**Distinguishing characters:** There are four main colour phases, scarlet, orange, apricot, and blackish green. Adult male, forehead, lores, through eye to ear-coverts and down side of neck black; crown to upper mantle grey; rest of upperparts green, including outer webs of flight feathers and tail, latter with yellow tips; below, chin to breast orange yellow, orange red, scarlet, black, apricot or buff, belly and under tail-coverts yellow, green or white; flanks olivaceous green or yellow; under wing-coverts and axillaries yellow (or green in the blackish green phase), edges of inner webs of flight feathers yellow. The female of the yellow phase is duller yellow; the females of the orange and scarlet phases have the chin to breast often more saffron, and the female of the apricot or buff-breasted phase has the flanks, belly and under tail-coverts greyer. The female of the black phase is unknown. Wing 83 to 95 mm. The immature bird has the forehead, lores, ear-coverts and side of neck grey; below, greenish yellow or buffish white in the apricot or buff-breasted phase, narrowly barred with dusky. The

# SHRIKES

young bird is similar to the immature but has the head, lores and ear-coverts green uniform with the mantle.

**General distribution:** Kenya Colony to south-eastern Belgian Congo, eastern Southern Rhodesia and the north-eastern Transvaal.

**Range in Eastern Africa:** Western and central Kenya Colony to the Zambesi River.

**Habits:** A most beautiful Shrike in any of its mutational phases. It is not confined to mountain forest in all localities, but is generally to be found therein and is more often heard than seen. Rare in Kenya Colony but further south not uncommon in its local habitats. It is often to be seen in company with a party of foraging Bulbuls among the creepers of secondary forest or forest edge, but is generally a bird of forest canopy not of undergrowth.

**Nest and Eggs:** Nest a shallow flimsy saucer of twigs and tendrils in creepers or dense vegetation at a considerable height from the ground. Eggs two, pale greenish heavily and evenly streaked with dark brown and chocolate on underlying streaks of mauve and bluish grey with a tendency to form a cap at the larger end; about $22 \cdot 5 \times 17$ mm.

**Recorded breeding:** Tanganyika Territory, October onwards, probably for most of the year. Nyasaland, July and August, also October and (breeding condition) December.

**Call:** A clear fluty whistle persistently repeated 'whu-koooo,' also a harsh scolding alarm note and a curious call of 'click-clack' like a lock shutting (Moreau). Zimmer reports also a creaking triple note which appears to be a breeding call but is also given by birds in immature plumage.

**1141  GREY-GREEN BUSH-SHRIKE.** *CHLOROPHONEUS BOCAGEI* (Reichenow). **Pl. 74.**

*Chlorophoneus bocagei jacksoni* (Sharpe). *Jackson's Bush-Shrike.*
*Dryoscopus jacksoni* Sharpe, Bull. B.O.C. 11, p. 57, 1901: Nandi, western Kenya Colony.

**Distinguishing characters:** Forehead, lores and stripe over eye white; crown to upper mantle, around eye, ear-coverts and down side of neck black; rest of mantle, rump and wings grey with an olive wash; tail black with narrow white tips to outer feathers; below, white, with a creamy or yellowish wash on breast, chest and

under wing-coverts. The sexes are alike. Wing 76 to 79 mm. The young bird has the top of the head greyish; mantle greenish; large pale tips to secondary coverts.

**Range in Eastern Africa:** Uganda and western Kenya Colony.

**Habits:** A bird of low forest or overgrown native plantations; very local and usually in pairs. It always appears to move slowly and deliberately about among branches. Not uncommon in Uganda, but rarer in Kenya Colony.

**Nest and Eggs:** Nest (of West African race) a shallow cup of vine tendrils. Eggs two, dull greenish white, smeared and blotched all over with olive brown and greyish stone; about 21 × 15·5 mm.

**Recorded breeding:** No records.

**Call:** Various. Van Someren records a long 'cheee,' a hunting call of 'twoo or twooi' as a duet, and a twanging note.

**Distribution of other races of the species:** West Africa and Angola, the nominate race being described from Cameroons.

**1142** FOUR-COLOURED BUSH-SHRIKE. *TELOPHORUS QUADRICOLOR* (Cassin).
*Telophorus quadricolor quadricolor* (Cass.). **Pl. 74.**
*Laniarius quadricolor* Cassin, Proc. Acad. Sci. Philad. p. 245, 1851: Durban, Natal, South Africa.

**Distinguishing characters:** Adult male, above, bright green; base of bill and over eye yellow or orange yellow; lores to under eye and down side of neck black; ear-coverts green; tail green or black and green; below, chin and neck rich scarlet sometimes tinged with orange; broad black band across chest joining up with black stripe from lores to neck; breast to under tail-coverts chrome or golden yellow, often with scarlet markings; flanks and under wing-coverts green. The female is usually less bright on breast and belly and has a slightly narrower chest band. In immature dress there is no black streak from lores to neck, the lores being yellow, and the black band across chest is narrower and often broken. Wing 78 to 85 mm. The young bird is yellowish green below as are the sides of the face; chin and throat washed with saffron; often faint barring on chest.

**General distribution:** Nyasaland to eastern Southern Rhodesia, Portuguese East Africa and Natal.

**Range in Eastern Africa:** Shiré River, Nyasaland.

**Habits:** Another beautiful but retiring species of thickets and dense coastal bush. Very little is recorded of its habits, but it has a loud distinctive melodious call uttered with nodding head which identifies it immediately.

**Nest and Eggs:** The nest is a loose saucer-shaped structure of twigs with a lining of leaf stalks, placed in a dense thicket. Eggs two or three, white or bluish white streaked and spotted with greyish brown and purplish grey; about 23 × 17 mm.

**Recorded breeding:** Nyasaland, April. Natal, October.

**Call:** A loud clear call described by Moreau as 'poo-poo-huery-huery' which is a very fair imitation, and the alarm note is a croaking growling sound. This refers actually to the following race, but there is probably little difference.

*Telophorus quadricolor nigricauda* (S. Clarke).
*Laniarius quadricolor nigricauda* Stephenson Clarke, Bull. B.O.C. 31, p. 32, 1913: Takaungu, eastern Kenya Colony.

**Distinguishing characters:** Differs from the nominate race in having more black in the tail. Wing 77 to 85 mm.

**Range in Eastern Africa:** Coastal area of Kenya Colony and Tanganyika Territory from the lower Tana River to the Rovuma River and inland to Taita and the Uluguru Mts.

**Habits:** As for the nominate race. Not uncommon in evergreen forests of foothills in eastern Tanganyika Territory.

**Recorded breeding:** No records.

**1143  DOHERTY'S BUSH-SHRIKE.   *TELOPHORUS DOHERTYI*** (Rothschild).
*Laniarius dohertyi* Rothschild, Bull. B.O.C. 11, p. 52, 1901: Kikuyu Escarpment, south-central Kenya Colony.

**Distinguishing characters:** Differs from the Four-coloured Bush-Shrike in having a broad band of scarlet on the forehead, the lores to over and under the eye and ear-coverts black, and a broader black band across chest. The female is similar to the male, but has a narrower black band across chest. Wing 79 to 81 mm. The young bird is yellowish green above, finely barred with blackish; upper tail-coverts barred with blackish; wing-coverts with yellowish tips; below, greenish yellow; throat to chest and flanks finely barred;

under tail-coverts dull red, tipped with yellowish and slightly barred; bill horn colour.

**General distribution:** Kivu area of eastern Belgian Congo to Uganda and Kenya Colony.

**Range in Eastern Africa:** Southern Uganda from the Ankole district to Mt. Elgon and Kenya Colony as far east as the Aberdare Mts. and Kikuyu.

**Habits:** A strikingly coloured species of dense cover at fairly high levels. Relatively common on Mt. Elgon at above 6,500 feet, living in dense cover of creepers and bamboos and spending much time on or near the ground. It is local but not uncommon in most of the Kenya Colony forests and is also found on the Kivu Volcanoes in similar cover.

**Nest and Eggs:** Undescribed.

**Recorded breeding:** No records.

**Call:** Fine and flute-like but no exact description given.

**1144** GREY-HEADED BUSH-SHRIKE. *MALACONOTUS BLANCHOTI* Stephens.
*Malaconotus blanchoti blanchoti* Steph. **Pl. 74.**
*Malaconotus blanchoti* Stephens, Gen. Zool. 13, p. 161, 1826: South Africa.

**Distinguishing characters:** Head, sides of face and upper mantle grey; lores white; rest of upperparts green with large pale yellow spots at ends of inner secondaries and tips of tail feathers; below, bright canary yellow; breast and chest saffron, or canary yellow; eye yellow; bill black. The sexes are alike. Wing 105 to 124 mm. The young bird has narrower yellow tips to the tail feathers and the bill is horn colour.

Rarely the whole upperside including wings and tail is grey with white spots at ends of inner secondaries and tips of tail feathers; edges of primaries white, not yellow; below, white; breast and chest buff; under wing-coverts and inner webs of flight feathers white, not yellow.

**General distribution:** Italian Somaliland to Nyasaland, eastern Cape Province and Natal.

**Range in Eastern Africa:** Southern Italian Somaliland to Kenya Colony, Tanganyika Territory and the Zambesi River.

**Habits:** A large striking-looking bird which yet manages to remain remarkably inconspicuous, though on occasions it will hawk from a perch like a true Shrike. Generally, however, it remains in the crown of a thick foliaged tree or bush and is only noticeable from its call. It appears to be somewhat migratory and Stoneham notes a definite movement through Uganda in July, the calls of the migrant birds being somewhat different from those of the local inhabitants.

**Nest and Eggs:** A rough platform of twigs with a rather flat inner cup of tendrils, or in places a rough mass of grass and leaves looking like an accumulation of rubbish, generally at a little height from the ground; often conspicuous but always firmly attached. Eggs two or three, greenish or pinkish white spotted and speckled with brown and lilac, forming a zone or cap at the larger end; about 29 × 20 mm.

**Recorded breeding:** Tanganyika Territory, October to December, also August. South-eastern Belgian Congo, September and October. Rhodesia, September or October. Nyasaland, September to January and April and May. Portuguese East Africa, August and September.

**Call:** Variable, a mellow whistle followed by a series of 'chaks' or a monotonous squealing pipe, or a rather ventriloquial three noted whistle. There is also a rattling call ending in a sort of sigh, and in this the bill starts by pointing straight upwards and is gradually depressed till it rests on the breast. Harsh rasping alarm note.

*Malaconotus blanchoti catharoxanthus* Neum.
*Malaconotus catharoxanthus* Neumann, J.f.O. p. 391, 1899: Bongo, Bahr-el-Ghazal, south-western Sudan.

**Distinguishing characters:** Differs from the nominate race in being rather larger. Wing 115 to 132 mm.

**General distribution:** Gambia and Sierra Leone to British Somaliland, Abyssinia, Uganda, south-eastern Belgian Congo, Angola and north-western Northern Rhodesia.

**Range in Eastern Africa:** The Sudan, Eritrea, Abyssinia and British Somaliland to Uganda; also Ukerewe Island, Lake Victoria.

**Habits:** As for the last race, with even more varied calls.

**Recorded breeding:** Nigeria, February, March, June and September. South-eastern Belgian Congo, September and October.

## SHRIKES

**1145 BLACK-CAP BUSH-SHRIKE.** *MALACONOTUS ALIUS* Friedmann.

*Malaconotus alius* Friedmann, Proc. N. Eng. Zool. Cl. 10, p. 5, 1927: Bagilo, Uluguru Mts., eastern Tanganyika Territory.

**Distinguishing characters:** Forehead to upper mantle, lores, round eyes, malar stripe and ear-coverts glossy blue-black: rest of upperparts including wings and tail dark green; below, sulphur-yellow; chest to under tail-coverts and flanks with an olivaceous wash; eyes brown; bill black. The female is more olivaceous below. Wing 104 to 107 mm. Juvenile plumage unrecorded.

**Range in Eastern Africa:** Bagilo, eastern Uluguru Mts., Tanganyika Territory.

**Habits:** Little known, except that it is usually found in the topmost branches of tall trees, often in company with the Grey-headed Bush Shrike.

**Nest and Eggs:** Undescribed.

**Recorded breeding:** No records.

**Call:** Described as a repeated 'ku-ku-kua-kua' (Andersen).

**1146 LAGDEN'S BUSH-SHRIKE.** *MALACONOTUS LAGDENI* (Sharpe).

*Laniarius lagdeni* Sharpe, P.Z.S. p. 54, pl. 5, 1884: Ashantee, Gold Coast Colony, West Africa.

**Distinguishing characters:** Differs from the Grey-headed Bush-Shrike in being generally darker; in having a heavier deeper bill; head, including lores and upper mantle, darker grey; the wing-coverts and subterminal ends of secondaries black; flanks green. The sexes are alike. Wing 114 mm. The young bird has a whiter chest.

**General distribution:** Gold Coast to western Uganda.

**Range in Eastern Africa:** Ruwenzori Mts., western Uganda.

**Habits:** Very little recorded, a forest species of tree tops found up to 9,000 feet on Mt. Ruwenzori.

**Nest and Eggs:** Undescribed.

**Recorded breeding:** No records.

**Call:** Unrecorded.

# SHRIKES

**1147 ROSY-PATCHED SHRIKE.** *RHODOPHONEUS CRUENTUS* (Hemprich & Ehrenberg).
*Rhodophoneus cruentus cruentus* (Hemp. & Ehr.). **Pl. 74.**
*Lanius cruentus* Hemprich & Ehrenberg, Symb. Phys. fol. c. pl. 3, fig. 1–3, 1828: Arkiko, near Massowa, Eritrea.

**Distinguishing characters:** Adult male, above, pale ashy brown with a very slight rosy wash; rump rosy; all except central tail feathers black with broad white ends; central tail feathers ashy brown; below, white with a rosy patch from centre of throat to centre of upper belly; flanks and under tail-coverts buffish. The female has the chin and throat white, a black gorget from the base of the lower mandible to breast and the rosy colour only from below the black gorget to the upper belly. Wing 87 to 98 mm. The young bird has buffish edges to the feathers of the upperparts, flight feathers, wing-coverts and tail; the young female has a dull blackish gorget.

**General distribution:** South-eastern Egypt at Gebel Elba to the Sudan and Eritrea.

**Range in Eastern Africa:** The Sudan from Kordofan to the Red Sea Province and Eritrea.

**Habits:** Birds of thorn scrub and aloes in desert country, usually in family parties and rather noisy. They spend a good deal of time on the ground and are extremely fast runners. Restless cheery inquisitive birds with a variety of ventriloquial notes.

**Nest and Eggs:** Nest of dry sticks and stems, very like a Dove's. Eggs two or three, pale greenish blue spotted and blotched with reddish brown and with lavender undermarkings; about $27 \times 19$ mm.

**Recorded breeding:** Red Sea Province Sudan, November.

**Call:** Normally a shrill piping chirp, but the birds include a very considerable variety of cries in their continual conversation to each other as they search a clump of bushes. Song uttered from a bush, a repeated 'do-me-so-ray' (Guichard).

*Rhodophoneus cruentus cathemagmenus* (Reichw.).
*Laniarius cathemagmenus* Reichenow, J.f.O. p. 63, 1887: Loeru, Kondoa-Irangi district, north-central Tanganyika Territory.

**Distinguishing characters:** Adult male, differs from the nominate race in being darker above with a distinct rosy wash, and the rosy colour of throat and breast a deeper shade; also this race usually has a black gorget in the male as well. The female can be distinguished from the nominate race by the darker upperside. Wing 89 to 98 mm.

**Range in Eastern Africa:** Southern Kenya Colony to eastern Tanganyika Territory as far south as the Kilosa area.

**Habits:** As for other races, locally common.

**Recorded breeding:** Tsavo, Kenya Colony (nestling), July.

*Rhodophoneus cruentus hilgerti* (Neum.).
*Pelicinius cruentus hilgerti* Neumann, O.M. p. 182, 1903: Sheik Hussein, British Somaliland.

**Distinguishing characters:** Adult male, differs from the nominate race in being darker above with a rosy wash, though not so pronounced as in the last race. In the female the black gorget extends a little further up the throat. Wing 88 to 102 mm.

**Range in Eastern Africa:** Eastern and southern Abyssinia to French and British Somalilands and central Kenya Colony.

**Habits:** As for the nominate race. Lort-Phillips records small parties performing antics in the evening, chasing each other round a stone with outstretched wings and tail and bobbing to one another on a bough while uttering a sharp metallic double note.

**Recorded breeding:** Southern Abyssinia, May, and November to January.

**1148** NICATOR. *NICATOR CHLORIS* (Valenciennes).
*Nicator chloris chloris* (Val.).
*Lanius chloris* Valenciennes, Dict. Sci. Nat. 40, p. 226, 1826: Galam, Senegal.

**Distinguishing characters:** Above, including side of face and tail green; a white spot above lores; wings dusky with green edges and large yellow spots to wing-coverts and secondaries; below, chin to chest slightly olivaceous grey; breast to belly greyish white; under wing-coverts, axillaries, inner webs of flight feathers, under tail-coverts and tips of tail feathers yellow. The sexes are alike, but the female sometimes lacks the white spot above the lores. Wing 85 to 114 mm. The young bird is rather paler above; has yellow tips to the primaries and rather narrower and more pointed outer tail feathers.

**General distribution:** Senegal to the Belgian Congo, the Sudan and Uganda.

**Range in Eastern Africa:** The Southern Sudan and Uganda.

**Habits:** A common forest species of undergrowth and the lower boughs of forest trees. Shy and difficult to see, the bird sits motionless in cover and suddenly gives a burst of song of considerable richness and power. Usually found singly, and except for its song and many different calls is entirely inconspicuous.

**Nest and Eggs:** The nest is a flat pad of stalks, tendrils and rootlets or a rough flat cup of grass in a tree or bush in dense cover. Eggs usually two, dull pinkish cream or putty-coloured with large blotches of brown and lilac in a zone at the larger end; about 25 × 16 mm.

**Recorded breeding:** Ashanti, February. Uganda, April, June and July.

**Call:** Very varied. Besides its normal song it has a quiet whispering sub-song, and many nasal and guttural notes, some of them remarkably Squirrel-like. Alarm note is a loud ringing 'zokh.'

*Nicator chloris gularis* Hartl. & Finsch.
*Nicator gularis* Hartlaub & Finsch, Van der Decken Reise, 4, Vög.
  Ostafr. p. 360, 1870: Shupanga, Zambesi River, east of the Shiré River mouth, Portuguese East Africa.

**Distinguishing characters:** Differs from the nominate race in having the top of the head olivaceous grey; a yellow spot above lores; sides of face grey or buffish grey; chin to chest buff and pale brown. The female often lacks the yellow spot above the lores. Wing 87 to 115 mm.

**General distribution:** Kenya Colony to Northern Rhodesia; Nyasaland and Zululand.

**Range in Eastern Africa:** Northern Kenya Colony and Tanganyika Territory to the Zambesi River, also Zanzibar Island.

**Habits:** As for the nominate race, but widely and more sparsely distributed. The song, which appears to increase in power in the eastern parts of its range, is loud and clear of beautiful liquid notes like that of a *Cossypha*. A bird of lower levels as a rule and notable for its curious nervous wing-flicking.

**Recorded breeding:** Portuguese East Africa, December to April. Zanzibar Island (breeding condition), January. Rhodesia, December and January.

## 1149 YELLOW-THROATED NICATOR. *NICATOR VIREO* Cabanis.

*Nicator vireo* Cabanis, J.f.O. p. 333, pl. 2, fig. 2, 1876: Chinchoxo, Portuguese Congo.

**Distinguishing characters:** Adult male, similar to the Nicator, but smaller and with a yellow throat. The sexes are alike. Wing 72 to 87 mm. The young bird has the forehead green not grey.

**General distribution:** Cameroons to northern Angola and Uganda.

**Range in Eastern Africa:** Bwamba area, western Uganda.

**Habits:** A species freely heard but less often seen, as it inhabits high leafy trees and dense creepers. It does not appear to differ from the preceding larger species in habits.

**Nest and Eggs:** Undescribed.

**Recorded breeding:** No records.

**Call:** A strikingly loud and pleasing song of half a dozen bugle-like notes, the middle ones rather higher. 'Chock-choi-choi-choi-chu-chu" (Van Someren).

Names in Sclater's *Syst. Av. Æthiop.* 2, 1930, which have been changed, or have become synonyms in this work:

*Eurocephalus rüppelli erlangeri* Zedlitz, treated as synonymous with *Eurocephalus anguitimens rüppelli* Bonaparte.
*Eurocephalus rüppelli deckeni* Zedlitz, treated as synonymous with *Eurocephalus anguitimens rüppelli* Bonaparte.
*Eurocephalus rüppelli böhmi* Zedlitz, treated as synonymous with *Eurocephalus anguitimens rüppelli* Bonaparte.
*Nilaus minor ruwenzorii* Bannerman, treated as synonymous with *Nilaus afer massaicus* Neumann.
*Nilaus afer erythreæ* Neumann, treated as synonymous with *Nilaus afer afer* (Latham).
*Nilaus afer hilgerti* Neumann, treated as synonymous with *Nilaus afer afer* (Latham).
*Lanius leucopygus* Heuglin, treated as synonymous with *Lanius elegans elegans* Swainson.
*Lanius excubitorius intercedens* Neumann, treated as synonymous with *Lanius excubitorius böhmi* Reichenow.
*Lanius antinorii antinorii* Salvadori, now *Lanius somalicus* Hartlaub.
*Lanius antinorii mauritii* Neumann, treated as synonymous with *Lanius somalicus* Hartlaub.
*Laniarius funebris rothschildi* Neumann, treated as synonymous with *Laniarius funebris* Hartlaub.
*Laniarius funebris degener* Hilgert, treated as synonymous with *Laniarius funebris* Hartlaub.
*Laniarius alboplagatus* (Jackson) treated as synonymous with *Laniarius funebris* Hartlaub.

Plate 76

Yellow Penduline Tit (p. 659)
*Anthoscopus parvulus parvulus*

Sennar Penduline Tit (p. 659)
*Anthoscopus punctifrons*

Southern Black Tit
*Parus niger* (p. 649)
Female     Male

Black Tit (p. 650)
*Parus leucomelas leucomelas*

Red-throated Tit (p. 654)
*Parus fringillinus*

African Penduline Tit (p. 655)
*Anthoscopus caroli sylviella*

African Penduline Tit (p. 656)
*Anthoscopus caroli roccatii*

Mouse-coloured Penduline Tit (p. 658)
*Anthoscopus musculus musculus*

Dusky Tit (p. 653)
*Parus funereus*

Cinnamon-breasted Tit (p. 654)
*Parus rufiventris pallidiventris*

Plate 77

Golden Oriole (p. 662)
*Oriolus oriolus oriolus*
Male            Female

Green-headed Oriole (p. 669)     Black-winged Oriole (p. 668)
*Oriolus chlorocephalus chlorocephalus*     *Oriolus nigripennis percivali*

Black-headed Forest Oriole     Black-headed Oriole (p. 665)
*Oriolus monacha monacha* (p. 664)     *Oriolus larvatus rolleti*

African Golden Oriole (p. 664)
*Oriolus auratus notatus*
Male            Female

# SHRIKES

*Laniarius nigerrimus* (Reichenow) a melanistic phase of and treated as synonymous with *Laniarius æthiopicus sublacteus* (Cassin).
*Laniarius ferrugineus somaliensis* Reichenow, treated as synonymous with *Laniarius æthiopicus erlangeri* Reichenow.
*Laniarius ruficeps cooki* Van Someren, treated as synonymous with *Laniarius ruficeps kismayensis* (Erlanger).
*Dryoscopus gambensis nyanzæ* Neumann, treated as synonymous with *Dryoscopus gambensis malzacii* (Heuglin).
*Tchagra australis congener* (Reichenow) treated as synonymous with *Tchagra australis minor* (Reichenow).
*Tchagra australis littoralis* (Van Someren) treated as synonymous with *Tchagra australis minor* (Reichenow).
*Tchagra senegala armena* (Oberholser) treated as synonymous with *Tchagra senegala senegala* (Linnæus).
*Tchagra senegala orientalis* (Cabanis) treated as synonymous with *Tchagra senegala senegala* (Linnæus).
*Tchagra senegala catholeuca* (Neumann) treated as synonymous with *Tchagra senegala senegala* (Linnæus).
*Tchagra senegala mozambica* (Van Someren) treated as synonymous with *Tchagra senegala senegala* (Linnæus).
*Tchagra senegala erlangeri* (Neumann) treated as synonymous with *Tchagra senegala habessinica* (Hemprich and Ehrenberg).
*Tchagra senegala warsangliensis* Clarke, treated as synonymous with *Tchagra senegala habessinica* (Hemprich and Ehrenberg).
*Chlorophoneus melamprosopus* (Reichenow) treated as synonymous with *Chlorophoneus multicolor batesi* Sharpe.
*Chlorophoneus rubiginosus münzneri* Reichenow, treated as synonymous with *Chlorophoneus nigrifrons* (Reichenow).
*Chlorophoneus nigrifrons elgeyuensis* Van Someren, treated as synonymous with *Chlorophoneus nigrifrons* (Reichenow).
*Chlorophoneus nigrifrons manningi* (Shelley) treated as synonymous with *Chlorophoneus nigrifrons* (Reichenow).
*Chlorophoneus nigrifrons conceptus* Hartert, treated as synonymous with *Chlorophoneus multicolor batesi* Sharpe.
*Chlorophoneus andaryæ* Jackson, treated as synonymous with *Chlorophoneus sulfureopectus similis* (Smith).
*Telophorus quadricolor intercedens* (Reichenow) treated as synonymous with *Telophorus quadricolor nigricauda* (Clarke).
*Malaconotus poliocephalus poliocephalus* (Lichtenstein) now *Malaconotus blanchoti blanchoti* Stephens.
*Malaconotus poliocephalus approximans* (Cabanis) treated as synonymous with *Malaconotus blanchoti blanchoti* Stephens.
*Malaconotus poliocephalus hypopyrrhus* Hartlaub, treated as synonymous with *Malaconotus blanchoti blanchoti* Stephens.
*Rhodophoneus cruentus kordofanicus* Sclater and Praed, treated as synonymous with *Rhodophoneus cruentus cruentus* (Hemprich and Ehrenberg).

Names introduced since 1930 and which have become synonyms in this work:

*Nilaus afer brevialatus* Grote, 1938, treated as synonymous with *Nilaus nigritemporalis* Reichenow.
*Corvinella corvina chapini* Friedmann and Bowen, 1933, treated as synonymous with *Corvinella corvina affinis* Hartlaub.
*Laniarius ferrugineus chyulu* Van Someren, 1939, treated as synonymous with *Laniarius æthiopicus ambiguus* Madarasq.
*Chlorophoneus sulfureopectus fricki* Friedmann, 1930, treated as synonymous with *Chlorophoneus sulfureopectus similis* (Smith).
*Chlorophoneus nigrescens* W. L. Sclater, 1931, treated as synonymous with *Chlorophoneus nigrifrons nigrifrons* (Reichenow).

## Addenda and Corrigenda

**1084** *C.q. martini.* General distribution: add North-eastern Northern Rhodesia.

**1088** *D.a. adsimilis.* Recorded breeding: add Eritrea and British Somaliland, June and July.

p. 581. *Malaconotus blanchoti* now *M. hypopyrrhus*.

**1098 & 1099** *Nilaus afer* and *Nilaus temporalis* are now treated as conspecific.

**1100** *Lanius elegans dubarensis.* Recorded breeding: add 'feeding young in March'.

**1114** *Lanius cristatus phœnicuroides* should be *Lanius cristatus speculigerus* Tscharowski, J.f.O., p. 322, 1874: Alt Tsaruchartui, eastern Siberia, Russia.

**1125** *Laniarius æthiopicus æthiopicus.* Delete from distribution the coastal area of British Somaliland.

**1133** Extends only to eastern Cape Province.

**1137** *Chlorophoneus multicolor.* It seems possible from recent research that this species is conspecific with 1140 *Chlorophoneus nigrifrons*.

**1139** *Chlorophoneus rubiginosus bertrandi* now *Chlorophoneus olivaceus bertrandi*.

**1144** For *Malaconotus blanchoti* read *Malacanotus hypopyrrhus* Hartl.
*Malaconotus hypopyrrhus hypopyrrhus* Hartl.
*Malaconotus hypopyrrhus* Hartlaub, Syst. Verz. Bremen, Ab. 1, Vög., p. 61, 1844: restricted type locality Durban, Natal. And delete from *General distribution* all areas north of southern Tanganyika Territory. The northern birds are *Malaconotus blanchoti approximans* Cabanis, in Van der Decken's Reise, 3, pt. 1, p. 27, 1869: Delaon River, Usambara.

**1147** *Rhodophoneus cruentus.* Recorded breeding: add Eritrea, January to March.
*R.c. hilgerti.* Recorded breeding: add British Somaliland, April and May.

p. 643. Delete *Malaconotus poliocephalus approximans*.

*Continued on p. 723*

## GREY HYPOCOLIUS

FAMILY—**HYPOCOLIIDÆ.**   Genus: *Hypocolius.*

A bird of Shrike-like habits but of uncertain affinities, usually found in small flocks, though not constantly so. The flight is undulating and at first sight it might in the field be mistaken for a Grey-backed Fiscal.

**1150  GREY HYPOCOLIUS.   *HYPOCOLIUS AMPELINUS*** Bonaparte.

*Hypocolius ampelinus* Bonaparte, Consp. 1, p. 336, 1850: River Tigris, Iraq.

**Distinguishing characters:** Adult male, general colour isabelline grey; around eye and ear-coverts to nape black; ear-coverts with a silvery wash; primaries black, with white or greyish and white ends; black ends to tail feathers. In immature dress the tips of the primaries are dusky. The female lacks the black around the eye and from the ear-coverts to nape; primaries grey with black subterminal ends and white tips. Wing 96 to 103 mm. The young bird differs from the adult female in having the primaries wholly isabelline grey and dusky.

**General distribution:** Iraq; in non-breeding season to Iran, India as far south as Bombay and Arabia to about lat. 21° N. Two records from Africa: Eritrea 1850 and Gebel Elba, south-eastern Egypt, 1943.

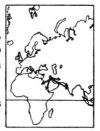

**Range in Eastern Africa:** One record, Massawa, Eritrea, as above.

**Habits:** A native of Iraq, and, as far as we are aware, there are only the above two records from Africa. In north-western India and Iran this species has been recorded in small flocks. The flight is Shrike-like, but it often perches in the centre of trees. Its food consists of insects, berries and dates, and it has a squeaking call. In Iraq it breeds among date-palms, making a cup-shaped nest and laying four dull white eggs with slate-coloured blotches; about 22 × 17 mm.

# FAMILY—**PARIDÆ**. TITS. Genera: *Parus* and *Anthoscopus*.

Thirteen species occur in Eastern Africa. Tits are found in all types of wooded country. They are very active little birds and their usual method of feeding is to search bark or branches for insects, and in doing so they assume a number of curious positions, as often as not upside down. They are always inquisitive and usually tame. The food is mainly insects and their larvæ, though small berries and fruits are also eaten.

Four species belong to the Penduline Tits, a group of small birds with short tails and short sharply pointed bills. Their nests and nesting habits are entirely different from the true Tits, but their general habits of feeding and sociability are the same.

## Key to the Adult Tits of Eastern Africa

| | | | |
|---|---|---|---|
| 1 | Size smaller, wing 58 mm. and under: | | 3–8 |
| 2 | Size larger, wing 63 mm. and over: | | 9–21 |
| 3 | Above, yellowish-green: | | 4–5 |
| 4 | Below, yellow: | YELLOW PENDULINE TIT *Anthoscopus parvulus* | **1163** |
| 5 | Below, warm creamy white: | SENNAR PENDULINE TIT *Anthoscopus punctifrons* | **1162** |
| 6 | Above, olive green or grey washed with olive green: | | 7–8 |
| 7 | Below, lower belly and under tail-coverts more or less buffish or tawny: | AFRICAN PENDULINE TIT *Anthoscopus caroli* | **1160** |
| 8 | Below, creamy white: | MOUSE-COLOURED PENDULINE TIT *Anthoscopus musculus* | **1161** |
| 9 | Mantle buffish white: | WHITE-BACKED BLACK TIT *Parus leuconotus* | **1156** |
| 10 | Mantle slate-grey with a greenish gloss: | DUSKY TIT *Parus funereus* | **1157** |

# TITS

| | | |
|---|---|---|
| 11 Mantle grey: | | 13–17 |
| 12 Mantle black: | | 18–21 |
| 13 Sides of head isabelline brown: | RED-THROATED TIT *Parus fringillinus* | 1159 |
| 14 Sides of head white: | GREY TIT *Parus afer* | 1151 |
| 15 Sides of face black uniform with rest of head: | | 16–17 |
| 16 Belly isabelline: | CINNAMON-BREASTED TIT *Parus rufiventris* | 1158 |
| 17 Belly greyish: | STRIPE-BREASTED TIT *Parus fasciiventer* | 1152 |
| 18 Breast to belly white: | WHITE-BREASTED TIT *Parus albiventris* | 1155 |
| 19 Breast and upper belly black: | | 20–21 |
| 20 Lower belly and under tail-coverts black: | BLACK TIT *Parus leucomelas* | 1154 |
| 21 Lower belly and under tail-coverts black and white: | SOUTHERN BLACK TIT *Parus niger* | 1153 |

**1151 GREY TIT. *PARUS AFER* Gmelin.**
*Parus afer griseiventris* Reichw.
*Parus griseiventris* Reichenow, J.f.O. p. 210, 1882: Kakoma, Tabora district, Tanganyika Territory.

**Distinguishing characters:** Top of head black, slightly glossy; patch on occiput, base of bill to ear-coverts and sides of neck white; mantle, rump and wing shoulders grey; median and greater wing-coverts and edges of flight feathers white; tail and upper tail-coverts black, former with outer web of outer feathers and tips of all feathers white; below, chin to neck slightly glossy black; rest of underparts pale buffish grey, with some black down the centre. The sexes are alike. Wing 74 to 83 mm. The young bird has the top of the head and chin to neck dull black and edges of flight feathers buffish.

**General distribution:** Tanganyika Territory to southern Belgian Congo, Northern Rhodesia, Nyasaland west of the Shiré River, eastern Bechuanaland and eastern Southern Rhodesia.

**Range in Eastern Africa:** Tanganyika Territory from the Tabora district to the south-western areas.

**Habits:** Generally common and well-distributed in small parties, scolding and chattering as they pass through the bush. Mostly found in open woodland in bush country and they are frequently members of mixed bird foraging parties. The habits of Tits are similar in whatever part of the world they are found.

**Nest and Eggs:** In a hole of a tree, bank or anthill, on a thick pad of soft material. Eggs usually three, white and rather glossy, spotted with brick red and with pale purplish grey underspotting; about 17·5 × 14 mm.

**Recorded breeding:** South-eastern Belgian Congo, September. Nyasaland, September and October. Southern Rhodesia, September to December.

**Call:** A 'chissik' often followed by three notes, with a more elaborate song-call of 'see-oo' repeated with variations.

*Note:* This species is said to be often parasitized by the Lesser Honey Guide.

*Parus afer thruppi* Shell.
*Parus thruppi* Shelley, Ibis, p. 406, pl. 11, fig. 2, 1885: Somaliland, east of long. 45° E. and north of lat. 5° N.

**Distinguishing characters:** Differs from the preceding race in having the black of the head and throat meeting at the sides of the neck. Wing 61 to 70 mm.

**Range in Eastern Africa:** Abyssinia and British and Italian Somalilands to Kenya Colony and north-eastern Tanganyika Territory.

**Habits:** As for the preceding race, but confined to dry bush veld and acacia scrub as a rule. Locally abundant wherever there are riverside trees in its desert habitat. It has several calls, among them a particularly harsh note, and a pleasant little warbling song.

**Recorded breeding:** Southern Abyssinia (breeding condition), February.

**Distribution of other races of the species:** South Africa, the nominate race being described from the Cape of Good Hope.

## TITS

**1152 STRIPE-BREASTED TIT.** *PARUS FASCIIVENTER* Reichenow.

*Parus fasciiventer* Reichenow, O.M. p. 31, 1893: Ruwenzori Mts.

**Distinguishing characters:** Very similar to the Grey Tit from which it differs in having the whole head and neck black. The sexes are alike. Wing 75 to 82 mm. The young bird has the head and chest brownish.

**General distribution:** Eastern Belgian Congo to Uganda.

**Range in Eastern Africa:** Ruwenzori Mts. area, western Uganda.

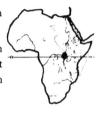

**Habits:** Of typical Tit-like habits, and usually tame and found in small parties, often with other species of birds. Locally abundant and found both in forest at 6,500 feet and up to the Tree-Heath zone at 11,000 feet.

**Nest and Eggs:** Undescribed.

**Recorded breeding:** Ruwenzori Mts. (nestling), December.

**Call:** A sharp 'chit-chit' with an occasional 'hee-i-pri chew' and a harsh scolding note (Van Someren).

**1153 SOUTHERN BLACK TIT.** *PARUS NIGER* Vieillot. **Pl. 76.**

*Parus niger* Vieillot, N. Dict. d'Hist. Nat. 20, p. 325, 1818: Sondag River, eastern Cape Province, South Africa.

**Distinguishing characters:** Adult male, general colour glossy blue black; rather slaty from breast to belly; edges of wing-coverts, flight feathers, edge of outer web of outer tail feathers and tips of all others white; lower belly and under tail-coverts black and white. The female is duller above and dull slaty grey below. Wing 74 to 88 mm. The young bird is similar to the adult female, but has a yellowish wash on the white of the wing-coverts and flight feathers.

**General distribution:** Northern Rhodesia as far west as Lusaka, Monze, Senanga and Sesheke to Nyasaland, Portuguese East Africa south of Lurio River, Damaraland, Bechuanaland, eastern Cape Province and Natal.

**Range in Eastern Africa:** Northern Portuguese East Africa from the Lurio River and Nyasaland to the Zambesi River.

**Habits:** Locally common in drier bush country and woodland in pairs and small parties. Typical Tit-like habits and consorts freely with other birds; rather noisy and always on the move. Possibly rather shyer than the Grey Tit.

**Nest and Eggs:** In a hole in a tree stump, or similar situation. Eggs three or four, creamy white well-spotted with bright rufous; about 18·5 × 14 mm.

**Recorded breeding:** Nyasaland, October and November. Rhodesia, September to January. Fairly often parasitized by Honey Guides.

**Call:** A considerable variety of notes including a trilling call, a rasping cry and a call of 'chip-wee.'

**1154 BLACK TIT.** *PARUS LEUCOMELAS* Rüppell.
*Parus leucomelas leucomelas* Rüpp. **Pl. 76.**
*Parus leucomelas* Rüppell, N. Wirbelt. Vög. p. 100, pl. 37, 1840: Halei Province, Temben, northern Abyssinia.

**Distinguishing characters:** Wholly black with a violet gloss, but with white outer webs to the greater wing-coverts and white edges to flight feathers; under wing-coverts black and white; narrow white edging and tips to outer tail feathers. The sexes are alike. Wing 74 to 92 mm. The young bird is duller than the adult with little or no gloss below and has the white on the wings washed with yellow.

**Range in Eastern Africa:** Eritrea, eastern Sudan in the Sennar area and Abyssinia to western Kenya Colony.

**Habits:** Locally plentiful in all types of suitable country from dry bush to dense woodland. Habits as for other Tits and is often with other species in a mixed bird party. Occasionally seen to take insects on the wing.

**Nest and Eggs:** In holes of trees or occasionally in buildings. Eggs typically Tit-like, up to five in number, pinkish white with rufous or brownish spots; about 20 × 14 mm.

**Recorded breeding:** Abyssinia, August.

**Call:** A great variety of calls including a clear bell-like piping call rather like some notes of a Thrush.

*Parus leucomelas insignis* Cab.
*Parus (Pentheres) insignis* Cabanis, J.f.O. p. 419, 1880: Malandje, northern Angola.

**Distinguishing characters:** Differs from the nominate race in having a more steel blue, not violet, gloss, and from the Southern Black Tit in having the lower belly and under tail-coverts black and only a narrow edging of white to the outer web of the outer tail feathers and tail tips, or tail wholly black. Wing 84 to 96 mm. The young bird has little or no yellow wash on the white of the wings.

**General distribution:** Portuguese Congo, Angola, Tanganyika Territory, Nyasaland and Damaraland.

**Range in Eastern Africa:** Tanganyika Territory from Iringa and Bagamoyo districts to south-western areas.

**Habits:** Common in open woodland in pairs or small parties, rather shyer than other Tits and keeping to taller trees. The call, according to Vincent, is a harsh 'twiddy-zeet-zeet-zeet' and occasionally a male gives a strange buzzing twitter ending on a shriller note 'zeu-zeu-zeu-twit.'

**Recorded breeding:** Tanganyika Territory highlands, September to November. Njombe, Tanganyika Territory, December. South-eastern Belgian Congo, October. Northern Rhodesia, October. Nyasaland, September to November. Portuguese East Africa (about to breed), March.

*Parus leucomelas guineensis* Shell.
*Parus leucomelas guineensis* Shelley, Bds. Afr. 2, p. 229, 1900: Volta River, Gold Coast, West Africa.

**Distinguishing characters:** Similar to the nominate race in colour but differs in having the under wing-coverts wholly white. Wing 71 to 88 mm.

**General distribution:** Senegal and Sierra Leone to French Equatorial Africa, the Sudan, northern Belgian Congo and Uganda.

**Range in Eastern Africa:** Western and southern Sudan to Uganda.

**Habits:** Widely distributed in woodland and forests, not usually common. Pairs or small parties usually keeping to higher trees and searching for food in crevices of bark; they also take insects on the wing. The call is described as a pleasant piping with several harsher jarring notes.

**Recorded breeding:** Darfur, April and May. Uganda, June, also March.

## 1155 WHITE-BREASTED TIT.  *PARUS ALBIVENTRIS*
Shelley. **Ph. xv.**

*Parus albiventris* Shelley, Ibis, p. 116, 1881: Ugogo, Dodoma district, central Tanganyika Territory.

**Distinguishing characters:** Adult male, general colour above similar to the Black Tit, but the breast to under tail-coverts wholly white and the black of the chin and chest extending down the flanks. The female is duller above and has the chin to chest dull sooty black. Wing 73 to 87 mm. The young bird has a yellow wash on the white edges to the flight feathers.

**General distribution:** Cameroons to the Sudan, Uganda, Kenya Colony and Tanganyika Territory.

**Range in Eastern Africa:** Southern half of the Sudan to Uganda, Kenya Colony and Tanganyika Territory as far south as the Ufipa Plateau and Iringa district.

**Habits:** Widely distributed in acacia steppe, woodland, forest and coastal scrub. Usually in pairs but quite large parties have been noticed. They have several times been recorded as visiting flowers, like Sunbirds, presumably for insects. Often seen in gardens, restless, active and noisy. On the Kenya Colony coast they are found among Mango trees, and are said to have quite a fine song. Roost in holes in trees.

**Nest and Eggs:** Nest a cup of hair, bark fibre and vegetable down placed in decaying stumps, holes or crevices in bark of trees usually at some height from the ground. Eggs three to five, white with rather sparse maroon spotting more concentrated at the larger end. No measurements available.

**Recorded breeding:** Kenya Colony highlands, March to June, also December. Tanganyika Territory (lower and middle uplands) October to December.

**Call:** A dry twittering trill, and a soft clicking feeding note. There is also a soft little song in the breeding season 'pee-pee-purr' (Couchman and Elliott).

TITS 653

**1156** WHITE-BACKED BLACK TIT. *PARUS LEUCONOTUS* Guérin.

*Parus leuconotus* Guérin, Rev. Mag. Zool. p. 162, 1843: Northern Abyssinia.

**Distinguishing characters:** Adult male, general colour, including wings and tail, glossy blue black; mantle buffish white. The female is duller than the male. Wing 71 to 81 mm. The young bird is a more dusky black than the adult female.

**Range in Eastern Africa:** Northern, central and eastern Abyssinia.

**Habits:** Little recorded. It is usually found in wooded mountain valleys and gorges from 6,000 feet upwards.

**Nest and Eggs:** Undescribed.

**Recorded breeding:** No records.

**Call:** Said to be like that of the Grey Tit.

**1157** DUSKY TIT. *PARUS FUNEREUS* (Verreaux). **Pl. 76.**
*Melanoparus funereus* J. & E. Verreaux, J.f.O. p. 104, 1855: Gabon.

**Distinguishing characters:** Adult male, general colour including wings and tail blackish slate grey with a greenish gloss; wing-coverts sometimes tipped with white. The female is greyer, less blackish. Wing 82 to 90 mm. The young bird is duller than the adult female, often with a brownish wash on throat to chest and sides of face.

**General distribution:** Cameroons and Gabon to Uganda, the Sudan and Kenya Colony.

**Range in Eastern Africa:** Uganda and south-eastern Sudan to western Kenya Colony.

**Habits:** Forest Tits of high tree tops usually in pairs or small parties. They would be frequently overlooked except for a piping call which is freely uttered.

**Nest and Eggs:** Undescribed.

**Recorded breeding:** No records.

**Call:** A churring alarm cry.

**1158** CINNAMON-BREASTED TIT. *PARUS RUFIVENTRIS*
Bocage.
*Parus rufiventris pallidiventris* Reichw. **Pl. 76.**
*Parus pallidiventris* Reichenow, J.f.O. p. 217, 1885: Kakoma, Tabora district, west-central Tanganyika Territory.

**Distinguishing characters:** Head and neck all round blue black; mantle to upper tail-coverts dark grey; wings blackish with white edges to coverts and flight feathers; tail black, with white edges to outer web of outer tail feathers and tips of all others; below, chest and breast grey; belly and under tail-coverts isabelline. The sexes are alike. The pale yellow iris is noticeable at close quarters. Wing 75 to 87 mm. The young bird is duller with a dull blackish head and a slight yellow wash on the white edges of the flight feathers.

**General distribution:** Tanganyika Territory to Nyasaland and eastern Southern Rhodesia.

**Range in Eastern Africa:** The Tabora district of Tanganyika Territory to Portuguese East Africa and Nyasaland.

**Habits:** A common species of open woodland at higher levels. Pairs or small parties, and often in mixed bird parties with typical Tit-like habits. Appear to be locally migratory.

**Nest and Eggs:** Nest of fine grass, fibre or felt-like material in a hole in a tree. Eggs four, whitish, well freckled with rufous; about 18 × 14 mm.

**Recorded breeding:** Nyasaland, October and probably later.

**Call:** A two-syllabled cry of 'chick-wee' also a rasping call.

**Distribution of other races of the species:** Angola and Damaraland to northern Nyasaland; the nominate race being described from Caconda, Angola.

**1159** RED-THROATED TIT. *PARUS FRINGILLINUS*
Fischer and Reichenow. **Pl. 76.**
*Parus fringillinus* Fischer & Reichenow, J.f.O. p. 56, 1884: Base of Mt. Meru, Arusha district, north-eastern Tanganyika Territory.

**Distinguishing characters:** Adult male, forehead, sides of head, neck all round and chin to breast isabelline brown; crown of head blackish with grey edges to the feathers; mantle to rump grey; wings black with white edges to wing-coverts and flight feathers; tail black with white edges to outer tail feathers and tips to all feathers; belly pale isabelline brown; flanks greyish. The female is slightly duller than the male. Wing 68 to 75 mm. The young bird has the crown of the head blackish with brown edges to the feathers, and the mantle slightly browner than the adult.

**Range in Eastern Africa:** South-central and south-western Kenya Colony to north-eastern Tanganyika Territory as far south as Mt. Gerui (Hanang) and the Masai Steppe at Kibaya.

**Habits:** A local species common only in southern Masailand and apparently of normal Tit-like habits.

**Nest and Eggs:** A nest described by Moreau was in a cavity behind the bark of a tree and was lined with down and fibres. The bird sat very tight. Eggs three, white with pale reddish spots and freckles; 17 × 13·5 mm.

**Recorded breeding:** Longido, Tanganyika Territory, January.

**Call:** A resonant 'Prttt-tchay-tchay-tchay—reminiscent of an English Willow-Tit' (Couchman & Elliott).

**1160 AFRICAN PENDULINE TIT.** *ANTHOSCOPUS CAROLI* (Sharpe).
*Anthoscopus caroli sylviella* Reichw. **Pl. 76.**
*Anthoscopus sylviella* Reichenow, O.M. p. 27, 1904: Usafua, Rungwe district, south-western Tanganyika Territory.

**Distinguishing characters:** Forehead buff or buffish white; rest of upperparts grey slightly washed with olivaceous; below, deepish buff, throat paler. The sexes are alike. Wing 52 to 58 mm. The young bird is similar to the adult.

**Range in Eastern Africa:** Tanganyika Territory from Longido to Kibaya, Iringa and the Rungwe area.

**Habits:** Small Tit-like birds usually seen searching bushes in family parties. They also search large flowers, and are generally quiet industrious little birds with a faint chirping note and very quick movements.

**Nest and Eggs:** A ball or pear-shaped nest of dense woolly material placed in or suspended from the bough of a bush. There is a well made entrance spout at the side which is barely large enough for the bird to pass through, and which is generally closed altogether with a bite or two from the bird's bill as it leaves; under this is often a little porch or platform. Eggs four to eight or more, pure white and oval; about 13·5 × 9·5 mm. There is some suspicion that more than one hen may lay in the same nest.

**Recorded breeding:** Tanganyika Territory, December and February, also building nest in July.

**Food:** Insects and their larvæ.

**Call:** A faint chirp or squeak, also a little squeaky two-syllabled whistle.

*Anthoscopus caroli sharpei* Hart.
*Anthoscopus sharpei* Hartert, Bull. B.O.C. 15, p. 75, 1905: Usambiro, Mwanza district, Tanganyika Territory.

**Distinguishing characters:** Differs from the preceding race in being generally darker; forehead tawny not pale buff. Wing 54 mm.

**Range in Eastern Africa:** Kenya Colony and Tanganyika Territory from the Kikuyu area to the Tabora district and west of Lake Natron.

**Habits:** As for other races.

**Recorded breeding:** South-western Kenya Colony, October.

*Anthoscopus caroli roccatii* Salvad. **Pl. 76.**
*Anthoscopus roccatii* Salvadori, Boll. Mus. Torino, 21, No. 542, p. 2, 1906: Entebbe, southern Uganda.

**Distinguishing characters:** Differs from the nominate race in being pale olive green above; forehead yellowish; below, dull yellowish; lower belly and under tail-coverts slightly buffish. Wing 52 to 56 mm.

**Range in Eastern Africa:** Southern half of Uganda.

**Habits:** Those of other races, in flight it looks like a little fluffy green ball. Occurs in woodland or occasionally in forest in small parties, but is not usually common.

**Recorded breeding:** Uganda, June. Mt. Elgon (breeding condition), December.

*Anthoscopus caroli rothschildi* Neum.
*Anthoscopus rothschildi* Neumann, J.f.O. p. 597, 1907: Simba, south-eastern Kenya Colony.

**Distinguishing characters:** Differs from the other races in being pale ochre yellow below. Wing 51 to 52 mm.

**Range in Eastern Africa:** The Simba and Kitui areas, Kenya Colony.

**Habits:** As for other races, seen in small parties among flowering sprays of acacias.

**Recorded breeding:** Southern Kenya Colony, March and April.

*Anthoscopus caroli robertsi* Haagn.
*Anthoscopus robertsi* Haagner, Ann. Trans. Mus. 1, p. 233, 1909: Villa Pereira, Boror, Portuguese East Africa.

**Distinguishing characters:** Differs from the other races in having the lower belly and under tail-coverts warmer buff. Wing 51 to 58 mm.

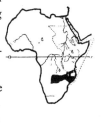

**General distribution:** Northern Rhodesia to Tanganyika Territory, Nyasaland and northern Portuguese East Africa.

**Range in Eastern Africa:** Tanganyika Territory at Liwale to the southern half of northern Portuguese East Africa and Nyasaland.

**Habits:** As for other races, widely distributed in open woodland.

**Recorded breeding:** Nyasaland, October to December.

*Anthoscopus caroli taruensis* V. Som.
*Anthoscopus rocatti taruensis* Van Someren, Bull. B.O.C. 41, p. 112, 1921: Samburu, eastern Kenya Colony.

**Distinguishing characters:** Similar to the last race but smaller. Wing 49 to 51 mm.

**Range in Eastern Africa:** Coastal area of Kenya Colony from the Tana River to Taru, Samburu and Changamwe, and to eastern Tanganyika Territory at Naberera 100 miles south of Mt. Kilimanjaro, Kijungu, Korogwe, Ngomeni and Morogoro.

**Habits:** As for other races.

**Recorded breeding:** No records.

*Anthoscopus caroli rhodesiæ* Scl.
*Anthoscopus ansorgei rhodesiæ* W. L. Sclater, Bull. B.O.C. 52, p. 143, 1932: Mt. Sunzu, near Abercorn, north-eastern Northern Rhodesia.

**Distinguishing characters:** Very similar to the last two races in colour above, but below, differs from them and from other preceding races in having the chin to breast dusky white; belly, lower flanks and under tail-coverts buffish; belly tinged with yellow. Wing 48 to 57 mm.

**General distribution:** South-western Tanganyika Territory to north-eastern Northern Rhodesia and south-eastern Belgian Congo.

**Range in Eastern Africa:** South-western Tanganyika Territory at the Ufipa Plateau.

**Habits:** As for other races.

**Recorded breeding:** No records.

**Distribution of other races of the species:** South Africa, the nominate race being described from Damaraland.

**1161 MOUSE-COLOURED PENDULINE TIT.** *ANTHOSCOPUS MUSCULUS* (Hartlaub).
*Anthoscopus musculus musculus* (Hartl.). **Pl. 76.**
*Ægithalus musculus* Hartlaub, O.C. p. 91, 1882: Lado, southern Sudan.

**Distinguishing characters:** Upperside mouse-grey; below creamy white; throat paler; lower flanks, lower belly and outer tail-coverts rather buffish. Wing 45 to 52 mm. The young bird is similar to the adult.

**Range in Eastern Africa:** Southern Sudan, the southern half of Abyssinia and British Somaliland, northern Uganda and western Kenya Colony.

**Habits:** Little recorded. A bird of rocky bush country in pairs or small parties searching thorn trees for food, and usually very tame.

**Nest and Eggs:** Nest very similar to those of other Penduline Tits and composed of plant down, placed in a low bush and usually easily visible. Eggs four, white; about 13 × 10 mm.

**Recorded breeding:** Southern Abyssinia, October.

**Food:** As for other species of the genus.

**Call:** A high thin 'dee-ee-dee-ee-dee' (Tomlinson).

*Anthoscopus musculus guasso* V. Som.

*Anthoscopus musculus guasso* Van Someren, Nov. Zool. 37, p. 359, 1932: Archer's Post, northern Guaso Nyiro, central Kenya Colony.

**Distinguishing characters:** Differs from the nominate race in being greyer above. Wing 48 to 50 mm.

**Range in Eastern Africa:** Northern, central and eastern Kenya Colony as far north as Marsabit, to southern Italian Somaliland and north-eastern Tanganyika Territory.

**Habits:** As for the nominate race.

**Recorded breeding:** No records.

## 1162 SENNAR PENDULINE TIT. *ANTHOSCOPUS PUNCTIFRONS* (Sundevall). Pl. 76.

*Ægithalus punctifrons* Sundevall, Œfv. K. Sv. Vet.-Akad. Förh. 7, p. 129, 1850: Sennar, eastern Sudan.

**Distinguishing characters:** Above, pale yellowish green; forehead yellow with black spots; below, varying from warm cream colour to very pale buff, with a paler throat. The sexes are alike. Wing 48 to 56 mm. Juvenile plumage unrecorded.

**General distribution:** The French Sahara to the Sudan.

**Range in Eastern Africa:** The Sudan from Darfur to the Sennar and Lake No area and central Eritrea.

**Habits:** Not uncommon unobtrusive little birds of plains and thorn bush. Generally plentiful where animals congregate to drink. Tame and always intently searching for food.

**Nest and Eggs:** Undescribed.

**Recorded breeding:** Darfur, probably September onwards.

**Food:** Insects and their larvæ.

**Call:** A high-pitched squeak.

## 1163 YELLOW PENDULINE TIT. *ANTHOSCOPUS PARVULUS* (Heuglin).

*Anthoscopus parvulus parvulus* (Heugl.). Pl. 76.

*Ægithalus parvulus* Heuglin, J.f.O. p. 260, 1864: Bongo, Bahr-el-Ghazal, southern Sudan.

**Distinguishing characters:** Above, bright yellowish green; forehead yellow with black spots; below, bright yellow. The sexes are alike. Wing 50 to 52 mm. Juvenile plumage unrecorded.

660 ORIOLES

**General distribution:** Lake Chad to the southern Sudan.

**Range in Eastern Africa:** Bahr-el-Ghazal to Rejaf, southern Sudan.

**Habits:** A rare and little known species whose habits do not seem to differ from others of the genus. It is described as a silent active little bird diligently searching for small caterpillars like a *Phylloscopus*.

**Nest and Eggs** (West African race): Nest a beautiful and elaborate construction of closely felted vegetable down, very strong and tough, rather like a closed purse in shape with an opening at the side of the top like a short flat tube, which is self-closing. Nest suspended from a bough or twig some feet from the ground. Eggs white without gloss; about 13 × 9·5 mm.

**Recorded breeding:** No records.

**Food:** As above.

**Call:** Unrecorded.

**Distribution of other races of the species:** West Africa.

Names in Sclater's *Syst. Av. Æthiop.* 2, 1930, which have been changed or have become synonyms in this work:

*Parus afer barakæ* Jackson, treated as synonymous with *Parus afer thruppi* Shelley.
*Parus afer parvirostris* Shelley, treated as synonymous with *Parus afer griseiventris* Reichenow.
*Parus niger purpurascens* Van Someren, treated as synonymous with *Parus leucomelas guineensis* Shelley.
*Parus niger lacuum* Neumann, treated as synonymous with *Parus leucomelas leucomelas* Rüppell.
*Parus albiventris curtus* Friedmann, treated as synonymous with *Parus albiventris* Shelley.
*Parus rufiventris rovumæ* Shelley, treated as synonymous with *Parus rufiventris pallidiventris* Reichenow.

FAMILY—**ORIOLIDÆ. ORIOLES.** Genus: *Oriolus*.

Seven species occur in Eastern Africa, one of which is a migrant from Europe and Asia during the non-breeding season. Orioles inhabit woodland and forest and their call and colour make them conspicuous, although as they more often settle in the centre of a tree than on the outside they are difficult to see and easily overlooked unless they call. Most Orioles will answer to a whistle.

## Key to the Adult Orioles of Eastern Africa

1. Head and neck green: GREEN-HEADED ORIOLE *Oriolus chlorocephalus* **1170**
2. Head and neck black: 4–9
3. Head and neck yellow: 10–11
4. Central tail feathers black or mainly black: BLACK-WINGED ORIOLE *Oriolus nigripennis* **1169**
5. Central tail feathers yellow green or mainly yellow green: 6–9
6. Edges of outer secondaries white or whitish: BLACK-HEADED ORIOLE *Oriolus larvatus* **1167**
7. Edges of outer secondaries grey: 8–9
8. Wing longer, over 128 mm.: BLACK-HEADED FOREST ORIOLE *Oriolus monacha* **1166**
9. Wing shorter, under 126 mm.: WESTERN BLACK-HEADED ORIOLE *Oriolus brachyrynchus* **1168**
10. Lores and wing-coverts black: GOLDEN ORIOLE, male *Oriolus oriolus* **1164**
11. Lores to behind eye black, wing-coverts broadly edged with yellow: AFRICAN GOLDEN ORIOLE, male *Oriolus auratus* **1165**
12. Top of head yellow green, throat white: 13–14
13. Darkish olivaceous stripe behind eye, broad yellow edges to wing-coverts: AFRICAN GOLDEN ORIOLE, female *Oriolus auratus* **1165**

14 No darkish olivaceous stripe behind eye, no broad yellow edges to wing-coverts:   GOLDEN ORIOLE, female *Oriolus oriolus*   1164

**1164 GOLDEN ORIOLE.** *ORIOLUS ORIOLUS* (Linnæus).
*Oriolus oriolus oriolus* (Linn.). **Pl. 77.**
*Coracias Oriolus* Linnæus, Syst. Nat. 10th ed. p. 107, 1758: Sweden.

**Distinguishing characters:** Adult male, above and below, golden yellow; lores, wings and tail black; broad yellow ends to primary coverts, and the apical quarter of all except central tail feathers yellow. The female is golden green above; rump and upper tail-coverts golden; wings blackish brown; tail blackish and golden, yellow tips narrow; below, chin to belly whitish streaked with black; flanks and under tail-coverts yellow; or else wholly yellow below with olive streaks; bill dark pink in both sexes. Wing 141 to 161 mm. The young bird has a dusky bill, and is more olivaceous golden yellow; wings and tail less black; below, chin to belly yellowish and white streaked with sepia.

**General distribution:** Central and southern Europe, southern Sweden to the Altai, Iran, Morocco and Algeria; in non-breeding season to Africa, Arabia, India and Ceylon.

**Range in Eastern Africa:** Throughout in non-breeding season.

**Habits:** Rather shy and prefers the cover of thick high trees. It is surprisingly inconspicuous for its size and colour, and unless its flute-like call is heard is quite easy to overlook. In the Sudan it is a passage migrant in September and October, repassing in March and April, but according to A. L. Butler, it is not much seen east of Gedaref. In Kenya Colony and Tanganyika Territory it is a passage migrant and also a visitor in small numbers occurring in both open and forested country and it reaches Rhodesia about November. It appears that usually the old males migrate to Europe some ten days before the females and return somewhat later than the females and young. The call is a loud flute-like whistle 'who-are-you.'

**Distribution of other races of the species:** India.

**1165 AFRICAN GOLDEN ORIOLE.** *ORIOLUS AURATUS* Vieillot.

*Oriolus auratus auratus* Vieill.
*Oriolus auratus* Vieillot, N. Dict. d'Hist. Nat. 18, p. 194, 1817: Gold Coast Colony, West Africa.

**Distinguishing characters:** Adult male, differs from the Golden Oriole in being richer golden yellow; black of lores extending to around and behind eye; very small yellow tips to primary coverts; other wing-coverts and secondaries broadly edged with golden yellow. The female has the upperparts more olivaceous yellow; blackish stripe through eye less extensive; less rich yellow below than the male, being more or less streaked with olive. Wing 132 to 154 mm. The immature male is similar to the adult, but has the mantle olivaceous, a dusky bill and some black streaks below. The young bird of both sexes can be distinguished from that of the Golden Oriole by the olivaceous stripe behind the eye and by the bright yellow or greenish-yellow edges to the lesser and secondary wing-coverts.

**General distribution:** Senegal to Abyssinia, Uganda and Kenya Colony.

**Range in Eastern Africa:** Central Sudan to Abyssinia, central Uganda and northern Kenya Colony.

**Habits:** A bird of open country or woodland in pairs or small parties, generally shy and not particularly common. Prefers the thick foliage of high trees. In northern Uganda apparently a non-breeding visitor only, occurring from June to February.

**Nest and Eggs:** The nest is a flimsy basket, often lined with a little fine grass but at times transparent, of stems and leaves bound to, or slung below, the fork of a bough and attached by cobwebs, etc. It is usually at some height and dependent from the outer branches of a tree. The opening is a wide gap at the top of the side. Eggs two, buffish pink with a few sparse red brown spots and streaks tending to form a zone at one end, and these markings are surrounded by a deep rose coloured suffusion; about 31 × 20 mm. Very beautiful eggs.

**Recorded breeding:** Nigeria, March and April.

**Food:** Largely caterpillars and fruit.

**Call:** A cheery whistled call 'fee-you-fee-you-fee-you,' and a wide range of mewing and squalling cries. The alarm call is a harsh 'mwah-mwah.'

*Oriolus auratus notatus* Peters. **Pl. 77.**
*Oriolus notatus* Peters, J.f.O. p. 132, 1868: Tete, Portuguese East Africa.

**Distinguishing characters:** Differs from the nominate race at all ages by having the yellow of the outer tail feathers extending to the base of the feather, at least on the inner web. Wing 132 to 154 mm.

**General distribution:** Angola to Tanganyika Territory, Damaraland and Southern Rhodesia.

**Range in Eastern Africa:** Tanganyika Territory to the Zambesi River, also Pemba, Zanzibar, Kwale and Mafia Islands; in non-breeding season to Uganda and Kenya Colony.

**Habits:** As for the nominate race, shy and difficult to see. It appears to visit southern Uganda in the non-breeding season, only occurring from April to August. It is relatively common in Rhodesia where it breeds from September onwards and leaves for the north in March. Breeds rarely in Nyasaland, and very probably in Tanganyika Territory also.

**Recorded breeding:** Rhodesia, September to November. Nyasaland, October to December. Portuguese East Africa (breeding condition), September.

**1166 BLACK-HEADED FOREST ORIOLE.** *ORIOLUS MONACHA* (Gmelin).
*Oriolus monacha monacha* (Gmel.). **Pl. 77.**
*Turdus monacha* Gmelin, Syst. Nat. 1, pt. 2, p. 824, 1799: Eritrea.

**Distinguishing characters:** Head and neck black; rest of upperparts olivaceous golden yellow; wings black with grey edges to flight feathers; outer webs of inner secondaries olivaceous yellow; inner webs of inner secondaries olivaceous yellow or dusky washed with olivaceous yellow; coverts of inner primaries edged with grey; broad white tips to primary coverts; rest of underparts golden yellow; tail, variable pattern of yellow green decreasing towards outer feathers, with broad golden yellow ends to all except the central tail feathers; outermost tail feather sometimes wholly golden yellow; bill dull red. The sexes are alike. Wing 128 to 145 mm. The immature bird is similar to the adult but has a darker coloured bill. The young bird has a black bill; head black as in the adult with some yellow streaks on chin and throat.

# ORIOLES

**Range in Eastern Africa:** Eritrea and northern Abyssinia as far south as the Shoa area.

**Habits:** Little definitely recorded. The distinction between the two species of Black-headed Oriole has only recently been realized and field notes are unreliable in the northern areas of their range.

**Nest, Eggs, Call:** See under the following races.

**Recorded breeding:** No records.

*Oriolus monacha meneliki* Blund. & Lovat.
*Oriolus meneliki* Blundell & Lovat, Bull. B.O.C. 10, p. 19, 1899: Burka, eastern Abyssinia.

**Distinguishing characters:** Differs from the nominate race in having a certain amount of black in the tail. Wing 130 to 150 mm.

**Range in Eastern Africa:** Northern to eastern and south-western Abyssinia.

**Habits:** Locally common in forest, woodland, or along belts of trees near water, gathering into flocks at certain times of year. Field notes are unreliable owing to the difficulty of distinguishing this species from the other Black-headed Oriole *Oriolus larvatus*, whose range overlaps.

**Nest and Eggs:** Undescribed.

**Recorded breeding:** No records.

**Food:** No information.

**Call:** Three main calls are described by C. W. Benson (1) three or four liquid whistles slurred together; (2) a harsh Shrike-like 'skaa-skaa'; (3) a rich loud 'li.'

*Note:* Intergrades with the nominate race between Lake Tana and the Shoa area.

**1167 BLACK-HEADED ORIOLE. *ORIOLUS LARVATUS***
Lichtenstein.
*Oriolus larvatus rolleti* Salvad. **Pl. 77. Ph. xv.**
*Oriolus rolleti* Salvadori, Atti. Acad. Torino, 7, p. 151, 1864: Lat. 7° N. on White Nile, southern Sudan.

**Distinguishing characters:** Differs from the Black-headed Forest Oriole in that the edges of the flight feathers are white or whitish, not grey; outer webs of inner secondaries black edged with olivaceous yellow; inner webs of inner secondaries black; coverts of

inner primaries edged with olivaceous yellow or grey. The sexes are alike. Wing 122 to 147 mm.; one from Abyssinia 115 mm. The young bird differs from the young bird of the last species in having the forehead and crown streaked yellow and black, also the black streaking of the chin and throat extends to the breast and the colour of the inner secondaries is as in the adult.

**General distribution:** The Sudan, Abyssinia, Italian Somaliland, and south to Angola, Bechuanaland, the Transvaal and Portuguese East Africa as far south as Barberton and Inhambane.

**Range in Eastern Africa:** Southern Sudan and southern and eastern Abyssinia, Italian Somaliland, Uganda, and Kenya Colony to the Zambesi River, also Ukerewe Island, Lake Victoria.

**Habits:** Birds of woodland, forest, bush or garden, usually in pairs but occasionally in flocks when fruit is ripe; not as shy as most species of Oriole and far more common. At times abundant but locally distributed owing to their preference for high shady trees. All Orioles have the habit of sitting motionless in a thick leafed tree, and if the call is not heard may be very easily overlooked in spite of their bright coloration. This species would be most difficult to distinguish from the Black-headed Forest Oriole in the field.

**Nest and Eggs:** The nest is a cup or basket of grass, or in some localities grey moss and tendrils, usually camouflaged to resemble its surroundings in colour, and attached underneath a slender fork or bough often at the tip of a branch and usually high above the ground. Eggs normally two in Equatorial Africa and up to four elsewhere; beautiful eggs, white, cream or pinkish with reddish-brown or sepia suffused spots and lavender undermarkings generally in a zone round the larger end; about 29 × 21 mm.

**Recorded breeding:** Uganda, January to May. Kenya Colony, March to June, also November. Tanganyika Territory, September to December. Nyasaland, September to November. Rhodesia, September to November. Portuguese East Africa, September.

**Food:** Caterpillars, seeds and fruit, and can be a nuisance among soft fruit.

**Call:** A flute-like 'hooi' or 'hueeoo' and quite a number of rasping clanging cries of which many renderings are given, and which at times incorporate the calls of other birds.

**Distribution of other races of the species:** South Africa, the nominate race being described from eastern Cape Province.

**1168** WESTERN BLACK-HEADED ORIOLE. *ORIOLUS BRACHYRYNCHUS* Swainson.
*Oriolus brachyrhynchus lætior* Sharpe.
*Oriolus lætior* Sharpe, Bull. B.O.C. 7, p. 17, 1897: Gabon.

**Distinguishing characters:** Similar to the Black-headed Forest Oriole but shorter winged. Wing 107 to 126 mm. The young bird differs from the young bird of that species in having the head yellow green with more or less distinct black streaks.

**General distribution:** Southern Nigeria to Gabon and Uganda.

**Range in Eastern Africa:** Uganda.

**Habits:** Abundant in the Semliki Valley among high trees. In Cameroons relatively common and found in tall trees in clearings or on the fringes of forest.

**Nest and Eggs:** Nest as for other Orioles, of lichens, cobwebs, etc. slung between horizontal twigs. Eggs undescribed.

**Recorded breeding:** Western Uganda (breeding condition), April.

**Food:** Mainly caterpillars.

**Call:** A variety of rich full throated notes, among them a loud whistling call described as 'or-iole' with the accent on the first syllable, also a loud squeaking call. The native name 'eja-koa' is also onomatopæic.

**Distribution of other races of the species:** West Africa, the nominate race being described from Sierra Leone.

**1169** BLACK-WINGED ORIOLE. *ORIOLUS NIGRIPENNIS* Verreaux.
*Oriolus nigripennis nigripennis* Verr.
*Oriolus (Barruffius) nigripennis* J. & E. Verreaux, J.f.O. p. 105, 1855: Gabon River.

**Distinguishing characters:** Differs from the other Black-headed Orioles in having the edges of the secondaries yellow green, not grey; tips of the primary coverts wholly black or with only an indication of white; and the central tail feathers wholly black, or rarely black and yellow green. Wing 111 to 131 mm. The immature bird differs from the adult in having the black head washed with green and the chest streaked black and yellow. The young bird has the mantle mottled with black; wing-coverts black with broad canary yellow edges; chest to upper belly and flanks streaked with black; bill slaty black.

**General distribution:** Fernando Po to Sierra Leone, the Sudan, Uganda and northern Angola.

**Range in Eastern Africa:** Southern Sudan to the Bwamba area, western Uganda.

**Habits:** Extremely difficult to tell in the field from the Western Black-headed Oriole, and having the same habits is easily overlooked. It may, therefore, be commoner than is supposed and in certain districts of west Uganda is quite plentiful in the forest canopy.

**Nest and Eggs:** Undescribed.

**Recorded breeding:** Ashanti, February. Cameroons (nestling), October. Southern Sudan (breeding condition), October and January.

**Food:** Fruit and caterpillars.

**Call:** A three-syllabled melodious call much softer than that of most Orioles. Van Someren describes it as 'tuulit' and it is often followed by other liquid notes to which the female replies with a mewing call.

*Oriolus nigripennis percivali* O. Grant. **Pl. 77.**
*Oriolus percivali* O. Grant, Bull. B.O.C. 14, p. 18, 1903: Kikuyu, south-central Kenya Colony.

**Distinguishing characters:** Differs from the nominate race in usually having white tips to the primary coverts and some white edging to the secondaries. It differs from the Black-headed Orioles in the black, or mainly black, central tail feathers and in the secondary coverts being darker with usually little or no grey colour. Wing 124 to 140 mm.

**Range in Eastern Africa:** Kenya Colony to Uganda and the Kungwe-Mahare Mts., western Tanganyika Territory.

**Habits:** As for the nominate race.

**Recorded breeding:** No records.

**1170 GREEN-HEADED ORIOLE.** *ORIOLUS CHLOROCEPHALUS* Shelley.

*Oriolus chlorocephalus chlorocephalus* Shelley. **Pl. 77.**

*Oriolus chlorocephalus* Shelley, Ibis, p. 183, pl. 4, 1896: Mt. Chiradzulu, southern Nyasaland.

**Distinguishing characters:** Whole head, neck, chin to chest, mantle and rump moss green; a canary yellow collar on hind neck; flight feathers and coverts with slate blue edging; inner secondaries edged with green; tail moss green with yellow ends to all except four central feathers; below, from breast to under tail-coverts canary yellow joining yellow collar on hind neck. The sexes are alike. Wing, male 133 to 141; female 129 to 132 mm. The young bird has a dull yellow green collar on hind neck; yellow tips to wing-coverts; chin and throat greenish yellow; green streaks on chest, breast and flanks.

**Range in Eastern Africa:** Cholo, Chiradzulu, Chikala and Soche Mts., southern Nyasaland to Chiperone Mt., Portuguese East Africa.

**Habits:** A relatively uncommon species, but now found to be more widespread than supposed. It often feeds in company with Black-headed Orioles when forest fruits are ripe.

**Nest and Eggs:** Nest a deep thick-walled pocket of beard-moss hanging from a tree usually on the edge of a clearing and some thirty feet from the ground. Eggs undescribed.

**Recorded breeding:** Nyasaland, August to October.

**Food:** Caterpillars, fruit and flowers, particularly Grevilleas.

**Call:** Said to be clearer and more liquid than that of the Black-headed Oriole, but otherwise similar, and only to be distinguished by an expert. Benson says that it has one distinctive call of its own, a nasal whining 'heee-eee-aaa.'

*Oriolus chlorocephalus amani* Benson.

*Oriolus chlorocephalus amani* Benson, Bull. B.O.C. 67, p. 26, 1946: Amani, north-eastern Tanganyika Territory.

**Distinguishing characters:** Differs from the nominate race in being smaller. Wing, male 123 to 134; female 123 to 127 mm.

**Range in Eastern Africa:** Lolkissale to the Usambara, Nguru and Uluguru Mts. and Rondo Plateau, Lindi district, Tanganyika Territory.

**Habits:** As for the nominate race.

**Recorded breeding:** Amani, Tanganyika Territory (nestling), November.

Names in Sclater's *Syst. Av. Æthiop.* 2, 1930, which have been changed or have become synonyms in this work:

*Oriolus monachus permistus* Neumann, treated as synonymous with *Oriolus monacha meneliki* Blundell and Lovat.
*Oriolus monacha kikuyuensis* Van Someren, treated as synonymous with *Oriolus larvatus rolleti* Salvadori.
*Oriolus larvatus reichenowi* Zedlitz, treated as synonymous with *Oriolus larvatus rolleti* Salvadori.

## FAMILY—CORVIDÆ. RAVENS, CROWS, ROOK, CHOUGH, PIAPIAC, and BUSH-CROW.

Genera: *Corvus, Corvultur, Rhinocorax, Coracia, Ptilostomus* and *Zavattariornis.*

Ten species occur in Eastern Africa. All are resident and conspicuous birds.

### KEY TO THE CORVIDÆ OF EASTERN AFRICA

1 Size smaller, wing 172 mm. and under: 3–4
2 Size larger, wing 255 mm. and over: 5–16
3 Tail short, general colour black and grey: STRESEMANN'S BUSH-CROW *Zavattariornis stresemanni* 1180
4 Tail long, general colour black: PIAPIAC *Ptilostomus afer* 1179

---

5 Bill heavy and deep: 6–7
6 White patch on occiput and base of neck joined by a white streak: THICK-BILLED RAVEN *Corvultur crassirostris* 1176
7 Broad white collar on hind neck: WHITE-NECKED RAVEN *Corvultur albicollis* 1175
8 Bill not heavy and deep: 9–16

# RAVENS, CROWS, ROOK, CHOUGH, etc.

9  Bill red:                            CHOUGH
                                        *Coracia pyrrhocorax*    **1178**
10 Bill black:                                                   11–16
11 General colour blue black
   and white:                           PIED CROW
                                        *Corvus albus*           **1172**
12 General colour blue black            INDIAN HOUSE-CROW
   and ashy grey:                       *Corvus splendens*       **1174**
13 General colour violet
   black and bronzy brown:              BROWN-NECKED RAVENS
                                        *Corvus corax ruficollis*
                                        *Corvus corax edithæ*    **1171**
14 General colour blue black:                                    15–16
15 Bill slender, nasal bristles
   not extending beyond
   nostrils:                            CAPE ROOK
                                        *Corvus capensis*        **1173**
16 Bill not slender, nasal
   bristles extending beyond
   nostrils:                            FAN-TAILED RAVEN
                                        *Rhinocorax rhipidurus*  **1177**

**1171 RAVEN.** *CORVUS CORAX* Linnæus.
*Corvus corax ruficollis* Less. *Brown-necked Raven.*
*Corvus ruficollis* Lesson, Traité d'Orn. p. 329, 1831: Cape Verde
Islands.

**Distinguishing characters:** Head, neck and underside dark bronzy brown; rest of upperparts, wings and tail glossy violet black. The sexes are alike. Wing 323 to 419 mm. The young bird is duller; head and underside dullish black, less bronzy brown; wings and tail as in adult. Albinism is not unknown.

**General distribution:** Cape Verde Islands and northern Africa to the Aral Sea, Russian Turkestan; also Arabia.

**Range in Eastern Africa:** Western and northern Sudan and Socotra Island.

**Habits:** A common bird of the sandy and gravelly deserts of the north-western Sudan, and found wherever there are villages and palm groves. At Khartoum it is abundant, as also in parts of Kordofan and on the Red Sea coast. It is often seen at dusk flighting

to some distant communal roosting place. Usually shy, even round towns.

**Nest and Eggs:** Nest in trees or on rocks, made of sticks, lined with camel hair or other soft material. Eggs two to six, dull bluish green spotted and blotched with brown and grey, rather dull and pale; about 42 × 32 mm.

**Recorded breeding:** Sudan, December to March.

**Food:** Probably omnivorous.

**Call:** A deep 'cronk.'

*Corvus corax edithæ* Phill. *Lesser Brown-necked Raven.*
*Corvus edithæ* Phillips, Bull. B.O.C. 4, p. 36, 1895: Dejamio. Hainwana Plain, British Somaliland.

**Distinguishing characters:** Differs from the preceding race in being usually smaller and having a smaller bill. Wing 311 to 367 mm.

**Range in Eastern Africa:** Abyssinia, British and Italian Somalilands and Kenya Colony.

**Habits:** A small desert Raven very similar in habits to the last race. Locally abundant, especially in Northern Kenya Colony and an inveterate scavenger, even invading huts or kitchens for food. It exists in the most arid localities, but seems to be particularly fond of water when it is available. Although classed as a race of the Raven, its appearance and call are much more like those of a Rook.

**Nest and Eggs:** Inland birds nest in trees, especially in slender sapling-like trees which are difficult to climb. Sea coast birds nest on cliffs or in bushes growing out of them. Exceptional nesting sites include the top of a wireless mast, and on telegraph wires against a pole. Nests are typical Crows' nests lined with hair, etc. Eggs three to five, pale blue, spotted or blotched with dark brown; about 42 × 30 mm.

**Recorded breeding:** Northern Kenya Colony, probably February. Italian Somaliland, April to June.

**Call:** Not a deep croak but a harsh 'caw' rather like that of an English Rook.

**Distribution of other races of the species:** Europe, Canary Islands, North Africa and India, the nominate race being described from Sweden.

**1172 PIED CROW. *CORVUS ALBUS* Müller. Ph. xv.**
*Corvus albus* P. L. S. Müller, Syst. Nat. Suppl. p. 85, 1776: Senegal.

**Distinguishing characters:** General colour black and white; broad white collar across base of hind neck; chest to upper belly white; rest of plumage glossy blue black. The sexes are alike. Wing 318 to 382 mm. The young bird has the black, especially on the underside, duller and with no gloss.

**General distribution:** Fernando Po and Senegal to British Somaliland and South Africa, also Aldabra and Comoro Islands and Madagascar.

**Range in Eastern Africa:** Central Sudan and British Somaliland to the Zambesi River, also Pemba and Zanzibar Islands.

**Habits:** Curiously local birds with a very wide general distribution and erratic migrational movements. These local movements are very difficult to classify, but this bird is definitely a common non-breeding visitor to Darfur from February to May, at which time it is also breeding freely in the rest of the Sudan. There is also a definite migration to the Tanganyika Territory highlands in November and December. These Crows are very common where they occur and are often numerous in towns and villages, being frequently found in company with the smaller Vultures, acting as scavengers. At times beneficial to agriculture, at others very definitely destructive. They occasionally gather into flocks of some hundreds and have a habit of circling high in the air in the heat of the day. It has been noticed that heat appears to distress them a good deal. Large numbers often roost together, but in spite of their abundance, nests are not commonly seen.

**Nest and Eggs:** Nest made of sticks and lined with any available material and placed either in trees or rocks, occasionally even on telegraph posts. Eggs three to six, bluish green spotted, blotched and streaked with olive brown or grey; about $44 \times 30$ mm., with considerable variation. Often parasitized by the Great Spotted Cuckoo.

**Recorded breeding:** Sudan, January to May. Abyssinia, March. Uganda and western Kenya Colony, January to June. Kenya Colony coast, November and December. Tanganyika Territory, September to November. Zanzibar Island, September, November also February. Rhodesia, August to October. Nyasaland, September to November.

**Food:** Omnivorous; among other things, garbage, shore offal, insects of all kinds, beetle larvæ, in search of which they do some damage to thatch, corn, maize, oil-palm nuts, wild fruits, in fact anything they can reach, and are also reported as doing much damage to ground-nut cultivation in West Africa, and to maize crops in Eastern Africa.

**Call:** A very hoarse guttural 'caw' at times almost a snore.

### 1173 CAPE ROOK. *CORVUS CAPENSIS* Lichtenstein.

*Corvus capensis kordofanensis* Laub.
*Corvus capensis kordofanensis* Laubmann, Verh. Orn. Ges. Bayern, 14, p. 103, 1919: Southern Kordofan, Sudan.

**Distinguishing characters:** Bill slender; nasal bristles short, not extending beyond the nostril; general colour black, glossed with steely blue. Worn birds are liable to be bronzy brown. The sexes are alike. Wing 298 to 363 mm. The young bird is duller with less gloss; a brownish wash on head and breast.

**Range in Eastern Africa:** Southern Sudan to Abyssinia, Eritrea, British Somaliland, Uganda and central Kenya Colony.

**Habits:** Locally common and is increasing in East Africa with white settlement. It is especially numerous at Nakuru, in the Rift Valley, and on the Uasingishu Plateau, Kenya Colony, at times to be seen in flocks of thousands. It has a curious habit when perched of puffing out the feathers of its head and neck so that they appear unduly large. In general a bird of villages and cultivated areas, and usually roosts in companies.

**Nest and Eggs:** Very neatly made nest of twigs, lined with hair or other material, and placed at the end of quite thin boughs not usually in a colony. Eggs two to four, usually three, salmon pink with purple and lilac spots and blotches; about 40 × 30 mm.

**Recorded breeding:** Southern Sudan, December to February. Abyssinia, April. Kenya Colony, March to July and October to December, somewhat irregularly.

**Food:** Largely insects, but probably omnivorous.

**Call:** A curiously harsh throaty call, almost a chuckle.

**Distribution of other races of the species:** South Africa, the nominate race being described from the Cape of Good Hope.

## 1174 INDIAN HOUSE-CROW. *CORVUS SPLENDENS* Vieillot.

*Corvus splendens splendens* Vieill.

*Corvus splendens* Vieillot, N. Dict. d'Hist. Nat. 8, p. 44, 1817: Bengal, India.

**Distinguishing characters:** Hinder part of crown, neck to upper mantle, chest and breast ashy grey with a slight vinous tone; forehead to crown, face, chin and throat, rest of mantle, wings and tail glossy purple, blue black and green; belly and under tail-coverts slightly glossy greyish, greenish and blue black. The sexes are alike. Wing 238 to 296 mm. The young bird is duller than the adult, and has practically no gloss on the forehead, crown, side of face and chin and throat.

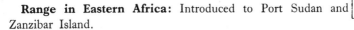

**General distribution:** India, except Kashmir to Sind; introduced to Zanzibar Island in the early nineties and to Port Sudan at a date unknown, also Arabia.

**Range in Eastern Africa:** Introduced to Port Sudan and Zanzibar Island.

**Habits:** This species has now been long enough established in Africa to be worthy of notice, in fact it has increased to a sufficient extent to become a nuisance, and seems not unlikely to spread. Fortunately it is pre-eminently a town bird and is not likely to disturb the balance of nature far afield in the countryside, but it has extended its range to most of the larger villages on the Island of Zanzibar.

**Nest and Eggs:** A substantial nest of sticks and other materials, usually in a coconut palm. Eggs up to six, pale blue, or bluish green, more or less heavily speckled and spotted with greyish or greenish brown; about 38 × 27 mm.

**Recorded breeding:** Red Sea Province, Sudan, February. Zanzibar Island, October to December.

**Food:** Anything it can pick up, growing crops, soft fruit, young poultry and small birds.

**Call:** A rather soft 'caw.'

**Distribution of other races of the species:** Kashmir, Sind, Ceylon, Burma, Malaya and southern China.

### 676 RAVENS, CROWS, ROOK, CHOUGH, etc.

**1175 WHITE-NECKED RAVEN.** *CORVULTUR ALBICOLLIS* (Latham).

*Corvus albicollis* Latham, Ind. Orn. 1, p. 151, 1790: Great Namaqualand, southern South West Africa.

**Distinguishing characters:** Adult male, bill heavy and deep with a white tip; head and neck bronzy brown; a broad white collar at base of hind neck; a few intermittent white feathers across chest; rest of plumage slightly glossy black. The female has a smaller bill. Wing 382 to 450 mm. The young bird is dull sooty black below; hinder part of white collar on hind neck streaked with black.

**General distribution:** Uganda and Kenya Colony to Namaqualand and Cape Province.

**Range in Eastern Africa:** Uganda and Kenya Colony to the Zambesi River.

**Habits:** Although ubiquitous in most districts of East Africa, it is essentially a bird of cliffs in the breeding season and also for roosting whenever possible. It generally appears in pairs at every camp and village, and is not usually gregarious, but at times flocks of hundreds have been noticed, and these were probably following a locust swarm. These birds are undoubtedly beneficial in this way, and in the number of mice and rats they catch, but they can be a plague to young poultry, though in many districts they appear never to touch them. They perform wonderful aerobatics and make much noise with their wings while doing so.

**Nest and Eggs:** Usually a large stick nest with a lining of soft material on the ledge of a cliff, but very occasionally nests are in trees. The nest is used year after year. Eggs up to six, blue-green and rather glossy with streaks and spots of dark green and brown tending to form a cap at one end; about 54 × 34 mm.

**Recorded breeding:** Ruwenzori Mts., February. Kenya Colony, October to December, also probably May and June. Tanganyika Territory, October. Nyasaland, September to November.

**Food:** Locusts, grasshoppers, grubs of all sorts, carrion, fruit, and whatever mammals, birds or reptiles they can catch.

**Call:** A curiously falsetto croak.

## RAVENS, CROWS, ROOK, CHOUGH, etc.

**1176 THICK-BILLED RAVEN.** *CORVULTUR CRASSIROSTRIS* (Rüppell).

*Corvus crassirostris* Rüppell, N. Wirbelt. Vög. p. 19, pl. 8, 1836: Halai, Eritrea.

**Distinguishing characters:** Differs from the White-necked Raven in having a heavier bill and in having a white patch on occiput joined by a narrow streak of white to a white patch at base of neck. The sexes are alike. Wing 427 to 472 mm. The young bird is dull black below.

**Range in Eastern Africa:** Eritrea and Abyssinia.

**Habits:** A common species in Abyssinia where it takes the place of the White-necked Raven and has similar habits. They usually perch on rocks or stumps and not on trees, and always appear for scraps when a camp is being moved or on any similar occasion. They will come miles to a slaughtered animal and probably possess a remarkable sense of smell. Occasionally gather into flocks and do some damage to crops, but are no doubt beneficial on balance. They perform extremely fine aerobatics in the breeding season.

**Nest and Eggs:** Nest in rocks or in trees where rocks are not available. Eggs undescribed.

**Recorded breeding:** Abyssinia, December to February.

**Food:** Omnivorous.

**Call:** Harsh and guttural.

**1177 FAN-TAILED RAVEN.** *RHINOCORAX RHIPIDURUS* (Hartert).

*Corvus rhipidurus* Hartert, Bull. B.O.C. 39, p. 21, 1918: Massawa, Eritrea.

**Distinguishing characters:** Wholly glossy blue black, with a purple and bronze sheen. The sexes are alike. Wing 342 to 406 mm. Liable to fade to bronzy brown, but can be distinguished from the Brown-necked Ravens by the shorter, stouter bill and shorter tail. The young bird is very similar to the adult, but duller in colour.

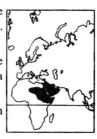

**General distribution:** Southern Sahara, Palestine, Arabia, the Sudan, Abyssinia, British and Italian Somalilands and Kenya Colony.

**Range in Eastern Africa:** The Sudan, Abyssinia, northern Kenya Colony and British and Italian Somalilands.

# 678  RAVENS, CROWS, ROOK, CHOUGH, etc.

**Habits:** A bird of rocks and cliffs which yet wanders very great distances for food. They can be recognized on the wing at once by the shortness of the tail, and they are generally in pairs, though at times they gather into flocks and soar high into the air till almost lost to sight. They have a curious habit of walking about with beaks open as if panting. Locally but widely distributed and are usually tame, often visiting camps and villages.

**Nest and Eggs:** Nests in inaccessible holes and crevices of cliffs, very rarely in buildings, made of twigs with any suitable lining. Breed in a colony as a rule. Eggs three or four, greenish blue speckled and blotched with olive brown and violet grey; about 46 × 32 mm.

**Recorded breeding:** Darfur, Sudan, May and June. Abyssinia, March. British Somaliland, April to June.

**Food:** Omnivorous.

**Call:** A falsetto croak with rather a cheery tone in it, and there is also a much lower note produced from deep down in the stomach only audible for a few yards.

**1178** CHOUGH. *CORACIA PYRRHOCORAX* (Linnæus).
*Coracia pyrrhocorax docilis* (Gmel.).
*Corvus docilis* Gmelin, Reise. Russ. 3, Theil. pl. 39, p. 365, 1774: Gilan, Iran.

**Distinguishing characters:** General colour glossy velvety black; wings and tail washed with purplish; bill and legs red. The sexes are alike. Wing 265 to 322 mm. The young bird is duller; wings and tail greener than adult; bill pinkish orange.

**General distribution:** Morocco to Algeria, Crete, Syria, the Caucasus, Iran, Afghanistan, Baluchistan, Arabia and Abyssinia.

**Range in Eastern Africa:** The Simen district, northern Abyssinia.

**Habits:** Common but local in the higher mountains of northern Abyssinia, roosting in cliffs in companies, and hovering and gliding in circles with outstretched wings in rocky places. They appear to be confined to one or two localities only, and are to be seen coming down to feed on pasture land from the cliffs above.

**Nest and Eggs:** On cliffs, though there is a report of their nesting in a village. A bulky nest of twigs, grass and roots lined with hair and rootlets. Eggs four to five, greenish white with rufous spots and blotches and faint purple undermarkings; about 36 × 26 mm.

**Recorded breeding:** Simen Mts. Abyssinia (breeding condition), May.

**Food:** Crustacea, worms and insects.

**Call:** Various calls of a rather chattering nature, sometimes almost Gull-like—with a sharper harsher alarm note.

**Distribution of other races of the species:** Europe and Asia, the nominate race being described from England.

**1179** PIAPIAC. *PTILOSTOMUS AFER* (Linnæus).
*Corvus afer* Linnæus, Syst. Nat. 12th ed. 1, p. 157, 1766: Senegal.

**Distinguishing characters:** Tail long and graduated; general colour black with slight gloss; wings and tail dusky brownish; underside of wings ashy; bill black. The sexes are alike. Wing 152 to 172 mm. The young bird is less glossy than the adult and has the bill violet, tip black.

**General distribution:** Senegal to Abyssinia and Uganda.

**Range in Eastern Africa:** Southern Sudan to southern Abyssinia and Uganda as far south as the Bwamba area.

**Habits:** Active lively birds usually seen in small parties playing follow-my-leader among palm trees, with which they are generally associated. Very noisy and carry out extensive aerial evolutions especially before roosting, which they do in companies. Usually shy and seem to have one or more sentinels always on the watch, but in West Africa are found in towns. General habits of a Starling rather than of a Crow. Extensively migratory, but movements most irregular. In many parts of the Sudan closely associated with domestic animals, feeding among them and perching on their backs.

**Nest and Eggs:** Nest in the tops of palm trees, of sticks, leaves and palm fibre, usually with some thorny boughs protecting the side entrance. Eggs three to seven, plain pale blue, or blue spotted and blotched with brown and pale grey; about 30 × 21 mm.

**Recorded breeding:** Nigeria, March and April. Sudan, March to June.

**Food:** Largely insects.

**Call:** A deep pipe. Alarm note a harsh scolding cry uttered with a bob of the head.

**1180 STRESEMANN'S BUSH-CROW.** *ZAVATTARIORNIS STRESEMANNI* Moltoni.

*Zavattariornis stresemanni* Moltoni, O.M. p. 80, 1938: Yavello, southern Abyssinia.

**Distinguishing characters:** Size small; upperside grey including lesser and median wing-coverts; forehead and upper tail-coverts whitish; wings and tail blue black, with a slight gloss; below, white; sides of chest and flanks grey; bare patch round eye black. The sexes are alike. Wing 137 to 150 mm. Juvenile plumage undescribed.

**Range in Eastern Africa:** The Boran district of southern Abyssinia.

**Habits:** It is curious that this bird has only recently been discovered, as it seems to be locally plentiful and quite conspicuous. It inhabits acacia thorn bush in parties of half a dozen or so during the non-breeding season and appears to be Starling-like in its habits. It does not breed in colonies, but more than two birds are frequently seen round a nest.

**Nest and Eggs:** The nest is a globular untidy looking structure of thorn twigs, inside which is a grass chamber with a thick vertical neck containing an entrance tube. It is placed in the top of a thorn tree. Eggs up to six, cream colour with pale lilac blotches tending to form a ring at the larger end; about 27 × 20 mm.

**Recorded breeding:** Yavello, southern Abyssinia, March.

**Food:** No information.

**Call:** A high-pitched 'chek.'

FAMILY—**STURNIDÆ. STARLINGS and OXPECKERS.** Genera: *Sturnus, Creatophora, Neocichla, Cinnyricinclus, Pholia, Speculipastor, Lamprocolius, Lamprotornis, Cosmopsarus, Onychognathus, Galeopsar, Pilorhinus, Pœoptera, Stilbopsar, Spreo* and *Buphagus*.

Thirty-seven species of Starlings and two species of Oxpecker are found in Eastern Africa. The European Starling is a very rare visitor and possibly only accidental. Starlings are found in all types of wooded country and are usually in flocks or parties and are noisy, conspicuous and gregarious birds. They have strong powers of flight and are, for the most part, highly beneficial to man in Africa. Their normal food is insects and their larvæ, with fruit when ripe, and in Africa the damage they do is negligible compared to the immense benefits they confer in the destruction of noxious insects. The velvety black spots found on the secondaries and wing-coverts of many of the Glossy Starlings contain an electric blue centre when seen in certain lights.

The Oxpeckers are also conspicuous birds haunting game and domestic animals: their economic value is a subject of some controversy.

### KEY TO THE ADULT STARLINGS OF EASTERN AFRICA

1 Bill yellow or yellow with a black tip:     2–3

2 Black patch at base of foreneck, outer web of secondaries white:    WHITE-WINGED BABBLING STARLING *Neocichla gutturalis*   1183

3 General colour iridescent green, bronze, violet and blue black:    EUROPEAN STARLING, breeding dress *Sturnus vulgaris*   1181

4 Bill yellow at base, red at tip:    YELLOW-BILLED OXPECKER *Buphagus africanus*   1217

5 Bill wholly red:    RED-BILLED OXPECKER *Buphagus erythrorhynchus*   1218

6 Bill white or whitish:     8–9

7 Bill black, blackish or dark brown, greyish brown or greenish brown:     10–72

8 Primaries chestnut:    WHITE-BILLED STARLING *Pilorhinus albirostris*   1207

9 Primaries black or blackish:    WATTLED STARLING *Creatophora cinerea*   1182

10 Speckled all over with buff and white:    EUROPEAN STARLING non-breeding dress *Sturnus vulgaris*\*    1181

11 Top of head white:    WHITE-CROWNED STARLING *Spreo albicapillus*    1212

12 Chest to belly golden yellow:    GOLDEN-BREASTED STARLING *Cosmopsarus regius*    1198

13 Below, white with black streaks and blobs:    VIOLET-BACKED STARLING, female *Cinnyricinclus leucogaster*    1184

14 Wholly dusky ash with a greenish gloss:    ASHY STARLING *Cosmopsarus unicolor*    1199

15 Below, wholly tawny and whitish:    SHARPE'S STARLING *Pholia sharpii*    1185

16 Chest to belly or breast to belly white:    17–21

17 White patch at base of primaries:    MAGPIE STARLING *Speculipastor bicolor*    1187

18 No white patch at base of primaries:    19–21

19 Upperside and neck all round metallic violet or ruby and violet:    VIOLET-BACKED STARLING, male *Cinnyricinclus leucogaster* 1184

20 Upperside, neck and chest pale ash grey:    FISCHER'S STARLING *Spreo fischeri*    1211

\* *Note:* Base of lower mandible dull yellowish.

| | | |
|---|---|---|
| 21 Upperside, neck, chest and breast black: | ABBOTT'S STARLING *Pholia femoralis* | 1186 |
| 22 Belly chestnut: | | 23–27 |
| 23 Axillaries and under tail-coverts white: | SUPERB STARLING *Spreo superbus* | 1216 |
| 24 Axillaries and under tail-coverts not white: | | 25–27 |
| 25 Head sooty brown, mantle and chin to breast glossy green: | CHESTNUT-BELLIED STARLING *Spreo pulcher* | 1213 |
| 26 Upperside and chin to breast glossy deep blue: | HILDEBRANDT'S STARLING *Spreo hildebrandti* | 1214 |
| 27 Above, richer violet blue; chestnut of breast to belly much darker: | SHELLEY'S STARLING *Spreo shelleyi* | 1215 |
| 28 Chest to belly glossy or metallic blue black, blue green or bronze: | | 29–72 |
| 29 Primaries chestnut: | | 30–54 |
| 30 Cushion of feathers on forehead: | BRISTLE-CROWNED STARLING *Galeopsar salvadorii* | 1206 |
| 31 Feathers of forehead normal: | | 32–54 |
| 32 Size smaller, wing under 106 mm.: | | 33–35 |
| 33 Wings black: | STUHLMANN'S STARLING, female *Stilbopsar stuhlmanni* | 1209 |

| | | |
|---|---|---|
| 34 Wings ashy: | NARROW-TAILED STARLING, female *Pœoptera lugubris* | 1208 |
| 35 Mantle glossy bronzy black: | KENRICK'S STARLING, female *Stilbopsar kenricki* | 1210 |
| 36 Size larger, wing over 115 mm. | | 37–54 |
| 37 Head and neck with grey markings: | | 38–43 |
| 38 Head and neck grey streaked with glossy green: | CHESTNUT-WING STARLING, female *Onychognathus fulgidus* | 1200 |
| 39 Head and neck greenish or bluish black spotted with grey; which extend over most of body plumage: | SLENDER-BILLED CHESTNUT-WING STARLING, female *Onychognathus tenuirostris* | 1205 |
| 40 Head and neck wholly grey: | SOMALI CHESTNUT-WING STARLING, female *Onychognathus blythii* | 1204 |
| 41 Head and neck grey streaked with glossy blue black: | | 42–43 |
| 42 Bill and tail short: | WALLER'S CHESTNUT-WING STARLING, female *Onychognathus walleri* | 1201 |
| 43 Bill and tail long: | RED-WING STARLING, female *Onychognathus morio* | 1203 |
| 44 Head and neck not grey: | | 45–54 |
| 45 Below washed with glossy green: | | 46–47 |

| | | |
|---|---|---|
| 46 Size larger, wing over 169 mm.; tail longer, bill shorter: | SOMALI CHESTNUT-WING STARLING, male *Onychognathus blythii* | **1204** |
| 47 Size smaller, wing under 156 mm.; tail shorter, bill longer: | SOCOTRA CHESTNUT-WING STARLING *Onychognathus frater* | **1202** |
| 48 Below, glossy blue black or violet: | | 49–54 |
| 49 Central tail feathers appreciably elongated: | SLENDER-BILLED CHESTNUT-WING STARLING, male *Onychognathus tenuirostris* | **1205** |
| 50 Central tail feathers not appreciably elongated: | | |
| 51 Bill shorter: | WALLER'S CHESTNUT-WING STARLING, male *Onychognathus walleri* | 51–54 **1201** |
| 52 Bill longer and stouter: | | 53–54 |
| 53 Sides of face and chin and throat glossy green: | CHESTNUT-WING STARLING, male *Onychognathus fulgidus* | **1200** |
| 54 Sides of face and chin and throat glossy blue black or violet black: | RED-WING STARLING, male *Onychognathus morio* | **1203** |
| 55 No chestnut in primaries: | | 56–72 |
| 56 Tail long: | | 57–60 |
| 57 Head and throat metallic dark blue and violet: | MEVE'S LONG-TAILED GLOSSY STARLING *Lamprotornis mevesii* | **1197** |

58 Head and throat metallic dull old gold: — 59–60

59 Mantle metallic dark blue and violet: — RÜPPELL'S LONG-TAILED GLOSSY STARLING *Lamprotornis purpuropterus* — 1196

60 Mantle metallic green with a bluish wash: — LONG-TAILED GLOSSY STARLING *Lamprotornis caudatus* — 1195

61 Tail normal: — 62–72

62 Above, and below, glossy blue black and violet: — STUHLMANN'S STARLING, male *Stilbopsar stuhlmanni* — 1209

63 Above and below, glossy bronzy and black: — KENRICK'S STARLING, male *Stilbopsar kenricki* — 1210

64 Top of head velvety violet: — PURPLE-HEADED GLOSSY STARLING *Lamprocolius purpureiceps* — 1193

65 Top of head metallic violet: — PURPLE GLOSSY STARLING *Lamprocolius purpureus* — 1191

66 Top of head metallic green or blue: — 67–72

67 Metallic coppery spot on side of neck, middle section of secondaries and tail velvety black: — SPLENDID GLOSSY STARLING *Lamprocolius splendidus* — 1192

68 Scapulars, rump and tail glossy violet, no black spots on wing-coverts: — BLACK-BREASTED GLOSSY STARLING *Lamprocolius corruscus* — 1194

## STARLINGS and OXPECKERS

69 Rump and tail metallic violet and blue, black spots on wing-coverts:    BRONZE-TAILED STARLING *Lamprocolius chalcurus*    **1189**

70 Rump and tail metallic golden green or bluish:    71–72

71 Longer tailed, inner webs of primaries notched:    BLUE-EARED GLOSSY STARLING *Lamprocolius chalybæus*    **1188**

72 Shorter tailed, no notch on inner webs of primaries:    LESSER BLUE-EARED GLOSSY STARLING *Lamprocolius chloropterus*    **1190**

## 1181 EUROPEAN STARLING. *STURNUS VULGARIS* Linnæus.

*Sturnus vulgaris vulgaris* Linn.
*Sturnis vulgaris* Linnæus, Syst. Nat. 10th ed. p. 167, 1758: Sweden.

**Distinguishing characters:** Adult male, iridescent green, bronze, violet and blue black, green predominating above and violet and blue black below; all body feathers tipped with buff or white giving a spotted appearance; in the breeding season the pale feather tips mainly or wholly disappear. Flight feathers and tail bronzy ash, edged and tipped with black and buff. The female and immature bird is rather duller and more broadly spotted. Wing 118 to 136 mm.

**General distribution:** Northern and central Europe to the valley of the Volga; in non-breeding season to western Asia, northern Africa and Arabia. Introduced into South Africa about 1898.

**Range in Eastern Africa:** One record from Dessie, northern Abyssinia, 12th June, 1936; the bird had been ringed in Silesia on 19th May, 1932.

**Habits:** Only known in Eastern Africa from one example, and that curiously enough a ringed one and taken in the summer when the bird should have been back in Europe. It seems unlikely that any considerable number wander so far south, but it is a bird that is extending its range and might easily start to colonize a suitable part of Eastern Africa, as it has done in South Africa where it was introduced.

**Distribution of other races of the species:** The Faroe Islands, Azores, other parts of Europe, Asia and India. Introduced into America, Australia and elsewhere where they are still spreading.

### 1182 WATTLED STARLING. *CREATOPHORA CINEREA* (Menschen).

*Rallus cinerea* Menschen, Ind. Syst. Mus. Geneve, pp. 40-41, No. 17, 1787: Foothills of the Cape of Good Hope, South Africa.

**Distinguishing characters:** Adult male, general colour grey, sometimes isabelline grey; lower rump usually paler; top of head, face, chin and throat bare and black, and with wattles on forehead, centre of crown and throat; bare skin on hinder crown and behind and below eye yellow; flight feathers and tail black with a greenish and bronzy gloss; primary coverts white, sometimes secondary coverts also white; bill white or fleshy white. In non-breeding dress the whole head is feathered with no wattles; bare black streaks on each side of throat. The adult female is similar to the adult male in non-breeding dress, but has the primary coverts black and the lower rump distinctly whitish. Wing 112 to 125 mm. The young male is similar to the adult female; the young female is much browner above and has the bill and the bare streaks on each side of throat dusky, yellowish or greenish yellow.

**General distribution:** The Sudan, Eritrea and British Somaliland to Angola, South Africa and south-western Arabia.

**Range in Eastern Africa:** Southern Sudan, Eritrea and British Somaliland to the Zambesi River, once recorded from Zanzibar Island.

**Habits:** A highly gregarious species with irregular local movements depending on the food supply. In South Africa it is one of the species called 'Locust Birds' from their habit of appearing in tens of thousands during a locust visitation. Sometimes they are found with other species, such as Oxpeckers and Glossy Starlings, and when in large flocks perform remarkable aerial evolutions. They prefer short grass country or open bush and are rather silent for Starlings.

**Nest and Eggs:** A mass of sticks and grass usually domed if single, but often piled on top of one another in a heap. If there are not enough bushes or trees in the locality, nests are made on the ground or in holes in the earth. Eggs two or three, but up to four or five are recorded in the south, pale blue, occasionally with black or brownish flecks; about $28 \times 21$ mm.

**Recorded breeding:** Abyssinia, irregularly, May and August. Kenya Colony, Turkana (breeding condition), July; Wajheir, May. Northern Tanganyika Territory, April and May.

**Food:** Caterpillars, grasshoppers, locusts, termites when swarming, and fruit in season. They follow a locust swarm closely and breed where the locust larvæ are, but if the locusts move too soon all the nests and young birds will be deserted.

**Call:** A rasping squeaking whistle, and a high-pitched squeak while feeding.

### 1183 WHITE-WINGED BABBLING STARLING. *NEOCICHLA GUTTURALIS* (Bocage).

*Neocichla gutturalis angustus* Fried.
*Neocichla gutturalis angustus* Friedmann, Journ. Wash. Acad. Sci. 20, p. 434, 1930: Muhulala, Kilimantinde, east-central Tanganyika Territory.

**Distinguishing characters:** Head and neck grey; mantle ashy brown, rump ashy; wing and tail bronzy black; edges and bases of secondaries white forming a distinct white patch; below, deep buff including under wing-coverts; centre of belly white; a black patch in centre of lower throat; bill black; eyes yellow. The sexes are alike. Wing 101 to 114 mm. The young bird has the top of the head blackish or dark brown; feathers of neck, mantle and wing-coverts blackish edged with buffish grey; below, buff with large black drop-like markings from chin to upper belly; bill yellow, usually with a blackish tip; eyes bluish grey.

**General distribution:** Central Tanganyika Territory to Nyasaland and Northern Rhodesia.

**Range in Eastern Africa:** Central to south-western Tanganyika Territory.

**Habits:** A bird of rather heavy flight, keeping to tree-tops as a rule, and looking not unlike a large *Prionops* on the wing. Somewhat shy and nowhere common. It usually occurs in flocks of ten to forty or so, and is a sociable Starling-like bird occasionally seen feeding on the ground.

**Nest and Eggs:** Undescribed.

**Recorded breeding:** No records.

**Food:** No particular information.

**Call:** Rather strident, and reminiscent of a Babbler, or even of a Parrot, but not so loud.

**Distribution of other races of the species:** Angola, the habitat of the nominate race.

### 1184 VIOLET-BACKED STARLING. *CINNYRICINCLUS LEUCOGASTER* (Boddært).

*Cinnyricinclus leucogaster leucogaster* (Bodd.).
*Turdus leucogaster* Boddært, Tabl. Pl. Enlum. p. 39, 1783: Whidah, Dahomey.

**Distinguishing characters:** Adult male, general colour metallic violet plum or cherry with bronzy and greenish reflections and having a scaly appearance; chest to under tail-coverts white. The female is mottled above with black and tawny with a slight iridescent blue-black wash on wing-coverts and inner secondaries; below, white, or buff and white, streaked and blobbed with black. Wing 90 to 114 mm. The young bird is similar to the adult female. Albinistic examples occur.

**General distribution:** Senegal to Eritrea, Abyssinia and Uganda; in non-breeding season to Kenya Colony and Tanganyika Territory.

**Range in Eastern Africa:** The Sudan, Eritrea, south-western Abyssinia and Uganda; in non-breeding season to Kenya Colony and western Tanganyika Territory as far south as the Kungwe-Mahare Mts.

**Habits:** Gregarious and beautiful birds subject to considerable movements but probably not so regularly migratory as the southern race. Commonly met with when fruit is ripe or when termites are swarming, but irregular in appearance. Their flight is fast and straight, but they are often seen hawking for insects in the air. Birds of forest and woodland, especially fond of *Podocarpus* fruit.

**Nest and Eggs:** In a hole in a tree at any height from the ground, lined with wool, hair or leaves. Eggs two or three, pale blue or bluish green, spotted with pale brown and with light purple undermarkings; about 24 × 17 mm.

**Recorded breeding:** Nigeria, March and April.

**Food:** Fruit and insects.

**Call:** A stuttering whistle often followed by a whine, also a twittering noise in flight.

*Cinnyricinclus leucogaster verreauxi* (Boc.).
*Pholidauges verreauxi* Bocage, in Finsch & Hartlaub, Vög. Ost. Afr. p. 867, 1870: Caconda, Angola.

**Distinguishing characters:** Adult male, differs from the nominate race in having white outer webs to the outer tail feathers in the male. The females and young birds of the two races are indistinguishable. Wing 91 to 113 mm.

**General distribution:** Kenya Colony, Tanganyika Territory, eastern and southern Belgian Congo, Angola, Damaraland, Northern Rhodesia, Portuguese East Africa, eastern Cape Province and Natal; in non-breeding season to the Sudan, Abyssinia, northern Belgian Congo, Uganda, Portuguese Congo and Angola.

**Range in Eastern Africa:** Kenya Colony and Tanganyika Territory to the Zambesi River, also Zanzibar and Mafia Islands; in non-breeding season to the Sudan and Abyssinia.

**Habits:** As for the northern race, but migratory with extensive northward movements in the non-breeding season, and less confined to tall trees. Eggs number up to four or five in the south.

**Recorded breeding:** Kenya Colony, February to April. Belgian Congo, October. Tanganyika Territory, October and November. Zanzibar Island, January to March. Nyasaland, September to December. Rhodesia, September to December. Natal, January (rare).

*Cinnyricinclus leucogaster arabicus* Grant & Praed.
*Cinnyricinclus leucogaster arabicus* C. Grant & Mackworth-Praed, Bull. B.O.C. 63, p. 7, 1942: Hajeilah, Yemen, southern Arabia.

**Distinguishing characters:** Differs from the nominate race in that the female is duller above, earth brown, not black, with the tawny edging to the feathers dull and sparse or totally absent. The

males are indistinguishable from those of the nominate race. Wing 94 to 112 mm. The young bird is similar to the adult female.

**General distribution:** The Sudan, Abyssinia and Arabia.

**Range in Eastern Africa:** Roseires area of the eastern Sudan to Abyssinia, except south-western area, and French and British Somalilands.

**Habits:** Indistinguishable from other races, but found in open country.

**Recorded breeding:** Arabia, June. British Somaliland, April and May.

**1185 SHARPE'S STARLING.** *PHOLIA SHARPII* (Jackson).
*Pholidauges sharpii* Jackson, Bull. B.O.C. 8, p. 22, 1898: Ravine, western Kenya Colony.

**Distinguishing characters:** Upperside including sides of face, and sides of chest metallic blue black with violet reflections; below, pale tawny, darker on belly and under tail-coverts. The sexes are alike. Wing 93 to 108 mm. The young bird has black arrow-shaped markings below.

**General distribution:** Eastern Belgian Congo to Abyssinia and Tanganyika Territory.

**Range in Eastern Africa:** Uganda to southern Abyssinia, Kenya Colony and Tanganyika Territory.

**Habits:** An inhabitant of high tree tops in forest country usually in small flocks and immediately recognizable by its call. Local but not uncommon.

**Nest and Eggs:** Once reported to be nesting with Swallows in holes of a cliff face, but little is recorded of its nesting habits, and it is more likely to breed in holes in trees as a rule. Eggs undescribed.

**Recorded breeding:** Northern Tanganyika Territory (breeding condition), October and January.

**Food:** Tree fruit and berries.

**Call:** Very thin and sharp, also a short song-phrase of four or five notes and a low sweet whistle.

## STARLINGS and OXPECKERS

**1186 ABBOTT'S STARLING.** *PHOLIA FEMORALIS* (Richmond).

*Pholidauges femoralis* Richmond, Auk, p. 160, 1897: Mt. Kilimanjaro, north-eastern Tanganyika Territory.

**Distinguishing characters:** Similar above to Sharpe's Starling, but bill longer, and chin to breast metallic blue black with violet reflections; belly and under tail-coverts buffish white. The sexes are alike. Wing 93 to 102 mm. The young bird is greyish black above with a slight greenish gloss; below, throat buff; chest and breast dusky brown; belly and under tail-coverts buffish white with broad black streaks.

**Range in Eastern Africa:** Southern Kenya Colony and north-eastern Tanganyika Territory from Escarpment to Mt. Kilimanjaro.

**Habits:** Another local forest species of rather curious distribution, shy and little known. Usually in flocks feeding on the fruits of trees. Believed to be extensive local migrants.

**Nest and Eggs:** Undescribed.

**Recorded breeding:** Kilimanjaro area, February and March.

**Food:** As above.

**Call:** A call of six whistled notes, three ascending and three descending in scale.

**1187 MAGPIE STARLING.** *SPECULIPASTOR BICOLOR* Reichenow.

*Speculipastor bicolor* Reichenow, O.C. p. 108, 1879: Kipini, mouth of Tana River, eastern Kenya Colony.

**Distinguishing characters:** Adult male, upperside and chin to chest glossy blue black; basal half of primaries and rest of underside including under wing-coverts creamy white. The female has the head and chin to chest dark grey; rest of upperparts less glossy than the male; a black glossy band across chest. Wing 105 to 122 mm. The young bird is dark greyish brown above; throat and upper breast paler.

**Range in Eastern Africa:** Southern Abyssinia, Kenya Colony and southern Italian Somaliland.

**Habits:** A species of irregular appearance in most localities and given to very wide wanderings. It is generally noticed in small

flocks at some height which can be identified by their shrill whistling cry, but it is sometimes abundant when figs or other fruit are ripe, and is usually shy.

**Nest and Eggs:** Nests as a rule are in holes in anthills, made of grass and green leaves in chambers at the end of tunnels. Eggs three or four, bluish green heavily speckled and spotted with rufous; about 27 × 20 mm.

**Recorded breeding:** Italian Somaliland, April and May.

**Food:** Fruit and insects.

**Call:** A peculiar shrill whistling cry in flight.

**1188** BLUE-EARED GLOSSY STARLING. *LAMPROCOLIUS CHALYBÆUS* (Hemprich & Ehrenberg).
*Lamprocolius chalybæus chalybæus* (Hemp. & Ehr.).
*Lamprotornis chalybæus* Hemprich & Ehrenberg, Symb. Phys. fol. y. pl. 10, 1828: Ambukol, Dongola, northern Sudan.

**Distinguishing characters:** General colour metallic golden green; hinder part of crown, nape, back and sides of neck metallic blue; throat washed with blue; black spot at tips of wing-coverts and inner secondaries; wing shoulder violet and blue; primaries notched on inner webs; rump and upper belly metallic blue; tail washed with blue; lower belly, under wing-coverts and axillaries violet. The sexes are alike, but the females are usually smaller than the males. Wing 127 to 156 mm. The young bird is sooty black washed with golden bottle green above, brighter on wings; tail bluish; below, a very slight wash of green. Albinistic examples occur.

**Range in Eastern Africa:** The Sudan and Eritrea, British Somaliland, northern Uganda and Kenya Colony, except southeastern area.

**Habits:** A common and noisy species, in places abundant. Generally seen in small flocks in open woodland feeding on the ground unless tree fruit happens to be ripe. Said to hop when on the ground, not to walk or run. Assemble in large flocks in the non-breeding season, and roost in large leafy trees or reed beds with a deafening chattering. Makes considerable local migrations. The wings make a loud and distinctive sound as a flock passes.

**Nest and Eggs:** Nest in a hole in a tree made of grass, feathers, etc., and occasionally the bulky nests of other species in the tops of thorn trees are made use of. It is doubtful if they ever make their own nests in trees. Eggs two to four, pale greenish blue, plain or spotted with dull rufous and grey; about 29 × 19 mm.

**Recorded breeding:** Darfur, April to June. Kenya Colony, March to July, also January.

**Food:** Insects, seeds and fruit.

**Call:** A musical whistle, also a variety of harsh notes and chatterings rather high pitched.

*Lamprocolius chalybæus sycobius* Hartl. **Pl. 78.**
*Lamprocolius sycobius* Hartlaub, J.f.O. p. 19, 1859: Tete, Portuguese East Africa.

**Distinguishing characters:** Differs from the nominate race in having the whole head and neck golden green; ear-coverts only blue. Wing 121 to 144 mm.

**General distribution:** Uganda and Kenya Colony to the Rhodesias, Bechuanaland and the eastern Transvaal.

**Range in Eastern Africa:** South-western Uganda and south-eastern Kenya Colony to the Zambesi River.

**Habits:** As for the nominate race, congregating into loose wavy flocks in the non-breeding season. These flocks often perch for a few seconds on a tree and then all fly off again. The call is a loud 'kwee-kwee' followed by a shrill rattling whistle. A flock utters a number of whistling, chirping and clicking notes, but the alarm call is a hysterical whistle of 'kwee-kwee' at which all conversation ceases and the flock takes wing (Vincent).

**Recorded breeding:** Tanganyika Territory, October to December. Nyasaland, September to December. Rhodesia, October to December.

**Distribution of other races of the species:** West Africa and Angola.

**1189** BRONZE-TAILED STARLING. *LAMPROCOLIUS CHALCURUS* (Nordmann).
*Lamprocolius chalcurus emini* Neum. **Pl. 78.**
*Lamprocolius chalcurus emini* Neumann, J.f.O. p. 81, 1920: Fatiko, northern Uganda.

**Distinguishing characters:** Similar to the Blue-eared Glossy Starling, but the ear-coverts deeper blue or violet; the central tail

feathers violet and blue, not green washed with blue; more violet on upper belly. The sexes are alike, but the female is usually smaller than the male. Wing 120 to 143 mm. The young bird is similar to that of the Blue-eared Glossy Starling.

**Range in Eastern Africa:** Southern Sudan, northern and eastern Uganda and western Kenya Colony as far south as Nyarondo.

**Habits:** Indistinguishable in the field from the last species as far as is known. It may have a distinctive behaviour, but no separate records have been made and it is possible that many of the records of the Blue-eared Glossy Starling really refer to this species.

**Nest and Eggs:** Nest (West African race) is a thick pad of grass and feathers in a natural hole in a tree. Eggs four, pale blue either plain or with light brown spots and speckles; about 27 × 20 mm.

**Recorded breeding:** No records.

**Food:** As for allied species.

**Call:** Nothing especially recorded.

**Distribution of other races of the species:** West Africa, the nominate race being described from Senegal.

**1190 LESSER BLUE-EARED GLOSSY STARLING.** *LAMPROCOLIUS CHLOROPTERUS* (Swainson).
*Lamprocolius chloropterus chloropterus* (Swain.). **Pl. 78.**
*Lamprotornis chloropterus* Swainson, An. Men. p. 359, 1838: West Africa.

**Distinguishing characters:** Similar to the Blue-eared Glossy Starling but smaller, with a shorter first primary, and lacking the distinct notches on the inner webs of the primaries; above, varying from metallic golden green to steel blue and violet; rump uniform with mantle; wing shoulder violet or violet and purple. The sexes are alike. Wing 109 to 133 mm. The young bird has the head and underside earth brown to ashy, with a slight iridescence on the crown; mantle, wings and tail duller than adult.

**General distribution:** Senegal to Abyssinia and Eritrea, Uganda and Kenya Colony.

**Range in Eastern Africa:** Central and southern Sudan, Eritrea, Abyssinia, Uganda and western Kenya Colony.

**Habits:** Not easy to distinguish in the field from either of the foregoing species, but is somewhat more lightly built and has a

## STARLINGS and OXPECKERS

rather different series of calls. Its habits are identical and it is often seen in very large flocks in the non-breeding season feeding on the fruit of trees. Like all Starlings will congregate at a locust or termite swarm.

**Nest and Eggs:** In the hole of a tree, an untidy mass of grass straw, leaves, etc. Eggs, as for the Blue-eared Glossy Starling but slightly smaller; about 27 × 16·5 mm.

**Recorded breeding:** No dates recorded.

**Food:** As above.

**Call:** Many chattering whistling calls but appreciably different from those of the two previous species to a trained ear. The male is said to have a pleasant seasonal song.

*Lamprocolius chloropterus elisabeth* Stres.
*Lamprocolius chloropterus elisabeth* Stresemann, O.M. p. 173, 1924: South Ufipa, south-western Tanganyika Territory.

**Distinguishing characters:** Adult similar to the preceding race. The young bird is russet brown below. Wing 110 to 126 mm.

**General distribution:** Kenya Colony to Tanganyika Territory, Northern Rhodesia, Nyasaland and Portuguese East Africa.

**Range in Eastern Africa:** South-eastern Kenya Colony and Tanganyika Territory to the Zambesi River.

**Habits:** As for other Glossy Starlings.

**Recorded breeding:** Nyasaland, September to December. Rhodesia, August and September.

**1191 PURPLE GLOSSY STARLING.** *LAMPROCOLIUS PURPUREUS* (Müller). **Pl. 78.**
*Turdus purpureus* P. L. S. Müller, Syst. Nat. Suppl. p. 143, 1776: Dahomey, West Africa.

**Distinguishing characters:** Forehead and sides of head metallic violet; nape metallic blue; mantle and wings metallic golden green; wing shoulder, rump and tail metallic blue, latter with some violet reflections; wing-coverts and innermost secondaries tipped with velvety black; below, wholly metallic violet or violet and blue. The sexes are alike. Wing 131 to 162 mm. The young bird has the head and nape dull metallic violet or bluish; mantle dull metallic golden green or blue; wings dull metallic green; rump

and tail dull metallic violet and blue; below, black with a violet wash.

**General distribution:** Senegal and Cameroons to the Sudan, Uganda and Kenya Colony.

**Range in Eastern Africa:** Western and southern Sudan, Uganda and western Kenya Colony.

**Habits:** Inhabits much the same country as the Blue-eared Glossy Starling and is often found with it. Its large bill and long neck, however, are a reasonable means of identification in the field. Stoneham notes its habit of stretching its neck and watching an intruder sideways. Sometimes occurs in enormous flocks among acacias in short grass country, and is also seen in small parties about rocky hills.

**Nest and Eggs:** Nest usually in the hole of a tree or in an old stump. Eggs two, pale blue freckled with reddish brown; about 27·5 × 20 mm.

**Recorded breeding:** Nigeria, April to July. Equatoria, April. Uganda, April.

**Food:** Fruit and insects.

**Call:** Varied calls, somewhat squeaky in character. A flock keep up a persistent rippling chattering while feeding.

**1192 SPLENDID GLOSSY STARLING.** *LAMPROCOLIUS SPLENDIDUS* (Vieillot).
*Lamprocolius splendidus splendidus* (Vieill.). **Pl. 78.**
*Turdus splendidus* Vieillot, Enc. Méth. 2, p. 653, 1822: Malimbe, Portuguese Congo.

**Distinguishing characters:** Adult male, forehead black; crown metallic golden green; nape metallic blue; upper mantle metallic golden green; lower mantle, scapulars, rump and upper tail-coverts metallic blue, latter with a green wash; lores and ear-coverts velvety black, latter with green reflections; spot on side of neck metallic copper; wing-coverts metallic blue and green, with subterminal velvety black spots; flight feathers velvety black, metallic green and blue; secondaries partly metallic purple; primaries deeply notched on inner webs; tail at base metallic blue and purple, middle section velvety black, ends of feathers metallic blue and green; below, chin to belly metallic violet and tarnished copper; sides of chest, lower belly and under tail-coverts metallic blue; centre of chest with blue reflections. The female is metallic blue below with some violet and tarnished copper. Wing 130 to 169 mm. The young male is duller above with only an indication of the black spots on the wing-coverts; no metallic copper spot on side of neck; below, blue black. The young female is blackish brown below with practically no blue-black gloss.

**General distribution:** Fernando Po and Nigeria to Abyssinia, Uganda, Kenya Colony, Tanganyika Territory, Angola and north-western Northern Rhodesia.

**Range in Eastern Africa:** Southern Sudan and western and southern Abyssinia to Uganda and western Kenya Colony, also Ukerewe Island, Lake Victoria.

**Habits:** Probably the finest of the short-tailed Glossy Starlings, identifiable by its large size and by the swishing noise produced by its wings in flight. Distinctly a tall woodland and secondary forest bird, frequently seen in the strips of 'gallery' forest that border the banks of rivers. Somewhat migratory, possibly regularly so, but its movements have not yet been worked out. When seen sideways in the air the back and breast are curiously humped. Generally shy.

**Nest and Eggs:** In holes in trees, of dry grass, etc. Eggs two or three, pale blue, freckled and spotted with reddish brown and with mauve undermarkings; about 31 × 22·5 mm.

**Recorded breeding:** Cameroons, August, (nestling) September. Uganda, February and March. Nandi, Kenya Colony, May to August. Katanga, Belgian Congo, September to November. Northern Rhodesia, a breeding visitor from August to November.

**Food:** Tree fruit, particularly wild figs, but probably almost omnivorous.

**Call:** Many nasal and guttural calls, also a series of whistles, the most usual call is a 'chak' or 'quank-quonk.'

**Distribution of other races of the species:** West Africa.

**1193** PURPLE-HEADED GLOSSY STARLING. *LAMPRO-COLIUS PURPUREICEPS* Verreaux. **Pl. 78.**

*Lamprocolius purpureiceps* J. & E. Verreaux, Rev. Mag. Zool. p. 418, 1851: Gabon.

**Distinguishing characters:** Top of head velvety violet; side of head, neck in part and chest metallic violet; rest of plumage metallic

violet-blue; mantle and belly greenish; tail with a coppery wash; no notch on inner webs of primaries, and no black spots on wing-coverts. The sexes are alike. Wing 109 to 132 mm. The young bird is duller above and dull black with some blue and green reflections below.

**General distribution:** Southern Nigeria and Gabon to south-western Belgian Congo and Uganda.

**Range in Eastern Africa:** Uganda.

**Habits:** A strictly forest species, inhabiting the tops of the highest trees in small parties. Locally migratory.

**Nest and Eggs:** Nest in holes in trees. Eggs undescribed.

**Recorded breeding:** Western Uganda, April.

**Food:** Mainly tree fruits.

**Call:** Described as a few decided notes followed by a series of half-tones, and then louder notes again.

**1194 BLACK-BREASTED GLOSSY STARLING.** *LAMPRO-COLIUS CORRUSCUS* (Nordmann).
*Lamprocolius corruscus mandanus* V. Som. **Pl. 78.**
*Lamprocolius corruscus mandanus* Van Someren, Bull. B.O.C. 41, p. 124, 1921: Manda Island.

**Distinguishing characters:** Adult male, top of head metallic blue or blue-green; neck all round, upper mantle and chin to chest metallic green, often with a violet or purple wash; ear-coverts violet; lower mantle metallic blue; scapulars, wing shoulder, rump, outer web of primaries and tail glossy violet; wing-coverts and outer webs of secondaries metallic blue with violet reflections; no black spots on wing-coverts; breast violet; belly black; lower flanks with metallic old gold reflections. The female has the belly sooty black. Wing 98 to 115 mm. The young bird is duller metallic green or blue and violet above; below, sooty black with dull green or blue reflections from chin to breast.

**Range in Eastern Africa:** Eastern Kenya Colony to southern Italian Somaliland, Tanganyika Territory and Portuguese East Africa, also Manda, Zanzibar and Mafia Islands.

**Habits:** Common and locally abundant in forest, forest clearings, and thick bush or in other forms of wooded country. Usually in fair sized flocks, and the noisy babbling sound made by a flock can be heard all day long. In the breeding season the male sings from a dead bough in the forest.

**Nest and Eggs:** In a hole in a tree or palm stump. Eggs three, pale bluish green; no measurements available.

**Recorded breeding:** Zanzibar Island, probably November and December. Mafia Island group, December and January.

**Food:** Berries, flowers, insects, etc.

**Call:** Very varied indeed and incorporates the notes of other species. Pakenham gives a list of nine or ten species whose calls are mimicked in this way. There is also a strong brief explosive warbling song.

*Lamprocolius corruscus vaughani* Bann.
*Lamprocolius corruscus vaughani* Bannerman, Bull. B.O.C. 46, p. 126, 1926: Pemba Island.

**Distinguishing characters:** Differs from the nominate race in having the top of the head violet. Wing 106 to 124 mm.

**Range in Eastern Africa:** Pemba Island.

**Habits:** A much more silent bird than the nominate race, living mainly in mangrove bush and among clove plantations. Usually in pairs or small parties, most active in the early mornings, rather skulking and silent by day. The favourite food appears to be the purple fruit of *Clausena anisata* (J. H. Vaughan).

**Recorded breeding:** November and December.

*Lamprocolius corruscus jombeni* V. Som.
*Lamprocolius corruscus jombeni* Van Someren, J.E.A. & U. Nat. Hist. Soc. No. 37, p. 5, 1931: Jombeni, north-east of Mt. Kenya, central Kenya Colony.

**Distinguishing characters:** Differs from the Manda Island race in being larger. Wing 116 to 121 mm.

**Range in Eastern Africa:** Mt. Kenya and the Jombeni Hills to the Garissa area, central Kenya Colony.

**Habits:** As for other races.

**Recorded breeding:** No records.

**Distribution of other races of the species:** South Africa, the nominate race being described from eastern Cape Province.

**1195 LONG-TAILED GLOSSY STARLING.** *LAMPROTORNIS CAUDATUS* (Müller).

*Turdus caudatus* P. L. S. Müller, Syst. Nat. Suppl. p. 144, 1776: Senegal.

**Distinguishing characters:** Tail long and graduated; head all round and throat metallic dull old gold; neck, mantle, wings, chest and breast metallic green with a bluish wash; flight feathers bluer; velvety black spots at end of wing-coverts and scapulars; rump metallic dark blue with violet reflections; tail and belly metallic dark blue and violet; centre of upper belly metallic coppery gold; some violet between old gold of sides of face and throat and green of neck. The female is usually smaller than the male. Wing 151 to 203 mm. The young bird is duller than the adult; head all round and throat more purplish and gold, and belly usually dull blackish.

**General distribution:** Senegal to the Sudan.

**Range in Eastern Africa:** Central and south-western Sudan.

**Habits:** A bird of open bush and light woodland, preferring high densely foliaged trees. Generally in small parties and rather shy, moving from tree to tree with harsh raucous cries. On the ground these birds hold their tails erect as a rule.

**Nest and Eggs:** In holes in trees, particularly Baobabs. Eggs two to four, pale blue, occasionally with light brown spots; about 29 × 21 mm.

**Recorded breeding:** Nigeria, September and October. Darfur, August to October.

**Food:** Fruit, insects, etc.

**Call:** Noisy birds with unpleasant raucous screeching cries and continual chattering in flight.

**1196 RÜPPELL'S LONG-TAILED GLOSSY STARLING.** *LAMPROTORNIS PURPUROPTERUS* Rüppell.

*Lamprotornis purpuropterus purpuropterus* Rüpp. **Pl. 78.**
*Lamprotornis purpuropterus* Rüppell, Syst. Uebers. pp. 64, 75, pl. 25, 1845: Shoa, central Abyssinia.

**Distinguishing characters:** Differs from the last species in having a shorter tail, the hind neck and rump violet, mantle dark blue and violet and no black spots on wing-coverts and scapulars. The sexes are alike. Wing 130 to 169 mm., tail 110 to 171 mm.

## STARLINGS and OXPECKERS

The young bird is generally duller; head black slightly violet on top; below, dull black with some metallic blue on chest and breast.

**Range in Eastern Africa:** Southern Sudan, central and eastern Abyssinia, Uganda, Kenya Colony and Tanganyika Territory.

**Habits:** Found in small flocks in any type of open country, often in native villages where it is quite fearless of man. It visits trees of gallery forest for fruit when ripe, but otherwise is not found in forest, and is commonly associated with Spreos or other Starlings. It has a pleasing song of simple but musical phrases of a rather repetitive character.

**Nest and Eggs:** A hole in a tree, usually a living tree, lined with any kind of material available. May also utilize old nests of other birds. Eggs two or three, pale blue or bluish green either plain or marked with rusty spots and blotches; about 29 × 21 mm.

**Recorded breeding:** Uganda, March and April, also probably August. North-western Kenya Colony, April to September.

**Food:** Almost omnivorous.

**Call:** A large variety of squeaky chatterings, with a harsh alarm call, and a varied Thrush-like song.

*Lamprotornis purpuropterus æneocephalus* Heugl.
*Lamprotornis æneocephalus* Heuglin, J.f.O. p. 22, 1863: Anseba, Eritrea.

**Distinguishing characters:** Differs from the nominate race in being usually longer-tailed. Wing 140 to 173 mm.; tail 159 to 211 mm.

**Range in Eastern Africa:** Eritrea to the Sennar and northern Kordofan areas of the Sudan.

**Habits:** As for the nominate race.

**Recorded breeding:** Blue Nile, Sudan, July and August.

**1197** MEVE'S LONG-TAILED GLOSSY STARLING. *LAMPROTORNIS MEVESII* (Wahlberg).
*Lamprotornis mevesii mevesii* (Wahlb.).
*Juida mevesii* Wahlberg, J.f.O. p. 1, 1857: Okavango River, Damaraland.

**Distinguishing characters:** Very similar in general appearance to Rüppell's Long-tailed Glossy Starling, but differs in having the

head, neck and throat dull metallic dark blue and violet (not old gold), and rump metallic coppery gold. Wing 130 to 160 mm.

**General distribution:** Southern Angola and the Limpopo River Valley to Damaraland, Bechuanaland and Nyasaland.

**Range in Eastern Africa:** Nyasaland to the Zambesi River.

**Habits:** Numerous and tame where found, generally a bird of Mopane belt at lower levels. Often seen in small flocks flying from one high tree to another or feeding on the ground. Said to roost in reed beds at night.

**Nest and Eggs:** In clefts of bark or holes of trees, often of Baobabs. Eggs pale greenish blue; about 28·5 × 20 mm.

**Recorded breeding:** Rhodesia, October and November. Nyasaland, February and probably October.

**Food:** Fruits, berries and insects, feeding largely on the ground.

**Call:** A loud whistle, also a harsh alarm call.

**Distribution of other races of the species:** Central Angola.

**1198** GOLDEN-BREASTED STARLING. *COSMOPSARUS REGIUS* Reichenow. **Pl. 78.**

*Cosmopsarus regius* Reichenow, O.C. p. 108, 1879: Massa, Tana River, eastern Kenya Colony.

**Distinguishing characters:** Tail long and steeply graduated; head and neck all round metallic green; ear-coverts, mantle, rump and wings metallic dark blue with a purple wash; tail metallic dull old gold with some blue and violet; below, patch on chest metallic violet, rest of underparts rich metallic golden yellow. The sexes are alike. Wing 117 to 142 mm. The young bird is duller than the adult; head, mantle and rump blackish with some green reflections; sides of head and chin to chest ashy brown.

**Range in Eastern Africa:** Southern Abyssinia, British and Italian Somalilands, eastern Kenya Colony and eastern Tanganyika Territory as far south as the Central Railway line.

**Habits:** In flocks in dry bush country, generally noticed on the top of a tree or leaving it with a chattering cry. In most places excessively shy, but in others is reported to be quite tame. One of the most beautiful of African birds.

**Nest and Eggs:** Nest in a hole or cleft of a tree, sparsely lined with straw, leaves or roots. Eggs two to four, in the northern part of its range up to six, elongated, pale greenish blue minutely speckled with reddish brown; about 27 × 17 mm.

**Recorded breeding:** Eastern Abyssinia, April and probably earlier. Northern Kenya Colony, November. Somaliland, April and May. North-eastern Tanganyika Territory, March.

**Food:** As for other Starlings, no particular information.

**Call:** A whistling chattering cry.

**1199** ASHY STARLING. *COSMOPSARUS UNICOLOR* Shelley.

*Cosmopsarus unicolor* Shelley, Ibis, p. 116, 1881: Ugogo, Dodoma district, central Tanganyika Territory.

**Distinguishing characters:** Tail long and steeply graduated; general colour dusky ash, with a green gloss above, brighter on the wings and tail. The sexes are alike. Wing 116 to 131 mm. The young bird is generally greyer and lighter than the adult.

**Range in Eastern Africa:** Tanganyika Territory from the Mwanza district to Dodoma, Kilosa and Tukuyu districts.

**Habits:** Fairly common in its restricted habitat, and found in small flocks among flat-topped acacias like the Golden-breasted Starling, in semi-arid country.

**Nest and Eggs:** Undescribed.

**Recorded breeding:** No records.

**Food:** Apparently largely insects.

**Call:** Unrecorded.

**1200** CHESTNUT-WING STARLING. *ONYCHOGNATHUS FULGIDUS* Hartlaub.

*Onychognathus fulgidus hartlaubii* Gray. **Pl. 78.**

*Onychognathus hartlaubii* Gray, P.Z.S. p. 291, 1858: Lower Niger River, southern Nigeria.

**Distinguishing characters:** Adult male, head and neck all round glossy green; crown with a bluish wash; rest of plumage deep glossy violet; primaries dark chestnut with black ends; secondaries and tail blue black, latter with glossy greenish edges. The female has the head and neck all round streaked grey and glossy green. Wing 117 to 132 mm. The young bird is black with a slight gloss.

**General distribution:** Sierra Leone to the Sudan.

**Range in Eastern Africa:** Southern Sudan.

**Habits:** A forest species inhabiting high trees, particularly found near hills in forest, usually seen in pairs or in small noisy parties.

**Nest and Eggs:** Undescribed.

**Recorded breeding:** Gold Coast, July.

**Food:** Fruit, seeds and insects.

**Call:** Various extremely raucous and Parrot-like whistles and squawks.

*Onychognathus fulgidus intermedius* Hart.
*Onychognathus intermedius* Hartert, No. Zool. 2, p. 56, 1895: Lukolele, lower Congo River, western Belgian Congo.

**Distinguishing characters:** Differs from the preceding race in having a longer bill. Wing 125 to 138 mm.

**General distribution:** Gabon to northern Angola, the Belgian Congo and Uganda.

**Range in Eastern Africa:** Uganda.

**Habits:** As for the nominate race.

**Recorded breeding:** No records.

**Distribution of other races of the species:** West Africa, the nominate race being described from São Thomé Island.

**1201 WALLER'S CHESTNUT-WING STARLING. *ONYCHOGNATHUS WALLERI* (Shelley).**
*Onychognathus walleri walleri* (Shell.). **Pl. 78.**
*Amydrus walleri* Shelley, Ibis, p. 335, pl. 8, 1880: Usambara Mts., north-eastern Tanganyika Territory.

**Distinguishing characters:** Adult male, differs from the Chestnut-wing Starling in having a shorter tail and a much smaller bill. The female has the top of the head and hind neck streaked grey and glossy blue; sides of face and chin to chest grey; a grey wash from breast to belly. Wing 121 to 140 mm. The young bird is black with a slight gloss.

**General distribution:** Kenya Colony to Tanganyika Territory and northern Nyasaland.

Plate 78

Black-breasted Glossy Starling (p. 700)
Blue-eared Glossy Starling (p. 695)
Bronze-tailed Starling (p. 695)
Golden-breasted Starling (p. 704)
Purple-headed Glossy Starling (p. 699)
Splendid Glossy Starling (p. 698)
Slender-billed Chestnut-wing Starling (p. 710)
Male   Female
Lesser Blue-eared Glossy Starling (p. 696)
Chestnut-wing Starling (p. 705)
Female   Male
Purple Glossy Starling (p. 697)
Waller's Chestnut-wing Starling (p. 706)
Female   Male
Rüppell's Long-tailed Glossy Starling (p. 702)
Red-wing Starling (p. 708)
Female   Male

Plate 79

D. M. Henry

Chestnut-bellied Starling (p. 717)     Shelley's Starling (p. 718)     Superb Starling (p. 719)

Stuhlmann's Starling (p. 714)
Male     Female     Hildebrandt's Starling (p. 718)     Kenrick's Starling (p. 714)
Female     Male

Pale White-eye (p. 733)     Green White-eye (p. 729)     Broad-ringed White-eye (p. 730)

Taita White-eye (p. 734)     Yellow White-eye (p. 726)     White-breasted White-eye (p. 734)

## STARLINGS and OXPECKERS

**Range in Eastern Africa:** Central Kenya Colony to Tanganyika Territory.

**Habits:** Arboreal birds of true forest, usually seen sitting on some bare branch on a tree top singly or in pairs. They are extremely pugnacious and attack any large birds or other Starlings which approach their nesting tree. Moreau records them actually settling on the backs of passing Hornbills in their eagerness to attack. Gather into small flocks in the non-breeding season and feed on fruit of forest trees. Usually noisy.

**Nest and Eggs:** In the hole of a branch of a high tree, generally inaccessible. Eggs undescribed.

**Recorded breeding:** Tanganyika Territory, August, also probably January. Nyasaland, October.

**Food:** Berries and tree fruit.

**Call:** A monotonous whistle of two or three notes, also a swearing noise at the nest.

*Onychognathus walleri elgonensis* (Sharpe).
*Amydrus elgonensis* Sharpe, Ibis, p. 242, 1891: Mt. Elgon.

**Distinguishing characters:** Differs from the nominate race in the female having the grey below confined to the chin and throat, which is streaked with blue. Males indistinguishable. Wing 116 to 134 mm.

**Range in Eastern Africa:** South-eastern Sudan, eastern Belgian Congo, Uganda and western Kenya Colony.

**Habits:** As for the nominate race.

**Recorded breeding:** Nandi, Kenya Colony, June.

**Distribution of other races of the species:** West Africa.

**1202 SOCOTRA CHESTNUT-WING STARLING.** *ONYCHOGNATHUS FRATER* (Sclater & Hartlaub).
*Amydrus frater* P. L. Sclater & Hartlaub, P.Z.S. p. 171, 1881: Socotra Island.

**Distinguishing characters:** Bill and wings longer than in the Chestnut-wing Starling; the chestnut in primaries not so dark and general colour glossy blue black with a greenish wash, especially below. The sexes are alike. Wing 152 to 163 mm. The young bird is sooty black, primaries similar to the adult.

**Range in Eastern Africa:** Socotra Island.
**Habits:** Little recorded.
**Nest and Eggs:** Undescribed.
**Recorded breeding:** No records.
**Food:** No information.
**Call:** A soft bell-like 'pee-hoo' also a harsh scream.

**1203** RED-WING STARLING. *ONYCHOGNATHUS MORIO* (Linnæus).

*Onychognathus morio morio* (Linn.). **Pl. 78.**
*Turdus morio* Linnæus, Syst. Nat. 12th ed. 1, p. 297, 1766: Cape of Good Hope.

**Distinguishing characters:** Adult male, bill long; general colour above glossy violet; ear-coverts greenish; below, glossy blue black; primaries chestnut with black tips. The female has the whole head and neck grey with violet and blue black streaks; upper chest grey, with much broader blue black streaks. Wing 137 to 165 mm. The young bird of both sexes is sooty black with some gloss; primaries as in adult, but the male is usually more glossy than the female.

**General distribution:** Nyasaland and Portuguese East Africa to South Africa.

**Range in Eastern Africa:** Nyasaland and Portuguese East Africa.

**Habits:** A common and widespread species inhabiting various types of country and always sociable, if not gregarious, even during the breeding season. In some places wild and wary inhabiting rocky gorges, in others breeding in the roofs of houses and very tame. They do not appear to migrate, but they range very widely for their food. Flight fast and dipping with a constant twittering call. They roost in companies among rocks when available and often cling to rocks with feet and tail.

**Nest and Eggs:** In caves or in holes in cliffs or buildings or the roofs of huts, occasionally in the crowns of palms; nest made of grass with a mud stiffening. Eggs three to five, bluish green with rufous spots; about 33 × 23 mm. but occasionally larger.

**Recorded breeding:** Rhodesia, October to April. Nyasaland, September to January. Portuguese East Africa, probably September and October.

# STARLINGS and OXPECKERS

**Food:** Varied, chiefly fruit and insects. They can be a serious pest to fruit-growers.

**Call:** Loud melodious whistles which also form a song, and a piping twittering call on the wing.

*Onychognathus morio rüppellii* (Verr.).
*Amydrus rüppellii* Verreaux, in Chenu's Encycl. d'Hist. Nat. Ois. 5, p. 166, 1856: Abyssinia.

**Distinguishing characters:** Differs from the nominate race in having a heavier bill with a more curved culmen. Wing 144 to 169 mm.

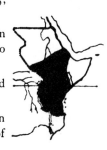

**Range in Eastern Africa:** Sudan (except western area) and Abyssinia to Tanganyika Territory.

**Habits:** As for the nominate race nesting both in villages and in high cliffs, common and widely spread. They have the habit of mobbing large birds.

**Recorded breeding:** Kenya Colony, January to April also October and November. Tanganyika Territory, October to December.

*Onychognathus morio neumanni* (Alex.).
*Amydrus neumanni* Alexander, Bull. B.O.C. 23, p. 41, 1908: Petti, northern Nigeria.

**Distinguishing characters:** Differs from the preceding races in having a shorter bill and more sharply curved culmen. Wing 158 to 174 mm.

**General distribution:** Northern Nigeria and northern Cameroons to French Equatorial Africa and the Sudan.

**Range in Eastern Africa:** Western Sudan.

**Habits:** As for other races.

**Recorded breeding:** No records.

*Onychognathus morio montanus* (Van Som.).
*Amydrus montanus* Van Someren, Bull. B.O.C. 40, p. 52, 1919: Mt. Elgon.

**Distinguishing characters:** Said to differ from the nominate and Abyssinian races in having a more slender bill. Wing 148 to 153 mm.

**Range in Eastern Africa:** Mt. Elgon, above 9,000 feet.
**Habits:** Common in fair sized flocks in its restricted habitat.
**Recorded breeding:** Mt. Elgon (breeding condition), January.
**Distribution of other races of the species:** West Africa.

**1204** SOMALI CHESTNUT-WING STARLING. *ONYCHOG-NATHUS BLYTHII* (Hartlaub).
*Amydrus blythii* Hartlaub, J.f.O. p. 32, 1859: Warsangeli country, eastern British Somaliland.

**Distinguishing characters:** Adult male, differs from the last species in having a longer and more graduated tail, and the female in having a wholly grey head and neck. Wing 149 to 181 mm. The young bird is duller than the adult; below, black with a slight metallic wash.

**Range in Eastern Africa:** Eritrea to eastern Abyssinia, British Somaliland and northern Italian Somaliland; also Abd-el-Kuri and Socotra Islands.

**Habits:** Similar to other Chestnut-winged Starlings haunting rocky places in flocks, but a more local and less common species than most. They roost in companies at night with a deafening chattering, generally in a ravine with overhung sides.

**Nest and Eggs:** Undescribed. Eggs, probably only one.
**Recorded breeding:** Socotra Island, October to December.
**Food:** Largely fruit.
**Call:** Loud and harsh, a sort of 'chee-chee-chee-chee-whoup' (O. Grant).

**1205** SLENDER-BILLED CHESTNUT-WING STARLING. *ONYCHOGNATHUS TENUIROSTRIS* (Rüppell).
*Onychognathus tenuirostris tenuirostris* (Rüpp.). **Pl. 78.**
*Lamprotornis tenuirostris* Rüppell, N. Wirbelt. Vög. p. 26, pl. 10, fig. 1, 1836: Eritrea.

**Distinguishing characters:** Adult male, bill slender; head and sides of face glossy green; feathers of head tipped with black; mantle and rump glossy violet; wings and tail black with a green gloss; primaries mainly chestnut with black ends; central tail feathers elongated; below, chin and throat glossy blue black; rest of underparts glossy violet. The female has grey tips to the head and body feathers. Wing 142 to 158 mm. The young bird is sooty black with some gloss, the male is usually more glossy than the female.

**Range in Eastern Africa:** Eritrea and Abyssinia to central Kenya Colony at Mt. Kenya.

**Habits:** Much as for the following race, rather uncommon among rocks or tall trees at high elevations, or along the bush fringes of precipices. Occasionally the male cruises high in the air with a piping song.

**Nest and Eggs:** As for the following race.

**Recorded breeding:** Mt. Kenya at 12,000 feet, August.

*Onychognathus tenuirostris theresæ* Meinertz.
*Onychognathus tenuirostris theresæ* Meinertzhagen, Bull. B.O.C. 57, p. 68, 1936: Northern Aberdare Mts., central Kenya Colony.

**Distinguishing characters:** Differs from the nominate race in having the top of the head violet. Wing 140 to 165 mm.

**General distribution:** Eastern Belgian Congo to Uganda, Kenya Colony, Tanganyika Territory and the Nyika Plateau, Nyasaland.

**Range in Eastern Africa:** Ruanda and Urundi, Uganda, central and south-eastern Kenya Colony and southern Tanganyika Territory.

**Habits:** A mountain species occurring from the lower forest zone upwards on most of the higher Eastern Africa mountains. Locally common on Mt. Kivu in parties and small flocks, feeding on the ground in forest clearings. Plentiful on Mt. Ruwenzori in large flocks in wooded valleys. Mountain forest species on the Aberdares coming to water in the evenings in flocks. Common in the Tanganyika Territory highlands round waterfalls, and often enter the water for food.

**Nest and Eggs:** Like a Thrush's nest of grass, moss, etc., stiffened with mud on ledges of rocks in caves or crevices and often under waterfalls. Eggs three or four, whitish with rufous spots and blotches. No measurements available.

**Recorded breeding:** Kenya Colony, August. Tanganyika Territory, probably October to December, but breeding season apparently irregular.

**Food:** Insects, molluscs and berries, particularly of *Podocarpus*. Olives are also eaten.

**Call:** A shrill cry. A flock makes a pleasing twittering noise on the wing.

### 1206 BRISTLE-CROWNED STARLING. *GALEOPSAR SALVADORII* Sharpe.

*Galeopsar salvadorii* Sharpe, Ibis, p. 241, pl. 4, 1891: Turquel, north-western Kenya Colony.

**Distinguishing characters:** Adult male, tail long and steeply graduated; central feathers considerably elongated; head and neck glossy violet; a cushion of soft feathers on forehead; rest of plumage glossy blue black; primaries chestnut with black tips. The female has a duller black cushion on the forehead, and a grey wash on the head and sides of face. Wing 151 to 166 mm. The young bird is duller than the adult with much less gloss and belly dull black or slightly greenish.

**Range in Eastern Africa:** British and Italian Somalilands, southern Abyssinia, eastern Uganda and northern Kenya Colony as far south as the Northern Guaso Nyiro.

**Habits:** Those of other Chestnut-wing Starlings, locally common in widely scattered colonies in rocky gorges.

**Nest and Eggs:** In Somaliland in holes or cracks in trees, in Turkana probably in inaccessible cliffs. Eggs two or three, rough chalky-white spotted and blotched with reddish brown and with lilac undermarkings (Sir G. Archer). No measurements available.

**Recorded breeding:** Turkana, Kenya Colony, probably March.

**Food:** No information.

**Call:** Unrecorded.

### 1207 WHITE-BILLED STARLING. *PILORHINUS ALBIROSTRIS* (Rüppell).

*Ptilonorhynchus (Kitta) albirostris* Rüppell, N. Wirbelt. Vög. p. 22, pl. 9, 1836: Taranta Mts., Acchele Guzai district, Eritrea.

**Distinguishing characters:** Adult male, bill white; tail square; general colour glossy violet and blue black; primaries chestnut with black tips. The female has the head, neck and chest washed with slightly glossy violet. Wing 151 to 165 mm. The young bird is much duller than the adult; some grey on the head and neck and a dusky white bill.

**Range in Eastern Africa:** Eritrea and Abyssinia as far south as the Arussi area.

**Habits:** Those of the Bristle-crowned Starling, haunting cliffs and old buildings in widely separated small colonies, and roosting communally at night among rocks. Common in the Blue Nile gorges.

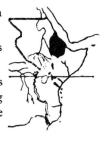

**Nest and Eggs:** Undescribed.

**Recorded breeding:** No records.

**Food:** No information.

**Call:** Described as loud and monotonous.

**1208** NARROW-TAILED STARLING. *PŒOPTERA LUGUBRIS* Bonaparte.

*Pœoptera lugubris major* Neum.

*Pœoptera lugubris major* Neumann, J.f.O. p. 82, 1920: Ituri, eastern Belgian Congo.

**Distinguishing characters:** Adult male, general colour glossy violet; wings dusky and ashy; tail long and graduated, feathers narrow. The female is slightly glossy blue black above and below; head washed with grey; primaries chestnut with dusky outer edges and ends; below, grey, slightly glossy. Wing 87 to 96 mm. The young bird is similar to the adult female.

**General distribution:** Northern Angola to western Uganda.

**Range in Eastern Africa:** Bwamba area, western Uganda.

**Habits:** A local species of the forests of western Uganda in flocks of up to one hundred or more, inhabiting high trees and behaving much like other Starlings. They roost in high trees and make aerial evolutions at dusk in a similar manner to the European Starling. They have been noted as hovering over the flowers of a flowering tree taking insects from the flowers.

**Nest and Eggs:** Undescribed. The nest of the West African race is usually in an old Barbet hole in a dead tree trunk, and the eggs three, pale bluish grey with scattered spots and speckles of brown at the larger end and with some gloss; about 22 × 16 mm.

**Recorded breeding:** No records.

**Food:** Largely fruit.

**Call:** A chirping note, uttered on the wing as well as when perched.

**Distribution of other races of the species:** West Africa, the nominate race being described from Gabon.

### 1209 STUHLMANN'S STARLING. *STILBOPSAR STUHLMANNI* Reichenow. Pl. 79.

*Stilbopsar stuhlmanni* Reichenow, O.M. p. 31, 1893: Badsua, south-west Lake Albert, north-eastern Belgian Congo.

**Distinguishing characters:** Adult male, size small; wholly glossy blue black and violet; wings and tail black. The female is greyish with a duller blue black and greenish gloss; primaries chestnut with black outer edges and ends. Wing 95 to 106 mm. The young bird is similar above to the adult male, but much duller, below sooty black. The primaries are black in the young male and chestnut in the young female.

**General distribution:** North-eastern Belgian Congo to Abyssinia and Kenya Colony.

**Range in Eastern Africa:** Southern Sudan to western and south-western Abyssinia, Uganda and central Kenya Colony.

**Habits:** A forest species frequenting the tops of high trees, of which little is recorded. Is said to be tame.

**Nest and Eggs:** Undescribed.

**Recorded breeding:** No records.

**Food:** Fruit.

**Call:** A loud shrill call like that of a Bee-eater.

### 1210 KENRICK'S STARLING. *STILBOPSAR KENRICKI* (Shelley).

*Stilbopsar kenricki kenricki* (Shell.). **Pl. 79.**

*Pœoptera kenricki* Shelley, Bull. B.O.C. 3, p. 42, 1894: Usambara, north-eastern Tanganyika Territory.

**Distinguishing characters:** Adult male, differs from Stuhlmann's Starling in being slightly glossy bronzy black, not blue black. The female has the head, neck and underside slate grey; mantle, wings and tail slightly glossy bronzy black; primaries chestnut with black outer-edges and ends. Wing, male 104 to 106, female 98 to 101 mm. The young bird is very similar above to the adult male, but much duller; below, sooty grey. The primaries are black in the young male, and chestnut in the young female.

**Range in Eastern Africa:** Mt. Meru, Mt. Kilimanjaro and the Usambara Mts., north-eastern Tanganyika Territory.

**Habits:** Small slender little Starlings of arboreal habits keeping to the tops of high forest trees, and noticeable by the musical babble kept up by a flock. Feed on tree fruit, especially that of *Ficus natalensis*.

**Nest and Eggs:** In a hole in the bough of a high tree. Eggs undescribed.

**Recorded breeding:** Tanganyika Territory, Mt. Kilimanjaro, February. Mt. Meru, April and possibly November. Amani, August.

**Food:** Berries and small fruits.

**Call:** A loud sweet 'peleep' and a babbling as above (Moreau).

*Stilbopsar kenricki bensoni* Van Someren.
*Stilbopsar kenricki bensoni* Van Someren, Bull. B.O.C. 46, p. 11, 1945: Meru, Mt. Kenya, Kenya Colony.

**Distinguishing characters:** Differs from the nominate race in being larger. Wing, male 105 to 111, female 102 mm.

**Range in Eastern Africa:** Mt. Kenya, Kenya Colony.

**Habits:** As for the nominate race.

**Recorded breeding:** Mt. Kenya, January.

**1211 FISCHER'S STARLING.** *SPREO FISCHERI* (Reichenow).

*Notauges fischeri* Reichenow, J.f.O. p. 54, 1884: Paré Mts., north-eastern Tanganyika Territory.

**Distinguishing characters:** Head and throat slightly glossy pale ash-grey; mantle, rump, wings and tail dark ash-grey with a green gloss, brighter on flight feathers and tail; below, chest and breast dark ash-grey; rest of underparts including under wing-coverts white. The sexes are alike. Wing 105 to 117 mm. The young bird has a few sandy tips to the mantle feathers.

**Range in Eastern Africa:** Southern Abyssinia at Dolo and southern Italian Somaliland to eastern half of Kenya Colony and northern Tanganyika Territory.

**Habits:** Relatively uncommon and found in small flocks in dry

acacia country. Habits little recorded but very much those of other Starlings, found in small parties and spending much time on the ground.

**Nest and Eggs:** A large round or pear-shaped grass structure with a side entrance hole and a lining of fine grass and wool or feathers, placed in a thorn tree. Eggs up to six, blue with sepia and lilac spots; about 24 × 19 mm.

**Recorded breeding:** Southern Abyssinia, north-eastern Kenya Colony and Italian Somaliland, April and May, also northern Kenya Colony, December.

**Food:** Largely caterpillars, beetles and other insects.

**Call:** 'A distinctive wheezy whirligig call, a repeated Creewee-creewoo' (Couchman and Elliott). A loud shrill whistle (Tomlinson).

### 1212 WHITE-CROWNED STARLING. *SPREO ALBI-CAPILLUS* Blyth.

*Spreo albicapillus* Blyth, J. As. Soc. Bengal, 24, p. 301, 1856: Warsangeli country, eastern British Somaliland.

**Distinguishing characters:** Top of head, outer webs of all except innermost secondaries, lower belly, under wing-coverts and under tail-coverts white; rest of upperparts slightly glossy green; wings and tail bluer green; sides of face and underside smoky brown streaked with white. The sexes are alike. Wing 143 to 163 mm. The young bird has the top of the head pale ashy-white, and the bill is yellow with a black tip, not wholly black as in the adult.

**Range in Eastern Africa:** Southern Abyssinia to British and Italian Somalilands.

**Habits:** Little or nothing recorded, except that they are assertive birds driving other species away from water. Rare and usually seen singly or in pairs.

**Nest and Eggs:** Nest a rounded grass structure with a side entrance placed in a tree and often surrounded with thorny twigs; usually found in scattered colonies. Eggs up to five, pale blue with scattered liver-coloured spots and occasionally deep mauve under-markings; about 30 × 20 mm.

**Recorded breeding:** Somaliland and south-eastern Abyssinia, March to May.

**Food:** No information.

**Call:** A harsh cry is recorded, but apparently very silent birds.

## 1213 CHESTNUT-BELLIED STARLING. *SPREO PULCHER* (Müller).

*Spreo pulcher rufiventris* (Rüpp.). **Pl. 79.**
*Lamprotornis rufiventris* Rüppell, N. Wirbelt. Vög. pp. 24 & 27, pl. 2, fig. 1, 1835: Northern Abyssinia.

**Distinguishing characters:** Head and sides of face sooty brown with a slight blue gloss; rest of upperparts including wings and tail dullish glossy green and blue; inner webs of primaries mainly whitish; below, chin to breast dullish glossy green; belly to under tail-coverts chestnut. The sexes are alike. Wing 111 to 120 mm. The young bird is duller and has the chin to breast ashy brown; lower mandible whitish, not wholly black as in adult.

**Range in Eastern Africa:** Central and eastern Sudan to Eritrea and northern Abyssinia.

**Habits:** Common and locally abundant in the western Sudan, usually associated with villages and cultivation, and found in small flocks feeding mainly on the ground. Local elsewhere and not usually tame, roosting in high trees in companies.

**Nest and Eggs:** A nest of dry grass, straw, etc. lined with softer materials, sometimes with an outer covering of thorny twigs. Entrance hole at side. Also uses old nests of *Bubalornis*. Eggs three to five, clear greenish blue spotted and blotched with various shades of rufous and with violet undermarkings; about 25·5 × 18·5 mm.

**Recorded breeding:** Sudan twice a year, September and October and February to April. Abyssinia, April and May, but nesting is irregular in any one district.

**Food:** Insects and various berries.

**Call:** Various rather pleasant calls and a harsh alarm note of 'kree,' also a weak song.

**Distribution of other races of the species:** West Africa, the nominate race being described from Senegal.

*Note:* In the Berber Province of the Sudan large numbers of thorny stick nests found in low thorn trees are believed to be the work of this bird. They are mainly utilized by Golden Sparrows and Lizards and it would be interesting to know if they are really built by this species.

### 1214 HILDEBRANDT'S STARLING. *SPREO HILDEBRANDTI* (Cabanis). Pl. 79.

*Notauges Hildebrandti* Cabanis, J.f.O. p. 233, pl. 3, 1878: Ukamba, south-central Kenya Colony.

**Distinguishing characters:** Above, glossy deep blue; green collar round hind neck; wings glossy green; black velvety spots at ends of wing-coverts; lesser wing-coverts blue; below, chin to chest glossy deep blue; breast to under tail-coverts chestnut, often paler on the breast. The sexes are alike. Wing 109 to 120 mm. The young bird is much duller above, lacks the green collar on the hind neck, and has the chin to chest dusky brown.

**Range in Eastern Africa:** Southern Kenya Colony to northern Tanganyika Territory south and east of Lake Victoria.

**Habits:** Similar to those of other Starlings, found in small scattered flocks in thorn-bush country. In some localities it roosts in holes in trees at night. Shy wary birds as a rule, feeding largely on the ground.

**Nest and Eggs:** Nests in holes in branches or in the trunks of trees, with a light pad of hair or fibre at the bottom. Eggs three or four, white and slightly glossy; about 25 × 18 mm.

**Recorded breeding:** Southern Kenya Colony, March and April.

**Food:** Insects and seeds, possibly fruit at some seasons.

**Call:** A low warbling 'chu-er—chu-er—chu-er—cher—cher—chule' (Van Someren). The flock call is a whistling 'chule.' Alarm call 'chu-ee.'

### 1215 SHELLEY'S STARLING. *SPREO SHELLEYI* Sharpe. Pl. 79.

*Spreo shelleyi* Sharpe, Cat. Bds. B.M. 13, p. 190, 1890: Somaliland.

**Distinguishing characters:** Differs from Hildebrandt's Starling in being richer violet blue above, including lesser wing-coverts and edges of primaries; chin to chest richer violet; breast to under tail-coverts much deeper and darker chestnut; the bill and first primary are also shorter than in Hildebrandt's Starling. In worn plumage more violet on the head, neck, upperside and chest, and the median wing-coverts and innermost secondaries lose most of the green gloss and fade to dusky. The sexes are alike. Wing 102 to 120 mm.

The young bird differs from that of Hildebrandt's Starling in having the head and mantle ash brown without any gloss; also in the more ashy, less blackish, undersides of the flight feathers and in the shorter first primary.

**Range in Eastern Africa:** British Somaliland to southern Abyssinia, south-eastern Sudan and eastern half of Kenya Colony.

**Habits:** Not uncommon but very local in flocks on the northern frontier of Kenya Colony. In Southern Abyssinia Erlanger described them as particularly tame and confiding.

**Nest and Eggs:** Undescribed.

**Recorded breeding:** No records.

**Food:** No information.

**Call:** Unrecorded.

### 1216 SUPERB STARLING. *SPREO SUPERBUS* (Rüppell).
Pl. 79.

*Lamprocolius superbus* Rüppell, Syst. Uebers, pp. 65, 75, pl. 26, 1845: Shoa, central Abyssinia.

**Distinguishing characters:** Head and sides of face black with a slight golden wash; neck, mantle and tail metallic deep blue; wings metallic green; black velvety spots at ends of wing-coverts; below, chin to breast metallic deep blue; belly and thighs chestnut; narrow band across breast, under tail-coverts, axillaries and under wing-coverts white. The sexes are alike. Wing 110 to 134 mm. The young bird has the head, neck and chin to breast dull black.

**Range in Eastern Africa:** Central Abyssinia to British Somaliland, southern Sudan, Uganda, Kenya Colony and Tanganyika Territory.

**Habits:** A common and beautiful Starling widely distributed in Eastern Africa; gregarious, tame and fearless of man, feeding on the ground in small parties. All the Starlings of this group are highly beneficial. This species has a form of courtship display, jumping about on the ground with drooping wings and outstretched neck.

**Nest and Eggs:** Usually a spherical ball of grass surrounded with thorny twigs and lined with feathers, placed low on thorn tree boughs or in bushes, but the birds often use holes in trees or cliffs, or the old nests of other birds. Eggs normally four, clear blue green; about 26 × 19 mm.

**Recorded breeding:** Somaliland, March. Kenya Colony, March to July, also October to December. Tanganyika Territory, April to July, also November to January.

**Food:** Insects, berries, etc.

**Call:** A cheerful chattering and a loud whining alarm call. Quite a pleasant varied whistling warbling song, often interspersed with mimicry of other birds.

### 1217 YELLOW-BILLED OXPECKER. *BUPHAGUS AFRICANUS* Linnæus.

*Buphagus africanus africanus* Linn.
*Buphaga africanus* Linnæus, Syst. Nat. 12th ed. 1, p. 154, 1766: Senegal.

**Distinguishing characters:** Base of lower mandible broad; above, including wings and tail dark earth brown; rump and upper tail-coverts buff brown; below, throat and neck dark earth brown; chest dusky; breast to under tail-coverts buff brown; bill, apical half bright red, basal half bright yellow. The sexes are alike. Wing 113 to 13 mm. The young bird has a dusky brown bill.

**General distribution:** Senegal to Eritrea, Abyssinia, Angola, Damaraland and Natal, but not eastern parts of Kenya Colony and Tanganyika Territory, nor Portuguese East Africa.

**Range in Eastern Africa:** Southern Sudan to Eritrea, south-western Abyssinia, Uganda, western halves of Kenya Colony and Tanganyika Territory and Nyasaland.

**Habits:** The Oxpeckers are among the most characteristic birds of Africa, attendant on all large animals, domesticated or wild. They climb all over the animal in every kind of posture and have a habit of hopping or dropping backwards as well as forwards. If alarmed they appear in a line along an animal's back with upraised bills and a hissing alarm note, and have earned the enmity of many a hunter by so doing. To domesticated animals they do serious damage, as they enlarge the wounds made by ticks to an unnecessary degree and actually feed on the animal's tissues, though they also destroy an immense amount of ticks. Regular dipping of animals causes them to disappear from a district if there is no large game in the neighbourhood.

## STARLINGS and OXPECKERS

**Nest and Eggs:** An untidy mass of grass or straw in the hole of a tree, generally lined or padded with hair. Eggs three to five, either very pale blue or white, occasionally spotted and blotched with pale brown and lilac; about 29 × 21 mm.

**Recorded breeding:** Nigeria, July. Sudan, May and June.

**Food:** Ticks as above.

**Call:** A hissing cackling sound used as an alarm note.

**Distribution of other races of the species:** The lower Congo River.

**1218 RED-BILLED OXPECKER. *BUPHAGUS ERYTHOR-HYNCHUS*** (Stanley).

*Buphagus erythrorhynchus erythrorhynchus* (Stan.).
*Tanagra erythroryncha* Stanley, in Salt's Travels Abyssinia, app. p. 59, 1814: Northern Abyssinia.

**Distinguishing characters:** Differs from the Yellow-billed Oxpecker in having a normal shaped lower mandible; the bill is wholly waxy red; there is a yellow ring round eye, and the upperside is paler earth brown; rump uniform with the mantle. The sexes are alike. Wing 105 to 121 mm. The young bird is more sooty brown above; chin to chest dusky brown; bill red at base, with apical third yellow and tip olive; the red bill is assumed before they moult to adult dress.

**Range in Eastern Africa:** Eritrea, Abyssinia (except southern area) and British Somaliland.

**Habits, Nest and Eggs:** As for the southern race.

**Recorded breeding:** Abyssinia, April and August.

*Buphagus erythrorhynchus caffer* Grote.
*Buphagus erythrorhynchus caffer* Grote, O.M. p. 13, 1927: Selala River, Transvaal, South Africa.

**Distinguishing characters:** Differs from the nominate race in being brown above, and warmer in tone of colour. Wing 107 to 130 mm.

**General distribution:** Abyssinia, the Sudan and Kenya Colony, to Southern Rhodesia and Natal.

*\*Note:* If original orthography is adhered to the spelling of this name should be *erythrorynchus*.

**Range in Eastern Africa:** Southern Abyssinia to southern Sudan, Uganda, Kenya Colony and the Zambesi River.

**Habits:** With the same main habits and characteristics as the Yellow-billed Oxpecker, this species is probably commoner and more widespread. Individual birds are reputed to return to individual animals each day. Native opinion is usually in favour of these birds, European the reverse, but the latter have, of course, other resources for getting rid of ticks. Moreau examined fifty-eight stomachs of these birds which contained 2,291 ticks. They roost in companies in reed beds or at times in holes in trees. The male has a chattering excited nuptial display which generally takes place on an animal's back.

**Nest and Eggs:** In the hole of a tree or the cleft of a rock, under eaves or in walls, an untidy mass of grass, etc. lined with hair. Eggs three to five, bluish white or else whitish and rather rough, freely spotted with various shades of brown; about 30 × 21·5 mm.

**Recorded breeding:** Southern Abyssinia, March. Uganda, June and July, also November and December. Kenya Colony, March to July. Tanganyika Territory, April and September.

**Food:** As above.

**Call:** A hissing 'tseee' and a warning rattle, also a shrill twittering in flight.

Names in Sclater's *Syst. Av. Æthiop.* 2, 1930, which have been changed or have become synonyms in this work.

*Creatophora carunculata* (Gmelin), now *Creatophora cinerea* (Menschen).
*Lamprocolius chloropterus cyanogenys* Sundevall, treated as synonymous with *Lamprocolius chloropterus chloropterus* (Swainson).
*Lamprocolius purpureus amethystinus* (Heuglin) treated as synonymous with *Lamprocolius purpureus* (Müller).
*Lamprocolius splendidus bailundensis* Neumann, treated as synonymous with *Lamprocolius splendidus splendidus* (Vieillot).
*Cosmopsarus regius magnificus* Van Someren, treated as synonymous with *Cosmopsarus regius* Reichenow.
*Onychognathus walleri nyasæ* (Shelley) treated as synonymous with *Onychognathus walleri walleri* (Shelley).
*Onychognathus morio shelleyi* (Hartert) treated as synonymous with *Onychognathus morio rüppellii* (Verreaux).

Names introduced since 1930 and which have become synonyms in this work:

*Cinnyricinclus leucogaster friedmanni* Bowen, 1930, treated as synonymous with *Cinnyricinclus leucogaster leucogaster* (Gmelin).
*Cinnyricinclus leucogaster lauragrayæ* Bowen, 1930, treated as synonymous with *Cinnyricinclus leucogaster verreauxi* (Bocage).

Plate 80

D. M. Henry

| Malachite Sunbird (p. 751) | Scarlet-tufted Malachite Sunbird (p. 752) | Tacazze Sunbird (p. 754) | Bronze Sunbird (p. 755) |
| Purple-breasted Sunbird (p. 757) | Red-chested Sunbird (p. 757) | Beautiful Sunbird (p. 759) | Pigmy Sunbird (p. 763) |
| Golden-winged Sunbird (p. 762) Female Male | | Superb Sunbird (p. 764) | Copper Sunbird (p. 765) |
| Splendid Sunbird (p. 766) | Shining Sunbird (p. 767) | Little Purple-banded Sunbird (p. 768) | Pemba Violet-breasted Sunbird (p. 771) |
| Mariqua Sunbird (p. 771) | White-bellied Sunbird (p. 776) | Variable Sunbird (p. 779) | Greater Double-collared Sunbird (p. 780) |

Plate 81

D. M. Henry

Southern Double-collared Sunbird (p. 781)  Eastern Double-collared Sunbird (p. 783)  Regal Sunbird (p. 788)  Loveridge's Sunbird (p. 790)
Amethyst Sunbird (p. 791)  Green-throated Sunbird (p. 793)  Scarlet-chested Sunbird (p. 794)  Mouse-coloured Sunbird (p. 798)
Green-headed Sunbird (p. 799)  Blue-throated Brown Sunbird (p. 801)  Olive Sunbird (p. 803)
  Male        Female
Collared Sunbird (p. 807)  Green Sunbird (p. 810)
  Female    Male                          Male            Female
       Banded Green Sunbird (p. 811)   Anchieta's Sunbird (p. 815)
Violet-backed Sunbird (p. 812)  Uluguru Violet-backed Sunbird (p. 814)  Amani Sunbird (p. 816)
                                                          Male      Female

# STARLINGS and OXPECKERS

*Onychognathus walleri keniensis* (Van Someren) 1931, treated as synonymous with *Onychognathus walleri walleri* (Shelley).

*Onychognathus tenuirostris raymondi* Meinertzhagen, 1936, treated as synonymous with *Onychognathus tenuirostris tenuirostris* (Rüppell).

*Spreo hildebrandti kellogorum* Neumann, 1944, treated as synonymous with *Spreo hildebrandti* (Cabanis).

## Addenda and Corrigenda

**1151** *Parus afer.* The East African Grey Tits are best regarded as specifically distinct from the South African *Parus afer.* They should be called the *Northern Grey Tit Parus griseiventris.*

**1153** *Parus niger.* It seems probable that there are two Black Tits distinguished by eye-colour, a yellow-eyed form extending from West Africa to the Sudan and northern Uganda, and a brown-eyed form extending from South Africa to Abyssinia. More work is necessary on the group before it can be decided that this difference is of specific value.

**1171** *Corvus corax edithæ.* Recorded breeding: British Somaliland, April. Eggs two.

**1172** *Corvus albus.* Recorded breeding: add Eritrea, February.

**1173** *Corvus capensis kordofanensis.* Recorded breeding: add British Somaliland, April to July.

**1174** *Corvus splendens.* Range in Eastern Africa: add Mombasa and adjacent mainland up to four miles north of the town: occurs also at Djibouti.

**1177** *Rhinocorax rhipidurus.* Range in Eastern Africa: add Eastern Uganda.

p. 680. *Pyrrhocorax pyrrhocorax baileyi* Rand & Vaurie, 1955, treated as synonymous with *Pyrrhocorax pyrrhocorax docilis* (Gmelin).

**1182** *Creatophora cinerea.* Recorded breeding: add British Somaliland, May.

**1187** *Speculipastor bicolor.* Range in Eastern Africa: add British Somaliland. Recorded breeding: add British Somaliland, April to June.

**1190** *Lamprocolius c. chloropterus.* Recorded breeding: add Eritrea (nestlings), June.

p. 701. *Note:* The Wedge-tailed Glossy Starling *Heteropsar acuticaudus* occurs on the Tanganyika Territory—Northern Rhodesian boundary and should probably be included in our area. It is a local species not easily identifiable in the field from any of the small Glossy Starlings, although a close examination would show the longer central tail feathers. Nest and eggs apparently undescribed and little is recorded of its habits.

**1196** *Lamprotornis purpuropterus æneocephalus.* Recorded breeding: add Eritrea, June and July.

**1200** *Onychognathus fulgidus hartlaubi*—the author is Hartlaub not Gray. Type locality is Fernando Po, which should be added to the *General distribution.*

**1201** *O.w. elgonensis. Nest and Eggs:* Eggs two, no description. *Recorded breeding:* Belgian Congo, December to February.

p. 706. *Onychognathus intermedius.* For No. read Nov.

**1204** *Onychognathus blythii. Nest and Eggs:* add nests in clefts in steep rock faces, occasionally under eaves of houses. *Recorded breeding:* add British Somaliland, April to October.

**1212** *Spreo albicapillus. Food:* Largely insects.

**1216** *Spreo superbus. Call:* add Often sings at night.

*Continued on p. 823*

# WHITE-EYES

FAMILY—**ZOSTEROPIDÆ**. **WHITE-EYES**. Genus: *Zosterops*.

Eight species occur in Eastern Africa, frequenting all types of bushed, wooded and forested country, though the different species keep more or less to their particular habitat. They are often seen in small flocks or parties and are very Warbler-like in appearance and actions. Their main food is insects, for which they search diligently with quick restless movements. They also feed on berries.

### KEY TO THE ADULT WHITE-EYES OF EASTERN AFRICA

| | | | |
|---|---|---|---|
| 1 | White ring round eye large and broad: | | 3–6 |
| 2 | White ring round eye not large and broad: | | 7–14 |
| 3 | Breast to belly grey: | TAITA WHITE-EYE *Zosterops silvanus* | 1225 |
| 4 | Breast to belly yellow: | | 5–6 |
| 5 | Forehead not distinctly yellow: | BROAD-RINGED WHITE-EYE *Zosterops eurycricotus* | 1222 |
| 6 | Forehead distinctly yellow: | KIKUYU WHITE-EYE *Zosterops kikuyuensis* | 1223 |
| 7 | Breast to belly white or grey: | | 8–9 |
| 8 | Above, pale olivaceous green or olivaceous yellow green: | WHITE-BREASTED WHITE-EYE *Zosterops abyssinicus* | 1226 |
| 9 | Above, moss green: | HEUGLIN'S WHITE-EYE *Zosterops pallidus* | 1224 |
| 10 | Breast to belly yellow: | | 11–14 |
| 11 | Above, more green than yellow, below less clear yellow, more washed with olive: | GREEN WHITE-EYE *Zosterops virens* | 1221 |

## WHITE-EYES

12  Above, more yellow than green, below clearer golden or canary yellow: 13–14
13  Forward part of crown and sides of face more yellow than green: PEMBA WHITE-EYE *Zosterops vaughani* 1220
14  Forward part of crown and sides of face more uniform with crown and mantle: YELLOW WHITE-EYE *Zosterops senegalensis* 1219

**1219 YELLOW WHITE-EYE.** *ZOSTEROPS SENEGALENSIS* Bonaparte.
*Zosterops senegalensis senegalensis* Bp.
*Zosterops senegalensis* Bonaparte, Consp. Av. 1, p. 399, 1850: Senegal.

**Distinguishing characters:** Bill small; above, greenish-canary yellow, with a mealy or powdery appearance; forehead yellow; below, clear canary yellow; a narrow ring of white feathers round eye; black usually confined to edge of mandible and under eye, little or none on lores. The sexes are alike. Wing 52 to 61 mm. The young bird is darker above than the adult.

**General distribution:** Senegal and Cameroons to Abyssinia, Eritrea, the Sudan and Uganda.

**Range in Eastern Africa:** Central and southern Sudan to Eritrea, Abyssinia (except southern and south-eastern areas) and north-western Uganda.

**Habits:** Insectivorous birds hunting bushes and trees in pairs or small parties, frequently with other species, especially Sunbirds. They prefer the denser foliaged trees and are usually in the tops of the trees, but are rendered noticeable by a constant twittering feeding call, almost a churring note. Usually very tame and particularly fond of water to bathe in. This species is usually found in thorn bush or more open woodland.

**Nest and Eggs:** A beautifully made little cup of grass or tendrils, covered with lichens on the outside and lined with hair. It is usually found in a thin upright fork of a tree or slung from a horizontal fork. Eggs two or three, clear blue or white; about 15 × 12 mm.

**Recorded breeding:** Northern Nigeria, June. Darfur, probably August to October. Uganda, April to June.

**Food:** Almost entirely insects, especially small caterpillars. Small fruit is also taken at times.

**Call:** A low piping whistle and the feeding call described above. There is also a curious short tinny note. The male has also a subdued song reminiscent of a Willow Warbler.

*Zosterops senegalensis anderssoni* Shell. **Pl. 79.**
*Zosterops anderssoni* Shelley, Bull. B.O.C. 1, p. 5, 1892: Elephant Vley, Damaraland, South West Africa.

**Distinguishing characters:** Above, darker than the nominate race, and below, a brighter tone of yellow. Wing 54 to 63 mm.

**General distribution:** Tanganyika Territory, Nyasaland, Damaraland, Bechuanaland, Northern Rhodesia, Portuguese East Africa, Southern Rhodesia, north-eastern Transvaal and northern Zululand.

**Range in Eastern Africa:** Southern areas of Tanganyika Territory from Kibungo, south-east of Uluguru Mts., Ubende area and Iringa to the Zambesi River.

**Habits:** Common upland and highland species among denser foliaged trees, particularly along water courses or at the edge of forest. Usually occurs in small loose flocks with constant low twittering calls. Some evidence of local migration or seasonal change of altitude.

**Recorded breeding:** Tanganyika Territory, November to January. South-eastern Belgian Congo, September. Rhodesia, November to March. Nyasaland, September to December, also March.

*Zosterops senegalensis flavilateralis* Reichw. **Ph. xv.**
*Zosterops flavilateralis* Reichenow, J.f.O. p. 193, 1892: Ndi, Taita, southern Kenya Colony.

**Distinguishing characters:** Above, rather less yellow than the nominate race; below, paler and clearer yellow but not so bright as in the preceding race; mantle greener than in the preceding race; black confined to edge of mandible and under eye, none on lores. Wing 52 to 59 mm.

# WHITE-EYES

**Range in Eastern Africa:** Kenya Colony, as far north as the Turkwel River and the Orr Valley to northern and central Tanganyika Territory as far south as Mpapwa and Morogoro; also Manda Island.

**Habits:** As for the nominate race, a common tame little bird with a twittering call, assiduously searching for insects in bush or woodland country. May be seen in small flocks in the non-breeding season.

**Recorded breeding:** Kenya Colony, March to June, also December and January.

*Zosterops senegalensis jubaensis* Erl.
*Zosterops jubaensis* Erlanger, O.M. p. 182, 1901: Damasso, lower Juba River, Italian Somaliland.

**Distinguishing characters:** Above much darker in tone than the nominate race and having the inner secondaries washed with ash colour. Wing 49 to 57 mm.

**Range in Eastern Africa:** Southern Abyssinia from Lake Rudolf, Omo River and upper Webi Shebeli areas to southern Italian Somaliland.

**Habits:** Of other races, a bird of dry thorn scrub and riverside tree belts.

**Recorded breeding:** No records.

**Distribution of other races of the species:** Sierra Leone.

**1220 PEMBA WHITE-EYE.** *ZOSTEROPS VAUGHANI* Bannerman.

*Zosterops vaughani* Bannerman, B.O.C. 44, p. 41, 1924: Pemba Island, eastern Africa.

**Distinguishing characters:** Bill long; lores black; above, yellowish green; forehead bright yellow; forward part of crown and sides of face more yellow than green; below, bright yellow; flanks faintly olivaceous. The sexes are alike. Wing 50 to 54 mm. Juvenile plumage unrecorded.

**Range in Eastern Africa:** Pemba Island.

**Habits:** One of the commonest birds of Pemba Island, found in all situations. They are usually seen in pairs or small parties searching for small berries and insects, and are tame delightful little birds.

**Nest and Eggs:** A small cup-shaped nest of grass or palm fibre a few feet up in the branches of a bush or small tree. Eggs two, plain pale blue. No measurements available.

**Recorded breeding:** Season indefinite, but mostly October to December, also March and July.

**Food:** As above, J. H. Vaughan notes that they are particularly fond of the fruits of the wild Mulberry and flower-heads of Chillies.

**Call:** A pleasing warbling song.

**1221 GREEN WHITE-EYE.** *ZOSTEROPS VIRENS* Sundevall.
*Zosterops virens stuhlmanni* Reichw.
*Zosterops stuhlmanni* Reichenow, J.f.O. p. 54, 1892: Bukoba, north-western Tanganyika Territory.

**Distinguishing characters:** Bill long; above, yellow green with a cinnamon wash; a narrow ring of white feathers round eye; lores and under eye-ring black; yellow on forehead narrow or almost absent; below, lemon yellow washed with olive green on cheeks, sides of chest and flanks. The sexes are alike. Wing 52 to 65 mm. The young bird is duller than the adult. Can be distinguished from the local race of the Yellow White-eye by its generally more olivaceous appearance, less bright upperside and less bright yellow underparts.

**General distribution:** Ruwenzori Mts. and Lake Edward areas, eastern Belgian Congo to Uganda and Tanganyika Territory.

**Range in Eastern Africa:** Uganda (but not the north-western and Mt. Elgon areas) to north-western Tanganyika Territory, also Nkose and Ukerewe Islands.

**Habits:** Indefatigable destroyers of insects, searching trees and bushes in a most thorough manner, these birds must be of great value among fruit trees. This species is somewhat more of a forest-edge bird than the last, and does not stray into dry country, but there is no hard and fast dividing line between their ranges. Quite large flocks may be seen in the non-breeding season, and it is locally common.

**Nest and Eggs:** A beautifully made deep cup of finely woven fibre and grasses decorated with spiders' webs or lichen and usually woven on to small twigs by its outer edges, so that it hangs from them. It is usually among the outer twigs of a bush or on the bough of a tree. Eggs two, rarely three, pale blue or bluish green, occasionally white; about 16 × 12 mm.

**Recorded breeding:** Uganda, March to August, also November to January.

**Food:** Insects and small fruit.

**Call:** A long series of high-pitched notes, clear and distinct, about fifteen notes in each burst, the whole forming almost a song (Belcher).

*Zosterops virens stierlingi* Reichw. **Pl. 79.**
*Zosterops stierlingi* Reichenow, J.f.O. p. 418, 1899: Iringa, south-central Tanganyika Territory.

**Distinguishing characters:** Differs from the preceding race in having the upperparts coldish yellow green with no cinnamon wash. Wing 53 to 66 mm.

**General distribution:** The Sudan to Kenya Colony, Tanganyika Territory, Portuguese East Africa, Nyasaland, south-eastern Belgian Congo as far north as Baraka, Northern Rhodesia and eastern Southern Rhodesia.

**Range in Eastern Africa:** Southern Sudan to Kenya Colony (except central area), Tanganyika Territory (except north-western and central areas) and the Zambesi River.

**Habits:** As for the last race, particularly partial to small patches of forest and keeping rather high up in the trees. Moreau describes the song as loud and clear and of a passionate quality.

**Recorded breeding:** Tanganyika Territory, probably November to January. Kungwe-Mahare Mts. (breeding condition), August. Nyasaland, October and November.

*Zosterops virens jacksoni* Neum.
*Zosterops jacksoni* Neumann, O.M. p. 23, 1899: Mau, Kenya Colony.

**Distinguishing characters:** Generally a distinctly darker tone than in the preceding race; above, sides of face and chest and flanks more olive green; yellow on forehead distinct and broad; eye ring rather larger than in other races, but never so large as in the Broad-ringed or Kikuyu White-eyes and yellow forehead not so distinct or extensive, more confined between bill and front of eye and not

extending to above or behind the eye; ear-coverts and below eye darker olive green. The female is often darker above than the male. Wing 57 to 63 mm.

**Range in Eastern Africa:** Central and south-western Kenya Colony to northern Tanganyika Territory west of Mt. Kilimanjaro, from about the eastern Trans-Nzoia, Kenya Colony, to Loliondo, Tanganyika Territory.

**Habits:** As for other races, common and gathering into flocks in the non-breeding season, and apparently remaining in flocks when already in breeding condition.

**Recorded breeding:** Kenya Colony, Eldama Ravine, February. Mau, September. Nairobi, January.

*Zosterops virens kaffensis* Neum.
*Zosterops kaffensis* Neumann, O.M. p. 10, 1902: Anderatscha, Kaffa, south-western Abyssinia.

**Distinguishing characters:** Differs from the Bukoba race in having a more distinct and broader yellow forehead, and in being slightly darker above, but not so olive green as the preceding race; yellow forehead not so bright nor so broad as in the preceding race; flanks less bright, paler olive green than in the Kikuyu White-eye, and eye ring smaller. Wing 55 to 66 mm.

**Range in Eastern Africa:** Abyssinia, but not western area around Goré, nor east of the Lakes.

**Habits:** As for other races, with a distinctly whining tone in the call.

**Recorded breeding:** Abyssinia, April to August (breeding condition); (fledgling) November.

**Distribution of other races of the species:** Fernando Po, Cameroons, French Equatorial Africa and northern Angola to South Africa, the nominate race being described from Natal.

### 1222 BROAD-RINGED WHITE-EYE. *ZOSTEROPS EURYCRICOTUS* Fischer & Reichenow. Pl. 79.

*Zosterops eurycricotus* Fischer & Reichenow, J.f.O. p. 55, 1884: Mt. Meru, Arusha district, northern Tanganyika Territory.

**Distinguishing characters:** Above, nearly moss green; a large white ring of feathers round eye which has a distinct sheen or silky

# WHITE-EYES

appearance, larger and perhaps more silky than in the Green White-eyes; below, olive yellow; less olive in centre of belly and on throat. The sexes are alike. Wing 58 to 63 mm. The young bird is duller and rather darker green than the adult. The nestlings have curious ear-like tufts of down above the eyes.

**Range in Eastern Africa:** The Monduli, Essimingor, Mt. Meru, Mt. Kilimanjaro, Lolkisale and Lossogonoi areas of northern Tanganyika Territory.

**Habits:** Much as for other species, but a bird of higher levels, quite common locally in flocks among trees, bushes and tree-heaths.

**Nest and Eggs:** A deep cup of fibre or lichen lined with fine fibres and attached or slung to large herbs or heaths. Eggs two, white, pink when fresh, rather long and narrow; about $18 \cdot 5 \times 12$ mm.

**Recorded breeding:** Mt. Meru, Tanganyika Territory, January. Mt. Kilimanjaro, March.

**Food:** Insects, also a large quantity of berries.

**Call:** Louder, but not so plaintive as that of the Green White-eye. It also has a loud clear warbling song with a cadence not unlike that of a Willow Warbler, and a twittering alarm cry.

## 1223 KIKUYU WHITE-EYE. *ZOSTEROPS KIKUYUENSIS* Sharpe.

*Zosterops kikuyuensis kikuyuensis* Sharpe. **Ph. xv.**
*Zosterops kikuyuensis* Sharpe, Ibis, p. 444, 1891: Kikuyu, Kenya Colony.

**Distinguishing characters:** Differs from the Broad-ringed White-eye in having a distinct yellow forehead. The sexes are alike. Wing 58 to 63 mm. The young bird is duller and rather darker green than the adult.

**Range in Eastern Africa:** North-western Uganda and Kenya Colony with the exception of the south-eastern area.

**Habits:** A forest species with the habits of others of the genus, common up to bamboo level on mountains. Often found in dark dripping surroundings in forest or forest edge in small cheerful flocks of up to twenty or more.

**Nest and Eggs:** The nest is a deep cup of woven beard-moss or lichen in a small tree or bush usually slung on to a horizontal fork. Eggs probably two, pale blue and elongated; about 19·5 × 11 mm.

**Recorded breeding:** Kenya Colony highlands, March to June, also more rarely November to January.

**Food:** Insects.

**Call:** Described as very similar to that of the Green White-eye, with a low short song, and a normal call of 'whii-tu.'

*Zosterops kikuyuensis mbuluensis* Scl. & Mor.
*Zosterops virens mbuluensis* Sclater & Moreau, Bull. B.O.C. 56, p. 13, 1935: Oldeani Forest, Arusha district, northern Tanganyika Territory.

**Distinguishing characters:** Differs from the Broad-ringed White-eye in having a slight but distinct yellow forehead, and from the nominate race in being lighter green above and with less yellow on the forehead. Wing 56 to 65 mm.

**Range in Eastern Africa:** Southern Kenya Colony at Oldonyo Erok and northern Tanganyika Territory at Nou, Ketumbeine, Mbagai twenty miles north-east of Ngorongoro, Longido, Oldeani, North Paré, Mt. Gerui (Hanang), Ufiome and Essimingor.

**Habits:** As for the other races, found abundantly in all suitable districts of mountain forest.

**Recorded breeding:** Northern Tanganyika Territory (breeding condition), January and March.

*Zosterops kikuyuensis chyuluensis* V. Som.
*Zosterops chyuluensis* Van Someren, J.E.A. & U. Nat. Hist. Soc. 14, p. 114, 1939: Chyulu Range, south-eastern Kenya Colony.

**Distinguishing characters:** Differs from the preceding races in being rich canary yellow below, with olivaceous flanks. Wing 58 to 65 mm.

**Range in Eastern Africa:** Chyulu Hills, south-eastern Kenya Colony.

**Habits:** As for other races, common at high levels in forest canopy or among Giant Lobelias in clearings. The nest is said to be very flimsily made of moss or lichens.

**Recorded breeding:** Chyulu Hills, probably December to March or later.

**1224 PALE WHITE-EYE. *ZOSTEROPS PALLIDUS* Swainson.**
*Zosterops pallidus poliogastra* Heugl. **Pl. 79.**
*Zosterops poliogastra* Heuglin, Ibis, p. 357, pl. 13, 1861: Highlands of northern Abyssinia.

**Distinguishing characters:** Forehead yellow; white eye ring; lores black; rest of upperparts and sides of neck and face moss green; below, chin to chest and under tail-coverts yellow; breast to belly whitish, with some yellow in centre of belly; flanks pale buff. The sexes are alike. Wing 58 to 67 mm. Juvenile plumage unrecorded.

**Range in Eastern Africa:** Northern, central, eastern and south-western Abyssinia, as far south as the Shoa, Lake Abaya and Gofa areas.

**Habits:** Very little recorded, but usually in pairs or small parties searching trees and bushes for insects with restless movements and little whistling cries.

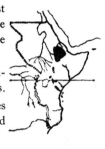

**Nest and Eggs:** Nest a very deep cup of fine grass in a bush or tree. Eggs two, pale sky blue; about 16 × 13 mm.

**Recorded breeding:** Abyssinia, April, also (breeding condition) December.

**Food:** Insects.

**Call:** A warbling rather like that of a Willow Warbler, and a Tit-like 'chee' while feeding.

*Zosterops pallidus winifredæ* Scl. & Mor.
*Zosterops winifredæ* W. L. Sclater & Moreau, Bull. B.O.C. 55, p. 14, 1934: Chome, Paré Mts., north-eastern Tanganyika Territory.

**Distinguishing characters:** Differs from the nominate race in having the flanks pale grey. Wing 58 to 60 mm.

**Range in Eastern Africa:** North-eastern Tanganyika Territory.
**Habits:** As for other races.
**Recorded breeding:** No records.

*Zosterops pallidus kulalensis* Will.
*Zosterops pallida kulalensis* Williams, Bull. B.O.C. 68, p. 101, 1948: Mt. Kulal, northern Kenya Colony.

**Distinguishing characters:** Similar to the Paré Mts. race but larger. The male has a lemon yellow streak down the centre of the breast to belly; the female has only a yellowish central wash. Wing 62 to 65 mm.

**Range in Eastern Africa:** Mt. Kulal, northern Kenya Colony.

**Habits:** As for other races.

**Recorded breeding:** No records.

**Distribution of other races of the species:** South-west and South Africa and Asia, the nominate race being described from South Africa.

**1225** TAITA WHITE-EYE. *ZOSTEROPS SILVANUS* Peters & Loveridge. **Pl. 79.**

*Zosterops silvanus* Peters & Loveridge, Proc. Biol. Soc. Wash. 58, p. 77, 1935: Mbololo, Taita, south-eastern Kenya Colony.

**Distinguishing characters:** Similar to the last species, but differs in having a large broad white ring of feathers round the eye; a bright yellow edge to the wing; chin and throat greener, less yellow. The sexes are alike. Wing 55 to 60 mm. The young bird is duller than the adult.

**Range in Eastern Africa:** Taita Hills, south-eastern Kenya Colony.

**Habits:** Little described, but not differing in any way from those of other White-eyes.

**Nest and Eggs:** Nest very similar to that of the Kikuyu White-eye, a deep cup of moss and lichens. Eggs undescribed.

**Recorded breeding:** Taita Hills, April to June.

**Food:** No information.

**Call:** Unrecorded.

**1226** WHITE-BREASTED WHITE-EYE. *ZOSTEROPS ABYSSINICUS* Guérin.

*Zosterops abyssinicus abyssinicus* Guér. **Pl. 79.**

*Zosterops abyssinica* Guérin, Rev. Zool. p. 162, 1843: Northern Abyssinia.

**Distinguishing characters:** Above, including sides of face, dull palish olivaceous green with a slight powdery appearance; below, chin to upper chest and under tail-coverts pale lemon yellow; chest to belly white, slightly greyish to buffish on flanks. The female is often duller than the male. Wing 53 to 61 mm. The young bird is rather duller than the adult female. The White-breasted White-eye can be distinguished from the Pale White-eye by the differently coloured upperside, lack of yellow on the forehead, paler yellow chin to upper chest, much whiter appearance below, and the horn-coloured, not blackish, bill.

**Range in Eastern Africa:** Red Sea Province of the Sudan and Eritrea to British Somaliland and Abyssinia from Lake Tana to Lake Zwai.

**Habits:** Of the other species of the genus, very common locally in any thick cover in the Abyssinian valleys, but does not appear to occur on the high plateau.

**Nest and Eggs:** A flimsy fairy-like cup-shaped nest decorated on the outside with cocoons. Eggs pale blue; no measurements available.

**Recorded breeding:** Red Sea coast, Sudan, March.

**Food:** No information.

**Call:** Unrecorded.

*Zosterops abyssinicus omoensis* Neum.
*Zosterops omoensis* Neumann, O.M. p. 162, 1904: Senti, between Ufa and Gofa, south-western Abyssinia.

**Distinguishing characters:** Differs from the nominate race in being more yellowish green above, but not the moss green of the Pale White-eye. Wing 54 to 59 mm.

**Range in Eastern Africa:** North-western, south-western and southern Abyssinia.

**Habits:** Of the other races.

**Recorded breeding:** North-western Abyssinia, June.

*Zosterops abyssinicus socotranus* Neum.
*Zosterops abyssinica socotrana* Neumann, Bull. B.O.C. 21, p. 59, 1908: Dahamis, Socotra Island.

**Distinguishing characters:** Differs from the nominate race in being rather brighter above, less olivaceous green. Wing 55 to 60 mm.

**Range in Eastern Africa:** Socotra Island.

**Habits:** Of the genus, plentiful in low bush-clad valleys, feeding in small flocks in a very Tit-like manner.

**Recorded breeding:** October and November.

**Distribution of other races of the species:** South-western Arabia.

Names in Sclater's *Syst. Av. Æthiop.* 2, 1930, which have been changed, or have become synonyms in this work:

*Zosterops senegalensis aurifrons* Heuglin, treated as synonymous with *Zosterops senegalensis senegalensis* Bonaparte.

*Zosterops superciliosa* Reichenow, treated as synonymous with *Zosterops senegalensis senegalensis* Bonaparte.

*Zosterops senegalensis niassæ* Reichenow, treated as synonymous with *Zosterops senegalensis anderssoni* Shelley.

*Zosterops senegalensis smithi* Neumann, treated as synonymous with *Zosterops senegalensis jubaensis* Erlanger.

*Zosterops virens usambaræ* Reichenow, treated as synonymous with *Zosterops virens stierlingi* Reichenow.

*Zosterops virens schoanus* Neumann, treated as synonymous with *Zosterops virens kaffensis* Neumann.

Names introduced since 1930 and which have become synonyms in this work:

*Zosterops virens sarmenticia* Bangs and Loveridge, 1931, treated as synonymous with *Zosterops virens stierlingi* Reichenow.

*Zosterops virens meruensis* Sclater and Moreau, 1935, treated as synonymous with *Zosterops eurycricotus* Fischer and Reichenow.

*Zosterops senegalensis australoabyssinicus* Benson, 1942, treated as synonymous with *Zosterops senegalensis jubaensis* Erlanger.

## Family—NECTARINIIDÆ. SUNBIRDS. Genera: *Nectarinia, Drepanorhynchus, Hedydipna, Cinnyris, Chalcomitra, Cyanomitra, Anthreptes, Hylia* and *Pholidornis*.

Fifty-six species are known to occur in Eastern Africa. Sunbirds are found in all types of country, and the males are often conspicuous objects, especially those species which inhabit the more open areas. They are subject to some movements in the breeding and non-breeding seasons, but in most cases this is a local one according to the flowering season of the trees, shrubs and plants to which they resort for the nectar and insect life. They extract the nectar both on the wing and from a perch, more usually from the latter. Some species puncture the larger tubular flowers which they cannot suck otherwise, and all devour a large quantity of small insects.

In some of the species the males assume a non-breeding dress, whilst in others there is no such dress, the males moulting from one full plumage to another at the end of the breeding period. It would appear that those species which breed only once a year have a non-breeding dress, but those that have two breeding seasons in the year, or of which there is sporadic breeding of individual pairs, have no non-breeding dress. Again, in some species the young males moult direct into a full breeding dress, whilst in others they assume an immature (intermediate) dress which in due course is discarded for the full breeding dress. In the field the situation is rather involved, as the young males are often assuming a full breeding dress at the same time that adults are assuming a non-breeding dress, and therefore, apparent adults in full breeding dress and adults in non-breeding dress can be seen in the same months. Although many species closely resemble each other they are distinguishable by size, colour, length of tail and shape of bill. The females, on the other hand, are not so easy to identify and in most cases in no way resemble the males in colouring. They are, however, clearly distinguishable one from the other, and the points to note are the size, colour of tail, and the length and curvature of the bill. The curvature of the bill agrees with that of the male.

The nest is suspended from a bough and is usually a long oval or pear-shaped structure with the opening at the top of the side, often with a little projecting porch over the opening. It is in many cases roughly made outside and may have a tail of loose material hanging from it. Eggs normally one or two, dull in colour and longitudinally streaked.

### Key to the Adult Male Sunbirds of Eastern Africa

1 Central tail feathers elongated:   3–16
2 Central tail feathers not elongated:   17–94
3 Chest to belly metallic violet: PURPLE-BREASTED SUNBIRD *Nectarinia purpureiventris* 1231
4 Breast to belly yellow: PIGMY SUNBIRD *Hedydipna platura* 1236
5 Breast to belly metallic green or mainly metallic green:   6–9

| | | |
|---|---|---|
| 6 | Tufts on side of chest red: | SCARLET-TUFTED MALACHITE SUNBIRD *Nectarinia johnstoni* **1228** |
| 7 | Tufts on side of chest yellow: | 8–9 |
| 8 | Red patch in centre of chest: | BEAUTIFUL SUNBIRD *Nectarinia pulchella* **1233** |
| 9 | No red patch in centre of chest: | MALACHITE SUNBIRD *Nectarinia famosa* **1227** |
| 10 | Breast metallic or black, belly black: | 11–16 |
| 11 | Edges of flight feathers and tail yellow: | GOLDEN-WINGED SUNBIRD *Drepanorhynchus reichenowi* **1235** |
| 12 | Edges of flight feathers and tail not yellow: | 13–16 |
| 13 | Throat metallic green, chest black: | BRONZY SUNBIRD *Nectarinia kilimensis* **1230** |
| 14 | Breast metallic ruby violet: | TACAZZE SUNBIRD *Nectarinia tacazze* **1229** |
| 15 | Indian red band across breast: | RED-CHESTED SUNBIRD *Nectarinia erythrocerca* **1232** |
| 16 | Orange red band across breast: | SMALLER BLACK-BELLIED SUNBIRD *Nectarinia nectarinioides* **1234** |
| 17 | Orange band across chest: | GREY-CHINNED SUNBIRD *Anthreptes rectirostris* **1272** |
| 18 | Breast to belly chocolate maroon: | SUPERB SUNBIRD *Cinnyris superbus* **1237** |

| | | |
|---|---|---|
| 19 | Breast to belly sooty black, sooty grey, or dusky grey: | 25–29 |
| 20 | Red band across chest: | 30–47 |
| 21 | No orange or red band across chest: | 48–94 |
| 22 | Belly white, greyish white or buffish white: | 48–58 |
| 23 | Belly various shades of yellow, orange yellow, red and yellow or green: | 59–74 |
| 24 | Belly black, blue black, smoky black or black washed with bronze: | 75–94 |
| 25 | Throat dingy grey: | GREEN HYLIA *Hylia prasina* **1281** |
| 26 | Throat metallic blue or green: | 27–29 |
| 27 | Mantle sooty brown: | BLUE-THROATED BROWN SUNBIRD *Cyanomitra cyanolæma* **1268** |
| 28 | Mantle moss green: | GREEN-HEADED SUNBIRD *Cyanomitra verticalis* **1266** |
| 29 | Mantle saffron: | BLUE-HEADED SUNBIRD *Cyanomitra alinæ* **1267** |
| 30 | Throat grey: | BANDED GREEN SUNBIRD *Anthreptes rubritorques* **1273** |
| 31 | Throat metallic green: | 32–47 |
| 32 | Belly black or smoky black: | 33–34 |
| 33 | Forehead and crown metallic purple: | SHINING SUNBIRD *Cinnyris habessinicus* **1240** |
| 34 | Forehead and crown metallic green: | SHELLEY'S DOUBLE-COLLARED SUNBIRD *Cinnyris shelleyi* **1246** |

35 Belly dusky olivaceous, yellowish olive, olivaceous green or pale dusky, with lower belly and under tail-coverts yellowish: 36–47

36 Size larger, wing 65 mm. and over: GREATER DOUBLE-COLLARED SUNBIRD *Cinnyris afer* **1252**

37 Size smaller, wing 65 mm. and under: 38–47

38 Red band across chest broader and extending to breast: 39–42

39 Upper tail-coverts metallic deep blue: NORTHERN DOUBLE-COLLARED SUNBIRD *Cinnyris reichenowi* **1256**

40 Upper tail-coverts green or bluish green: 41–43

41 Under wing-coverts dusky: OLIVE-BELLIED SUNBIRD *Cinnyris chloropygius* **1257**

42 Under wing-coverts white: TINY SUNBIRD *Cinnyris minullus* **1258**

43 Red band across chest narrower and more confined to chest: 44–47

44 Tail blue black:

45 Pectoral tufts sharply defined: EASTERN DOUBLE-COLLARED SUNBIRD *Cinnyris mediocris* **1254**

46 Yellow of pectoral tufts extending down side of chest band: MOREAU'S SUNBIRD *Cinnyris moreaui* **1255**

# SUNBIRDS

47 Tail dusky blackish, less glossy:  
SOUTHERN DOUBLE-COLLARED SUNBIRD  
*Cinnyris chalybeus*    1253

---

48 Throat and neck below uniform with belly:  
MOUSE-COLOURED SUNBIRD  
*Cyanomitra veroxii*    1265

49 Chin and upper throat dusky grey:  
SOCOTRA SUNBIRD  
*Cyanomitra balfouri*    1270

50 Chin black, upper throat metallic violet:    51–52

51 Sides of face brownish:  
ULUGURU VIOLET-BACKED SUNBIRD  
*Anthreptes neglectus*    1275

52 Sides of face black:  
WESTERN VIOLET-BACKED SUNBIRD  
*Anthreptes longuemarei*    1274

53 Chin to chest metallic green and violet, violet and steel blue or bottle green:    54–58

54 Tufts on side of chest yellow:  
WHITE-BELLIED SUNBIRD  
*Cinnyris talatala*    1249

55 Tufts on side of chest orange or orange and yellow:    56–58

56 Maroon band across breast:  
ANGOLA WHITE-BELLIED SUNBIRD  
*Cinnyris oustaleti*    1250

57 Throat metallic bottle green:  
AMANI SUNBIRD  
*Anthreptes pallidigaster*    1278

58 Throat metallic violet and steel blue: WHITE-BELLIED VARIABLE SUNBIRD
*Cinnyris venustus albiventris* **1251**

| | | |
|---|---|---|
| 59 Mantle smoke grey: | ANCHIETA'S SUNBIRD *Anthreptes anchietæ* | **1276** |
| 60 Mantle streaked: | TIT-HYLIA (both sexes) *Pholidornis rushiæ* | **1282** |
| 61 Mantle green: | | 62–67 |
| 62 Head grey: | GREY-HEADED SUNBIRD *Anthreptes axillaris* | **1280** |
| 63 Head green: | | 64–67 |
| 64 Forehead and throat metallic blue black: | YOKANA SUNBIRD *Anthreptes reichenowi* | **1279** |
| 65 Forehead and throat green: | | 66–67 |
| 66 Bill and tail long: | OLIVE SUNBIRD *Cyanomitra olivacea* | **1269** |
| 67 Bill and tail short: | LITTLE GREEN SUNBIRD *Anthreptes seimundi* | **1277** |
| 68 Mantle metallic green with a golden wash or green with a blue wash: | | 69–74 |
| 69 Forehead metallic purple violet: | YELLOW AND ORANGE-BELLIED VARIABLE SUNBIRDS *Cinnyris venustus* races | **1251** |
| 70 Forehead metallic green: | | 71–74 |
| 71 Breast and belly yellow: | COLLARED SUNBIRD *Anthreptes collaris* | **1271** |
| 72 Breast and belly red and yellow: | | 73–74 |
| 73 Bill shorter, central tail feathers slightly elongated: | REGAL SUNBIRD *Cinnyris regius* | **1259** |

| | | |
|---|---|---|
| 74 Bill longer, central tail feathers not elongated: | LOVERIDGE'S SUNBIRD *Cinnyris loveridgei* | 1260 |
| 75 Mantle velvety black or bronzy velvety black: | | 76–80 |
| 76 Chin and throat metallic rosy purple: | AMETHYST SUNBIRD *Chalcomitra amethystina* | 1261 |
| 77 Chin and throat metallic green: | GREEN-THROATED SUNBIRD *Chalcomitra rubescens* | 1262 |
| 78 Chin metallic green, neck to chest red: | SCARLET-CHESTED SUNBIRD *Chalcomitra senegalensis* | 1263 |
| 79 Chin black, neck to chest red: | HUNTER'S SUNBIRD *Chalcomitra hunteri* | 1264 |
| 80 Mantle metallic golden copper: | COPPERY SUNBIRD *Cinnyris cupreus* | 1238 |
| 81 Mantle various shades metallic green: | | 82–94 |
| 82 Forehead metallic purple: | | 83–85 |
| 83 Chest and breast mixed metallic purple and red, tufts on side of chest yellow: | SPLENDID SUNBIRD *Cinnyris coccinigaster* | 1239 |
| 84 Breast violet, tufts on side of chest orange and yellow: | NORTHERN ORANGE-TUFTED SUNBIRD *Cinnyris oseus* | 1248 |
| 85 Breast chocolate, tufts on side of chest orange and yellow: | ORANGE-TUFTED SUNBIRD *Cinnyris bouvieri* | 1247 |

86 Forehead metallic green: 87–94
87 Maroon band across the chest: 88–89
88 Size larger, wing 64 mm. and over: MARIQUA SUNBIRD
*Cinnyris mariquensis* **1245**
89 Size smaller, wing 61 mm. and under: LITTLE PURPLE-BANDED SUNBIRD
*Cinnyris bifasciatus* **1241**
90 Violet band across chest: 91–94
91 Mantle metallic blue and green: PEMBA VIOLET-BREASTED SUNBIRD
*Cinnyris pembæ* **1244**
92 Mantle metallic green: 93–94
93 Bill and wing longer: VIOLET-BREASTED SUNBIRD
*Cinnyris chalcomelas* **1243**
94 Bill and wing shorter: TSAVO PURPLE-BANDED SUNBIRD
*Cinnyris tsavoensis* **1242**

## KEY TO THE ADULT FEMALE SUNBIRDS OF EASTERN AFRICA

1 Some metallic colouring in plumage: 3–9
2 No metallic colouring in plumage except perhaps on upper tail-coverts and tail: 10–92
3 Head and chin to chest metallic violet blue: BLUE-HEADED SUNBIRD
*Cyanomitra alinæ* **1267**
4 Top of head metallic green, chin to chest dusky grey: GREEN-HEADED SUNBIRD
*Cyanomitra verticalis* **1266**

## SUNBIRDS

5 Whole upperside except neck metallic blue and violet:     ULUGURU VIOLET-BACKED SUNBIRD *Anthreptes neglectus* **1275**

6 Wing shoulders, tail and upper tail-coverts metallic violet:     WESTERN VIOLET-BACKED SUNBIRD *Anthreptes longuemarei* **1274**

7 Whole upperside metallic green:     8–9

8 Below, yellow:     COLLARED SUNBIRD *Anthreptes còllaris* **1271**

9 Below, greyish with a yellow wash:     BANDED GREEN SUNBIRD *Anthreptes rubritorques* **1273**

---

10 Red patch in centre of breast and upper belly:     ANCHIETA'S SUNBIRD *Anthreptes anchietæ* **1276**

11 Edges of flight feathers and tail golden yellow:     GOLDEN-WINGED SUNBIRD *Drepanorhynchus reichenowi* **1235**

12 No red patch below and edges of flight feathers and tail not golden yellow:     13–92

13 Tail washed with glossy green:     17–21

14 Tail non-glossy green or dusky, edged with non-glossy green:     22–36

15 Tail blackish or brownish, with a bronzy gloss:     37–44

16 Tail glossy blue black:     45–92

17 Below, yellowish, breast streaked:     SPLENDID SUNBIRD *Cinnyris coccinigaster* **1239**

18 Below, darker olivaceous yellow mottled with darker colour: MALACHITE SUNBIRD *Nectarinia famosa* **1227**

19 Below, yellow, no streaks or mottling: BEAUTIFUL SUNBIRD *Nectarinia pulchella* **1233**

20 Below, chin and upper throat dusky grey, belly white: SOCOTRA SUNBIRD *Cyanomitra balfouri* **1270**

21 Below, greyish white: MOUSE-COLOURED SUNBIRD *Cyanomitra veroxii* **1265**

---

22 Top of head grey: GREY-HEADED SUNBIRD *Anthreptes axillaris* **1280**

23 Top of head mottled dusky and grey: PURPLE-BREASTED SUNBIRD *Nectarinia purpureiventris* **1231**

24 Top of head not grey nor mottled dusky and grey: 25–36

25 Bill long and curved: 26–29

26 Below, chin and neck white or pale brown, chest mottled: BLUE-THROATED BROWN SUNBIRD *Cyanomitra cyanolæma* **1268**

27 Below, plain olivaceous or yellowish green: 28–29

28 Under tail-coverts olivaceous green: OLIVE SUNBIRD *Cyanomitra olivacea* **1269**

29 Under tail-coverts orange: SUPERB SUNBIRD *Cinnyris superbus* **1237**

| | | |
|---|---|---|
| 30 | Bill short and practically straight: | 31–36 |
| 31 | Broad greenish white stripe over eye: | GREEN HYLIA<br>*Hylia prasina* **1281** |
| 32 | No broad stripe over eye: | 33–36 |
| 33 | Throat whitish: | YOKANA SUNBIRD<br>*Anthreptes reichenowi* **1279** |
| 34 | Throat green: | 35–36 |
| 35 | Below, olivaceous green: | GREY-CHINNED SUNBIRD<br>*Anthreptes rectirostris* **1272** |
| 36 | Below, yellow green: | LITTLE GREEN SUNBIRD<br>*Anthreptes seimundi* **1277** |

| | | |
|---|---|---|
| 37 | Underside plain olivaceous and pale yellow: | SOUTHERN DOUBLE-COLLARED SUNBIRD<br>*Cinnyris chalybeus* **1253** |
| 38 | Underside streaked: | 39–44 |
| 39 | Throat dusky blackish: | 40–41 |
| 40 | Edges of primary coverts white: | SCARLET-CHESTED SUNBIRD<br>*Chalcomitra senegalensis* **1263** |
| 41 | No white edges to primary coverts: | GREEN-THROATED SUNBIRD<br>*Chalcomitra rubescens* **1262** |
| 42 | Throat olivaceous: | 43–44 |
| 43 | Bill longer, white edges to primary coverts: | HUNTER'S SUNBIRD<br>*Chalcomitra hunteri* **1264** |
| 44 | Bill shorter, no white edges to primary coverts: | AMETHYST SUNBIRD<br>*Chalcomitra amethystina* **1261** |

| | | |
|---|---|---|
| 45 | Orange red tufts on side of chest: | SCARLET-TUFTED MALACHITE SUNBIRD<br>*Nectarinia johnstoni* **1228** |

46 No tufts on side of chest: 47–92
47 Below, dusky olivaceous: 48–49
48 Size larger, wing 67 mm. and over: TACAZZE SUNBIRD
   *Nectarinia tacazze* **1229**
49 Size smaller, wing 53 mm. and under: NORTHERN ORANGE-TUFTED SUNBIRD
   *Cinnyris oseus* **1248**
50 Below, yellow, yellowish or almost white with no streaking: 51–57
51 Bill long: 52–56
52 Upper tail-coverts edged with metallic green: 53–56
53 Belly distinctly yellowish: WHITE, YELLOW AND ORANGE-BELLIED VARIABLE SUNBIRDS
   *Cinnyris venustus* **1251**
54 Belly almost white: WHITE-BELLIED SUNBIRD
   *Cinnyris talatala* **1249**
55 Upper tail-coverts plain olivaceous green: COPPER SUNBIRD
   *Cinnyris cupreus* **1238**
56 Above, pale olive: PEMBA VIOLET-BREASTED SUNBIRD
   *Cinnyris pembæ* **1244**
57 Bill short: PYGMY SUNBIRD
   *Hedydipna platura* **1236**
58 Below, yellow or yellowish with underlying or indistinct streaking: 59–67
59 Size larger, wing 63 mm. and over: BRONZE SUNBIRD
   *Nectarinia kilimensis* **1230**
60 Size smaller, wing 55 mm. and under: 61–67
61 Above, grey brown: ANGOLA WHITE-BELLIED SUNBIRD
   *Cinnyris oustaleti* **1250**

| | | |
|---|---|---|
| 62 Above, dark olivaceous: | | 63–64 |
| 63 Under wing-coverts dusky: | OLIVE-BELLIED SUNBIRD *Cinnyris chloropygius* | 1257 |
| 64 Under wing-coverts white: | TINY SUNBIRD *Cinnyris minullus* | 1258 |
| 65 Above, pale olivaceous: | | 66–67 |
| 66 Bill narrower at base, no light edges to primary coverts: | ORANGE-TUFTED SUNBIRD *Cinnyris bouvieri* | 1247 |
| 67 Bill wider at base, light edges to primary coverts: | VIOLET-BREASTED SUNBIRD *Cinnyris chalcomelas* | 1243 |
| 68 Underside yellow with distinct streaking: | | 69–78 |
| 69 Tail rather graduated and central tail feathers elongated: | RED-CHESTED SUNBIRD *Nectarinia erythrocerca* | 1232 |
| 70 Tail square: | | 71–78 |
| 71 Size smaller, wing 56 mm. and under; bill shorter and smaller: | | 72–75 |
| 72 Above, more dusky olivaceous: | LITTLE PURPLE-BANDED SUNBIRD *Cinnyris bifasciatus* | 1241 |
| 73 Above, more olivaceous green, less dusky in tone: | | 74–75 |
| 74 Under wing-coverts heavily mottled: | SMALLER BLACK-BELLIED SUNBIRD *Nectarinia nectarinioides* | 1234 |
| 75 Under wing-coverts clearer whitish, not heavily mottled: | TSAVO PURPLE-BANDED SUNBIRD *Cinnyris tsavoensis* | 1242 |

76 Size larger, wing 56 mm. and over; bill longer and stouter: 77–78

77 Tail rather longer, 40 mm. and over: MARIQUA SUNBIRD
*Cinnyris mariquensis* **1245**

78 Tail rather shorter, 39 mm. and under: SHELLEY'S DOUBLE-COLLARED SUNBIRD
*Cinnyris shelleyi* **1246**

79 Underside green: 80–92

80 A blue-grey sheen on upperparts: 81–82

81 Sides of face grey, size larger, wing 55 mm. and over: LOVERIDGE'S SUNBIRD
*Cinnyris loveridgei* **1260**

82 Sides of face green, size smaller, wing 54 mm. and under: MOREAU'S SUNBIRD
*Cinnyris moreaui* **1255**

83 Top of head and sides of face uniform with upperparts; no sheen on upperparts: 84–92

84 Size larger, wing 60 mm. and over; bill larger: GREATER DOUBLE-COLLARED SUNBIRD
*Cinnyris afer* **1252**

85 Size smaller, wing 54 mm. and under; bill shorter: 86–92

86 Above, bright green: REGAL SUNBIRD
*Cinnyris regius* **1259**

87 Above, duller olivaceous green: 88–92

88 Bill longer: EASTERN DOUBLE-COLLARED SUNBIRD
*Cinnyris mediocris* **1254**

| | | |
|---|---|---|
| 89 Bill shorter: | NORTHERN DOUBLE-COLLARED SUNBIRD *Cinnyris reichenowi* | **1256** |
| 90 Underside ash grey or whitish: | | 91–92 |
| 91 Bill shorter and almost straight, a greenish gloss on upperside: | AMANI SUNBIRD *Anthreptes pallidigaster* | **1278** |
| 92 Bill longer and curved, no gloss on upperside: | SHINING SUNBIRD *Cinnyris habessinicus* | **1240** |

**1227** MALACHITE SUNBIRD. *NECTARINIA FAMOSA* (Linnæus).
*Nectarinia famosa cupreonitens* Shell. **Pls. 80, 82. Ph. xv.**
*Nectarinia cupreonitens* Shelley, Mon. Nect. p. 17, pl. 6, 1876: Abyssinia.

**Distinguishing characters:** Adult male, upperside, head and neck all round metallic green with a golden wash; upper tail-coverts metallic emerald green; flight feathers and tail black, latter and inner secondaries edged with green or blue; central tail feathers elongated; below, chest to under tail-coverts metallic deep blue, golden green at some angles; tuft on sides of chest bright yellow. The non-breeding dress is similar to the dress of the adult female, but the black flight feathers and tail with the elongated central feathers are retained, as also are the metallic wing and upper tail-coverts, and in some cases sparse metallic tips occur on the body. The female has the upperside and sides of face slightly olivaceous brown including wings; a more or less distinct superciliary stripe; tail blackish and slightly metallic; outermost feather edged and tipped with white, remainder, except central pair which are not elongated, tipped with white; below, dusky yellow mottled with darker colour. Wing 63 to 75 mm. The young bird is similar to the adult female, but rather greener above and rather yellower below.

**General distribution:** Eritrea to Abyssinia, the Sudan, eastern Belgian Congo, Tanganyika Territory, Northern Rhodesia and Nyasaland.

**Range in Eastern Africa:** Eritrea, Abyssinia, southern Sudan and Uganda to the Zambesi River.

**Habits:** A highland species usually found at levels above 5,500 feet, and frequenting the moorlands, scrub or bamboo zones of most of the higher ranges. Occurs in Abyssinia from 7,000 to 14,000 feet and is commonest on Mt. Kenya at about the same levels. In most areas it appears to be locally migratory, but the fact that it has a non-breeding dress may account for its presumed disappearance in part. Usually seen in pairs or small parties among flowers, Kniphofia, large thistles, Leonotis, Heaths, etc., and often locally common. Takes no notice of cold or strong winds.

**Nest and Eggs:** Spherical or pear-shaped often bulky nest of beard-moss, grass and spiders' webs, lined with down, fur or feathers when available and often conspicuous, suspended from the outside of a bush, usually three or four feet from the ground among large herbs or flowering plants. Eggs one rarely two, pale cream with streaks and spots of grey or greyish brown; or closely mottled with slate grey all over; about 21 $\times$ 13·5 mm.

**Recorded breeding:** Abyssinia, July to September. Kenya Colony, May to December, mostly June and October. Tanganyika Territory in many months, possibly most often in March and April.

**Call:** A shrill harsh call uttered on the wing; the male in the breeding season has a jingling song of a rapid succession of 'chees,' the female's call is a harsh 'tsi-tsee.'

**Distribution of other races of the species:** South Africa, the nominate race being described from the Cape of Good Hope.

**1228** SCARLET-TUFTED MALACHITE SUNBIRD. *NECTARINIA JOHNSTONI* Shelley.
*Nectarinia johnstoni johnstoni* Shell. **Pls. 80, 82.**
*Nectarinia johnstoni* Shelley, P.Z.S. p. 227, pl. 14, 1885: Mt. Kilimanjaro.

**Distinguishing characters:** Adult male, general colour metallic moss green with a wash of blue on chest, flanks and upper belly; flight feathers and tail bluish black; central tail feathers greatly elongated and edged with metallic green; tufts on side of chest bright red; belly black. In non-breeding dress, the top and sides of the head, mantle, and chin to chest are dusky olivaceous; belly and flanks olivaceous. The female is similar to that of the Tacazze Sunbird, but has orange-red tufts on side of chest and is more dusky from chin to breast and has a square tail. Wing 75 to 85 mm. The young bird is similar to the young bird of the Tacazze Sunbird and has no tufts on side of chest.

**Range in Eastern Africa:** Mt. Kenya and Aberdare Mts., central Kenya Colony to north-eastern Tanganyika Territory from Mt. Olosirwa to Mt. Kilimanjaro.

**Habits:** An abundant species in the Alpine zones of the higher mountains, occurring from about 5,500 feet to the limit of plant growth. Fond of the flowers of Giant Senecios, Heaths and Lobelias, and of Aloe flowers at lower altitudes. On Mt. Kenya confined to open moorland and very tame, but usually local in a small area. They are said to roost in holes in the matted clusters of dead leaves on the Giant Senecios. A considerable portion of the food consists of Diptera, and the birds are capable of catching them on the wing like a Flycatcher.

**Nest and Eggs:** Nest oval, of plant down, fur, and dried stems, or rootlets in a Tree-Heath or low shrub. Eggs one or two, usually one, whitish or creamy white, thickly streaked with pinkish brown or scrawled and dotted all over with brown and with lilac undermarkings; about 20 × 12 mm.

**Recorded breeding:** Kenya Colony, Mt. Kenya, January, July and August. Odd nests at other seasons. Tanganyika Territory, Mt. Kilimanjaro, December and January, also September and October. Mt. Olosirwa, January.

**Call:** The song uttered from the top of a tree, rock or tall plant, is 'tsik-tsik-tsik-tsik-tsik-tree-tree-tree-tsirrurr' (Elliott). Call a sharp 'tsik.'

*Nectarinia johnstoni salvadorii* Shell.
*Nectarinia salvadorii* Shelley, Bull. B.O.C. 13, p. 61, 1903: Kachere, Nyika Plateau, Nyasaland.

**Distinguishing characters:** Adult male, differs from the nominate race in having the chest, flanks and upper belly metallic moss green with no blue wash and a shorter bill. Wing 67 to 80 mm.

**General distribution:** Tanganyika Territory to Nyasaland.

**Range in Eastern Africa:** South-western Tanganyika Territory.

**Habits:** As for the nominate race, but two eggs are as usual as one.

**Recorded breeding:** Western Nyasaland, October and November.

*Nectarinia johnstoni dartmouthi* O. Grant.

*Nectarinia dartmouthi* O. Grant, Bull. B.O.C. 16, p. 117, 1906: Mubuku Valley, Ruwenzori Mts., western Uganda.

**Distinguishing characters:** Adult male, differs from the nominate race in having the top of the head and mantle more metallic emerald green with blue reflections in certain lights; upper tail-coverts, flanks and upper belly metallic violet blue. This race has no non-breeding dress. The female is darker than that of the nominate race. Wing 72 to 86 mm.

**General distribution:** Eastern Belgian Congo and western Uganda.

**Range in Eastern Africa:** Ruwenzori Mts., western Uganda.

**Habits:** As for the nominate race, common in the Lobelia and Groundsel zone. Very pugnacious and with a short cheerful burring song. The male has a short soaring flight in the breeding season.

**Recorded breeding:** Ruwenzori and Kivu ranges, November to January, but irregularly at other seasons also.

**1229** TACAZZE SUNBIRD. *NECTARINIA TACAZZE* (Stanley). **Pls. 80, 82.**

*Certhia tacazze* Stanley, in Salt's Abyss. App. p. 58, 1814: Tacazze River, northern Abyssinia.

**Distinguishing characters:** Adult male, head and neck all round metallic bronzy green; mantle to upper tail-coverts and wing-coverts metallic ruby violet; wings and tail black, central tail feathers elongated; below, chest and breast metallic ruby violet; rest of underparts black. The non-breeding dress is similar to the dress of the adult female, but the black flight feathers and tail with the elongated central feathers are retained, also the metallic wing and upper tail-coverts, and in some cases sparse metallic tips occur on the body feathers. There is evidence that in the highlands of Kenya Colony there is no non-breeding dress in many cases. The female differs from that of the Malachite Sunbird in being more dusky below from chin to chest, and in having the tail longer and the central tail feathers lightly elongated. Wing 66 to 85 mm. The young bird of both sexes is similar to the adult female, but is rather yellower.

## SUNBIRDS

**Range in Eastern Africa:** Southern Sudan, Eritrea, Abyssinia, Kenya Colony and north-eastern Tanganyika Territory.

**Habits:** A common and often abundant species of the higher levels, particularly common on the Abyssinian plateau. Often occurs in company with other Sunbirds and may be seen in rough and cold weather in the most exposed situations. Appears to be locally migratory, but records are somewhat unreliable owing to its change of plumage.

**Nest and Eggs:** Nest of fibres and lichen plastered with spiders' webs and often decorated with bark or other material, pear-shaped and suspended by a woven rope of spiders' webs to a branch or twig. Egg usually one, of pale ground colour heavily mottled and scrawled with wavy marks of brown and sepia; about 21 × 14 mm.

**Recorded breeding:** Abyssinia, probably mostly May and June. Kenya Colony, May to December, mostly June, October and November, also recorded in February. Northern Tanganyika Territory, January and February.

**Call:** A high-pitched short note is the only call recorded.

### 1230 BRONZE SUNBIRD. *NECTARINIA KILIMENSIS* Shelley.

*Nectarinia kilimensis kilimensis* Shell. **Pls. 80, 82. Ph. xvi.**
*Nectarinia kilimensis* Shelley, P.Z.S. p. 555, 1884: Mt. Kilimanjaro.

**Distinguishing characters:** Adult male, above, including head and neck dull metallic green, often with bronzy old gold reflections; mantle, rump and wing-coverts dull metallic bronzy old gold; wings and tail purplish black; central tail feathers greatly elongated and narrowly edged with dull metallic bronzy old gold; below, chin to chest dull metallic bronzy old gold; rest of underparts purplish black. There is no non-breeding dress. The female differs from the female of the Malachite Sunbird in having a more downward curved bill; the central tail feathers slightly elongated; tail washed with purplish not blue; chin whitish; below, much brighter yellow and streaked, not mottled. Wing 64 to 80 mm. The young bird is similar to the adult female, but the young male has the throat feathers black tipped with white.

**General distribution:** North-eastern Belgian Congo, Uganda, Kenya Colony and Tanganyika Territory.

**Range in Eastern Africa:** Uganda and Kenya Colony to central and western areas of Tanganyika Territory as far south as the Ufipa Plateau.

**Habits:** Common and widespread in Kenya Colony and Uganda in gardens, scrub, or open woodlands at levels below 6,000 feet, being particularly in evidence at Erythrina flowers. Visits also Leonotis, Crotalaria, etc. Appears to be partially migratory, visiting north-central Tanganyika Territory only in February and March.

**Nest and Eggs:** Nest of grass fibre, lichen, etc. bound with cobwebs and attached to the tip of a twig a few feet from the ground; it often has somewhat of a projecting porch. Egg usually one, pale bluish cream spotted and streaked with ashy brown or whitish with heavy sepia spots and blotches; about 20 × 13 mm.

**Recorded breeding:** Uganda, April to August. Kenya Colony, January to August, also November and December. Northern Tanganyika Territory, October.

**Call:** A loud shrill 'pee-yoo-peeyoo,' also a double piping call.

*Nectarinia kilimensis arturi* Scl.
*Nectarinia arturi* P. L. Sclater, Bull. B.O.C. 19, p. 30, 1906: Wolverhampton, south Melsetter district, Southern Rhodesia.

**Distinguishing characters:** Adult male differs from the nominate race in having the head, neck and mantle washed with metallic copper. There is no non-breeding dress. The female is similar to that of the nominate race. Wing 63 to 78 mm.

**General distribution:** Tanganyika Territory to north-eastern Northern Rhodesia, Nyasaland and eastern Southern Rhodesia.

**Range in Eastern Africa:** Southern areas of Tanganyika Territory as far north as Njombe and Kilosa.

**Habits:** As for the nominate race, but also found commonly on open grassland. Visits flowers of Loranthus, Buffalo Bean, etc. and appears to be migratory at least in the south of its range.

**Recorded breeding:** Tanganyika Territory uplands, probably March and April. Nyasaland, March, June and July. Southern Rhodesia, September to May.

**Distribution of other races of the species:** Angola.

## 1231 PURPLE-BREASTED SUNBIRD. *NECTARINIA PURPUREIVENTRIS* (Reichenow). Pls. 80, 82.

*Cinnyris purpureiventris* Reichenow, O.M. p. 61, 1893: Migere, south of Lake Edward, eastern Belgian Congo.

**Distinguishing characters:** Adult male, forehead, front part of crown, sides of face and chin and throat dull velvety metallic violet; hinder part of crown and occiput, sides of neck and lower throat metallic blue, green and bronzy old gold; hind neck and wing shoulder metallic purplish violet; mantle to rump, rest of wing-coverts and edges of inner secondaries metallic bronzy old gold; edges of inner secondaries and secondary coverts sometimes greenish; flight feathers and tail black; central tail feathers greatly elongated, with narrow metallic bronzy old gold or greenish edges; chest to belly deep metallic violet. The non-breeding dress is similar to the dress of the adult female, but the black flight feathers and tail with the elongated central feathers are retained, also the metallic wing and upper tail-coverts. The female has the head dusky with grey tips to the feathers giving a scaly appearance; rest of upperparts olive green; below, chin and throat black with white tips to feathers, rest of underparts dull yellowish green. Wing 61 to 69 mm. The young bird is similar to the adult female.

**General distribution:** Kivu area, eastern Belgian Congo to Mt. Ruwenzori.

**Range in Eastern Africa:** Ruwenzori area, western Uganda.

**Habits:** Locally common, but irregular in appearance in certain valleys of Mt. Ruwenzori. Generally noticed at Erythrina flowers.

**Nest and Eggs:** Undescribed.

**Recorded breeding:** No records.

**Call:** Unrecorded.

## 1232 RED-CHESTED SUNBIRD. *NECTARINIA ERYTHROCERCA* Hartlaub. Pls. 80, 82. Ph. xvi.

*Nectarinia erythrocerca* Hartlaub, Syst. Orn. West Afr. p. 270, 1857: While Nile, south of lat. 8° N., southern Sudan.

**Distinguishing characters:** Adult male, head and neck all round and mantle metallic green; lower rump, upper tail-coverts, some of the wing-coverts and band across chest metallic violet; wings and

tail black, latter narrowly edged with metallic green; central tail feathers elongated; breast Indian red with some metallic violet tips; rest of underparts black. There is no non-breeding dress. The female is olivaceous brown above; flight feathers edged dull yellow; tail blue black slightly washed with bronze and green, and tipped and edged with whitish; central tail feathers slightly elongated; cheeks and underside yellowish, with dark centres to the feathers, more profuse on chin and throat; occasionally a trace of red on the chest. Wing 54 to 65 mm. The young male is more olive green above than the adult female and has a wholly black chin and throat. The young female is similar to the adult female.

**General distribution:** Eastern Belgian Congo to the Sudan, Uganda, Kenya Colony and Tanganyika Territory.

**Range in Eastern Africa:** Southern Sudan, Uganda, western Kenya Colony and northern Tanganyika Territory from the Bukoba to the Mwanza and Kahama areas.

**Habits:** Locally common both in gardens and open scrub, probably the commonest Sunbird in Uganda. Very pugnacious, attacking any species of bird which comes near, but tame and confiding with man. Visits all kinds of flowers.

**Nest and Eggs:** Nest of grass fibre and dry leaves neatly bound with spider webs and ornamented with leaves or seed heads, and with an inner lining of vegetable down and feathers, in shrubs or low bushes, often by roadsides. Eggs usually one, rarely two, white or bluish white with longitudinal streaks and blotches of grey, or occasionally grey all over; about $17 \times 11 \cdot 5$ mm.

**Recorded breeding:** Uganda, every month recorded except August and September, commonest in April. Western Kenya Colony, June, July, August, December and January.

**Call:** A repeated 'trink-trink,' sharp and distinctive.

**1233** BEAUTIFUL SUNBIRD. *NECTARINIA PULCHELLA* (Linnæus).
*Nectarinia pulchella pulchella* (Linn.). **Pl. 82.**
*Certhia pulchella* Linnæus, Syst. Nat. 12th ed. 1, p. 187, 1766: Senegal.

**Distinguishing characters:** Adult male, general colour including belly metallic golden green; flight feathers and tail blue black, central pair of tail feathers elongated and edged with metallic green; red patch in centre of chest; sides of chest mixed yellow and metallic golden green. The non-breeding dress is similar to the dress of the adult female, but the black flight feathers and tail with elongated central feathers are retained, also the metallic wing and upper tail-coverts; occasionally sparse metallic tips occur on the body feathers. The female is pale ashy olive above including sides of face, washed with faint yellow; tail blackish with a faint greenish wash and narrow metallic green edges, white edges to outer webs of outer feathers and white ends to all the feathers; below, chin and throat white or yellowish white; rest of underparts pale yellow. Wing 47 to 65 mm. The young bird is similar to the adult female, but the young male has the chin and throat black.

**General distribution:** Senegal and Sierra Leone to the Sudan.

**Range in Eastern Africa:** The Darfur and Kordofan areas, western Sudan.

**Habits, Nest and Eggs:** See under other races.

**Recorded breeding:** French Sudan, July. Nigeria, June to October, also February. Darfur, May and June.

*Nectarinia pulchella melanogastra* Fisch. & Reichw. **Pl. 80.**
*Nectarinia melanogastra* Fischer & Reichenow, J.f.O. p. 181, 1884: Nguruman, north end of Lake Natron, southern Kenya Colony.

**Distinguishing characters:** Adult male, differs from the nominate race in having the belly black, not metallic green. Wing 48 to 65 mm. Can be distinguished from the Red-chested Sunbird in having the sides of the chest yellow. There is no non-breeding dress in Kenya Colony.

**Range in Eastern Africa:** Western, central and southern Kenya Colony to north-eastern Tanganyika Territory as far as the Ufipa and Iringa areas.

**Habits:** As for other races, a common species in western Kenya Colony.

**Recorded breeding:** Southern Kenya Colony, April and May. Tanganyika Territory, May and June, also probably March and April and possibly December.

*Nectarinia pulchella lucidipectus* Hart.
*Nectarinia pulchella lucidipectus* Hartert, Nov. Zool. 28, p. 123, 1921: Wad Medani, Blue Nile, eastern Sudan.

**Distinguishing characters:** Adult male, differs from the nominate race in having the sides of chest wholly yellow, not yellow and metallic golden green; some black on the lower belly. Wing 52 to 62 mm.

**Range in Eastern Africa:** South-western and central Sudan to Eritrea, Abyssinia, Uganda and western Kenya Colony.

**Habits:** Common among gardens and cultivated land, usually in small flocks and generally near water, but very widely distributed and locally abundant. Visits many kinds of flowers but normally feeds among Acacia blossom and Aloe flowers. The males are very pugnacious and fine songsters.

**Nest and Eggs:** A typical elongated Sunbird's nest suspended from a twig, compactly made of grass or bark fibres, tendrils and leaves, bound tightly with cobwebs and with the entrance at the side of the top. Eggs two, occasionally one, glossy white densely mottled all over with ash grey and ochreous brown tending to darken into streaks or a zone at the larger end; about $17 \cdot 5 \times 11 \cdot 5$ mm., but smaller eggs are recorded from Turkana, about $15 \times 10 \cdot 5$ mm. Plain white eggs are also recorded, but they may be those of some parasitical species.

**Recorded breeding:** Nile Valley, June to September. Equatoria, November. Abyssinia, probably July and August. Turkana, Kenya Colony, May to September. Uganda (nestling), November.

**Call:** A low pleasant warbling song. The call is persistent and monotonous, usually of one syllable only.

## 1234 SMALLER BLACK-BELLIED SUNBIRD. *NECTARINIA NECTARINIOIDES* (Richmond).

*Nectarinia nectarinioides nectarinioides* (Rich.). **Pl. 82.**
*Cinnyris nectarinioides* Richmond, Auk p. 158, 1897: Plains east of Mt. Kilimanjaro, south-eastern Kenya Colony.

**Distinguishing characters:** Adult male, not unlike the Nguruman race of the Beautiful Sunbird but differs by the orange red, not bright red, chest. There is no non-breeding dress. The female is olivaceous brown above, including sides of face; below, pale yellowish

# SUNBIRDS

narrowly streaked with dusky; tail bluish, edges and tips of outer feathers whitish; central tail feathers slightly longer than others; occasionally a trace of red on the chest. Wing 48 to 57 mm. The young bird differs from the adult female in having a pale streak over the eye; whitish streak down side of throat; throat black merging into black and olive mottling on upper breast.

**Range in Eastern Africa:** Northern Guaso Nyiro, Kenya Colony to north-eastern Tanganyika Territory.

**Habits:** Very much those of the Beautiful Sunbird, with which it can be easily confused in the field. Quite common locally, feeding largely on insects at Acacia, Loranthus and Baobab flowers, and usually keeping to the higher branches of trees.

**Nest and Eggs:** Nest of fine grass and seed heads, plastered with spiders' webs and ornamented with bark and leaves, and with a lining of feathers or vegetable down. The nest has a tapering support above it and usually a porch and a tail below it. Eggs one or two, pale olive grey almost entirely covered with light chocolate with darker streaks and flecks; about $16 \times 11 \cdot 5$ mm.

**Recorded breeding:** South-eastern Kenya Colony, July and August.

**Call:** A falling jangling little song, containing a double 'tee-tss, tee-tss' and a high pitched call of three or four notes (Williams).

*Nectarinia nectarinioides erlangeri* Reichw.
*Nectarinia erlangeri* Reichenow, Vög. Afr. 3, p. 496, 1905: Dolo, southern Abyssinia.

**Distinguishing characters:** Adult male differs from the nominate race in having the breast more crimson red. Wing 49 to 55 mm.

**Range in Eastern Africa:** North-eastern Kenya Colony to the Juba River Valley area of southern Abyssinia and southern Italian Somaliland.

**Habits:** As for the nominate race.

**Recorded breeding:** Southern Abyssinia, April. Juba River, June and July, also probably January.

## 1235 GOLDEN-WINGED SUNBIRD. *DREPANORHYNCHUS REICHENOWI* Fischer. Pl. 80.

*Drepanorhynchus reichenowi* Fisher, J.f.O. p. 56, 1884: Lake Naivasha, southern Kenya Colony.

**Distinguishing characters:** Adult male, above, head, mantle to upper tail-coverts and wing shoulder metallic bronzy old gold with green reflections; wings and tail black with broad golden yellow edges; central tail feathers elongated; below, chin to chest metallic copper and ruby red; rest of underparts deep black. In non-breeding dress the whole head to mantle and chin to chest is dull black uniform with rest of underparts; wings with metallic wing shoulders and tail, as in breeding dress. The female is olivaceous above and yellowish mottled with olivaceous below; flight feathers and tail edged with golden yellow. Wing 66 to 85 mm. The young female is similar to the adult female, but the young male is dull blackish below usually with some olivaceous feathers.

**General distribution:** Eastern Belgian Congo to Uganda and Tanganyika Territory.

**Range in Eastern Africa:** Southern half of Uganda to central Kenya Colony and north-eastern Tanganyika Territory.

**Habits:** Common at all levels above 5,000 feet, and locally abundant in open country, bush or cultivated land. It has a loud whirring flight and visits many species of flowers, notably Leonotis, Crotalaria, sweet potatoes, and various *Acanthaceæ* and *Leguminosæ*. Occurs up to at least 10,500 feet.

**Nest and Eggs:** Wide mouthed nest of fine grass blades and thistle down or other vegetable down in a small bush, often conspicuous but soundly attached. Eggs, usually one, whitish ground colour streaked or heavily mottled with grey and brown, frequently capped at the larger end; about 21 × 13 mm.

**Recorded breeding:** Kenya Colony, May to December, commonest June and October, November, also occasionally in other months. Northern Tanganyika Territory, January to March.

**Call:** A sibilant 'zweet' and a sharp 'chink' are the most common calls.

**1236 PYGMY SUNBIRD.** *HEDYDIPNA PLATURA* (Vieillot).
*Hedydipna platura platura* (Vieill.). **Pls. 80, 82.**
*Cinnyris platurus* Vieillot, N. Dict. d'Hist. Nat. 31, p. 501, 1819: Senegal.

**Distinguishing characters:** Adult male, above, metallic golden green, including sides of face and wing-coverts; rump and upper tail-coverts violet; wings dusky with narrow light edges to flight feathers; innermost secondaries and tail blue black; central tail feathers greatly elongated; below, chin and neck metallic golden green; chest to under tail-coverts golden yellow. The non-breeding dress is similar to the dress of the adult female, including the flight feathers and tail, but the metallic upper wing-coverts and black under wing-coverts and axillaries are retained and in some sparse metallic tipped feathers occur on the body. The female is ashy above and yellow below. Wing 53 to 61 mm. The young bird is similar to the adult female, but the young male has a dusky patch on chin and throat.

**General distribution:** Senegal to the Sudan, Uganda and Kenya Colony.

**Range in Eastern Africa:** The Darfur and Kordofan areas of the Sudan as far east as Melit to northern Uganda and north-western Kenya Colony.

**Habits:** Over much of its range the commonest Sunbird locally, frequenting the higher boughs of Acacias and other trees of the bush, and not usually found near water. It is locally migratory and arrives in the Welle district from the Sudan in December to breed, leaving again in April. It does not seem to be dependent on flowers, as it leaves its breeding ground as the main flowering season approaches. Occurs in flocks of twenty to thirty in the non-breeding season.

**Nest and Eggs:** An oval nest of fine grass, spiders' webs, etc., with a porched side entrance, decorated outside with leaves or lichens and suspended from a twig. It has a thick felted lining of down. Eggs two or three, white, pinkish when fresh, occasionally with small rufous speckling; about $15 \times 10 \cdot 5$ mm.

**Recorded breeding:** Nigeria, December to March. Welle River, Belgian Congo, January to April. North-western Uganda, October and December.

**Call:** A soft silvery trilling song, also a call of 'cheek' or 'cheek-cheek.'

*Hedydipna platura metallica* (Lichtenstein).
*Nectarinia metallica* Lichtenstein, Verz. Doubl. p. 15, 1823: Dongola, northern Sudan.

**Distinguishing characters:** Adult male, differs from the nominate race in having a metallic violet band across the upper chest and the rump and upper tail-coverts darker violet; rarely the violet band across the chest is almost absent, and the violet blue confined to the upper tail-coverts. The non-breeding dress is assumed in the same way as in the nominate race; the female and young bird are indistinguishable from those of that race. Wing 51 to 61 mm.

**General distribution:** Egypt and the Sudan, to British Somaliland and Kenya Colony, also south-western Arabia.

**Range in Eastern Africa:** The Sudan as far south and west as Meidob, Bara, El Obeid, eastern Darfur and Kordofan areas to Eritrea, Abyssinia, British Somaliland and northern Kenya Colony.

**Habits:** As for the nominate race, but a bird of the desert edge and generally common wherever Sodom-Apple grows. Restless little birds with swift darting flight, but very tame.

**Nest and Eggs:** Nest oval, of fine fibres, down and rootlets closely woven and held together with spiders' webs, and with a dense lining of down or feathers. Nest either in a fork or attached to a thin twig. Side entrance. Eggs two to four, elongated, white with a rosy flush when fresh, finely speckled or spotted with rufous and with larger grey undermarkings.

**Recorded breeding:** Sudan, March to September and even later.

**Call:** A few soft twittering notes in the breeding season.

**1237** SUPERB SUNBIRD. *CINNYRIS SUPERBUS* (Shaw). Pls. 80, 82.

*Cinnyris superbus superbus* (Shaw).
*Certhia superba* Shaw, Gen. Zool. 8, p. 193, 1811: Malimba, Portuguese Congo.

**Distinguishing characters:** Adult male, above, top of head metallic blue, with greenish and violet reflections in certain lights; rest of upperparts metallic golden green; wings and tail velvety black; below, chin to chest metallic ruby red and deep blue; breast to belly chocolate maroon; some black on lower belly and under tail-coverts. There is no non-breeding dress. The female is olivaceous, with a pale stripe over the eye; below, yellower, edges of primaries yellowish or orange; under tail-coverts orange and sometimes orange spots on breast. Wing, male 73 to 83, female 69 to 75 mm. The young bird is similar to the female, but is less green and has the under tail-coverts olivaceous yellow.

**General distribution:** Cameroons and Angola, to eastern Belgian Congo and Uganda.

**Range in Eastern Africa:** Western and southern Uganda.

**Habits:** This species is somewhat aberrant from the rest of the genus in that it makes a nest twice as large as that of any other species. It is a restless bird of forest or forest clearings, feeding among the flowers of high forest trees. The males breed before assuming full plumage in some cases.

**Nest and Eggs:** Nest very large and untidy, either a handful of dry grass and a few leaves, or a mass of lichen. In either case it is suspended from a bough and has a rough tail of material and is very easily passed over as an accumulation of rubbish. It has a lining of fine hair-like fibre. Eggs one or two, pale cream or pale bluish white and glossy, with sparse blotches and speckles of dark grey or purplish black; about $21 \cdot 5 \times 15$ mm.

**Recorded breeding:** Uganda, March to May.

**Call:** A sharp chirp. The male has a little jingling song.

**Distribution of other races of the species:** French Guinea to Nigeria.

**1238 COPPER SUNBIRD.** *CINNYRIS CUPREUS* (Shaw).
*Cinnyris cupreus cupreus* (Shaw). **Pls. 80, 82.**
*Certhia cuprea* Shaw, Gen. Zool. 8, p. 201, 1811: Malimba, Portuguese Congo.

**Distinguishing characters:** Adult male, head and neck all round to upper chest and upper mantle metallic golden copper with purple and ruby red reflections; lower mantle, rump, upper tail-coverts and lesser wing-coverts metallic purple; flight feathers and tail blue black; rest of underparts black. The non-breeding dress is similar to the dress of the adult female, but the black flight feathers and tail are retained, also the metallic wing and upper tail-coverts and in some cases sparse metallic tips appear on the body feathers. The female is olivaceous green above and yellow below, lighter on the throat;

tail blue black with light edges and tips to the outer feathers. Wing, male 54 to 67, female 48 to 62 mm. The young bird is similar to the adult female, but the young male has a dusky throat.

**General distribution:** Senegal and Portuguese Congo to Abyssinia, the Sudan, Belgian Congo, Uganda, Kenya Colony, Tanganyika Territory, Northern Rhodesia, Nyasaland and Portuguese East Africa.

**Range in Eastern Africa:** Abyssinia to the Sudan, Uganda and Kenya Colony to the Zambesi River.

**Habits:** A common species of most of Eastern Africa in bush, open woodland or forest clearings, though in Abyssinia it is not found freely above 6,000 feet. There is undoubtedly considerable local migration. Rather restless bird with jerky erratic flight, but tame and often nesting near huts or buildings. This species has been noticed in parties of considerable size which gather to roost in a tree.

**Nest and Eggs:** Nest of dry grass lined with seed heads and with a sort of beard below the lip of the entrance, generally at the end of a low bough. It is neatly and compactly built and often has a lining of vegetable down. Sir Charles Belcher noted that many nests overhung holes in anthills or in the ground. Nest built by female alone. Eggs two, pale yellowish white to deep cream, or pale brown with obscure cloudings of grey and with streaks of brownish black; about 17 × 12 mm.

**Recorded breeding:** Nigeria, August to October. Uganda, March to September, also October and November. Tanganyika Territory, probably March onwards. Southern Belgian Congo, December to April. Nyasaland, February and March.

**Call:** A sharp little chipping song with an occasional explosive 'pst.' A roosting flock makes a continual reedy warbling.

**Distribution of other races of the species:** Angola.

**1239 SPLENDID SUNBIRD.** *CINNYRIS COCCINIGASTER* (Latham). **Pls. 80, 82.**

*Certhia coccinigaster* Latham, Suppl. Ind. Orn. in Gen. Synop. Suppl. p. 35, 1801: Senegal.

**Distinguishing characters:** Adult male, head, sides of face and chin to chest metallic purple; mantle to upper tail-coverts and lesser wing-coverts metallic emerald green with slight golden wash; occiput and upper tail-coverts washed with metallic blue; wings and tail black washed with metallic green, the latter with metallic green edging; chest and breast mixed metallic purple and bright red; under tail-coverts metallic blue; rest of underparts black; yellow tufts on side of chest. There is no non-breeding dress. The immature dress is similar to the adult female, but more ash-coloured above, with chin to chest as in the adult male. The female is olivaceous above; tail blackish with a slight green gloss; pale yellowish below streaked with dusky except on belly, throat paler; occasionally some reddish spots on underside. Wing 61 to 75 mm. The young female is similar to the adult female, but has the chin to upper chest black. The young male also has the chin to upper chest black, but is rather darker, less green above than the young female. The adult and young female may be distinguished from that of the Copper Sunbird by the larger size and by the shorter green-washed tail.

**General distribution:** Senegal and Gabon to the Sudan.

**Range in Eastern Africa:** South-western Sudan.

**Habits:** A common West African species which just reaches our area. In the Gold Coast it is the commonest Sunbird. It lives among flowering trees in forest or in coconut groves and is somewhat migratory, keeping chiefly to the Savanna belt.

**Nest and Eggs:** Nest oval, of fine grass with side entrance, dependent from creepers or boughs. Eggs undescribed.

**Recorded breeding:** West Africa, January to May. North-eastern Belgian Congo (nestling), August.

**Call:** A distinctive melodious song, at first loud and clear and then dying away and getting slower as if the bird was becoming sleepy (Bates).

**1240 SHINING SUNBIRD.** *CINNYRIS HABESSINICUS* (Hemprich & Ehrenberg).

*Cinnyris habessinicus habessinicus* (Hemp. & Ehr.). **Pls. 80, 82.**
*Nectarinia (Cinnyris) habessinicus* Hemprich & Ehrenberg, Symb. Phys. fol. a, pl. 4, 1828: Eilet, Eritrea.

**Distinguishing characters:** Adult male, forehead and crown metallic purple; head, neck all round, mantle to rump and wing-coverts metallic green with an old gold wash; rump bluer; a broad red band across chest and breast, some feathers tipped with metallic blue; yellow tufts on side of chest; wings, tail and belly slightly glossy blue black. There is no non-breeding dress. The immature dress

is similar to the adult female, but the chin to neck is metallic green and the red band across the chest is more or less indicated. The female is ash coloured, rather paler below; tail slightly glossy blue black with whitish edges to outer tail feathers and white tips to all. Wing 57 to 73 mm. The young bird is similar to the adult female, but is yellower below inclining to orange on chest; the young male has a black throat.

**Range in Eastern Africa:** Eastern Sudan, the greater part of Abyssinia and British Somaliland.

**Habits:** A common species in Abyssinia and British Somaliland at all elevations up to 12,000 feet and locally migratory.

**Nest and Eggs:** Nest of fibres, spiders' webs and cocoons suspended from the end of a branch. Eggs usually one only, dull white with heavy ash-coloured streaking; about $18 \cdot 5 \times 12$ mm.

**Recorded breeding:** Red Sea coast, April. Eastern Abyssinia, April.

**Call:** A rasping little note.

*Cinnyris habessinicus turkanæ* van Som.
*Cinnyris habessinicus turkanæ* van Someren, Bull. B.O.C. 40, p. 94, 1920: Kohua River, Lake Rudolf, Kenya Colony.

**Distinguishing characters:** Differs from the nominate race in lacking the metallic blue band below the red chest band. Wing 59 to 68 mm.

**Range in Eastern Africa:** Southern Abyssinia to Kenya Colony from Lake Rudolf to the Turkana area.

**Habits:** As for the nominate race.

**Recorded breeding:** Northern Kenya Colony, December.

**Distribution of other races of the species:** Arabia.

**1241 LITTLE PURPLE-BANDED SUNBIRD.** *CINNYRIS BIFASCIATUS* (Shaw).
*Cinnyris bifasciatus microrhynchus* Shell. **Pls. 80, 82.**
*Cinnyris microrhynchus* Shelley, Mon. Nect. p. 219, pl. 67, 1876: Dar-es-Salaam, eastern Tanganyika Territory.

**Distinguishing characters:** Adult male, head and neck all round, mantle to rump and lesser wing-coverts metallic green with a golden wash; band across chest metallic violet; band across breast maroon with some metallic violet tips to the feathers; belly black;

wings and tail black with a slight blue black gloss; edge of wings slightly metallic green. The non-breeding dress is similar to that of the adult female, but the black wings and tail and metallic wing-coverts and upper tail-coverts are retained; throat blackish. The female is slightly olivaceous ash colour above; below, pale yellowish with black streaks; throat dusky to black often with whitish tips to feathers; tail black with a slight blue black gloss, whitish edges to outer feathers and whitish tips to all except central pair. In non-breeding dress the upperparts are brighter olivaceous. Wing, male 51 to 61, female 48 to 56 mm. The young bird is similar to the adult female, but rather more olivaceous, less ashy grey, above.

**General distribution:** Uganda to Kenya Colony, south-eastern Belgian Congo, Northern Rhodesia, eastern Southern Rhodesia, eastern Transvaal and north-eastern Zululand.

**Range in Eastern Africa:** Uganda and Kenya Colony (but not southern areas from the middle Tana River to Athi, Taita and Voi) to the Zambesi River; also Zanzibar and Mafia Islands.

**Habits:** A common and often abundant species of bush country or coastal mangrove swamp, with considerable local movements. It also occurs in gardens and cultivated land. Partial to flowering Acacias and usually appears when they come into flower.

**Nest and Eggs:** Nest a typical Sunbird's of grass fibres bound with spiders' webs and with lichen or dead leaves stuck on as if to give a deliberately ragged appearance. Strongly made with or without a lining of feathers, and attached to a bough a few feet from the ground. Eggs two, purplish grey or slate grey all over with occasional spots and streaks of black; about $17 \times 11 \cdot 5$ mm.

**Recorded breeding:** Zanzibar Island, May, July to September, November and December. Mombasa, March to May, also November. Tanganyika Territory, August. Mafia Island (breeding condition), August. South-eastern Belgian Congo, August. Northern Nyasaland, May to August and (breeding condition) October. Southern Nyasaland, June. Southern Rhodesia, October. Portuguese East Africa, October and November.

**Call:** Song a high pitched little falling trill. The normal call of the male is 'b-r-r-zi' answered by the female with 'br-r-r' (Moreau).

**Distribution of other races of the species:** Gabon to Angola and Nyasaland, the nominate race being described from the Portuguese Congo.

# SUNBIRDS

**1242** TSAVO PURPLE-BANDED SUNBIRD. *CINNYRIS TSAVOENSIS* V. Som.

*Cinnyris bifasciatus tsavoensis* Van Someren, Nov. Zool. 29, p. 196, 1922: Tsavo, south-eastern Kenya Colony.

**Distinguishing characters:** Adult male, differs from the preceding species in having usually a narrower and more violet than maroon band on the breast. There is no non-breeding dress. The female and young bird are indistinguishable from that of the preceding species. Wing, male 54 to 59; female 49 to 52 mm.

**Range in Eastern Africa:** Southern Italian Somaliland to eastern Kenya Colony and eastern Tanganyika Territory, but not the coastal areas.

**Habits:** As for the last species and locally abundant at Aloe flowers.

**Nest and Eggs:** The nest is attached direct to a branch and is compact with no tail, and usually covered with lichens. Eggs similar to those of the last species, but smaller; about $15 \cdot 5 \times 10 \cdot 5$ mm.

**Recorded breeding:** Tana River, Kenya Colony, October.

**Call:** Unrecorded.

**1243** VIOLET-BREASTED SUNBIRD. *CINNYRIS CHALCOMELAS* Reichenow. **Pl. 82.**

*Cinnyris chalcomelas* Reichenow, Vög. Afr. 3, p. 482, 1905: Kismayu, southern Italian Somaliland.

**Distinguishing characters:** Adult male, similar in general appearance to the Tsavo Purple-banded Sunbird, but larger; bill longer and rather straighter; wings, tail and belly darker, more glossy blue black; band on breast even deeper violet; edge of wing metallic green. There is no non-breeding dress. The female is also larger and longer billed than the female of the last species and is only faintly streaked below. Wing, male 59 to 63, female 55 to 56 mm. The young bird is very similar to the adult female.

**Range in Eastern Africa:** Southern Italian Somaliland to the Mombasa area of Kenya Colony and as far west as Tsavo.

**Habits:** No information.

**Nest and Eggs:** Undescribed.

**Recorded breeding:** No records.

**Call:** Unrecorded.

## 1244 PEMBA VIOLET-BREASTED SUNBIRD. *CINNYRIS PEMBÆ* Reichenow. **Pls. 80, 82.**

*Cinnyris pembæ* Reichenow, O.M. p. 180, 1905: Pemba Island.

**Distinguishing characters:** Adult male, head and neck all round and mantle to upper tail-coverts metallic blue and green; lesser wing-coverts metallic violet and green; greater wing-coverts edged with metallic green; wings and tail deep slightly glossy blue black; edge of wing metallic violet; tail feathers edged with metallic blue; a metallic violet band across chest; belly blue black. There is no non-breeding dress. The immature dress is similar to that of the adult female, but the throat and neck is metallic green and in some cases there is a partial metallic bluish violet band on the chest, and sparse metallic tips to head and body feathers. The female is pale olive above; a distinct whitish superciliary stripe; below, pale yellow; tail blue black with white tips and white edges to outer tail feathers. Wing, male 50 to 56, female 48 to 50 mm. The young bird is similar to the adult female but perhaps slightly more yellowish olive above, and the young male has a blackish throat.

**Range in Eastern Africa:** Lamu, eastern Kenya Colony and Pemba Island.

**Habits:** A locally plentiful species of normal Sunbird habits, attending Hibiscus flowers and being particularly fond of the Millingtonia trees. The males are very pugnacious when breeding.

**Nest and Eggs:** A typical Sunbird's nest rather strongly built of normal materials and usually lined with seed down; the nest has no streamers attached and hangs from a bough some four to eight feet from the ground. Eggs two, greenish white marbled all over with grey and ashy brown. No measurements available.

**Recorded breeding:** Chiefly May, June and July, but in any month from May to January.

**Call:** A low, rapid and rather poor song (Vaughan).

## 1245 MARIQUA SUNBIRD. *CINNYRIS MARIQUENSIS* Smith.

*Cinnyris mariquensis suahelicus* Reichw. **Pls. 80, 82.**

*Cinnyris suahelicus* Reichenow, J.f.O. p. 161, 1891: Tabora district, central Tanganyika Territory.

**Distinguishing characters:** Adult male, similar to the Little Purple-banded Sunbird, but much larger and belly a more smoky

black. There is no non-breeding dress. The female also resembles that of the Little Purple-banded Sunbird but is much larger and brighter yellow below. Wing, male 64 to 70, female 61 to 65 mm.; tail, female 40 to 44 mm. The young bird is very similar to the adult female, but the young male has a black throat.

**General distribution:** Uganda and Kenya Colony to north-eastern Northern Rhodesia at Mwenzo and Tanganyika Territory.

**Range in Eastern Africa:** Central Uganda to central Kenya Colony and the Iringa district of southern Tanganyika Territory.

**Habits:** Local but not uncommon in scrub and Acacia country, usually found in open bush where the grass is short. Visits any kind of flower, and if flowers are not available searches grassheads for insects. Generally seen in pairs.

**Nest and Eggs:** Nest of cottony vegetable down bound with spider webs and lined with feathers, usually dependent from the branch of a shrub. Eggs two, cream, greyish white, or pale greenish white, with brown streaks and specklings at the larger end; about 17 × 13 mm.

**Recorded breeding:** Kenya Colony, March and April. Uganda, September. Tanganyika Territory uplands, probably March and April.

**Call:** A vigorous rather monotonous little song.

*Cinnyris mariquensis osiris* (Finsch).
*Nectarinia osiris* Finsch, Trans. Zool. Soc. 7, p. 230, 1872: Senafé; Acchele Guzai district, Eritrea.

**Distinguishing characters:** Adult male, differs from the preceding race in having the breast band darker with more extensive violet coloration; and the belly black. Wing, male 64 to 71, female 58 to 64 mm.

**Range in Eastern Africa:** Eritrea and British Somaliland to south-eastern Sudan, northern Uganda and northern Kenya Colony.

**Habits:** As for the last race.

**Recorded breeding:** Turkana, Kenya Colony (breeding condition), June.

**Distribution of other races of the species:** South Africa, the nominate race being described from north of Kurrichane, Bechuanaland.

## 1246 SHELLEY'S DOUBLE-COLLARED SUNBIRD. *CINNYRIS SHELLEYI* Alexander.

*Cinnyris shelleyi shelleyi* Alex. **Pl. 82.**
*Cinnyris shelleyi* Alexander, Bull. B.O.C. 8, p. 54, 1899: Zambesi River, 60 miles east of Kafue-Zambesi rivers junction.

**Distinguishing characters:** Adult male, similar to the Mariqua Sunbird in size, shape of bill and general colour, but has a broad crimson band across breast. There is no non-breeding dress. The female is also similar to that of the Mariqua Sunbird but is shorter tailed and has paler and greener upper tail-coverts. Wing, male 63 to 68, female 56 to 59 mm.; tail, female 37 to 39 mm.; bill from base, male 20 to 23 mm. The plumage of the young bird is not recorded.

**General distribution:** Northern Rhodesia to Nyasaland, Tanganyika Territory, Portuguese East Africa and Southern Rhodesia.

**Range in Eastern Africa:** South-western Tanganyika Territory to Portuguese East Africa and Nyasaland.

**Habits:** Common, widely distributed in open woodland on the central plateau of Portuguese East Africa, the males perching on bare tree tops and uttering a quick chittering call. Particularly fond of Loranthus flowers.

**Nest and Eggs:** The only nests so far described were in small trees about six feet from the ground and were very small with a thin rather untidy roof, made of grass and cobwebs and lined with feathers. Eggs two, pale olive with smudged black spots above lighter slate coloured smudges; about 15 × 11 mm.

**Recorded breeding:** Rhodesia, November. Portuguese East Africa, probably March and April.

**Call:** A rapidly repeated 'didi-didi.' Song a nasal 'chibbee-cheeu-cheeu' (Vincent).

*Cinnyris shelleyi hofmanni* Reichw.

*Cinnyris hofmanni* Reichenow, O.M. p. 91, 1915: Magogoni, Ruvu River, eastern Tanganyika Territory.

**Distinguishing characters:** Differs from the nominate race in having a shorter bill. The female is dull olive green above; tail as in adult male; a narrow but distinct yellowish stripe over eye; yellowish moustachial streak; chin and throat dusky; rest of underparts pale yellow, with dusky centres to feathers of chest, breast and flanks; a variable broken orange-red band across breast. Wing, male 58 to 63, female 56 to 57; bill from base, male 18 to 20, female 19; tail, male 35 to 38, female 31 to 34 mm. The young bird is similar to the adult female, but has a black throat and the young male has a crimson band across the breast.

**Range in Eastern Africa:** Ruvu River area to Morogoro, eastern Tanganyika Territory.

**Habits:** As for the nominate race.

**Recorded breeding:** Morogoro (breeding condition), November.

## 1247 ORANGE-TUFTED SUNBIRD. *CINNYRIS BOUVIERI* Shelley. Pl. 82.

*Cinnyris bouvieri* Shelley, Mon. Nect. p. 227, pl. 70, 1877: Landana, Portuguese Congo.

**Distinguishing characters:** Adult male, very similar to the Mariqua Sunbird, but has a violet forehead; chin and upper throat non-metallic black; a narrow deep chocolate band on breast below the blue and violet band; orange and yellow tufts on the sides of the chest. There is no non-breeding dress. The female is olivaceous above, with a superciliary stripe over and behind the eye; below, pale yellow indistinctly streaked with dusky; throat dusky; tail blue black with white edges to outer feathers and white tips to others. Wing, male 52 to 61, female 52 to 54 mm. The young bird is similar to the adult female, but the young male has a darker throat.

**General distribution:** Northern Cameroons to Portuguese Congo, eastern Belgian Congo and Uganda.

**Range in Eastern Africa:** Western and southern Uganda.

**Habits:** Rare and little known in Uganda, generally seen at Erythrina trees. Sir F. Jackson records a number seen in the

Botanical Gardens at Entebbe between July and September. Serle remarks that in the Cameroon highlands it also inhabits open hillsides far from trees.

**Nest and Eggs:** Nest, a typical domed Sunbirds' nest of fine grass and gossamer with a lining of plant down. Eggs undescribed.

**Recorded breeding:** Cameroons (building), October.

**Call:** Unrecorded.

**1248** NORTHERN ORANGE-TUFTED SUNBIRD. *CINNYRIS OSEUS* Bonaparte.
*Cinnyris oseus decorsei* Oust. **Pl. 82.**
*Cinnyris decorsei* Oustalet, Bull. Mus. Paris, 10, p. 536, 1904: Lake Chad region.

**Distinguishing characters:** Adult male, very similar to the Orange-tufted Sunbird, but bill shorter and the chin and throat metallic blue; it also lacks the chocolate breast band. The non-breeding dress is similar to that of the adult female, but the black wings and tail and metallic wing-coverts are retained; sparse metallic tipped feathers occur on the body. The female differs from the female of the Orange-tufted Sunbird in being shorter billed, more dusky, less yellow below, and having the upper tail-coverts darker with metallic bottle green edges; edges of tail feathers metallic bottle green. Wing, male 50 to 55, female 48 to 54 mm. The young bird is very similar to that of the Orange-tufted Sunbird but has a shorter bill.

**General distribution:** Lake Chad area to the Sudan.

**Range in Eastern Africa:** Western and southern Sudan.

**Habits:** Migratory over much of its range and not well known. It appears to visit the high ground of the Jebel Marra to breed from September to March; on the Welle River, Belgian Congo, it appears in October and stays till February, leaving when about to breed.

**Nest and Eggs:** Nest a typical Sunbird's of dry stems and thistle leaves lined with down and suspended from a low twig or frond. Eggs two, dull white, speckled and clouded with very small spots of grey or greyish brown—chiefly in a zone at the larger end; about 15 × 11 mm.

**Recorded breeding:** Jebel Marra, Sudan (high elevations), November to February, probably a migrant from the south.

**Call:** Unrecorded.

**Distribution of other races of the species:** Syria, Palestine and south-western Arabia, the nominate race being described from Palestine.

## 1249 WHITE-BELLIED SUNBIRD. *CINNYRIS TALATALA* Smith.

*Cinnyris talatala anderssoni* (Strick.). **Pls. 80, 83.**
*Nectarinia anderssoni* Strickland, Cont. Orn. p. 153, 1852: Damaraland.

**Distinguishing characters:** Adult male, upperside and sides of face metallic green washed with old gold and sometimes with blue reflections; upper tail-coverts bluer; wings sooty black; tail blue black with metallic green edges, underside with lighter edges to outermost feathers and light tips to all; below, neck metallic blue green; chin sometimes dull black; chest band metallic violet followed by a blackish band; rest of underparts white or white tinged yellow; yellow tufts at sides of chest. In non-breeding dress similar to the adult female, but the blue black tail and the metallic wing-coverts and upper tail-coverts are retained, with sometimes sparse metallic tipped feathers on head and mantle. The female is ashy brown above; tail bluish black with metallic green edges and metallic green edges to upper tail-coverts, these being bright green and not bottle green as in the female of the Northern Orange-tufted Sunbird; below, dusky white. Wing, male 52 to 58, female 48 to 54 mm. The young bird is similar to the adult female, but is usually more olivaceous above and yellower below.

**General distribution:** Southern Angola to Damaraland, the Rhodesias, Nyasaland, Tanganyika Territory, Portuguese East Africa and Zululand.

**Range in Eastern Africa:** Southern Tanganyika Territory to Nyasaland and Portuguese East Africa.

**Habits:** Common and locally abundant birds, generally seen among Acacias in pairs and always very active. The male has a fine Canary-like song uttered from a bush top.

**Nest and Eggs:** Typical Sunbird's nest suspended from a thorny bush a few feet from the ground. Eggs two, grey or greyish white, spotted, speckled and clouded with slaty brown or purplish brown; about 16 × 11·5 mm.

**Recorded breeding:** Nyasaland, April and July to October. Rhodesia, September to November. Portuguese East Africa, February to August.

**Call:** Song as above, powerful for the size of the bird, also a short penetrating 'zitting' call and at times a whistle.

**Distribution of other races of the species:** South Africa.

**1250** ANGOLA WHITE-BELLIED SUNBIRD. *CINNYRIS OUSTALETI* (Bocage).
*Nectarinia oustaleti* Bocage, Jorn. Lisboa 6, p. 254, 1878: Caconda, Benguella, southern Angola.

**Distinguishing characters:** Adult male, differs from the White-bellied Sunbird in having a shorter bill; mixed orange and yellow tufts at side of chest; violet band across the breast tipped with maroon. The non-breeding dress is dark ashy above, but the wings and tail, metallic wing-coverts and upper tail-coverts are as in the breeding dress, and there are a few sparse metallic tipped feathers on the head and mantle; below, throat dusky with some metallic tips; chest with some metallic tips to feathers; tufts at sides of chest present. The female is grey brown above; upper tail-coverts black with bronzy and green edges; below, chin to chest dusky with indistinct darker streaks; belly creamy white. It can be distinguished from the female of the White-bellied Sunbird in being darker above, with no olivaceous wash; below, chin to chest more dusky, with indistinct darker streaks; belly creamy white. Wing, male 53 to 58, female 55 mm. Juvenile plumage unrecorded.

**General distribution:** Southern Angola to north-eastern Northern Rhodesia.

**Range in Eastern Africa:** Not yet recorded, but may occur between south end of Lake Tanganyika and north end of Lake Nyasa as it has been taken at Mwenzo, about twenty miles south of the boundary.

**Habits:** No information.

**Nest and Eggs:** Undescribed.

**Recorded breeding:** No records.

**Call:** Unrecorded.

**1251** VARIABLE SUNBIRD. *CINNYRIS VENUSTUS* (Shaw & Nodder).

*Cinnyris venustus albiventris* (Strick.).    White-bellied Variable Sunbird.

*Nectarinia albiventris* Strickland, Contr. Orn. p. 42, pl. 86, 1852: Ras Hafun, northern Italian Somaliland.

**Distinguishing characters:** Adult male, similar to the White-bellied Sunbird, but has a violet forehead, orange and yellow tufts at sides of chest and throat steel blue and violet, not green. In non-breeding dress similar to the adult female, but the metallic wing and upper tail-coverts are retained and there are often a considerable number of metallic tips to the body feathers. The female is ashy brown above, below, white; no metallic edging to upper tail-coverts and only faint green or no green edging to tail feathers. Wing, male 52 to 58, female 48 to 53 mm. The young bird is very similar to the adult female, but as a rule lacks the metallic edges to the upper tail-coverts; the young male has a dusky to blackish throat.

**Range in Eastern Africa:** British and Italian Somaliland to southern Abyssinia and eastern half of Kenya Colony, also Manda Island.

*Note:* Intermediates between this race, the Sudan race and the Kenya Colony race occur wherever they meet.

**Habits:** Not uncommon locally in Somaliland among Aloes and Acacias, and abundant on Manda Island among dense bush on sand hills. Tame fearless little birds, locally common.

**Nest and Eggs:** See under other races.

**Recorded breeding:** Southern Abyssinia, April and May. Manda Island, May.

*Cinnyris venustus fazoqlensis* (Heugl.).    Northern Yellow-bellied Variable Sunbird.

*Nectarinia fazoqlensis* Heuglin, Orn. Nordost Afr. 2, pt. 2, p. 70, 1871: South of Fazogli Mt., Sennar district, eastern Sudan.

**Distinguishing characters:** Adult male, differs from the preceding race in having the breast and belly yellow; throat metallic green. The female is more olivaceous above; below, yellow or washed with yellow with rarely some indistinct olivaceous streaking. Wing, male 50 to 57, female 46 to 51 mm. The female is very similar to that of the Beautiful Sunbird and can only be distinguished by the tail feathers being blue black not black with a greenish wash, and there is no metallic edging to the upper tail-coverts. The young bird can be distinguished from the young bird of the Beautiful Sunbird by the same characters in the tail feathers.

**General distribution:** Eritrea and Abyssinia.

**Range in Eastern Africa:** Eritrea and Abyssinia (except south-western areas).

**Habits:** As for other races.

**Recorded breeding:** Abyssinia, May, also August to October.

*Cinnyris venustus falkensteini* Fischer & Reichw. *Yellow-bellied Variable Sunbird.* **Pls. 80, 83. Ph. xvi.**
*Cinnyris falkensteini* Fischer & Reichenow, J.f.O. p. 56, 1884: Lake Naivasha, Kenya Colony.

**Distinguishing characters:** Adult male, similar to the last race but with the throat steel blue and violet. The female is indistinguishable from that of the last race. Wing, male 47 to 57, female 47 to 52 mm.

**General distribution:** The Sudan, Abyssinia, Kenya Colony, Tanganyika Territory, Nyasaland, Portuguese East Africa and the Rhodesias.

**Range in Eastern Africa:** South-eastern Sudan and south-western Abyssinia, Kenya Colony except the coastal areas, Tanganyika Territory, Nyasaland and Portuguese East Africa.

**Habits:** A common and widespread species, in several districts the commonest of the Sunbirds. It inhabits plantations and cultivated country, woodland or forest edge, and long grass country with plenty of bushes, and occurs from the low levels up to the higher plateaux of Abyssinia and Nyasaland.

**Nest and Eggs:** Nest oval, usually rather lightly woven of grass and fibre and lined with feathers and vegetable down, suspended from a bush creeper or large plant at any height from the ground. Eggs two, white or greyish white, densely speckled with brown and ash grey to form a cap or zone at the larger end; about 15 × 11 mm. Occasionally parasitized by Klaas's Cuckoo.

**Recorded breeding:** Kenya Colony apparently in most months of the year, but chiefly March to June. Tanganyika Territory,

January to April, also October. Nyasaland, March to August. Rhodesia at any time of year. Portuguese East Africa (nestlings), July.

**Call:** A pleasant short twittering warbling song. Sharp alarm call of 'cheer-cheer' and a variety of chirping notes.

*Cinnyris venustus igneiventris* Reichw. Orange-bellied Variable Sunbird.
*Cinnyris igneiventris* Reichenow, O.M. p. 171, 1899: Karagwe, north-western Tanganyika Territory.

**Distinguishing characters:** Adult male, differs from the Yellow-bellied races in having the breast to belly mixed orange and yellow. Wing, male 50 to 55, female 47 to 49 mm.

**General distribution:** Eastern Belgian Congo to Uganda and Tanganyika Territory.

**Range in Eastern Africa:** Western Uganda to north-western Tanganyika Territory.

**Habits:** As for other races.

**Recorded breeding:** Belgian Congo, July and August.

**Distribution of other races of the species:** West Africa, the nominate race being described from Sierra Leone.

**1252** GREATER DOUBLE-COLLARED SUNBIRD. *CINNYRIS AFER* (Linnæus).
*Cinnyris afer stuhlmanni* Reichw. **Pls. 80, 83.**
*Cinnyris stuhlmanni* Reichenow, O.M. p. 61, 1893: Nsaugani, Ukondju, eastern Belgian Congo.

**Distinguishing characters:** Adult male, head and neck all round, lesser wing-coverts, mantle and rump metallic moss-green with a slight golden wash; upper tail-coverts metallic violet; wings blackish; flight feathers and greater coverts edged with olive green; tail blue black, outermost feathers dusky black; narrow band of violet across chest; broad scarlet band across breast; tufts on sides of chest bright yellow; belly and under tail-coverts dusky olivaceous. The non-breeding dress is very similar to the adult female, but the darker wings and metallic wing-coverts are retained. The female is olivaceous green, paler on the belly; wings rather less blackish than the male; under wing-coverts yellowish white, not dusky; tail as in male, but shorter and with some white on ends of outer feathers. Wing, male 65 to 67; female 60 mm. The young bird is very similar to the adult female; but the young male may have a dusky throat.

**General distribution:** Eastern Belgian Congo and western Uganda.

**Range in Eastern Africa:** Ruwenzori Mts., western Uganda.

**Habits:** A local species occurring on Mt. Ruwenzori in a belt from about 10,000 to 11,500 feet mostly in the upper Bamboo and Tree Heath zones, but at lower elevations also in the eastern Belgian Congo. The male has a courtship display with drooping quivering wings and yellow plumes raised fanwise. Locally common.

**Nest and Eggs:** An oval moss nest usually placed among the moss dependent from the ends of Tree-Heath branches, and therefore difficult to detect. Egg normally one only, ground colour white but usually obscured with dark olive, heavily freckled with dark ash grey; about 19 × 13 mm.

**Recorded breeding:** Ruwenzori Mts., December and January.

**Call:** The male has a short cheerful song, rather harsh in tone. Alarm call a loud chirp.

**Distribution of other races of the species:** South Africa, the nominate race being described from the Cape of Good Hope.

**1253** SOUTHERN DOUBLE-COLLARED SUNBIRD. *CINNYRIS CHALYBEUS* (Linnæus).
*Cinnyris chalybeus intermedius* (Boc.). **Pls. 81, 83.**
*Nectarinia intermedia* Bocage, Jorn. Lisboa 6, p. 120, 1878: Caconda, Benguella, Angola.

**Distinguishing characters:** Adult male, head and neck all round, mantle and lesser wing-coverts golden metallic green; rump olivaceous; upper tail-coverts olivaceous or olivaceous more or less tipped with metallic blue, copper or green, rarely with violet; wings and tail dusky blackish with an olivaceous wash; flight feathers edged with olive; below, a narrow metallic blue band across chest; a broad scarlet band across the breast; tufts at sides of chest yellow; belly dusky; lower belly and under tail-coverts yellowish. There is no non-breeding dress. The female is olivaceous above and olivaceous and pale yellow below; tail browner, with a slight bronzy wash. Wing, male 59 to 64, female 56 to 57 mm.; culmen, male 20 to 22, female 20 to 21 mm. Juvenile plumage unrecorded.

**General distribution:** Angola to south-eastern Belgian Congo, Tanganyika Territory and Nyasaland.

**Range in Eastern Africa:** Central and southern Tanganyika Territory.

**Habits:** A common species of Brachystegia-wooded uplands, visiting all kinds of flowers and usually tame. In certain cases extends into mountain forest.

**Nest and Eggs:** Relatively large nest of dry grass lined with seed down and feathers, with a large entrance hole with a porch over it, hanging from a bush or tree at any height from the ground. Eggs usually two, uniformly mottled and freckled with grey, pinkish grey, or purplish grey, occasionally with sparse dark spots and scrawls; about $16 \cdot 5 \times 11 \cdot 5$ mm.

**Recorded breeding:** Tanganyika Territory, May to October.

**Call:** A little twitter. The male has a really fine song in the nesting season.

*Cinnyris chalybeus graueri* Neum.
*Cinnyris afra graueri* Neumann, Bull. B.O.C. 21, p. 55, 1908: western Kivu Volcanoes, eastern Belgian Congo.

**Distinguishing characters:** Adult male, differs from the preceding race in having the upperparts washed with metallic blue. There is no non-breeding dress. The immature dress is similar to that of the adult but has mixed olivaceous and metallic feathers above, the metallic feathers being greener than in the adult and having olivaceous not blackish bases; flight feathers as in the adult female; below, red chest band mixed with olivaceous. The female is darker olivaceous green below than the female of the preceding race. Wing, male 64 to 65, female 58 to 62 mm.

**General distribution:** Eastern Belgian Congo and Uganda.

**Range in Eastern Africa:** Ruanda, eastern Belgian Congo and south-western Uganda at Mt. Muhavura.

**Habits:** Locally abundant and the commonest Sunbird of the Birunga Volcanoes.

**Recorded breeding:** Kivu area, September to November, also March and April.

*Cinnyris chalybeus bractiatus* Vinc.

*Cinnyris chalybeus bractiatus* Vincent, Bull. B.O.C. 53, p. 146, 1933: Fort Chiququa, Mashonaland, Southern Rhodesia.

**Distinguishing characters:** The adult male differs from the preceding race in being longer billed, with paler belly, and upper tail-coverts always blue. The female is similar to that of the preceding race. Wing, male 59 to 65, female 57 to 62 mm.; culmen, male 23 to 25·5, female 21 to 23 mm.

**General distribution:** Fort Jameson area, Northern Rhodesia to southern Nyasaland, northern Portuguese East Africa and Southern Rhodesia.

**Range in Eastern Africa:** Southern part of northern Portuguese East Africa and Nyasaland.

**Habits:** As for other races.

**Recorded breeding:** Nyasaland, April to November. Rhodesia, August to October, also May.

**Distribution of other races of the species:** South Africa, the nominate race being described from the Cape of Good Hope.

**1254** EASTERN DOUBLE-COLLARED SUNBIRD. *CINNYRIS MEDIOCRIS* Shelley.

*Cinnyris mediocris mediocris* Shell. **Pls. 81, 83.**

*Cinnyris mediocris* Shelley, P.Z.S. p. 228, 1885: Mt. Kilimanjaro, north-eastern Tanganyika Territory.

**Distinguishing characters:** Adult male, head and neck all round, lesser wing-coverts, mantle and rump metallic golden green with often a bluish wash; upper tail-coverts metallic blue; wings blackish with olive edges to flight feathers; tail blue black, outer feathers whitish on outer edges and tips; a narrow metallic blue band across chest; a broad scarlet band on breast; tufts on sides of chest yellow; belly to under tail-coverts yellowish olive. There is no non-breeding dress. The female is olivaceous moss-green above, wings duller black than in male and the olive edges to flight feathers more distinct; tail as in male; below, olive yellow, flanks and throat more olivaceous moss-green. Wing, male 51 to 58, female 48 to 52 mm. The young bird is similar to the adult female.

**Range in Eastern Africa:** Kenya Colony to north-eastern Tanganyika Territory as far south as Mt. Gerui (Hanang), but not Paré and Usambara Mts.

**Habits:** A species of high mountain forest and moorland, often occurring in cold desolate country among Kniphofias. It also inhabits forest and is found among high trees frequently in mixed bird parties.

**Nest and Eggs:** Spherical nest, with opening near the top, of lichens, grass and rootlets warmly lined with vegetable down. It is placed in shrubs, brambles or trees at any height from the ground. Eggs two, greenish white or pale pinkish buff heavily freckled or marbled with greyish brown or purplish brown; about 16·5 × 11·5 mm.

**Recorded breeding:** Kenya Colony, most months from May to December, mainly June and November. Northern Tanganyika Territory (breeding condition), October to January.

**Call:** Alarm note a chattering interspersed with whistles. A rather jingling but melodious little song.

*Cinnyris mediocris usambaricus* Grote.
*Cinnyris mediocris usambaricus* Grote, O.M. p. 86, 1922: Mlalo, near Wilhelmstal, Usambara, north-eastern Tanganyika Territory.

**Distinguishing characters:** Differs from the preceding races in having a narrower scarlet band on breast; upper tail-coverts violet. The female has the throat more distinctly streaked. Wing, male 53 to 56, female 48 to 50 mm.

**Range in Eastern Africa:** Paré and Usambara Mts., north-eastern Tanganyika Territory.

**Habits:** As for the nominate race.

**Recorded breeding:** No records.

*Cinnyris mediocris bensoni* Will.
*Cinnyris mediocris bensoni* Williams, Bull. B.O.C. 73, p. 10, 1953: Dedza Mt., Nyasaland.

**Distinguishing characters:** Adult male differs from the Iringa race in having the belly to under tail-coverts more dusky olivaceous green. The female is rather darker than that of the Iringa race. Wing, male 55 to 59, female 50 to 53 mm.

**General distribution:** Eastern Northern Rhodesia, Nyasaland and Portuguese East Africa.

**Range in Eastern Africa:** Namuli and Chiperone Mts., Portuguese East Africa.

**Habits:** As for the nominate race.

**Recorded breeding:** No records.

*Cinnyris mediocris fülleborni* Reichw.
*Cinnyris fülleborni* Reichenow, O.M. p. 7, 1899: Kalinga, Iringa district, south-central Tanganyika Territory.

**Distinguishing characters:** Adult male, differs from the nominate race in having the belly to under tail-coverts olivaceous green; upper tail-coverts violet. The female is darker, more moss-green below. Wing, male 53 to 59, female 50 to 54 mm.

**General distribution:** Tanganyika Territory to Portuguese East Africa, north-eastern Northern Rhodesia, and Nyasaland.

**Range in Eastern Africa:** Central to south-western Tanganyika Territory from the Dabaga and Njombe areas to northern Portuguese East Africa and Nyasaland.

**Habits:** Birds of mountain forest and scrub from 5,000 feet upwards. Very tame and usually common.

**Recorded breeding:** Tanganyika Territory, July, also irregularly October to December. Nyasaland, June to August.

**1255** MOREAU'S SUNBIRD. *CINNYRIS MOREAUI* Sclater.
*Cinnyris mediocris moreaui* W. L. Sclater, Ibis p. 214, 1933: Maskati, Nguru Mts., east central Tanganyika Territory.

**Distinguishing characters:** Differs from the Eastern Double-collared Sunbird in having the yellow of the pectoral tufts extending the whole breadth and slightly below the red chest band; chest band more orange red, less scarlet; rump plain olive green; upper tail-coverts green and violet; belly paler yellowish olive. There is no non-breeding dress. The female has a metallic greenish grey wash on head and mantle. Wing, male 55 to 57, female 50 to 54 mm. The young bird is duller than the adult female.

**General distribution:** Tanganyika Territory to Nyankhowa Mt., northern Nyasaland.

**Range in Eastern Africa:** South-west Masai, Mpapwa and Kilosa areas and the Nguru Mts. to south-western areas, Tanganyika Territory.

**Habits, Nest and Eggs:** Not distinguished from those of the Eastern Double-collared Sunbird, of which this species has hitherto been regarded as a race.

**Recorded breeding:** See above.

### 1256 NORTHERN DOUBLE-COLLARED SUNBIRD. *CINNYRIS REICHENOWI* Sharpe.

*Cinnyris reichenowi reichenowi* Sharpe. **Pl. 83.**

*Cinnyris reichenowi* Sharpe, Ibis p. 444, 1891: Sotik, western Kenya Colony.

**Distinguishing characters:** Adult male, very similar to the Eastern Double-collared Sunbird, but has a shorter bill; above, not so golden metallic green; under wing-coverts more dusky, less white or yellowish white. It is an almost exact small replica of the Ruwenzori race of the Greater Double-collared Sunbird. There is no non-breeding dress. The female and young bird are similar to those of the Eastern Double-collared Sunbird but have shorter bills. Wing, male 52 to 56, female 50 to 51 mm.

**General distribution:** Southern Sudan to eastern Belgian Congo, Uganda and the western half of Kenya Colony.

**Range in Eastern Africa:** Southern Sudan and Uganda to Kenya Colony as far east as Mt. Urguess, Embu and Chuka.

**Habits:** Not uncommon in western Kenya Colony and Uganda frequenting flowers of aloes, thistles, Salvias, etc. Fearless pugnacious little birds.

**Nest and Eggs:** Typical Sunbird's nest of grass and lichens, lined with down and feathers, usually in moss-covered branches and difficult to find. Eggs two, mauvish cream or claret brown, clouded with dark pinkish grey and with a few blackish specks; about 17 × 12 mm. (West African race).

**Recorded breeding:** North-eastern Belgian Congo, April. Uganda, July.

**Call:** A loud cheerful jingling song of short duration.

**Distribution of other races of the species:** West Africa.

Plate 82

Females of

| | | | |
|---|---|---|---|
| Malachite Sunbird (p. 751) | Scarlet-tufted Malachite Sunbird (p. 752) | Tacazze Sunbird (p. 754) | Bronze Sunbird (p. 755) |
| Purple-breasted Sunbird (p. 757) | Red-chested Sunbird (p. 757) | Smaller Black-bellied Sunbird (p. 760) | Beautiful Sunbird (p. 758) |
| Superb Sunbird (p. 764) | Pigmy Sunbird (p. 763) | Copper Sunbird (p. 765) | Splendid Sunbird (p. 766) |
| Shining Sunbird (p. 767) | Little Purple-banded Sunbird (p. 768) | Violet-breasted Sunbird (p. 770) | Pemba Violet-breasted Sunbird (p. 771) |
| Mariqua Sunbird (p. 771) | Shelley's Double-collared Sunbird (p. 773) | Orange-tufted Sunbird (p. 774) | Northern Orange-tufted Sunbird (p. 775) |

Plate 83

White-bellied Sunbird (p. 776)
Variable Sunbird (p. 779)
Eastern Double-collared Sunbird (p. 783)
Northern Double-collared Sunbird (p. 786)
Loveridge's Sunbird (p. 790)
Blue-throated Brown Sunbird (p. 801)
Plain-backed Sunbird (p. 817)
Banded Green Sunbird (p. 811)
Hunter's Sunbird (p. 797)
Amethyst Sunbird (p. 791)

Females of
Southern Double-collared Sunbird (p. 781)
Greater Double-collared Sunbird (p. 780)
Regal Sunbird (p. 788)
Olive-bellied Sunbird (p. 787)
Uluguru Violet-backed Sunbird (p. 814)
Violet backed Sunbird (p. 812)
Scarlet-chested Sunbird (p. 796)
Little Green Sunbird (p. 816)
Scarlet-chested Sunbird (p. 794)
Green-throated Sunbird (p. 793)

## SUNBIRDS

**1257** OLIVE-BELLIED SUNBIRD. *CINNYRIS CHLORO-PYGIUS* (Jardine).

*Cinnyris chloropygius orphogaster* Reichw. **Pl. 83.**
*Cinnyris chloropygius orphogaster* Reichenow, O.M. p. 169, 1899: Bukoba, north-western Tanganyika Territory.

**Distinguishing characters:** Adult male, similar in size to the Eastern Double-collared Sunbird, but bill rather straighter; upperparts and under wing-coverts similar to the Northern Double-collared Sunbird, but upper tail-coverts metallic green or bluish green, not deep metallic blue; only faint narrow bluish band across chest. There is no non-breeding dress. The female is similar to that of the Yellow-bellied Variable Sunbird but is darker, more olive green above; tail duller blue black; upper tail-coverts olive without metallic green edges; below, dull yellow with some olivaceous streaking. Wing, male 50 to 55, female 47 to 50 mm. The young bird is similar to the adult female, but more ashy, less green, above.

**General distribution:** The Belgian Congo, the south-western Sudan, Uganda and Tanganyika Territory.

**Range in Eastern Africa:** South-western Sudan to Uganda and western Tanganyika Territory at Biharamulo.

**Habits:** Found in scrub and forest outskirts, more numerous in the west of Uganda and quite common in the Semliki Valley.

**Nest and Eggs:** A well-made nest of fine grass blades ornamented with lichen and thickly lined with vegetable down, usually dependent from a twig some feet from the ground and often with a beard of pendent material. Eggs two, pale greyish or greenish white with dark grey or brown longitudinal streaks or blotches tending to form a cap at the larger end; about $15 \times 10 \cdot 5$ mm.

**Recorded breeding:** Central Belgian Congo, September to December. Uganda, April to November.

**Call:** A thin not unpleasant sibilant 'si-si-si-si-si-si.'

*Cinnyris chloropygius bineschensis* Neum.
*Cinnyris chloropygius bineschensis* Neumann, O.M. p. 185, 1903: Detschabassa, Binescho, Kaffa, south-western Abyssinia.

**Distinguishing characters:** Adult male, differs from the preceding race in having the under wing-coverts and belly darker. Wing measurement unrecorded.

**Range in Eastern Africa:** South-western Abyssinia.

**Habits:** No information.

**Recorded breeding:** No records.

**Distribution of other races of the species:** Fernando Po and West Africa, the nominate race being described from the Niger River.

**1258 TINY SUNBIRD.** *CINNYRIS MINULLUS* Reichenow.
*Cinnyris minullus marginatus* O. Grant.
*Cinnyris marginatus* O. Grant, Bull. B.O.C. 19, p. 106, 1907: Between Kasongo and Ponthierville, eastern Belgian Congo.

**Distinguishing characters:** Adult male, differs from the Olive-bellied Sunbird in being smaller; metallic blue tips to scarlet breast band, and white, not dusky, under wing-coverts. There is no non-breeding dress. The female also has white, not dusky, under wing-coverts. Wing, male 45 to 49, female 44 to 46 mm. The young bird is similar to the adult female but has a horn-coloured bill.

**General distribution:** Eastern Belgian Congo and Uganda.

**Range in Eastern Africa:** The Bwamba area of western Uganda.

**Habits:** Indistinguishable in the field from the Olive-bellied Sunbird and apparently identical in habits.

**Nest and Eggs** (Cameroons race): Nest a pocket of rootlets and moss without a hanging beard of loose material. Eggs two, white spotted with grey and dark brown; about $14 \cdot 5 \times 10$ mm.

**Recorded breeding:** No records.

**Call:** Unrecorded.

**Distribution of other races of the species:** Cameroons, the habitat of the nominate race.

**1259 REGAL SUNBIRD.** *CINNYRIS REGIUS* Reichenow.
*Cinnyris regius regius* Reichw. **Pls. 81, 83.**
*Cinnyris regius* Reichenow, O.M. p. 32, 1893: Ruwenzori Mts., western Uganda.

**Distinguishing characters:** Adult male, head and neck all round, lesser wing-coverts, mantle and rump metallic deep green with a golden wash; upper tail-coverts metallic violet and purple; wings blackish with distinct olivaceous yellow edges to flight feathers; tail glossy blue black, with a violet wash; central feathers slightly elongated; below, a narrow band of metallic violet across chest; a broad stripe of orange red from breast down centre of belly to under tail-coverts; sides of body and flanks bright yellow. There is no non-breeding dress. The immature dress is parti-coloured, being a mixture of adult male and female dress; wings, including lesser wing-coverts, upper and under tail-coverts and tail as in adult male. The female is similar to that of the Northern Double-collared Sunbird, but has a longer bill and is brighter green above. Wing, male 53 to 56, female 49 to 52 mm.; culmen from base, male 20 to 22, female 18 to 20 mm. The young bird is duller and rather darker than the adult female.

**General distribution:** Eastern Belgian Congo and Uganda.

**Range in Eastern Africa:** Western Uganda.

**Habits:** Numerous in the forest zone of Mt. Ruwenzori from 7,000 to 9,000 feet. Usually rather shy birds, but little information available.

**Nest and Eggs:** A typical Sunbird's nest of fibres, moss and tendrils with a thick lining of plant down or feathers. They are often attached to fern fronds. Eggs undescribed.

**Recorded breeding:** Western Uganda, July and August, but apparently an irregular breeder.

**Call:** A sweet sparkling little song often heard in dense mist and dripping surroundings.

*Cinnyris regius kivuensis* Schout.
*Cinnyris regius kivuensis* Schouteden, Rev. Zool. Bot. 30, p. 166, 1937: Kivu, eastern Belgian Congo.

**Distinguishing characters:** Differs from the nominate race in having a shorter bill. Wing, male 51 to 54, female 48 to 49 mm.; culmen from base, male 18 to 19 mm.

**General distribution:** Uganda to the Kivu area, eastern Belgian Congo.

**Range in Eastern Africa:** Behungi and south-west Kigezi areas, south-western Uganda.

**Habits:** Local in scrub and bamboo forest.

**Recorded breeding:** No records.

*Cinnyris regius anderseni* Williams.

*Cinnyris regius anderseni* Williams, Ibis p. 644, 1950: Kungwe-Mahare Mts., western Tanganyika Territory.

**Distinguishing characters:** Size and length of bill as in the Kivu race, but bill broader at base. The female is washed with grey above. Wing, male 52 to 55, female 48 to 50 mm.; culmen from base, male 20 mm.

**Range in Eastern Africa:** Kungwe-Mahare area, western Tanganyika Territory.

**Habits:** Little noted, found in mixed forest and bamboo.

**Recorded breeding:** No records.

**1260 LOVERIDGE'S SUNBIRD.** *CINNYRIS LOVERIDGEI* Hartert. **Pls. 81, 83.**

*Cinnyris loveridgei* Hartert, Bull. B.O.C. 42, p. 49, 1922: Uluguru Mts., eastern Tanganyika Territory.

**Distinguishing characters:** Adult male, upperparts, wings and tail as in the Regal Sunbird, but central tail feathers not elongated; size larger and bill very much longer; differs from the Regal Sunbird below in being yellowish olive from metallic chest band to under tail-coverts, the orange red being confined to the breast and upper belly. There is no non-breeding dress. The female has the top of the head and sides of face grey; rest of upperparts, wings and tail as in the female of the Regal Sunbird but with a distinct glossy blue grey sheen which is also noticeable on head, sides of face and lesser wing-coverts; below, chin grey, rest of underparts slightly olivaceous yellow. Wing, male 56 to 60, female 55 to 57 mm. The young bird is similar to the adult female, but has the throat and upper breast pale grey; rest of underparts pale greyish green.

**Range in Eastern Africa:** Uluguru Mts., eastern Tanganyika Territory.

**Habits:** Locally common in rain forest, but habits little noted. Occurs from 2,500 to 7,500 feet.

**Nest and Eggs:** A rather elongated nest of grass, moss and dead leaves lined with fine rootlets and hair-like mycelium, placed among hanging moss and most difficult to see. Eggs undescribed.

**Recorded breeding:** Tanganyika Territory, September to February.

**Call:** A soft jingling song not unlike that of the Eastern Double-collared Sunbird but less musical. Also a loud squeak (Moreau).

**1261** AMETHYST SUNBIRD. *CHALCOMITRA AMETHYSTINA* (Shaw).
*Chalcomitra amethystina kirkii* (Shell.). **Pls. 81, 83.**
*Cinnyris kirkii* Shelley, Mon. Nect. p. 273, pl. 85, 1876: Shupanga, Shiré River, southern Nyasaland.

**Distinguishing characters:** Adult male, forehead to crown of head metallic green with a slight golden wash; chin and throat metallic rosy purple; rest of plumage velvet black, slightly browner above; wings and tail glossed with bronze; wing shoulder metallic violet and blue. Rarely there are lemon yellow tufts on sides of chest. There is no non-breeding dress. The female is olivaceous ashy green above; flight feathers blackish glossed with bronze; tail bronzy black edged with olive and tipped with white; below, olive yellow with dusky streaks. Wing, male 65 to 71, female 60 to 65 mm. The immature male is similar to the adult female but has a metallic chin and throat as in the adult male. The young bird is very similar to the adult female, but is yellower above and has yellow edges to flight feathers and a dusky throat; the young male has the chin and throat blacker. It may be distinguished from the young bird of the Scarlet-chested Sunbird by the absence of barring on the underside and by the absence of white edges to the primary coverts, except occasionally on the outermost feather.

**General distribution:** Tanganyika Territory to Nyasaland, Portuguese East Africa and eastern Southern Rhodesia.

**Range in Eastern Africa:** Tanganyika Territory (but not coastal areas nor north of a line from Kigoma to about Pangani) to the Zambesi River.

**Habits:** Common and widely distributed in bush and open woodland. Very pugnacious little birds attacking other species freely. Not usually a bird of gardens but seems to prefer wild country, and is often found in large timber.

**Nest and Eggs:** Nest at some height from the ground and dependent from a twig, oval or pear-shaped, of dry grass bound with spider webs and usually ornamented with lichens, but there is some evidence that the birds will adopt and adapt any bunch of hanging material. The nest is warmly lined with vegetable down and has a porch and often a hanging tail. Eggs two, cream, buff, or pale brown with dark brown or slate cloudy edged blotches at the larger end; about 18·5 × 12·5 mm.

**Recorded breeding:** Tanganyika Territory, February to April or later. Nyasaland, September to December, occasionally January to April. Portuguese East Africa, January. Rhodesia, August to November.

**Call:** A sharp ticking alarm note, also a penetrating 'cheep-cheep' and a stuttering twitter while feeding. A fine breeding song of the male has a rather Robin-like quality.

*Chalcomitra amethystina kalckreuthi* (Cab.).

*Cinnyris* (*Chalcomitra*) *kalckreuthi* Cabanis, J.f.O. pp. 205, 227, 1878: Mombasa, eastern Kenya Colony.

**Distinguishing characters:** Similar to the nominate race, but shorter winged. Very rarely the upper tail-coverts are tipped with metallic purple and blue. Wing, male 58 to 65, female 58 to 62 mm.

**Range in Eastern Africa:** Southern Italian Somaliland to coastal areas of Tanganyika Territory as far west as Dolo and North Paré Hills and the Mamboio, Morogoro district, also Kwale and Juani Islands, south of Dar-es-Salaam.

**Habits:** Locally common, with the habits of the preceding race, but occasionally found commonly in towns, even nesting on verandahs of houses.

**Recorded breeding:** Lamu, March to July, also October and November. Southern Kenya Colony, March and April.

*Chalcomitra amethystina doggetti* (Sharpe). **Ph. xvi.**

*Cinnyris doggetti* Sharpe, Ibis p. 116, 1902: Ravine, western Kenya Colony.

**Distinguishing characters:** Adult male, differs from the two preceding races in having the forehead and crown washed with metallic blue. The longer bill and bronzy tail distinguishes the female from that of the Mariqua Sunbird. Wing, male 65 to 70, female 61 to 65 mm.

**Range in Eastern Africa:** South-eastern Sudan to western and central Kenya Colony as far east as the Chyulu Hills and northern Tanganyika Territory.

**Habits:** Locally common with the habits of other races, mostly a bird of gardens or forest edges in Kenya Colony.

**Recorded breeding:** South-eastern Sudan (near breeding condition), May. Kenya Colony highlands, March to June, also November and December. Northern Tanganyika Territory (breeding condition), October, also April.

**Distribution of other races of the species:** Angola and Northern Rhodesia to South Africa, the nominate race being described from the Cape of Good Hope.

### 1262 GREEN-THROATED SUNBIRD. *CHALCOMITRA RUBESCENS* (Vieillot).

*Chalcomitra rubescens rubescens* (Vieill.). **Pls. 81, 83.**
*Cinnyris rubescens* Vieillot, N. Dict. d'Hist. Nat. 31, p. 506, 1819: Malimba, Portuguese Congo.

**Distinguishing characters:** Adult male, forehead black; front half of crown and neck in front metallic green; a band of metallic violet behind green on crown and a narrow band of metallic violet and blue across chest, occasionally with maroon tips to feathers; rest of upperside including wings and tail velvety brownish black washed with bronze; rest of underparts velvety brownish black. There is no non-breeding dress. The immature bird is similar to the adult female, but has a metallic green chin and throat. The female is dusky brown above with an olivaceous wash; slight bronzy green wash on wings and tail; below, chin and throat whitish; rest of underparts yellowish, broadly streaked with dusky. Wing 59 to 72 mm. The young bird is similar to the adult female, but has a darker chin and throat and dusky blobs and spots on the underside.

**General distribution:** Fernando Po to Cameroons, Gabon, the Sudan, Uganda, Tanganyika Territory and northern Angola.

**Range in Eastern Africa:** Southern Sudan to Uganda and north-western Tanganyika Territory.

**Habits:** Much more of a forest species, locally plentiful, but visiting more open country when certain flowers appear, notably those of Erythrina.

**Nest and Eggs:** Nest of usual type with a porch, largely composed of lichens and lined with plant down. Eggs two, creamy white thickly covered with dark brown or lavender, usually in longitudinal markings; about 17 × 12 mm.

**Recorded breeding:** Uganda, October to December, also young in June. Central and south-western Belgian Congo, September to November.

**Call:** Unrecorded.

*Chalcomitra rubescens kakamegæ* (V. Som.).

*Cinnyris angolensis kakamegæ* Van Someren, Bull. B.O.C. 41, p. 113, 1921: Kakemega, Nandi, western Kenya Colony.

**Distinguishing characters:** Differs from the nominate race in being generally darker, and the female more heavily streaked below. Wing 70 to 72 mm.

**Range in Eastern Africa:** Northern Kavirondo to Nandi, western Kenya Colony.

**Habits:** As for the nominate race.

**Recorded breeding:** No records.

**1263 SCARLET-CHESTED SUNBIRD.** *CHALCOMITRA SENEGALENSIS* (Linnæus).

*Chalcomitra senegalensis gutturalis* (Linn.). **Pls. 81, 83.**
*Certhia gutturalis* Linnæus, Syst. Nat. 12th ed. 1, p. 186, 1766: South-eastern Cape Province, South Africa.

**Distinguishing characters:** Adult male, forehead to crown, chin and upper neck in front metallic green; lower neck in front to chest crimson red or rarely orange, with metallic subterminal blue tips to feathers; wing shoulder metallic violet; rest of plumage velvety black; wings and tail browner with a bronzy gloss. There is no non-breeding dress. The female is olivaceous dusky above; flight feathers and tail washed with bronze; below, olivaceous yellow with broad dusky streaks; throat more uniform dusky. In worn dress the general appearance is considerably darker. There is occasionally a coloured feather or so on the chest. Wing, male 67 to 80, female 62 to 73 mm.; culmen from base, male 25 to 32, female 24 to 30 mm. The immature male is similar to the adult female but has the chin to chest as in the adult male and in some cases sparse metallic tipped feathers on forehead and blackish feathers on upper-

## SUNBIRDS

parts. The young bird has a black or blackish chin and neck in front; the rest of the underparts more or less distinctly barred and blobbed with black. The female and young bird can be distinguished from those of the Amethyst Sunbird by the white edging to the primary coverts.

**General distribution:** Kenya Colony to Tanganyika Territory, Angola, Damaraland, eastern Cape Province and Natal.

**Range in Eastern Africa:** South-eastern Kenya Colony from Ukamba and Lamu to Tanganyika Territory and the Zambesi River, also Zanzibar Island, and rarely south Pemba Island.

**Habits:** A common and conspicuous species, locally abundant, and found in all types of country including gardens. This race is, in part at least, migratory, and appears to visit the Tanganyika Territory uplands to breed. The food includes flying ants and other larger insects.

**Nest and Eggs:** Nest very varied as to materials and neatness of construction, generally large, oval and without a tail, usually placed fairly low down and dependent from a branch; often a pair nest year after year in the same tree. There are many records of association with hornets' nests. Eggs usually two, white, cream or bluish white, heavily streaked and spotted with brown or grey; about 19 × 13·5 mm.

**Recorded breeding:** Kenya Colony, April to July, also September and October. Tanganyika Territory, March and April, also July. Zanzibar Island, January and February, and June to October. South-eastern Belgian Congo, August, September and December. Rhodesia, August to March, mostly August to October. Nyasaland, March to May and August to December. Portuguese East Africa, May to December.

**Call:** A variety of 'cheeps' and 'ticks' with a distinctive squeaky trilling song. The most usual call consists of three clear notes uttered slowly in a descending scale.

*Chalcomitra senegalensis cruentata* (Rüpp.).
*Nectarinia cruentata* Rüppell, Syst. Uebers p. 26, pl. 9, 1845: Simen
 Province, northern Abyssinia.

**Distinguishing characters:** Adult male, similar to the South African race, but has the chin or chin and upper throat black; metallic colouring on wing shoulder confined to base. The female

differs from that of the South African race in lacking the white edges to the primary coverts. Wing, male 68 to 80, female 63 to 71 mm.

**Range in Eastern Africa:** The Sennar and Boma areas of the eastern Sudan to Eritrea, Abyssinia and northern Kenya Colony.

**Habits:** As for the South African race and at least partially migratory. It appears to be a resident on the high plateau of Abyssinia, but is stated to visit Bogosland, Eritrea, only from May to July. Its diet includes flying ants, and it has been noticed coming regularly to a swarm.

**Recorded breeding:** Abyssinian plateau, April to October.

*Chalcomitra senegalensis acik* (Hartm.). **Pl. 83.**
*Nectarinia acik* Hartmann, J.f.O. p. 205, 1866: Djur, Bahr-el-Ghazal, south-western Sudan.

**Distinguishing characters:** Adult male, differs from the South African race in having the wing shoulder uniform with rest of wing, usually not metallic. The female is more dusky and more densely marked below and often has no white edging to the primary coverts. Albinistic examples are known. The females without white on the primary coverts can be distinguished from the female of the Amethyst Sunbird by the very much darker coloration both above and below. From the female of the Green-throated Sunbird with which it agrees in size, length and shape of bill and general colour above, including wings and tail, it can be distinguished by the much darker dusky olive yellow underparts with rather indistinct markings as against the lighter underparts with distinct streaks of the female of the Green-throated Sunbird. Wing, male 65 to 71, female 60 to 63 mm.; culmen from base, male 22 to 25, female 22 to 25 mm. The young bird can be distinguished from the young bird of the Green-throated Sunbird by the blacker throat which extends on to the chest, and the darker ochre yellow underparts with distinctly black markings as against the more dusky throat and paler underparts with dusky, not black markings, of the young bird of the Green-throated Sunbird. The female and young bird of the Green-throated Sunbird have no white edging to the primary coverts.

**General distribution:** Northern Cameroons to the Sudan, north-eastern Belgian Congo and Uganda.

**Range in Eastern Africa:** Western and southern Sudan, as far east as Torit and the Dongotona Mts., to northern Uganda.

**Habits:** As for other races, and is also partially migratory, noticed arriving in the Bahr-el-Ghazal in February and leaving at the end of April. Locally very common. Eggs are very variable with green or pale blue ground instead of white, but always longitudinally streaked with grey and brown.

**Recorded breeding:** Darfur, September. Northern Uganda, April to July, also October to January.

*Chalcomitra senegalensis lamperti* (Reichw.). **Ph. xvi.**
*Cinnyris senegalensis lamperti* Reichenow, J.f.O. p. 196, 1897: Moshi, north-eastern Tanganyika Territory.

**Distinguishing characters:** Adult male, similar to the last race but larger and longer billed; indications of metallic colouring at base of wing shoulder. The female is not so dusky and is less olive yellow and less densely marked below. Wing, male 69 to 81, female 64 to 70 mm.; culmen from base, male 28 to 31; female 25 to 29 mm.

**General distribution:** Eastern Belgian Congo, Uganda and the southern Sudan to Kenya Colony and Tanganyika Territory.

**Range in Eastern Africa:** Imatong Mts. and Laboni Forest, southern Sudan to Uganda, Kenya Colony as far east as the Chyulu Hills and northern Tanganyika Territory from the Bukoba and Mwanza districts to Loliondo, Moshi and Mt. Kilimanjaro.

**Habits:** As for other races, very widely distributed in many types of country.

**Recorded breeding:** Kenya Colony and northern Tanganyika Territory, December to March, but may be found breeding in any month.

**Distribution of other races of the species:** West Africa to Damaraland, the nominate race being described from Senegal.

**1264** HUNTER'S SUNBIRD. *CHALCOMITRA HUNTERI* (Shelley). **Pl. 83.**
*Cinnyris hunteri* Shelley, P.Z.S. p. 365, pl. 41, fig. 2, 1889: Useri River, east of Mt. Kilimanjaro, Kenya Colony-Tanganyika Territory boundary.

**Distinguishing characters:** Adult male, above, similar to the Scarlet-chested Sunbird but with the lower rump and upper tail-coverts metallic violet and purple; the metallic violet wing shoulder

extends to the adjoining side of the neck; throat black; below, the scarlet neck and chest has a narrow subterminal yellow bar between the scarlet tips and the black base, and there are only sparse metallic blue tips. There is no non-breeding dress. The immature male is similar to the adult female but has the chin to chest as in the adult male. The female is similar to that of the Amethyst Sunbird, but is longer billed and has white edges to the primary coverts. Wing, male 66 to 76, female 63 to 66 mm. The young bird is similar to that of the Scarlet-chested Sunbird, but is longer billed and has a darker, less distinctly bronzy tail.

**Range in Eastern Africa:** British and Italian Somaliland to southern Abyssinia, northern Kenya Colony, as far west as Lake Rudolf, Kitui and Tsavo, also Mt. Kilimanjaro and Mkomasi, but not the coastal areas of north-eastern Tanganyika Territory.

**Habits:** Locally common and conspicuous among thorn-bush and Sansieveria, but little is recorded of its habits.

**Nest and Eggs:** Nest elliptical with a suspending streamer and a long tail below, of fine grass, rootlets, leaves, etc., compacted with spiders' webs and with a dense lining of feathers. It has a definite porch. Egg one only as a rule, olive grey with copious streaks and longitudinal blotches of brown and grey; about $19 \times 11 \cdot 5$ mm.

**Recorded breeding:** Garissa, Tana River, October.

**Call:** Said to be indistinguishable from that of the Scarlet-chested Sunbird.

**1265** MOUSE-COLOURED SUNBIRD. *CYANOMITRA VEROXII* (Smith).
*Cyanomitra veroxii fischeri* (Reichw.). **Pl. 81.**
*Cinnyris fischeri* Reichenow, J.f.O. p. 142, 1880: Mozambique, northern Portuguese East Africa.

**Distinguishing characters:** Adult male, above, grey with a slight greenish gloss; tail dull blue black with some greenish edges to outer webs; below, pale greyish white; tufts at side of chest mixed red and yellow. There is no non-breeding dress. The female is smaller than the male, otherwise similar. Wing, male 61 to 63, female 55 to 56 mm. The young bird has the mantle washed with olivaceous; underside washed with yellowish.

**General distribution:** Italian Somaliland to Portuguese East Africa, as far south as Coguno, Inhambane district.

**Range in Eastern Africa:** Coastal areas of Italian Somaliland to Kenya Colony, Tanganyika Territory and Portuguese East Africa from Kismayu to the Zambesi River.

**Habits:** Scarce, local and shy species of which very little is recorded. It inhabits low hot coastal bush, and forest fringes at low levels. The male is noticeable for its cheerful song from a prominent perch in the morning and evening.

**Nest and Eggs:** The nest is a rough untidy ragged structure with a pendent tail attached to a creeper, piece of bark or even wire in densely shaded places. It has been noted in a deserted hut. The eggs are two, white but so heavily mottled as to be almost entirely purplish brown; about 16·5 × 12 mm.

**Recorded breeding:** Malindi, Kenya Colony, May.

**Call:** A loud and beautiful, though brief, warble (Moreau). Pakenham describes the song of the Zanzibar race as 'chyo-chyo-tsi' or 'tsitsitsi jujujuju' with variations.

*Cyanomitra veroxii zanzibarica* Grote.
*Chalcomitra veroxii zanzibarica* Grote, Ibis p. 350, 1932: Zanzibar Island.

**Distinguishing characters:** Differs from the preceding race in being smaller and paler, with less gloss on upperparts. Wing, male 59 to 61, female 55 mm.

**Range in Eastern Africa:** Zanzibar Island.

**Habits:** Not uncommon locally, but shy and difficult to observe with rapid and often prolonged flight. Habits similar to those of the nominate race. The male has a display which brings the pectoral tufts into prominence.

**Recorded breeding:** Probably December to April.

**Distribution of other races of the species:** South Africa, the habitat of the nominate race.

**1266** GREEN-HEADED SUNBIRD. *CYANOMITRA VERTI-CALIS* (Latham).
*Cyanomitra verticalis viridisplendens* (Reichw.). **Pl. 81.**
*Cinnyris viridisplendens* Reichenow, J.f.O. p. 54, 1892: Bukoba, north-western Tanganyika Territory.

**Distinguishing characters:** Adult male, head, neck and sides of face metallic green; rest of upperparts moss green; below, chin to

chest metallic greenish blue; rest of underparts dusky grey; tufts on sides of chest cream-coloured. There is no non-breeding dress. The female has the chin to chest dusky grey and lacks the tufts on side of chest. Wing 60 to 71 mm. The young bird has the forehead and crown black; chin to chest black; rest of underparts olivaceous; feathers on breast yellowish.

**General distribution:** Eastern Belgian Congo and the southern Sudan, Uganda, Kenya Colony, Tanganyika Territory, north-eastern Northern Rhodesia and northern Nyasaland.

**Range in Eastern Africa:** Southern Sudan to central Kenya Colony and Tanganyika Territory.

**Habits:** Locally common and fairly widespread species of forest or forest edge and thickly wooded river banks. Inhabits undergrowth as well as tree tops and is usually seen searching for insects like a Warbler; found up to 7,000 feet or more.

**Nest and Eggs:** Nest, suspended from a bush, of banana bark, fibre, etc., lined with grass and with long streamers hanging from the lip of the entrance hole, and often from the sides as well. Nest often well concealed. Eggs two, pinkish white or buff, spotted, blotched and marbled with various shades of brown up to deep chocolate; about 19·5 × 14 mm.

**Recorded breeding:** Uganda, April to September. Tanganyika Territory (breeding condition), December.

**Call:** A distinctive 'tweezee' or 'cheerick,' also a breeding season song by the male.

**Distribution of other races of the species:** West Africa to Angola and the Belgian Congo, the nominate race being described from Senegal.

**1267 BLUE-HEADED SUNBIRD.** *CYANOMITRA ALINÆ* Jackson.

*Cyanomitra alinæ alinæ* Jacks.
*Cyanomitra alinæ* Jackson, Bull. B.O.C. 14, p. 94, 1904: Ruwenzori Mts., western Uganda.

**Distinguishing characters:** Adult male, differs from the Green-headed Sunbird in having the mantle saffron; the head and chin to chest metallic violet blue and the breast to belly sooty black; lower flanks and under tail-coverts yellowish. There is no non-breeding dress. The female is similar to the male, but has usually a less black underside, and can be distinguished from the male of the Green-headed Sunbird by the colour of the head and neck. Wing 59 to 68 mm. The young bird can be distinguished from the young bird of the Green-headed Sunbird by the saffron colour of the mantle and the more uniform dusky coloration from chin to belly.

**Range in Eastern Africa:** Ruwenzori Mts. to Kigezi, western Uganda.

**Habits:** Plentiful in and below the forest line on Mt. Ruwenzori feeding at flowers of Erythrina. R. B. Woosnam noted that the females were much less numerous than the males at these trees, about one to eight.

**Nest and Eggs:** Undescribed.

**Recorded breeding:** No records.

**Call:** Noisy little birds with a rather distinctive call of 'tcii-tcii-tcii-yehu' (Van Someren).

**Distribution of other races of the species:** The Belgian Congo.

**1268** BLUE-THROATED BROWN SUNBIRD. *CYANO-MITRA CYANOLÆMA* (Jardine).

*Cyanomitra cyanolæma cyanolæma* (Jardine). **Pls. 81, 83.**

*Nectarinia cyanolæma* Jardine, Con. Orn. p. 154, 1851: Fernando Po.

**Distinguishing characters:** Adult male, forehead and forepart of crown metallic steel blue; rest of upperparts sooty brown; below, chin and neck metallic steel blue; rest of underparts sooty grey, darker on chest; tufts at sides of chest pale yellow. There is no non-breeding dress. The female is yellowish green above; light stripe above and below eye; below, chin and neck white or pale brown; rest of underparts more or less whitish, grey and yellow green; lower flanks and under tail-coverts yellower green. Wing 62 to 75 mm. The young male is sooty brown above including whole head and chin to chest; wings and tail with yellow green edging; breast and belly streaked sooty brown and whitish. The young female is more olive green above than the adult; below, paler olive green; throat slightly dusky.

**General distribution:** Gold Coast to Uganda and northern Angola; also Fernando Po.

**Range in Eastern Africa:** Western and southern Uganda.

**Habits:** A forest species inhabiting clearings and more open places, but also found in the tops of high forest trees.

**Nest and Eggs:** The nest is described as a three foot long bundle of fibrous material with the egg chamber as an enlargement below the middle lined with soft rootlets, and may be placed at some height in a tree. Eggs two, bluish or brownish white with light brown mottling and with conspicuous spots of dark purplish brown; about 18 × 12·5 mm.

**Recorded breeding:** West Africa, June and July. Uganda, April, also August and September.

**Call:** Described as a continuous chirping which betrays the bird's presence in the forest canopy.

**Distribution of other races of the species:** French Guinea and Sierra Leone.

**1269** OLIVE SUNBIRD. *CYANOMITRA OLIVACEA* (Smith).
*Cyanomitra olivacea olivacina* (Peters).
*Nectarinia olivacina* Peters, J.f.O. p. 50, 1881: Inhambane, Portuguese East Africa.

**Distinguishing characters:** Adult male, above, olive green; wings and tail dusky with olive green edges; below, olivaceous green; tufts at sides of chest yellow; bill wholly black. There is no non-breeding dress. The female also has the yellow tufts at the side of the chest but has a shorter tail than the male at all ages. Wing, male 57 to 61, female 51 to 55 mm.; tail, male 48 to 50, female 39 to 42 mm. The young bird has the chin to chest orange yellow and both sexes have the yellow tufts at the sides of the chest.

**General distribution:** Portuguese East Africa from about the Rovuma River to Inhambane.

**Range in Eastern Africa:** Coastal area of Portuguese East Africa around the mouth of the Rovuma River.

**Habits:** As for the following race, widely spread but not common, usually seen low down in trees or bushes.

**Nest and Eggs and Call:** See under the following race.

**Recorded breeding:** No records.

Plate 84

Parrot-billed Sparrow (p. 880)  
Chestnut Sparrow (p. 882)  
Grey-headed Sparrow (p. 877)  
Swahili Sparrow (p. 879)  
Sudan Golden Sparrow (p. 881)  
Female  Male  
Arabian Golden Sparrow (p. 882)  
Female  Male  
Somali Sparrow (p. 875)  
Female  Male  
Kenya Rufous Sparrow (p. 873)  
Female  Male  
Rufous Sparrow (p. 872)  
Female  Male  
House Sparrow (p. 869)  
Female  Male  
Spanish Sparrow (p. 874)  
Female  Male  
Desert Sparrow (p. 876)  
Female  Male

Plate 85

Pale Rock-Sparrow (p. 887)  Bush Petronia (p. 886)  Yellow-spotted Petronia (p. 884)
　　　　　　　　　　　　　　　Female　　　　　Male
Heuglin's Masked Weaver  Tanganyika Masked Weaver (p. 895)  Baglafecht Weaver (p. 901)
　(p. 896)　　　　　　　　　　　　　　　　　　　　　　　　　Female　　　　Male
Black-headed Weaver (p. 890)　　Vitelline Masked Weaver (p. 899)  Chestnut Weaver (p. 907)
　　　　　　　　　　　　　　　Male—N.-b.d.　　Male—B.d.
Layard's Black-headed Weaver  Little Weaver (p. 903)  Masked Weaver (p. 898)
　(p. 891)　　　　　　　　　　　　　　　　　　　Male—N.-b.d.　　Male—B.d.
Speke's Weaver　　Northern Masked Weaver  Yellow-backed Weaver  Golden-backed Weaver
　(p. 892)　　　　　(p. 894)　　　　　　　(p. 904)　　　　　　　(p. 906)

# SUNBIRDS

**Cyanomitra olivacea neglecta** Neum. **Pl. 81.**

*Cyanomitra obscura neglecta* Neumann, J.f.O. p. 297, 1900: Kibwesi, Ukamba, south-eastern Kenya Colony.

**Distinguishing characters:** Differs from the preceding race in having a more curved bill, and in being paler below. Both sexes have yellow tufts on the sides of the chest. Wing, male 61 to 65, female 55 to 59 mm.; tail, male 47 to 55, female 41 to 47 mm.

**Range in Eastern Africa:** Central and north-eastern Kenya Colony east of the Mau, to north-eastern Tanganyika Territory at Monduli and Mts. Meru and Kilimanjaro.

**Habits:** Usually a forest species preferring damp places, but often found in the open also, particularly when certain trees are in flower, or on adjoining moorland. At other times searches trees or even house verandahs like a Tit for insects. The male has a little fluttering display flight.

**Nest and Eggs:** In any situation, but usually in damp evergreen forest. The nest looks like a collection of rubbish, and is rough and untidy with a tail several inches long, and is lined with seed down. Eggs two, white, buff, or coffee-colour, speckled or marbled all over with sepia and brownish purple; about $17 \cdot 5 \times 12$ mm.

**Recorded breeding:** South-eastern Kenya Colony, probably December to March. Tanganyika Territory, September to December, also probably July.

**Call:** A harsh churring and ticking alarm note, also a high-pitched double squeak. The male has a loud and clear song, something like that of a Willow Warbler, which varies somewhat locally but is always distinctive.

*Cyanomitra olivacea changamwensis* (Mearns).
*Cinnyris changamwensis* Mearns, Smiths. Misc. Coll. 56, No. 14, p. 4, 1910: Changamwe, near Mombasa, eastern Kenya Colony.

**Distinguishing characters:** Differs from the last race in being more yellowish olive above. Both sexes have yellow tufts on the sides of the chest. Wing, male 57 to 64, female 55 to 59 mm.; tail, male 44 to 55, female 40 to 45 mm.

**Range in Eastern Africa:** Coastal areas of Italian Somaliland, Kenya Colony and north-eastern Tanganyika Territory from the Juba River to Pangani and as far west as the Taita Hills.

**Habits:** Common in wooded country of the coastal belt.

**Recorded breeding:** No records.

*Cyanomitra olivacea alfredi* Vinc.

*Cyanomitra olivacea alfredi* Vincent, Ibis p. 90, 1934: Namuli Mt., northern Portuguese East Africa.

**Distinguishing characters:** Similar to the Inhambane race but larger. Wing, male 60 to 67, female 55 to 59 mm.; tail, male 49 to 58, female 42 to 45 mm. Both sexes have yellow tufts on sides of chest.

**General distribution:** Tanganyika Territory to eastern Northern Rhodesia, Nyasaland, and northern Portuguese East Africa.

**Range in Eastern Africa:** Tanganyika Territory from Mpapwa and the Uluguru Mts. to Portuguese East Africa and Nyasaland.

**Habits:** A species of mountain rain forest, locally common, also found along the wooded banks of rivers. The nest is as for other races and usually built under substantial cover of some sort, with the entrance hole rather low in the side.

**Recorded breeding:** Nyasaland, September to January, also (breeding condition) May.

*Cyanomitra olivacea puguensis* V. Som.

*Cyanomitra olivacea puguensis* Van Someren, Bull. B.O.C. 59, p. 86, 1939: Kilindoni Forest, Mafia Island, eastern Tanganyika Territory.

**Distinguishing characters:** Similar to the Mombasa race but larger, darker above and bill stouter; differs from the Inhambane race in being darker above and less greenish below. Both sexes have yellow tufts on sides of chest. Wing, male 57 to 64, female 55 to 57 mm.; tail, male 45 to 54, female 39 to 42 mm.

**Range in Eastern Africa:** Eastern Tanganyika Territory from the Pugu Hills to the Rufigi River and the Mahenge and Lindi districts, also Mafia Island.

**Habits:** As for other races.

**Recorded breeding:** No records.

# SUNBIRDS

*Cyanomitra olivacea ragazzii* (Salvad.).

*Eleocerthia ragazzii* Salvadori, Ann. Mus. Civ. Genova 26, p. 247, 1888: Ferkerie-ghem Forests, Shoa, central Abyssinia.

**Distinguishing characters:** Differs from the preceding races in having the base of the lower mandible white, and in the female having no yellow tufts on sides of chest. The female is smaller than the male. Wing, male 59 to 65, female 50 to 58 mm.; tail, male 46 to 56, female 38 to 46 mm.

**Range in Eastern Africa:** Boma Plateau, eastern Sudan to Abyssinia.

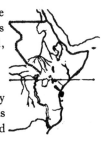

**Habits:** As for other races, and found both in forest and in relatively open country.

**Recorded breeding:** No records.

*Cyanomitra olivacea granti* Vin.

*Cyanomitra olivacea granti* Vincent, Ibis p. 91, 1934: Pemba Island.

**Distinguishing characters:** Similar to the Mafia Island race but bill straighter and rather longer. The female has no yellow tufts on sides of chest. Wing, male 55 to 62, female 52 to 56 mm.; tail, male 42 to 53, female 39 to 43 mm.

**Range in Eastern Africa:** Pemba and Zanzibar Islands.

**Habits:** As for other races, locally common. The nest is of dry grass and fibre with a long streamer at the bottom. The male has a display which often includes fluttering and falling in the air round the top of a bush.

**Recorded breeding:** Pemba Island, September, November to January and March.

*Cyanomitra olivacea lowei* Vinc.

*Cyanomitra olivacea lowei* Vincent, Ibis p. 91, 1934: Kafulafuta River, north-eastern Northern Rhodesia.

**Distinguishing characters:** Nearest to the Namuli race but differs in being paler, less green, below, and in the female having no tufts on sides of chest; the base of the lower mandible is often pale, not wholly black. Wing, male 62 to 67, female 57 to 61 mm.; tail, male 50 to 59, female 41 to 46 mm.

**General distribution:** Tanganyika Territory to south-eastern Belgian Congo and Northern Rhodesia.

**Range in Eastern Africa:** Western Tanganyika Territory from Kabogo and the Kungwe-Mahare to the upper waters of the Nyamanzi River and the Ufipa Plateau.

**Habits:** Mostly a forest species of higher ground.

**Recorded breeding:** No records.

*Cyanomitra olivacea vincenti* Grant & Praed.
*Cyanomitra olivacea vincenti* C. Grant & Mackworth-Praed, Bull. B.O.C. 64, p. 18, 1943: Kapenguria, west Suk, north-western Kenya Colony.

**Distinguishing characters:** Similar to the Shoa race but larger. The female has no yellow tufts on sides of chest. Wing, male 65 to 73, female 57 to 65 mm.; tail, male 46 to 57, female 40 to 50 mm.

**General distribution:** The Sudan and north-eastern Belgian Congo to Uganda, Kenya Colony and Tanganyika Territory.

**Range in Eastern Africa:** Southern Sudan, Uganda, western Kenya Colony, north-western Tanganyika Territory and Ukerewe Island.

**Habits:** Locally common on Mts. Elgon and Ruwenzori with fast darting flight among undergrowth accompanied by a zipping cry. Also common locally in forested areas. The nest is occasionally found under the eaves of houses.

**Recorded breeding:** Uganda, February to April, also September and October.

**Distribution of other races of the species:** West and South Africa, the nominate race being described from Durban, Natal.

**1270** SOCOTRA SUNBIRD.   *CYANOMITRA BALFOURI* (Sclater & Hartlaub).
*Cinnyris balfouri* P. L. Sclater & Hartlaub, P.Z.S. p. 169, pl. 15, fig. 2, 1881: Socotra Island.

**Distinguishing characters:** Adult male, streaked with dusky and whitish above, streaks broader on mantle; malar streak white; below, chin and upper throat dusky grey; lower throat to breast mottled dusky grey and white; belly white; tufts on sides of chest yellow. There is no non-breeding dress. The female lacks the yellow tufts at side of chest. Wing, male 64 to 65, female 58 to 60 mm. Juvenile plumage unrecorded.

# SUNBIRDS

**Range in Eastern Africa:** Socotra Island.
**Habits:** No information.
**Nest and Eggs:** Undescribed.
**Recorded breeding:** No records.
**Call:** Unrecorded.

### 1271 COLLARED SUNBIRD. *ANTHREPTES COLLARIS* (Vieillot).

*Anthreptes collaris hypodilus* (Jard.).
*Nectarinia hypodila* Jardine, Contr. Orn. p. 153, 1851: Clarence, Fernando Po.

**Distinguishing characters:** Adult male, head all round, upperside, lesser wing-coverts, chin and throat metallic green with a golden wash; wings black with green edges; tail blue black with metallic green edges; below, a band of metallic violet across chest; rest of underparts yellow, slightly dusky on chest and flanks. There is no non-breeding dress. The female has the chin and neck in front dusky yellow. Wing, male 50 to 55, female 47 to 52 mm. The young bird is similar to the adult female, but lacks the golden wash above and the chin and neck in front are more dusky.

**General distribution:** Fernando Po, Cameroons and northern Angola to northern Belgian Congo and the Sudan.

**Range in Eastern Africa:** Southern Sudan west of the Nile.

**Habits:** As for other races.

**Nest, Eggs, Food and Call:** As for the following race, but the nest is said usually to have a curved porch over the entrance hole.

**Recorded breeding:** No records.

*Anthreptes collaris zambesiana* (Shell.). **Pl. 81.**
*Anthodiæta zambesiana* Shelley, Mon. Nect. 2, p. 243, pl. 3, 1880: Shupanga, Shiré River, southern Nyasaland.

**Distinguishing characters:** Differs from the preceding race in being brighter yellow below, less dusky, in both sexes. Wing, male 50 to 58, female 48 to 55 mm.

**General distribution:** Southern Belgian Congo and Northern Rhodesia to Tanganyika Territory, Nyasaland, Portuguese East Africa as far south as Inhambane and Bechuanaland.

**Range in Eastern Africa:** South-western Tanganyika Territory, Portuguese East Africa and Nyasaland.

**Habits:** Common and widespread in well watered wooded country searching foliage restlessly in pairs and small parties, with an incessant twittering chirp and a good deal of wing flickering. This species appears to be little attracted to flowers and its habits are much more those of a small Warbler.

**Nest and Eggs:** The nest is round or elliptical, lightly but strongly made of grass fibre and stems woven together with spiders' webs, often at some height from the ground. Usually there is a lining of plant down or feathers, and the nest is suspended from a bush or shrub some feet from the ground. Eggs two or three, greenish white or buffish white with blotches, spots and speckles of various shades of brown and lilac grey; about 15 × 11 mm.

**Recorded breeding:** Southern Belgian Congo, October. Nyasaland, February, April, May and September to December. Portuguese East Africa, March to May, also probably later in year.

**Food:** Almost entirely insects and their larvæ.

**Call:** As above, a sharp chirping 'preep' or 'tsiwu' repeatedly uttered, also a song in the breeding season.

*Anthreptes collaris elachior* Mearns.
*Anthreptes collaris elachior* Mearns, Smiths, Misc. Coll. 56, No. 14, p. 5, 1910: Changamwe, near Mombasa, eastern Kenya Colony.

**Distinguishing characters:** Adult male, differs from the preceding race in being slightly paler yellow below, and the female has a yellower, less dusky throat. Wing, male 48 to 56, female 46 to 53 mm.

**Range in Eastern Africa:** Eastern Kenya Colony as far west as Thika, Tsavo and Taveta to eastern Tanganyika Territory as far west as Mt. Kilimanjaro, Kilosa and Mahenge, also Manda, Zanzibar and Mafia Islands.

**Habits:** As for the last race. Moreau notes that its nest has more than once been found in the immediate vicinity of bees' nests. The nest is compactly built and usually well hidden, occasionally it has a small porch and a streamer of grass dependent from the lower lip.

**Recorded breeding:** Kenya Colony coast, April to June, also August and November. Zanzibar Island, April, May and July to November. Mafia Island, August, also (breeding condition) June.

*Anthreptes collaris garguensis* Mearns. **Ph. xvi.**
*Anthreptes collaris garguensis* Mearns, Proc. U.S. Nat. Mus. 48, p. 389, 1915: Mt. Gargues, north-central Kenya Colony.

**Distinguishing characters:** Differs from the last race in the slightly brighter yellow of the male, and in the female having the chin to upper chest dusky yellow. Wing, male 51 to 58, female 50 to 55 mm.

**Range in Eastern Africa:** Southern Sudan east of the Nile to Uganda, western and central Kenya Colony and northern Tanganyika Territory as far east as Monduli, Oldeani and Esimingor, also Kome and Ukerewe Islands, Lake Victoria.

**Habits:** As for other races, common in gardens and open forest and found up to considerable elevations on the Ruwenzori Mts.

**Recorded breeding:** Uganda, May to July, also October to December. Kenya Colony highlands, April to June, also December. Northern Tanganyika Territory (breeding condition), December.

*Anthreptes collaris jubaensis* Van Som.
*Anthreptes collaris jubaensis* van Someren, Nov. Zool. 37, p. 358, 1932: Hollesheid, Juba River, southern Italian Somaliland.

**Distinguishing characters:** Adult male, differs from the other races in being brighter yellow below and with less dusky flanks. The female is also brighter below and less dusky on chin to chest. Wing, male 51 to 54, female 48 to 52 mm.

**Range in Eastern Africa:** Southern Abyssinia and southern Italian Somaliland.

**Habits:** As for other races.

**Recorded breeding:** Italian Somaliland, April.

**Distribution of other races of the species:** West and South Africa, the nominate race being described from the Gamtoos River, Cape Province.

**1272** GREEN SUNBIRD. *ANTHREPTES RECTIROSTRIS* (Shaw).

*Anthreptes rectirostris tephrolæma* (Jard. & Fras.). Grey-chinned Sunbird. **Pl. 81.**

*Nectarinia tephrolæmus* Jardine & Fraser, Con. Orn. p. 154, 1851: Fernando Po.

**Distinguishing characters:** Adult male, above, including lesser wing-coverts, metallic golden green; lower rump and upper tail-coverts green; wings and tail dusky edged with green; below, chin and upper throat grey; side of face and neck metallic golden green; an ochre yellow band across chest; breast pale grey; belly, under tail-coverts and flanks pale yellow; bright yellow tufts at sides of chest. There is no non-breeding dress. The female is olive green above, rarely with a slight metallic gloss, and paler olive green below; throat duller and lower belly and under tail-coverts yellower; no tufts at sides of chest. Wing 54 to 64 mm. The young bird is similar to the adult female, but rather greener above and yellower below, and the young male has some metallic golden green feathers on the upperparts and side of face, and metallic golden green lesser wing-coverts.

**General distribution:** Fernando Po to Cameroons, Gabon and northern Angola to Uganda, the Sudan and Kenya Colony.

**Range in Eastern Africa:** Southern Sudan, Uganda and western Kenya Colony.

**Habits:** A comparatively rare and little known species, mostly of forest margins or more open spaces hunting among foliage for insects.

**Nest and Eggs:** Nest, hanging from a bough or creeper, made of fibres, lichens and leaves bound with cobwebs and lined with plant down. Eggs (of the nominate race) two, ashy grey with a slight violet wash and with irregular dark spots and markings. Measurements unrecorded.

**Recorded breeding:** No records.

**Food:** Insects and small fruit.

**Call:** A faint sibilant little cry freely uttered.

**Distribution of other races of the species:** West Africa, the nominate race being described from Gambia.

## SUNBIRDS

**1273** BANDED GREEN SUNBIRD. *ANTHREPTES RUBRI-TORQUES* Reichenow. **Pls. 81, 83.**

*Anthreptes rubritorques* Reichenow, O.M. p. 181, 1905: Mlalo, Usambara, north-eastern Tanganyika Territory.

**Distinguishing characters:** Adult male, above, including lesser wing-coverts, rump and upper tail-coverts metallic bluish green with a golden wash; wings and tail blackish with very narrow greenish edges; below, chin to breast grey with a narrow red band across chest; belly, under tail-coverts and flanks mixed grey and yellow; tufts on sides of chest orange and yellow. There is no non-breeding dress. The female is duller above than the male; below, as in male with no red band; chin to chest uniform with flanks and upper belly; no tufts at sides of chest. Wing, male 59 to 63, female 52 mm. The young bird is blackish olive green above, below, yellowish olive.

**Range in Eastern Africa:** Usambara to Nguru Mts., north-eastern Tanganyika Territory.

**Habits:** An inhabitant of the canopy of mountain forest, easily overlooked. It also occurs in lowland forest but also keeps always to the tree tops. Not uncommon at Amani where it occurs round the Institute.

**Nest and Eggs:** A neat pear-shaped nest of fine stems and lichen with side top entrance and lined with plant down, made by the female only, and placed at some height in creepers or thick cover. Eggs undescribed.

**Recorded breeding:** Tanganyika Territory, October and November.

**Food:** Insects and small berries.

**Call:** The male gives a string of insistent chirps from a tree top, at times quickening into a song.

**1274** VIOLET-BACKED SUNBIRD. *ANTHREPTES LONGUEMAREI* (Lesson).

*Anthreptes longuemarei haussarum* Neum.

*Anthreptes longmari haussarum* Neumann, O.M. p. 6, 1906: Agowe, Tongwe, Togoland.

**Distinguishing characters:** Adult male, above, including tail, metallic violet, inclining to bronzy in worn dress; occasionally some metallic blue feathers on rump; sides of face and flight feathers black; wing shoulder metallic violet; metallic green near edge of wing; below, chin black; throat metallic violet; rest of underparts white; tufts on sides of chest yellow. There is no non-breeding dress. The immature dress is similar to that of the adult female. The female is sooty grey above often with a violet wash; superciliary stripe white; wing shoulder plain or violet; tail as in male; below, chin to breast white; belly and flanks yellow; there are no yellow tufts on sides of chest. Wing 63 to 81 mm. The young bird is more olivaceous above, with no violet on wing shoulder, and below, from chin to belly yellow.

**General distribution:** Liberia to Nigeria, Cameroons, the Sudan and Uganda.

**Range in Eastern Africa:** South-western Sudan and north-western Uganda.

**Habits:** Usually seen in small parties in bush or fringing forest hunting for insects in a restless manner like a flock of Tits.

**Nest, Eggs, etc.:** See under the following race.

**Recorded breeding:** No records.

*Anthreptes longuemarei orientalis* Hartl. **Pls. 81, 83.**
*Anthreptes orientalis* Hartlaub, J.f.O. p. 213, 1880: Lado, southern Sudan.

**Distinguishing characters:** Adult male, above, more metallic blue and violet than the preceding race; a metallic green patch on rump and more metallic green on the wing shoulder. There is no non-breeding dress. The female has no metallic violet wash on upperparts; wing shoulder usually plain but sometimes metallic violet; below, the belly is whiter only faintly washed with yellow; tail as in male. Wing 59 to 72 mm.

**Range in Eastern Africa:** Central Abyssinia to southern Sudan, British and Italian Somalilands, northern Uganda, Kenya Colony and Tanganyika Territory as far south as Kidete, Kilosa district.

**Habits:** Locally common in dry thorn bush country, with the same general habits as other species of the genus. There appears to be evidence of a consistent nesting association with wasps. Often feeds on after dark.

**Nest and Eggs:** Nest of fine grass and down felted together into a compact wall, neatly made and with a small porch, usually low on a thorn tree. It often has something of a tail below it. Eggs two, greenish grey closely speckled with grey or sometimes with dark brown and black; about 17·5 × 12 mm.

**Recorded breeding:** Southern Sudan, November. Southern Abyssinia, February. Northern Kenya Colony, May and October to January. Tana River, Kenya Colony, October. Tanganyika Territory, February and November.

**Food:** Insects.

**Call:** A slow twittering song and a sharp alarm note of 'skee.' There is also a high-pitched chipping note.

*Anthreptes longuemarei angolensis* Neum.
*Anthreptes longmari angolensis* Neumann, O.M. p. 6, 1906: Duque de Braganza, northern Angola.

**Distinguishing characters:** Adult male, differs from the Sudan race in being washed with buff below. The female has the belly, flanks and under tail-coverts yellow. Wing 64 to 84 mm.

**General distribution:** Angola to southern Belgian Congo, Tanganyika Territory, northern and north-eastern Northern Rhodesia, Portuguese East Africa north of the Zambesi River and west of the Shiré River and western Nyasaland.

**Range in Eastern Africa:** South-western Tanganyika Territory.

**Habits:** As for other races. Eggs are somewhat larger; about 19·5 × 13 mm.

**Recorded breeding:** Southern Belgian Congo, September to November. Nyasaland, September to January.

*Anthreptes longuemarei nyassæ* Neum.
*Anthreptes longmari nyassæ* Neumann, O.M. p. 7, 1906: Lahengula, Zomba, southern Nyasaland.

**Distinguishing characters:** Adult male, differs from the Sudan and Angolan races in being metallic violet above, not blue and violet; though often with some metallic blue green feathers on rump. The female is similar to that of the last race, often with a violet wash above and some violet on wing shoulders, but less yellow on belly. Wing 68 to 85 mm.

**General distribution:** Tanganyika Territory to Portuguese East Africa and eastern Southern Rhodesia.

**Range in Eastern Africa:** Eastern and southern Tanganyika Territory as far north as the Dar-es-Salaam area to Portuguese East Africa.

**Habits:** A common species of the open woodland and bush country, feeding among small branches like an Eremomela, in pairs or small parties, and searching for food like Tits. Very often members of mixed bird parties. The flight is direct and buoyant. The nests are usually high up in trees at the ends of branches and are described as large and covered outside with moss or lichens.

**Recorded breeding:** No records.

*Anthreptes longuemarei neumanni* Zedl.
*Anthreptes longmari neumanni* Zedlitz, J.f.O. p. 73, 1916: Afgoi, southern Italian Somaliland.

**Distinguishing characters:** Differs from the nominate race in being smaller. Wing 61 to 64 mm.

**Range in Eastern Africa:** Juba River Valley, southern Italian Somaliland.

**Habits:** As for other races.

**Recorded breeding:** No records.

**Distribution of other races of the species:** West Africa, the nominate race being described from Senegal.

### 1275 ULUGURU VIOLET-BACKED SUNBIRD. *ANTHREPTES NEGLECTUS* Neumann. **Pls. 81, 83.**

*Anthreptes longuemarei neglectus* Neumann, O.M. p. 13, 1922: Uluguru Mts., eastern Tanganyika Territory.

**Distinguishing characters:** Adult male, differs from the preceding species in being usually more metallic blue and violet; the flight feathers edged with yellow green; below, buffish white. There is no non-breeding dress. The female is similar above to the male, including wings and tail; below, greyish white, including chin and throat; lower belly and under tail-coverts yellow. Wing 64 to 75 mm. The young bird is dark grey above with a metallic wash; wing shoulder metallic as in adult; wings and tail as in adult; below, as adult female but less grey.

**Range in Eastern Africa:** North-eastern Tanganyika Territory to Portuguese East Africa.

**Habits:** Not common and usually if not exclusively found in evergreen forest. Habits as for other species but little recorded.

**Nest and Eggs:** The only nest known was seen some twenty-five feet up, hanging from the bough of a forest tree. It appeared to be typical of the genus, but very compact and neatly made.

**Recorded breeding:** No records.

**Food:** Insects.

**Call:** A loud persistent squeak.

## 1276 ANCHIETA'S SUNBIRD. *ANTHREPTES ANCHIETÆ* (Bocage). Pl. 81.

*Nectarinia anchietæ* Bocage, Jorn. Lisboa 6, p. 208, 1878: Caconda, Benguella, Angola.

**Distinguishing characters:** Adult male, forehead, crown and chin to chest metallic blue black; rest of upperparts sooty brown; sides of breast and upper belly yellow; centre of breast, centre of upper belly and under tail-coverts red; lower belly grey or grey and yellowish. There is no non-breeding dress. The female has the chin and throat duller metallic blue black; less red on the breast to upper belly, and a few metallic tips on forehead feathers. Wing 60 to 67 mm. The young bird has the chin and throat dull dusky; rest of underparts olivaceous with no red or yellow colour.

**General distribution:** Angola to Northern Rhodesia, Tanganyika Territory, Portuguese East Africa and Nyasaland.

**Range in Eastern Africa:** Western Tanganyika Territory.

**Habits:** Common in parts of Rhodesia and northern Nyasaland, and apparently subject to wide seasonal movements. Noted at flowers of Tecoma bushes and Proteas.

**Nest and Eggs:** Typical Sunbird's nest of grass felted with plant down suspended from a bough some feet from the ground. Eggs two, bluish white with black spots and scrawls and grey undermarkings; about 18 × 11·5 mm.

**Recorded breeding:** Northern Nyasaland, September and October, but probably double-brooded.

**Food:** Apparently more often seen at flowers than most *Anthreptes*, but nothing is recorded of its food.

**Call:** A weak quavering song and a penetrating monosyllabic call note.

**1277 LITTLE GREEN SUNBIRD.** *ANTHREPTES SEIMUNDI* (O. Grant).

*Anthreptes seimundi minor* Bates. **Pl. 83.**

*Anthreptes seimundi minor* Bates, Bull. B.O.C. 46, p. 107, 1926: Sannaga, Cameroons.

**Distinguishing characters:** Above green; below, paler yellow green. There is no non-breeding dress. The sexes are alike. Wing 48 to 57 mm. The young bird is duller in colour; more olivaceous above and below.

**General distribution:** Cameroons to Gabon, and northern Angola to Uganda.

**Range in Eastern Africa:** Uganda.

**Habits:** A species of forest clearings and the tangled growth of old cultivations, of which little is recorded within our area.

**Nest and Eggs:** Nests of the usual Sunbird type, abundantly lined with plant down and neatly finished off. Eggs two, cream densely freckled all over with brown; about $14 \cdot 5 \times 11$ mm.

**Recorded breeding:** West Africa, December, April and May.

**Food:** Insects.

**Call:** A squeaky little call of 'tuip-tuip-tuip' (Van Someren).

**Distribution of other races of the species:** West Africa, the nominate race being described from Fernando Po.

**1278 AMANI SUNBIRD.** *ANTHREPTES PALLIDIGASTER* Sclater & Moreau. **Pl. 81.**

*Anthreptes pallidigaster* W. L. Sclater & Moreau, Bull. B.O.C. 56, p. 17, 1935: Sigi Valley, four miles east of Amani, north-eastern Tanganyika Territory.

**Distinguishing characters:** Adult male, above, including lesser wing-coverts and upper tail-coverts, metallic bottle green; rump grey; wings black; tail glossy blue black with lightish ends to outer feathers; below, sides of face and chin to chest metallic bottle green, sometimes with violet reflections; rest of underparts white; tufts on sides of chest orange. There is no non-breeding dress. The female is grey above with slight metallic reflections; tail as in male; below, wholly creamy white and with no tufts on sides of chest. Wing 50 to 54 mm. The young bird is paler grey above than the adult female, with little metallic reflection.

# SUNBIRDS

**Range in Eastern Africa:** Sokoke forest, eastern Kenya Colony to Amani, Usambara Mts., north-eastern Tanganyika Territory.

**Habits:** Apparently a very local species indeed, inhabiting the tree tops of lowland forest and coming to flowering trees when in bloom, particularly Erythrinas. Generally seen in small flocks, and may easily be taken for a *Spermestes* at a distance.

**Nest and Eggs:** Undescribed.

**Recorded breeding:** No records.

**Food:** Small spiders and larvæ of insects.

**Call:** A minute jingling sibilant song, hardly audible.

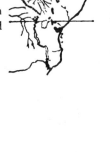

### 1279 PLAIN-BACKED SUNBIRD. *ANTHREPTES REICHENOWI* Gunning.

*Anthreptes reichenowi yokanæ* Hart. **Pl. 83.**

*Anthreptes yokanæ* Hartert, Bull. B.O.C. 41, p. 63, 1921: Rabai, north of Mombasa, eastern Kenya Colony.

**Distinguishing characters:** Adult male, forehead and chin to chest metallic blue black; rest of plumage green; yellower in centre of breast and belly; tufts at sides of chest yellow. There is no non-breeding dress. The female has the forehead green and the chin and neck in front yellowish white. Wing 51 to 57 mm. The immature male has the metallic crown and throat patches duller and smaller. The young bird is similar to the adult female, but is more olivaceous above.

**Range in Eastern Africa:** Coastal areas of Kenya Colony and Tanganyika Territory from Sokoke and Arabuko to Tanga.

**Habits:** A little known bird of lowland forest tree tops, generally confined to evergreen forest below 1,500 feet. Very often a member of a mixed bird party.

**Nest and Eggs:** A rather flimsy nest of skeleton leaves, bark and stems with a porch, hanging from a bough. Eggs two, white, spotted and freckled with dull red brown and with slight mauve under-spotting often forming a band; about $15 \cdot 5 \times 11$ mm.

**Recorded breeding:** Kenya Colony coast, April and May, and August to November.

**Food:** Spiders, termites and other insects.

**Call:** A little rambling song reminiscent of the sub-song of a Willow Warbler.

**Distribution of other races of the species:** Portuguese East Africa, the habitat of the nominate race.

### 1280 GREY-HEADED SUNBIRD. *ANTHREPTES AXILLARIS* (Reichenow).

*Camaroptera axillaris* Reichenow, O.M. p. 32, 1893: Uvamba (= Toro), western Uganda.

**Distinguishing characters:** Adult male, head grey; rest of upperparts green; sides of face and throat greyish white; rest of underparts yellow green; tufts on sides of chest orange; bill horn, lower mandible paler. There is no non-breeding dress. The female has no tufts on sides of chest. Wing 61 to 71 mm. The young bird has the head and throat green uniform with the rest of the plumage.

**General distribution:** Eastern Belgian Congo as far west as Bima to Uganda.

**Range in Eastern Africa:** Uganda.

**Habits:** Not uncommon in the larger Uganda forests, and abundant in the Semliki Valley among creeper covered trees in old clearings and secondary forest. Also found in undergrowth of true forest, but not in the tree tops.

**Nest and Eggs:** Undescribed.

**Recorded breeding:** Western Uganda (breeding condition), July, and August.

**Food:** No information.

**Call:** Unrecorded.

### 1281 GREEN HYLIA. *HYLIA PRASINA* (Cassin).

*Hylia prasina prasina* (Cass.).
*Sylvia prasina* Cassin, Proc. Acad. Sci. Philad. 7, p. 325, 1855: Moonda River, Gabon.

**Distinguishing characters:** Above, dark olive green; a broad greenish white stripe from base of bill to over and behind eye; lores and ear-coverts dark olive green; below, dusky grey washed with olivaceous; bill black; feet and toes olive green. There is no non-

## SUNBIRDS

breeding dress. The sexes are alike. Wing 56 to 72 mm. The young bird has a yellow bill and yellow feet.

**General distribution:** Portuguese Guinea and Gabon to the Sudan, Uganda and Kenya Colony.

**Range in Eastern Africa:** Southern Sudan to Uganda and western Kenya Colony.

**Habits:** Common forest species in Uganda, usually in tree tops but also in undergrowth. Often a member of a mixed bird party and appears to search foliage like an *Anthreptes*. Has a habit of freely flirting its wings.

**Nest and Eggs:** A large flattish domed nest, covered with scraps of vegetable material and cobwebs and very rough in appearance, generally low down at the ends of boughs of forest trees, or in bushes or large herbs, built in, not suspended. Eggs one or two, white; about 19 × 14·5 mm.

**Recorded breeding:** Cameroons (nestling), February and April. Western Uganda (breeding condition), July and August.

**Food:** Scale insects and wax.

**Call:** A double note like a saw being filed (Bates), also a call somewhat reminiscent of a Honey Guide's 'chuzz.' The song is a full-throated warble with occasional harsh notes in it.

**Distribution of other races of the species:** Fernando Po.

**1282** TIT-HYLIA. *PHOLIDORNIS RUSHIÆ* (Cassin).
*Pholidornis rushiæ denti* O. Grant.
*Pholidornis denti* O. Grant, Bull. B.O.C. 19, p. 41, 1907: Avakubi, north-eastern Belgian Congo.

**Distinguishing characters:** Head, mantle and chin to breast streaked dusky brown and pale buff; superciliary stripe whitish; wings and tail blackish; wing-coverts edged with pale buff; rump, upper tail-coverts, belly and under tail-coverts yellow. The sexes are alike, but the eye is red in the male and grey in the female. Wing 49 mm. Juvenile plumage unrecorded.

**General distribution:** Eastern Belgian Congo and Uganda.

**Range in Eastern Africa:** Uganda.

**Habits:** A little known species of the tops of tall trees in forest, usually seen in company with Sunbirds.

**Nest and Eggs:** The nest of this species is described as like that of a Weaver with the entrance hole underneath, large for the bird, and composed of thick felted down. It is usually attached to vines or creepers. Eggs two, white; no measurements available.

**Recorded breeding:** No records.

**Food:** No information; probably insects.

**Call:** A scarcely audible little twitter.

**Distribution of other races of the species:** West Africa to Angola, the nominate race being described from Gabon.

Names in Sclater's *Syst. Av. Æthiop.* 2, 1930, which have been changed; or have become synonyms in this work:

*Nectarinia famosa æneigularis* Sharpe, treated as synonymous with *Nectarinia famosa cupreonitens* Shelley.
*Nectarinia famosa centralis* Van Someren, treated as synonymous with *Nectarinia famosa cupreonitens* Shelley.
*Nectarinia famosa subfamosa* Salvadori, treated as synonymous with *Nectarinia famosa cupreonitens* Shelley.
*Cinnyris mariquensis hawkeri* Neumann, treated as synonymous with *Cinnyris mariquensis osiris* (Finsch.)
*Cinnyris leucogaster* Vieillot, now *Cinnyris talatala* A. Smith.
*Cinnyris venustus niassæ* Reichenow, treated as synonymous with *Cinnyris venustus falkensteinii* Fischer & Reichenow.
*Cinnyris chalybeus manoensis* Reichenow, treated as synonymous with *Cinnyris chalybeus intermedius* (Bocage).
*Cinnyris chalybeus gertrudis* Grote, treated as synonymous with *Cinnyris chalybeus intermedius* (Bocage).
*Chalcomitra angolensis* (Lesson) now *Chalcomitra rubescens* (Vieillot).
*Chalcomitra senegalensis æquatorialis* (Reichenow) treated as synonymous with *Chalcomitra senegalensis lamperti* (Reichenow).
*Chalcomitra senegalensis inæstimata* (Hartert) treated as synonymous with *Chalcomitra senegalensis gutturalis* (Linnæus).
*Anthreptes collaris ugandæ* Van Someren, treated as synonymous with *Anthreptes collaris garguensis* Mearns.
*Anthreptes tephrolæma elgonensis* Van Someren, treated as synonymous with *Anthreptes rectirostris tephrolæma* (Jardine & Fraser).

Names introduced since 1930 and which have become synonyms in this work:

*Nectarinia nectarinioides beveni* Van Someren, 1930, treated as synonymous with *Nectarinia erythrocerca* Hartlaub.
*Cinnyris superbus buvuma* Van Someren, 1932, treated as synonymous with *Cinnyris superbus superbus* Shaw.
*Cinnyris cupreus septentrionalis* Vincent, 1936, treated as synonymous with *Cinnyris cupreus cupreus* (Shaw).
*Cinnyris venustus sukensis* Van Someren, 1932, treated as synonymous with *Cinnyris venustus fazoqlensis* (Heuglin).
*Cinnyris chalybeus zonarius* Vincent, 1933, treated as synonymous with *Cinnyris chalybeus bractiatus* Vincent.
*Cinnyris chalybeus namwera* Vincent, 1933, treated as synonymous with *Cinnyris chalybeus bractiatus* Vincent.
*Chalcomitra senegalensis erythrinæ* Stoneham, 1933, treated as synonymous with *Chalcomitra senegalensis lamperti* Reichenow.

## CREEPERS

*Cyanomitra olivacea pembæ* Granvik, February 1934, now *Cyanomitra olivacea granti* Vincent, January 1934.
*Cyanomitra olivacea chyulu* Van Someren, 1939, treated as synonymous with *Cyanomitra olivacea neglecta* Neumann.
*Anthreptes collaris djamdjamensis* Benson, 1942, treated as synonymous with *Anthreptes collaris jubaensis* Van Someren.
*Anthreptes orientalis barbouri* Friedmann, 1931, treated as synonymous with *Anthreptes longuemarei orientalis* Hartlaub.

FAMILY—**CERTHIIDÆ. CREEPERS.** Genus: *Salpornis*.

One member of this family occurs in Eastern Africa. It is a bird of the open woodlands, running about on the trunks and limbs of trees, after the manner of a Wryneck or Woodpecker, in search of its insect food. Its remarkable coloration causes it to be easily recognized, but it is by no means common.

**1283** SPOTTED CREEPER. *SALPORNIS SPILONOTA* (Franklin).

*Salpornis spilonota salvadori* (Boc.).

*Hylypsornis salvadori* Bocage, Jorn. Lisboa 6, p. 198, 1878: Cacondo, Benguella, Angola.

**Distinguishing characters:** General colour blackish, spotted above and below with white; underparts washed with buff; wings spotted and tipped with white; tail barred blackish and white. The sexes are alike. Wing 89 to 102 mm. The young bird is similar to the adult, but the markings below are less sharp and there is less buff wash.

**General distribution:** Kenya Colony to Angola and Southern Rhodesia.

**Range in Eastern Africa:** Kenya Colony to Portuguese East Africa and Nyasaland.

**Habits:** Usually seen climbing a tree trunk starting from the bottom and carefully searching the bark as it makes its way up, very much in the manner of the palæarctic Tree-Creepers. Emin remarks that the first specimen obtained was climbing a corn stalk. Not uncommon but local in open woodland and frequently a member of a mixed bird party. The flight is fast and dipping like that of a Woodpecker.

**Nest and Eggs:** The nest, which has been very rarely found, is a neat cup of stems, bark and other materials, beautifully decorated with lichens, and is saddled on a horizontal branch generally at a fork, sometimes at considerable height from the ground. Eggs up to three, pale turquoise blue, heavily zoned with lavender and with irregular black dots and occasional brown smudges; about 18 × 13 mm.

**Recorded breeding:** Kedowa, Kenya Colony, May. Northern Rhodesia, August and September. Nyasaland, September and October. Elizabethville, Southern Belgian Congo, October.

**Food:** Insects and their larvæ.

**Call:** A shrill thin whistle 'sweepy-swip-swip-swip' (Vincent), also described as a Coot-like 'kek-kek-kek-kek.'

*Salpornis spilonota emini* Hartl.
*Salpornis emini* Hartlaub, P.Z.S. p. 415, pl. 37, 1884: Langomeri, West Nile, north-western Uganda.

**Distinguishing characters:** Differs from the preceding race in having a longer bill, and the white spots below much smaller. Wing 85 to 96 mm.

**General distribution:** Portuguese Guinea and Sierra Leone to the Sudan and Uganda.

**Range in Eastern Africa:** The Sudan and northern Uganda.

**Habits:** As for the last race.

**Recorded breeding:** Cameroons, March.

*Salpornis spilonota erlangeri* Neum.
*Salpornis spilonota erlangeri* Neumann, O.M. p. 52, 1907: Anderatscha, Kaffa, south-western Abyssinia.

**Distinguishing characters:** Similar to the southern race but bill shorter. Wing 90 to 95 mm.

**Range in Eastern Africa:** Western Abyssinia from Dangila southwards.

**Habits:** As for other races, locally not uncommon on the high plateau.

**Recorded breeding:** Abyssinia, feeding young in May.

**Distribution of other races of the species:** India, the habitat of the nominate race.

## Addenda and Corrigenda

p. 724. Genus *Zosterops*. From further information and research it appears that the whole group is in a very fluid state and it may well be that yellow-bellied and grey-bellied birds are really conspecific. The latest work on the subject is that by R. E. Moreau in the British Museum (Natural History) Bulletin, Aves 2, 1957, but the subject is still hampered by insufficient information.

**1226** *Zosterops abyssinicus abyssinicus.* *Recorded breeding:* add British Somaliland, June. Eggs three.

**1227** *Nectarinia famosa.* *General Distribution:* and *Range in Eastern Africa:* delete Eritrea.

**1231** *Nectarinia purpureiventris.* Add *Nest and Eggs:* Nest attached to a branch of a bush or shrub a few feet from the ground and made of moss, lichens, flower-heads and poppies. Eggs normally one, occasionally two, more or less heavily speckled with olive-grey in a zone round the larger end; about 19·5 × 13 mm. *Recorded breeding:* Belgian Congo, May.

**1235** *Drepanorhynchus reichenowi.* For Fisher read Fischer.

**1240** *Cinnyris habessinicus.* *Recorded breeding:* add British Somaliland, February to July and October and November (double brooded). *Call:* add 'a whistle up and down on two notes'. Alarm call 'chit-chit-chit-chit, also a similar squeaky song with the accent on the first syllable'.

**1245** *Cinnyris mariquensis* (p. 773). *Recorded breeding:* add Hargeisa, April to July. Egg one.

**1253** *Cinnyris chalybeus intermedius.* *General distribution:* delete Nyasaland. For culmen male 20 to 22 read 19–22.
*Cinnyris chalybeus graueri* should be *Cinnyris afer graueri.*

p. 783. For *Cinnyris chalybeus bractiatus* read *Cinnyris chalybeus manoënsis* Reichenow, and for 'preceding' read 'Angolan'. Culmen, male 23–27 mm. *Range in Eastern Africa:* add southern Tanganyika Territory.

**1256** Northern Double-collared Sunbird. For *Cinnyris reichenowi* Sharpe, read *Cinnyris ludovicensis* Bocage, and *Cinnyris ludovicensis reichenowi* for *Cinnyris reichenowi reichenowi.* *Distribution of other races of the species:* add 'the nominate race being described from Angola'.

**1259** *Cinnyris regius.* *Nest and Eggs:* Egg one, grey with numerous darker markings round the larger end; about 17 × 12 mm. (Kivu race).

**1261** *Cinnyris amethystina kirkii.* *Range in Eastern Africa:* In north-western Tanganyika Territory the range extends as far north as Kibondo and the Urundi boundary.

**1262** *C.r. rubescens.* *General Distribution:* add north-western Northern Rhodesia.

p. 820. Delete reference to *Cinnyris chalybeus manoënsis.*

*Continued on* p. 1104

## FAMILIES—PASSERIDÆ and PLOCEIDÆ. SPARROWS, WEAVERS, WAXBILLS, WIDOW-BIRDS, etc.

One hundred and sixty-three species are known from Eastern Africa. They are found in all types of country, and are represented in all parts of the area. Many have in the male a different dress in the breeding and non-breeding season, a fact which increases the difficulty of identification in the field. Their nesting habits differ considerably, some being gregarious. The food is grain, seeds and insects, and some are undoubtedly a nuisance to the cultivator, especially those species which flock in the non-breeding season. The young, however, are probably always fed on insects and at certain times of year they are beneficial, just as are Sparrows in Europe.

KEY TO THE ADULT WEAVER BIRDS, INCLUDING THE SPARROWS, WEAVERS, MALIMBES, ANAPLECTES, QUELEAS, BISHOPS, WIDOW-BIRDS, MANNIKINS, SILVER-BILLS, NEGRO-FINCHES, TWIN-SPOTS, CRIMSON-WINGS, SEED-CRACKERS, CUT-THROATS, QUAIL-FINCHES, LOCUST-FINCHES, PYTILIAS, FIRE-FINCHES, WAXBILLS, OLIVE-BACKS, CORDON-BLEUS, GRENADIERS, INDIGO-BIRDS AND WHYDAHS OCCURRING IN EASTERN AFRICA

1 Orange or red orange spot on forehead: FIRE-FRONTED BISHOP male in breeding dress *Euplectes diadematus* 1368

2 No orange or red orange spot on forehead: 3–416

3 Rump and upper tail-coverts white: 4–7

4 Central tail feathers elongated: PIN-TAILED WHYDAH male in breeding dress *Vidua macroura* 1441

5 Central tail feathers normal length: 6–7

6 Top of head mottled pale brown and dusky: DONALDSON-SMITH'S SPARROW WEAVER *Plocepasser donaldsoni* 1289

| | | |
|---|---|---|
| 7 Top of head black: | STRIPE-BREASTED SPARROW-WEAVER *Plocepasser mahali* | **1287** |
| 8 Rump and upper tail-coverts not white: | | 9–416 |
| 9 General colour mottled brownish and dusky; inner webs of flight feathers and tail tawny chestnut: | RUFOUS-TAILED WEAVER *Histurgops ruficauda* | **1290** |
| 10 General colour above, black and red; below, red and white: | RED-HEADED WEAVER male *Anaplectes melanotis* | **1359** |
| 11 General colour above olivaceous brown indistinctly streaked, cheeks, chin and throat black; below, yellow: | COMPACT WEAVER male and female in breeding dress *Pachyphantes pachyrhynchus* | **1350** |
| 12 Some spotting and barring in plumage: | | 34–82 |
| 13 No spotting or barring in plumage: | | 83–416 |
| 14 Mantle green: | | 83–87 |
| 15 Mantle buff, below, mainly blue: | | 88–97 |
| 16 Not blue below: | | 98–416 |
| 17 Mantle orange red or orange brown: | | 98–103 |
| 18 Mantle various shades of grey, various shades of brown, russet, sandy, sooty black, or ashy: | | 104–130 |
| 19 General colour grey: | | 131–132 |

20 General colour olive green and greenish yellow: 133–134
21 General colour olive and red: 135–142
22 General colour claret red or rose pink: 143–146
23 General colour mainly chocolate brown, dark maroon, chestnut and black or chestnut above and below or mantle only, but not merely top of head or rump: 147–152
24 General colour mainly black and crimson, or black and orange red: 153–165
25 General colour mainly blue black washed with violet, purple or green: 166–176
26 General colour black and white, black, black and grey, blue black or black and buff, with or without red or yellow wing shoulders or red or yellow band on foreneck: 177–205
27 General colour slate and red, brown and red, or olive and red: 206–216
28 General colour above streaked black and grey, or black and greenish, crown and nape black; below, chin to breast bright yellow, belly whitish: 217–218
29 General colour mainly bright yellow, either above or below or both: 219–287

30 General colour above various shades of yellow, but not bright yellow, mantle streaked; below, yellow, no black or chestnut on head:     288–302

31 Outer edges of flight feathers yellow; usually some palish yellow in plumage mainly confined to head, throat and chest; mantle streaked:     303–343

32 No yellow edging to flight feathers, head and mantle streaked:     344–389

33 Mantle streaked, head plain:     390–416

---

34 Top of head and mantle grey, wing-coverts spotted white: GREY-HEADED NEGRO-FINCH *Nigrita canicapilla* **1386**

35 Mantle spotted:     36–37

36 Mantle spotted white: LOCUST-FINCH *Ortygospiza locustella* **1404**

37 Mantle spotted black: CUT-THROAT, female *Amadina fasciata* **1402**

38 Inner secondaries only barred: BLACK-CHEEKED WAXBILL *Estrilda erythronotos* **1427**

39 All flight feathers except inner secondaries barred on outer webs:     40–41

40 Mantle chestnut: RUFOUS-BACKED MANNIKIN *Spermestes nigriceps* **1381**

41 Mantle blue black: BLACK AND WHITE MANNIKIN *Spermestes poensis* **1380**

42 Large spots below: 43-47
43 Mantle and chest green: GREEN-BACKED TWIN-SPOT
*Mandingoa nitidula* **1407**
44 Mantle dusky brown, chest spotted: BROWN TWIN-SPOT
*Clytospiza monteiri* **1389**
45 Mantle crimson, chest spotted: DUSKY TWIN-SPOT
*Clytospiza dybowskii* **1390**
46 Mantle brown, chest crimson: PETER'S TWIN-SPOT
*Hypargos niveoguttatus* **1406**
47 Mantle slate, chest red: RED-HEADED BLUE-BILL female
*Spermophaga ruficapilla* **1391**
48 Small spots more confined to sides of chest and flanks: 49-57
49 Mantle brownish slate or brownish grey: 50-51
50 Sides of face and throat black: BLACK-FACED FIRE-FINCH male
*Lagonosticta larvata* **1416**
51 Sides of face and throat slate: BARAKA DUSKY FIRE-FINCH
*Lagonosticta cinereovinacea* **1415**
52 Mantle russet or earth brown washed with red or vinous: 52-54
53 Under tail-coverts brown: RED-BILLED FIRE-FINCH
*Lagonosticta senegala* **1413**
54 Under tail-coverts black: AFRICAN FIRE-FINCH male
*Lagonosticta rubricata* **1411**
55 Mantle earth brown not washed with red or vinous: 56-57

| | | | |
|---|---|---|---|
| 56 | Sides of face slate: | BLACK-FACED FIRE-FINCH female *Lagonosticta larvata* | **1416** |
| 57 | Sides of face washed with crimson: | AFRICAN FIRE-FINCH female *Lagonosticta rubricata* | **1411** |
| 58 | Below, more or less barred and spotted: | | 59–68 |
| 59 | Upper tail-coverts blue: | PURPLE GRENADIER female *Granatina ianthinogaster* | **1433** |
| 60 | Upper tail-coverts ashy: | QUAIL-FINCH *Ortygospiza atricollis* | **1403** |
| 61 | Upper tail-coverts whitish, barred with dusky: | BRONZE MANNIKIN *Spermestes cucullatus* | **1379** |
| 62 | Upper tail-coverts red: | | 63–68 |
| 63 | Edges of flight feathers and wing-coverts red: | RED-WINGED PYTILIA *Pytilia phœnicoptera* | **1408** |
| 64 | Edges of flight feathers and wing-coverts orange: | ORANGE-WINGED PYTILIA *Pytilia afra* | **1409** |
| 65 | Edges of flight feathers and wing-coverts green: | GREEN-WINGED PYTILIA *Pytilia melba* | **1410** |
| 66 | Edges of flight feathers and wing-coverts earth brown, uniform with rest of upperparts: | | 67–68 |
| 67 | Below, yellow or orange yellow, flanks barred: | ZEBRA WAXBILL *Estrilda subflava* | **1421** |
| 68 | Below, crimson wine colour, narrow scattered bars on chest: | BAR-BREASTED FIRE-FINCH *Lagonosticta rufopicta* | **1414** |
| 69 | Mantle barred: | | 70–71 |

70 Top of head ashy brown uniform with mantle: SILVER-BILL
*Euodice malabarica* **1383**

71 Top of head barred: CUT-THROAT, male
*Amadina fasciata* **1402**

72 Top of head and round eyes black: 73-74

73 Breast to under tail-coverts dusky to blackish: BLACK-HEADED WAXBILL
*Estrilda atricapilla* **1424**

74 Breast to under tail-coverts whitish: BLACK-CROWNED WAXBILL
*Estrilda nonnula* **1425**

75 Top of head brownish, red streak through eye: 76-79

76 Wing-coverts crimson: CRIMSON-RUMPED WAXBILL
*Estrilda rhodopyga* **1420**

77 Wing-coverts uniform with mantle: 78-79

78 Upper tail-coverts and tail black: BLACK-RUMPED WAXBILL
*Estrilda troglodytes* **1419**

79 Upper tail-coverts uniform with mantle: WAXBILL
*Estrilda astrild* **1418**

80 No red streak through eye, upper tail-coverts crimson: 81-82

81 Mantle green: YELLOW-BELLIED WAXBILL
*Coccopygia melanotis* **1417**

82 Mantle brown: FAWN-BREASTED WAXBILL
*Estrilda paludicola* **1422**

---

83 Head and tail black: WHITE-COLLARED OLIVE-BACK, male
*Nesocharis ansorgei* **1429**

84 Top of head grey, tail green: GREY-HEADED OLIVE-BACK
*Nesocharis capistrata* **1428**

| | | |
|---|---|---|
| 85 | Top of head and tail green: | 86–87 |
| 86 | Below, whitish, washed with yellow on chest; edges inner webs of flight feathers yellow: | UGANDA ORANGE WEAVER, female *Ploceus aurantius* 1334 |
| 87 | Below, yellow, belly white; inner webs of flight feathers buffish: | MASKED WEAVER, female and male in non-breeding dress *Ploceus intermedius* 1319 |

| | | |
|---|---|---|
| 88 | Red patch on side of head: | RED-CHEEKED CORDON-BLEU, male *Uræginthus bengalus* 1431 |
| 89 | No red patch on side of head: | 90–97 |
| 90 | Below, bright violet blue: | PURPLE GRENADIER, male *Granatina ianthinogaster* 1433 |
| 91 | Below, pale blue: | 92–97 |
| 92 | Top of head blue: | BLUE-CAPPED CORDON-BLEU, male *Uræginthus cyanocephalus* 1432 |
| 93 | Forehead only blue: | 94–97 |
| 94 | Blue continuous from ear-coverts to side of neck: | CORDON-BLEU *Uræginthus angolensis* 1430 |
| 95 | Brown of sides of neck extends forwards behind ear-coverts: | 96–97 |
| 96 | Brighter blue, second primary notched: | RED-CHEEKED CORDON-BLEU, female *Uræginthus bengalus* 1431 |

| | | |
|---|---|---|
| 97 | Paler blue, second primary not notched: | BLUE-CAPPED CORDON-BLEU, female *Uræginthus cyanocephalus* **1432** |
| 98 | Forehead and top of head orange red: | **99–100** |
| 99 | Tail short and brown: | ZANZIBAR RED BISHOP male in breeding dress *Euplectes nigroventris* **1363** |
| 100 | Tail longer and black: | BLACK-WINGED RED BISHOP male in breeding dress *Euplectes hordeacea* **1365** |
| 101 | Forehead and top of head black: | **102–103** |
| 102 | Upper tail-coverts long and orange red: | RED BISHOP male in breeding dress *Euplectes orix* **1363** |
| 103 | Upper tail-coverts normal length and black: | BLACK BISHOP male in breeding dress *Euplectes gierowii* **1366** |
| 104 | Mantle grey, outer edges of flight feathers red or yellow: | RED-HEADED WEAVER female *Anaplectes melanotis* **1359** |
| 105 | Mantle dusky brown, upper tail-coverts red; outer edges of flight feathers not red or yellow: | WHITE-HEADED BUFFALO-WEAVER *Dinemellia dinemelli* **1286** |
| 106 | Mantle pale brown, bill ivory: | BLACK-CAPPED SOCIAL WEAVER *Pseudonigrita cabanisi* **1292** |

107 Mantle earth brown, bill black:     WHITE-BREASTED NEGRO-FINCH *Nigrita fusconota* **1388**

108 Mantle earth brown, or washed with russet, top of head chestnut:     CHESTNUT-CROWNED SPARROW-WEAVER *Plocepasser superciliosus* **1288**

109 Mantle dusky sandy, top of head black speckled with white:     SPECKLE-FRONTED WEAVER *Sporopipes frontalis* **1311**

110 Top of head not chestnut, nor black speckled with white:     111–130

111 Mantle isabelline grey, chin and throat black:     DESERT SPARROW, male *Passer simplex* **1299**

112 Mantle earth brown, ashy brown, russet brown, or ashy grey, yellow spot on lower neck in front:     YELLOW-SPOTTED PETRONIA *Petronia xanthosterna* **1308**

113 Chin and throat not black, no yellow spot on lower neck in front:     114–130

114 Rump chestnut:     115–120

115 Bill stout, mantle tawny or dusky brown:     PARROT-BILLED SPARROW *Passer gongonensis* **1303**

116 Bill normal:     117–120

117 Mantle tawny or dusky grey:     118–119

118 Mantle tawny, below, pale, belly white:     GREY-HEADED SPARROW *Passer griseus* **1300**

119 Mantle dusky grey, below, dusky:    SWAINSON'S SPARROW *Passer swainsoni* **1301**

120 Mantle earth or dusky brown:    SWAHILI SPARROW *Passer suahelicus* **1302**

121 Rump not chestnut:    122–130

122 Tail black:    123–126

123 Top of head grey:    124–126

124 Mantle biscuit brown, outer tail feathers only with light tips:    GREY-HEADED SILVER-BILL *Odontospiza caniceps* **1384**

125 Mantle greyish buff, tail feathers with broad greyish ends:    GREY-HEADED SOCIAL WEAVER *Pseudonigrita arnaudi* **1291**

126 Mantle sooty brown, blackish brown or sooty black, base of primaries white, usually some white on forehead:    GROSBEAK WEAVER, male *Amblyospiza albifrons* **1358**

127 Tail not black:    128–130

128 Mantle and general colour above sandy:    DESERT SPARROW, female *Passer simplex* **1299**

129 Mantle and general colour above ashy:    ARABIAN GOLDEN SPARROW female *Auripasser euchlorus* **1305**

130 Mantle and general colour above isabelline brown:    PALE ROCK-SPARROW *Carpospiza brachydactyla* **1310**

---

131 Head grey, bill blue grey:    LAVENDER WAXBILL *Estrilda perreini* **1423**

| | | | |
|---|---|---|---|
| 132 | Head black and white, bill pink: | JAVA SPARROW *Padda orizivora* | **1385** |
| 133 | Inner webs of flight feathers pale buff: | VIEILLOT'S BLACK WEAVER female *Melanopteryx nigerrimus* | **1344** |
| 134 | Inner webs of flight feathers yellow: | WEYN'S WEAVER, female *Melanopteryx weynsi* | **1345** |
| 135 | Bill red: | SHELLEY'S CRIMSON-WING male *Cryptospiza shelleyi* | **1401** |
| 136 | Bill black: | | 137–142 |
| 137 | Red patch round eye: | RED-FACED CRIMSON-WING male *Cryptospiza reichenovii* | **1398** |
| 138 | No red patch round eye: | | 139–142 |
| 139 | Mantle crimson: | | 140–142 |
| 140 | Size larger, wing over 66 mm.: | SHELLEY'S CRIMSON-WING female *Cryptospiza shelleyi* | **1401** |
| 141 | Size smaller, wing under 60 mm.: | REL-FACED CRIMSON-WING female *Cryptospiza reichenovii* | **1398** |
| 142 | Mantle more brick red in tone: | ABYSSINIAN CRIMSON-WING *Cryptospiza salvadorii* | **1399** |
| 143 | Breast to under tail-coverts black: | BLACK-BELLIED FIRE-FINCH male *Estrilda rara* | **1426** |
| 144 | Lower belly and under tail-coverts black: | | 145–146 |

## SPARROWS, WEAVERS, WAXBILLS, etc.

145 Bill black, lower mandible pink:  
    BLACK-BELLIED FIRE-FINCH female  
    *Estrilda rara*    1426

146 Bill wholly slate:  
    JAMESON'S FIRE-FINCH  
    *Lagonosticta jamesoni*    1412

147 Above, and below, chocolate brown:  
    CHESTNUT SPARROW, male  
    *Sorella eminibey*    1306

148 Above, dusky, below, dark maroon:  
    CHESTNUT-BREASTED NEGRO-FINCH  
    *Nigrita bicolor*    1387

149 Chest to belly chestnut, rest of plumage black, yellow band across upper mantle:  
    YELLOW-MANTLED WEAVER, male  
    *Melanoploceus tricolor*    1353

150 Mainly chestnut above and below:    151–152

151 Wings edged with yellow, tail greenish yellow:  
    CINNAMON WEAVER male in breeding dress  
    *Ploceus badius*    1328

152 Wings edged with buffish, tail dusky:  
    CHESTNUT WEAVER male in breeding dress  
    *Ploceus rubiginosus*    1327

153 Head all round crimson:    154–162  
154 Tail black:    155–158  
155 Bill shorter and deeper, iridescent blue:    156–157  
156 Head red to nape:  
    RED-HEADED BLUE-BILL male  
    *Spermophaga ruficapilla*    1391

157 Forehead red, crown and nape black:  
    GRANT'S BLUE-BILL, male  
    *Spermophaga poliogenys*    1392

| | | |
|---|---|---|
| 158 | Bill longer and less deep, black: | CRESTED MALIMBE *Malimbus malimbicus* **1355** |
| 159 | Tail black and crimson: | 160–162 |
| 160 | Width of bill at base 18–19 mm.: | LARGE-BILLED SEED-CRACKER, male *Pirenestes maximus* **1395** |
| 161 | Width of bill at base 14–17 mm.: | BLACK-BELLIED SEED-CRACKER, male *Pirenestes ostrinus* **1393** |
| 162 | Width of bill at base 12–13 mm.: | ROTHSCHILD'S SEED-CRACKER, male *Pirenestes rothschildi* **1394** |
| 163 | Top of head orange red, ear-coverts, chin and throat black: | 164–165 |
| 164 | Chest to belly black: | RED-HEADED MALIMBE *Malimbus rubricollis* **1356** |
| 165 | Chest to belly red: | RED-BELLIED MALIMBE *Malimbus erythrogaster* **1357** |

| | | |
|---|---|---|
| 166 | Central tail feathers long; under wing-coverts white: | STEEL-BLUE WHYDAH male in breeding dress *Vidua hypocherina* **1442** |
| 167 | Central tail feathers normal length; under wing-coverts brown and white: | 168–176 |
| 168 | Wing and tail blackish: | 169–171 |
| 169 | General colour dull blue black with a green wash: | GREEN INDIGO-BIRD male in breeding dress *Hypochera ænea* **1434** |

| | | |
|---|---|---|
| 170 | General colour glossy violet blue black: | PURPLE INDIGO-BIRD male in breeding dress *Hypochera ultramarina* **1435** |
| 171 | General colour dull violet blue black: | SOUTH AFRICAN INDIGO-BIRD male in breeding dress *Hypochera amauropteryx* **1436** |
| 172 | Wings and tail ash-coloured: | 173–176 |
| 173 | General colour dull velvety violet purple: | DUSKY INDIGO-BIRD male in breeding dress *Hypochera funerea* **1437** |
| 174 | General colour dull velvety black: | BLACK INDIGO-BIRD male in breeding dress *Hypochera nigerrima* **1438** |
| 175 | General colour dull violet blue black: | CAMEROON INDIGO-BIRD male in breeding dress *Hypochera camerunensis* **1439** |
| 176 | General colour glossy blue black: | ALEXANDER'S INDIGO-BIRD male in breeding dress *Hypochera nigeriæ* **1440** |
| 177 | General colour black and white: | MAGPIE MANNIKIN *Amauresthes fringilloides* **1382** |
| 178 | General colour mainly black: | 179–205 |
| 179 | Wing shoulder yellow: | 180–186 |
| 180 | Rump and mantle black: | 181–184 |
| 181 | Flight feathers black to base: | 182–183 |

# SPARROWS, WEAVERS, WAXBILLS, etc.

182 Bill bluish white:     MARSH WIDOW-BIRD male in breeding dress *Coliuspasser hartlaubi* **1374**

183 Bill black:     YELLOW-SHOULDERED WIDOW-BIRD male in breeding dress *Coliuspasser macrocercus* **1371**

184 Flight feathers white at base:     WHITE-WINGED WIDOW-BIRD male in breeding dress *Coliuspasser albonotatus* **1373**

185 Mantle yellow:     YELLOW-MANTLED WIDOW-BIRD male in breeding dress *Coliuspasser macrourus* **1372**

186 Rump yellow:     YELLOW BISHOP male in breeding dress *Euplectes capensis* **1367**

187 Wing shoulder tawny:     JACKSON'S WIDOW-BIRD male in breeding dress *Drepanoplectes jacksoni* **1378**

188 Wing shoulder red:     189–190

189 Size larger, central tail feathers greatly elongated:     LONG-TAILED WIDOW-BIRD male in breeding dress *Coliuspasser progne* **1377**

190 Size smaller, central tail feathers not elongated:     FAN-TAILED WIDOW-BIRD male in breeding dress *Coliuspasser axillaris* **1370**

191 Wing shoulder black:     192–205

192 White bases to head and body feathers:     BUFFALO-WEAVER *Bubalornis albirostris* **1284** RED-BILLED BUFFALO-WEAVER *Bubalornis niger* **1285**

193 No white bases to head and body feathers, mantle black: 194–205

194 Central tail feathers normal length: 195–199

195 Colour wholly black: 196–197

196 Bill heavy, feet brown: VIEILLOT'S BLACK WEAVER male
*Melanopteryx nigerrimus* 1344

197 Bill much less heavy, feet blackish: MAXWELL'S BLACK WEAVER
*Melanoploceus albinucha* 1354

198 Yellow band across upper mantle: YELLOW-MANTLED WEAVER, female
*Melanoploceus tricolor* 1353

199 Forehead and sides of face yellow: BLACK-BILLED WEAVER
*Heterhyphantes melanogaster* 1346

200 Central tail feathers greatly elongated: 201–205

201 Central tail feathers buff: FISCHER'S WHYDAH male in breeding dress
*Vidua fischeri* 1443

202 Central tail feathers black: 203–205

203 With or without red or orange band across neck in front: RED-COLLARED WIDOW-BIRD male in breeding dress
*Coliuspasser ardens* 1375

204 Crown of head to nape and a band across neck in front red: RED-NAPED WIDOW-BIRD male in breeding dress
*Coliuspasser laticuada* 1376

| | | |
|---|---|---|
| 205 | Chestnut band across chest: | PARADISE WHYDAH male in breeding dress *Steganura paradisæa* **1444** BROAD-TAILED PARADISE WHYDAH male in breeding dress *Steganura orientalis* **1445** |

| | | |
|---|---|---|
| 206 | General colour slate and red: | 207–208 |
| 207 | Mantle red, below, slate: | DUSKY CRIMSON-WING *Cryptospiza jacksoni* **1400** |
| 208 | Mantle slate, below, chin to chest and flanks red: | GRANT'S BLUE-BILL female *Spermophaga poliogenys* **1392** |
| 209 | General colour brown, rump red: | 210–216 |
| 210 | General colour warm brown: | 211–213 |
| 211 | Width of bill at base 18–19 mm.: | LARGE-BILLED SEED-CRACKER, female *Pirenestes maximus* **1395** |
| 212 | Width of bill at base 14–17 mm.: | BLACK-BELLIED SEED-CRACKER, female *Pirenestes ostrinus* **1393** |
| 213 | Width of bill at base 12–13 mm.: | ROTHSCHILD'S SEED-CRACKER, female *Pirenestes rothschildi* **1394** |
| 214 | General colour colder brown: | 215–216 |
| 215 | Width of bill at base 13·5–16 mm.: | URUNGU SEED-CRACKER *Pirenestes frommi* **1397** |

216 Width of bill at base
9–12 mm.:      NYASALAND SEED-CRACKER
*Pirenestes minor*    **1396**

217 Forehead and forecrown yellow:    EMIN'S WEAVER, male
*Othyphantes emini*    **1347**

218 Forehead and forecrown black:    EMIN'S WEAVER, female
*Othyphantes emini*    **1347**

219 Mantle chestnut:    SUDAN GOLDEN SPARROW male
*Auripasser luteus*    **1304**

220 Mantle plain green:    STRANGE WEAVER
*Hyphanturgus alienus*    **1339**

221 Mantle mainly black or brownish black, heavily streaked with black:    **222–236**

222 Forehead or forehead and top of head yellow:    **223–224**

223 Chin and throat black:    BLACK-NECKED WEAVER male
*Hyphanturgus nigricollis*    **1336**

224 Chin and throat yellow:    REICHENOW'S WEAVER male
*Othyphantes reichenowi*    **1348**

225 Forehead, top of head and mantle black or dark coloured:    **226–236**

226 Chin to chest black:    **227–228**

227 Edges of inner webs of flight feathers yellow:    WEYN'S WEAVER, male
*Melanopteryx weynsi*    **1345**

228 Edges of inner webs of flight feathers buff:    CLARKE'S WEAVER
*Ploceus golandi*    **1320**

229 Chin and throat black or dark coloured:    **230–232**

| | | | |
|---|---|---|---|
| 230 | Chest yellow, bill white: | DARK-BACKED WEAVER male *Symplectes bicolor* | **1335** |
| 231 | Chest chestnut, bill black: | USAMBARA WEAVER *Hyphanturgus nicolli* | **1340** |
| 232 | Sides of face yellow, black stripe through eye: | BLACK-NECKED WEAVER female *Hyphanturgus nigricollis* | **1336** |
| 233 | Sides of face black: | | 234–236 |
| 234 | Chin and throat dusky or dusky and black: | DARK-BACKED WEAVER *Symplectes bicolor* | **1335** |
| 235 | Chin and throat yellow: | REICHENOW'S WEAVER female *Othyphantes reichenowi* | **1348** |
| 236 | Mantle green, heavily streaked: | STUHLMANN'S WEAVER *Othyphantes stuhlmanni* | **1349** |
| 237 | Mantle yellow: | | 238–240 |
| 238 | Top of head chestnut: | BROWN-CAPPED WEAVER male *Phormoplectes insignis* | **1351** |
| 239 | Top of head black: | | 240–243 |
| 240 | Wings and tail black: | BROWN-CAPPED WEAVER female *Phormoplectes insignis* | **1351** |
| 241 | Mantle mottled yellow and black: | | 242–243 |
| 242 | Nape black: | LAYARD'S BLACK-HEADED WEAVER male in breeding dress *Ploceus nigriceps* | **1313** |
| 243 | Nape not black: | BLACK-HEADED WEAVER male in breeding dress *Ploceus cucullatus* | **1312** |
| 244 | Top of head yellow: | | 245–248 |

245 Chest to belly black: YELLOW-CROWNED BISHOP
male in breeding dress
*Euplectes afra*   **1369**

246 Chest to belly yellow:   247–248

247 Wing 81 to 89 mm.: SPEKE'S WEAVER, male
*Ploceus spekei*   **1314**

248 Wing 73 to 76 mm.: FOX'S WEAVER, male
*Ploceus spekeoides*   **1315**

249 Mantle yellow, greenish yellow, or olive yellow:   250–279

250 Head mainly green: OLIVE-HEADED GOLDEN WEAVER
*Hyphanturgus olivaceiceps*   **1338**

251 Some chestnut or orange on head, but no black:   252–260

252 Head orange, some chestnut on lower neck in front: GOLDEN PALM-WEAVER
male
*Ploceus bojeri*   **1331**

253 Head yellow, nape chestnut: TAVETA GOLDEN WEAVER
male
*Ploceus castaneiceps*   **1333**

254 Head all round faintly washed with orange: GOLDEN WEAVER
male in breeding dress
*Xanthophilus subaureus*   **1341**

255 Top of head uniform with mantle: GOLDEN WEAVER
male in non-breeding dress
*Xanthophilus subaureus*   **1341**

256 Forehead and chin washed with chestnut: RÜPPELL'S WEAVER, male in non-breeding dress
*Ploceus galbula*   **1329**

257 Forehead and chin distinctly chestnut:  RÜPPELL'S WEAVER
male in breeding dress
*Ploceus galbula*     1329

258 Lores, sides of face, below eye and chin to throat dark chestnut:     259–260

259 Primaries canary yellow, ends dusky: SOUTHERN BROWN-THROATED WEAVER
male in breeding dress
*Ploceus xanthopterus*     1332

260 Primaries dusky, edged with olive green: NORTHERN BROWN-THROATED WEAVER, male
*Ploceus castanops*     1330

261 Some black on head:     262–281

262 Narrow black streak through eye: SPECTACLED WEAVER
*Hyphanturgus ocularis*     1337

263 Lores black: ORANGE WEAVER, male
*Ploceus aurantius*     1334

264 Broad black patch from lores to sides of face and ear-coverts: BAGLAFECHT WEAVER
male in breeding dress
*Ploceus baglafecht*     1323

265 Lores, round eye, ear-coverts, sides of face and chin to lower neck black: HEUGLIN'S MASKED WEAVER
male in breeding dress
*Ploceus heuglini*     1318

266 Lores to round eyes, chin and upper neck black, top of head orange, black band round back of crown: BERTRAM'S WEAVER, male
*Xanthoploceus bertrandi*     1352

267 Black band across forehead: 268–274

268 Lower neck to flanks chestnut: TANGANYIKA MASKED WEAVER, male *Ploceus reichardi* 1317

269 Chin to upper throat black: VITELLINE MASKED WEAVER male in breeding dress *Ploceus vitellinus* 1321

270 Black of chin and throat extending to a point on lower neck: 271–280

271 Ear-coverts yellow: NORTHERN MASKED WEAVER male in breeding dress *Ploceus tæniopterus* 1316

272 Ear-coverts black: 273–281

273 Size larger, wing over 72 mm.: SOUTHERN MASKED WEAVER male in breeding dress *Ploceus velatus* 1322

274 Size smaller, wing under 67 mm.: LITTLE WEAVER male in breeding dress *Ploceus luteolus* 1324

275 Black extending to crown and level with back of eyes: 276–277

276 Size larger, wing over 69 mm.: MASKED WEAVER male in breeding dress *Ploceus intermedius* 1319

277 Size smaller, wing under 65 mm.: SLENDER-BILLED WEAVER male *Icteropsis pelzelni* 1343

278 Black extending to occiput: 279–281

## SPARROWS, WEAVERS, WAXBILLS, etc.

| | | |
|---|---|---|
| 279 Mantle yellow, underside chestnut: | GOLDEN-BACKED WEAVER male in breeding dress *Ploceus jacksoni* | **1326** |
| 280 Mantle golden olive, underside washed with chestnut from neck to breast: | YELLOW-BACKED WEAVER male in breeding dress *Ploceus capitalis* | **1325** |
| 281 Mantle olive green, underside olive yellow: | BERTRAM'S WEAVER, female *Xanthoploceus bertrandi* | **1352** |
| 282 No black on head and no chestnut in plumage: | | 283–287 |
| 283 Edges of flight and tail feathers white: | ARABIAN GOLDEN SPARROW male *Auripasser euchlorus* | **1305** |
| 284 Edges of flight and tail feathers yellow: | | 285–287 |
| 285 Size small, wing under 65 mm.: | SLENDER-BILLED WEAVER female *Icteropsis pelzelni* | **1343** |
| 286 Size medium, wing 66 to 74 mm.: | GOLDEN WEAVER female in breeding dress *Xanthophilus subaureus* | **1341** |
| 287 Size large, wing over 82 mm.: | HOLUB'S GOLDEN WEAVER *Xanthophilus xanthops* | **1342** |
| 288 Bill short and conical: | PARASITIC WEAVER, male *Anomalospiza imberbis* | **1405** |
| 289 Bill normal: | | 290–302 |
| 290 Mantle not boldly streaked: | | 291–297 |
| 291 Mantle yellow green or yellow green and ashy grey: | | 292–295 |

848   SPARROWS, WEAVERS, WAXBILLS, etc.

292 Size larger, wing 78 mm. and over:     293–294

293 First primary broader:    BLACK-HEADED WEAVER female in breeding dress *Ploceus cuculiatus*    1312

294 First primary narrower:    LAYARD'S BLACK-HEADED WEAVER female in breeding dress *Ploceus nigriceps*    1313

295 Size smaller, wing 76 mm. or less:    FOX'S WEAVER female *Ploceus spekeoides*    1315

296 Mantle mustard yellow:    GOLDEN PALM-WEAVER female *Ploceus bojeri*    1332

297 Mantle olive yellow:    GOLDEN WEAVER female in non-breeding dress *Xanthophilus subaureus*    1341

298 Mantle boldly streaked:     299–302

299 Mantle olive green:    GOLDEN-BACKED WEAVER female and male in non-breeding dress *Ploceus jacksoni*    1326

300 Mantle yellow green:     301–302

301 Rump yellow, underside richer yellow:    MASKED WEAVER female in breeding dress *Ploceus velatus*    1322

302 Rump paler yellowish green, underside paler yellow:    TAVETA GOLDEN WEAVER female *Ploceus castaneiceps*    1333

303 General colour above buff or tawny, broadly streaked with black:     304–309

| | | |
|---|---|---|
| 304 | Bill red: | 305–306 |
| 305 | Head brownish, finely streaked: | RED-BILLED QUELEA female and male in non-breeding dress *Quelea quelea* **1360** |
| 306 | Head red, or black and red, or pink or buff: | RED-BILLED QUELEA male in breeding dress *Quelea quelea* **1360** |
| 307 | Bill black: | 308–309 |
| 308 | Whole head red, chin and throat blackish: | RED-HEADED QUELEA male in breeding dress *Quelea erythrops* **1361** |
| 309 | Whole head to chest red: | CARDINAL QUELEA male in breeding dress *Quelea cardinalis* **1362** |
| 310 | Mantle green or yellow green: | 311–316 |
| 311 | Sides of face and throat brownish: | NORTHERN BROWN-THROATED WEAVER female *Ploceus castanops* **1330** |
| 312 | Sides of face and throat yellow: | 313–320 |
| 313 | Top of head more olive green: | 314–317 |
| 314 | Edges of flight feathers and tail light yellow: | 315–317 |
| 315 | Inner webs of flight feathers brighter yellow: | TANGANYIKA MASKED WEAVER female *Ploceus reichardi* **1317** |

316 Inner webs of flight feathers paler or buffish yellow: HEUGLIN'S MASKED WEAVER female and male in non-breeding dress *Ploceus heuglini* **1318**

317 Mantle brown or buff, little or no green or olive: 318–343

318 Top of head bright olive green: 319–320

319 Bill larger, first primary broader: BLACK-HEADED WEAVER female and male in non-breeding dress *Ploceus cucullatus* **1312**

320 Bill smaller, first primary narrower: LAYARD'S BLACK-HEADED WEAVER, female and male in non-breeding dress *Ploceus nigriceps* **1313**

321 Top of head washed with dull olive or brown: 322–343

322 Top of head and mantle grey brown: BAGLAFECHT WEAVER female and male in non-breeding dress *Ploceus baglafecht* **1323**

323 Top of head dull olive, finely streaked: 324–343

324 Mantle heavily and broadly streaked with black: 325–330

325 Tail brown: NORTHERN MASKED WEAVER female and male in non-breeding dress *Ploceus tæniopterus* **1316**

326 Tail olive green: 327–343

327 Rump pale brown: 328–329

| | | |
|---|---|---|
| 328 | Chest yellow: | SOUTHERN BROWN-THROATED WEAVER female and male in non-breeding dress *Ploceus xanthopterus* **1332** |
| 329 | Chest buff: | CINNAMON WEAVER, female and male in non-breeding dress *Ploceus badius* **1328** |
| 330 | Rump olivaceous: | VITELLINE MASKED WEAVER female and male in non-breeding dress *Ploceus vitellinus* **1321** |
| 331 | Top of head brown broadly streaked: | 332-335 |
| 332 | Bill larger: | RED-HEADED QUELEA, female and male in non-breeding dress *Quelea erythrops* **1361** |
| 333 | Bill smaller: | 334-335 |
| 334 | Feet larger: | FIRE-FRONTED BISHOP female and male in non-breeding dress *Euplectes diademata* **1368** |
| 335 | Feet smaller: | CARDINAL QUELEA, female and male in non-breeding dress *Quelea cardinalis* **1362** |
| 336 | Mantle less heavily and broadly streaked with dusky: | 337-343 |
| 337 | Flanks and under tail-coverts yellowish: | 338-339 |
| 338 | Size larger, wing over 76 mm.: | SPEKE'S WEAVER, female *Ploceus spekei* **1314** |
| 339 | Size smaller, wing under 65 mm.: | LITTLE WEAVER, female and male in non-breeding dress *Ploceus luteolus* **1324** |

852    SPARROWS, WEAVERS, WAXBILLS, etc.

340 Flanks and under tail-coverts white:     341–343

341 Upper tail-coverts buff: YELLOW-BACKED WEAVER female and male in non-breeding dress *Ploceus capitalis*    **1325**

342 Upper tail-coverts brighter olive yellow: SOUTHERN MASKED WEAVER female and male in non-breeding dress *Ploceus velatus*    **1322**

343 Upper tail-coverts duller olive yellow: RÜPPELL'S WEAVER, female *Ploceus galbula*    **1329**

---

344 Head and mantle heavily and broadly streaked:     345–386

345 Rump yellow: YELLOW BISHOP, female and male in non-breeding dress *Euplectes capensis*    **1367**

346 Rump not yellow:     347–389

347 Under wing-coverts black or blackish:     348–356

348 Sides of face and underside washed with yellow:     349–353

349 Tail elongated, bill smaller: RED-NAPED WIDOW-BIRD male in non-breeding dress *Coliuspasser laticauda*    **1376**

350 Tail longer but not elongated, bill larger: BLACK BISHOP, female and male in non-breeding dress *Euplectes gierowii*    **1366**

351 Tail shorter:     352–353

352 Bill larger: BLACK-WINGED RED BISHOP female and male in non-breeding dress *Euplectes hordeacea*    **1365**

# SPARROWS, WEAVERS, WAXBILLS, etc.

| | | |
|---|---|---|
| 353 | Bill smaller: | RED-COLLARED WIDOW-BIRD* female and male in non-breeding dress *Coliuspasser ardens* **1375** |
| 354 | Sides of face and underside with no yellow wash: | 355–389 |
| 355 | First primary longer: | LONG-TAILED WIDOW-BIRD female and male in non-breeding dress *Coliuspasser progne* **1377** |
| 356 | First primary shorter: | MARSH WIDOW-BIRD, female and male in non-breeding dress *Coliuspasser hartlaubi* **1374** |
| 357 | Under wing-coverts white, central streak along top of head: | 358–360 |
| 358 | White on inner webs of tail feathers: | PIN-TAILED WHYDAH female and male in non-breeding dress *Vidua macroura* **1441** |
| 359 | White in tail confined to outer web and tips: | STEEL-BLUE WHYDAH female and male in non-breeding dress *Vidua hypocherina* **1442** |
| 360 | Under wing-coverts tawny: | FAN-TAILED WIDOW-BIRD female and male in non-breeding dress *Coliuspasser axillaris* **1370** |
| 361 | Under wing-coverts buff, buffish and dusky or brown and white: | 362–389 |
| 362 | Sides of chest and flanks heavily streaked with black: | 363–370 |

\* The Red-naped Widow-Bird, female, is similar but larger.

# SPARROWS, WEAVERS, WAXBILLS, etc.

363 Size larger, wing over 78 mm.: JACKSON'S WIDOW-BIRD female and male in non-breeding dress *Drepanoplectes jacksoni* **1378**

364 Size smaller, wing under 74 mm.: 365–370

365 Superciliary stripe, sides of face and below washed with yellow: YELLOW-CROWNED BISHOP female in breeding dress *Euplectes afra* **1369**

366 Superciliary stripe, sides of face and chin to chest buffish: 367–370

367 Below, distinctly streaked: YELLOW-CROWNED BISHOP female and male in non-breeding dress *Euplectes afra* **1369**

368 Below, indistinctly streaked: 369–370

369 Under wing-coverts and inner edges to flight feathers rather darker buff: RED BISHOP female and male in non-breeding dress *Euplectes orix* **1363**

370 Under wing-coverts and inner edges of flight feathers rather paler buff: ZANZIBAR RED BISHOP female and male in non-breeding dress *Euplectes nigroventris* **1364**

371 Sides of chest and flanks not heavily streaked: 372–389

## SPARROWS, WEAVERS, WAXBILLS, etc.

372 Crown tawny with black streaks:     FISCHER'S WHYDAH, female and male in non-breeding dress
*Vidua fischeri*     **1443**

373 Centre of crown creamy white, bordered with black:     374–375

374 Size larger, wing over 73 mm.:     PARADISE WHYDAH, female and male in non-breeding dress
*Steganura paradisæa*     **1444**
BROAD-TAILED PARADISE WHYDAH, female and male in non-breeding dress
*Steganura orientalis*     **1445**

375 Size smaller, wing under 66 mm.:     INDIGO BIRDS, female and male in non-breeding dress
*Hypochera* species     **1434–1440**

376 Centre of crown streaked with black:     377–386

377 Bill short and conical:     PARASITIC WEAVER, female
*Anomalospiza imberbis*     **1405**

378 Bill normal:     379–386

379 Tail shorter:     YELLOW-CROWNED BISHOP male in non-breeding dress
*Euplectes afra*     **1369**

380 Tail longer:     381–386

381 Tail square:     382–383

382 Top of head and rump distinctly streaked, bill smaller:     WHITE-WINGED WIDOW-BIRD, female and male in non-breeding dress
*Coliuspasser albonotatus*     **1373**

# SPARROWS, WEAVERS, WAXBILLS, etc.

383 Top of head finely streaked, rump plain, bill larger: CHESTNUT WEAVER, female and male in non-breeding dress *Ploceus rubiginosus* **1327**

384 Tail graduated: 385–386

385 Under wing-coverts and edge of wing buffish white: YELLOW-MANTLED WIDOW-BIRD, female and male in non-breeding dress *Coliuspasser macrourus* **1372**

386 Under wing-coverts buff, edge of wing yellow: YELLOW-SHOULDERED WIDOW-BIRD, female and male in non-breeding dress *Coliuspasser macrocercus* **1371**

387 Mantle not heavily and broadly streaked: 388–389

388 Top of head blackish, broad buff stripe over eye, below, plain brown: COMPACT WEAVER, female and male in non-breeding dress *Pachyphantes pachyrhynchus* **1350**

389 Top of head streaked brown, no stripe over eye, below, streaked white and dusky black, bill large: GROSBEAK WEAVER, female *Amblyospiza albifrons* **1358**

---

390 Chin and throat black: 391–399

391 Top of head chestnut: SPANISH SPARROW, male *Passer hispaniolensis* **1297**

392 Top of head tawny: SOMALI SPARROW, male *Passer castanopterus* **1298**

| | | |
|---|---|---|
| 393 | Top of head grey or earth brown: | 394–416 |
| 394 | Stripe over eye chestnut: | HOUSE SPARROW, male *Passer domesticus* **1293** |
| 395 | Stripe over eye tawny: | 396–401 |
| 396 | Rump grey: | SOCOTRA SPARROW, male *Passer insularis* **1296** |
| 397 | Rump tawny: | 398–401 |
| 398 | Ear-coverts white, bordered with black: | RUFOUS SPARROW, male *Passer motitensis* **1294** |
| 399 | Ear-coverts grey: | KENYA RUFOUS SPARROW male *Passer rufocinctus* **1295** |
| 400 | Chin and throat chestnut: | CHESTNUT SPARROW, female *Sorella eminibey* **1306** |
| 401 | Throat spotted with black: | SPANISH SPARROW, female *Passer hispaniolensis* **1297** |
| 402 | Chin and throat dusky: | 403–406 |
| 403 | Stripe over eye buffish white: | 404–406 |
| 404 | Rump tawny: | KENYA RUFOUS SPARROW female *Passer rufocinctus* **1295** |
| 405 | Rump grey: | SOCOTRA SPARROW, female *Passer insularis* **1296** |
| 406 | Stripe over eye tawny: | RUFOUS SPARROW, female *Passer motitensis* **1294** |
| 407 | No black or dusky on chin and throat: | 408–416 |
| 408 | Chin and throat washed with yellowish: | SUDAN GOLDEN SPARROW female *Auripasser luteus* **1304** |
| 409 | Yellow spot at base of neck in front: | 410–413 |

# SPARROWS, WEAVERS, WAXBILLS, etc.

410 Mantle streaked: 411–412
411 Size larger, wing 82 mm. and over: YELLOW-THROATED PETRONIA
*Petronia superciliaris* 1307
412 Size smaller, wing 82 mm. and under: BUSH PETRONIA, female
*Petronia dentata* 1309
413 Yellow spot at base of neck in front, mantle plain: BUSH PETRONIA, male
*Petronia dentata* 1309
414 No yellow on chin and throat or at base of neck: 415–416
415 Wing shoulder chestnut: HOUSE SPARROW, female
*Passer domesticus* 1293
416 Wing shoulder earthy brown: SOMALI SPARROW, female
*Passer castanopterus* 1298

**BUFFALO-WEAVERS, SPARROW-WEAVERS, RUFOUS-TAILED WEAVER, SOCIAL WEAVERS.** FAMILY—PLOCEIDÆ.
Genera: *Bubalornis, Dinemellia, Plocepasser, Histurgops* and *Pseudonigrita*.

There are three species of Buffalo-Weaver, three Sparrow-Weavers, one Rufous-tailed Weaver and two Social Weavers in Eastern Africa. Buffalo Weavers were so named by the early travellers in South Africa who saw them following the herds of Buffalo.

**1284 BUFFALO-WEAVER.** *BUBALORNIS ALBIROSTRIS* (Vieillot).
*Coccothraustes albirostris* Vieillot, N. Dict. d'Hist. Nat. 13, p. 535, 1817: Senegambia.

**Distinguishing characters:** Adult male, wholly black with white bases to the feathers; narrow white edges to outer web of primaries; no white at base of flight feathers; bill white in the breeding season, black in the non-breeding season. The female is similar to the male but has a blackish bill. Wing 110 to 131 mm. The young bird is dull black or brownish black above; mottled or streaked dull black and white below; bill horn-colour.

**General distribution:** Senegal and Portuguese Guinea to Eritrea, Abyssinia and Uganda.

**Range in Eastern Africa:** The western Sudan to Eritrea, northern Abyssinia and Uganda.

**Habits:** Very Starling-like in appearance, feeding in flocks of ten or twelve on the ground, especially in cultivated areas. The flight is rather slow and hovering with deliberate wing movements, but the birds run fast on the ground and climb quickly in trees. They are chiefly notable for their noisiness and for the enormous communal nests they make, at times filling up the whole of a thorn tree till it looks like an umbrella. Nests are also built round those of large birds occasionally, and Dr. J. G. Myers records a number of small colonies almost touching that of a White-backed Vulture. The nests are kept in repair throughout the year and may be used for many seasons.

**Nest and Eggs:** The nest is a large mass of thorns and sticks common to the colony, in which are the individual domed grass nests of each pair. The latter are lined with fine grass and rootlets and are entered from below. Colonies may be long distances apart. Eggs in Darfur are given as three or four pale blue and rough; about 30 × 21 mm. In Karamoja, Uganda, also, Granvik noted blue eggs; about 26 × 18 mm. Over the whole of the rest of Africa this species lays Sparrow-like eggs of a whitish or pale green ground colour, blotched, mottled and streaked with ashy grey and brown, of measurements which fall between the two given above. The clutch is normally two or three. Is it possible that the blue eggs were really laid by some intruding species, possibly a Starling?

**Recorded breeding:** Nigeria, September. Darfur, Sudan, July to September. Equatoria, Sudan, August and September. Abyssinia and Somaliland, May.

**Food:** Seeds, grain, locusts and other insects.

**Call:** Noisy birds, keeping up a falsetto croaking call 'ee-stiguwi-stiguwi,' both in flight and when feeding and with a variety of chattering calls at the nest.

**1285** RED-BILLED BUFFALO-WEAVER. *BUBALORNIS NIGER* Smith.

*Bubalornis niger intermedius* (Cab.).

*Textor intermedius* Cabanis, J.f.O. p. 413, 1868: Kisuani, Usambara, north-eastern Tanganyika Territory.

**Distinguishing characters:** Differs from the Buffalo-Weaver in being a rather deeper black tone in general colour, but with white bases to the feathers; some white at base of flight feathers; bill yellow, scarlet, crimson or vermilion. The female is duller, often more brownish above; below, mottled dull black and white; bill blackish. Wing 108 to 127 mm. The young bird is similar to the adult female, but the young male has the bill red, orange or yellowish orange; the young female horn, tip orange.

**Range in Eastern Africa:** Southern Sudan, southern Abyssinia, British Somaliland, Kenya Colony and northern Tanganyika Territory.

**Habits:** Those of the preceding species; locally common but with very wide intervals between colonies.

**Nest and Eggs:** Apparently indistinguishable from those of the preceding species.

**Recorded breeding:** British Somaliland, March. Northern Kenya Colony, January. Western Kenya Colony, May to September, mostly July and August. Southern Kenya Colony, March to May. North-western Tanganyika Territory, March. Usambara Mts., Tanganyika Territory. a migrant to breed from March to June.

**Food:** Seeds and insects.

**Call:** A loud 'churr,' but probably many calls similar to those of the preceding species.

**Distribution of other races of the species:** Angola and South Africa, the nominate race being described from the Transvaal.

**1286** WHITE-HEADED BUFFALO-WEAVER. *DINEMELLIA DINEMELLI* (Rüppell).

*Dinemellia dinemelli dinemelli* (Rüpp.).

*Textor dinemelli* Rüppell, Syst. Uebers. p. 72, pl. 30, 1845: Shoa, central Abyssinia.

**Distinguishing characters:** Head and neck all round and underparts white; or sometimes top of head pale greyish white; mantle, wings, and tail dusky brown; basal half of primaries white; lower rump, upper and under tail-coverts and spot on bend of wing orange red. The sexes are alike. Wing 107 to 128 mm. The young bird is similar to the adult but the lower rump, upper and under tail-coverts and spot on bend of wing are orange. More or less albinistic individuals are not uncommon.

**Range in Eastern Africa:** The Sudan to Abyssinia, British Somaliland, Uganda and Kenya Colony.

**Habits:** Noticeable birds with a loud call, usually in small scattered flocks feeding largely on the ground among thorn trees. They fly ahead of one, settling on the lower boughs of the trees and do not look in the least like Weaver-Birds in the field; their call and habits are almost Parrot-like, and they are usually shy.

**Nest and Eggs:** The nests are hung from the boughs of thorn trees, often low down, and are usually in a scattered colony, though occasionally clustered together. They are rough untidy structures of thorny twigs, retort-shaped with the entrance from below, and with a warm inner lining of grass and a few feathers. Eggs three or four, greyish white or greyish green, streaked, marbled and blotched with olive, brown, grey and sepia or black, often capped at the larger end; about $25 \times 19$ mm.

**Recorded breeding:** Equatoria, Sudan, August and September. British Somaliland, March and April. Italian Somaliland, March and May. Southern Abyssinia, May. North-western Kenya Colony, May to September. Southern Kenya Colony, March and April.

**Food:** Seeds, fruit and insects.

**Call:** A harsh trumpet-like cry, freely uttered, also a bubbling twittering call.

*Dinemellia dinemelli boehmi* Reichw.
*Dinemellia boehmi* Reichenow, J.f.O. p. 372, 1885: Kakoma, Tabora district, central Tanganyika Territory.

**Distinguishing characters:** Differs from the nominate race in having the mantle, wings and tail black. Wing 126 to 136 mm.

**Range in Eastern Africa:** South-eastern Kenya Colony and Tanganyika Territory.

**Habits:** As for the nominate race.

**Recorded breeding:** Tanganyika Territory, November.

## 862 SPARROWS, WEAVERS, WAXBILLS, etc.

*Dinemellia dinemelli ruspolii* Salvad.
*Dinemellia ruspolii* Salvadori, Mem. Ac. Torino (2), 44, p. 558, 1894: Banan, about 110 miles west of Dolo, southern Abyssinia.

**Distinguishing characters:** Differs from the nominate race in having broad white edges to the outer webs of the secondaries and scapulars. Wing 107 mm.

**Range in Eastern Africa:** Southern Abyssinia and Italian Somaliland.

**Habits:** As for the nominate race.

**Recorded breeding:** No records.

### 1287 STRIPE-BREASTED SPARROW-WEAVER. *PLOCEPASSER MAHALI* Smith.

*Plocepasser mahali pectoralis* (Pet.).
*Philagrus pectoralis* Peters, J.f.O. p. 133, 1868: Inhambane, southern Portuguese East Africa.

**Distinguishing characters:** Top of head blackish; white superciliary streak broadening behind eye; sides of face and mantle russet earth-brown; lower rump and upper tail-coverts white; wings and tail blackish; broad white edges to wing-coverts; flight feathers edged with brown; tail edged and tipped with white; moustachial streak black; below, white; chest feathers with brown centres. The sexes are alike. Wing 90 to 108 mm. The young bird is similar to the adult, but the bill is blackish and horn-colour, not wholly black.

**General distribution:** Kenya Colony, Tanganyika Territory, Northern Rhodesia, Nyasaland and Portuguese East Africa.

**Range in Eastern Africa:** Eastern Kenya Colony and eastern and southern Tanganyika Territory to the Zambesi River.

**Habits:** Noisy chattering Sparrow-like birds with a white rump noticeable in flight, common in dry bush and desert country, but also found round villages and native crops, or in short grass plains. Local, but abundant where they occur, and their shrill harsh whistling cries render them conspicuous. They feed on the ground in flocks, and keep about their nesting colonies throughout the year.

**Nest and Eggs:** The nest is a large untidy globe of dry grass with a neck pointing downwards, but there is frequently a second opening on the other side which is used as a roosting chamber. Nests are often in large colonies usually on one side of a group of trees, but occasionally covering them. Eggs two, rarely three, cream or pinkish white with reddish freckles or spots and greyish undermarkings often in a zone at the larger end; about 24 × 15 mm.

**Recorded breeding:** Central Kenya Colony, June. Tanganyika Territory, September, possibly also May. Nyasaland, August, also October and November. Rhodesia, November to January, also at the end of March.

**Food:** Seeds, grain and insects.

**Call:** A chorus of harsh chattering and bubbling whistles in a colony. A loud constantly repeated alarm note. The male has a pleasing varied song in the breeding season, often singing after dark.

*Plocepasser mahali melanorhynchus* Bp.
*Plocepasser melanorhynchus* Bonaparte, Consp. Gen. Av. 1, p. 444, 1850: Shoa, central Abyssinia.

**Distinguishing characters:** Differs from the preceding race in having the mantle earth brown; sides of face blackish and brown centres to chest feathers absent or confined to sides of chest. Wing 90 to 105 mm.

**Range in Eastern Africa:** Central and eastern Abyssinia and the southern Sudan to Uganda and Kenya Colony, but not eastern areas.

**Habits:** As for the last race, very local but abundant where found. Granvik notes that the nest has only one entrance when there are eggs, but another is opened on the other side when the young hatch, so that there is a curved tube through the nest. MacInnes says that each nest he found with eggs was below a roosting nest. Tomlinson says that it would be impossible to lose direction in northern Kenya Colony as all the nests of this species are consistently on the west side of the trees.

**Recorded breeding:** Equatoria, Sudan, June, September and February. Uganda, April and August. Suk and Turkana, northwestern Kenya Colony, June and July. Kenya Colony highlands, August and September.

## SPARROWS, WEAVERS, WAXBILLS, etc.

*Plocepasser mahali propinquatus* Shelley.

*Plocepasser propinquatus* Shelley, Ibis p. 6, 1887: Bardera, Juba River, southern Italian Somaliland.

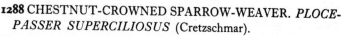

**Distinguishing characters:** Differs from the last race in being smaller. Wing 80 to 87 mm.

**Range in Eastern Africa:** Southern Italian Somaliland.

**Habits:** As for other races.

**Recorded breeding:** No records.

**Distribution of other races of the species:** Angola and South Africa, the nominate race being described 'from between the Orange River and the Tropic.'

**1288** CHESTNUT-CROWNED SPARROW-WEAVER. *PLOCEPASSER SUPERCILIOSUS* (Cretzschmar).

*Plocepasser superciliosus superciliosus* (Cretz.).

*Ploceus superciliosus* Cretzschmar, in Rüppell's Atlas, Vögel, p. 24, pl. 15, 1827: Kordofan, Sudan.

**Distinguishing characters:** Top of head and sides of face chestnut; superciliary stripe white; malar stripe black; mantle and rump earth brown washed with russet; wing-coverts tipped with white; below, throat white; rest of underparts greyish white. The sexes are alike. Wing 82 to 96 mm. The young bird is similar to the adult, but is more fluffy in appearance and slightly paler chestnut on the head.

**General distribution:** Senegal to the Sudan.

**Range in Eastern Africa:** The Sudan as far east as Roseires and Fazogli.

**Habits:** Rather inconspicuous birds in parties of four or five in woodland or high bush, local and not usually common. They may be recognized by their call, which is a Bunting-like sibilant trill. Little is recorded of their habits.

**Nest and Eggs:** A rather coarse domed nest of grass and leaves lined with feathers placed along or at the end of an Acacia branch, but not dependent from it. Eggs two, pinkish cream or reddish white thickly spotted or blotched with pinkish or mauve brown, and with pale lilac undermarkings; about $23 \times 15$ mm.

**Recorded breeding:** Nigeria, April to July. Sudan, February and June.

**Food:** No information.

**Call:** As above, a skirling trill.

*Plocepasser superciliosus brunnescens* Grote.
*Plocepasser superciliosus brunnescens* Grote, J.f.O. p. 399, 1922: Bosum, Ubangi-Shari, French Equatorial Africa.

**Distinguishing characters:** Differs from the nominate race in having a warmer wash of russet on the mantle. Wing 81 to 95 mm.

**General distribution:** French Equatorial Africa to the Sudan, Abyssinia and Uganda.

**Range in Eastern Africa:** South-western Sudan to Walamo and Gofa, western Abyssinia and north-western Uganda.

**Habits:** As for the nominate race.

**Recorded breeding:** Uganda, September, also a nestling in May.

*Plocepasser superciliosus bannermani* Grant & Praed.
*Plocepasser superciliosus bannermani* C. Grant & Mackworth-Praed, Bull. B.O.C. 64, p. 18, 1943: Gomit River, Big Abbai, 60 miles south of Lake Tana, northern Abyssinia.

**Distinguishing characters:** Differs from the nominate race in having the mantle earth brown. Wing 86 to 97 mm.

**Range in Eastern Africa:** Eritrea to eastern and southern Abyssinia, south-eastern Sudan and north-western Kenya Colony.

**Habits:** As for the nominate race, uncommon and local; Cheesman notes that it looks like a red-headed Sparrow.

**Recorded breeding:** Abyssinia, May. North-western Kenya Colony, August.

**Distribution of other races of the species:** Cameroons.

**1289 DONALDSON SMITH'S SPARROW-WEAVER.** *PLOCEPASSER DONALDSONI* Sharpe.

*Plocepasser donaldsoni* Sharpe, Bull. B.O.C. 5, p. 14, 1895: near Lasamis, between Lake Rudolf and the northern Guaso Nyiro, northern Kenya Colony.

**Distinguishing characters:** Above, pale brown; top of head mottled with dusky; lower rump and upper tail-coverts white; ear-

coverts blackish and pale brown; narrow black moustachial stripe; below buffish white with brownish mottling from chest to flanks. The sexes are alike. Wing 89 to 95 mm. The young bird is browner above with pale tawny edges to the feathers of the mantle and flight feathers.

**Range in Eastern Africa:** South-western Abyssinia and northern Kenya Colony, as far south as Isiolo.

**Habits:** A scarce and little known bird inhabiting the hottest spots of rocky desert country and of very local occurrence. Usually in scattered flocks and associates little with other species.

**Nest and Eggs:** Grass nests with short entrance tubes and lined with feathers, in colonies identical with those of the Stripe-breasted Sparrow-Weaver. Eggs pinkish or greyish brown, very finely and diffusely speckled with mauve and reddish brown; about 24 × 15 mm.

**Recorded breeding:** North-western frontier of Kenya Colony, November and December, also June and July.

**Food:** Mainly grass seeds.

**Call:** A soft double cluck, a loud Parrot-like 'chink-chink,' and the male has a pleasant little song.

**1290 RUFOUS-TAILED WEAVER.** *HISTURGOPS RUFICAUDA* Reichenow.

*Histurgops ruficauda* Reichenow, J.f.O. p. 67, 1887: Wembere Steppe, central Tanganyika Territory.

**Distinguishing characters:** General colour mottled brownish and dusky grey; inner webs of flight feathers, except ends, tawny; tail tawny chestnut, central feathers blackish brown. The sexes are alike. Wing 120 to 128 mm. The young bird is rather browner, especially below, than the adult.

**Range in Eastern Africa:** Northern Tanganyika Territory from the Mwanza to the Arusha, Tabora and Mkalama districts.

**Habits:** Not uncommon in many parts of northern Tanganyika Territory where it occurs in noisy sociable flocks. Feeds on the ground and walks slowly and deliberately. So far little is recorded of it, but it prefers wooded hills and associates with Fischer's Love-Bird.

**Nest and Eggs:** A roughly made round grass nest in the fork of a bough or hanging from it, and there is an entrance spout at the

**Food:** No information.

**Call:** Unrecorded.

*Note:* From the absence of any further specimens from this well-known area, it seems doubtful if this can be a species. It is most likely a partially melanistic phase of the Masked Weaver, *Ploceus intermedius intermedius* Rüppell.

## 1321 VITELLINE MASKED WEAVER. *PLOCEUS VITELLINUS* (Lichtenstein).

*Ploceus vitellinus vitellinus* (Licht.). **Pls. 85, 88.**
*Fringilla vitellina* Lichtenstein, Verz. Doubl. p. 23, 1823: Senegambia.

**Distinguishing characters:** Adult male in breeding dress very similar to the Northern Masked Weaver, but with a smaller bill and the black confined to the chin and upper throat, not extending down the front of the neck; inner webs of flight feathers yellow. In non-breeding dress the whole upperside is olivaceous green with blackish streaks; below, chin to breast and under tail-coverts pale yellow; belly white; bill horn-colour. The female in breeding dress has the upperparts olivaceous or yellowish green; mantle and scapulars streaked with blackish; below, pale yellow, including under tail-coverts; chest and flanks buff and yellow; centre of belly white; bill horn. In non-breeding dress the mantle is less olivaceous; breast to under tail-coverts white; little or no buff on chest and flanks; bill horn-colour. Wing, male 68 to 75, female 66 to 70 mm. The young bird is similar to the adult female in non-breeding dress, but the upperparts are less olivaceous and the sides of throat and chest are buff not yellow.

**General distribution:** Senegal to the Sudan.

**Range in Eastern Africa:** The Sudan from Darfur to Sennar.

**Habits:** Common and abundant species of wild country, nesting in small colonies in Acacias, and often found in thick bush. The nests are often built by males in immature dress. Occasionally two females share a nest.

**Nest and Eggs:** Nests are heart-shaped, or blunt pear-shaped, with a short entrance funnel from below at the stalk of the pear, made of grass. They are built by the males only and are suspended from an Acacia twig, often three to four on one twig. They make numerous 'cocks' nests also. Eggs two to four, extremely variable.

the main types being white or pinkish white with brown or rufous spots and blotches, and greenish blue spotted with brown and chocolate with mauve undermarkings; about 20 × 13·5 mm.

**Recorded breeding:** Nigeria, June and July. Air, September, Sudan, June to August and November.

**Food:** Seeds, worms and insects.

**Call:** The male has a long drawn wheezy call note.

*Ploceus vitellinus uluensis* (Neum.).
*Hyphantornis vitellinus uluensis* Neumann, J.f.O. p. 282, 1900: Ulu Mts., Machakos, Kenya Colony.

**Distinguishing characters:** Adult male, differs from the nominate race in having more black on the forehead. Wing, male 68 to 75, female 61 to 71 mm.

**Range in Eastern Africa:** Abyssinia and British Somaliland to southern Sudan, Uganda, Kenya Colony and northern Tanganyika Territory as far south as Shinyanga, Monduli and Bagamoyo.

**Habits:** As for the nominate race, but the nests have usually a longer entrance funnel, almost a tail of green grass, and the colonies may be large.

**Recorded breeding:** Eastern Abyssinia, April and May. British Somaliland (nestling), September. Turkana, Kenya Colony, June and July. Southern Kenya Colony, March to May. Northern Tanganyika Territory, May.

**1322 SOUTHERN MASKED WEAVER. *PLOCEUS VELATUS*** Vieillot.
*Ploceus velatus tahatali* Smith. **Pl. 87, 89.**
*Ploceus tahatali* A. Smith, Rep. Exp. C. Afr. p. 50, 1836: Marico River, western Transvaal, South Africa.

**Distinguishing characters:** The adult male in breeding dress has a narrow black band on forehead; sides of face including ear-coverts, chin and throat black, with black of throat extending on to lower neck; top of head more or less saffron to chestnut; nape and underparts golden yellow; mantle yellow green streaked with dusky; tail dusky green; chest washed with saffron; edges of inner webs of flight feathers yellow; bill black. In non-breeding dress the head, sides of face and mantle are olive-yellow, the latter streaked with dusky; wings edged with pale yellow; rump and tail yellow, less bright than in breeding dress; below, chin to chest pale yellow; belly white; bill horn-colour. The female in breeding dress is yellow green above, mantle streaked with dusky; superciliary stripe narrow and yellow; below, bright yellow, deeper in tone on chest; bill horn. In non-breeding dress the female is similar to the male in non-breeding dress, but is rather more olive and ashy above and paler yellow from chin to chest. Wing, male 72 to 80, female 64 to 74 mm. The young bird is similar to the adult female in non-breeding dress.

**General distribution:** Angola to Nyasaland, the Rhodesias and the western Transvaal.

**Range in Eastern Africa:** Nyasaland.

**Habits:** An abundant species of the southern half of Rhodesia, less plentiful in the north, but locally common in Nyasaland and Portuguese East Africa. It is gregarious at all times, active and usually tame, most frequently found in low lying country.

**Nest and Eggs:** Nest in colonies in any sort of situation. Nest strong and with a warm lining compactly made of coarse grass lined with finer grass and seed heads, and without a spout, rather triangular or kidney-shaped in section. Eggs two or three, variable, white, blue or pale blue, also cream or greenish blue spotted with brown or red; about 21 × 14 mm.

**Recorded breeding:** Nyasaland, January to April. Portuguese East Africa, March. Rhodesia, September to March.

**Food:** Common on grain crops, and also eat much fruit, the feathers of the underside being often stained pinkish with fruit juice.

**Call:** Noisy chattering at the breeding colonies, and the male has a hissing note.

**Distribution of other races of the species:** South Africa, the nominate race being described from Namaqualand.

**1323** BAGLAFECHT WEAVER. *PLOCEUS BAGLAFECHT* (Daudin).
*Ploceus baglafecht baglafecht* (Daud.). **Pl. 85.**
*Loxia baglafecht* Daudin, in Buffon, Hist. Nat. (Didot's ed.), Quad. 14, p. 245, 1802: Eritrea.

**Distinguishing characters:** Adult male in breeding dress has the forehead and forecrown yellow; crown and nape yellowish green; mantle green with faint dark streaks; rump green; tail brownish washed with green; wings edged yellowish green; lores and round eye to ear-coverts black; chin sometimes black; side of neck and underside yellow; belly often whitish; edges of inner webs of flight feathers buff or buffish yellow; bill black. In non-breeding dress the upperside and sides of face are ashy with a buffish wash, some dull black on the forehead; lores, around eyes, ear-coverts and mantle streaked with dusky black; tail green; wing edging greenish; below, buff; bill black. The female in breeding dress differs from the male in having the forehead and forecrown green; lores and round eye to ear-coverts duller black with a greenish wash. The non-breeding dress is similar to that of the male. Wing, male 75 to 84, female 73 to 81 mm. The young male has the upperside and sides of face green; mantle streaked with dusky; below, yellow, buffish on lower flanks, lower belly whitish; the young female has the top of the head olivaceous; mantle brown streaked with dusky; below, buff; chin yellow; centre of belly whitish.

**Range in Eastern Africa:** Eritrea, Abyssinia and the Boma Plateau of the south-eastern Sudan.

**Habits:** A common resident of the high plateau of Abyssinia, nesting singly and generally near water. The cock makes a separate nest to sleep in.

**Nest and Eggs:** Nest, flask or retort-shaped, of green grass, rather large with a short spout, strongly made and with a pad of grass heads or feathers inside. Eggs two, pinkish white, or more commonly bluish green, handsomely spotted and blotched with rufous or purplish brown and with lilac undermarkings; about 22 × 17 mm.

**Recorded breeding:** Abyssinia, June to September.

**Food:** No information.

**Call:** A vigorous wheezy song by the male in the breeding season is the only call recorded.

*Ploceus baglafecht eremobius* (Hartl.).
*Symplectes eremobius* Hartlaub, Zool. Jahrb. 2, p. 320, 1887: Chor Mabrué, at lat. 4° 3′ N.; long. 29° 25′ E. north-eastern Belgian Congo.

**Distinguishing characters:** Adult male, differs from the nominate race in being smaller and having the lower breast to under tail-coverts white. Female plumage undescribed. Wing 74 to 77 mm.

# SPARROWS, WEAVERS, WAXBILLS, etc.

**General distribution:** North-eastern Belgian Congo to the Sudan and Uganda.

**Range in Eastern Africa:** South-western Sudan at Li Rangu and western Uganda at Unyoro.

**Habits:** As for other races.

**Recorded breeding:** No records.

**Distribution of other races of the species:** Banso Mts., Cameroons.

## 1324 LITTLE WEAVER. *PLOCEUS LUTEOLUS* (Lichtenstein).

*Ploceus luteolus luteolus* (Licht.). **Pls. 85, 88.**
*Fringilla luteola* Lichtenstein, Verz. Doubl. p. 23, 1823: Senegal.

**Distinguishing characters:** Adult male in breeding dress, forehead, forecrown, sides of face, ear-coverts and chin to throat black; crown, sides of neck, upper tail-coverts and underparts canary yellow; mantle greenish yellow more or less streaked; wings and tail dusky edged with yellow; bill black. In non-breeding dress the top of the head is green, mantle buff streaked with dusky; sides of face and chin to chest and flanks buff; breast to belly white; bill dusky horn-colour. The female in breeding dress is yellow below; forehead green uniform with top of head; sides of face and chin to chest pale yellow; bill black. In non-breeding dress the head is dull olivaceous; mantle ashy streaked with dusky; sides of face and under tail-coverts yellowish; rest of underparts white washed with buff on chest and flanks; bill horn. Wing, male 58 to 65, female 58 to 65 mm. The young bird is similar to the adult female in non-breeding dress.

**General distribution:** Senegal to Eritrea, Abyssinia, the Sudan and Kenya Colony.

**Range in Eastern Africa:** Central and southern Sudan to Eritrea, Abyssinia and north-western Kenya Colony.

**Habits:** A common and gregarious species of thorn-scrub, woodland or gardens, the commonest Weaver of the red sandstone country of Kordofan, gathering into very large flocks on cultivated ground. The nests are suspended from thorn trees and there is a strong tendency toward a wasp association—in some localities all the nests are in trees which also contain wasps' nests.

**Nest and Eggs:** Nests shaped like a spherical pipe with the stem pointing downwards, the entrance funnel running up towards the top of the bowl, and there is a sort of gate inside to prevent the eggs falling down the tube as the nest sways in the wind. It is constructed of delicate grass lattice work, in the Sudan usually unlined or with a fine grass lining, in Abyssinia with a lining of hair or other soft material. Eggs two or three, white and oval; about 18 × 13 mm.

**Recorded breeding:** Timbuktu, French Sudan, August. Kano, Nigeria, June to August. Darfur and other parts of Sudan, July to November. Abyssinia, August.

**Food:** No information.

**Call:** The male has a varied little song in the breeding season, interspersed with jarring notes.

*Ploceus luteolus kavirondensis* (V. Som.).
*Sitagra luteola kavirondensis* Van Someren, Bull. B.O.C. 41, p. 123, 1921: Soronko River, Mt. Elgon, Uganda.

**Distinguishing characters:** Differs from the nominate race in being rather greener above with more pronounced streaking, and with less yellow on occiput. Wing, male 58 to 67, female 58 to 65 mm.

**Range in Eastern Africa:** Uganda, western Kenya Colony and northern Tanganyika Territory.

**Habits:** As for the nominate race.

**Recorded breeding:** Western Kenya Colony, July to September.

**1325 YELLOW-BACKED WEAVER.** *PLOCEUS CAPITALIS* (Latham).
*Ploceus capitalis dimidiatus* (Ant. & Salvad.). **Pls. 85, 88.**
*Hyphantornis dimidiatus* Antinori & Salvadori, Atti. R. Accad. Torino 8, p. 360, 1873: Kassala, eastern Sudan.

**Distinguishing characters:** Adult male in breeding dress, head and neck all round black; nape and rump yellow; mantle and wing-coverts olivaceous yellow; wings edged yellow; tail olivaceous; below, chest chestnut; rest of underparts yellow, with more or less chestnut on breast and flanks; edges of inner webs of flight feathers bright yellow; bill black. In non-breeding dress upperside brown; head olivaceous yellow contrasting with mantle; superciliary stripe yellow; mantle and scapulars broadly streaked with black; below, white; chest and flanks buff; chin, throat and chest sometimes washed with yellow; bill, upper mandible blackish, lower horn-colour. The female in non-breeding dress is similar to the male in the same dress but smaller. In breeding dress the top of the head is less yellowish and the sides of the face buff. Wing, male 68 to 82, female 65 to 76 mm. The young bird is similar to the adult female but has the top of the head slightly more olivaceous, the male being usually larger than the female.

**General distribution:** Eritrea, Abyssinia, the Sudan, eastern Belgian Congo, Uganda, Kenya Colony and Tanganyika Territory.

**Range in Eastern Africa:** Western Eritrea, north-western Abyssinia to the southern Sudan, Uganda, western Kenya Colony and the Mwanza district of northern Tanganyika Territory.

**Habits:** Common and abundant species in Uganda, very gregarious and nesting in colonies with other species of Weavers, usually over or near water. They often start a colony and desert it suddenly after a few days.

**Nest and Eggs:** The nests, which are roughly kidney-shaped and without a funnel, are built mainly by the male, the female helping with the lining, and are suspended from trees, bushes, elephant-grass, or papyrus. They are of grass lined with broad reed leaves, with an inner lining of fine grass. Eggs usually two, most variable, white, pink, brown, chocolate, liver colour, terra cotta, or various shades of green, with or without brown or purplish spotting, and rather hard shelled; about 22 × 14·5 mm.

**Recorded breeding:** Uganda, February to July, October and November. Northern Tanganyika Territory, January to April and irregularly later in the year.

**Food:** Grass seeds, with some insect food at times.

**Call:** Described as somewhat harsher than those of other species usually found with it.

*Ploceus capitalis dichrocephalus* (Salvad.). **Pl. 86.**
*Hyphantornis dichrocephalus* Salvadori, Ann. Mus. Genova (2) 16, p. 45, 1896: Italian Somaliland.

**Distinguishing characters:** Adult male, differs from the preceding race in having the head and sides of face blackish chestnut, top of head darker; nape and sides of neck chestnut not yellow; below, chin and throat dusky saffron. The female has the chin to chest and sides of face and neck pale yellowish; flanks slightly buffish. Wing, male 71 to 79, female 66 to 68 mm. The young bird is similar to that of the preceding race.

**Range in Eastern Africa:** Southern Abyssinia, southern Italian Somaliland and north-eastern Kenya Colony.

**Habits:** As for the preceding race.

**Recorded breeding:** Southern Abyssinia, March.

**Distribution of other races of the species:** Nigeria to French Equatorial Africa, the nominate race being described from Nigeria.

**1326** GOLDEN-BACKED WEAVER. *PLOCEUS JACKSONI* Shelley. **Pls. 85, 88.**

*Ploceus jacksoni* Shelley, Ibis p. 293, pl. 7, 1888: Lake Jipe, north-eastern Tanganyika Territory.

**Distinguishing characters:** The adult male in breeding dress is very similar to the Yellow-backed Weaver but differs in that the black of the head extends on to the nape; yellow of mantle more golden, less olivaceous; chestnut much darker and deeper in tone. In non-breeding dress the upperparts are more olivaceous than the non-breeding dress of the Yellow-backed Weaver; below, wholly canary yellow, lower belly white. The female is similar to the male in non-breeding dress but smaller. Xanthochroic examples are occasionally found. Wing, male 71 to 81, female 67 to 72 mm. The young bird has the chest and flanks buffish.

**Range in Eastern Africa:** South-eastern Sudan, Uganda, western Kenya Colony, Tanganyika Territory as far south as the Lugufu River and Iringa and Urundi, Belgian Congo.

**Habits:** A river swamp or lake-side species, nesting in long grass, reeds, Ambatch or Acacias over water in small colonies. They are very numerous in some localities and in common with other Weavers gather into flocks that are a formidable pest on native crops.

**Nest and Eggs:** The nest, built by the cock, is compact and strongly made of grass lined with fine grass, suspended from boughs or reeds, and oval or spherical in shape. Eggs two or three, turquoise blue with spots and blotches of reddish brown and with pale mauve undermarkings; about 21 × 14·5 mm.

**Recorded breeding:** Lake Victoria, January to May, also November and December in wet seasons. Kenya Colony, in many months. Elgeyu, July and August. Karamoja, June and July. Tanganyika Territory, January to May, also September onwards.

**Food:** Grain and weed seeds, insects mainly in breeding season.

**Call:** Nothing particularly recorded.

**1327 CHESTNUT WEAVER.  *PLOCEUS RUBIGINOSUS*** Rüppell.

*Ploceus rubiginosus rubiginosus* Rüpp. **Pls. 85, 89.**
*Ploceus rubiginosus* Rüppell, N. Wirbelt. Vög. p. 93, pl. 33, fig. 1, 1840: Temben province, northern Abyssinia.

**Distinguishing characters:** Adult male, head and neck all round black; rest of plumage chocolate; wings and tail black edged with buff; edges of inner webs of flight feathers pale buff; bill black. The non-breeding dress is buffish brown above, broadly streaked, except on rump, with black; wings including wing-coverts edged with buffish white; stripe over eye and sides of face buff; chest and flanks pale tawny; throat and belly white; edges of inner webs of flight feathers pale buff; bill dark horn. In immature dress the whole of the head, throat and chest are pale chestnut; mantle pale chestnut with black streaks. The female is similar to the male in non-breeding dress, but is smaller and the chest and flanks are buff. Wing, male 81 to 86, female 77 to 81 mm. The young bird is similar to the adult male in non-breeding dress.

**Range in Eastern Africa:** Eritrea, Abyssinia, British and Italian Somaliland, south-eastern Sudan, Kenya Colony as far west as Mt. Maroto and Nairobi and Tanganyika Territory as far south as the Dodoma district.

**Habits:** Very local and possibly migratory species, common in the Acacia bush of Turkana, Kenya Colony. The males are not present at the breeding colonies, and appear to live in bachelor parties among high Acacias elsewhere. The breeding colonies of females are exceedingly noisy.

**Nest and Eggs:** Nests retort-shaped, stout and very untidy with spikes of protruding grass, in enormous numbers in closely packed colonies. Eggs three or four, turquoise blue, pale greenish blue, or white; about 22 × 15 mm.

**Recorded breeding:** Central Abyssinia, October. Kenya Colony May to July. Northern Tanganyika Territory, March to June.

**Food:** Small seeds and insects.

**Call:** Nothing noted except the intense chattering of the breeding colonies.

**Distribution of other races of the species:** Angola and South West Africa.

**1328** CINNAMON WEAVER. *PLOCEUS BADIUS* (Cassin).
*Ploceus badius badius* (Cass.). **Pls. 86, 89.**
*Hyphantornis badius* Cassin, Prov. Acad. Sci. Philad. 5, p. 57, 1850: Fazogli, Blue Nile, eastern Sudan.

**Distinguishing characters:** Adult male in breeding dress, head and neck in front to upper chest black; mantle chestnut; rump mixed chestnut and golden yellow; wings edged with golden yellow; tail olivaceous edged on inner webs with yellow; below, chest and breast chestnut; belly and under tail-coverts yellow; edges of inner webs of flight feathers and axillaries golden yellow; thighs yellow; bill black. In non-breeding dress brown above; head green; superciliary stripe yellow; mantle and scapulars broadly streaked with black; wing-coverts broadly edged with green; below, white; chest and flanks buff; bill blackish, lower mandible horn. The female is similar to the male in non-breeding dress, but is smaller and has a more olivaceous green head. It is not dissimilar in general appearance to the female of the Chestnut Weaver from which it can be distinguished by the green top to the head, the yellow outer and inner webs to the flight feathers and the yellow edged tail. Wing, male 75 to 78, female 65 to 70 mm. The young bird is similar to the adult in non-breeding dress.

*Note:* The male in non-breeding dress and the adult female can only be distinguished from the Yellow-backed Weaver in similar dress by the clear buff brown of the chest, which is tinged with yellowish in the Yellow-backed Weaver.

**Range in Eastern Africa:** The Blue Nile Valley, Sudan.

**Habits:** Nothing has been noted of this species except that it occurs in large flocks and feeds on grassy plains. They assume breeding plumage in May and June.

**Nest and Eggs:** Undescribed.

**Recorded breeding:** No records.

**Food:** No information.

**Call:** Unrecorded.

*Ploceus badius axillaris* (Heugl.).

*Hyphantornis axillaris* Heuglin, J.f.O. p. 381, 1867: Kidj-Neger district, near Shambé, southern Sudan.

**Distinguishing characters:** Adult male, differs from the nominate race in having the black extending on to the hind neck and the belly and under tail-coverts chestnut. The female has a duller green head than the female of the nominate race and usually blacker streaks on the mantle. Wing, male 70 to 73, female 63 to 65 mm.

**Range in Eastern Africa:** Southern Sudan.

**Habits:** No information.

**Recorded breeding:** No records.

**1329 RÜPPELL'S WEAVER.** *PLOCEUS GALBULA* Rüppell.
**Pls. 86, 89.**

*Ploceus galbula* Rüppell, N. Wirbelt. Vög. p. 92, pl. 32, fig. 2, 1840: Modat Valley, Eritrea.

**Distinguishing characters:** Adult male in breeding dress, forehead, sides of face and chin chestnut; rest of head and underside golden yellow; mantle and rump yellowish green, former streaked with dusky; wings and tail dusky black edged with yellow; inner webs of flight feathers edged with yellow; bill black. In non-breeding dress the crown and nape are yellowish green uniform with the mantle; bill less black. The female is ashy brown above streaked with dusky; rump and tail olivaceous; flight feathers edged with yellow; below, chin to chest buff; rest of underparts white; flanks ashy; bill horn-colour. Wing, male 66 to 79, female 65 to 75 mm. The young bird is similar to the adult female, but is more buffish yellow on the chest.

**General distribution:** Sudan to Eritrea, Abyssinia, British Somaliland and western Arabia.

**Range in Eastern Africa:** Central (between Shendi and El Duem) to eastern Sudan, Eritrea, Abyssinia and British Somaliland.

**Habits:** A species which congregates into the most enormous flocks of hundreds of thousands of birds which look like a locust swarm in the distance. Abundant and often swarming in every little patch of bush where they occur. They would do very great damage to crops in a cultivated area.

**Nest and Eggs:** Pear-shaped nests with side openings, generally hanging from Acacias, preferably in the neighbourhood of water if available. Eggs two to four, pinkish white, pale green or blue, either plain or spotted with brown, black or violet; about 21 × 15 mm.

**Recorded breeding:** Gedaref, Sudan, from May onwards, main body not till July or August. Red Sea coast, March to May. Southern Arabia, January and February. Anseba Valley, Eritrea, August. Eastern Abyssinia, February, May, June and August. British Somaliland, February and March.

**Food:** No information, but probably mainly grass seeds.

**Call:** A wheezy little song like a creaky hinge.

**1330** NORTHERN BROWN-THROATED WEAVER. *PLOCEUS CASTANOPS* Shelley. **Pls. 86, 89.**

*Ploceus castanops* Shelley, P.Z.S. p. 35, 1888: Wadelai, north-western Uganda.

**Distinguishing characters:** Adult male, very similar to Rüppell's Weaver, but bill less heavy and more sharply pointed; chestnut on chin extending down front of neck and first primary much longer; inner webs of flight feathers much paler yellow; bill black. There is no non-breeding dress. The female has the head and mantle streaked brown and black; lores and around eye blackish; wings blackish, edged with yellow; tail rather darker than the female of Rüppell's Weaver; below, darkish yellowish buff; edges of inner webs of flight feathers as in male; bill black. Wing, male 70 to 77, female 66 to 69 mm. The young bird is similar to the adult female, but has a warmer buff tone on the chest and a horn-coloured bill. The young male is larger than the female.

**Range in Eastern Africa:** Uganda and Belgian Ruanda.

**Habits:** A bird of very different habits to the normal run of Weavers. It nests in colonies in reed beds, elephant grass or low scrub and is often seen feeding over the water and searching lily pads for food. In the non-breeding season it betakes itself to the forest and creeps about among boughs, feeding on insects like a Warbler.

**Nest and Eggs:** Nests are made of broad grass strips with a lining of fine grass and feathers, rather round, and usually with a distinct porch over the entrance. There is, however, some discrepancy in the descriptions of these nests. Eggs two or three, very variable, pinkish white or pale blue, either plain or with any degree of reddish brown spotting; about 22 × 14·5 mm.

**Recorded breeding:** Uganda, February to August, also November and December.

**Food:** Largely insects.

**Call:** Unrecorded.

### 1331  GOLDEN PALM-WEAVER.  *PLOCEUS BOJERI* (Cabanis). Pls. 86, 89.

*Hyphantornis bojeri* Cabanis, in van der Decken's Reise. Vög. 3, p. 32, 1869: Mombasa, eastern Kenya Colony.

**Distinguishing characters:** Adult male, general colour golden yellow; whole of head and throat orange; lower neck in front saffron; mantle, wings and tail washed with olivaceous; bill black. There is no non-breeding dress. The female is mustard yellow above with indistinct dusky streaks on mantle; wings dusky with yellow edges to feathers; tail olivaceous yellow; below, bright yellow; chest more chrome yellow; bill horn-colour, upper mandible darker. Wing, male 74 to 76, female 66 to 71 mm. The young bird is similar to the adult female but is paler yellow below.

**Range in Eastern Africa:** Southern Italian Somaliland, eastern Kenya Colony and north-eastern Tanganyika Territory, but confined to the coastal areas further south; also Manda Island.

**Habits:** Much confused in the past with the Golden Weaver, and notes are not reliable in most cases, but according to various observers it has certain characteristics of breeding habits and call which will identify a colony in the field. Noisy and gregarious but rather local species.

**Nest and Eggs:** Nests are roundish ovals without porch or spout, and are hung from boughs or from leaves of palm trees. Eggs two or three, rich glossy dark olive green with indistinct dark mottling; about 21·5 × 14 mm.

**Recorded breeding:** Kenya Colony coast freely, but dates need checking owing to confusion mentioned above. Nests of this species have been recorded from March to May and August to November.

**Food:** No information.

**Call:** Little exactly recorded beyond the sparrowy chattering of the breeding colony, but see under the Taveta Golden Weaver.

**1332** SOUTHERN BROWN-THROATED WEAVER. *PLOCEUS XANTHOPTERUS* (Finsch & Hartlaub). **Pls. 86, 89.**

*Hyphantornis xanthopterus* Finsch & Hartlaub, Vög. Ost. Afr. p. 399, 1870: Shupanga, Shiré River, Nyasaland.

**Distinguishing characters:** Adult male in breeding dress, general colour golden yellow; sides of face and chin to neck chestnut; mantle sometimes greenish; some variable dusky markings in wings and tail; bill black. In non-breeding dress the top of head and neck are olivaceous; mantle and rump buffish brown, former with black streaks; wings and tail as in the male in breeding dress; faint superciliary stripe; sides of face and chin to chest buff with a yellow wash; flanks buff; belly white; bill horn-colour, upper mandible blackish. The female in breeding dress is olivaceous above washed with yellow; rump buffish brown; wings dusky edged with bright yellow above and below, including upper wing-coverts; tail olivaceous with yellow edges; sides of face and chin to chest yellow, brighter than in non-breeding dress; flanks buff; belly white; bill as for male in non-breeding dress. The female in non-breeding dress is similar to the male in non-breeding dress, but is smaller; wings dusky with yellow edges to feathers; tail olivaceous with yellow edges; bill unchanged. Wing, male 70 to 77, female 61 to 67 mm. The young bird is similar to the adult female but has the sides of the face, body and flanks brownish.

**General distribution:** Nyasaland and northern Portuguese East Africa, Bechuanaland, Southern Rhodesia, the Transvaal, Natal and Zululand.

**Range in Eastern Africa:** Northern Portuguese East Africa to the Zambesi River.

**Habits:** Little recorded, and probably differing little from those of allied species with which it often consorts, and breeds in close proximity.

**Nest and Eggs:** Small well-woven nests without porch or spout, of grass lined with fine grass and seed heads, and generally over or near water. Nests often in reeds, but also dependent from bushes or trees over water. Eggs usually two, are normally pale green mottled dark brown or greyish brown. Other variations are pale brick colour, bluish green, or reddish white, plain or blotched and speckled with rufous or brown; about 21 × 14 mm.

**Recorded breeding:** Nyasaland, November to February. Tanganyika Territory, November to February.

**Food:** No information.

**Call:** Unrecorded.

### 1333 TAVETA GOLDEN WEAVER. *PLOCEUS CASTANEICEPS* (Sharpe). Pls. 86, 89.

*Hyphantornis castaneiceps* Sharpe, Cat. Bds. Brit. Mus. 13, p. 448, pl. 13, fig. 5, 1890: Taveta, south-eastern Kenya Colony.

**Distinguishing characters:** Adult male, very similar to the Southern Brown-throated Weaver, but the sides of the face and chin to chest are golden yellow, and the first primary is longer; the occiput is chestnut and this extends in a narrow circle behind the ear-coverts to the lower neck; bill black. There is no non-breeding dress. The female is very similar to the female of the Golden Palm-Weaver but is more olivaceous green above with more distinct dusky streaks. Wing, male 75 to 78, female 67 to 73 mm. The young bird is rather browner above than the adult female; below, sides of face and throat pale yellow; chest and flanks buffish; lower belly white.

**Range in Eastern Africa:** South-eastern Kenya Colony from Tsavo, Taveta and Lake Jipe to eastern Tanganyika Territory as far south as Kilosa, Morogoro and Ifakara.

**Habits:** Notes are not always reliable owing to confusion with other species, but this is a common waterside Weaver in most of coastal Eastern Africa. Noisy and gregarious species, which may abruptly desert its breeding colony for no apparent reason, and return after an absence of weeks or months.

**Nest and Eggs:** Oval, unlined nests of green grass usually over water, attached to two or several stems of reeds or grass. Eggs two or three, glossy dark olive green, often with indistinct dark mottling; about 23 × 15 mm.

**Recorded breeding:** See remarks under the Golden Palm-Weaver.

**Food:** Mainly corn and grass seeds.

**Call:** The usual chattering at the breeding colony, but the chattering is lower pitched and more sparrowy than that of the Golden Palm-Weaver, and is interspersed with a peculiar ugly creaking song 'eee-urr-twee-twee-twee.'

**1334 ORANGE WEAVER.** *PLOCEUS AURANTIUS* (Vieillot).
*Ploceus aurantius rex* Neum. **Pls. 86, 89.**
*Ploceus aurantius rex* Neumann, Bull. B.O.C. 23, p. 12, 1908: Entebbe, southern Uganda.

**Distinguishing characters:** Adult male, head and underside orange golden yellow; black spot in front of eye sometimes extending to lores; chin and throat, mantle, rump and tail olivaceous yellow; wings edged with yellow; bill horn-colour. There is no non-breeding dress. The female is pale green above with indistinct dusky streaks on mantle; superciliary stripe greenish yellow; wings edged pale yellow; sides of face and chin to chest more or less pale yellow; breast and flanks dusky; belly to under tail-coverts white. Wing, male 74 to 78, female 68 to 71 mm. The young bird is similar to the adult female.

**Range in Eastern Africa:** Uganda to north-western Tanganyika Territory.

**Habits:** In the non-breeding season lives with other Weavers in large flocks and roosts in reed beds. In the nesting season it lives mainly by the shores of lakes nesting in reeds or Ambatch. Sir F. Jackson notes that the male, who does the building, is not at all industrious and the nest takes a long time to make. Nests are usually in a small colony often in association with colonies of other species.

**Nest and Eggs:** Nest constructed of rough-edged grass and not particularly well built, somewhat rougher in appearance than that of allied species, and with a small porch over the entrance. It is usually low down in reeds or Ambatch, not more than four feet up as a rule, and is lined with broad bladed grass. Eggs usually two, pale brown, pinkish white, blue or green, with rufous markings of any intensity from freckles to broad blotches; about $21 \times 14 \cdot 5$ mm.

**Recorded breeding:** Lake Victoria, February to May, also November.

**Food:** No information.

**Call:** Unrecorded.

**Distribution of other races of the species:** West Africa to the Belgian Congo, the nominate race being described from the Portuguese Congo.

Plate 88

Females of

Black-headed Weaver (p. 890)
Breeding dress    Non-breeding dress
Speke's Weaver    Northern Masked Weaver
(p. 892)          (p. 894)
Heuglin's Masked Weaver (p. 896)
Breeding dress    Non-breeding dress
Little Weaver (p. 903)
Breeding dress    Non-breeding dress

Layard's Black-headed Weaver (p. 891)
Breeding dress    Non-breeding dress
Tanganyika Masked Weaver    Masked Weaver
(p. 895)                     (p. 898)
Vitelline Masked Weaver (p. 899)
Breeding dress    Non-breeding dress
Yellow-backed Weaver    Golden-backed Weaver
(p. 904)                (p. 906)

Plate 89

**Females of**
Fox's Weaver (p. 893)   Chestnut Weaver (p. 907)   Cinnamon Weaver (p. 908)   Rüppell's Weaver (p. 909)
Northern Brown-throated Weaver (p. 910)   Golden Palm-Weaver (p. 911)   Southern Brown-throated Weaver (p. 912)
Breeding dress   Non-breeding dress
Taveta Golden Weaver (p. 913)   Orange Weaver (p. 914)   Southern Masked Weaver (p. 900)
Breeding dress   Non-breeding dress
Golden Weaver (p. 922)   Vieillot's Black Weaver (p. 926)   Weyn's Weaver (p. 927)
Breeding dress   Non-breeding dress

## 1335 DARK-BACKED WEAVER. *SYMPLECTES BICOLOR* (Vieillot).

*Symplectes bicolor kersteni* (Finsch. & Hartl.). **Pl. 86.**
*Sycobrotus kersteni* Finsch & Hartlaub, Vög. Ostafr. p. 404, pl. 6, 1870: Zanzibar Island.

**Distinguishing characters:** Above, wholly deep black including whole of head and throat; below, golden yellow; under wing-coverts white. There is no non-breeding dress. The sexes are alike. Wing 86 to 99 mm. The young bird is similar to the adult, but has the chin and throat yellow with some black markings.

**Range in Eastern Africa:** Coastal areas of southern Italian Somaliland and Kenya Colony to Tanganyika Territory as far south as Dar-es-Salaam and as far west as the Kilosa districts, also Zanzibar Island.

**Habits:** An insectivorous non-gregarious species inhabiting forest trees or thick cover, especially trees covered with creepers. Feeds like a Tit, searching foliage and often hanging on head downwards. They are noticeable noisy bright-coloured birds found mostly in pairs or in parties of six to ten in the non-breeding season.

**Nest and Eggs:** Nest an inverted retort of tendrils or grass, with a long entrance funnel, looking very much like a stocking suspended by the foot, usually at some height, from the branches of forest trees or occasionally from telegraph wires. Eggs two, buffish, pinkish, or greenish white, spotted or blotched with brown and rufous and with slight blue-grey undermarkings; about 21 × 15 mm.

**Recorded breeding:** Coastal Kenya Colony, Tanganyika Territory and Zanzibar Island, breeding season indefinite.

**Food:** Exclusively insects.

**Call:** A curious call of five notes punctuated by clicks is the song and it sounds rather like the squeaking of a hinge. Many other calls, notably a loud thin squeak and a loud 'wheet-wheet.'

*Symplectes bicolor amaurocephalus* (Cab.).
*Sycobrotus amaurocephalus* Cabanis, J.f.O. p. 349, pl. 21, fig. 1, 1880: Malange, Angola.

**Distinguishing characters:** Differs from the preceding race in having the mantle and rump greyish, contrasting with the brownish black on top of the head; below, paler golden yellow; bill less heavy

and deep. Wing 75 to 92 mm. The young bird has the chin and throat dusky white.

**General distribution:** Angola to southern Belgian Congo, north-eastern Northern Rhodesia and Tanganyika Territory.

**Range in Eastern Africa:** Western Tanganyika Territory.

**Habits:** As for the last race.

**Recorded breeding:** No records.

*Symplectes bicolor stictifrons* (Fisch. & Reichw.).
*Symplectes stictifrons* Fischer & Reichenow, J.f.O. p. 373, 1885: Lindi, south-eastern Tanganyika Territory.

**Distinguishing characters:** Adult male, differs from the Zanzibar race in having the upperside from the top of the head to rump dusky brown; forehead and throat feathers tipped with white. The female is greyer above than the male. Wing 79 to 91 mm. The young bird has the chin and throat yellowish white.

**General distribution:** Tanganyika Territory to southern Nyasaland, eastern Southern Rhodesia and Portuguese East Africa as far south as Inhambane.

**Range in Eastern Africa:** Southern Tanganyika Territory to the Zambesi River.

**Habits:** As for other races, common but local at any altitude but always in dense cover, often in mixed bird parties.

**Recorded breeding:** Nyasaland highlands, September to November, later in the lowlands, February and March. Portuguese East Africa, September to October.

*Symplectes bicolor mentalis* Hartl.
*Symplectes mentalis*, Hartlaub, J.f.O. p. 314, 1891: Buguera, west of Lake Albert, north-eastern Belgian Congo.

**Distinguishing characters:** Head black as in the Zanzibar race, but yellow below paler and extending up the centre of the throat; mantle and rump similar to the Angolan race. Wing 81 to 88 mm.

**General distribution:** The Sudan, north-eastern Belgian Congo, Uganda and Kenya Colony.

**Range in Eastern Africa:** Southern Sudan, Uganda and western Kenya Colony.

**Habits:** As for other races, but relatively uncommon.

**Recorded breeding:** Western Uganda (breeding condition), July and October.

**Distribution of other races of the species:** Fernando Po, Cameroons and South Africa, the nominate race being described from eastern Cape Province, South Africa.

## 1336 BLACK-NECKED WEAVER. *HYPHANTURGUS NIGRICOLLIS* (Vieillot).

*Hyphanturgus nigricollis nigricollis* (Vieill.).
*Malimbus nigricollis* Vieillot, Ois. Chant. p. 74, pl. 45, 1805: Malimba, Portuguese Congo.

**Distinguishing characters:** Adult male, forehead to nape, sides of face and chest yellow washed with chestnut; a stripe through eye and chin to neck in front black; mantle, wings and tail brownish black; flight feathers and tail narrowly edged with greenish; rump greenish; breast to under tail-coverts golden yellow; bill black. There is no non-breeding dress. The female has the forehead to nape black; mantle washed with green; superciliary stripe golden yellow; chin and neck in front golden yellow uniform with rest of underparts; bill black. Wing 72 to 81 mm. The young bird is similar to the adult female but is olive green above and paler yellow below; bill horn-colour.

**General distribution:** Cameroons to Abyssinia, Angola, Uganda and Tanganyika Territory.

**Range in Eastern Africa:** Southern Sudan to western Abyssinia, Uganda, and north-western Tanganyika Territory.

**Habits:** A widely distributed but never numerous species of forest or damp low-lying country, usually seen hunting for insects like a Tit. They are said to be rather noisy and inquisitive.

**Nest and Eggs:** Nest usually solitarily, but occasionally breed in a small colony. The nest, which is often well hidden, is usually made of wiry grass and has a long rather narrow entrance funnel pointing downwards. It is usually at least ten feet from the ground. Eggs two rarely three, very variable, pale blue, white, greenish, pink or reddish white with a variety of grey, brown, or mauve markings; about 21 × 14 mm.

**Recorded breeding:** Cameroons (nestlings), March, April and November. Uganda, March to June. Tana River, Kenya Colony, July.

**Food:** Mainly insects, with some green vegetable matter at times.

**Call:** A remarkable double note, soft and musical, described by Woosnam as like the noise made by two or three glass finger-bowls meeting.

*Hyphanturgus nigricollis melanoxanthus* Cab. Pl. 86.
*Hyphanturgus melanoxanthus* Cabanis, J.f.O. pp. 205, 232, 1878: Mombasa, south-eastern Kenya Colony.

**Distinguishing characters:** Adult male, differs from the nominate race in having the occiput and the whole upperside black. The female is also black above. Wing 71 to 80 mm.

**Range in Eastern Africa:** Southern Abyssinia from the Omo River to Kenya Colony and the Dodoma district of Tanganyika Territory.

**Habits:** As for the nominate race.

**Recorded breeding:** Malindi, Kenya Colony, May. Southern Kenya Colony, March to May. Tanganyika Territory, July.

**Distribution of other races of the species:** Fernando Po and Senegal to French Equatorial Africa.

**1337 SPECTACLED WEAVER.** *HYPHANTURGUS OCULARIS* (A. Smith).
*Hyphanturgus ocularis crocatus* (Hartl.). Pl. 86.
*Hyphantornis crocata* Hartlaub, Abh. Nat. Ver. Bremen 7, p. 100, 1881: Magungo, north-eastern end of Lake Albert, north-western Uganda.

**Distinguishing characters:** Adult male, forehead and crown golden yellow; rest of upperparts including wings and tail yellowish green; below, similar to the Black-headed Weaver; bill black. There is no non-breeding dress. The female differs from the male in having the chin and neck in front golden yellow washed with chestnut. Wing 67 to 80 mm. The young bird is similar to the adult female in having the forehead, chin and neck in front yellow; bill horn-colour.

# SPARROWS, WEAVERS, WAXBILLS, etc.

**General distribution:** Cameroons to north-eastern Belgian Congo, the Sudan, Abyssinia, northern Angola, Kenya Colony and Tanganyika Territory.

**Range in Eastern Africa:** The southern Sudan to southern Abyssinia, Uganda, western Kenya Colony and northern and western Tanganyika Territory as far as the Bukoba district, Loliondo, Oldeani and Monduli, also Ukerewe Island.

**Habits:** As for the following race, and often seen creeping about among foliage searching for insects.

**Recorded breeding:** Uganda, February to August, also November and December. South-eastern Belgian Congo, September to February.

*Hyphanturgus ocularis suahelicus* (Neum.). **Ph. xvii.**
*Ploceus ocularius suahelicus* Neumann, J.f.O. p. 339, 1905: Lewa, Usambara, north-eastern Tanganyika Territory.

**Distinguishing characters:** Adult male, differs from the preceding race in having a distinct chestnut wash on the forehead, sides of face and neck in both sexes. Wing 66 to 80 mm.

**General distribution:** Eastern Kenya Colony to north-eastern Northern Rhodesia, Nyasaland and the Zambesi River.

**Range in Eastern Africa:** Eastern Kenya Colony to the Zambesi River.

**Habits:** A non-gregarious and usually shy species, common and widely distributed, but rarely abundant. They occur at all levels in open country or in any form of woodland or scrub, or in clearings of the forest, and are often members of a mixed bird party. Conspicuous birds and largely insectivorous.

**Nest and Eggs:** A beautifully woven unlined nest of fine grass and fibre, usually green in colour, with a long thin entrance spout from below. Nests are usually attached to the ends of boughs or palm fronds, frequently over water and low down. The length of the entrance funnel varies, and occasionally there are two, or none. Eggs two, rarely three, white or pale bluish green, spotted or speckled with reddish brown and violet grey; about 22 × 15 mm.

**Recorded breeding:** Tana River, Kenya Colony, August. Kenya Colony highlands, June, July, November and December.

Tanganyika Territory, December to March. Nyasaland, October to April, mostly October and November. Rhodesia, October to December, occasionally March and April. Portuguese East Africa, September to December.

**Food:** Mainly insects, but also berries and flowers.

**Call:** A shrill tremulous warbling song rather like that of a Willow Warbler, but much louder. The normal call is a liquid whistling chirp, and the alarm note a throaty chattering.

*Note:* The Uganda and Usambara races meet at Monduli.

**Distribution of other races of the species:** South Africa, the habitat of the nominate race.

**1338** OLIVE-HEADED GOLDEN WEAVER. *HYPHAN-TURGUS OLIVACEICEPS* (Reichenow). **Pl. 86.**

*Symplectes olivaceiceps* Reichenow, O.M. p. 7, 1899: Songea, south-western Tanganyika Territory.

**Distinguishing characters:** Adult male, forehead and crown of head golden yellow; rest of upperparts moss green; nape and upper tail-coverts washed with yellow; wings and tail slate with narrow yellow edging; sides of face, chin and throat moss green; lower neck and chest chestnut; breast to under tail-coverts golden yellow; bill black. There is no non-breeding dress. The female has the forehead and crown uniform moss green with the mantle, and the chestnut of the lower neck and chest is paler. Wing 79 to 85 mm. The young bird has the top of the head green, uniform with the mantle; wing-coverts and flight feathers edged with yellow; sides of face, chin and throat yellow uniform with rest of underparts; a wash of saffron on chest; bill pinkish horn.

**General distribution:** Tanganyika Territory to Nyasaland and Portuguese East Africa.

**Range in Eastern Africa:** South-western Tanganyika Territory to Portuguese East Africa.

**Habits:** An insectivorous species, solitary or in pairs and nowhere common. It is found in open Brachystegia woodland, particularly in woodland bordering open grassland and is frequently a member of a mixed bird party. It is recorded as creeping about like the preceding species searching among boughs for food.

**Nest and Eggs:** Undescribed.

**Recorded breeding:** Nyasaland, September and October. Portuguese East Africa, probably March and April.

**Food:** Insects.

**Call:** A loud chattering call note.

**1339** STRANGE WEAVER. *HYPHANTURGUS ALIENUS* (Sharpe). Pl. 86.

*Sitagra aliena* Sharpe, Bull. B.O.C. 13, p. 21, 1902: Ruwenzori Mts., western Uganda.

**Distinguishing characters:** Adult male, whole of head and neck black; rest of upperside including tail and outer webs of flight feathers green; below, chest and sides of lower neck chestnut; rest of underparts golden yellow; flanks greenish yellow; bill black. There is no non-breeding dress. The female has the chestnut extending up the neck in front with less black, sometimes mixed black and chestnut and sometimes black confined to chin. Wing 66 to 76 mm. The young bird has the head and sides of face green uniform with the mantle; below, chin to chest yellow green washed with chestnut or dusky green; breast and under tail-coverts variable whitish and yellow or wholly yellow; bill dusky brown. The young male in immature dress has the chin and throat as in the adult female; bill blackish horn.

**General distribution:** Eastern Belgian Congo and Uganda.

**Range in Eastern Africa:** Western and south-western Uganda.

**Habits:** Occurs in forest and open country at from 5,500 to 8,500 feet on Mt. Ruwenzori. Local and not uncommon where found in pairs and family parties. They creep about and feed among thick cover and creepers like Tits or Warblers.

**Nest and Eggs:** Nest composed of tendrils with a few grass strips, suspended from a bough at no great height. Eggs two, long and oval, creamy white thickly speckled with brick red and with lilac grey underspots; about 23 × 15 mm.

**Recorded breeding:** Mt. Ruwenzori, January.

**Food:** Mainly insects and a few berries.

**Call:** Unrecorded.

**1340 USAMBARA WEAVER.** *HYPHANTURGUS NICOLLI* (Sclater). **Pl. 86.**

*Ploceus (Symplectes) nicolli* W. L. Sclater, Bull. B.O.C. 52, p. 26, 1931: Amani, Usambara Mts., north-eastern Tanganyika Territory.

**Distinguishing characters:** Adult male, forehead dull yellow; chin to nape and sides of face and throat dusky olivaceous with some dull yellow; mantle dull black with a slight yellowish wash; upper tail-coverts dull black and yellow; wings and tail slate black; chest chestnut; breast to under tail-coverts bright yellow; bill black. There is no non-breeding dress. The female has the whole head dusky brown; mantle black. Wing 83 to 87 mm. Juvenile plumage unrecorded.

**Range in Eastern Africa:** Usambara Mts., north-eastern Tanganyika Territory.

**Habits:** A little known species seen usually at the edge of forest. A rather silent bird with a soft call and probably insectivorous.

**Nest and Eggs:** Undescribed, though a nest rather like that of the Dark-backed Weaver may belong to this species.

**Recorded breeding:** No records.

**Food:** Probably insects.

**Call:** A soft little call of 'sur-swee-ee.'

**1341 GOLDEN WEAVER.** *XANTHOPHILUS SUBAUREUS* (Smith).

*Xanthophilus subaureus aureoflavus* (Smith). **Pls. 86, 89.**

*Ploceus aureo-flavus* A. Smith, Ill. Zool. S. Afr. Aves, text to pl. 30, 1839: Zanzibar Island.

**Distinguishing characters:** Adult male in breeding dress, wholly canary yellow with a greenish wash above; forehead, crown, sides of face, and chin to throat saffron; bill black. In non-breeding dress the top of the head and the mantle are greenish yellow; the sides of face, chin and throat yellow with very slight indication of the saffron colour; bill horn-colour. The female in breeding dress is very similar to the adult male in non-breeding dress; above yellowish, including edges of wing feathers and tail, with indistinct streaks on mantle; below, similar to male; flight feathers dusky with inner webs edged with yellow. In non-breeding dress the upperparts are greener with distinct dark streaks on mantle and scapulars; belly white or yellow and white; bill horn-colour. Wing, male 69 to 80, female 66 to 74 mm. The young male has the head olivaceous green; mantle olivaceous distinctly streaked with dusky; below, chin to under tail-coverts pale yellow; centre of belly whitish. The young female is similar to the young male, but has the breast to under tail-coverts white contrasting with the pale yellow chin to chest.

**General distribution:** Kenya Colony, Tanganyika Territory, Nyasaland and Portuguese East Africa.

**Range in Eastern Africa:** Kenya Colony, Tanganyika Territory as far west as Embu, Meru, Mt. Kilimanjaro, Kilosa and the Tukuyu district to Nyasaland and Portuguese East Africa; also Zanzibar Island.

**Habits:** Freely confused with the Golden Palm-Weaver in the past, and with very similar habits. A highly gregarious species with noisy chattering breeding colonies, which may be in trees, bushes or reeds. The males do most, if not all, the nest building, and there is some evidence of polygamy. Abundant locally and congregates in very large flocks in the non-breeding season.

**Nest and Eggs:** Nest neat and strong, a double ball of grass basket-work with entrance at the bottom, usually attached to a stalk or twig by a single support. Eggs two, three or four, pale blue or white, either plain or with a few dark spots; about 22 × 14·5 mm.

**Recorded breeding:** Kenya Colony, at any time from March to January, probably two broods in the year. Zanzibar Island, mainly March to June and October to January. Northern Portuguese East Africa, February. Nyasaland, October to February.

**Food:** Seeds, grain and insects.

**Call:** Many chattering calls.

**Distribution of other races of the species:** South Africa.

**1342** HOLUB'S GOLDEN WEAVER. *XANTHOPHILUS XANTHOPS* (Hartlaub).

*Xanthophilus xanthops xanthops* (Hartl.). **Pl. 86.**
*Hyphantornis xanthops* Hartlaub, Ibis p. 342, 1862: Angola.

**Distinguishing characters:** Adult male, forehead, sides of face and underside golden yellow; neck in front washed with orange; nape to mantle and tail golden green; wings edged golden green;

edges of inner webs of flight feathers buffish yellow; bill black. There is no non-breeding dress. The female has the whole upperside including forehead and sides of face duller golden green; below, yellow with little or no orange on neck in front; bill black. Wing, male 86 to 97, female 82 to 92 mm. The young bird is more olivaceous green above than the adult female, with indistinct streaks on mantle; below, chest and flanks washed with buff; bill dusky horn-colour. The young bird can be distinguished from the young bird of the Black-headed Weaver and Layard's Black-headed Weaver by the whole underside being yellow, and by the edges of the inner webs of the flight feathers being buffish yellow not yellow.

**General distribution:** Portuguese Congo to Angola, Belgian Congo, Uganda, Kenya Colony, Tanganyika Territory, Nyasaland and northern Portuguese East Africa.

**Range in Eastern Africa:** Uganda, Kenya Colony and northern and western Tanganyika Territory to the Zambesi River.

**Habits:** Usually found in pairs or small scattered parties in open bush country near water or marshes. Little is recorded of their habits, but they seem to be shy birds as a rule, often seen in company with Babblers or Laniarius Shrikes. They appear to keep near the nesting site for most of the year and often use the nest for roosting or play.

**Nest and Eggs:** Breed two or three pairs together but not in colonies. The nest is the largest of all the Weavers, rough and loosely woven of coarse grass with a finer lining, without porch or entrance funnel, conspicuous and bulky, usually over water or in patches of reeds. Eggs two, rarely three, very variable, white, cream, blue or greenish, plain or with reddish brown and grey spots, speckles or blotches; about 25 × 16 mm.

**Recorded breeding:** Uganda, February to June, also October to December. Kenya Colony highlands, February to May, also December. Southern Belgian Congo, September to November and February to April. Tanganyika Territory, September to January. Nyasaland, September to April. Portuguese East Africa, October to April.

**Food:** Insects and wild fruit, but do considerable damage locally to coffee plantations by eating the ripe berries.

**Call:** A harsh chirp.

**Distribution of other races of the species:** Southern Africa.

**1343 SLENDER-BILLED WEAVER. ICTEROPSIS PELZELNI** (Hartlaub). **Pl. 87.**

*Sitagra pelzelni* Hartlaub, Zool. Jahrb. 2, p. 343, pl. 14, figs. 9, 10 1887: Magungo, north-eastern end of Lake Albert, north-western Uganda.

**Distinguishing characters:** Adult male, forehead, forecrown, sides of face, ear-coverts and chin and throat black, extending to a point on the upper chest; hind crown, nape, side of neck and below golden yellow; mantle green; wings dusky edged with yellow; edges of inner webs of flight feathers buff; tail olivaceous edged with yellow; bill black and rather long and narrow. There is no non-breeding dress. The male in immature dress is similar to the adult female but the bill is horn-colour. The female has the forehead, forecrown, sides of face and chin and throat yellow; hind crown and nape green; bill black. Wing 56 to 65 mm. The young bird is olivaceous green above; mantle streaked with dusky; superciliary stripe pale yellow; flight feathers edged with yellow; wing-coverts and innermost secondaries edged with buff; below, buffish yellow; bill horn-colour.

**General distribution:** Eastern Belgian Congo to Uganda, Kenya Colony and Tanganyika Territory.

**Range in Eastern Africa:** Uganda to western Kenya Colony and Tanganyika Territory at south end of Lake Victoria, also Kaserazi Island, south-west Lake Victoria.

**Habits:** A locally common species normally confined to swampy ground or damp woodlands. At certain places, notably at Entebbe, Uganda, it has adapted itself to changed surroundings and now nests in trees all over the town. It feeds largely on insects and examines bark and tree boughs like a Tit or a Warbler, often hanging on head downwards.

**Nest and Eggs:** Normally in papyrus swamps, a small roughly built grass nest lined with feathers, usually concealed to some extent and very loosely woven. Eggs two or three, white or pinkish, occasionally spotted with pink or brown; about $18 \cdot 5 \times 13$ mm.

**Recorded breeding:** Uganda February to May, possibly October and November also. Western Kenya Colony, July to September or later.

**Food:** Insects.

**Call:** Nothing particularly recorded.

### 1344 VIEILLOT'S BLACK WEAVER. *MELANOPTERYX NIGERRIMUS* (Vieillot). Pl. 89.

*Ploceus nigerrimus* Vieillot, N. Dict. d'Hist. Nat. 34, p. 130, 1819: Portuguese Congo.

**Distinguishing characters:** Adult male, wholly black including bill. There is no non-breeding dress. The female is olivaceous above streaked with blackish on head, mantle and scapulars; streaks on head finer and narrower; below, olivaceous yellow; edges of inner webs of flight feathers buffish; bill horn-colour. Wing 77 to 92 mm. The immature male is similar to the adult female but the streaks on the mantle are broader and blacker. The young bird is similar to the adult female, but is browner above especially on mantle and rump.

**General distribution:** Southern Nigeria and Cameroons to northern Angola, the Sudan, Uganda, Kenya Colony and Tanganyika Territory.

**Range in Eastern Africa:** Southern Sudan to Uganda, western Kenya Colony and western Tanganyika Territory.

**Habits:** A noisy, energetic and conspicuous species, breeding in large colonies, frequently in tall trees. They are rather prone to nest round the nests of Birds of Prey or other large birds. Pugnacious birds, often tearing down each other's nests or those of other species. Abundant in forest clearings and open spaces near villages. The cocks breed before they come into full plumage.

**Nest and Eggs:** The nest is composed of broad strips of banana leaf or other similar material, the bird seizing a strip and flying off till it comes away. The nest is built by the cock and is round, with a large semi-circular entrance tube pointing downwards. Eggs two or three, blue, pale blue or greenish blue, often with darker clouding at one end; about 26 × 16·5 mm.

**Recorded breeding:** Uganda, at all times of the year. Mt. Elgon, December and January. Tanganyika Territory, August and September.

**Food:** Seeds, and a considerable quantity of insect food, particularly when feeding young.

**Call:** Colonies are described as noisy, but no individual calls have been noted.

# SPARROWS, WEAVERS, WAXBILLS, etc. 927

**1345 WEYNS'S WEAVER.   *MELANOPTERYX WEYNSI*** Dubois. **Pls. 87, 89.**

*Melanopteryx weynsi* Dubois, O.M. p. 69, 1900: Bumba, Congo River, between long. 22° and 23° E., northern Belgian Congo.

**Distinguishing characters:** Adult male, whole head to chest and mantle black; flight feathers edged with golden yellow above and below; rump black and green; tail green; rest of underparts bright canary yellow; flanks chestnut; bill black. There is no non-breeding dress. The female is olivaceous green above; mantle and scapulars streaked with dusky; wings and tail as in male; below, greenish yellow; flanks green; centre of belly whitish; bill black. The adult female has a very close general resemblance to the adult female of Vieillot's Black Weaver, but it has a black bill, green tail and yellow on inner webs of flight feathers. Rarely the female has a mixed male and female dress. Wing, male 80 to 83, female 76 to 80 mm. The young bird is duller than the adult female; chin to breast more olivaceous yellow; bill horn-colour.

**General distribution:** Northern Belgian Congo to Uganda.

**Range in Eastern Africa:** Uganda.

**Habits:** A rare forest species probably nesting in colonies. They have been noted as feeding on wild figs in forest in July and August, and were noisy while doing so.

**Nest and Eggs:** Undescribed.

**Recorded breeding:** Uganda (breeding condition), July.

**Food:** As above.

**Call:** Unrecorded.

**1346 BLACK-BILLED WEAVER.   *HETERHYPHANTES MELANOGASTER*** (Shelley).

*Heterhyphantes melanogaster stephanophorus* Sharpe.
*Heterhyphantes stephanophorus* Sharpe, Ibis p. 117, pl. 6, fig. 2, 1891: Mau, south-western Kenya Colony.

**Distinguishing characters:** Adult male, forehead, crown and sides of face yellow; streak through eye and rest of plumage black; bill black. There is no non-breeding dress. The female has the chin and neck in front yellow uniform with the sides of the face and forehead. Wing 68 to 79 mm. In immature dress both sexes have the forehead and sides of face yellow; rest of upperparts dullish black;

below, sooty brown; chin and throat dull yellow, washed with chestnut; chest brighter brown; bill blackish. The young bird has the forehead, forecrown, sides of face, chin and throat greenish golden yellow; rest of upperparts dull sooty black with a slight yellow wash; below, greenish olive yellow; chest brighter; bill horn-colour.

**Range in Eastern Africa:** Southern Sudan, Belgian Congo, Uganda and western Kenya Colony.

**Habits:** Not uncommon locally singly or in pairs in impenetrable forest undergrowth, among creepers and thorny plants. Rarely flies, but hops about in thick cover keeping low down, and is not found in the overhead canopy.

**Nest and Eggs:** The only nest known was among lianas in deep forest on the outermost twigs of a small bush. It contained two eggs, white with rufous and grey spots, but they were destroyed by some bird or animal before they could be measured.

**Recorded breeding:** Mt. Elgon, June. Belgian Congo, April.

**Food:** No information.

**Call:** Unrecorded.

**Distribution of other races of the species:** West Africa, the nominate race being described from Cameroon Mt.

**1347** EMIN'S WEAVER. *OTHYPHANTES EMINI* (Hartlaub). Pl. 87.

*Sycobrotus emini* Hartlaub, O.C. p. 92, 1882: Agaru, east of Nimule, southern Sudan.

**Distinguishing characters:** Adult male in breeding dress, forehead and forecrown golden yellow; lores, round eye, sides of face and ear-coverts black; rest of crown to mantle black, latter with green or grey edges to feathers; rump grey; upper tail-coverts and tail green; chin to chest golden yellow; breast to belly white; wing border and edges of inner webs of flight feathers pale yellow; bill black. In non-breeding dress the hinder crown to mantle and scapulars is grey with black streaks. The female in breeding dress is similar to the male in breeding dress, but has the forehead and fore-crown black. In non-breeding dress the mantle and scapulars are grey with black streaks. Wing 72 to 83 mm. In immature dress both sexes have the forehead and fore-crown blackish, rest of upperparts grey with black streaks on mantle and scapulars; rump and

# SPARROWS, WEAVERS, WAXBILLS, etc.

upper tail-coverts grey; below, creamy white, chest buffish; chin slightly yellow; wings and tail as in adult; bill black. The young bird has the head and sides of face yellow green; mantle and rump pale brown, former streaked with dusky; chin to chest chrome yellow; breast to belly creamy white; bill horn-colour. The immature dress is very similar to the non-breeding dress of the Baglafecht Weaver, but the upperside is usually clearer grey, not ashy with a buff wash.

**General distribution:** Abyssinia to the Sudan, eastern Belgian Congo and Uganda.

**Range in Eastern Africa:** Abyssinia to southern Sudan and western and central Uganda.

**Habits:** Local and little recorded, usually seen in marshy country or in wide clearings.

**Nest and Eggs:** Nests oval and without spouts, thickly built of dry grass and lined with grass heads, with entrance hole low on one side. They may be at the end of boughs of high trees, or more usually at the end of a palm frond. Eggs two or three, variable, bluish green with occasional dark spots, or white or sea-green well marked with pale brown, or white with light reddish brown markings; about 22 × 15 mm.

**Recorded breeding:** Central Abyssinia, July. Harar, April to June. Southern Sudan, July and August.

**Food:** No information.

**Call:** An attractive rather musical chattering song.

**1348 REICHENOW'S WEAVER.** *OTHYPHANTES REICHENOWI* (Fischer).

*Othyphantes reichenowi reichenowi* (Fisch.). **Pl. 87. Ph. xvii.**

*Sycobrotus reichenowi* Fischer, J.f.O. p. 180, 1884: Great Arusha, northern Tanganyika Territory.

**Distinguishing characters:** Adult male, forehead, fore-crown, behind ear-coverts and underside yellow; lores, round eyes, ear-coverts, and rest of upperparts black; wings edged with golden yellow; tail olive green; lower rump and upper tail-coverts mixed black and yellow; edges of inner webs to flight feathers pale yellow; bill black. There is no non-breeding dress. The female is similar to the male but has the forehead, fore-crown, and behind ear-coverts black. Wing 75 to 86 mm. In immature dress the male is similar

to the adult female but the whole upperside is mottled and streaked black and green; bill horn-colour, upper mandible blackish. The young bird is green above, streaked with blackish; bill horn-colour, sometimes with upper mandible dusky. Occasionally xanthochroic examples occur, which are wholly yellow with a whitish bill and scattered normally coloured feathers.

**Range in Eastern Africa:** Mega district of southern Abyssinia to central and southern Kenya Colony and northern Tanganyika Territory.

**Habits:** Found in pairs or small parties in various types of country, nesting near water at any height from the ground, fairly common in woodland or forest clearings.

**Nest and Eggs:** A stoutly made nest of green grass lined with seed heads, usually bound on to a twig or sewn on to the middle rib of a palm leaf. There is usually only one occupied nest on a tree, but a number of cocks' nests or forsaken nests may be present. Eggs two to three, usually two, variable, blue, white or greenish, plain or spotted with rufous, dark brown, or purplish brown; about 22 × 16 mm.

**Recorded breeding:** Kenya Colony highlands, March to July, also November to January. Northern Tanganyika Territory, January.

**Food:** Grass seeds, berries, insects, almost omnivorous.

**Call:** A little rambling unmelodious song. The call is a sharp chirp and the alarm note a raucous 'twee-up' (Moreau).

*Othyphantes reichenowi fricki* Mearns.
*Othyphantes fricki* Mearns, Smiths Misc. Coll. 61, No. 14, p. 1, 1913:
 Aletta=Alata, 26 miles east of Lake Abaya, southern Abyssinia.

**Distinguishing characters:** Adult male, similar to the nominate race, but mantle black and green, and black of sides of face usually joined to black of hind neck; differs from the following race in having green on the mantle and a much narrower black band joining the face and hind-neck. The female differs from the nominate race in having the mantle streaked black and green, not black only. Wing 81 to 84 mm.

**Range in Eastern Africa:** Southern Abyssinia from Alata to Alghe.

# SPARROWS, WEAVERS, WAXBILLS, etc.

**Habits:** Locally common on the outskirts of evergreen forest in a few localities in southern Abyssinia.

**Recorded breeding:** No records.

*Othyphantes reichenowi nigrotemporalis* Granv.

*Othyphantes reichenowi nigrotemporalis* Granvik, O.M. p. 40, 1922: Mt. Elgon, Uganda-Kenya Colony boundary.

**Distinguishing characters:** Adult male, differs from the nominate race in having the black from lores, around eyes and ear-coverts joining up with the black of the upperparts. The female is indistinguishable from that of the preceding race. Wing 78 to 85 mm.

**Range in Eastern Africa:** Yavello district of southern Abyssinia to north-western and western Kenya Colony.

**Habits:** As for the nominate race but confined to higher altitudes, locally common. Occurs on Mt. Elgon from 6,500 to 9,000 feet.

**Recorded breeding:** Mt. Elgon, December and January, also June and July.

*Note:* Intermediates between this race and the nominate one occur in the Elgon and Nandi areas.

## 1349 STUHLMANN'S WEAVER. *OTHYPHANTES STUHLMANNI* (Reichenow).

*Othyphantes stuhlmanni stuhlmanni* (Reichw.). **Pl. 87.**

*Symplectes stuhlmanni* Reichenow, O.M. p. 29, 1893: Bukoba, north-western Tanganyika Territory.

**Distinguishing characters:** Adult male, top of head, sides of face including ear-coverts and malar stripe black; rest of upperparts golden green; mantle and scapulars broadly streaked with black; wings edged with golden green; tail olive green; below, bright chrome yellow; edges of inner webs of flight feathers chrome yellow; bill black. There is no non-breeding dress. The female is similar to the male but usually has the occiput mixed black and green. Wing 74 to 84 mm. The immature dress of both sexes is similar to that of the adult but the top of the head is green; forehead, sides of face and ear-coverts more or less blackish; bill dusky or black. The young bird has the top of the head and sides of face moss-green; rest of upperparts brighter yellowish green, including tail; mantle and scapulars streaked with dusky; wings edged with bright yellowish green; underside buffish or chrome yellow; bill horn-colour.

# 932 SPARROWS, WEAVERS, WAXBILLS, etc.

**General distribution:** Eastern Belgian Congo to Uganda and Tanganyika Territory.

**Range in Eastern Africa:** Central Uganda to western Tanganyika Territory as far south as the Kungwe-Mahare Mts.

**Habits:** Occurs not uncommonly in bush or more open forest country singly or in pairs, but gather in larger flocks in the evening to roost in elephant grass or reeds. Nest singly or in a small scattered colony.

**Nest and Eggs:** Nest suspended from a bough and often somewhat concealed by pendent leaves. Eggs two, white, spotted or blotched with rufous and with pale grey underspots; about 22 × 15 mm.

**Recorded breeding:** Uganda, March to June, and September to December; (nestling) January.

**Food:** No information.

**Call:** Reported to have quite a pleasant song.

*Othyphantes stuhlmanni sharpii* Shell.
*Otyphantes sharpii* Shelley, Ibis p. 557, 1898: South-western Tanganyika Territory.

**Distinguishing characters:** Adult male, differs from the nominate race in having the golden yellowish green of the upperparts replaced by green, and the underside more lemon yellow. Wing 76 to 82 mm.

**Range in Eastern Africa:** Southern Tanganyika Territory from the Ufipa Plateau to Iringa district.

**Habits:** As for the nominate race.

**Recorded breeding:** Tanganyika Territory, December and January.

**Distribution of other races of the species:** Nyika Plateau, Nyasaland.

**1350 COMPACT WEAVER.** *PACHYPHANTES PACHY-RHYNCHUS* (Reichenow). **Pl. 87.**

*Ploceus pachyrhynchus* Reichenow, O.M. p. 29, 1893: Kerevia, Semliki River Valley, eastern Belgian Congo.

**Distinguishing characters:** Adult male in breeding dress, forehead chestnut; crown and over eye yellow; nape green or dusky brown; mantle and scapulars green streaked with dusky; rump brown; wings and tail dusky edged with buffish white; sides of face, chin and throat black; breast, chest and flanks yellow; belly and under tail-coverts brown; edges of inner webs of flight feathers buffish white; bill slate, short and thick. In non-breeding dress the whole plumage is brown; upperparts brown or ashy brown streaked with dusky; top of head and streak through eye blackish; stripe over eye, sides of face and underside brown or buffish brown; bill slaty horn or slate. The female in breeding dress differs from the male in having the forehead and crown of head mixed blackish green and yellow; bill slate. In non-breeding dress the female is similar to the male in the same dress. Wing 61 to 72 mm. The young bird is similar to the adult in non-breeding dress, but has the stripe over the eye, sides of neck and underside washed with pale yellow; bill horn-colour.

**General distribution:** Sierra Leone to Abyssinia, Uganda, Kenya Colony, Portuguese Congo and northern Angola.

**Range in Eastern Africa:** Southern Sudan and south-western Abyssinia, to Uganda and western Kenya Colony.

**Habits:** Locally common or even abundant in open park-like country in flocks. Also found in patches of long grass.

**Nest and Eggs:** Nest oval and very neat and compact of fine grass lined with plant down, usually slung from tall grass stems. Entrance about one-third of the way from the top of the side, and in fact very like the nest of a Grosbeak Weaver. Eggs three, occasionally four, rather small, white, pale blue, or stone grey, sometimes with grey spotting; about $19\cdot5 \times 13\cdot5$ mm.

**Recorded breeding:** Uganda, April to July.

**Food:** Insects and grass seeds.

**Call:** A rather melodious 'cheewery-cheewery-cheewery,' also a single 'chee' (Marchant).

**1351 BROWN-CAPPED WEAVER.** *PHORMOPLECTES INSIGNIS* (Sharpe).

*Phormoplectes insignis insignis* (Sharpe). **Pl. 87.**

*Sycobrotus insignis* Sharpe, Ibis p. 117, pl. 6, fig. 1, 1891: Mt. Elgon, Uganda.

**Distinguishing characters:** Adult male, forehead to nape chestnut; sides of face, chin and throat, scapulars, wings and tail black;

rest of plumage golden yellow; bill black. There is no non-breeding dress. The female has the forehead to nape black; bill black. Wing 81 to 90 mm. The young male has the forehead to nape and sides of face green flecked with black; chin and throat yellow; wings edged with yellow and buff; bill horn-colour. The young female differs from the young male in having the forehead to nape dull black flecked with yellow.

**General distribution:** Cameroons highlands to the Sudan, Kenya Colony and Tanganyika Territory.

**Range in Eastern Africa:** Imatong Mts., southern Sudan, below 6,000 feet to Uganda, Kenya Colony and western and northern Tanganyika Territory as far south as the Kungwe-Mahare Mts.

**Habits:** Locally common in forest glades and forest bush with rich vegetation. Always in pairs and found in high trees where it can be seen hanging in any position in search of insects and climbing about like a Tit.

**Nest and Eggs:** Nest of creepers or convolvulus tendrils with a long entrance funnel which is woven on to the underside of a bough, not suspended (F. J. Jackson). Eggs two, pale blue, occasionally with faint brown spots; about 22 × 16 mm.

**Recorded breeding:** Nandi, Kenya Colony, May.

**Food:** Insects.

**Call:** A silent species, with an occasional rasping note, but in the breeding season the male has a clear whistling rather nasal song.

**Distribution of other races of the species:** Fernando Po.

**1352 BERTRAM'S WEAVER.  *XANTHOPLOCEUS BER-TRANDI* (Shelley). Pl. 87.**

*Hyphantornis bertrandi* Shelley, Ibis p. 23, pl. 2, 1893: Plains near Milanji, southern Nyasaland.

**Distinguishing characters:** Adult male, forehead and forecrown saffron; patch on hind crown and sides of face black; rest of upperparts green, including tail; hind neck yellowish green; wings blackish edged with green; chin and throat black; rest of underparts yellow with a saffron wash on neck and chest; edges of inner webs of flight feathers buffish yellow; bill black. There is no non-breeding dress. The female has the whole top of head and sides of face black; chin and throat usually black, but sometimes mainly yellow; bill black. Wing 76 to 85 mm. The young bird has the top of the head green with some black, and sides of face mixed yellow and black; bill blackish horn.

**General distribution:** Central Tanganyika Territory to Nyasaland and Tete Province, Portuguese East Africa.

**Range in Eastern Africa:** Tanganyika Territory from Mpapwa, the Nguru and Uluguru Mts. and Mahenge to Nyasaland.

**Habits:** Fairly common along stream beds in the Tanganyika Territory highlands, and on hill sides with sparse trees and long grass in Nyasaland.

**Nest and Eggs:** A round nest of broad-leaved grass without porch or funnel built among the lower boughs of riverside trees. Eggs two, green blotched and spotted with dull rusty red; about 22·5 × 16 mm.

**Recorded breeding:** Tanganyika Territory (about to breed), December. Nyasaland, August to November.

**Food:** No information.

**Call:** Unrecorded.

### 1353 YELLOW-MANTLED WEAVER. *MELANOPLOCEUS TRICOLOR* (Hartlaub).

*Melanoploceus tricolor interscapularis* (Reichw.).

*Ploceus interscapularis* Reichenow, O.M. p. 29, 1893: Budoko, Ituri district, eastern Belgian Congo.

**Distinguishing characters:** Adult male, whole head, mantle, rump, wings and tail black; a bright yellow collar across upper mantle; chest to belly chestnut; under tail-coverts black; bill black. There is no non-breeding dress. The female is wholly black except for the yellow collar across the mantle; chin to belly more sooty black; bill black. Wing 84 to 87 mm. The young bird is dull blackish; forehead to mantle dull rufous; bill horn-colour.

**General distribution:** Eastern Belgian Congo to Uganda and northern Angola.

**Range in Eastern Africa:** Uganda.

**Habits:** A rare forest Weaver found among high trees, the trunks of which it climbs in a spiral manner like a Woodpecker. Silent birds easily overlooked.

**Nest and Eggs:** An untidy loosely woven nest of rootlets and grass at the end of a branch, usually out of sight and high above the ground. Occasionally nest in a small colony but usually singly. Eggs of this species are reputed to be white or pinkish; about 23 × 16 mm.

**Recorded breeding:** Uganda, probably June to October.

**Food:** Largely insects.

**Call:** The only call described is a short 'cheret.'

**Distribution of other races of the species:** West Africa, the nominate race being described from Sierra Leone.

### 1354 MAXWELL'S BLACK WEAVER. *MELANOPLOCEUS ALBINUCHA* (Bocage).

*Melanoploceus albinucha maxwelli* (Alex.).
*Melanopteryx maxwelli* Alexander, Bull. B.O.C. 13, p. 54, 1903: Ribola wa Moka, Fernando Po.

**Distinguishing characters:** Colour wholly black, including bill and feet; eye white. The sexes are alike. Wing 77 to 86 mm. The young bird is duller black above with some grey or olivaceous on head; below, variable sooty grey, whitish sooty grey or slightly olivaceous; bill and feet dusky horn-colour.

**General distribution:** Fernando Po to southern Nigeria, Cameroons, north-eastern Belgian Congo and Uganda.

**Range in Eastern Africa:** Bwamba area of western Uganda.

**Habits:** Little recorded. Usually seen in flocks of some size coming to roost in trees or long grass.

**Nest and Eggs:** Undescribed, but nests in small colonies in the tops of thickly foliaged forest trees.

**Recorded breeding:** No records.

**Food:** No information.

**Call:** A running chatter at the breeding colonies, not particularly described.

**Distribution of other races of the species:** Sierra Leone to Gold Coast, the nominate race being described from Sierra Leone.

# SPARROWS, WEAVERS, WAXBILLS, etc.

## MALIMBES, GROSBEAK WEAVER, RED-HEADED WEAVER and QUELEAS. FAMILY—PLOCEIDÆ. Genera: *Malimbus, Amblyospiza, Anaplectes* and *Quelea*.

Three species of Malimbes, one Grosbeak Weaver, one Red-headed Weaver and three Queleas occur in Eastern Africa. The Malimbes are forest dwellers, the Grosbeak Weaver is a low country species, and the Red-headed Weaver is a bird of light woodlands. The Queleas or Diochs are birds of open country, and in the non-breeding season congregate into large flocks and are a nuisance to the farmer.

### 1355 CRESTED MALIMBE. *MALIMBUS MALIMBICUS* (Daudin).

*Malimbus malimbicus crassirostris* Hart.
*Malimbus malimbicus crassirostris* Hartert, Nov. Zool. 26, p. 140, 1919: Budongo Forest, Unyoro, western Uganda.

**Distinguishing characters:** Adult male, forehead, round and behind eye and chin black; crown, elongated crest, ear-coverts, sides of face and throat to chest glossy crimson red; rest of plumage glossy black; wings and tail duller; bill black. The female is similar to the male but lacks the elongated crest. Wing, male 90 to 93, female 82 to 89 mm. The young bird is similar to the adult female, but duller in colour; more sooty brown below; crimson of head and neck much duller and intermixed with black or blackish; bill horn-colour.

**General distribution:** Eastern Belgian Congo and western Uganda.

**Range in Eastern Africa:** Western Uganda.

**Habits:** A species of dense bush or lower evergreen forest, usually seen as a member of a mixed bird party, but of which little is recorded. It may be found at any height in the forest from undergrowth to top canopy.

**Nest and Eggs:** An untidy nest of tendrils, fibre or strips of palm leaves roughly woven together with a short entrance half-tube or porch underneath and attached to a hanging frond in forest undergrowth. Eggs two, greenish or pinkish white heavily mottled and shaded with light brown and pale grey; about 24 × 16 mm.

**Recorded breeding:** Western Uganda, April to August.

# SPARROWS, WEAVERS, WAXBILLS, etc.

**Food:** Believed to be entirely insects in spite of its heavy bill.

**Call:** Various harsh chirping calls with a few more musical notes at times.

**Distribution of other races of the species:** West Africa and Angola, the nominate race being described from Malimba, Portuguese Congo

### 1356 RED-HEADED MALIMBE. *MALIMBUS RUBRICOLLIS* (Swainson).

*Malimbus rubricollis centralis* Reichw.
*Malimbus rubricollis centralis* Reichenow, O.M. p. 30, 1893: Nduluma, Ituri district, eastern Belgian Congo.

**Distinguishing characters:** Adult male, forehead to hind-neck and sides of neck orange red; rest of plumage slightly glossy black; bill black. The female has the forehead and fore-crown also black. Wing 93 to 105 mm. The young bird is duller than the adult; orange of head and sides of neck paler; bill horn-colour.

**General distribution:** Eastern Belgian Congo to the Sudan, Uganda and Kenya Colony.

**Range in Eastern Africa:** Southern Sudan, Uganda and western Kenya Colony.

**Habits:** Relatively common in the forests of Uganda among the tops of tall trees, where it climbs about and frequently hangs on underneath branches upside down. It is also said to climb up boughs like a Woodpecker. Not gregarious and usually seen in pairs.

**Nest and Eggs:** Large untidy retort-shaped nests of creepers and grass, with a wide spout, usually dependent from the top branches of high trees. Eggs are said to be white, but probably vary considerably in ground colour. No measurements available.

**Recorded breeding:** Uganda, April to August.

**Food:** Probably mainly insects but berries also.

**Call:** A harsh chirp and a low wheezy call.

**Distribution of other races of the species:** Fernando Po and West Africa to Angola, the nominate form being described from Malimba, Portuguese Congo.

# SPARROWS, WEAVERS, WAXBILLS, etc. 939

**1357 RED-BELLIED MALIMBE.** *MALIMBUS ERYTHROGASTER* Reichenow.

*Malimbus erythrogaster fagani* O. Grant.

*Malimbus fagani* O. Grant, Bull. B.O.C. 21, p. 15, 1907: Fort Beni, eastern Belgian Congo.

**Distinguishing characters:** Similar to the Red-headed Malimbe, but has the red of the chest and belly joining up with the sides of neck and top of head; the black chin and throat patch forming a V towards the chest. The female is similar but has a black forehead. Wing 90 mm. The young bird has the throat red with white and black bases to the feathers; flanks and under tail-coverts brownish black.

**General distribution:** Eastern Belgian Congo to Uganda.

**Range in Eastern Africa:** Bwamba area, western Uganda.

**Habits:** Little recorded. An uncommon species of forest. Chapin records that this species frequently builds close to nests of Birds of Prey or other large birds.

**Nest and Eggs:** In high forest trees, apparently undescribed.

**Recorded breeding:** Congo Forest, March to September.

**Food:** No information.

**Call:** Unrecorded.

**Distribution of other races of the species:** West Africa and Belgian Congo, the nominate race being described from Cameroons.

**1358 GROSBEAK WEAVER.** *AMBLYOSPIZA ALBIFRONS* (Vigors).

*Amblyospiza albifrons albifrons* (Vig.).

*Pyrrhula albifrons* Vigors, P.Z.S. p. 92, 1831: Algoa Bay, Cape Province, South Africa.

**Distinguishing characters:** Bill notably heavy and deep; adult male, forehead usually white, but not constantly so; rest of head and neck all round to upper mantle and upper chest rather dark sooty brown; rest of upperparts including wings and tail brownish black; wings with white bases to flight feathers, showing as a white patch on primaries both above and below; rest of underparts sooty grey; in fresh dress all the feathers are tipped and edged with brown and white; bill blackish horn. The female is wholly brown above

with lighter edges to feathers; below, white or creamy white heavily streaked with blackish; no white in wing; bill yellowish horn. Wing, male 90 to 104, female 85 to 95 mm. The young bird is similar to the adult female, bill yellow.

**General distribution:** South-eastern Belgian Congo to Tanganyika Territory, Nyasaland, the Rhodesias, Cape Province and Natal.

**Range in Eastern Africa:** South-western Tanganyika Territory and Nyasaland.

**Habits:** Large gregarious Weavers of low swampy country, usually common. Noticeable birds with a conspicuous white wing bar in flight, which is dipping like that of a Woodpecker; but mostly remarkable for their nests which are beautifully woven of fine materials.

**Nest and Eggs:** Nest a sphere or oval, very finely woven of thin grass strips, usually fixed on to two or three upright stems of reeds or grass in swampy places. There is a small entrance hole at the side, but the presence or absence of a porch varies with the locality. In Natal there is usually a porch, in Nyasaland no porch. Nests are usually in a colony and in this race are said to be built almost entirely by the male, though the female reshapes the inside. Eggs usually three, cream or creamy pink with reddish brown spots often suffused; about $22 \cdot 5 \times 15 \cdot 5$ mm.

**Recorded breeding:** Nyasaland, October to April. Rhodesia, February and March. Natal, October and January to March.

**Food:** Seeds of grasses and sedges, berries when available. The young are fed by regurgitation, not with insects.

**Call:** Various loud calls not particularly described, the male also has a pleasant little song.

*Amblyospiza albifrons melanota* (Heugl.).
*Coryphegnathus melanotus* Heuglin, J.f.O. p. 21, 1863: Upper White Nile, Sudan.

**Distinguishing characters:** Adult male, differs from the nominate race in being blacker, less brownish, on the mantle, rump, wings and tail. The female and young bird are rather warmer brown above than those of the nominate race. Wing, male 92 to 100, female 84 to 92 mm.

# SPARROWS, WEAVERS, WAXBILLS, etc.

**Range in Eastern Africa:** Abyssinia, southern Sudan, Uganda, eastern and northern Kenya Colony, southern Italian Somaliland, and western Tanganyika Territory as far south as the Kungwe-Mahare Mts.

**Habits:** As for the nominate race. Nests in a small colony of five to six pairs. The nest was seen to be constructed by both birds in fourteen days (Woosnam). In the non-breeding season the birds are often found in forest undergrowth.

**Recorded breeding:** Abyssinia, May, June and October. Lado, November. Uganda, February to August.

*Amblyospiza albifrons unicolor* (Fisch. & Reichw.).
*Pyrenestes unicolor* Fischer & Reichenow, O.C. p. 88, 1878: Zanzibar Island.

**Distinguishing characters:** Adult male, differs from the preceding races in being generally sooty black including head and neck. The female and young bird are similar to those of the nominate race. Wing, male 87 to 92, female 78 to 90 mm.

**Range in Eastern Africa:** Eastern Kenya Colony and eastern Tanganyika Territory from Lamu to Bagamoyo and as far west as Mt. Kilimanjaro, Monduli and the Nguru Hills; also Pemba, Zanzibar and Mafia Islands.

**Habits:** As for the nominate race. Nests in colonies in swamps and along stream beds. The nests are not laid in at once and are often deserted. They are said to be made mainly by the male. 'Cock's nests' are also made for roosting purposes and usually have a larger aperture. The male has a pleasant little trilling song which he sings as he works at the nests.

**Recorded breeding:** Eastern Tanganyika Territory, March to May. Zanzibar Island, May to July, with a partial second brood in November. Pemba Island, December, also other months.

*Amblyospiza albifrons montana* V. Som.
*Amblyospiza albifrons montana* Van Someren, Bull. B.O.C. 41, p. 122, 1921: Fort Hall, south-central Kenya Colony.

**Distinguishing characters:** Adult male, similar to the last race but larger, and bill often rather stouter. The female is similar to the female of the last race. Wing, male 93 to 97, female 84 to 90 mm.

## 942  SPARROWS, WEAVERS, WAXBILLS, etc.

**Range in Eastern Africa:** Central and south-western Kenya Colony to Tanganyika Territory from Lake Victoria to Mt. Kilimanjaro and as far south as Kilosa; also Kome and Ukerewe Islands, southern Lake Victoria.

**Habits:** As for the nominate race; common in suitable localities. No porch to nest, incubation by hen only.

**Recorded breeding:** Kenya Colony highlands, March to June, also December.

**Distribution of other races of the species:** West Africa and Angola.

**1359 RED-HEADED WEAVER.** *ANAPLECTES MELANOTIS*
(Lafresnaye).
*Anaplectes melanotis melanotis* (Lafres.). **Pl. 90.**
*Ploceus melanotis* Lafresnaye, Rev. Zool. p. 20, 1839: Senegal.

**Distinguishing characters:** Adult male, top of head, neck all round and throat to breast scarlet red, varying in extent from chest to upper belly; near nostrils, lores, round eyes and ear-coverts to chin black; rest of upperside including wings and tail grey; mantle black and red; outer edges of flight and tail feathers scarlet; rest of underparts white; bill red or orange red. In non-breeding dress the mantle is grey or brownish grey. The female is wholly grey above; wholly white below; a grey wash on chest; edges of flight and tail feathers paler scarlet red and occasionally yellowish. Wing 74 to 89 mm. The young bird has the head olivaceous or saffron; sides of face and chest buffish yellow; edges of flight feathers and tail orange or reddish orange; bill dusky.

**General distribution:** Senegal to British Somaliland and south to Tanganyika Territory and southern Belgian Congo.

**Range in Eastern Africa:** Abyssinia, southern Sudan, British Somaliland, Uganda, Kenya Colony, and Tanganyika Territory as far south as the Ufipa and Iringa areas and Dar-es-Salaam.

**Habits:** Birds of bush country or open woodland, preferring the tops of fairly large trees, usually not uncommon but rather shy, and found in pairs or small parties. They feed like Tits hanging on in all positions from small branches. They are believed to be polygamous and nests are usually found two or three together. They are constructed by the male but the female gives some attention to the inside.

**Nest and Eggs:** Rough but elaborate retort-shaped nests made of tendrils or leaf midribs with a long dependent spout. Easy to see but usually placed high at the ends of thin boughs and inaccessible. Eggs two or three, either plain pale blue, sometimes with a zone of deeper cloudings, or pale blue with occasional spots or speckles of purplish brown or black; about 21 × 13 mm.

**Recorded breeding:** Sudan, August to October. Eastern Abyssinia, February. Southern Abyssinia, July, October and February. Kenya Colony, March and April. Tanganyika Territory, January to March.

**Food:** Largely insects.

**Call:** A soft liquid 'clink' (Elliott).

*Note:* Intermediates between this race and the following occur in western Nyasaland and central Tanganyika Territory.

*Anaplectes melanotis rubriceps* (Sund.).
*Ploceus (Hyphantornis) rubriceps* Sundevall, Œfv. K. Sv. Vet.-Ak. Förh. 7, p. 97, 1850: Mohapoani, Witfontein Mts., w. Transvaal.

**Distinguishing characters:** Adult male, differs from the nominate race in having little or no black on head and edges of flight and tail feathers yellow; in breeding dress the mantle is black and red; grey in non-breeding dress. The female has the top of the head saffron or olivaceous yellow; sides of face and chin to upper chest yellow; edges of flight feathers and tail as in male. In immature dress the male differs from the adult female in usually having the top of the head warmer coloured and often tinged with orange; below, often a few red tipped feathers on chin, throat and chest. Wing 75 to 87 mm.

**General distribution:** Southern Belgian Congo, Tanganyika Territory, Nyasaland, Bechuanaland and northern, western and eastern Transvaal.

**Range in Eastern Africa:** Tanganyika Territory, Nyasaland and Portuguese East Africa.

**Habits:** As for the nominate race, nesting in small colonies of three or four pairs. Nests are occasionally attached to telegraph wires. Almost certainly polygamous, the male building several nests for several wives. They appear to make some seasonal movements.

**Recorded breeding:** Nyasaland, September to December. Rhodesia, September to December.

*Anaplectes melanotis jubaensis* V. Som.
*Anaplectes jubaensis* Van Someren, Bull. B.O.C. 40, p. 94, 1920: South-west of Juba River, southern Italian Somaliland.

**Distinguishing characters:** Adult male, differs from the nominate race in having the underside wholly scarlet red. Wing about 81 mm.

**Range in Eastern Africa:** Southern Italian Somaliland.

**Habits:** As for the other races.

**Recorded breeding:** No records.

**1360** RED-BILLED QUELEA. *QUELEA QUELEA* (Linnæus).
*Quelea quelea lathamii* (Smith). **Pl. 91.**
*Loxia lathamii* A. Smith, Rep. Exp. C. Afr. p. 51, 1836: near Kurrichane, western Transvaal, South Africa.

**Distinguishing characters:** Adult male, colour and extent of head markings variable; forehead, sides of face, chin and throat either blackish or brownish black, or a rosy, buffish or tawny colour which sometimes extends to the sides of neck and chest; sometimes chin and throat and ear-coverts whitish; rest of upperparts streaked with black and buff; edges of flight and tail feathers yellowish; below, whitish or buffish white with rosy or tawny on breast spreading individually more or less to belly and often with some dark mottling on breast and flanks; bill red. The female has the top of the head brownish or greyish brown finely streaked with dusky; superciliary stripe and ear-coverts dusky, chin and throat white or buffish white; bill red. Wing 61 to 71 mm. The male in non-breeding dress and the young bird are similar to the adult female.

**General distribution:** Angola to the Rhodesias, northern Portuguese East Africa and South Africa.

**Range in Eastern Africa:** Nyasaland and southern half of northern Portuguese East Africa.

**Habits:** As for the following race, congregating into vast flocks, and nesting in enormous numbers in a very small area. Nests are said to be usually in trees. It is so local in the breeding season that very few colonies have been found.

**Nest and Eggs:** See under the following race, but eggs measure about 19 × 14 mm.

**Recorded breeding:** Southern Rhodesia, September to December. Cape Province, April.

*Quelea quelea æthiopica* (Sund.).
*Ploceus æthiopicus* Sundevall, Œfv. K. Sv. Vet.-Akad. Förh. 7, p. 126, 1850: Sennar, eastern Sudan.

**Distinguishing characters:** Adult male, differs from the preceding race in having the colour on the head extending to the occiput. There is the same variation of head colours. The female and young bird are similar to those of the last race but the bill is slightly more robust. Wing 62 to 73 mm.

**General distribution:** The Sudan to British Somaliland, northeastern Belgian Congo and Tanganyika Territory.

**Range in Eastern Africa:** The Sudan, Abyssinia, Eritrea and British Somaliland to Tanganyika Territory.

**Habits:** A gregarious species congregating into immense flocks and doing very great damage to crops. Locally migratory. Emin noted that in the non-breeding season they frequented the reed beds of the Sobat River and spent the breeding season in the tablelands of Kordofan and Sennar. Large flocks also occur in Kenya Colony and in Uganda, and it is a major pest of rice in eastern Tanganyika Territory. Vast flocks can be seen going to roost in reed beds and forming wreaths like smoke in the sky.

**Nest and Eggs:** Extremely few breeding colonies have ever been found in spite of its locust-like numbers. Nests in a Kenya Colony highland colony were neat but flimsy networks of fine green grass woven on to reed or grass stems. Eggs two to four or five, pale blue or bluish white, plain or evenly spotted or freckled with dusky brown; about 18 × 12 mm.

**Recorded breeding:** Sudan, September. Kenya Colony, Nairobi plains, May; southern areas, March and April.

**Food:** Seeds, corn and other crops to which it can be very destructive.

**Call:** A noisy chattering while in flocks. Alarm call a sharp 'chak-chak.'

**Distribution of other races of the species:** West Africa, the nominate race being described from Senegal.

### 1361 RED-HEADED QUELEA. *QUELEA ERYTHROPS* (Hartlaub). Pl. 91.

*Ploceus erythrops* Hartlaub, Rev. Zool. p. 109, 1848: São Thomé Island, West Africa.

**Distinguishing characters:** Adult male, differs from the Red-billed Quelea in having the whole head and throat scarlet red, often with a blackish chin and centre of throat; tail shorter; bill black in breeding dress. The female has a horn-coloured bill; general colour darker especially below than the females of the last species; forehead, superciliary stripe and sides of face distinctly washed with yellow. The male in non-breeding dress is similar to the adult female; rarely there is a wash of red on the sides of the face and forehead. Wing 52 to 66 mm. The young bird is similar to the adult female, but has pale sandy edges to the feathers of the mantle, wing-coverts and inner secondaries.

**General distribution:** Senegal and Cameroons to Abyssinia, and south to eastern Angola, Northern Rhodesia, Nyasaland, south-eastern Cape Province and Natal; also Princes and São Thomé Islands.

**Range in Eastern Africa:** The Sudan and Abyssinia to the Zambesi River.

**Habits:** A species of very similar habits to the last but not so numerous, continually chattering and quite tame. Also a pest of rice fields in certain localities. Locally migratory in large flocks in the non-breeding season.

**Nest and Eggs:** Nest in compact colonies, and nests are within a few feet of each other, attached to two or three stems of grass or stout herbs. Nests are spherical, made of grass with a side entrance. Eggs usually two, olive green with dusky spots or uniform pale greenish blue; about 18 × 13 mm.

# SPARROWS, WEAVERS, WAXBILLS, etc.

**Recorded breeding:** Nyasaland, March and April.

**Food:** Grass seeds, small corn, etc.

**Call:** Nothing noted except the continuous chattering of the flocks.

## 1362 CARDINAL QUELEA. *QUELEA CARDINALIS* (Hartlaub).

*Quelea cardinalis cardinalis* (Hartl.). **Pl. 91.**

*Hyphantica cardinalis* Hartlaub, J.f.O. p. 325, 1880: Lado, southern Sudan.

**Distinguishing characters:** Adult male, head and chin to chest crimson red, occiput and nape streaked tawny and blackish suffused with crimson red; rest of upperparts streaked tawny and blackish; below, breast to under tail-coverts creamy or buffish white; flanks streaked with pale brown; bill black. It differs from the Red-headed Quelea in its usually smaller size, smaller bill and in having the red extending on to the chest. The female has the top of the head and nape narrowly streaked with tawny and black; superciliary streak buff; chin and throat pale yellowish buff; bill horn-colour. The male in non-breeding dress is similar to the adult female, but usually has some crimson markings or wash on the head, chin and throat. It may be distinguished from the female of the Red-headed Quelea, not only by its usually smaller size and distinctly smaller bill, but by the light streaking above having an olivaceous or yellowish tinge. Wing 57 to 63 mm. The young bird is warmer buff than the adult female, especially below, with some dusky specks on chest.

**Range in Eastern Africa:** Southern Abyssinia, south-eastern Sudan and Uganda to western Kenya Colony and north-western Tanganyika Territory.

**Habits:** Locally abundant and often in vast flocks in grass country or on cultivated land, with the habits of other Queleas; migratory, and irregular in breeding both as to season and locality.

**Nest and Eggs:** Nests of 'watch pocket' type, *i.e.* semi-domed with large entrance hole at the side of the top. Often the nest is among low plants and a leaf is drawn over to form the top. The nest takes about a day and a half to construct and both birds work at it (Granvik). It is made of fine grass without lining, suspended between stems of grass or herbs. Eggs two or three, pale bluish white densely mottled all over with reddish brown or greyish mauve with occasional darker scrawls, or heavily spotted and blotched with the same colours, or with pinkish or bluish green ground colour more or less heavily freckled; about 17 × 12·5 mm.

**Recorded breeding:** Uganda, August. North-western Kenya Colony, June to September. Southern Kenya Colony, March to June. Northern Tanganyika Territory, April and May.

**Food:** Grass seeds, corn, etc., at times a pest of crops.

**Call:** Noisy birds in the breeding season, the males making what Van Someren describes as a 'sizzling' call.

*Quelea cardinalis pallida* Friedm.
*Quelea cardinalis pallida* Friedmann, Proc. Biol. Soc. Washington 44, p. 119, 1931: Indunumara Mts., northern Kenya Colony.

**Distinguishing characters:** Differs from the nominate race in being paler tawny above with narrower dark streaks. Wing 56 to 60 mm.

**Range in Eastern Africa:** Northern Kenya Colony.

**Habits:** As for the nominate race.

**Recorded breeding:** No records.

*Quelea cardinalis rhodesiæ* Grant & Praed.
*Quelea cardinalis rhodesiæ* C. Grant & Mackworth-Praed, Bull. B.O.C. 64, p. 65, 1944: near Molito's, Petauke, east Luangwa district, north-eastern Northern Rhodesia.

**Distinguishing characters:** Adult male, differs from the nominate race in having the tawny and black streaked occiput clearly demarcated from the crimson crown and with no red suffusion; upperparts in both sexes darker and with broader and blacker streaks. Wing 56 to 63 mm.

**General distribution:** Kenya Colony to Tanganyika Territory and Northern Rhodesia.

**Range in Eastern Africa:** South central Kenya Colony to Tanganyika Territory.

**Habits:** As for the nominate race, breeding in large open colonies, nests a few yards apart, frequently in association with other Weavers, in long grass in swampy places.

**Recorded breeding:** Kenya Colony highlands, May to July. Tanganyika Territory, January to March.

# SPARROWS, WEAVERS, WAXBILLS, etc.

**BISHOPS and WIDOW-BIRDS.** FAMILY—PLOCEIDÆ. Genera: *Euplectes, Coliuspasser* and *Drepanoplectes.*

Seven species of Bishops and nine Widow-Birds occur in Eastern Africa. The males have a breeding and non-breeding dress, and in the Widow-Birds the females are appreciably smaller than the males. Mostly gregarious at some time of year. The Bishops, in particular, congregate into flocks in the non-breeding season when the males are not distinguishable in the field from the females. At certain times of the year all may be a nuisance in cultivated areas.

**1363 RED BISHOP.** *EUPLECTES ORIX* (Linnæus).
*Euplectes orix franciscana* (Isert). **Pls. 87, 90.**
*Loxia franciscana* Isert, Schriff. Ges. Nat. Fr. Berlin 9, p. 332, pl. 9, 1789: Accra, West Africa.

**Distinguishing characters:** Adult male, top of head to occiput, sides of face, ear-coverts and breast to belly black; rest of plumage except wings and tail red, occasionally orange yellow; mantle slightly brownish red; wings and tail dusky with lighter edges; upper and under tail-coverts usually equal with end of tail but not constantly so; bill black. The female is broadly streaked above with brownish buff and black; superciliary stripe buff; below, buffish; belly whiter; chest and flanks streaked with dark brown; bill horn-colour. The male in non-breeding dress is similar to the adult female, but larger. Wing, male 60 to 66, female 53 to 60 mm. The young bird is similar to the adult female, but has broader and paler buff edges to feathers of upperparts, wings and tail.

**General distribution:** Senegal to Abyssinia, Uganda and Kenya Colony.

**Range in Eastern Africa:** The Sudan to northern Abyssinia, Uganda and Kenya Colony as far south as Ravine and Lake Hannington.

**Habits:** Abundant in most localities and very conspicuous during the breeding season. The male hovers with a purring noise of the wings and with feathers much puffed out. Each male has a very definite territory. Definitely polygamous, and nests are not found in compact colonies but three or four within a small area. They have a habit of roosting in the nests of other species of birds which has given rise to some confusion. In the non-breeding season these

# SPARROWS, WEAVERS, WAXBILLS, etc.

birds gather into Sparrow-like flocks of large size in company with other species, and raid crops when ripe. Locally migratory.

**Nest and Eggs:** Nest in loose colonies in which each male has his own pitch, and probably three or four wives. Nests are oval of fine grass with entrance hole at the side near the top. Eggs two to four, usually three, pale turquoise blue; about $16 \cdot 5 \times 12 \cdot 5$ mm.

**Recorded breeding:** Sudan, Darfur, May to July. Kassala, August to November. Equatoria, September to November. Northern Abyssinia, June and July. Uganda, June to October.

**Food:** Seeds, grain, etc.

**Call:** Much noisy chattering, but little definite recorded.

*Euplectes orix sundevalli* Bp.
*Euplectes sundevalli* Bonaparte, Consp. Gen. Av. 1, p. 446, 1850: Eastern Transvaal, South Africa.

**Distinguishing characters:** Adult male, differs from the preceding race in being larger and in having the black confined to the forehead and usually also a black chin. The female is similar to that of the last race but larger. Wing, male 63 to 74, female 54 to 64 mm.

**General distribution:** Northern Rhodesia to southern Nyasaland, northern Portuguese East Africa, eastern half of Southern Rhodesia and the eastern Transvaal.

**Range in Eastern Africa:** Nyasaland and southern parts of northern Portuguese East Africa.

**Habits:** As for the preceding race, breeding in tall grass in low lying land. Nests woven on to upright stems and made of coarse grass outside with a closely woven inner lining which projects in a sort of porch; eggs somewhat larger, about $18 \times 14$ mm.

**Recorded breeding:** Nyasaland, January to March. Portuguese East Africa and Southern Rhodesia, January to March.

*Euplectes orix nigrifrons* (Böhm).
*Pyromelana nigrifrons* Böhm, J.f.O. p. 177, 1884: Karema, Ubende, western Tanganyika Territory.

**Distinguishing characters:** Adult male, differs from the last race in being more orange red. The female is similar to that of the last race. Wing, male 63 to 70; female 65 to 69 mm. The young bird has some brown streaks on the chest and flanks.

General distribution: Eastern Belgian Congo to Uganda, Kenya Colony, Tanganyika Territory and western Nyasaland.

Range in Eastern Africa: Central Uganda to Kenya Colony as far east as Machakos and western and south-central Tanganyika Territory.

Habits: As for other races, but usually not in colonies and often nesting some way from water.

Recorded breeding: Kenya Colony, mostly May to August. Tanganyika Territory, May, June and January to March.

*Euplectes orix pusilla* (Hart.).
*Pyromelana franciscana pusilla* Hartert, Bull. B.O.C. 11, p. 71, 1901: Lake Stephanie, southern Abyssinia.

Distinguishing characters: Differs from the West African race in being paler orange red and occasionally orange yellow; upper and under tail-coverts not extending to tip of tail. The female is paler than the female of the West African race; above streaked with buffish and black not brownish buff and black. Wing, male 61 to 66, female 55 to 60 mm.

Range in Eastern Africa: Southern and eastern Abyssinia, central Eritrea and the Somalilands.

Habits: As for other races.

Recorded breeding: Abyssinia, August and September onwards.

Distribution of other races of the species: Angola and South Africa, the nominate race being described from Angola.

### 1364 ZANZIBAR RED BISHOP. *EUPLECTES NIGROVENTRIS* Cassin. Pl. 90.

*Euplectes nigroventris* Cassin, Proc. Acad. Sci. Philad. p. 66, 1848: Zanzibar Island.

Distinguishing characters: Adult male, differs from the South African and Tanganyika Territory races of the Red Bishop in being smaller, and in having the top of the head deep orange and the whole underside black except for buffish thighs and orange under tail-coverts; neck in front occasionally red, or mixed red and black. The female is similar to that of the paler races of the last species, but the under wing-coverts and inner edges to the flight feathers are paler buff. The male in non-breeding dress is similar to the adult

## 952  SPARROWS, WEAVERS, WAXBILLS, etc.

female, but is larger. Wing, male 57 to 65, female 52 to 59 mm. The young bird is similar to the adult female but has paler buff edges to the feathers of the upperparts, wings and tail.

**Range in Eastern Africa:** Eastern Kenya Colony as far north as Lamu and as far west as Voi to eastern Tanganyika Territory as far west as Kilosa and northern Portuguese East Africa as far south as Mocuba; also Manda, Zanzibar and Kwale Islands.

**Habits:** A common bird of coastal Eastern Africa in small parties. In the breeding season it nests almost in colonies, but each male has its own territory and is definitely polygamous. Very noisy while breeding.

**Nest and Eggs:** Nests are long ovals of coarse grass lined with finer grass, sometimes with a porch, attached to grass, bushes, or reed stems in swampy places. The nest looks rather frail and transparent and is often conspicuous. Eggs two or three, pale blue occasionally spotted with dusky brown; about 17 × 12 mm.

**Recorded breeding:** Kenya Colony, Witu, May; Mombasa, April and July. Zanzibar Island, May to July, also October to December.

**Food:** Mainly grass seeds and rice in season.

**Call:** Noisy birds, but nothing particular recorded.

**1365 BLACK-WINGED RED BISHOP.** *EUPLECTES HORDEACEA* (Linnæus).

*Euplectes hordeacea hordeacea* (Linn.). **Pl. 90.**

*Loxia hordeacea* Linnæus, Syst. Nat. 10th ed. p. 173, 1758: Senegal.

**Distinguishing characters:** Adult male, very similar to the South African race of the Red Bishop and to the Zanzibar Red Bishop, colour varies from red to orange with only a narrow band of black on the forehead; wings black, first primary longer; tail longer and black; under tail-coverts buff; bill black. Occasionally albinistic or with general colour very pale and wings and tail white; bill and eye remain black. The female is broadly streaked with brownish buff and black above; superciliary stripe yellowish; below, yellowish buff with some narrow streaking on chest and flanks; wings, including wing-coverts, and tail dusky black; bill horn-colour. Occasionally assumes partial male breeding dress. The blackish wings, under wing-coverts and tail, longer first primary and yellowish wash below, distinguish the female from the other Red

Bishops. The male in non-breeding dress differs from the adult female in being larger, with broader black streaks above; wings and tail black; no yellowish wash below. Wing, male 71 to 86, female 64 to 75 mm. The young bird is similar to the adult female, but has broader sandy edges to the feathers of the mantle, wing-coverts, and inner secondaries, the male being larger than the female.

**General distribution:** Senegal to the Sudan, and south to Angola, South West Africa, the Zambesi River and eastern Southern Rhodesia.

**Range in Eastern Africa:** The western Sudan, Uganda, central and southern Kenya Colony and Tanganyika Territory, and to the Zambesi River, also Pemba and Zanzibar Islands.

**Habits:** A widespread, conspicuous and common species of grass country. The males make display flights over the grass tops with rump feathers puffed out and twittering calls, and very beautiful they are. They do not breed in their first season and are definitely polygamous. In the non-breeding season these birds occur in flocks with other Weavers and Bishops. The non-breeding plumage is acquired by a very rapid moult after breeding.

**Nest and Eggs:** Nests are domed purses of green grass woven on to two or three upright grass stems, or occasionally bushes or herbs, and have a porch of grass heads. Eggs two to four, pale turquoise blue, occasionally with spots of black, brown or violet; about 17·5 × 13 mm.

**Recorded breeding:** Sudan, August to October. Kenya Colony, mostly May and June. Tanganyika Territory, March to June. Pemba and Zanzibar Islands, May to August with a partial second brood in December and January. Nyasaland and Portuguese East Africa, February to May. Rhodesia, April and May.

**Food:** Grass seeds, corn and rice.

**Call:** Various twittering chattering cries.

*Euplectes hordeacea craspedoptera* (Bp.).
*Ploceus craspedoptera* Bonaparte, Consp. Gen. Av. 1, p. 446, 1850: Abyssinia.

**Distinguishing characters:** Adult male, differs from the nominate race in usually having the under tail-coverts white, often with black streaks. Wing, male 72 to 80, female 65 to 71 mm.

# SPARROWS, WEAVERS, WAXBILLS, etc.

**Range in Eastern Africa:** Southern Sudan and Abyssinia to Uganda and northern Kenya Colony as far south as the Nandi area.

**Habits:** As for the nominate race.

**Recorded breeding:** North-western Kenya Colony, June to August.

## 1366 BLACK BISHOP. *EUPLECTES GIEROWII* Cabanis.

*Euplectes gierowii friederichseni* Fisch. & Reichw. **Pl. 90.**

*Euplectes friederichseni* Fischer & Reichenow, J.f.O. p. 54, 1884: Nguruman, north of end Lake Natron, southern Kenya Colony.

**Distinguishing characters:** Adult male, occiput, sides of neck and collar on lower neck in front and chest, mantle and upper rump orange; rest of plumage black; bill black. It differs from the Black-winged Bishop in having the forehead and fore-crown, lower rump and upper tail-coverts black and a narrower collar on lower neck and chest. The female is broadly streaked above with buff and black; wings and tail dusky black edged with buff; below, yellowish buff, with black spots on chest; bill horn-colour. It is larger and has a larger bill than the female of the Black-winged Red Bishop. The male in non-breeding dress has the head, mantle and upper rump black edged with buff; superciliary stripe buff; sides of face mottled black and buff; below, buff with black streaks on chest. Wing, male 80 to 83, female 70 mm. The young bird is similar to the adult female but has smaller spots on the chest; the young male is larger than the young female.

**Range in Eastern Africa:** Southern Kenya Colony and northern Tanganyika Territory, south and east of Lake Victoria.

**Habits:** Little noted, but apparently those of other Bishops. Very local in open bush and rough cultivations.

**Nest and Eggs:** Nest made of coarse grass loosely woven and lined with grass seed heads, placed in the fork of a shrub or in among coarse grass and herbs by a river bank. Eggs dull greenish blue; about 20 × 15 mm.

**Recorded breeding:** Northern Tanganyika Territory, June, also (breeding condition) November.

**Food:** Grass seeds.

**Call:** Unrecorded.

## SPARROWS, WEAVERS, WAXBILLS, etc.

*Euplectes gierowii ansorgei* (Hart.).
*Pyromelana ansorgei* Hartert, in Ansorge's Under Afr. Sun. p. 344, pl. 2, 1899: Masindi, Unyoro, western Uganda.

**Distinguishing characters:** Adult male, differs from the preceding race in having the black of the throat extending further down, leaving a narrow collar on lower neck; above, the upper mantle only is lemon or orange yellow. The male in non-breeding dress, female and young bird are similar to those of the preceding race. Wing, male 82 to 95, female 72 to 77 mm.

**General distribution:** Abyssinia to the Sudan, north-eastern Belgian Congo and Uganda.

**Range in Eastern Africa:** Central Abyssinia to the Sudan and Uganda.

**Habits:** Local and uncommon, found in elephant grass in swampy places, usually one male to three or four females.

**Recorded breeding:** Uganda, June to October, also December.

**Distribution of other races of the species:** Angola, the habitat of the nominate race.

**1367 YELLOW BISHOP.** *EUPLECTES CAPENSIS* (Linnæus).
*Euplectes capensis xanthomelas* Rüpp. **Pls. 87, 90. Ph. xvii.**
*Euplectes xanthomelas* Rüppell, N. Wirbelt, Vög. p. 94, 1840: Temben, northern Abyssinia.

**Distinguishing characters:** Adult male, wholly black with bright lemon yellow wing shoulders, lower back, and rump; under wing-coverts buff; light edges to flight feathers; bill, upper mandible blackish slate, lower whitish. The female is broadly streaked above with brownish and black; rump mustard brown or mustard yellow; wing shoulder yellow with black centres to feathers; underside paler brown with narrower dusky streaks; throat and centre of breast to belly whitish or pale buffish; bill horn-colour. The male in non-breeding dress differs from the adult female in being larger and in having blacker streaks above; wing shoulders, lower back and rump lemon yellow; flight feathers and tail black. Wing, male 67 to 81, female 60 to 71 mm. The male in immature dress is similar to the adult female, but is larger. The young bird is streaked above with tawny and black including the wing shoulder; below, paler tawny; chest more spotted than streaked with dusky brown.

## 956 SPARROWS, WEAVERS, WAXBILLS, etc.

**General distribution:** Abyssinia and the Sudan to Angola and the north-eastern Transvaal.

**Range in Eastern Africa:** Abyssinia and the Sudan to the Zambesi River.

**Habits:** A common and widespread species of grassland, found both in open scrub or on grassy plains. It has definite breeding areas at which the males appear some time before the females. In courtship the male flies toward the female with vibrating wings and then drops into grass or bush. Collect in the usual flocks with other species in the non-breeding season. Believed to be polygamous.

**Nest and Eggs:** The nest is a semi-domed oval with a reinforced circular side of top entrance, made of grass and usually woven on to stems two of which support either side of the entrance. It is compact and stoutly made with coarse grass outside and usually placed low down in herbage. Eggs two or three, greyish white thickly spotted, streaked, or blotched with ashy grey and dark brown or sepia, but the ground colour may also be olive brown or greyish green; about 20 × 15 mm.

**Recorded breeding:** South-eastern Sudan, October. Abyssinia, July to September. Uganda, May and June, also October and November. Kenya Colony, April to August, also October and January. Tanganyika Territory, March to May, also January. Nyasaland, December to May.

**Food:** Grass seeds and corn when available, the young are fed on insects.

**Call:** Rather silent birds, but the call is a low chirp, and there is a feeble twittering song.

**Distribution of other races of the species:** West and South Africa, the nominate race being described from Cape of Good Hope.

### 1368 FIRE-FRONTED BISHOP. *EUPLECTES DIADEMATA* Fischer & Reichenow. Pl. 90.

*Euplectes diadematus* Fischer & Reichenow, O.C. p. 88, 1878: Malindi, eastern Kenya Colony.

**Distinguishing characters:** Adult male, forehead orange or reddish orange; rest of head, neck all round and underside black; mantle and rump golden yellow, former streaked with black; wings

## SPARROWS, WEAVERS, WAXBILLS, etc.

blackish; wing-coverts and inner secondaries edged with buff; primaries edged with yellow; tail ashy, under tail-coverts golden yellow; bill black. The female is streaked above with buff and black; below, chin to chest and flanks buff, more or less streaked with darker buff on chest and flanks; breast to belly white; bill horn-colour. It is very similar to the females of the Red Bishops, but may be distinguished by the yellow edges to the flight feathers which are not found in those species. The females of the Red-headed and Cardinal Quelea also have yellow edges to the flight feathers, but the female of the Red-headed Quelea is normally larger and rather heavier billed, and the female of the Cardinal Quelea is more densely marked above with a yellow washed chin and throat. The male in non-breeding dress is similar to the adult female, but is usually larger. Wing, male 60 to 62, female 57 to 60 mm. The young bird is very similar to the adult female.

**Range in Eastern Africa:** Eastern areas of Kenya Colony to Tanganyika Territory from Lamu to the Pangani River, and as far west as Marsabit, Chanler's Falls, Tsavo, Bura and Mkomasi.

**Habits:** Unrecorded as far as can be traced, but presumably those of other Bishops. It is common in rice fields of the coastal strip or in grass patches, but is not recorded as congregating into large flocks and appears to be scarce and local.

**Nest and Eggs:** Undescribed.

**Recorded breeding:** Taita, Kenya Colony (breeding condition), April.

**Food:** No information.

**Call:** A sharp 'ze-ze' and a grasshopper-like sizzling call (Moreau).

**1369 YELLOW-CROWNED BISHOP.** *EUPLECTES AFRA* (Gmelin).

*Euplectes afra afra* (Gmel.).
*Loxia afra* Gmelin, Syst. Nat. 1, pt. 2, p. 857, 1789: Senegal.

**Distinguishing characters:** Adult male, above bright lemon yellow; upper mantle more or less black; narrow band on forehead to ear-coverts, chin, neck in front and breast to belly black; band across chest, flanks and under tail-coverts bright lemon yellow; wings and tail blackish edged with buffish white; chestnut marks in centre of chest band; bill black. The female in breeding dress is

streaked above with buff and dusky black; edges of wing and tail feathers pale buff; broad superciliary stripe white or pale yellow and extending well behind eye; underside whitish washed with yellow and with some dusky streaks on chest and flanks; bill horn-colour. The male in non-breeding dress is similar to the adult female in non-breeding dress, but has blacker streaks above. The female in non-breeding dress is buff below, more distinctly streaked on chest and flanks. Wing, male 55 to 61, female 54 to 58 mm. The young bird is similar to the adult female but is browner above with buff edges to the feathers of the upperparts, wing-coverts and innermost secondaries; below, chin to breast buff; streaks on chest and flanks narrow.

**General distribution:** Fernando Po and Senegal to the Sudan.

**Range in Eastern Africa:** Western Sudan.

**Habits:** Absolutely confined to swamps in the western Sudan, and with the habits of the genus. The males erect the yellow feathers of the back into a sort of puff as they fly about and fight on the breeding grounds, and look like little golden balls buzzing about with rapid changes of direction. Probably polygamous.

**Nest and Eggs:** Neat compact nests with entrance at the side near the top, made of grass lined with fine grass heads usually placed in low tangled vegetation well out in a swamp over water. Eggs two to four, white and glossy with small intensely black spots and speckles; about 16·5 × 12·5 mm.

**Recorded breeding:** Nigeria, July to September. Darfur, Sudan, September and October.

**Food:** Grass seeds.

**Call:** Described as monotonous.

*Euplectes afra stricta* Hartl. **Pl. 90.**
*Euplectes strictus* Hartlaub, Syst. Orn. West-Afr. p. 129, 1857: Simen, northern Abyssinia.

**Distinguishing characters:** Adult male, differs from the nominate race in being larger and having the chin to belly uniform black. The female is also larger and in breeding dress is much browner than the female of the nominate race, has blacker stripes above, and broader blacker stripes on chest and flanks; chest to belly washed with yellowish. In non-breeding dress it is similar to the females of the other races but is larger. The male in non-breeding dress is similar to the female in breeding dress but is larger and has no yellowish wash below. Wing, male 68 to 72, female 64 to 68 mm.

**Range in Eastern Africa:** Eastern Sudan and Abyssinia.

**Habits:** A common resident of the high plateau of Abyssinia, nesting over running water and collecting into large flocks when not breeding. The flight is straight and fast at all times. Eggs normally one or two to the clutch.

**Recorded breeding:** Kassala, August and September. Abyssinia, January.

*Euplectes afra ladoensis* Reichw.
*Euplectes ladoensis* Reichenow, J.f.O. p. 218, 1885: Lado, southern Sudan.

**Distinguishing characters:** Adult male, differs from the nominate race in having the chest black, uniform with chin to belly; often yellowish bases to chest feathers. The female is similar to that of the nominate race. The male in non-breeding dress is similar to the adult female. Wing, male 56 to 61, female 55 to 61 mm.

**General distribution:** Eastern Belgian Congo to the Sudan, Kenya Colony and Tanganyika Territory.

**Range in Eastern Africa:** The southern Sudan to western and central Kenya Colony, Urundi and eastern Tanganyika Territory as far south as 150 miles south of Mt. Kilimanjaro.

**Habits:** As for other races, nesting in long grass in damp places. Eggs are bluish white sparsely spotted with reddish and dusky brown or even black at the larger end, two or three to the clutch; about 17·5 × 13 mm.

**Recorded breeding:** Central Kenya Colony, April, June and September. Tanganyika Territory (nestling), June.

*Euplectes afra niassensis* Meise.
*Euplectes afra niassensis* Meise, Mitt. Zool. Mus. Berlin 22, p. 150, 1937: Mitimone, Rovuma River, southern Tanganyika Territory.

**Distinguishing characters:** Adult female very similar in general colour above to the nominate race, but below, chest, breast and flanks buff with narrow brown streaks. Wing 54 mm. The male, seasonal and juvenile plumages are at present unknown.

**Range in Eastern Africa:** The type locality. This race is only known from one adult female.

**Habits:** No information.

**Recorded breeding:** No records.

**Distribution of other races of the species:** Portuguese Guinea, Angola and South Africa.

### 1370 FAN-TAILED WIDOW-BIRD. *COLIUSPASSER AXILLARIS* (Smith).

*Coliuspasser axillaris axillaris* (Smith). **Pl. 90.**
*Vidua axillaris* A. Smith, Ill. Zool. S. Afr. Aves, pl. 17, 1838: Eastern Cape Province.

**Distinguishing characters:** Adult male, wholly black including wings and tail, with an orange red wing shoulder; lesser wing-coverts and under wing-coverts brown; edges of greater wing-coverts and edges of flight feathers buff; bill bluish. The female is broadly streaked above with buff and black; superciliary stripe buff; wings and tail dusky with buff edges;·feathers of wing shoulder black with orange edging; below, buffish white; chest and flanks darker with some streaking; under wing-coverts brown; bill horn-colour. The male in non-breeding dress is similar to the adult female; wings as in breeding dress but the wing shoulder is more bright orange than red orange. Wing, male 81 to 95, female 68 to 78 mm. The immature male is similar to the adult female, but is larger. The young bird is rather browner above than the adult female and has paler under wing-coverts; feathers of the wing shoulder black edged with buff; the male is larger than the female. It may be distinguished from the young bird of the Yellow Bishop by the russet brown under wing-coverts and plain buff chest.

**General distribution:** Nyasaland and northern Portuguese East Africa, to eastern Cape Province and Natal.

**Range in Eastern Africa:** South-western areas of northern Portuguese East Africa and Nyasaland.

**Habits:** A common and widespread bird of swampy grass patches or stream sides. Its flight is noticeably heavy in comparison with other Weavers. Noisy birds but shy, and while breeding, very shy. The courtship flight is a slow flapping over the grass tops with rounded wings ending with a sudden backward jerk to land on a grass stem. In the non-breeding season congregate into considerable flocks.

**Nest and Eggs:** Nests are made of coarse grass lined with finer grass, thickly walled and semi-domed usually with living grass woven into the dome. They are placed low over water among tangled vegetation or close to the ground in grass tufts and are not easy to find, though at times almost in colonies. Eggs two to four, glossy greenish grey clouded and marbled with purplish brown and ash grey, and with darker purple spots and scrawls; about 19 × 13·5 mm.

**Recorded breeding:** Nyasaland, December to April. Portuguese East Africa, May. Natal, November to February.

**Food:** Corn, grass seeds and insects.

**Call:** Various twittering cries.

*Coliuspasser axillaris phœniceus* Heugl.
*Coliuspasser phœniceus* Heuglin, J.f.O. p. 304, 1862: Sobat River, eastern Sudan.

**Distinguishing characters:** Adult male, differs from the nominate race in having the wing shoulder more orange yellow, not so deep orange red, and in having the primary and secondary coverts wholly brown. Rarely the wing shoulder and wing-coverts are black with some brown feathers. The female is indistinguishable from that of the nominate race. Wing, male 78 to 95, female 65 to 77 mm.

**General distribution:** Abyssinia to Uganda, Kenya Colony, eastern Belgian Congo, Tanganyika Territory and northern Nyasaland.

**Range in Eastern Africa:** Southern Abyssinia to central and western Uganda, Kenya Colony and Tanganyika Territory, in non-breeding season to the Sudan.

**Habits:** As for other races, the call in the Sudan is described as a melancholy flute-like pipe.

**Recorded breeding:** Uganda in most months, commonest March to May, and October to December. Western Kenya Colony, April to July. Western Tanganyika Territory, January to March.

*Note:* Intermediates between this race and the Shoa race occur in the Kambata area of Abyssinia.

# 962  SPARROWS, WEAVERS, WAXBILLS, etc.

*Coliuspasser axillaris zanzibarica* (Shell.).
*Urobrachya zanzibarica* Shelley, P.Z.S. p. 586, 1881: Malindi, eastern Kenya Colony.

**Distinguishing characters:** Adults differ from the last race in having a stouter bill. Occasionally the female assumes male dress with a horn-white bill. Wing, male 81 to 96, female 70 to 74 mm. The young male has orange edging to the feathers of the wing shoulder and is paler above than the young male of the nominate race.

**Range in Eastern Africa:** Eastern Kenya Colony and Tanganyika Territory as far west as Kilosa and Iringa, also the Mafia Islands group, but not Zanzibar Island in spite of its name.

**Habits:** As for other races.

**Recorded breeding:** Mafia Island, May.

*Coliuspasser axillaris traversii* (Salvad.).
*Urobrachia traversii* Salvadori, Ann. Mus. Civ. Genova 26, p. 287, 1888: Sutta, Shoa, central Abyssinia.

**Distinguishing characters:** Adult male, differs from other races in having the wing shoulder orange yellow. The female is not distinguishable from other races. Wing, male 78 to 94, female 72 to 75 mm.

**Range in Eastern Africa:** Northern and central Abyssinia.

**Habits:** As for other races.

**Recorded breeding:** No records.

**Distribution of other races of the species:** Cameroons to Angola and Northern Rhodesia.

**1371 YELLOW-SHOULDERED WIDOW-BIRD. *COLIUSPASSER MACROCERCUS* (Lichtenstein). Pl. 90.**
*Fringilla macrocerca* Lichtenstein, Verz. Doubl. p. 24, 1823: Abyssinia.

**Distinguishing characters:** Adult male, general colour wholly black; wing shoulder and edge of wing yellow; light edging to flight feathers; under wing-coverts buff or buff and black; bill black; tail long and graduated. The female is broadly streaked above with buff and black; superciliary stripe, sides of face and throat yellowish

## SPARROWS, WEAVERS, WAXBILLS, etc.

buff; feathers of wing shoulder and bend of wing edged with chrome yellow; rest of underparts buff. It may be distinguished from the female of the Yellow Bishop, which also has yellow edges to the feathers of the wing shoulder, by its larger size and buff coloured rump streaked with black. The male in non-breeding dress differs from the female in being larger with broader markings above; wings as in male in breeding dress with the yellow shoulder. Wing, male 80 to 94, female 71 to 77 mm. The male in immature dress is similar to the adult female, but is larger. Occasionally the female is subject to melanism. The young bird is pale tawny; above broadly streaked with black; rather narrower streaks on top of head; wings and wing-coverts blackish edged and tipped with pale fawn colour; below, tawny; bill and feet pale horn.

**Range in Eastern Africa:** Southern Eritrea and northern and central Abyssinia to Uganda and western Kenya Colony.

**Habits:** A common resident of the high plateaux of Eritrea and Abyssinia, feeding in large flocks and roosting in reed beds. They fly low and fast and their wings make a swishing noise. The male in display is said to make a curious noise like rustling grass. Prefer grassy land near marshes.

**Nest and Eggs:** Flimsy transparent nests of fine grass with a side entrance and with living grass woven on to it. They are well hidden in a grass tuft, or slung on to longer grass, and there is a percentage of 'cocks' nests also. Eggs two to four, blue ground colour mottled all over with brownish olive, not unlike a small Crow's egg; about 20·5 × 14 mm.

**Recorded breeding:** Eritrea, August. Northern Abyssinia (nestlings), August. Yala River, Kavirondo, Kenya Colony, July.

**Food:** No information.

**Call:** A melancholy pipe.

1372 YELLOW-MANTLED WIDOW-BIRD. *COLIUSPASSER MACROURUS* (Gmelin).
*Coliuspasser macrourus macrourus* (Gmel.). **Pl. 90.**
*Loxia macroura* Gmelin, Syst. Nat. I, pt. 2, p. 845, 1789: Dahomey, West Africa.

**Distinguishing characters:** Adult male, differs from the Yellow-shouldered Widow-Bird in having the mantle as well as wing

shoulder yellow. Rarely melanic examples occur having the mantle and wing shoulders black. The female can be distinguished from the female of the Yellow-shouldered Widow-Bird by the duller black streaks on the upperparts. It can be distinguished from the female of the Yellow-Bishop by the brown, not mustard-coloured rump. The male in non-breeding dress is similar to the Yellow-shouldered Widow-Bird in the same dress but may be distinguished by the slightly duller black streaking of the upperparts. The female in non-breeding dress is suffused with yellow below. Wing, male 76 to 88, female 63 to 77 mm. The male in immature dress is similar to the adult female, but is larger and may be distinguished from the Yellow-shouldered Widow-Bird in immature dress by the duller black streaks on the upperparts. Rarely the female assumes partial male dress. The young bird has pale buff or fawn-coloured edges to the feathers of the upperparts and flight feathers and is buff below washed with yellow.

**General distribution:** Senegal and the Sudan to northern Angola, the Rhodesias, Nyasaland and Portuguese East Africa.

**Range in Eastern Africa:** The southern Sudan, Uganda and Kenya Colony to the Zambesi River.

**Habits:** Birds of open grassy plains and swampy low lying country, usually seen flitting about over the grass and settling on the tallest grass head. Widespread but local, congregating into large flocks to roost in the non-breeding season.

**Nest and Eggs:** A more or less spherical nest of grass, rather flimsy and transparent, but with much living grass woven on to it, usually in short grass in swampy places. Eggs two or three, blue or greenish blue, but almost covered with a fine dense speckling and scrolling of brown or grey brown; about 20 × 15 mm.

**Recorded breeding:** Nigeria, August to October. Southern Sudan, September and October. Uganda, August and September. Tanganyika Territory, March onwards, rather a late breeder. Southern Belgian Congo, January to April. Rhodesia, January to April. Nyasaland and Portuguese East Africa, January to April.

**Food:** No information.

**Call:** A thin 'z-e-e-e.'

## SPARROWS, WEAVERS, WAXBILLS, etc.

*Coliuspasser macrourus conradsi* Berger.
*Coliuspasser macroura conradsi* Berger, J.f.O. p. 487, 1908: Ukerewe Island, Lake Victoria, Tanganyika Territory.

**Distinguishing characters:** Adult male, differs from the nominate race in having a longer tail. Wing 78 to 86 mm.

**Range in Eastern Africa:** Ukerewe Island, southern Lake Victoria, Tanganyika Territory.

**Habits:** As for other races.

**Recorded breeding:** Ukerewe Island, June.

**1373** WHITE-WINGED WIDOW-BIRD.  *COLIUSPASSER ALBONOTATUS* (Cassin).
*Coliuspasser albonotatus albonotatus* (Cass.). **Pl. 90.**
*Vidua albonotata* Cassin, Proc. Acad. Sci. Philad. 4, p. 65, 1848: Durban, Natal, South Africa.

**Distinguishing characters:** Adult male, very similar to the Yellow-shouldered Widow-Bird, but smaller, with a shorter squarer tail; bases of the flight feathers, under wing-coverts and the greater part of the primary and secondary coverts white. The female is also very similar to the female of the Yellow-shouldered Widow-Bird, but is smaller, has light bases to the flight feathers below and very pale buffish white under wing-coverts. The male in non-breeding dress differs from the adult female in being larger and having the wings and shoulder patch of the male in breeding dress. Wing, male 72 to 81, female 62 to 72 mm. The young bird has the feathers of the upperparts edged with fawn colour; below, buff; chest and flanks darker; under wing-coverts and bases of flight feathers as in adult female.

**General distribution:** South-eastern Belgian Congo, Tanganyika Territory, Rhodesia and Nyasaland to Bechuanaland, the eastern Transvaal and Natal.

**Range in Eastern Africa:** Southern Tanganyika Territory to the Zambesi River.

**Habits:** Conspicuous birds locally common on open ground with long grass, usually seen sitting on grass heads and spreading their long split tails with a cheerful twittering. Definitely polygamous, the males having small territories about fifty yards square. They are said to fly faster and higher than most Widow-Birds.

**Nest and Eggs:** Nest spherical with a large side entrance made of woven grass blades, with an inner lining of fine thin grass heads coiled round with a few projecting as a sort of porch. Eggs two or three, pale greenish blue spotted and streaked with brown and grey; about 19 × 14 mm.

**Recorded breeding:** Tanganyika Territory, January to March. South-eastern Belgian Congo, April and May. Nyasaland, January to March. Portuguese East Africa, March. Rhodesia, January to March.

**Food:** Grass and other seeds.

**Call:** Nothing noted except the cheerful twittering mentioned above.

*Coliuspasser albonotatus eques* (Hartl.).
*Vidua eques* Hartlaub, P.Z.S. p. 106, pl. 15, 1863: Tabora, central Tanganyika Territory.

**Distinguishing characters:** Adult male, differs from the nominate race in having the wing shoulder cinnamon brown. Wing, male 74 to 85, female 63 to 72 mm.

**Range in Eastern Africa:** Western Sudan to southern Abyssinia, Uganda, Kenya Colony and Tanganyika Territory as far south as the Iringa area.

**Habits:** As for the nominate race.

**Recorded breeding:** Darfur, September, probably a migrant to breed. Kenya Colony, March to May, also December and January. Tanganyika Territory, January to March.

*Note:* In the Iringa area both yellow and cinnamon brown shouldered birds occur and in the Mzimba district of Nyasaland some cinnamon brown colour occurs in the yellow shoulder.

**Distribution of other races of the species:** São Thomé Island, Gabon and Angola.

**1374 MARSH WIDOW-BIRD.  COLIUSPASSER HARTLAUBI** (Bocage).
*Coliuspasser hartlaubi psammocromius* (Reichw.). **Pl. 90.**
*Penthetria psammocromia* Reichenow, O.M. p. 39, 1900: Tandala, Tukuyu district, south-western Tanganyika Territory.

**Distinguishing characters:** Adult male, general colour black with a long tail; wing shoulder yellow; primary and secondary

## SPARROWS, WEAVERS, WAXBILLS, etc.

coverts pale buff; under wing-coverts pale buff and black; bill bluish white. The female is broadly streaked above with black and brown; feathers of wing shoulder edged with dull yellow; below, buffish brown with brown streaks on chest and flanks; under wing-coverts mainly black; bill horn-colour. The male in non-breeding dress is larger than the female, more broadly streaked above and has the wings as in breeding dress. Wing, male 100 to 113, female 77 to 92 mm. The young bird has the feathers of the upperparts edged with tawny, underside browner than the adult female with no distinct streaks; bill flesh-coloured.

**General distribution:** Nyasaland and Tanganyika Territory.

**Range in Eastern Africa:** Southern Tanganyika Territory.

**Habits:** Locally common in sedgy swampy valleys. Habits little noted, apparently not differing from those of other Widow-Birds, but in the breeding season confined to patches of dense wiry grass about fifteen inches high.

**Nest and Eggs:** A domed nest of fine grass loosely woven, with a side entrance, placed among living grass which is bent down over it in swampy places in highland country. Eggs two; pale green freckled with brown and with pale lilac undermarkings; about 22 × 16 mm.

**Recorded breeding:** Tanganyika Territory, January, February and May. Nyika Plateau, Nyasaland, October and November.

**Food:** Seeds, berries and insects.

**Call:** Unrecorded.

*Coliuspasser hartlaubi humeralis* (Sharpe).
*Penthetriopsis humeralis* Sharpe, Bull. B.O.C. 11, p. 57, 1901: Nandi, western Kenya Colony.

**Distinguishing characters:** Adult male, differs from the preceding race in having the pale buff confined to the lesser wing-coverts, and the tail feathers shorter and broader. The female has broader tail feathers than the female of the preceding race. Wing, male 95 to 101, female 79 to 90 mm.

**General distribution:** Cameroons to Kenya Colony.

**Range in Eastern Africa:** Uganda and western Kenya Colony.

**Habits:** As for the nominate race, local and uncommon.

**Recorded breeding:** Uganda, May to August.

**Distribution of other races of the species:** Angola to Belgian Congo and Northern Rhodesia, the nominate race being described from Angola.

### 1375 RED-COLLARED WIDOW-BIRD. *COLIUSPASSER ARDENS* (Boddært).

*Coliuspasser ardens ardens* (Bodd.). **Pl. 90.**
*Fringilla ardens* Boddært, Tabl. Pl. Enl. p. 39, 1783: Eastern Cape Province, South Africa.

**Distinguishing characters:** Adult male, general colour black; band across lower neck in front yellow, orange, rusty orange or red; whitish edges to flight feathers and wing-coverts; tail long and graduated, central tail feather shortest; bill black. The female is streaked above with black and tawny; superciliary stripe yellowish buff; below, buff; belly whitish; chin and throat washed with yellow; chest deeper buff; under wing-coverts dusky brown and buff; feathers of wing shoulder edged with buff or tawny; bill horn-colour; occasionally assumes partial male dress. The male in non-breeding dress is larger, with broader black streaks above, and has a slightly longer tail than the female; wings and tail black; bill dusky horn. The male in immature dress is similar to the adult female, but is larger. Wing, male 68 to 81, female 62 to 72 mm.; tail, female 37 to 47 mm. The young bird usually has rather broader and paler tawny streaks above. The female, male in non-breeding dress, and the young bird can be distinguished from the Black-winged Red Bishop by the smaller bill and longer tail and the warmer tone of colour, especially below, the male is also more heavily streaked on the rump. Occasionally the male in non-breeding dress has elongated dusky brown tail feathers.

**General distribution:** The Sudan, Uganda and Kenya Colony to Angola, eastern Cape Province and Natal.

**Range in Eastern Africa:** Southern Sudan, Uganda and Kenya Colony to the Zambesi River.

**Habits:** Common in patches of grass in suitable country, feeding in a closely packed flock and making small local movements. In the breeding season the male has a noticeable display flight in conjunction with a curious ticking song. Tame, and roost in large flocks in elephant grass.

**Nest and Eggs:** The nest is a semi-transparent affair of woven grass lined with a well woven layer of dry grass and with projecting flower heads forming a porch. It is usually slung on to grass stems or placed in the top of a grass tuft with green living grass woven down to form the dome. Eggs two to four, bluish or greyish green, usually heavily blotched and speckled with greyish brown and dark umber brown, often forming a cap at the larger end; about 18·5 × 13·5 mm.

**Recorded breeding:** Uganda, April to June, also September. Kenya Colony highlands, April. Southern Belgian Congo, January to May. Rhodesia, January to May. Tanganyika Territory, January to May. Nyasaland, January to May.

**Food:** Seeds and insects.

**Call:** A plaintive squealing chirp, and a metallic grasshopper-like ticking song in the breeding season.

*Coliuspasser ardens concolor* (Cass.).
*Vidua concolor* Cassin, Proc. Acad. Sci. Philad. p. 66, 1848: Senegal, West Africa.

**Distinguishing characters:** Adult male, differs from the nominate race in having no band across the lower neck in front. The female, male in non-breeding dress and young bird are similar to the nominate race. Wing, male 72 to 82, female 59 to 71 mm.

**General distribution:** Senegal and the Sudan, Uganda, northern Angola, Tanganyika Territory, Nyasaland and Portuguese East Africa.

**Range in Eastern Africa:** South-western Sudan to Uganda, western, south-central and south-western Tanganyika Territory and Portuguese East Africa.

**Habits:** As for the nominate race.

**Recorded breeding:** Uganda, April onwards. Tanganyika Territory, January to March. Portuguese East Africa, February and March.

*Note:* In the eastern Belgian Congo, Uganda, western, south-central, and south-western Tanganyika Territory, Nyasaland, and northern Portuguese East Africa intermediates between this and the nominate race occur.

**1376** RED-NAPED WIDOW-BIRD. *COLIUSPASSER LATI-CAUDA* (Lichtenstein).

*Coliuspasser laticauda laticauda* (Licht.). **Ph. xvii.**

*Fringilla laticauda* Lichtenstein, Verz. Doubl. p. 24, 1823: Northern Abyssinia.

**Distinguishing characters:** Adult male, differs from the preceding species in having the crown of the head to the nape red, joining up on the sides of the neck with a red band on lower neck in front; the mantle feathers edged with tawny. The female and young bird are similar to the last species but they have longer tails. They are also very similar to the young birds of the Yellow Bishop, but may be distinguished by the more uniform dusky, less mottled wing shoulder and wholly dusky, not buff, under wing-coverts. The male in non-breeding dress is larger than the female, is more broadly streaked above and more thickly streaked below, and has the black wings and long tail of the breeding dress edged with tawny; bill horn-colour. The immature male is similar to the adult female, but is larger. Wing, male 75 to 90, female 66 to 71 mm.; tail, female 49 to 60 mm.

**Range in Eastern Africa:** Abyssinia to south-eastern Sudan.

**Habits:** Very much those of the Red-collared Widow-Bird. This species is a common but local resident of the high plateau of Abyssinia occurring on grassy slopes in lightly wooded country in small colonies. The male has a display flight in an almost vertical position, ending by a flapping glide on to a bush with outspread wings, while making a tinkling, ticking noise.

**Nest and Eggs:** The nest is started by weaving living grass into a ring, and a flimsy well-woven fine grass nest is then built into it and woven on. It is roughly spherical with entrance hole at the side. Eggs two or three, pale blue with streaks and blotches of pale sepia and greyish or purplish brown, and with pale grey or purple undermarkings; about $19 \cdot 5 \times 14$ mm.

**Recorded breeding:** Abyssinia, April and May, also September.

**Food:** Corn, seeds and insects, including termites.

**Call:** The song of the male is much like the noise of winding a clock; there is also a grasshopper-like trilling, and a noise like a toy balloon deflating.

*Coliuspasser laticauda suahelica* (V. Som.).

*Penthetria laticauda suahelica* Van Someren, Bull. B.O.C. 41, p. 121, 1921: Nairobi River, Kenya Colony.

**Distinguishing characters:** Adult male, similar to the nominate race but longer tailed. The females, males in non-breeding dress, and young birds of the two races are identical. Wing, male 73 to 86, female 65 to 74 mm.

**Range in Eastern Africa:** Kenya Colony, but not the coastal areas, and north-eastern Tanganyika Territory.

**Habits:** As for the nominate race, common in the Kenya Colony highlands from about 5,000 to 8,000 feet. Believed to be polygamous.

**Recorded breeding:** Kenya Colony, March to July, also December and January.

**1377** LONG-TAILED WIDOW-BIRD or SAKABULA. *COLIUSPASSER PROGNE* (Boddaert).
*Coliuspasser progne delamerei* Shell. **Pl. 90.**
*Coliuspasser delamerei* Shelley, Bull. B.O.C. 13, p. 73, 1903: Ngari Mossor, north of Mt. Kenya, Kenya Colony.

**Distinguishing characters:** Adult male, general colour wholly black including under wing-coverts; wing shoulder orange red; median wing-coverts white; tail long, central pair of feathers very long; bill bluish white. The female is streaked above with buff or tawny and black; below, streaked with pale buff on chest, breast and flanks; under wing-coverts black; tail feathers rather narrow and pointed; bill horn-colour. The male in non-breeding dress is larger than the female, is more broadly streaked above and below and has the wings and wing shoulder as in breeding dress; bil unchanged. Rarely the male in non-breeding dress has elongated brownish black tail feathers. Wing, male 127 to 147, female 93 to 110 mm. The male in immature dress is similar to the adult female, but is larger. The young bird is very similar to the adult female, but the male is larger and rather warmer coloured on the chest.

**Range in Eastern Africa:** Western and central Kenya Colony.

**Habits:** Local in swampy grassy places in flocks consisting of one or two males and a number of females. The males fly with tail drooping and somewhat spread, with slow regular movements of wings. In wet weather they cannot fly at all and are easily caught by hand. Congregate in the non-breeding season into flocks and roost in reed beds.

**Nest and Eggs:** Nests as for other species, placed in tussocky grass usually towards the top of a tussock, large domed structures of grass with a lining of seedheads. Eggs one to three, normally two, pale bluish green freckled and streaked with brown; about 23·5 × 16·5 mm.

**Recorded breeding:** Kenya Colony, February to July, mainly March and April, but depending on the rains.

**Food:** No information.

**Call:** A sharp loud alarm call.

**Distribution of other races of the species:** Angola and South Africa, the nominate race being described from the Cape of Good Hope.

**1378 JACKSON'S WIDOW-BIRD.** *DREPANOPLECTES JACKSONI* Sharpe. Pl. 90. Ph. xvii.

*Drepanoplectes jacksoni* Sharpe, Ibis p. 246, pl. 5, 1891: near Lake Nakuru, western Kenya Colony.

**Distinguishing characters:** Adult male, general colour black; wing shoulder wholly light brown or light brown or yellowish brown with black centres to feathers; rest of wing feathers edged with pale brown; under wing-coverts deep buff; tail long, feathers broad and curved towards ends; bill bluish or greenish and black. The female is similar to that of the Marsh Widow-Bird, but has buff, not mainly black, under wing-coverts; bill horn-colour. The male in non-breeding dress is larger and browner than the female; wings as in the male in breeding dress but with broad tawny brown edges to the feathers. Wing, male 87 to 95, female 78 to 86 mm. The immature male is similar to the adult female, but is larger. The young bird is very similar to the adult female, but the male is larger, chest warmer in colour and with less streaking.

**Range in Eastern Africa:** Western and central Kenya Colony to northern Tanganyika Territory at Loliondo and the Crater Highlands.

# SPARROWS, WEAVERS, WAXBILLS, etc. 973

**Habits:** Common and conspicuous species of highland grass country, especially where there are tufts of grass surrounded by short turf. The cocks make rings in the grass as playgrounds from which they dance up and down in the air most of the day. These rings have a tuft of grass in the centre, in which are usually two recesses. The bird springs about two feet with head thrown back, feet hanging, and tail curved up to touch the nape except for two feathers which point outwards and downwards. The playground ring is worn almost bare of grass and a female often sits on the tuft of grass in the centre. The species is believed to be polygamous as a rule. Large flocks in the non-breeding season on cultivated land, roosting in companies in reeds or long grass.

**Nest and Eggs:** A flimsy fine grass nest lined with seed heads with a side entrance, placed low in grass with growing grass woven on to the roof, not usually close to the dancing rings. Eggs two or three, pale blue or pale greyish green spotted, blotched and streaked all over with pale brown and greyish brown; about 22 × 15 mm.

**Recorded breeding:** Kenya Colony, Nandi, August and September. Nairobi, April to June. Northern Tanganyika Territory, January to March.

**Food:** Seeds and insects, very addicted to the seeds of *Crambe kilimandscharica* (Elliott).

**Call:** A soft 'tu' while flying and a little song of 'si-si-si-glip-glip-glip-glip.' The call in display is a little dull bell-like sound, and in the breeding season there are various clicks and whistles, as well as an extraordinary sound exactly like that made by an animal cropping grass.

## Addenda and Corrigenda

p. 824. For Families *Passeridæ* and *Ploceidæ* read Family *Ploceidæ*.

p. 837. For *Pirenestes maximus* read *Pirenestes frommi*.

p. 841. For *Pirenestes maximus* read *Pirenestes frommi*. Delete item 215.

**1286** *Dinemellia dinemelli*. *Recorded breeding:* add British Somaliland, April to June. Eggs up to six.

**1288** *Plocepasser superciliosus*. Probably best treated binomially and the two following races sunk in it.

*Plocepasser superciliosus brunnescens*. Regarded as synonymous with the nominate race (p. 865).

*Plocepasser superciliosus bannermani*. Regarded as synonymous with the nominate race (p. 865).

*Continued on p. 974*

## SPARROWS, WEAVERS, WAXBILLS, etc.

### Addenda and Corrigenda

**1293** *Passer domesticus* races. *P.d. indicus. General distribution:* delete Indo-China and Wady Halfa. The birds in the Wady Halfa area have been named by Meinertzhagen *P.d. halfæ*, Bull. B.O.C. 41, p. 67, 1921. Some specimens have a superficial resemblance to *P.d. indicus.* They are doubtfully separable from Egyptian birds.

**1294** *Passer motitensis shelleyi. Recorded breeding:* add British Somaliland, March and April.

**1298** *Passer castanopterus. Nest and Eggs:* add A typical Sparrow's nest under the eaves of a building. *Recorded breeding:* add British Somaliland, April to July.

**1301** *Passer swainsoni. Recorded breeding:* add Eritrea, January to March, also May to November. British Somaliland, April to July.

**1302** *Passer suahelicus. General distribution:* delete Northern Rhodesia.

**1305** *Auripasser euchlorus. Recorded breeding:* British Somaliland coast, April to June. Nests at any height in trees. Locally migratory and disappears in the non-breeding season.

**1308** *Petronia xanthosterna. Recorded breeding:* add Eritrea, January and February. *Habits:* add Said to be very destructive to young pine and cypress seedlings.

**1317** *Ploceus reichardi. Distinguishing characters:* add. Male in non-breeding dress olive-green above; mantle streaked with blackish; wings, tail and upper tail-coverts as in breeding dress; below, bright yellow; wash of saffron on chest; bill, upper mandible dusky, lower pale horn colour. Wing, male, 66 to 71 mm. *Nest and Eggs:* add Nest, short retort or kidney shaped, attached by one side to a single stem of elephant grass or to an Ambatch bush. Entrance hole below, no tube. Nest of grass lined with finer grass. Eggs vary from pale olive to pale blue in ground colour. *Recorded breeding:* add Rukwa Valley, January to April.

**1318** *Ploceus heuglini.* Now *P. atrogularis atrogularis* Heuglin, J.f.O., p. 245, 1864: Bongo, Bahr.-el; -Ghazal, south-western Sudan.

**1319** *Ploceus intermedius. Nest and Eggs:* add It has been observed several times that this bird only adds the spout to its nest when the young have already hatched out and just before they grow large enough to fall out through efforts to get to the food first. Also it is possible that this spout is added to prevent the nest from being easily plundered by birds of prey (Tribe).

**1320** *Ploceus golandi.* A second adult male of this species was collected in 1955 from a small and shy flock of birds feeding in forest tree-tops some 70 miles north of Mombasa. *Distinguishing characters:* add Eye rich brown; bill black; feet and toes flesh pink. Wing 71 to 74 mm. Stomach contents largely insects.

**1324** *Ploceus luteolus. Recorded breeding:* add Eritrea, June and July.

**1325** *Ploceus capitalis dimidiatus.* This race is confined to the eastern Sudan and Abyssinia. Southern birds from Uganda, Kenya Colony and Tanganyika Territory should be called *P.c. fischeri* Reichw.

**1330** *Ploceus castanops. Recorded breeding:* add Bukoba, western Lake Victoria, May.

**1338** *Hyphanturgus olivaceiceps. Nest and Eggs:* add Nest of Usnea lichen attached to the branch of a tree 25 ft. from the ground among lichens and very inconspicuous. Nest round, with entrance hole at bottom, without spout. Eggs two, bright turquoise blue; about 20 × 15 mm.

*Continued on p.* 1104

# SPARROWS, WEAVERS, WAXBILLS, etc.

## MANNIKINS, SILVER-BILLS, JAVA SPARROW and NEGRO-FINCHES. FAMILY—PLOCEIDÆ. Genera: *Spermestes, Amauresthes, Euodice, Odontospiza, Padda* and *Nigrita*.

Four species of Mannikin, two Silver-bills, the Java Sparrow, which is an introduced species, and three Negro-Finches occur in Eastern Africa.

### 1379 BRONZE MANNIKIN. *SPERMESTES CUCULLATUS* Swainson.

*Spermestes cucullatus cucullatus* Swain.
*Spermestes cucullata* Swainson, Bds. W. Afr. 1, p. 201, 1837: Senegal.

**Distinguishing characters:** Top of head, sides of face, wing shoulder and patch on flanks glossy bottle green; chin to upper chest glossy bronzy black with a wash of bottle green; hind neck, mantle and wings ash-coloured; lower rump and upper tail-coverts barred ashy and black; tail black; rest of underparts white; flanks and under tail-coverts barred with black; bill black and blue grey. The sexes are alike. Wing 45 to 52 mm. The young bird is wholly pale brown; paler below; tail black; bill blackish horn.

**General distribution:** Principé and São Thomé Islands, Senegal and Gabon, to Uganda and the Sudan. Introduced to Porto Rico Island, Greater Antilles.

**Range in Eastern Africa:** Uganda and the southern Sudan.

**Habits:** Tame, cheerful little birds, usually abundant, feeding on the ground like miniature Sparrows. Common on cultivated land, in villages, and in any type of open land or bush up to the edge of forest. Congregate into large flocks and follow their food supply as the various grasses or millet ripen. They roost communally, frequently in a densely packed cluster in an old nest, and often on top of one another, and have been known to build a nest for roosting purposes only. Enormously prolific, but the young are preyed on by very many species of birds, Kingfishers, Coucals, etc.

**Nest and Eggs:** Large untidy grass nests with a thick lining of feathers, and usually a porch of grass heads, placed in bushes, forks of branches or in thatch of huts, but in wild country very frequently in close association with wasps' or hornets' nests. Occasionally utilize the old nests of Weavers. Eggs four to eight, white; about 13·5 × 10 mm.

**Recorded breeding:** Nigeria, June to December. Southern Sudan, September. Uganda in many months, mainly April to August. Several broods in the year.

**Food:** Grass, rice and millet seeds, often feeding hanging on upside down.

**Call:** A sharp alarm note, and much wheezy twittering.

*Spermestes cucullatus scutatus* Heugl. **Ph. xviii.**
*Spermestes scutatus* Heuglin, J.f.O. p. 18, 1863: Lake Tana, northern Abyssinia.

**Distinguishing characters:** Differs from the nominate race in lacking the glossy bottle green patch on the flanks. Wing 45 to 54 mm.

**General distribution:** Abyssinia and the Sudan, to the Portuguese Congo, Angola, eastern Cape Province and Natal.

**Range in Eastern Africa:** Abyssinia and the Sudan to the Zambesi River, also Pemba, Zanzibar and Mafia Islands.

**Habits:** As for the nominate race.

**Recorded breeding:** Abyssinia, August and September. Kenya Colony, throughout the year. Tanganyika Territory, February to May. Zanizbar Island, all the year round. Nyasaland, August to May. Rhodesia, January to April. In most places there are several broods in the year.

*Note:* Intermediates between this and the nominate race occur in the south-western Sudan and Uganda, notably in the southern areas.

**1380 BLACK AND WHITE MANNIKIN.** *SPERMESTES POENSIS* (Fraser).

*Spermestes poensis poensis* Fras.
*Amadina Poensis* Fraser, P.Z.S. 1842, p. 145: Clarence, Fernando Po.

**Distinguishing characters:** Head to chest, mantle, wings and tail glossy black washed with violet; outer webs of flight feathers, rump and upper tail-coverts barred black and white; chest to under tail-coverts white; flanks black with white edges to feathers; bill blue grey. The sexes are alike. Wing 49 to 54 mm. The young bird is dusky above; chin to breast grey; rest of underparts buffish

## SPARROWS, WEAVERS, WAXBILLS, etc.

white; flanks dusky; outer webs of flight feathers faintly barred with brown; bill slate.

**General distribution:** Fernando Po and Cameroons to northern Angola, Belgian Congo, Abyssinia, Kenya Colony and Tanganyika Territory.

**Range in Eastern Africa:** South-western Sudan to south-western Abyssinia, Uganda, western Kenya Colony and north-western Tanganyika Territory.

**Habits:** Mainly a species of forest or forest edge in dense secondary growth, but has otherwise the same habits as the Bronze Mannikin with which it frequently consorts in mixed flocks. Locally migratory according to seasonal food supply.

**Nest and Eggs:** Large untidy grass nests often high in shady trees, but also occasionally in association with wasps. Eggs three to six, white; about 14 × 11 mm.

**Recorded breeding:** Uganda, April to August. Belgian Congo, December.

**Food:** As for the Bronze Mannikin.

**Call:** A plaintive chirp, not particularly described.

**Distribution of other races of the species:** West Africa.

### 1381 RUFOUS-BACKED MANNIKIN. *SPERMESTES NIGRICEPS* Cassin.

*Spermestes nigriceps nigriceps* Cass. **Ph. xviii.**
*Spermestes nigriceps* Cassin, Proc. Acad. N. Sci. Philad. p. 185, 1852: Zanzibar Island.

**Distinguishing characters:** Differs from the Black and White Mannikin in having the mantle, wing-coverts and innermost secondaries chestnut, sometimes washed with black; bill blue grey. The sexes are alike. Wing 46 to 51 mm. The young bird is slightly chestnut brown above, with faint brown bars on outer webs of flight feathers; below, dingy white; flanks brown; bill blue grey. The mantle has a chestnut wash and is much warmer brown than in the young bird of the Bronze Mannikin.

**General distribution:** Kenya Colony to Nyasaland, Southern Rhodesia, Portuguese East Africa and Natal; also southern Arabia.

**Range in Eastern Africa:** Central and eastern Kenya Colony to the Zambesi River, also Pemba and Zanzibar Islands.

**Habits:** Locally common species in palm groves, thick forest, bush or open woodland, especially in gardens or plantations where there is water. Their habits are essentially those of the Bronze Mannikin but they are somewhat more arboreal, and more partial to the vicinity of swamps and streams. They have been noted as flying into a white wall in strong sunlight and killing themselves. Inland a bird of middle or higher levels, but also abundant on the coast.

**Nest and Eggs:** Nest a grass ball like that of the Bronze Mannikin, with the entrance a mere slit in the side, but at times with a porch of grass heads and usually placed in trees at some height from the ground. Four or five birds have been noticed working at one nest. Eggs four to six, white; about 14 × 11 mm.

**Recorded breeding:** Kenya Colony highlands, March to July, also November to February. Kenya Colony coast, October to December. Tanganyika Territory, February and March. Zanzibar Island, October, December and March. Nyasaland, February to August.

**Food:** Mainly grass seeds but also insects, and have been recorded catching flying Termites.

**Call:** Not yet distinguished from that of the last species.

*Spermestes nigriceps minor* Erl.

*Spermestes nigriceps minor* Erlanger, O.M. p. 22, 1903: Fanole, lower Juba River, southern Italian Somaliland.

**Distinguishing characters:** Differs from the nominate race in being smaller. Wing 43 to 45 mm.

**Range in Eastern Africa:** Southern Italian Somaliland.

**Habits:** As for the nominate race.

**Recorded breeding:** No records.

1382 MAGPIE MANNIKIN. *AMAURESTHES FRINGILLOIDES* (Lafresnaye).

*Ploceus fringilloides* Lafresnaye, Mag. Zool. pl. 48, 1835: Liberia, West Africa.

**Distinguishing characters:** Bill large; head and neck all round, patch on sides of chest, tail, rump and upper tail-coverts glossy blue black; mantle brown with black centres to feathers and some white

Plate 90

Females of

| Red Bishop (p. 949) | Zanzibar Red Bishop (p. 951) | Black-winged Red Bishop (p. 952) | Black Bishop (p. 954) |
| Yellow Bishop (p. 955) | Fire-fronted Bishop (p. 956) | Yellow-crowned Bishop (p. 958) | Fan-tailed Widow-Bird (p. 960) |
| Yellow-shouldered Widow-Bird (p. 962) | White-winged Widow-Bird (p. 965) | Yellow-mantled Widow-Bird (p. 963) | Red-collared Widow-Bird (p. 968) |
| Marsh Widow-Bird (p. 966) | Jackson's Widow-Bird (p. 972) | Long-tailed Widow-Bird (p. 971) | Red-headed Weaver (p. 942) |

Plate 91

|  | Parasitic Weaver (p. 1000) |  |  |
|---|---|---|---|
| Male breeding dress | Male non-breeding dress | Female | Young |
| Red-billed Quelea (p. 944) | | Red-headed Quelea (p. 946) | |
| Male | Female | Male | Female |
| Black-bellied Seed-cracker (p. 989) | Lesser Seed-cracker (p. 990) | | Cardinal Quelea (p. 947) |
| Male | Male | Female | |
| Female | | | |
| | Red-faced Crimson-Wing (p. 992) | | Abyssinian Crimson-Wing (p. 993) |
| | Male | Female | |
| Warsangli Linnet (p. 1085) | Shelley's Crimson-Wing (p. 995) | | Dusky Crimson-Wing (p. 994) |
| | Male | Female | |

shaft stripes; wings dusky; rest of underparts white, lower belly and under tail-coverts wash_d with buff; flanks with black and brown markings; inner webs of flight feathers buffish white; bill, upper mandible black, lower grey. The sexes are alike. Wing 57 to 66 mm. The young bird is dusky brown above; lower rump, upper tail-coverts and tail bluish black; below pale buff; bill blackish horn.

**General distribution:** Senegal to the Sudan, Uganda, Northern Rhodesia, eastern Transvaal and Natal.

**Range in Eastern Africa:** Southern Sudan, Uganda, and Kenya Colony to the Zambesi River; also Zanzibar Island, but there uncommon.

**Habits:** Lively dainty little birds mostly seen feeding on the ground in flocks in cultivated land or gardens. Frequently associate with other Mannikins. Prolific, and the young assume adult plumage in about eight weeks and probably breed soon afterwards.

**Nest and Eggs:** Untidy grass nests with a few leaves and stems plastered on, often of green grass and lined with finer grass. They are usually placed in trees or bamboos about twelve or fifteen feet from the ground. Eggs up to six, white; measurements variable, about 17 × 12 mm.

**Recorded breeding:** Uganda, April to October. Tanganyika Territory, all the year round. Portuguese East Africa, February and March. Nyasaland, March and October.

**Food:** Grass seeds and millet, particularly fond of 'hill-rice.'

**Call:** A loud shrill chirruping 'pee-oo-pee-oo' uttered in flight, also a thin cheeping alarm note.

**1383** SILVER-BILL. *EUODICE MALABARICA* (Linnæus).
*Euodice malabarica cantans* (Gmel.).
*Loxia cantans* Gmelin, Syst. Nat. 1, pt. 2, p. 859, 1789: Senegal.

**Distinguishing characters:** Above ashy brown; forehead edged with paler colour, giving a faint scaly appearance; inner secondaries ashy brown with faint bars; remainder of flight feathers, upper tail-coverts and tail bronzy black, latter graduated and central tail feathers pointed; underparts white, but chin to chest and flanks suffused with buff; bill blue grey. The sexes are alike. Wing 52 to 57 mm. The young bird has faint buff edges to the feathers of the upperparts, and brown edges to the tail feathers; bill blue-grey.

# 980   SPARROWS, WEAVERS, WAXBILLS, etc.

**General distribution:** Senegal to the Sudan.

**Range in Eastern Africa:** The Sudan.

**Habits:** Common, tame and sociable little birds perching in trees in dense flocks almost touching one another. Abundant in most parts of the Sudan round villages, but are not particularly active birds and are recorded as sitting huddled together for long periods. They also extend far into desert country.

**Nest and Eggs:** Sparrow-like nest of grass, twigs, etc., lined with dry grass and with a pad of hair or feathers, placed in bushes, trees or thatch, but so often in Weavers' old nests that there is doubt as to whether they build nests for themselves at all in some areas. In their own nests there is a vertical side entrance and often some attempt at a porch over it. One or two instances of wasps' nests association have been recorded. Eggs up to twelve, normally five to six, white, without gloss; about 16 × 11 mm.

**Recorded breeding:** French West Africa, October and November. Sudan, all the year round but mostly September to May.

**Food:** Mainly grass and weed seeds.

**Call:** A whispering little double trill with a sharper chirping alarm note.

*Euodice malabarica orientalis* (Lor. & Hellm.).
*Aidemosyne orientalis* von Lorenz & Hellmayr, O.M. p. 39, 1901: Yeshbum, south-western Arabia.

**Distinguishing characters:** Differs from the nominate race in being browner above and with upperparts more distinctly barred. Wing 50 to 56 mm.

**General distribution:** The Sudan to Eritrea, Abyssinia and British Somaliland, to Uganda, Kenya Colony and Tanganyika Territory, also Arabia.

**Range in Eastern Africa:** Red Sea Province, Sudan, Eritrea, Abyssinia, British Somaliland, Uganda, Kenya Colony and north-western Tanganyika Territory.

**Habits:** As for the nominate race. Nests are described as substantial and bristly with large side entrances and porches, but probably the greater number are in the old nests of various species of Weavers.

## SPARROWS, WEAVERS, WAXBILLS, etc.

**Recorded breeding:** Kenya Colony highlands, May to July. Northern Tanganyika Territory, July.

**Distribution of other races of the species:** India and Ceylon.

### 1384 GREY-HEADED SILVER-BILL. *ODONTOSPIZA CANICEPS* (Reichenow).

*Pitylia caniceps* Reichenow, O.C. p. 139, 1879: Massa, Tana River, Kenya Colony.

**Distinguishing characters:** Adult male, head all round grey; sides of face and throat speckled with white; mantle and underside slightly vinous biscuit brown; rump and upper tail-coverts white; wings and tail black; wing-coverts ashy; bill blue black. The sexes are alike. Wing 55 to 62 mm. The young bird has no white speckles on the sides of the face and throat.

**Range in Eastern Africa:** Southern Abyssinia to Uganda, and Kenya Colony to the Iringa area of Tanganyika Territory.

**Habits:** Little known, except as birds of thorn bush country, coming to water holes or streams to drink in small flocks.

**Nest and Eggs:** Nests are large untidy structures of fine grass-heads lined with feathers placed at the ends of boughs of moderately high trees, or among the upgrowing shoots of pollarded trees. Eggs four or five, rarely six, white; about 16 × 12 mm.

**Recorded breeding:** Kenya Colony, May and June.

**Food:** Grass and other seeds.

**Call:** Unrecorded.

### 1385 JAVA SPARROW. *PADDA ORYZIVORA* (Linnæus).

*Loxia oryzivora* Linnæus, Syst. Nat. 10th ed. p. 173, 1758: Java, Dutch East Indies.

**Distinguishing characters:** Top of head, chin, upper throat, band behind ear-coverts, upper tail-coverts and tail black; sides of face to ear-coverts and under tail-coverts white; rest of plumage grey; belly and lower flanks vinous; bill pink. The sexes are alike. Wing 66 to 72 mm. The young bird is brownish and grey above; top of head greyer; below, brown; sides of face and ear-coverts brown, with a dusky band behind the latter.

**General distribution:** Java, Bali and Sumatra, introduced to St. Helena, the Seychelles, Pemba and Zanzibar Islands and the coast of Tanganyika Territory.

**Range in Eastern Africa:** Introduced and established in Pemba and Zanzibar Islands and on the coast of Tanganyika Territory.

**Habits:** Now fully established in Zanzibar Island, but does not appear to spread very fast.

**Nest and Eggs:** Nests are Sparrow-like grass structures in colonies in trees, walls, or roofs of houses. Eggs up to seven or eight, white; about 20 × 13·5 mm.

**Recorded breeding:** Zanzibar Town and northern part of Zanzibar Island, May to August.

**Food:** Corn and seeds.

**Call:** A soft metallic whistling song usually ending in a trill. Normal call a sharp 'chyup' with an occasional grating sound.

### 1386 GREY-HEADED NEGRO-FINCH. *NIGRITA CANICAPILLA* (Strickland).

*Nigrita canicapilla canicapilla* (Strickl.).
*Æthiops canicapilla* Strickland, P.Z.S. p. 30, 1841: Fernando Po.

**Distinguishing characters:** Forehead and sides of face black, bordered with whitish grey; top of head and mantle grey; rump whitish grey; wings and tail black; wing-coverts with white spots; underside black continuous with forehead and sides of face; under wing-coverts white; bill black. The sexes are alike. Wing 64 to 75 mm. The young bird is dull sooty black; faint white spots on wing-coverts; under wing-coverts white; bill blackish horn.

**General distribution:** Fernando Po to southern Nigeria, Portuguese Congo, Belgian Congo and Uganda.

**Range in Eastern Africa:** Bwamba area, western Uganda.

**Habits, Nest, Eggs, etc.:** See under the following race.

**Food:** No information, but the food in West Africa is largely palm-nut husk.

**Call:** A plaintive whistle of three notes, soft and only audible a few yards 'a-ee-o.'

*Nigrita canicapilla schistacea* Sharpe.
*Nigrita schistacea* Sharpe, Ibis p. 118, 1891: Sotik, western Kenya Colony.

**Distinguishing characters:** Differs from the nominate race in having the head and mantle darker grey. Wing 64 to 73 mm.

**Range in Eastern Africa:** Southern Sudan to Uganda, Kenya Colony and northern Tanganyika Territory.

**Habits:** Tame and inquisitive little birds found in pairs or small parties in the open parts of forests usually near water. Locally common and may be found at any level up to the tops of tall trees, often in company with Sunbirds.

**Nest and Eggs:** Large untidy nests of fibres, dead leaves, and banana leaf strips, with a side entrance, at any height from the ground in a tree. Eggs usually four, white, pinkish when fresh; about 17 × 13 mm.

**Recorded breeding:** Uganda, March to July. Tanganyika Territory (breeding condition), January.

*Nigrita canicapilla diabolica* (Reichw. & Neum.).
*Atopornis diabolicus* Reichenow & Neumann, O.M. p. 74, 1895: Kifinika, Mt. Kilimanjaro, north-eastern Tanganyika Territory.

**Distinguishing characters:** Differs from the preceding races in having the mantle blackish slate, a greyer border to black of forehead and sides of face, and a blackish grey rump; innermost secondaries sometimes tipped with white. Wing 63 to 67 mm.

**Range in Eastern Africa:** Mt. Kilimanjaro to Oldeani, northern Tanganyika Territory.

**Habits:** As for the preceding race but confined to mountain forest.

**Recorded breeding:** No records.

*Nigrita canicapilla candida* Mor.
*Nigrita canicapilla candida* Moreau, Bull. B.O.C. 62, p. 43, 1942: Ujamba, Kungwe-Mahare Mts., western Tanganyika Territory.

**Distinguishing characters:** Differs from the other races in having the hind crown, nape and upper mantle as well as rump whitish. Wing 70 mm.

984  SPARROWS, WEAVERS, WAXBILLS, etc.

**Range in Eastern Africa:** Kungwe-Mahare Mts., western Tanganyika Territory.

**Habits:** Unrecorded.

**Recorded breeding:** No records.

**Distribution of other races of the species:** West Africa and Angola.

### 1387 CHESTNUT-BREASTED NEGRO-FINCH. *NIGRITA BICOLOR* (Hartlaub).

*Nigrita bicolor brunnescens* Reichw. **Pl. 92.**

*Nigrita bicolor brunnescens* Reichenow, O.M. p. 173, 1902: Principé Island, Gulf of Guinea, West Africa.

**Distinguishing characters:** Above, sooty greyish brown; forehead with a varying amount of dark maroon; tail black; underside, dark maroon; bill black. The sexes are alike. Wing 58 to 64 mm. The young bird is browner above and pale chestnut brown below; sides of face, chin and throat brown.

**General distribution:** Principé Island, Nigeria, Cameroons, Gabon, and northern Angola to Uganda.

**Range in Eastern Africa:** Uganda.

**Habits:** Small parties in old forest clearings, little known.

**Nest and Eggs:** A large Estrilda-like nest of dry leaves lined with grass and grass-tops, in the fork of a bough. Eggs up to five, white; about 16 × 11·5 mm.

**Recorded breeding:** Southern Cameroons, May and June.

**Food:** Largely insects, but in West Africa palm-nut husks also.

**Call:** Males are said to have a sweet little song 'ki-yu-ki-yu-weh-weh-weh' (Bates).

**Distribution of other races of the species:** West Africa, the nominate race being described from the Gold Coast.

### 1388 WHITE-BREASTED NEGRO-FINCH. *NIGRITA FUSCONOTA* Fraser.

*Nigrita fusconota fusconota* Fras.

*Nigrita fusconotus* Fraser, P.Z.S. p. 145, 1842: Clarence, Fernando Po.

**Distinguishing characters:** Forehead to nape, sides of face, wing shoulder, lower rump, upper tail-coverts and tail glossy blue black; rest of upperparts earth brown; underside white or buffish white; bill black. The sexes are alike. Wing 50 to 56 mm. The young bird has the forehead to nape and sides of face dusky earth brown.

**General distribution:** Fernando Po, southern Nigeria, Cameroons and Gabon to northern Angola, Uganda and Kenya Colony.

**Range in Eastern Africa:** Uganda and western Kenya Colony.

**Habits:** Not uncommon in pairs or small parties in the more open parts of the forest. Usually seen hunting for insects and searching leaves like small Warblers or Bulbuls.

**Nest and Eggs:** A domed nest made of bark shreds, bast or palm-fibre, with a side entrance usually in a small tree or in the angle of a palm frond. Eggs up to six, white; about 14 × 11 mm.

**Recorded breeding:** Cameroons, June. Western Uganda, July and August.

**Food:** Insects, and like the preceding species is said to feed to some extent on the husk of the nuts of the Oil-palm.

**Call:** A high-pitched sizzling call in a descending scale 'te-te-te-te-teeeee' often repeated (Van Someren).

**Distribution of other races of the species:** Gold Coast Colony.

**TWIN-SPOTS, BLUE-BILLS, SEED-CRACKERS, CRIMSON-WINGS, CUT-THROAT, QUAIL and LOCUST-FINCHES, and PARASITIC WEAVER.** FAMILY—**PLOCEIDÆ.** Genera: *Clytospiza, Spermophaga, Pirenestes, Cryptospiza, Amadina, Ortygospiza, Anomalospiza, Hypargos* and *Mandingoa*.

Four species of Twin-spots, two Blue-bills, five Seed-crackers, four Crimson-wings, one Cut-throat, one Quail Finch, one Locust-Finch and one Parasitic Weaver occur in Eastern Africa.

**1389 BROWN TWIN-SPOT.** *CLYTOSPIZA MONTEIRI* (Hartlaub).
*Clytospiza monteiri ugandensis* (V. Som.). **Pl. 92.**
*Hypargus monteiri ugandensis* Van Someren, Bull. B.O.C. 41, p. 115, 1921: Masindi, western Uganda.

**Distinguishing characters:** Adult male, head grey; mantle and wings dusky brown; upper tail-coverts crimson; tail bronzy black;

below, chin and throat grey with a crimson stripe down centre of throat; rest of underparts chestnut thickly spotted with white; bill black. The female has a buffish white stripe down the centre of the chin and throat. Wing 55 to 61 mm. The young bird is rather browner above; chin and throat uniform grey; rest of underparts russet brown with faint indications of spotting.

**General distribution:** Eastern Cameroons to the Sudan and Uganda.

**Range in Eastern Africa:** Southern Sudan and Uganda.

**Habits:** Little known species of forest clearings or fringes, usually in pairs. Occasionally in cultivated lands or grass land outside forest.

**Nest and Eggs:** Believed to use the old nests of various species of Weavers, relining them with hair or fine fibres. J. G. Myers records small loosely made domed grass nests in the forks of trees. Eggs four to six, white; no measurements available.

**Recorded breeding:** Equatoria, September. Uganda, June and October.

**Food:** No information, except that they are fond of Termites.

**Call:** Nothing recorded except a constant twittering.

**Distribution of other races of the species:** West Africa to Angola, the nominate race being described from Bembe, Angola.

**1390** DUSKY TWIN-SPOT. *CLYTOSPIZA DYBOWSKII* (Oustalet). Pl. 92.

*Lagonosticta dybowskii* Oustalet, Le Nat. p. 231, 1892: Upper Kemo, Ubangi River, French Equatorial Africa.

**Distinguishing characters:** Adult male, head and neck all round, upper mantle and chest slate grey; mantle to upper tail-coverts crimson; wings dusky brown, with some white tips to coverts; breast to under tail-coverts black with white spots on breast and flanks; tail black; bill black. The female has the whole underside grey with white spots on breast and flanks. Wing 50 to 55 mm. Juvenile plumage unrecorded.

**General distribution:** Sierra Leone to northern Cameroons, French Equatorial Africa, northern Belgian Congo and the Sudan.

**Range in Eastern Africa:** South-western Sudan.

# SPARROWS, WEAVERS, WAXBILLS, etc.

**Habits:** Little known, in West Africa a bird of high wet grassland and rocks.

**Nest and Eggs:** Undescribed.

**Recorded breeding:** No records.

**Food:** Insects and seeds.

**Call:** Unrecorded.

**1391 RED-HEADED BLUE-BILL.** *SPERMOPHAGA RUFICAPILLA* (Shelley).

*Spermophaga ruficapilla ruficapilla* (Shell.).
*Spermospiza ruficapilla* Shelley, P.Z.S. p. 30, 1888: Bellima, Welle district, north-eastern Belgian Congo.

**Distinguishing characters:** Adult male, head all round, chin to breast, flanks and upper tail-coverts crimson; rest of plumage black; bill iridescent blue, cutting edges and tip red. The female has the mantle dark slate grey; belly grey densely spotted with white; bill as in the male. Wing 68 to 76 mm. The young bird is sooty black above, rather browner below; upper tail-coverts crimson; bill as in adult.

**General distribution:** Northern Angola to eastern Belgian Congo, the Sudan, Kenya Colony and Tanganyika Territory.

**Range in Eastern Africa:** South-eastern Sudan to Uganda, western Kenya Colony and western Tanganyika Territory at the Kungwe-Mahare Mts.

**Habits:** Uncommon, though possibly commoner than supposed, shy skulking silent little birds of the undergrowth of secondary evergreen forest in swampy places. Little is known of their habits, but they flick their wings and tail repeatedly. Moreau notes that they never appear to stand upright, but always adopt a flat crouching position.

**Nest and Eggs:** The nest is rather large and untidy, a dome of grass or leaves lined with fine grass and often covered by hanging fern fronds. Eggs three or four, white; no measurements available.

**Recorded breeding:** Uganda, April to July, also probably October to February.

**Food:** Probably mainly grass seeds.

**Call:** A scarcely audible squeak, and a clacking Chat-like call. There is also a low song composed of chuckling clinking noises.

*Spermophaga ruficapilla cana* (Friedm.).
*Spermospiza ruficapilla cana* Friedmann, Proc. New Engl. Zool. Cl. 10, p. 7, 1927: Amani, north-eastern Tanganyika Territory.

**Distinguishing characters:** Adult male, differs from the nominate race in having the black replaced by dark slate grey. The female differs from the female of the nominate race in having the mantle paler grey. Wing 70 to 72 mm. The young bird is browner than that of the nominate race.

**Range in Eastern Africa:** Usambara district, north-eastern Tanganyika Territory.

**Habits:** As for the nominate race, locally common in the Amani forests in eastern Tanganyika Territory. The normal pose is horizontal not upright, and there is constant flicking of wings and twitching of tail. Moreau has observed what appeared to be a breeding display of the male in September, and which he describes as very like that of a tame pigeon.

**Recorded breeding:** Amani, January.

**1392 GRANT'S BLUE-BILL.**   *SPERMOPHAGA POLIOGENYS* (O. Grant).
*Spermospiza poliogenys* O. Grant, Bull. B.O.C. 19, p. 32, 1906: Twenty miles north of Fort Beni, eastern Belgian Congo.

**Distinguishing characters:** Adult male, differs from that of the Red-headed Blue-bill in having the red on the top of the head confined to the forehead; rest of top of head black; mantle more glossy. The female has the top and sides of head to mantle dark grey; chin to breast red; rest of underparts spotted with white. Wing 65 to 73 mm. The young bird is grey or greyish brown; upper tail-coverts red; faint indications of spots below in female.

**General distribution:** Eastern Belgian Congo to Uganda.

**Range in Eastern Africa:** Bwamba area of western Uganda.

**Habits:** A little known inhabitant of heavy primary forest, usually skulking in undergrowth near the ground.

**Nest and Eggs:** Undescribed.

**Recorded breeding:** No records.

**Food:** No information.

**Call:** Unrecorded.

## SPARROWS, WEAVERS, WAXBILLS, etc.

**1393 BLACK-BELLIED SEED-CRACKER. *PIRENESTES OSTRINUS*** (Vieillot). **Pl. 91.**

*Loxia ostrina* Vieillot, Ois. Chant p. 79, pl. 48, 1805: Southern Gabon coast.

**Distinguishing characters:** Adult male, head all round, chin to breast, flanks, upper tail-coverts, central pair and outer webs of tail feathers crimson; rest of plumage black; bill blue black. The female has the black replaced by brown and the crimson is less bright. Wing 64 to 72 mm.; width of lower mandible at base 14 to 17 mm. The young bird is wholly brown with a duller red tail.

**General distribution:** Interior of Gold Coast to Gabon, the Belgian Congo, Uganda and Angola.

**Range in Eastern Africa:** Uganda.

**Habits:** Nowhere abundant, but not uncommon in small numbers in forest clearings or in dense scrub near water. Usually shy.

**Nest and Eggs:** A large untidy nest of coarse grass with loose ends hanging out and with a fine grass lining, usually at some height from the ground. Eggs three, white without gloss; about 18 × 14 mm. The male assists in the incubation.

**Recorded breeding:** Uganda, April.

**Food:** Hard weed seeds.

**Call:** Unrecorded.

**1394 ROTHSCHILD'S SEED-CRACKER. *PIRENESTES ROTHSCHILDI*** Neumann.

*Pyrenestes ostrinus rothschildi* Neumann, J.f.O. p. 528, 1910: Warri, lower Nigeria.

**Distinguishing characters:** Differs from the Black-bellied Seed-cracker in being smaller. Wing 60 to 65 mm.; width of lower mandible at base 12 to 13 mm.

**General distribution:** Gold Coast and Cameroons to Uganda and northern Angola.

**Range in Eastern Africa:** Uganda.

**Habits:** As for the last species, but little known. The relationship of the birds of this group is obscure, and whether they are species or races is still a matter of some controversy.

**Nest and Eggs and Call:** Undescribed.

**Recorded breeding:** No records.

### 1395 LARGE-BILLED SEED-CRACKER. *PIRENESTES MAXIMUS* Chapin.

*Pyrenestes ostrinus maximus* Chapin, Am. Mus. Nov. No. 56, p. 7, fig. 5, 1923: Faradje, north-eastern Belgian Congo.

**Distinguishing characters:** Differs from the two preceding species in having an appreciably larger and heavier bill. Wing 70 to 74 mm.; width of lower mandible at base 18 to 19 mm.

**General distribution:** Togoland and Northern Nigeria, to the Belgian Congo, the Sudan and Uganda.

**Range in Eastern Africa:** South-western Sudan and Uganda.

**Habits:** Inhabits savannah country and other patches of woodland but little recorded.

**Nest and Eggs:** Nest in fork of small tree in dense shade, oval, of fine grass without lining, decorated with ferns and long grass blades. Eggs four, dull white; about $18 \cdot 5 \times 14 \cdot 5$ mm. (E. T. M. Reid).

**Recorded breeding:** Equatoria, September.

**Food:** Hard seeds.

**Call:** Unrecorded.

### 1396 LESSER SEED-CRACKER. *PIRENESTES MINOR* Shelley. Pl. 91.

*Pyrenestes minor* Shelley, Ibis p. 20, 1894: near Zomba, southern Nyasaland.

**Distinguishing characters:** Adult male, forehead, fore-crown, sides of face, chin to chest, upper tail-coverts, central pair and outer webs of other tail feathers crimson; rest of plumage earth brown; bill black. The female differs from the male in having the lower throat and chest earth brown. Wing 58 to 62 mm.; width of lower mandible at base 9 to 12 mm. The young bird has the whole head and chin to chest earth brown; bill blackish horn.

# SPARROWS, WEAVERS, WAXBILLS, etc.

**General distribution:** Tanganyika Territory to Nyasaland and Portuguese East Africa as far south as the Beira area.

**Range in Eastern Africa:** Eastern Tanganyika Territory at the Uluguru Mts. and Pugu Hills to the Zambesi River.

**Habits:** Inhabits thick woodland along streams, or the edges of forest, usually in localities with a high rainfall. Generally seen low down among branches and undergrowth, but not in the densest cover, and usually in pairs.

**Nest and Eggs:** Little recorded, but said to make a large untidy domed nest with a small side entrance in a tree. Eggs five or six, white; about 17 × 13 mm. (Reichenow).

**Recorded breeding:** Nyasaland, January, March and December. Portuguese East Africa (apparently breeding), March.

**Food:** Seeds.

**Call:** A Sparrow-like 'tzeet' and a sharp clicking alarm note. There is a soft chittering trilling song.

## 1397 URUNGU SEED-CRACKER. *PIRENESTES FROMMI*
Kothe.

*Pyrenestes ostrinus frommi* Kothe, O.M. p. 70, 1911: Kitungulu, Urungu, Ufipa district, south-western Tanganyika Territory.

**Distinguishing characters:** Differs from the Lesser Seed-cracker in having an appreciably larger and heavier bill. Wing 61 to 69 mm.; width of lower mandible at base 13·5 to 16 mm.

**General distribution:** Tanganyika Territory to Chinteche, northern Nyasaland and northern Portuguese East Africa at Furancungo and Mocuba.

**Range in Eastern Africa:** Tanganyika Territory from the Uluguru Mts. to south-western areas and the Mocuba district of Portuguese East Africa.

**Habits:** Inhabits thick patches of bushy cover along waterways and edges of forest.

**Nest and Eggs:** Undescribed.

**Recorded breeding:** No records.

**Food:** No information.

**Call:** Unrecorded.

## 1398 RED-FACED CRIMSON-WING. *CRYPTOSPIZA REICHENOVII* (Hartlaub).

*Cryptospiza reichenovii australis* Shell. **Pl. 91.**

*Cryptospiza australis* Shelley, Ibis p. 184, 1896: Mt. Chiradzulu, southern Nyasaland.

**Distinguishing characters:** Adult male, lores and patch round eyes crimson red; head to upper mantle dark olive; rest of mantle to upper tail-coverts, wing-coverts and outer webs of innermost secondaries dark reddish crimson; wings and tail blackish; below, paler olive than head; flanks dark crimson; bill black. The female has the lores and round the eyes olive; below, paler olive than head, but chest to belly sometimes greyish olive. Wing 51 to 58 mm. The young bird has the mantle olive brown, with less crimson on the rump and upper tail-coverts than in the adult; bill black.

**General distribution:** Eastern Belgian Congo, Uganda and Tanganyika Territory to Nyasaland, eastern Southern Rhodesia and western Portuguese East Africa.

**Range in Eastern Africa:** Uganda and Tanganyika Territory to the Zambesi River.

**Habits:** Quite common and often abundant in forest or bamboo at all elevations, but shy silent birds easily overlooked and rarely seen outside dense shade. Usually seen in small parties feeding on the seeds of a grass that is found by stream sides in forest, and seldom fly more than a few yards. Will come out of forest to millet or other crops if they are near enough.

**Nest and Eggs:** Nest egg-shaped, of fine grass or skeleton leaves lined with feathers, grass seed-heads and horsehair-like mycelium, with a wide porch over the side entrance. Nests placed in bushes, tree ferns, etc., but particularly in the prickly heads of a forest shrub (*Cyclomorpha*), and on some occasions alongside the nests of a vicious red ant. Eggs three, white; about 17·5 × 13 mm.

**Recorded breeding:** Amani, Tanganyika Territory, apparently in any month of the year. Nyasaland, February to April. Portuguese East Africa, February and March.

**Food:** Grass and sedge seeds, millet if available.

**Call:** A high pitched sharp 'tzeet' also a little falling song of four notes followed by a chirp.

**Distribution of other races of the species:** Fernando Po and West Africa, the nominate race being described from Cameroons.

### 1399 ABYSSINIAN CRIMSON-WING. *CRYPTOSPIZA SALVADORII* Reichenow.

*Cryptospiza salvadorii salvadorii* Reichw. **Pl. 91.**
*Cryptospiza salvadorii* Reichenow, J.f.O. p. 187, 1892: Sciolitat, Shoa, central Abyssinia.

**Distinguishing characters:** Adult male, differs from the Red-faced Crimson-wing in having the lores and round the eye olive; general colour rather paler and the crimson redder in tone. The female is rather paler and the crimson redder than in the female of the Red-faced Crimson-wing. Wing 55 to 60 mm. The young bird has the mantle olive brown, with less crimson on the rump and upper tail-coverts than in the adult.

**Range in Eastern Africa:** Central and southern Abyssinia and northern Kenya Colony.

**Habits:** As for other races.

**Nest and Eggs:** See under other races.

**Recorded breeding:** South-western Abyssinia, August.

*Cryptospiza salvadorii ruwenzori* Scl.
*Cryptospiza salvadorii ruwenzori* W. L. Sclater, Bull. B.O.C. 46, p. 45, 1925: Mubuku Valley, Ruwenzori Mts., western Uganda.

**Distinguishing characters:** Differs from the nominate race in being much paler below. From the Red-faced Crimson-wing it can be distinguished by the duller crimson and the much paler underparts, the male having the lores and round the eyes olive. Wing 56 to 59 mm.

**General distribution:** Eastern Belgian Congo and Uganda.

**Range in Eastern Africa:** Western Uganda.

**Habits:** As for the following race, not uncommon in forest.

**Recorded breeding:** No records.

*Cryptospiza salvadorii kilimensis* Scl.
*Cryptospiza salvadorii kilimensis* W. L. Sclater, Bull. B.O.C. 55, p. 13, 1934: Ngare-Nairobi, west of Mt. Kilimanjaro, north-eastern Tanganyika Territory.

**Distinguishing characters:** Differs from the preceding race in being darker, rather more greyish olive below, and from the Red-

faced Crimson-wing in the male lacking the red on lores and round eyes, and in having a greyer tone on head and underside. Wing 54 to 59 mm.

**Range in Eastern Africa:** Southern Sudan and eastern Uganda, western and central Kenya Colony and north-eastern Tanganyika Territory.

**Habits:** In forest or dense riverside scrub at all elevations, shy skulking little birds which slip back into cover at any alarm. Locally not uncommon and come out to feed on grass seeds or crops when situated close to thick cover.

**Nest and Eggs:** Nest an oval ball of grass and tendrils covered with moss, with the opening near the top of the side, placed in dense creepers or in the fork of a small tree in forest. Eggs four or five, white, occasionally, if not always, with minute grey spots; about 17 × 12 mm.

**Recorded breeding:** Mt. Elgon, probably June and July, also December and January. Kenya Colony highlands, August to October, also probably March and April. Mt. Kilimanjaro, February.

**Food:** Seeds and small corn.

**Call:** A low 'chip-chip.'

**1400** DUSKY CRIMSON-WING. *CRYPTOSPIZA JACKSONI* Sharpe. Pl. 91.

*Cryptospiza jacksoni* Sharpe, Bull. B.O.C. 13, p. 8, 1902: Ruwenzori Mts., western Uganda.

**Distinguishing characters:** Head, sides of face, sides of neck, mantle to upper tail-coverts and flanks reddish crimson; collar on hind neck dark slate grey, extending more or less to centre of occiput and joining grey of underparts; below, dark slate grey; wings and tail blackish; bill black. The sexes are alike, but the female usually has the dark slate grey extending on to the crown of the head. Wing 55 to 60 mm. The young bird has the whole head dark slate grey and the crimson of the mantle to upper tail-coverts duller.

**General distribution:** Eastern Belgian Congo and Uganda.

**Range in Eastern Africa:** Ruwenzori Mts. to south-western Uganda.

Plate 92

Green-backed Twin-Spot (p. 1004)
Male   Female
African Fire-Finch (p. 1009)
Male   Female
Black-bellied Waxbill (p. 1030)
Male   Female
Orange-winged Pytilia (p. 1006)
Male
Brown Twin-Spot (p. 985)
Male   Female

Locust-Finch (p. 999)
Male   Female
Jameson's Fire-Finch (p. 1011)
Male   Female
Green-winged Pytilia (p. 1007)
Male   Female
Green-winged Pytilia (p. 1009)
Male
Dusky Twin-Spot (p. 986)
Male

Red-winged Pytilia (p. 1005)
Male
Chested-breasted Negro-Finch
Male   (p. 984)

Plate 93

Red-billed Fire-Finch (p. 1014)  
  Male  
  Female  
Black-headed Waxbill (p. 1028)  
Black-crowned Waxbill (p. 1029)  
  Red-cheeked Cordon-Bleu (p. 1035)  
  Male    Female  

Waxbill (p. 1020)  
Bar-breasted Fire-Finch (p. 1015)  
Fawn-breasted Waxbill (p. 1025)  
Yellow-bellied Waxbill (p. 1018)  
Cordon-Bleu (p. 1033)  
  Female    Male  

Zebra Waxbill (p. 1024)  
  Male  
  Female  
Black-faced Fire-Finch (p. 1016)  
  Male    Female  
Blue-capped Cordon-Bleu (p. 1037)  
  Male  
  Female

**Habits:** A bird of dense forest undergrowth, very shy and difficult to see. It occurs on Mt. Ruwenzori at between 6,000 and 8,000 feet, and comes out to feed on millet or grass seeds, disappearing into the forest again if disturbed. Said to be not uncommon near Mt. Kivu in the Belgian Congo.

**Nest and Eggs:** Undescribed.

**Recorded breeding:** No records.

**Food:** Small seeds.

**Call:** Unrecorded.

**1401 SHELLEY'S CRIMSON-WING.** *CRYPTOSPIZA SHELLEYI* Sharpe. Pl. 91.

*Cryptospiza shelleyi* Sharpe, Bull. B.O.C. 13, p. 21, 1902: Ruwenzori Mts., western Uganda.

**Distinguishing characters:** Adult male, head, sides of face, mantle, rump and upper tail-coverts dark red; wings and tail blackish; below, olive; belly and under tail-coverts black; lower flanks rufous; bill red. The female has the head and sides of face olive; bill, upper mandible black, lower red. The young bird is similar to the adult female. Wing 66 mm.

**General distribution:** Virunga Volcanoes of the eastern Belgian Congo to the Ruwenzori Mts., Uganda.

**Range in Eastern Africa:** Ruwenzori Mts., western Uganda.

**Habits:** Another shy species of dense forest undergrowth, rarely seen and diving into impenetrable jungle as soon as it is disturbed. Nothing is known of its life history.

**Nest and Eggs:** Undescribed.

**Recorded breeding:** No records.

**Food:** Small seeds.

**Call:** A series of twittering notes rising and falling 'tü-tü-tü-ti-ti-ti.'

**1402 CUT-THROAT.** *AMADINA FASCIATA* (Gmelin).

*Amadina fasciata fasciata* (Gmel.).

*Loxia fasciata* Gmelin, Syst. Nat. 1, pt. 2, p. 859, 1789: Senegal.

**Distinguishing characters:** Adult male, above, greyish fawn colour; head, rump and upper tail-coverts barred with black; mantle, scapulars and innermost secondaries more sparsely and irregularly

barred with black; below, a band of reddish crimson from ear-coverts to lower throat; chin buff; some black barring on lower neck in front; rest of underparts vinous fawn colour; centre of belly chocolate; upper belly and flanks irregularly barred with black. The female has the ear-coverts and lower throat fawn colour, speckled black; centre of belly white or with a paler chestnut patch. Wing 64 to 70 mm. The young male has a paler red band across the throat and the centre of the belly paler; the young female is similar to the adult female.

**General distribution:** Senegal and northern Nigeria to the Sudan and Uganda.

**Range in Eastern Africa:** The western and south-western Sudan and northern Uganda.

**Habits:** Birds of dry bush or open country, extending far into the desert, but very common round villages and cultivation. Breed in single pairs, but are highly gregarious in the non-breeding season and gather into dense packs. Said to come to drink in flocks in the middle of the day. General habits very Sparrow-like.

**Nest and Eggs:** A ball of dry grass with often a short funnel entrance, in a bush or in the hole of a tree, and lined with feathers. They also make extensive use of the old nests of other birds, mainly Ploceine Weavers. Eggs up to nine, normally four to seven, dull chalky white and oval; about 17 × 13·5 mm.

**Recorded breeding:** French Sudan, November. Sudan, January, February, August and September.

**Food:** Corn, seeds, insects, etc.

**Call:** Many chirruping calls of a Sparrow-like nature.

*Amadina fasciata alexanderi* Neum.
*Amadina fasciata alexanderi* Neumann, Bull. B.O.C. 23, p. 43, 1908: Waram, Hawash River, eastern Abyssinia.

**Distinguishing characters:** Differs from the nominate race in having broader barring both above and below. Wing 63 to 70 mm.

**Range in Eastern Africa:** Eritrea, Abyssinia and the Somalilands to south-eastern Sudan, Kenya Colony and eastern Tanganyika Territory.

**Habits:** As for the nominate race, a bird of dry country.

# SPARROWS, WEAVERS, WAXBILLS, etc. 997

**Recorded breeding**: Danakil, Abyssinia, May, November and December, in the nests of the White-headed Buffalo-Weaver. Kenya Colony, Turkana (breeding condition), July; highlands, May to August.

*Amadina fasciata meridionalis* Neun.
*Amadina fasciata meridionalis* Neunzig, J.f.O. p. 198, 1910: Northern Rhodesia.

**Distinguishing characters**: Similar to the Abyssinian race but differs in having a rather smaller bill. Wing 63 to 67 mm.

**General distribution**: Northern Rhodesia to southern Nyasaland, Southern Rhodesia, Bechuanaland except Ngamiland, and the Transvaal.

**Range in Eastern Africa**: Nyasaland.

**Habits**: As for the nominate race, often in flocks with Queleas and other Weavers. Appear to make general use of other birds' nests for nesting.

**Recorded breeding**: Nyasaland, March to May onwards.

**Distribution of other races of the species**: Ngamiland, northwestern Bechuanaland.

**1403  QUAIL-FINCH.    *ORTYGOSPIZA ATRICOLLIS*** (Vieillot).

*Ortygospiza atricollis atricollis* (Vieill.).
*Fringilla atricollis* Vieillot, N. Dict. d'Hist. Nat. 12, p. 182, 1817: Senegal.

**Distinguishing characters**: Adult male, forehead, sides of face and throat black; rest of upperparts including wings greyish brown; below, chin white or black; chest, breast and flanks barred earth brown and white; centre of lower breast and upper belly russet brown; lower belly white; under tail-coverts russet brown and black; tail blackish tipped with white; bill in breeding season red; in non-breeding season upper mandible black, lower red. The female has the forehead and sides of face earth brown; chin white, throat grey. Wing 47 to 55 mm. The young bird is rather paler above with lighter edges to the feathers; below, chest and breast brown unbarred; flanks barred dirty white.

**General distribution**: Senegal and Gambia to the Sudan and Uganda.

**Range in Eastern Africa:** Southern Sudan and north-western Uganda.

**Habits:** Unmistakable little birds of swamps and tussocky bogs, flying round with quick spasmodic flight and suddenly dropping to earth. In small parties or flocks in the non-breeding season, feeding on grassy plains or crops. They make an insistent metallic chirruping on the wing which is quite distinctive. In the courtship flight the male towers high into the air and falls like a stone to earth making the same sort of clicking noise.

**Nest and Eggs:** A pear-shaped rather rough nest of grass lined with feathers or soft material, placed on or near the ground among tussocky grass. It often has a rough porch over the entrance. Eggs usually four to six, white; about 15 × 11 mm.

**Recorded breeding:** Nigeria, September to December. Northern Uganda, June to November.

**Food:** Probably entirely grass seeds.

**Call:** As above, and occasionally the males give a rather harsh subdued warble while on the ground.

*Ortygospiza atricollis fuscocrissa* Heugl.
*Ortygospiza fuscocrissa* Heuglin, J.f.O. p. 18, 1863: Lake Tana, northern Abyssinia.

**Distinguishing characters:** Differs from the nominate race in being generally darker; white streaks on lores and round eyes; flanks usually barred black and white. Wing 52 to 59 mm.

**Range in Eastern Africa:** Eritrea, Abyssinia and northern Kenya Colony as far south as the Likipia Escarpment and the northern Uaso Nyiro.

**Habits:** As for the nominate race.

**Recorded breeding:** No records.

*Ortygospiza atricollis mülleri* Zedl. **Ph. xvii.**
*Ortygospiza atricollis mülleri* Zedlitz, J.f.O. p. 604, 1911: Simbiti, Wembere Steppe, north-central Tanganyika Territory.

**Distinguishing characters:** Similar to the last race, but above greyer in tone. Wing 53 to 56 mm.

**General distribution:** Kenya Colony, Tanganyika Territory and northern Nyasaland at Karonga.

**Range in Eastern Africa:** Kenya Colony from Baringo to Tanganyika Territory as far south as Njombe and south-western areas

**Habits:** As for other races.

**Recorded breeding:** Kenya Colony highlands, April to August, also January. Tanganyika Territory, northern areas, February, May, June and October. Dar-es-Salaam, July.

*Ortygospiza atricollis dorsostriata* V. Som.
*Ortygospiza atricollis dorsostriata* Van Someren, Bull. B.O.C. 41, p. 115, 1921: Southern Ankole, south-western Uganda.

**Distinguishing characters:** Similar to the Abyssinian race but has much less or no white above lores and round eyes, and little or no white on chin. It is darker and browner above, less grey, than the preceding race. The female has the chin grey. Wing 50 to 53 mm.

**Range in Eastern Africa:** Western and southern Uganda.

**Habits:** As for other races.

**Recorded breeding:** No records.

**Distribution of other races of the species:** West and South Africa and the Rhodesias.

**1404 LOCUST-FINCH.** *ORTYGOSPIZA LOCUSTELLA* (Neave).
*Ortygospiza locustella locustella* (Neave). **Pl. 92.**
*Paludipasser locustella* Neave, Bull. B.O.C. 25, p. 25, 1909: Upper Luansenshi River, north-east of Lake Bangweolo, north-eastern Northern Rhodesia.

**Distinguishing characters:** Adult male, forehead, sides of face, chin to chest, and all except central upper tail-coverts vinous crimson; edges of flight feathers pale vinous crimson; fore-crown to nape streaked black and brown; wing-coverts pale vinous crimson with whitish tips; innermost secondaries black spotted with white; rest of plumage dull black; mantle and rump spotted with white; under wing-coverts buff; bill red, in non-breeding season upper mandible black, lower red. The female has the forehead and sides of face black, chin to belly buffish white; flanks barred black and white. Wing 44 to 47 mm. The young bird has the whole of the upperside streaked black and brown with some white spotting on mantle and rump; flight feathers and wing-coverts edged with brown; below, browner than the adult female; bill dusky black.

**General distribution:** Northern Rhodesia as far west as Ndola and Kalomo to Nyasaland, Tanganyika Territory, the Tete area of Portuguese East Africa and Southern Rhodesia.

**Range in Eastern Africa:** Southern Tanganyika Territory from the Iringa district to the Northern Rhodesia boundary.

**Habits:** Very like those of the Quail-Finch but more local, inhabiting wet grassland and bogs of fine wiry grass. Locust-like little birds in dense flocks which only travel a few yards at a time, but fly straight and fast with rapid dipping flight. Individuals lie closely and then rise quickly to some height before descending to the ground again.

**Nest and Eggs:** A domed nest of fine soft grass lined with a few feathers among matted grass on the ground. Eggs believed to be five or six but clutches of two are also recorded, plain white; about 13 × 10 mm.

**Recorded breeding:** Sao, Iringa district, Tanganyika Territory, probably March to May. Northern Nyasaland, January to March and October. Rhodesia, May, probably January to May. Portuguese East Africa, March.

**Food:** Grass seeds.

**Call:** A querulous 'pink-pink.'

**Distribution of other races of the species:** Belgian Congo.

**1405 PARASITIC WEAVER.** *ANOMALOSPIZA IMBERBIS* (Cabanis).
*Anomalospiza imberbis imberbis* (Cab.). **Pl. 91.**
*Crithagra imberbis* Cabanis, J.f.O. p. 412, 1868: Mombasa, eastern Kenya Colony.

**Distinguishing characters:** This is a species of very distinctive parasitical breeding habits, and also the young bird is very dissimilar to the adult, having a pattern of plumage very like a young *Cisticola*. Adult male, forehead, sides of face and underside canary yellow; rest of upperparts greenish yellow streaked with black; wings and tail dusky edged with greenish yellow; lower flanks streaked with dusky; bill almost black. In fresh dress the forehead to rump is more olive green with black streaks and greyish edges to feathers of mantle and scapulars; general tone duller and darker than in worn

## SPARROWS, WEAVERS, WAXBILLS, etc.

breeding dress; below, yellow slightly washed with olivaceous and with paler edges to the feathers; bill dusky horn-colour. In immature dress the forehead to rump is darkish olive green, streaks dull blackish and feathers edged with greyish; general tone duller and darker than the adult; head and hind neck variously streaked or plain; below, dull yellow with pale edges to the feathers, giving a sort of silvery appearance; bill dusky or horn-colour. The female is buffish tawny above, broadly streaked with black; superciliary stripe buff; dusky stripe through eye; below, chin, throat and belly whitish; chest, breast and flanks buffish brown streaked darker brown; sides of face, superciliary stripe, and chin and throat usually washed with yellowish. The immature female has the sides of face and chest more dusky; breast and flanks distinctly streaked. Wing 63 to 73 mm. The young bird is tawny or sandy buff above, broadly streaked with black; inner secondaries and tail black with broad sandy buff edges; sides of face and chin and throat rather more tawny; rest of underparts pale buffish white, with some black streaks on lower flanks; bill, upper mandible black, lower straw coloured with extreme tip black.

**General distribution:** Abyssinia and the Sudan to the Belgian Congo and Tanganyika Territory.

**Range in Eastern Africa:** Abyssinia, the Sudan, Uganda, Kenya Colony and Tanganyika Territory; also Pemba and Zanzibar Islands.

**Habits:** Locally common species, congregating into large flocks in the non-breeding season, usually in bush, cultivated land or open grass country. Definitely parasitic, but otherwise with the habits of any other small ground-feeding Weaver-Finch.

**Nest and Eggs:** Parasitic on several species, particularly on Cisticolas, Prinias, Quail-Finches and possibly on some Sparrows. Eggs almost certainly of this species are thick shelled, pale blue or bluish white with sparse blackish spots, and violet undermarkings; about 17 × 13 mm., usually larger than those of the host.

**Recorded breeding:** Abyssinia (nestlings), August and November. Tanganyika Territory (nestlings), May. Pemba Island, September to January.

**Food:** Probably largely grass seeds.

**Call:** The song is a squeaky 'tsileu-tsileu-tsileu' and there is much chattering during flight.

*Anomalospiza imberbis rendalli* (Trist.).

*Crithagra rendalli* Tristram, Ibis p. 130, 1895: Barberton, Transvaal, South Africa.

**Distinguishing characters:** Adult male, in worn breeding dress similar to the nominate race; in fresh dress it differs in being brighter above, greener and with narrower greyish edges to the feathers; below, brighter, yellower and with narrower light edges to the feathers; bill sepia, keel and base of lower mandible pallid flesh; *i.e.* blackish in dried skin and not horn-colour. Female and young bird indistinguishable from the female and young bird of the nominate race. Wing 64 to 73 mm.

**General distribution:** South-eastern Belgian Congo, Rhodesias and Nyasaland to Portuguese East Africa, Damaraland and the Transvaal.

**Range in Eastern Africa:** Portuguese East Africa at Ile to the Zambesi River.

**Habits:** As for the nominate race.

**Recorded breeding:** Nyasaland and Portuguese East Africa (breeding condition), January and February. South-eastern Belgian Congo, December. Southern Rhodesia, February.

**Distribution of other races of the species:** Sierra Leone to Cameroons.

**1406 PETER'S TWIN-SPOT.** *HYPARGOS NIVEOGUTTA-TUS* (Peters).

*Spermophaga niveoguttata* Peters, J.f.O. p. 133, 1868: Inhambane, Portuguese East Africa.

**Distinguishing characters:** Adult male, top of head greyish brown; mantle to rump and wings russet brown, mantle sometimes washed with crimson; hind neck washed with deep crimson; upper tail-coverts deep crimson; tail blackish and deep crimson; below, lores, over eyes, sides of face, sides of neck and chin to chest deep crimson; rest of underparts black; sides of chest and flanks spotted with white. The female has the top of the head and sides of face greyer; chin to chest buff and paler crimson; rest of underparts grey or buffish grey spotted as in male with each spot encircled in black. Wing 52 to 60 mm. The young bird is on the upperside similar to, but paler than, the adult female; chin buff; rest of underparts russet brown; centre of breast to belly black.

## SPARROWS, WEAVERS, WAXBILLS, etc.

**General distribution:** Kenya Colony, Tanganyika Territory, the Belgian Congo, Northern Rhodesia, eastern Southern Rhodesia and Portuguese East Africa.

**Range in Eastern Africa:** Eastern Kenya Colony and Tanganyika Territory to the Zambesi River.

**Habits:** Quiet little birds found in pairs or small parties along the heavily bushed sides of streams, or the edges of forest, or in thick tangled bush where there is long grass. Tame and inquisitive and usually seen feeding on the ground on paths through bush near water.

**Nest and Eggs:** A rather large domed nest grass with a lining of feathers, either on the ground or in a low bush. Some evidence of occasional association with hornets' nests. Eggs three, white; about 16 × 12·5 mm.

**Recorded breeding:** Kilindini, Kenya Colony, July. Northern Tanganyika Territory (breeding condition), April. Rhodesia, April and May, also probably December. Nyasaland, January to May. Portuguese East Africa, January to March.

**Food:** Grass and weed seeds.

**Call:** A hissing sound. The song is a feeble stuttering trill and there are also faint grasshopper-like squeaks.

### 1407 GREEN-BACKED TWIN-SPOT. *MANDINGOA NITIDULA* (Hartlaub).

*Mandingoa nitidula nitidula* (Hartl.).

*Estrilda nitidula* Hartlaub, Ibis p. 269, 1865: Natal, South Africa.

**Distinguishing characters:** Adult male, lores and round eyes to chin tomato red; whole upperside, sides of neck, throat to chest and under tail-coverts olive green; rump golden olive; breast to belly black densely spotted with white; wings dusky edged with olive green; tail black edged with olive green, central tail feathers wholly olive green. The female is paler than the male; lores and round eyes to chin pale orange. Wing 50 to 53 mm. The young bird is duller olive green; rump olive green; lores and round eyes to chin buff; below, grey.

**General distribution:** Portuguese East Africa to eastern Transvaal and Natal.

**Range in Eastern Africa:** Ribaue, Portuguese East Africa, to the Zambesi River.

**Habits:** As for other races.

**Recorded breeding:** Portuguese East Africa (about to breed), May.

*Mandingoa nitidula schlegeli* (Sharpe).
*Pytelia schlegeli* Sharpe, Ibis p. 482, pl. 14, 1870: Fantee, Gold Coast Colony.

**Distinguishing characters:** Adult male, differs from the nominate race in being larger billed; lores and round eyes to chin rich tomato red; throat and chest orange red. The female has the throat to chest golden olive. Wing 48 to 55 mm.

**General distribution:** Fernando Po, Sierra Leone and Cameroons to northern Angola and Uganda.

**Range in Eastern Africa:** Western Uganda.

**Habits:** As for other races, rare and little seen.

**Recorded breeding:** Cameroons (nestling), November.

*Mandingoa nitidula chubbi* (O. Grant). **Pl. 92.**
*Pytelia chubbi* O. Grant, Bull. B.O.C. 29, p. 64, 1912: Marsabit, northern Kenya Colony.

**Distinguishing characters:** Adult male, differs from the nominate race in having the throat to chest washed with golden olive or more rarely with orange red; bill smaller than in the preceding race. The young bird is brighter olive green above than the young bird of the nominate race and washed with olive, less grey below. Wing 46 to 54 mm.

**General distribution:** Abyssinia and Kenya Colony to north-eastern Northern Rhodesia, Nyasaland and eastern Southern Rhodesia.

**Range in Eastern Africa:** Southern Abyssinia to Kenya Colony, Ukerewe Island, Tanganyika Territory and Nyasaland; also Pemba and Zanzibar Islands, where it is rare.

**Habits:** Silent inconspicuous little birds of forest edge or dense thickets. On Pemba Island live in thickets in the clove plantations and are much prized as cage-birds and heavily trapped in consequence. In Nyasaland widely distributed in evergreen forest.

Curiously prone to flying into telegraph wires, etc. (Moreau). Feed largely on the ground.

**Nest and Eggs:** A domed grass nest, rather large and untidy in the branches of trees. Eggs white; about 16·5 × 11·5 mm.

**Recorded breeding:** No records.

**Food:** Mainly grass seeds.

**Call:** A chirping 'tzeet' and various squeaky noises. There is also a sweet little subdued song.

## PYTILIAS, FIRE-FINCHES, WAXBILLS, OLIVE-BACKS, CORDON-BLEUS and GRENADIER. FAMILY—PLOCEIDÆ.

Genera: *Pytilia, Lagonosticta, Coccopygia, Estrilda, Nesocharis, Uræginthus* and *Granatina.*

Three species of Pytilias; six Fire-Finches; eleven Waxbills; two Olive-backs; three Cordon-bleus and one Grenadier occur in Eastern Africa.

Except for the Olive-backs all are common and widespread species, many of which are plentiful around villages and farm lands, and spend much of their time on the ground. They feed mainly on weed seeds and do little harm agriculturally.

**1408** RED-WINGED PYTILIA. *PYTILIA PHŒNICOPTERA* Swainson.

*Pytilia phœnicoptera lineata* Heugl. **Pl. 92.**

*Pytelia lineata* Heuglin, J.f.O. p. 17, 1863: Lake Tana, northern Abyssinia.

**Distinguishing characters:** Adult male, general colour dark grey; outer edges of flight feathers, upper wing-coverts, lower rump, upper tail-coverts, central tail feathers and outer edges of rest of tail feathers crimson; breast to under tail-coverts barred with white; eye brown; bill red. The female is rather browner above. Wing 59 to 61 mm. The young bird is browner; the crimson colour much duller; bill red brown.

**Range in Eastern Africa:** Abyssinia.

**Habits:** Noted in small flocks with other Weavers in the Blue Nile district.

**Nest, Eggs, etc.:** See under the following race.

**Recorded breeding:** No records.

**Pytilia phœnicoptera emini** Hart.

Pytilia phœnicoptera emini Hartert, Nov. Zool. 6, p. 413, 1899: Lado, southern Sudan.

**Distinguishing characters:** Differs from the preceding race in being greyer, less brown, above; eye red; bill black. Wing 56 to 60 mm.

**Range in Eastern Africa:** Southern Sudan and Uganda.

**Habits:** A shy and scarce species of bush and tall grass country, preferring fairly dense cover, usually in pairs, but feeding on the ground with various species of Waxbills.

**Nest and Eggs:** The nest of the nominate race is a globular mass of grass tops loosely put together and thickly lined with feathers with a round hole at the side, placed in the top of a bush or similar situation. Eggs four, white and large for the bird; about 15·5 × 12·5 mm.

**Recorded breeding:** No records.

**Food:** Seeds and insects.

**Call:** An occasional chirping call.

**Distribution of other races of the species:** West Africa, the nominate race being described from Senegal.

**1409 ORANGE-WINGED PYTILIA.** *PYTILIA AFRA* (Gmelin). Pl. 92.

Fringilla afra Gmelin, Syst. Nat. 1, pt. 2, p. 905, 1789: Angola.

**Distinguishing characters:** Adult male, forehead, sides of face, round eyes to ear-coverts, chin, lower rump, upper tail-coverts, central tail feathers and outer webs of rest of tail feathers crimson; crown to mantle and rump, lesser wing-coverts and innermost secondaries olive green; edges of flight feathers and greater wing-coverts orange; below, throat and neck grey; chest olive green more or less washed with orange; rest of underparts olive green barred with whitish; centre of belly white; bill scarlet. The female has the forehead olive green; sides of face, round eyes to ear-coverts, chin and underside olivaceous grey barred with white, but chin and neck often without barring; belly buffish; bill sepia, lower mandible orange. Wing 55 to 64 mm. The young bird is similar to the adult female, but has the lower rump rather more orange red.

**General distribution:** The Sudan and Abyssinia to the Portuguese Congo, northern Angola, Northern Rhodesia and Portuguese East Africa.

**Range in Eastern Africa:** Southern Sudan and southern Abyssinia to the Zambesi River, also Zanzibar Island.

**Habits:** Widely distributed and locally common in scattered flocks or small parties in bush or semi-open country, feeding on the ground.

**Nest and Eggs:** A very frail spherical nest of grass in the fork of a bush or plant. Eggs three to five, white; about $16 \cdot 5 \times 12 \cdot 5$ mm.

**Recorded breeding:** Southern Abyssinia (breeding condition), June. Tanganyika Territory, April to June. Zanzibar Island, April to June. South-eastern Belgian Congo, April and May. Northern Rhodesia, January to May. Nyasaland, March to June.

**Food:** Grass and other small seeds.

**Call:** A piping whistle of two notes, but more often a single flat 'saaa' not particularly distinctive.

### 1410 GREEN-WINGED PYTILIA. *PYTILIA MELBA* (Linnæus).

*Pytilia melba melba* (Linn.). **Pl. 92.**

*Fringilla melba* Linnæus, Syst. Nat. 10th ed. p. 180, 1758: Angola.

**Distinguishing characters:** Adult male, forehead, malar stripe, chin and throat scarlet; crown, round eyes, lores to upper mantle and sides of neck grey; mantle to rump and wings, including edges of flight feathers, olive green with a golden wash; upper tail-coverts, central tail feathers and edges of rest of tail feathers crimson; below, chest olivaceous golden green, but sometimes red of throat extends on to chest; rest of underparts barred white and black or white and olive green, barring sometimes broken up into spots; lower belly and under tail-coverts buff, latter often barred; bill and eyes red. The female has the red of the head replaced by grey; chin and throat often barred. Wing 54 to 62 mm. The young bird is browner than the adult female; no barring below, but has often some crescent-shaped darker markings giving a scaly appearance; bill sepia. It can be distinguished from the young bird of the Orange-winged Pytilia by the brownish olive green, not orange, edging to the flight feathers.

**General distribution:** Angola, Damaraland, Belgian Congo, the Rhodesias, Bechuanaland, Tanganyika Territory, Nyasaland, Transvaal and Natal.

**Range in Eastern Africa:** Southern Tanganyika Territory from the Ufipa Plateau and Iringa to south-western areas.

**Habits:** Inconspicuous birds of thorny thickets and undergrowth, creeping about and feeding low down or on the ground rather like Hedge-Sparrows in Europe, but usually in pairs or family parties. Locally common and very silent, rarely flying more than a few yards.

**Nest and Eggs:** A roughly made oval nest of dry grass sometimes lined with feathers or vegetable down with side entrance and a rough porch in low bushes. Eggs three to seven, white and rather round; about 15 × 12 mm. Incubation is believed to be largely by the male.

**Recorded breeding:** Tanganyika Territory, April. Northern Rhodesia, January to April. Nyasaland and Portuguese East Africa, February to June.

**Food:** Grass and other small seeds.

**Call:** Usual call a low single 'wick,' also a long plaintive whistle and a monotonous song in the breeding season.

*Pytilia melba grotei* Reichw.

*Pytelia melba grotei* Reichenow, J.f.O. p. 227, 1919: Kionga, at mouth of Rovuma River, Portuguese East Africa.

**Distinguishing characters:** Adult male, differs from the nominate race in having the scarlet of the throat extending on to the chest. The females are indistinguishable. Wing 55 to 61 mm.

**General distribution:** Belgian Congo and Uganda to southern Nyasaland.

**Range in Eastern Africa:** Uganda, Tanganyika Territory and Portuguese East Africa to Nyasaland.

**Habits:** As for other races.

**Recorded breeding:** No records.

SPARROWS, WEAVERS, WAXBILLS, etc. 1009

*Pytilia melba citerior* Strickl.
*Pytelia citerior* Strickland, Con. Orn. p. 151, 1852: Kordofan, Sudan.

**Distinguishing characters**: Adult male, differs from the two preceding races in having the lores and round the eyes red; in being much paler with golden yellow chest and palish olivaceous barring. The female is also much paler with pale barring more confined to the breast and flanks. Wing 55 to 61 mm.

**General distribution**: Senegal to the Sudan.

**Range in Eastern Africa**: Western and central Sudan.

**Habits**: As for other races, tame, inquisitive, silent little birds of bushy thickets.

**Recorded breeding**: Nigeria, August and September. Sudan, May to July, also November and February.

*Pytilia melba soudanensis* (Sharpe). **Pl. 92.**
*Zonogastris soudanensis* Sharpe, Cat. Bds. B.M. 13, p. 298, 1890: Upper White Nile, southern Sudan.

**Distinguishing characters**: Adult male, differs from the preceding race in having the red of the lores and round the eyes not extending behind the eye; generally darker and barring below darker. The female is rather darker than the female of the preceding race with more and darker barring below. Wing 54 to 63 mm.

**Range in Eastern Africa**: The Sudan to Eritrea, the Somalilands, Kenya Colony and north-eastern Tanganyika Territory.

**Habits**: As for other races.

**Recorded breeding**: Sudan, October. Abyssinia, May and June. Uganda and Kenya Colony, March to May.

**1411 AFRICAN FIRE-FINCH.** *LAGONOSTICTA RUBRICATA* (Lichtenstein).
*Lagonosticta rubricata rhodopareia* Heugl. **Pl. 92. Ph. xviii.**
*Lagonosticta rhodopareia* Heuglin, J.f.O. p. 16, 1868: Keren, Eritrea.

**Distinguishing characters**: Adult male, top of head, mantle and wings dark earth brown with a vinous wash; upper tail-coverts claret red; tail black, edges of basal half of feathers claret red; round eyes, sides of face and underside claret red with a slight bloom; lower belly, thighs and under tail-coverts black; a few white spots on sides of chest and flanks; bill slate grey; second primary notched. The female is clearer earth brown above; below, vinous brown; wings, tail, lower belly, thighs and under tail-coverts as in the male. Wing 45 to 53 mm. The young bird has the sides of face and the underside buffish brown; lower belly dusky; bill, upper mandible dark brown or blackish, lower light horn-colour.

**General distribution:** Eritrea, Abyssinia, eastern Belgian Congo, Uganda and Kenya Colony.

**Range in Eastern Africa:** Eritrea, Abyssinia, south-eastern Sudan, Uganda and Kenya Colony.

**Habits:** Widely distributed but not very common species with the usual habits of the genus, picking about in grass in bushy places. Tame and seen usually in pairs or small parties feeding on or near the ground, or making short quick flights to another feeding place. The black tail is noticeable in flight.

**Nest and Eggs:** A circular rather flimsy grass nest lined with feathers with entrance at the side. It is usually placed in a grass tuft or in a bush hidden by grass. Eggs three to six, normally four, white; about 15 × 11·5 mm.

**Recorded breeding:** Kenya Colony, April to June.

**Food:** Small seeds and insects.

**Call:** The call is a sharp 'chit-chit-chit-chit-chit' and there is a low chirrupy wavering song. There are also various squeaky whistles.

*Lagonosticta rubricata congica* Sharpe.
*Lagonosticta congica* Sharpe, Cat. Bds. B.M. 13, p. 280, 1890: Kasongo, eastern Belgian Congo.

**Distinguishing characters:** Differs from the preceding race in being darker, less earth brown above. Wing 47 to 51 mm.

**General distribution:** The Sudan to Uganda, eastern Belgian Congo and Tanganyika Territory.

**Range in Eastern Africa:** Yambio area, south-western Sudan to the Ruwenzori and the Bwamba area of western Uganda and to the Kungwe-Mahare Mts. in western Tanganyika Territory.

**Habits:** As for other races.

**Recorded breeding:** No records.

# SPARROWS, WEAVERS, WAXBILLS, etc.

*Lagonosticta rubricata hæmatocephala* Neum.
*Lagonosticta rubricata hæmatocephala* Neumann, O.M. p. 168, 1907: Songea, south-western Tanganyika Territory.

**Distinguishing characters:** Adult male, differs from the Eritrean race in being above rather clearer earth brown, top of head to nape distinctly washed with claret red. The female differs from the female of the Eritrean race in having the top of the head to nape distinctly washed with claret red; below, more clearly washed with claret red, less brownish. Wing 46 to 51 mm.

**General distribution:** Tanganyika Territory and south-eastern Belgian Congo to Rhodesia and Portuguese East Africa.

**Range in Eastern Africa:** Tanganyika Territory from Mt. Kilimanjaro and the Usambara Mts. to the Zambesi River.

**Habits:** As for the Eritrean race but usually commoner, and seen freely in native villages.

**Recorded breeding:** Tanganyika Territory, March. South-eastern Belgian Congo, February to April. Nyasaland, February to June. Portuguese East Africa, March to May.

**Distribution of other races of the species:** West and South Africa, the nominate race being described from eastern Cape Province.

### 1412 JAMESON'S FIRE-FINCH. *LAGONOSTICTA JAMESONI* Shelley.

*Lagonosticta jamesoni jamesoni* Shell. **Pl. 92.**
*Lagonosticta jamesoni* Shelley, Ibis p. 355, 1882: Tatin River, Matabeleland, Southern Rhodesia.

**Distinguishing characters:** Adult male, differs from the African Fire-Finch in being paler above and washed with rose pink; below, rose pink; lores and chin deeper rose pink; second primary not notched. The female is also paler than the female of the African Fire-Finch and has no notch on the second primary. Wing 45 to 50 mm. The young bird is earth brown above; russet brown below; upper and under tail-coverts and tail as in adult. It can only be distinguished from the young bird of the African Fire-Finch by the absence of the notch on the second primary.

**General distribution:** Tanganyika Territory to the Rhodesias, Bechuanaland, the Transvaal and Zululand.

**Range in Eastern Africa:** Tanganyika Territory.

**Habits:** Not uncommon but rather local species usually found in pairs or family parties on wooded hills where they feed among shrubs or bushes overgrown with grass. Definitely a country bird, preferring low thick cover, and is not likely to be met with in towns or villages.

**Nest and Eggs:** A loosely-made spherical nest of grass stems lined with seed heads and a few feathers, and with entrance at the side of the top, usually low in small shrubs amongst grass. Eggs three or four, white; about $14 \cdot 5 \times 11$ mm.

**Recorded breeding:** Tanganyika Territory, January to March. Southern Rhodesia, February to May. Nyasaland, March and June.

**Food:** Grass seeds.

**Call:** The only call recorded is a clicking alarm note.

*Lagonosticta jamesoni taruensis* V. Som.
*Lagonosticta jamesoni taruensis* Van Someren, Bull. B.O.C. 40, p. 54, 1919: Tsavo, southern Kenya Colony.

**Distinguishing characters:** Rather brighter in general colour than the nominate race. Wing 45 to 50 mm.

**Range in Eastern Africa:** Eastern Kenya Colony from Lamu to Mombasa and as far west as South Ukamba, Tsavo and Taita.

**Habits:** As for the nominate race.

**Recorded breeding:** No records.

**1413** RED-BILLED FIRE-FINCH. *LAGONOSTICTA SENEGALA* (Linnæus).

*Lagonosticta senegala brunneiceps* Sharpe. **Ph. xviii.**
*Lagonosticta brunneiceps* Sharpe, Cat. Bds. B.M. 13, p. 277, 1890: Maragaz, Eritrea.

**Distinguishing characters:** Adult male, centre of crown to mantle and rump pale russet brown slightly washed with cerise red; forehead, round eyes, sides of face and upper tail-coverts cerise red; flight feathers brown; tail black edged with red; below, cerise red; lower belly and under tail-coverts pale brown; usually a few very small white spots on sides of chest; second primary notched; bill pinkish red, ridge of culmen black. In first adult dress the mantle is less washed with red and the underside duller cerise red. The female has the lores cerise red; whole of upperparts earth brown; over eye, sides of face and chin sometimes washed with red; flight feathers, tail and upper tail-coverts as in male; below, pale

# SPARROWS, WEAVERS, WAXBILLS, etc.

buffish brown with small whitish spots on chest, sides of chest and upper flanks; bill as in male. Wing 46 to 54 mm. The young bird is very similar to the adult female but has no red on lores and no spots on chest, sides of chest or flanks; bill dusky.

**Range in Eastern Africa:** The Sudan, except the south-western area, Eritrea and all Abyssinia, except the eastern areas.

**Habits:** In one or other of its numerous racial forms, this is one of the most noticeable and widespread of African birds. It may be seen feeding mainly on the ground in all towns and villages and often enters houses. It has been well named as the 'Animated Plum.' It is probably the commonest and tamest bird in Africa, usually in small flocks when not breeding.

**Nest and Eggs:** The nest is a loose ball of dry grass, rags or any other material lined with a quantity of feathers, placed in thatch, low walls, bushes or any suitable position. The entrance is at the side of the top, and there is often a sort of porch above it. Eggs three to six, normally four, white; about 13·5 × 10·5 mm. Incubation is by both sexes.

**Recorded breeding:** Sudan irregularly, mostly July and August. Abyssinia, September and October.

**Food:** Grass seeds or anything else suitable.

**Call:** A piping whistle.

*Lagonosticta senegala somaliensis* Salvad.
*Lagonisticta somaliensis* Salvadori, Mem. Accad. Torino (2), 44, p. 557, 1894: Ogaden, eastern Abyssinia.

**Distinguishing characters:** Adult male, differs from the preceding race in being paler cerise red. The female is similar to the female of the preceding race. Wing 47 to 51 mm.

**Range in Eastern Africa:** Eastern Abyssinia to Italian Somaliland.

**Habits:** As for other races.

**Recorded breeding:** No records.

*Lagonosticta senegala rendalli* Hart.
*Lagonosticta senegala rendalli* Hartert, Nov. Zool. 5, p. 72, 1898: Upper Shiré River, southern Nyasaland.

**Distinguishing characters:** Adult male, differs from the Eritrean race in having the pale brown of the belly and under tail-coverts extending well up towards the breast, and more extensive

# 1014 SPARROWS, WEAVERS, WAXBILLS, etc.

small white spots on sides of chest as well as on the chest. The female is similar to that of the Eritrean race. Wing 46 to 52 mm.

**General distribution:** Southern Belgian Congo to Rhodesia, southern Nyasaland, Bechuanaland and the Transvaal.

**Range in Eastern Africa:** Nyasaland.

**Habits:** As for other races.

**Recorded breeding:** Nyasaland, February to October.

*Lagonosticta senegala ruberrima* Reichw. **Pl. 93.**
*Lagonosticta brunneiceps ruberrima* Reichenow, O.M. p. 24, 1903: Bukoba, north-western Tanganyika Territory.

**Distinguishing characters:** Adult male, differs from the Eritrean race in being darker wine red and with lower belly and under tail-coverts dusky brown; whole upperparts more distinctly washed with wine red. The female usually has a wash of wine red on the upperparts and a faint wash of this colour from chin to chest, but some are merely rather darker in general colour than the female of the Eritrean race. Albinistic specimens occur. Wing 45 to 52 mm. The young bird is darker than that of the Eritrean race.

**General distribution:** Eastern Belgian Congo to Uganda, Kenya Colony, Tanganyika Territory and western Nyasaland.

**Range in Eastern Africa:** Uganda to central Kenya Colony and Tanganyika Territory.

**Habits:** As for other races, the female usually closes the entrance of the nest with feathers on leaving.

**Recorded breeding:** Uganda and Kenya Colony throughout the year. Tanganyika Territory, March to July.

*Lagonosticta senegala rhodopsis* (Heugl.).
*Estrelda rhodopsis* Heuglin, J.f.O. p. 166, 1863: near Wau, Bahr-el-Ghazal, south-western Sudan.

**Distinguishing characters:** Adult male, differs from the Eritrean race in being paler russet brown above, and paler cerise red below; whole belly and under tail-coverts pale brown. The female is paler than that of the Eritrean race. Wing 47 to 52 mm.

**General distribution:** Northern Nigeria and northern Cameroons to the Sudan.

# SPARROWS, WEAVERS, WAXBILLS, etc. 1015

**Range in Eastern Africa:** South-western Sudan.

**Habits:** As for other races.

**Recorded breeding:** Nigeria, November to January. Southern Sudan, August.

**Distribution of other races of the species:** West Africa to Damaraland, the nominate race being described from Senegal.

### 1414 BAR-BREASTED FIRE-FINCH. *LAGONOSTICTA RUFOPICTA* (Fraser).

*Lagonosticta rufopicta lateritia* Heugl. **Pl. 93.**
*Lagonosticta (Estrelda) lateritia* Heuglin, J.f.O. p. 262, 1864: Djur, Bahr-el-Ghazal, south-western Sudan.

**Distinguishing characters:** Not unlike the Red-billed Fire-Finch, but dark earth brown above; upper tail-coverts crimson; tail deeper black; forehead, above eyes, sides of face and underside from chin to breast crimson wine colour; feathers of chest and breast tipped with white, forming more or less broken bars; belly dusky; under tail-coverts buff; second primary not notched; bill crimson, ridge of culmen black. The sexes are alike. Wing 47 to 52 mm. The young bird is earth brown with a wash of crimson from chest to breast; upper tail-coverts dull crimson; bill dusky.

**General distribution:** North-eastern Belgian Congo to the Sudan and Uganda.

**Range in Eastern Africa:** Southern Sudan and north-western Uganda.

**Habits:** Another town dwelling Fire-Finch, found freely in villages and equally tame and fearless, with the habits of the Red-billed species.

**Nest and Eggs:** Nest a spherical ball of grass with an inner layer of grass tops and a lining of feathers, rather frail, with side entrance. Eggs four, white; about 14 × 11 mm.

**Recorded breeding:** Southern Sudan and the Nile Province of Uganda, June to October.

**Food:** Grass seeds.

**Call:** A musical twittering song.

**Distribution of other races of the species:** West Africa to French Equatorial Africa, the nominate race being described from Cape Coast Castle, Sierra Leone.

**1415 DUSKY FIRE-FINCH.** *LAGONOSTICTA CINEREO-VINACEA* Sousa.

*Lagonosticta cinereovinacea graueri* Roths.

*Lagonosticta graueri* Rothschild, Bull. B.O.C. 23, p. 102, 1909: Forest near Baraka, eastern Belgian Congo.

**Distinguishing characters:** Head all round to upper rump and chest brownish slate; lower rump and upper tail-coverts maroon red; wings rather brownish slate; tail, breast and belly to under tail-coverts black; flanks maroon red with white spots; eyes red; bill violet black. The sexes are alike, but the eye is brown in the female. Wing about 52 mm. The young bird is dusky brown; upper tail-coverts dull maroon red; chin to breast more slaty; flanks washed with maroon red; bill black.

**General distribution:** Eastern Belgian Congo to Uganda.

**Range in Eastern Africa:** South-western Uganda.

**Habits:** No information.

**Nest and Eggs:** Undescribed.

**Recorded breeding:** No records.

**Food:** No information.

**Call:** Unrecorded.

**Distribution of other races of the species:** Angola, the habitat of the nominate race.

**1416 BLACK-FACED FIRE-FINCH.** *LAGONOSTICTA LARVATA* (Rüppell).

*Lagonosticta larvata larvata* (Rüpp.). **Pl. 93.**

*Amadina larvata* Rüppell, N. Wirbelt. Vög. p. 97, pl. 36, fig. 1, 1840: Simien, northern Abyssinia.

**Distinguishing characters:** Adult male, near nostrils, sides of face, above eyes to ear-coverts and chin and throat black; top of head dark brownish grey often clearly washed with maroon red; nape, sides of neck to chest and breast maroon red; mantle and wings dark brownish grey; upper tail-coverts and edges of tail feathers maroon red; flanks sometimes washed with maroon red; small white spots on flanks; belly and under tail-coverts grey and black or black; bill slate grey. The female has the upperside and sides of face dusky earth brown; upper tail-coverts and tail as in male; below, pale brown, more or less dusky on flanks, with a slight vinous wash; small white spots on flanks; bill slate grey. Wing 51 to 54 mm. The young bird is similar to the adult female.

**Range in Eastern Africa:** The eastern Sudan and Abyssinia.

**Habits:** Shy, silent species seen among bamboos at 3,000 to 5,000 feet, or among grass along the banks of wooded streams, in small parties, occasionally in large flocks.

**Recorded breeding:** No records.

*Lagonosticta larvata nigricollis* Heugl.
*Lagonosticta nigricollis* Heuglin, J.f.O. p. 273, 1863: Djur River, Bahr-el-Ghazal, south-western Sudan.

**Distinguishing characters:** Adult male, differs from the nominate race in being grey above and paler below with a vinous wash. The female is paler than the female of the nominate race. Wing 52 to 54 mm.

**General distribution:** The Shari Ubangi area of French Equatorial Africa to the Sudan and Uganda.

**Range in Eastern Africa:** South-western Sudan and northern Uganda.

**Habits:** Shy little birds in small flocks in thick high grass. The male has a loud clear song in May.

**Recorded breeding:** No records.

*Lagonosticta larvata togoensis* (Neum.).
*Estrilda larvata togoensis* Neumann, O.M. p. 167, 1907: Kete Kratschi, Togoland, West Africa.

**Distinguishing characters:** Adult male, slightly paler grey above than in the nominate race. Female is less grey, more earth brown. Wing 50 to 55 mm.

**General distribution:** Gold Coast and Togoland to the Sudan.

**Range in Eastern Africa:** Darfur, western Sudan.

**Habits:** In small numbers in woodlands, probably a non-breeding migrant to western Darfur. Usually seen feeding on the ground.

**Nest and Eggs:** The nest is a loose ball of withered grass lined with grass tops and feathers, and with the entrance at the side. It is placed low in bushes or in similar situations. Eggs three or four, white; about 14 × 11 mm.

**Recorded breeding:** Nigeria, July and August.

**Food:** No information, presumably small seeds.

**Call:** Unrecorded.

**Distribution of other races of the species:** Senegal, Gambia and Portuguese Guinea.

### 1417 YELLOW-BELLIED WAXBILL. *COCCOPYGIA MELANOTIS* (Temminck).
*Coccopygia melanotis quartinia* (Bp.).
Estrelda quartinia Bonaparte, Consp. Gen. Av. 1, p. 461, 1850: Abyssinia.

**Distinguishing characters:** Forehead to nape and sides of face grey; mantle and wings green with fine indistinct barring; rump and upper tail-coverts scarlet; tail black; below, chin to breast pale grey; rest of underparts greenish yellow; centre of belly brighter yellow; bill, upper mandible black, lower red. The sexes are alike. Wing 45 to 51 mm. The young bird is similar to the adult, but has the rump and upper tail-coverts orange and the bill black.

**Range in Eastern Africa:** Abyssinia, Eritrea and the southern Sudan.

**Habits:** As for the following race.

**Recorded breeding:** Eritrea and central Abyssinia, September and October. Southern Abyssinia, June.

*Coccopygia melanotis kilimensis* Sharpe. **Pl. 93. Ph. xviii.**
*Coccopygia kilimensis* Sharpe, Cat. Bds. B.M. 13, p. 307, 1890: Mt. Kilimanjaro, north-eastern Tanganyika Territory.

**Distinguishing characters:** Differs from the preceding race in being slightly smaller and slightly more olivaceous on mantle. Wing 41 to 48 mm.

**General distribution:** Eastern Belgian Congo to Uganda, Tanganyika Territory, north-eastern Northern Rhodesia, Nyasaland and eastern Southern Rhodesia.

**Range in Eastern Africa:** Uganda and Tanganyika Territory to the Zambesi River.

**Habits:** Particularly beautiful little birds found in small parties in open parts of forest or among thick low cover or bracken outside forest. Tame and rather conspicuous, although of retiring habits. On mountains common among low bush and Solanum along streams. Widespread and usually common in suitable localities.

**Nest and Eggs:** Nest a long ovoid of dry grass thickly lined with feathers and seed down, rather flimsy and not too securely attached, built in forks of shrubs or trees and often conspicuous. Eggs three or four, white; about 13 × 10 mm.

**Recorded breeding:** Uganda and Kenya Colony in many months, mainly March to June. Tanganyika Territory, breeds all the year round, particularly February to April. Nyasaland, February to May. Portuguese East Africa, April.

**Food:** Small seeds, particularly fond of seeds of a large thistle (Woosnam).

**Call:** A sharp high pitched metallic call uttered usually in flight.

**Distribution of other races of the species:** Angola and South Africa, the nominate race being described from eastern Cape Province.

**1418 WAXBILL.** *ESTRILDA ASTRILD* (Linnæus).
*Estrilda astrild minor* (Cab.).
*Habropyga minor* Cabanis, J.f.O. p. 229, 1878: Voi River, southeastern Kenya Colony.

**Distinguishing characters:** Adult male, above, pale brown, closely and narrowly barred; top of head slightly greyer; lores and streak through eye to top of ear-coverts crimson red; below, sides of face, chin and throat white, sometimes faintly washed with pink; rest of underparts closely and narrowly barred with dusky white and brown; centre of breast and belly pale crimson red; lower belly and under tail-coverts black; bill red. The female has rather less red and black on the belly. Wing 42 to 47 mm. The young bird has the streak through eye more orange red and the barring below more suffused, less clear; bill blackish.

**Range in Eastern Africa:** Coastal areas of Kenya Colony and Tanganyika Territory from Lamu to about the Rufigi River and as far west as Tsavo and Lake Jipe, also Zanzibar and Mafia Islands.

**Habits:** As for other races, common in grassy country near cultivation.

**Nest and Eggs:** See under the following race.

**Recorded breeding:** Zanzibar Island, May to October.

*Estrilda astrild cavendishi* Sharpe. **Pl. 93. Ph. xix.**

*Estrilda cavendishi* Sharpe, Ibis p. 110, 1900: Mapicuti, Cheringoma district, southern Portuguese East Africa.

**Distinguishing characters:** Adult male, above, browner than the last race; below, wholly suffused with pink; broader bright crimson red patch in centre of breast to belly and a larger black patch on lower belly. The female is less suffused below with pink, has little or no crimson red on centre of breast to belly, and less black on lower belly. Wing 44 to 52 mm.

**General distribution:** Kenya Colony, Tanganyika Territory, south-eastern Belgian Congo, Nyasaland, Rhodesia and Portuguese East Africa.

**Range in Eastern Africa:** Kenya Colony and Tanganyika Territory to the Zambesi River, but not north-western Tanganyika Territory nor coastal areas from the Tana River to about the Rufigi River.

**Habits:** A common and ubiquitous species, occasionally occurring in very large flocks indeed, found in villages, cultivated land, and grass country, or even in open places in forest. Typical Waxbill habits and completely fearless of man; feeding on or near the ground.

**Nest and Eggs:** A large grass nest shaped rather like a pear with the stalk inclined somewhat downwards, and the entrance tube is at the stalk end. Very frequently there is a sort of canopy or super-structure on top of the nest presumably used as a roosting place. Nest in low cover or in grassy banks. Eggs four to six as a rule; white, flushed pink when fresh; about $13 \times 10$ mm. In common with other small Waxbills, very liable to be parasitized by the Pin-tailed Whydah.

**Recorded breeding:** Kenya Colony at most seasons, mainly from March to July, also November to January. Tanganyika Territory, February onwards. South-eastern Belgian Congo, March. Nyasaland, January to April. Portuguese East Africa, February to April.

**Food:** Mainly grass seeds.

**Call:** A constant reedy twittering while flying in a flock.

*Estrilda astrild peasei* Shell.
*Estrilda peasei* Shelley, Bull. B.O.C. 13, p. 75, 1903: Jeffi Dunsa, thirty miles east of Adis Ababa, central Abyssinia.

**Distinguishing characters:** Very similar to the preceding race but above browner including top of head; below, lacking the bright crimson red patch in centre of breast to belly. The sexes are alike. Wing 45 to 53 mm.

**Range in Eastern Africa:** Central and southern Sudan and Abyssinia.

**Habits:** As for other races.

**Recorded breeding:** Central and eastern Abyssinia, May and June.

*Estrilda astrild nyanzæ* Neum.
*Estrilda astrild nyanzæ* Neumann, J.f.O. p. 596, 1907: Bukoba, north-western Tanganyika Territory.

**Distinguishing characters:** Adult male, similar to the preceding race, but the top of the head washed with grey; centre of belly palish crimson. The female is less suffused with crimson below. Wing 45 to 50 mm.

**General distribution:** Eastern Belgian Congo, Uganda and Tanganyika Territory.

**Range in Eastern Africa:** Uganda and north-western Tanganyika Territory.

**Habits:** As for other races.

**Recorded breeding:** Uganda, March to May, also October.

**Distribution of other races of the species:** West and South Africa, also Cape Verde, St. Helena and other islands, the nominate race being described from Cape Town.

**1419 BLACK-RUMPED WAXBILL.** *ESTRILDA TROGLO-DYTES* (Lichtenstein).
*Fringilla troglodytes* Lichtenstein, Verz. Doubl. p. 26, 1823: Senegambia.

**Distinguishing characters:** Very similar to the Abyssinian race of the Waxbill, but above, greyish tawny with very indistinct close narrow barring; upper tail-coverts and tail black, latter with some

white edging to outer feathers; below, buffish with a pink wash; very indistinct close and narrow barring on flanks; a small crimson patch in centre of lower belly; under tail-coverts buffish white; bill red. The sexes are alike. Wing 44 to 50 mm. The young bird lacks the crimson streak from lores through eye to top of ear-coverts and has a black bill.

**General distribution:** Senegal to Abyssinia, south-western Eritrea, the Sudan and Uganda.

**Range in Eastern Africa:** Central Sudan to north-western Abyssinia and north-western Uganda.

**Habits:** Locally common in swampy places at low levels and along the banks of streams, with the habits of other Waxbills.

**Nest and Eggs:** Nest pear-shaped of dry grass on the ground or in a grass tuft with a porch at the side. Eggs three to six, white and glossy; about 13 × 11 mm.

**Recorded breeding:** Nigeria, July and August. Sudan, June to August. Uganda, August to November.

**Food:** No information.

**Call:** Unrecorded.

**1420 CRIMSON-RUMPED WAXBILL. *ESTRILDA RHODO-PYGA* Sundevall.**
*Estrilda rhodopyga rhodopyga* Sund.
*Estrilda rhodopyga* Sundevall, Œfv. K. Sv. Vet.-Akad. Förh. 7, p. 126, 1850: Sennar, eastern Sudan.

**Distinguishing characters:** Above, very similar to the Abyssinian race of the Waxbill but barring closer and narrower and more indistinct; upper tail-coverts and edges of central tail feathers crimson; wing-coverts and inner secondaries edged with crimson; below, buff with indistinct barring on flanks; under tail-coverts often barred; bill blackish or dark sepia. The sexes are alike. Wing 45 to 50 mm. The young bird has no crimson stripe through eye.

**Range in Eastern Africa:** Central Sudan to Eritrea, Abyssinia and western British Somaliland.

**Habits:** Common locally in small parties feeding on grass heads and extremely tame.

**Nest and Eggs:** Undescribed.

# SPARROWS, WEAVERS, WAXBILLS, etc. 1023

**Recorded breeding:** Sudan, probably November.
**Food:** Insects and their larvæ and small seeds.
**Call:** Unrecorded.

*Estrilda rhodopyga centralis* Kothe.
*Estrilda rhodopyga centralis* Kothe, O.M. p. 70, 1911: Kisenyi, Lake Kivu, Ruanda, Belgian Congo.

**Distinguishing characters:** Differs from the nominate race in being generally darker in tone. Wing 45 to 50 mm.

**General distribution:** The Sudan, Uganda, Ruanda, Kenya Colony, Tanganyika Territory and Nyasaland as far south as thirty miles north of Nkata, also Likoma Island.

**Range in Eastern Africa:** Southern Sudan, Uganda and Ruanda to Kenya Colony and Tanganyika Territory as far south as Iringa; also Rusinga Island, Lake Victoria.

**Habits:** Common at lower levels in open country with scattered clumps of bush, and in old cultivation.

**Recorded breeding:** No records.

**1421 ZEBRA WAXBILL.** *ESTRILDA SUBFLAVA* (Vieillot).
*Estrilda subflava subflava* (Vieill.).
*Fringilla subflava* Vieillot, N. Dict. d'Hist. Nat. 30, p. 575, 1819: Senegal.

**Distinguishing characters:** Adult male, above, earth brown including wings; a crimson stripe from lores to over and behind eye; upper tail-coverts reddish crimson; tail blackish, outer webs of outer tail-feathers on underside white; sides of face olivaceous; below, chin and throat bright yellow; rest of underparts rich orange with some bright yellow; flanks olivaceous barred with darker olivaceous and pale yellow; bill red; ridge of culmen and under side of lower mandible black. The female is pale yellow below, slightly dusky on chin to chest; under tail-coverts orange; barring on flanks less sharp; no crimson stripe over eye. Wing 41 to 49 mm. The young bird is wholly earth brown above; upper tail-coverts rather paler; no crimson stripe over eye; below, buff; flanks not barred; bill blackish.

**General distribution:** Senegal to Abyssinia, eastern Belgian Congo and Uganda.

**Range in Eastern Africa:** Central Sudan, Abyssinia and Uganda.

**Habits:** As for the following race.

**Recorded breeding:** Nigeria, November. Sudan, many broods in the year, mostly in June to September.

*Estrilda subflava clarkei* (Shell.). **Pl. 93. Ph. xix.**
*Coccopygia clarkei* Shelley, Bull. B.O.C. 13, p. 75, 1903: Richmond Road (=Thornhill junction), Natal, South Africa.

**Distinguishing characters:** Adult male, differs from the nominate race in being more extensively bright yellow, with the rich orange colour more confined to the chest and under tail-coverts. The female is yellower below. Wing 42 to 49 mm.

**General distribution:** Kenya Colony to Angola, eastern Southern Rhodesia and Natal.

**Range in Eastern Africa:** Central Kenya Colony to the Zambesi River; also Pemba, Zanzibar and Mafia Islands.

**Habits:** Widespread species of grassland and cultivations. Usually in small flocks and often near water. Tame and conspicuous little birds, feeding largely on grass seeds and with a habit of sliding down a grass stem to look for fallen seeds.

**Nest and Eggs:** A barrel-shaped nest of grass lined with feathers placed a foot or so from the ground in coarse herbage, but more frequently adopts the old nests of other birds and relines them. Noted as using old nests of Bishop-birds, Cisticola, Amblyospiza, Ploceus, Coliuspasser, etc. Incubation is largely by the male. Eggs four or five; white; about $14 \times 10$ mm.

**Recorded breeding:** Uganda, January to March, July, October and December. Kenya Colony, April to July. Zanzibar Island, May to September, and probably October. Tanganyika Territory, March to May. South-eastern Belgian Congo, April to June. Rhodesia, April and May. Nyasaland, April to August.

**Food:** Largely grass seeds.

**Call:** A metallic twittering.

**1422** FAWN-BREASTED WAXBILL. *ESTRILDA PALUDICOLA* Heuglin.

*Estrilda paludicola paludicola* Heugl. **Pl. 93.**

*Estrelda paludicola* Heuglin, J.f.O. p. 166, 1863: Middle course of Bahr-el-Ghazal, south-western Sudan.

**Distinguishing characters:** Above, russet brown, very finely, narrowly and indistinctly barred; top of head usually greyer; sides of face grey; lower rump and upper tail-coverts claret red; tail black, outer tail feathers edged with whitish; below, creamy white; often a wash of wine red on lower belly; bill red. The sexes are alike. Wing 45 to 50 mm. The young bird is similar to the adult but has a black bill.

**General distribution:** Eastern Belgian Congo to the Sudan, Uganda and Kenya Colony.

**Range in Eastern Africa:** Southern Sudan to Uganda as far south as the Semliki River and the Entebbe area, and western Kenya Colony as far east as Mau.

**Habits:** Locally common in small parties in swampy grass land near rivers, with the habits of other Waxbills, either in open woodland country or occasionally in forest.

**Nest and Eggs:** A spherical nest with a superstructure and sometimes a short entrance funnel compactly built of fine grass and placed either on or just off the ground among grass. Eggs up to ten, white; about 13 × 10 mm.

**Recorded breeding:** Southern Sudan, July and August. Northern Uganda, January to May, also August and October and November.

**Food:** Grass seeds.

**Call:** A pleasing little song.

*Estrilda paludicola roseicrissa* Reichw.

*Estrilda roseicrissa* Reichenow, J.f.O. p. 47, 1892: Bukoba, north-western Tanganyika Territory.

**Distinguishing characters:** Differs from the nominate race in being rather warmer russet brown above, including top of head which is never grey; below, whiter, with usually more wine colour on lower belly. Wing 45 to 49 mm.

**General distribution:** Eastern Belgian Congo, Uganda and Tanganyika Territory.

**Range in Eastern Africa:** Uganda as far north as south end of Ruwenzori Mts., and about eighty miles west of Entebbe to north-western Tanganyika Territory.

**Habits:** Plentiful in parties of up to thirty or more, often with other Waxbills.

**Recorded breeding:** Tanganyika Territory, probably March and April.

*Estrilda paludicola ochrogaster* Salvad.
*Estrilda ochrogaster* Salvadori, Boll. Zool. Anat. Torino 12, no. 287, p. 4, 1897: Northern Abyssinia.

**Distinguishing characters:** Differs from the nominate race in being bright ochreous yellow below. Wing 44 to 51 mm.

**Range in Eastern Africa:** Eastern Sudan and Abyssinia.

**Habits:** Birds of uplands, but not found on the high plateau, with the habits of other races.

**Recorded breeding:** No records.

*Estrilda paludicola marwitzi* Reichw.
*Estrilda marwitzi* Reichenow, O.M. p. 40, 1900: Malangali, Uhehe, Iringa district, Tanganyika Territory.

**Distinguishing characters:** Nearest to the Bukoba race but differs in being warmer russet brown above; below, whiter, often with a greyish wash from chin to chest. Wing 45 to 49 mm.

**Range in Eastern Africa:** Kahama to the Iringa and Njombe areas, Tanganyika Territory.

**Habits:** As for other races.

**Recorded breeding:** Njombe, Tanganyika Territory, February to April.

**Distribution of other races of the species:** Nigeria to the Belgian Congo and Angola.

## SPARROWS, WEAVERS, WAXBILLS, etc. 1027

**1423 LAVENDER WAXBILL. *ESTRILDA PERREINI*** (Vieillot).

*Estrilda perreini perreini* (Vieill.).
*Fringilla perreini* Vieillot, N. Dict. d'Hist. Nat. 12, p. 179, 1817: Malimbe, Portuguese Congo.

**Distinguishing characters:** General colour dusky grey; lores, narrow stripe through eye and chin black; lower rump and upper tail-coverts red; tail black; lower belly and under tail-coverts blackish slate; bill blue grey. The sexes are alike. Wing 47 to 55 mm. The young bird has the lores and chin only black; lower rump and upper tail-coverts duller and darker red.

**General distribution:** Portuguese Congo and northern Angola to Northern Rhodesia and Tanganyika Territory.

**Range in Eastern Africa:** Western and south-western Tanganyika Territory as far north as the Kungwe-Mahare Mts.

**Habits:** An uncommon species of woodland and open spaces of forest where thick bushes occur among grass. Inconspicuous birds with unobtrusive and little recorded habits.

**Nest and Eggs:** Retort-shaped nests with a short tubular entrance sloping downwards from the top of the nest. Rather large nests of strong grass lined with soft seed heads and placed in shrubs or small trees at several feet from the ground. Eggs up to four or more, white; about 14·5 × 11 mm.

**Recorded breeding:** South-eastern Belgian Congo, March and April.

**Food:** Small seeds—no particular information.

**Call:** A thin whistling 'pseeu-pseeu.'

*Estrilda perreini incana* Sund.
*Estrilda incana* Sundevall, Œfv. K. Sv. Vet. Akad. Förh. 7, p. 98, 1850: Durban, Natal, South Africa.

**Distinguishing characters:** Differs from the nominate race in being generally paler. Wing 45 to 50 mm.

**General distribution:** Southern Nyasaland to eastern Southern Rhodesia and Natal.

**Range in Eastern Africa:** Nyasaland.

**Habits:** An apparently uncommon species of thick cover and dense undergrowth of which little is recorded. Shy retiring little birds seldom seen.

**Recorded breeding:** Nyasaland (breeding condition), April. Portuguese East Africa, May and June.

**Distribution of other races of the species:** Sao Thomé Island and Angola.

### 1424 BLACK-HEADED WAXBILL. ESTRILDA ATRICAPILLA Verreaux.

*Estrilda atricapilla graueri* Neum. Pl. 93.

*Estrilda atricapilla graueri* Neumann, Bull. B.O.C. 21, 1908, p. 55: Sabinyo Volcano, Kivu district, eastern Belgian Congo.

**Distinguishing characters:** Adult male, head to nape and round eye black; mantle, scapulars, wing-coverts and innermost secondaries narrowly barred dark grey and black; rump and upper tail-coverts claret red; tail black; sides of face and neck and chin to upper belly greyish white; lower belly and under tail-coverts black; flanks claret red; bill black. The female is browner on the mantle. Wing 44 to 49 mm. The young bird has the mantle and scapulars dark dusky brown and from chest to under tail-coverts dusky blackish with no red on flanks; bill black.

**General distribution:** South-central and eastern Belgian Congo, Uganda and Kenya Colony.

**Range in Eastern Africa:** Western and southern Uganda to Kenya Colony as far east as Mt. Kenya.

**Habits:** A mountain species, not uncommon up to about 8,000 feet, found in flocks in glades and open spaces in forest. Habits little recorded, and in some cases this bird has been confused with the next species, so that records are unreliable.

**Nest and Eggs:** The nest is much like those of other Waxbills but it is always lined with feathers, and has no superstructure. It is placed in leafy shrubs on forest outskirts or in thick cover along streams. Eggs four or five, pinkish white when fresh; about 14 × 10 mm.

**Recorded breeding:** Uganda (nestling), February. Kenya Colony highlands, November.

**Food:** Grass seeds.

**Call:** Said to be a faint twittering note. It would be interesting to know if it can be distinguished from that of the following species.

**Distribution of other races of the species:** West Africa and the Belgian Congo, the nominate race being described from Gabon.

## 1425 BLACK-CROWNED WAXBILL. *ESTRILDA NONNULA* Hartlaub.

*Estrilda nonnula nonnula* Hartl. **Pl. 93.**

*Astrilda nonnula* Hartlaub, J.f.O. p. 425, 1883: Kudurma, Bahr-el-Ghazal, south-western Sudan.

**Distinguishing characters:** Adult male, above similar to the Black-headed Waxbill but barring of mantle and scapulars paler grey and duller black; sides of face and underside white with a slight greyish wash from chin to chest and on flanks and under tail-coverts; flanks paler red; belly white; under tail-coverts greyish white; bill black and red. The female is rather paler grey on mantle and scapulars. Wing 44 to 50 mm. The young bird differs from the young bird of the Black-headed Waxbill in being paler buffish brown below including the under tail-coverts; bill black.

**General distribution:** Cameroons to eastern Belgian Congo, the Sudan, Uganda, Tanganyika Territory and Kenya Colony.

**Range in Eastern Africa:** Uganda, south-western Sudan, north-western Tanganyika Territory and Kenya Colony as far east as Molo.

**Habits:** Common and often abundant species of open country congregating in thousands on millet crops. In other areas and at other times of the year they are found among bushes eating buds as Bullfinches do in Europe. They occur up to fairly high elevations and are also found in gardens, woodland glades, or even forest.

**Nest and Eggs:** Oval nest of fresh grass lined with down and feathers, with a side entrance, and usually a sort of second nest on the top in which the male roosts. It is placed in bushes, trees, or any sort of cover. Eggs four to six, white; about 13·5 × 10·5 mm.

**Recorded breeding:** Uganda, January to May, also September and October, but probably in all months of the year. From the time nest building starts to the time the young leave the nest is less than a month.

**Food:** As above, also any form of small seeds.

**Call:** An excited twittering cry, not particularly described.

**Distribution of other races of the species:** Fernando Po.

**1426 BLACK-BELLIED WAXBILL.  ESTRILDA RARA** (Antinori).

*Estrilda rara rara* (Ant.). Pl. 92.

*Habropyga rara* Antinori, Cat. Ucc. p. 72, 1864: between White Nile and Ghazal River, southern Sudan.

**Distinguishing characters:** Adult male, very similar to the African Fire-Finch, but differs in having the black of the belly and under tail-coverts extending up to the breast; the red below is more vinous and the mantle is washed with vinous red; no white spots on sides of chest and flanks; bill, upper mandible black, lower crimson not slate grey. The female differs from the female of the African Fire-Finch in having the sides of the face grey; the chin and throat washed with grey and the upper mandible of the bill black, lower pink not slate grey; a crimson spot in front of eye. Wing 46 to 52 mm. The young bird is sooty brown; upper tail-coverts crimson; the young male being more or less washed with vinous all over.

**General distribution:** Cameroons to the Sudan, Uganda and Kenya Colony.

**Range in Eastern Africa:** Southern Sudan, Uganda and western Kenya Colony.

**Habits:** Little recorded, but apparently a fairly common species in Uganda though rare elsewhere. Usually in small flocks in grass among Acacia scrub.

**Nest and Eggs:** Nest a sphere of loose grass lined with rootlets and grass fibres in bushes or trees or occasionally in the thatch of a hut. Eggs three or four, white and slightly glossy; about 14 × 11 mm.

**Recorded breeding:** Equatoria, August to November.

**Food:** Insects and presumably seeds.

**Call:** Unrecorded.

**Distribution of other races of the species:** Nigeria.

**1427 BLACK-CHEEKED WAXBILL. *ESTRILDA ERYTHRONOTOS* (Vieillot).**

*Estrilda erythronotos charmosyna* (Reichw.).
*Habropyga charmosyna* Reichenow, O.C. p. 78, 1881: Bardera, Juba River, southern Italian Somaliland.

**Distinguishing characters:** Adult male, head, mantle and scapulars pale vinaceous; head rather paler; forehead whitish; mantle and scapulars narrowly and indistinctly barred; rump and upper tail-coverts claret red; tail black with a bronzy wash; wing-coverts and innermost secondaries barred greyish white and black; sides of face, round eye to ear-coverts and chin black; a whitish mark behind ear-coverts; below, pinkish vinaceous, narrowly and very indistinctly barred; flanks washed with claret red; bill blue grey, tip black. The female is less vinaceous both above and below. Wing 51 to 56 mm. The young bird is similar to the adult female.

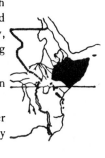

**Range in Eastern Africa:** Abyssinia and British and Italian Somalilands to northern Kenya Colony.

**Habits:** Is said to prefer rocky scrub country and to be rather noisy with a loud sweet whistle in two cadences. It has also a pretty warbling song.

**Nest and Eggs:** See under the following race.

**Recorded breeding:** No records.

*Estrilda erythronotos delamerei* Sharpe.
*Estrilda delamerei* Sharpe, Bull. B.O.C. 10, p. 102, 1900: Athi River, Kenya Colony.

**Distinguishing characters:** Adult male, differs from the preceding race in being darker and washed with grey above and below; chin and upper throat black; belly and under tail-coverts black. The female has the chin only black, and is paler below with the belly and under tail-coverts uniform in colour with the rest of the underparts. Wing 50 to 55 mm. The young bird is similar to the adult female, the young male having some blackish on belly and under tail-coverts and more washed with claret red on flanks.

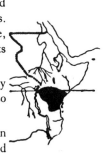

**Range in Eastern Africa:** Southern Uganda to Kenya Colony and Tanganyika Territory as far south as the Iringa district; also Ukerewe Island, Lake Victoria.

**Habits:** Locally common and rather more a bird of trees than scrub. Habits little recorded, but unlike the preceding race, is said to be silent and secretive. In the Chyulu Hills, Van Someren records it as a species of grass and *Erythrina* association, nesting low in bushes amongst grass.

**Nest and Eggs:** A rather large grass nest with side entrance and a porch over it, often untidy outside and lined with fine grass, in a thickly foliaged tree. Eggs about four, white; about 14·5 × 11·5 mm.

**Recorded breeding:** Iringa, Tanganyika Territory, February to April.

**Food:** Grass seeds.

**Call:** See under the preceding race.

**Distribution of other races of the species:** South Africa, the nominate race being described from the western Transvaal.

**1428 GREY-HEADED OLIVE-BACK.** *NESOCHARIS CAPISTRATA* (Hartlaub).

*Pytelia capistrata* Hartlaub, J.f.O. p. 259, 1861: Bissao, Portuguese Guinea.

**Distinguishing characters:** Narrow band on forehead and sides of face white; top of head to nape grey; rest of upperparts and tail olive green; below, chin and throat to behind ear-coverts black; rest of underparts grey; flanks yellow; bill black. The sexes are alike. Wing 55 to 60 mm. The young bird has the forehead, top of head to nape, and sides of face darkish grey; bill whitish, tip greyish.

**General distribution:** Gambia and Portuguese Guinea to the Sudan, Uganda and north-eastern Belgian Congo.

**Range in Eastern Africa:** South-western Sudan to north-western Uganda.

**Habits:** A rare species of which little is known. It is said to feed among tall trees searching for insects like a Tit or a Warbler.

**Nest and Eggs:** Undescribed.

**Recorded breeding:** Sierra Leone (nestling), October. Uganda, September.

**Food:** Believed to be almost entirely insectivorous.

**Call:** Unrecorded.

### 1429 WHITE-COLLARED OLIVE-BACK. *NESOCHARIS ANSORGEI* (Hartert).

*Pytelia ansorgei* Hartert, Bull. B.O.C. 10, p. 26, 1899: Luimi River, Toro, western Uganda.

**Distinguishing characters:** Adult male, head all round black; collar on hind neck grey; rest of upperparts golden olive green; tail black; below, collar on neck white; chest and breast golden olive green; belly and under tail-coverts grey; bill black. The female has a less pronounced collar on hind neck. Wing 50 mm. The young bird is said to be similar to the adult female.

**General distribution:** Eastern Belgian Congo and Uganda.

**Range in Eastern Africa:** Western Uganda.

**Habits:** A little known species of dense herbaceous scrub.

**Nest and Eggs:** Undescribed.

**Recorded breeding:** No records.

**Food:** No information.

**Call:** Unrecorded.

### 1430 CORDON-BLEU. *URÆGINTHUS ANGOLENSIS* (Linnæus).

*Uræginthus angolensis niassensis* Reichw. **Pl. 93.**

*Uræginthus angolensis niassensis* Reichenow, Mitt. Zool. Mus. Berlin 5, p. 228, 1911: Songea, southern Tanganyika Territory.

**Distinguishing characters:** Adult male, above, earth brown; rump, upper tail-coverts and tail blue; round eye, sides of face and underside blue; centre of breast to under tail-coverts buff brown; flanks blue; second primary slightly notched on inner web; bill rose madder or mauve with black tip and cutting edges. The female is paler than the male; bill blue grey or dark rose madder. Wing 47 to 55 mm. The young bird is similar to the adult female but paler; bill blackish.

**General distribution:** Kenya Colony, Tanganyika Territory, Northern Rhodesia, Nyasaland, Southern Rhodesia, Portuguese East Africa, the Transvaal and Natal.

**Range in Eastern Africa:** South-eastern Kenya Colony and Tanganyika Territory to the Zambesi River, also Zanzibar Island (probably escapes from captivity).

**Habits:** Widely distributed and locally plentiful species of lower levels, usually seen in flocks, pairs or small parties on the ground or among low herbage. The flocks rise with a sort of hissing twitter. Tame and freely found round human habitations.

**Nest and Eggs:** Nest of grass lined with fine grass and a few feathers, spherical or oval with side entrance. Nest placed in any sort of situation, bushes, trees or walls of huts, but when in a bush or tree is very often in close contact with a nest of wasps or hornets. Also utilize Weavers' old nests occasionally, and in some localities nest on iron telegraph poles. Eggs four to six, white; about 14 × 11 mm.

**Recorded breeding:** Tanganyika Territory, February to April. Nyasaland, January to May. Portuguese East Africa, February to June. Southern Rhodesia, March to May.

**Food:** Mainly grass seeds but young are fed on insects.

**Call:** A sharp little whistle 'sweet-sweet-sweet' and a harsh rattling alarm call.

**Distribution of other races of the species:** Angola and the Belgian Congo, the nominate race being described from Angola.

### 1431 RED-CHEEKED CORDON-BLEU. *URÆGINTHUS BENGALUS* (Linnæus).

*Uræginthus bengalus bengalus* (Linn.).
*Fringilla bengala* Linnæus, Syst. Nat. 12th ed. 1, p. 323, 1766: Senegal.

**Distinguishing characters:** Adult male, differs from the Cordon-bleu in having the ear-coverts red; second primary slightly notched on inner web; bill pearly pink, tip black. The female is paler and has the ear-coverts blue. Wing 48 to 56 mm. The young bird is similar to the adult female but has the blue below confined to the chin and chest; rest of underparts buff brown.

**General distribution:** Senegal and Portuguese Guinea to the Sudan.

**Range in Eastern Africa:** Western Sudan.

**Habits:** A tame and abundant species in all types of country, particularly noticeable round villages and in cultivated areas, usually in small parties feeding on or near the ground.

**Nest and Eggs:** Nest a sphere or oval of dry grass, rather roughly made, with a lining of fine grass tops and feathers; occasionally there is an outer network of stems and rootlets; entrance hole at the side rather low down. Nests may be anywhere, in bushes, trees, thatch or quite frequently in Weavers' old nests, but this race does not seem to affect the vicinity of Hornets' nests to the same degree as others. Eggs about four to five, white and oval with faint gloss; about 14·5 × 11 mm.

**Recorded breeding:** Nigeria, June to August. Western Sudan, August to December.

**Food:** Grass seeds, etc.

**Call:** A plaintive little pipe or mouse-like squeak, and a series of chattering alarm notes. A very soft little song of three syllables, the last being longer than the others.

*Uræginthus bengalus schoanus* Neum. **Pl. 93.**
*Uræginthus bengalus schoanus* Neumann, J.f.O. p. 350, 1905: Ejere, Shoa, central Abyssinia.

**Distinguishing characters:** Differs from the nominate race in being brighter and more tawny in colour above and with more blue on the belly. Wing 49 to 57 mm.

**General distribution:** Eritrea, Abyssinia and the Sudan to north-eastern Belgian Congo, Uganda, Kenya Colony and Tanganyika Territory.

**Range in Eastern Africa:** Eritrea, Abyssinia, eastern and southern Sudan, Uganda, northern and western Kenya Colony and northern and western Tanganyika Territory as far south as Kasulu and Monduli.

**Habits:** As for the nominate race, but in many localities the nests are often in close association with hornets' nests. Abundant species, but subject to temporary total disappearances.

**Recorded breeding:** Sudan, Upper Nile, November; Kassala, December; Equatoria, September to November. Abyssinia, April to November. Uganda, June to August.

*Uræginthus bengalus brunneigularis* Mearns. **Ph. xviii.**
*Uræginthus bengalus brunneigularis* Mearns, Smiths Misc. Coll. 56, No. 20, p. 6, 1911: Wambugu, near Mt. Kenya, Kenya Colony.

**Distinguishing characters:** Male similar to the nominate race; female differs from the nominate race in having the sides of the face, chin and throat brown or mixed brown and blue. Wing 49 to 55 mm.

**Range in Eastern Africa:** South-central Kenya Colony from the Mt. Kenya area to the coast.

**Habits:** As for other races; no regular hornets' nest association at breeding site.

**Recorded breeding:** Kenya Colony, March to August, also December.

*Uræginthus bengalus ugogoensis* Reichw.
*Uræginthus bengalus ugogoensis* Reichenow, Mitt. Zool. Mus. Berlin 5, p. 228, 1911: Seke, Ugogo, Dodoma district, central Tanganyika Territory.

**Distinguishing characters:** Nearest to the Shoa race but paler tawny brown above, less warm in tone. The female differs from the male and female of the Cordon-bleu in having the brown of the sides of the neck extending more forward below the ear-coverts, and not with continuous blue from the ear-coverts down the side of the neck as in the Cordon-bleu; the buff brown of the underside rather more extensive on the breast and belly. Wing 52 to 55 mm.

**Range in Eastern Africa:** Tanganyika Territory from the Dodoma and Iringa districts to the coast.

**Habits:** As for other races. The call is a double mouse-like squeak.

**Recorded breeding:** Tanganyika Territory, January to April, also probably October to December.

*Uræginthus bengalus katangæ* Vinc.
*Uræginthus bengalus katangæ* Vincent, Bull. B.O.C. 54, p. 174, 1934: Elisabethville, south-eastern Belgian Congo.

**Distinguishing characters:** Differs from the Abyssinian and Dodoma races in being darker above, more earth brown, less tawny. Wing 50 to 55 mm.

**General distribution:** Central and southern Belgian Congo to Tanganyika Territory and Northern Rhodesia.

**Range in Eastern Africa:** Central and western Tanganyika Territory from Kasulu, Loliondo and Monduli to the south-western areas.

**Habits:** As for other races, though not usually common.

**Recorded breeding:** Southern Belgian Congo, March and May.

*Note:* This race and the Abyssinian race meet at Kasulu and Monduli.

**Distribution of other races of the species:** Angola and Northern Rhodesia.

**1432 BLUE-CAPPED CORDON-BLEU. *URÆGINTHUS CYANOCEPHALUS* (Richmond). Pl. 93.**

*Estrilda cyanocephala* Richmond, Auk. p. 157, 1897: Useri, near Mt. Kilimanjaro, Kenya Colony.

**Distinguishing characters:** Adult male, similar to the Cordon-bleu but has the forehead to nape blue uniform with the sides of the face and underparts; second primary not notched on inner web; bill red, deep pink or dull beetroot, tip black. The female differs from the male in having the blue of the top of the head confined to the forehead, or absent altogether; it differs from the female Cordon-bleu in being paler above and below; buff colour more confined to breast and flanks and less blue on flanks; second primary not notched. Wing 53 to 59 mm. The young bird has the forehead, sides of face and chin and throat much paler blue; breast buffish not blue as in the young of the Cordon-bleu; bill dusky.

**Range in Eastern Africa:** Kenya Colony to southern Italian Somaliland and Tanganyika Territory as far south as the Dodoma district.

**Habits:** Birds of dry desert country, locally common in pairs or small parties with the habits of the Red-cheeked Cordon-bleu but rather shyer.

**Nest and Eggs:** Nest of grass, oval or barrel shaped, with entrance hole low down at the side; they also freely adopt the old nests of Weavers. Commonly nest in association with hornets in low Acacia bushes. Eggs four to six at least, white; about 14·5 × 11 mm.

**Recorded breeding:** Northern Kenya Colony (nestling), February. Northern Tanganyika Territory, November to January, and also June.

**Food:** Probably mainly grass seeds but also said to feed freely on termites.

**Call:** Has a simple but pleasing song. The normal call is a twittering piping 'see-pee' and there is also a thin squeak.

**1433** PURPLE GRENADIER. *GRANATINA IANTHINO-GASTER* (Reichenow).

*Granatina ianthinogaster ianthinogaster* (Reichw). **Pl. 94.**
*Uræginthus ianthinogaster* Reichenow, O.C. p. 114, 1879: Massa, Tana River, eastern Kenya Colony.

**Distinguishing characters:** Adult male, head and neck all round and chin to throat russet brown, violet blue round eyes; mantle and wings earth brown with a wash of russet; upper tail-coverts bright violet blue; tail black; rest of underparts bright violet blue with some russet brown markings; bill red, culmen and tip blackish. The female is like the male but is whitish blue round the eyes; below, russet brown with dull white spots and barring; centre of lower belly white. Wing 56 to 63 mm. The young bird is similar to the adult female but has no ring round eye and no white spots or barring below; upper tail-coverts bright blue; tail black; bill reddish horn.

**Range in Eastern Africa:** Central, eastern and southern Kenya Colony from the Tana River to Tanganyika Territory as far south as the Iringa area.

**Habits:** Locally common or even abundant in thick thorn scrub and aloes, and in places in open bush country. Usually secretive but tame, and inhabit the lower parts of bushes among grass and aloes, feeding on the ground. Frequently 'twinkles' wings and tail. This species is known to be parasitized by Fischer's Whydah.

**Nest and Eggs:** Nest a loosely woven sphere of fine grass lined with feathers in the forks of low bushes. Eggs three to five, white and glossy; about 15·5 × 12·2 mm.

**Recorded breeding:** Southern Kenya Colony, March and April. Northern Tanganyika Territory, December to February.

**Food:** Grass and other small seeds.

**Call:** A thin trilling reeling note, and a weak Canary-like tinkling song.

*Granatina ianthinogaster hawkeri* Phill.
*Granatina hawkeri* Phillips, Bull. B.O.C. 8, p. 23, 1898: Bari, Webi Shebeli River, south-eastern Abyssinia.

**Distinguishing characters:** Adult male, differs from the nominate race in having the mantle and wings more russet, less earth brown. The female also has the head and mantle paler brown; round eye almost white. Wing 52 to 58 mm.

**Range in Eastern Africa:** South-eastern Sudan and southern Abyssinia to British and Italian Somalilands, north-eastern Uganda and northern Kenya Colony.

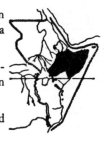

**Habits:** As for other races. Erlanger notes that the males continue to bring feathers to the nest all through the incubation period.

**Recorded breeding:** South-eastern Abyssinia, April to June and November. British Somaliland (nestling), June.

*Granatina ianthinogaster roosvelti* Mearns.
*Granatina ianthinogaster roosvelti* Mearns, Smiths Misc. Coll. 61, No. 9, p. 3, 1913: Southern Guaso Nyiro, Kenya Colony.

**Distinguishing characters:** Adult male, differs from the nominate race in having the mantle and wings darker, washed with dusky. The female has the mantle and wings earth brown, and the blue round eye less whitish. Wing 57 to 63 mm.

**Range in Eastern Africa:** Western Kenya Colony; also Rusinga Island, Lake Victoria.

**Habits:** As for other races.

**Recorded breeding:** Islands of eastern Lake Victoria, January.

## INDIGO-BIRDS and WHYDAHS. Family—PLOCEIDÆ.
Genera: *Hypochera*, *Vidua* and *Steganura*.

Seven species of Indigo-Birds, three Whydahs and two Paradise Whydahs occur in Eastern Africa. All are conspicuous species in breeding dress, and very easily overlooked at other times.

Indigo-Birds are plentiful around villages and farm lands and spend much of their time on the ground. The males in breeding dress of the different species which occur in the same area are difficult to distinguish in the field, unless the comparative characters have been studied in skins, and in non-breeding dress they would look very much alike. Owing to the difficulty of identification in the field, we can give no definite field notes for each species. Some are rare and some are common, but when all is said, very little is definitely known about them. They are usually tame, fearless little birds and enter houses freely. They are found in flocks in any sort of open country or in clearings, but do not breed in colonies.

Frequently the males sit on dead trees or telegraph wires, and they are believed to be polygamous, while in some species, and possibly in all, there is considerable suspicion of parasitism on Waxbills. The males utter a twittering, slightly metallic 'chi-chi-chi-chit' reminiscent of a European Linnet. They have a quick shuffling movement when feeding on the ground. Much is still to be learnt about their breeding habits, and it would well repay observers to study them. It seems almost incredible that of two species that can hardly be distinguished in the hand, one should be parasitic and one not. The percentage of males in breeding dress is apparently a small one. Food small seeds and insects.

The male Whydahs in breeding dress are very distinctive. They flock in the non-breeding season and are somewhat of a nuisance to the farmer.

**1434** GREEN INDIGO-BIRD. *HYPOCHERA ÆNEA* Hartlaub.
*Hypochera ænea codringtoni* Neave. **Pl. 94.**
*Hypochera codringtoni* Neave, Mem. Lit. Phil. Soc. Manchester, 51, no. 10, p. 94, 1907: Molilo's, Loangwa Valley, north-eastern Northern Rhodesia.

**Distinguishing characters:** Adult male, general colour dull blue black with a bottle green wash; wings and tail blackish; white patch on sides of rump; bill white; feet salmon pink. In non-breeding dress the top of the head is blackish with a broad central buff stripe; sides of face and stripe over eye buff; broad blackish patch behind eye; mantle streaked blackish and buff; rump plain earth brown, indistinctly mottled with dusky; wings and tail blackish with buffish edging; below, chin buff; centre of belly and under tail-coverts white; outer edges of flight feathers buff; under wing-coverts mottled brown and white. The female is similar to the male in non-breeding dress but smaller. Wing, male 66 to 70, female 64 mm. Juvenile plumage unrecorded.

**General distribution:** Tanganyika Territory to Northern Rhodesia and Nyasaland.

**Range in Eastern Africa:** Tanganyika Territory as far north as Iringa.

**Habits:** A relatively scarce species.

**Nest and Eggs:** Undescribed.

**Recorded breeding:** Tanganyika Territory uplands, January to March (breeding condition), 'probably parasitic.'

**Food:** No information.

**Call:** Unrecorded.

**Distribution of other races of the species:** West Africa, the nominate race being described from Senegambia.

### 1435 PURPLE INDIGO-BIRD. *HYPOCHERA ULTRAMARINA* (Gmelin).

*Hypochera ultramarina ultramarina* (Gmel.).
*Fringilla ultramarina* Gmelin, Syst. Nat. 1, pt. 2, p. 927, 1789: Abyssinia.

**Distinguishing characters:** Adult male, general colour glossy violet blue black; patch of white on sides of rump; flight feathers and tail black or blackish; bill white or pinkish white; feet coral or orange red. In non-breeding dress the top of the head is blackish with a broad central buff stripe; sides of face and stripe over eye buff; blackish streak behind eye; mantle streaked brown and black; rump less distinctly streaked; wings and tail blackish with brownish edging; below, buff; centre of belly and under tail-coverts white; outer edges of flight feathers buff, under wing-coverts mottled brown and white. The female is similar to the male in non-breeding dress. It is browner and rather less clearly streaked above than the female of the Green Indigo-Bird. Wing, male 58 to 68, female 60 to 65 mm. The young bird is browner than the adult female, less distinctly streaked on mantle and has no central stripe on head.

**Range in Eastern Africa:** The Sudan, south western Eritrea and Abyssinia.

**Habits:** Abundant and tame in many villages often in company with Fire-Finches.

**Nest and Eggs:** Said to make Sparrow-like nests of all sorts of material, and eggs are given as three to five, bluish white; about 15 × 11·5 mm., but see remarks under the group heading.

**Recorded breeding:** Sudan, probably August to March, and often breeding in immature plumage.

**Food:** Small seeds.

**Call:** The male has a fine song, energetic and melodious.

*Hypochera ultramarina neumanni* Alex.

*Hypochœra neumanni* Alexander, Bull. B.O.C. 23, p. 33, 1908: Lake Chad, northern Nigeria.

**Distinguishing characters:** Adult male, general colour less violet in tone than in the nominate race. Female rather paler in general colour than that of the nominate race. Wing, male 61 to 64, female 60 to 66 mm.

**General distribution:** Northern Nigeria to the Sudan.

**Range in Eastern Africa:** Western Sudan.

**Habits:** As for the genus.

**Recorded breeding:** Darfur, probably October to January, not in colonies.

*Hypochera ultramarina orientalis* Reichw. **Pl. 94.**

*Hypochera ultramarina orientalis* Reichenow, Vög. D.O.A. p. 188, 1894: Paré Mts., north-eastern Tanganyika Territory.

**Distinguishing characters:** Adult male, differs from the nominate race in being more violet in general colour and very slightly less glossy. The female has the sides of the crown rather blacker than the female of the nominate race. Wing, male 63 to 71, female 63 to 64 mm.

**General distribution:** Uganda, Kenya Colony, the Belgian Congo and Tanganyika Territory.

**Range in Eastern Africa:** Uganda, Kenya Colony and Tanganyika Territory; also Buvuma and Rusinga Islands, Lake Victoria.

**Habits:** As for the genus, the male is said to hover over the female during the courtship flight.

**Recorded breeding:** Kisumu, Kenya Colony, probably May and June. Northern Uganda, September. Northern Tanganyika Territory (breeding condition), January.

**1436 SOUTH AFRICAN INDIGO-BIRD.** *HYPOCHERA AMAUROPTERYX* Sharpe.

*Hypochœra amauropteryx* Sharpe, Cat. Bds. B.M. 13, p. 309, 1890: Rustenburg, western Transvaal.

**Distinguishing characters:** Adult male, general colour dull darkish violet blue black; wings and tail blackish; white patch on sides of rump; bill and feet salmon, salmon pink or yellow orange.

## SPARROWS, WEAVERS, WAXBILLS, etc.    1043

In non-breeding dress it is darker brown and more broadly streaked than the Purple Indigo-Bird. The female is similar to the male in non-breeding dress. Wing, male 64 to 70, female 64 to 65 mm. Juvenile plumage unrecorded.

**General distribution:** Tanganyika Territory to Nyasaland, Damaraland and the Transvaal.

**Range in Eastern Africa:** Central Tanganyika Territory from Kahama and Mkalama to south-western areas.

**Habits:** Locally common in open country or large clearings.

**Nest and Eggs:** Parasitic on the Red-billed Fire-Finch and other Waxbills, eggs white, no measurements available.

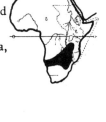

**Recorded breeding:** Nyasaland and Portuguese East Africa, February to June.

**Food:** No information.

**Call:** Unrecorded.

**1437  DUSKY INDIGO-BIRD.    *HYPOCHERA FUNEREA*** (Tarragon).

*Hypochera funerea funerea* (Tarrag.).    **Pl. 94.**
*Fringilla funerea* de Tarragon, Rev. Zool. Paris p. 180, 1874: Natal, South Africa.

**Distinguishing characters:** Adult male, general colour dullish velvety deep violet purple; wings and tail ash-coloured; patches of white on the sides of the rump; bill white or pinkish white; feet coral red, reddish orange, flesh, pinkish white, pale pink or white. The female, male in non-breeding dress, and young bird are similar to the South African Indigo-Bird but more clearly streaked above, and warmer brown above than the Green Indigo-Bird. Wing, male 65 to 70, female 62 to 66 mm.

**General distribution:** Tanganyika Territory to Nyasaland, Portuguese East Africa, the Rhodesias and Natal.

**Range in Eastern Africa:** Southern Tanganyika Territory from the Rovuma River to Nyasaland and Portuguese East Africa.

**Habits:** Of the genus, but distinctly local and generally in large clearings or open country.

**Nest and Eggs:** Present information unreliable, but believed to be parasitic.

3x

**Recorded breeding:** Rovuma River, southern Tanganyika Territory (nestling), August. Nyasaland and Portuguese East Africa (breeding condition), February to July.

**Food:** No information.

**Call:** A sustained twittering, almost a song.

*Hypochera funerea wilsoni* Hart.
*Hypochœra wilsoni* Hartert, Nov. Zool. 8, p. 342, 1901: Yelwa, middle Niger River, Nigeria.

**Distinguishing characters:** Differs from the nominate race in being smaller. The female is more sandy coloured above. Wing, male 63 to 67, female 60 to 61 mm.

**General distribution:** Senegal, Portuguese Guinea, Nigeria and Cameroons to the Sudan.

**Range in Eastern Africa:** Southern Sudan to the Turkwel River area of north-western Kenya Colony.

**Habits:** Of the genus.

**Recorded breeding:** No records.

**1438** BLACK INDIGO-BIRD. *HYPOCHERA NIGERRIMA* Sharpe. Pl. 94.

*Hypochœra nigerrima* Sharpe, P.Z.S. p. 133, 1871: Lucala River, northern Angola.

**Distinguishing characters:** Adult male, general colour dull matt velvety black with a deep dark violet-purple wash; wings and tail ash-colour; a white patch on sides of rump; bill white or pinkish white; feet white or very pale flesh. The dull blackish appearance with very dark violet purple sheen distinguishes this species from the Purple and Dusky Indigo-Birds. The non-breeding dress is very much browner both above and below than the other species. The female is similar to the male in non-breeding dress. Wing, male 65 to 70, female 63 mm. The young bird is similar to the male in non-breeding dress, but lacks the central stripe on the crown.

**General distribution:** Angola to Tanganyika Territory, Nyasaland and northern Portuguese East Africa.

**Range in Eastern Africa:** Southern half of Tanganyika Territory to Portuguese East Africa and Nyasaland.

**Habits:** Of the genus, often associating with the Pin-tailed Whydah.

**Nest and Eggs:** Undescribed.

**Recorded breeding:** Tanganyika Territory (breeding condition), January.

**Food:** No information.

**Call:** Unrecorded.

**1439 CAMEROONS INDIGO-BIRD.** *HYPOCHERA CAMERUNENSIS* Grote.

*Hypochera chalybeata camerunensis* Grote, J.f.O. p. 398, 1922: between Nola and Mbaiki, south-eastern Cameroons.

**Distinguishing characters:** Adult male, general colour dull violet blue black; wings and tail ash-colour; white patches on sides of rump; bill whitish; feet light brownish plum. The dull colour and ash-coloured wings and tail distinguish this species from the Purple Indigo-Bird. The non-breeding dress, adult female and young bird are very similar to the Purple Indigo-Bird, but darker brown above than the Chad race of that species. Wing, male 62 to 67, female 61 to 63 mm.

**General distribution:** Senegal to Gambia, Sierra Leone and Abyssinia.

**Range in Eastern Africa:** Central and south-western Sudan to northern Abyssinia.

**Habits:** Of the genus, little recorded.

**Nest and Eggs:** Undescribed. Method of breeding unknown.

**Recorded breeding:** No records.

**Food:** No information.

**Call:** The male has quite a sweet little song.

**1440 ALEXANDER'S INDIGO-BIRD.** *HYPOCHERA NIGERIÆ* Alexander.

*Hypochera nigeriæ* Alexander, Bull. B.O.C. 23, p. 33, 1908: Kiri, Gongola River, northern Nigeria.

**Distinguishing characters:** Adult male, general colour glossy blue black with a greenish wash; wings and tail ash-colour; white patches on sides of rump; bill white or pale flesh; feet white or flesh coloured. The slight greenish wash distinguishes this species from the Cameroons Indigo-Bird, and this colour and the ash-coloured wings and tail distinguishes it from the Purple Indigo-Bird. The non-breeding dress, adult female and young bird are very similar to the Chad race of the Purple Indigo-Bird. Wing, male 61 to 66; female 62 mm.

**General distribution:** Sierra Leone and northern Nigeria to north-eastern Belgian Congo and the Sudan.

**Range in Eastern Africa:** Western Sudan.

**Habits:** Of the genus, migratory in the western Sudan.

**Nest and Eggs:** Undescribed. Method of breeding unknown.

**Recorded breeding:** Darfur, Sudan, September.

**Food:** No information.

**Call:** A sweet song is recorded in the breeding season.

**1441 PIN-TAILED WHYDAH. *VIDUA MACROURA* (Pallas). Pl. 94.**

*Fringilla macroura* Pallas, in Vroeg's Cat. Adum. No. 144, p. 3, 1764: Angola.

**Distinguishing characters:** Adult male, head, mantle and wings black; white collar on hind neck; lesser and secondary wing-coverts white forming a patch on the wing; rump and upper tail-coverts white with some black streaks; tail black with white on inner webs; four central feathers black and considerably elongated; sides of face and underside white; chin black; under wing-coverts and inner webs of flight feathers white; bill red, scarlet, orange scarlet, rich orange chrome, orange coral or crimson; feet grey black, brownish black, black, pale cold sepia or slate. In non-breeding dress centre of crown tawny; broad black streaks on each side of crown from base of bill to nape; rest of upperparts broadly streaked black and tawny; flight feathers black narrowly edged whitish; innermost secondaries and wing-coverts black broadly edged with tawny; tail as in breeding dress but central tail feathers only very slightly longer than the others; sides of face buffish white with black moustachial streak and black streak behind eye; below, white with more or less buffish tawny on chest and flanks; some streaks on sides of chest; bill and feet as in breeding dress. The female is similar to the male in non-breeding dress, but is smaller and less boldly streaked above; towards the end of the breeding season practically all the tawny colour has faded out; bill in breeding season, upper mandible dark brown or blackish, lower greenish pink, dark madder, dusky or black; in non-breeding season red, dull red, or pale red. Very rarely the female in the breeding season has two blackish brown elongated central tail feathers. The immature dress of both sexes is similar to the adult in non-breeding dress, but with the tail of the young bird. The young bird is earth brown above with some indistinct streaks on mantle; below, buff; no white in tail; bill black to reddish; feet brown. Wing, male 66 to 78, female 61 to 70 mm.

**General distribution:** Fernando Po, St. Thomé Island and Senegal to Abyssinia, Cape Province and Natal, also Mayotte Island in the Comoros.

**Range in Eastern Africa:** The Sudan, Eritrea and Abyssinia to the Zambesi River; also Zanzibar and Mafia Islands.

**Habits:** Delightful little birds with curious little jerky movements at all times. Widely distributed and usually common all over Eastern Africa with local seasonal movements. Pugnacious in the breeding season, but gather into large flocks at other times of year and roost in companies in trees like Starlings. The male has a curious habit of wiggling his long tail about which has earned him in Uganda the name of Tadpole Wagtail, and he has also a dancing courtship flight before the female, which, as the breeding season is a prolonged one, is often seen. Almost certainly polygamous.

**Nest and Eggs:** Parasitic, chiefly on Estrilda Waxbills and Cisticolas, even laying the first egg in the nest. Eggs are white or pale cream; about 14·5 × 11 mm.

**Recorded breeding:** Abyssinia, August to November. Sudan, April to July, also September and November. Uganda, March to November. Kenya Colony, at any time of year. Tanganyika Territory, irregular. Zanzibar Island, December to May. Portuguese East Africa, March to June. Rhodesia, February to May. Nyasaland, December to April.

**Food:** Mainly grass and other small seeds.

**Call:** Alarm note a sharp chirp, a twittering little shrill laughing song in the breeding season.

1048   SPARROWS, WEAVERS, WAXBILLS, etc.

**1442 STEEL-BLUE WHYDAH.   VIDUA HYPOCHERINA**
Verreaux.   Pl. 94.

*Vidua hypocherina* J. & E. Verreaux, Rev. Mag. Zool. p. 260, pl. 16, 1856: East Africa.

**Distinguishing characters:** Adult male, general colour violet blue black; under wing-coverts and edges of inner webs of flight feathers white; white patch on each side of rump; tail black with narrow white tips, four central feathers greatly elongated; bill grey flesh; feet dusky flesh. Specimens which have lost the elongated tail feathers can be at once distinguished from the Indigo-Birds by the white under wing-coverts and inner webs of the flight feathers. In non-breeding dress it closely resembles the Pin-tailed Whydah in similar dress, but the bill is brownish and the white in the tail is confined to the edge and tips. The female is similar to the male in non-breeding dress. Wing, male 63 to 67, female 60 to 65 mm. The young bird is paler ash-brown above than the young bird of the Pin-tailed Whydah, and has a smaller bill.

**Range in Eastern Africa:** Eastern Abyssinia and British Somaliland to Kenya Colony and Tanganyika Territory as far south as the Dodoma and Morogoro districts, also Kome Island, Lake Victoria.

**Habits:** Locally common in dry thorn country or in open cultivated lands associating in flocks with other small Weavers, particularly with Indigo-Birds. Little is known of its habits but it is presumed to be parasitic.

**Nest and Eggs:** Undescribed.

**Recorded breeding:** Kavirondo, Kenya Colony (breeding condition), June. Tanganyika Territory (breeding condition), February.

**Food:** Small seeds.

**Call:** Unrecorded.

**1443 FISCHER'S WHYDAH.   VIDUA FISCHERI (Reichenow).**
Pl. 94.

*Linura fischeri* Reichenow, O.C. p. 91, 1882: Usegua, eastern Tanganyika Territory.

**Distinguishing characters:** Adult male, forehead, crown and breast to under tail-coverts buff; sides of face, chin to chest, neck all round and mantle black; wings dusky black; under wing-coverts dusky black and white; edges of inner webs of flight feathers pale buffish; lower rump and upper tail-coverts buff streaked with black;

## SPARROWS, WEAVERS, WAXBILLS, etc.

tail dusky, four central tail feathers buff, narrow and greatly elongated; bill red; feet orange. In non-breeding dress the top of the head and sides of face are pale tawny, the former streaked with dusky; superciliary stripe tawny; rest of upperparts deep buff streaked with black; below, chin, neck and flanks deep buff; rest of underparts buffish white; under wing-coverts dusky buff; bill red. The female is similar to the male in non-breeding dress. Wing, male 65 to 71, female 61 to 65 mm. The young bird has the whole head to breast and flanks tawny; mantle, rump, wings and tail earth brown; mantle faintly streaked with dusky; belly and under tail-coverts white; bill black or slate.

**Range in Eastern Africa:** Eastern Abyssinia and British Somaliland to Kenya Colony and Tanganyika Territory as far south as the Morogoro district.

**Habits:** Only known as birds of dry thorn bush and desert scrub, not usually common, shy, and in small flocks. They have a dancing courtship flight.

**Nest and Eggs:** Parasitic on small Weavers, notably the Purple Grenadier. Eggs white; about 15·5 × 12·5 mm.

**Recorded breeding:** Southern Abyssinia (probable), June. Tsavo, Kenya Colony (breeding condition), April. Northern Tanganyika Territory (breeding condition), March to May.

**Food:** Insects and larvæ, grass and other small seeds.

**Call:** A short Bunting-like song (Elliott). A small sweet rather lisping Serin-like song (Moreau).

*Note:* As this Whydah is parasitic on the Purple Grenadier the young will be found with the foster parents, but they are at once distinguishable from the young of that species by not having blue upper tail-coverts nor a black tail; they are also longer winged and smaller billed.

**1444 PARADISE WHYDAH.** *STEGANURA PARADISÆA*
(Linnæus). **Pl. 94.**
*Emberiza paradisæa* Linnæus, Syst. Nat. 12th ed. p. 32, 1766: Angola.

**Distinguishing characters:** Adult male, head all round, throat mantle, rump, upper and under tail-coverts, wing-coverts, innermost secondaries and tail black; broad band on hind neck golden buff; chest chestnut; breast to belly buff; centre of belly whiter; central pair of tail feathers broad and short with bare shafts elongated; next pair broad, greatly elongated and tapering towards tips, outer webs much narrower than inner webs towards tips; other tail feathers normal with very narrow white edging to tips and inner webs; bill black. In non-breeding dress a creamy broad streak from forehead to occiput streaked with black, bordered by broad black streaks; sides of face and throat creamy white with a black streak behind eye; mantle and scapulars tawny streaked black; innermost secondaries, rump and upper tail-coverts black and broadly bordered with ashy grey; wing-coverts black broadly bordered with tawny; below, chest and flanks deep tawny with black streaks; centre of breast to belly and under tail-coverts white. The female has similar markings to the male in non-breeding dress but is much duller, more earth brown than tawny, with chest and flanks buffish brown. Wing, male 75 to 85, female 73 to 83 mm. The young bird is earth brown above including wings and tail; below, chin to chest pale earth brown; rest of underparts white; bill horn-colour.

**General distribution:** Eritrea, Abyssinia, the Sudan to Angola and South Africa.

**Range in Eastern Africa:** Eritrea, Abyssinia, south-eastern Sudan, north-eastern Uganda and Kenya Colony to the Zambesi River.

**Habits:** As this species and the next have only recently been separated, field notes may refer to either, and as far as is known there is no difference in habits, though further observation may produce facts of interest. Widely distributed and not uncommon, the males attract attention by their dipping flight and their habit of soaring into the air and dropping perpendicularly: the flight is also noticeably noisy with a rustling sound.

**Nest and Eggs:** Parasitic on the Green-winged and other Pytilias; eggs white and rather larger than the hosts' eggs; about $18 \cdot 5 \times 13 \cdot 5$ mm.

**Recorded breeding:** Sudan, breeding condition, September. Western Tanganyika Territory, March to June, and November and December. Northern Tanganyika Territory, March to August, and October to December. Central and southern Tanganyika Territory, January to April, and October to December. Nyasaland (breeding condition), April and May.

**Food:** Grass seeds.

**Call:** Nothing particularly recorded.

## SPARROWS, WEAVERS, WAXBILLS, etc.

**1445 BROAD-TAILED PARADISE WHYDAH.** *STEGANURA ORIENTALIS* (Heuglin).

*Steganura orientalis orientalis* (Heugl.).

*Vidua paradisea orientalis* Heuglin, Orn. Nordost. Afr. 2, p. 583, 1871: Keren, Eritrea.

**Distinguishing characters:** Adult male, differs from the Paradise Whydah in having the elongated pair of tail feathers broad to their tips, with both webs of about equal width; broad band on hind neck golden buff and chestnut. The non-breeding dress is paler tawny than that of the Paradise Whydah and the streaks on the mantle are narrower. The female is paler tawny above than the female of the Paradise Whydah and the streaks on mantle are narrower. Wing, male 74 to 80, female 72 to 75 mm. The young bird is paler earth brown than the young bird of the Paradise Whydah.

**General distribution:** Lake Chad to Eritrea and the Sudan.

**Range in Eastern Africa:** Eritrea to western and southern Sudan.

**Habits:** See under the preceding species.

**Nest and Eggs:** Undescribed.

**Recorded breeding:** No records.

**Food:** No information.

**Call:** Unrecorded.

*Steganura orientalis obtusa* Chap. **Pl. 94.**

*Steganura aucupum obtusa* Chapin, Am. Mus. Nov. 43, p. 6, 1922: Luchenya, southern Nyasaland.

**Distinguishing characters:** Adult male, differs from the nominate race in having the elongated pair of tail feathers shorter and much broader. Wing, male 81 to 87 mm. The female and juvenile plumages are unrecorded.

**General distribution:** Eastern and southern Belgian Congo to Kenya Colony, Tanganyika Territory, Angola, eastern Southern Rhodesia and southern Portuguese East Africa as far south as Gazaland and Beira.

**Range in Eastern Africa:** South-central Kenya Colony to western Tanganyika Territory and Nyasaland.

**Habits:** Of the nominate race.

1052  SPARROWS, WEAVERS, WAXBILLS, etc.

**Recorded breeding:** Nyasaland (breeding condition), April and May. Portuguese East Africa, March to July.

**Distribution of other races of the species:** West Africa.

Names in Sclater's *Syst. Av. Æthiop.*, 2, 1930, which have been changed or have become synonyms in this work:

*Bubalornis albirostris nyanzæ* (Neumann), treated as synonymous with *Bubalornis niger intermedius* (Cabanis).

*Passer griseus eitreæ* Zedlitz, treated as synonymous with *Passer griseus neumanni* Zedlitz.

*Passer griseus mosambicus* Van Someren, treated as synonymous with *Passer griseus ugandæ* Reichenow.

*Sorella eminibey guasso* Van Someren, treated as synonymous with *Sorella eminibey* Hartlaub.

*Gymnoris pyrgita reichenowi* Zedlitz, treated as synonymous with *Petronia xanthosterna pyrgita* (Heuglin).

*Sporopipes frontalis abyssinicus* Mearns, treated as synonymous with *Sporopipes frontalis frontalis* (Daudin).

*Ploceus insignis ornatus* Granvik, treated as synonymous with *Phormoplectes insignis insignis* (Sharpe).

*Ploceus emini budongoensis* (Van Someren) treated as synonymous with *Othyphantes emini* (Hartlaub).

*Ploceus nigriceps graueri* Hartert, treated as synonymous with *Ploceus nigriceps* (Layard).

*Ploceus cucullatus femininus* (O. Grant) treated as synonymous with *Ploceus cucullatus bohndorffi* Reichenow.

*Ploceus pachyrhynchus omoensis* (Neumann) treated as synonymous with *Pachyphantes pachyrhynchus* (Reichenow).

*Ploceus ocularius abayensis* Neumann treated as synonymous with *Hyphanturgus ocularis crocatus* (Hartlaub).

*Xanthophilus xanthops camburni* (Sharpe) treated as synonymous with *Xanthophilus xanthops xanthops* Hartlaub.

*Icteropsis pelzelni futa* Bangs & Phillips, treated as synonymous with *Icteropsis pelzelni* (Hartlaub).

*Ploceus flavissimus* Neumann, treated as synonymous with *Ploceus galbula* Rüppell.

*Amblyospiza albifrons æthiopica* Neumann, treated as synonymous with *Amblyospiza albifrons melanota* (Heuglin).

*Quelea quelea centralis* Van Someren, treated as synonymous with *Quelea quelea æthiopica* (Sundevall).

*Euplectes orix wertheri* (Reichenow) treated as synonymous with *Euplectes orix nigrifrons* (Böhm).

*Euplectes rufigula* (Van Someren) treated as synonymous with *Euplectes nigroventris* Cassin.

*Euplectes hordeacea sylvatica* (Neumann) treated as synonymous with *Euplectes hordeacea hordeacea* (Linnæus).

*Euplectes hordeacea changamwensis* (Mearns) treated as synonymous with *Euplectes hordeacea hordeacea* (Linnæus).

*Euplectes capensis crassirostris* (O. Grant) treated as synonymous with *Euplectes capensis xanthomelas* Rüppell.

*Euplectes taha intercedens* (Erlanger) treated as synonymous with *Euplectes afra stricta* Hartlaub.

*Coliuspasser ardens teitensis* (Van Someren) treated as synonymous with *Coliuspasser ardens ardens* (Boddaert).

*Spermestes bicolor stigmatophorus* Reichenow, treated as synonymous with *Spermestes poensis poensis* (Fraser).

*Euodice cantans inornata* (Mearns) treated as synonymous with *Euodice malabarica cantans* (Gmelin).

# SPARROWS, WEAVERS, WAXBILLS, etc. 1053

*Euodice cantans meridionalis* (Mearns) treated as synonymous with *Euodice malabarica orientalis* (Lorenz & Hellmayr).
*Nigrita bicolor saturatior* Reichenow, treated as synonymous with *Nigrita bicolor brunnescens* Reichenow.
*Cryptospiza reichenovii ocularis* Sharpe, treated as synonymous with *Cryptospiza reichenovii australis* Shelley.
*Cryptospiza salvadorii borealis* Percival, treated as synonymous with *Cryptospiza salvadorii salvadorii* Reichenow.
*Pirenestes ostrinus centralis* Neumann, treated as synonymous with *Pirenestes rothschildi* Neumann.
*Pytelia melba belli* O. Grant, treated as synonymous with *Pytilia melba melba* (Linnæus).
*Pytelia melba jessei* Shelley, treated as synonymous with *Pytilia melba soudanensis* (Sharpe).
*Lagonosticta rubricata hildebrandti* Neumann, treated as synonymous with *Lagonosticta rubricata rhodopareia* Heuglin.
*Lagonosticta senegala flavodorsalis* Zedlitz, treated as synonymous with *Lagonosticta senegala rhodopsis* Heuglin.
*Lagonosticta senegala abayensis* Neumann, treated as synonymous with *Lagonosticta senegala brunneiceps* Sharpe.
*Lagonosticta senegala kikuyuensis* Van Someren, treated as synonymous with *Lagonosticta senegala ruberrima* Reichenow.
*Lagonosticta cinereovinacea rudolfi* (Hartert) treated as synonymous with *Lagonosticta cinereovinacea graueri* Rothschild.
*Coccopygia melanotis nyansæ* (Neumann) treated as synonymous with *Coccopygia melanotis kilimensis* Sharpe.
*Estrilda astrild massaica* Neumann, treated as synonymous with *Estrilda astrild cavendishi* Sharpe.
*Estrilda astrild macmillani* O. Grant, treated as synonymous with *Estrilda astrild peasei* Shelley.
*Estrilda xanthophrys* W. L. Sclater, treated as synonymous with *Estrilda troglodytes* (Lichtenstein).
*Estrilda atricapilla kandti* Reichenow, treated as synonymous with *Estrilda nonnula nonnula* Hartlaub.
*Estrilda charmosyna kiwanukæ* Van Someren, treated as synonymous with *Estrilda erythronotos delamerei* Sharpe.
*Uræginthus bengalus ugandæ* Zedlitz, treated as synonymous with *Uræginthus bengalus schoanus* Neumann.
*Granatina ianthinogaster ugandæ* Van Someren, treated as synonymous with *Granatina ianthinogaster hawkeri* Phillips.
*Hypochera ultramarina purpurascens* Reichenow, treated as synonymous with *Hypochera nigerrima* Sharpe.

Names introduced since 1930, which have become synonyms in .this work:

*Passer griseus turkanæ* Granvik, 1934, treated as synonymous with *Passer gongonensis gongonensis* (Oustalet).
*Passer griseus tertale* Benson, 1942, treated as synonymous with *Passer gongonensis gongonensis* (Oustalet).
*Petronia superciliaris oraria* Grote, 1930, treated as synonymous with *Petronia superciliaris* (Blyth).
*Gymnoris pyrgita dankali* Thesiger & Meynell, 1935, treated as synonymous with *Petronia xanthosterna pyrgita* (Heuglin).
*Ploceus jacksoni jucundus* Friedmann, 1931, treated as synonymous with *Ploceus jacksoni* Shelley.
*Ploceus aureoflavus pallidiceps* Vincent, 1933, treated as synonymous with *Xanthophilus subaureus aureoflavus* (Smith).
*Ploceus aureoflavus reicherti* Meise, 1934, treated as synonymous with *Xanthophilus subaureus aureoflavus* (Smith).
*Euplectes zavattarii* Moltoni, 1943, treated as synonymous with *Euplectes orix pusilla* (Hartert).

*Cryptospiza reichenovii sanguinolenta* Vincent, 1933, treated as synonymous with *Cryptospiza reichenovii australis* Shelley.

*Cryptospiza salvadorii chyuluensis* Van Someren, 1939, treated as synonymous with *Cryptospiza salvadorii kilimensis* Sclater.

*Amadina fasciata albitorquata* Vincent, 1933, treated as synonymous with *Amadina fasciata meridionalis* Neunzig.

*Paludipasser iriscæ* Roberts, 1933, treated as synonymous with *Ortygospiza locustella locustella* Neave.

*Anomalospiza imberbis nyassæ* Benson, 1938, treated as synonymous with *Anomalospiza imberbis rendalli* (Tristam).

*Pytilia melba jubaensis* Van Someren, 1932, treated as synonymous with *Pytilia melba soudanensis* (Sharpe).

*Estrilda ianthinogaster somereni* Delacour, 1943, treated as synonymous with *Granatina ianthinogaster hawkeri* Phillips.

*Steganura aucupum kaduliensis* Bowen, 1931, treated as synonymous with *Steganura orientalis orientalis* Heuglin.

## Addenda and Corrigenda

**1379-1381** The genus *Spermestes* should now be *Lonchura* (Fraser).

**1383** *Euodice malabarica orientalis*. Recorded breeding: add Eritrea, February and March.

**1395** Delete *Pirenestes maximus* which is a synonym of 1397 *Pirenestes frommi*.

**1397** *Pirenestes frommi* was described from a young bird. The bill measurements should be 18 to 19 mm.

**1398** *Cryptospiza reichenowi*. The Uganda birds agree best with the nominate race, not with *C.r. australis*. They were originally separated as a species under the name *Cryptospiza ocularis* Sharpe.

**1400** *Cryptospiza jacksoni*. Add Nest and Eggs: Nest in a shrub or small tree, oval, mainly of grass with a little moss, lined with feathers or hair. Eggs two, white; about 17 × 12 mm. Recorded breeding: Belgian Congo, July to September, probably also in April and May.

**1403** *Ortygospiza atricollis fuscocrissa*. Recorded breeding: add Eritrea, June.

*O.a. dorsostriata*. General distribution: Kasai district, Belgian Congo to Uganda.

*Ortygospiza atricollis dorsostriata* is not a race of *O. atricollis* but a race of, or identical with, *Ortygospiza gabonensis* Lynes, Bull. B.O.C. 33, p. 131, 1914: Gabon.

**1407** *Mandingoa n. schlegeli*. Range in Eastern Africa: add south-central Uganda.

**1410** *Pytilia m. grotei* should be *P.m. belli* O. Grant, Bull. B.O.C. 21, p. 14, 1907: Mokia, Toro, Uganda.

*P.m. belli*. Recorded breeding: add Uganda, October.

**1429** *Nesocharis ansorgei*. General distribution: add South-central Belgian Congo. Range in Eastern Africa: add Katera forest, Masaka, Uganda. Nest and Eggs: add Believed to utilise the old nests of Ploceid Weavers.

**1431** *Uræginthus bengalus schoanus*. Recorded breeding: add Eritrea, October.

**1441** *Vidua macroura*. Add to map and distribution Angola.

**1442** *Vidua hypocherina*. Distinguishing characters: add Male in non-breeding dress has no long central tail feathers.

**1443** *Vidua fischeri*. Distinguishing characters: as in note under 1442.

*Continued on p. 1105*

# FINCHES

FAMILY—**FRINGILLIDÆ. FINCHES.** Genera: *Bucanetes, Rhynchostruthus, Serinus, Linurgus, Carduelis* and *Warsanglia*.

Twenty-one species occur in Eastern Africa. Their habits do not differ from those of Finches elsewhere, and they are found in all types of country throughout Eastern Africa.

We can see no justification for upholding the genus *Poliospiza* as distinct from *Serinus*.

### KEY TO THE ADULT FINCHES OCCURRING IN EASTERN AFRICA

1 Some yellow or olivaceous yellow or green in plumage: 3–35
2 No yellow or olivaceous yellow or green in plumage except on rump: 36–49
3 Some black or dusky black on head, chin and throat: 5–13
4 No clear black or dusky black on head, chin and throat yellow washed with yellow or olive green: 14–35
5 Head and throat black, collar on hind neck bright yellow, mantle green, underside olivaceous and yellow: ORIOLE-FINCH, male *Linurgus olivaceus* **1463**
6 No collar on hind neck: 7–35
7 Plumage dull olive green, top of head dusky black: BLACK-HEADED SISKIN male *Carduelis nigriceps* **1465**
8 Head and throat black, ear-coverts white, edges of secondaries, wing-coverts and tail bright yellow: GOLDEN-WINGED GROSBEAK *Rhynchostruthus socotranus* **1447**

1056 FINCHES

9 Black or dusky black confined to forehead, sides of face and chin and upper throat: 10–13
10 Forehead band white or earth brown: THICK-BILLED SEED-EATER *Serinus burtoni* **1462**
11 Forehead band black: 12–13
12 Band behind black band on forehead light yellow: BLACK-FACED CANARY male *Serinus capistratus* **1451**
13 Whole of head behind black band on forehead uniform with rest of upperparts: AFRICAN CITRIL male *Carduelis citrinelloides* **1464**
14 Chin and throat yellow or olivaceous or washed with yellow: 15–35
15 Remainder of plumage brown or buff heavily streaked above and below: STREAKY SEED-EATER *Serinus striolatus* **1461**
16 Remainder of plumage green or yellow or olivaceous not heavily streaked below: 17–35
17 Above, green not streaked: ORIOLE-FINCH female *Linurgus olivaceus* **1463**
18 More or less streaked above: 19–35
19 Rump yellow contrasting with rest of upperparts: 20–23
20 No yellow band on forehead: 21–22

| | | |
|---|---|---|
| 21 | Lower rump and upper tail-coverts yellow: | GROSBEAK CANARY male *Serinus donaldsoni* **1452** |
| 22 | Lower rump and upper tail-coverts green: | KENYA GROSBEAK CANARY *Serinus buchanani* **1453** |
| 23 | Yellow band on forehead: | 24–25 |
| 24 | Belly white: | WHITE-BELLIED CANARY *Serinus dorsostriatus* **1449** |
| 25 | Belly yellow: | YELLOW-FRONTED CANARY *Serinus mozambicus* **1448** |
| 26 | Rump green or yellowish green not contrasting with rest of upperparts: | 27–35 |
| 27 | Forehead and forecrown golden yellow: | YELLOW-CROWNED CANARY, male *Serinus flavivertex* **1454** |
| 28 | Yellow band on forehead: | BRIMSTONE CANARY *Serinus sulphuratus* **1450** |
| 29 | Forehead and forecrown uniform with rest of head: | 31–35 |
| 30 | Forehead to sides of face and chin and throat slightly dusky: | BLACK-HEADED SISKIN female *Carduelis nigriceps* **1465** |
| 31 | Forehead to sides of face and chin and throat as rest of plumage: | 32–35 |
| 32 | Flight feathers and tail dusky: | BLACK-FACED CANARY female *Serinus capistratus* **1451** |
| 33 | Flight feathers and tail black: | 34–35 |

| | | |
|---|---|---|
| 34 General colour duller, more olivaceous: | YELLOW-CROWNED CANARY female *Serinus flavivertex* | 1454 |
| 35 General colour brighter green and yellow: | AFRICAN CITRIL female *Carduelis citrinelloides* | 1464 |

| | | |
|---|---|---|
| 36 Rump white: | WHITE-RUMPED SEED-EATER *Serinus leucopygius* | 1458 |
| 37 Rump yellow: | | 38–41 |
| 38 Bill large: | GROSBEAK CANARY female *Serinus donaldsoni* | 1452 |
| 39 Bill not large: | | 40–41 |
| 40 Throat to chest yellow: | YELLOW-THROATED SEED-EATER *Serinus flavigula* | 1460 |
| 41 No yellow on throat to chest: | YELLOW-RUMPED SEED-EATER *Serinus atrogularis* | 1459 |
| 42 Rump chestnut: | WARSANGLI LINNET *Warsanglia johannis* | 1466 |
| 43 Rump not white or yellow or chestnut, but uniform with rest of upperparts: | | 44–49 |
| 44 Some rose colour in plumage: | TRUMPETER-BULLFINCH *Bucanetes githagineus* | 1446 |
| 45 No rose colour in plumage: | | 46–49 |
| 46 Sides of face and ear-coverts black: | BLACK-EARED SEED-EATER *Serinus mennelli* | 1457 |
| 47 Sides of face and ear-coverts ashy brown: | | 48–49 |

Plate 94

Purple Grenadier (p. 1038)
Female     Male
Pin-tailed Whydah (p. 1046)
Female     Male
Fischer's Whydah (p. 1048)
Female     Male
Dusky Indigo-Bird (p. 1043)
Female     Male
Black Indigo-Bird (p. 1044)

Paradise Whydah (p. 1049)
Male     Female
Broad-tailed Paradise Whydah (p. 1051)
Steel-blue Whydah (p. 1048)
Male     Female
Purple Indigo-bird (p. 1042)
Male     Female
Green Indigo-Bird (p. 1040)

Plate 95

Yellow-fronted Canary (p. 1061)  
Black-headed Siskin (p. 1084)  
Yellow-crowned Canary (p. 1068)  
Brown-rumped Seed-eater (p. 1070)  
Cabanis' Bunting (p. 1088)  
Somali Golden-breasted Bunting (p. 1091)  

African Citril (p. 1083)  
Male    Female  
Brimstone Canary (p. 1065)  
White-bellied Canary (p. 1064)  
Streaky-headed Seed-eater (p. 1070)  
Golden-breasted Bunting (p. 1090)  
Brown-rumped Bunting (p. 1092)  

Grosbeak Canary (p. 1067)  
Black-faced Canary (p. 1066)  
Male    Female  
Oriole Finch (p. 1080)  
Male    Female  
Trumpeter Bullfinch (p. 1059)  
Ortolan (p. 1092)  
Three-streaked Bunting (p. 1089)

| 48 Underside ashy brown, throat streaked: | BROWN-RUMPED SEED-EATER *Serinus tristriatus* | 1455 |
|---|---|---|
| 49 Underside white, throat and chest streaked: | STREAKY-HEADED SEED-EATER *Serinus gularis* | 1456 |

**1446 TRUMPETER-BULLFINCH.** *BUCANETES GITHAGINEUS* (Lichtenstein).

*Bucanetes githagineus githagineus* (Licht.). **Pl. 95.**

*Fringilla githaginea* Lichtenstein, Verz. Doubl. p. 29, 1823: Dera, Upper Egypt.

**Distinguishing characters:** Adult male, forehead, crown, sides of face and chin to chest pale grey washed with rose colour; mantle earth brown with a slight rosy wash; wings dusky, ends black, tips white; edges of wing-coverts and flight feathers rose pink; tail dusky, blacker towards ends of feathers, tips white, outer edges rosy; upper tail-coverts and rest of underparts rose colour; bill orange red. The female has less rose colour, and is paler earth brown above including head; below, buffish brown; bill pale orange. Wing 77 to 88 mm. The young bird is pale tawny buff above and below.

**General distribution:** Egypt and the Sudan.

**Range in Eastern Africa:** The Anglo-Egyptian Sudan as far south as Dongola and the Red Sea Province.

**Habits:** Locally plentiful in the northern Sudan along the desert edge raiding crops on cultivated land. Usually in small parties running along the ground like Larks and are extremely difficult to see when stationary.

**Nest and Eggs:** Nest a cup of roots and grass lined with any kind of soft material, usually placed under cover of a stone or rock. Eggs four or five, pale greenish white with rufous spots usually in a zone at the larger end; about 20 × 15 mm.

**Recorded breeding:** Ahaggar Mts., French Sudan, February. Egypt, March.

**Food:** Small seeds and insects.

**Call:** A trumpeting call, not very loud, also a double call of 'ter-ter.'

**Distribution of other races of the species:** Morocco to Tunisia and Arabia.

### 1447 GOLDEN-WINGED GROSBEAK. *RHYNCHOSTRUTHUS SOCOTRANUS* Sclater & Hartlaub.

*Rhynchostruthus socotranus socotranus* Scl. & Hartl.
*Rhyncostruthus socotranus* P. L. Sclater & Hartlaub, P.Z.S. p. 171, pl. 17, fig. 2, 1881: Goehel Valley, Socotra Island.

**Distinguishing characters:** Adult male, head all round to upper chest black; ear-coverts white; mantle and rump earth brown; wings and tail black; wing-coverts, edges of secondaries and tail feathers bright yellow; below, ashy grey, paler on belly; bill black. The female has less black on the head; nape and upper chest brown. Wing 82 to 90 mm. The young bird has the head ashy brown; streaked with brown both above and below; yellow on the wings and tail paler.

**Range in Eastern Africa:** Socotra Island.

**Habits:** No information.

**Nest and Eggs:** Undescribed.

**Recorded breeding:** No records.

**Food:** No information, see under the following race.

**Call:** Unrecorded.

*Rhynchostruthus socotranus louisæ* Phil.
*Rhynchostruthus louisæ* Phillips, Bull. B.O.C. 6, p. 47, 1897: Sheikh, British Somaliland.

**Distinguishing characters:** Adult male, differs from the nominate race in having the top of the head brown; black confined to sides of face and chin to neck; ear-coverts greyish with a whitish edging to black of face. The female has black on face and chin only. Wing 78 to 86 mm.

**Range in Eastern Africa:** British Somaliland.

**Habits:** The only note we have is that birds were seen feeding on the buds of an Euphorbia.

**Recorded breeding:** No records.

**Distribution of other races of the species:** Arabia.

### 1448 YELLOW-FRONTED CANARY. *SERINUS MOZAMBICUS* (Müller).

*Serinus mozambicus mozambicus* (Müll.). **Pl. 95.**

*Fringilla mozambica* P. L. S. Müller, Syst. Nat. Suppl. p. 163, 1776: Mozambique, northern Portuguese East Africa.

**Distinguishing characters:** Adult male, forehead, broad streak over eye and lower rump yellow; upper tail-coverts green; rest of upperparts dull green streaked with black; wings dusky edged with yellowish green including wing-coverts; tail dusky edged with yellowish green and tipped with white; lores through eye to ear-coverts blackish; cheeks yellow; moustachial stripe black; chin to under tail-coverts yellow; sides of neck and chest dusky; bill, upper mandible brown, lower whitish. The female is usually slightly browner on head and mantle. Wing 61 to 72 mm. The young bird is duller than the adult female with less yellow on forehead; chest spotted with dusky; some streaking on sides of chest and flanks.

**General distribution:** Tanganyika Territory to Angola, Damaraland, Bechuanaland, eastern Cape Province and Natal, introduced into Mauritius, Réunion and Amirante Islands.

**Range in Eastern Africa:** Eastern Tanganyika Territory to the Zambesi River, also Mafia Island, and a sight-record from Zanzibar Island.

**Habits:** Common and often abundant in gardens, cultivated land, or any form of lightly wooded country. Very freely kept as cage birds owing to their tameness and sweet song. Gregarious in the non-breeding season and usually seen feeding on the ground in small flocks.

**Nest and Eggs:** Nest a shallow compact cup of grass and weed stems bound with cobwebs and lined with rootlets, usually saddled on the fork of a small tree or shrub. Eggs three or four, greenish white with sparse spotting of reddish brown and occasionally with pale purple blotches; about $16 \cdot 5 \times 12 \cdot 5$ mm.

**Recorded breeding:** Nyasaland in most months, chiefly January to August. Portuguese East Africa, March to May.

**Food:** Grass and weed seeds, but at times of year buds and flowers of trees.

**Call:** A sweet song reminiscent of a Canary but not as powerful.

*Serinus mozambicus barbatus* (Heugl.).
*Crithagra barbata* Heuglin, J.f.O. p. 248, 1864: Djur, Bahr-el-Ghazal, south-western Sudan.

**Distinguishing characters:** Differs from the nominate race in having the head and mantle much brighter green; chin sometimes white. The sexes are alike. Wing 61 to 71 mm.

**General distribution:** Southern French Equatorial Africa to northern Belgian Congo from about long. 19° E., the Sudan, Uganda and Kenya Colony.

**Range in Eastern Africa:** Western and south-western Sudan, Uganda and central and eastern Kenya Colony, also Lamu Island.

**Habits:** As for other races, common among leafy trees, gardens, and cultivation.

**Recorded breeding:** Western Sudan, July to October. Uganda, March to June, also November and December. Kenya Colony in many months, often June to August.

*Serinus mozambicus pseudobarbatus* V. Som.
*Serinus pseudobarbatus* Van Someren, Bull. B.O.C. 40, p. 56, 1919: Fort Ternan, Kavirondo, western Kenya Colony.

**Distinguishing characters:** Differs from the nominate race in being more yellow green above, and from the preceding race in being darker yellow green above and rather more heavily streaked. Wing 67 to 73 mm.

**General distribution:** Kenya Colony, Tanganyika Territory and the south-eastern Belgian Congo.

**Range in Eastern Africa:** Western Kenya Colony to western and south-central Tanganyika Territory at Kigoma and Iringa.

**Habits:** As for other races.

**Recorded breeding:** Tanganyika Territory in several months, commonly February to April.

*Serinus mozambicus grotei* Scl. & Praed.
*Serinus mozambicus grotei* W. L. Sclater & Mackworth-Praed, Ibis p. 581, 1931: Sennar, eastern Sudan.

**Distinguishing characters:** Very similar above to the nominate race, but yellow of forehead, over eye and below, much paler. Wing 61 to 70 mm.

**Range in Eastern Africa:** Eastern and south-eastern Sudan and western Abyssinia.

**Habits:** As for other races.

**Recorded breeding:** No records.

*Serinus mozambicus gommaensis* Grant & Praed.
*Serinus mozambicus gommaensis* C. Grant & Mackworth-Praed, Bull. B.O.C. 66, p. 18, 1945: Gomma, southern Abyssinia.

**Distinguishing characters:** General colour similar to the nominate race but usually less yellow on forehead; stripe over eye not extending so far back; darker dull green above than the preceding race. Wing 64 to 71 mm.

**Range in Eastern Africa:** Abyssinia, except the Wallega area near the Sudan border and central Eritrea.

**Habits:** As for other races.

**Recorded breeding:** No records.

**Distribution of other races of the species:** Senegal to Angola, and São Thomé Island.

### 1449 WHITE-BELLIED CANARY. *SERINUS DORSOSTRIATUS* (Reichenow).

*Serinus dorsostriatus dorsostriatus* (Reichw.).
*Crithagra dorsostriatus* Reichenow, J.f.O. p. 72, 1887: Kagehi, Mwanza district, Tanganyika Territory.

**Distinguishing characters:** Adult male, size, shape, bill and general colour similar to the Yellow-fronted Canary, but the tail is longer and the belly and lower flanks white, the latter with some black streaks; bill horn-colour, lower mandible paler. The female is earth brown above streaked with dusky; lower rump yellow; wings and tail dusky black, edged with whitish and yellowish white; superciliary stripe white; below, pale ashy brown, lighter on belly and under tail-coverts; lores to ear-coverts and moustachial stripe dusky black; some blackish spots on upper chest. Wing 74 to 75 mm. The young bird is similar to the adult female, but has little indication of a moustachial streak and the throat to breast and flanks are more streaked than spotted with dusky. The young male is washed with yellow.

**Range in Eastern Africa:** Biharamulo district to the Tabora and Mwanza districts, northern Tanganyika Territory.

**Habits:** As for other races.

**Nest and Eggs and Call:** See under the following race.

**Recorded breeding:** Northern Tanganyika Territory (breeding condition), February.

*Serinus dorsostriatus maculicollis* Sharpe. **Pl. 95.**

*Serinus maculicollis* Sharpe, Bull. B.O.C. 4, p. 41, 1895: Milmil, British Somaliland.

**Distinguishing characters:** Adult male, differs from the nominate race in being smaller, and the mantle is less yellow, more sage-green with broader streaking; under tail-coverts white not yellow. The female has the chest and breast yellow with black spots on the former; chin and throat either yellow or white. Wing 66 to 74 mm.

**Range in Eastern Africa:** Southern Abyssinia, British Somaliland, south-eastern Sudan to Uganda and northern Kenya Colony as far south as Lake Baringo and the Northern Guaso Nyiro.

**Habits:** A typical Canary with the habits of other small Finches, occurring in small flocks in patches of bush in open country or along lake shores. Locally common.

**Nest and Eggs:** A small nest of fibres and rootlets often with a twig foundation, lined with hair and plant down and also usually patched outside with down, placed in the fork of a tree or bush. Eggs two to four, pale blue, either plain or spotted and streaked with black and sepia; about 16 × 13 mm.

**Recorded breeding:** Abyssinia, April. Uganda, June and July, also November to January.

**Food:** Grass seeds.

**Call:** A shrill 'whee,' and a sweet quite loud song.

*Serinus dorsostriatus harterti* Zedl.

*Serinus dorsostriatus harterti* Zedlitz, J.f.O. p. 47, 1916: Afgoi, Webi Shebeli River, southern Italian Somaliland.

**Distinguishing characters:** Similar to the preceding race but smaller. Wing 62 to 67 mm.

**Range in Eastern Africa:** Italian Somaliland.

**Habits:** As for other races.

**Recorded breeding:** No records.

*Serinus dorsostriatus taruensis* V. Som.
*Serinus maculicollis taruensis* Van Someren, Bull. B.O.C. 41, p. 114, 1921: M'buyuni, south-eastern Kenya Colony.

**Distinguishing characters:** Adult male, darker green above, less yellow green than the two preceding races; below, less clear yellow; darker and slightly olivaceous on sides of chest. The female is similar to that of the preceding race. Wing 67 to 78 mm.

**Range in Eastern Africa:** Southern Kenya Colony to north-eastern and central Tanganyika Territory, as far south as Dodoma and as far west as Mt. Gerui (Hanang).

**Habits:** Locally common in thorn bush country.

**Recorded breeding:** North-eastern Tanganyika Territory, Longido, May and June, also January; Mkomasi, June.

## 1450 BRIMSTONE CANARY. *SERINUS SULPHURATUS* (Linnæus).
*Serinus sulphuratus sharpii* Neum. **Pl. 95. Ph. xix.**
*Serinus sharpii* Neumann, J.f.O. p. 287, 1900: Marangu, Mt. Kilimanjaro, north-eastern Tanganyika Territory.

**Distinguishing characters:** Adult male, above, yellowish olive green, streaked with dusky, upper tail-coverts not streaked; lores to ear-coverts and moustachial stripe yellowish olive green; wings and tail black edged with yellowish olive green; narrow band on forehead, stripe over eye, stripe between eye and moustachial stripe and underside rich canary yellow; bill horn-colour, lower mandible paler. The female is more heavily streaked above. Wing 72 to 84 mm. The young bird is similar to the adult female but is more olive green above and less bright yellow below.

**General distribution:** Eastern Belgian Congo, Uganda and Kenya Colony to Angola and eastern Southern Rhodesia.

**Range in Eastern Africa:** Uganda and central Kenya Colony to the Zambesi River.

**Habits:** A common and widespread species particularly partial to gardens and cultivated land, but also found more sparingly in any type of open bush or light woodland. The male sings from a prominent perch and the song is sweet and varied. They do not seem to congregate into flocks as much as other Canaries, but at times they do a good deal of damage in gardens.

**Nest and Eggs:** Nest of grass, rootlets and fibres, well lined with plant down and placed in the fork of a tree or bush, or not uncommonly in a bunch of bananas. Eggs normally three, pale green or very pale blue with sparse dark brown or black spots or blotches; about 20 × 13 mm.

**Recorded breeding:** Uganda in any month, most commonly April to June and October to December. Kenya Colony and Tanganyika Territory, mainly October to March. South-eastern Belgian Congo, May and September. Nyasaland, January and March to October. Portuguese East Africa, May and June.

**Food:** Insects, seeds, and very largely young shoots and buds of trees and shrubs.

**Call:** Song as above; the normal call is a throaty chirrup.

**Distribution of other races of the species:** South Africa, the nominate race being described from the Cape of Good Hope.

**1451 BLACK-FACED CANARY.** *SERINUS CAPISTRATUS* (Finsch).

*Serinus capistratus capistratus* (Finsch). Pl. 95.

*Crithagra capistratus* Finsch, in Finsch & Hartlaub, Van der Decken Reisen, 4, Vög. Ostafr. p. 458, 1870: Golungo Alto, northern Angola.

**Distinguishing characters:** Adult male, in general appearance a small Brimstone Canary with a narrow black band on forehead in front of the yellow band; lores to ear-coverts and chin black. The female also in general appearance resembles a small Brimstone Canary, but has dusky olive streaks from chin to chest. Wing 60 to 68 mm. The young bird is more heavily streaked above and below.

**General distribution:** Portuguese Congo to northern Angola, eastern and southern Belgian Congo and Northern Rhodesia.

**Range in Eastern Africa:** Western Urundi.

**Habits:** A scarce species of which little has been recorded in any part of its range.

**Nest and Eggs:** Unknown.

**Recorded breeding:** No records.

**Food:** No information.

**Call:** Unrecorded.

Plate 96

South African Black Flycatcher (p. 185)
*Melaenornis pammelaina pammelaina*

Square-tailed Drongo (p. 566)
*Dicrurus ludwigii sharpei*

Fülleborn's Black Boubou
*Laniarius fülleborni* (p. 610)

Mountain Sooty Boubou (p. 609)
*Laniarius poensis holomelas*

Velvet-mantled Drongo (p. 563)
*Dicrurus modestus coracinus*

Slate-coloured Boubou
*Laniarius funebris* (p. 608)

Sooty Boubou (p. 609)
*Laniarius leucorhynchus*

Black Flycatcher (p. 184)
*Melaenornis edolioides lugubris*

Drongo (p. 564)
*Dicrurus adsimilis adsimilis*

Black Cuckoo-Shrike (p. 556)
*Campephaga sulphurata*

Tropical Boubou (p. 614)
*Laniarius aethiopicus erlanger*

*Serinus capistratus koliensis* Grant & Praed.
*Serinus capistratus koliensis* C. Grant & Mackworth-Praed, Bull. B.O.C. 72, p. 1, 1952: Onyulu's, Koli River, Lango, central Uganda.

**Distinguishing characters:** The male differs from the nominate race in lacking the black on the forehead, sides of face and chin, and being in general appearance very similar to the female of the nominate race, but has darker and more distinct streaks on the head and mantle, and the streaks on the chin to chest darker and much sharper. Wing 61 to 67 mm. The sexes are practically alike.

**Range in Eastern Africa:** Central and southern Uganda to western Kenya Colony and Ruanda.

**Habits:** Recorded as breeding in a swamp at Kisumu.

**Recorded breeding:** Uganda and Kenya Colony, July and August.

**1452 GROSBEAK CANARY.** *SERINUS DONALDSONI* Sharpe. Pl. 95.

*Serinus donaldsoni* Sharpe, Bull. B.O.C. 4, p. 41, 1895: Darde River, Sheikh Hussein, Arussi, south-central Abyssinia.

**Distinguishing characters:** Adult male, top of head and mantle green with blackish streaks; wings and tail dusky with light edges; sides of face greenish; lower rump, upper tail-coverts, superciliary stripe and chin to under tail-coverts yellow; sides of chest and flanks streaked with blackish; bill heavy and horn-colour. The female is ashy brown above, streaked with black; rump and upper tail-coverts yellowish green or yellowish orange; white streaks above and below eyes; rest of sides of face ashy brown; white patch at base of lower mandible; slight yellowish edges to wing and tail feathers; underside white, streaked with black, except on the belly. Wing 76 to 86 mm. The young bird is similar to the adult female.

**Range in Eastern Africa:** British Somaliland to northern Italian Somaliland, eastern and southern Abyssinia and Kenya Colony as far south as Garsen on the lower Tana River.

**Habits:** Local in several localities of southern Abyssinia, and the males sing from a prominent perch. The song is described by Benson as 'hiki-hiki-hiki-hiki-hiki-hiki-hirrer' with a curious silvery tone.

**Recorded breeding:** Southern Abyssinia (breeding condition), August.

**1453 KENYA GROSBEAK CANARY.** *SERINUS BUCHANANI* Hart.

*Serinus buchanani* Hartert, Bull. B.O.C. 39, p. 50, 1919: Maktau, south-eastern Kenya Colony.

**Distinguishing characters:** Differs from the Grosbeak Canary in being greener above and paler below, more greenish yellow. The female differs from the male in having the streaks on the sides of the chest and flanks extending across the chest. Wing 81 to 92 mm. The young bird is similar to the adult female.

**Range in Eastern Africa:** Southern Kenya Colony to central Tanganyika Territory.

**Habits:** Normally seen singly or in pairs and not in flocks, feeding among the upper branches of trees in an agile Tit-like manner, and found in open bush or thorn country.

**Nest and Eggs:** A rather flat nest of twigs, rootlets and vegetable down in the fork of a tree lined with fine rootlets or fibre. Eggs three, pale blue with dots and lines of purplish black; about $19 \cdot 5 \times 15$ mm.

**Recorded breeding:** South-eastern Kenya Colony, February to April, also September. Northern Tanganyika Territory (approaching breeding condition) April.

**Food:** Grass seeds and other vegetable matter, including Acacia fruit.

**Call:** Silent birds, an occasional long-drawn squeak of alarm has been noted and very occasionally a short song with a higher final note.

**1454 YELLOW-CROWNED CANARY.** *SERINUS FLAVIVERTEX* (Blanford).
*Serinus flavivertex flavivertex* (Blanf.). **Pl. 95.**
*Crithagra flavivertex* Blanford, Ann. Mag. N.H. (4) 4, p. 330, 1869: Adigrat, northern Abyssinia.

**Distinguishing characters:** Adult male, forehead to crown golden yellow; occiput and mantle green streaked with black; rump and upper tail-coverts yellow green, streaked with black; wings black; wing shoulder, edges of secondaries and secondary coverts yellow, forming two yellow bands along wing; primaries narrowly edged with yellow on middle sections; tail black edged with yellow;

lores through eyes to ear-coverts olivaceous; sides of face and underside greenish yellow; centre of lower belly white; bill horn-colour. The female is less bright; more distinctly streaked on head and mantle; below, some streaking on throat, chest and flanks. Wing 73 to 84 mm. The young bird is brown or buffish brown above, heavily streaked with black; yellow bands along wings buffish yellow, or greenish yellow; tail as in adult; below, yellowish buff or pale yellow, heavily streaked except for lower belly; bill horn-colour.

**Range in Eastern Africa:** Eritrea, Abyssinia, Kenya Colony and north-eastern Tanganyika Territory as far south as Mt. Gerui (Hanang).

**Habits:** A highland or mountain species found both above and below the forest belt, usually seen in small parties, but occasionally in very large flocks. The males sing from a tree top.

**Nest and Eggs:** A typical Canary nest of rootlets and fibres lined with soft materials in a tree or bush. Eggs two to four, usually three, pale greenish blue or bluish white with a few dark spots; about 18 × 13 mm.

**Recorded breeding:** Abyssinia, June. Kenya Colony highlands, May to September, also February. Tanganyika Territory, January.

**Food:** Weed seeds.

**Call:** A soft tinkling song, described as reminiscent of a European Goldfinch.

*Serinus flavivertex sassii* Neum.
*Serinus flavivertex sassii* Neumann, O.M. p. 13, 1922: Chingogo Forest, Ruanda, eastern Belgian Congo.

**Distinguishing characters:** Differs from the nominate race in having the tail pale yellow with a black line along the shaft of each feather and their outer edges yellow green; broader black centres to central feathers. Wing 71 to 80 mm.

**General distribution:** Eastern Belgian Congo to Uganda, Tanganyika Territory, north-eastern Northern Rhodesia and Nyasaland.

**Range in Eastern Africa:** Ruanda and western Uganda to south-western Tanganyika Territory.

**Habits:** As for other races, locally common at high elevations.

**Recorded breeding:** Nyasaland, October and November.

**Distribution of other races of the species:** Angola.

## FINCHES

**1455 BROWN-RUMPED SEED-EATER.** *SERINUS TRIS-TRIATUS* Rüppell.

*Serinus tristriatus tristriatus* Rüpp. **Pl. 95.**

*Serinus tristriata* Rüppell, N. Wirbelt. Vög. p. 97, pl. 35, fig. 2, 1840: Taranta Pass, Acchele Guzai district, Eritrea.

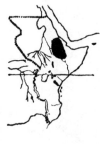

**Distinguishing characters:** General colour ashy brown, distinctly streaked with blackish on head; more indistinctly streaked on mantle; superciliary stripe and chin white, latter with blackish spots on sides; bill horn-colour. The sexes are alike. Wing 64 to 70 mm. The young bird is more clearly streaked on mantle and the underside is streaked throughout.

**Range in Eastern Africa:** Eritrea and Abyssinia.

**Habits:** Locally common and even abundant in the highlands of Abyssinia, usually singly or in pairs, occasionally in flocks.

**Nest and Eggs:** A very neatly made cup nest of stems, grass, wool and spiders' webs in trees and bushes at any height from the ground. Eggs three or four, pale greenish blue with brown, violet or blackish spots; about 19 × 14 mm.

**Recorded breeding:** Abyssinia, April to October.

**Food:** No information.

**Call:** A Sparrow-like chirp and a clear Canary-like song.

*Serinus tristriatus pallidior* (Phil.).

*Poliospiza pallidior* Phillips, Ibis p. 398, 1898: Wagga, British Somaliland.

**Distinguishing characters:** Differs from the nominate race in being paler below, especially on belly. Wing 64 to 70 mm.

**Range in Eastern Africa:** Central and eastern Eritrea to British Somaliland.

**Habits:** No information.

**Recorded breeding:** No records.

**1456 STREAKY-HEADED SEED-EATER.** *SERINUS GULARIS* (Smith).

*Serinus gularis reichardi* (Reichw.). **Pl. 95.**

*Poliospiza reichardi* Reichenow, J.f.O. p. 209, 1882: Kakoma, Tabora district, Tanganyika Territory.

**Distinguishing characters:** Top of head streaked ashy brown, blackish and white; superciliary stripe white; rest of upperparts, sides of face, wings and tail ashy brown; mantle faintly streaked with dusky; below, white; chest distinctly streaked with dusky; sides of chest, breast and flanks less distinctly streaked; head and throat sometimes washed with yellowish; bill horn-colour. The sexes are alike. Wing 76 to 80 mm. The young bird is more heavily streaked on the mantle and more clearly streaked on the breast and flanks.

**General distribution:** Tanganyika Territory to northern Portuguese East Africa, Nyasaland, southern Belgian Congo and Northern Rhodesia.

**Range in Eastern Africa:** Tanganyika Territory to Portuguese East Africa from the Tabora district to Unangu.

**Habits :** A common upland species of gardens, cultivated land or Brachystegia woodland, congregating into large flocks and doing damage to both corn and soft fruit at times. At other periods they are rather inconspicuous birds, but the male has a butterfly-like courtship flight like an English Greenfinch (Lynes).

**Nest and Eggs:** Nest of rootlets and stems lined with soft material usually at some height in a tree. They are often in a little colony of four or five pairs. Eggs two to four, faint bluish white with a few dots and streaks of purplish black; about $20 \times 15$ mm.

**Recorded breeding:** Tanganyika Territory, January to March. Nyasaland, April and May, also October.

**Food:** As above.

**Call:** A quiet Canary-like song, very pleasing and often varied with mimicry of other birds. Ordinary call 'see-e-ee.'

*Serinus gularis striatipectus* (Sharpe).
*Poliospiza striatipectus* Sharpe, Ibis p. 258, 1891: Elgeyu, west-central Kenya Colony.

**Distinguishing characters:** Differs from the preceding race in being browner above; below, streaks absent or very indistinct. Wing 77 to 84 mm.

**Range in Eastern Africa:** South-western Sudan, Uganda and western and central Kenya Colony.

**Habits:** As for the preceding race.

**Recorded breeding:** Kenya Colony highlands, July.

*Serinus gularis erlangeri* (Reichw.).
*Poliospiza erlangeri* Reichenow, O.M. p. 146, 1905: Ladsho, Arussi-Gallaland, south-central Abyssinia.

**Distinguishing characters:** Differs from the preceding race in being dark ash brown above with more distinct streaks on mantle; below, heavily streaked on chest and flanks. Wing 75 to 82 mm.

**Range in Eastern Africa:** South-eastern Sudan to southern half of Abyssinia.

**Habits:** Common in the highlands of southern Abyssinia.

**Recorded breeding:** Southern Abyssinia (breeding condition), September.

**Distribution of other races of the species:** West and South Africa, the nominate race being described from Bechuanaland.

### 1457 BLACK-EARED SEED-EATER. *SERINUS MENNELLI* (Chubb).

*Poliospiza mennelli* E. C. Chubb, Bull. B.O.C. 21, p. 62, 1908: Tjokos' Kraal, Shangani River, Southern Rhodesia.

**Distinguishing characters:** Adult male, differs from the Streaky-headed Seed-eater in being ashy grey above; streaks on head distinctly black and white; sides of face black; below, whiter and with streaking sharper; bill brown. The female has the streaks on head and sides of face duller black. Wing 80 to 86 mm. The young bird is brown with paler brown tips to the wing-coverts and innermost secondaries.

**General distribution:** Eastern Angola to south-eastern Belgian Congo, Nyasaland, Rhodesia and Portuguese East Africa.

**Range in Eastern Africa:** Lake Amaramba area of Portuguese East Africa.

**Habits:** A highland species of woodland, or wooded rocky hillsides, not uncommon but very local. In the breeding season the males have a courtship flight consisting of a mount to some height and then a series of swerving dives interspersed with butterfly-like flaps.

**Nest and Eggs:** Nest a cup, mainly of lichens, at some height in the fork of a tree. Eggs two or three, pale greenish, spotted or freckled with black and with lilac grey undermarkings; about 17·5 × 13·5 mm. Much larger measurements are recorded by Priest.

**Recorded breeding:** South-eastern Belgian Congo, January. Northern Rhodesia, February onwards. Nyasaland and Portuguese East Africa, December to May.

**Food:** Seeds with some fruit at times.

**Call:** A series of nasal bleats, also in the breeding season a series of uneven twittering whistles, together with a sweet song by the male from some prominent perch.

**1458** WHITE-RUMPED SEED-EATER. *SERINUS LEUCO-PYGIUS* (Sundevall).
*Serinus leucopygius leucopygius* (Sund.).
*Crithagra leucopygia* Sundevall, Œfv. K. Sv. Vet. Akad. Handl. 7, p. 127, 1850: Sennar Province, eastern Sudan.

**Distinguishing characters:** A small bird, ashy grey above, spotted and streaked with dusky; rump white; below, sides of face and chin to breast ashy grey; rest of underparts white; flanks streaked dusky; bill brown. The sexes are alike. Wing 63 to 70 mm. The young bird is more streaked above and spotted on chest, breast and flanks.

**Range in Eastern Africa:** Central, eastern and southern Sudan, western Eritrea and north-western Abyssinia.

**Habits:** A common species of the Sudan and of the lower levels of north-western Abyssinia, the birds being noticeable in flight by their white rumps. They occur both in gardens and cultivated land, and also in dry open bush, and their habits are those of other small Finches.

**Nest and Eggs:** A very neat cup in the outer fork of a bough, or lashed between two twigs, made of fibre, plant stems, and hair, lined with feathers or vegetable down. Eggs three or four, pale bluish white, or pale greenish grey sparingly spotted with brown or black; about 16 × 12 mm.

**Recorded breeding:** Sudan, August onwards.

**Food:** Seeds, millet, etc.

**Call:** The song is clear but not loud. Emin Pasha describes them as 'warbling their varied stanzas among the acacias.'

*Note:* Intermediates between this race and the next occur along the White Nile from Fashoda to Mongalla and the Nuba Hills, Kordofan.

*Serinus leucopygius riggenbachi* Neum.
*Serinus leucopygius riggenbachi* Neumann, Bull. B.O.C. 21, p. 44, 1908: Thiés, near Dakar, Senegal.

**Distinguishing characters:** Differs from the nominate race in having the throat whiter, less dusky; chin to chest spotted with dusky, the spots being larger than in the young bird of the nominate race. Wing 63 to 70 mm.

**General distribution:** Senegal and Gambia to the Sudan.

**Range in Eastern Africa:** Western and south-western Sudan.

**Habits:** As for other races.

**Recorded breeding:** Nigeria, July and August. Western Sudan, September to November.

**1459 YELLOW-RUMPED SEED-EATER.** *SERINUS ATROGULARIS* (A. Smith).
*Serinus atrogularis xanthopygius* Rüpp.
*Serinus xanthopygius* Rüppell, N. Wirbelt Vög. p. 96, pl. 35, fig. 1, 1840: Schoada Valley, Simen, northern Abyssinia.

**Distinguishing characters:** Above ash brown streaked with dusky; rump yellow; upper tail-coverts edged with yellow; sides of face and underside ashy or ashy brown; chin and throat white or buffish white; a broken dusky band across lower neck; belly white; bill horn-colour. The sexes are alike. Wing 67 to 72 mm. The young bird is spotted on the chest and breast.

**Range in Eastern Africa:** Eritrea to Abyssinia as far south as the Gofa area.

**Habits:** As for other races, found among junipers in stony country.

**Nest, Eggs and Call:** See under the following race.

**Recorded breeding:** No records.

*Serinus atrogularis reichenowi* Salvad.
*Serinus reichenowi* Salvadori, Ann. Civ. Mus. Genova 26, p. 272, 1888: Cialalaka, Shoa, central Abyssinia.

**Distinguishing characters:** Differs from the preceding race in being browner above with streaking more distinct; a white superciliary stripe; cheeks white and dusky; whitish ends to wing-coverts,

# FINCHES

flight feathers and tail; below, white or buffish white with more distinct streaks on flanks and a more distinct broken band across lower neck. Wing 60 to 71 mm. The young bird is more heavily spotted on chest, breast and flanks than the young bird of the preceding race.

**Range in Eastern Africa:** Central Abyssinia to south-eastern Sudan, eastern Uganda, Kenya Colony and Tanganyika Territory, also Rusinga Island, eastern Lake Victoria.

**Habits:** A bird of irregular distribution, in places not uncommon in bush or woodland near water; partially migratory locally, and usually seen at lower levels feeding on the seed-heads of flowers.

**Nest and Eggs:** A very small cup nest in a bush, or on the branch of a tree, made of rootlets and grass stems, lined with plant down, and usually bound over with cobwebs. Eggs normally three, bluish white, freckled and spotted with brown or black, and often with a few dull mauve undermarkings; about 16 × 12 mm.

**Recorded breeding:** Abyssinia, May. Western Kenya Colony, May to September, also December and January. Southern Kenya Colony, March and April. Tanganyika Territory, January to March.

**Food:** Seeds of flowers, particularly Compositæ, also noted eating unripe seeds of blue Lupins.

**Call:** A clear Canary-like song in the breeding season.

*Serinus atrogularis somereni* Hart.
*Serinus angolensis somereni* Hartert, Bull. B.O.C. 29, p. 63, 1912: Toro, eastern Uganda.

**Distinguishing characters:** Differs from the preceding race in being darker both above and below; more black on throat and lower neck and underside usually washed with brown or dusky brown. Wing 65 to 72 mm.

**General distribution:** Belgian Congo to Uganda.

**Range in Eastern Africa:** Western Uganda.

**Habits:** As for other races.

**Recorded breeding:** Uganda, May.

*Serinus atrogularis hilgerti* Zedl.
*Serinus angolensis hilgerti* Zedlitz, O.M. p. 76, 1912: Afgoi, southern Italian Somaliland.

**Distinguishing characters:** Differs from the preceding race in being much paler in general colour. Wing 60 to 63 mm.

**Range in Eastern Africa:** Southern Italian Somaliland and eastern Kenya Colony as far south as Mombasa.

**Habits:** As for other races.

**Recorded breeding:** No records.

**Distribution of other races of the species:** Angola, South Africa, and south-western Arabia, the nominate race being described from the western Transvaal.

**1460 YELLOW-THROATED SEED-EATER.** *SERINUS FLAVIGULA* Salvad.
*Serinus flavigula* Salvadori, Ann. Civ. Mus. Genova, 26, p. 272, 1888: Ambokana, Shoa, central Abyssinia.

**Distinguishing characters:** Above, similar to the northern Abyssinian race of the Yellow-rumped Seed-eater but greyer; sides of face grey; primaries darker with light edges; below, chin white; throat to chest yellow; a blackish band across lower neck; sides of chest and flanks grey; belly white. Apparently the sexes are alike. Wing 64 to 70 mm. Juvenile plumage unrecorded.

**Range in Eastern Africa:** Abyssinia from Shoa to Yavello.

**Habits:** Undescribed, only very few individual specimens are known, and appear to have been met with casually among other Seed-eaters.

**Nest and Eggs:** Undescribed.

**Recorded breeding:** No records.

**Food:** No information.

**Call:** Unrecorded.

**1461 STREAKY SEED-EATER.** *SERINUS STRIOLATUS* (Rüppell).
*Serinus striolatus striolatus* (Rüpp.). **Ph. xix.**
*Pyrrhula striolata* Rüppell, N. Wirbelt. Vög. p. 99, pl. 37, fig. 1, 1840: Halai, Acchele Guzai district, Eritrea.

**Distinguishing characters:** Above, streaked buff, brown and black; flight and tail feathers edged with green, superciliary stripe buff or whitish; round eyes to ear-coverts and moustachial stripe blackish brown; below, chin and throat buff, sometimes washed with faint yellow; rest of underparts buff, heavily streaked on chest, breast and flanks, rarely slightly yellowish; bill horn-colour. The sexes are alike. Wing 65 to 75 mm. The young bird has rather narrower streaks below.

**General distribution:** Eritrea to Abyssinia, the Sudan, eastern Belgian Congo, Uganda, Kenya Colony and Tanganyika Territory.

**Range in Eastern Africa:** Eritrea, Abyssinia, the south-eastern Sudan, southern Uganda, Ruanda, Urundi, Kenya Colony and north-eastern Tanganyika Territory.

**Habits:** Common highland and mountain species in any sort of habitat up to 14,000 feet. Found freely in gardens where it is reported to do a certain amount of damage, and locally known as 'Sparrow.'

**Nest and Eggs:** Nest of rootlets, twigs, grass and moss lined with hair and feathers, a rather flat cup placed usually in a low bush or creeper but may be at any height from the ground. Eggs three or four, occasionally five, creamy or greenish white, with dark brown or chocolate scrawls and speckles; about 20 × 14·5 mm.

**Recorded breeding:** Abyssinia, April to October. Kenya Colony, March to August, also November and December. Northern Tanganyika Territory, April.

**Food:** Mainly seeds.

**Call:** A soft call of three notes, the first higher pitched (Moreau). Also quite a good songster with a pleasant tinkling song, reminiscent of a Bulbul.

*Serinus striolatus whytii* Shell.
*Serinus whytii* Shelley, Ibis p. 528, pl. 11, 1897: Nyika Plateau, western Nyasaland.

**Distinguishing characters:** Differs from the nominate race in having the top of the head and sides of face streaked lemon yellow and black; superciliary stripe and throat washed with bright lemon yellow; chin blackish and yellow. Wing 65 to 70 mm.

**General distribution:** Tanganyika Territory to Nyasaland as far south as the Vipya Plateau.

**Range in Eastern Africa:** The Dabaga and Njombe areas of south-central Tanganyika Territory to northern end Lake Nyasa.

**Habits:** Local and relatively scarce at edge of evergreen forest at higher levels. The nest is described as a cup of fine twigs lined with very fine grass and a few feathers. Eggs pale blue with the same markings as other races; about 19 × 14 mm.

**Recorded breeding:** Tanganyika Territory (about to breed), February. Northern Nyasaland, October and November.

*Serinus striolatus graueri* Hart.
*Serinus striolatus graueri* Hartert, Bull. B.O.C. 19, p. 84, 1907: Mt. Ruwenzori, eastern Belgian Congo.

**Distinguishing characters:** Differs from the nominate race in being distinctly warmish buff below. Wing 64 to 71 mm.

**Range in Eastern Africa:** Ruwenzori Mts., Uganda.

**Habits:** As for other races, mainly a mountain species, noted as feeding on the seeds of creepers in deep forest valleys.

**Recorded breeding:** No records.

**1462 THICK-BILLED SEED-EATER.** *SERINUS BURTONI* (Gray).
*Serinus burtoni albifrons* (Sharpe).
*Crithagra albifrons* Sharpe, Ibis p. 118, 1891: Kikuyu, central Kenya Colony.

**Distinguishing characters:** Bill large; above dark earth brown, sometimes with a faint wash of yellowish, with some dusky streaking, especially on head; band of white on forehead; sometimes some white feathers on dull black sides of face and throat; edges of wing-coverts, flight feathers and tail olive green; tips of wing-coverts and innermost secondaries buffish white; below, chin and upper throat dull black; rest of underparts earthy or buffish brown streaked with dusky; bill horn-colour. The sexes are alike. Wing 81 to 87 mm. The young bird is rather browner above; tips of wing-coverts and innermost secondaries more buffish; bill darker.

**Range in Eastern Africa:** Highlands of central Kenya Colony.

**Habits:** A species of mountain bush or forest, clumsy-looking silent birds usually seen in pairs or small parties, local and not usually common. Their normal habitat is a wind-swept mountain side, and although the species has a wide range in its various racial forms, very little has been recorded of it, except that all observers have described this unfortunate bird as dull and uninteresting.

**Nest and Eggs:** Believed to make a light frail nest, but at present nothing definite has been recorded of either the nest or eggs.

**Recorded breeding:** No records.

**Food:** Hard seeds, mainly of trees or shrubs.

**Call:** Said to be a low squeaking note.

*Serinus burtoni kilimensis* (Richm.).
*Crithagra kilimensis* Richmond, Auk. 14, p. 155, 1897: Mt. Kilimanjaro.

**Distinguishing characters:** Differs from the preceding race in having little or no white on forehead and in being darker above. Wing 81 to 89 mm.

**Range in Eastern Africa:** Southern Kenya Colony to northeastern and western Tanganyika Territory as far south as the Kungwe-Mahare Mts.

**Habits:** As for other races, found at forest edge from 4,000 to 8,500 feet.

**Recorded breeding:** Northern Tanganyika Territory, August, also (breeding condition), April.

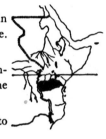

*Serinus burtoni melanochrous* Reichw.
*Serinus melanochrous* Reichenow, O.M. p. 122, 1900: Tandala, Ukinga, Tukuyu district, south-western Tanganyika Territory.

**Distinguishing characters:** Differs from the preceding race in having the throat whitish or yellowish; underside broadly and heavily streaked with black. Wing 82 to 86 mm.

**Range in Eastern Africa:** South-central to south-western Tanganyika Territory from Njombe to Ukinga.

**Habits:** Large dull silent birds in small parties mainly found inside forest jungle.

**Recorded breeding:** Tanganyika Territory, probably October to December.

*Serinus burtoni tanganjicæ* Granv.

*Serinus albifrons tanganjicæ* Granvik, J.f.O. (Sond.) p. 191, 1923: Forest west of Lake Tanganyika, eastern Belgian Congo.

**Distinguishing characters:** Differs from the Kilimanjaro race in being distinctly browner below. Wing 82 to 87 mm.

**General distribution:** Eastern Belgian Congo to Uganda and the Lake Kivu area.

**Range in Eastern Africa:** Ruwenzori Mts. to Lake Kivu.

**Habits:** An uncommon mountain species, habits as for other races.

**Recorded breeding:** No records.

*Serinus burtoni gurneti* (Gyld.).

*Poliospiza burtoni gurneti* Gyldenstolpe, Ark. Zool. 19, A, no. 1, p. 18, 1926: Mt. Elgon.

**Distinguishing characters:** Similar to the preceding race, but larger. Wing 87 to 91 mm.

**Range in Eastern Africa:** Eastern Uganda and western Kenya Colony.

**Habits:** A mountain forest species, habits as for other races.

**Recorded breeding:** No records.

**Distribution of other races of the species:** West Africa, the nominate race being described from Cameroon Mt.

**1463** ORIOLE-FINCH. *LINURGUS OLIVACEUS* (Fraser).
*Linurgus olivaceus kilimensis* (Reichw. & Neum.). **Pl. 95.**
*Hyphantospiza kilimensis* Reichenow & Neumann, O.M. p. 74, 1895: Mt. Kilimanjaro.

**Distinguishing characters:** Adult male, head all round to nape and chest black; usually, but not constantly, a golden yellow band round hind neck; sides of neck golden yellow; rest of upperparts rich moss-green or olivaceous moss-green; wings black with pale yellow tips and golden yellow edges to wing-coverts and secondaries; inner webs of flight feathers sharply edged with white; tail pale olivaceous moss-green, tips golden yellow; below, golden yellow; sides of chest and flanks olivaceous green; bill orange yellow. The female is wholly olive green, paling to greenish yellow on belly; wings as in male but with yellow edging; bill yellowish orange or yellowish horn. Wing 71 to 78 mm. The young bird is duller than the adult female and has pale greenish tips to the wing-coverts.

**General distribution:** Kenya Colony, Tanganyika Territory and northern Nyasaland.

**Range in Eastern Africa:** Mt. Kenya to north-eastern and south-western Tanganyika Territory.

**Habits:** A shy and little known mountain forest species, usually seen feeding in small scattered flocks at the forest edge, or among the boughs of large trees.

**Nest and Eggs:** Nest a rather shallow cup of moss, lichens and rootlets, lined with soft lichen or plant down. They are usually in low bushes. Eggs (of nominate race) two, white, sparsely speckled with reddish brown and occasional hair-lines. No measurements available.

**Recorded breeding:** Northern Tanganyika Territory (breeding condition), July and October.

**Food:** Noted as grass seeds, tree seeds, the seeds of a saprophytic Orchid and occasional caterpillars.

**Call:** A soft plaintive 'twee' (Moreau), or 'sit-sit' (Couchman) with a musical whistling song, and a thin squeaking alarm note.

*Linurgus olivaceus elgonensis* V. Som.
*Linurgus elgonensis* Van Someren, Nov. Zool. 25, p. 283, 1918: Mt. Elgon.

**Distinguishing characters:** Differs from the preceding race in being more golden yellow both above and below. Wing 70 to 72 mm.

**Range in Eastern Africa:** Imatong Mts. south-eastern Sudan to eastern Uganda and western Kenya Colony.

**Habits:** Rare and found among bamboos and dense bush up to 10,000 feet, usually shy and wary.

**Recorded breeding:** No records.

*Linurgus olivaceus keniensis* V. Som.
*Linurgus keniensis* Van Someren, Bull. B.O.C. 43, p. 154, 1923: Meru Forest, northern Mt. Kenya, central Kenya Colony.

**Distinguishing characters:** Differs from the preceding race in having the mantle, scapulars and rump darker, more greenish. Wing 73 to 77 mm.

**Range in Eastern Africa:** Central Kenya Colony.

**Habits:** As for other races.

**Recorded breeding:** No records.

**Distribution of other races of the species:** West Africa, the nominate race being described from Fernando Po.

**1464 AFRICAN CITRIL.** *CARDUELIS CITRINELLOIDES* (Rüppell).

*Carduelis citrinelloides citrinelloides* (Rüpp.).
*Serinus citrinelloides* Rüppell, N. Wirbelt. Vög. p. 95, pl. 34, fig. 1, 1840: Simen, northern Abyssinia.

**Distinguishing characters:** Adult male, with or without a narrow black band on forehead; superciliary stripe yellow, or absent; rest of upperparts yellow green with narrow black streaks, or darker, less yellow green with wider, more distinct, black streaks; lores to ear-coverts, and chin and upper throat black or dusky black; wing-coverts, flight feathers and tail black edged with dull or greenish yellow; below, golden yellow to lemon yellow, with, in some, olivaceous flanks and streaks on chest and flanks; bill whitish horn.\*
From the Black-faced Canary it can be distinguished by having no yellow band behind the black forehead band and by the black, not dusky, flight and tail feathers. The female is similar to the dark-backed male; superciliary stripe absent; lores to ear-coverts, chin and upper throat not black or slightly dusky greenish; below, distinctly striped on throat, chest and flanks. Wing 62 to 72 mm. The young bird is browner above than the adult female, and much paler yellow below with sometimes a wash of chrome. It can be distinguished from the young bird of the Yellow-crowned Canary by the smaller size, less black wings and tail, more olive, less brown appearance, and green rump and upper tail-coverts, not brown or buff.

**Range in Eastern Africa:** Eritrea and Abyssinia to the south-eastern Sudan.

**Habits:** A common species at all levels, usually seen in flocks and with no particularly distinctive habits.

\**Note:* These two rather different-looking adult dresses are found in the same month.

**Nest and Eggs:** A neat little cup nest of dry grass and rootlets decorated with cobwebs and cocoons and lined with down or soft material. It is placed in the fork of a tree, shrub, or large plant. Eggs two or three, dull white or cream with reddish or brownish purple spots, speckles and blotches; about 17 × 12·5 mm.

**Recorded breeding:** Abyssinia, September to December.

**Food:** Mainly seeds, also noted taking flying ants, and searching leaves for insects.

**Call:** A soft clicking cheep. Song notable, a clean piping whistle of some power.

*Carduelis citrinelloides hypostictus* (Reichw.). **Pl. 95.**
*Spinus citrinelloides hypostictus* Reichenow, Vög. Afr. 3, p. 275, 1904: Moshi, north-eastern Tanganyika Territory.

**Distinguishing characters:** Differs from the nominate race in having the dusky black of the lores, round eyes, cheeks, chin and throat washed with whitish, giving a hoary appearance. The female has the lores to cheeks and chin grey. Wing 60 to 74 mm.

**General distribution:** Kenya Colony and Tanganyika Territory to Nyasaland, northern Portuguese East Africa and north-eastern Northern Rhodesia.

**Range in Eastern Africa:** South-eastern Kenya Colony and north-eastern Tanganyika Territory to the Zambesi River.

**Habits:** As for other races, the male sings sweetly from a tree, or tall herbage, but they appear to remain in flocks even in the breeding season. A common and widespread species of lower and middle uplands.

**Recorded breeding:** Chyulu Hills, south-eastern Kenya Colony, March to May. Tanganyika Territory, January to April. Nyasaland, January to March, and September. Portuguese East Africa, June to September.

*Carduelis citrinelloides frontalis* (Reichw.).
*Spinus citrinelloides frontalis* Reichenow, Vög. Afr. 3, p. 275, 1904: Lake Kivu, eastern Belgian Congo.

**Distinguishing characters:** Differs from the nominate race in having a yellow band across the forehead. From the Black-faced Canary the male and female can be distinguished only by the longer

sharply pointed bill; the young bird has also a longer and sharper bill and is browner, especially above, than the young bird of the Black-faced Canary. Wing 58 to 70 mm. Xanthochroic examples are known.

**General distribution:** Eastern Belgian Congo to Uganda and Tanganyika Territory.

**Range in Eastern Africa:** Uganda to western Tanganyika Territory as far south as the Kasulu area.

**Habits:** As for other races, locally plentiful and especially fond of cultivations and banana plantations.

**Recorded breeding:** Uganda, March to July, also November and December.

*Carduelis citrinelloides kikuyuensis* (Neum.). **Pl. xix.**
*Spinus citrinelloides kikuyuensis* Neumann, J.f.O. p. 356, 1905: Kikuyu, central Kenya Colony.

**Distinguishing characters:** Adult male, similar to the nominate race in the darker dress, but eye stripe more distinct. The female is similar to that of the nominate race. Wing 62 to 70 mm.

**Range in Eastern Africa:** Northern and central Kenya Colony.

**Habits:** As for the other races, usually common and widespread occurring freely in clearings where flower heads are in seed in small twittering flocks. Common inhabitants of gardens, with a low but sweet song.

**Recorded breeding:** Kenya Colony highlands, May to July, also December.

**1465 BLACK-HEADED SISKIN.** *CARDUELIS NIGRICEPS* (Rüppell). **Pl. 95.**
*Serinus nigriceps* Rüppell, N. Wirbelt. Vög. p. 96, pl. 34, fig. 2, 1840: Simen Province, northern Abyssinia.

**Distinguishing characters:** Adult male, head and neck all round dull black; rest of plumage dull olive green; wing shoulder and rump brighter; lower belly whitish; wing-coverts, wings and tail black, edged and tipped with white and pale olive green; bill blackish. The female has the head and neck dull olive green washed with blackish on forehead, crown, sides of head and chin to throat. From the female of the Yellow-crowned Canary it can be distinguished by the dusky forehead, sides of face, chin and throat. Wing 74 to 80 mm. The young bird is similar to the adult female but is browner above, streaked with dusky.

**Range in Eastern Africa:** Northern and central Abyssinia.

**Habits:** Common but little known bird of the higher levels of Abyssinia, always noted in flocks.

**Nest and Eggs:** Nest placed low down in bushes or on hanging boughs, well made and compact with a deep cup, composed of rootlets and stems with a finer lining. Eggs two or three, bluish white with a few brown spots; indistinguishable from those of some of the Seed-eaters; about $18 \times 13$ mm.

**Recorded breeding:** Central Abyssinia, May and June, September and October.

**Food:** No information.

**Call:** Unrecorded.

**1466 WARSANGLI LINNET.** *WARSANGLIA JOHANNIS* Stephenson Clarke. **Pl. 91.**

*Warsanglia johannis* Stephenson Clarke, Bull. B.O.C. 40, p. 48, 1919: Mush Haled, eastern British Somaliland.

**Distinguishing characters:** Adult male, forehead and superciliary stripe white; rest of top of head, mantle, scapulars, upper tail-coverts, sides of neck and sides of chest dusky grey; a blackish grey stripe through eye; rump and lower flanks chestnut; wings and tail black; bases of flight feathers and of all except central tail feathers white; sides of face and rest of underparts white; a slight blackish grey streak below ear-coverts; bill blackish horn; feet black. Wing 80 mm. Female and juvenile plumages unknown.

**Range in Eastern Africa:** Warsangli area, eastern British Somaliland.

**Habits:** Unknown, the type and only specimen was secured from a small flock at a water hole. A remarkably distinct species of unknown affinities.

**Nest and Eggs:** Unknown.

**Recorded breeding:** No records.

**Food:** No information.

**Call:** Unrecorded.

Names in Sclater's *Syst. Av. Æthiop.*, 2, 1930, which have been changed or have become synonyms in this work:

*Serinus mozambicus aurifrons* Sclater & Praed, now *Serinus mozambicus grotei* Sclater & Praed.
*Serinus sulphuratus shelleyi* Neumann, treated as synonymous with *Serinus sulphuratus sharpii* Neumann.
*Poliospiza angolensis* (Gmelin), now *Serinus atrogularis* (A. Smith).
*Poliospiza collaris* Reichenow, treated as synonymous with *Serinus flavigula* (Salvadori).
*Poliospiza dimidiata* Madarasz, treated as synonymous with *Serinus flavigula* (Salvadori).
*Poliospiza pachyrhyncha* Reichenow, treated as synonymous with *Serinus donaldsoni* Sharpe.
*Poliospiza striolata ugandæ* Van Someren, treated as synonymous with *Serinus striolatus striolatus* (Rüppell).

Names introduced since 1930 and which have become synonyms in this work:

*Serinus mozambicus gertrudis* Grote, 1934, treated as synonymous with *Serinus mozambicus mozambicus* (Müller).
*Serinus flavivertex elgonensis* Granvik 1934, treated as synonymous with *Serinus flavivertex flavivertex* (Blanford).
*Linurgus kilimensis rungwensis* Bangs & Loveridge, 1931, treated as synonymous with *Linurgus olivaceus kilimensis* (Reichenow & Neumann).
*Spinus citrinelloides chyulu* Van Someren, 1939, treated as synonymous with *Carduelis citrinelloides hypostictus* Reichenow.

Family—**EMBERIZIDÆ**. BUNTINGS. Genera: *Emberiza* and *Fringillaria*.

Twelve species occur in Eastern Africa, of which three are migrants in the non-breeding season from Europe and Asia. They are mainly ground feeding birds and more or less gregarious in the non-breeding season.

### Key to the Adult Buntings of Eastern Africa

1 No distinct stripe over eye:    3–6

2 Distinct white stripe over eye:    7–21

3 Head earth brown or olivaceous, mantle earth brown, underside yellowish olive:    CINEREOUS BUNTING *Emberiza cineracea*    **1474**

4 Head grey, mantle warm brown:    5–6

# BUNTINGS

5 Throat tawny, chest grey:
  CRETZSCHMAR'S BUNTING
  *Emberiza cæsia* **1473**

6 Throat pale yellow, chest yellowish olivaceous, grey or buff:
  ORTOLAN
  *Emberiza hortulana* **1472**

7 No white stripe under eye:
  8–11

8 Ends of tail feathers white, underside yellow:
  9–11

9 No stripe down centre of crown:
  CABANIS' BUNTING
  *Emberiza cabanisi* **1467**

10 Distinct stripe down centre of crown:
   THREE-STREAKED BUNTING
   *Emberiza orientalis* **1468**

11 Tail uniform in colour, underside brown:
   HOUSE-BUNTING
   *Fringillaria striolata* **1477**

12 White stripe under eye:
   13–21

13 Ends of tail feathers white, underside yellow:
   14–17

14 No white on wing-coverts:
   BROWN-RUMPED BUNTING
   *Emberiza forbesi* **1471**

15 White tips to wing-coverts:
   16–17

16 Mantle feathers edged with white or buffish white:
   SOMALI GOLDEN-BREASTED BUNTING
   *Emberiza poliopleura* **1470**

17 Mantle feathers edged with sandy:
   GOLDEN-BREASTED BUNTING
   *Emberiza flaviventris* **1469**

18 Tail all one colour:
   19–21

## BUNTINGS

19 Underside sooty grey, chin whitish:     SOUTHERN ROCK-BUNTING *Fringillaria capensis* **1475**

20 Underside tawny chestnut and whitish, chin whitish:     SOCOTRA MOUNTAIN-BUNTING *Fringillaria socotrana* **1478**

21 Underside warm russet, throat black:     CINNAMON-BREASTED ROCK-BUNTING *Fringillaria tahapisi* **1476**

**1467 CABANIS' BUNTING.** *EMBERIZA CABANISI* (Reichenow). **Pl. 95.**

*Polymitra (Fringillaria) cabanisi* Reichenow, J.f.O. p. 233, pl. 2, figs. 2, 3, 1875: Cameroons.

**Distinguishing characters:** Adult male, above, blackish brown with darker streaking; white stripe from base of bill, over eye and down side of neck; sides of face black; ends of wing-coverts white; apical third of outer tail feathers and ends of all others except central pairs white; below, chin to neck white; rest of underparts canary yellow; under tail-coverts dusky at base, white at ends. The female is browner above, including sides of face; stripe over eye and chin to throat buffish white. Wing 70 to 86 mm. The young bird is somewhat more russet brown above and has dark buff ends to the wing-coverts.

**General distribution:** Cameroons to the Sudan and Uganda.

**Range in Eastern Africa:** Southern Sudan and western Uganda.

**Habits:** Habits little recorded, but believed to be those of allied species of Buntings; it is a widely distributed but not common species, often associated with water.

**Nest and Eggs:** Nest a shallow cup of dried leaves and dead stems lined with fibres in a low bush or creeper. Eggs two, white with scrawls and blotches of pale umber brown and grey; about 22 × 15·5 mm.

**Recorded breeding:** Cameroons and Nigeria, June to September.

# BUNTINGS

**Food:** Insects and small seeds.

**Call:** A low soft call of 'tweee' and a song of thin clear notes 'twee-twee-sheshesheshe' (Marchant).

## 1468 THREE-STREAKED BUNTING. *EMBERIZA ORIENTALIS* (Shelley).

*Emberiza orientalis orientalis* (Shell.). **Pl. 95.**

*Fringillaria orientalis* Shelley, P.Z.S. p. 308, 1882: Mamboio, Morogoro district, eastern Tanganyika Territory.

**Distinguishing characters:** Differs from Cabanis' Bunting in being brown or grey and brown above, with distinct black streaks; a white or grey streak down centre of crown; lower flanks grey. The female has usually a less distinct, or very faint, streak down centre of crown; sides of crown and face dusky or brownish. Wing 75 to 89 mm. The young bird is russet brown above streaked with blackish; superciliary stripe pale brown; below, paler yellow; chest and often flanks streaked with dusky brown.

**General distribution:** Tanganyika Territory to the Belgian Congo, Nyasaland, Northern Rhodesia, eastern Southern Rhodesia and Portuguese East Africa.

**Range in Eastern Africa:** Tanganyika Territory to the Zambesi River.

**Habits:** A not uncommon species of light bush country, woodlands or fairly open ground, feeding in pairs or family parties on the ground. Its habits are very much those of the Golden-breasted Bunting, and it is widely distributed through much of the range of that species. The birds are fond of patches of bare ground.

**Nest and Eggs:** Nest a well made cup, somewhat stouter than the nest of the Golden-breasted Bunting, of strips of grass lined with finer grass, and placed in the fork of a small shrub, or some similar situation. Eggs normally three, light green with scrawls of umber and sepia, and with lilac undermarkings; about 20·5 × 15 mm.

**Recorded breeding:** Northern Tanganyika Territory, October to March. Nyasaland, November to January.

**Food:** Seeds and insects taken from the ground.

**Call:** A piercing sweetly modulated whistling song 'wee-chidder-chidder-chidder-wee' rather variable (Moreau).

**Distribution of other races of the species:** Angola.

**1469 GOLDEN-BREASTED BUNTING.** *EMBERIZA FLAVIVENTRIS* Stephen.

*Emberiza flaviventris flaviventris* Steph. **Pl. 95. Ph. xix.**
*Emberiza flaviventris* Stephen, Gen. Zool. 9, pt. 2, p. 374, 1815: Cape of Good Hope, South Africa.

**Distinguishing characters:** Adult male, differs from Cabanis' Bunting in having the lores and a stripe from base of bill under eye to side of neck white, forming a white stripe over and below eye; black stripes behind eye and along moustache; mantle and edges of innermost secondaries chestnut variously streaked with greyish tawny or dusky; lower flanks white. The female is duller than the male; black of head more dusky and mantle sometimes brownish. Wing 71 to 91 mm. The young bird has a buff streak down centre of crown; below, paler yellow.

**General distribution:** The Sudan to Uganda, Kenya Colony, Angola and South Africa.

**Range in Eastern Africa:** South-eastern Sudan to Uganda, Kenya Colony and the Zambesi River.

**Habits:** A very common, widespread and noticeable species of gardens, woodland, or open country. Generally singly or in pairs, or three or four together, feeding mostly on the ground. Belcher describes this bird as taking the place of the Chaffinch in England. The song of the male in the early morning is a very characteristic sound.

**Nest and Eggs:** Nest a delicate and frail cup of grass and fibre, a few feet from the ground in the fork of a shrub, often quite exposed; in some areas twigs are used as a foundation. Eggs two or three, greyish white, greenish or bluish, with a zone of black and sepia spots and scrawls and violet undermarkings; about $19 \cdot 5 \times 14$ mm.

**Recorded breeding:** Kenya Colony, in many months and probably three or four broods in the year. Tanganyika Territory, November to April. Nyasaland, October to January. Portuguese East Africa, November and December.

**Food:** Seeds, insects, etc. mainly picked up from the ground.

**Call:** A typical Bunting-like 'zizi-zizi.' A pleasant little song of 'chwee-chi-it-twee' (Belcher) and a trilling note. Alarm note a grating 'treech.'

Fischer's Sparrow-Lark (p. 38)
*Eremopterix leucopareia*
Red-capped Lark (p. 42)
*Calandrella cinerea saturatior*
Rosy-breasted Long-claw (p. 80)
*Macronyx ameliae wintoni*

Rufous-naped Lark (p. 11)
*Mirafra africana athi*
Richard's Pipit (p. 67)
*Anthus novaeseelandiae lacuum*
Yellow-throated Long-claw (p. 76)
*Macronyx croceus*

Dark-capped Bulbul (p. 113)
*Pycnonotus tricolor fayi*
Fischer's Greenbul (p. 129)
*Phyllastrephus fischeri placidus*
Yellow-whiskered Greenbul (p. 150)
*Stelgidocichla latirostris eugenia*

Zanzibar Sombre Greenbul (p. 147)
*Andropadus importunus fricki*
Swamp Flycatcher (p. 168)
*Alseonax aquaticus infulatus*
Dusky Flycatcher (p. 165)
*Alseonax adustus pumilus*

White-eyed Slaty Flycatcher (p. 180)
*Dioptrornis fischeri fischeri*
Yellow Flycatcher (p. 191)
*Chloropeta natalensis massaica*
Paradise Flycatcher (p. 219)
*Tchitrea viridis ferreti*

Chin-spot Puff-back Flycatcher (p. 202)
*Batis molitor molitor*
Olive Thrush (p. 243)
*Turdus olivaceous abyssinicus*
Sooty Chat (p. 284)
*Myrmecocichla nigra*

Ph. x

Stonechat (p. 288)
*Saxicola torquata axillaris*
White-browed Robin-Chat (p. 291)
*Cossypha heuglini heuglini*
Robin-Chat (p. 300)
*Cossypha caffra iolaema*

Spotted Morning Warbler (p. 315)
*Cichladusa guttata guttata*
Red-backed Scrub-Robin (p. 317)
*Erythropygia zambesiana brunneiceps*
Brown-backed Scrub-Robin (p. 322)
*Erythropygia hartlaubi kenia*

Greater Swamp-Warbler (p. 379)
*Calamocichla gracilirostris nilotica*
Abyssinian Black Wheatear (p. 268)
*Oenanthe lugubris schalowi*
Black-collared Apalis (p. 406)
*Apalis pulchra pulchra*

White-browed Crombec (p. 429)
*Sylvietta leucophrys leucophrys*
Grey-capped Warbler (p. 421)
*Eminia lepida*
Buff-bellied Warbler (p. 422)
*Phyllolais pulchella*

Red-faced Crombec (p. 426)
*Sylvietta whytii jacksoni*
Black-breasted Apalis (p. 407)
*Apalis flavida flavocincta*
Grey-backed Camaroptera (p. 441)
*Camaroptera brevicaudata brevicaudata*

Hunter's Cisticola (p. 478)
*Cisticola hunteri prinioides*
Red-faced Cisticola (p. 482)
*Cisticola erythrops sylvia*
Singing Cisticola (p. 481)
*Cisticola cantans pictipennis*

Ph. xiii

Winding Cisticola (p. 486)
*Cisticola galactotes nyansae*
Stout Cisticola (p. 489)
*Cisticola robusta ambigua*
Tawny-flanked Prinia (p. 506)
*Prinia subflava tenella*

Banded Martin (p. 544)
*Riparia cincta suahelica*
African Sand Martin (p. 543)
*Riparia paludicola ducis*
Black Rough-winged Swallow (p. 549)
*Psalidoprocne holomelaena holomelaena*

Black Cuckoo-Shrike (p. 556)
*Campephaga sulphurata*
Fiscal (p. 593)
*Lanius collaris humeralis*
Tropical Boubou (p. 614)
*Laniarius aethiopicus ambiguus*

Black-backed Puff-Back (p. 617)
*Dryoscopus cubla hamatus*
Black-headed Bush-Shrike (p. 623)
*Tchagra senegala senegala*
Brown-headed Bush-Shrike (p. 626)
*Tchagra australis emini*

Ph. xv

White-breasted Tit (p. 652)
*Parus albiventris*
Black-headed Oriole (p. 665)
*Oriolus larvatus rolleti*
Pied Crow (p, 673)
*Corvus albus*

Yellow-White-eye (p. 726)
*Zosterops senegalensis flavilateralis*
Kikuyu White-eye (p. 731)
*Zosterops kikuyuensis kikuyuensis*
Malachite Sunbird (p. 751)
*Nectarinia famosa cupreonitens*

Bronze Sunbird (p. 755)
*Nectarinia kilimensis kilimensis*
Red-chested Sunbird (p. 757)
*Nectarinia erythrocerca*
Yellow-bellied Variable Sunbird (p. 779)
*Cinnyris venustus falkensteini*

Amethyst Sunbird (p. 792)
*Chalcomitra amethystina doggetti*
Scarlet-chested Sunbird (p. 797)
*Chalcomitra senegalensis lamperti*
Collared Sunbird (p. 809)
*Anthreptes collaris garguensis*

Spectacled Weaver (p. 919)
*Hyphanturgus ocularis suahelicus*
Reichenow's Weaver (p. 929)
*Othyphantes reichenowi reichenowi*
Yellow Bishop (p. 955)
*Euplectes capensis xanthomelas*

Red-naped Widow-Bird (p. 970)
*Coliuspasser laticauda laticauda*
Jackson's Widow-Bird (p. 972)
*Drepanoplectes jacksoni*
Quail-Finch (p. 998)
*Ortygospiza atricollis mülleri*

Ph. xviii

Bronze Mannikin (p. 976)
*Spermestes cucullatus scutatus*
Rufous-backed Mannikin (p. 977)
*Spermestes nigriceps nigriceps*
Red-billed Fire-Finch (p. 1012)
*Lagonosticta senegala brunneiceps*

African Fire-Finch (p. 1009)
*Lagonosticta rubricata rhodopareia*
Yellow-bellied Waxbill (p. 1018)
*Coccopygia melanotis kilimensis*
Red-cheeked Cordon-Bleu (p. 1035)
*Uraeginthus bengalus brunneigularis*

Waxbill (p. 1020)
*Estrilda astrild cavendishi*
Zebra Waxbill (p. 1024)
*Estrilda subflava clarkei*
Brimstone Canary (p. 1065)
*Serinus sulphuratus sharpii*

Streaky Seed-eater (p. 1076)
*Serinus striolatus striolatus*
African Citril (p. 1084)
*Carduelis citrinelloides kikuyuensis*
Golden-breasted Bunting (p. 1090)
*Emberiza flaviventris flaviventris*

*Emberiza flaviventris flavigastra* Cretz.
*Emberiza flavigastra* Cretzschmar, in Rüpp. Atlas, Vög. p. 38, pl. 25, 1828: Kordofan, Sudan.

**Distinguishing characters:** Differs from the nominate race in being generally paler and with more extensive white on flanks. Wing 72 to 83 mm.

**General distribution:** French Sudan to the Sudan and Eritrea.

**Range in Eastern Africa:** Central Sudan from Darfur to Sennar and Eritrea.

**Habits:** As for the nominate race, common in bush and open woodland, with a light silvery song. Eggs are two in number, white, copiously scribbled with thin black lines (Lynes).

**Recorded breeding:** West Africa, June and July. Western Sudan, June to September.

**1470 SOMALI GOLDEN-BREASTED BUNTING.** *EMBERIZA POLIOPLEURA* (Salvadori). **Pl. 95.**
*Fringillaria poliopleura* Salvadori, Ann. Mus. Civ. Genova 26, p. 269, 1888: Soddé, Shoa, central Abyssinia.

**Distinguishing characters:** Differs from the Golden-breasted Bunting in having the mantle feathers edged with white, or pale buffish white, giving the upperside a brighter appearance; rump paler grey, feathers with black subterminal spots. The female has the sides of the crown and face dusky black. Wing 66 to 79 mm. The young bird has little yellow below; chest streaked with black.

**Range in Eastern Africa:** Abyssinia and the Somalilands to south-eastern Sudan, Kenya Colony and north-eastern Tanganyika Territory as far west as Moroto and the Mt. Kilimanjaro area.

**Habits:** A more or less desert species of dry thorn scrub, usually common and tame where it occurs. It takes the place of the Golden-breasted Bunting and its habits are similar.

**Nest and Eggs:** Nest a rough cup of stiff grass stems lined with finer grasses in the fork of a bush. Eggs two or three, glossy white or whitish with a zone of blackish dots and scrawls and pale undermarkings; about $18 \cdot 5 \times 13$ mm.

**Recorded breeding:** South-eastern Sudan (breeding condition), April. Abyssinia and Italian Somaliland, April to June. Tanganyika Territory, November.

Food: As for allied species.

Call: Similar to those of other Buntings, the song is a sort of 'tizekh-tizekh' repeated several times.

### 1471 BROWN-RUMPED BUNTING. *EMBERIZA FORBESI* Hartlaub.

*Emberiza forbesi forbesi* Hart. Pl. 95.

*Emberiza forbesi* Hartlaub, O.C. p. 92, 1882: Langomeri, north-western Uganda.

**Distinguishing characters:** Differs from the Golden-breasted and Somali Golden-breasted Buntings in having the rump chestnut brown, not grey; tips of wing-coverts buff, not white. Wing 70 to 75 mm.

**Range in Eastern Africa:** Abyssinia to Uganda and the southern Sudan.

**Habits:** A relatively uncommon species of which little is recorded. Widely distributed in semi-arid country and usually noticeable by the loud song of the males in the breeding season.

**Nest and Eggs:** Apparently undescribed.

**Recorded breeding:** No records.

**Food:** No information.

**Call:** A rather pleasant and quite loud song.

**Distribution of other races of the species:** West Africa.

### 1472 ORTOLAN. *EMBERIZA HORTULANA* Linnæus. Pl. 95.

*Emberiza Hortulana* Linnæus, Syst. Nat. 10th ed. p. 177, 1758: Sweden.

**Distinguishing characters:** Adult male, head all round to nape and chest greyish olivaceous yellow; often some olivaceous streaks on chest; eyelid, moustachial stripe and chin and throat pale yellow; rest of upperparts warmish brown streaked with black; rump not streaked; rest of underparts tawny chestnut; some white in outer tail feathers. The female is duller than the male and often more distinctly streaked on chest. The immature dress is similar to the adult female. Wing 81 to 98 mm. The young bird is more earthy brown above, streaked with blackish including head; sides of face and underside buff or pale brown, sometimes with a slight yellowish wash; sides of face, throat, chest and flanks streaked blackish.

**General distribution:** Europe and Asia; in non-breeding season to Africa as far south as Rio-de-Oro, Darfur and northern Kenya Colony, also south-western Arabia.

**Range in Eastern Africa:** As far south as the western Sudan, Eritrea and northern Kenya Colony in non-breeding season.

**Habits:** A common often abundant palæarctic passage migrant or winter visitor, in pairs or small parties, occasionally in flocks. It does not appear to reach much south of the Sudan and Abyssinia, and there are few records as yet from Kenya Colony. Usually seen on cultivated ground or in short grass country.

**1473** CRETZSCHMAR'S BUNTING. *EMBERIZA CÆSIA* Cretzschmar.

*Emberiza cæsia* Cretzschmar, in Rüpp. Atlas, Vög. p. 17, pl. 10, 1827: Kurgos Island, Berber district, Sudan.

**Distinguishing characters:** Adult male, differs from the Ortolan in having the head, neck, sides of face and chest grey; chin and throat tawny. The female has the top of the head streaked with brown and blackish; a narrow blackish moustachial stripe; chest less grey and spotted with blackish. Wing 76 to 90 mm. The young bird is warmer brown above than the young of the Ortolan, and is pale tawny chestnut below.

**General distribution:** South-eastern Europe and western Asia; in non-breeding season to the Sudan, Eritrea and southern Arabia.

**Range in Eastern Africa:** Eritrea and the Sudan as far south as the White Nile and Sennar in non-breeding season.

**Habits:** A common palæarctic winter migrant from the north to the Nile Valley, usually in small flocks, from September to March, but it has not been noted much south of Kawa. It frequents cultivated land and gardens, particularly on the desert edge.

**1474** CINEREOUS BUNTING. *EMBERIZA CINERACEA* Brehm.

*Emberiza cineracea semenowi* Sar.
*Emberiza (Hypocentor) semenowi* Sarudny, Orn. Jahrb. 15, p. 217, 1904: Gebel Tuve, Arabistan, south-western Iran.

**Distinguishing characters:** Adult male, top of head to nape and sides of face yellowish olive; rest of upperparts earth brown; mantle streaked with dusky; edges of wing-coverts and innermost secondaries browner; below, yellow, olivaceous on sides of chest and

flanks; some white in outer tail feathers. The female has the head more earthy brown streaked with dusky and slightly washed with olivaceous; below, duller and some dusky streaks on chest. Wing 84 to 90 mm. Juvenile plumage unrecorded.

**General distribution:** Syria and Iran; in non-breeding season to the Sudan, Eritrea and south-western Arabia.

**Range in Eastern Africa:** The Red Sea Province of the Sudan and Eritrea, in non-breeding season.

**Habits:** An uncommon palæarctic winter migrant to the north-eastern part of our area, particularly the Red Sea coast.

**Distribution of other races of the species:** Asia Minor to the Persian Gulf, the nominate race being described from Smyrna.

**1475 SOUTHERN ROCK-BUNTING.** *FRINGILLARIA CAPENSIS* (Linnæus).
*Fringillaria capensis vincenti* Lowe.
*Fringillaria capensis vincenti* P. R. Lowe, Bull. B.O.C. 52, p. 144, 1932: Zobué, north of Tete, northern Portuguese East Africa.

**Distinguishing characters:** Centre of crown grey with black markings; sides of crown black; white stripe from base of bill, over eye to above ear-coverts; black stripe from lores, through eye to ear-coverts; white stripe from gape to below eye and ear-coverts; black moustachial stripe, often joining up behind white stripe below eye to black stripe through eye; rest of upperparts sooty brown indistinctly streaked with blackish; wing shoulder and edges of wing-coverts chestnut; below, chin and throat white; rest of underparts sooty grey; no white in tail. The sexes are alike. Wing 72 to 82 mm. The young bird has the black of the sides of the crown less defined.

**General distribution:** Nyasaland and northern Portuguese East Africa.

**Range in Eastern Africa:** Portuguese East Africa and Nyasaland.

**Habits:** An uncommon species living among grass tufts, aloes and boulders, on the rocky slopes of hills. It is usually found with the Cinnamon-breasted Rock-Bunting, but appears larger and plumper in outline. Very local.

**Nest and Eggs:** Nest a cup of fine grass and fine rootlets at the base of a grass tuft among moss and lichens on rocks. Eggs three, bluish white, heavily freckled and blotched with reddish brown and with a few pale lilac undermarkings at the larger end; about 20·5 × 15 mm.

**Recorded breeding:** Nyasaland and Portuguese East Africa, September, also March and April.

**Food:** Seeds and insects.

**Call:** A small Sparrow-like chirping note. The song of the male is a Canary-like whistle 'tree-re-reet' with the second syllable highest (Vincent).

**Distribution of other races of the species:** South Africa, the nominate race being described from the Cape of Good Hope.

## 1476 CINNAMON-BREASTED ROCK-BUNTING. *FRINGILLARIA TAHAPISI* A. Smith.

*Fringillaria tahapisi tahapisi* Smith.
*Fringillaria tahapisi* A. Smith, Rep. Exp. C. Afr. p. 48, 1836: South-eastern Transvaal, South Africa.

**Distinguishing characters:** Adult male, head and chin to chest black; white markings on top of head; white stripes over and under eye and along moustache; feathers of chin to chest tipped with white, wearing to black; rest of upperparts streaked russet and black; wing-coverts and flight feathers edged with tawny; tail dull black; underside warm russet; edges of inner webs of secondaries often russet. The female has the top of the head tawny streaked with black; chin and throat duller black. Wing 70 to 83 mm. The young bird has the whole head streaked tawny and black; stripes on head tawny not white; chin and throat dusky; tawny edges of wing-coverts and innermost secondaries broader.

**General distribution:** Gabon to Angola, Abyssinia, British Somaliland, the Sudan and South Africa, but not western and southern Cape Province.

**Range in Eastern Africa:** Central Abyssinia to British Somaliland, south-eastern Sudan and south to the Zambesi River.

**Habits:** A widespread, but not usually numerous species, of dry rocky ground in the north of its range, but in the south is found in many types of open country, and is fond of dry bare patches of ground. Usually very tame and has a sort of shuffling hopping gait on the ground.

**Nest and Eggs:** Nest a small frail cup of grass lined with rootlets on the ground at the base of a stone or stunted bush. Eggs usually three, pale bluish white, often thickly spotted and blotched with various shades of brown, but without scrawls; about 18 × 14 mm.

**Recorded breeding:** Southern Kenya Colony, March to June, also January. Tanganyika Territory, January to May. South-eastern Belgian Congo, May. Rhodesia, May and June. Nyasaland, May to August.

**Food:** Seeds and insects.

**Call:** A low bleating alarm note. The song is a pleasant little skirl reminiscent of an English Yellow-hammer.

*Fringillaria tahapisi insularis* Grant & Forbes.
*Fringillaria insularis* O. Grant & Forbes, Bull. Liv. Mus. 2, p. 2, 1899: Adho Dimellus, Socotra Island.

**Distinguishing characters:** General colour paler than the nominate race. Wing 69 to 77 mm.

**Range in Eastern Africa:** Socotra Island.

**Habits:** Found at lower levels, habits unrecorded.

**Recorded breeding:** January (nestling).

*Fringillaria tahapisi septemstriata* (Rüpp.).
*Emberiza septemstriata* Rüppell, N. Wirbelt. Vög. p. 86, pl. 30, 1840: Gondar, northern Abyssinia.

**Distinguishing characters:** Differs from the nominate race in having inner primaries, secondaries and inner webs of the outer primaries russet, ends dusky. Wing 72 to 80 mm.

**Range in Eastern Africa:** Eastern Sudan, Eritrea and northern Abyssinia.

**Habits:** A local resident of rocky woodland at lower levels, with a short sweet song.

**Recorded breeding:** Northern Abyssinia, December.

*Fringillaria tahapisi goslingi* Alex.
*Fringillaria goslingi* Alexander, Bull. B.O.C. 16, p. 124, 1906: Kina, northern Nigeria.

**Distinguishing characters:** Similar to the preceding race in the amount of russet in the flight feathers, but general colour paler. Wing 70 to 80 mm.

**General distribution:** Sierra Leone, Gold Coast Colony and northern Nigeria to the Sudan.

## BUNTINGS

**Range in Eastern Africa:** Western and south-western Sudan.

**Habits:** Locally common among grass and rocks at lower levels, in the Jebel Marra it occurs in great numbers and assembles in large flocks to drink. The nest is very frail and may be in any situation on the ground, a furrow in a field of corn has been recorded, and the eggs often have a greenish ground.

**Recorded breeding:** West Africa, October to February, also July. Western Sudan, November to January.

**Distribution of other races of the species:** Southern Arabia.

**1477 HOUSE-BUNTING.** *FRINGILLARIA STRIOLATA* (Lichtenstein).

*Fringillaria striolata striolata* (Licht.).
*Fringilla striolata* Lichtenstein, Verz. Doubl. p. 24, 1823: Ambukol, Berber district, Sudan.

**Distinguishing characters:** Adult male, head streaked white and black; superciliary stripe white; mantle to upper tail-coverts tawny with dusky streaks; wing shoulder and edges of flight feathers warm tawny; sides of face buff streaked with black; more or less distinct whitish streaks below eye and along moustache; below, chin and throat blackish with white edges to the feathers, sometimes extending to breast; rest of underparts buff to pale tawny; inner webs of flight feathers tawny; tail dusky with edges of outer webs tawny. The female has the head streaked buff and duller black; chin and throat less black. Wing 69 to 83 mm. The young bird is similar to the adult female.

**General distribution:** The Sudan, Eritrea, Abyssinia, Palestine, Arabia and India.

**Range in Eastern Africa:** The Berber and Red Sea provinces of the Sudan to eastern Abyssinia, Eritrea and British Somaliland.

**Habits:** Birds of stony desert country, hiding among stones.

**Recorded breeding:** No records.

*Fringillaria striolata saturatior* Sharpe.
*Fringillaria saturatior* Sharpe, Bull. B.O.C. 11, p. 47, 1901: Lake Stephanie, southern Abyssinia.

**Distinguishing characters:** Differs from the nominate race in having the throat greyer, with the black confined to streaks; chest to belly deeper russet brown. Wing 71 to 85 mm.

**Range in Eastern Africa:** Western Sudan to south-western Abyssinia and north-eastern Kenya Colony.

**Habits:** Locally common or even abundant congregating into flocks to feed or drink. In the lower parts of its range it becomes the local House-Bunting, and is found in cultivated areas; but on the Jebel Marra has all the habits of a true Rock-Bunting.

**Nest and Eggs:** A cup nest well concealed in a hollow of the ground, slightly made of dry grass and rootlets with fine rootlets and a few hairs. Eggs two or three, pale greenish white blotched and spotted with sepia or chocolate, and with lavender undermarkings; about 19 × 14·5 mm.

**Recorded breeding:** October to December, also February.

**Food:** Small seeds.

**Call:** A Bunting-like chirp and a small wheezy tri-syllabic song.

**1478** SOCOTRA MOUNTAIN-BUNTING. *FRINGILLARIA SOCOTRANA* Grant & Forbes.

*Fringillaria socotrana* O. Grant & Forbes, Bull. Liv. Mus. 2, p. 2, 1899: Adho Dimellus, Socotra Island.

**Distinguishing characters:** Stripe down centre of crown white, bordered by a broad black stripe; superciliary stripe and stripe below eye white; stripe through eye and along moustache black; mantle tawny streaked with black; rump white with some black spots; wing shoulder and edges of flight feathers tawny chestnut; tail dusky with tawny edges to outer feathers; below, chin and throat white; chest and breast tawny chestnut; belly and under tail-coverts whitish. The female is rather duller than the male. Wing 67 to 71 mm. Juvenile plumage unrecorded.

**Range in Eastern Africa:** Socotra Island.

**Habits:** Little recorded, but found at higher levels, above 4,000 feet.

**Nest and Eggs:** Undescribed.

**Recorded breeding:** No records.

**Food:** No information.

**Call:** Unrecorded.

# BUNTINGS

Names in Sclater's *Syst. Av. Æthiop.*, 2, 1930, which have been changed or have become synonyms in this work:

*Emberiza affinis* Heuglin, now *Emberiza forbesi* Hartlaub.
*Fringillaria striolata jebelmarræ* Lynes treated as synonymous with *Fringillaria striolata saturatior* Sharpe.

Names introduced since 1930 which have become synonyms in this work:
*Fringillaria striolata dankali* Thesiger & Meynell, 1934, treated as synonymous with *Fringillaria striolata striolata* Lichtenstein.

## Addenda and Corrigenda

**1447** *Rhynchostruthus socotranus louisæ*. *Distinguishing characters:* Wing should be 78 to 89 mm.

**1459** (p. 1075). Type locality of *Serinus angolensis somereni* is in western not eastern Uganda.

**1463** *Linurgus olivaceus kilimensis*. *General distribution:* delete Kenya Colony.

p. 1086. *Serinus flavivertex elgonensis* Granvik 1934. The name is preoccupied by *Serinus gularis elgonensis* (Grant) 1912.

**1466** *Warsanglia johannis* S. Clarke, pl. 91. Delete 'female and juvenile plumages unknown', and add to *Distinguishing characters:* The female has less white on the forehead; top of head, mantle, scapulars and upper tail-coverts paler, slightly brownish-grey streaked with dusky; stripe through eye; sides of neck and sides of chest paler grey; tips of secondaries and secondary coverts white; rather less chestnut on rump; eye brown; bill greyish horn; feet and toes blackish horn. Wing 72 to 80 mm. The young bird is brownish above spotted with dusky, including rump; below, white spotted and streaked with blackish, including lower flanks. *Habits:* Found in high juniper forest, singly or in small parties, flight fast (A. R. Tribe). *Food:* Grass seeds, also seen to feed on *Salvia* sp. *Call:* A high-pitched 'tssp' or 'chip'.

**1468** *Emberiza orientalis* is best treated as a race of *Emberiza cabanisi* (1467).

**1469** *Emberiza flaviventris*. For Stephen read Stephens.

*Note:* In conformity with the original work we have retained the names of countries or areas as they then were, e.g. British and Italian Somaliland.
The second edition of Volume 2 was closed for the addition of Addenda and Corrigenda and handed to the Printers on 1 January 1959.

# ADDENDA

**671** *Alæmon hamertoni altera.* For Warsangeli read Haber Toojaala.

**672** *Ammomanes c. kinneari.* Delete *General distribution:* add *Range in Eastern Africa:* Omdurman and fifty miles to the south.

**673** *Ammomanes d. akeleyi.* For Elliott read D. G. Elliot.
*A.d. samharensis. Recorded breeding:* add Eritrea nestlings in May.
*A.d. erythrochroa.* Add *General distribution:* North-eastern French Equatorial Africa to central Sudan.

**675** *Galerida c. isabellina.* Add 'extends to Ennedi, north-eastern French Equatorial Africa.
*G.c. altirostris. Recorded breeding:* add Eritrea, December to March.

**676** *Galerida m. prætermissa. Recorded breeding:* add Eritrea, nestlings in June and August.
*G.m. elliotti. Elliotti* should be *ellioti.*

**678** *Pseudalæmon f. fremantlii. Habits:* add 'local on light grey stony soil, feeding much by digging' (G. Clarke).
*P.f. megaënsis. Range in Eastern Africa:* add 'to Marsabit in northern Kenya Colony.

**680** *Eremopterix n. melanauchen. Recorded breeding:* add Eritrea, January, March and September.

**689** *Aëthocorys personata. Call:* Various slurred whistling calls of a somewhat plaintive nature, the commonest being a soft 'pee-ay'.

p. 46. Add: *Mirafra albicauda rukwensis* White, 1956, included in *Mirafra albicauda.*
Add: *Calandrella cinerea williamsi* Clancey, 1952, treated as synonymous with *Calandrella cinerea saturatior* Reichenow.

**775** *Eurillas virens.* Delete description of young bird and substitute: The young bird is very similar to the adult, but has slight russet edges to the wing-coverts, and a pale lower mandible.

p. 152. Add: *Ixonotus guttatus ngoma* Rand 1955, treated as synonymous with *Ixonotus guttatus.*
*Phyllastrephus albigula itoculo* is synonymous with *Ph. fischeri münzneri* Reichenow, not *Ph. f. fischeri.*

pp. 157 & 158. Delete *Parisoma böhmi* and *P. lugens* which are now treated as Warblers.

p. 161. *Parisoma plumbeum* now *Myioparus plumbeus.*

**779** *Muscicapa semitorquata.* For Hofmeyer read Homeyer.

**783** *A. seth-smithi.* Belgian Congo eggs given as clear brown with darker markings; about 16 × 10 mm. (Prigogine).

**785** *Alseonax c. cinereolus. Range in Eastern Africa:* delete 'eastern half' before Tanganyika Territory.

**787** *Parisoma böhmi.* Transferred to Warblers, next to *Sylvia.*

**788** Now *Myioparus plumbeus.*
*M.p. plumbeus.* For young read immature, and add, the young bird is spotted above with buff and blackish, and mottled below with blackish.

**789** *Parisoma lugens* transferred to Warblers.

# ADDENDA

**790** *Lioptilornis* should be *Lioptilus*. Add: *General distribution:* Mt. Musokolo, eastern Belgian Congo and Ruanda.

**799** *Melænornis p. pammelaina*. Add: *General distribution:* Lower Congo. River and Angola.

**809** *Erythrocercus l. livingstonei* is extra limital. The race should be *E.l. francisi* W. L. Sclater, Bull. B.O.C. 7, p. 60, 1898: Inhambane, Portuguese East Africa. *General distribution:* should read Portuguese East Africa to Zambesi River and southern Nyasaland from Liwonde and Netia.

**812** *Bias musicus*. The breeding record from western Uganda under *B.m. changamwensis* should refer to the preceding race.

**818** *Batis o. orientalis*. *Recorded breeding:* add Eritrea, July.

**825** *Dyaphorophyia jamesoni* should be considered a race of the Red-cheeked Wattle-eye *Dyaphorophyia blissetti* Sharpe, Ann. Mag. N.H. (10), p. 451, 1872: Wassau, Gold Coast.

**827** For *Erannornis* read *Elminia*.
*Erannornis longicauda kivuensis* should be *Elminia albicauda kivuensis*, and the range should extend to south-eastern Tanganyika Territory.

**832** *Tchitrea v. ferreti*. *Recorded breeding:* add British Somaliland, April to June.

p. 226. *Erythrocercus livingstonei monapo* is a synonym of *E.l. francisi* Sclater, not of *E.l. livingstonei*.

p. 237. For 901 read 900.

**868** *Œnanthe p. livingstonei*. For Murchison Flats read Murchison Falls. Add: *Recorded breeding:* Rukwa Valley, south-western Tanganyika Territory, a migrant to breed from June to November. Add: *Call:* Sings much at night.

**869** *Cercomela m. lypura*. Add: *Recorded breeding:* Eritrea, January and February, and British Somaliland, March and April.

**885** *Cossypha semirufa*. *Recorded breeding:* Add:· Eritrea, March to August.

**890** *Cossypha natalensis natalensis* is extra-limital. The present race should be: *Cossypha natalensis intensa* Mearns, Smiths, Misc. Coll. 61, p. 2, 1932: Taveta, Kenya Colony.

**900** *S. erythrothorax*. *Distinguishing characters:* add: breast, under tail-coverts, also yellowish-white, or rarely cadmium yellow.

**901** *Alethe castanea*. Add: *Nest and Eggs:* On ground in buttress of tree in forest, of dead leaves; cup of moss-lined Mycelium and fine blackish rootlets. Eggs probably normally two, bluntly ovate, pale olive-greenish, densely covered in fine markings of dull reddish and pale chestnut, a few underlying markings of mauve and lilac, zoned at larger end; about 20 × 14 mm. *Recorded breeding:* Uganda, March.

**903** *Alethe p. ufipæ*. Sambawanga should be Sumbawanga.

**910** *Erythropygia zambesiana*. This and its races are now considered conspecific with the White-browed Scrub Robin, *Erythropygia leucophrys*, of South Africa: *Sylvia leucophrys* Vieillot, N. Dict. d'Hist. Nat. 11, p. 191, 1817: Gamtoos River, Cape Province.
The races should therefore be: *E. leucophrys zambeziana*, *E.l. brunneiceps* and *E.l. sclateri*.

**911** *Erythropygia leucoptera pallida* should be *E.l. eluta* Bowen. *E.l. eluta* Bowen, Proc. Biol. Soc. Wash. 47, p. 159, 1934: Kismayu, Italian Somaliland.
*E.l. pallida* Benson becomes a synonym.

# ADDENDA 1103

p. 332. *Turdus libonyanus cinerascens.* Delete from synonyms, see under 839. *T.l. niassæ* and *T.l. costæ* are synonyms of *T.l. cinerascens.*

p. 333. *Turdus libonyanus tropicalis* Peters, now considered a synonym of *Turdus libonyanus libonyanus* (Smith).
*Erythropygia leucophrys jungens* Bowen 1934, treated as synonymous with *Erythropygia zambesiana zambesiana* Sharpe.
*Erythropygia leucophrys pallida* Benson is synonymous with *Erythropygia leucophrys eluta* Bowen.
Delete from synonym: *Psophocichla guttata belcheri.* Cf. above, under 846.

**957** *Calamocichla leptorhyncha leptorhyncha, C.l. macrorhyncha, C.l. jacksoni* and *C.l. tsanæ* are all races of *Calamocichla gracilirostris.*

**967** *Calamonastes f. stierlingi.* Distinguishing characters: young bird, delete 'faintly' before 'washed with yellow'. Recorded breeding: add Tabora, Tanganyika Territory, nestling January.

**968** *Calamonastes s. simplex.* Recorded breeding: add British Somaliland, February to April.

p. 395. *Sphenœcus* should be *Sphenœacus.*

**970** *Apalis murina, A.m. youngi, A.m. whitei* and *A.m. fuscigularis* are best treated as races of *Apalis thoracica* (Shaw & Nodder).
*Motacilla thoracica* Shaw & Nodder, Nat. Misc. 22, p. 969, 1811: Oliphants River, Cape Province.

**971** *Apalis f. flavigularis, A.f. griceiceps, A.f. uluguru* and *A.f. lynesi* are best treated as races of *Apalis thoracica* (Shaw & Nodder).

**972** *Apalis denti* is the female of 976 *A. nigrescens* of which it is a synonym, q.v. in addenda.

**974** *Apalis alticola brunneiceps* is a synonym of *A. alticola* (Shelley). *Cisticola alticola* Shelley, Bull. B.O.C. 8, p. 35, 1899: Old Fife, Tanganyika Territory–Nyasaland boundary.

**976** *Apalis nigrescens* should be treated as a race of 972 *A. rufogularis.* Distinguishing characters: delete 'the sexes are alike'. Wing 45 to 52 mm. General distribution: add Kayoyo, southern Belgian Congo and north-western Northern Rhodesia.

**987** *Apalis rufifrons rufifrons.* Recorded breeding: add Eritrea, March.

**995** *Phyllolais pulchella.* Recorded breeding: add Eritrea, nestling July.

**999** *Sylvietta isabellina.* For Elliott read D. G. Elliot.

**1000** Add *Distribution of other races of the species:* Belgian Congo and South Africa, the nominate race being described from Angola.

**1003** *Eremomela icteropygialis griseoflava. Range in Eastern Africa:* for lat. 9° N. read lat. 4° N. *E.i. bello.* Wing 58 not 68 mm.

**1004** *Eremomela flavicrissalis.* For p. 48 read p. 481.

**1005** *Eremomela canescens canescens.* Add March before 1864. *E.c. elegans.* Add July before 1864.

**1007** *Eremomela turneri* must now be regarded as a species distinct from *E. badiceps.* Not only do they breed in the same area, but *E. turneri* is distinguished by its weaker feet and slender claws.

**1007 & 1012** 'Fraser' should be in brackets.

**1008** *E.u. rensi.* General distribution should include Fort Jameson area, Northern Rhodesia.

**1012** *C.s. ugandæ. Distinguishing characters:* add Eye orange-brown; bill black; bare skin on side of neck cobalt blue; feet and toes pale brownish flesh colour.

# ADDENDA

**1016** For Bechuanaland read 'near Zeerust, south-western Transvaal'.

**1018** *Cisticola aridula lavendulæ.* Recorded breeding: add Eritrea (fledglings), March.

**1025** *C. lateralis antinorii.* General distribution: add French Congo. North-eastern should be northern Belgian Congo.

**1029** *Cisticola chubbi* should be *Cisticola chubbi chubbi.* Add to Distribution of other races of the species: Eastern Belgian Congo.

**1030** *Cisticola hunteri hypernephala.* Elliott should be D. G. Elliot.

**1032** *Cisticola erythrops.* 'Hartlaub' should be in brackets.

**1037** *Cisticola brachyptera.* For 1890 read 1870.

**1045** *Prinia subflava desertæ* should be *Prinia subflava pallescens* Madarasz, Ann. Mus. Hung., p. 593, pl. ii, fig. 3,|1914: Senga = Singa, Blue Nile, eastern Sudan. *P.s. desertæ* becomes a synonym.

**1046** *Prinia somalica somalica.* Range in Eastern Africa: add Harar, eastern Abyssinia.

**1047** *Prinia gracilis carlo.* Recorded breeding: add Eritrea, December to May.

**1053** *Bathmocercus rufus jacksoni* should be *Bathmocercus rufus vulpinus* Reichenow, Nov. Zool., p. 160, 1895: Aruwimi River, north-eastern Belgian Congo. *B.r. jacksoni* becomes a synonym.

p. 518. *Calamocichla leptorhyncha nuerensis* Lynes, treated as synonymous with *Calamocichla leptorhyncha jacksoni* Neumann (now *Calamocichla gracilirostris jacksoni*).

**1267** *Cyanomitra alinæ.* Nest and Eggs: Nest attached to a bush a few feet from the ground and composed of green moss, rootlets and dead leaves lined with plant down. Egg one, reddish-brown with darker markings; about 20 × 14·5 mm. (Belgian Congo race).

**1268** *Cyanomitra cyanolæma cyanolæma* should probably be *C. cyanolæma octaviæ* Amadon, Bull. Amer. Mus. Nat. Hist. 100, p. 427, 1953: Cameroon, but more material is needed.

**1271** *Anthreptes collaris.* The race *A.c. hypodilus* is confined to Fernando Po. Mainland birds under that heading should be called *A.c. somereni* Chapin.

**1274** (p. 814). *Anthreptes longuemarei neumanni.* For nominate race read Lado race.

**1278** *Anthreptes pallidigaster.* Recorded breeding: add Sokoke Forest, May.

**1346** *H.m. stephanophorus.* Eggs also described as white; about 22 × 16 mm.

**1350** *Pachyphantes pachyrhynchus.* Now *Pachyphantes superciliosus* (Shelley).

*Hyphantornis superciliosus* Shelley, Ibis, p. 140, 1873: Accra, Gold Coast.

**1360** *Quelea quelea.* Distinguishing characters: add Forehead band normally black. Recorded breeding: add Breeds in South Africa, January to April.

p. 945. *Quelea quelea æthiopica.* Distinguishing characters: add Forehead often of same colour as crown of head, not black, more extensive rosy below. Range in Eastern Africa: add At least whole of Tanganyika Territory is an area of intermediates between this race and South African race.

**1363** (p. 951). *Euplectes orix pusilla.* Recorded breeding: add Eritrea, July and August.

# ADDENDA

**1376** *C.l. laticauda.* Range in Eastern Africa: add central Eritrea.

**1377** *Coliuspasser progne delamerei.* Distinguishing characters: add after 'bill unchanged', tail feathers as in female.

**1378** *Drepanoplectes jacksoni.* Distinguishing characters: add 'to male in non-breeding dress', tail feathers as in female.

**1444** *Steganura paradisea.* Distinguishing characters: as in note under 1442.

p. 1053. *Euplectes zavattarii* Moltoni 1943. Gandarobe. Sazam, Omo. 31.5.39. Treated by us as a melanistic example of *E. orix pusilla*, Bull. B.O.C., p. 60, 1938. The type is a mounted bird with underside as in *Euplectes nigriventris* but rather faded. The crown and nape show an extraordinary mixture of red and black feathers. It is only understandable as a throw-back to a common ancestor of *E. orix* and *E. nigriventris*.

Add *Euplectes capensis litoris* Neunzig 1928, treated as synonymous with *Euplectes capensis xanthomelas* Rüpp.

Delete *Pytelia melba belli*, replace by *Pytelia melba grotei* Reichenow, treated as synonymous with *Pytelia melba belli* Grant, cf. No. 1410 above.

Since Volume 2 was published 11 species and 14 subspecies have been newly reported or recognised from Eastern Africa. They are as follows:

## Additions to the East African List since 31 December 1953

**660A MARSABIT LARK.** *MIRAFRA WILLIAMSI* Macdonald.

*Mirafra williamsi* Macdonald, Bull. B.O.C. 76, p. 70, 1956: Marsabit, northern Kenya Colony.

**Distinguishing characters:** Above, dark russet brown with narrow dusky streaks; innermost secondaries more blackish with buff edges; outer tail feathers mainly white, penultimate feather with white outer web; buff stripe over eye; below, pale buff; chin and throat white; chest russet with blackish-brown spots; bill fairly thick and heavy; eye brown; bill greyish-brown; feet and toes pinkish. The sexes are alike. Wing 83 to 84 mm. The young bird has not been described.

*Note:* In this species the first primary is similar to that of the Northern White-tailed Bush-Lark and the Kordofan Bush-Lark, but it differs from the former in the pattern of the tail feathers, length of innermost secondaries and size of bill, and from the latter in size of bill, toes and second primary.

**Range in Eastern Africa:** Marsabit area, northern Kenya Colony.

**Habits:** Flight said to be slower than that of the Flappet-Lark, and the bird is less skulking and more easily seen. Occurs on same ground as the Flappet-Lark, in rather open, dry, overgrazed areas.

**Nest and Eggs:** Unknown.

**Recorded breeding:** No records.

**Call:** Nothing recorded.

**672 *Ammomanes cinctura arenicolor* Sund.**

*Alauda arenicolor* Sundevall, Œfv. K. Vet. Akad. Förh., p. 128, 1850: Lower Egypt.

**Distinguishing characters:** Above, has a distinctive greyish tone. Wing 85 to 97 mm.

**General distribution:** North Africa, Egypt, Arabia, Palestine, Sudan.

**Range in Eastern Africa:** Northern Sudan.

**Habits:** As for other races.

**672 *Ammomanes cinctura pallens* Le R.**

*Ammomanes phœnicura pallens* Le Roi, O.M., p. 6, 1912: Bir Sani, Bajuda Steppe, eastern Dongola, Sudan.

**Distinguishing characters:** Paler above than the Omdurman race and lacking the grey wash of the Egyptian race. Wing 85 to 90 mm.

**General distribution:** Air and Ennedi, French West and Equatorial Africa to Sudan.

**Range in Eastern Africa:** Central Sudan from Dongola province to Port Sudan and Sinkat.

**Habits:** As for other races.

*Note:* The above two races have been separated from *Ammomanes cinctura kinneari*, of which the *General Distribution* should be deleted and its *Range in Eastern Africa* should be altered to 'Omdurman and about 50 miles to the south'.

**685 *Calandrella rufescens vulpecula* White.**

*Calandrella rufescens vulpecula* White, Bull. B.O.C. 75, p. 3, 1955: 10 miles west of Bohotleh, southern British Somaliland.

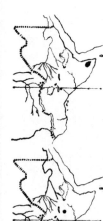

**Distinguishing characters:** Differs from other races in having the edges of the feathers of the upperside warm foxy rufous; below, more vinous pink. Wing 86 to 92 mm.

**Range in Eastern Africa:** Bohotleh area, southern British Somaliland.

**Habits:** Said to differ markedly from the Somali race, being much wilder and more solitary.

**689 *Aëthocorys personata mcchesneyi* Will.**

*Aëthocorys personata mcchesneyi* Williams, Bull. B.O.C. 77, p. 157, 1957: Plateau 6 miles north of Mt. Marsabit.

**Distinguishing characters:** Paler than other races; below, sandier, reddish brown and more clearly marked above. Wing 84 to 87 mm.

**Range in Eastern Africa:** Marsabit Plateau.

**Habits:** As for other races.

# ADDENDA

**703** *Anthus similis schoutedeni* Chap.

*Anthus similis schoutedeni* Chapin, Rev. Zool. Bot. Afr. 29, p. 345, 1937: Kwamouth, middle Congo River, western Belgian Congo.

**Distinguishing characters:** A dark-coloured race with little rufous tinge; well spotted on breast and distinctly white on throat, abdomen and tail-coverts. Wing 83 to 92 mm.

**General distribution:** Western Belgian Congo to south-western Tanganyika Territory.

**Range in Eastern Africa:** South-western Tanganyika Territory.

**Habits:** Not particularly recorded.

**703** *Anthus similis hallæ* White.

*Anthus similis hallæ* White, Bull. B.O.C. 77, p. 30, 1957: Lake Karange, Ankole, Uganda.

**Distinguishing characters:** Rather smaller than the Abyssinian race, and greyish not brownish above and with less pronounced streaking. Wing 86 to 97 mm.

**General distribution:** Eastern Belgian Congo and western Uganda.

**Range in Eastern Africa:** Western Uganda.

**Habits:** Not particularly recorded.

**739A** CAPUCHIN BABBLER. *PHYLLANTHUS ATRIPENNIS* (Swainson).

*Phyllanthus atripennis bohndorffi* Sharpe.

*Crateropus bohndorffi* Sharpe, Journ. Linn. Soc. London, Zool. 17, p. 422, 1884: Sassa, Niam Niam country, Belgian Congo.

**Distinguishing characters:** A chocolate-coloured bird with a grey scaly feathered head; dark lores and frontal band; dark wings and tail; throat as body; eye brownish red; bill yellow; feet and legs dull greenish grey. The sexes are alike. Wing 108 to 115 mm.

**General distribution:** Southern Welle, Aruwimi and Ituri districts, the Semliki Valley and western Uganda.

**Range in Eastern Africa:** Bwamba Forest, western Uganda.

**Habits:** A rarely seen bird of dense patches of secondary forest, found generally in small parties, with normal Babbler habits. A recent discovery in our area, birds being caught in mist nets in forest clearings.

**Nest and Eggs:** Undescribed.

**Recorded breeding:** Probably throughout the year.

**Food:** Insects, molluscs and probably fruit.

**Call:** Hoarse scolding notes rising to a high chatter (Chapin). Normal feeding calls a succession of squeaky quacking noises (C. S. Webb). Occasionally a plaintive whistle.

**Distribution of other races of the species:** West Africa.

## ADDENDA

**758** *Phyllastrephus fischeri münzneri* Reichw.

*Phyllastrephus placidus münzneri* Reichenow, O.M. p. 181, 1916: Sanyi, Mahenge, south-eastern Tanganyika Territory.

**Distinguishing characters:** Differs from the nominate race in being greener on the mantle and rather larger. Wing, male 93 to 97 mm., as against 79 to 89 in the nominate race; female 82 to 86 mm., as against 75 to 80 in the nominate race; bill from base, male 24 to 27 mm., as against 20 to 25 in the nominate race; female 21 to 23 mm., as against 20 mm. in the nominate race.

**Range in Eastern Africa:** Sigi Valley, east Usambara Mts., Uluguru Mts., Morogoro, Pugu Hills, Dar-es-Salaam, Mahenge, eastern Tanganyika Territory to Netia, eastern Portuguese East Africa.

**770** *Chlorocichla lætissima schoutedeni* Prig.

*Chlorocichla lætissima schoutedeni* Prigogine, Rev. Zool. Bot. Afr. 49, p. 348, 1954: Kabobo, eastern Belgian Congo, also Kasangu, south-western Tanganyika Territory.

**Distinguishing characters:** Brighter green than the nominate race above, and green with only a yellowish wash below. Wing 100 to 112 mm.

**General distribution:** Eastern Belgian Congo to south-western Tanganyika Territory.

**Range in Eastern Africa:** South-western Tanganyika Territory.

**Habits:** As for the nominate race.

**772A** ANSORGE'S GREENBUL. *ANDROPADUS ANSORGEI* Hartert.

*Andropadus ansorgei kavirondensis* V. Som.

*Charitillas kavirondensis* Van Someren, Bull. B.O.C. 40, p. 95, 1920: Kakamega Forest, Kavirondo, Kenya Colony.

**Distinguishing characters:** Above, olive-brown; edges of flight feathers olivaceous; below, chin to belly greyish with a slight yellow wash on belly; flanks slightly darker; eye brown; bill blackish brown; feet and toes olive. The sexes are alike. Wing 69 to 80 mm.

**Range in Eastern Africa:** Western Kenya Colony.

**Habits:** As for other Greenbuls, little recorded.

**Nest and Eggs:** Undescribed.

**Recorded Breeding:** No records.

**Call:** No reliable information.

*Note:* This bird was previously treated as a race of *Andropadus gracilis* Cab. from which it only differs in the grey colour of its underside.

**799A** YELLOW-EYED BLACK FLYCATCHER. *MELÆNORNIS ARDESIACA* Berlioz.

*Melænornis ardesiaca* Berlioz, Bull. Mus. Hist. Paris, ser. 2, vol. 8,

# ADDENDA

p. 329, 1936: Mbwahi, west of Lake Kivu.

**Distinguishing characters:** Whole plumage dark grey with a bluish tinge, becoming blackish grey round the bill and eyes. Wings rather rounded, 82 to 94 mm. The sexes are alike. Eye yellow, not brown, as in the Black Flycatcher; bill strong and distinctly broad. The young bird has pale spots on the throat and breast.

**General distribution:** Imperfectly known owing to confusion with other species.

**Range in Eastern Africa:** Western Uganda.

**Habits:** Very little known. Described as an inhabitant of dense undergrowth at the edge of almost impenetrable forest which is usually seen in the early morning. The bright yellow eye is quite noticeable.

**Nest and Eggs:** Nest, a cup in the forks of small trees, not particularly described. Eggs unknown.

**Recorded breeding:** Belgian Congo, April, but also earlier.

**Call:** None recorded.

**964** *Seicercus umbrovirens williamsi* (Clan.).

*Phylloscopus umbrovirens williamsi* Clancey, Bull. B.O.C. 76, p. 10, 1956: 10 miles north of Erigavo, eastern British Somaliland.

**Distinguishing characters:** Differs from the nominate race in being darker, warmer brown above and rather paler below. Wing 55 mm.

**Range in Eastern Africa:** Erigavo area, eastern British Somaliland.

**Habits:** As for other races.

**1002A SOMALI SHORT-BILLED CROMBEC.** *SYLVIETTA PHILIPPÆ* Williams.

*Sylvietta philippæ* Williams, Ibis, p. 582, 1955: Galkayu (= Rocca Littoria), central Italian Somaliland.

**Distinguishing characters:** Bill short; above grey, including innermost secondaries and tail; rest of flight feathers dusky; white stripe over eye; lores, round eyes and ear-coverts dusky grey; chin and throat white, rest of underparts pale yellow; lower flanks inclining to brownish; eye orange-brown; bill grey; feet and toes pinkish flesh. The sexes are alike. Wing 51 to 55 mm.

**Range in Eastern Africa:** Central British Somaliland to central Italian Somaliland.

**Habits:** Inhabits the denser parts of semi-desert bush on rocky or red sand soil. In life appears very like *Eremomela flavicrissalis* with which it occurs, but is distinguished by its red legs. Usually tame.

**Nest and Eggs:** Unknown.

**Breeding records:** None.

**Call:** A short soft 'tsssp' rarely uttered.

**1065A LARGER STRIPED SWALLOW.** *HIRUNDO CUCULLATA* (Boddaert).

*Hirundo cucullata* Boddaert, Tabl. Pl. Enl.-p. 45, 1783: Cape of Good Hope.

**Distinguishing characters:** Top of head to nape chestnut with narrow glossy blue streaks; mantle, scapulars and wing-coverts glossy steel blue; flight feathers and tail more dusky, latter with white spots on inner webs; rump tawny; sides of face, ear-coverts and rest of underparts white or buffish white streaked with dusky; eye brown; bill black; feet and toes dusky brown. The sexes are alike. Wing 116 to 130 mm. The young bird has black mixed with chestnut on top of the head, and tawny tips to the innermost secondaries and wing-coverts.

**General distribution:** South Africa, in non-breeding season to the Belgian Congo, Northern Rhodesia and Tanganyika Territory.

**Range in Eastern Africa:** South-western Tanganyika Territory in non-breeding season.

**Habits:** A common and familiar migratory species of southern Africa, tame and friendly, breeding in houses, verandahs, caves or culverts. Only recently noted within our area but apparently a regular non-breeding visitor in small numbers.

**1146A FIERY-BREASTED BUSH-SHRIKE.** *MALACONOTUS CRUENTUS* (Lesson).

*Malaconotus cruentus adolfi-frederici* Reichenow, O.M., p. 191, 1908: Bwamba Forest, northern Uganda.

**Distinguishing characters:** Forehead, lores and round eyes white; crown, cheeks, ear-coverts, neck and part of mantle blue-grey; back, rump and upper tail-coverts olive-green; wing-coverts and inner webs of wing feathers black, except lesser wing-coverts which are green, and the inner secondaries which are green with a black patch; outer webs of wing feathers green, inner feathers with broad yellow tips, and a touch of yellow on the outer ones; tail olive-green tipped yellow with a black subterminal band. Underside bright yellow heavily washed with vermilion red; bill black; legs blue. The sexes are alike. Wing 105 to 115 mm. (in other races). Immature and young with a brownish wash on head and neck, also on rump; wings, tail and back much as in adult; below, pale yellow with greyish breast soon becoming apricot yellow.

**Range in Eastern Africa:** Only recorded from western Uganda.

**Habits:** A very bright-coloured and striking-looking bird of the denser parts of forest or woodland, more often heard than seen. It is a widespread species of the West African forest which only just reaches our area.

**Nest and Eggs:** Very little known and descriptions vary to such a degree that more knowledge is necessary.

**Recorded breeding:** No local records.

**Call:** A succession of short 'coos' and another call of 'kick-ik-ik' and rasping notes, possibly from the other bird of a pair.

**1235A** *Drepanorhynchus reichenowi lathburyi* Will.

*Nectarinia reichenowi lathburyi* Williams, Bull. B.O.C. 76, p. 137, 1956: Mt. Gouguez (Uraguess), Mathews Range, northern Kenya Colony.

**Distinguishing characters:** Similar to the nominate race in colour but smaller and with a more sharply curved bill. Female shows the same differences and is also slightly darker above and deeper yellow below than the nominate race. Wing 67 to 80 mm., culmen 22·5 to 24 against 25 to 28 in the nominate race.

**Range in Eastern Africa:** Confined to the Mathews Range and Mt. Nyiro in northern Kenya Colony.

**Habits:** As for the nominate race.

**Recorded breeding:** Estimated breeding season prolonged, but mainly February to April.

**1313** Layard's Spot-backed Weaver *Ploceus nigriceps* is best treated as a race of 1312 *Ploceus cucullatus* and should be:

*Ploceus cucullatus nigriceps* Layard, Bds. S. Afr. 1st ed., p. 180, 1867: Bulawayo.

**General distribution:** Southern Italian Somaliland, Kenya Colony and eastern Tanganyika Territory, south to eastern Cape Province.

**Range in Eastern Africa:** delete South-western Uganda.

add:

*Ploceus cucullatus graueri* Hartert, Bull. B.O.C. 29, p. 24, 1911: Usambara, north of Lake Tanganyika.

**Distinguishing characters:** More cinnamon colour on breast. Wing 85 to 88 mm.

**General distribution:** Eastern Belgian Congo to Urundi.

**Range in Eastern Africa:** Urundi.

**Habits:** As for other races.

**1335** *Symplectes bicolor*. It seems now quite probable that *S. bicolor* and *S. kersteni* are different species. The English name for the latter would be the *East African Dark-backed Weaver*.

p. 917. Add *Symplectes bicolor kigomaënsis* Grant & Praed.

*Symplectes bicolor kigomaënsis* C. Grant & Mackworth-Praed, Bull. B.O.C. 76, p. 33, 1956: Kazinga, near Kigoma, western Tanganyika Territory.

**Distinguishing characters:** Differs from the preceding races in the top of the head and nape being brownish black, contrasting with the mantle. Wing 81 to 95 mm.

**General distribution:** Southern Belgian Congo, Northern Rhodesia and Tanganyika Territory.

**Range in Eastern Africa:** Western Tanganyika Territory.

*Note: Symplectes bicolor amaurocephalus* (Cabanis) is confined to Angola.

### 1357A BLUE-BILLED MALIMBE. *MALIMBUS NITENS* (Gray).

*Malimbus nitens microrhynchus* Reichenow, O.M., p. 160, 1908: Cenda River, Ituri district.

**Distinguishing characters:** A wholly black bird with a red patch on the lower throat and chest and a blue bill. The female is similar but somewhat duller black. Eye pinkish brown to red; bill bluish horn to bright blue; legs and feet dark greyish slate. Wing 79 to 93 mm.

**General distribution:** North-eastern Belgian Congo to western Uganda.

**Range in Eastern Africa:** Bwamba forest, western Uganda.

**Habits:** Birds of thick forest, usually shy except when breeding, and easily overlooked.

**Nest and Eggs:** The nominate race appears to nest almost invariably over a forest pool or water of some sort. Nest suspended from branches or palm leaves, of grass or fibre, rather untidy and roughly made. Eggs two, pale stone or greenish grey, blotched or freckled with brown and mauve; about 25 × 16 mm. (nominate race).

**Recorded breeding:** Bwamba forest, March onwards.

**Food:** Mainly insects.

**Call:** A long-drawn 'ze-e-e-e-e' and a chirping call.

**Distribution of other races of the species:** West Africa, the nominate race being described from Sierra Leone.

### 1367 *Euplectes capensis xanthomelas* Rüpp.

The distribution of this race should stop at northern Nyasaland and Tanganyika Territory. Birds from south of that are *Euplectes capensis zambesiensis* Roberts, Ann. Trans. Mus. 8, p. 266, 1922: Bowi, Portuguese East Africa, in which the females and non-breeding males are less heavily streaked and have paler yellow rumps.

### 1387A PALE-FRONTED NEGRO-FINCH. *NIGRITA LUTEIFRONS* Verreaux.

*Nigrita luteifrons luteifrons* J. & E. Verreaux, Rev. Mag. Zool. Ser. 2, vol. 3, p. 20, 1851: Gabon.

**Distinguishing characters:** A grey-backed bird with a pale yellowish buff forehead. Lores, wings, tail and whole of underside black. Female similar but duller and greyer. Eye pale grey; bill black; legs and feet flesh colour. Wing 55 to 61 mm. The immature dress is lead colour with no pale forehead, and dark grey washed with brownish buff below.

# ADDENDA

**General distribution:** Southern Nigeria to northern Angola and eastwards to the Semliki Valley and western Uganda.

**Range in Eastern Africa:** Bwamba Forest, western Uganda.

**Habits:** Little known active birds feeding in the tops of trees in forest and behaving very like Sunbirds.

**Nest and Eggs:** Undescribed.

**Recorded breeding:** No records.

**Food:** Largely insects.

**Call:** Nothing on record.

**1396A VINCENT'S SEED-CRACKER.** *PIRENESTES VINCENTI* (Benson).

*Pirenestes vincenti* C. W. Benson, Bull. B.O.C. 75, p. 110, 1955: 64 miles west-north-west of Mocuba, Portuguese East Africa.

**Distinguishing characters:** This is a name proposed with some hesitation by Mr. Benson for the large-billed form of *Pirenestes* which occurs with *P. minor*, and is only distinguishable from it in that respect. Width of lower mandible 14 to 15 mm.

**Range in Eastern Africa:** Southern Tanganyika Territory, Nyasaland and Portuguese East Africa.

**Habits:** Indistinguishable from those of *P. minor* and more work must be done on this group before a decision can be made as to the real standing of this bird.

**1431A** *Uræginthus bengalus kigomaënsis* Grant & Praed.

*Uræginthus bengalus kigomaënsis* C. Grant & Mackworth-Praed, Bull. B.O.C. 76, p. 34, 1956: Kigoma, western Tanganyika Territory.

**Distinguishing characters:** Differs from the Shoa race in being darker earth-brown above, and from the Ugogo race in being brighter blue below; ear-coverts claret red. Wing about 55 mm.

**Range in Eastern Africa:** Western to north-eastern Tanganyika Territory.

*Note:* *Uræginthus bengalus katangæ* Vincent does not occur in Eastern Africa.

# INDEX

*An index to groups of birds under their English names is given after the Introduction on p. vii*

## A

abayensis, Lagonosticta senegala, 1053
—, Ploceus ocularius, 1052
—, Sylvietta whytii, 426
abdominalis, Eremomela flaviventris, 432
—, — icteropygialis, 432
aberdare, Cisticola robusta, 490
aberrans njombe, Cisticola, 466
abessinica, Camaroptera brevicaudata, 443
—, — griseoviridis, 443
abietinus, Phylloscopus collybita, 384
abyssinica bannermani, Hirundo, 539
—, Eremomela canescens, 434
—, — elegans, 434
—, Hirundo abyssinica, 537
—, Turdoides leucocephala, 84
— unitatis, Hirundo, 538
abyssinicus atriceps, Pseudoalcippe, 105
—, Bradypterus babœcala, 369
— chyulu, Pseudoalcippe, 106
— omoensis, Zosterops, 735
—, Ploceus cucullatus, 889
—, Pseudoalcippe abyssinicus, 104, 106
— socotranus, Zosterops, 735
—, Sporopipes frontalis, 1052
— stictigula, Pseudoalcippe, 106
— stierlingi, Pseudoalcippe, 105
—, Turdus olivaceus, 243, 332, 333
—, Zosterops abyssinicus, 734
acaciæ, Argya fulva, 93
acholiensis, Sheppardia æquatorialis, 307
acik, Chalcomitra senegalensis, 796
acredula, Phylloscopus trochilus, 383, 517
Acrocephalus arundinaceus arundinaceus, 363
— — zarudnyi, 364
— bæticatus cinnamomeus, 366
— — suahelicus, 367
— griseldis, 364
— palustris, 366
— schœnobænus, 368
— scirpaceus crassirostris, 517

Acrocephalus scirpaceus fuscus, 365
— — griseus, 517
— — scirpaceus, 365
adsimilis, Dicrurus adsimilis, 564, 567
— divaricatus, Dicrurus, 567
— jubaensis, Dicrurus, 565
adustus chyulu, Alseonax, 167
— fülleborni, Alseonax, 166
— marsabit, Alseonax, 167
— minimus, Alseonax, 164, 225
— murinus, Alseonax, 165
— pumilus, Alseonax, 165, 226
— roehli, Alseonax, 166
— subadustus, Alseonax, 166
Ægithalus *see* Anthoscopus
ægra, Œnanthe leucopyga, 265
ænea codringtoni, Hypochera, 1040
æneigularis, Nectarinia famosa, 820
æneocephalus, Lamprotornis purpuropterus, 703
æquatorialis acholiensis, Sheppardia, 307
—, Apalis flavida, 518
—, Chalcomitra senegalensis, 820
—, Megabyas flammulatus, 196
—, Schœnicola brevirostris, 519
—, Sheppardia æquatorialis, 306
—, Urolestes melanoleucus, 605
æthiopica, Amblyospiza albifrons, 1052
—, Hirundo, 528
—, Platysteira cyanea, 209
—, Quelea quelea, 945, 1052
æthiopicus ambiguus, Laniarius, 614, 643
— erlangeri, Laniarius, 614, 643
—, Laniarius æthiopicus, 611
—, major, Laniarius, 612
— mossambicus, Laniarius, 613
— sublacteus, Laniarius, 613, 643
æthiops cryptoleuca, Myrmecocichla, 285
— sudanensis, Myrmecocichla, 286
Aëthocorys personata intensa, 45
— — personata, 45
— — yavelloensis, 45
afer barakæ, Parus, 660
— brevialatus, Nilaus, 642
— erythreæ, Nilaus, 642
— grauerti, Cinnyris, 781
— griseiventris, Parus, 647, 660

# INDEX

afer hilgerti, Nilaus, 642
— massaicus, Nilaus, 586
— minor, Nilaus, 586
—, Nilaus afer, 584, 642
— parvirostris, Parus, 660
—, Ptilostomus, 679
— stuhlmanni, Cinnyris, 780
— thruppi, Parus, 648, 660
affinis, Apalis porphyrolæma, 518
—, Corvinella corvina, 604, 643
—, Dryoscopus affinis, 618
—, Emberiza, 1099
—, Prinia subflava, 506
— senegalensis, Dryoscopus, 619
afra, Euplectes afra, 957
— ladoensis, Euplectes, 959
— niassensis, Euplectes, 959
—, Pytilia, 1006
— stricta, Euplectes, 958, 1052
africana athi, Mirafra, 11, 46
— dohertyi, Mirafra, 46
— zuluensis, Mirafra, 12
— harterti, Mirafra, 13
— kurræ, Mirafra, 13
—, Luscinia megarhynchos, 331
— nigrescens, Mirafra, 11
— ruwenzoria, Mirafra, 46
— tropicalis, Mirafra, 11, 46
africanoides alopex, Mirafra, 17
— intercedens, Mirafra, 17, 46
— longonotensis, Mirafra, 46
— macdonaldi, Mirafra, 46
africanus, Buphagus africanus, 720
Agrobates galactotes familiaris, 357
— — galactotes, 356
— — hamertoni, 358
— — minor, 357
— — syriacus, 357
aguimp vidua, Motacilla, 49
Aidemosyne see Euodice
airensis, Cercomela melanura, 272
akeleyæ, Alethe poliocephala, 311
akeleyi, Ammomanes deserti, 26
Alæmon alaudipes desertorum, 23
— — meridionalis, 24
— — hamertoni altera, 25
— — hamertoni, 25
— — tertia, 25
alaudipes desertorum, Alæmon, 23
— meridionalis, Alæmon, 24
Alaudula somalica, 41
alba forwoodi, Motacilla, 48
—, Motacilla alba, 48
alberti, Prionops, 572
albicapilla omoensis, Cossypha, 298
albicapillus, Spreo, 716
albicauda kivuensis, Elminia, 215
—, Mirafra, 7
albiceps, Psalidoprocne, 553
albicollis, Corvultur, 676
—, Muscicapa, 164

albicrissalis, Bradypterus alfredi, 372
albifrons æthiopica, Amblyospiza, 1052
—, Amblyospiza albifrons, 939
— clericalis, Pentholæa, 278
—, Eremopterix nigriceps, 36
— melanota, Amblyospiza, 940, 1052
— montana, Amblyospiza, 941
— pachyrhyncha, Pentholæa, 279
—, Pentholæa albifrons, 278
—, Serinus burtoni, 1078
— tanganjicæ, Serinus, 1080
— unicolor, Amblyospiza, 941
albigula, Phyllastrephus debilis, 127
— shimbanus, Phyllastrephus, 152
albigularis, Bessonornis, 333
—, Hirundo albigularis, 527
—, Phyllastrephus, 131
— porotoensis, Bessonornis, 333
albinucha maxwelli, Melanoploceus, 936
albipectus, Malacocincla, 102, 106
—, Turdinus, 102
albirostris, Bubalornis, 858
— nyanzæ, Bubalornis, 1052
—, Pilorhinus, 712
albiscapulata, Saxicola, 281
—, Thamnolæa cinnamomeiventris, 281
albistriata, Sylvia cantillans, 354
albitorquata, Amadina fasciata, 1054
albiventris, Camaroptera brevicaudata, 519
—, Cinnyris venustus, 778
— curtus, Parus, 660
—, Parus, 652, 660
albofasciata, Saxicola torquata, 287
albonotatus, Coliuspasser albonotatus, 965
— eques, Coliuspasser, 966
—, Trochocercus albonotatus, 217
alboplagatus, Laniarius, 642
albus, Corvus, 673
Alethe anomala gurué, 304
— castanea woosnami, 308
— choloensis choloensis, 309
— — namuli, 310
— fülleborni fülleborni, 312
— — usambaræ, 313
— lowei, 314
— macclounii njombe, 333
— montana, 313
— poliocephala akeleyæ, 311
— — carruthersi, 310
— — kungwensis, 311
— — ufipæ, 311
— poliophrys, 312
— poliothorax, 103
alexanderi, Amadina fasciata, 996
—, Eremomela flaviventris, 432
—, — icteropygialis, 432

# INDEX

alexinæ, Schœnicola brevirostris, 395, 519
alfredi albicrissalis, Bradypterus, 372
—, Cyanomitra olivacea, 804
— itoculo, Phyllastrephus, 152
— kungwensis, Bradypterus, 372
—, Phyllastrephus, 126
— sjostedti, Bradypterus, 518
alienus, Hyphanturgus, 921
alinæ, Cyanomitra alinæ, 800
alius, Malaconotus, 638
alopex, Mirafra africanoides, 17
alpestris, Turdus torquatus, 239
alpinus, Seicercus umbrovirens, 390
Alseonax adustus chyulu, 167
— — fülleborni, 166
— — marsabit, 167
— — minimus, 164, 225
— — murinus, 165
— — pumilus, 165, 226
— — roehli, 166
— — subadustus, 166
— aquaticus aquaticus, 168
— — infulatus, 168, 225
— — ruandæ, 225
— cassini, 169
— cinereus cinereus, 170
— — cinereolus, 171, 225, 226
— — kikuyuensis, 225
— flavipes, 225
— gambagæ, 162
— griseigularis griseigularis, 171
— minimus chyulu, 167
— — djamdjamensis, 225
— — interpositus, 226
— — marsabit, 167
— murinus rœhli, 166
— pumila, 165
— seth-smithi, 169, 225, 226
altera, Alæmon hamertoni, 25
alticola brunneiceps, Apalis, 401
altirostris, Galerida cristata, 29
altumi, Bradypterus, 518
altus, Artisornis metopias, 518
Amadina fasciata albitorquata, 1054
— — alexanderi, 996
— — fasciata, 995
— — meridionalis, 997, 1054
amani, Dioptrornis fischeri, 226
—, Oriolus chlorocephalus, 669
—, Phyllastrephus orostruthus, 133
Amauresthes fringilloides, 978
amaurocephalus, Symplectes bicolor, 915
amauropteryx, Hypochera, 1042
amauroura, Melocichla mentalis, 514, 519
ambigua, Cisticola robusta, 489
ambiguus, Laniarius æthiopicus, 614, 643

Amblyospiza albifrons æthiopica, 1052
— — albifrons, 939
— — melanota, 940, 1052
— — montana, 941
— — unicolor, 941
ameliæ wintoni, Macronyx, 80
amethystina doggetti, Chalcomitra, 792
— kalckreuthi, Chalcomitra, 792
— kirkii, Chalcomitra, 791
amethystinus, Lamprocolius purpureus, 722
Ammomanes cinctura kinneari, 25
— deserti akeleyi, 26
— — assabensis, 27
— — erythrochroa, 27
— — samharensis, 27
ampelinus, Hypocolius, 645
Amydrus see Onychognathus
Anaplectes melanotis jubaensis, 944
— — melanotis, 942
— — rubriceps, 943
anchietæ, Anthreptes, 815
—, Bocagia minuta, 629
andaryæ, Chlorophoneus, 643
anderseni, Cinnyris regius, 790
anderssoni, Cinnyris talatala, 776
—, Tephrocorys cinerea, 46
—, Zosterops senegalensis, 726, 736
Andropadus curvirostris curvirostris, 147
— gracilis gracilis, 145, 152
— — kavirondensis, 145
— importunus fricki, 147
— — hypoxanthus, 146
— — insularis, 145, 151
— — insularis kilimandjaricus, 151
— — somaliensis, 151
angolensis, Anthreptes longuemarei, 813
— arcticincta, Hirundo, 527
—, Chalcomitra, 820
— hilgerti, Serinus, 1076
—, Hirundo angolensis, 526
— kakamegæ, Cinnyris, 794
— kungwensis, Dryoscopus, 622
— mandensis, Dryoscopus, 622
—, Monticola, 258
— niassensis, Uræginthus, 1033
—, Poliospiza, 1086
— somereni, Serinus, 1075
anguitimens rüppelli, Eurocephalus, 583, 642
angusticauda, Cisticola, 500
angustus, Neocichla gutturalis, 689
ankole, Cisticola brachyptera, 497
annæ, Anthus novæseelandiæ, 68
anomala, Dessonornis anomala, 302
— grotei, Dessonornis, 303, 333
— gurué, Dessonornis, 304

# INDEX

anomala macclounii, Dessonornis, 303, 333
— mbuluensis, Dessonornis, 304
Anomalospiza imberbis imberbis, 1000
— — nyassæ, 1054
— — rendalli, 1002, 1054
ansorgei, Euplectes gierowii, 955
—, Nesocharis, 1033
— nilotica, Calamocichla, 379
Anthoscopus caroli rhodesiæ, 658
— — robertsi, 657
— — roccatii, 656
— — rothschildi, 657
— — sharpei, 656
— — sylviella, 655
— — taruensis, 657
— musculus guasso, 659
— — musculus, 658
— parvulus parvulus, 659
— punctifrons, 659
Anthreptes anchietæ, 815
— axillaris, 818
— collaris djamdjamensis, 821
— — elachior, 808
— — garguensis, 809, 820
— — hypodilus, 807
— — jubaensis, 809, 821
— — ugandæ, 820
— — zambesiana, 807
— longuemarei angolensis, 813
— — haussarum, 811
— — neglectus, 814
— — neumanni, 814
— — nyassæ, 813
— — orientalis, 812, 821
— pallidigaster, 816
— rectirostris tephrolæma, 810, 820
— reichenowi yokanæ, 817
— rubritorques, 811
— seimundi minor, 816
— tephrolæma elgonensis, 820
Anthus brachyurus leggei, 72
— caffer australoabyssinicus, 73
— — blayneyi, 72
— campestris campestris, 61, 81
— — griseus, 81
— cervinus, 73
— gouldi turneri, 81
— latistriatus, 69
— leucophrys bohndorffi, 65
— — omoensis, 64
— — zenkeri, 65, 81
— lineiventris, 74
— melindæ, 71
— nicholsoni chyuluensis, 81
— novæseelandiæ, 81
— novæseelandiæ annæ, 68
— — cinnamomeus, 67
— — lacuum, 67
— — lichenya, 69

Anthus novæseelandiæ lynesi, 68
— richardi, 81
— similis, 81
— — hararensis, 62, 81
— — jebelmarræ, 64
— — neumannianus, 81
— — nivescens, 62
— — nyassæ, 63
— — sokotræ, 63
— sokokensis, 71
— sordidus, 81
— — longirostris, 81
— trivialis trivialis, 70
— vaalensis goodsoni, 66
— — saphiroi, 66
antinorii, Cisticola lateralis, 471
—, Lanius antinorii, 642
— mauritii, Lanius, 642
—, Psalidoprocne, 552
Apalis alticola brunneiceps, 401
— argentea, 419
— bamendæ bensoni, 519
— — chapini, 415
— — strausæ, 416, 519
— binotata binotata, 411
— — personata, 411
— caniceps caniceps, 409
— — neglecta, 410
— — tenerrima, 410
— chariessa, 413, 519
— — macphersoni, 519
— chirindensis lightoni, 404
— cinerea cinerea, 400
— flavida æquatorialis, 518
— — flavocincta, 407, 518
— — golzi, 408
— — malensis, 409
— — viridiceps, 408
— flavigularis flavigularis, 398
— — griseiceps, 398, 519
— — lynesi, 399
— — uluguru, 399
— griseiceps chyulu, 519
— jacksoni jacksoni, 412
— karamojæ, 418
— melanocephala ellinoræ, 404
— — fuliginosa, 303
— — lightoni, 404
— — melanocephala, 402
— — moschi, 403
— — muhuluensis, 405, 519
— — nigrodorsalis, 402
— — songeaensis, 519
— — tenebricosa, 403
— moreaui, 418
— — sousæ, 418
— murina fuscigularis, 398
— — murina, 396
— — whitei, 397
— — youngi, 397
— nigrescens, 405

# INDEX

Apalis nigriceps collaris, 413
— porphyrolæma affinis, 518
— — porphyrolæma, 414, 518
— — vulcanorum, 518
— pulchra pulchra, 406
— ruficeps, 518
— rufifrons rufidorsalis, 417
— — rufifrons, 416
— — smithii, 417
— rufogularis denti, 400
— ruwenzorii, 407
— thoracica injectiva, 519
— — youngi, 397
approximans, Malaconotus poliocephalus, 643
aquaticus, Alseonax aquaticus, 168
— infulatus, Alseonax, 168, 225
— ruandæ, Alseonax, 225
arabica, Ptyonoprogne obsoleta, 546
arabicus, Cinnyricinclus leucogaster, 691
arboreus, Passer domesticus, 870
archeri, Cossypha, 294
—, Eremomela griseoflava, 518
—, Heteromirafra ruddi, 22
arcticincta, Hirundo angolensis, 527
ardens, Coliuspasser ardens, 968, 1052
— concolor, Coliuspasser, 969
— teitensis, Coliuspasser, 1052
argentata, Thamnolæa, 332
argentea, Apalis, 419
—, Cisticola natalensis, 493
Argya amauroura, 514
— aylmeri aylmeri, 96
— — boranensis, 97
— — keniana, 96
— — loveridgei, 96
— — mentalis, 97
— fulva acaciæ, 93
— rubiginosa emini, 95
— — heuglini, 95
— — rubiginosa, 94
— — sharpii, 95
aridula, Cisticola aridula, 457
— lavendulæ, Cisticola, 458
— tanganyika, Cisticola, 458
Arizelocichla masukuensis kakamegæ, 141
— — masukuensis, 140
— — roehli, 140
— milanjensis chyulu, 152
— — milanjensis, 138
— — striifacies, 139, 152
— nigriceps chlorigula, 138
— — fusciceps, 136
— — kungwensis, 138
— — neumanni, 137
— — nigriceps, 136
— — percivali, 151
— — usambaræ, 137, 151

Arizelocichla tephrolæma kikuyuensis, 135
armena, Tchagra senegala, 643
arnaudi australoabyssinicus, Pseudonigrita, 868
— dorsalis, Pseudonigrita, 868
—, Pseudonigrita arnaudi, 867
arnotti collaris, Thamnolæa, 332
— leucolæmia, Thamnolæa, 332
—, Thamnolæa arnotti, 283, 332
arquata, Cichladusa, 314
arsinoe, Pycnonotus barbatus, 115
— somaliensis, Pycnonotus, 116
Artisornis metopias, 419, 518
— — altus, 518
Artomyias fuliginosa minuscula, 190
arturi, Nectarinia kilimensis, 756
arundinaceus, Acrocephalus arundinaceus, 363
— zarudnyi, Acrocephalus, 364
aschani, Camaroptera brevicaudata, 519
—, Hirundo senegalensis, 554
asmaraensis, Calandrella cinerea, 43
assabensis, Ammomanes deserti, 27
astrild cavendishi, Estrilda, 1020, 1053
— macmillani, Estrilda, 1053
— massaica, Estrilda, 1053
— minor, Estrilda, 1019
— nyanzæ, Estrilda, 1021
— peasei, Estrilda, 1021, 1053
athensis, Calandrella rufescens, 41
athi, Mirafra africana, 11, 46
Atimastillas flavicollis pallidigula, 151
— flavicollis shelleyi, 151
atricapilla dammholzi, Sylvia, 351
— graueri, Estrilda, 1028
— kandti, Estrilda, 1053
—, Sylvia atricapilla, 350
atricauda, Melocichla mentalis, 515
atriceps, Pseudoalcippe abyssinicus, 105
atricollis dorsostriata, Ortygospiza, 997, 1054
— fuscocrissa, Ortygospiza, 998
— mulleri, Ortygospiza, 998
—, Ortygospiza atricollis, 997
atrocærulea, Hirundo atrocærulea, 530
— lynesi, Hirundo, 531
atrogularis hilgerti, Serinus, 1076
—, Œnanthe deserti, 332
— reichenowi, Serinus, 1074
—, Serinus, 1086
— somereni, Serinus, 1075
— xanthopygius, Serinus, 1074
aucheri, Lanius elegans, 589
aucupum kaduliensis, Steganura, 1054

# INDEX

aucupum obtusa, Steganura, 1051
aurantiigula, Macronyx, 79, 81
— subocularis, Macronyx, 81
aurantius rex, Ploceus, 914
auratus notatus, Oriolus, 664
—, Oriolus auratus, 663
aureoflavus pallidiceps, Ploceus, 1053
— reicherti, Ploceus, 1053
—, Xanthophilus subaureus, 922, 1053
aurifrons, Serinus mozambicus, 1086
—, Zosterops senegalensis, 736
Auripasser euchlorus, 882
— luteus, 881
aussæ, Cercomela melanura, 273
australis congener, Tchagra, 643
—, Cryptospiza reichenovii, 992, 1053, 1054
— emini, Tchagra, 626
—, Hyliota australis, 199, 226
— inornata, Hyliota, 226
— littoralis, Tchagra, 643
— minor, Tchagra, 626, 643
—, Stelgidocichla latirostris, 152
— usambaræ, Hyliota, 200
australoabyssinica, Mirafra pœcilosterna, 20
—, Anthus caffer, 73
—, Pseudonigrita arnaudi, 868
—, Turdus tephronotus, 333
—, Zosterops senegalensis, 736
awemba, Cisticola robusta, 491
axillaris, Anthreptes, 818
—, Coliuspasser axillaris, 960
— phœniceus, Coliuspasser, 961
—, Ploceus badius, 909
—, Saxicola torquata, 288
—, traversii, Coliuspasser, 962
— zanzibarica, Coliuspasser, 962
aylmeri, Argya aylmeri, 96
— boranensis, Argya, 97
— keniana, Argya, 96
— loveridgei, Argya, 96
— mentalis, Argya, 97
ayresii, Cisticola ayresii, 459
— entebbe, Cisticola, 460
— imatong, Cisticola, 461
— mauensis, Cisticola, 460

## B

babœcala abyssinicus, Bradypterus, 369
—, Bradypterus babœcala, 517
— centralis, Bradypterus, 369
— elgonensis, Bradypterus, 369
— moreaui, Bradypterus, 370
— sudanensis, Bradypterus, 370
bactrianus, Passer domesticus, 871
badiceps, Eremomela badiceps, 437

badiceps latukæ, Eremomela, 438
— turneri, Eremomela, 437
badius axillaris, Ploceus, 909
—, Ploceus badius, 908
Bæopogon indicator chlorosaturata, 121
bæticatus cinnamomeus, Acrocephalus, 366
— suahelicus, Acrocephalus, 367
bafirawari, Bradornis, 180
baglafecht eremobius, Ploceus, 902
—, Ploceus baglafecht, 901
bailundensis, Lamprocolius splendidus, 722
bairdii melanops, Prinia, 511
balfouri, Cyanomitra, 806
bamendæ bensoni, Apalis, 519
— chapini, Apalis, 415
— strausæ, Apalis, 416, 519
bangsi, Sheppardia cyornithopsis, 333
bannermani, Hirundo abyssinica, 539
—, Plocepasser superciliosus, 865
baraka, Sylvietta virens, 429
—, Turdus olivaceus, 244
barakæ, Illadopsis rufipennis, 106
—, Parus afer, 660
—, Sathrocercus lopezi, 377
barbarus mufumbiri, Laniarius, 607
barbata, Erythropygia, 319
— greenwayi, Erythropygia, 321
— rovumæ, Erythropygia, 333
barbatus arsinoe Pycnonotus, 115
— schoanus, Pycnonotus, 115
—, Serinus mozambicus, 1062
— somaliensis, Pycnonotus, 116
barbouri, Anthreptes orientalis, 821
barbozæ, Hyliota flavigaster, 198
bartteloti, Cossypha cyanocampter, 296
batesi, Chlorophoneus multicolor, 629, 643
Bathmocercus rufus jacksoni, 517
Batis capensis dimorpha, 201
— — mixta, 200
— — reichenowi, 201
— diops, 202
— fratrum, 207
— minor erlangeri, 206, 226
— — minor, 206
— — nyanzæ, 226
— — suahelica, 207
— molitor molitor, 202, 226
— — puella, 226
— — soror, 203
— mystica, 226
— orientalis bella, 226
— — chadensis, 205
— — lynesi, 205
— — minor, 206

# INDEX

Batis orientalis orientalis, 204, 226
— perkeo, 205
— puella soror, 203
beema, Budytes flavus, 53
belcheri, Psophocichla guttata, 333
belli, Cisticola cantans, 481
—, Eremomela icteropygialis, 432
—, Pytilia melba, 1053
bengalus brunneigularis, Uræginthus, 1035
— katangæ, Uræginthus, 1036
— schoanus, Uræginthus, 1035, 1053
— ugandæ, Uræginthus, 1053
— ugogoensis, Uræginthus, 1036
—, Uræginthus bengalus, 1034
bensoni, Apalis bamendæ, 519
—, Cinnyris mediocris, 784
—, Phyllastrephus terrestris, 123
—, Stilbopsar kenricki, 715
bertrandi, Chlorophoneus rubiginosus, 631
—, Xanthoploceus, 934
Bessonornis albigularis, 333
— — porotoensis, 333
beveni, Nectarinia nectarinioides, 820
Bias feminina, 197
— musicus changamwensis, 197
— — femininus, 197
bicolor amaurocephalus, Symplectes, 915
— brunnescens, Nigrita, 984, 1053
— kersteni, Symplectes, 915
— mentalis, Symplectes, 916
— saturatior, Nigrita, 1053
—, Speculipastor, 693
— stictifrons, Symplectes, 916
— stigmatophorus, Spermestes, 1052
bifasciatus microrhynchus, Cinnyris, 768
— tsavoensis, Cinnyris, 770
bimaculata, Melanocorypha bimaculata, 39
— rufescens, Melanocorypha, 39
bineschensis, Cinnyris chloropygius, 787
binotata, Apalis binotata, 411
— personata, Apalis, 411
bivittatus, Trochocercus cyanomelas, 215
blanchoti catharoxanthus, Malaconotus, 637
—, Malaconotus blanchoti, 636, 643
blanfordi, Psalidoprocne pristoptera, 552
—, Sylvia leucomelæna, 352
—, Tephrocorys, 44
blayneyi, Anthus caffer, 72
Bleda alfredi, 126
— eximia ugandæ, 117
— syndactyla woosnami, 117
blythii, Onychognathus, 710
blythi, Sylvia curruca, 348
bocagei jacksoni, Chlorophoneus, 633
Bocagia minuta anchietæ, 629
— — minuta, 628
bodessa, Cisticola chiniana, 469
—, — subruficapilla, 469
boehmi, Dinemellia dinemelli, 861
bogdanowi, Lanius, 603
böhmi, Eurocephalus rüppelli, 642
—, Lanius excubitorius, 592, 642
— marsabit, Parisoma, 173
—, Myopornis, 188
—, Parisoma böhmi, 172
— somalicum, Parisoma, 173
bohndorffi, Anthus leucophrys, 65
—, Ploceus cucullatus, 890, 1052
bojeri, Ploceus, 911
bonelli orientalis, Phylloscopus, 385
boranensis, Argya aylmeri, 97
borbonica, madagascariensis, Phedina, 540
borealis, Cryptospiza salvadorii, 1053
borin, Sylvia, 350
bororensis, Camaroptera brachyura, 439
bottæ frenata, Œnanthe, 269
bouvieri, Cinnyris, 774
bowdleri, Bradornis pallidus, 178, 225
brachydactila, Calandrella brachydactila, 39
— hermonensis, Calandrella, 40
— longipennis, Calandrella, 40
brachydactyla, Carpospiza, 887
brachyptera ankole, Cisticola, 497
—, Cisticola brachyptera, 495
— hypoxantha, Cisticola, 496
— isabellina, Cisticola, 496
— katonæ, Cisticola, 496
— kericho, Cisticola, 498
— reichenowi, Cisticola, 497
— zedlitzi, Cisticola, 497
brachypterus, Bradypterus brachypterus, 517
— centralis, Bradypterus, 369
— fraterculus, Bradypterus, 518
— moreaui, Bradypterus, 370
brachyrhynchus lætior, Oriolus, 667
brachyura bororensis, Camaroptera, 439
— dilutior, Sylvietta, 424
— fuggles - couchmani, Camaroptera, 440
— hilgerti, Sylvietta, 518
— leucopsis, Sylvietta, 424, 518
— micrura, Sylvietta, 518
— pileata, Camaroptera, 439
—, Sylvietta brachyura, 423, 518
brachyurus leggei, Anthus, 72

# INDEX

bractiatus, Cinnyris chalybeus, 783, 820
Bradornis bafirawari, 180
— griseus erlangeri, 225
— — neumanni, 178
— — ukamba, 226
— microrhynchus, 179, 226
— pallidus bowdleri, 178, 225
— — chyuluensis, 226
— — granti, 225
— — griseus, 178, 225, 226
— — leucosoma, 226
— — neumanni, 178, 225
— — pallidus, 177
— — sharpei, 225
— — suahelicus, 225
— — subalaris, 177, 226
— pumilus, 179, 225
— taruensis, 225
Bradypterus alfredi albicrissalis, 372
— — kungwensis, 372
— — sjöstedti, 518
— altumi, 518
— babœcala abyssinicus, 369
— — babœcala, 517
— — centralis, 369
— — elgonensis, 369
— — moreaui, 370
— — sudanensis, 370
— brachypterus brachypterus, 517
— — centralis, 369
— — fraterculus, 518
— — moreaui, 370
— carpalis, 371, 518
— graueri, 371
— roehli, 518
— usambaræ, 376
— yokanæ, 518
brevialatus, Nilaus afer, 643
brevicaudata abessinica, Camaroptera, 443
— albiventris, Camaroptera, 519
— aschani, Camaroptera, 519
—, Camaroptera brevicaudata, 441, 519
— erlangeri, Camaroptera, 444, 519
— griseigula, Camaroptera, 443
— noomei, Camaroptera, 518
— sharpei, Camaroptera, 444, 518
— tincta, Camaroptera, 442
brevirostris æquatorialis, Schœnicola, 519
— alexinæ, Schœnicola, 395, 519
— chyulu, Schœnicola, 519
—, Schœnicola brevirostris, 394
brunneiceps, Apalis alticola, 401
—, Erythropygia zambesiana, 317
—, Lagonosticta senegala, 1012, 1053
— ruberrima, Lagonosticta, 1014

brunneigularis, Uræginthus bengalus, 1035
brunnescens cinnamomea, Cisticola, 462
—, Cisticola brunnescens, 461
— hindii, Cisticola, 462
— nakuruensis, Cisticola, 463
—, Nigrita bicolor, 984, 1053
—, Plocepasser superciliosus, 865
— wambera, Cisticola, 463
Bubalornis albirostris, 858
— — nyanzæ, 1052
— niger intermedius, 860, 1052
Bucanetes githagineus githagineus, 1059
buchanani, Serinus, 1068
bucolica, Heliocorys modesta, 32
budongoensis, Ploceus emini, 1052
—, Seicercus, 391
Budytes feldegg, 56, 58
— flavus beema, 53
— — campestris, 58
— — dombrowskii, 53
— — flavus, 53
— — melanogriseus, 58
— — rayi, 58
— leucocephalus, 57
— luteus flavissimus, 55, 58
— — luteus, 54, 58
— perconfusus, 54
— superciliaris, 57
— thunbergi cinereocapillus, 56
— — thunbergi, 55
Buphagus africanus africanus, 720
— erythrorhynchus caffer, 721
— — erythrorhynchus, 721
burigi, Lanius souzæ, 599
burtoni albifrons, Serinus, 1078
— gurneti, Serinus, 1080
— kilimensis, Serinus, 1079
— melanochrous, Serinus, 1079
— tanganjicæ, Serinus, 1080
buryi, Lanius elegans, 589
buvuma, Cinnyris superbus, 820

## C

cabanisi, Emberiza, 1088
—, Lanius, 596
—, Phyllastrephus fischeri, 130, 151
—, Pseudonigrita, 868
— sucosus, Phyllastrephus, 151
cabanisii, Ploceus intermedius, 898
cæsia, Emberiza, 1093
— pura, Coracina, 561
caffer australoabyssinicus, Anthus, 73
— blayneyi, Anthus, 72
—, Buphagus erythrorhynchus, 721
caffra iolæma, Cossypha, 300
—, Saxicola torquata, 287, 333

# INDEX

Calamocichla ansorgei nilotica, 379
— foxi, 379
— gracilirostris nilotica, 379
— — parva, 378
— leptorhyncha jacksoni, 381
— — leptorhyncha, 380, 518
— — macrorhyncha, 381
— — tsanæ, 381
Calamonastes fasciolatus stierlingi, 392
— simplex simplex, 393
— — undosus, 394
Calamonastides gracilirostris, 382
— gracilirostris zuluensis, 518
— palustris, 518
Calandrella blanfordi blanfordi, 44
— brachydactila brachydactila, 39
— — hermonensis, 40
— — longipennis, 40
— cinerea asmaraensis, 43
— — cinerea, 46
— — erlangeri, 43, 46
— — saturatior, 42
— rufescens athensis, 41
— — megaensis, 42
— — somalica, 41
Calendula dunni, 28
caliginosa, Corvinella corvina, 605
calurus ndussumensis, Trichophorus, 116
Camaroptera axillaris, 818
— brachyura bororensis, 439
— — fuggles-couchmani, 440
— — pileata, 439
— brevicaudata albiventris, 519
— — abessinica, 443
— — aschani, 519
— — brevicaudata, 441, 519
— — erlangeri, 444, 519
— — griseigula, 443
— — noomei, 518
— — sharpei, 444, 518
— tincta, 442
— caniceps, 409
— chloronota toroensis, 441
— dorcadichroa, 389
— superciliaris ugandæ, 445
camburni, Xanthophilus xanthops, 1052
camerunensis, Hypochera chalybeata, 1045
Campephaga flava, 562
— hartlaubi, 562
— petiti, 558
— phœnicea, 558, 562
— quiscalina martini, 559
— — münzneri, 560
— sulphurata, 556, 562
— xanthornoides, 562
campestris, Anthus campestris, 61, 81
—, Budytes flavus, 58

campestris griseus, Anthus, 81
cana, Spermophaga ruficapilla, 988
candida, Mirafra, 46
—, Nigrita canicapilla, 983
canescens abyssinica, Eremomela, 434
— elegans, Eremomela, 434, 518
—, Eremomela canescens, 433, 518
canicapilla candida, Nigrita, 983
— diabolica, Nigrita, 983
—, Nigrita canicapilla, 982
— schistacea, Nigrita, 983
caniceps, Apalis caniceps, 409
— mentalis, Sigmodus, 572 `
— neglecta, Apalis, 410
—, Odontospiza, 981
— tenerrima, Apalis, 410
cantans belli, Cisticola, 481
—, Cisticola cantans, 479
— concolor, Cisticola, 480
—, Euodice malabarica, 979, 1052
— inornata, Euodice, 1052
— meridionalis, Euodice, 1053
— munzneri, Cisticola, 481
— pictipennis, Cisticola, 481
cantillans, albistriata, Sylvia, 354
— chadensis, Mirafra, 7
— marginata, Mirafra, 6, 46
capensis crassirostris, Euplectes, 1052
— dimorpha, Batis, 201
— kordofanensis, Corvus, 674
— mixta, Batis, 200
— reichenowi, Batis, 201
— suahelicus, Phyllastrephus, 122
— vincenti, Fringillaria, 1094
— wellsi, Motacilla, 51
— xanthomelas, Euplectes, 955, 1052
capistrata, Nesocharis, 1032
capistratus koliensis, Serinus, 1067
—, Serinus capistratus, 1066
capitalis dichrocephalus, Ploceus, 905
— dimidiatus, Ploceus, 904
cardinalis pallida, Quelea, 948
—, Quelea cardinalis, 947
— rhodesiæ, Quelea, 948
Carduelis citrinelloides citrinelloides, 1082
— — frontalis, 1083
— — hypostictus, 1083, 1086
— — kikuyuensis, 1084
— nigriceps, 1084
carlo, Prinia gracilis, 509
carnapi dilutior, Sylvietta, 424
caroli rhodesiæ, Anthoscopus, 658
— robertsi, Anthoscopus, 657
— roccatii, Anthoscopus, 656
— rothschildi, Anthoscopus, 657
— sharpei, Anthoscopus, 656
— sylviella, Anthoscopus, 655

# INDEX

caroli taruensis, Anthoscopus, 657
carpalis, Bradypterus, 371, 518
Carpospiza brachydactyla, 887
carruthersi, Alethe poliocephala, 310
—, Cisticola, 487
carunculata, Creatophora, 722
cassini, Alseonax, 169
castanea, Dyaphorophyia castanea, 211
— woosnami, Alethe, 308
castaneiceps, Ploceus, 913
castanops, Ploceus, 910
castanopsis, Heliolais, 519
castanopterus fulgens, Passer, 876
—, Passer castanopterus, 875
catharoxanthus, Malaconotus blanchoti, 637
cathemagmenus, Rhodophoneus cruentus, 639
catholeuca, Tchagra senegala, 643
caudatus, Lamprocolius, 702
cavei, Eremopterix signata, 46
—, Sathrocercus cinnamomeus, 374
cavendishi, Estrilda astrild, 1020, 1053
centralis, Bradypterus babœcala, 369
—, — brachypterus, 369
—, Chlorocichla flaviventris, 142, 152
—, Estrilda rhodopyga, 1023
—, Malimbus rubricollis, 938
—, Nectarinia famosa, 820
—, Pirenestes ostrinus, 1053
—, Quelea quelea, 1052
—, Turdus pelios, 242
Cercomela dubia, 276, 332
— familiaris falkensteini, 274, 332
— — gambagæ, 332
— — omoensis, 275
— — sennaarensis, 332
— melanura airensis, 272
— — aussæ, 273
— — lypura, 272
— scotocerca enigma, 332
— — furensis, 274
— — scotocerca, 273
— — spectatrix, 273
— — turkana, 274
Cercotrichas podobe podobe, 325
Certhilauda meridionalis, 24
— somalica, 23
cerviniventris, Phyllastrephus, 130
cervinus, Anthus, 73
chadensis, Batis orientalis, 205
—, Mirafra cantillans, 7
chagwensis, Stelgidillas gracilirostris, 143
chalcomelas, Cinnyris, 770
Chalcomitra amethystina doggetti, 792

Chalcomitra amethystina kalckreuthi, 792
— — kirkii, 791
— angolensis, 820
— hunteri, 797
— rubescens kakamegæ, 794
— — rubescens, 793, 820, 823
— senegalensis acik, 796
— — æquatorialis, 820
— — cruentata, 795
— — erythrinæ, 820
— — gutturalis, 794, 820
— — inæstimata, 820
— — lamperti, 797, 820
chalcurus emini, Lamprocolius, 695
chalybæus, Lamprocolius chalybæus, 694
— sycobius, Lamprocolius, 695
chalybea, Psalidoprocne, 550
chalybeata camerunensis, Hypochera, 1045
chalybeus bractiatus, Cinnyris, 783, 820
— gertrudis, Cinnyris, 820
— graueri, Cinnyris, 782
— intermedius, Cinnyris, 782, 820
— manoensis, Cinnyris, 820
— namwera, Cinnyris, 820
— zonarius, Cinnyris, 820
changamwensis, Bias musicus, 197
—, Cyanomitra olivacea, 803
—, Euplectes hordeacea, 1052
chapini, Apalis bamendæ, 415
—, Corvinella corvina, 643
chariessa, Apalis, 413, 519
— macphersoni, Apalis, 519
Charitillas gracilis kavirondensis, 145
— minor, 152
charmosyna, Estrilda erythronotos, 1031
— kiwanukæ, Estrilda, 1053
cheniana schillingsi, Mirafra, 46
chiniana bodessa, Cisticola, 469
— emendata, Cisticola, 471
— fischeri, Cisticola, 468
— heterophrys, Cisticola, 469
— humilis, Cisticola, 468
— mocuba, Cisticola, 519
— procera, Cisticola, 470, 519
— simplex, Cisticola, 467
— ukamba, Cisticola, 470
— victoria, Cisticola, 470
chiridensis lightoni, Apalis, 404
chlorigula, Arizelocichla nigriceps, 138
chloris gularis, Nicator, 641
—, Nicator gularis, 640
chlorocephalus amani, Oriolus, 669
—, Oriolus chlorocephalus, 669
Chlorocichla flaviventris centralis, 142, 152

# INDEX

Chlorocichla flaviventris chyuluensis, 152
— — occidentalis, 141
— lætissima, 142
chloronota, Sylvietta leucophrys, 430
— toroensis, Camaroptera, 441
Chloropeta natalensis massaica, 191
— — natalensis, 190
— similis, 192
Chloropetella holochlora, 192, 226
— holochlorus suahelica, 226
Chlorophoneus andaryæ, 643
— bocagei jacksoni, 633
— melamprosopus, 643
— multicolor batesi, 629, 643
— nigrescens, 643
— nigrifrons, 632, 643
— — conceptus, 643
— — elgeyuensis, 643
— — manningi, 643
— rubiginosus bertrandi, 631
— — münzneri, 643
— sulfureopectus fricki, 643
— — similis, 630, 643
chloropterus cyanogenys, Lamprocolius, 722
— elisabeth, Lamprocolius, 697
—, Lamprocolius chloropterus, 696, 722
chloropygius bineschensis, Cinnyris, 787
— orphogaster, Cinnyris, 787
chlorosaturata, Bæopogon indicator, 121
chocolatinus, Dioptrornis chocolatinus, 183
— reichenowi, Dioptrornis, 183
choloensis, Alethe choloensis, 309
— namuli, Alethe, 310
chrysopygia, Œnanthe, 266
chubbi, Cisticola, 477
—, Mandingoa nitidula, 1004
—, Sylvietta ruficapilla, 428
chuka, Geokichla gurneyi, 249
chyulu, Alseonax adustus, 167
—, Apalis griseiceps, 519
—, Arizelocichla milanjensis, 152
—, Cyanomitra olivacea, 821
—, Geokichla gurneyi, 249
—, Laniarius ferrugineus, 643
—, Melocichla mentalis, 519
—, Pseudoalcippe abyssinicus, 106
—, Pycnonotus tricolor, 152
—, Schœnicola brevirostris, 519
—, Seicercus umbrovirens, 518
—, Spinus citrinelloides, 1086
chyuluensis, Anthus nicholsoni, 81
—, Bradornis pallidus, 226
—, Chlorocichla flaviventris, 152
—, Cryptospiza salvadorii, 1054

chyuluensis, Phyllastrephus fischeri, 152
—, Sathrocercus cinnamomeus, 518
—, Turdus olivaceus, 333
—, Zosterops kikuyuensis, 732
Cichladusa arquata, 314
— guttata guttata, 315
— — rufipennis, 316
cincta erlangeri, Riparia, 544
— suahelica, Riparia, 544
cinctura kinneari, Ammomanes, 25
cineracea semenowi, Emberiza, 1093
cinerascens, Sporopipes frontalis, 888
—, Turdus libonyanus, 332
cinerea anderssoni, Tephrocorys, 46
—, Apalis cinerea, 400
— asmaraensis, Calandrella, 43
—, Calandrella cinerea, 46
—, Creatophora, 688, 722
— erlangeri, Calandrella, 43, 46
— fuertesi, Tephrocorys, 46
—, Motacilla cinerea, 52
— ruficeps, Tephrocorys, 46
— saturatior, Calandrella, 42
—, Turdoides plebeja, 85
cinereocapillus, Budytes thunbergi, 56
cinereola, Cisticola cinereola, 502
— schillingsi, Cisticola, 502
cinereolus, Alseonax cinereus, 171, 225, 226
cinereovinacea graueri, Lagonosticta, 1016, 1053
— rudolfi, Lagonosticta, 1053
cinereus, Alseonax cinereus, 170
— cinereolus, Alseonax, 171, 225, 226
— kikuyuensis, Alseonax, 225
—, Cisticola brunnescens, 462
cinnamomeiventris albiscapulata, Thamnolæa, 281
— subrufipennis, Thamnolæa, 280, 332
cinnamomeus, Acrocephalus bæticatus, 366
—, Anthus novæseelandiæ, 67
— cavei, Sathrocercus, 374
— chyuluensis, Sathrocercus, 518
— macdonaldi, Sathrocercus, 375
— nyassæ, Sathrocercus, 373
— rufoflavidus, Sathrocercus, 374, 518
—, Sathrocercus cinnamomeus, 373
— ufipæ, Sathrocercus, 375
Cinnyricinclus leucogaster arabicus, 691
— — friedmanni, 722
— — lauragrayæ, 722
— — leucogaster, 690, 722
— — verreauxi, 691, 722

Cinnyris afer graueri, 781
— — stuhlmanni, 780
— bifasciatus microrhynchus, 768
— — tsavoensis, 770
— bouvieri, 774
— chalcomelas, 770
— chalybeus bractiatus, 783, 820
— — gertrudis, 820
— — graueri, 782
— — intermedius, 782, 820
— — manoensis, 820
— — namwera, 820
— — zonarius, 820
— chloropygius bineschensis, 787
— — orphogaster, 787
— coccinigaster, 766
— cupreus cupreus, 765, 820
— — septentrionalis, 820
— habessinicus habessinicus, 767
— — turkanæ, 768
— leucogaster, 820
— loveridgei, 790
— mariquensis hawkeri, 820
— — osiris, 772, 820
— — suahelicus, 771
— mediocris bensoni, 784
— — fülleborni, 785
— — mediocris, 783
— — moreaui, 785
— — usambaricus, 784
— minullus marginatus, 788
— moreaui, 785
— oseus decorsei, 775
— oustaleti, 777
— pembæ, 771
— regius anderseni, 790
— — kivuensis, 789
— — regius, 788
— reichenowi reichenowi, 786
— shelleyi hofmanni, 774
— — shelleyi, 773
— superbus buvuma, 820
— — superbus, 764, 820
— talatala, 820
— — anderssoni, 776
— tsavoensis, 770
— venustus albiventris, 778
— — falkensteini, 779, 820
— — fazoqlensis, 778, 820
— — igneiventris, 780
— — niassæ, 820
— — sukensis, 820
Cisticola aberrans njombe, 466
— angusticauda, 500
— aridula aridula, 457
— — lavendulæ, 458
— — tanganyika, 458
— ayresii ayresii, 459
— — entebbe, 460
— — imatong, 461
— — mauensis, 460

Cisticola brachyptera ankole, 497
— — brachyptera, 495
— — hypoxantha, 496
— — isabellina, 496
— — katonæ, 496
— — kericho, 498
— — reichenowi, 497
— — zedlitzi, 497
— brunnescens brunnescens, 461
— — cinnamomea, 462
— — hindii, 462
— — nakuruensis, 463
— — wambera, 463
— cantans belli, 481
— — cantans, 479
— — concolor, 480
— — münzneri, 481
— — pictipennis, 481
— carruthersi, 487
— chiniana bodessa, 469
— — emendata, 471
— — fischeri, 468
— — heterophrys, 469
— — humilis, 468
— — mocuba, 519
— — procera, 470, 519
— — simplex, 467
— — ukamba, 470
— — victoria, 470
— chubbi, 477
— cinereola cinereola, 502
— — schillingsi, 502
— emini emini, 474
— — lurio, 475
— — petrophila, 474
— — teitensis, 475
— erythrops nilotica, 482
— — nyasa, 483
— — pyrrhomitra, 483
— — sylvia, 482
— eximia eximia, 463
— fulvicapilla muelleri, 499
— galactotes galactotes, 484
— — hæmatocephala, 485
— — lugubris, 485
— — marginata, 486
— — nyansæ, 486
— — suahelica, 486
— — zalingei, 487
— hæsitata, 456
— hunteri hunteri, 478
— — hypernephala, 479
— — masaba, 479
— — prinioides, 478
— juncidis perennia, 456
— — terrestris, 454
— — uropygialis, 455
— lais distincta, 465
— — mariæ, 467
— — semifasciata, 464
— lateralis antinorii, 471

# INDEX

Cisticola lugubris nyansæ, 486
— — suahelica, 486
— nana, 501
— natalensis argentea, 493
— — inexpectata, 493
— — kapitensis, 493
— — katanga, 494
— — littoralis, 494
— — matengorum, 519
— — natalensis, 491, 519
— — tonga, 494
— — valida, 492
— nigriloris, 476
— njombe mariæ, 467
— — njombe, 466
— robusta aberdare, 490
— — ambigua, 489
— — awemba, 491
— — nuchalis, 489
— — omo, 490
— — robusta, 488
— — schraderi, 490
— ruficeps mongalla, 504
— — ruficeps, 503
— — scotoptera, 504
— strangei kapitensis, 493
— subruficapilla bodessa, 469
— terrestris mauensis, 460
— — nakuruensis, 463
— tinniens oreophila, 504
— troglodytes ferruginea, 499
— — troglodytes, 498
— woosnami lufira, 473
— — schusteri, 473
— — woosnami, 472
citerior, Pytilia melba, 1009
citrinelloides, Carduelis citrinelloides, 1082
— chyulu, Spinus, 1086
— frontalis, Carduelis, 1083
— hypostictus, Carduelis, 1083, 1086
— kikuyuensis, Carduelis, 1084
citriniceps, Eremomela scotops, 436
clamans, Spiloptila, 516
clamosa, Turdoides melanops, 97
clara, Motacilla clara, 50
—, Parisoma lugens, 175
— torrentium, Motacilla, 50
clarkei, Estrilda subflava, 1024
—, Turdoides leucopygia, 91
clericalis, Pentholæa albifrons, 278
Clytospiza dybowskii, 986
— monteiri ugandensis, 985
coccinigaster, Cinnyris, 766
Coccopygia melanotis kilimensis, 1018, 1053
— — nyansæ, 1053
— — quartinia, 1018
codringtoni, Hypochera ænea, 1040
Coliuspasser albonotatus albonotatus, 965
— — eques, 966
— ardens ardens, 968, 1052
— — concolor, 969
— — teitensis, 1052
— axillaris axillaris, 960
— — phœniceus, 961
— — traversii, 962
— — zanzibarica, 962
— hartlaubi humeralis, 967
— — psammocromius, 966
— laticauda laticauda, 970, 973
— — suahelica, 971
— macrocercus, 962
— macrourus conradsi, 965
— — macrourus, 963
— progne delamerei, 971
collaris, Apalis nigriceps, 413
— djamdjamensis, Anthreptes, 821
— elachior, Anthreptes, 808
— garguensis, Anthreptes, 809, 820
— humeralis, Lanius, 593
— hypodilus, Anthreptes, 807
— jubaensis, Anthreptes, 890, 821
—, Mirafra, 18
—, Poliospiza, 1086
— smithii, Lanius, 594
—, Thamnolæa arnotti, 332
— ugandæ, Anthreptes, 820
— zambesiana, Anthreptes, 807
collurio, Lanius, 600
collybita abietinus, Phylloscopus, 384
—, Phylloscopus collybita, 384
comitatus stuhlmanni, Pedilorhynchus, 189
communis icterops, Sylvia, 349
— rubicola, Sylvia, 350
—, Sylvia communis, 349
conceptus, Chlorophoneus nigrifrons, 643
concinnata, Prionops cristata, 571
concolor, Cisticola cantans, 480
—, Coliuspasser ardens, 969
—, Macrosphenus, 447
concreta graueri, Dyaphorophyia, 213
— kungwensis, Dyaphorophyia, 213
— silvæ, Dyaphorophyia, 213
congener, Tchagra australis, 643
congensis, Andropadus gracilirostris, 144
—, Stelgidillas gracilirostris, 144
congica, Lagonosticta rubricata, 1010
congicus, Erythrocercus mccallii, 193
conradsi, Coliuspasser macrourus, 965
cooki, Laniarius ruficeps, 643
Coracia pyrrhocorax docilis, 678
Coracina cæsia pura, 561
— pectoralis, 560

## INDEX

coracinus, Dicrurus modestus, 563, 567
corax edithæ, Corvus, 672
— ruficollis, Corvus, 671
cordofanica, Mirafra, 8
cordofanicus, Passer motitensis, 872
coronata kordofanensis, Thamnolæa, 283
—, Thamnolæa coronata, 282
corruscus jombeni, Lamprocolius, 701
— mandanus, Lamprocolius, 700
— vaughani, Lamprocolius, 701
corvina affinis, Corvinella, 604, 643
— caliginosa, Corvinella, 605
— chapini, Corvinella, 643
— togoensis, Corvinella, 605
Corvinella corvina affinis, 604, 643
— — caliginosa, 605
— — chapini, 643
— — togoensis, 605
Corvultur albicollis, 676
— crassirostris, 677
Corvus albus, 673
— capensis kordofanensis, 674
— corax edithæ, 672
— — ruficollis, 671
— splendens splendens, 675
Cosmopsarus regius, 704, 722
— — magnificus, 722
— unicolor, 705
Cossypha albicapilla omoensis, 298
— archeri, 294
— caffra iolæma, 300
— cyanocampter bartteloti, 296
— gutturalis, 329
— heuglini euronota, 333
— — heuglini, 291
— — intermedia, 292, 333
— insulana granti, 333
— — kungwensis, 296, 333
— natalensis, 297
— niveicapilla, 299, 333
— — melanoptera, 333
— polioptera kungwensis, 296
— — polioptera, 295
— semirufa donaldsoni, 294
— — intercedens, 293
— — semirufa, 293
Cotyle see Riparia
craspedoptera, Euplectes hordeacea, 953
crassirostris, Acrocephalus scirpaceus, 517
—, Corvultur, 677
—, Euplectes capensis, 1052
—, Malimbus malimbicus, 937
—, Sylvia hortensis, 351
Crateropus see Turdoides
— plebeius emini, 87
— smithi omoensis, 91

crawfurdi, Eremomela griseoflava, 518
Creatophora carunculata, 722
— cinerea, 688, 722
Criniger verreauxi ndussumensis, 116
cristata altirostris, Galerida, 29
— concinnata, Prionops, 571
— isabellina, Galerida, 28
— melanoptera, Prionops, 576
— omoensis, Prionops, 576
—, Prionops cristata, 570
— somaliensis, Galerida, 29
— zalingei, Galerida, 30
cristatus isabellinus, Lanius, 601
—, Lanius, 570
— phœnicuroides, Lanius, 602
crocatus, Hyphanturgus ocularia, 918, 1052
croceus, Macronyx, 76
cruentata, Chalcomitra senegalensis, 795
cruentus cathemagmenus, Rhodophoneus, 639
— hilgerti, Rhodophoneus, 640
— kordofanicus, Rhodophoneus, 643
—, Rhodophoneus cruentus, 639, 643
cryptoleuca, Myrmecocichla æthiops, 285
Cryptolopha see Seicercus
Cryptospiza jacksoni, 994
— reichenovii australis, 992, 1053, 1054
— — ocularis, 1053
— — sanguinolenta, 1054
— salvadorii borealis, 1053
— — chyuluensis, 1054
— — kilimensis, 993, 1054
— — ruwenzori, 993
— — salvadorii, 993, 1053
— shelleyi, 995
cubla hamatus, Dryoscopus, 617
cucullatus abyssinicus, Ploceus, 889
— bohndorffi, Ploceus, 890, 1052
— femininus, Ploceus, 1052
— scutatus, Spermestes, 976
—, Spermestes cucullatus, 975
cupreonitens, Nectarinia famosa, 751, 820
cupreus, Cinnyris cupreus, 765, 820
— septentrionalis, Cinnyris, 820
curruca blythi, Sylvia, 348
—, Sylvia curruca, 348
curtus, Parus albiventris, 660
curvirostris, Andropadus curvirostris, 147
cyanea æthiopica, Platysteira, 209
— nyanzæ, Platysteira, 208
cyanecula, Cyanosylvia svecica, 328
cyanocampter bartteloti, Cossypha, 296
cyanocephala, Uræginthus, 1037

# INDEX

cyanogenys, Lamprocolius chloropterus, 722
cyanolæma, Cyanomitra cyanolæma, 801
cyanomelas bivittatus, Trochocercus, 215
— kikuyuensis, Trochocercus, 216
— vivax, Trochocercus, 216
Cyanomitra alinæ alinæ, 800
— balfouri, 806
— cyanolæma cyanolæma, 801
— olivacea alfredi, 804
— — changamwensis, 803
— — chyulu, 821
— — granti, 805, 821
— — lowei, 805
— — neglecta, 803, 821
— — olivacina, 802
— — pembæ, 821
— — puguensis, 804
— — ragazzii, 805
— — vincenti, 806
— veroxii fischeri, 798
— — zanzibarica, 799
— verticalis viridisplendens, 799
Cyanosylvia svecica cyanecula, 328
— — magna, 329
— — svecica, 328
— wolfi magna, 329
cyornithopsis bangsi Sheppardia, 333
— lopezi, Sheppardia, 307
cypriaca, Œnanthe leucomela, 264

## D

dammholzi, Sylvia atricapilla, 351
dankali, Fringillaria striolata, 1099
—, Gymnoris pyrgita, 1053
dartmouthi, Nectarinia johnstoni, 754
daurica domicella, Hirundo, 534
— emini, Hirundo, 534
— melanocrissa, Hirundo, 533
— rufula, Hirundo, 532
— scullii, Hirundo, 533
debilis albigula, Phyllastrephus, 127
—, Phyllastrephus debilis, 127
— rabai, Phyllastrephus, 128, 152
deckeni, Eurocephalus rüppelli, 642
—, Turdus olivaceus, 243
decorsei, Cinnyris oseus, 775
degener, Laniarius funebris, 642
degeni, Mirafra fischeri, 46
delamerei, Coliuspasser progne, 971
—, Estrilda erythronotos, 1031, 1053
—, Pseudalæmon fremantlii, 33
Delichon urbica urbica, 548
dentata, Petronia dentata, 886
denti, Apalis rufogularis, 400
—, Philodornis rushiæ, 819
desertæ, Prinia subflava, 507

deserti akeleyi, Ammomanes, 26
— assabensis, Ammomanes, 27
— atrogularis, Œnanthe, 332
— erythrochroa, Ammomanes, 27
—, Œnanthe deserti, 262, 332
— oreophila, Œnanthe, 262
— samharensis, Ammomanes, 27
desertorum, Alæmon alaudipes, 23
Dessonornis anomala anomala, 302
— — grotei, 303, 333
— — gurué, 304
— — macclounii, 303, 333
— — mbuluensis, 304
diabolica, Nigrita canicapilla, 983
diademata, Euplectes, 956
dichrocephalus, Ploceus capitalis, 905
Dicrurus adsimilis adsimilis, 564, 567
— — divaricatus, 567
— — jubaensis, 565
— elgonensis, 567
— ludwigii ludwigii, 565
— — münzneri, 566
— — sharpei, 566, 567
— modestus coracinus, 563, 567
— — ugandensis, 567
diffusus ugandæ, Passer, 878
dilutior, Sylvietta brachyura, 424
—, — carnapi, 424
dimidiata marwitzi, Hirundo, 529
—, Poliospiza, 1086
dimidiatus, Ploceus capitalis, 904
dimorpha, Batis capensis, 201
dinemelli boehmi, Dinemellia, 861
—, Dinemellia dinemelli, 860
— ruspolii, Dinemellia, 862
Dinemellia dinemelli boehmi, 861
— — dinemelli, 860
— — ruspolii, 862
diops, Batis, 202
Dioptrornis chocolatinus chocolatinus, 183
— — reichenowi, 183
— fischeri amani, 226
— — fischeri, 180
— — nyikensis, 181
— — toruensis, 182
— — ufipæ, 182
distans, Malacocincla rufipennis, 101
distincta, Cisticola lais, 465
divaricatus, Dicrurus adsimilis, 567
djamdjamensis, Alseonax minimus, 225
—, Anthreptes collaris, 821
docilis, Coracia pyrrhocorax, 678
dodsoni, Pycnonotus, 114, 152
doggetti, Chalcomitra amethystina, 792
dohertyi, Mirafra africana, 46
—, Telophorus, 635

# INDEX

dombrowski, Budytes flavus, 53
domesticus arboreus, Passer, 870
— bactrianus, Passer, 871
— indicus, Passer, 869
— niloticus, Passer, 871
domicella, Hirundo daurica, 534
donaldsoni, Cossypha semirufa, 294
—, Plocepasser, 865
—, Serinus, 1067, 1086
dorcadichroa, Camaroptera, 389
dorcadichrous, Seicercus umbrovirens, 389
dorsalis, Fiscus, 596
—, Pseudonigrita arnaudi, 868
dorsostriata, Ortygospiza atricollis, 999, 1054
— harterti, Serinus, 1064
— maculicollis, Serinus, 1064
—, Serinus dorsostriatus, 1063
— taruensis, Serinus, 1065
Drepanoplectes jacksoni, 972
Drepanorhynchus reichenowi, 762
Dryodromas rufidorsalis, 417
— smithii, 417
Dryoscopus affinis affinis, 618
— — senegalensis, 619
— angolensis kungwensis, 622
— — nandensis, 622
— cubla hamatus, 617
— gambensis erwini, 621
— — erythreæ, 621
— — malzacii, 620, 643
— — nyanzæ, 643
— holomelas, 609
— jacksoni, 633
— malzacii erythreæ, 621
— pringlii, 619
dubarensis, Lanius elegans, 590
—, — excubitor, 590
dubia, Cercomela, 276, 332
ducis, Riparia paludicola, 543
dunni, Ammomanes dunni, 28
Dyaphorophyia castanea castanea, 211
— concreta graueri, 213
— — kungwensis, 213
— — silvæ, 213
— jamesoni, 212
dybowskii, Clytospiza, 986

E

edithæ, Corvus corax, 672
edolioides lugubris, Melænornis, 184, 226
— schistacea, Melænornis, 184
— ugandæ, Melænornis, 226
elachior, Anthreptes collaris, 808
elæica, Hippolais pallida, 360
elegans abyssinica, Eremomela, 434
— aucheri, Lanius, 589

elegans buryi, Lanius, 589
— dubarensis, Lanius, 590
—, Eremomela canescens, 434, 518
— lahtora, Lanius, 588
—, Lanius elegans, 587, 642
— pallidirostris, Lanius, 588
— uncinatus, Lanius, 589
elgeyuensis, Chlorophoneus nigrifrons, 643
elgonensis, Anthreptes tephrolæma, 820
—, Bradypterus babœcala, 369
—, Dicrurus, 567
—, Eremomela pusilla, 518
—, Linurgus olivaceus, 1081
—, Onychognathus walleri, 707, 723
—, Pogonocichla stellata, 324
—, Serinus flaviventris, 1086
—, Turdus olivaceus, 332
elisabeth, Lamprocolius chloropterus, 697
ellinoræ, Apalis melanocephala, 404
elliotti, Galerida malabarica, 30
Elminia albicauda kivuensis, 215
— teresita, 214
Emberiza affinis, 1099
— cabanisi, 1088
— cæsia, 1093
— cineracea semenowi, 1093
— flaviventris flavigastra, 1091
— — flaviventris, 1090
— forbesi forbesi, 1092, 1099
— hortulana, 1092
— orientalis orientalis, 1089
— poliopleura, 1091
emendata, Cisticola chiniana, 471
emini, Argya rubiginosa, 95
— budongoensis, Ploceus, 1052
—, Cisticola emini, 474
—, Hirundo daurica, 534
—, Lamprocolius chalcurus, 695
— lurio, Cisticola, 475
—, Othyphantes, 928, 1052
— petrophila, Cisticola, 474
—, Pytilia phœnicoptera, 1006
—, Salpornis spilonota, 822
—, Sporopipes frontalis, 888
—, Tchagra australis, 626
—, Tchitrea nigriceps, 224
— teitensis, Cisticola, 475
—, Turdoides jardinei, 87
Eminia lepida, 421
eminibey, guasso, Sorella, 1052
—, Sorella, 882, 1052
Empidornis semipartitus kavirondensis, 187
— — orleansi, 187
— — semipartitus, 187
enigma, Cercomela scotocerca, 332
entebbe, Cisticola ayresii, 460

epulatus seth-smithi, Pedilorhynchus, 169
eques, Coliuspasser albonotatus, 966
Erannornis longicauda kivuensis, 215
—— teresita, 214
eremobius, Ploceus baglafecht, 902
Eremomela badiceps badiceps, 437
—— latukæ, 438
—— turneri, 437
— canescens abyssinica, 434
—— canescens, 433, 518
—— elegans, 434, 518
— elegans abyssinica, 434
— flavicrissalis, 433
— flaviventris abdominalis, 432
—— alexanderi, 432
— griseoflava, 518
—— archeri, 518
—— belli, 432
—— crawfurdi, 518
—— karamojensis, 518
— icteropygialis, 518
—— abdominalis, 432
—— alexanderi, 432
—— belli, 432
—— griseoflava, 431, 518
—— polioxantha, 431
— pusilla elgonensis, 518
—— tessmani, 518
— scotops citriniceps, 436
—— kikuyuensis, 519
—— occipitalis, 436, 519
—— pulchra, 436
—— scotops, 435
— usticollis rensi, 438, 519
Eremopterix leucopareia, 38
— leucotis leucotis, 34
—— madaraszi, 35
—— melanocephala, 35
— nigriceps albifrons, 36
—— melanauchen, 36
— signata cavei, 46
—— harrisoni, 38, 46
—— signata, 37
ericetorum philomelos, Turdus, 239
eritreæ, Passer griseus, 1052
erlangeri, Batis minor, 206, 226
—, Bradornis griseus, 225
—, Calandrella cinerea, 43, 46
—, Camaroptera brevicaudata, 444, 519
—, Erythropygia quadrivirgata, 321
—, Eurocephalus rüppelli, 642
—, Laniarius æthiopicus, 614, 643
—, Nectarinia nectarinioides, 761
—, Pinarochroa sordida, 332
—, Prinia somalica, 508
—, Riparia cincta, 544
—, Salpornis spilonota, 822
—, Serinus gularis, 1072
—, Tchagra senegala, 643

ernesti, Pinarochroa sordida, 277
erwini, Dryoscopus gambensis, 621
erythreæ, Dryoscopus gambensis, 621
—, — malzacii, 621
—, Nilaus afer, 642
—, Seicercus umbrovirens, 517
erythrinæ, Chalcomitra senegalensis, 820
erythrocerca, Nectarinia, 757, 820
Erythrocercus holochlorus, 192
— livingstonei livingstonei, 194, 226
—— monapo, 226
—— thomsoni, 194
— mccallii congicus, 193
erythrochroa, Ammomanes deserti, 27
erythrogaster fagani, Malimbus, 939
—, Laniarius, 606
erythronotos charmosyna, Estrilda, 1031
— delamerei, Estrilda, 1031, 1053
erythrops nilotica, Cisticola, 482
— nyasa, Cisticola, 483
— pyrrhomitra, Cisticola, 483
—, Quelea, 946
— sylvia, Cisticola, 482
erythroptera jodoptera, Heliolais, 512, 518
— kavirondensis, Heliolais, 518
— kirbyi, Heliolais, 518
— major, Heliolais, 513
— rhodoptera, Heliolais, 512, 518, 519
Erythropygia barbata, 319
—— greenwayi, 321
—— rovumæ, 333
— hamertoni, 358
— hartlaubi hartlaubi, 321
—— kenia, 322
— leucophrys ruficauda, 333
—— soror, 333
—— vansomereni, 333
— leucoptera leucoptera, 318, 333
—— pallida, 319
— quadrivirgata erlangeri, 321
—— greenwayi, 321
—— quadrivirgata, 320, 333
— vulpina, 333
— zambesiana brunneiceps, 317
—— sclateri, 318
—— zambesiana, 316, 333
erythropygia, Pinarocorys, 21
erythrorhynchus, Buphagus erythrorhynchus, 721
— caffer, Buphagus, 721
erythrothorax mabiræ, Stiphrornis, 307
Estrilda astrild cavendishi, 1020, 1053
—— macmillani, 1053

# INDEX

Estrilda astrild massaica, 1053
— — minor, 1019
— — nyanzæ, 1021
— — peasei, 1021, 1053
— atricapilla graueri, 1028
— — kandti, 1053
— charmosyna kiwanukæ, 1053
— erythronotos charmosyna, 1031
— — delamerei, 1031, 1053
— nonnula nonnula, 1029, 1053
— paludicola marwitzi, 1026
— — ochrogaster, 1026
— — paludicola, 1025
— — roseicrissa, 1025
— perreini incana, 1027
— — perreini, 1027
— rara rara, 1030
— rhodopyga centralis, 1023
— — rhodopyga, 1022
— subflava clarkei, 1024
— — subflava, 1023
— troglodytes, 1021, 1053
— xanthophrys, 1053
euchlorus, Auripasser, 882
eugenia, Stelgidocichla latirostris, 150, 152
Euodice cantans inornata, 1052
— — meridionalis, 1053
— malabarica cantans, 979, 1052
— — orientalis, 980, 1053
Euplectes afra afra, 957
— — ladoensis, 959
— — niassensis, 959
— — stricta, 958, 1052
— capensis crassirostris, 1052
— — xanthomelas, 955, 1052
— diademata, 956
— gierowii ansorgei, 955
— — friederichseni, 954
— hordeacea changamwensis, 1052
— — craspedoptera, 953
— — hordeacea, 952, 1052
— — sylvatica, 1052
— nigroventris, 951, 1052
— orix franciscana, 949
— — nigrifrons, 950, 1052
— — pusilla, 951, 1053
— — sundevalli, 950
— — wertheri, 1052
— rufigula, 1052
— taha intercedens, 1052
— zavattarii, 1053
Eurillas virens shimba, 152
— — virens, 148
— — zanzibaricus, 149
— — zombensis, 149, 152
Eurocephalus anguitimens ruppelli, 583, 642
— rüppelli böhmi, 642
— — deckeni, 642
— — erlangeri, 642

euronota, Cossypha heuglini, 333
eurycricotus, Zosterops, 730, 736
eversmani, Phylloscopus trochilus, 517
excubitor dubarensis, Lanius, 590
excubitorius böhmi, Lanius, 592, 642
— intercedens, Lanius, 642
—, Lanius excubitorius, 591
eximia, Cisticola eximia, 463
— ugandæ, Bleda, 117

## F

fagani, Malimbus erythrogaster, 939
falkensteini, Cercomela familiaris, 274, 332
—, Cinnyris venustus, 779, 820
familiaris, Agrobates galactotes, 357
— falkensteini, Cercomela, 274, 332
— gambagæ, Cercomela, 332
— omoensis, Cercomela, 275
— sennaarensis, Cercomela, 332
famosa æneigularis, Nectarinia, 820
— centralis, Nectarinia, 820
— cupreonitens, Nectarinia, 751, 820
— subfamosa, Nectarinia, 820
fasciata albitorquata, Amadina, 1054
— alexanderi, Amadina, 996
—, Amadina fasciata, 995
— meridionalis, Amadina, 997, 1054
fasciiventer, Parus, 649
fasciolatus stierlingi, Calamonastes, 392
fayi, Pycnonotus tricolor, 113
fazoqlensis, Cinnyris venustus, 778, 820
feldegg, Budytes, 56, 58
femininus, Bias musicus, 197
—, Ploceus cucullatus, 1052
femoralis, Pholia, 693
ferreti, Tchitrea viridis, 219, 226
ferruginea, Cisticola troglodytes, 499
ferrugineus chyulu, Laniarius, 643
— somaliensis, Laniarius, 643
fischeri amani, Dioptrornis, 226
— cabanisi, Phyllastrephus, 130, 151
— chyuluensis, Phyllastrephus, 152
—, Cisticola chiniana, 468
—, Cyanomitra veroxii, 798
— degeni, Mirafra, 46
—, Dioptrornis fischeri, 180
—, Linura, 1048
— marsabit, Phyllastrephus, 152
—, Mirafra rufocinnamomea, 14, 46
— natalicus, Turdus, 333
— nyikensis, Dioptrornis, 181
—, Phyllastrephus fischeri, 128, 152
— placidus, Phyllastrephus, 129, 152
—, Psophocichla guttata, 252
—, Spreo, 715

# INDEX

fischeri toruensis, Dioptrornis, 182
— ufipæ, Dioptrornis, 182
— Vidua, 1048
— zombæ, Mirafra, 46
Fiscus dorsalis, 596
flammulatus æquatorialis, Megabyas, 196
flava, Campephaga, 562
flavicans hypochondriacus Macrosphenus, 446
flavicollis flaviguia, Pyrrhurus, 119, 151
—, Macronyx, 78
— pallidigula, Atimastillas, 151
— shelleyi, Atimastillas, 151
— soror, Pyrrhurus, 120
flavicrissalis, Eremomela, 433
flavida æquatorialis, Apalis, 518
— flavocincta, Apalis, 407, 518
— golzi, Apalis, 408
— malensis, Apalis, 409
— tenerrima, Apalis, 410
— viridiceps, Apalis, 408
flavigaster barbozæ, Hyliota, 198
—, Hyliota flavigaster, 198
flavigastra, Emberiza flaviventris, 1091
flavigula, Pyrrhurus flavicollis, 119, 151
—, Serinus, 1076, 1086
flavigularis, Apalis flavigularis, 398
— griseiceps, Apalis, 398, 519
— lynesi, Apalis, 399
— uluguru, Apalis, 399
flavilateralis, Zosterops senegalensis, 726
flavipes, Alseonax, 225
flavissimus, Budytes luteus, 55, 58
—, Ploceus, 1052
flaviventris abdominalis, Eremomela, 432
— alexanderi, Eremomela, 432
— centralis, Chlorocichla, 142, 152
— chyuluensis, Chlorocichla, 152
—, Emberiza flaviventris, 1090
— flavigastra, Emberiza, 1091
— occidentalis, Chlorocichla, 141
flavivertex elgonensis, Serinus, 1086
— sassii, Serinus, 1069
—, Serinus flavivertex, 1068, 1086
flavocincta, Apalis flavida, 407, 518
flavodorsalis, Lagonosticta senegala, 1053
flavostriatus kungwensis, Phyllastrephus, 126
— litoralis, Phyllastrephus, 152
— olivaceo-griseus, Phyllastrephus, 125
— tenuirostris, Phyllastrephus, 124, 152
— vincenti, Phyllastrephus, 125

flavus beema, Budytes, 53
—, Budytes flavus, 53
— campestris, Budytes, 58
— dombrowskii, Budytes, 53
— melanogriseus, Budytes, 58
— rayi, Budytes, 58
fluviatilis, Locustella, 362
forbesi, Emberiza forbesi, 1092, 1099
forwoodi, Motacilla alba, 48
foxi, Calamocichla, 379
franciscana, Euplectes orix, 949
fraseri intermedia, Stizorhina, 226
— vulpina, Stizorhina, 195, 226
Fraseria ocreata ocreata, 186
—, Onychognathus, 707
fraterculus, Bradypterus brachypterus, 518
fratrum, Batis, 207
fremantlii delamerei, Pseudalæmon, 33
— megaensis, Pseudalæmon, 34
—, Pseudalæmon fremantlii, 33
frenata, Œnanthe bottæ, 269
fricki, Andropadus importunus, 147
—, Chlorophoneus sulfureopectus, 643
—, Othyphantes reichenowi, 930
friederichseni, Euplectes gierowii, 954
friedmanni, Cinnyricinclus leucogaster, 722
Fringillaria capensis vincenti, 1094
— socotrana, 1098
— striolata dankali, 1099
— — jebelmarræ, 1099
— — saturatior, 1097, 1099
— — striolata, 1097, 1099
— tahapisi goslingi, 1096
— — insularis, 1096
— — septemstriata, 1096
— — tahapisi, 1095
fringillinus, Parus, 654
fringilloides, Amauresthes, 978
frommi, Pirenestes, 991
frontalis abyssinicus, Sporopipes, 1052
—, Carduelis citrinelloides, 1083
— cinerascens, Sporopipes, 888
— emini, Sporopipes, 888
—, Sporopipes, frontalis, 887, 1052
fuertesi, Tephrocorys cinerea, 46
fuggles-couchmani, Camaroptera brachyura, 440
—, Seicercus umbrovirens, 391
fulgens, Passer castanopterus, 876
fulgidus hartlaubii, Onychognathus, 705
— intermedius, Onychognathus, 706
fuliginosa, Apalis melanocephala, 403
— minuscula, Artomyias, 190

fuligula pusilla, Ptyonoprogne, 548
— rufigula, Ptyonoprogne, 547, 554
fülleborni, Alethe fülleborni, 312
—, Alseonax adustus, 166
—, Cinnyris mediocris, 785
—, Laniarius, 610
—, Macronyx, 77
— usambaræ, Alethe, 313
fulva acaciæ, Argya, 93
fulvescens ugandæ, Malacocincla, 100
fulvicapilla muelleri, Cisticola, 499
funebris degener, Laniarius, 642
—, Laniarius, 608, 642
— rothschildi, Laniarius, 642
funerea, Hypochera funerea, 1043
— wilsoni, Hypochera, 1044
funereus, Parus, 653
furensis, Cercomela scotocerca, 274
—, Mirafra rufocinnamomea, 16
—, Ploceus tæniopterus, 895
fusca, Locustella luscinioides, 362
fusciceps, Arizelocichla nigriceps, 136
fuscigularis, Apalis murina, 398
fusciventris, Ptyonoprogne rufigula, 554
fuscocrissa, Ortygospiza atricollis, 998
fusconota, Nigrita fusconota, 984
fuscus, Acrocephalus scirpaceus, 365
futa, Icteropsis pelzelni, 1052

## G

gabunensis, Neocossyphus rufus, 255
galactotes, Agrobates galactotes, 356
—, Cisticola galactotes, 484
— familiaris, Agrobates, 357
— hæmatocephala, Cisticola, 485
— hamertoni, Agrobates, 358
— lugubris, Cisticola, 485
— marginata, Cisticola, 486
— minor, Agrobates, 357
— nyansæ, Cisticola, 486
— suahelica, Cisticola, 486
— syriacus, Agrobates, 357
— zalingei, Cisticola, 487
galbula, Ploceus, 909, 1052
Galeopsar salvadorii, 712
Galerida cristata altirostris, 29
— — isabellina, 28
— — somaliensis, 29
— — zalingei, 30
— malabarica, 46
— elliotti, 30
— — huriensis, 31
— — prætermissa, 30
— theklæ, 46
galinieri, Parophasma, 176
gallarum, Mirafra hypermetra, 9

galtoni omoensis, Saxicola, 275
gambagæ, Alseonax, 162
—, Cercomela familiaris, 332
—, Muscipapa striata, 162, 225
— somaliensis, Muscicapa, 225
gambensis erwini, Dryoscopus, 621
— erythreæ, Dryoscopus, 621
— malzacii, Dryoscopus, 620, 643
— nyanzæ, Dryoscopus, 643
garguensis, Anthreptes collaris, 809, 820
Geokichla gurneyi chuka, 249
— — chyulu, 249
— — gurneyi, 248, 332
— — kilimensis, 250
— — otomitra, 332
— — raineyi, 248, 332
— — usambaræ, 332
— piaggiæ hadii, 251
— — kilimensis, 250, 332
— — piaggiæ, 249
— — rowei, 250
— — williamsi, 251
gertrudis, Cinnyris chalybeus, 820
—, Serinus mozambicus, 1086
gierowii ansorgei, Euplectes, 955
— friederichseni, Euplectes, 954
giffardi, Heliocorys modesta, 32
gilletti, Mirafra, 19
githagineus, Bucanetes githagineus, 1059
golandi, Ploceus, 898
golzi, Apalis flavida, 408
golzii, Luscinia megarhyncha, 333
gommaensis, Serinus mozambicus, 1063
gongonensis jubaensis, Passer, 881
—, Passer gongonensis, 880, 1053
goodsoni, Anthus leucophrys, 66
—, — vaalensis, 66
gordoni, Hirundo semirufa, 536
goslingi, Fringillaria tahapisi, 1096
gouldi turneri, Anthus, 81
gracilirostris, Calamonastides, 382
— chagwensis, Stelgidillas, 143
— congensis, Stelgidillas, 144
— nilotica, Calamocichla, 379
— parva, Calamocichla, 378
— percivali, Stelgidillas, 143
— zuluensis, Calamornis, 518
gracilis, Andropadus gracilis, 145, 152
— carlo, Prinia, 509
— kavirondensis, Andropadus, 145
—, Prinia gracilis, 509
graculinus, Sigmodus retzii, 574, 576
Granatina ianthinogaster hawkeri, 1038, 1053, 1054
— — ianthinogaster, 1038
— — roosvelti, 1039
— — ugandæ, 1053

# INDEX

grandis, Melocichla mentalis, 513
granti, Bradornis pallidus, 225
—, Cossypha insulana, 333
—, Cyanomitra olivacea, 805, 821
—, Sathrocercus mariæ, 377
—, Tchitrea suahelica, 223, 226
granviki, Melocichla mentalis, 515
Graucalus *see* Coracina
graueri, Bradypterus, 371
—, Cinnyris chalybeus, 782
—, Dyaphorophyia concreta, 213
—, Estrilda atricapilla, 1028
—, Lagonosticta cinereovinacea, 1016, 1053
—, Ploceus nigriceps, 1052
—, Serinus striolatus, 1078
—, Turdus olivaceus, 246, 333
greenwayi, Erythropygia barbata, 321
—, — quadrivirgata, 321
griseiceps, Apalis flavigularis, 398, 519
— chyulu, Apalis, 519
—, Suaheliornis kretschmeri, 134
— uluguru, Apalis, 399
griseigula, Camaroptera brevicaudata, 443
griseigularis, Alseonax griseigularis, 171
griseiventris, Parus afer, 647, 660
griseldis, Acrocephalus, 364
griseoflava archeri, Eremomela, 518
— belli, Eremomela, 432
— crawfurdi, Eremomela, 518
—, Eremomela, 518
—, — icteropygialis, 431, 518
— karamojensis, Eremomela, 518
griseopyga, Hirundo griseopyga, 539
griseoviridis abessinica, Camaroptera, 443
— sharpei, Camaroptera, 444
grisescens, Mirafra africana, 12
griseus, Acrocephalus scirpaceus, 517
—, Anthus campestris, 81
—, Bradornis pallidus, 178, 225, 226
— eritreæ, Passer, 1052
— erlangeri, Bradornis, 225
— jubaensis, Passer, 881
— mosambicus, Passer, 1052
— neumanni, Bradornis, 178
— —, Passer, 878, 1052
—, Passer griseus, 877
— tertale, Passer, 1053
— turkanæ, Passer, 1053
— ugandæ, Passer, 878, 1052
— ukamba, Bradornis, 226
grotei, Dessonornis anomala, 303, 333
—, Pytilia melba, 1008
—, Serinus mozambicus, 1062, 1086

guasso, Anthoscopus musculus, 659
—, Sorella eminibey, 1052
gubernator, Lanius, 601
guineensis, Parus leucomelas, 651, 660
gularis erlangeri, Serinus, 1072
—, Nicator chloris, 641
— reichardi, Serinus, 1070
— striatipectus, Serinus, 1071
gunningi sokokensis, Sheppardia, 304
gurneti, Serinus burtoni, 1080
gurneyi chuka, Geokichla, 249
— chyulu, Geokichla, 249
—, Geokichla gurneyi, 248, 332
— keniensis, Geokichla, 332
— kilimensis, Geokichla, 250
— otomitra, Geokichla, 332
— raineyi, Geokichla, 248, 332
— usambaræ, Geokichla, 332
gurué, Dessonornis anomala, 304
guttata belcheri, Psophocichla, 333
—, Cichladusa guttata, 315
— fischeri, Psophocichla, 252
—, Psophocichla guttata, 251, 333
— rufipennis, Cichladusa, 316
guttatus, Ixonotus, 122
guttifer, Pogonocichla stellata, 322, 333
gutturalis angustus, Neocichla, 689
—, Chalcomitra senegalensis, 794, 820
—, Irania, 329
Gymnoris pyrgita dankali, 1053
— — reichenowi, 1052

## H

habessinica, Tchagra senegala, 624, 643
habessinicus, Cinnyris habessinicus, 767
— turkanæ, Cinnyris, 768
hadii, Geokichla piaggiæ, 251
hæmatocephala, Cisticola galactotes, 485
—, Lagonosticta rubricata, 1011
hæsitata, Cisticola, 456
hafizi, Luscinia megarhynchos, 331, 333
hamatus, Dryoscopus cubla, 617
hamertoni, Agrobates galactotes, 358
—, Alæmon hamertoni, 25
— altera, Alæmon, 25
—, Erythropygia, 358
— tertia, Alæmon, 25
hararensis, Anthus similis, 62, 81
harrisoni, Eremopterix signata, 38, 46
harterti, Mirafra africana, 13
—, Serinus dorsostriatus, 1064

hartlaubi, Campephaga, 562
—, Erythropygia hartlaubi, 321
— humeralis, Coliuspasser, 967
— kenia, Erythropygia, 322
— psammocromius, Coliuspasser, 966
hartlaubii, Onychognathus fulgidus, 705
—, Turdoides leucopygia, 90
haussarum, Anthreptes longuemarei, 811
hawkeri, Cinnyris mariquensis, 820
—, Granatina ianthinogaster, 1038, 1053, 1054
Hedydipna platura metallica, 764
— — platura, 763
Heliocorys modesta bucolica, 32
— — giffardi, 32
— modesta modesta, 31
Heliolais castanopsis, 519
— erythroptera jodoptera, 512, 518
— — kavirondensis, 518
— — kirbyi, 518
— — major, 513
— — rhodoptera, 512, 518, 519
helleri, Turdus, 246
hemileucus, Passer insularis, 874
hemprichii, Saxicola torquata, 290
hermonensis, Calandrella brachydactila, 40
Heterhyphantes melanogaster stephanophorus, 927, 973
Heteromirafra ruddi, 22
— — archeri, 22
heterophrys, Cisticola chiniana, 469
heuglini, Argya rubiginosa, 95
—, Cossypha heuglini, 291
— euronota, Cossypha, 333
— intermedia, Cossypha, 292, 333
—, Œnanthe, 270
—, Ploceus heuglini, 896
hildebrandti kellogorum, Spreo, 723
—, Lagonosticta rubricata, 1053
—, Spreo, 718, 723
hilgerti, Nilaus afer, 642
—, Rhodophoneus cruentus, 640
—, Serinus atrogularis, 1076
—, Sylvietta brachyura, 518
hindei, Turdoides, 93
hindii, Cisticola brunnescens, 462
Hippolais icterina, 359
— languida, 359
— olivetorum, 359
— pallida elæica, 360
— — pallida, 360
Hirundo abyssinica abyssinica, 537
— — bannermani, 539
— — unitatis, 538
— æthiopica, 528
— albigularis albigularis, 527
— angolensis angolensis, 526

Hirundo angolensis arcticincta, 527
— atrocærulea atrocærulea, 530
— — lynesi, 531
— daurica domicella, 534
— — emini, 534
— — melanocrissa, 533
— — rufula, 532
— — scullii, 533
— dimidiata marwitzi, 529
— griseopyga griseopyga, 539
— megaensis, 530
— rustica rothschildi, 525
— — rustica, 524
— — transitiva, 525
— semirufa gordoni, 536
— senegalensis aschani, 554
— — hybrida, 554
— — monteiri, 536, 554
— — senegalensis, 535, 554
— smithii smithii, 531
hispanica melanoleuca, Œnanthe, 263
hispaniolensis, Passer hispaniolensis, 874
Histurgops ruficauda, 866
hofmanni, Cinnyris shelleyi, 774
holochlora, Chloropetella, 192, 226
— suahelica, Chloropetella, 226
holomelæna massaica, Psalidoprocne, 554
—, Psalidoprocne holomelæna, 549, 554
— ruwenzori, Psalidoprocne, 550
holomelas, Dryoscopus, 609
—, Laniarius poensis, 609
hordeacea changamwensis, Euplectes, 1052
— craspedoptera, Euplectes, 953
—, Euplectes hordeacea, 952, 1052
— sylvatica, Euplectes, 1052
hortensis crassirostris, Sylvia, 351
hortulana, Emberiza, 1092
humeralis, Coliuspasser hartlaubi, 967
—, Lanius collaris, 593
humilis, Cisticola chiniana, 468
hunteri, Chalcomitra, 797
—, Cisticola hunteri, 478
— hypernephala, Cisticola, 479
— masaba, Cisticola, 479
— prinioides, Cisticola, 478
huriensis, Galerida malabarica, 31
hybrida, Hirundo senegalensis, 554
Hylia prasina prasina, 818
Hyliota australis australis, 199, 226
— — inornata, 226
— — usambaræ, 200
— flavigaster barbozæ, 198
— — flavigaster, 198
— slatini, 226
Hypargos niveoguttatus, 1002

# INDEX

hypermetra gallarum, Mirafra, 9
— kathangorensis, Mirafra, 10
— kidepoensis, Mirafra, 10
—, Mirafra, hypermetra, 9
hypernephala, Cisticola hunteri, 479
Hyphanturgus alienus, 921
— nicolli, 922
— nigricollis melanoxanthus, 918
— — nigricollis, 917
— ocularis crocatus, 918, 1052
— — suahelicus, 919
— olivaceiceps, 920
Hypochera ænea codringtoni, 1040
— amauropteryx, 1042
— camerunensis, 1045
— chalybeata camerunensis, 1045
— funerea funerea, 1043
— — wilsoni, 1044
— nigeriæ, 1045
— nigerrima, 1044, 1053
— ultramarina neumanni, 1042
— — orientalis, 1042
— — purpurascens, 1053
— — ultramarina, 1041
hypocherina, Vidua, 1048
hypochloris, Phyllastrephus, 131
hypochondriacus, Macrosphenus flavicans, 446
Hypocolius ampelinus, 645
hypodilus, Anthreptes collaris, 807
hypoleuca kilosa, Turdoides, 97
— —, Turdoides, 92, 97
—, Turdoides hypoleuca, 92
— semitorquata, Muscicapa, 163
hypopyrrhus, Malaconotus poliocephalus, 643
hypospodia, Pinarochroa sordida, 276
hyposticus, Carduelis citrinelloides, 1083, 1086
hypoxantha, Cisticola brachyptera, 496
— reichenowi, Cisticola, 497
hypoxanthus, Andropadus importunus, 146

## I

ianthinogaster, Granatina ianthinogaster, 1038
— hawkeri, Granatina, 1038, 1053, 1054
— roosvelti, Granatina, 1039
— somereni, Estrilda, 1054
— ugandæ, Granatina, 1053
icterina, Hippolais, 359
icterinus sethsmithi, Phyllastrephus, 133
icterops, Sylvia communis, 349
Icteropsis pelzelni, 925, 1052
— — futa, 1052

icteropygialis abdominalis, Eremomela, 432
— alexanderi, Eremomela, 432
— belli, Eremomela, 432
—, Eremomela, 518
— griseoflava, Eremomela, 431, 518
— polioxantha, Eremomela, 431
igneiventris, Cinnyris venustus, 780
Illadopsis rufipennis barakæ, 106
— — puguensis, 101
imatong, Cisticola ayresii, 461
imberbis, Anomalospiza imberbis, 1000
— nyassæ, Anomalospiza, 1054
— rendalli, Anomalospiza, 1002, 1054
immutabilis, Prinia mistacea, 518
importunus fricki, Andropadus, 147
— hypoxanthus, Andropadus, 146
— insularis, Andropadus, 145, 151
inæstimata, Chalcomitra senegalensis, 820
Incana incana, 446
incana, Drymocichla, 422
—, Estrilda perreini, 1027
indicator chlorosaturata, Bæopogon, 121
indicus, Passer domesticus, 869
inexpectata, Cisticola natalensis, 492
infulatus, Alseonax aquaticus, 168, 225
injectiva, Apalis thoracica, 519
inornata, Euodice cantans, 1052
—, Hyliota australis, 226
insignis ornatus, Ploceus, 1052
—, Parus leucomelas, 651
—, Phormoplectes insignis, 933, 1052
insulana granti, Cossypha, 333
— kungwensis, Cossypha, 296, 333
insularis, Andropadus importunus, 145, 151
—, Fringillaria tahapisi, 1096
— hemileucus, Passer, 874
— kilimandjaricus, Andropadus, 151
—, Passer insularis, 874
— somaliensis, Andropadus, 151
intensa, Aëthocorys personata, 45
—, Pogonocichla, 333
intercedens, Cossypha semirufa, 293
—, Euplectes taha, 1052
—, Lanius excubitorius, 642
—, Mirafra africanoides, 17, 46
—, Telophorus quadricolor, 643
intermedia, Cossypha heuglini, 292, 333
—, Stizorhina fraseri, 226
intermedius, Bubalornis niger, 860, 1052
— cabanisii, Ploceus, 898
—, Cinnyris chalybeus, 782, 820
—, Onychognathus fulgidus, 706

intermedius, Ploceus intermedius, 897
—, Sigmodus retzii, 576
interpositus, Alseonax minimus, 226
interscapularis, Melanoploceus tricolor, 935
iloæma, Cossypha caffra, 300
Irania gutturalis, 329
irisæ, Paludipasser, 1054
isabellina, Cisticola brachyptera, 496
—, Galerida cristata, 28
—, Œnanthe, 261
—, Sylvietta, 427
isabellinus, Lanius cristatus, 601
itoculo, Phyllastrephus alfredi, 152
Ixonotus guttatus, 122

## J

jacksoni, Apalis jacksoni, 412
—, Bathmocercus rufus, 517
—, Calamocichla leptorhyncha, 381
—, Chlorophoneus bocagei, 633
—, Cryptospiza, 994
—, Drepanoplectes, 972
— jucundus, Ploceus, 1053
—, Parisoma lugens, 175
—, Platysteira peltata, 226
—, Ploceus, 906, 1053
—, Sylvietta whytii, 426
—, Zosterops virens, 729
jamesi kismayensis, Tchagra, 627
— mandanus, Tchagra, 627
—, Tchagra jamesi, 626
jamesoni, Dyaphorophyia, 212
—, Lagonosticta jamesoni, 1011
— taruensis, Lagonosticta, 1012
jardinei emini, Turdoides, 87
— kirki, Turdoides, 86
jebelmarræ, Anthus similis, 64
—, Fringillaria striolata, 1099
—, Lanius, 590
—, Saxicola torquata, 289
jessei, Pytilia melba, 1053
jodoptera, Heliolais erythroptera, 512, 518
johannis, Warsanglia, 1085
johnstoni dartmouthi, Nectarinia, 754
—, Nectarinia johnstoni, 752
—, Pogonocichla stellata, 333
— salvadorii, Nectarinia, 753
—, Seicercus ruficapillus, 387
jombeni, Lamprocolius corruscus, 701
jubaensis, Anaplectes melanotis, 944
—, Anthreptes collaris, 809, 821
—, Dicrurus adsimilis, 565
—, Passer gongonensis, 881
—, Pytilia melba, 1054
—, Turdoides squamulata, 86
—, Zosterops senegalensis, 727, 736

jucundus, Ploceus jacksoni, 1053
juncidis perennia, Cisticola, 456
— terrestris, Cisticola, 454
— uropygialis, Cisticola, 455

## K

kaduliensis, Steganura aucupum, 1054
kaffensis, Zosterops virens, 730, 736
kakamariæ, Petronia xanthosterna, 885
kakamegæ, Arizelocichla masukuensis, 141
—, Chalcomitra rubescens, 794
kalckreuthi, Chalcomitra amethystina, 792
kandti, Estrilda atricapilla, 1053
kapitensis, Cisticola natalensis, 493
—, — strangei, 493
karamojæ, Apalis, 418
karamojensis, Eremomela griseoflava, 518
katanga, Cisticola natalensis, 494
katangæ, Uræginthus bengalus, 1036
kathangorensis, Mirafra hypermetra, 10
katonæ, Cisticola, brachyptera, 496
kavirondensis, Andropadus gracilis, 145
—, Chlorocichla gracilis, 145
—, Empidornis semipartitus, 187
—, Heliolais erythroptera, 518
—, Mirafra rufocinnamomea, 16
—, Ploceus luteolus, 904
kellogorum, Spreo hildebrandti, 723
kenia, Erythropygia hartlaubi, 322
keniana, Argya aylmeri, 96
keniensis, Geokichla gurneyi, 332
—, Linurgus olivaceus, 1081
—, Onychognathus walleri, 723
—, Sigmodus scopifrons, 575
kenricki bensoni, Stilbopsar, 715
—, Stilbopsar kenricki, 714
kericho, Cisticola brachyptera, 498
kersteni, Symplectes bicolor, 915
kibaliensis, Trochocercus nigromitratus, 226
kidepoensis, Mirafra hypermetra, 10
kikuyuensis, Alseonax cinereus, 225
—, Arizelocichla tephrolæma, 135
—, Carduelis citrinelloides, 1084
—, chyuluensis, Zosterops, 732
—, Eremomela scotops, 519
—, Lagonosticta senegala, 1053
— mbuluensis, Zosterops, 732
—, Oriolus monacha, 670
—, Trochocercus bivittatus, 261
—, — cyanomelas, 216
—, Zosterops kikuyuensis, 732

# INDEX

kilimandjaricus, Andropabus insularis, 151
kilimensis arturi, Nectarinia, 756
—, Coccopygia melanotis, 1018, 1053
—, Cryptospiza salvadorii, 993, 1054
—, Geokichla gurneyi, 250
—, — piaggiæ, 250, 332
—, Linurgus olivaceus, 1080, 1086
—, Nectarinia kilimensis, 755
— rungwensis, Linurgus, 1086
—, Serinus burtoni, 1079
kilosa, Turdoides hypoleuca, 97
kinneari, Ammomanes cinctura, 25
kirbyi, Heliolais erythroptera, 518
kirki, Sigmodus scopifrons, 575
—, Turdoides jardinei, 86
kirkii, Chalcomitra, 791
kismayensis, Laniarius ruficeps, 616, 643
—, Tchagra jamesi, 627
kivuensis, Cinnyris regius, 789
—, Elminia albicauda, 215
—, Erannornis longicauda, 215
—, Malacocincla pyrrhoptera, 103
kiwanukæ, Estrilda charmosyna, 1053
koliensis, Serinus capistratus, 1067
kordofanensis, Corvus capensis, 674
—, Thamnolæa coronata, 283
kordofanicus, Rhodophoneus cruentus, 643
kretschmeri griseiceps, Suaheliornis, 134
—, Suaheliornis kretschmeri, 134
kulalensis, Zosterops pallidus, 733
kungwensis, Alethe poliocephala, 311
—, Arizelocichla nigriceps, 138
—, Bradypterus alfredi, 372
—, Cossypha insulana, 296, 333
—, — polioptera, 296
—, Dryoscopus angolensis, 622
—, Dyaphoropygia concreta, 213
—, Phyllastrephus flavostriatus, 126
kurræ, Mirafra africana, 13

## L

lacuum, Anthus novæseelandiæ, 67
—, Parus niger, 660
ladoensis, Euplectes afra, 959
lætior, Oriolus brachyrhynchus, 667
lætissima, Chlorocichla, 142
lætus, Seicercus, 392
lagdeni, Malaconotus, 638
Lagonosticta cinereovinacea graueri, 1016, 1053
— — rudolfi, 1053
— jamesoni jamesoni, 1011
— — taruensis, 1012
— larvata larvata, 1016
— — nigricollis, 1017

Lagonosticta larvata togoensis, 1017
— rubricata congica, 1010
— — hæmatocephala, 1011
— — hildebrandti, 1053
— — rhodopareia, 1009, 1053
— rufopicta lateritia, 1015
— senegala abayensis, 1053
— — brunneiceps, 1012, 1053
— — flavodorsalis, 1053
— — kikuyuensis, 1053
— — rendalli, 1013
— — rhodopsis, 1014, 1053
— — ruberrima, 1014, 1053
— — somaliensis, 1013
lahtora, Lanius elegans, 588
lais distincta, Cisticola, 465
— mariæ, Cisticola, 467
— semifasciata, Cisticola, 464
lamperti, Chalcomitra senegalensis, 797, 820
Lamprocolius caudatus, 702
— chalcurus emini, 695
— chalybæus chalybæus, 694
— — sycobius, 695
— chloropterus chloropterus, 696, 722
— — cyanogenys, 722
— — elisabeth, 697
— corruscus jombeni, 701
— — mandanus, 700
— — vaughani, 701
— purpureiceps, 699
— purpureus, 697, 722
— — amethystinus, 722
— splendidus bailundensis, 722
— — splendidus, 698, 722
— superbus, 719
Lamprotornis mevesii mevesii, 703
— purpuropterus æneocephalus, 703
— — purpuropterus, 702
languida, Hippolais, 359
Laniarius æthiopicus æthiopicus, 611
— — ambiguus, 614, 643
— — erlangeri, 614, 643
— — major, 612
— — mossambicus, 613
— — sublacteus, 613, 643
— alboplagatus, 642
— barbarus mufumbiri, 607
— bertrandi, 631
— erythrogaster, 606
— ferrugineus chyulu, 643
— — somaliensis, 643
— fülleborni, 610
— funebris, 608, 642
— — degener, 642
— — rothschildi, 642
— leucorhynchus, 609
— lühderi, 616
— nigerrimus, 643

# INDEX

Laniarius poensis, holomelas, 609
— ruficeps cooki, 643
— — kismayensis, 616, 643
— — ruficeps, 614
— — rufinuchalis, 615
Lanius antinorii antinorii, 642
— — mauritii, 642
— bogdanowi, 603
— cabanisi, 596
— collaris humeralis, 593
— — smithii, 594
— collurio, 600
— cristatus, 570
— — isabellinus, 601
— — phœnicuroides, 602
— elegans aucheri, 589
— — buryi, 589
— — dubarensis, 590
— — elegans, 587, 642
— — lahtora, 588
— — pallidirostris, 588
— — uncinatus, 589
— excubitor dubarensis, 590
— excubitorius böhmi, 592, 642
— — excubitorius, 591
— — intercedens, 642
— gubernator, 601
— jebelmarræ, 590
— leucopygus, 642
— mackinnoni, 598
— marwitzi, 594
— minor, 592
— nubicus, 597
— poliocephalus, 568
— senator niloticus, 603
— somalicus, 595, 642
— souzæ burigi, 599
larvata, Lagonosticta larvata, 1016
— nigricollis, Lagonosticta, 1017
— togoensis, Lagonosticta, 1017
larvatus reichenowi, Oriolus, 670
— rolleti, Oriolus, 665, 670
lateralis antinorii, Cisticola, 471
lateritia, Lagonosticta rufopicta, 1015
lathamii, Quelea quelea, 944
laticauda, Coliuspasser laticauda, 970, 973
— suahelica, Coliuspasser, 971
latirostris australis, Stelgidocichla, 152
— eugenia, Stelgidocichla, 150, 152
latistriatus, Anthus, 69
latukæ, Eremomela badiceps, 438
lauragrayæ, Cinnyricinclus leucogaster, 722
lavendulæ, Cisticola aridula, 458
layardi fayi, Pycnonotus, 113
— micrus, Pycnonotus, 112
—, Pycnonotus xanthopygos, 111
leggei, Anthus brachyurus, 72
lepida, Eminia, 421

leptorhyncha, Calamocichla leptorhyncha, 380, 518
— jacksoni, Calamocichla, 381
— macrorhyncha, Calamocichla, 381
— tsanæ, Calamocichla, 381
leucocephala abyssinica, Turdoides, 84
—, Budytes, 57
—, Turdoides leucocephala, 83
leucogaster arabicus, Cinnyricinclus, 691
—, Cinnyricinclus leucogaster, 690, 722
—, Cinnyris, 820
— friedmanni, Cinnyricinclus, 722
— lauragrayæ, Cinnyricinclus, 722
— verreauxi, Cinnyricinclus, 691, 722
leucolæmia, Thamnolaea arnotti, 332
leucomela cypriaca, Œnanthe, 264
—, Œnanthe leucomela, 263
leucomelæna blanfordi, Sylvia, 352
— somaliensis, Sylvia, 352
leucomelas guineensis, Parus, 651, 660
— insignis, Parus, 651
—, Parus leucomelas, 650, 660
leuconotus, Parus, 653
leucopareia, Eremopterix, 38
leucophrys bohndorffi, Anthus, 65
— chloronota, Sylvietta, 430
— goodsoni, Anthus, 66
— omoensis, Anthus, 64
— ruficauda, Erythropygia, 333
— saphiroi, Anthus, 66
— soror, Erythropygia, 333
—, Sylvietta leucophrys, 429
— vansomereni, Erythropygia, 333
— zenkeri, Anthus, 65, 81
leucopleura, Thescelocichla, 118
leucopogon reichenowi, Prinia, 510
leucopsis, Sylvietta brachyura, 424, 518
leucoptera, Erythropygia leucoptera, 318, 333
— pallida, Erythropygia, 319
— sclateri, Erythropygia, 318
leucopyga ægra, Œnanthe, 265
—, Œnanthe leucopyga, 264
leucopygia clarkei, Turdoides, 91
— hartlaubii, Turdoides, 90
— limbata, Turdoides, 90
— omoensis, Turdoides, 91
— smithii, Turdoides, 91
—, Turdoides leucopygia, 89
leucopygius riggenbachi, Serinus, 1074
—, Serinus leucopygius, 1073
leucopygus, Lanius, 642
leucorhynchus, Laniarius, 609
leucosoma, Bradornis pallidus, 226

# INDEX

leucotis, Eremopterix leucotis, 34
— madaraszi, Eremopterix, 35
— melanocephala, Eremopterix, 35
libanotica, Œnanthe œnanthe, 260
libonyanus cinerascens, Turdus, 332
— costæ, Turdus, 332
— niassæ, Turdus, 332
— tropicalis, Turdus, 240, 332
lichenya, Anthus novæseelandiæ, 69
lightoni, Apalis chirindensis, 404
— — melanocephala, 404
limbata, Turdoides leucopygia, 90
lineata, Pytilia phœnicoptera, 1005
lineiventris, Anthus, 74
Linura fischeri, 1048
Linurgus kilimensis rungwensis, 1086
— olivaceus elgonensis, 1081
— — keniensis, 1081
— — kilimensis, 1080, 1086
Lioptilornis rufocinctus, 176
litoralis, Phyllastrephus flavostriatus, 152
litsipsirupa simensis, Psophocichla, 252
— —, Psophocichla, 253
littoralis, Cisticola natalensis, 494
—, Tchagra australis, 643
livingstonei, Erythrocercus livingstonei, 194, 226
— monapo, Erythrocercus, 226
— thomsoni, Erythrocercus, 194
livingstonii, Œnanthe pileata, 270
Locustella fluviatilis, 362
— luscinioides fusca, 362
— — luscinioides, 361
— nævia straminea, 363
locustella, Ortygospiza locustella, 999, 1054
longicauda kivuensis, Erannornis, 215
— teresita, Erannornis, 214
longipennis, Calandrella brachydactila, 40
longirostris, Anthus sordidus, 81
—, Monticola solitaria, 257
longonotensis, Mirafra africanoides, 46
longuemarei angolensis, Anthreptes, 813
— haussarum, Anthreptes, 811
— neglectus, Anthreptes, 814
— neumanni, Anthreptes, 814
— nyassæ, Anthreptes, 813
— orientalis, Anthreptes, 812, 821
lopezi barakæ, Sathrocercus, 377
—, Sheppardia cyornithopsis, 307
loveridgei, Argya aylmeri, 96
—, Cinnyris, 790
lowei, Alethe, 314
—, Cyanomitra olivacea, 805
lucidipectus, Nectarinia pulchella, 760

ludoviciæ, Turdus, 240
ludwigii, Dicrurus ludwigii, 565
— münzneri, Dicrurus, 566
— sharpei, Dicrurus, 566, 567
lufira, Cisticola woosnami, 473
lugens clara, Parisoma, 175
— jacksoni, Parisoma, 175
—, Parisoma lugens, 174
— persica, Œnanthe, 266
lugubris, Cisticola galactotes, 485
— major, Pœoptera, 713
—, Melænornis edolioides, 184, 226
— nyansæ, Cisticola, 486
—, Œnanthe lugubris, 268
— schalowi, Œnanthe, 268
— suahelica, Cisticola, 486
— vauriei, Œnanthe, 269
lühderi, Laniarius, 616
lurio, Cisticola emini, 475
Luscinia luscinia, 332
— megarhyncha golzii, 333
— megarhynchos africana, 331
— — hafizi, 331, 333
— — megarhynchos, 330
luscinioides fusca, Locustella, 362
—, Locustella luscinioides, 361
lusitana erythrochroa, Ammomanes, 27
luteola kavirondensis, Sitagra, 904
luteolus kavirondensis, Ploceus, 904
—, Ploceus luteolus, 903
luteus, Auripasser, 881
—, Budytes luteus, 54, 58
— flavissimus, Budytes, 55, 58
lynesi, Anthus novæseelandiæ, 68
—, — rufulus, 68
—, Apalis flavigularis, 399
—, Batis orientalis, 205
—, Hirundo atrocærulea, 531
—, Mirafra, rufa, 18
lypura, Cercomela melanura, 272

## M

mabiræ, Stiphrornis erythrothorax, 307
macarthuri, Pogonocichla stellata, 324
macclounii, Dessonornis anomala, 303, 333
— mbuluensis, Bessonornis, 304
— njombe, Alethe, 333
macdonaldi, Mirafra africanoides, 46
—, Sathrocercus cinnamomeus, 375
mackenzianus, Seicercus umbrovirens, 389, 518
mackinnoni, Lanius, 598
macmillani, Estrilda astrild, 1053
macphersoni, Apalis chariessa, 519
macrocercus, Coliuspasser, 962
Macronyx ameliæ wintoni, 80
— aurantiigula, 79, 81

# INDEX

Macronyx aurantiigula subocularis, 81
— croceus, 76
— flavicollis, 78
— fülleborni, 77
— sharpei, 78
macrorhyncha, Calamocichla leptorhyncha, 381
Macrosphenus concolor, 447
— flavicans hypochondriacus, 446
macrourus, Coliuspasser macrourus, 963
— conradsi, Coliuspasser, 965
maculicollis, Serinus dorsostriatus, 1064
— taruensis, Serinus, 1065
madagascariensis, Phedina borbonica, 540
madaraszi, Eremopterix leucotis, 35
magna, Cyanosylvia svecica, 329
—, — wolfi, 329
magnificus, Cosmopsarus regius, 722
mahali melanorhynchus, Plocepasser, 863
— pectoralis, Plocepasser, 862
— propinquatus, Plocepasser, 864
major, Heliolais erythroptera, 513
—, Laniarius æthiopicus, 612
—, Pœoptera lugubris, 713
malabarica cantans, Euodice, 979, 1052
— elliotti, Galerida, 30
—, Galerida, 46
— huriensis, Galerida, 31
— orientalis, Euodice, 980, 1053
— prætermissa, Galerida, 30
Malacocincla albipectus, 102, 106
— fulvescens ugandæ, 100
— pyrrhoptera kivuensis, 103
— — pyrrhoptera, 102
— rufipennis distans, 101
— — puguensis, 101
— — rufipennis, 100
Malaconotus alius, 638
— blanchoti blanchoti, 636, 643
— — catharoxanthus, 637
— lagdeni, 638
— poliocephalus approximans, 643
— — hypopyrrhus, 643
— — poliocephalus, 643
malensis, Apalis flavida, 409
malimbicus crassirostris, Malimbus, 937
Malimbus erythrogaster fagani, 939
— malimbicus crassirostris, 937
— nigricollis, 917
— rubricollis centralis, 938
malzacii, Dryoscopus gambensis, 620, 643
— erythreæ, Dryoscopus, 621

mandanus, Lamprocolius corruscus, 700
—, Tchagra jamesi, 627
Mandingoa nitidula chubbi, 1004
— — nitidula, 1003
— — schlegeli, 1004
manningi, Chlorophoneus, nigrifrons, 643
manoensis, Cinnyris chalybeus, 820
marginata, Cisticola galactotes, 486
—, Mirafra cantillans, 6, 46
marginatus, Cinnyris minullus, 788
mariæ, Cisticola lais, 467
—, — njombe, 467
— granti, Sathrocercus, 377
—, Sathrocercus mariæ, 375, 518
— usambaræ, Sathrocercus, 376, 518
mariquensis hawkeri, Cinnyris, 820
— osiris, Cinnyris, 772, 820
— suahelicus, Cinnyris, 771
marsabit, Alseonax adustus, 167
—, — marsabit, 167
—, Parisoma böhmi, 173
—, Phyllastrephus fischeri, 152
martini, Campephaga quiscalina, 559
marwitzi, Estrilda paludicola, 1026
—, Hirundo dimidiata, 529
—, Lanius, 594
masaba, Cisticola hunteri, 479
massaica, Chloropeta natalensis, 191
—, Estrilda astrild, 1053
—, Gymnoris pyrgita, 885
—, Mirafra pœcilosterna, 46
—, Petronia xanthosterna, 885
—, Psalidoprocne holomelæna, 554
massaicus, Nilaus afer, 586, 642
masukuensis, Arizelocichla masukuensis, 140
— kakamegæ, Arizelocichla, 141
— roehli, Arizelocichla, 140
matengorum, Cisticola natalensis, 519
mauensis, Cisticola ayresii, 460
—, — terrestris, 460
mauritii, Lanius antinorii, 642
maximus, Pirenestes, 990
maxwelli, Melanoploceus albinucha, 936
mbololo, Seicercus ruficapilla, 518
mbuluensis, Dessonornis anomala, 304
—, Zosterops kikuyuensis, 732
mccallii congicus, Erythrocercus, 193
mediocris bensoni, Cinnyris, 784
—, Cinnyris mediocris, 783
— fülleborni, Cinnyris, 785
— moreaui, Cinnyris, 785
— usambaricus, Cinnyris, 784
Megabyas flammulatus æquatorialis, 196

# INDEX

megaensis, Calandrella rufescens, 42
—, Hirundo, 530
—, Pseudalæmon fremantlii, 34
megarhyncha golzii, Luscinia, 333
megarhynchos africana, Luscinia, 331
— hafizi, Luscinia, 331, 333
—, Luscinia megarhynchos, 330
melæna, Pentholaea, 279
Melænornis edolioides lugubris, 184, 226
— — schistacea, 184
— — ugandæ, 226
— pammelaina pammelaina, 185
— — tropicalis, 186
melamprosopus, Chlorophoneus, 643
melba belli, Pytilia, 1053
— citerior, Pytilia, 1009
— grotei, Pytilia, 1008
— jessei, Pytilia, 1053
— jubaensis, Pytilia, 1054
—, Pytilia melba, 1007, 1053
— soudanensis, Pytilia, 1009, 1053, 1054
melanauchen, Eremopterix nigriceps, 36
melanocephala, Apalis melanocephala, 402
— ellinoræ, Apalis, 404
—, Eremopterix leucotis, 35
— fuliginosa, Apalis, 403
— lightoni, Apalis, 404
— momus, Sylvia, 353
— moschi, Apalis, 403
— muhuluensis, Apalis, 405, 519
— mystacea, Sylvia, 354
— nigrodorsalis, Apalis, 402
— songeænsis, Apalis, 519
— tenebricosa, Apalis, 403
melanochrous, Serinus burtoni, 1079
Melanocorypha bimaculata bimaculata, 39
— — rufescens, 39
melanocrissa, Hirundo daurica, 533
melanogaster stephanophorus, Heterhyphantes, 927, 973
melanogastra, Nectarinia pulchella, 759
melanogriseus, Budytes flavus, 58
melanoleuca, Œnanthe hispanica, 263
melanoleucus æquatorialis, Urolestes, 605
Melanoploceus albinucha maxwelli, 936
— tricolor interscapularis, 935
melanops, clamosa, Turdoides, 97
—, Prinia bairdii, 511
— sharpei, Turdoides, 88, 97
— vepres, Turdoides, 88
melanoptera, Cossypha niveicapilla, 333
—, Prionops cristata, 576

Melanopteryx nigerrimus, 926
— weynsi, 927
melanorhynchus, Plocepasser mahali, 863
melanota, Amblyospiza albifrons, 940, 1052
melanotis, Anaplectes melanotis, 942
— jubaensis, Anaplectes, 944
— kilimensis, Coccopygia, 1018, 1053
— nyansæ, Coccopygia, 1053
— quartinia, Coccopygia, 1018
— rubriceps, Anaplectes, 943
melanoxanthus, Hyphanturgus, nigricollis, 918
melanura airensis, Cercomela, 272
— aussæ, Cercomela, 273
— lypura, Cercomela, 272
melindæ, Anthus, 71
Melocichla mentalis amauroura, 514, 519
— — atricauda, 515
— — chyulu, 519
— — grandis, 513
— — granviki, 515
— — orientalis, 515
meneliki, Oriolus monacha, 665, 670
mennelli, Serinus, 1072
mentalis amauroura, Melocichla, 514, 519
—, Argya aylmeri, 97
— atricauda, Melocichla, 515
— chyulu, Melocichla, 519
— grandis, Melocichla, 513
— granviki, Melocichla, 515
— orientalis, Melocichla, 515
—, Platysteira peltata, 211, 226
—, Sigmodus caniceps, 572
—, Symplectes bicolor, 916
meridionalis, Alæmon alaudipes, 24
—, Amadina fasciata, 997, 1054
—, Eudoice cantans, 1053
meruensis, Zosterops virens, 736
metallica, Hedydipna platura, 764
metopias altus, Artisornis, 519
—, Artisornis, 419, 518
mevesii, Lamprotornis mevesii, 703
microrhynchus, Bradornis, 179, 226
—, Cinnyris bifasciatus, 768
micrura, Sylvietta brachyura, 518
micrus, Pycnonotus layardi, 112
milanjensis, Arizelocichla milanjensis, 152
— chyulu, Arizelocichla, 152
— striifacies, Arizelocichla, 139, 152
—, Turdus olivaceus, 244
minima, Sylvietta whytii, 426
minimus, Alseonax adustus, 164, 225
— chyulu, Alseonax, 167
— djamdjamensis, Alseonax, 225

# INDEX

minimus interpositus, Alseonax, 226
— marsabit, Alseonax, 167
minor, Agrobates galactotes, 357
—, Anthreptes seimundi, 816
—, Batis minor, 206
—, — orientalis, 206
—, Charitillas, 152
— erlangeri, Batis, 206, 226
—, Estrilda astrild, 1019
—, Lanius, 592
—, Nilaus afer, 585
— nyanzæ, Batis, 226
—, Pirenestes, 990
—, Pycnonotus tricolor, 151
—, Riparia paludicola, 542
— ruwenzorii, Nilaus, 642
— schoensis, Riparia, 543
—, Spermestes nigriceps, 978
— suahelica, Batis, 207
—, Tchagra australis, 625, 643
minullus marginatus, Cinnyris, 788
—, Seicercus ruficapillus, 386, 518
minuscula, Artomyias fuliginosa, 190
minuta anchietæ, Bocagia, 629
—, Bocagia minuta, 628
Mirafra africana athi, 11, 46
— — dohertyi, 46
— — zuluensis, 12
— — harterti, 13
— — kurræ, 13
— — nigrescens, 11
— — ruwenzoria, 46
— — tropicalis, 11, 46
— africanoides alopex, 17
— — intercedens, 17, 46
— — longonotensis, 46
— — macdonaldi, 46
— albicauda, 7
— candida, 46
— cantillans chadensis, 7
— — marginata, 6, 46
— cheniana schillingsi, 46
— collaris, 18
— cordofanica, 8
— fischeri degeni, 46
— — zombæ, 46
— gilletti, 19
— hypermetra gallarum, 9
— — hypermetra, 9
— — kathangorensis, 10
— — kidepoensis, 10
— pœcilosterna australoabyssinica, 20
— — massaica, 46
— — pœcilosterna, 20, 46
— pulpa, 46
— rufa lynesi, 18
— — rufa, 18
— rufocinnamomea fischeri, 14, 46
— — furensis, 16
— — kavirondensis, 16

Mirafra rufocinnamomea omoensis, 16
— — rufocinnamomea, 13, 46
— — sobatensis, 15
— — tigrina, 15
— — torrida, 15
— sharpei, 10
mistacea immutabilis, Prinia, 518
—, Prinia mistacea, 518
— superciliosa, Prinia, 518
mixta, Batis mixta, 200
mocuba, Cisticola chiniana, 519
modesta bucolica, Heliocorys, 32
— giffardi, Heliocorys, 32
—, Heliocorys modesta, 31
modestus coracinus, Dicrurus, 563, 567
— ugandensis, Dicrurus, 567
Modulatrix stictigula pressa, 302
— — stictigula, 301
molitor, Batis molitor, 202, 226
— puella, Batis, 226
— soror, Batis, 203
momus, Sylvia melanocephala, 353
monacha, Œnanthe, 267
— kikuyuensis, Oriolus, 670
— meneliki, Oriolus, 665, 670
—, Oriolus monacha, 664
— permistus, Oriolus, 670
monapo, Erythrocercus livingstonei, 226
mongalla, Cisticola ruficeps, 504
montana, Alethe, 313
—, Amblyospiza albifrons, 941
montanus, Onychognathus, 709
monteiri, Hirundo senegalensis, 536, 554
— ugandensis, Clytospiza, 985
Monticola angolensis, 258
— rufocinerea rufocinerea, 258, 333
— saxatilis, 256
— solitaria longirostris, 257
— — solitaria, 257
moreaui, Apalis moreaui, 418
—, Bradypterus babœcala, 370
—, — brachypterus, 370
—, Cinnyris mediocris, 785
— sousæ, Apalis, 418
morio montanus, Onychognathus, 709
— neumanni, Onychognathus, 709
—, Onychognathus morio, 708
— rüppellii, Onychognathus, 709, 722
— shelleyi, Onychognathus, 722
mosambicus, Passer griseus, 1052
moschi, Apalis melanocephala, 403
mossambicus, Laniarius æthiopicus, 613
Motacilla agiump vidua, 49
— alba alba, 48

# INDEX

Motacilla alba forwoodi, 48
— capensis wellsi, 51
— cinerea cinerea, 52
— cinereocapilla, 56
— clara clara, 50
— — torrentium, 50
— feldegg, 56
— flava, 53
— flavissima, 55
— thunbergi, 55
motitensis cordofanicus, Passer, 872
— shelleyi, Passer, 872
mozambica, Tchagra senegala, 643
mozambicus aurifrons, Serinus, 1086
— barbatus, Serinus, 1062
— gertrudis, Serinus, 1086
— grotei, Serinus, 1062, 1086
— gommaensis, Serinus, 1063
— pseudobarbatus, Serinus, 1062
—, Serinus mozambicus, 1061, 1086
muelleri, Cisticola fulvicapilla, 499
mufumbiri, Laniarius barbarus, 607
muhuluensis, Apalis melanocephala, 405, 519
mülleri, Ortygospiza atricollis, 998
multicolor batesi, Chlorophoneus, 629, 643
münzneri, Campephaga quiscalina, 560
—, Chlorophoneus rubiginosus, 643
—, Cisticola cantans, 481
—, Dicrurus ludwigii, 566
murina, Apalis murina, 396
— fuscigularis, Apalis, 396
— whitei, Apalis, 397
— youngi, Apalis, 397
murinus, Alseonax adustus, 165
— roehli, Alseonax, 166
Muscicapa albicollis, 164
— gambagæ somaliensis, 225
— hypoleuca semitorquata, 163
— striata gambagæ, 162, 225
— — neumanni, 162
— — striata, 161
musculus, Anthoscopus musculus, 658
— guasso, Anthroscopus, 659
musicus changamwensis, Bias, 197
— femininus, Bias, 197
Myopornis böhmi, 188
Myrmecocichla æthiops cryptoleuca, 285
— — sudanensis, 286
— nigra, 284
— mystacea, Sylvia melanocephala, 354
mystica, Batis, 226

## N

nævia straminea, Locustella, 363
nakuruensis, Cisticola brunnescens, 463
—, — terrestris, 463
namuli, Alethe choloensis, 310
namwera, Cinnyris chalybeus, 820
nana, Cisticola, 501
—, Sylvia nana, 355
nandensis, Dryoscopus angolensis, 622
natalensis argentea, Cisticola, 493
—, Chloropeta natalensis, 190
—, Cisticola natalensis, 491, 519
—, Cossypha, 297
— inexpectata, Cisticola, 493
— kapitensis, Cisticola, 493
— katanga, Cisticola, 494
— littoralis, Cisticola, 494
— massaica, Chloropeta, 191
— matengorum, Cisticola, 519
— tonga, Cisticola, 494
— valida, Cisticola, 492
natalicus, Turdus fischeri, 333
naumanni, Pycnonotus tricolor, 152
ndussumensis, Trichophorus calurus, 116
Nectarinia acik, 796
— erythrocerca, 757, 820
— famosa æneigularis, 820
— — centralis, 820
— — cupreonitens, 751, 820
— — subfamosa, 820
— johnstoni dartmouthi, 754
— — johnstoni, 752
— — salvadorii, 753
— kilimensis arturi, 756
— — kilimensis, 755
— metallica, 764
— nectarinioides beveni, 820
— — erlangeri, 761
— — nectarinioides, 760
— pulchella lucidipectus, 760
— — melanogastra, 759
— — pulchella, 758
— purpureiventris, 757
— tacazze, 754
nectarinioides beveni, Nectarinia, 820
— erlangeri, Nectarinia, 761
—, Nectarinia nectarinioides, 760
neglecta, Apalis caniceps, 410
—, Cyanomitra olivacea, 803, 821
neglectus, Anthreptes longuemariæ, 814
Neocichla gutturalis angustus, 689
Neocossyphus poensis præpectoralis, 255
— rufus gabunensis, 255
— — rufus, 254

Neolestes torquatus, 150
Nesocharis ansorgei, 1033
— capistrata, 1032
neumanni, Anthreptes longuemarei, 814
—, Arizelocichla nigriceps, 137
—, Bradornis pallidus, 178, 225
—, Hypochera ultramarina, 1042
—, Muscicapa striata, 162
—, Onychognathus, 709
—, Passer griseus, 878, 1052
—, Sigmodus retzii, 576
neumannianus, Anthus similis, 81
niassæ, Cinnyris venustus, 820
—, Turdus libonyanus, 332
—, Zosterops senegalensis, 736
niassensis, Euplectes afra, 959
—, Uræginthus angolensis, 1033
Nicator chloris chloris, 640
— — gularis, 641
—˙vireo, 642
nicolli, Hyphanturgus, 922
nicholsoni chyuluensis, Anthus, 81
— hararensis, Anthus, 62
— nyassæ, Anthus, 63
niger intermedius, Bubalornis, 860, 1052
— lacuum, Parus, 660
—, Parus, 649
— purpurascens, Parus, 660
nigeriæ, Hypochera, 1045
nigerrima, Hypochera, 1044, 1053
nigerrimus, Laniarius, 643
—, Melanopteryx, 926
nigra, Myrmecocichla, 284
nigrescens, Apalis, 405
—, Chlorophoneus, 643
—, Mirafra africana, 11
nigricans, Pinarocorys, 21
nigricauda, Telophorus quadricolor, 635, 643
nigriceps albifrons, Eremopterix, 36
—, Arizelocichla nigriceps, 136
— chlorigula, Arizelocichla, 138
— collaris, Apalis, 413
— emini, Tchitrea, 224
— fusciceps, Arizelocichla, 136
— grauerii, Ploceus, 1052
— kungwensis, Arizelocichla, 138
— melanauchen, Eremopterix, 36
— minor, Spermestes, 978
— neumanni, Arizelocichla, 136
— percivali, Arizelocichla, 151
nigriceps, Ploceus, 891, 1052
—, Serinus, 1084
— somereni, Tchitrea, 225
—, Spermestes nigriceps, 977
— usambaræ, Arizelocichla, 137, 151
nigricollis, Hyphanturgus nigricollis, 917
—, Lagonosticta larvata, 1017

nigricollis melanoxanthus, Hyphanturgus, 918
nigrifrons, Chlorophoneus, 632, 643
— conceptus, Chlorophoneus, 643
— elgeyuensis, Chlorophoneus, 843
—, Euplectes orix, 950, 1052
— manningi, Chlorophoneus, 843
—, Pyromelana, 950
nigriloris, Cisticola, 476
nigripennis, Oriolus nigripennis, 667
— percivali, Oriolus, 668
Nigrita bicolor brunnescens, 984, 1053
— — saturatior, 1053
— canicapilla candida, 983
— — canicapilla, 982
— — diabolica, 983
— — schistacea, 983
— fusconota fusconota, 984
nigritemporalis, Nilaus, 586, 643
nigrodorsalis, Apalis melanocephala, 402
nigromitratus kibaliensis, Trochocercus, 228
— toroensis, Trochocercus, 219, 226
nigrotemporalis, Othyphantes reichenowi, 931
nigroventris, Euplectes, 971, 1052
Nilaus afer afer, 584, 642
— — brevialatus, 643
— — erythreæ, 642
— — hilgerti, 642
— — massaicus, 586, 642
— — minor, 585
— — ruwenzorii, 642
— — nigritemporalis, 586, 643
nilotica, Calamocichla gracilirostris, 379
—, Cisticola erythrops, 482
niloticus, Lanius senator, 603
—, Passer domesticus, 871
nisoria, Sylvia, 355
nitens, Trochocercus nitens, 218
nitidula chubbi, Mandingoa, 1004
—, Mandingoa nitidula, 1003
— schlegeli, Mandingoa, 1004
niveicapilla, Cossypha, 299, 333
— melanoptera, Cossphya, 333
niveoguttatus, Hypargos, 1002
nivescens, Anthus similis, 62
njombe aberrans, Cisticola, 466
—, Alethe macclounii, 333
—, Cisticola njombe, 466
— mariæ, Cisticola, 467
nonnula, Estrilda nonnula, 1029, 1053
noomei, Camaroptera brevicaudata, 518
notatus, Oriolus auratus, 664
novæseelandiæ annæ, Anthus, 68
—, Anthus, 81
— cinnamomeus, Anthus, 67

# INDEX

novæseelandiæ lacuum, Anthus, 67
— lichenya, Anthus, 69
— lynesi, Anthus, 68
nubicus, Lanius, 597
nuchalis, Cisticola robusta, 489
nyansæ, Cisticola galactotes, 486
—, Coccopygia melanotis, 1053
nyanzæ, Batis minor, 226
—, Bubalornis albirostris, 1052
—, Dryoscopus gambensis, 643
—, Estrilda astrild, 1021
—, Platysteira cyanea, 208
nyasa, Cisticola erythrops, 483
nyassæ, Onychognathus walleri, 722
nyassæ, Anomalospiza imberbis, 1054
—, Anthreptes longuemarei, 813
—, Anthus similis, 63
—, Sathrocerus cinnamomeus, 373
nyikæ, Turdus olivaceus, 245, 332
nyikensis, Dioptrornis fischeri, 181

## O

obbiensis, Spizocorys, 44
obsoleta arabica, Ptyonoprogne, 546
—, Ptyonoprogne obsoleta, 545
— reichenowi, Ptyonoprogne, 546
obtusa, Steganura aucupum, 1051
—, — orientalis, 1051
occidentalis, Chlorocichla flaviventris, 141
occipitalis, Eremomela scotops, 436, 519
ochrogaster, Estrilda paludicola, 1026
ochrogularis, Seicercus ruficapillus, 388
ochruros phœnicuroides, Phœnicurus, 327
ocreata, Fraseria ocreata, 186
ocularis crocatus, Hyphanturgus, 918, 1052
— suahelicus, Hyphanturgus, 919
ocularis, Cryptospiza reichenovii, 1053
ocularius abayensis, Ploceus, 1052
Odontospiza caniceps, 981
Œnanthe bottæ frenata, 269
— chrysopygia, 266
— deserti atrogularis, 332
— — deserti, 262, 332
— — oreophila, 262
— heuglini, 270
— hispanica melanoleuca, 263
— isabellina, 261
— leucomela cypriaca, 264
— — leucomela, 263
— leucopyga ægra, 265
— — leucopyga, 264
— lugens persica, 266
— lugubris lugubris, 268

Œnanthe lugubris schalowi, 268
— — vauriei, 269
— monacha, 267
— nigra, 284
— œnanthe libanotica, 260
— — œnanthe, 259
— phillipsi, 261
— pileata livingstonii, 270
— xanthroprymna, 265
œnanthe libanotica, Œnanthe, 260
oldeani, Turdus olivaceus, 245
oleaginea, Psalidoprocne orientalis, 551
olimotiensis, Pinarochroa sordida, 277
olivacea alfredi, Cyanomitra, 804
— changamwensis, Cyanomitra, 803
— chyulu, Cyanomitra, 821
— granti, Cyanomitra, 805, 821
— lowei, Cyanomitra, 805
— neglecta, Cyanomitra, 803, 821
— olivacina, Cyanomitra, 802
— pembæ, Cyanomitra, 821
— puguensis, Cyanomitra, 804
— ragazzi, Cyanomitra, 805
— vincenti, Cyanomitra, 806
olivaceiceps, Hyphanturgus, 920
olivaceo-griseus, Phyllastrephus flavostriatus, 125
olivaceus abyssinicus, Turdus, 243, 332, 333
— baraka, Turdus, 244
— chyuluensis, Turdus, 333
— deckeni, Turdus, 243
— elgonensis, Linurgus, 1081
— elgonensis, Turdus, 332
— graueri, Turdus, 246, 333
— keniensis, Linurgus, 1081
— kilimensis, Linurgus, 1080, 1086
— milanjensis, Turdus, 244
— nyikæ, Turdus, 245, 332
— oldeani, Turdus, 245
— roehli, Turdus, 245
— uluguru, Turdus, 332
olivacina, Cyanomitra olivacea, 802
olivetorum, Hippolais, 359
omo, Cisticola robusta, 490
omoensis, Anthus leucophrys, 64
—, Cercomela familiaris, 275
—, Cossypha albicapilla, 298
—, Mirafra rufocinnamomea, 16
—, Ploceus pachyrhynchus, 1052
—, Prionops cristata, 576
—, Seicercus umbrovirens, 390
—, Turdoides leucopygia, 91
—, Zosterops abyssinicus, 735
Onychognathus blythii, 710
— frater, 707
— fulgidus hartlaubii, 705
— — intermedius, 706
— morio montanus, 709

## INDEX

Onychognathus morio morio, 708
— — neumanni, 709
— — rüppellii, 709, 722
— — shelleyi, 722
— tenuirostris raymondi, 723
— — tenuirostris, 710, 723
— — theresæ, 711
— walleri elgonensis, 707, 723
— — keniensis, 723
— — nyasæ, 722
—'— walleri, 706, 722, 723
oraria, Petronia superciliaris, 1053
oreophila, Cisticola tinniens, 504
—, Œnanthe deserti, 262
orientale, Parisoma plumbeum, 174
orientalis, Anthreptes longuemarei, 812, 821
— barbouri, Anthreptes, 821
—, Batis orientalis, 204, 226
— bella, Batis, 226
— chadensis, Batis, 205
—, Emberiza, 1089
—, Euodice malabarica, 980, 1053
—, Hypochera ultramarina, 1042
— lynesi, Batis, 205
—, Melocichla mentalis, 515
— minor, Batis, 206
— obtusa, Steganura, 1051
— oleaginea, Psalidoprocne, 551
—, Phylloscopus bonelli, 385
—, Pogonocichla stellata, 323, 333
—, Psalidiprocne orientalis, 550
—, — petiti, 550
—, Pyrrhurus scandens, 120
—, Saxicola torquata, 333
—, Steganura orientalis, 1051, 1054
—, Tchagra senegala, 643
—, Vidua paradisea, 1051
Oriolus auratus auratus, 663
— — notatus, 664
— brachyrhynchus lætior, 667
— chlorocephalus amani, 669
— — chlorocephalus, 669
— larvatus reichenowi, 670
— — rolleti, 665, 670
— monacha kikuyuensis, 670
— — meneliki, 665, 670
— — monacha, 664
— — permistus, 670
— nigripennis nigripennis, 667
— — percivali, 668
— oriolus oriolus, 662
orix franciscana, Euplectes, 949
— nigrifrons, Euplectes, 950, 1052
— pusilla, Euplectes, 951, 1053
— sundevalli, Euplectes, 950
— wertheri, Euplectes, 1052
orleansi, Empidornis semipartitus, 187
ornatus, Ploceus insignis, 1052

orostruthus amani, Phyllastrephus, 133
—, Phyllastrephus orostruthus, 132
orphogaster, Cinnyris chloropygius, 787
Orthotomus *see* Heliolais
Ortygospiza atricollis atricollis, 997
— — dorsostriata, 999, 1054
— — fuscocrissa, 998
— — mülleri, 998
— locustella locustella, 999, 1054
oryzivora, Padda, 981
oseus decorsei, Cinnyris, 775
osiris, Cinnyris mariquensis, 772, 820
ostrinus centralis, Pirenestes, 1053
—, Pirenestes, 989
Othyphantes emini, 928, 1052
— reichenowi fricki, 930
— — nigrotemporalis, 931
— — reichenowi, 929
— stuhlmanni sharpii, 932
— — stuhlmanni, 931
otomitra, Geokichla gurneyi, 332
oustaleti, Cinnyris, 777

## P

Pachyphantes pachyrhynchus, 932, 1052
Pachyprora *see* Batis
pachyrhyncha, Pentholæa albifrons, 278
—, Poliospiza, 1086
pachyrhynchus omoensis, Ploceus, 1052
—, Pachyphantes, 932, 1052
Padda oryzivora, 981
pallida, elæica, Hippolais, 360
—, Erythropygia leucoptera, 319
—, Hippolais pallida, 360
—, Petronia xanthosterna, 885
—, Quelea cardinalis, 948
—, Sylvietta rufescens, 427
pallidiceps, Ploceus aureoflavus, 1053
pallidigaster, Anthreptes, 816
pallidigula, Antimastillas flavicollis, 151
pallidior, Serinus tristriatus, 1070
pallidirostris, Lanius elegans, 588
pallidiventris, Parus rufiventris, 654, 660
pallidus bowdleri, Bradornis, 178, 225
—, Bradornis pallidus, 177
—, chyuluensis, Bradornis, 226
— granti, Bradornis, 225
— griseus, Bradornis, 178, 225, 226
— kulalensis, Zosterops, 733
— leucosoma, Bradornis, 226
— neumanni, Bradornis, 178, 225

# INDEX

pallidus poliogastra, Zosterops, 733
— sharpei, Bradornis, 225
— suahelicus, Bradornis, 225
— subalaris, Bradornis, 177, 226
— winifredæ, Zosterops, 733
paludicola ducis, Riparia, 543
—, Estrilda paludicola, 1025
— marwitzi, Estrilda, 1026
— minor, Riparia, 542
— ochrogaster, Estrilda, 1026
—, Riparia paludicola, 542
— roseicrissa, Estrilda, 1025
— schoensis, Riparia, 543
Paludipasser see Ortygospiza
palustris, Acrocephalus, 366
—, Calamornis, 518
pammelaina, Melænornis pammelaina, 185
— tropicalis, Melænornis, 186
paradisæa, Steganura, 1049
Parisoma böhmi böhmi, 172
— — marsabit, 173
— — somalicum, 173
— lugens clara, 175
— — jacksoni, 175
— — lugens, 174
— plumbeum orientale, 174
— — plumbeum, 173
Parophasma galinieri, 176
Parus afer barakæ, 660
— — griseiventris, 647, 660
— — parvirostris, 660
— — thruppi, 648, 660
— albiventris, 652, 660
— — curtus, 660
— fringillinus, 654
— funereus, 653
— fusciiventer, 649
— leucomelas guineensis, 651, 660
— — insignis, 651
— — leucomelas, 650, 660
— leuconotus, 653
— niger, 649
— — lacuum, 660
— — purpurascens, 660
— rufiventris pallidiventris, 654, 660
— — rovumæ, 660
parva, Calamocichla gracilirostris, 378
parvirostris, Parus afer, 660
parvulus, Anthoscopus parvulus, 659
Passer castanopterus castanopterus, 875
— — fulgens, 876
— domesticus arboreus, 870
— — bactrianus, 871
— — indicus, 869
— — niloticus, 871
— gongonensis gongonensis, 880, 1053
— — jubaensis, 881

Passer griseus eritreæ, 1052
— — griseus, 877
— — jubaensis, 881
— — mosambicus, 1052
— — neumanni, 878, 1052
— — tertale, 1053
— — turkanæ, 1053
— — ugandæ, 878, 1052
— hispaniolensis hispaniolensis, 874
— insularis hemileucus, 874
— — insularis, 874
— motitensis cordofanicus, 872
— — shelleyi, 872
— rufocinctus, 873
— simplex, 876
— suahelicus, 879
— swainsonii, 879, 973
peasei, Estrilda astrild, 1021, 1053
pectoralis, Coracina, 560
—, Plocepasser, 862
Pedilorhynchus comitatus stuhlmanni, 189
— epulatus seth-smithi, 169
pelios centralis, Turdus, 242
— schuetti, Turdus, 241
—, Turdus pelios, 241
— ubendeensis, Turdus, 333
peltata jacksoni, Platysteira, 226
— mentalis, Platysteira, 211, 226
—, Platysteira peltata, 210
pelzelni futa, Icteropsis, 1052
—, Icteropsis, 925, 1052
pembæ, Cinnyris, 771
—, Cyanomitra olivacea, 821
Pentholaea albifrons albifrons, 278
— — clericalis, 278
— — pachyrhyncha, 279
— melæna, 279
percivali, Arizelocichla nigriceps, 151
—, Oriolus nigripennis, 668
—, Stelgidillas gracilirostris, 143
perconfusus, Budytes, 54
perennia, Cisticola juncidis, 456
perkeo, Batis, 205
permistus, Oriolus monachus, 670
perreini, Estrilda perreini, 1027
— incana, Estrilda, 1027
persica, Œnanthe lugens, 266
personata, Aëthocorys personata, 45
—, Apalis binotata, 411
— intensa, Aëthocorys, 45
— yavelloensis, Aëthocorys, 45
perspicillata, Tchitrea, 226
— ruwenzoriæ, Tchitrea, 222
— ungujaensis, Tchitrea, 226
petiti, Campephaga, 558
— orientalis, Psalidoprocne, 550
Petronia brachydactyla, 887
— dentata dentata, 886
— superciliaris, 883, 1053
— — oraria, 1053

# INDEX

Petronia xanthosterna kakamariæ, 885
—— massaica, 885
—— pallida, 885
—— pyrgita, 884, 1052, 1053
Petrophila rufocinerea tenis, 333
petrophila, Cisticola emini, 474
Phedina borbonica madagascariensis, 540
phillipsi, Œnanthe, 261
philomelos, Turdus ericetorum, 239
phœnicea, Campephaga, 558, 562
phœniceus, Coliuspasser axillaris, 961
phœnicoptera emini, Pytilia, 1006
— lineata, Pytilia, 1005
phœnicuroides, Lanius cristatus, 602
—, Phœnicurus ochruros, 327
Phœnicurus ochruros phœnicuroides, 327
— phœnicurus phœnicurus, 326
—— samamisicus, 327
phœnicurus samamiscus, Phœnicurus, 327
Pholia femoralis, 693
— sharpii, 692
Pholidornis rushiæ denti, 819
Phormoplectes insignis insignis, 933, 1052
Phyllastrephus albigula shimbanus, 152
— albigularis, 131
— alfredi, 126
—— itoculo, 152
—— cabanisi sucosus, 151
— capensis suahelicus, 122
— cerviniventris, 130
— debilis albigula, 127
—— debilis, 127
—— rabai, 128, 152
— fischeri cabanisi, 130, 151
—— chyuluensis, 152
—— fischeri, 128, 152
—— marsabit, 152
—— placidus, 129, 152
— flavostriatus kungwensis, 126
—— litoralis, 152
—— olivaceo-griseus, 125
—— tenuirostris, 124, 152
—— vincenti, 125
— hypochloris, 131
— icterinus sethsmithi, 133
— kretschmeri, 134
— orostruthus amani, 133
—— orostruthus, 132
— strepitans, 123
— sucosus sylvicultor, 151
— tephrolæmus usambaræ, 137
— terrestris bensoni, 123
—— suahelicus, 122
— xavieri sethsmithi, 133

Phyllolais pulchella, 422
Phylloscopus bonelli orientalis, 385
— collybita abietinus, 384
—— collybita, 384
— sibilatrix, 385
— trochilus acredula, 383, 517
—— eversmani, 517
—— trochilus, 382
piaggiæ, Geokichla piaggiæ, 249
— hadii, Geokichla, 251
— kilimensis, Geokichla, 250, 332
— rowei, Geokichla, 250
— williamsi, Geokichla, 251
pictipennis, Cisticola cantans, 481
pileata, Camaroptera brachyura, 439
— livingstonii, Œnanthe, 270
Pilorhinus albirostris, 712
Pinarochroa sordida erlangeri, 332
—— ernesti, 277
—— hypospodia, 276
—— olimotiensis, 277
—— schoana, 332
—— sordida, 276, 332
Pinarocorys erythropygia, 21
— nigricans, 21
Pirenestes frommi, 991
— maximus, 990
— minor, 990
— ostrinus, 989
—— centralis, 1053
— rothschildi, 989, 1053
placidus, Phyllastrephus fischeri, 129, 152
platura, Hedydipna platura, 763
— metallica, Hedydipna, 764
Platysteira cyanea æthiopica, 209
—— nyanzæ, 208
— peltata jacksoni, 226
—— mentalis, 211, 226
—— peltata, 210
plebeja cinerea, Turdoides, 85
—, Turdoides plebeja, 84
Plocepasser donaldsoni, 865
— mahali melanorhynchus, 863
—— pectoralis, 862
—— propinquatus, 864
— superciliosus bannermani, 865
—— brunnescens, 865
—— superciliosus, 864
Ploceus æthiopicus, 945
— aurantius rex, 914
— aureoflavus pallidiceps, 1053
—— reicherti, 1053
— badius axillaris, 909
—— badius, 908
— baglafecht baglafecht, 901
—— eremobius, 902
— bojeri, 911
— capitalis dichrocephalus, 905
—— dimidiatus, 904
— castaneiceps, 913

# INDEX

Ploceus castanops, 910
— cucullatus abyssinicus, 889
— — bohndorffi, 890, 1052
— — femininus, 1052
— flavissimus, 1052
— galbula, 909, 1052
— golandi, 898
— heuglini heuglini, 896
— intermedius cabanisii, 898
— — intermedius, 897
— jacksoni, 906, 1053
— — jucundus, 1053
— luteolus kavirondensis, 904
— — luteolus, 903
— nigriceps, 891, 1052
— — graueri, 1052
— reichardi, 895
— rubiginosus rubiginosus, 907
— spekei, 892
— spekeoides, 893
— tæniopterus furensis, 895
— — tæniopterus, 894
— velatus tahatali, 900
— vitellinus uluensis, 900
— — vitellinus, 899
— xanthopterus, 912
plumata poliocephala, Prionops, 568
—  vinaceigularis, Prionops, 569, 576
plumbeiceps violacea, Tchitrea, 223
plumbeum orientale, Parisoma, 174
—, Parisoma plumbeum, 173
podobe, Cercotrichas podobe, 325
pœcilosterna, australoabyssinica, Mirafra, 20
— massaica, Mirafra, 46
—, Mirafra pœcilosterna, 20, 46
poensis, holmelas, Laniarius, 609
— præpectoralis, Neocossyphus, 255
—, Spermestes poensis, 976, 1052
Pœoptera see Stilbopsar
Pogonocichla intensa, 333
— stellata elgonensis, 324
— — guttifer, 322, 333
— — johnstoni, 333
— — macarthuri, 324
— — orientalis, 323, 333
— — ruwenzorii, 324
poliocephala akeleyæ, Alethe, 311
— carruthersi, Alethe, 310
— kungwensis, Alethe, 311
—, Prionops plumata, 568
— ufipæ, Alethe, 311
poliocephalus approximans, Malaconotus, 643
— hypopyrrhus, Malaconotus, 643
—, Malaconotus poliocephalus, 643
poliogastra, Zosterops pallidus, 733
poliogenys, Spermophaga, 988
poliolopha, Prionops, 571
poliophrys, Alethe, 312
poliopleura, Emberiza, 1091

polioptera, Cossypha polioptera, 295
— kungwensis, Cossypha, 296
Poliospiza collaris, 1086
— dimidiata, 1086
— pachyrhyncha, 1086
— striolata ugandæ, 1086
poliothorax, Alethe, 103
—, Tchitrea, 226
polioxantha, Eremomela icteropygialis, 431
porotoensis, Bessonornis albigularis, 333
porphyrolæma affinis, Apalis, 518
—, Apalis porphyrolæma, 414, 518
— vulcanorum, Apalis, 518
præpectoralis, Neocossyphus poensis, 255
prætermissa, Galerida malabarica, 30
prasina, Hylia prasina, 818
Pratincola axillaris, 288
pressa, Modulatrix stictigula, 302
pringlii, Dryoscopus, 619
Prinia bairdii melanops, 511
— gracilis carlo, 509
— — gracilis, 509
— leucopogon reichenowi, 510
— mistacea immutabilis, 518
— — mistacea, 518
— — superciliosa, 518
— somalica erlangeri, 508
— — somalica, 508
— subflava affinis, 506
— — desertæ, 507
— — subflava, 505, 518
— — tenella, 506, 518
prinioides, Cisticola hunteri, 478
Prionops alberti, 572
— concinnatus, 571
— cristata concinnata, 571
— — cristata, 570
— — melanoptera, 576
— — omoensis, 576
— plumata poliocephala, 568
— — vinaceigularis, 569, 576
— poliolopha, 571
pristoptera blanfordi, Psalidoprocne, 552
—, Psalidoprocne pristoptera, 551
procera, Cisticola chiniana, 470, 519
progne delamerei, Coliuspasser, 971
promiscua, Saxicola torquata, 289
propinquatus, Plocepasser mahali, 864
Psalidoprocne albiceps, 553
— antinorii, 552
— chalybea, 550
— holomelæna holomelæna, 549, 554
— — massaica, 554
— — ruwenzori, 550

# INDEX

Psalidoprocne orientalis oleaginea, 551
—— orientalis, 550
— petiti orientalis, 550
— pristoptera blanfordi, 552
—— pristoptera, 551
psammocromius, Coliuspasser hartlaubi, 966
Pseudalæmon fremantlii delamerei, 33
—— fremantlii, 33
—— megaensis, 34
Pseudoalcippe abyssinicus abyssinicus, 104, 106
—— atriceps, 105
—— chyulu, 106
—— stictigula, 106
—— stierlingi, 105
pseudobarbatus, Serinus mozambicus, 1062
Pseudonigrita arnaudi arnaudi, 867
—— australoabyssinicus, 868
—— dorsalis, 868
— cabanisi, 868
Psophocichla guttata belcheri, 333
—— fischeri, 252
—— guttata, 261, 333
— litsipsirupa simensis, 252
—— stierlingi, 253
Ptilonorhynchus albirostris, 712
Ptilostomus afer, 679
Ptyonoprogne fuligula pusilla, 548
—— rufigula, 547, 554
— obsoleta arabica, 546
—— obsoleta, 545
—— reichenowi, 546
— rufigula fusciventris, 554
—— rupestris, 545
Ptyrticus turdinus turdinus, 99
puella, Batis molitor, 226
— soror, Batis, 203
— unitatis, Hirundo, 538
puguensis, Cyanomitra olivacea, 804
—, Illadopsis rufipennis, 101
pulchella lucidipectus, Nectarinia, 760
— melanogastra, Nectarinia, 759
—, Nectarinia pulchella, 758
—, Phyllolais, 422
pulcher rufiventris, Spreo, 717
pulchra, Apalis pulchra, 406
—, Eremomela scotops, 436
pulpa, Mirafra, 46
pumilus, Alseonax adustus, 165, 226
—, Bradornis, 179, 225
punctifrons, Anthoscopus, 659
pura, Coracina cæsia, 561
purpurascens, Hypochera ultramarina, 1053
—, Parus niger, 660
purpureiceps, Lamprocolius, 699

purpureiventris, Nectarinia, 757
purpureus amethystinus, Lamprocolius, 722
—, Lamprocolius, 697, 722
purpuropterus æneocephalus, Lamprotornis, 703
—, Lamprotornis purpuropterus, 702
pusilla elgonensis, Eremomela, 518
—, Euplectes orix, 951, 1053
—, Ptyonoprogne fuligula, 548
— tessmani, Eremomela, 518
Pycnonotus barbatus arsinoe, 115
—— schoanus, 115
—— somaliensis, 116
— dodsoni, 114, 152
— tricolor chyulu, 152
—— fayi, 113
—— minor, 151
—— naumanni, 152
—— tricolor, 112, 151
— xanthopygos layardi, 111
—— micrus, 112, 152
—— spurius, 112
pyrgita dankali, Gymnoris, 1053
—, Petronia xanthosterna, 884, 1052, 1053
— reichenowi, Gymnoris, 1052
Pyromelana see Euplectes
pyrrhocorax docilis, Coracia, 678
pyrrhomitra, Cisticola erythrops, 483
pyrrhoptera kivuensis, Malacocincla, 103
—, Malacocincla pyrrhoptera, 102
Pyrrhulauda see Eremopterix
Pyrrhurus flavicollis flavigula, 119, 151
—— soror, 120
— scandens orientalis, 120
Pytilia afra, 1006
— melba belli, 1053
—— citerior, 1009
—— grotei, 1008
—— jessei, 1053
—— jubaensis, 1054
—— melba, 1007, 1053
—— soudanensis, 1009, 1053, 1054
— phœnicoptera emini, 1006
—— lineata, 1005

## Q

quadricolor intercedens, Telophorus, 643
— nigricauda, Telophorus, 635, 643
—, Telophorus quadricolor, 634
quadrivirgata erlangeri, Erythropygia, 321
—, Erythropygia quadrivirgata, 320, 333
— greenwayi, Erythropygia, 321

# INDEX

quartinia, Coccopygia melanotis, 1018
Quelea cardinalis cardinalis, 947
— — pallida, 948
— — rhodesiæ, 948
— erythrops, 946
— quelea æthiopica, 945, 1052
— — centralis, 1052
— — lathami, 944
quelimanensis, Seicercus ruficapillus, 387
quiscalina martini, Campephaga, 559
— münzneri, Campephaga, 560

## R

rabai, Phyllastrephus debilis, 128, 152
ragazzii, Cyanomitra olivacea, 805
raineyi, Geokichla gurneyi, 248, 332
rara, Estrilda rara, 1030
rayi, Budytes flavus, 58
raymondi, Onychognathus tenuirostris, 723
rectirostris tephrolæma, Anthreptes, 810, 820
regius anderseni, Cinnyris, 790
—, Cinnyris regius, 788
—, Cosmopsarus, 704, 722
— kivuensis, Cinnyris, 789
— magnificus, Cosmopsarus, 722
reichardi, Ploceus, 895
—, Serinus gularis, 1070
reichenovii australis, Cryptospiza, 992, 1053, 1054
— ocularis, Cryptospiza, 1053
— sanguinolenta, Cryptospiza, 1054
reichenowi, Batis capensis, 201
—, Cinnyris reichenowi, 786
—, Cisticola brachyptera, 497
—, Dioptrornis chocolatinus, 183
—, Drepanorhynchus, 762
— fricki, Othyphantes, 930
—, Gymnoris pyrgita, 1052
— nigrotemporalis, Othyphantes, 913
—, Oriolus larvatus, 670
—, Othyphantes reichenowi, 929
—, Prinia leucopogon, 510
—, Ptyonoprogne obsoleta, 546
—, Serinus atrogularis, 1074
— yokanæ, Anthreptes, 817
reicherti, Ploceus aureoflavus, 1053
remigialis, Tchagra senegala, 624
rendalli, Anomalospiza imberbis, 1002, 1054
—, Lagonosticta senegala, 1013
rensi, Eremomela usticollis, 438, 519
restricta, Tchitrea viridis, 221
retzii graculinus, Sigmodus, 574, 576

retzii intermedius, Sigmodus, 576
— neumanni, Sigmodus, 576
— tricolor, Sigmodus, 573
rex, Ploceus aurantius, 914
Rhinocorax rhipidurus, 677
rhipidurus, Rhinocorax, 677
rhodesiæ, Anthoscopus caroli, 658
—, Quelea cardinalis, 948
rhodopareia, Lagonosticta rubricata, 1009, 1053
Rhodophoneus cruentus cathemagmenus, 639
— — cruentus, 639, 643
— — hilgerti, 640
— — kordofanicus, 643
rhodopsis, Lagonosticta senegala, 1014, 1053
rhodoptera, Heliolais erythroptera, 512, 518, 519
rhodopyga centralis, Estrilda, 1023
—, Estrilda rhodopyga, 1022
Rhynchostruthus socotranus louisæ, 1060
— — socotranus, 1060
richardi, Anthus, 81
riggenbachi, Serinus leucopygius, 1074
Riparia arabica, 546
— cincta erlangeri, 544
— — suahelica, 544
— — paludicola ducis, 543
— — minor, 542
— — paludicola, 542
— — schoensis, 543
— riparia riparia, 541
— — reichenowi, 546
robertsi, Anthoscopus caroli, 657
robusta aberdare, Cisticola, 490
— ambigua, Cisticola, 489
— awemba, Cisticola, 491
—, Cisticola robusta, 488
— nuchalis, Cisticola, 489
— omo, Cisticola 490
—, Saxicola torquata, 333
— schraderi, Cisticola, 490
roccatii, Anthoscopus caroli, 656
roehli, Alseonax adustus, 166
—, — murinus, 166
—, Arizelocichla masukuensis, 140
—, Bradypterus, 518
—, Turdus olivaceus, 245
rolleti, Oriolus larvatus, 665, 670
roosvelti, Granatina ianthinogaster, 1039
roseicrissa, Estrilda paludicola, 1025
rothschildi, Anthoscopus caroli, 657
—, Hirundo rustica, 525
—, Laniarius funebris, 642
—, Pirenestes, 989, 1053
rovumæ, Erythropygia barbata, 333
—, Parus rufiventris, 660

# INDEX

rowei, Geokichla piaggiæ, 250
ruandæ, Alseonax aquaticus, 225
ruberrima, Lagonosticta senegala, 1014, 1053
rubescens, Chalcomitra rubescens, 793, 820, 823
— kakamegæ, Chalcomitra, 794
rubetra, Saxicola rubetra, 290
rubicola caffra, Saxicola, 287
—, Sylvia communis, 350
rubiginosa, Argya rubiginosa, 94
— emini, Argya, 95
— heuglini, Argya, 95
— sharpii, Argya, 95
rubiginosus bertrandi, Chlorophoneus, 631
— münzneri, Chlorophoneus, 643
—, Ploceus rubiginosus, 907
rubricata congica, Lagonosticta, 1010
— hæmatocephala, Lagonosticta, 1011
— hildebrandti, Lagonosticta, 1053
— rhodopáreia, Lagonosticta, 1009, 1053
rubriceps, Anaplectes melanotis, 943
rubricollis centralis, Malimbus, 938
rubritorques, Anthreptes, 811
ruddi archeri, Heteromirafra, 22
—, Heteromirafra, 22
rudolfi, Lagonosticta cinereovinacea, 1053
rufa lynesi, Mirafra, 18
—, Mirafra rufa, 18
rufescens athensis, Calandrella, 41
— megaensis, Calandrella, 42
—, Melanocorypha bimaculata, 39
— pallida, Sylvietta, 427
— somalica, Calandrella, 41
ruficapilla cana, Spermophaga, 988
— chubbi, Sylvietta, 284
—, Spermophaga ruficapilla, 987
ruficapillus johnstoni, Seicercus, 387
— mbololo, Seicercus, 518
— minullus, Seicercus, 386, 518
— ochrogularis, Seicercus, 388
— quelimanensis, Seicercus, 387
ruficauda, Erythropygia leucophrys, 333
—, Histurgops, 866
ruficeps, Apalis, 518
—, Cisticola ruficeps, 503
— cooki, Laniarius, 643
— kismayensis, Dryoscopus, 616, 643
— —, Laniarius, 616
— mongalla, Cisticola, 504
— rufinuchalis, Laniarius, 615
— scotoptera, Cisticola, 504
—, Tephrocorys cinerea, 46
ruficollis, Corvus corax, 671
rufidorsalis, Apalis rufifrons, 417

rufifrons, Apalis rufifrons, 416
— rufidorsalis, Apalis, 417
— smithii, Apalis, 417
rufigula, Euplectes, 1052
— fusciventris, Ptyonoprogne, 554
—, Ptyonoprogne fuligula, 547, 554
rufinuchalis, Laniarius ruficeps, 615
rufipennis barakæ, Illadopsis, 106
—, Cichladusa guttata, 316
— distans, Malacocincla, 101
—, Malacocincla rufipennis, 100
— puguensis, Malacocincla, 101
rufiventer somereni, Terpsiphone, 225
rufiventris pallidiventris, Parus, 654, 660
— rovumæ, Parus, 660
—, Spreo pulcher, 717
rufocinctus, Lioptilornis, 176
—, Passer, 873
rufocinerea, Monticola rufocinerea, 258, 333
rufocinnamomea fischeri, Mirafra, 14, 46
— furensis, Mirafra, 16
— kavirondensis, Mirafra, 16
—, Mirafra rufocinnamomea, 13, 46
— omoensis, Mirafra, 16
— sobatensis, Mirafra, 15
— tigrina, Mirafra, 15
— torrida, Mirafra, 15
rufoflavidus, Sathrocercus cinnamomeus, 374, 518
rufogularis denti, Apalis, 400
rufopicta lateritia, Lagonosticta, 1015
rufuensis, Turdoides hypoleuca, 92, 97
rufula, Hirundo daurica, 532
rufulus lynesi, Anthus, 68
rufus gabunensis, Neocossyphus, 255
—, jacksoni, Bathmocercus, 517
—, Neocossyphus rufus, 254
rungwensis, Linurgus kilimensis, 1086
rupestris, Ptyonoprogne, 545
— reichenowi, Riparia, 546
rüppelli böhmi, Eurocephalus, 642
— deckeni, Eurocephalus, 642
— erlangeri, Eurocephalus, 642
—, Eurocephalus anguitimens, 583, 642
—, Sylvia, 353
rüppellii, Onychognathus morio, 709, 722
rushiæ denti, Pholidornis, 819
ruspolii, Dinemellia dinemelli, 862
rustica, Hirundo rustica, 524
— rothschildi, Hirundo, 525
— transitiva, Hirundo, 525

# INDEX

ruwenzori, Cryptospiza salvadorii, 993
—, Psalidoprocne holomelæna, 550
ruwenzoria, Mirafra africana, 46
ruwenzoriæ, Tchitrea perspicillata, 222
—, Tchitrea suahelica, 222
ruwenzorii, Apalis, 407
—, Nilaus minor, 642
—, Pogonocichla stellata, 324

## S

Salpornis spilonota emini, 822
— — erlangeri, 822
— — salvadori, 821
salvadori, Salpornis spilonota, 821
salvadorii borealis, Cryptospiza, 1053
— chyuluensis, Cryptospiza, 1054
—, Cryptospiza salvadorii, 993, 1053
—, Galeopsar, 712
— kilimensis, Cryptospiza, 993, 1054
—, Nectarinia johnstoni, 753
— ruwenzori, Cryptospiza, 993
samamisicus, Phœnicurus phœnicurus, 327
samharensis, Ammomanes deserti, 27
sanguinolenta, Cryptospiza reichenovii, 1054
saphiroi, Anthus leucophrys, 66
—, — vaalensis, 66
sarmenticia, Zosterops virens, 736
sassii, Serinus flavivertex, 1069
Sathrocercus cinnamomeus cavei, 374
— — chyuluensis, 518
— — cinnamomeus, 373
— — macdonaldi, 375
— — nyassæ, 373
— cinnamomeus rufoflavidus, 374, 518
— — ufipæ, 375
— lopezi barakæ, 377
— mariæ granti, 377
— — mariæ, 375, 518
— — usambaræ, 376, 518
saturatior, Calandrella cinerea, 42
—, Fringillaria striolata, 1097, 1099
—, Nigrita bicolor, 1053
saxatilis, Monticola, 256
Saxicola rubetra rubetra, 290
— torquata albofasciata, 287
— — axillaris, 288
— — caffra, 287, 333
— — hemprichii, 290
— — jebelmarræ, 289
— — orientalis, 333
— — promiscua, 289
— — robusta, 333
— — stonei, 333
— — variegata, 286

scandens orientalis, Pyrrhurus, 120
Scepomycter winifredæ, 420
schalowi, Œnanthe lugubris, 268
schillingsi, Cisticola cinereola, 502
—, Mirafra cheniana, 46
schistacea, Melænornis edolioides, 184
—, Nigrita canicapilla, 983
schlegeli, Mandingoa nitidula, 1004
schoana, Pinarochroa sordida, 332
schoanus, Pycnonotus barbatus, 115
—, Uræginthus bengalus, 1035, 1053
—, Zosterops virens, 736
Schœnicola brevirostris æquatorialis, 519
— — alexinæ, 395, 519
— — brevirostris, 394
— — chyulu, 519
schœnobænus, Acrocephalus, 368
schoensis, Riparia paludicola, 543
schraderi, Cisticola robusta, 490
schuetti, Turdus pelios, 241
schusteri, Cisticola woosnami, 473
scirpaceus, Acrocephalus scirpaceus, 365
—, crassirostris, Acrocephalus, 517
— fuscus, Acrocephalus, 365
— griseus, Acrocephalus, 517
sclateri, Erythropygia leucoptera, 318
—, — zambesiana, 318
scopifrons keniensis, Sigmodus, 575
— kirki, Sigmodus, 575
—, Sigmodus scopifrons, 574
scotocerca, Cercomela scotocerca, 273
— enigma, Cercomela, 332
— furensis, Cercomela, 274
— spectatrix, Cercomela, 273
— turkana, Cercomela, 274
scotops citriniceps, Eremomela, 436
—, Eremomela scotops, 435
— kikuyuensis, Eremomela, 519
— occipitalis, Eremomela, 436, 519
— pulchra, Eremomela, 436
scotoptera, Cisticola ruficeps, 504
scullii, Hirundo daurica, 533
scutatus, Spermestes cucullatus, 976
Seicercus budongoensis, 391
— lætus, 392
— ruficapillus johnstoni, 387
— — minullus, 386, 518
— — mbololo, 518
— — ochrogularis, 388
— — quelimanensis, 387
— umbrovirens alpinus, 390
— — chyulu, 518
— — dorcadichrous, 389
— — erythreæ, 517
— — fuggles-couchmani, 391
— — mackenzianus, 389, 518

# INDEX

Seicercus umbrovirens omoensis, 390
—— umbrovirens, 388, 517
—— wilhelmi, 390
seimundi minor, Anthreptes, 816
semenowi, Emberiza cineracea, 1093
semifasciata, Cisticola lais, 464
semipartitus, Empidornis semipartitus, 187
— kavirondensis, Empidornis, 187
— orleansi, Empidornis, 187
semirufa, Cossypha semirufa, 293
— donaldsoni, Cossypha, 294
— gordoni, Hirundo, 536
— intercedens, Cosspyha, 293
—, Thamnolæa, 281
semitorquata, Muscicapa hypoleuca, 163
senator niloticus, Lanius, 603
senegala abayensis, Lagonosticta, 1053
— armena, Tchagra, 643
— brunneiceps, Lagonosticta, 1012, 1053
— catholeuca, Tchagra, 643
— erlangeri, Tchagra, 643
— flavodorsalis, Lagonosticta, 1053
— habessinica, Tchagra, 624, 643
— kikuyuensis, Lagonosticta, 1053
— mozambica, Tchagra, 643
— orientalis, Tchagra, 643
— remigialis, Tchagra, 624
— rendalli, Lagonosticta, 1013
— rhodopsis, Lagonosticta, 1014, 1053
— ruberrima, Lagonosticta, 1014, 1053
— somaliensis, Lagonosticta, 1013
—, Tchagra senegala, 623, 643
— warsangliensis, Tchagra, 643
senegalensis acik, Chalcomitra, 796
— æquatorialis, Chalcomitra, 820
— anderssoni, Zosterops, 726, 736
— aschani, Hirundo, 554
— aurifrons, Zosterops, 736
— australoabyssinicus, Zosterops, 736
— cruentata, Chalcomitra, 795
—, Dryoscopus affinis, 619
— erythrinæ, Chalcomitra, 820
— flavilateralis, Zosterops, 726
— gutturalis, Chalcomitra, 794, 820
—, Hirundo senegalensis, 535, 554
— hybrida, Hirundo, 554
— inæstimata, Chalcomitra, 820
— jubaensis, Zosterops, 727, 736
— lamperti, Chalcomitra, 797, 820
— monteiri, Hirundo, 536, 554
— niassæ, Zosterops, 736
— smithi, Zosterops, 736
—, Zosterops senegalensis, 725, 736

sennaarensis, Cercomela familiaris, 332
septemstriata, Fringillaria tahapisi, 1096
septentrionalis, Cinnyris cupreus, 820
Serinus atrogularis, 1086
—— hilgerti, 1076
—— reichenowi, 1074
—— somereni, 1075
—— xanthopygius, 1074
— buchanani, 1068
— burtoni albifrons, 1078
—— gurneti, 1080
—— kilimensis, 1079
—— melanochrous, 1079
—— tanganjicæ, 1080
— capistratus capistratus, 1066
—— koliensis, 1067
— donaldsoni, 1067, 1086
— dorsostriatus dorsostriatus, 1063
—— harterti, 1064
—— maculicollis, 1064
—— taruensis, 1065
— flavigula, 1076, 1086
— flavivertex elgonensis, 1086
—— flavivertex, 1068, 1086
—— sassii, 1069
— gularis erlangeri, 1072
—— reichardi, 1070
—— striatipectus, 1071
— leucopygius leucopygius, 1073
—— riggenbachi, 1074
— mennelli, 1072
— mozambicus aurifrons, 1086
—— barbatus, 1062
—— gertrudis, 1086
—— grotei, 1062, 1086
—— gommaensis, 1063
—— mozambicus, 1061, 1086
—— pseudobarbatus, 1062
— nigriceps, 1084
— striolatus graueri, 1078
—— striolatus, 1076, 1086
—— whytii, 1077
— sulphuratus sharpii, 1065, 1086
—— shelleyi, 1086
— tristriatus pallidior, 1070
—— tristiatus, 1070
seth-smithi, Pedilorhynchus epulatus, 169
sethsmithi, Phyllastrephus icterinus, 133
—, — xavieri, 133
sharpei, Anthoscopus caroli, 656
—, Bradornis pallidus, 225
—, Camaroptera brevicaudata, 444, 518
—, Dicrurus ludwigii, 566, 567
—, Macronyx, 78
—, Mirafra, 10

# INDEX

sharpei, Sheppardia sharpei, 305, 333
—, Turdoides melanops, 88, 97
— usambaræ, Sheppardia, 306
sharpii, Argya rubiginosa, 95
—, Othyphantes stuhlmanni, 932
—, Pholia, 692
—, Serinus sulphuratus, 1065, 1086
shelleyi, Atimastillas flavicollis, 151
—, Cinnyris shelleyi, 773
—, Cryptospiza, 995
— hofmanni, Cinnyris, 774
—, Onychognathus morio, 722
—, Passer motitensis, 872
—, Serinus sulphuratus, 1086
—, Spreo, 718
Sheppardia æquatorialis acholiensis, 307
— — æquatorialis, 306
— cyornithopsis bangsi, 333
— — lopezi, 307
— gunningi sokokensis, 304
— sharpei sharpei, 305, 333
— — usambaræ, 306
shimba, Eurillas virens, 152
shimbanus, Phyllastrephus albigula, 152
sibilatrix, Phylloscopus, 385
Sigmodus caniceps mentalis, 572
— retzii graculinus, 574, 576
— — intermedius, 576
— — neumanni, 576
— — tricolor, 573
— scopifrons keniensis, 575
— — kirki, 575
— — scopifrons, 574
signata cavei, Eremopterix, 46
—, Eremopterix signata, 37
— harrisoni, Eremopterix, 38, 46
silvæ, Dyaphorophyia concreta, 213
—, — graueri, 213
silvanus, Zosterops, 734
simensis, Psophocichla litsipsirupa, 252
similis, Anthus, 81
—, Chloropeta, 142
—, Chlorophoneus sulfureopectus, 630, 643
— hararensis, Anthus, 62, 81
— jebelmarræ, Anthus, 64
— neumannianus, Anthus, 81
— nivescens, Anthus, 62
— nyassæ, Anthus, 63
— sokotræ, Anthus, 63
simplex, Calamonastes simplex, 393
—, Cisticola chiniana, 467
—, Passer, 876
— undosus, Calamonastes, 394
Sitagra aliena, 921
sjöstedti, Bradypterus alfredi, 518
slatini, Hyliota, 226
slatini, Zosterops senegalensis, 736
smithii, Apalis rufifrons, 417
—, Hirundo smithii, 531
—, Lanius collaris, 594
—, Turdoides leucopygia, 91
sobatensis, Mirafra rufocinnamomea, 15
socotrana, Fringillaria, 1098
socotranus louisæ, Rhynchostruthus, 1060
—, Rhynchostruthus socotranus, 1060
—, Zosterops, abyssinicus, 735
sokokensis, Anthus, 71
—, Sheppardia gunningi, 304
sokotræ, Anthus similis, 63
solitaria longirostris, Monticola, 257
—, Monticola solitaria, 257
somalica, Calandrella rufescens, 41
—, Certhilauda, 23
— erlangeri, Prinia, 508
— megaensis, Calandrella, 42
—, Prinia somalica, 508
somalicum, Parisoma böhmi, 173
somalicus, Lanius, 595, 642
somaliensis, Andropadus insularis, 151
—, Galerida cristata, 29
—, Lagonosticta senegala, 1013
—, Laniarius ferrugineus, 643
—, Muscicapa gambagæ, 225
—, Pycnonotus arsinoe, 116
—, — barbatus, 115
—, Sylvia leucomelæna, 352
somereni, Estrilda ianthinogaster, 1054
—, — atrogularis, 1075
—, Tchitrea nigriceps, 225
—, Terpsiphone rufiventer, 225
songeaensis, Apalis melanocephala, 519
sordida erlangeri, Pinarochroa, 332
— ernesti, Pinarochroa, 277
— hypospodia, Pinarochroa, 276
— olimotiensis, Pinarochroa, 277
—, Pinarochroa sordida, 276, 332
— schoana, Pinarochroa, 332
sordidus, Anthus, 81
Sorella eminibey, 882, 1052
— — guasso, 1052
soror, Batis molitor, 203
—, — puella, 203
—, Erythropygia leucophrys, 333
—, Pyrrhurus flavicollis, 120
soudanensis, Pytilia melba, 1009, 1053, 1054
sousæ, Apalis moreaui, 418
souzæ burigi, Lanius, 599
speciosa, Tchitrea viridis, 220
spectatrix, Cercomela scotocerca, 273

# INDEX

Speculipastor bicolor, 693
spekei, Ploceus, 892
spekeoides, Ploceus, 893
Spermestes bicolor stigmatophorus, 1052
— cucullatus cucullatus, 975
— — scutatus, 976
— nigriceps minor, 978
— — nigriceps, 977
— poensis poensis, 976, 1052
Spermophaga poliogenys, 988
— ruficapilla cana, 988
— — ruficapilla, 987
spilonota emini, Salpornis, 822
— erlangeri, Salpornis, 822
— salvadori, Salpornis, 821
Spiloptila clamans, 516
Spinus see Carduelis
Spizocorys athensis, 41
— obbiensis, 44
splendens, Corvus splendens, 675
splendidus bailundensis, Lamprocolius, 722
—, Lamprocolius splendidus, 698, 722
Sporopipes frontalis abyssinicus, 1052
— — cinerascens, 888
— — emini, 888
— — frontalis, 887, 1052
Spreo albicapillus, 716
— fischeri, 715
— hildebrandti, 718, 723
— — kellogorum, 723
— pulcher rufiventris, 717
— shelleyi, 718
— superbus, 719
spurius, Pycnonotus xanthopygos, 112
squamulata jubaensis, Turdoides, 86
—, Turdoides squamulata, 85
Steganura aucupum kaduliensis, 1054
— — obtusa, 1051
— orientalis obtusa, 1051
— — orientalis, 1051, 1054
— paradisæa, 1049
Stelgidillas gracilirostris chagwensis, 143
— — congensis, 144
— — percivali, 143
Stelgidocichla latirostris australis, 152
— latirostris eugenia, 150, 152
stellata elgonensis, Pogonocichla, 324
— guttifer, Pogonocichla, 322, 333
— johnstoni, Pogonocichla, 333
— macarthuri, Pogonocichla, 324
— orientalis, Pogonocichla, 323, 333
— ruwenzorii, Pogonocichla, 324

Stenostira plumbea, 173
stephanophorus, Heterhyphantes melanogaster, 927, 973
stictifrons, Symplectes bicolor, 916
stictigula, Modulatrix stictigula, 301
— pressa, Modulatrix, 302
—, Pseudoalcippe abyssinicus, 106
stierlingi, Calamonastes fasciolatus, 392
—, Pseudoalcippe abyssinicus, 105
—, Psophocichla litsipsirupa, 253
—, Zosterops virens, 729, 736
stigmatophorus, Spermestes bicolor, 1052
Stilbopsar kenricki bensoni, 715
— — kenricki, 714
— stuhlmanni, 714
Stiphrornis erythrothorax mabiræ, 307
Stizorhina fraseri intermedia, 226
— — vulpina, 195, 226
straminea, Locustella nævia, 363
strangei kapitensis, Cisticola, 493
strausæ, Apalis bamendæ, 416, 519
strepitans, Phyllastrephus, 123
stresemanni, Zavattariornis, 680
striata gambagæ, Muscicapa, 162, 225
—, Muscicapa striata, 161
— neumanni, Muscicapa, 162
striatipectus, Serinus gularis, 1071
stricta, Euplectes afra, 958, 1052
striifacies, Arizelocichla milanjensis, 139, 152
striolata dankali, Fringillaria, 1099
—, Fringillaria striolata, 1097, 1099
— jebelmarræ, Fringilla, 1099
— saturatior, Fringillaria, 1097, 1099
— ugandæ, Poliospiza, 1086
striolatus graueri, Serinus, 1078
—, Serinus striolatus, 1076, 1086
— whytii, Serinus, 1077
stuhlmanni, Cinnyris afer, 780
—, Othyphantes stuhlmanni, 931
—, Pedilorhynchus comitatus, 189
— sharpii, Othyphantes, 932
—, Stilbopsar, 714
—, Zosterops virens, 728
Sturnus vulgaris vulgaris, 687
suahelica, Batis minor, 207
—, Chloropetella holochlorus, 226
—, Cisticola galactotes, 486
—, Coliuspasser laticauda, 971
— granti, Tchitrea, 223, 226
—, Riparia cincta, 544
— ruwenzoriæ, Tchitrea, 222
—, Tchitrea suahelica, 221, 226
suahelicus, Acrocephalus bæticatus, 367
—, Batis minor, 207
—, Bradornis pallidus, 225

# INDEX

suahelicus, Cinnyris mariquensis, 771
—, Hyphanturgus ocularis, 919
—, Passer, 879
—, Phyllastrephus terrestris, 122
Suaheliornis kretschmeri griseiceps, 134
— — kretschmeri, 134
subadustus, Alseonax adustus, 166
subalaris, Bradornis pallidus, 177, 226
subaureus aureoflavus, Xanthophilus, 922, 1053
subfamosa, Nectarinia famosa, 820
subflava affinis, Prinia, 506
— clarkei, Estrilda, 1024
— desertæ, Prinia, 507
—, Estrilda subflava, 1023
—, Prinia subflava, 505, 518
— tenella, Prinia, 506, 518
sublacteus, Laniarius æthiopicus, 613, 643
subruficapilla bodessa, Cisticola, 469
subrufipennis, Thamnolæa cinnamomeiventris, 280, 332
sucosus, Phyllastrephus cabanisi, 151
— sylvicultor, Phyllastrephus, 151
sudanensis, Bradypterus baboecala, 370
—, Myrmecocichla æthiops, 286
sukensis, Cinnyris venustus, 820
sulfureopectus fricki, Chlorophoneus, 643
— similis, Chlorophoneus, 630, 643
sulphurata, Campephaga, 556, 562
sulphuratus sharpii, Serinus, 1065, 1086
— shelleyi, Serinus, 1086
sundevalli, Euplectes orix, 950
superbus buvuma, Cinnyris, 820
—, Cinnyris superbus, 764, 820
—, Spreo, 719
superciliaris, Budytes, 57
— oraria, Petronia, 1053
— ugandæ, Camaroptera, 445
superciliosa desertæ, Prinia, 507
—, Prinia mistacea, 518
—, Zosterops, 736
superciliosus bannermani, Plocepasser, 865
— brunnescens, Plocepasser, 865
—, Plocepasser superciliosus, 864
svecica cyanecula, Cyanosylvia, 328
—, Cyanosylvia svecica, 328
— magna, Cyanosylvia, 329
swainsonii, Passer, 879, 973
sycobius, Lamprocolius chalybæus, 695
sylvatica, Euplectes hordeacea, 1052
Sylvia atricapilla atricapilla, 350
— — dammholzi, 351
— borin, 350

Sylvia cantillans albistriata, 354
— communis communis, 349
— — icterops, 349
— — rubicola, 350
— curruca blythi, 348
— — curruca, 348
— hortensis crassirostris, 351
— leucomelæma blanfordi, 352
— — somaliensis, 352
— melanocephala momus, 353
— — mystacea, 354
— nana nana, 355
— nisoria, 355
— rüppelli, 353
sylvia, Cisticola erythrops, 482
sylvicultor, Phyllastrephus sucosus, 151
sylviella, Anthoscopus caroli, 655
Sylvietta brachyura brachyura, 423, 518
— — dilutior, 424
— — hilgerti, 518
— — leucopsis, 424, 518
— — micrura, 518
— isabellina, 427
— leucophrys chloronota, 430
— — leucophrys, 429
— rufescens pallida, 427
— ruficapilla chubbi, 428
— virens baraka, 429
— whytii abayensis, 426
— — jacksoni, 426
— — minima, 426
— — whytii, 425
Symplectes bicolor amaurocephalus, 915
— — kersteni, 915
— — mentalis, 916
— — stictifrons, 916
syndactyla woosnami, Bleda, 117
syriaca, Curruca galactotes, 357
syriacus, Agrobates galactotes, 357

## T

tacazze, Nectarinia, 754
tæniopterus furensis, Ploceus, 895
—, Ploceus tæniopterus, 894
taha intercedens, Euplectes, 1052
tahapisi, Fringillaria tahapisi, 1095
— goslingi, Fringillaria, 1096
— insularis, Fringillaria, 1096
— septemstriata, Fringillaria, 1096
tahatali, Ploceus velatus, 900
talatala anderssoni, Cinnyris, 776
—, Cinnyris, 820
tanganjicæ, Serinus burtoni, 1080
tanganyika, Cisticola aridula, 458
Tarsiger see Pogonocichla
taruensis, Anthoscopus caroli, 657
—, Bradornis, 225

# INDEX

taruensis, Lagonosticta jamesoni, 1012
—, Serinus dorsostriatus, 1065
Tchagra australis congener, 643
— — emini, 626
— — littoralis, 643
— — minor, 625, 643
— jamesi jamesi, 626
— — kismayensis, 627
— — mandanus, 627
— senegala armena, 643
— — catholeuca, 643
— — erlangeri, 643
— — habessinica, 624, 643
— — mozambica, 643
— — orientalis, 643
— — remigialis, 624
— — senegala, 623, 643
— — warsangliensis, 643
Tchitrea nigriceps emini, 224
— — somereni, 225
— perspicillata, 226
— — ungujaensis, 226
— poliothorax, 226
— plumbeiceps violacea, 223
— suahelica granti, 223, 226
— — ruwenzoriæ, 222
— — suahelica, 221, 226
— viridis ferreti, 219, 226
— — restricta, 221
— — speciosa, 220
teitensis, Cisticola emini, 475
—, Coliuspasser ardens, 1052
Telephonus see Tchagra
Telophorus dohertyi, 635
— quadricolor intercedens, 643
— — nigricauda, 635, 643
— — quadricolor, 634
tenebricosa, Apalis melanocephala, 403
tenebrosa, Turdoides tenebrosa, 89
tenella, Prinia subflava, 506, 518
tenellus, Tmetothylacus, 75
tenerrima, Apalis caniceps, 410
—, — flavida, 410
tenis, Petrophila rufocinerea, 333
tenuirostris, Onychognathus tenuirostris, 710, 723
—, Phyllastrephus flavostriatus, 124, 152
— raymondi, Onychognathus, 723
— theresæ, Onychognathus, 711
Tephrocorys see Calandrella
tephrolæma, Anthreptes rectirostris, 810, 820
— elgonensis Anthreptes, 820
— kikuyuensis, Arizelocichla, 135
— kungwensis, Arizelocichla, 138
tephrolæmus, usambaræ, Phyllastrephus, 137

tephronotus australoabyssinicus, Turdus, 247, 333
teresita, Elminia, 214
—, Erannornis longicauda, 214
Terpsiphone rufiventer somereni, 225
terrestris bensoni, Phyllastrephus, 123
—, Cisticola juncidis, 454
— mauensis, Cisticola, 460
— nakuruensis, Cisticola, 463
— suahelicus, Phyllastrephus, 122
tertale, Passer griseus, 1053
tertia, Alæmon hamertoni, 25
tessmani, Eremomela pusilla, 518
Thamnolæa argentata, 332
— arnotti arnotti, 283, 332
— — collaris, 332
— — leucolæmia, 332
— cinnamomeiventris albiscapulata, 281
— — subrufipennis, 280, 332
— coronata coronata, 282
— — kordofanensis, 283
— semirufa, 281
theklæ, Galerida, 46
— huriensis, Galerida, 31
theresæ, Onychognathus tenuirostris, 711
Thescelocichla leucopleura, 118
thomsoni, Erythrocercus livingstonei, 194
thoracica injectiva, Apalis, 519
— youngi, Apalis, 397
thruppi, Parus afer, 647, 660
thunbergi, Budytes thunbergi, 55
— cinereocapillus, Budytes, 56
tigrina, Mirafra rufocinnamomea, 15
tincta, Camaroptera brevicaudata, 442
tinniens oreophila, Cisticola, 504
Tmetothylacus tenellus, 75
togoensis, Corvinella corvina, 605
—, Lagonosticta larvata, 1017
tonga, Cisticola, natalensis, 494
toroensis, Camaroptera chloronota, 441
—, Trochocercus nigromitratus, 219, 226
toruensis, Dioptrornis fischeri, 182
torquata albofasciata, Saxicola, 287
— axillaris, Saxicola, 288
— caffra, Saxicola, 287, 333
— hemprichii, Saxicola, 290
— jebelmarræ, Saxicola, 289
— orientalis, Saxicola, 333
— promiscua, Saxicola, 289
— robusta, Saxicola, 333
— stonei, Saxicola, 333
— variegata, Saxicola, 286
torquatus alpestris, Turdus, 239

# INDEX

torquatus, Neolestes, 150
torrentium, Motacilla clara, 50
torrida, Mirafra rufocinnamomea, 15
transitiva, Hirundo rustica, 525
traversii, Coliuspasser axillaris, 962
Trichophorus calurus ndussumensis, 116
— flavigula, 119
tricolor chyulu, Pycnonotus, 152
— fayi, Pycnonotus, 113
— interscapularis, Melanoploceus, 935
— minor, Pycnonotus, 151
— naumanni, Pycnonotus, 152
—, Pycnonotus tricolor, 112, 151
—, Sigmodus retzii, 573
tristiatus pallidior, Serinus, 1070
—, Serinus tristriatus, 1070
trivialis, Anthus trivialis, 70
trochilus acredula, Phylloscopus, 383, 517
— eversmani, Phylloscopus, 517
—, Phylloscopus trochilus, 382
Trochocercus albonotatus albonotatus, 217
— cyanomelas bivittatus, 215
— — kikuyuensis, 216
— — vivax, 216
— nigromitratus kibaliensis, 226
— nigromitratus toroensis, 219, 226
— nitens nitens, 218
troglodytes, Cisticola troglodytes, 498
—, Estrilda, 1021, 1053
— ferruginea, Cisticola, 499
tropicalis, Melænornis pammelaina, 186
—, Mirafra africana, 11, 46
—, Turdus libonyanus, 240, 332
tsanæ, Calamocichla leptorhyncha, 381
tsavoensis, Cinnyris, 770
Turdinus albipectus, 102
— pyrrhopterus kivuensis, 103
— rufipennis distans, 101
— ugandæ, 100
turdinus, Ptyrticus turdinus, 99
Turdoides hindei, 93
— hypoleuca hypoleuca, 92
— — kilosa, 97
— — rufuensis, 92, 97
— jardinei emini, 87
— — kirki, 86
— leucocephala abyssinica, 84
— — leucocephala, 83
— leucopygia clarkei, 91
— — hartlaubii, 90
— — leucopygia, 89
— — limbata, 90
— — omoensis, 91
— — smithii, 91

Turdoides melanops clamosa, 97
— — sharpei, 88, 97
— — vepres, 88
— plebeja cinerea, 85
— — plebeja, 84
— squamulata jubaensis, 86
— — squamulata, 85
— tenebrosa tenebrosa, 89
Turdus ericetorum philomelos, 239
— fischeri natalicus, 333
— helleri, 246
— libonyanus cinerascens, 332
— — costæ, 332
— — niassæ, 332
— libonyanus tropicalis, 240, 332
— ludoviciæ, 240
— olivaceus abyssinicus, 243, 332, 333
— — baraka, 244
— — chyuluensis, 333
— — deckeni, 243
— — elgonensis, 332
— — graueri, 246, 333
— — milanjensis, 244
— — nyikæ, 245, 332
— — oldeani, 245
— — roehli, 245
— — uluguru, 332
— pelios centralis, 242
— — pelios, 241
— — schuetti, 241
— — ubendeensis, 333
— tephronotus, 247, 333
— — australoabyssinicus, 233
— torquatus alpestris, 239
turkana, Cercomela scotocerca, 274
turkanæ, Cinnyris habessinicus, 768
—, Passer griseus, 1053
turneri, Anthus gouldi, 81
—, Eremomela badiceps, 437

## U

ubendeensis, Turdus pelios, 333
ufipæ, Alethe poliocephala, 311
—, Dioptrornis fischeri, 182
—, Sathrocercus cinnamomeus, 375
ugandæ, Anthreptes collaris, 820
—, Bleda eximia, 117
—, Camaroptera superciliaris, 445
—, Granatina ianthinogaster, 1053
—, Malacocincla fulvescens, 100
—, Melænornis edolioides, 226
—, Passer griseus, 878, 1052
—, Poliospiza striolata, 1086
—, Uræginthus bengalus, 1053
ugandensis, Clytospiza monteiri, 985
—, Dicrurus modestus, 567
ugogoensis, Uræginthus bengalus, 1036
ukamba, Bradornis griseus, 226

# INDEX

ukamba, Cisticola chiniana, 470
ultramarina, Hypochera ultramarina, 1041
— neumanni, Hypochera, 1042
— orientalis, Hypochera, 1042
— purpurascens, Hypochera, 1053
uluensis, Ploceus vitellinus, 900
uluguru, Apalis flavigularis, 399
—, Turdus olivaceus, 332
umbrovirens alpinus, Seicercus, 390
— chyulu, Seicercus, 518
— dorcadichrous, Seicercus, 389
— erythreæ, Seicercus, 517
— fuggles-couchmani, Seicercus, 391
— mackenzianus, Seicercus, 389, 518
— omoensis, Seicercus, 390
—, Seicercus umbrovirens, 388, 517
— wilhelmi, Seicercus, 390
uncinatus, Lanius elegans, 589
undosus, Calamonastes simplex, 394
ungujaensis, Tchitrea perspicillata, 226
unicolor, Amblyospiza albifrons, 941
—, Cosmopsarus, 705
unitatis, Hirundo abyssinica, 538
Uræginthus angolensis niassensis, 1033
— bengalus bengalus, 1034
— — brunneigularis, 1035
— — katangæ, 1036
— — schoanus, 1035, 1053
— — ugandæ, 1053
— — ugogoensis, 1036
— cyanocephalus, 1037
urbica, Delichon urbica, 548
Urolestes melanoleucus æquatorialis, 605
uropygialis, Cisticola juncidis, 455
usambaræ, Alethe fülleborni, 313
—, Arizelocichla nigriceps, 137, 151
—, Bradypterus, 376
—, Geokichla gurneyi, 382
—, Hyliota australis, 200
—, Phyllastrephus tephrolæmus, 137
—, Sathrocercus mariæ, 376, 518
—, Sheppardia sharpei, 306
—, Zosterops virens, 736
usambaricus, Cinnyris mediocris, 784
usticollis rensi, Eremomela, 438, 519

## V

vaalensis goodsoni, Anthus, 66
— saphiroi, Anthus, 66
valida, Cisticola natalensis, 492
vansomereni, Erythropygia leucophrys, 333
variegata, Saxicola torquata, 286
vaughani, Lamprocolius corruscus, 701
—, Zosterops, 727
vauriei, Œnanthe lugubris, 269
velatus tahatali, Ploceus, 900
venustus albiventris, Cinnyris, 778
— falkensteini, Cinnyris, 779, 820
— fazoqlensis, Cinnyris, 778, 820
— igneiventris, Cinnyris, 780
— niassæ, Cinnyris, 820
— sukensis, Cinnyris, 820
vepres, Turdoides melanops, 88
veroxii fischeri, Cyanomitra, 798
— zanzibarica, Cyanomitra, 799
verreauxi, Cinnyricinclus leucogaster, 691, 722
— ndussumensis, Criniger, 116
verticalis viridisplendens, Cyanomitra, 799
victoria, Cisticola chiniana, 470
Vidua fischeri, 1048
— hypocherina, 1048
— macroura, 1046
vidua, Motacilla aguimp, 49
vinaceigularis, Prionops plumata, 569, 576
vincenti, Cyanomitra olivacea, 806
—, Fringillaria capensis, 1094
—, Phyllastrephus flavostriatus, 125
violacea, Tchitrea plumbeiceps, 223
virens baraka, Sylvietta, 429
—, Eurillas virens, 148
— jacksoni, Zosterops, 729
— kaffensis, Zosterops, 730, 736
— meruensis, Zosterops, 736
— mbuluensis, Zosterops, 732
— sarmenticia, Zosterops, 736
— schoanus, Zosterops, 736
— shimba, Eurillas, 152
— stierlingi, Zosterops, 729, 736
— stuhlmanni, Zosterops, 728
— usambaræ, Zosterops, 736
— zanzibaricus, Eurillas, 149
— zombensis, Eurillas, 149, 152
vireo, Nicator, 642
viridiceps, Apalis flavida, 408
viridis ferreti, Tchitrea, 219, 226
— restricta, Tchitrea, 221
— speciosa, Tchitrea, 220
viridisplendens, Cyanomitra verticalis, 799
vitellinus, Ploceus vitellinus, 899
— uluensis, Ploceus, 900
vivax, Trochocercus cyanomelas, 216
vulcanorum, Apalis porphyrolæma, 518
vulgaris, Sturnus vulgaris, 687
vulpina, Erythropygia, 333
—, Stizorhina fraseri, 195, 226

# INDEX

## W

walleri elgonensis Onychognathus, 707, 723
— keniensis, Onychognathus, 723
— nyasæ, Onychognathus, 722
—, Onychognathus walleri, 706, 722, 723
wambera, Cisticola brunnescens, 463
Warsanglia johannis, 1085
warsangliensis, Tchagra senegala, 643
wellsi, Motacilla capensis, 51
wertheri, Euplectes orix, 1052
weynsi, Melanopteryx, 927
whitei, Apalis murina, 397
whytii abayensis Sylvietta, 426
— jacksoni, Sylvietta, 426
— minima, Sylvietta, 426
—, Serinus striolatus, 1077
—, Sylvietta whytii, 425
wilhelmi, Seicercus umbrovirens, 390
williamsi, Geokichla piaggiæ, 251
wilsoni, Hypochera funerea, 1044
winifredæ, Scepomycter, 420
—, Zosterops pallidus, 733
wintoni, Macronyx ameliæ, 80
wolfi magna, Cyanosylvia, 329
woosnami, Alethe castanea, 308
—, Bleda syndactyla, 117
—, Cisticola woosnami, 472
— lufira, Cisticola, 473
— schusteri, Cisticola, 473

## X

xanthomelas, Euplectes capensis, 955, 1052
Xanthophilus subaureus aureoflavus, 922, 1053
— xanthops camburni, 1052
— — xanthops, 923, 1052
xanthophrys, Estrilda, 1053
Xanthoploceus bertrandi, 934
xanthoprymna, Œnanthe, 265
xanthops camburni, Xanthophilus, 1052
—, Xanthophilus xanthops, 923, 1052
xanthopterus, Ploceus, 912
xanthopygius, Serinus atrogularis, 1074
xanthopygos layardi, Pycnonotus, 111
— micrus, Pycnonotus, 112, 152
— spurius, Pycnonotus, 112
xanthornoides, Campephaga, 562
xanthosterna kakamariæ, Petronia, 885
xanthosterna massaica, Petronia, 885
— pallida, Petronia, 885
— pyrgita, Petronia, 884, 1052, 1053
xavieri sethsmithi, Phyllastrephus, 133

## Y

yavelloensis, Aëthocorys personata, 45
yokanæ, Anthreptes reichenowi, 817
—, Bradypterus, 518
youngi, Apalis murina, 397
—, Apalis thoracica, 397

## Z

zalingei, Cisticola galactotes, 487
—, Galerida cristata, 30
zambesiana, Anthreptes collaris, 807
— brunneiceps, Erythropygia, 317
—, Erythropygia zambesiana, 316, 333
— sclateri, Erythropygia, 318
zanzibarica, Coliuspasser axillaris, 962
—, Cyanomitra veroxii, 799
zanzibaricus, Eurillas virens, 149
zarudnyi, Acrocephalus arundinaceus, 364
zavattarii, Euplectes, 1053
Zavattariornis stresemanni, 680
zedlitzi, Cisticola brachyptera, 497
zenkeri, Anthus leucophrys, 65, 81
zombæ, Mirafra fischeri, 46
zombensis, Eurillas virens, 149, 152
zonarius, Cinnyris chalybeus, 820
Zosterops abyssinicus abyssinicus, 734
— — omoensis, 735
— — socotranus, 735
— — eurycricotus, 730, 736
— kikuyuensis chyuluensis, 732
— — kikuyuensis, 731
— — mbuluensis, 732
— — pallidus kulalensis, 733
— — poliogastra, 733
— — winifredæ, 733
— senegalensis anderssoni, 726, 736
— — aurifrons, 736
— — australoabyssinicus, 736
— — flavilateralis, 726
— — jubaensis, 727, 736
— — niassæ, 736
— — senegalensis, 725, 736
— — smithi, 736
— silvanus, 734
— superciliosa, 736
— vaughani, 727
— virens jacksoni, 729

# INDEX

Zosterops virens kaffensis, 730, 736
— — meruensis, 736
— — sarmenticia, 736
— — schoanus, 736
— — stierlingi, 729, 736

Zosterops virens stuhlmanni, 728
— — usambaræ, 736
zuluensis, Calamornis gracilirostris, 518
—, Mirafra africana, 12

Addenda and Corrigenda

# INDEX TO ADDENDA AND CORRIGENDA IN SECOND EDITION OF VOLUME TWO

## A

abyssinica, Hirundo abyssinica, 554
abyssinicus, Zosterops abyssinicus, 823
Acrocephalus griseldis, 519
acuticaudus, Heteropsar, 723
adolfi-frederici, Malaconotus cruentus, 1110
adsimilis, Dicrurus adsimilis, 644
æneocephalus, Lamprotornis purpuropterus, 723
æthiopica, Hirundo, 554
—, Quelea quelea, 1104
æthiopicus, Laniarius æthiopicus, 644
Aëthocorys personata, 1101
— — mcchesneyi, 1106
afer graueri, Cinnyris, 823
—, Nilaus, 644
—, Parus, 723
africanoides macdonaldi, Mirafra, 46
Agrobates galactotes galactotes, 519
— — minor, 519
akeleyi, Ammomanes deserti, 1101
Alæmon alaudipes desertorum, 46
— hamertoni altera, 1101
Alauda arenicolor, 1106
alaudipes desertorum, Alæmon, 46
albicapillus, Spreo, 723
albicauda kivuensis, Elminia, 1102
—, Mirafra, 46
— rukweriensis, Mirafra, 1101
albigula itoculo, Phyllastrephus, 1101
albigularis, Phyllastrephus abigularis, 152
albus, Corvus, 723
Alethe castanea, 1102
— poliocephala ufipae, 1102
alinæ, Cyanomitra, 1104
Alseonax cinereus cinereolus, 1101
— seth-smithi, 1101
altera, Alæmon hamertoni, 1101
alticola, Apalis alticola, 1103
— brunneiceps, Apalis, 1103
—, Cisticola, 1103
altirostris, Galerida cristata, 1101
amaurocephalus, Symplectes bicolor, 1112

amethystina kirkii, Chalcomitra, 823
Ammomanes cinctura arenicolor, 1106
— — kinneari, 1101, 1106
— — pallens, 1106
— deserti akeleyi, 1101
— — erythrochroa, 1101
— — samharensis, 1101
— phœnicura pallens, 1106
Andropadus ansorgei, 152
— — kavirondensis, 1108
— gracilis, 152
— importunus fricki, 152
angolensis somereni, Serinus, 1099
ansorgei, Andropadus, 152
— kavirondensis, Andropadus, 1108
—, Nesocharis, 1054
Anthreptes collaris collaris, 1104
— — hypodilus, 1104
— — somereni, 1104
— longuemarei neumanni, 1104
— pallidigaster, 1104
Anthus leucophrys zenkeri, 81
— novæseelandiæ lacuum, 81
— similis hallæ, 1107
— — schoutedeni, 1107
— vaalensis goodsoni, 81
antinorii, Cisticola lateralis, 1104
Apalis alticola alticola, 1103
— — brunneiceps, 1103
— denti, 1103
— flavigularis flavigularis, 1103
— — griseiceps, 1103
— — lynesi, 1103
— — uluguru, 1103
— murina fuscigularis, 1103
— — murina, 1103
— — whitei, 1103
— — youngi, 1103
— nigrescens, 1103
— rufifrons rufifrons, 1103
— rufogularis, 1103
— thoracica, 1103
approximans, Malaconotus blanchoti, 644
— — poliocephalus, 644
ardesiaca, Melænornis, 1108
arenicolor, Alauda, 1106
— Ammomanes cinctura, 1106
aridula lavendulæ, Cisticola, 1104

# INDEX TO ADDENDA AND CORRIGENDA

atricollis dorsostriata, Ortygospiza, 1054
— fuscocrissa, Ortygospiza, 1054
atripennis bohndorffi, Phyllanthus, 1107
atrogularis, Ploceus atrogularis, 974
Auripasser euchlorus, 974
australis, Cryptospiza reichenovii, 1054

## B

badiceps, Eremomela, 1103
baileyi, Pyrrhocorax pyrrhocorax, 723
bannermani, Plocepasser superciliosus, 973
barbatus schoanus, Pycnonotus, 152
— somaliensis, Pycnonotus, 152
Bathmocercus rufus jacksoni, 1104
— — vulpinus, 1104
Batis orientalis orientalis, 1102
belcheri, Psophocichla guttata, 334, 1103
belli, Pytilia melba, 1054, 1105
bengalus katangæ, Uræginthus, 1113
— kigomænsis, Uræginthus, 1113
bengalus schoanus, Uræginthus, 1054
bensoni, Phyllastrephus terrestris, 152
bertrandi, Chlorophoneus olivaceus, 644
— — rubiginosus, 644
Bias musicus changamwensis, 1102
— — musicus, 1102
bicolor amaurocephalus, Symplectes, 1112
— kigomænsis, Symplectes, 1111
—, Speculipastor, 723
—, Symplectes bicolor, 1111
blanchoti approximans, Malaconotus, 644
—, Malaconotus, 644
blissetti, Dyaphorophyia, 1102
blythii, Onychognathus, 723
bocagei, Peliocichla, 334
—, Turdus pelios, 334
böhmi, Parisoma, 1101
bohndorffi, Crateropus, 1107
—, Phyllanthus atripennis, 1107
brachyptera, Cisticola, 1104
bractiatus, Cinnyris chalybeus, 823
Bradypterus rufescens, 519
brunneiceps, Apalis alticola, 1103
—, Erythropygia leucophrys, 1102

brunnescens, Plocepasser superciliosus, 973
Budytes feldegg, 81

## C

Calamocichla gracilirostris gracilirostris, 519, 1103
— — jacksoni, 519, 1104
— — leptorhyncha, 519
— — macrorhyncha, 519
— — nilotica, 519
— — parva, 519
— — tsanæ, 519
— leptorhyncha jacksoni, 1103, 1104
— — leptorhyncha, 1103
— — macrorhyncha, 1103
— — nuerensis, 1104
— — tsanæ, 1103
— rufescens rufescens, 519
— — foxi, 519
— — nilotica, 519
Calamoherpe gracilirostris, 519
Calamonastes fasciolatus stierlingi, 1103
— simplex simplex, 1103
Calandrella cinerea saturatior, 1101
— — williamsi, 1101
— rufescens vulpecula, 1106
calurus emini, Criniger, 152
Camaroptera superciliaris ugandæ, 1103
Campephaga quiscalina martini, 644
canescens elegans, Eremomela, 1103
—, Eremomela canescens, 1103
cantillans chadensis, Mirafra, 46
— marginata, Mirafra, 46
capensis kordofanensis, Corvus, 723
— litoris, Euplectes, 1105
— xanthomelas, Euplectes, 1105, 1112
— zambesiensis, Euplectes, 1112
capitalis dimidiatus, Ploceus, 974
— fischeri, Ploceus, 974
carlo, Prinia gracilis, 1104
caspica, Motacilla, 81
castanea, Alethe, 1102
castanops, Ploceus, 974
castanopterus, Passer, 974
Cercomela melanura lypura, 1102
chadensis, Mirafra cantillans, 46
Chalcomitra amethystina kirkii, 823
— rubescens rubescens, 823

# INDEX TO ADDENDA AND CORRIGENDA

chalybeus bractiatus, Cinnyris, 823
— graueri, Cinnyris, 823
— intermedius, Cinnyris, 823
— mancensis, Cinnyris, 823
changamwensis, Bias musicus, 1102
Charitillas gracilis kavirondensis, 152
— kavirondensis, 1108
Chlorocichla lætissima, lætissima, 152
— — schoutedeni, 1108
Chlorophoneus multicolor, 644
— nigrifrons, 644
— olivaceus bertrandi, 644
— rubiginosus bertrandi, 644
chloropterus, Lamprocolius, 723
chubbi, Cisticola chubbi, 1104
cinctura arenicolor, Ammomanes, 1106
— kinneari, Ammomanes, 1101, 1106
— pallens, Ammomanes, 1106
cinerascens, Turdus libonyanus, 334, 1103
cinera, Creatophora, 723
— saturatior, Calandrella, 1101
— williamsi, Calandrella, 1101
cinereolus, Alseonax cinereus, 1101
cinereus cinereolus, Alseonax, 1101
Cinnyris afer graueri, 823
— chalybeus bractiatus, 823
— — graueri, 823
— — intermedius, 823
— — mancensis, 823
— habessinicus, 823
— ludovicensis ludovicensis, 823
— — reichenowi, 823
— mariquensis, 823
— regius, 823
— reichenowi, 823
Cisticola alticola, 1103
— aridula lavendulæ, 1104
— brachyptera, 1104
— chubbi chubbi, 1104
— erythrops, 1104
— hunteri hypernephala, 1104
— lateralis antinorii, 1104
Coliuspasser laticauda laticauda, 1105
— progne delamerei, 1105
collaris, Anthreptes collaris, 1104
— hypodilus, Anthreptes, 1104
— somereni, Anthreptes, 1104
Coracia pyrrhocorax docilis, 723
corax edithæ, Corvus, 723
Corvus albus, 723
— capensis kordofanensis, 723

Corvus corax edithæ, 723
— splendens, 723
Cossypha natalensis intensa, 1102
— — natalensis, 1102
— semirufa, 1102
costæ, Turdus libonyanus, 1103
Crateropus bohndorffi, 1107
Creatophora cinerea, 723
Cringer calurus emini, 152
cristata altirostris, Galerida, 1101
— isabellina, Galerida, 1101
cristatus phœnicuroides, Lanius, 644
— speculigerus, Lanius, 644
cruentus adolfi-frederici, Malaconotus, 1110
— hilgerti, Rhodophoneus, 644
Cryptospiza jacksoni, 1054
— ocularis, 1054
— reichenovii, 1054
— — australis, 1054
cucullata, Hirundo, 1110
cucullatus graueri, Ploceus, 1111
— nigriceps, Ploceus, 1111
cyanolæma, Cyanomitra cyanolæma, 1104
— octaviæ, Cyanomitra, 1104
Cyanomitra alinæ, 1104
— cyanolæma cyanolæma, 1104
— — octaviæ, 1104

D

delamerei, Coliuspasser progne, 1105
denti, Apalis, 1103
desertæ, Prinia subflava, 1104
deserti akeleyi, Ammomanes, 1101
— erythrochroa, Ammomanes, 1101
— samharensis, Ammomanes, 1101
desertorum, Alæmon alaudipes, 46
Dicrurus adsimilis adsimilis, 644
dimidiatus, Ploceus capitalis, 974
dinemelli, Dinemellia, 973
Dinemellia dinemelli, 973
docilis, Coracia pyrrhocorax, 723
dodsoni, Pycnonotus, 152
domesticus halfæ, Passer, 974
— indicus, Passer, 974
dorsostriata, Ortygospiza atricollis, 1054
Drepanoplectes jacksoni, 1105
Drepanorhynchus reichenowi, 823
— — lathburyi, 1111
dubarensis, Lanius elegans, 644
Dyaphorophyia blissetti, 1102
— jamesoni, 1102

# INDEX TO ADDENDA AND CORRIGENDA

## E

edithæ, Corvus corax, 723
elegans dubarensis, Lanius, 644
—, Eremomela canescens, 1103
elgonensis, Serinus flavivertex, 1099
—, Serinus gularis, 1099
ellioti, Galerida praetermissa, 1101
Elminia albicauda kivuensis, 1102
eluta, Erythropygia leucoptera, 1102
Emberiza flaviventris, 1099
— orientalis, 1099
emini, Criniger calurus, 152
Erannornis longicauda kivuensis, 1102
Eremomela badiceps, 1103
— canescens canescens, 1103
— — elegans, 1103
— flavicrissalis, 1103
— icteropygialis griseoflava, 1103
— turneri, 1103
— usticollis rensi, 1103
Eremopterix nigriceps melanauchen, 1101
Erythrocercus livingstonei francisi, 1102
— — livingstonei, 1102
— — monapo, 1102
erythrochroa, Ammomanes deserti, 1101
erythrops, Cisticola, 1104
Erythropygia leucophrys brunneiceps, 1102
— — jungens, 1103
— — sclateri, 1102
— — zambesiana, 1102
— leucoptera eluta, 1102
— — pallida, 1102
— zambesiana zambesiana, 1102, 1103
erythrothorax, Stiphrornis, 1102
euchlorus, Auripasser, 974
Euodice malabarica orientalis, 1054
Euplectes capensis litoris, 1105
— — xanthomelas, 1105, 1112
— — zambesiensis, 1112
— nigriventris, 1105
— orix pusilla, 1104, 1105
— zavattarii, 1105
Eurillas virens, 1101

## F

famosa, Nectarinia, 823
fasciolatus stierlingi, Calamonastes, 1103
feldegg, Budytes, 81

fischeri münzneri, Phyllastrephus, 1101, 1108
—, Phyllastrephus fischeri, 152, 1101
—, Ploceus capitalis, 974
—, Vidua, 1054
flavicrissalis, Eremomela, 1103
flavigularis, Apalis flavigularis, 1103
— griseiceps, Apalis, 1103
— lynesi, Apalis, 1103
— uluguru, Apalis, 1103
flaviventris, Emberiza, 1099
flavivertex elgonensis, Serinus, 1099
foxi, Calamocichla rufescens, 519
francisi, Erythrocercus livingstonei, 1102
fremantlii megænsis, Pseudalæmon 1101
—, Pseudalæmon fremantlii, 1101
fricki, Andropadus importunus, 152
frommi, Pirenestes, 973, 1054
fulgidus hartlaubi, Onychognathus, 723
fuscigularis, Apalis murina, 1103
fuscocrissa, Ortygospiza atricollis, 1054

## G

gabonensis, Ortygospiza, 1054
galactotes, Agrobates galactotes, 519
— minor, Agrobates, 519
Galerida cristata altirostris, 1101
— — isabellina, 1101
— — malabarica ellioti, 1101
— — prætermissa, 1101
golandi, Ploceus, 974
goodsoni, Anthus vaalensis, 81
gracilirostris, Calamoherpe, 519
— Calamocichla, gracilirostris, 519, 1103
— jacksoni, Calamocichla, 519, 1104
— — leptorhyncha, Calamocichla, 519
— macrorhyncha, Calamocichla, 519
— nilotica, Calamocichla, 519
— parva, Calamocichla, 519
— tsanæ, Calamocichla, 519
gracilis, Andropadus, 152
— carlo, Prinia, 1104
— kavirondensis, Charitillas, 152
graueri, Cinnyris afer, 823
— — chalybeus, 823

# INDEX TO ADDENDA AND CORRIGENDA

graueri, Ploceus cucullatus, 1111
griseiceps, Apalis flavigularis, 1103
griseiventris, Parus, 723
griseldis, Acrocephalus, 519
griseoflava, Eremomela icteropygialis, 1103
grotei, Pytilia melba, 1054, 1105
gularis elgonensis, Serinus, 1099
guttata belcheri, Psophocichla, 334, 1103
—, Psophocichla guttata, 334
guttatus, Ixonotus guttatus, 1101
— ngoma, Ixonotus, 1101

## H

habessinicus, Cinnyris, 823
halfæ, Passer domesticus, 974
hallæ, Anthus similis, 1107
hamertoni altera, Alæmon, 1101
hartlaubi, Onychognathus fulgidus, 723
Heterhyphantes melanogaster stephanophorus, 1104
Heteropsar acuticaudus, 723
heuglini, Œnanthe, 334
—, Ploceus, 974
hilgerti, Rhodophoneus cruentus, 644
Hirundo abyssinica abyssinica, 554
— æthiopica, 554
— cucullata, 1110
hunteri hypernephala, Cisticola, 1104
hypernephala, Cisticola hunteri, 1104
Hyphantornis superciliosus, 1104
Hyphanturgus olivaceiceps, 974
hypocherina, Vidua, 1054
hypodilus, Anthreptes collaris, 1104
hypopyrrhus, Malaconotus, 644

## I

icteropygialis griseoflava, Eremomela, 1103
importunus fricki, Andropadus, 152
indicus, Passer domesticus, 974
intensa, Cossypha natalensis, 1102
intermedius, Cinnyris chalybeus, 823
—, Onychognathus, 723
— Ploceus, 974
isabellina, Galerida cristata, 1101
—, Sylvietta, 1103
itoculo, Phyllastrephus albigula, 1101

Ixonotus guttatus guttatus, 1101
— — ngoma, 1101

## J

jacksoni, Bathmocercus rufus, 1104
—, Calamocichla gracilirostris, 519, 1104
—, — leptorhyncha, 1103, 1104
— Cryptospiza, 1054
—, Drepanoplectes, 1105
jamesoni, Dyaphorophyia, 1102
johannis, Warsanglia, 1099
jungens, Erythropygia leucophrys, 1103

## K

katangæ, Uræginthus bengalus, 1113
kavirondensis, Andropadus ansorgei, 1108
—, Charitillas, 1108
—, — gracilis, 152
kersteni, Symplectes, 1111
kigomænsis, Symplectes bicolor, 1111
—, Uræginthus bengalus, 1113
kilimensis, Linurgus olivaceus, 1099
kinneari, Ammomanes cinctura, 1101, 1106
kirkii, Chalcomitra amethystina, 823
kivuensis, Elminia albicauda, 1102
— Erannornis longicauda, 1102
kordofanensis, Corvus capensis, 723

## L

lacuum, Anthus novæseelandiæ, 81
lætissima, Chlorocichla lætissima, 152
— schoutedeni, Chlorocichla, 1108
Lamprocolius chlopterus, 723
Lamprotornis purpuropterus æneocephalus, 723
Laniarius æthiopicus æthiopicus, 644
Lanius cristatus phœnicuroides, 644
— — speculigerus, 644
— elegans dubarensis, 644
lateralis antinorii, Cisticola, 1104
lathburyi, Drepanorhynchus reichenowi, 1111
—, Nectarinia reichenowi, 1111

# INDEX TO ADDENDA AND CORRIGENDA

laticauda, Coliuspasser laticauda, 1105
lavendulæ, Cisticola aridula, 1104
leptorhyncha, Calamocichla gracilirostris, 519
— — leptorhyncha, 1103
— jacksoni, Calamocichla, 1103, 1104
— macrorhyncha, Calamocichla, 1103
— nuerensis, Calamocichla, 1104
— tsanæ, Calamocichla, 1103
leucomela, Œnanthe, 334
leucophrys brunneiceps, Erythropygia, 1102
— jungens, Erythropygia, 1103
— sclateri, Erythropygia, 1102
— zambesiana, Erythropygia, 1102
— zenkeri, Anthus, 81
leucoptera eluta, Erythropygia, 1102
— pallida, Erythropygia, 1102
libonyanus cinerascens, Turdus, 334, 1103
— costæ, Turdus, 1103
— niassæ, Turdus, 1103
— tropicalis, Turdus, 334, 1103
—, Turdus libonyanus, 1103
— verreauxi, Turdus, 1103
Linurgus olivaceus kilimensis, 1099
Lioptilornis, 1102
Lioptilus, 1102
litoris, Euplectes capensis, 1105
litsipsirupa, Psophocichla litsipsirupa, 334
livingstonei, Erythrocercus livingstonei, 1102
— francisi, Erythrocercus, 1102
— monapo, Erythrocercus, 1102
livingstonii, Œnanthe pileata, 1102
Lonchura, 1054
longicauda kivuensis, Erannornis, 1102
longuemarei neumanni, Anthreptes, 1104
lousiæ, Rhynchostruthus socotranus, 1099
ludovicensis, Cinnyris ludovicensis, 823
— reichenowi, Cinnyris, 823
ludoviciæ, Turdus, 334
lugens, Parisoma, 1101
lugubris, Œnanthe lugubris, 334
— vaurei, Œnanthe, 334
luteifrons, Nigritra luteifrons, 1112
luteolus, Ploceus, 974
lynesi, Apalis flavigularis, 1103
lypura, Cercomela melanura, 1102

## M

macdonaldi, Mirafra africanoides, 46
macrorhyncha, Calamocichla gracilirostris, 519
—, Calamocichla leptorhyncha, 1103
macroura, Vidua, 1054
malabarica ellioti, Galerida, 1101
— orientalis, Euodice, 1054
— prætermissa, Galerida, 1101
Malacocincla, 152
Malaconotus blanchoti, 644
— — approximans, 644
— cruentus adolfi-frederici, 1110
— hypopyrrhus, 644
— poliocephalus approximans, 644
Malimbus nitens microrhynchus, 1112
Mandingoa nitidula schlegeli, 1054
manœnsis, Cinnyris chalybeus, 823
marginata, Mirafra cantillans, 46
mariquensis, Cinnyris, 823
martini, Campephaga quiscalina, 644
maximus, Pirenestes, 973, 1054
mcchesneyi, Aëthocorys personata, 1106
megænsis, Pseudalæmon fremantlii, 1101
Melænornis ardesiaca, 1108
— pammelaina, 1102
melanauchen, Eremopterix nigriceps, 1101
melanogaster stephanophorus, Heterhyphantes, 1104
melanura lypura, Cercomela, 1102
melba belli, Pytilia, 1054, 1105
— grotei, Pytilia, 1054, 1105
microrhynchus, Malimbus nitens, 1112
minor, Agrobates galactotes, 519
—, Pirenestes, 1113
Mirafra africanoides macdonaldi, 46
— albicauda, 46
— — rukweriensis, 1101
— cantillans marginata, 46
— — chadensis, 46
— rufa, 46
— sharpei, 46
— williamsi, 1105
monacha, Œnanthe, 334
monapo, Erythrocercus livingstonei, 1102
Monticola rufocinerea rufocinerea, 334
Motacilla caspica, 81
— pleschanka, 334

# INDEX TO ADDENDA AND CORRIGENDA

motitensis shelleyi, Passer, 974
multicolor, Chlorophoneus, 644
münzneri, Phyllastrephus fischeri, 1101, 1108
murina, Apalis murina, 1103
— fuscigularis, Apalis, 1103
— whitei, Apalis, 1103
— youngi, Apalis, 1103
Muscicapa semitorquata, 1101
musicus, Bias musicus, 1102
— changawensis, Bias, 1102
Myioparus plumbeus, 1101

## N

natalensis, Cossypha natalensis, 1102
— intensa, Cossypha, 1102
Nectarinia famosa, 823
— purpureiventris, 823
— reichenowi lathburyi, 1111
Nesocharis ansorgei, 1054
neumanni, Anthreptes longuemarei, 1104
ngoma, Ixonotus guttatus, 1101
niassæ, Turdus libonyanus, 1103
niger, Parus, 723
nigrescens, Apalis, 1103
nigriceps melanauchen, Eremopterix, 1101
—, Ploceus cucullatus, 1111
nigrifrons, Chlorophoneus, 644
Nigrita luteifrons luteifrons, 1112
nigriventris, Euplectes, 1105
Nilaus afer, 644
— temporalis, 644
nilotica, Calamocichla gracilirostris, 519
—, — rufescens, 519
nitens microrhynchus, Malimbus, 1112
novæseelandiæ lacuum, Anthus, 81
nuerensis, Calamocichla leptorhyncha, 1104

## O

octaviæ, Cyanomitra cyanolæma, 1104
ocularis, Cryptospiza, 1054
Œnanthe heuglini, 334
— leucomela, 334
— lugubris lugubris, 334
— — vaurei, 334
— monacha, 334
— phillipsi, 334
— pileata livingstonii, 1102
— pleschanka pleschanka, 334

olivaceiceps, Hyphanturgus, 974
olivaceus bertrandi, Chlorophoneus, 644
— kilimensis, Linurgus, 1099
Onychognathus blythii, 723
— fulgidus hartlaubi, 723
— intermedius, 723
orientalis, Batis orientalis, 1102
—, Emberiza, 1099
—, Euodice malabarica, 1054
orix pusilla, Euplectes, 1104, 1105
Ortygospiza atricollis dorsostriata, 1054
— — fuscocrissa, 1054
— gabonensis, 1054

## P

Pachyphantes pachyrhynchus, 1104
— superciliosus, 1104
pachyrhynchus, Pachyphantes, 1104
pallens, Ammomanes cinctura, 1106
—, — phœnicura, 1106
pallescens, Prinia subflava, 1104
pallida, Erythropygia leucoptera, 1102
pallidigaster, Anthreptes, 1104
pammelaina, Melænornis, 1102
paradisea, Steganura, 1105
Parisoma böhmi, 1101
— lugens, 1101
— plumbeum, 1101
Parus afer, 723
— griseiventris, 723
— niger, 723
parva, Calamocichla gracilirostris, 519
Passer castanopterus, 974
— domesticus halfæ, 974
— — indicus, 974
— motitensis shelleyi, 974
— suahelicus, 974
— swainsoni, 974
Peliocichla bocagei, 334
— schuetti, 334
pelios bocagei, Turdus, 334
— schuetti, Turdus, 334
— Turdus, pelios, 334
Pelocichla, 334
personata, Æthocorys, 1101
— mcchesneyi, Aëthocorys, 1106
Petronia xanthosterna, 974
philippæ, Syvietta, 1109
phillipsi, Œnanthe, 334
phœnicura pallens, Ammomanes, 1106

# INDEX TO ADDENDA AND CORRIGENDA

phœnicuroides, Lanius cristatus, 644
Phyllanthus atripennis bohndorffi, 1107
Phyllastrephus albigula itoculo, 1101
— albigularis albigularis, 152
— fischeri fischeri, 152, 1101
— — münzneri, 1101, 1108
— terrestris bensoni, 152
Phyllolais pulchella, 1103
Phylloscopus umbrovirens williamsi, 1109
pileata livingstonii, Œnanthe, 1102
Pirenestes frommi, 973, 1054
— maximus, 973, 1054
— minor, 1113
— vincenti, 1113
pleschanka, Œnanthe pleschanka, 334
Plocepasser superciliosus, 973
— — bannermani, 973
— — brunnescens, 973
Ploceus atrogularis atrogularis, 974
— capitalis dimidiatus, 974
— — fischeri, 974
— castanops, 974
— cucullatus graueri, 1111
— — nigriceps, 1111
— golandi, 974
— heuglini, 974
— intermedius, 974
— luteolus, 974
— reichardi, 974
plumbeum, Parisoma, 1101
plumbeus, Myioparus, 1101
poliocephala ufipae, Alethe, 1102
poliocephalus approximans, Malaconotus, 644
prætermissa, Galerida malabarica, 1101
Prinia gracilis carlo, 1104
— somalica somalica, 1104
— subflava desertæ, 1104
— — pallescens, 1104
progne delamerei, Coliuspasser, 1105
Pseudalæmon fremantlii fremantlii, 1101
— — magænsis, 1101
Psophocichla guttata belcheri, 334, 1103
— — guttata, 334
— litsipsirupa litsipsirupa, 334
pulchella, Phyllolais, 1103
purpureiventris, Nectarinia, 823
purpuropterus æneocephalus, Lamprotornis, 723
pusilla, Euplectes orix, 1104, 1105

Pycnonotus barbatus schoanus 152
— — somaliensis, 152
Pycnonotus dodsoni, 152
pyrrhocorax docilis, Coracia, 723
Pyrrhocorax pyrrhocorax baileyi, 723
Pytilia melba belli, 1054, 1105
— — grotei, 1054, 1105

## Q

Quelea quelea æthiopica, 1104
quiscalina martini, Campephaga, 644

## R

regius, Cinnyris, 823
reichardi, Ploceus, 974
reichenovii australis, Cryptospiza, 1054
—, Cryptospiza reichenovii, 1054
reichenowi Cinnyris, 823
— Cinnyris ludovicensis, 823
— Drepanorhynchus, 823
— lathburyi, Drepanorhynchus, 1111
— —, Nectarinia, 1111
rensi, Eremomela usticollis, 1103
Rhinocorax rhipidurus, 723
rhipidurus, Rhinocorax, 723
Rhodophoneus cruentus hilgerti, 644
Rhynchostruthus socotranus lousiæ, 1099
rubescens, Chalcomitra rubescens, 823
rubiginosus bertrandi, Chlorophoneus, 644
rufa, Mirafra, 46
rufescens, Bradypterus, 519
—, Calamocichla rufescens, 519
— foxi, Calamocichla, 519
— nilotica, Calamocichla, 519
— vulpecula, Calandrella, 1106
rufifrons, Apalis rufifrons, 1103
rufocinerea, Monticola rufocinerea, 334
rufogularis, Apalis, 1103
rufus jacksoni, Bathmocercus, 1104
— vulpinus, Bathmocercus, 1104
rukweriensis, Mirafra albicauda, 1101

## S

samharensis, Ammomanes deserti, 1101

# INDEX TO ADDENDA AND CORRIGENDA

saturatior, Calandrella cinerea, 1101
schlegeli, Mandingoa nitidula, 1054
schoanus, Pycnonotus barbatus, 152
—, Uræginthus bengalus, 1054
schoutedeni, Anthus similis, 1107
—, Chlorocichla lætissima, 1108
schuetti, Peliocichla, 334
—, Turdus pelios, 334
sclateri, Erythropygia leucophrys, 1102
Seicercus umbrovirens williamsi, 1109
semirufa, Cossypha, 1102
semitorquata, Muscicapa, 1101
senegala, Tchagra sengala, 644
Serinus angolensis somereni, 1099
— flavivertex elgonensis, 1099
— gularis elgonensis, 1099
seth-smithi, Alseonax, 1101
sharpei, Mirafra, 46
shelleyi, Passer motitensis, 974
similis hallæ, Anthus, 1107
— schoutedeni, Anthus, 1107
simplex, Calamonastes simplex, 1103
socotranus lousiæ, Rhynchostruthus, 1099
somalica, Prinia somalica, 1104
somaliensis, Pycnonotus barbatus, 152
somereni, Anthreptes collaris, 1104
—, Serinus angolensis, 1099
speculigerus, Lanius cristatus, 644
Speculipastor bicolor, 723
Spermestes, 1054
Sphenœacus, 1103
splendens, Corvus, 723
Spreo albicapillus, 723
— superbus, 723
Steganura paradisea, 1105
stephanophorus, Heterhyphantes melanogaster, 1104
stierlingi, Calamonastes fasciolatus, 1103
Stiphrornis erythrothorax, 1102
suahelicus, Passer, 974
subflava desertæ, Prinia, 1104
— pallescens, Prinia, 1104
superbus, Spreo, 723
superciliosus bannermani, Plocepasser, 973
— brunnescens, Plocepasser, 973
—, Hyphantornis, 1104
—, Pachyphantes, 1104
—, Plocepasser superciliosus, 973
swainsoni, Passer, 974

Sylvietta isabellina, 1103
— philippæ, 1109
Symplectes bicolor amaurocephalus, 1112
— — bicolor, 1111
— — kigomænsis, 1111
— kersteni, 1111

## T

Tchagra senegala senegala, 644
temporalis, Nilaus, 644
terrestris bensoni, Phyllastrephus, 152
thoracica, Apalis, 1103
Trichastoma, 152
Tricophorus, 152
Trichostoma, 152
tropicalis, Turdus libonyanus, 334, 1103
tsanæ, Calamocichla gracilirostris, 519
—, — leptorhyncha, 1103
Turdus libonyanus cinerascens, 334, 1103
— — costæ, 1103
— — libonyanus, 1103
— — niassæ, 1103
— — tropicalis, 334, 1103
— — verreauxi, 334
— ludoviciæ, 334
— pelios bocagei, 334
— — pelios, 334
— — schuetti, 334
turneri, Eremomela, 1103

## U

ufipæ, Alethe poliocephala, 1102
ugandæ, Camaroptera superciliaris, 1103
uluguru, Apalis flavigularis, 1103
umbrovirens williamsi, ·Phlloscopus, 1109
— —, Seicercus, 1109
Uræginthus bengalus katangæ, 1113
— — kigomænsis, 1113
— — schoanus, 1054
usticollis rensi, Eremomela, 1103

## V

vaalensis goodsoni, Anthus, 81
vaurei, Œnanthe lugubris, 334
verreauxi, Turdus libonyanus, 334
Vidua fischeri, 1054

# INDEX TO ADDENDA AND CORRIGENDA

Vidua hypocherina, 1054
— macroura, 1054
vincenti, Pirenestes, 1113
virens, Eurillas, 1101
vulpecula, Calandrella rufescens, 1106
vulpinus, Bathmocercus rufus, 1104

## W

Warsanglia johannis, 1099
whitei, Apalis murina, 1103
williamsi, Calandrella cinerea, 1101
—, Mirafra, 1105
—, Phylloscopus umbrovirens, 1109
—, Seicercus umbrovirens, 1109

## X

xanthomelas, Euplectes capensis, 1105, 1112
xanthosterna, Petronia, 974

## Y

youngi, Apalis murina, 1103

## Z

zambesiana, Erythropygia leucophrys, 1102
—, — zambesiana, 1102, 1103
zambesiensis, Euplectes capensis 1112
zavattarii, Euplectes, 1105
zenkeri, Anthus leucophrys, 81
Zosterops, 823
— abyssinicus abyssinicus, 823

# Index of English Group Names in Volumes One and Two

## A

Akalats, ii, 304
Alethes, ii, 308
Apalis Warblers, ii, 396
Avocet, i, 368

## B

Babblers, ii, 82
—, Hill-, ii, 104
—, Thrush-, ii, 99
Barbets, i, 697, 731
Bat-eating Buzzard, i, 165
Bateleur, i, 188
Bee-eaters, i, 584
Birds of Prey, i, 116
Bishops, ii, 949
Bitterns, i, 54
Blackcap, ii, 348
Blue-bills, ii, 987
Bluethroat, ii, 328
Boobys, i, 19
Boubous, ii, 608
Bristle-bills, ii, 117
Broadbills, i, 792
Brownbuls, ii, 122
Brubrus, ii, 584
Buffalo-Weavers, ii, 858
Bulbuls, ii, 107
Buntings, ii, 1086
Bush-Crow, ii, 680
Bush-Robins, ii, 322
Bush-Shrikes, ii, 614–616, 623–638
Bustards, i, 311
Button-Quails, i, 442
Buzzards, i, 193

## C

Canaries, ii, 1061
Cape Hen, i, 13
Catbird, Abyssinian, ii, 176
Chanting-Goshawks, i, 209
Chats, ii, 271
—, Cliff-, ii, 280
—, Robin-, ii, 291
Chatterers, ii, 93
Chiff-Chaff, ii, 384
Chough, ii, 678
Chukor, i, 226

Cisticolas, ii, 448
Citril, African, ii, 1082
Cliff-Chats, ii, 280
Colies, i, 684
Coots, i, 303
Cordon-Bleus, ii, 1033
Cormorants, i, 24
Coucals, i, 513
Coursers, i, 396
Crakes, i, 287
Cranes, i, 306
Creepers, ii, 821
Crimson-wings, ii, 992
Crombecs, ii, 423
Crows, ii, 670
Cuckoo-Falcon, i, 158
Cuckoo Shrikes, ii, 555
Cuckoos, i, 492
Curlew, i, 392
— Sandpiper, i, 379
—, Stone, i, 328
Cut-throat, ii, 995

## D

Darters, i, 28
Diochs, ii, 937
Doves, i, 470
Drongos, ii, 562
Ducks, i, 83–109
Dunlin, i, 379

## E

Eagles, i, 167
—, Fish, i, 187
—, Harrier, i, 182
Egrets, i, 42–45
Eremomelas, ii, 430

## F

Falcon, Cuckoo, i, 158
Falcons, i, 138
Finch, Oriole-, ii, 1080
—, Locust, ii, 999
—, Negro-, ii, 982
—, Quail-, ii, 997
Finches, ii, 1055
—, Fire-, ii, 1009
Finfoot, i, 305

# INDEX OF ENGLISH GROUP NAMES

Fire-Finches, ii, 1009
Fiscals, ii, 591, 593–596
Flamingoes, i, 80
Flycatchers, ii, 153
Forest-Robin, ii, 307
Forest-Warbler, ii, 419
Francolins, i, 227
Frigate-Birds, i, 29

## G

Gallinules, i, 299
Game Birds, i, 221
Gannets, i, 19
Geese, i, 110
Glass-eyes, ii, 439
Go-away-birds, i, 536
Godwits, i, 391
Gonoleks, ii, 606
Goshawk, Chanting-, i, 209
—, Gabar, i, 209
Goshawks, i, 199
Grasshopper Buzzard, i, 187
Grebes, i, 3
Greenbuls, ii, 116
Greenshank, i, 390
Grenadier, Purple, ii, 1038
Guinea-Fowls, i, 270
Gulls, i, 413

## H

Hadada, i, 74
Hammerkop, i, 60
Harrier-Eagles, i, 182
Harrier-Hawk, i, 217
Harriers, i, 214
Hawk-Eagles, i, 174–179
Hawk, Long-tailed, i, 213
Helmet-Shrikes, ii, 567
Hemipodes, i, 442
Herons, i, 34
Hill-Babblers, ii, 104
Hobbies, i, 144
Honey-birds, i, 746
Honey Buzzard, i, 165
Honey-guides, i, 738
Hoopoes, i, 629
—, Wood, i, 632
Hornbills, i, 604
Hylias, ii, 818
Hypocolius, Grey, ii, 645

## I

Ibis, Wood-, i, 70
Ibises, i, 71
Illadopses, ii, 100
Indigo-birds, ii, 1039

## J

Jabiru, i, 68
Jacanas, i, 333

## K

Kestrels, i, 149
Kingfishers, i, 566
Kites, i, 160
Knot, i, 381

## L

Lammergeyer, i, 192
Lapwing, Long-toed, i, 365
Lapwings, i, 351
Larks, ii, 1
Leaf-Loves, ii, 119
Lily Trotters, i, 333
Lizard-Buzzard, i, 180
Locust-Finch, ii, 999
Longbills, ii, 446
Long-claws, ii, 76
Lovebirds, i, 550

## M

Malimbes, ii, 937
Mannikins, ii, 975
Man-o'-war Birds, i, 29
Marabou, i, 69
Martins, ii, 521
Moorhens, i, 301
Morning Warblers, ii, 314
Mousebirds, i, 684

## N

Negro-Finches, ii, 982
Nicators, ii, 640
Nightingale, ii, 330
Nightjars, i, 663
Noddy, i, 440

## O

Olive-backs, ii, 1032
Open-Bill, i, 67
Oriole-Finch, ii, 1080
Orioles, ii, 660
Ortolan, ii, 1092
Osprey, i, 219
Ostrich, i, 1
Owls, i, 642
Oxpeckers, ii, 720
Oyster-Catcher, i, 367

# INDEX OF ENGLISH GROUP NAMES

## P

Parrakeets, i, 549
Parrots, i, 539
Partridge, Sand, i, 226
—, Stone-, i, 268
Pelicans, i, 32
Petrels, i, 7
Petronias, ii, 869
Phalaropes, i, 395
Piapiac, ii, 679
Pigeons, i, 458
—, Green, i, 485
Pipits, ii, 58
Pittas, i, 796
Plantain-eater, i, 535
Plover, Crab, i, 409
—, Egyptian, i, 408
—, Quail-, i, 444
—, Ringed, i, 339
Plovers, i, 335
Pratincoles, i, 403
Prinias, ii, 505
Puff-backs, ii, 617
Pytilias, ii, 1005

## Q

Quail-Finches, ii, 997
Quail-Plover, i, 444
Quails, i, 264
—, Button-, i, 442
Queleas, ii, 944

## R

Rails, i, 285
Ravens, ii, 670
Redshanks, i, 387
Redstarts, ii, 326
Ringed Plover, i, 339
Robin, Forest-, ii, 307
—, White-throated, ii, 329
Robins, Bush, ii, 322
—, Scrub-, ii, 316
Robin-Chats, ii, 291
Rock-Chats, ii, 272
Rock-Sparrow, ii, 887
Rollers, i, 556
Rook, Cape, ii, 674
Ruff, i, 383

## S

Saddle-bill, i, 68
Sanderling, i, 382
Sandgrouse, i, 445
Sandpiper, Curlew, i, 379
—, Marsh, i, 389
Sandpipers, i, 384

Scimitar-bills, i, 639
Scrub-Robins, ii, 316
Secretary Bird, i, 128
Seed-crackers, ii, 989
Seed-eaters, ii, 1070
Seesee, i, 225
Shearwaters, i, 7
Shikra, i, 205
Shrikes, ii, 576
—, Cuckoo, ii, 555
—, Helmet-, ii, 567
—, Red-billed, ii, 567
Silver-bills, ii, 979
Silver-bird, ii, 187
Siskin, Black-headed, ii, 1084
Skimmer, i, 441
Skuas, i, 411
Snipe, i, 371
—, Painted, i, 370
Socotra Warbler, ii, 439
Sparrow-Hawks, i, 199
Sparrow, Java, ii, 981
—, Rock-, ii, 887
Sparrow-Weavers, ii, 862
Sparrows, ii, 869
Spinetails, i, 789
Spoonbills, i, 78
Spot-throat, ii, 301
Sprosser, ii, 332
Spurfowl, i, 256
Starlings, ii, 680
Stilt, i, 369
Stints, i, 380
Stonechats, ii, 286
Stone Curlew, i, 328
Stone-Partridge, i, 268
Storks, i, 62
Sunbirds, ii, 736
Swallows, ii, 521
Swifts, i, 776

## T

Terns, i, 422
Thicknees, i, 327
Thrush-Babbler, ii, 99
Thrushes, ii, 238
—, Ant-, ii, 254
—, Rock-, ii, 256
Tinker-birds, i, 722
Tit-Hylia, ii, 819
Tits, ii, 646
—, Penduline, ii, 655
Trogons, i, 693
Tropic Birds, i, 16
Trumpeter-Bullfinch, ii, 1059
Turacos, i, 521
Turnstone, i, 384
Turtle-Dove, i, 469
Twin-Spots, ii, 985, 1002

# INDEX OF ENGLISH GROUP NAMES

## V

Vulture, Bearded, i, 192
—, Palm-nut, i, 190
Vultures, i, 129

## W

Waders, i, 376
Wagtails, ii, 48
—, Yellow, ii, 52
Waldrapp, i, 73
Warblers, ii, 335
—, Apalis, ii, 396
—, Camaroptera, ii, 439
—, Cisticola, ii, 448
—, Eremomela, ii, 430
—, Grass, ii, 448
—, Morning, ii, 314
—, Prinia, ii, 505
Wattle-eyes, ii, 208
Waxbills, ii, 1018
Weaver, Parasitic, ii, 1000
—, Rufous-tailed, ii, 866
Weaver, Speckle-fronted, ii, 887
Weavers, Buffalo-, ii, 858
—, Grosbeak, ii, 939
—, Ploceine, ii, 889
—, Red-headed, ii, 942
—, Social, ii, 867
—, Sparrow, ii, 862
Whale-Bird, i, 14
Wheatears, ii, 259
Whimbrel, i, 394
Whinchat, ii, 290
White-eyes, ii, 724
Whitethroats, ii, 348
Whydahs, ii, 1046
Widow-Birds, ii, 960
Willow Warblers, ii, 382
Wood-Hoopoes, i, 632
Woodpeckers, i, 749
Wren-Warblers, ii, 392
Wrynecks, i, 773

## Y

Yellow-bill, i, 519